Environmental Earth Sciences

Series editor

James W. LaMoreaux, Tuscaloosa, USA

More information about this series at http://www.springer.com/series/8394

Vladimir F. Krapivin · Costas A. Varotsos
Vladimir Yu. Soldatov

New Ecoinformatics Tools in Environmental Science

Applications and Decision-making

Springer

Vladimir F. Krapivin
Informatics
Kotelnikov Institute of Radioengineering
Fryazino
Russia

Vladimir Yu. Soldatov
Informatics
Kotelnikov Institute of Radioengineering
Fryazino
Russia

Costas A. Varotsos
Environmental Physics and Meteorology
University of Athens
Athens
Greece

ISSN 2199-9155 ISSN 2199-9163 (electronic)
Environmental Earth Sciences
ISBN 978-3-319-13977-7 ISBN 978-3-319-13978-4 (eBook)
DOI 10.1007/978-3-319-13978-4

Library of Congress Control Number: 2014957398

Springer Cham Heidelberg New York Dordrecht London

Printed on acid-free paper

Springer International Publishing AG Switzerland is part of Springer Science+Business Media (www.springer.com)

Preface

The design problem of geoinformation monitoring systems needs the solution of a wide spectrum of tasks incoming to the competence of many areas of knowledge. The complex character of this problem is stipulated by the assembly of heterogeneous and differently systematic and applied research that is realized in many countries in the framework of national and international environmental programs. The basic goal of all similar researches consists in the attempt to answer the unique basic question: What structure and working regime of the observation system are to be in order to guarantee the reliable assessments of its current state and the prognosis of its evolution in the nearest and perspective future? Unfortunately, present science does not give an affirmative reply to this question. Ecoinformatics tries to answer this question by developing new informational technologies and producing a theoretical base for a new stage of the environmental science developing.

Ecoinformatics is the science of information in ecology and environmental science that integrates environmental and information sciences. Ecoinformatics tries to create the tools for the assessment and analysis of natural systems that exist under different conditions. In this context, ecoinformatics develops computer technologies relevant to the management of ecological data and information delivered existing and planned geoinformational systems. As a result ecoinformatics proposes new tools based on the coupled use of ecological, geophysical, and mathematical knowledge to integrate information, data, methods, algorithms, and computer technologies for providing ecological data to scientific or policy making processes.

Ecoinformatics is studying the problems that arise due to use of informatics methods for solution of environment control tasks. It is developed areas such as elaboration of new informational technologies for monitoring data processing, making and development of algorithms for spatial–temporal interpolation, analysis of correlation functions in the nature–anthropogenic systems taking into account hierarchy of spatial and temporal scales, and search for the most efficient methods for the synthesis of monitoring system structures.

Ecoinformatics devotes priority to technologies in the framework in which the informatics methods are used for research of dynamic characteristics of environmental systems, assessment of large-scale consequences of anthropogenic activity, and prognosis of biogeochemical processes from local to global scales. This field of study also includes a creation and application of mathematical models for natural and anthropogenic processes, search of criteria for reliable assessment of human living condition quality, and detection of causes stimulating the disturbances of sanitary epidemiological conditions in the environment.

Ecoinformatics studies theoretical and applied tasks that are aimed at understanding the role of mathematical modeling methods, theory of complex systems, nonlinear programming, computer cartography, remote sensing technology, and expert systems in the study of processes happening in the environment.

Ecoinformatics joins knowledge from such areas as physics, mathematics, biology, chemistry, sociology, ecology, economics, and law that solve environmental problems. It helps to find answers to many questions on the conflicts between nature and human society. Most experts suggest that an efficient way to resolve these conflicts would be creation of a unified planetary-scale adaptive geoinformation monitoring system, which should be based on knowledge bases and global datasets that are constantly updated. The adaptive nature of such a system should be provided by continuously correcting the data acquisition mode and by varying the parameters and structure of the global model.

This book develops ideas and technologies that could help solve many environmental problems and describes a simulation system based on sets of computer algorithms that process data from global and regional monitoring. The book has ten chapters that consider various aspects of the environmental problems and develops information-modeling technologies for operational diagnostics of environmental processes including stressful natural phenomena. In particular, the book discusses the following topics:

- Natural catastrophes as a dynamic category of environmental phenomena;
- Remote sensing and mathematical modeling for effective forecast of large-scale land territories and water areas behavior;
- Practical applications of microwave radiometric technologies along with other remote sensing technologies in different situations in hydrology for underground water mapping and for fire hazard determination in forested and peat bog environments;
- Decision making in complicated conditions;
- Evolution modeling in ecology and hydrometeorology;
- Arctic Basin pollution dynamics under intensive anthropogenic forcing due to a variety of industrial activities;
- Target-oriented models that may be effective in socioeconomic areas; and
- Regional and global ecological monitoring functioning based on the fundamentals of the noosphere paradigm.

The book discusses the global implications of environmental degradation—a model for predicting anthropogenic influences on global environmental change and for incorporating monitoring data into the predictions. Global problems of the nature–society system dynamics are considered and the key problems of ensuring its sustainable development are studied. Emphasis is placed on global geoinformation monitoring, which could provide a reliable control for the development of environmental processes by obtaining prognostic estimates of the consequences of the realization of anthropogenic projects. Ecoinformatics tools developed in this book offer a new approach to the study of global environmental changes, focusing on it as an area of study involving many scientific disciplines. These tools provide a unique insight into the social context of global changes in biogeochemical cycles, and is a timely contribution to the current debate into global warming and sustainable development of the nature–society system.

The book is aimed at specialists dealing with the development of information-modeling technologies to protect the natural world. Global modeling, climate change, problems inherent in relationships between society and nature, geopolitics, international relations, and methodology of interdisciplinary studies are studied in-depth. It is of special interest to designers and users of information-modeling technologies in the field of population protection from natural disasters.

Contents

About the Authors

Vladimir F. Krapivin was educated at the Moscow State University as mathematician in 1959. He received his Ph.D. in Radiophysics from the Kotelnikov Institute of Radioengineering and Electronics (KIRE) of Russian Academy of Sciences (RAS) in 1966. The Doctoral Dissertation in Geophysics was defended by him at the Moscow Institute of Oceanology of RAS in 1973. He is leader of Informatics Department in KIRE RAS. His responsibilities include all environmental and research activities of the Institute, including the design and construction of monitoring systems. He became Full Professor of Radiophysics in 1987 at KIRE RAS.

The title "Honoured Worker of Science" was given to him in 1999 by the Russian Government.

He was appointed Grand Professor in 2003 at the World University for Development of Science, Education and Society, and in 2012 at the Nguyen Tat Thanh University (Ho Chi Minh City, Vietnam). Vladimir Krapivin is a full member of the Russian Academy of Natural Sciences and Balkan Academy of Sciences, New Culture and Sustainable Development. He has published 32 books and 450 papers in the field of ecoinformatics, remote sensing, and global modeling. He has specialized in investigating global environmental change by the application of modeling technology.

Costas A. Varotsos received his B.Sc. in Physics at the Athens University in 1980 and his Ph.D. in Atmospheric Physics in 1984. He is Full Professor in Atmospheric Physics at the Department of Environmental Physics and Meteorology of the Faculty of Physics and Dean of the School of Sciences of the National and Kapodistrian University of Athens. Since 1989 he teaches Atmospheric and Environmental Physics and Chemistry, which are also the main topics of his research interests. He is Honorary Professor of the Russian Academy of Natural Sciences, Fellow (elected) of the Royal Meteorological Society (Oxford, UK), and Full Member (elected) of the European Academy of Natural Sciences (Hanover, Germany).

He is a member of several scientific societies including the American Meteorological Society, the American Geophysical Union, and the EGU. He has been appointed as Specialty Chief Editor of "Frontiers in Environmental Science" of the NATURE Publishing Group, Editor of the "International Journal of Remote Sensing" of T&F, Editor of the "Remote Sensing Letters" of T&F, Editor of "The Scientific World Journal" of Hindawi Publ. Co., Advisor of the "Environmental Science and Pollution Research", Guest Editor of a number of Journals, and member of the Editorial board in several International Journals indexed in WoS. He is also serving as reviewer of various Journals, EU, UN, and US Scientific Proposals and Reports. He has published more than 250 papers and 20 books in the fields of atmospheric physics, atmospheric chemistry, and environmental change.

Vladimir Yu. Soldatov was born on April 17, 1986 in Fryazino of Moscow Region, Russian Federation. He educated as an Engineer in Applied Mathematics at the Moscow Technological University "Stankin." He received his Ph.D. in Experimental Physics from the Kotelnikov Institute of Radioengineering and Electronics of Russian Academy of Sciences in 2011. He has edited monographs and has authored over 120 scientific articles, reports, and book chapters in the field of experimental hydrophysics, decision making, prognosis theory, and mathematical ecology. At present, Vladimir Soldatov is studying the problem of prognosis and control of hydrophysical, hydrological, and hydrochemical processes. His major research interests include a creation of new information-modeling technologies for environmental monitoring systems.

Vladimir Soldatov took part in many scientific projects funded by Russian Fund for Basic Research, Russian Academy of Sciences, and International Science and Technology Center. He proposed a new technology for diagnosis of tropical hurricanes beginning in oceans with remote sensing methods and mathematical modeling.

Summary

The greatly enhanced awareness today of environmental problems has led to a resurgence of interest in all aspects of global environmental monitoring. The need to reduce pollution of the hydrosphere and to reduce the risk of damage from extreme environmental processes has led to a large increase in the number of theoretical and applied investigations in the field of global ecoinformatics. The growth of needs in practically all spheres of remote environmental monitoring has resulted in numerous problems arising from the many unsolved tasks including the assessment of important characteristics of the soil–plant formations and oceans. It has become a priority to develop new concepts and approaches for assessing and dealing with natural and man-made system dynamics. A major priority is global problems connected with creation of effective information technologies to data processing within the environment studies.

The complexity, importance, and interdisciplinary nature of ecoinformatics allow the solution of many environmental problems by means of combined use of environmental monitoring systems and mathematical models. The main objective of the book lies in the working information technology for the combined use of modeling technology and remote sensing measurements in the assessment of environmental system states as well as to illustrate this technology with computer calculations for various environmental problems. Different tasks related to the assessment and prognosis of natural systems' dynamics are solved based on the remote sensing measurements using mobile platforms. New information-modeling technology is proposed to be used for optimization of remote sensing monitoring systems. This technology is based on sets of computer algorithms for comprehensive analysis of data from global and regional monitoring systems.

The chapters in the theoretical part of the book contain descriptions of rigorous algorithms and environmental models. The applied part considers specific problems of environmental dynamics in areas of different countries where *on-site* experiments were realized to test new information-modeling technology that enhances the GIS to GIMS.

This book is the result of a detailed study of the environment by means of computer algorithms and simulation models. It describes the sets of algorithms and models for comprehensive analysis of data from global and regional monitoring systems. The main purpose of this book is to develop a universal information-modeling technology to estimate the characteristics of environmental subsystems both at global and regional scales. Applied mathematicians, hydrologists, geophysicists, radiophysicists, and other researchers of the environment will find a wealth of information in this book.

This book should prove useful to students in a variety of disciplines since it provides basic information in an interdisciplinary area and encourages students, through the references supplied, to seek out and to read the original monographs and articles cited. It should also be of interest to professionals in academia, in consulting offices, and government agencies, and to others responsible for the investigation, evaluation, and solution of complicated environmental issues.

Abbreviations and Acronyms

AAMU	Alabama Agricultural and Mechanical University
AARS	Asian Association on Remote Sensing
ABL	Atmospheric Boundary Layer
ABPCV	Archive of Biometric and Production Characteristics of Vegetation
ACCSV	Arctic Climate Change, Substrate, and Vegetation
ACES	Aral–Caspian Expert System
ACIA	Arctic Climate Impact Assessment
ACRS	Asian Conference on Remote Sensing
ACSYS	Arctic Climate SYstem Study
ADEOS	ADvanced Earth Observation Satellite
AEPS	Arctic Environmental Protection Strategy
AGCM	Atmosphere Global Climate Model
AGIR	Asociatia Generala a Inginerilor din Romania
AIMES	Analysis, Integration, and Modeling of the Earth System
AIMS	Adaptive Information-Modeling System
AIP	American Institute of Physics
AMAP	Arctic Monitoring and Assessment Programme
AMMA	African Monsoon Multidisciplinary Analysis
AMSR	Advanced Microwave Scanning Radiometer
ANWAP	Arctic Nuclear Waste Assessment Programme
AO	Arctic Oscillations
APDA	Arctic Precipitation Data Archive
APS	Atmosphere–Plant–Soil
ARCSS	ARCtic System Science Programme
ARCUS	Arctic Research Consortium of United States
ARDB	Arctic Run-off Data Base
ARF	Aerosol Radiative Forcing
ARISTI	All-Russian Institute for Scientific and Technical Information
ARM	Atmosphere Radiation Measurement
ASI	Advanced Science Institutes

ASP	Arctic Studies Program
ASPC	Atlantic Storm Prediction Center
ASTER	Advanced Spaceborne Thermal Emission and reflection Radiometer
ATLAS	Arctic Transitions in the Land–Atmosphere System
AVHRR	Advanced Very High Resolution Radiometer
AVSS	Atmosphere–Vegetation–Soil System
AYRS	Angara–Yenisei River System
AYRSSM	AYRS Simulation Model
BARC	Burned Area Reflectance Classification
BC	Black Carbon
BCSS	Biosphere/Climate/Society System
BDMS	Base of the Decision Making System
BEXCO	Busan EXhibition and COnvention center
BSAP	Baltic Sea Action Plan
BSS	Biosphere/Society System
CA	Climate Agenda
CAD	Computer-Aided Design
CALRS	Cluster Analysis focused on the account of Local Reading of Sensors
CANO	Central Administration of Navigation and Oceanography
CART	Cloud And Radiation Testbed
CASRS	Cluster Analysis focused on sings Space of Remote Sensors
CATCH	Computer And Technology Crime High
CBS	Climate–Biosphere–Society
CCM	Community Climate Model
CCN	Cloud Concentration Nucleii
CCSS	Common Core State Standards
CDR	Climate Data Record
CEAI	Control Engineering and Applied Informatics
CEOP	Coordinated Enhanced Observing Period
CF	Cloud Feedback
CFL	Circumpolar Flaw Lead
CHAMP	CHAllenging Minisatellite Payload
CG	Cloud–Ground
CGCM	Canadian Global Climate Model
CHC	Canadian Hurricane Center
CIESIN	Center for International Earth Science Information Network
CliC	Climate and Criosphere
CLIVAR	CLImate VARiability and predictability
CLRTAP	Convention on Long-Range Transboundary Air Pollution
CM	Climate Model
CMAQ	Congestion Mitigation and Air Quality model
CMIP	Coupled Model Intercomparison Project
CNES	Centre National d'Etudes Spatiales (Space Agency of France)

COADS	Comprehensive Ocean–Atmosphere Data Set
COLA	Center for Ocean–Land–Atmosphere studies
COP	Conference Of the Parties
COPES	Coordinated Observations and Prediction of the Earth System
COST	Cooperation in Science and Technology
CRED	Centre for Research on the Epidemiology of Disasters
CRC	Chemical Rubber Company
CSCS	Control Systems and Computer Science
CSI	Canopy Structure Index
CSIRO	Commonwealth Scientific and IndustRial Organization
CSL	Climatic Snow Line
DDC	Data Distribution Center
DEKLIM	DEutsches KLIMaforschung's program
DIC	Dissolved Inorganic Carbon
DLR	Deutsches zentrum für Luft- und Raumfahrte.V. (German Aerospace Center)
DMS	DiMethylSulfide
DMSP	Decision Making Sequent Procedure
DNA	DeoxyriboNucleic Acid
DO	Dansgaard–Oeschger
DOC	Dissolved Organic Carbon
DPIR	Drivers, Pressures, Impact, Response
EASE	Equal-Area Scalable Earth
ECHAM	European Center HAMburg
ECMWF	European Center for Medium-range Weather Forecast
EDB	Extendable Data Base
EES	Ecologo-Economic System
EIA	Environmental Impact Assessment
EIER	Engineering Institute for the Electronics and Radiotechnics
ELA	Equilibrium Line Altitude
ENSO	El Niño–Southern Oscillation
ENVISAT	ENVIronmental SATellite
EMW	ElectroMagnetic Waves
EOF	Empirical Orthogonal Function
EOS	Earth Observing System
EOSDIS	EOS Data and Information System
EPA	Environmental Protection Agency
ERB	Earth's Radiation Balance
ESA	European Space Agency
ESA	Environmental Systems Analysis
ESAIEP	Expert System for an Adaptive Identification of the Environmental Parameters
ESDIS	Earth Science Data and Information System
ESECEZ	Expert System for Ecological Control of Estuary Zone
ESM	Earth System Model

ESPAP	Expert System for Physics of Atmospheric Pollution
ESTAR	Electronically Scanned Thin Array Radiometer
ETM+	Enhanced Thematic Mapper Plus
EUP	Enterprise Unified Process
EUSAR	EUropean conference on Synthetic Aperture Radar
EVI	Enhanced Vegetation Index
EWT	Equivalent Water Thickness
FAO	Food and Agriculture Organization
FASIR	Fourier-Adjusted, Solar zenith angle-corrected, Interpolated, and Reconstructed data
FCCC	Framework Climate Change Convention
FGGE	First GARP Global Experiment
FMC	Fuel Moisture Content
FPAR	Fraction of PAR
FRA	Forest Resources Assessment
FTP	File Transfer Protocol
GARP	Global Atmospheric Research Program
GATOR	Gas-Aerosol-TranspOrt-Radiation-chemistry model
GCCM	Global Carbon Cycle Model
GCM	Global Climate Model
GCMM	General Circulation and Mesoscale Meteorological model
GCN	Global Core Network
GCOM	Global Change Observation Mission
GCOS	Global Climate Observation System
GCP	Global Carbon Project
GDP	Gross Domestic Product
GEF	Global Ecological Fund
GEO	Group on Earth Observations
GEOS	Global Earth Observation System
GEOSS	Global Earth Observation System of Systems
GEWEX	Global Energy and Water cycle Experiment
GFDL	Geophysical Fluid Dynamics modeL
GHG	GreenHouse Gas
GHOST	Global HOlocene Spatial and Temporal climate variability
GIGAS	GEOS, INSPIRE, and GMES an Action in Support
GIMS	Global Information-Modeling System
GIMSAF	GeoInformation Monitoring System of Agricultural Function
GIS	Geographical Information System
GISP2	Greenland Ice Sheet Project Two
GISS	Goddard Institute for Space Studies
GLOBEC	GLOBal ocean ECosystem dynamics
GLOSS	Global Sea Level Observing System
GMES	Global Monitoring for Environment and Security
GMNSS	Global Model of the NSS
GNSS	Global Navigation Satellite System

GOES	Geostationary Operational Environmental Satellite
GOOS	Global Ocean Observing System
GPM	Global Precipitation Measurement
GPS	Global Positioning System
GSM	Global Simulation Model
GST	Global Surface Temperature
GTOS	Global Terrestrial Observation System
GWEM	Global Wildland fire Emission Model
GWP	Global Warming Potential
HARC	Human Dimensions of the Arctic System
HCN	Hydrogen CyaNide
HDI	Human Development Index
HELCOM	HELsinki COMmission
HSCaRS	Hydrology, Soil Climatology and Remote Sensing
IA	Integral Assessment
IACCA	Inter-Agency Committee on the Climate Agenda
IASC	International Arctic Science Committee
IASTED	International Association of Science and Technology for Development
IBCAO	International Bathymetric Chart of the Arctic Ocean
IBP	International Biological Program
ICGGM	International Center on the Global Geoinformation Monitoring
ICID	International Commission on Irrigation and Drainage
ICLIPS	Integrated assessment of CLImate Protection Strategies
ICLRT	International Center for Lightning Research and Testing
ICSU	International Council of Scientific Unions
IE	Interactive Ensemble
IEA	International Energy Agency
IEEE	Institute of Electrical and Electronics Engineers
IFG	International Forum on Globalization
IGAC	International Global Atmospheric Chemistry
IGARSS	International Geoscience And Remote Sensing Symposium
IGBP	International Geosphere–Biosphere Program
IGCM	Intermediate General Circulation Model
IGOS	Integrated Global Observing Strategy
IHDP	International Human Dimensions Program
IIASA	International Institute for Applied Systems Analysis
IIC	Information Interface with Computer
IISGGM	International Integrated System for Global Geoinformation Monitoring
IMAGE	Integrated Model to Assess the Greenhouse Effect
IMARC	Intelligent Multifrequency Airborne polarimetric Radar Complex
IMS	Ice Mapping System
IMTEM	Information-Modeling Technology for the Environmental Monitoring

INSPIRE	INfrastructure for SPatial InfoRmation in the European community
IPAB	International Program for Antarctic Buoys
IPCC	Intergovernmental Panel on Climate Change
IPSL	Institute Pierre Simon Laplace
IPY	International Polar Year
IR	InfraRed
ISAC	International Study of Arctic Change
ISGGM	Integrated System for Global Geoinformation Monitoring
ISPRS	International Society for Photogrammetry and Remote Sensing
ISRS	International Symposium on Remote Sensing
ITBL	Internal Thermal Boundary Layer
ITCZ	Inter-Tropical Convergence Zone
ITEX	International Tundra EXperiment
ITHEA	Institute if information THEories and Applications
JASL	Joint Archive for Sea Level
JERS	Japanese Earth Resources Satellite
JGOFS	Joint Global Ocean Flux Study
KIREE	Kotelnikov Institute of RadioEngineering and Electronics
KIT	Karlsrucher Instituts für Technologie
KP	Kyoto Protocol
LAI	Leaf Area Index
LAII	Land–Atmosphere–Ice Interactions
LLGHG	Long-Lived GreenHouse Gas
LMDZ	Laboratoire de Météorologie Dynamique Zoom
LOSAC	L-band Ocean Salinity Airborne Campaign
LPJ-DGVM	Lund–Potsdam–Jena Dynamic Global Vegetation Model
LRTAP	Long-Range Transport of Air Pollution
LSAT	Land Surface Air Temperature
LUT	LookUp Table
LWRF	LongWave Radiative Forcings
MACS	Microwave Autonomous Copter System
MAT	Moist Acidic Tundra
MCBS	Magnetosphere–Climate–Biosphere–Society
MCRM	Markov Chain Reflectance Model
MDA	Model-Driven Architecture
MEA	Millennium Ecosystems Assessment
MEM	Microwave Emission Model
MGBC	Model of the Global Biogeochemical Cycle
MGC	Minor Gas Components
MGCDC	Model of Global Carbon Dioxide Cycle
MGSC	Model of Global Sulfur Cycle
MIROC	Model for Interdisciplinary Research On Climate
MISR	Multiangle Imaging SpectroRadiometer
MMTS	Maximum–Minimum Temperature System
MNT	Moist Nonacidic Tundra

MODIS	MODerate resolution Imaging Spectroradiometer
MR	Microwave Radiometer
MSAVI	Modified Soil-adjusted Vegetation Index
MSD	Mean Squared Displacement
MSP	Multichannel SpectroPolarimeter
MSSA	Multichannel Singular Spectral Analysis
MTLE	Model of Transmission Line with Exponential current decay
MTVZA	Module of Temperature and Vlazhnost (humidity) Zonding of the Atmosphere
MUE	Model of the Upwelling Ecosystem
MVI	Microwave Vegetation Index
NAM	Northern hemisphere Annular Mode
NAO	North Atlantic Oscillation
NASA	National Aeronautical and Space Administration
NASDA	National Space Development Agency (Japan)
NATO	North Atlantic Treaty Organization
NAWQA	National Water-Quality Assessment Program
NBR	Normalized Burn Ratio
NCAR	National Center for Atmospheric Research
NCDC	National Climatic Data Center
NDBC	National Data Buoy Center NOAA
NDVI	Normalized Difference Vegetation Index
NDWI	Normalized Difference Water Index
NEE	Net Ecosystem Exchange
NEW	North-East Water
NGDC	National Geophysical Data Center
NH	Northern Hemisphere
NHC	National Hurricane Center
NMAT	Nocturnal Marine Air Temperature
NMD	Normalized Mean Difference
NME	Normalized Mean Error
NMHC	Non-Methane HydroCarbons
NN	Neural Network
NOAA	National Oceanic and Atmospheric Administration
NODC	National Oceanographic Data Center
NPOESS	National Polar-orbiting Operational Environmental Satellite System
NPP	Net Primary Productivity
NSIDC	National Snow and Ice Data Center
NSS	Nature–Society System
NTU	Nephelometric Turbidity Units
NVR	Nadir Viewing Rada
OAII	Ocean/Atmosphere/Ice Interactions
OAS	Ocean–Atmosphere System
OC	Organic Carbon

OECD	Organization for Economic Cooperation and Development
OGCM	Oceanic General Circulation Model
OM	Organic Matter
ONR	Office of Naval Research
OSE	Okhotsk Sea Environment
PAGES	Pilot Analysis of the Global EcoSystems
PALE	Paleoclimates of Arctic Lakes and Estuaries
PALS	Passive and Active L and S band system
PAR	Photosynthetically Active Radiation
PARCS	Paleoenvironmental ARCtic Sciences
PBL	Planetary Boundary Layer
PCE	Peruvian Current Ecosystem
PCEM	Peruvian Current Ecosystem Model
PCP	Packet of Computer Programs
PEHP	Production of Electric power and Heat for the Population
PEL	Petroleum Economics Limited
PGT	Pasquill-Gifford-Turner
PIRA	Petroleum Industry Research Associates
PIRATA	PIlot Research moored Array in the Tropical Atlantic
PLC	Pollution Load Compilation
PMR	Passive Microwave Radar
PNAS	Proceedings of the National Academy of Sciences
POC	Particulate Organic Carbon
POLDER	POLarization and Directionality of Earth's Reflectances
PORSEC	Pan Ocean Remote Sensing Conference
PSC	Polar Stratospheric Clouds
PWI	Plant Water Indexes
QSAR	Quantitative Structure–Activity Relationships
RAISE	Russian–American Initiative on Shelf–land Environments
RAS	Russian Academy of Sciences
RCA	Russian Custom Academy
RF	Radiative Forcing
RGGI	Regional Greenhouse Gas Initiative
RIBAMOD	RIver BAsin MODeling
RIMS	Regional Information-Modeling System
RNA	RIver BAsin MODeling
RSMC	Regional Specialized Meteorological Center
RT	Radiative Transfer
SAC-C	Satélite de Aplicaciones Científicas-C
SALLJEX	South American Low-Level Jet Experiment
SASS	Subsonic ASSessment program
SAM	Southern hemisphere Annual Mode
SAR	Synthetic Aperture Radar
SAS	Surface Active Substance
SAT	Surface Air Temperature

SATVI	Soil-adjusted Total Vegetation Index
SAVI	Soil-adjusted Vegetation Index
SBI	Shelf–Basin Interactions
SCE	Snow Cover Extent
SCOPE	Scientific Committee On Problems of the Environment
SDS	Scott Data System
SEA	Strategic Environmental Assessment
SEATCH	Study of Environmental ArcTic CHange
SEDAAR	Strategic Environmental Distributed Active Archive Resource
SGM	Spatial Global Model
SGPE	Southern Great Plains Experiment
SH	Southern Hemisphere
SHEBA	Surface HEat Budget of the Arctic
SHF	Super High Frequency
SHIPS	Statistical Hurricane Intensity Prediction Scheme
SIMS	Synthesis, Integration, and Modelling Studies
SLAR	Side-Looking Airborne Radar
SMEX	Soil Moisture Experiments
SMMR	Scanning Multichannel Microwave Radiometer
SMOS	Soil Moisture and Ocean Salinity
SMOSE	Simulation Model of the Okhotsk Sea Environment
SONEX	SAAS Ozone and Nitrogen oxide Experiment
SPARC	Stratospheric Processes And their Role in Climate
SPF	Soil–Plant Formation
SPM	Summary for Policy Makers
SPOT	Systéme Probatoire d'Observation de la Terre
SPREP	Secretariat of the Pacific Regional Environment Programme
SPRINTARS	SPectral RadIatioN-Transport model for AeRosol Species
SR	Simple Ratio
SRB	Surface Radiation Budget
SRVI	Simple Ratio Vegetation Index
SSAPP	Simulation System for the Atmosphere Pollution Physics
SSHHI	Simulation System for Hydrophysical and Hydrochemical Investigations
SSM/I	Spatial Sensor Microwave/Imager
SSMAE	Spatial Simulation Model of the Arctic Ecosystem
SSMIS	Special Sensor Microwave Imager/Sounder
SST	Sea Surface Temperature
STOCHEM	UK Meteorological Office Global Three-Dimensional Lagrangian Chemistry model
SVI	Spectral Vegetation Index
SWR	Short-Wave Radiation
SWRF	Short-Wave Radiative Forcings
TAO	Tropical Atmosphere Ocean
TAR	Third Assessment Report

TCWC	Tropical Cyclone Warning Center
TERC	Terrestrial Environment Research Center
THORPEX	THe Observing system Research and Predictability EXperiment
TIR	Thermal Infrared Radiometer
TL	Transmission Line
TM	Thematic Mapper
TOA	Top-Of-Atmosphere
TOGA	Tropical Ocean Global Atmosphere
TOMS	Total Ozone Mapping Spectrometer
TORRO	TORnado and storm Research Organization
TRITON	TRIangle Trans-Ocean buoy Network
TRMM	Tropical Rainfall Measuring Mission
TTP	Technology Transfer and Promotion
TVDI	Temperature Vegetation Dryness Index
UHI	Urban Heat Island
UHSLC	University of Hawaii Sea Level Center
UML	Unified Modeling Language
UNCSD	United Nations Conference on Sustainable Development
UNEP	United Nations Environment Programme
UNESCO	United Nations Educational, Scientific and Cultural Organization
UNDP	United Nations Development Programme
UNFCCC	United Nations Framework Convention on Climate Change
UQHL	Upper Quasi-Homogeneous Layer
USA	United States of America
USCRN	U.S. Climate Reference Network
USDA	United States Department of Agriculture
USGS	United States Geological Survey
USSR	Union of Soviet Socialistic Republics
UV	UltraViolet
VAMOS	Variability of the American Monsoon Systems
VDI	Vegetation Dryness Index
VHF	Very High Frequency
VI	Vegetation Indices
VWC	Vegetation Water Content
WAM	West African Monsoon
WB	World Bank
WCP	World Climate Programme
WCRP	World Climate Research Programme
WCSDG	World Commission on the Social Dimension of Globalization
WDI	Water Deficit Index
WEO	World Energy Outlook
WHO	World Health Organization
WISE	Web-based Inquiry Science Environment
WIT	Wessex Institute Transactions
WMO	World Meteorological Organization

WOCE	World Ocean Circulation Experiment
WR	Wood Remains
WRI	World Resources Institute
WTF	Wet Tropical Forests
WWRP	World Weather Research Programme
ZEUS	Zentren EUropaischen Supercomputings

Chapter 1
Information-Modeling Technology for the Environmental Monitoring

1.1 The Principal Conception of the Information-Modeling Technology

The methods of the local environment diagnostics do not provide complex assessment for an environmental element (i.e., natural objects or process states), especially in the case when this environmental element occupies extensive space. Any technical means for the data collection of the environment allow production of information that is characterized as episodic in time and fragmentary in space. In particular, the microwave remote-sensing systems that are widely used to equip the flying laboratories and the natural-resources satellites supply the data sets that are geographically distributed. Reconstruction of the information obtained is possible only by employing methods of spatial-temporal interpolation for the development of which several techniques and algorithms of simulation modeling have been used in the past (e.g., Cakaj 2010; Cracknell et al. 2009a; Krapivin and Shutko 2012). For example, the spatial modeling has been widely used for the interpretation of the data obtained by environmental instruments, which monitor the lithosphere, cryosphere, hydrosphere, vegetation interfaces and urban environments (Kelly et al. 2005; Polishchuk and Polishchuk 2013; Soldatov 2011; Kareiva 2001; Le Corbusie 1977; Lu and Li 2003; Safai et al. 2004).

For the accomplishment of the complex task of the environmental diagnostics, it is necessary to built up a system that incorporates such functions, as well as data collection (by means of remote sensing and in-situ methods), their analysis and processing. A system of this type is capable to conduct systematic observations and assessments of the environmental state, to predetermine forecasting diagnostics of changes of the environmental elements (due to anthropogenic impacts) and to analyze the evolution of environmental processes, taking into account the anthropogenic scenarios. One of the system functions is to provide warnings about undesirable changes in the environment. Attainment of such functions for

V.F. Krapivin et al., *New Ecoinformatics Tools in Environmental Science*,
Environmental Earth Sciences, DOI 10.1007/978-3-319-13978-4_1

environmental monitoring is feasible with the use of simulation methods allowing the development of a model for the investigation of the natural subsystem.

The development of models for the description of environmental, biogeochemical, hydrophysical, climatic and socio-economic processes (that provide the synthesis of images for its subsystems), inevitably demands the establishment of monitoring systems with processing and formation of suitable databases. As has already been demonstrated by many authors there exist several balanced criteria for the information selection, taking into account the hierarchy in the cause-effect correlations within the biosphere. These include a co-ordination of the acceptance and the depth in the spatial discretisation, in order the atmosphere, the land ecosystems and the hydrosphere to be fully described (Krapivin 2000b; Mustard 2001; Smith 2008).

As it has already been shown by numerous studies, an application of mathematical modeling for the satellite monitoring systems may yield practical outcome only with the creation of a unique data network that should be consolidated to the model of nature-society system (NSS). In general, the reconstruction of a spatial image of a specific territory of the ecosystem by utilizing the multi-channel information from different sensors requires the accomplishment of the optimization task. For the latter, it is necessary to determine the optimal structure for the monitoring system with high reliability of the results obtained.

To accomplish this task the Global Information—Modeling System (GIMS) that has been developed by Krapivin (1993, 2000d) could be employed. The GIMS tool considers each environmental sub-system as a part of the global nature-society system. In this case, a specific model of the ecosystem has boundary conditions that are defined by GIMS. The development and use of the GIMS is mainly based on the Information-Modeling Technology for the Environmental Monitoring (IMTEM) that optimizes the monitoring system structure at the expense of its adaptive regime. A conceptual scheme of an adaptive regime for environmental monitoring premises the assumption of architecture of such monitoring system, which brings together knowledge from different disciplines and enables to manage the supply of this knowledge. This is possible through the combined use of GIS technology, simulation modeling and methods for expert systems (Maguire et al. 1991).

The global information system (GIS) tool guarantees the geographical data processing, the relation with existing databases and the topological representation for studied territories. With the help of technology modeling, the GIS can be extended to GIMS, modifying some functions of the user interface for computer cartographic systems, including forecasting assessments of conditions for environmental system functioning within predefined scenarios. Furthermore, these parametric changes can be used both for the evaluation of model coefficients and for the prognostic assessment of environmental dynamics based on the evolutionary modeling. Such a modification requires the support of combining mathematical modeling with remote sensing to solve the following problems (Krapivin and Shutko 2012):

- Forecast of the onset and the dangerous level of natural disasters, the emergency situations and the manmade catastrophes.
- Control of possible disasters and catastrophes (including complex meteorological conditions) with the provision of information for decision making.
- Impact assessment of disasters and catastrophes for cities, agricultural and forest areas, sea and coastal flora and fauna.
- Delivery plans for rescue operations.

The broadening of the GIS functions to the GIMS by means of the IMTEM technology enables to solve key work packages of large-scale industrial centers, like the following:

- Investigation of seasonal parameters for municipal and suburban landscapes, geophysical fields and local anomalies of various types, detection of laws for phenomenological and topological characteristics of disasters, representation of the investigation results in the form of thematic maps of standardized scales.
- Development of the proper methodology for the assessment of ecological and sanitary state of living, industrial, forest-park and suburban zones, reservoirs and rivers, main heating and gasoline lines, transport and electrical nets.
- Investigation of seasonal and daily dynamics of characteristics for areas where municipal and industrial wastes are stores, as well as for pollution sources of land covers, atmosphere and reservoirs.
- Solution of inverse tasks and elaboration of statistical similarity criteria with regard to the local anthropogenic and geophysical features of municipal and suburban territories, low atmosphere, cloudiness and ozone layer, pollution dynamics and their features.

Monitoring systems based on IMTEM technology can be carried out using the data delivered from the following levels of information with appropriate techniques, methodical and metrology bases:

(1) Measurements by means of satellite sensors.

- estimations of global ecological characteristics for studied region (taking into account the monitoring needs for lower levels).
- determination of the optical transparency, the state of gas and aerosol constituents in the Earth's atmosphere.
- principal characteristics of anthropogenic pollution of the atmosphere (e.g., state of the ozone layer) and water basins.
- state of land covers and architectural objects.

(2) Special investigations performed by the boards of multi-purpose and special flying laboratories about the:

- ecological characteristics of regional environment (taking into account the third and fourth level needs).
- energetic and polarization characteristics of microwave fields that characterize the urban and other landscapes.

- radio-thermal and radar images of urban regions, megacities, forests, etc.
- images and parameters of heat field of landscape in near and far IR-ranges.
- optical power of transparency, specific humidity and other atmospheric characteristics observed in nadir and zenith directions.

(3) Helicopter and light-aviation laboratories that supply the eco-monitoring data in the height region 80–500 m about the:

- regional ecological characteristics (taking into account the needs for the ground-based stations and flying observational platforms).
- surface air pollution by gases and solid aerosols.
- background level and anomalies of the environmental radio-activity and heat fields (in the IR and SHF ranges).
- manmade air pollution in residential areas.

(4) Observational laboratories based on automobiles and ships as well as in-situ stations. The main target of these laboratories and stations is the detection of the anomalies in the local environment.

(5) This level is supported by GIS, IMTEM and GIMS technologies:

- collection, annotation and accumulation of data supplied by the four levels of eco-monitoring, bringing them to unique spatial-temporal co-ordinates.
- express analysis and processing of the experimental and calculated estimations for the state of the environmental elements with allocation of their features and their connection to the specific points on the area and map.
- identification of pollutant sources and other disturbances in the environment, assessment of dynamics and forecasting of the consequences from pollutant impacts on ecological and sanitary conditions, detection of the ways for the pollutants migration and determination of concentration of unhealthy contaminants on regional territory.
- preparation of the consumer's information in the standardized formats that provide the reconstruction of objective image for ecological and sanitary situation, as well as substantiation of ways for rational solution of present and perspective nature-conservative problems.

The hierarchical structure of environmental monitoring systems optimizes the use of financial resources with high quality of final results. This is the basic argument for the creation of geoinformation monitoring systems using the technology of hierarchical synthesis. In particular, GIS is the most developed technology of environmental monitoring. GIS technology is used in many countries with very high economic efficiency and lies at the interfaces between databases and remote sensing. The main components of GIS are the computer-assisted network, database, network communication and system representation of real situation on the display. Numerous examples of the GIS synthesis show that GIS technology provides a convenient tool for the population to control the monitoring regime and it is an efficient mechanism for the integration of multiple-factor information about the object under investigation. However, GIS has serious limitations when monitoring complex problems must be solved to form the dynamical image of environmental sub-system under the conditions of incomplete information. Basic flaw of the GIS technology is that it is not oriented to the multifaceted forecast of the state of the monitoring object. In this regard, Krapivin and Varotsos (2007, 2008) proposed the GIS modification by means of a combined use of GIS and modeling technologies. The result of this attempt is the development of a new GIMS technology based on the formula GIMS = GIS U IMTEM. This GIMS technology eliminates many of the above-said GIS flaws and allows the establishment of monitoring systems that display prognosis functions. The GIMS architecture is shown in Fig. 1.1.

In principle, the required information may be obtained by using in-situ and remote sensing measurements and by getting access to a-prior knowledge-based data in the GIS databases. However, the following problems must be solved:

- what kind of instruments should be used to conduct the so-called ground-truth and remote sensing measurements?
- what is the cost to be paid for the in-situ and remote sensing information?
- what kind of balance should be taken into account for the information content of the in-situ and remote sensing observations along with their corresponding cost? and

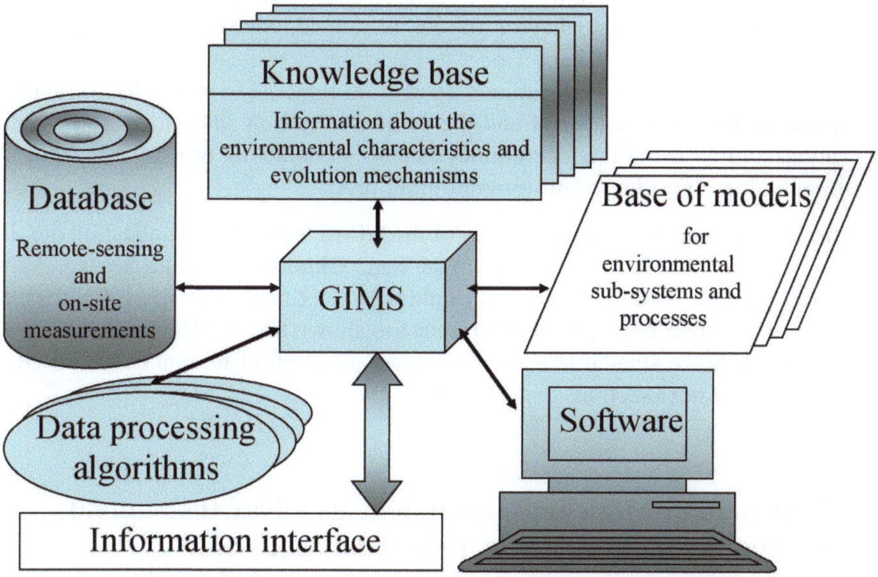

Fig. 1.1 Conceptual diagram showing the definition of the GIMS technology structure

- what kind of mathematical models must be used for both the interpolation and the extrapolation of data, in terms of time and space in order to reduce the frequency and thus the cost of the observations and to increase the reliability of forecasting the environmental behavior of the observed items?

These and other problems are often solved by using a monitoring system, which is based on the combination of the functions of environmental data acquisition, control and analysis, as well as forecasting the characteristics of the most important processes in the environment. In other words, this unification forms the new information technology called the GIMS-technology. The relationship between GIMS and GIS technologies was defined above by the formula GIMS = GIS ∪ IMTEM that increases the spectrum of the solved environmental problems (Krapivin and Shutko 2012; Shutko and Krapivin 2011).

The GIMS technology was successfully tested and applied in many countries in the World. It has been proven that the advantages of GIMS application consist, first of all, in the following capabilities (Shutko et al. 2010):

- to plan in situ and remote parts of experiment, optimizing the customer requirements for a quality of data and its cost;
- to simulate the data in the form of maps having in fact the data measured along the flight line by collecting the in situ measurements and the remote observations without using scanning technologies which are extremely cost-effective;
- to forecast/predict the behavior of the examined environmental/geophysical systems; and
- to expand the scale of the examined areas from regional to global by combining optimally the volume/amount and the essence of the in situ and remote observations and by using this operational data set in modeling process according to the required algorithmic and programming tasks.

Three of the demonstrative examples of the GIMS technology application are the creation of the GIMS for Bulgaria, Aral Sea, Okhotsk Sea and Sea of Azov (Kondratyev et al. 2002; Krapivin and Shutko 1989; Krapivin et al. 2007c, 2009d; Shutko and Krapivin 2010). This experience has shown that GIMS technology must be developed by the standardization and universalization of its functions, which is described in the next section.

1.2 Information-Modeling Technology as a Key Instrument for the Solution of Environmental Problems

1.2.1 Standardization and Universalization of the Functions of the Global Information-Modeling System

The GIMS includes a set of items whose functions are determined by a number of standard tasks. These items form the informational architecture of the monitoring system, which includes the basis of models that describe the environmental subsystems that enable environmental diagnosis. A GIMS—oriented system shell is formed to formalize the input information. An additional database level is synthesized to have multiple assignments of semantic structures with real environmental subsystems with variable dimensions. Such GIMS completion is formed through IMTEM that offers additional functions, like:

- maximum utilization of the monitoring data about the environmental objects in presence of unremovable information;
- conducting experiments from ground locations, mobile platforms and aircraft/satellite combining the advantages of microwave radiometers, optical sensors and other means;
- incorporation into the remotely sensed data flow of the GPS-positioned in situ measurements and prior knowledge-based information; and
- adaptation of the GIMS structure and functions to the environmental sub-system under the study.

Let us designate semantic structures by the matrix symbol $A_i = \|a_{i1,\ldots,is}\|$, where element $a_{i1,\ldots,is}$ matches the object, process, phenomenon, event or other environmental bifurcation. Actually, matrix A_i is the parametrical image of the real environmental subsystem with its specific features. Parameter s reflects the dimension of the informational type for the subsystem section. The structures $\{A_i\}$ identify both the spatial distributionof the subsystem components and their types and parameters. The basic structures of $\{A_i\}$ have four dimensions: i_1 by latitude, i_2 by longitude, i_3 by height, and i_4 by time. Other structures of $\{A_i\}$ determine model coefficients, types of soil-plant formations (SPFs), precipitation, temperature, radiation, etc.

The semantic structures $\{A_i\}$, called identifiers, are used by the basic models for the formation of initial fields, validation of model output, and for preparation of the final or intermediate reports. Land cover classification is the main function of the GIMS. The identifier of basic land cover classes provides a correspondence between different types of SPFs and their parameters and spatial structure. An example of such an identifier is given in Fig. 1.2. Each of the identifier elements can have a vector structure connected with the description of various classifications of land cover and allowing the formation of global land cover classification maps by means of interpolation and extrapolation through the analysis of satellite data.

The GIMS database together with the structures of the $\{A_i\}$ consists of information about the model coefficients and a set of scenario fragments. The structures $\{A_i\}$ link up the knowledge base with the database. Each symbol of A_i is decoded in conformity with the depth hierarchy and reflects just how reliable the description of the environmental subsystem is, both qualitatively and quantitatively.

Identification procedure allows formal description of the environmental subsystem image with fixation of geographical coordinates. According to scheme of Fig. 1.2, series of identifiers are placed to the GIMS database. These identifiers

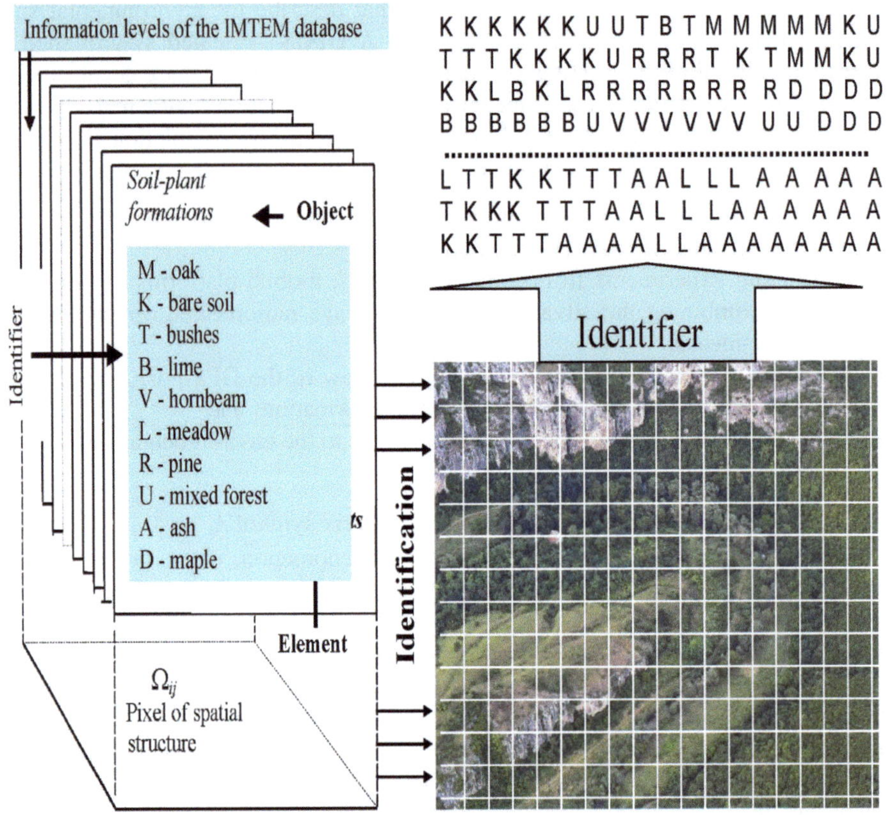

Fig. 1.2 Scheme of cartographic identification of the environmental subsystem realized in the IMTEM technology for the monitoring case of natural park Rusenski Lom in Bulgaria (Soldatov 2011; Krapivin and Shutko 2012)

form studied subsystem image and allows classification of spatial pixels both contoured by subsystem territory and no owned. In particular, such identifier can represented in the form of matrix $A_k = \left\| a_{ij}^k \right\|$ where

$$a_{ij}^k = \begin{cases} 0, \text{ if pixel } \Xi_{ij}^k \text{ lies beyond the bounds of studied territory } \Xi, \\ \alpha, \text{ if pixel } \Xi_{ij}^k \text{ belongs to studied territory } \Xi \text{ and object } \alpha \text{ is situated there.} \end{cases}$$

As a result, the topological procedure in the studied environmental subsystem located in territory Ξ of the existing database can be performed automatically taking into account its spatial pixel hierarchy. This procedure is schematically shown in Fig. 1.3.

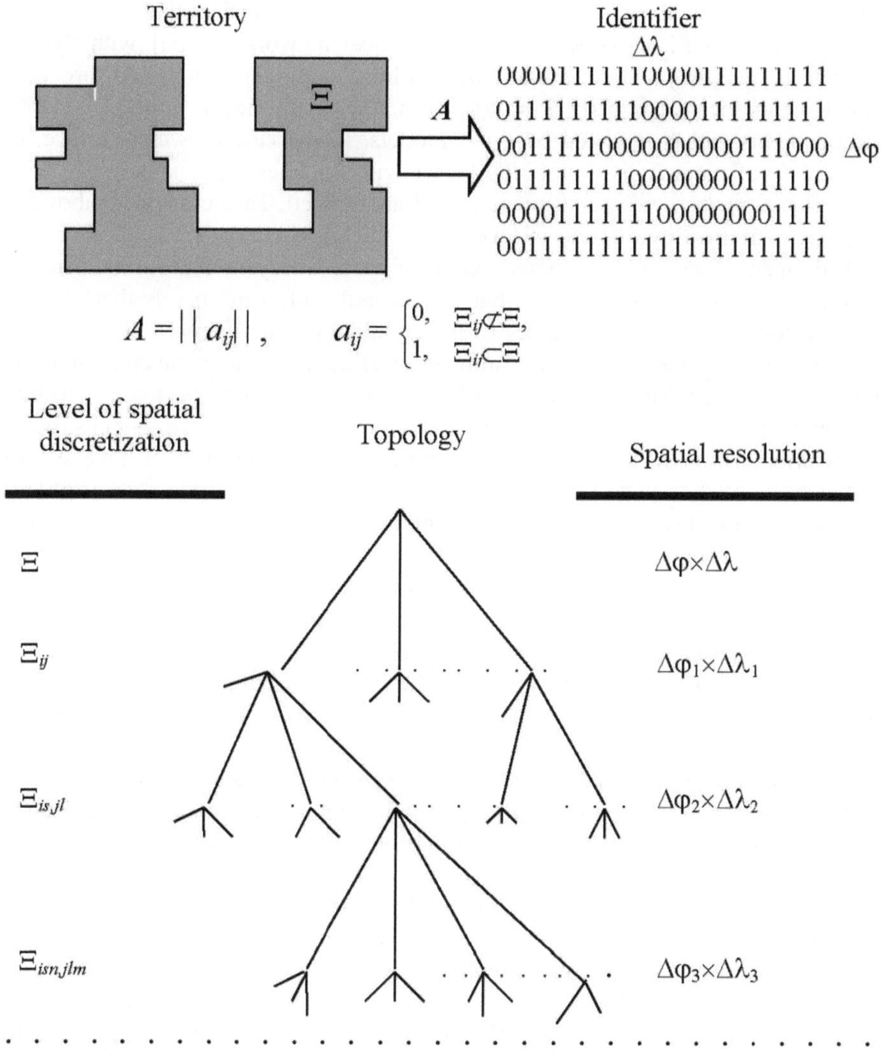

Fig. 1.3 Procedure of topological coordination of environmental subsystem structure with the GIMS database

1.2.2 Simulation Experiment Based on the Global Information-Modeling System

Application of the GIMS technology for the solution of specific environmental tasks demands development of required models and use of proper algorithms for the evaluation of parameters for these models. Realization of simulation experiment by means of the GIMS starts from isolation of area Ξ, where the studied environmental subsystem is functioning. It is realized by means of the procedure that is depicted in Fig. 1.3. The area Ξ is covered by geographical network $\{\varphi_i, \lambda_j\}$ with the discretization steps $\Delta\varphi_i$ and $\Delta\lambda$ by latitude and longitude, respectively. All processes and elements located on pixel $\Xi_{ij} = \{(\varphi, \lambda): \varphi_i \leq \varphi \leq \varphi_i + \Delta\varphi_i; \lambda_j \leq \lambda \leq \lambda_j + \Delta\lambda_j\}$ is homogeneous and is modeled by point models. In the case of water surface, the water body of pixel Ξ_{ij} is divided by layers with thickness Δz_k, namely 3D pixels $\Xi_{ijk} = \{(\varphi, \lambda, z): (\varphi, \lambda) \in \Xi_{ij}, z_k \leq z \leq z_k + \Delta z_k\}$ are defined. The atmosphere above the pixel Ξ_{ij} is divided by altitude in levels with thickness Δh_s.

The interactions in the studied environmental subsystem are considered as interactions between its elements that are located within the pixels that are base network for the numerical schemes of the simulation experiment.

The pixels Ξ_{ij} and Ξ_{ijk} are not uniform considering their parameters and their functional characteristics. In other words, the model binding to database is realized through an heterogeneity. Moreover, to avoid the redundancy in the model structure it is a priori supposed that all elements and processes of territory Ξ have specific spatial discretization. The variability of the spatial discretizations within different items of the model is neutralized in algorithmic level by coordinating the data flows from the monitoring system. As a result, the model structure is irrespective from the database structure and therefore it does not change when database structure is changed. Similar independence is realized between the model items. It is achieved by means of basic information pathway that controls the data exchange between items through inputs and outputs only. In the case of disconnection of some items, their outputs are replaced by the available database outputs. It is schematically shown in the Fig. 1.4.

The user forms a spatial image of the environmental subsystem and a management regime of the simulation experiment that is supported by the existing database and the already structured knowledge base. As a result, the formalized structure of the algorithmic and modelled procedures is generated, in such a manner as to be oriented to the climatic zone and the socio-economic conditions of its operation. This structure has features of intellectual information system that help to realize the simulation experiments without additional processing of algorithms and models. The user may choose algorithms and models and their modifications from the existing knowledge base of GIMS reaching the required level of adequacy in the description of the studied environmental subsystem.

The basic principles of the GIMS technology are:

(1) Integration, unification and coordination of the existing environmental monitoring systems based on the unique organizational and science-methodic base.
(2) Optimization of equipment and economic resources for synthesis of the environmental monitoring systems, their functioning and modifications.
(3) Coordination and compatibility of information fluxes in the monitoring system by using the unique coordinate-time system, common system for classification, coding, formats and data structure.

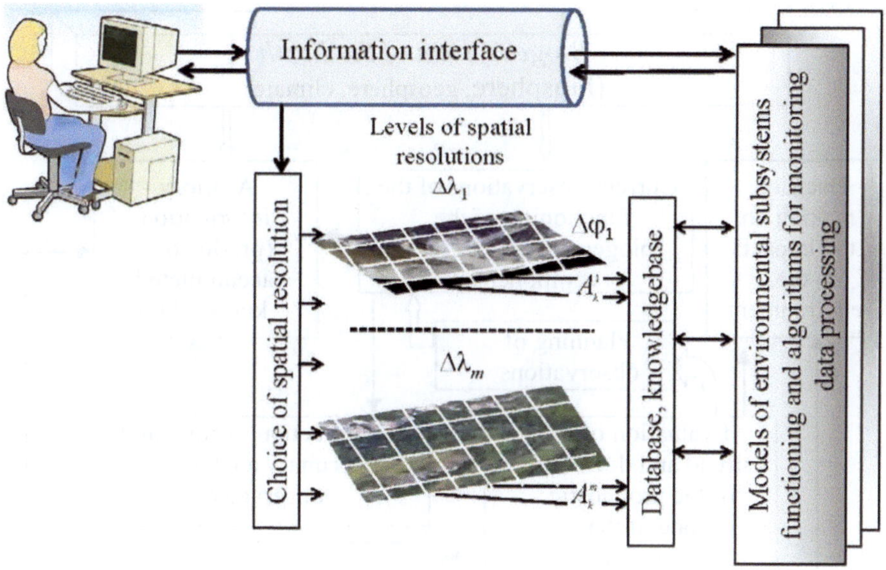

Fig. 1.4 Scheme for simulation experiment management within the dialog regime with structural formation of models and algorithms. *Notations* $\Delta\varphi_i$ and $\Delta\lambda_j$ are dimensions of geographical grid by latitude and longitude on *i*th level, respectively; A_k^i are the semantic identifiers supporting the formation of spatial and subject image of *k*th modelled environment on *i*th level of spatial discretization

(4) Focus of the admittance to information through international informational networks with maximum extension of user service.
(5) Independence of the environmental monitoring by the disparity of state borders and ecosystem boundaries.

Realization of these principles provides simulation modeling experiments of different spatial scales, i.e., from local to global. In this case, the functioning of GIMS is presented in Fig. 1.5. This scheme depicts the support of the simulation experiment in its adaptive regime, when it is regularly performing the model modernization and the monitoring system structure management. Thus, the GIMS technology optimizes the information fluxes to achieve effective result from all possible sensors. The latter is schematically explained in Fig. 1.6.

Thus, the GIMS technology considers any environmental subsystem as an element of nature interacting through biospheric, climatic, and socio-economic processes. A model is then created describing these interactions and the functioning of various levels of the space-time hierarchy of the whole combination of processes in the subsystem. The model encompasses characteristic features for typical elements of the natural and anthropogenic processes and its development is based on the existing information base. The model structure is directed towards the adaptive regime of its use (Fig. 1.7).

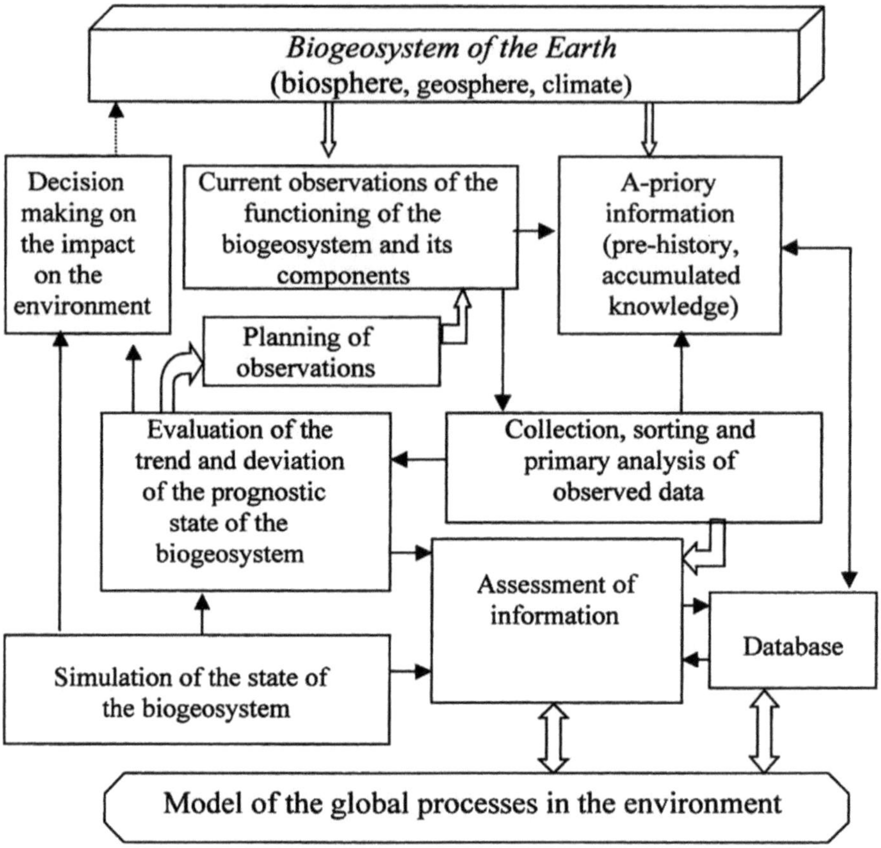

Fig. 1.5 Concept of environmental monitoring that is based on the GIMS technology

Fig. 1.6 Optimal integration of environmental data delivered by different monitoring systems and other data sources

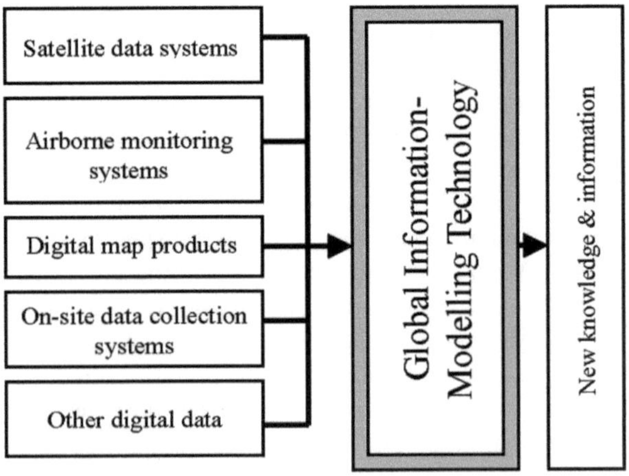

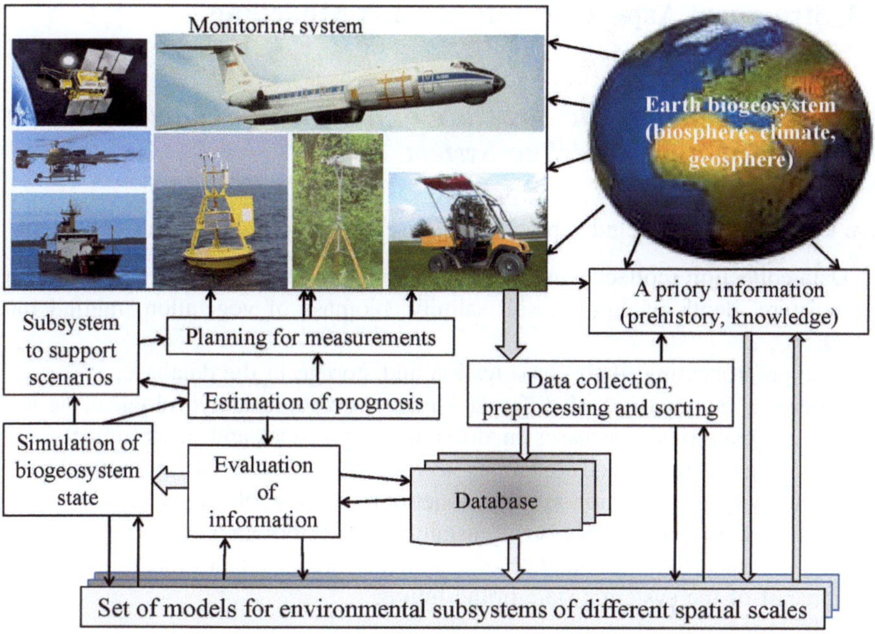

Fig. 1.7 Conceptual block-diagram of geoinformation monitoring using GIMS technology

The combination of the environmental information acquisition system, the model of the typical geoecosystem functioning, the computer cartography system and the tools of artificial intelligence will result in the building of the geoinformation monitoring system of the typical element, capable of implementing the following tasks:

- evaluation of the global change effects on the environment of the typical element of the nature/society system;
- evaluation of the role of environmental change occurring in the typical element due to climatic and biospheric changes on the Earth and its territories;
- evaluation of the environmental state of the atmosphere, hydrosphere and soil-plant;
- formation and renewal of information structures based on ecological, climatic, demographic and economic parameters;
- operative cartography of the situation of the landscape;
- forecast of the ecological consequences from the implementation of anthropogenic scenarios;
- designation of land covers, natural phenomena, populated landscapes, surface contaminations of landscapes, hydrological systems and forests;
- safety evaluation of population.

The construction of the GIMS is associated with the biosphere, climate and social medium which are characterized by a given level of spatial hierarchy.

1.3 Structural Aspects of Information-Modelling Technology

1.3.1 Description of the Structure of a Common Global Information-Modeling System

The GIMS-structure includes blocks which serve the following main operations:

- Data collection (collects current information about the system soil/canopy: soil moisture, depth of aquifer, soil salinity, biomass of vegetation, rainfall rate, others);
- Data preprocessing, their classification and storage in the database;
- Modeling (simulation) of different kinds of ecological, hydrological, agricultural, climatologic processes in different geophysical and environmental systems. These are blocks containing a variety of models of agricultural crops productivity, of irrigation systems functioning, geo-ecology and epidemiological models of certain deceases transmitted by vectors, and others;
- Assessment of the current state of a particular geophysical system;
- Forecast of the system's state in the future;
- A feedback support.

The data collection block consists of two main sub-blocks, namely:

- a sub-block for collecting prior knowledge-based information;
- a sub-block for collecting current data of the in situ measurements and the remote observations from mobile, aircraft and satellite platforms.

The first of these two sub-blocks plays a very important role in GIMS by constituting a model of the observed geo-ecosystem. It includes the relationships describing different geo-ecosystems functioning as well as the relationships between the geo-ecosystem and the surrounding media.

The second sub-block consists of three boxes providing the following operations:

- In situ measurements that include sampling (probes) and utilizing current information accessible from the meteorological stations and from other accessible sources.
- Remote observations from the aircraft (piloted and unmanned). There are different types of sensors used for remote sensing observations. Among them are the sensors operating in optical, infrared, and microwave bands of electromagnetic wavelengths. The data measured by all of them can potentially be used in GIMS. In the majority of cases we have considered in this box the data collected by passive microwave radiometers and partly by the infrared radiometers have been used. These sensors provide the measurements of the following parameters:

(a) soil moisture,
(b) depth to a shallow aquifer,
(c) biomass of vegetation above a water surface (rice crop, wetland vegetation) or above a wet ground,
(d) the temperature assessments of a water surface, land and dense vegetation canopy.

- Satellite observations of the Earth/Atmosphere/Ocean/Biosphere. Optical sensors provide crop type and land use classification with a ground resolution of about 1–50 m. Infrared sensors provide measurements of the surface temperature of land, water and upper boundary of vegetative canopy with a ground resolution of 10–100 m. Microwave radiometers provide estimates of soil moisture, general condition of vegetation, temperature variations of ocean surface and of the surface of dry land with a spatial resolution of around 20–50 km.

Algorithmic and programming maintenance of the GIMS implement the following functions:

- Parameterization of the current and prognostic weather and climatic conditions based on the in situ and remote observations and by running the required models.
- Providing the cluster analysis.
- Modeling of a number of agricultural crops productivity.
- Modeling of typical irrigation systems functioning.
- Collection of in situ, aircraft and spacecraft observations with adequate classification and storage capabilities.
- Estimates of trends of the developed geo-systems and evaluation of the discrepancies between the measured data and model predictions.
- Current service in the data flow control and parameter assessments.
- Data processing (interpolation, extrapolation) using different methods.
- Estimates of soil moisture profile and a total water content in 1-m soil layer using the data of mobile and aircraft observations at microwaves and a prior knowledge-based information about the soil.
- Modeling of the soil concentration variations.
- Modeling of the humus characteristics variations.
- Modeling of the soil/canopy system growth for 30 plant types.
- Mapping of the modeling results, measurements and inter/extrapolation.
- Visualization of the data in the form of Figures, Graphics, Tables, etc.

There are other computer programs and models used for global change studies, e.g., the model of carbon circulation, the model of oxygen circulation and others (Krapivin and Varotsos 2007, 2008).

As mentioned above, the construction of the GIMS is connected with consideration of the biosphere, climate and social medium that are characterized by the given level of spatial hierarchy. Table 1.1 gives general description of the items of the GIMS technology.

Table 1.1 Basic items of the GIMS technology

Item	Item function
PADAS	Planning and analysis of the data acquisition systems
IDPDA	Initial data processing and data acquisition
CM	Computer mapping
ESA	Evaluation of the state of the atmosphere
ESSPC	Evaluation of the state of the soil-plant cover
ESWM	Evaluation of the state of the water medium
REES	Risk evaluation of ecological safety
ICESD	Identification of causes of ecological and sanitary disturbances
IS	Intelligent support

The item PADAS implements the task of the experimental planning by performing the analysis of the structure of the environmental data acquisition system, making use of data derived by satellites, flying laboratories, and movable and stationary ground observation facilities. The laboratories are equipped with the necessary software and hardware tools to allow determination of the degree of the environmental contamination, of the ecological situation, mapping of the characteristic geological formations, detection of soil subsurface centers of ecological defects, performing the all-weather land-cover classification and detection of permafrost disturbances, oil spills, forest states and pollution of bodies of water.

The item IDPDA accomplishes methods and algorithms for synchronous analysis of aero-space information and ground measurements using space-time interpolation procedures. Retrieval of the data and their reduction to the conventional time scale is performed. Model parameters are determined. Thematic classification of the data is carried out, in order to perform space-time combination of images in the optical, IR and microwave ranges and of trace measurements obtained from devices of various types.

The algorithms that are developed in the item CM are suitable for the drawing of computer maps with characteristic marks for evaluating the ecological situation. Multilevel scaling and fragmentation of the territory is envisaged. The output maps with the information required by the user is provided through the user interface.

Models of atmospheric pollution developed due to evaporation and burning of oil products, natural gas and other outputs of industrial enterprises are employed within the item ESA. In particular, the problem of the evaluation of the atmosphere dust content is solved. The gas and aerosol composition of the near-earth atmospheric layer are provided and forecasting maps of their distribution over the earth's surface are drawn.

Item ESSPC focuses on the following tasks:

- classification of the floristic background taking into account the microrelief, soil type and its salinity, humidification and the degree of the soil brine mineralization;
- disclosure of the micro- and macro-relief peculiarities and subsurface anomalies;

- determination of the structural topology of the land cover;
- emergence of forests, swamps, agricultural crops and pastures.

In the item ESWMA the complex simulation model of the territory is formulated taking into account the seasonal changes of surface and river runoff, the influence of snow cover and permafrost and the regime of precipitation and evapotranspiration. A model is developed by employing the water quality dynamics for the hydrologic network of the territory.

The item REES accomplishes the algorithms for the evaluation of the damage to nature, economic stability and population health, depending on the environmental changes that are connected with the natural trends of meteorological, biogeochemical, biogeocenotic, micro-biologic, radiologic and other natural processes as well as the enhancement of the manmade environmental stress.

The task of revealing the sources of environmental pollution is implemented in the item ICESD. This subsystem determines the source coordinates, the magnitude and the possible time of the non-planned incoming of polluting substances. The dynamic characteristics of the pollution sources are herein given. Additionally, the a priori unknown pollution sources are revealed and the directions of possible trans-boundary transport of pollutants are determined.

The software-mathematical algorithms are formulating in the item IS in order to provide the user with intelligent support in performing the complex analysis of objective information formed in the framework of the simulation experiment. The necessary information for the objective dialogue with the global model of the nature/society system model is provided in a convenient form for the user. The input of the data processing corrections is also provided. The knowledge base of the anthropogenic, demographic and socio-economic processes on the territory is formed.

1.3.2 The Subsystems of the Global Information-Modeling System

The items PADAS and IDPDA ensure the fulfillment of such GIMS technology functions as systematic collection, generalization, storage, utilization and distribution of information about the environmental parameters in the form that is adapted to the users. The technical realization of these functions is implemented by using fiber-optical sensors and measuring systems (Burkov and Krapivin 2009). The use of the optical fiber technology for data transmission within the environmental monitoring systems improves their reliability and the survivability of the information-measuring systems. These effects are achieved as a result of the optical fiber measuring systems that have not interference effects and their functioning does no depend on the electromagnetic radiation and other physical fields. Under this, synchronization of the ground fiber optical sensors with the satellite microwave radiometers is achieved without noises deduced from the electromagnetic fields, high temperatures and aggressive mediums (Burkov et al. 2008).

As mentioned above the item IDPDA analyses the environmental information and in that way helps to assess real state of the nature/society subsystems within concrete space-temporal boundaries. Here it is necessary to note that measurements conducted by satellites or flying laboratories deliver small volumes of statistically reliable samples. Moreover, these data are usually non-stationary. The formation of the input information in the GIMS based on monitoring data that received episodically in time and fragmentarily in space is a problem, which can be solved by using specific algorithms that are included in the item IDPDA. These algorithms are described in Chap. 2. Choice of one of them is realized automatically by item IDPDA depending on the type of treated task. Large-scale tasks of cartographic representation of monitoring data are carried out using spline or differential interpolation. In the case, when un-removable information uncertainty exists, then the evolutionary modeling method is employed.

The method of differential approximation has universal character and as unique method of the theory of approximation functions is often used for the analysis of dynamic information. In the case when the task of data reconstruction within the inter-trass space is performed, the method of differential approximation reduces this task to that of the measurement data reduction to unique time. Moreover, coupled use of the differential approximation and spline-approximation methods raises the precision of task solution when there exist cross-trasses of satellites and flying laboratories.

The study of the natural objects and processes using remote sensing methods has promoted the evolutionary technology paradigm as a knowledge technology directed towards the restored adaptation (Bukatova et al. 1991). This paradigm lies in the GIMS technology base. Succession for this paradigm needs the improvement of "super-intellect" taking into consideration of un-removable information uncertainty that is practically always present in the global ecoinformatics. The evolutionary modeling is such a super-intellect. It is a method of adaptive structural identification of objects, based on the synthesis of structured models using the simulation mechanisms of natural evolution and their self-organization.

The evolutionary intellectual technology creates the model for nature-society subsystem that appears an unexpected temporal change. This model guarantees the elimination of information uncertainty at any time of monitoring. Figure 1.8 illustrates the main scheme for this technology. In the usual case, there are two processes that are permanently alternated—the process of structural adaptation and the process of usage. The arbitrary stage of adaptation is characterized by synthesis of models that are used for forecasting, interpolation and other operations with fragmentary data. The most efficient model is the choice from stage to stage of adaptation. The procedure of evolutionary selection between models provides the GIMS functioning that is practically unlimited in time when an un-removable informational uncertainty is present.

The item CM performs informational filling of the final maps according to user's requirements and prepares operative informational bulletins following the changing situations at all spatial scales. Cartographic identification of the environmental objects is realized in accordance to the schemes depicted in Figs. 1.2 and 1.3.

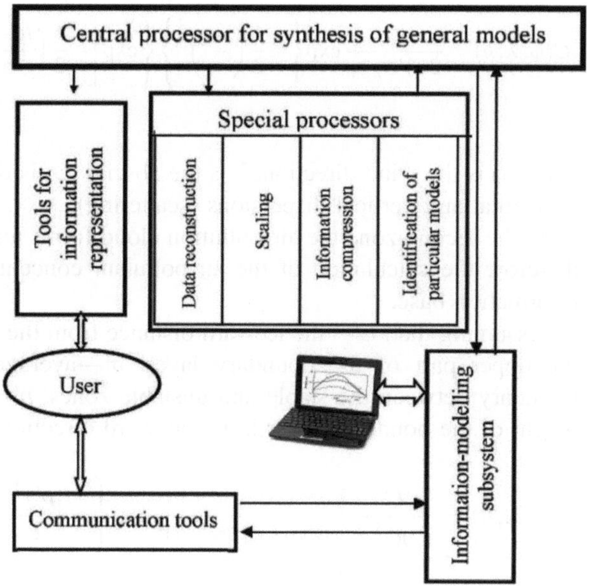

Fig. 1.8 Principal structural scheme of evolutionary technology use in the information-modeling monitoring system

The GIMS using mainly the item ESA develops models that describe processes of the air-pollution spread due to different anthropogenic activity. The change in the concentration of every atmospheric pollutant C is described by the following equation:

$$\partial C(t, \varphi, \lambda, h)/_{\partial t} + \nabla \cdot \vec{V} C = \nabla D \cdot \nabla C + R \tag{1.1}$$

where $\vec{V}\left(V_{\varphi}, V_{\lambda}, V_h\right)$ is the wind velocity, φ and λ are latitude and longitude, respectively; h is the altitude, t is the time, D is the molecular diffusion coefficient, R is the changes due to atmospheric turbulence, emission and mixing.

The item ESA refers to model of Gauss-type as an essential procedure for the dispersion of the atmospheric pollutants. This model requires a preliminary input information about the height of pollution source and dispersion characteristics within the pollutant cloud. The different approximations of the Gauss stream in the neighborhood of the pollutant sources depend on the stability parameters of the atmospheric surface layer. The model configuration depends on geographical coordinates. For example, clusters with stable and unstable air-pollutant fluxes are considered primarily within the boundaries between the areas land and water. With this approach, three specific zones can be selected: the zone with undisturbed dispersion, the precipitation zone and the delay zone. Each zone has specific equations for the parameterization of the air-pollution dispersion.

In the zone with undisturbed dispersion, the pattern of distribution of the air-pollutants fluxes is influenced by an homogeneous and stable layer, i.e.,:

$$C(\varphi,\lambda,h)=\frac{Q}{2\pi V\sigma_{\lambda}\sigma_{h}}\exp\left\{-\frac{1}{2}\left(\frac{\lambda}{\sigma_{\lambda}}\right)^{2}\right\}\left\{\exp\left[-\frac{1}{2}\left(\frac{h-Z}{\sigma_{h}}\right)^{2}\right]+\exp\left[-\frac{1}{2}\left(\frac{h+Z}{\sigma_{h}}\right)\right]\right\} \quad (1.2)$$

where φ is the wind direction, λ is the abscissa, h is the ordinate, σ_{φ} and σ_{p} are the horizontal and vertical dispersions (scattering).

In the second zone the air-pollution cloud forms under instability conditions and therefore the calculation of the air-pollutant concentrations requires a broadened information base.

Assuming that: φ'—the leeward distance from the intersection of the cloud with the upper part of the boundary layer, σ'—average value of scattering at the boundary between the stable and unstable zones, $p' = [L(\varphi) - Z]/\sigma_h$, $L(\varphi)$—the height of the boundary layer in the leeward direction φ, then:

$$\begin{aligned}C(\varphi,\lambda,h)=&\frac{Q}{(2\pi)^{3/2}}\int_{0}^{\varphi}\frac{1}{\sigma'\sigma_{h}(V,\varphi,\varphi')V}\exp\left\{-\frac{(p')^{2}}{2}-\frac{1}{2}\left(\frac{\lambda}{\sigma'}\right)^{2}\right\}\\&\times\left\{\exp\left[\frac{-(h-Z)^{2}}{2\sigma_{h}^{2}(V,\varphi,\varphi')}\right]+\exp\left[-\frac{(h+Z)^{2}}{2\sigma_{h}^{2}(V,\varphi,\varphi')}\right]\right\}\frac{\partial p'}{\partial\varphi'}d\varphi'\end{aligned} \quad (1.3)$$

The third zone is quite specific enough due to the conditions of cloud formation. This cloud is completely inside the atmospheric boundary layer (ABL), and the concentration of pollutants within it can be calculated with the formula:

$$\begin{aligned}C(\varphi,\lambda,h)=&\frac{Q}{2\pi V\sigma_{\lambda}\sigma_{h}}\exp\left[-\frac{1}{2}\left(\frac{\lambda}{\sigma_{h}}\right)^{2}\right]\\&\times\left\{\exp\left[-\frac{1}{2}\left(\frac{h-Z}{\sigma_{h}}\right)^{2}\right]+\exp\left[-\frac{1}{2}\left(\frac{h+Z}{\sigma_{h}}\right)^{2}\right]\right\}\end{aligned} \quad (1.4)$$

for $\sigma_h < 0.47L(\varphi)$, and

$$C(\varphi,\lambda,h)=\frac{Q}{\pi V\sigma_{\lambda}L(\varphi)}\exp\left[-\frac{1}{2}\left(\frac{\lambda}{\sigma_{\lambda}}\right)^{2}\right] \quad (1.5)$$

for $\sigma_h > 0.8L(\varphi)$. Within the interval $0.47 \leq L(\varphi) \leq 0.8L(\varphi)$ the concentration C is calculated with the interpolation formula between the points $\sigma_h = 0.47L(\varphi)$ and $\sigma_h = 0.8L(\varphi)$.

The Gauss-type model describes well the fields of air-pollutants over territories at scales up to 100 km. For large territories, other types of models are used. For example, Eulerian and Lagrangian models are used when Gaussian model appears high errors. The item ESA employs these models depending on the territory area and taking into account the availability of monitoring data.

```
  W   150   120    90    60    30     0     30    60    90   120   150    E
N ..................****. *********.......................................... N
 ..................AAA. . *********......**...............A...............
 ............AAA. AAAA.....********..............C......AAAAA.....A........
 ....CCCC.....CCA. A. AAAC...*****.........LLL.....C. CCCCCMMMMCAAAAAAAA.....
 MC. FFFMFMFFFFCCCCCAA. . MM. . ***. . . L......DDDFF. FFFFCLLLFFFMFFFFFMAAAMMMMMMM
 ....LLFFFFCCCCCCCC...C.....L..........RG. DDDDDDDDDFFFFDDDDDDDDDDMMFFMCCM. .
 .....L...FFFFGGGFF...CFC...........+...R. RGGGGGGGDGGGGGGGGDDDDDDM...L....
 ...L......+RGGGGGDDF. FFFF..........++. ++RRRRXXXXGXXXXGGGGGGDDGDDD...L....
50...........RRXWXXGGDDDDD. D............+++++XXWWVVVVVVVGGGGGGGGGDDD........50
 ............RXWWX++. +++G............+++++XXWWV. S. @@QQSGVVVVGXRR..........
 ............BBBXXX++++..............++. . P++. . WQ. KS&QQQSSSSSSXX++. . +.......
 ............RBUWXP+PP..............PP....P. WWWW&K&&QQSSSSSQXX. +. . P........
 .............@UUUPPPP..............&&&&.....&&WUUUUUQBBQQ+++...P.........
30.............@@@UP. . P.............HHHHHHHHHHHHHHTHHHJPQQQQPPPP............30
 ..............@@@..................HHHHHHHHH. HH. TTTHNNYZZPPPP.............
 ...............U@...J............TTTTTTTTTTTTTTT...NNI. YYYI..............
 ................YYZ...J..........TTTTTTTTTTT. TT....JN. . YYY...............
 ..................YZ.............JJJJJTJJJNTT....NN...YY. . Y............
10...................Z. JJ..........JJJJJJJINNTT.....N....Y...............10
 ....................ZJJZ..........YJZYJJJJJT..........Z................
 ....................ZZZZJ............ZYYYYJNN..........Z. . Z.............
 ...................QZZZZJZZ..........YYZYJJJ............Z. Z.............
 ...................WZZZZJJJJJ.........JYYNNJ...................ZZ.......
10....................ZZZZJJJJ..........JNNNN......................Z......10
 ....................WWZZZJJY..........JNNNN. . Z...............NN.........
 .....................WWNZYJY..........JJJNJ. J...............NNNNN.......
 .....................TNNYYY...........TJJJJ. N.............JJTTJJJ.......
 .....................TNNYP.............HHJE................JJTTTTNN......
30.....................UUEY..............HEE.................HHTTTJJ......30
 .....................UUEI..............E..................U...UUP......
 .....................UEE...........................................P.
 ....................+VV..................................P....+. .
 ....................+V...........................................
50....................+E...........................................50
 .....................+...........................................
S ................................................................... S
  W   150   120    90    60    30     0     30    60    90   120   150    E
```

Fig. 1.9 Spatial distribution of soil-plant formations in the biosphere in geographical zones of 4° × 5°. The list of symbols and quantitative characteristics of the types of land vegetation formations which are given in Tables 1.2 and 1.3

The GIMS structure includes tools for the measurement of the land cover characteristics, algorithms and software, as well as technologies for the synthesis of the soil-plant formation images based on the monitoring data and other information containing in the environmental databases. The GIMS database has series of spatial images of the soil-plant covers that were formed in the framework of different international and national environmental programs. Figures 1.9, 1.10 and 1.11 show various examples of spatial images for soil-plant formations.

Table 1.2 Identifiers adopted in the GIMS for designation of the soil-plant formations: σ is the area of the soil-plant formation (10^6 km^2) and P is the production of soil-plant formation (kg C m^{-2} year^{-1})

Type of the soil-plant formation	Identifier
Arctic deserts and tundras	A
Alpine deserts	B
Tundras	C
Mid-taiga forests	D
Pampas and grass savannas	E
North-taiga forests	F
South-taiga forests	F
Sub-tropical deserts	G
Sub-tropical and tropical grass-tree thickets of the tugai type	I
Tropical savannas	J
Saline lands	K
Forest-tundra	L
Mountain tundra	M
Tropical xerophytic open woodlands	N
Aspen-birch sub-taiga forests	O
Sub-tropical broad-leaved and coniferous forests	P
Alpine and sub-alpine meadows	Q
Broad-leaved coniferous forests	R
Sub-boreal and saltwort deserts	S
Tropical deserts	T
Xerophytic open woodlands and shrubs	U
Dry steppes	V
Moderately arid and arid (mountain including) steppes	W
Forest-steppes (meadow steppes)	X
Variably-humid deciduous tropical forests	Y
Humid evergreen tropical forests	Z
Broad-leaved forests	+
Sub-tropical semi-deserts	&
Sub-boreal and wormwood deserts	@
Mangrove forests	#
Lack of vegetation	*

The item ESSPC carries out the following tasks:

- classification of floristic treatments with consideration of micro-relief, soil type and its salinity, moist and level of the soil brine mineralization;
- identification of features of the anomalies in micro- and macro-relief, and subsurface;
- definition of the topological structure of the land covers; and
- indication of the state of forests, marshes, agricultural fields, pasture lands, and other biomes.

Table 1.3 Quantitative characteristics of the types of the land vegetation formations

Type of vegetation formation (see Table 1.2 for explanations)	Formation area (10^6 km^2)	Annual production (kg/m^2/year)	Phytomass (kg/m^2)	Dead organic matter (kg/m^2)
A	2.55	0.17	0.4	1.3
C	2.93	0.36	1.9	5.6
M	2.33	0.38	1.9	5.5
L	1.55	0.65	3.8	9.0
F	5.45	0.54	10.0	8.1
D	5.73	0.63	22.5	10.8
G	6.60	0.65	23.5	14.5
R	2.12	0.87	25.0	25.1
+	7.21	1.25	45.0	24.8
P	5.75	1.72	43.0	22.2
U	3.91	0.56	3.8	15.0
X	3.72	0.74	1.9	38.0
W	4.29	0.79	1.9	33.0
E	1.66	1.11	3.8	21.0
V	2.66	0.38	0.8	12.6
#	2.08	0.45	0.4	12.1
S	2.69	0.25	0.2	8.1
&	1.99	0.35	0.8	8.8
H	7.16	0.12	0.1	1.4
B	1.15	0.47	0.8	16.9
Q	3.54	0.76	1.9	24.0
Z	10.4	3.17	60.0	21.6
Y	7.81	2.46	60.0	20.5
N	9.18	1.42	10.0	15.1
J	17.1	1.35	0.1	2.0
T	11.5	0.18	0.4	4.9
K	0.38	0.18	45.0	21.6
I	0.9	1.96	45.0	21.6
*	14.6	0	0	0

Bazilevich and Rodin (1987)

A main subsystem of the GIMS is the complex simulation model of the regional water balance that takes into consideration the changeability in surface and river runoffs, impacts of snow cover, precipitation regime and evapotranspiration. Figures 1.12, 1.13 and 1.14 illustrate schematic structures that are realized in the item ESWM of GIMS. The model of water quantity is synthesized properly to be a tool for assessing the dynamic characteristics of hydrologic regime taking into

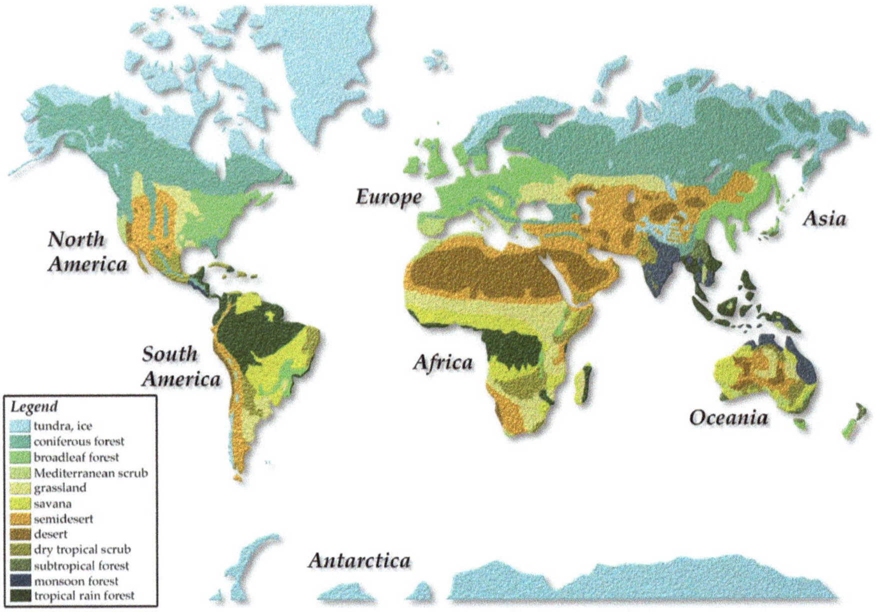

Fig. 1.10 Map of major types of the land covers (http://landcovers.usgs.gov/classes.php)

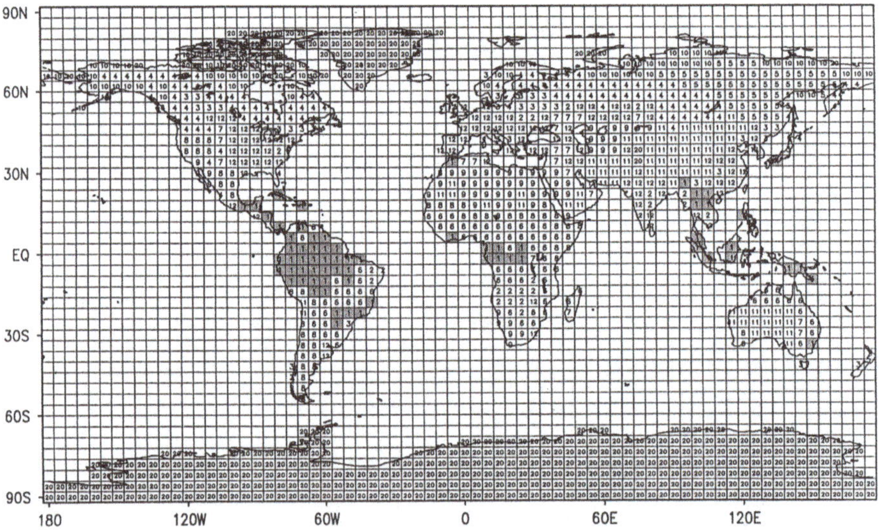

Fig. 1.11 Fragment of the GIMS database that characterizes global distribution of land cover types. *Notation 1* broad-leaved evergreen trees (tropical humid forest), *2* deciduous broad-leaved trees, *3* mixed deciduous broad-leaved and coniferous evergreen trees, *4* coniferous evergreen trees, *5* coniferous deciduous trees, *6* grass, *7* meadows, *8* broad-leaved brushes that are located separately or as moderate tangles, *9* broad-leaved deciduous brushes that are located separately or on bare soil, *10* tundra, *11* desert (bare soil), *12* tillage for winter wheat cultivation only, *20* long-time or glacier ice

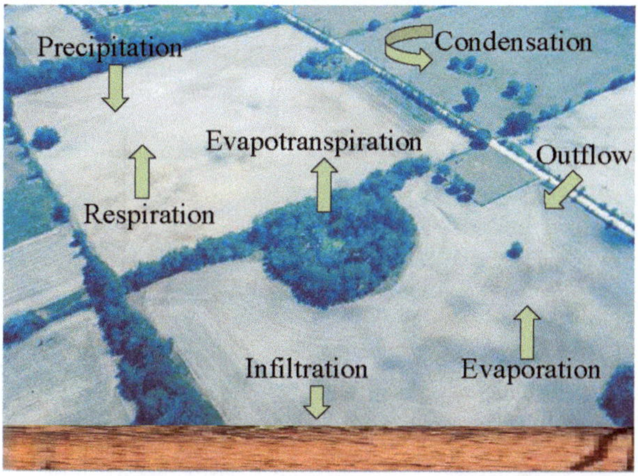

Fig. 1.12 Components of the water balance for a restricted region

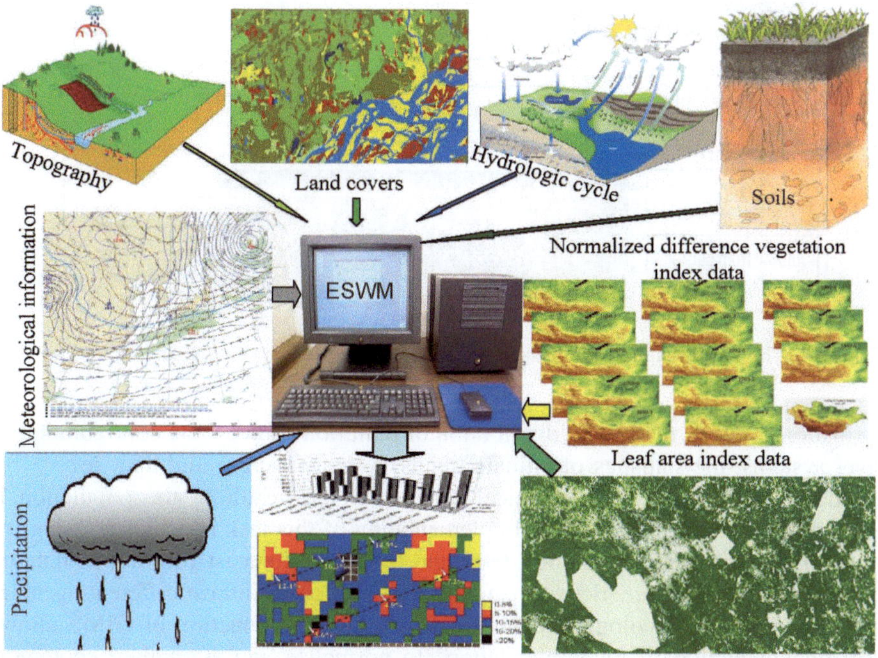

Fig. 1.13 A few parts of the information base of the regional water balance model

consideration all the possible sources of water pollution. The ESWM component provides interim information on other items that parameterize the functions of the soil-vegetation systems and detects disturbances in ecological and sanitary conditions. The initial information about the item ESWM consists of data on soil-plant

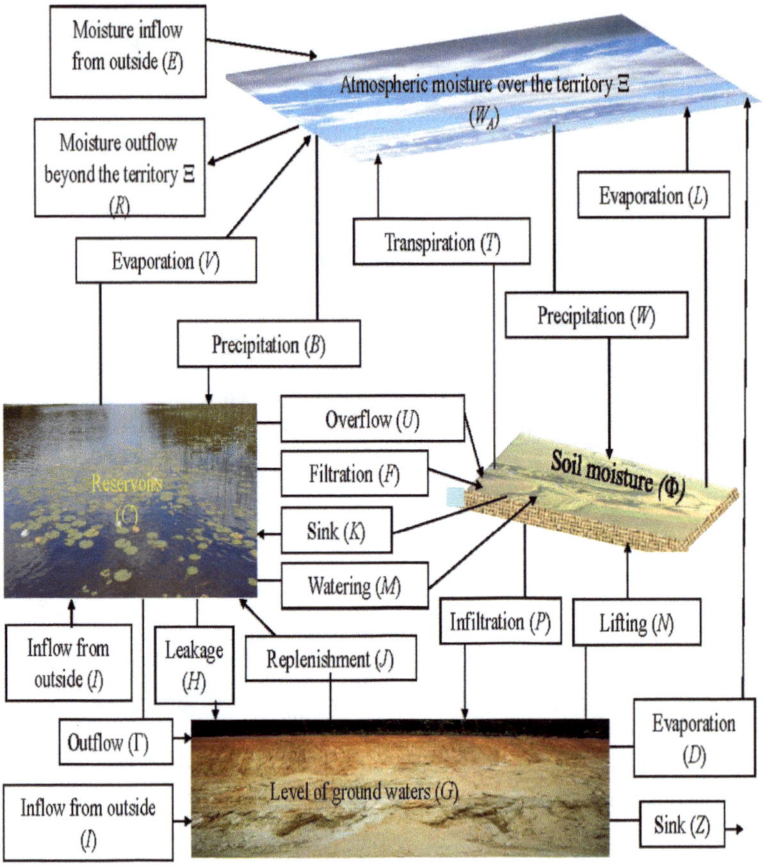

Fig. 1.14 The block-diagram of the sample model of water balance in a restricted territory

formations, reliefs, the spatial distribution of pollution sources, the configuration of river systems, the statistics of climate trends, water resources and their use.

Risk evaluation of the ecological safety and the health of the population is important function of the GIMS. Implementation of this task is achieved using the item REES which allow for short-term and long-term forecasts, to deliver data on environmental characteristics to management services, to inform different agencies about disasters and ecologically dangerous events. This function of GIMS allows the implementation of recommendations to normalize and strengthen ecological situations, and the perfection of ecological safety systems. It develops the algorithms that help to assess the environmental damage, economic stability shock, and the deterioration of population health depending on the regional natural and man-made changes of meteorological trends, biogeochemical, microbiological and radiological processes. Indicators of biocomplexity, biodiversity, survivability, and bioavailability are assessed along with it.

An important structural feature of the GIMS is that it is constituted by the global simulation model for nature/society system that controls information fluxes between global databases and the GIMS database protected from the exaggeration and uncertainty at the same time. A truly global model plays the role of context information for the GIMS workspace. Hierarchical structure of this workspace is determined by series of directories and files stored in the configuration that is identical to structure shown in Figs. 1.3 and 1.4. Subsystem of intelligent support helps user to overcome isolated virtually unsolved questions displaying all the questionable features of the studied problem. One of such unsolved question is the choice of an adequate model. Choice of the model type done at the first stage of the GIMS synthesis is often a poorly formalized procedure. In this case, it is required to run the following stages:

- primary informal description of the monitoring object;
- formulation of the objectives for investigation;
- substantial analysis of the a priori information about the environmental subsystem;
- development of the mathematical (formalized) model for the subsystem functioning;
- set-up of the modelling algorithm;
- synthesis of software package for the model run in a computer;
- model verification and evaluation of its parameters;
- choice of the alternative versions of the model to adapt to the structure of the GIMS database;
- elaboration of scenarios on simulation experiments that reflect the objectives of the research model; and
- utilization of the model in an operational monitoring system.

Synthesis of the model as part of the GIMS item requires setting boundaries of the object monitoring. This stage includes this model and models of more high levels that parameterize the environment beyond those boundaries. The number of the model levels depends on the spatial resolution of the monitoring system. The model must be equipped with the following functions:

1. The measuring function that uses similarity of the model and the object to be modeled.
2. The descriptive function that characterizes the object features in its various states.
3. The interpretative function that describes the power of the model adaptability and the applications of its results.
4. The explanatory function which represents the model ability for the interpretation of the monitoring data in accordance with the terms of formal tools of mathematical methods used in the model.
5. The prognostic function which is associated with the probability to forecast the evolution of the environmental subsystem with given precision and for conditions under which it is not observed.

6. The criterion function of the model, which allows the verification of the correctness of the environmental knowledge and suggests the ability to model the experiment for the purpose of receiving new information.

1.3.3 The Global Information-Modeling System-Based on Monitoring Systems

Structural filling of the real information-modeling system depends on several circumstances. The owner of the system may have requirements that are not performed as part of the basic version of GIMS. In this case the additional items are synthesized and basic version of the GIMS is built by connecting the new items to the central information line. Some experience of such procedures is presented in Table 1.4 that features a series of regional information-modeling systems (RIMS) synthesized during the last time.

The composition of real GIMS starts with analysing all the data on the monitoring subsystem. Usually these data are used to develop the methodology of sharing the remotely sensed data, the in situ measurements and information permanently collected from the existing meteorological stations which provided non-stop measurements of precipitation, soil moisture, soil and atmospheric temperature at different levels, and other data. In the case of RIMS-B, the availability of all this instrumentation and corresponding data allows the development of a management technology for non-stop monitoring of soil moisture and moisture related parameters, such as dryness index, risk of water shortage in local areas, drought, and others.

As seen from Table 1.14 the GIMS technology has been successfully tested and applied in many regions. It has shown that the advantages of applying GIMS provide firstly the following capabilities:

- to design the in-situ and remote parts of the experiment optimizing customer requirements for data quality and cost;
- to simulate the actual measurements along the flight line in the form of maps, with the collection of the in-situ measurements and remote observations not using scanning technologies, which is extremely cost-effective;
- to forecast/predict the behavior of the examined environmental/geophysical subsystems; and
- to expand the scale of the examined areas from regional to global by combining optimally the volume/amount and the essence of the in situ and remote observations and by using these operational data in the modeling process according to the required algorithmic and computer programming tasks.

Each implementation of GIMS technology has theoretical and experimental section which has both its general concept and also the development of its specific spacecraft, aircraft and ground-based platforms. Therefore, the actual composition

Table 1.4 Short characteristics of regional information-modelling systems which have been developed with the GIMS technology

RIMS version	RIMS functions	References
Synthesized RIMS-AYR in the frame of the agreement between the U.S. Department of Energy and the Institute of Ecoinformatics of the Russian Academy of Natural Sciences	Assessment of the influence of the Angara-Yenisei River (AYR) system to the pollution level of the Arctic Basin with radionuclides, heavy metals and oil hydrocarbons	Krapivin et al. (1997a, 1998a), Krapivin and Phillips (2001a), Phillips et al. (1997)
Synthesized RIMS-ACS in the framework of scientific collaboration between the Georgetown University (Washington, USA) and Institute of Ecoinformatics of Russian Academy of Natural Sciences	Assessment of the characteristics of the hydrological regime of the Aral-Caspian system (ACS) and accomplishment of the forecast of these characteristics in the framework of climate change scenarios and anthropogenic intervention to the functioning of this system	Kondratyev et al. (2001), Krapivin and Potapov (2002), Chukhlantsev et al. (2004)
Synthesized RIMS-GCC in the frame of the collaboration between the Institute of Physics of Vietnamese Academy of Sciences and Technologies (Hochiminh City) and the Kotelnikov Institute of Radioengineering and Electronics of the Russian Academy of Sciences	Evaluation of the sinks of atmospheric carbon in hydrosphere, on the land and in geosphere taking into consideration the spatial distribution of the soil-plant formations, the heterogeneity of the World Ocean ecosystem characteristics and geospheric processes. These data are used as input information for global carbon cycle (GCC) model	Burkov and Krapivin (2005), Krapivin and Kondratyev (2002), Kondratyev et al. (2002)
Synthesized RIMS-AB in the framework of the agreement between the U.S. Department of Energy and the Institute of Ecoinformatics of the Russian Academy of Natural Sciences	Calculation of the spatial distribution of pollutants in the Arctic Basin (AB) taking into account the river outflow from all continents and the atmospheric transport from distant territories	Kondratyev et al. (1990b, 2004a), Krapivin and Kondratyev (2002)
Combined RIMS-SA as a result of the scientific cooperation of the Kotelnikov Institute of Radioengineering and the Electronics of the Russian Academy of Sciences with the Institute of Market Problems and Economic-Ecological Researches (Ukraine)	Supply of the Geoecological information-modelling system for the Sea of Asov (SA) with systematic collection, generalization, storage, usage and distribution of information about the environmental characteristics of the Sea of Azov zone in the forms that are much adopted to practical use	Krapivin et al. (2009d)

(continued)

Table 1.4 (continued)

RIMS version	RIMS functions	References
Combined RIMS-TC as a result of the accomplishment of the ISTC Project No. 3827	Detection of the tropical cyclone (TC) onset in its earlier phase based on the meteorological information delivered by buoy stations of the TAO/TRITON/PIRATA/ RAMA system	Bondur et al. (2009), Soldatov (2009a), Soldatov et al. (2010a), Sukov et al. (2008a)
Combined RIMS-OS in the frame of the collaboration between the Kotelnikov Institute of Radioengineering and Electronics of RAS with the Ice Laboratory of Hokkaido University (Mombetsu, Hokkaido, Japan)	Calculation of the biocomplexity distribution for total area of sea by using the simulation model of the Okhotsk Sea based on the existing database and the fragmentary information deduced from observations	Aota et al. (1991a, b, 1992, 1993), Krapivin and Mkrtchyan (2006a, 2007c)
Combined RIMS-B in the frame of the national Bulgarian Project: "Development in Bulgaria of Geo-Information Monitoring System (GIMS), for emergency mapping and risk assessment of the areas with water seepage through levees, of the zones with dangerously high groundwater level and other hydrological, ecological and agricultural phenomena"	Development of a powerful centralized information and management service in Bulgaria using the GIMS technology. It has been achieved by developing and adapting to the real Bulgarian environments of the IMTEM technology for the land surface and water areas and using remote sensing tools	Krapivin and Shutko (2012), Shutko et al. (2010), Verba et al. (2013), Nitu et al. (2013)

of GIMS is possible in the context of cooperation between specialists from various fields of science and engineering.

Table 1.5 shows the stages for the RIMS-B synthesis. The implementation of these stages allowed the demonstrational flights of the aircraft laboratory Miramap (The Netherlands) equipped with advanced radiometric system (Shutko et al. 2010). A comparison of in-situ, remote sensing and modeling results leads to the following conclusions:

- for the practical use of the GIMS technology in the Bulgarian conditions it is necessary to create the archive of characteristics for soil-plant formations that reflect the spatial distributions of soil types and their features, as well as the biometrical and production characteristics of vegetation cover; and
- it is necessary to implement studies and to establish timely warnings about the quasi-stationary character of the climate and geophysical parameters both in separate regions and on a national basis.

Table 1.5 Specification of theoretical and experimental studies the realization of which given possibility to synthesis of the RIMS-B

Stage	Performed tasks
1. RIMS-B database formation	• Collection and generalization of cartographic data about the spatial structure of the soil-plant formations and their types • Determination of the biometrical characteristics of the land covers • Identification of the landscape structures with designation of their morphology and other characteristics that are required for the indication of facies • Definition of the cartographic scheme for the structure of the river basins and isolated reservoirs
2. Development of typical model of hydrological regime for a restricted territory	• Analysis of the existing models of the regional water balance and selection of the model that is most informative for the Bulgarian environments • Elaboration of models for the Bulgarian river systems and specification of the elements of the regional water balance model • Choice of the algorithm for coordination of remote sensing data using selected models with the determination of the model parameters when a disagreement between the measured data and modeling results is derived
3. Elaboration of algorithms for the coordination of the in-situ and remote sensing measurements of the ecological characteristics	• Choice of the size of the spatial pixels that support the required precision of the assessment of the state of the hydrological processes and systems • Development of mnemonic language for database coordination with the model items of hydrological regime for a selected territory • Creation of algorithm for the spatial image reconstruction of the regional hydrological regime based on the remote sensing and in-situ measurements delivered episodically in time and fragmentary in space • Development of the method to overcome the non-stationarities in the delivered data of the soil moisture and the ground water level

(continued)

Table 1.5 (continued)

Stage	Performed tasks
4. Elaboration of the algorithm for the management of decision making under the selection of measurement strategy or modernization of the typical model of the regional water balance	• Choice of scale and criterion for assessment of discrepancy between observational data and modeling results • Determination of decision making procedure concerning the monitoring regime or hydrological model correction depending on the results of comparison of forecasting results with observational data • Document preparation for the regulation of the RIMS-B user operations in the case when the system detects the onset of a dangerous environmental event (flood, soil over-moistening, leakage etc.)
5. Determination of the reporting forms about the results obtained from the RIMS-B functioning	• Choice of the reporting forms that provide the requirements of services to make a decision concerning the realization of actions about the management of the hydrological system and/or defense of the population • Computer implementation of the selected reporting forms that fulfill the informational requirements of various services
6. Formulation of synoptic scenarios	• Establishment of the instruction about the interaction of RIMS-B operator with the regional and national meteorological services • Synthesis of typical synoptic scenarios for separate regions of Bulgaria based on the study of the existing global and European climate change scenarios, as key elements of RIMS-B which will be used in regime "by default" • Analysis of climatic trends in the Bulgarian territory and elaboration of forecasts in evolution of these trends • Synthesis of an evolutionary model for synoptic trend towards an early distortion of the regional synoptic situation using the available meteorological information

1.4 The Evolutionary Modeling as an Item of the Information-Modeling Technology

1.4.1 The Evolutionary Modeling Technology

The composition of an intelligent support subsystem of the GIMS is based on methods that simulate evolution processes (Bukatova 1992; Bukatova and Makrusev 2003; Bukatova et al. 1991; Gulyaev et al. 1987, 1989, 2009; Norris and

McCulloch 2003; Pollak 1990). These methods were the foundation for a new science called "evoinformatics" (Krapivin and Potapov 2002; Nitu et al. 2004). This science lies at the boundary of disciplines such as neurocybernetics, cognitive psychology, artificial intelligence, theory of systems, theory of survivability, and systemology.

Recent advances in informatics and information technology have enabled mathematical modeling and computer technology applications to enter such domains as ecology, biophysics, and medicine. The concept of experimental model has lost its novelty since long ago and has been replaced by the term of "computing experiment", which is used in many studies covering a vast range of subjects including the biospheric experiments (Kondratyev et al. 2003b, 2004b). All such works imply a priori availability of a more or less adequate model implemented as an array of tools in an algorithmic language. To manipulate a model to carry out a series of specific computing experiments one needs a general purpose computer. It is at this point that the researcher may be challenged with insurmountable difficulties caused by constraints on the computer's memory and speed of operation.

Many of such endeavors indicate that modern hardware can handle relatively complex models. Yet the same experience prompts the need for constant improvement of modeling techniques, because the researcher runs into conflict between his desire to enhance the accuracy of a model and limited capabilities of the computer. Building a model that is completely representative of a real world entity is clearly not feasible: on the one hand, taking account of all the parameters of the entity leads to the "evil of multi-dimensionality" and, on the other hand, simple models that can only cope with a small number of parameters are simply not up to the task of simulating the complex entities under consideration. Besides, such projects involve ocean physics, geophysics, global ecology, socio-economics, etc. The building of an adequate model is in principle impossible because of unattainability of information completeness. Such systems can only deal with applied problems from the domains of global ecology, biophysics, and medicine. Furthermore, difficulties in these subject areas arise at the early stages of research (e.g., when attempting to formulate a model).

What is to be done when the currently available knowledge does not allow for synthesis of a mathematical model of an entity or a process? The answer on this question gives a theory of learning computers of the evolutionary type. We shall retain the term "model" although it is used here in a somewhat different sense. What it implies here is the description of entities changing over time in an unpredictable manner and, by virtue of this, ensuring the irremovability of information uncertainty at any moment. Such are natural systems studied e.g. in global ecology, geophysics, biophysics and medicine. Consequently, a model treated in a broad sense must provide for continuous adaptation to the changing behavior and structure of the observed entity. It is clear that universal models can be built only through synthesis of particular models. Models of this kind are implemented for problems of recognition and prediction.

Thus, let a real-world object *A* has some unknown algorithm of operation, only some previous history of operation of finite length been known. We need to efficiently simulate the functioning of A by using models built for some system operating in the real time. It is necessary to build a sequence of improving models $\{A_k\}$, where k = 1, 2, …. Cybernetics began to ponder how this could be attained virtually from the moment the first, primitive by today's standard, machines emerged. It was then that the idea of creating artificial intelligence and designing "thinking machines" appeared, giving rise to the development of robotics. Being aware that a computer is an obedient executor of some program, cybernetics doggedly worked on the possibility of imparting to the machine a measure of unpredictable behavior with some facets of "innovation". Treating intelligence as the ability to correctly respond to a novel situation, scientists came to the conclusion that machines capable of adapting to the level of individual components and their structural organization are in fact feasible.

The idea of model-free learning for computer is discussed during last years. The basic problem is how to teach a computer by changing the structure of relations between its elements. This analogy with the neural operation of the brain established in neurophysiology yielded fruit and helped advance far along the scale of artificial intelligence capabilities. Fogel et al. (1966) gave an impetus to an entirely new cybernetic trend.

Evolutionary modeling on the whole can be represented by a hierarchical two-tier procedure (Fig. 1.15). At the first tier there are two constantly alternating processes conventionally termed the structural adaptation and utilization processes. At the kth stage of adaptation at the time of operation of the structural adaptation algorithm, models of the sequence $\{A_{s,i}\}$, $i = 1, \ldots, M_s$ are synthesized. Special rules form memory from the most effective models $<A_s^1, \ldots, A_s^k>$. At the stage of utilization following the stage of adaptation, the system selects the most efficient model. A schematic diagram of the ith step of kth stage of adaptation of models is represented in Fig. 1.16. The "object" block here denotes that the real-world object is defined by some previous history.

The procedure of the evolutionary selection of models provides for a virtually time-unlimited operation of the system under irremovable information uncertainty. Apart from the previous history which, as a rule, does not meet the requirements of traditional statistical analysis, the researcher has no other available information. Under such conditions one obviously has to use the maximum of available information, in particular, that on the operation of the adaptation and utilization stages.

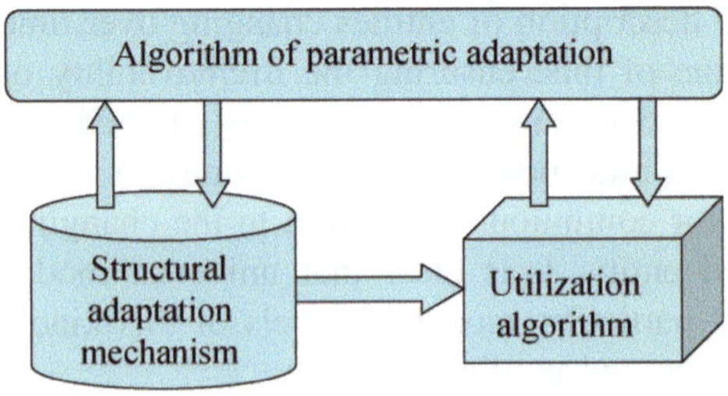

Fig. 1.15 Schematic representation of the concept of evolutionary modeling (Bukatova et al. 1991)

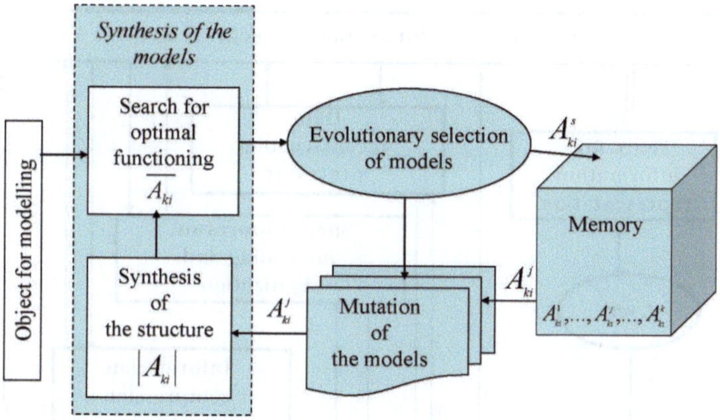

Fig. 1.16 Conceptual flow-diagram of the *i*th step at the *k*th stage of the adaptation process

In the parametric adaptation algorithm, this information is used as early as the adaptation of the first-tier parameters, i.e. the characteristic number for each mode of changes; multiplicity of the change mode utilization; length of the change mode list; the distribution of change modes in the list; the distribution of *k*-models of A_k^i, memory volume at the stage of adaptation, the prehistory length, etc.

The endeavors to search for a building block base to synthesize network with a variable structure in an attempt to create special flexible hardware based on the new principles of information processing have offered up some results. Recent advances in microelectronics have helped solve the problem of selecting components for structures with variable fields of relations (Bukatova 1992).

The unification and specialization are characteristic features of the evolutionary software as a consequence of the minimum of a priori information, effective mechanisms of adaptation and the modular principle of realization. With the orientation to up-to-date personal computer engineering and diverse range of application problems, these specific features have made it possible to work out evolutionary computation technologies, in which for an active dialogue with a user, a set of software modules is realized, as well as the adjustment of the evolutionary facility to the specificity of the problem being solved (Bukatova et al. 1991). Regardless of the field of application, evolutionary technology with software support is characterized by adaptability, flexibility, dynamism, self-correction. At the same time, the main distinction consists in the high effectiveness and adjustability under the conditions of the maximum informative uncertainty, including that of an irremovable character. This approach permits to realize the adaptive procedure in Fig. 1.5. Evolutionary modeling technology really does give a new structure of the AIMS (Fig. 1.17).

Evolutionary modeling is effective at bringing about an adaptation process within the AIMS technology to correct the functional and parametric structure of the basic model. Search of correlations for the description local function in framework of common adaptation procedure other methods can use. For example, Fang and Liang (2003) studied two LAI retrieval schemes with the neural network algorithm:

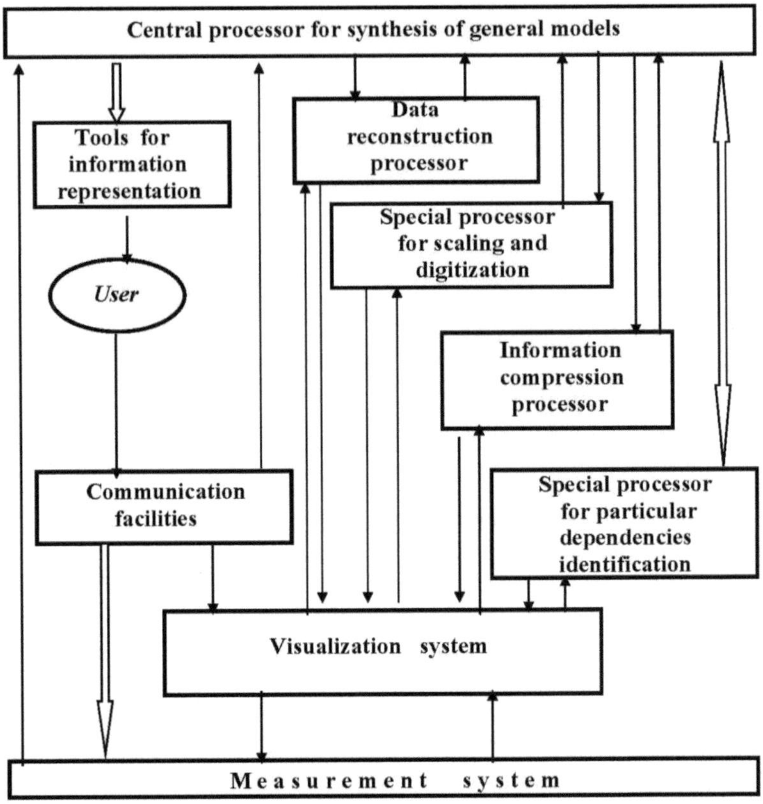

Fig. 1.17 Structure of the geoinformation monitoring system and its evolutionary devices

- extraction of LAI from atmospherically corrected surface reflectance;
- extraction of LAI from top-of-atmosphere (TOA) raw radiances detected by the Enhanced Thematic Mapper Plus (ETM+) sensor.

LAI is estimated through the inversion of an RT model basing on the Landsat ETM+ reflectance and radiance data. The first step in this procedure is the creating of an appropriate Lookup Table (LUT) method. Then, the Markov Chain Reflectance Model (MCRM) is used.

1.4.2 Tools of Evolutionary Modeling Technology

Realization of type model screening in framework of evolution procedure needs to define model class that will be used. There are two possible alternatives for model type choice:

- analytical or simulation/computational type that has not a relation to the subject area for study; or
- mathematical equations describing the subject area.

In the firs case a model type can be one of the following:

- self-organizing model (Ivachnenko et al. 1984);
- model of neural computation (Jordan and Sejnowski 2001; Obermayer and Sejnowski 2001);
- genetic model (Neale and Cardon 1992); and
- graphical model (Koller and Friedman 2009).

In the second case mathematical equations are formed depending on the subject area that is considered. Selection of model type in many respects is no simple task. Expert experience plays important role.

One of simple methods is method of self-organizing models that used effectively for the problem solution of the spatial-temporal recovery of the data of monitoring. The idea of this approach is based on traditional functions approximation theory.

Let an object or process be described by the model $\Psi = f(a_1, \ldots, n)$ where parameters $\{a_i\}$ reflect the quantitative, functional, and structural sections of the phenomenon under study. A multitude of possible types of the function f is determined on basis of an expert estimation with consideration of a priori information and heuristic set of partial descriptions of the phenomenon. The training sequence $\{f_i\}$ is constructed which serves the basis for multi-row selection to choose the model of an optimal complexity and acceptable accuracy. The first level of selection consists in calculation of the row $\{y_s\}$, where $y_i = g(a_{i-1}, a_i)$ $\left(s = 1, \ldots, L = C_n^2; i = 1, \ldots, n\right)$. The second level of selection gives the row $\{z_p\}$, where $z_p = g\left(y_{j-1}, y_j\right)$ $(p = 1, \ldots, C_L^2; j = 1, \ldots, L)$. The process of selection is continued till the most regular mathematical description of the phenomenon under study is obtained. Estimation of the accuracy of the obtained model and the choice of the moment of the end of the process of selection depend on the chosen criterion of discrepancy between theoretic and empirical image of phenomenon. The root mean square deviation criterion is most often used, and polynomial serves as the function f. The procedure of the models selection consists in a gradual complication of the polynomial approximation.

Method of the group consideration of arguments has been described in detail in the work of Ivachnenko et al. (1984), where its various modifications are given with examples of the efficient use in solving the applied problems.

Models of neural computation are usually applied for neuro-biological computation. For example, the Hodgkin–Huxley model, widely regarded as one of the great achievements of 20th-century biophysics, describes how action potentials in neurons are initiated and propagated in axons via voltage-gated ion channels. It is a set of nonlinear ordinary differential equations that that approximates the electrical characteristics of excitable cells such as neurons and cardiac myocytes. These equations were introduced by Hodgkin and Huxley (1952) to explain the results of voltage clamp experiments on the squid giant axon. Analytic solutions do not exist, but the Levenberg–Marquardt algorithm (Marquardt 1963), a modified

Gauss–Newton algorithm, is often used to fit these equations to voltage-clamp data. This algorithm, known as the damped least squares method, provides a numerical solution to the problem of minimizing a function, generally nonlinear, over a space of parameters of the function.

Subsequent development of these approaches to the solution of tasks related to biological evolution simulation is based on parameterization of basic evolution factors—hereditary mutability, competition, and selection. Progress in this direction was achieved due to the introduction of structured model (C-model) into evolutionary schemes (Bukatova and Makrusev 2003, 2004a, b).

Bukatova and Matveev (2004b) developed an idea of neuron networks for tasks when the processing images are needed. In this case there are three possibilities for neuron network selection:

1. Neuron network as finite-automat stratified model or Petri net (René and Hassane 2005).
2. Modification of NeuroShell 2, Neural Analyzer, and BrainMaker Pro depending on the concrete neuro-paradigm (Lowd 1977).
3. Integrity-evolutionary integration medium that includes tools for structural search of effective configuration of neuron net.

Each of case includes the tools for structural search of efficient configuration for neuron network. Let input measurements $x_c(t)$ from finite-dimensional space **X** are transformed to the solutions $y_c(t)$ from finite-dimensional space **Y** by means of neuron network that realizes unknown transformer **F**. In this case it is needed to synthesis **F: X** → **Y**. This operation can be realized by means of evolutionary synthesis algorithm represented in Figs. 1.15 and 1.16. This algorithm uses neurons and connections between them. Quantity criterion defines the compromise between network complexity and precision results. This compromise is achieved in framework of structural evolutionary synthesis that is based on stochastic search of C-models. Variety of possible situations is considered by Bukatova and Makrusev (2004a, b). Particularly, Krapivin et al. (2008a) considering the problem of information-measuring system synthesize use a modern technologies and element base of microelectronics and optoelectronics jointly with adaptive-evolutionary technology. Gulyaev et al. (1987, 1989) developed technology of evolutionary computer that is based on the optical electronics element base including discrete logical elements.

1.5 A Global Model as Unit of the Information-Modeling Technology

1.5.1 Principal Structure of Global Model

Solution of global environmental problems is possible by means of the GIMS technology that parameterizes processes in the nature/society system (NSS) of global scale. There exist various approaches to the synthesis of a global model to describe all aspects of man's interactions with environmental bodies and with their physical, biological and chemical systems. One such way has its origin in the studies of the Computer Centre of the Russian Academy of Sciences in Moscow (Krapivin et al. 1982). A global model of this type is formed on basis of the detailed description of the climate system with the consideration of a small set of biospheric components. This strategy of global modeling is adhered to in the Potsdam Institute for Climate Impact Research studies (Boysen 2000) where the Moscow Global Model prototypes are developed. More than 30 climate models are being developed in different countries as attempts to bring forth new trends in the science of global change (Demirchian, Kondratyev 2004; Claussen et al. 1999). Unfortunately global- and regional-scale studies on the processes and impacts of global change using this approach have not produced results that are enough satisfactory. That is why another approach to the global modeling problem has been developed by many authors (Krapivin 1993; Sellers et al. 1996; Kondratyev et al. 2002; Degermendzhi et al. 2009). This approach is known as evolutionary modeling.

The traditional approaches to building a global model encounter some difficulties of the algorithmic description with respect to many socio-economic, ecological and climatic processes so that one has to deal with information uncertainty. These approaches to global modeling simply ignore such uncertainty and consequently the structure of the resultant models does not adequately reflect the real processes. Evolutionary modeling makes it possible to remove this drawback by the synthesis of a combined model whose structure is subject to adaptation against the background of the history of a system of the biosphere and climate components. The implementation of such a model can also be combined in various classes of models using conventional software and hardware and special-purpose processors of the evolutionary type. The form of such combination is diverse, depending on the spatial-temporal completeness of the databases.

The experience in global modeling abounds in examples of insolvable problems encountered when looking for ways to describe the scientific and technological advances and human activity in its diverse manifestations. No lesser difficulties arise in modeling climate described by a superimposition of processes with different temporal variability rates. As to completeness of description in the global model, it is impossible to clearly delineate the bounds of information availability and the extent of the required spatial and structural detail. Therefore, without going into natural-philosophical analysis of global problems, and skirting the issue of the ultimate solution to global modeling, we will confine ourselves to the discussion of only one of the possible approaches. This approach will demonstrate in which way evolutionary modeling implemented on special processors can help overcome computing, algorithmic and other difficulties of global modeling. All of this does imply that a search for effective models of the traditional type can apt be considered perspective. At present, the building of global biogeocenotic models is not seen as difficult. Many such models have been created, and the gathering of information to support them is under way. The history of the interaction of the biosphere with the climate system and human society is not known sufficiently, which is one of the hurdles, e.g. in the description of climatic cycles. To build a global model accounting for the interaction of the biosphere, climate system, magnetosphere, etc. it is necessary to apply the evolutionary approach which helps to overcome the uncertainties in the description of this interaction. As a result of the adjustment of such a model to the history of the prescribed cycle, we will obtain a model implicitly tracing various regularities of the dynamics of the biosphere in the past and allowing for forecast assessments to be made in the same temporal cycle. A special-processor version of this model completely removes all the existing algorithmic and computing hurdles arising from the large dimensionality of the global model and the conditions of irreducible nonparametrical uncertainty.

Figure 1.18 shows the conceptual diagram of this new type of global model. The data archive is formed here as two structures. Data of the first type for the computer models of the biosphere processes are stored as climatic maps and as tables of the model equation coefficients. It is necessary to fill in all cells of the schematic maps. Data of the second type are represented as fragments recorded disparately (possibly irregularly) in time and space, i.e. CO_2 concentration, temperature, precipitation, pressure, population numbers, availability of resources, etc. Data of this type are used to adjust the evolutionary processor to the given class of models, e.g. finite automata. As a result of this procedure the model is adapted to the history of the prescribed time cycle. As has been shown by Bukatova et al. (1991) a stable

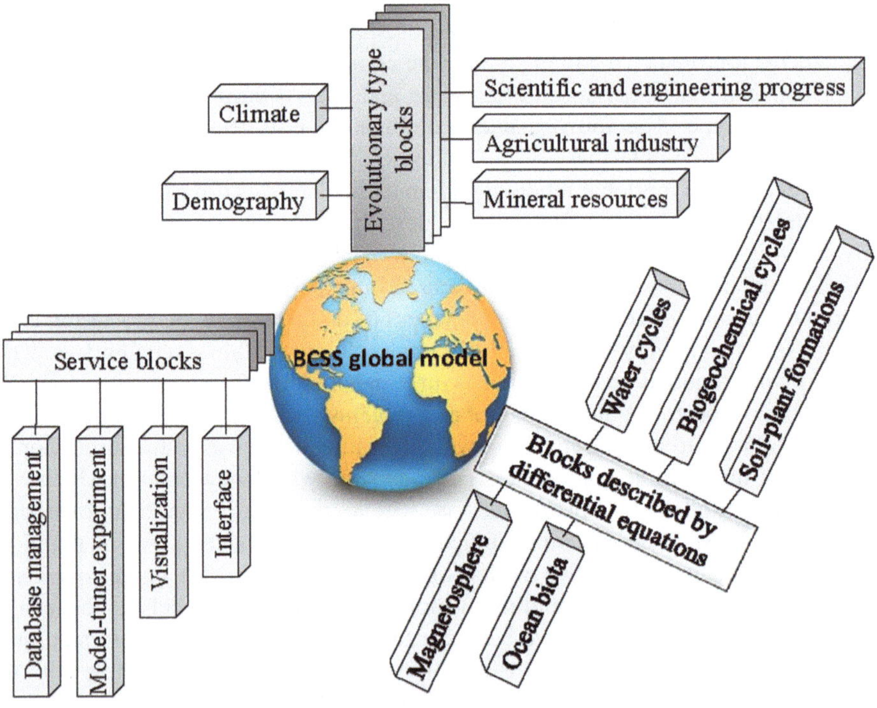

Fig. 1.18 Structure of global model using the evolutionary blocks

forecast is produced with 75–95 % reliability covering several temporal steps. The extent of a forecast is determined by the depth of the history against the background of the saturation of the effect. Given the need for a forecast under the conditions of change in the trends of human economic activities, an evolutionary processor is adjusted to the assigned scenario, thus automatically providing for simulation of the corresponding response of the biosphere to this change.

The suggested structure of the global model thus ensures a flexible combination of models of the traditional and evolutionary types. The proposed approach helps escape the need to model unstable processes (climatic, socio-economic, demographic, etc.) and provides for overcoming uncertainty. A model of this new type makes it possible to go from learning experiments to the assessment of the viability of the biosphere with regard to actual trends of anthropogenic stresses in all regions of the globe.

A departure from the established global modeling techniques based on the new information technology makes it possible to proceed to creating a global system of monitoring with the global model as a portion of the support for the system. The structure of such a system is represented in Fig. 1.19. Application of the evolutionary computer technology provides for categorization of the whole system by a class of subsystems with variable structure and for making it adaptable to changes in the natural process or entity under observation. Furthermore, it becomes possible to detail heterogeneously the natural systems under study in the space of phase variables, and to select nonuniform geographical grids in a sampling analysis of the planetary surface, i.e. arbitrary insertion of significant regularities at the regional level becomes possible.

The automatic system for processing of global information is aimed at the acquisition of combined models reflected in the real-time scale of the climatic and anthropogenic changes in the biosphere and based on the known history (or, rather, its simulation). The system relies on the set of models of biosphere processes and, using software of other units with the help of the scenario of anthropogenic behavior formulated at input, provides for prompt assessment of the environmental state and for forecast assessments within the framework of this scenario. The automatic system for processing of global information also consists of the advantage in that it formulates the entry and bound conditions for particular built-in models in the study of regional systems and virtually substitutes for field measurements of those conditions.

The first version of the global model (Krapivin et al. 1982) was oriented towards rigid spatial-temporal detailing and therefore required a large quantity of information. The subsequent development of an automatic system for processing of global information has made it possible, owing to the evolutionary technology, to reject the generally accepted regular geophysical grid in archive development and to solve this problem using algorithms for the recovery of spatial-temporal information.

Global model aims to parameterize the main processes that take place in the biosphere/climate/society system. The existing global knowledge and databases

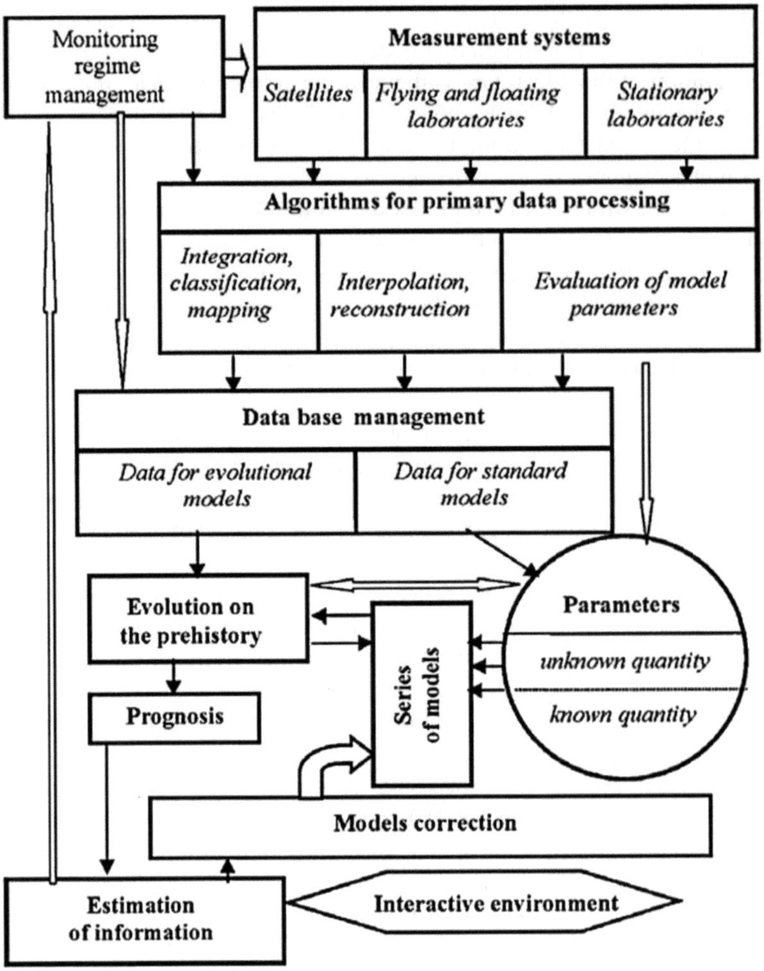

Fig. 1.19 Structure of the learning monitoring system of the environment implementing evolutionary computer technology and relying on the set of models of biosphere processes

enables the synthesis model describing the interactions between areas and elements of this system represented in Figs. 1.20, 1.21 and 1.22.

1.5.2 Global Biogeochemical Cycles as Units of Global Model

Interactions between the abiotic factors of the environment and living organisms of the biosphere are accompanied by a continuous matter cycle in nature. Different species of living organisms assimilate substances needed for their growth and life

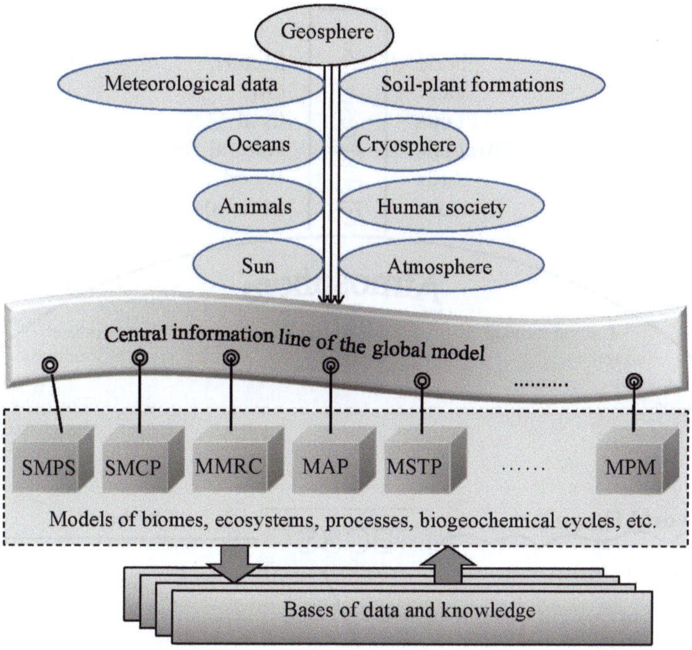

Fig. 1.20 The information-functional structure of the global model. The notation is given in Table 1.6

support emitting to the environment products of metabolism and other complex mineral and organic compounds of chemical elements in the form of non-assimilated food or dead biomasses. As a result of the biosphere evolution, a stable chain of global biogeochemical cycles has been formed whose violation in the second half of the 20th century has made humankind face many principal problems such as an unpredicted climate change due to the greenhouse effect, a decrease of biodiversity, progressing desertification, and many others. Indeed, questions about what is the matter with the Earth's climate and what are the consequences for the ozone layer depletion remain unanswered despite huge economic expenditures on their study.

Now it is clear that these and other nature-protection questions cannot be answered without developing an effective global monitoring system based on the global model of the biosphere/climate/society system as GIMS item, one of the basic units of which is the unit simulating the biogeochemical cycles of basic chemical elements of the biosphere (Zhu and Anderson 2002; Kondratyev et al. 2002). It is this approach under conditions of implementation of the Kyoto Protocol (KP) ideas that will make it possible to assess the anthropogenic fluxes of pollutants and to estimate permissible emissions of carbon, chlorine, sulfur, fluorine, methane, and other chemical elements to the environment as well as to regulate the problems of the GHGs emissions market (Kalb et al. 2004; Pan 2005).

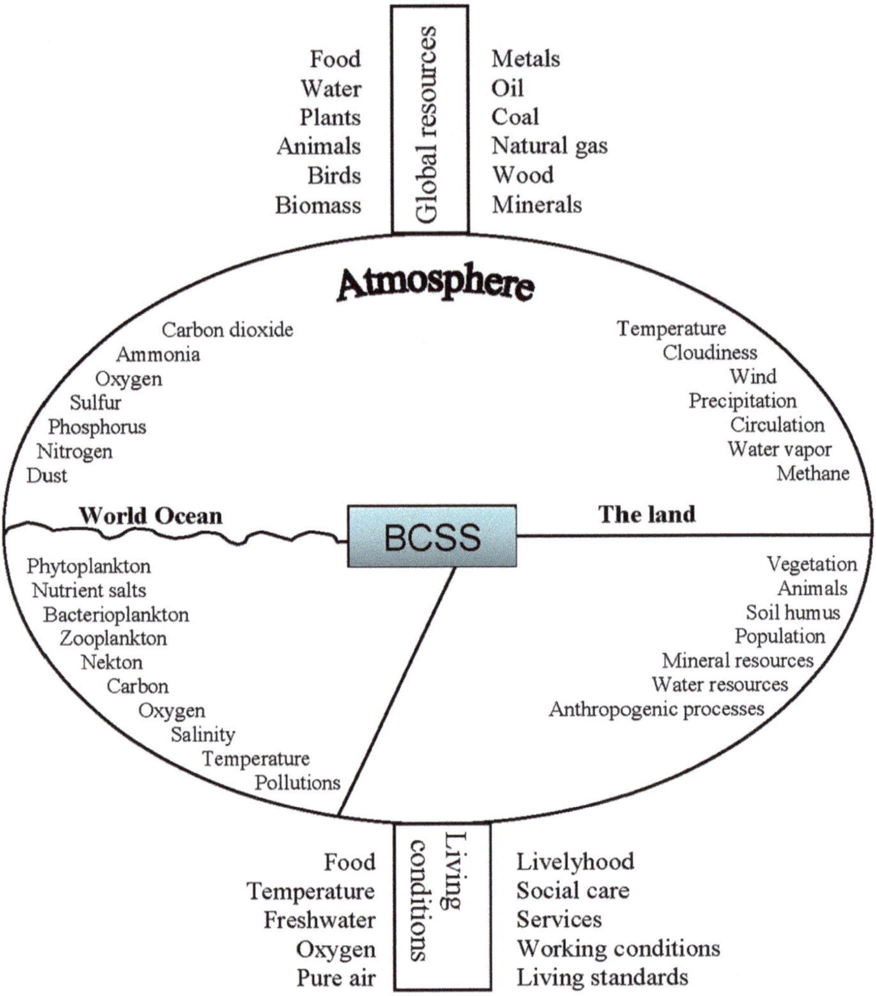

Fig. 1.21 Key elements of the biosphere/climate/society system (BCSS) and energy components to be taken into account on the global ecodynamics forecast within the framework of global model use as the external informational source for regional GIMS

Fundamental connections between characteristics of the biological state of the environment, such as biodiversity in ecosystems, the state and dynamics of food chains, and interactions of the biosystem with the cycle of biogenic elements have been poorly studied, both in land and water ecosystems. Among numerous questions appearing in studies of the global biogeochemical processes the following are of key importance:

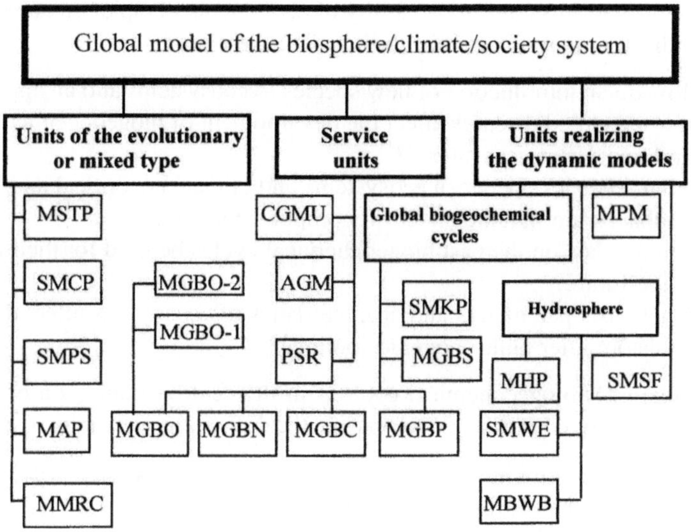

Fig. 1.22 The block scheme of the BCSS global model. The notation is given in Table 1.6

(1) What physical, biological, chemical and social processes are the basic ones in regulation of the cycles of carbon, nitrogen, sulfur, water and other elements both in space and in time?

- What mathematical relations are determinants in the parameterization of biological processes in the computer models of biogeochemical cycles?
- What are dependences between biodiversity, structure of ecological chains, and biogeochemical cycles in land and water ecosystems?
- What processes are determinants in the transport of biogenic salts and pollutants in space in general and between various ecosystems, in particular?
- What are mechanisms that relate one biogeochemical cycle to another, and do general principles of parameterization of these relations exist or they depend on the type of chemical elements and ecosystems under consideration?

(2) What are forms and ways of anthropogenic interference to global biogeochemical cycles?

- How do humans influence the biogeochemical cycles, change their rates and spatial distributions of chemical elements, forming inputs and outputs of numerical models, and what are consequences of this interference?
- How does a change in the land use strategy affect the re-distribution of chemical elements in space and in time?
- What anthropogenic pollutants are important for the biogeochemical influence on ecosystems and how to predict them?

(3) What mechanisms control the ability of ecosystems to rapidly restore and what are indicators that reflect this ability of ecosystems?

- How do an introduction of new species to ecosystems and an appearance of new unstudied diseases affect the development of biogeochemical cycles in land and water ecosystems?
- What feedbacks between ecosystems and climate are critical and how these feedbacks are parameterized in computer models?
- Can the data on the past biogeochemical cycles be used for their prediction in the future?
- What basic parameters and characteristics of ecosystems affect their ability to restore after anthropogenic forcings?

The global CO_2 biogeochemical cycle is in the centre of attention of scientists. Specialists of many countries try to answer the following questions:

(i) What concentrations of CO_2 can be expected in future with present or predicted rates of organic fuels burning?
(ii) What climate changes can result from increased concentrations of CO_2?
(iii) What are consequences of climate changes for the biosphere?
(iv) What can the mankind undertake in order to either reduce negative consequences of climate change or prevent them?

Clearly, according to rough model estimates, the industrial civilization should now search for new sources of energy which would decrease the rates of organic fuel burning and, hence, reduce the external forcings on natural biogeochemical cycled. The atmosphere is one of the important reservoirs taking part in formation of these cycles. On the whole, chemistry and physics of the atmospheric processes suffer changes, without a study of which reliable assessment of the state of the atmosphere and the dynamics and photochemical processes in it is impossible (Brasseur 2005).

During the last decade the words "greenhouse effect" could be seen in numerous publications on the problems of global climate change on Earth (Ichikawa 2004). This term implies a totality of descriptions of the effects appearing in the climate system and connected with a number of natural and anthropogenic processes. On the whole, the notion of the greenhouse effect refers to an explanation of changes in the atmospheric thermal regime, as a result of the impact of some gases on the process of solar radiation absorption. Many gases are characterized by a high stability and long residence in the atmosphere (Table 1.7). Carbon dioxide is one of them. As for the role of CO_2, more than a century ago, Arrhenius (1896) was the first to draw the conclusion that its emission in fuels burning can lead to a climate warming. In subsequent decades this sagacious conclusion turned out to be a weird foresight and sorrowful forecast. After all, in the global historical long-range perspective, CO_2 content in the atmosphere had been changing stably with variations of about 20 ppm, at least, during 11,000 years before the industrial epoch. In this long-term context the anthropogenic increase of atmospheric CO_2 by $\sim$100 ppm for the last 200 years is a dramatic change in the global carbon cycle. This increase

Table 1.6 Characteristic of the BCSS global model units

Identifier of the unit in Figs. 1.20, 1.21 and 1.22	Characteristic of the unit functions
SMPS	A set of models of the population size dynamics with regard to the age structure (Logofet 2002; Caswell 2006). Matrix population models are considered as a powerful tool for investigating the population dynamic consequences of the structure of the life cycle and the vital rates
SMCP	A set of models of climatic processes with differently detailed consideration of parameters and their correlations (Kiehl and Gent 2004; Dessler 2011). Simple climate models are the bases for the BCSS model
MMRC	Model of the mineral resources control (Nitu et al. 2004; Krapivin and Varotsos 2007, 2008; Krapivin 2008b). The knowledge base of the BCSS model is filled with information about the present distribution of the energy sources and other natural resources listed in Fig. 1.21
MAP	Model of agricultural production (Ashraf et al. 2012; Konrad and Joachim 2013). The simple predictive agricultural production models are based on the data of regional crops
MSTP	Model of scientific-technical progress (Krapivin and Varotsos 2007, 2008). Model reflects the tendency in the use of global resources (Fig. 1.21) and economical growth
CGMU	Control of the global model units and database interface. Coordination of inputs and outputs of all global model items is provided
AGM	Adjustment of the global model to the simulation experiment conditions and its control. Delimitation of functional regions between global model and regional GIMS is installed
PSR	Preparation of simulation results to visualization or other forms of account. Visualization panels and tabs help to explore computational results. The results include predictions, models, accuracy metrics and plots
MBWB	Model of the biospheric water balance (Krapivin and Kondratyev 2002; Krapivin and Shutko 2012)
MGBC	Model of the global biogeochemical cycle of carbon dioxide (Kondratyev et al. 2003c; Kondratyev and Krapivin 2004a, b)
MGBS	Model of the global biogeochemical cycle of sulfur compounds (Krapivin and Varotsos 2008)
MGBO	Model of the global biogeochemical cycle of oxygen and ozone (Krapivin 2000a)
MGBN	Model of the global biogeochemical cycle of nitrogen (Krapivin 2000c)
MGBP	Model of the global biogeochemical cycle of phosphorus (Krapivin and Kondratyev 2002)
SMKP	A set of models of kinetics of some types of pollutants in different media (Kondratyev et al. 2006a; Krapivin and Potapov 2006a, b)

(continued)

Table 1.6 (continued)

Identifier of the unit in Figs. 1.20, 1.21 and 1.22	Characteristic of the unit functions
SMWE	A set of models of water ecosystems in different climatic zones (Degermendzi 1987)
MHP	Model of hydrodynamic processes (Kondratyev et al. 2002)
SMSF	A set of models of soil-plant formations (Kondratyev et al. 2004a; Burkov and Krapivin 2009)
MPM	Model of processes in magnetosphere (Korgenevsky et al. 1989)

Table 1.7 Characteristic of the most important greenhouse gases

Gas	Life time in the atmosphere (years)	Anthropogenic emission, MTC (%)	Average concentration	D_a (%)	I_s (%)	Percent of total (%)
CO_2	3–5	1585.7 (84)	362 ppmv	76	0.5	99.438
NO_x	100–150	97.5 (5)	308 ppbv	6	0.25	0.471
CH_4	11	175.8 (9)	1815 ppbv	13	1.0	0.084
HCFC, HFC, PFC	75–111	31.4 (2)	0.34–0.54 ppbv	5	7	0.007

Notation: *ppmv* parts per million by volume, *ppbv* parts per billion by volume, D_a is distribution in the atmosphere, I_s is increase in speed

is connected with emissions to the atmosphere of ~400 PgC during this period mainly due to deforestation and fossil fuel burning.

Numerous long-term observations in various latitudinal belts show a high level of correlation between temperature and CO_2 content. The atmosphere-ocean interaction contributes most into this dependence. Though the atmosphere and the ocean are in equilibrium with respect to CO_2 exchange, still this equilibrium is regularly violated. The most serious causes of this violation are:

(1) SST variations;
(2) changes in the ocean volume; and
(3) changes in the regime of the vertical circulation of the ocean.

In general, an efficiency of these causes can be characterized by the following ratio of the forcing on CO_2 concentrations in the atmosphere. The first cause contributes about 65 % into the change of the CO_2 partial pressure in the atmosphere (p_a). The remaining 35 % are contributed by the second and third causes. Quantitatively, this relationship is characterized by a 6 % increase of the atmospheric CO_2 partial pressure per 1 °C increase of the temperature of the ocean's upper layer. Also, a 1 % decrease of the ocean volume raises p_a by 3 %. On the whole, as Perry (2001) notes, understanding the role of the atmosphere-ocean system in global changes requires a study of its dynamics with consideration of various kinds of information for a long historical period. Of course, it is important here to reasonably use the respective models and data from the paleocenographic record. This is only possible with coordination of various programs on study of the atmosphere-ocean system.

An assessment of the greenhouse effect requires a complex consideration of the interaction of all processes of energy transformation on Earth. However, in the diversity of processes (from astronomical to biological) that affect the climate system on various time scales, there exists a hierarchy in their significance. But this hierarchy cannot be constant, since the role of some processes can vary in a wide range of their significance for climatic variations. Consideration of one factor separately simplifies an analysis of its impact on climate. In fact, the impact of the greenhouse effect is determined by an exceeding of surface temperature T_L over the effective temperature T_e. The Earth's surface temperature T_L is a function of surface emissivity κ. The effective temperature T_e is a function of emissivity α of the atmosphere-land-ocean system. In general, the parameters κ and α depend on many factors, in particular, on the CO_2 concentration in the atmosphere. There are a lot of simple and complicated numerical models where attempts have been made to parameterize these dependences. Unfortunately, there is not a single model that can meet the requirements of adequacy and reliably describe the prehistory of the climatic trends on Earth. Nevertheless, one can state that the greenhouse effect depends non-linearly on the difference $T_L - T_e$, that is, on atmospheric turbidity, especially in the long-wave region. The more CO_2 is in the atmosphere, the stronger is the atmospheric turbidity. The strongest effect of CO_2 on the atmospheric turbidity is in the long-wave region 12–18 μm. This effect is weaker in the wavelength intervals 7–8, 9–10, 2.0, 2.7, and 4.3 μm. It is clear that with the increasing partial pressure of CO_2 in the atmosphere the role of various bands of CO_2 will grow, and it means that with the intensified CO_2 absorption bands the upward long-wave radiation flux will decrease. At the same time, the downward long-wave radiation flux on the Earth surface will increase. From the available estimates, a reduction of the upward and increase of the downward fluxes are estimated at 2.5 and 1.3 W m^{-2}, respectively.

Thus, to estimate the level of the greenhouse effect due to CO_2 and other GHGs (Table 1.8), it is necessary to know how to predict their concentration in the atmosphere, with all feedbacks in their global biogeochemical cycle taken into account (Watson et al. 2000; Krapivin and Potapov 2007). This problem touches upon several spheres of science—biogeochemistry, geochemistry, soil science, ecology, agrochemistry, geology, oceanology, physiology, and radiochemistry. The present methods of the global ecoinformatics enable one to combine knowledge accumulated in these fields.

Of course, the global cycle of chemical elements should be studied not only to be able to assess the climatic consequences of the anthropogenic activity but also to understand the prospects of the environmental dynamics from the viewpoint of its quality and possibility of life. Since the cycles of chemical elements in nature are closely connected with living substance activity, one can single out the geological,

Table 1.8 Greenhouse gases and global warming potentials (GWPs)

Gas	100 year GWP	ΔGWP (%)	Gas	100 year GWP	ΔGWP (%)
Carbon dioxide (CO_2)	1	55	HFC-227ea (C_3HF_7)	2,900	0.69
Methane (CH_4)	21	17	HFC-236fa ($C_3H_2F_6$)	6,300	0.75
Nitrous oxide (N_2O)	310	5	HFC-43-10mee ($C_5H_2F_{10}$)	1,300	0.75
HFC-23 (CHF_3)	11,700	0.96	Perfluoromethane (CF_4)	6,500	1.15
HFC-125 (C_2HF_5)	2,800	0.75	Perfluoroethane (C_2F_6)	9,200	0.75
HFC-134a (CH_2FCF_3)	1,300	0.34	Perfluorobutane (C_4F_{10})	7,000	0.87
HFC-143a (CF_3CH_3)	3,800	0.75	Perfluorohexane (C_6F_{14})	7,400	0.75
HFC-152a (CH_3CHF_2)	140	0.28	Sulphur hexafluoride (SF_6)	23,900	0.3

biogenic, and biological cycles of this rotation. The biogenic cycle includes sub-cycles, such as biogeochemical, biogeocenotic, and geochemical.

Land ecosystems play an important role in the biogeochemical cycles' dynamics on Earth. The anthropogenic changes in vegetation covers affect the biogeochemical cycles and, thereby, other processes, climate included. The most well-known and important biogeochemical cycles, for example, include cycles of carbon, nitrogen, oxygen, phosphorus, and water. Biogeochemical cycles always involve equilibrium states: a balance in the cycling of the element between land surface compartments. Chemical elements participate in the processes of photosynthesis and respiration of plants, as well as their dying-off, through which their exchange is formed between compartments and elements of land ecosystems. The most characteristic features of biogeochemical cycles of individual chemical elements are as follows:

- The nitrogen cycle is a much more complicated biogeochemical cycle but also cycles through living parts and nonliving parts including the water, land, and air. Nitrogen is a very important element in that it is part of both proteins, present in the composition of the amino acids that make up proteins, as well as nucleic acids such as DNA and RNA, present in nitrogenous bases. The largest reservoir of nitrogen is the atmosphere, in which about 78 % of nitrogen is contained as nitrogen gas (N_2). Nitrogen gas is "fixed," in a process called nitrogen fixation. Nitrogen fixation combines nitrogen with oxygen to create nitrates (NO_3).
- The oxygen cycle is the biogeochemical cycle that describes the movement of oxygen within and between its three main reservoirs: the atmosphere, the biosphere, and the lithosphere. The main driving factor of the oxygen cycle is photosynthesis, which is responsible for the modern Earth's atmosphere and life.
- The carbon cycle is the biogeochemical cycle by which carbon is exchanged between the biosphere, geosphere, hydrosphere, and atmosphere of the Earth. The cycle is usually thought of as four major reservoirs of carbon interconnected by pathways of exchange. The reservoirs are the atmosphere, the terrestrial biosphere (which usually includes freshwater systems and non-living organic material, such as soil carbon), the oceans, and the sediments.
- The phosphorus cycle is the biogeochemical cycle that describes the movement of phosphorus through the lithosphere, hydrosphere, and biosphere. Unlike

many other biogeochemicals, the atmosphere does not play a significant role in the movements of phosphorus, because phosphorus and phosphorus-based compounds are usually solids at the typical ranges of temperature and pressure found on Earth.

- The essential steps of the sulfur cycle are:

  (1) Mineralization of organic sulfur to the inorganic form, hydrogen sulfide: (H_2S).
  (2) Oxidation of sulfide and elemental sulfur (S) and related compounds to sulfate (SO_4^{2-}).
  (3) Reduction of sulfate to sulfide.
  (4) Microbial immobilization of the sulfur compounds and subsequent incorporation into the organic form of sulfur.

- The water cycle is the continuous circulation of water within the Earth's hydrosphere. As water moves through the cycle, it changes state between liquid, solid, and gas phases.
- Hydrogen is one of the constituents of water. It recycles as in the other biogeochemical cycles. It is actively involved with the other cycles like the carbon cycle, nitrogen cycle and sulfur cycle as well.

A detailed description of biogeochemical cycles of carbon, nitrogen, phosphorus, sulfur, and water in land ecosystems has been given in a works of Kondratyev et al. (2003b, c). The main reservoirs of these elements are biomass and soil, between which a matter exchange takes place through respiration of plants, their photosynthesis and dying-off. Modeling of this exchange requires knowledge of the spatial structure of vegetation cover and its classification.

- Population density;
- Potential natural vegetation;
- Cropland extent from 1700 to present;
- Grazing land extent;
- Built-up land extent;
- The major crops extent;
- Land suitability for cultivation.

Of course, an accurate assessment of the fluxes of chemical elements in the atmosphere-vegetation-soil system is only possible with a detailed inventory of land covers. For instance, Fang et al. (2001) have undertaken such an inventory for seven time periods over the territory of China, marking planted and natural forests. It was shown that a maximum rate (0.035 PgC $year^{-1}$) of carbon assimilation from the atmosphere was observed between 1989 and 1993, with the periods of maximum for different types of forests not coinciding. This confirms the fact that for an accurate and reliable calculation of carbon fluxes in the atmosphere-vegetation-soil system, one should have characteristics of vegetation covers of different types distributed in space and time. And since there is no such concentrated data, all available estimates of CO_2 sinks on land cannot be considered reliable. This is confirmed by the data of the structural analysis of the forest ecosystems' biodiversity in South Asia, Far East, and Japan performed by the Institute for Global Environmental Strategies (IGES) and in which estimates of the rates of forests' degradation are given (Inoue and Isozaki 2003). As Austin et al. (2004) have shown, the sporadic nature of water availability in arid and deserted territories is the cause of great shifts in the C/N ratio and, hence, considerable heterogeneities in the biogeochemical cycles of these territories.

From estimates of Stoll-Kleemann and O'Riordan (2004), about 70 % of land surface are anthropogenically affected causing changes in biodiversity thousands times faster than it takes place naturally. Global biodiversity cannot be maintained without changing the strategy of human behaviour in the sphere of the environmental protection. Therefore, one should expect a crisis in biodiversity, unless the international cooperation on its protection becomes effective.

Global scales of variability of the biogeochemical cycles of many elements raises a problem of control of the state of water ecosystems not only with local sources of pollution taken into account, but also, and to a greater extent, distant transports of chemical matter and biological pollutions. Input of various substances to water ecosystems leads to a degradation of fish populations and a change in sanitary conditions for population in adjacent regions. The ways of penetration of undesired substances to water ecosystems are diverse, including river and shore runs-off as basic highways of pollutants propagation. Therefore, protection of water ecosystems under present-day conditions of globalization requires technologies and systems to control additional fluxes of nitrogen and phosphorus which provide a minimum of the oxygen balance violation and preserve natural trends of the living elements' biomass. As Fourie et al. (2004) noted, it is especially important for water ecosystems in many regions of Africa, where only the atmosphere is an external source of additional biogenic elements.

Inland water ecosystems are divided into fresh-water and salted ecosystems. The simplest scheme of life organization in these ecosystems consists in interactions of living elements with abiotic components (penetration of light, water currents, dissolved nutrient concentrations, and suspended solids). The producers supply O_2 to the aquatic systems through photosynthesis. This O_2 is then used by the producers, consumers and decomposers through aerobic respiration. The CO_2 enters an aquatic system from the atmosphere and through aerobic respiration by producers, consumers, and decomposers and it's removed by photosynthesizing producers. The concentration of oxygen in water depends on the amount of pollutants entering the ecosystem. These pollutants, depending on their type, can affect directly the aquatic organisms and indirectly through the process of eutrophication. As a result, the input of pollutants to the water ecosystemleads to a change of its role in the gas exchange with the atmosphere.

There are not less than 1,500 substances as pollutants in freshwater ecosystems. Among them are the following:

- *Acids and alkalies*. Most freshwater lakes, streams, and ponds have a natural pH in the range of 6–8. Acid deposition has many harmful ecological effects when the pH of most aquatic systems falls below 6 and especially below 5.
- *Anions*. The most toxic form of cyanide is free cyanide, which includes the cyanide anion itself and hydrogen cyanide, HCN, either in a gaseous or aqueous state. One teaspoon of a 2 % cyanide solution can kill a person.
- *Detergents*. There are two kinds of detergents with different characteristics: phosphate detergents and surfactant detergents. Detergents that contain phosphates are highly caustic, and surfactant detergents are very toxic.
- *Gases*. Some gases that can harm aquatic freshwater life are gases such as chlorine, ammonia and methane.
- *Heat*. Respiration and growth rates may be changed and these may alter the feeding rates of organisms. The reproduction period may be brought forward and development may be speeded up. Parasites and diseases may also be affected. An increase of temperature also means a decrease in oxygen solubility.
- *Heavy metals*. The most common heavy metal pollutants are arsenic, cadmium, chromium, copper, nickel, lead and mercury. Some metals, such as manganese, iron, copper, and zinc are essential micronutrients. Each type of heavy metals in its own way affects the water ecosystem's biochemistry and can accumulate in bottom deposits and in the biomass of living elements.
- *Nutrients*. Too many nutrients stimulate the rapid growth of plants and algae, clogging waterways and sometimes creating blooms of toxic blue-green algae. This process is called eutrophication.
- *Organic pollution*. Organic pollution occurs when large quantities of organic compounds, which act as substrates for microorganisms, are released into watercourses. Organic pollutants consist of proteins, carbohydrates, fats and nucleic acids in a multiplicity of combinations. Organic pollution affects the organisms living in a stream by lowering the available oxygen in the water.
- *Pathogens*. A pathogen is an organism that produces a disease.

The process of eutrophication is the most widespread phenomenon in the life of water ecosystems. An excess amount of nitrogen and phosphorus leads to a rapid growth of phytoplankton or algae biomass, and as a result, the content of oxygen in the water decreases sharply, and the mortality of living organisms grows substantially. The gas exchange with the atmosphere is violated. From available estimates, the share of eutrophicated lakes on different continents constitutes: Asia—54 %, Europe—53 %, North America—48 %, South America—41 %, and Africa—28 %. In the present-day world, it is difficult to differentiate the anthropogenic process of eutrophication from natural because of globalized biogeochemical cycles and difficult control of the fluxes of chemical elements through the atmosphere and river run-off.

Stability of the biosphere as a global ecosystem and its self-regulating ability are determined by the cyclic character of the processes of exchange with matter, energy, and information taking place in it and realized due to the incoming solar energy and activity of living substance. These processes manifest themselves in the following three basic forms:

(1) *Biological cycle* covers all biophyllic elements and vitally important microelements and is characterized by selection of light-weight isotopes of carbon, hydrogen, nitrogen, and sulfur from heavier forms.
(2) *Water cycle in the biosphere* determines the planetary transitions of various components such as aerosols, microorganisms, dissolved and suspended substances.
(3) *Processes of erosion, chemical denudation, transition, sedimentation, and accumulation* of mechanical and chemical deposits on land and in the ocean provide the global circulation of matter and energy.

Therefore a discussion of the greenhouse effect problem cannot by constructive without a complex consideration of feedbacks of the CO_2 cycle with the biogeochemical processes in the presence of other elements such as nitrogen, sulfur, phosphorus, methane, ozone, water, and others (Fasham 2003; Melillo et al. 2003; Wang et al. 2005). The processes of CO_2 assimilation from the atmosphere are affected by numerous natural and anthropogenic factors manifested through a long chain of cause-and-effect bonds. For instance, acid rains affect the state of the vegetation cover and the latter affects the CO_2 exchange at the atmosphere-land boundary. The use of nitrogen ($\approx 90 \times 10^6$ t N year^{-1}) and phosphorus fertilizers in agriculture changes the role of cultural plants in assimilation of atmospheric CO_2 and affects the rates of decomposition of the soil organic matter. Besides, in agriculture manure is an important source of greenhouse gases CH_4 and N_2O. So, Park et al. (2006) studied the fluxes of CH_4 and N_2O from supplies of liquid pig manure under cold climate conditions at an annual mean temperature below 10 °C at the farms of Ontario (Canada) for the period 2000–2002. At the annual mean air temperature 8.4 °C the manure temperature was, on the average, higher by 4 °C, and an average content of dry matter in manure and decomposition potential ranged within 0.6–3 % and $-232 \div 333$ mV. Average fluxes of N_2O changed depending on the firm from zero to 337.6 μg m^{-2} s^{-1} in summer and to 101.8 μg m^{-2} s^{-1} in winter. Monthly mean fluxes of CH_4 ranged between 4.6×10^{-3} and 1.05 mg m^{-2} s^{-1}. The presence of such data for the globe would make it possible to specify the structure of the biogeochemical cycle of GHGs.

However, the complexity of the biogeochemical cycles of GHGs and estimates of how well they are understood indicate the necessity of caution when predicting global changes and the development of new information technologies to study these cycles in correlation with other global processes. The developed countries spend huge resources to create an information base which would provide reliable predictions of climate change. But, practically all international programs targeting this are investigating parts of the overall scheme. For instance, the scientific priorities of the Joint Global Ocean Flux Study (JGOFS) program include:

- Determine how changes in basin-scale forcing affect the dynamics of the North Pacific Drift Current and how these dynamics affect the nutrient and carbon trapping capacity of the California Current System.
- Understand the imbalance between nitrogen fixation and denitrification (the marine nitrogen cycle) and its relationship to the ability of the oceanic biological pump to sequester anthropogenic carbon dioxide.
- Quantify how regime shifts interact with seasonal and stochastic variability to produce extreme events such as the recent coccolithophorid bloom in the Bering Sea and the basin-scale hypoxia (Murata and Takizawa 2002; Weeks et al. 2004).

Unfortunately, even a complex program such as the GCP cannot resolve the problem of assessing enough information for reliable prediction of global change. One of the technologies capable to constructively resolve this problem is the GIMS-technology.

On the whole, many chemical elements, especially GHGs, getting to the environment from anthropogenic sources, become an object of not only biogeochemical analysis but also of economic consideration. Such a multi-purpose analysis in connection with CH_4 was carried out at the Second International Conference in Novosibirsk in 2000 (Bazhin 2000; Byakola 2000). Such connections should be thoroughly systematized and parameterized. Otherwise it is impossible to speak about any reliable assessment of the role of the biosphere in assimilation of excess CO_2 from the atmosphere. Complex studies in this direction are being carried out, for instance, in several laboratories in the USA and Europe (Friedrich 2001). Measurements of spatial and temporal distributions of gases related to the global CO_2 cycle are made in the zones of the functioning observatories with the use of flying laboratories and specialized stationary platforms. Accumulation of such data will make it possible to reveal dependences needed for the global model. However, the USA take their irreconcilable stand with respect to the Kyoto Protocol despite the fact that CO_2 emissions from their territory reach almost 25 % (541×10^7 t CO_2 year^{-1}) of all its anthropogenic sources over the globe. In March 2001, President Bush said he wouldn't be ratifying the Kyoto Protocol because it could significantly damage the country financially. He was also concerned about the pressure on "industrialized" countries to cut back on carbon dioxide, while developing countries weren't expected to cut theirs back too. Emissions in America have continued to rise and are now 11 % higher than in 1990, even though when they did temporarily sign up to Kyoto, they promised a 6 % reduction.

All this confirms the fact that fragmentary studies of the global carbon cycle (i.e., not based on a complex such as that described in Krapivin and Kondratyev (2002) will always raise doubts. For global conclusions, like those made in the Kyoto Protocol recommendations, we need to be sure that the predicted global consequences are accurate. Nevertheless, such conclusions and assessments are necessary. Unfortunately, most of the international programs on the subject considered are not aimed at the development of the global modeling technology and do not concentrate the efforts of specialists on deriving numerical NSS models.

Existing global models are simple and inadequately provided with data bases. Three directions in the global modeling have been described in the works of Kondratyev et al. (2002) and Boysen (2000). In each of them one or several components are absent, but on the whole, conceptually they are mutually additively. This makes it possible to combine them and, hence, to derive a global model which would take into account the most important processes in the nature-society system. One of them is the gas exchange between the atmosphere and vegetation cover described in global models at a very high level. Nevertheless, models of land ecosystem have recently appeared, such as Biome-GCP, Forest-GCP or TsuBiMo (Wang et al. 2005; Alexandrov et al. 2005) which simulate the supplies and fluxes of energy, water, hydrogen, and nitrogen in the vegetation cover, litter, and soil, which enables us to specify the role of land in regulating radiation balance of the atmosphere-plant-soil system. In particular, Wang et al. (2005) have studied the fluxes of H_2O, CO_2, and nitrogen in the plains of North China from the data on vegetation and soils for 2002 and showed that the model Biome-GCP provides a reliable assessment of the response of land ecosystems to the anthropogenic interference into the natural balance of water, carbon, and nitrogen. Without any interference these ecosystems are in a balanced state with regard to these elements, but interference intensifies the fluxes of CO_2 and H_2O, and excess nitrogen gets into the soil and neighboring water basins, contaminating them.

Clearly, a systematization of the models of land ecosystems and their introduction to GIMS as alternative units, whose application is regulated by available data bases, will make it possible to markedly reduce the level of uncertainty in the estimates of CO_2 sinks and sources. Such a study has been exemplified in the work of Alexandrov et al. (2005) where a calibration of TsuBiMo by the data base of the Osnabrueck Centre for Environment and Technology (Germany) is demonstrated. The proposed scheme of calibration consists of four stages:

(1) Reduction of the number of unspecified parameters of the model by introducing a generalized parameter.
(2) Evaluation of generalized parameters from available data base.
(3) Formation of the empirical model relating generalized parameters to climate.
(4) Establishing relationship between the global multitude of generalized parameters and global fields of climatic variables.

Applying this scheme of calculation to the TsuBiMo model, Alexandrov et al. (2005), from measurements of CO_2 fluxes in the neighborhood of Takayama (Japan), constructed an empirical model to calculate monthly mean temperatures and showed that the accuracy of calculations of the inter-annual and intra-annual variability of biomes productivity can be substantially increased on a local level.

The photochemical system of the atmosphere has been poorly studied and is not considered in global models. Knowledge of the laws of changes in the incoming radiation intensity in connection with the effects of its assimilation by gases and aerosols will make it possible to raise the accuracy of the greenhouse effect estimates. And first of all, of importance here is the role of molecular nitrogen, ozone, water vapour, nitric oxide, sulfur dioxide, nitrogen dioxide, CH_4, CO_2, and other gases. Greenhouse gases, other than CO_2, can in the future play an unpredictable role in formation of the Earth radiation balance. Therefore some models are proposed here which enable one to parameterize part of the indicated elements.

1.6 Block Schemes of Models for Biogeochemical Cycles

1.6.1 Schemes of the Global Carbon Cycle

The reliability of the assessment of the role of CO_2 in the greenhouse effect formation depends on a detailed consideration of the global biogeochemical carbon cycle dynamics in the models and on the accuracy of the assessment of its characteristics. There are dozens of diagrams of the global carbon cycle in the form of CO_2 changes. Consider some of them to demonstrate their principal features and to understand the limit for necessary details of the simulation of the carbon cycle compounds, beyond which it is impossible to obtain an additional knowledge about this cycle and, hence, about the greenhouse effect due to CO_2. Note that all known diagrams of the global CO_2 cycle are divided into two classes: point wise (globally averaged) and spatial (locally averaged). All the diagrams are similar in that the biosphere is divided into the atmosphere, oceans, and land ecosystems. Many diagrams divide carbon into organic and inorganic forms. As a rule, the time step of averaging all processes and reservoirs of carbon is assumed to be 1 year, and therefore the atmospheric reservoir is considered as homogeneously mixed-up (pointwise). The World Ocean and surface ecosystems are considerably detailed. This detailing is based on global data bases for these reservoirs of carbon. As a rule, the final results of the diagrams studies are either of methodical character or they predict the atmospheric CO_2 concentrations within the limits of a certain scenario of the anthropogenic activity.

The schemes in Figs. 1.23 and 1.24 give an idea about the amounts of carbon supplies for its basic reservoirs. The estimates shown in this scheme differ drastically from the estimates by other authors. Nevertheless, their relationships and orders of magnitude coincide in most cases. As seen, the largest carbon supply is concentrated in the World Ocean. A minimum of it is in the atmosphere.

Natural processes determining the global carbon cycle dynamics have various time scales. Some of them, such as burying the dead organic matter at the bottom of the oceans, have characteristic time scales of hundreds and thousands years. Others, for instance the biological carbon cycle on land, have a period of several tens of years. Therefore a consideration of the timescales of the carbon cycle in the biosphere is an important stage of studies of the dynamics of the CO_2 content in the atmosphere. Here it is also important to consider the fact that the characteristic time of the complete mixing of the atmosphere constitutes from several months to 2 years. Regular measurements at different stations of monitoring have shown that the CO_2 concentration varies considerably during 1 year. The difference between maximum and minimum estimates of the atmospheric CO_2 concentrations varies

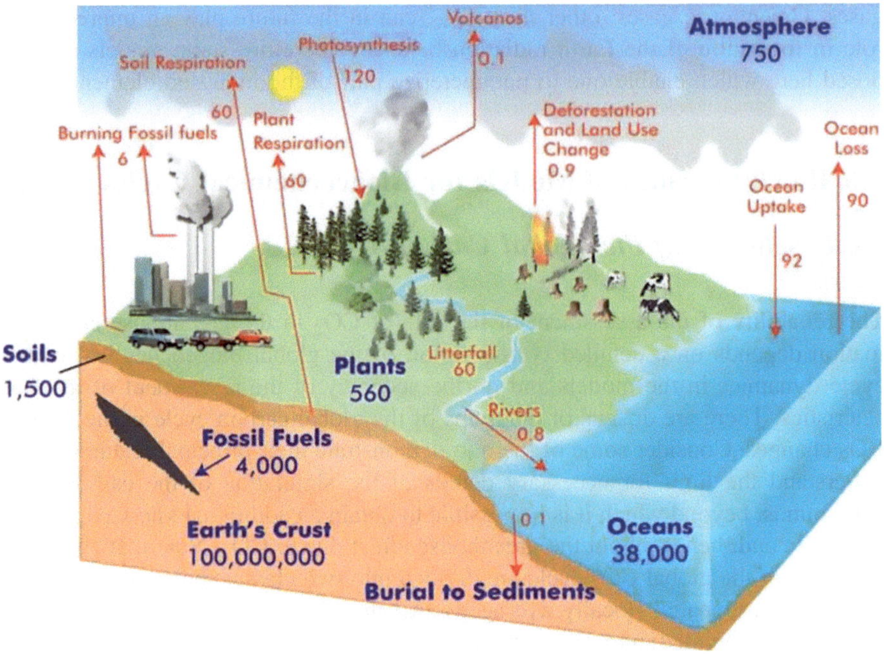

Fig. 1.23 A simplified diagram of the global carbon cycle. Pool sizes, shown in *blue*, are given in petagrams (Pg) of carbon. Fluxes, shown in *red*, are in Pg per year (http://www.globe.gov/projects/carbon)

from 10 ppm at the South Pole to 15 × 10 ppm at the NH high-latitudes. This spatial non-uniformity is explained by the presence in the Northern Hemisphere of large seasonally photosynthesizing vegetation communities.

Diagram of the fast carbon cycle represented in Fig. 1.25 shows the movement of carbon between land, atmosphere, and oceans in billions of tons of carbon per year. Yellow numbers are natural fluxes, red are human contributions in billions of tons of carbon per year. White numbers indicate stored carbon. As it follows from Figs. 1.23, 1.24 and 1.25 latitudinal distribution of CO_2 in the atmosphere has non-linear character. An increased CO_2 content in the NH atmosphere is closely connected with the impact of human activity through direct CO_2 emissions and due to the impact on vegetation cover. Almost 90 % of total carbon emissions due to organic fuel burning fall on the zone 30°N–60°N. It follows from the data in Table 1.9 that the conceptual schemes of the global biogeochemical cycle of carbon dioxide should also consider the spatial non-uniformity of atmospheric processes (Kaminski et al. 2001).

An important constituent of most of the conceptual schemes of the global carbon cycle is the structure of carbon fluxes in the World Ocean. As follows from Table 1.9, there is a certain information possibility to select in the oceans several layers by depths and to distinguish between the spatial non-uniformities in the structure of the ocean surface. Most of the authors consider the vertical structure of

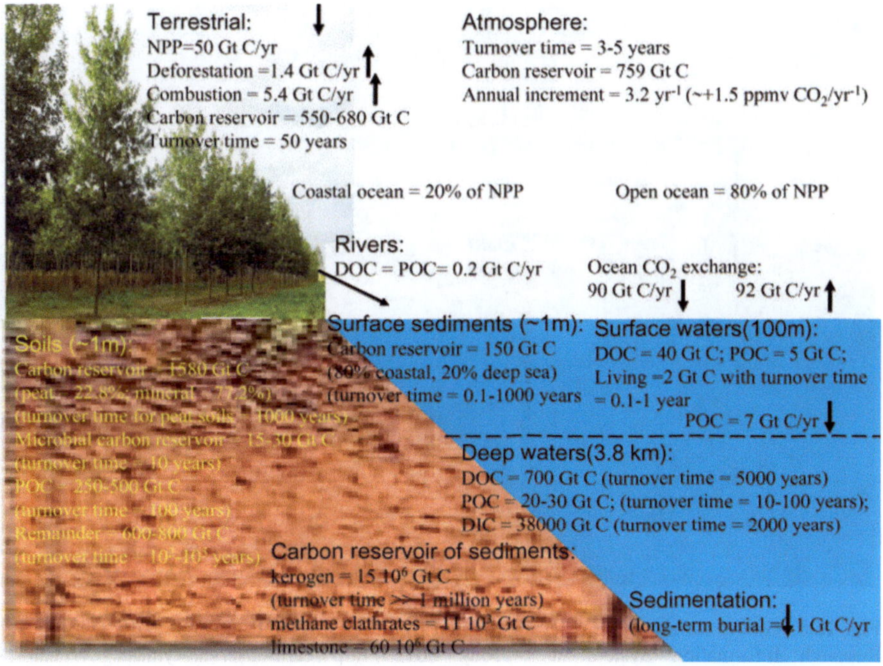

Fig. 1.24 Global carbon reservoirs, fluxes, and turnover times. Pool sizes and fluxes are given in Gt (10^{15} g) C and Gt C year^{-1}, respectively. Turnover times (reservoir divided by largest flux to or from reservoir) are in parentheses. To convert Gt C to moles C, multiply by 8.3×10^{10} (http://www.ess.uci.edu/~reeburgh/fig1.html)

the oceans in the form of two or three layers covering the photic layer and deep layers. Within the photic layer, the layers can be selected above and beneath the thermocline. The spatial non-uniformity of the ocean is demonstrated through upwelling and latitudinal zones, with different rates and directions of the oceans-atmosphere CO_2 exchange. More detailed schemes of the global carbon cycle take into account non-uniformities in the carbonate system of the oceans, which makes it possible to considerably increase the accuracy of the respective models. Two of the widely used schemes of this type are shown in Figs. 1.26 and 1.27.

The role of the World Ocean in the global CO_2 cycle is mainly manifested through the process of its exchange on the atmosphere-ocean border. The intensity of gas exchange between the ocean and the atmosphere is determined by the dynamic characteristics of the turbulent layers of water and air near the interface. Here, numerous physical schemes appear that reflect the situations of sea wave formation, as well as formation of foam and various films. As a result, carbon dioxide either dissolves in the ocean providing thereby the inflow of CO_2 needed for photosynthesis or is emitted from the ocean into the atmosphere. The cause of this binary situation on the air-water border is the difference between the partial pressures of CO_2 in the atmosphere and CO_2 dissolved in the water. Actually, this

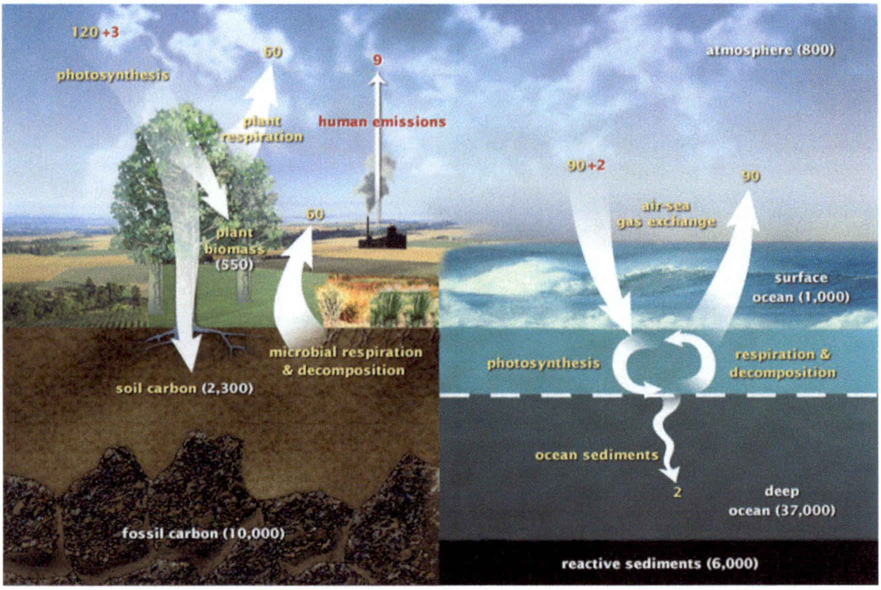

Fig. 1.25 Global carbon cycle diagram adapted from U.S. DOE, Biological and Environmental Research Information System at 6 May 2012. http://earthobservatory.nasa.gov/Features/CarbonCycle/

Table 1.9 Global carbon reservoirs (Gt)

Reservoir	Amount (C)
Atmosphere	720
World Ocean	38,400
Total inorganic carbon	37,400
Surface layer	670
Deep layers	36,730
Total organic carbon	1,000
Lithosphere	
Carbonate sedimentary rocks	>60,000,000
Kerogens	15,000,000
Land biosphere	2,000
Living biomass	600–1,000
Dead biomass	1,200
Biosphere of inland waters	1–2
Burnt fuel	4,130
Coal	3,510
Oil	230
Natural gas	140
Others (peat, etc.)	250

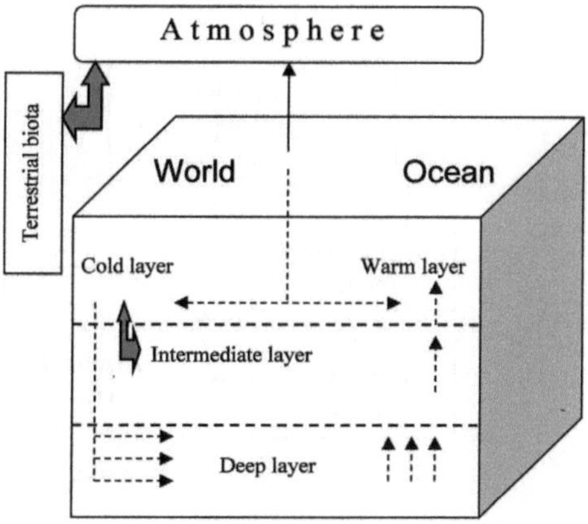

Fig. 1.26 A scheme of the carbon reservoirs and fluxes in the models with the vertical structure of the World Ocean

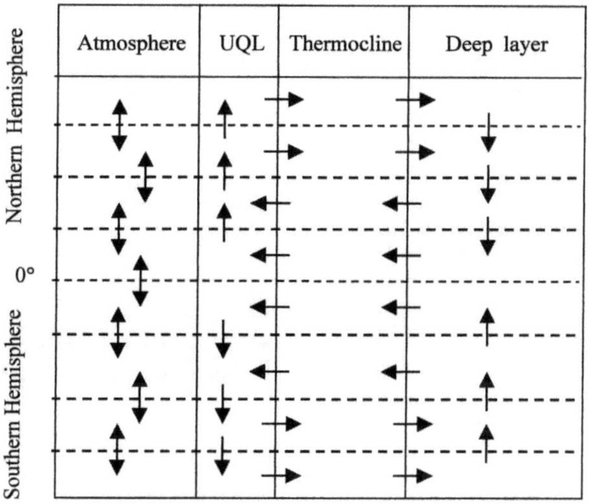

Fig. 1.27 A scheme of the World Ocean global circulation as the basis for the description of carbon cycles. *UQL* upper quasi-homogeneous layer (Tarko 2001, 2003)

directed transport of CO_2 on the border "atmosphere-ocean" is more complicated. Its study requires expensive field experiments and a detailed classification of both synoptic and physico-geographical situations on the oceans' surface.

In the programmes of studies of the global carbon cycles the emphasis is placed on the role of surface ecosystems in its formation (Houghton et al. 2001). In the process of photosynthesis, plants assimilate carbon dioxide and, on the contrary, the decomposing dead plants emit carbon dioxide into the atmosphere. Thus a continuous CO_2 exchange between living and dead organic matter and the atmosphere takes place in the land biosphere. There are many conceptual diagrams, which formalize this exchange and serve the basis for global models of the carbon cycle. Two examples of such diagrams are given in Figs. 1.28 and 1.29.

Clearly, the accuracy of estimates of carbon fluxes in the terrestrial part of the biosphere is the function of a detailed quantization of the types of soil-plant

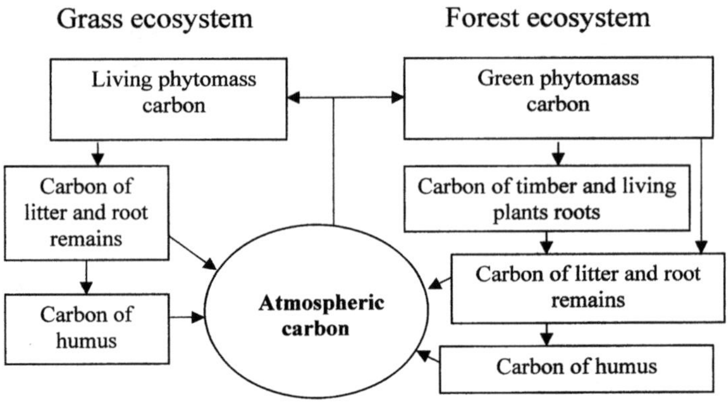

Fig. 1.28 The diagram of the carbon fluxes in the system atmosphere-plant-soil with the grass and forest ecosystems as an example

Fig. 1.29 The scheme of the global carbon cycle by Watson et al. (2000). *Designations*: *GPP* gross primary productivity, *NPP* net primary productivity, *NEP* net ecosystem productivity, *NBP* net biome productivity

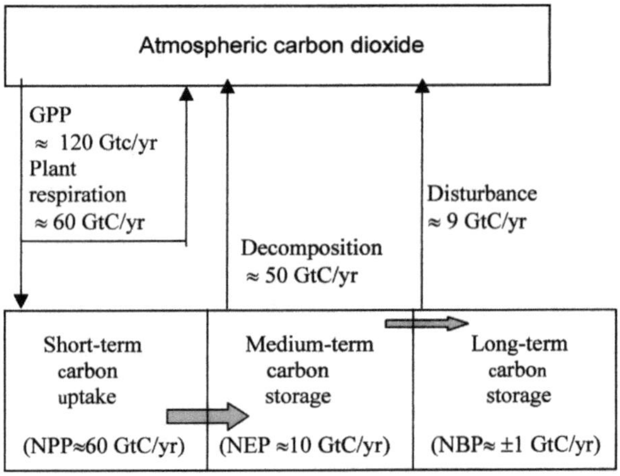

formations and accuracy of the parameterization of the biocenotic processes. In this regard the world maps of vegetation and soils have been drawn, their areas estimated, and data on the vital functions of soil microorganisms have been accumulated, and technologies of an operational monitoring of landscapes have been worked out. Unfortunately, there are no reliable enough estimates of the limits for considering details of the soil-plant formations, which would provide the needed accuracy of the carbon fluxes estimation. To solve this problem, a global model of carbon cycle is needed which would reflect a hierarchy of important, by their expert estimates, constituents of the biosphere and the processes of carbon transformation. An exemplary scheme of carbon flux in this model is characterized in Fig. 1.30. Numerical experiments with this model will make it possible to obtain comparative assessments of the consequences of complicating its various constituents and thereby to approach sequentially the level of an optimal spatial quantization.

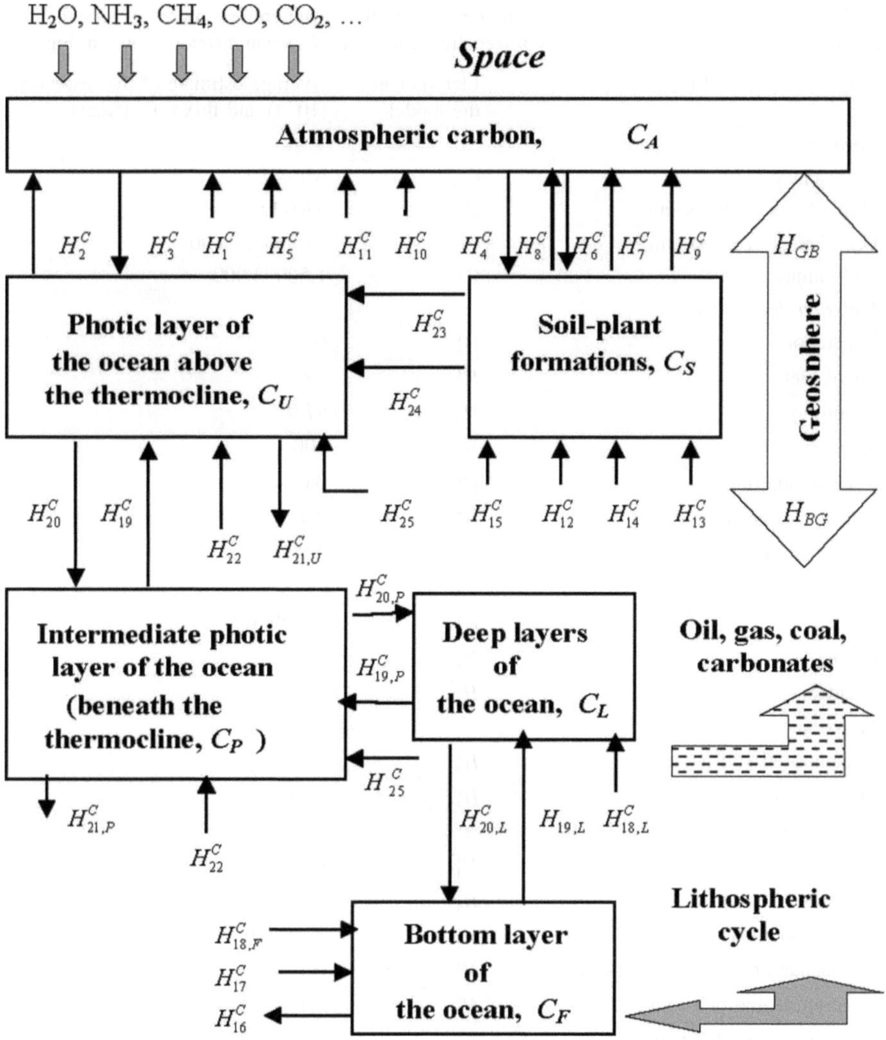

Fig. 1.30 The block-diagram of the global biogeochemical cycle of carbon dioxide in the system atmosphere-land-ocean. The CO_2 reservoirs and fluxes are described in Table 1.10 (Varotsos et al. 2014)

1.6.2 Conceptual Schemes of the Nitrogen Cycle in Nature

The global cycle of nitrogen as one of the nutrient elements is a mosaic structure of local processes of its compounds formed due to water migration and atmospheric processes. The present-day nitrogen cycle is strongly subject to anthropogenic forcings manifested through interference into the nitrogen cycle both directly and via the influence on the related processes. Therefore the construction of an adequate

Table 1.10 Reservoirs and fluxes of carbon as CO_2 in the biosphere in a model of global carbon dioxide cycle (MGCDC) of the global biogeochemical cycle of carbon dioxide shown in Fig. 1.30

CO_2 reservoirs and fluxes	Identifier in the model	Average estimate of the reservoir (10^9 t) and flux (10^9 t/year)
Carbon		
Atmosphere	C_A	650–750
Photic layer of the ocean	C_U	580–1,020
Deep layers of the ocean	C_L	34,500–37,890
Soil humus	C_S	1,500–3,000
Emission due to burning		
Vegetation	H_8^C	6.9
Fossil fuel	H_1^C	3.6
Desorption	H_2^C	97.08
Sorption	H_3^C	100
Rock weathering	H_4^C	0.04
Volcanic emanations	H_5^C	2.7
Assimilation by land vegetation	H_6^C	224.4
Respiration		
Plants	H_7^C	50–59.3
People	H_{10}^C	0.7
Animals	H_{11}^C	4.1
Emission		
Decomposed soil humus	H_9^C	139.5
Plants' roots	H_{15}^C	56.1
Vital functions		
Population	H_{12}^C	0.3
Animals	H_{13}^C	3.1
Vegetation decay	H_{14}^C	31.5–50
Sedimentation to bottom deposits	H_{16}^C	0.1–0.2
Solution of marine sediments	H_{17}^C	0.1
Decomposition of detritus		
Photic layer	H_{22}^C	35
Deep layers of the ocean	H_{18}^C	5
Uplifting with deep waters	H_{19}^C	45

(continued)

Table 1.10 (continued)

CO_2 reservoirs and fluxes	Identifier in the model	Average estimate of the reservoir (10^9 t) and flux (10^9 t/year)
Lowering with surface waters and due to gravitational sedimentation	H_{20}^C	40
Photosynthesis	H_{21}^C	69
Underground sink	H_{23}^C	0.5
Surface sink	H_{24}^C	0.5–0.6
Breathing of living organisms in the ocean	H_{25}^C	25
Geospheric sink of carbon	H_{BG}	70
Geospheric source of carbon	H_{GB}	17

model of the nitrogen cycle in nature should be based on the description of the whole complex of natural processes and those initiated by humans. A general description of supplies and fluxes of nitrogen is schematically given in Figs. 1.31, 1.32, 1.33, 1.34 and 1.35.

The natural sources of nitrogen oxides are connected with the vital functions of bacteria, volcanic eruptions, as well as several atmospheric phenomena (e.g., lightening discharges). The biogeochemical cycle of nitrogen includes the processes

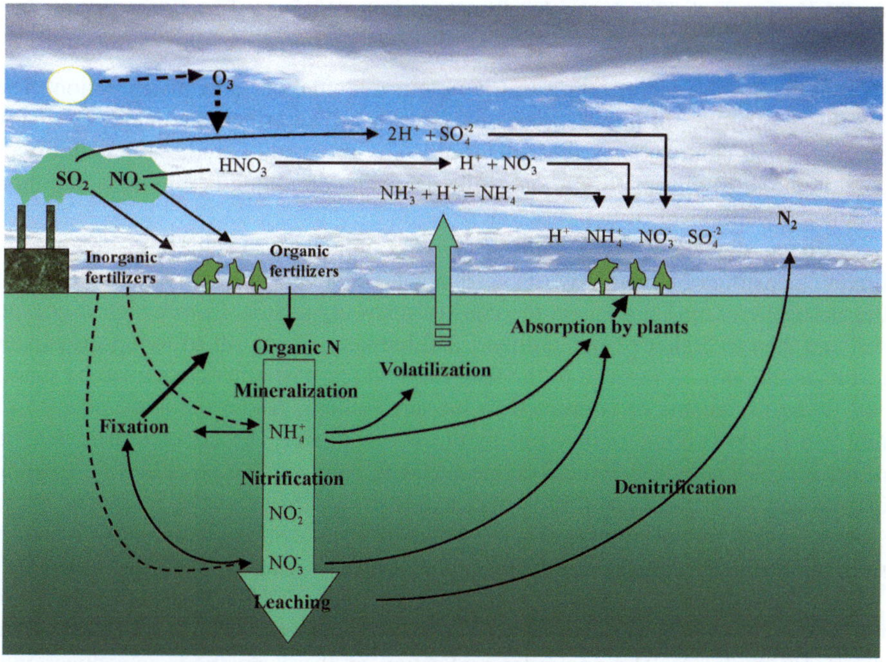

Fig. 1.31 A scheme for the circulation of sulphur and nitrogen with the formation of acid precipitation

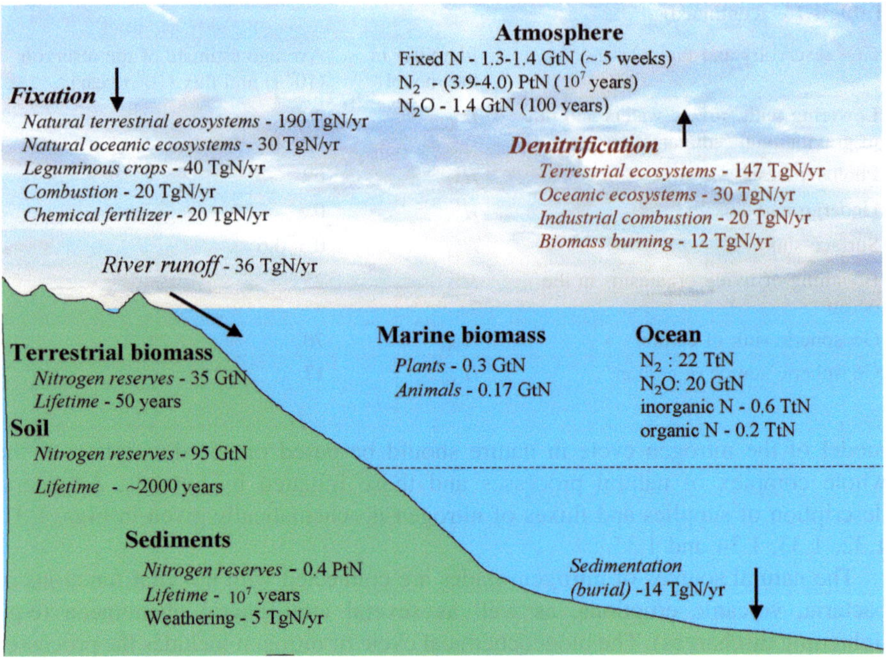

Fig. 1.32 Reserves, fluxes, and cycling times of nitrogen in the atmosphere-biosphere-geosphere system (Harrison et al. 2005; Vitousek 2004). *Notation* Pt = 10^{15} tons, Tg = 10^{12} g, Gt = 10^9 tons

such as fixation, mineralization, nitrification, assimilation, and dissimilation. The structural schemes of these processes have been described in detail by many authors. Their complexity level is determined by the goal of studies, availability of data on the rates of transformation of the nitrogen-containing compounds and their supplies, by the level of detailing, etc.

Nitrogen moves in the biosphere by a complicated meandering structure of fluxes consisting of a hierarchy of cycles at various levels of life organization. From the atmosphere, nitrogen enters the cells of microorganisms, from which it goes to soil and then to higher plants, animals, and humans. Dying off of living organisms results in the turn of nitrogen to the soil, from which it is either consumed by plants and living organisms or is emitted to the atmosphere. About the same scheme of nitrogen oxide cycling is inherent to the hydrosphere. The characteristic feature of these cycles is their openness connected with the available processes of removal of nitrogen from the biospheric balance into rocks, from where it returns much slower. Taking into account the nature of the nitrogen cycle in the biosphere and its reservoir structure enables one to formulate a global scheme of nitrogen fluxes.

To simplify the calculation scheme presented in Fig. 1.35, advection processes in the balance equations of nitrogen cycle can be described by a superposition of the fluxes H_{14}^N and H_{15}^N. The computer realization of these equations introduces into them some corrections for the agreement between the dimensionalities of the

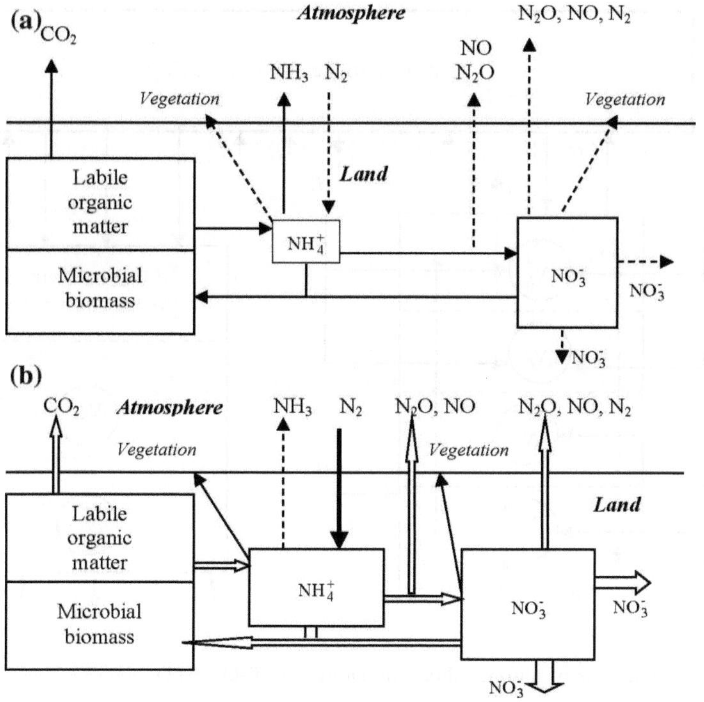

Fig. 1.33 Block-diagram of biogeochemical cycles of C and N in the water-limited ecosystems (Austin et al. 2004). **a** Dry season; **b** wet season

Fig. 1.34 The scheme of nitrogen fluxes in the marine medium

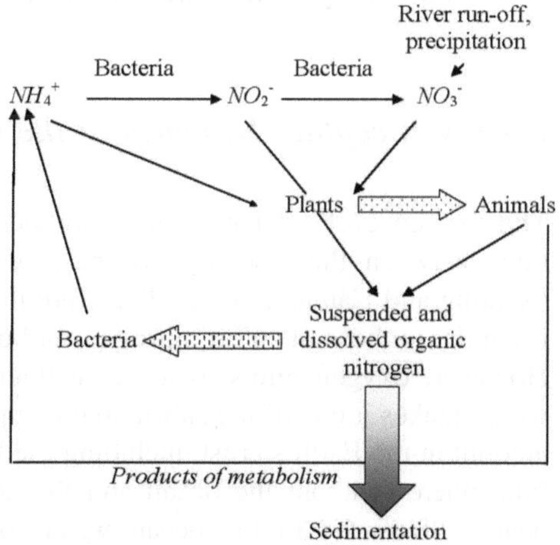

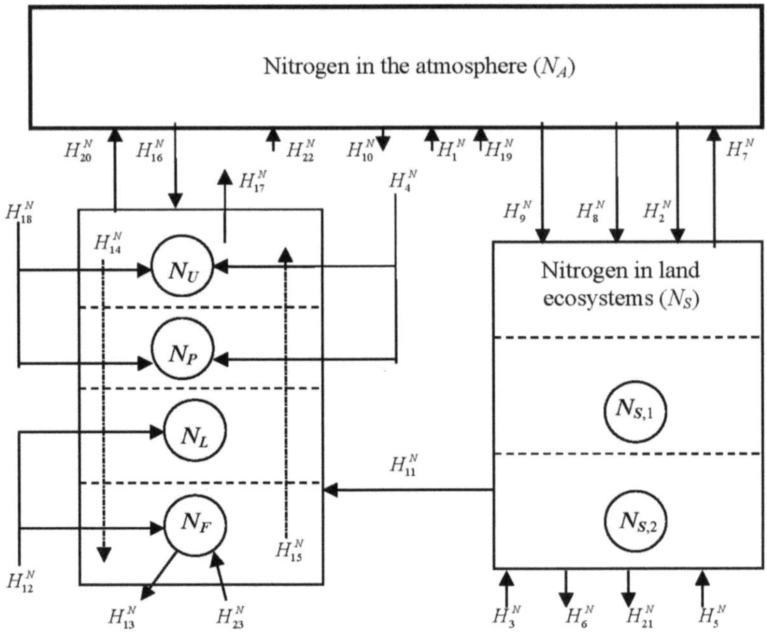

Fig. 1.35 The scheme of nitrogen fluxes in nature (see Table 1.11)

variables in conformity with the spatial digitization of specific territory. Therefore the estimates of the fluxes H_i^N given below for their consideration in the GIMS should be corrected following this criterion. Detailed description of the nitrogen cycle as unit of the GIMS was given by Nitu et al. (2000a).

1.6.3 Conceptual Schemes of the Global Oxygen Cycle

The oxygen cycle in nature is composed of characteristic biogeochemical transitions between the reservoirs of basic constituents circulating in the biosphere (Kasting and Canfield 2012). Therefore the block-scheme of the oxygen exchange is similar to those of sulfur, nitrogen, carbon, and phosphorus (Figs. 1.36 and 1.37). However, oxygen refers to the constituents spread over the globe most widely, which makes it one of the substantial components of the biogeochemical cycles. Its amount in the Earth's crust, including the hydrosphere, reaches 49 % by mass. The lithosphere (without the ocean and the atmosphere) contains 47.2 % of oxygen, water—88.89 %. In the ocean water, oxygen constitutes 85.82 %, the living substance contains 65 % by mass. These estimates testify to a significance of oxygen for the biosphere, an appearance and existence of which are determined by the presence of oxygen. Now about 39×10^{14} t O_2 circulate in the biosphere,

Table 1.11 Characteristics of reservoirs and fluxes of nitrogen in the biosphere (Fig. 1.35)

Reservoirs (Gt) and fluxes (10^6 t/year)	Identifier	Estimate
Nitrogen supplies		
Atmosphere	N_A	39×10^5
Soil	N_S	280
Photic and intermediate layer of the ocean	$N_U + N_P$	2,800
Deep and bottom layer of the ocean	$N_L + N_F$	36,400
Natural sources of the hydrosphere	H_1^N	0.392
Technogenic accumulation		
Fuel burning	H_2^N	22.8
Fertilizers production	H_2^N	41.8
Input due to dead organisms		
On land	H_3^N	42.2
In upper layers of the World Ocean	H_{18}^N	5
In deep layers of the World Ocean	H_{12}^N	7.8
Input due to organisms functioning		
On land	H_5^N	0.1
In the World Ocean	H_4^N	0.3
Biological fixation		
On land	H_6^N	20.3
In the World Ocean	H_{17}^N	10
In the atmosphere	H_{10}^N	40
Denitrification		
On land	H_7^N	52
In the World Ocean	H_{20}^N	49.8
Atmospheric fixation		
Over land	H_8^N	4
Over the World Ocean	H_{16}^N	3.6
Run-off from land into the World Ocean	H_{11}^N	38.6
Precipitation	H_{13}^N	0.5
Vertical exchange processes in the oceans		
Descending	H_{14}^N	0.2
Lifting	H_{15}^N	7.5
Anthropogenic emissions to the atmosphere	H_{19}^N	15
Removal of nitrogen from the cycle due to sedimentation	H_{21}^N	0.2
Input of nitrogen to the atmosphere during rocks weathering	H_{22}^N	0.217
Input of nitrogen to the water medium with dissolving sediments	H_{23}^N	0.091

37×10^{18} mol O_2 reside in the atmosphere, the oceans and the long-lived biota contain 219×10^{15} and 180×10^{15} mol O_2, respectively. The periods of complete cycle of oxygen vary from 3×10^6 years for the atmosphere to 22 days for the

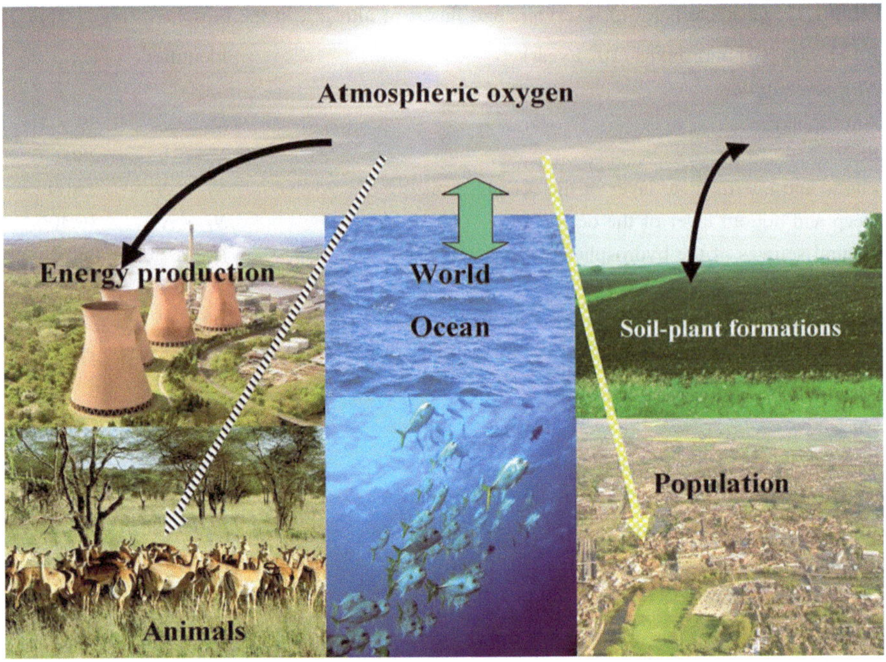

Fig. 1.36 Simplified scheme of the biogeochemical oxygen cycle in the biosphere

surface waters of the World Ocean. On the whole, for the oceans and long-lived biota, oxygen completes its cycle for 500 and 50 years, respectively.

Oxygen is present in the biosphere in the form of molecular oxygen (O_2), ozone (O_3), atomic oxygen (O), and as a constituent of various oxides. On the one hand, oxygen maintains life on the Earth due to the process of respiration and formation of the ozone layer, and on the other hand, it itself is the product of the organisms' functioning. This fact hinders a description of its cycle, since it requires a synthesis of the descriptions of various processes. Here an attempt has been made to describe the cycle and to derive a model of the global biogeochemical oxygen cycle (MGBO) as unit of the BCSS global model.

Many authors believe that in the nearest future nothing threatens the stability of the global biogeochemical cycle of oxygen. This statement is not valid for ozone, whose concentration and spatial distribution suffered serious changes during the last decades. According to Kondratyev and Varotsos (2000), the available observations of the vertical profile of atmospheric ozone show a very complicated spatial and temporal variability that depends on many characteristics of the nature-society system. The MGBO unit at a parameterization of ozone fluxes follows a numerical model by Aloyan (2004) and Arutiunian et al. (2004), with a needed correlation taken into account. This correction consists in the substitution of some functional dependence for scenarios reflecting the dynamics of changes in concentrations of the chemicals not described in the global model of the carbon cycle.

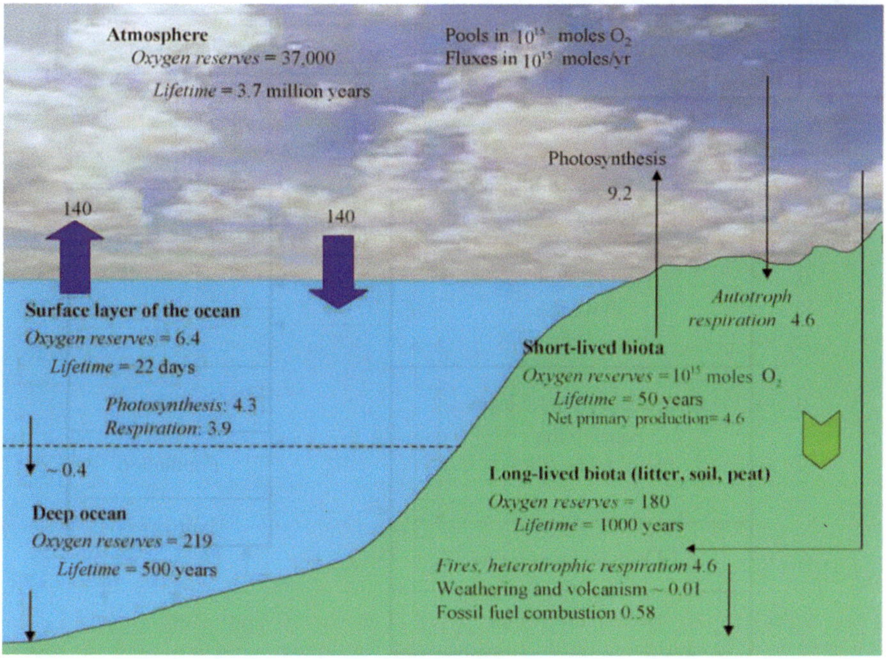

Fig. 1.37 Reserves, fluxes, and living times of the oxygen in its basic reservoirs

Now and in the geological past there have been two sources of oxygen—endogenic and photosynthetical. Without dwelling upon the respective scientific discussions and all existing concepts, we shall try to describe the sources of oxygen in the present biosphere, following numerous studies in this field.

The basic source of atomic oxygen is the photosynthesis of plants, whose equation has the form:

$$2H_2O + CO_2 \xrightarrow[\text{chlorophyll}]{\text{light}} COH + H_2O + O_2$$

Photosynthesis produces annually above 50×10^9 t of oxygen, that is, of an order of 3.3×10^{14} % of its supplies in the atmosphere. Hence, we see that only due to photosynthesis can the oxygen supplies in the atmosphere be totally renewed during a time period of 300 thousand years. About 80 % of the total amount of oxygen produced by photosynthesis result from phytoplankton functioning, and land vegetation communities produce only 20 %.

Global scheme of oxygen cycle in the BCSS was proposed by Krapivin and Varotsos (2008) that is represented in Fig. 1.38.

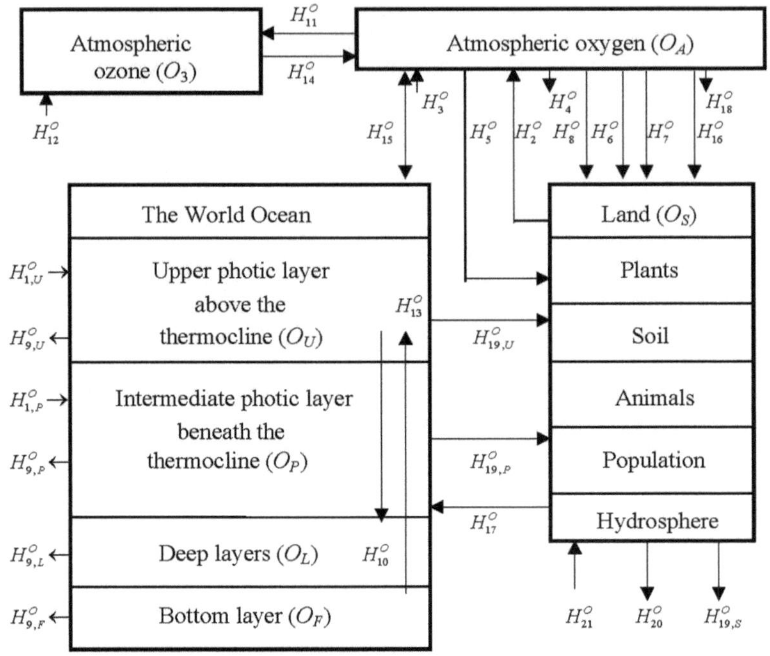

Fig. 1.38 Oxygen fluxes in the BCSS realized as the MGBO unit. Notation is given in Table 1.12

1.6.4 Biogeochemical Cycle of Methane

The SMKP item of the BCSS model includes additional descriptions of the methane cycle components that are not taken into account in other biogeochemical cycles. Methane like carbon dioxide belongs to greenhouse gases. The spectrum of its natural and anthropogenic sources is wide, and its greenhouse effect exceeds 20 times that of CO_2 though its concentration ($\sim$1.6 ppm) in the atmosphere is about 200 times less than that of CO_2 (Polischuk and Polischuk 2013). By the greenhouse effect, it is now second to carbon dioxide. Methane also affects the ozone content in the stratosphere and plays a key role in transformation of chemically active Cl into less active HCl. Before the human interference, the natural cycle of methane had been balanced with respect to climate. With extracting natural combustible gases consisting by 90–95 % of methane, humankind has contributed to this cycle factors of instability and uncertainty. In the whole, during the last 200 years the anthropogenic contribution to the input of CH_4 to the atmosphere has doubled. A situation has occurred when the difference between methane concentrations at Poles reaches 150 ppb. Most of the authors estimated the level of the global emission of methane into the atmosphere at 535×10^6 t CH_4 year^{-1}, of which 375×10^6 t CH_4 year^{-1} is of anthropogenic origin (50×10^6 t CH_4 year^{-1} being from the rice fields).

Table 1.12 Estimates of the reservoirs and fluxes of oxygen and ozone in the MGBO unit used in its adjustment (Krapivin and Varotsos 2008)

Reservoirs (t/km^2), fluxes $\left(\text{t km}^{-2}\,\text{year}^{-1}\right)$	Identifier	Estimate
Oxygen in the upper photic layer of the World Ocean	O_U	0.8×10^8
Oxygen in the transition layer of the ocean	O_P	0.7×10^9
Oxygen in deep ocean	O	3×10^4
Oxygen in the bottom layer of the ocean	O_F	9×10^3
Oxygen in the atmosphere	O_A	0.24×10^7
Oxygen in the surface part of the hydrosphere	O_S	0.6×10^8
Ozone	O_3	0.23
Photosynthesis in the ocean	H_1^0	108–388
Photosynthesis on land	H_2^0	70–100
Photodecomposition of water in the atmosphere	H_3^0	0.008
Oxidation processes in the atmosphere	H_4^0	0.009
Respiration of plants	H_5^0	0.07–0.1
Respiration of animals	H_6^0	50–60
Respiration of humans	H_7^0	70–80
Oxidation-restoration processes in soil	H_8^0	1
Oxidation processes in the World Ocean	H_9^0	164
Descending of the oxygen-saturated waters	H_{10}^0	190
Decomposition and destruction of O_2 in the atmosphere	H_{11}^0	0.76–1.32
Formation of O_3 from NO_2	H_{12}^0	0.23–22.2
Lifting of dissolved oxygen in the upwelling zones	H_{13}^0	36
Decomposition and destruction of ozone in the atmosphere	H_{14}^0	1.48–1.66
Exchange on the border atmosphere-ocean	H_{15}^0	18–140
Exchange on the border atmosphere-inland water bodies	H_{16}^0	18–140
Transport of oxygen into the ocean by the river run-off	H_{17}^0	50
Anthropogenic consumption of oxygen	H_{18}^0	60–90
Expenditures of O_2 on metabolism of aquatic animals	H_{19}^0	0.2
Oxidation processes in the continental water bodies	H_{20}^0	90–200
Photosynthesis in the continental water bodies	H_{21}^0	100–400

The anthropogenic input of methane is expected to grow within the next 20–30 years, though in some developed regions, measures are taken to reduce the anthropogenic emissions of methane into the atmosphere. Nevertheless, the concentration of methane in the present atmosphere increases 7 times faster than the growth of CO_2 concentration, so that its amount increases annually by 2 %, that is, by the year 2020 the amount of methane in the atmosphere can double compared to 2000, which, from numerous estimates, will lead to a global warming by 0.2–0.4 °C. Like in the case of CO_2, these estimates will remain rather doubtful and contradictory until the global model mentioned above is synthesized. However, with present knowledge, only first steps are possible in modeling all these features of the global cycle of CH_4.

The sources of methane are oil, sedimentary and ejected rocks, bottom sediments of lakes, seas, oceans, and other objects of the hydrosphere, as well as soil, peat-bogs, rice plantations, and others:

- natural gas and oil extracting and refining systems;
- vital functions of animals;
- heaps of solid waste;
- coal mines functioning;
- processing of stock-breeding waste;
- sewage processing;
- anaerobic decomposition of organic matter in flooded rice fields;
- fossil fuel burning at stationary and mobile systems;
- burning of agricultural waste, biomass, rubbish, and fires in savannahs; and
- various industrial processes.

The basic reaction of methane formation due to anaerobic fermentation or mineralization of organic matter is as follows:

$$C_6H_{12}O_6 \rightarrow 3CO_2 + 3CH_4.$$

Sinks for tropospheric methane are:

- reaction with hydroxyl radical ($\sim$90 %);
- transport to the stratosphere ($\sim$5 %); and
- oxidation in dry soil ($\sim$5 %).

Ways of methane transformation, as shown in Table 1.13, include numerous processes by 70–80 % of biogenic origin mostly affected by humans. Of course, the significance of these processes varies depending on many natural and anthropogenic parameters. Relationships of individual elements of correlation between the cycles of CO_2, CH_4, and other chemicals vary, too. At any rate, it is clear that depending on the strategy adopted for the nature-society system, in due course the composition of the terrestrial atmosphere will change substantially. Suffice it to say that only burning of 1 m^3 of methane extracts from the atmosphere 2 m^3 O_2. From the open dust-heaps and municipal and industrial sewage, the atmosphere receives annually about 2 % of anthropogenic methane ($\sim$270–460 $\times$ 10^6 t C). These integral estimates do not permit one to calculate an actual distribution of CH_4 flux in the atmosphere.

A certain contribution to detailing of the spatial distribution of the sources of methane was made at the Second International Conference on the problems of methane held in Novosibirsk in 2000. Proceedings of this Conference contain concrete data on the sources of methane in many regions of the globe. For instance, according to Byakola (2000), within the framework of the international UNEP/GDP project, an inventory of the sources and sinks of CO_2 and CH_4 has been made for the territory of Uganda (236 $\times$ 10^3 km^2). In Uganda, basic anthropogenic sources of

Table 1.13 Sources of the input of CH_4 into the terrestrial atmosphere

Source of CH_4	Area of the source (10^6 km^2)	Rate of CH_4 formation (g m^{-2} $year^{-1}$)	Average rate of CH_4 formation (10^6 t/year)
Rice fields	1.35	206	280
Marshes	2.6	50–100	130–260
Fresh-water lakes	2.5	50–100	1.25–25
Arid soils	30	0.44	10
Woodlands	44	0.01–0.09	0.4
Tundra	8	10	0.8–8
Oceans	361	0.012	4–6.7
Shelf zones	1.4	5–10	0.07–1.4
Animals			101–220
Termites			150
Fossil fuels			100
Dust-heaps			20–40
Sewage			30–40

methane are agriculture, municipal sewage, and biomass burning. In 1990, the stock-breeding and rice fields in Uganda contributed to the atmosphere 205.45×10^3 t and 23.45×10^3 t CH_4, respectively. The agricultural waste burning added 3.55×10^3 t CH_4.

Naturally, Uganda needs to reduce the GHGs emissions, but the threshold at which such emissions must not be exceeded is unknown. Of course, the stock-breeding and rice production in Uganda will develop in future, increasing thereby the volumes of CH_4 emitted to the atmosphere. Hence, a balanced correlation should be sought of the economy in the country and the state of the environment. This problem can be solved with the use of new technologies of nature use. In particular, one of the ways to reduce CH_4 emissions is a secondary utilization of organic waste, for instance, in paper production. In Uganda, up to 16 % of the urban wastes are used in paper production.

The gas transport systems are one of the powerful anthropogenic sources of CH_4. The work of Coconea et al. (2000) contains information about methane emissions from pipelines in the territory of Romania, the country that signed the Lisbon Protocol in 1994 and now supports the Kyoto Protocol. Romania is the first country in Europe where in 1917 a 50-km pipeline was laid to transport natural gas. At present, natural gas constitutes 36 % of energy resources of the country, the share of oil and coal constituting 32.6 and 15.2 %, respectively. Therefore the problem of anthropogenic input of CH_4 from the territory of Romania into the atmosphere is rather urgent. Here, like in Uganda, the saving technologies play an important role, reducing by 38.9 % the leakage of methane from the pipelines during the last 20 years constituting 55.35 % in 1994 with respect to the leakage in 1987. On the whole, both extraction and distribution of coal, oil, and gas in the territory of

Romania give 56 % of the total amount of CH_4 emitted from this territory. Agriculture takes the second place (29 %).

One of the significant sources of CH_4 is the territory of Russia, which contributes to the atmosphere about 47×10^6 t CH_4 year^{-1}, and this flux is expected to reach 78×10^6 t CH_4 year^{-1} by 2025. This increase will be caused by the developing infrastructure of the gas, oil, and coal industry. On global scales, these trends will be practically observed in all countries. In Table 1.13 the contribution of the coal industry into CH_4 production is estimated for various global regions. These estimates are determined by technologies used in the coal industry. On the average, the contributions of various sources into the coal industry itself constitute: 70 %—underground ventilation in the coal-mines, 20 %—underground drainage; 5 %—surface loading and unloading operations; 4 %—opencast mining of deposits; 1 %—derelict mines.

The global cycle of methane has been studied inadequately, and therefore its modeling faces a lot of unsolved problems. The CH_4 fluxes from the waterlogged territories have been studied best. These fluxes constitute about 20 % of the total input of methane to the atmosphere from all sources (Tables 1.14 and 1.15). Note that almost 80 % of the sources of methane are of the biological nature, so that the anthropogenic interference into its natural cycle is also possible through violation of various biospheric processes. In particular, on the waterlogged territories, methane forms only due to biological processes.

The hydrospheric sources of methane can be presented by a multi-layer model (Figs. 1.39, 1.40 and 1.41). This scheme describes the vertical structures of most of the water bodies. Methane forms in the layer of bottom deposits due to bacteria functioning, and in the zone with oxygen, methane is partially oxidized giving carbon dioxide $CH_4 + 2O_2 \rightarrow CO_2 + 2H_2O + E$. Bacteria taking part in methane

Table 1.14 Emissions of methane by the coal industry in various countries (Gale and Freund 2000; IEA 2007a, b)

Country	Coal reserves (million tonnes)	Coal production (million tonnes oil equivalent/year)	CH_4 emissions (10^6 t/year)	Specific rate of CH_4 emision (kg CH_4/ton of coal)
Australia	78,500	203.1	0.8	3.5
England	220	11.3	0.5	7.4
Germany	6,739	50.3	1	3.6
India	93,445	209.7	0.4	1.5
China	114,500	1212.3	7.7	6.7
Poland	14,000	67.0	0.6	3
Russia	157,010	144.5	4.5	8.3
USA	246,643	595.1	4.3	5
Czechoslovakia	5,552	23.7	0.3	3.4
South Africa	48,750	144.8	1	0.5
Total World	*909,064*	*3079.7*	*21.7*	*4.9*

Table 1.15 Methane emissions from different sources recalculated for carbon equivalent (EPA 2001)

Anthropogenic source of methane	1990	1995	2000	2005	2010
Output of natural gas and oil	181.1	177.0	185.2	186.9	190.2
Animals' vital functions	157.4	143.4	144.0	149.5	150.8
Solid waste heaps	136.9	131.4	133.9	134.4	135.5
Coal mines functioning	82.8	62.6	59.0	59.3	59.0
Processing of stock-breeding waste	27.9	26.8	28.1	29.2	29.8
Sewage processing	10.1	9.8	9.8	10.4	10.4
Other sources connected with agriculture	7.9	8.2	8.5	8.2	8.2
Industrial and municipal sectors	15.0	13.9	13.7	13.9	14.8

Estimates of methane fluxes are given in 10^6 t C year^{-1}

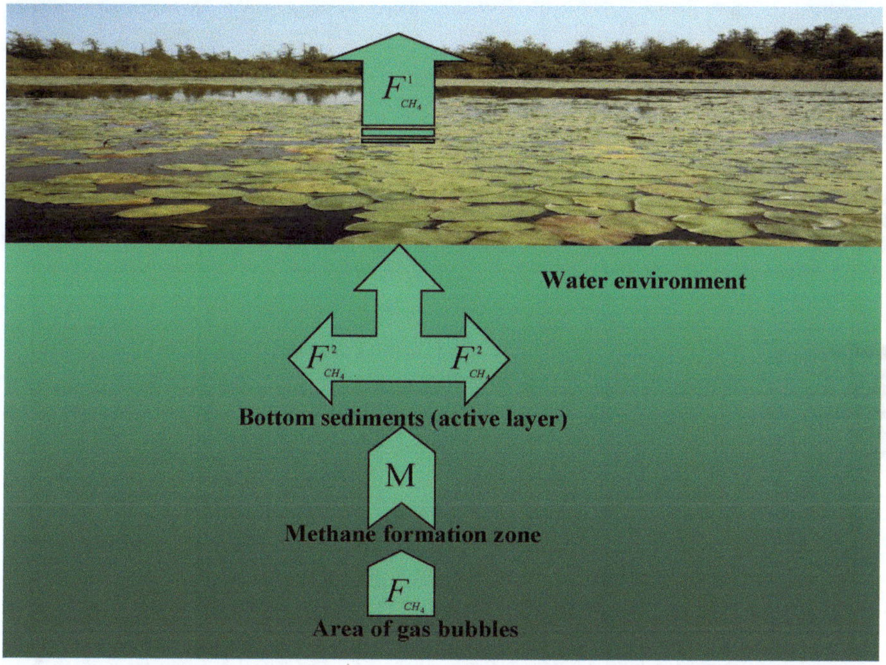

Fig. 1.39 Block-diagram for formation and transport of methane in the waterlogged country. *Notation*: $F^1_{CH_4}$ is the methane flux across the atmosphere/water body interface; $F^2_{CH_4}$ is the oxidation of methane in aerobic zone; F_{CH_4} is the intensity of methane source; M is the methane concentration

oxidation use the released energy E for the organic matter synthesis. The remaining methane gets to the atmosphere and, in contrast to CO_2, practically does not return to the water medium. This is somehow connected with the fact that the CH_4 solubility in water is almost 40 times lower than that of CO_2. The lifetime τ_H of

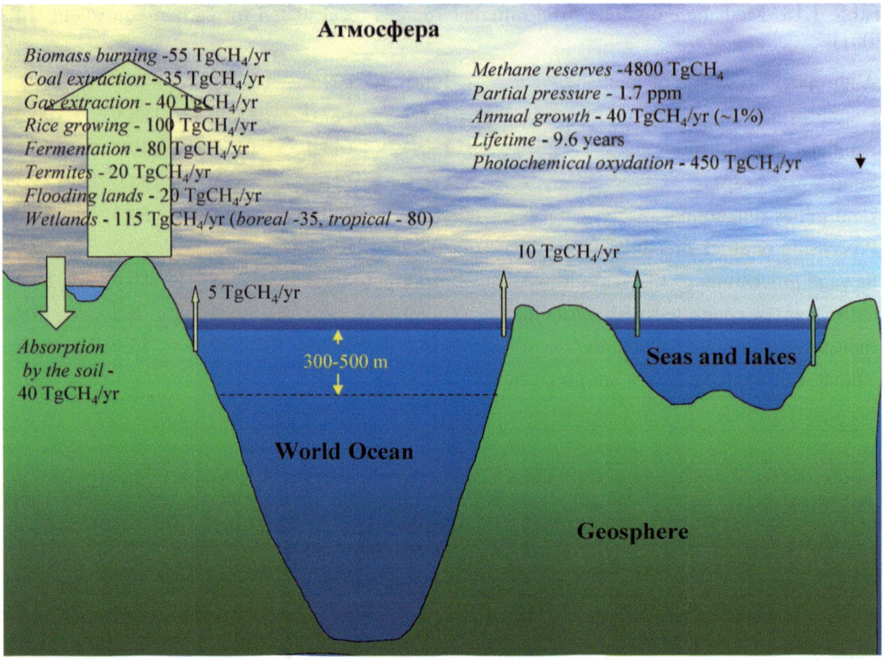

Fig. 1.40 Reserves and fluxes of the methane in the atmosphere-ocean-land system (Fung et al. 1991). *Notation* $Tg = 10^{12}$ g

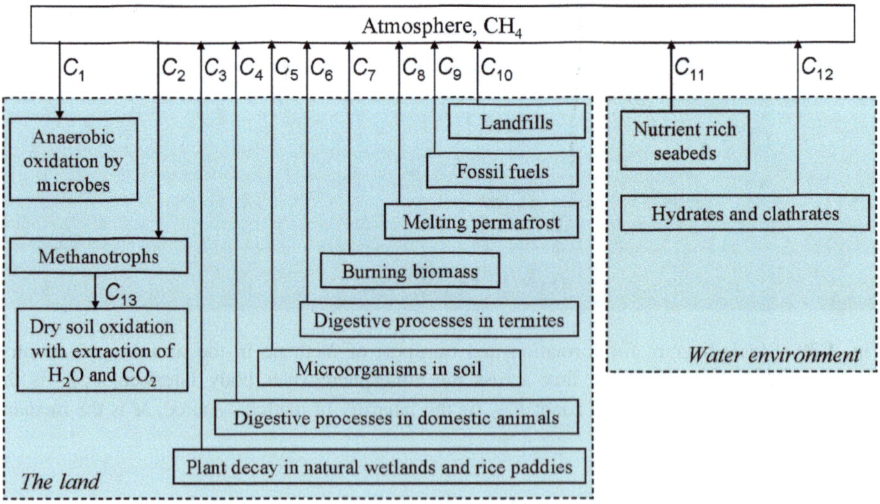

Fig. 1.41 Global methane cycle in the atmosphere-ocean-land system

methane in the atmosphere is estimated at about 5 years. Its extraction from the atmosphere takes place due to the participation of methane in photochemical reactions, resulting in methane oxidation first to CO, and then to CO_2. The cycle CO–OH–CP_4 plays an important role in the cycle of methane:

$$OH + CH_4 \rightarrow CH_3 + H_2O, \quad OH + CO \rightarrow CO_2 + H$$

The participating OH-radicals form in the atmosphere during the water vapor photolysis. As a result, atmospheric methane is oxidized.

Methane fluxes C_i (i = 1–13) are functions of temperature, microbial population, geographical coordinates and other environmental characteristics that are specific for pixels Ξ_{ij}. For example, Panikov and Dedysh (2000) developed the model for methane emission to the atmosphere from snow-covered bogs in West Siberia. This approach can be used for the modeling of anaerobic formation of CH_4 in pixels with the permafrost. In this case flux C_8 can be described by the equation:

$$dC_8/dt = \mu X(1 - Y)/Y,$$

where μ is specific growth rate of respective microbial population with biomass X, Y is biomass yield per unit of substrate consumed. Microbial biomass X is formed with the following law:

$$dX/dt = \mu X,$$

where $\mu = \mu_{max} SR/(S + K_S)$, S is catabolic substrate concentration, R is the function of physiological state, K_S is saturation constant, numerically equal to that substrate concentration at which microbial specific growth rate attains the half of maximal value ($\mu = 0.5\mu_{max}$).

Natural wetlands and rice paddies deliver to the atmosphere more 30 % of global CH_4 emission. Flux C_3 can be parameterized by the following equation:

$$C_3 = H_r f_1(T_s) f_2(h) f_3(\text{pH}) f_4(r_p),$$

where H_r is heterotrophic respiration, T_s is soil temperature, h is water table position, r_p is redox potential, functions f_i (i = 1–4) parameterized the CH_4 emission rates. Fluxes C_1 and C_2 that characterize major atmospheric CH_4 sinks are mainly parameterized by its reaction with hydroxyl (OH) radical. These fluxes depend on the OH levels and reaction rate. Under this, it is known that increase in methane leads to positive feedback (Xu et al. 2007; Krapivin and Varotsos 2008).

The human interference into the processes described by this diagram breaks the natural stability of the balance CH_4/CO/CO_2. In particular, the reclaiming of marshes is one of such destabilizing factors. For instance, the drainage of 20 % of marshes leads o a natural reduction of CH_4 emissions from the marshes by 20 %, and on the whole, the amount of methane is reduced by 4 %, which practically does not influence climate, but causes changes in the biogeochemical cycles of ozone and carbon dioxide with unpredicted consequences. These estimates are important for the final conclusion about the level of the integral greenhouse effect. However, the solution of this problem is connected with many factors, neglecting of which leads to serious errors. For instance, the CH_4 flux on the border atmosphere-marsh depends on the vertical profile of the temperature in the marsh body. In a simplest case, if we denote as $T_W(z, t)$ the temperature at a time moment t at a depth z and write the equation of heat conductivity

$$\frac{\partial T_W(z,t)}{\partial t} = a^2 \frac{\partial^2 T_W(z,t)}{\partial z^2} \tag{1.6}$$

where $a^2 = Kc^{-1}\rho^{-1}$, K is the coefficient of heat conductivity, c is specific heat capacity, ρ is the medium density, then an estimation of the flux $F^1_{CH_4}$ as a time function becomes dependent of the multitude of the poorly assessed characteristics of the environment.

Let the marsh surface temperature vary cyclically with frequency ω and amplitude A, decreasing with depth: $T_W(0, t) = A(z) \cdot \cos(\omega t)$, where

$$A(z) = A(0) \cdot \exp\left(-\sqrt{\frac{\omega}{2a}} z\right) \tag{1.7}$$

The solution to Eq. (1.7) enables us to trace the temperature variations $T_W(z, t)$ and suggests the conclusion that in this case these variations weakly depend on $T_W(0, t)$. Even if $T_W(0, t)$ increases by 2 °C, then according to (1.7), the amplitude of temperature changes with depth will rapidly decrease to 0.97, 0.33, and 0.01 °C at depths 40 cm, 2 m, and 3 m, respectively. Hence, with a 2 °C increase in the average global atmospheric temperature, flux $F^1_{CH_4}$ will increase by no more than 1.4 %.

Comparing the global significance of the CO_2 and CH_4 cycles in the system atmosphere-marshes, note that the CO_2 cycle promotes a climatic stabilization, whereas the CH_4 cycle intensifies the climate changes. With a climate warming the marshes assimilate part of CO_2 from the atmosphere and reduce thereby the greenhouse effect. On the contrary, when the climate warms due to increasing $F^1_{CH_4}$, the greenhouse effect intensifies.

The western Siberian region of Russia is characterized by numerous intensive natural and anthropogenic sources of methane formation. These are marshes, tundra, permafrost, oil and gas deposits. In this region the flux $F^1_{CH_4}$ varies widely both during a year and shorter time periods. From measurements carried out by Jagovkina et al. (2000) at the coastline of Yamal in June 1996, the CH_4 concentration in the atmosphere at a height of 2 m varied from 1.83 ppmv in June, 18 to 1.98 ppmv in June, 23, with an average daily value of 0.032 ppmv.

The peatbogs of Siberia are quite special in the global cycle of methane. They play a unique role in the biogeochemical cycles of methane and carbon dioxide. On the one hand, they are a non-anthropogenic source of CH_4 and CO_2, but on the other hand, they are intensive assimilators of carbon from the atmosphere and extract it from the natural cycle for a long time. The marshes of West Siberia, for instance, contain 20–30 % of the global carbon supplies. The intensity of CH_4 emissions from the marshes is, on the average, almost 2,000 times weaker than that of CO_2. 35–50 % of all methane emitted from the territory of Russia fall on the marshes. The West Siberian marshes emit to the atmosphere not more than 1.7×10^6 t CH_4 year^{-1}, which does not exceed 1 % of the global CH_4 flux. The spatial heterogeneity of the flux $F^1_{CH_4}$ is high, which is determined by diverse characteristics of the marsh ecosystems. In particular, the upper oligotrophic coniferous-shrubby sphagnous swamps emit 0.9–10 mg C m^{-2} h^{-1} (Dementjeva 2000). This estimate is rather approximate, since the scattering of such estimates by various authors constitutes hundreds of percent. For instance, a drained sphagnous swamp of the transitional type can emit 142–204 g C m^{-2} h^{-1}, and the rush-sphagnous bogs—83.5–309 mg C m^{-2} h^{-1}.

The main mechanism for the formation of methane in a marsh is connected with the functioning of special groups of microorganisms. Part of methane due to diffusion is emitted to the atmosphere, but most of methane remains in the peat layer and is gradually emitted to the atmosphere.

Remaking the nature, humankind interferes into the natural biogeochemical balance of greenhouse gases practically all over the world. One of the aspects of this remaking is a reduction of the areas of marshes and their transformation into agricultural fields. Diverse human agricultural activity adds to the atmosphere 20 % of all the anthropogenic flux of greenhouse gases. For instance, in the USA it is 30 %. The stockbreeding contributes considerably to this flux. In California and Wisconsin each hectare of pastures emits annually 502 kg CH_4 (or 10,511 kg CO_2) and 134 kg CH_4 (or 2,814 kg CO_2), respectively. In New Zealand such emissions of CH_4 are estimated at 291 kg CH_4 (or 6,110 kg CO_2) (Johnson and Ulyatt 2000).

Among the Kyoto Protocol signatories, England takes the ninth place by the volume of reduced emissions of greenhouse gases. The decreasing trend of methane emissions is part of the general reduction of emissions of six greenhouse gases (CO_2, CH_4, N_2O, hydrofluorocarbons, perfluorocarbons, sulfur hexafluoride) from 1990. In 2000 the greenhouse emissions decreased by 15 % compared to 1990. By 2010, the CH_4 emissions will constitute 20,134 t year^{-1}. This reduction will be reached mainly due to new technologies in processing the wastes and in the coal industry. On the whole, in England, according to the developed scenario, emissions of methane by 2010 will decrease by 14 % in agriculture, by 82 % in the coal industry, by 29 % in the oil and gas industry, and by 73 % in wastes processing. The possibility of realization of this scenario is confirmed by the CH_4 decreasing trend in 1998 compared to 1990. For instance, during this period, emissions of methane in the coal industry have decreased by 64 %, and in wastes processing by 29 %. In 1990, the share of wastes processing in England constituted 32 % of all CH_4 emissions, only 3 % of these emissions being connected with sewage processing.

In agriculture, emissions of CH_4 in England constituted $1{,}037 \times 10^3$ t in 1990 and 998×10^3 t in 1998. The scenario of reduction of the flux $F^1_{CH_4}$ from the territory of England due to improved technologies in agriculture foresees emissions of 902–983 $\times 10^3$ t CH_4 in 2010. The CH_4 emission from burning the agricultural wastes is completely excluded, and in the stockbreeding, emissions of methane are reduced by 8 % compared to 1990.

In the coal industry in England, emissions of methane in 1990 constituted 819×10^3 t with the main contribution to this flux made by underground operations. This constituted 24 % of the whole flux of methane from the territory of England. In 1998 the flux $F^1_{CH_4}$ decreased to 264×10^3 t and by 2010 it should decrease to 218×10^3 t. A similar decreasing trend of methane emissions from the territory of England remains in the oil and gas industry, too. According to the scenario, the contribution of these sectors of energy production into the flux $F^1_{CH_4}$ will decrease from 540×10^3 t in 1990 to 349–464×10^3 t in 2010 (Meadows 2000).

1.6.5 Block-Schemes of Global Sulfur Cycle

Very likely the major sulfur fluxes of both natural and anthropogenic origin have been described by Ivanov (1981) and Ivanov and Freney (1983). An increase of intensity and spatial distribution of anthropogenic processes during the last decades has also manifested itself through a change of the scale of propagation of sulfur compounds in the biosphere. This effect is confirmed by that compared with the pre-industrial period, sedimentation of sulfur over the continents and oceans has increased by 162.5 and 24.6 %, respectively. Emissions of sulfur to the atmosphere reached 93×10^6 tS year^{-1}. The anthropogenic flux of sulfur in the form of SO_2 is easily estimated, provided 3.1×10^{12} kg of coal are globally burnt every year, with an average content of sulfur in it 2.5 % (by weight) (Berndt et al. 2004).

An improvement of the global model of the biosphere is connected with an extension of biogeochemical cycles considered in it. A necessity to include in the model a unit describing sulfur fluxes in nature systems is dictated by dependence of biotic processes on the content of sulfur in the biosphere compartments. Available data on supplies and fluxes of sulfur compounds in the atmosphere, soils, vegetation cover, and hydrosphere, enable one to formulate mathematical relationships to describe the global sulfur cycle.

Sulfur compounds affect strongly the quality of the environment and change its role in regulation of the greenhouse effect. For instance, in December 1952 a fog consisting of a mixture of smoke and coal dust covered London. As a result, during 1 week more than 2,000 people died of air pollution. These events had happed before, but had not been recorded as in this case. Measurements carried out in the St. Bartholomew hospital have shown that the concentration of particles of smoke and SO_2 exceeded several milligrams per m^3. In general, London at that time depended on the use of coal for space heating and energy production, and therefore after that event attempts have been undertaken to remove sulfur from coal before its burning. But nevertheless, in 1962 the tragedy repeated with 800 victims due to smog.

Beginning from 1970, in the OECD countries the problem of air quality has become the subject of studies at many scientific centers. Oil from the counties of the Middle East becomes the main source of energy. The content of sulfur in oil constitutes 2.5–3 %. In 1985 some European countries signed the CLRTAP protocol on a 30 % reduction of sulfur emissions. As a result, now SO_2 emissions decreased by more than 50 % compared to 1980. Of course, this was possible largely due to transition of Europe to the use of Russian gas.

It should be mentioned that along with the formation of acid rains, sulfur compounds directly affect a decrease of the greenhouse effect. For instance, sulfate ion has an opposite effect on a change of air temperature compared to CO_2 and, hence, reduces the effect of climate warming. Photochemistry of sulfur cycle has two the most important reactions:

- reaction from source gases to SO_2, and
- reactions leading from SO_2 to H_2SO_4.

The global sulfur cycle consists of the mosaic structure of local fluxes of its compounds with other elements formed due to water migration and atmospheric processes. It is schematically shown in Figs. 1.42 and 1.43. The conceptual schemes of the global and regional cycles of sulfur have been described in detail by many authors (Nitu et al. 2000b; Krapivin and Varotsos 2008). However, the existing

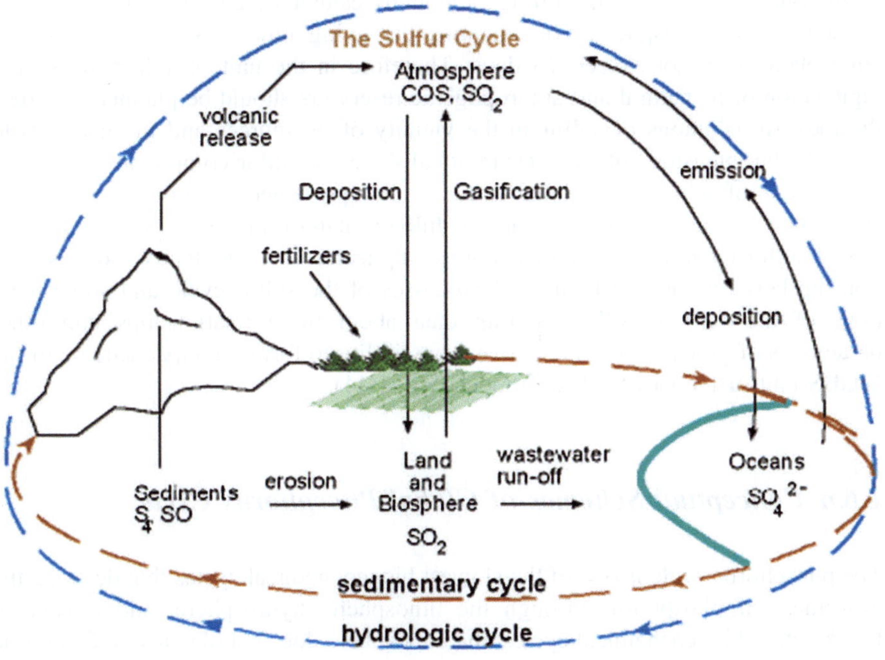

Fig. 1.42 Principal scheme of global sulfur cycle (http://www.enviroliteracy.org)

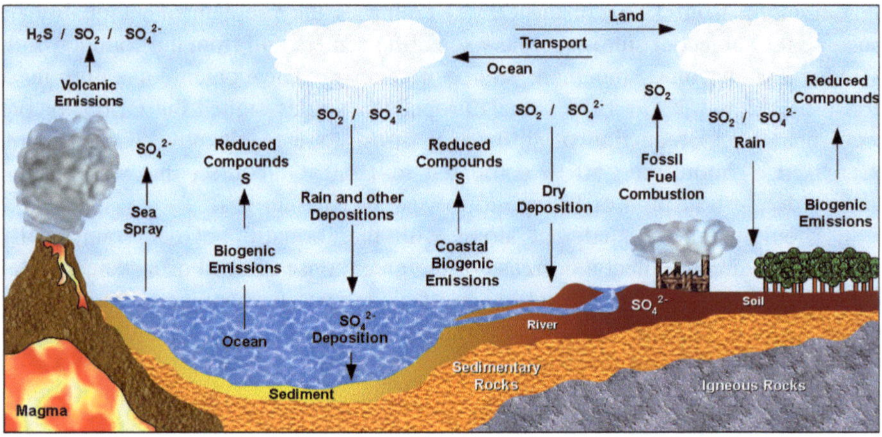

Fig. 1.43 Reservoirs and fluxes of global sulfur cycle (http://www.meas.ncsu.edu/airquality)

models have been developed for autonomous functioning and usage, which makes is difficult to include them into the global model without substantial changes in their parametric space. The model of the global sulfur cycle proposed here has been derived in the form of a unit with inputs and outputs, which enables one to match it with other units of the global model via their inputs and outputs.

In contrast to hydrogen, sulfur compounds cannot be attributed to long-lived elements of the biosphere. For example, the living time of sulfur oxide in the atmosphere does not exceed 15 days. Therefore in the unit of sulfur the spatial digitization of its natural and anthropogenic reservoirs should be planned to reflect the local distributions of sulfur in the vicinity of its sources and to enable us to estimate the intensities of the inter-regional fluxes of sulfur compounds.

Analysis of schemes of global sulfur cycle represented in Figs. 1.41 and 1.42 shows that the quantitative estimate of different natural processes involved in the global sulfur cycle appears much more complicated. It is evidently there is relationship between general biological processes of the sulfur cycle and some reactions of the carbon cycle. Existing data about these relationships and other parameters of global sulfur cycle give a possibility to form principal scheme of the MGBS unit of the BCSS global model (Fig. 1.44).

1.6.6 Conceptual Schemes of Global Phosphorus Cycle

The phosphorus cycle is one of the slowest biogeochemical cycles that describe the movement of phosphorus through the lithosphere, hydrosphere, and biosphere. Unlike other biogeochemical cycles, the atmosphere does not play a significant role

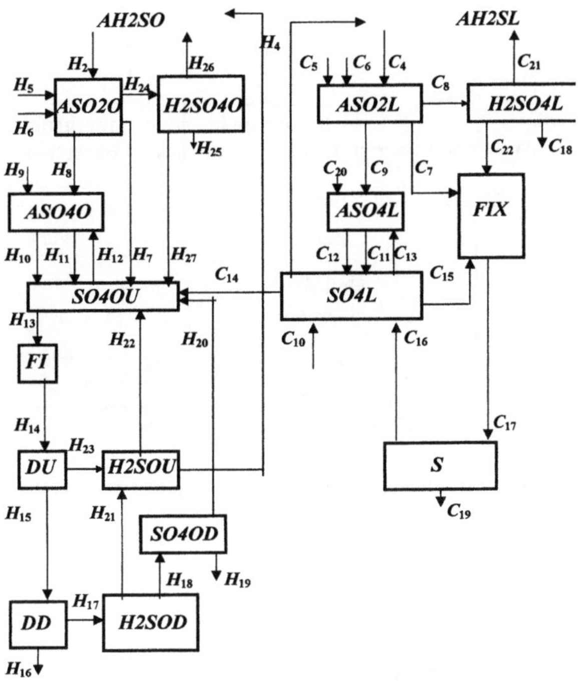

Fig. 1.44 The scheme of sulfur fluxes in the environment (MGBS unit of the BCSS global model). The notation is given in Tables 1.16 and 1.17

in the movement of phosphorus. In contrast to nitrogen, the main reservoir of phosphorus in the biosphere is not the atmosphere but the rocks and other deposits formed in the past geological epochs, which, being subject to erosion, emit phosphates. Besides, there are other mechanisms of the return of phosphorus to the biospheric cycle, but, as a rule, they are not that efficient. One of these mechanisms is fish catching, returning to land from the hydrosphere about 60×10^3 t P/year, as well as an extraction of phosphorus-containing rocks estimated at $1–2 \times 10^6$ t P/year. The present cycle of phosphorus is closed by its fluxes to the bottom deposits in the World Ocean to which it gets with sewage, as well as with the coast and river run-off.

Schematic diagrams of global phosphorus cycle are represented in Figs. 1.45 and 1.46. They reflect main processes of phosphorus in the environment:

- phosphorus is an essential nutrient for plants and animals which use of it to build nuclei acid, phosphorus bilayer, and adenosine triphosphate;
- phosphorus minerals are used to build strong bones and teeth;
- phosphorus comes from dead organic matter and from the weathering/erosion of phosphates from rocks; and
- plants get it from the soil and animals get it from eating plants.

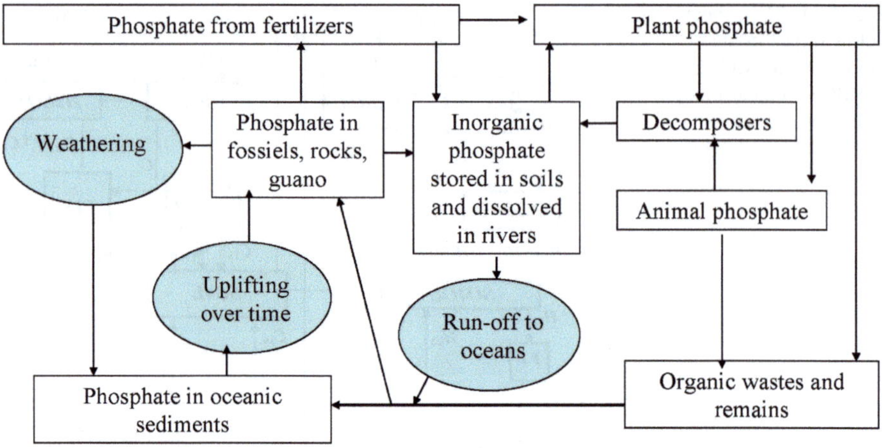

Fig. 1.45 A schematic representation of global phosphorus cycle

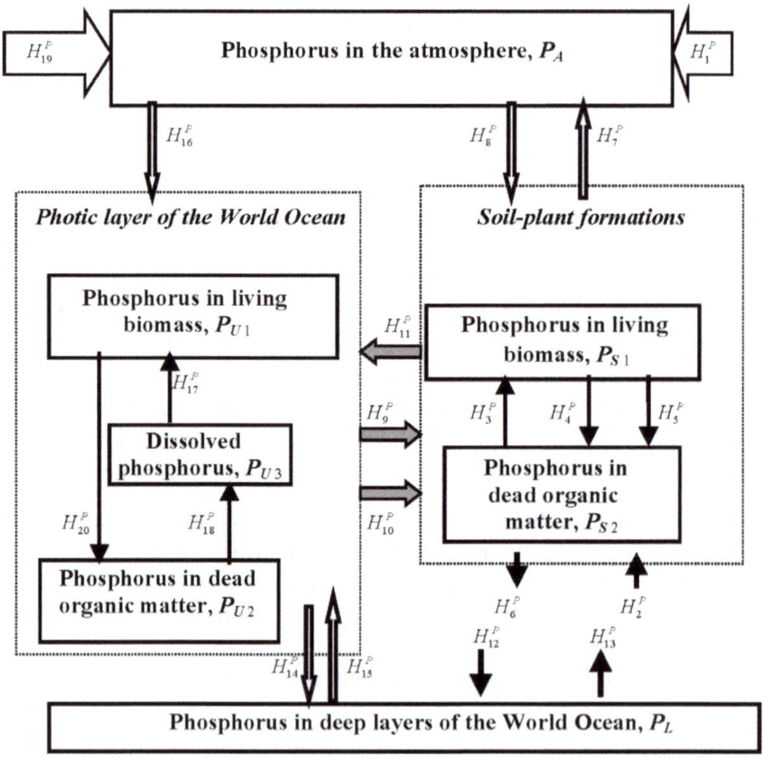

Fig. 1.46 The scheme of phosphorus fluxes in the BCSS. Notation is given in Table 1.18

1.7 Models of Global Biogeochemical Cycles

1.7.1 Key Aspects of Global Biogeochemical Cycles

Key aspects of global biogeochemical cycles were discussed by Krapivin and Varotsos (2008). Global model of the biosphere/climate/society system is basic unit of the GIMS that parameterizes main biogeochemical cycles providing boundary information conditions for RIMS. Figures 1.30, 1.35, 1.38, 1.43 and 1.45 give a possibility to write balance equations for each biogeochemical cycle.

Fundamental connections between characteristics of the biological state of the environment, such as biodiversity in ecosystems, the state and dynamics of food chains, and interactions of the biosystem with the cycle of biogenic elements have been poorly studied, both in land and water ecosystems. Among numerous questions appearing in studies of the global biogeochemical processes the following are of key importance:

(1) What physical, biological, chemical and social processes are the basic ones in regulation of the cycles of carbon, nitrogen, sulfur, water and other elements both in space and in time?

- What mathematical relations are determinants in the parameterization of biological processes in the computer models of biogeochemical cycles?
- What are dependences between biodiversity, stricture of ecological chains, and biogeochemical cycles in land and water ecosystems?
- What processes are determinants in the transport of biogenic salts and pollutants in space in general and between various ecosystems, in particular?
- What are mechanisms that relate one biogeochemical cycle to another, and do general principles of parameterization of these relations exist or they depend on the type of chemical elements and ecosystems under consideration?

(2) What are forms and ways of anthropogenic interference to global biogeochemical cycles?

- How do humans influence the biogeochemical cycles, change their rates and spatial distributions of chemical elements, forming inputs and outputs of numerical models, and what are consequences of this interference?
- How does a change in the land use strategy affect the re-distribution of chemical elements in space and in time?
- What anthropogenic pollutants are important for the biogeochemical influence on ecosystems and how to predict them?

(3) What mechanisms control the ability of ecosystems to rapidly restore and what are indicators that reflect this ability of ecosystems?

- How do an introduction of new species to ecosystems and an appearance of new unstudied diseases affect the development of biogeochemical cycles in land and water ecosystems?
- What feedbacks between ecosystems and climate are critical and how these feedbacks are parameterized in computer models?
- Can the data on the past biogeochemical cycles be used for their prediction in the future?
- What basic parameters and characteristics of ecosystems affect their ability to restore after anthropogenic forcings?

The interactions of biogeochemical cycles influence and maintain our climate system. Land use and fossil fuel emissions are currently impacting the biogeochemical cycles of carbon, nitrogen and sulfur on land, in the atmosphere, and in the oceans (Schulze et al. 2001). Our understanding of biogeochemical cycles is directly related to the research and analyses that are brought to them, and to the methods that are used. Jacobson (2000) notes that the study of cycles as a model for the earth's changing climate has become a new science. Earth Systems Science is the basis for understanding all aspects of anthropogenic global change, such as chemically forced global climate change. Earth Systems Science is an integrated discipline that has been rapidly developing over the last two decades.

Biogeochemical cycles are characterized by the variety of pathways which are passed by chemical elements. Ecosystems have many biogeochemical cycles operating as a part of the global or regional system, for example the water cycle, the carbon cycle, the nitrogen cycle, etc. All chemical elements occurring in organisms are part of biogeochemical cycles. In addition to being a part of living organisms, these chemical elements also cycle through abiotic factors of ecosystems such as water (hydrosphere), land (lithosphere), and/or the air (atmosphere). All of this performs the biogeochemical cycle modeling as very ambiguous task.

1.7.2 The Carbon Cycle Modeling

The carbon cycle is the biogeochemical cycle by which carbon is exchanged among the biosphere, pedosphere, geosphere, hydrosphere, and atmosphere of the Earth. Figures 1.23, 1.24, 1.25, 1.26, 1.27, 1.28 and 1.29 present a diversity of possible scheme for modeling.

Considering the notations in Fig. 1.30, the balance equations to describe the global carbon cycle are written as:

$$\frac{\partial C_A}{\partial t} = H_{GB} - H_{BG} + H_5^C - V_\varphi \frac{\partial C_A}{\partial \varphi} - V_\lambda \frac{\partial C_A}{\partial \lambda} + \begin{cases} -H_2^C + H_3^C, & (\varphi, \lambda) \in \Xi_O; \\ H_1^C + H_4^C - H_6^C + \sum_{i=7}^{11} H_i^C, & (\varphi, \lambda) \in \Xi \backslash \Xi_O; \end{cases}$$

$$\frac{\partial C_{S1}}{\partial t} = H_6^C - H_7^C - H_8^C - H_{14}^C - H_{15}^C;$$

$$\frac{\partial C_{S2}}{\partial t} = \sum_{i=1}^{4} H_{i+11}^C - H_4^C - H_9^C - H_{23}^C - H_{24}^C;$$

$$\frac{\partial C_U}{\partial t} + v_\varphi^U \frac{\partial C_U}{\partial \varphi} + v_\lambda^U \frac{\partial C_U}{\partial \lambda} = Q_U + H_{22}^C + H_{25}^C - H_{21,U}^C - H_3^C + H_2^C + H_{19,U}^C - H_{20,U}^C;$$

$$\frac{\partial C_P}{\partial t} + v_\varphi^P \frac{\partial C_P}{\partial \varphi} + v_\lambda^P \frac{\partial C_P}{\partial \lambda} = H_{19,P}^C - H_{19,U}^C + H_{20,U}^C - H_{20,P}^C - H_{21,P}^C + H_{22}^C + H_{25}^C;$$

$$\frac{\partial C_L}{\partial t} + v_\varphi^L \frac{\partial C_L}{\partial \varphi} + v_\lambda^L \frac{\partial C_L}{\partial \lambda} = H_{18,L}^C + H_{19,L}^C - H_{19,P}^C + H_{20,P}^C - H_{20,L}^C;$$

$$\frac{\partial C_F}{\partial t} + v_\varphi^F \frac{\partial C_F}{\partial \varphi} + v_\lambda^F \frac{\partial C_F}{\partial \lambda} = H_{17}^C - H_{16}^C + H_{18,F}^C - H_{19,L}^C + H_{20,L}^C$$

The flux Q_U is formed from H_{23}^C and H_{24}^C. Let $Q_U = 0$ for the pelagic regions Ω_{OP} of the World Ocean. Describe the formation of Q_U on the shelf by a simple algorithm with the supposed uniform distribution of the sink from the *K*th region to the *M*th water basin:

$$Q_U = \begin{cases} 0, & (\varphi, \lambda) \in \Xi_{OP}, \\ (H_{23}^C + H_{24}^C)\sigma_{OP}/\sigma_L, & (\varphi, \lambda) \in \Xi_O \backslash \Xi_{OP}, \end{cases}$$

where σ_{OP} and σ_L are the areas of the water basins Ξ_{OP} and Ξ_L, respectively; Ξ is the biosphere area, Ξ_O is the World Ocean area.

Fluxes H_i^C are described in Krapivin and Varotsos (2008). In particular, main significance is given to the H_3^C flux of CO_2 dissolved in the upper layer of the ocean that can be calculated by the formula

$$H_3^C = \psi(T_L)\sqrt{p_a}/(1 + 0.5C_S)$$

where $\psi(T_L)$ is the function of the effect of temperature on CO_2 solubility, C_S is the water salinity. An expression of the ψ function has been studied inadequately. There are only some experimental estimates (Alekseev et al. 1992). Of course, a lot of simplified approaches to the determination of the direction of CO_2 fluxes on the atmosphere-ocean border prevail here. In a simplest case, the fluxes H_2^C and H_3^C can be considered as mutually exclusive. That is, a valve is supposed to exist on the atmosphere-ocean border. Its switching is determined by the pH parameter. A critical level of pH is about 8.11. At $pH \leq 8.11$ the ocean assimilates CO_2, and at $pH > 8.11$ the ocean emits CO_2.

The MGBC item of the BCSS global model is correlated with the MGBO, MGBN, MGBP and MGBS items. Each of them is described by the system of differential equations that reflect the balance relationships between corresponding reservoirs. For example, oxygen and ozone cycles depend on the photosynthesis both in oceans and on the land.

1.7.3 Nitrogen Cycle Modeling

A comparative analysis of the model schemes of the flux diagrams of nitrogen compounds in nature proposed by various experts makes it possible to construct a block-scheme presented in Fig. 1.35. Here the atmosphere, soil, lithosphere, and hydrosphere are considered as nitrogen reservoirs. Three first reservoirs are described by 2-D models, and the hydrosphere is described by a 3-D multi-layer model. The characteristics of nitrogen fluxes between these reservoirs are given in Table 1.11. The equations of the model are written as

$$\frac{\partial N_A}{\partial t} + V_\varphi \frac{\partial N_A}{\partial \varphi} + V_\lambda \frac{\partial N_A}{\partial \lambda} = H_1^N + \begin{cases} H_{20}^N - H_{16}^N, & (\varphi, \lambda) \in \Omega_O \\ H_7^N + H_{19}^N - H_8^N - H_9^N + H_{22}^N - H_2^N - H_{10}^N, & (\varphi, \lambda) \in \Omega / \Omega_O \end{cases}$$

$$\frac{\partial N_{S1}}{\partial t} = H_8^N + H_6^N - H_3^H$$

$$\frac{\partial N_{S2}}{\partial t} = H_2^N + H_3^N + H_5^N + H_9^N - H_6^N - H_7^N - H_{11}^N - H_{21}^N$$

$$\frac{\partial N_U}{\partial t} + v_\varphi \frac{\partial N_U}{\partial \varphi} + v_\lambda \frac{\partial N_U}{\partial \lambda} = H_{16}^N + H_{4,U}^N + H_{18,U}^N + H_{11}^N - H_{17,U}^N - H_{20}^N - H_{14,UP}^N - H_{15,UP}^N$$

$$\frac{\partial N_P}{\partial t} + v_\varphi \frac{\partial N_P}{\partial \varphi} + v_\lambda \frac{\partial N_P}{\partial \lambda} = H_{18,P}^N + H_{4,P}^N + H_{14,UP}^N + H_{15,PL}^N - H_{17,P}^N - H_{14,PL}^N - H_{15,UP}^N$$

$$\frac{\partial N_L}{\partial t} = Q_L + H_{12,L}^N + H_{14,PL}^N + H_{15,LF}^N - H_{14,LF}^N - H_{15,PL}^N$$

$$\frac{\partial N_F}{\partial t} = Q_F + H_{12,F}^N + H_{23}^N + H_{14,LF}^N - H_{13}^N - H_{15,LF}^N$$

where $V(V_\varphi, V_\lambda)$ is the wind speed, $v(v_\varphi, v_\lambda)$ is the current velocity in the ocean, Q_L and Q_F are functions describing the mixing of the deep waters of the ocean.

To simplify the calculation scheme presented in the diagram in Fig. 1.35, advective processes in the equations of the MGBN item can be described by a superposition of the fluxes H_{14}^N and H_{15}^N. The computer realization of the equations of the MGBN item introduces into the equations of the MGBN item some corrections for the agreement between the dimensionalities of the variables in conformity with the spatial digitization of Ξ. Therefore the estimates of the fluxes H_i^N given below for their consideration in the MGBN should be corrected following this criterion.

Different versions of the MGBN unit of the BCSS model can be synthesized depending on the parameterization of fluxes H_i^N. Krapivin and Varotsos (2008) the flux H_{16}^N of nitrogen fixed in the atmosphere over any ocean basin described by the relationship:

$$H_{16}^N = [\lambda_1(\theta_1)^{\Delta T} + \lambda_2 R_W]N_A,$$

where ΔT is the atmospheric temperature variation, θ_1 is the indicator of the temperature dependence of the rate of atmospheric fixation of nitrogen, R_W is precipitation, λ_1 and λ_2 are the coefficients.

The equation of atmospheric fixation over a land site Ξ_{ij} is written by analogy to H_{16}^N:

$$H_{8,ij}^N = [\lambda_3(\theta_1)^{\Delta T} + \lambda_4 R_{W,ij}]N_A,$$

where λ_3 and λ_4 are coefficients.

To estimate the coefficients λ_i (i = 1, …, 4), as a first approximation one can use average data on nitrogen fluxes and precipitation. If we assume $H_{16}^N = 9.96 \times 10^{-3}$ t km^{-2} year^{-1}, $H_8^N = 0.027$ t km^{-2} year^{-1}, estimate local precipitation over the ocean and land at 1.01 and 0.24 m year^{-1}, respectively, and the convective precipitation over the ocean and land at 0.19 and 0.116 m year^{-1}, respectively, we obtain $\lambda_1 = 0.00498$, $\lambda_2 = 0.00458$, $\lambda_3 = 0.0135$, and $\lambda_4 = 0.0285$. These estimates are easily specified with account of local data at a fixed time moment for smaller regions and water bodies.

The flux of nitrogen H_1^N is determined by the geothermal activity of the Earth. Its estimates testify to a necessity of consideration of this constituent in the global model. In particular, for instance, in the nitrogen fumaroles of Vesuvius the content of nitrogen by weight constitutes 98 %, in gases of the lavas of the Hawaiian volcanoes there are only 5.7 % of nitrogen, and over the globe an input of juvenile nitrogen averages 0.4×10^6 t year^{-1}. Let H_1^N be a function of time approximating a statistical series of observations. A more strict account of this flux of nitrogen in the model can be realized by using the algorithms of parameterization of random processes, for instance, with the use of the evolutionary modeling. However, within the global model, orientated toward describing the processes in time steps of decades, it is enough to use average annual data.

The flux H_1^N can be, to some extent, interpreted as a compensation for the fluxes H_{13}^N and H_{21}^N.

The nitrogen supplies on land consist of the assimilable nitrogen in soil $N_{S2} \approx 0.19 \times 10^4$ t km^{-2}, in plants (12 × 10^9 t), and living organisms (0.2 × 10^9 t). A diversity of nitrogen fluxes is formed here of the processes of nitrifucation, denitrification, ammonification, fixation, and river run-off. The intensities of these fluxes depend on climatic conditions, temperature regime, moistening, as well as chemical and physical properties of soil. Many qualitative and quantitative characteristics of these dependences have been described in the literature (Hellebrand et al. 2003). Let us consider some of them.

Nitrification is the biological oxidation of ammonia with oxygen into nitrite followed by the oxidation of these nitrites into nitrates. Nitrification is an important step in the nitrogen cycle in soil. Nitrification involves the oxidation of nitrogen by specialized bacteria (*Nitrosomonas, Nitrobacter,* etc.). The return of nitrogen to the cycle due to living micro-organisms is one of the stabilizing natural processes. Simplifying the whole process of the transformation of ammonia salts into nitrates, present the activity of heterotrophic micro-organisms and saprophags as a generalized process of the organic matter decomposition. The rate of the organic matter decomposition and nitrification increases with an increasing temperature, reaching its optimal value at T_a = 34.5 °C. Therefore for the flux H_3^N an approximation $H_3^N = \lambda_{N\,\kappa} M_\kappa$ can be assumed, where M_κ is the rate of the dying-off of the component κ, $\lambda_{N\kappa}$ is the content of nitrogen in the component κ.

Denitrification process exists in anoxic environments when nitrate and nitrite act as electron acceptors (oxidizers) and the nitrification reactions reverse: $NO_3^- \Rightarrow NO_2^- \Rightarrow NO$. The processes of denitrification $\left(H_7^N\right)$ on land are important channels of the input of nitrogen into the atmosphere. The intensity of these processes depends on temperature, humidity, pollution of soils with poisonous chemicals, and pH. The quantitative and functional characteristics of these dependences have been well studied. Within the global model it is possible to take into account only the factors of temperature and humidity:

$$H_7^N = \lambda_6 \theta_2^{\Delta T} W_S \frac{N_S}{k_1 + N_S},$$

where W_S is soil moisture, θ_2 is the temperature coefficient, λ_6 and k_1 are empirical parameters. If we assume $H_7^N = 0.318$ t/km^2/year, then $\lambda_6 = 0.496$, $k_1 = 0.556$.

In the biological cycle of nitrogen of importance are the processes of its fixation by microorganisms and plants whose intensity is estimated at 148 × 10^6 t year^{-1}. The rate of fixation, depending on the character of the medium, can vary reaching 3 × 10^9 t year^{-1} in highly productive regions. The nitrogen flux H_{10}^N depends on the distribution of vegetation cover and can be described by the equation $H_{10}^N = \sigma_\kappa \lambda_\kappa R_\kappa / \sigma_{ij}$, where σ_κ is the area under vegetation of the κ type on the territory Ξ_{ij} of the area σ_{ij}, R_κ is the productivity of plants of the κ type, λ_κ is the coefficient.

Fixation of nitrogen by plants directly from the soil via the root systems (flux H_6^N) occupies a principle place in the cycle of nitrogen, especially on the territories with cultural vegetation. For instance, an increase of the share of legumes in agriculture can raise H_6^N up to 35 t km^{-2} year^{-1}. Therefore a consideration of this flux in the model is necessary and can be realized in the following form:

$$H_6^N = \sigma_\kappa R_\kappa \mu_\kappa / \sigma_{ij}$$

where μ_κ is the constant.

The rate of assimilation of nitrogen by the roots of plants is known to depend much on the soil temperature regime, decreasing a little with the temperature lowering to 8–10 °C and dropping at temperatures below 5–6 °C. The motion of nitrogen from the roots to the upper parts of the plants slows down, too. The formula for H_6^N reflects this regularity through the respective reactions of the plants' productivity with the time lag neglected.

On land, plants assimilate annually about 30×10^6 t N from the atmosphere and more than 5.3×10^6 t N directly from the soil. Approximate estimates of the productivity of various types of vegetation average R_κ = 710–3243 t km^{-2} year^{-1}. Hence, we have $\mu_\kappa = 0.134 \times 10^{-5}$ to 0.506×10^{-4}.

On global scales, the ways of nitrogen migration include the transport of its compounds between land and oceans due to water run-off. The annual input of nitrogen from land into the World Ocean is estimated at 38.6×10^6 t. If the total sink to the ocean from land is described by the function W_{SO}, then the nitrogen flux H_{11}^N can be approximated by the expression

$$H_{11}^N = \lambda_N N_{S2}[1 - exp(-k_N W_{SO})],$$

where λ_N and k_N are the coefficients. The functional form foresees that the nitrogen flux from land to the ocean is equal to zero in the absence of the run-off and its stabilization at a level λ_N, with the run-off volume considerably increasing. To estimate the parameters λ_N and k_N, it is necessary to take into account the spatial heterogeneity of the types of soil-vegetation formations, relief, and other geophysical parameters. In particular, the content of nitrogen compounds in water differs as a function of the run-off territory. The river waters in the forest regions with the temperate climate contain 0.4 mg ℓ^{-1} of nitrates, for the arid climate this value is 1.45 mg ℓ^{-1}. The concentration of nitrates increases sharply in drainage waters of the irrigation systems (5.5 mg ℓ^{-1}), in the river waters of thickly populated regions (25 mg ℓ^{-1}), and reaches a maximum in the soil solutions of the salted irrigates soils (200 mg ℓ^{-1}). Ground waters contain from 10 to 100 mg ℓ^{-1} of nitrates. The total run-off of water into the World Ocean reaches 50×10^3 km^3, 30 % of which being the underground run-off, hence, the total flux of nitrogen per unit area of the ocean will be 0.107 t year^{-1}. Assuming W_{SO} = 0.337 m and that a 95 % level of the sink saturation is reached at a fivefold increase of W_{SO}, we obtain k_N = 1.367, λ_N = 0.708.

The surface part of the nitrogen cycle contains a constant process of removal of nitrogen from the biosphere into deposits, in particular, as a result of accumulation of saltpetre on the Earth surface through erosion and alkalifying. From the available estimates, $H_{21}^N \approx 3.9 \times 10^{-4}$ t/km^2/year, with $H_{21}^N < H_{22}^N$, but $H_{21}^N + H_{13}^N \cong H_1^N + H_{22}^N + H_{23}^N$. This relationship follows from the fact that during the geological period the loss of nitrogen had been balanced by its input. Of course, in the present biosphere, with the changing intensity of most of the fluxes enumerated in Table 1.11, this balance has started breaking due to increasing fluxes H_9^N and H_{19}^N.

Finally, one should mention the fact that in connection with the persistent C/N ratio for different types of soils and climatic zones for nitrogen fluxes it is important to find correlations between factors of biogeochemical cycles' regulation on a regional level. Quite special here are arid regions where soils are poor in microorganisms, and the regime of moisture cycle is determined by division into dry and wet seasons. In this connection, Austin et al. (2004) shown that the episodic nature of water availability in arid and semiarid ecosystems has significant consequences on belowground carbon and nutrient cycling. Pulsed water events directly control the C/N of microbially available substrate. A level of this control depends on the spatial and temporal heterogeneity of vegetation cover, topographic position and soil texture. Seasonal distribution of water pulses finally lead to the change in biogeochemical cycling in water-limited ecosystems. Schematic outline of biogeochemical cycles of C and N in arid and semiarid ecosystems under dry season, and after a rainfall pulse is given in Fig. 1.33.

In the seawater, nitrogen is present in the form of dissolved gas, ions of ammonium NH_4^+, nitrite NO_2^-, nitrate NO_3^-, and in the form of various organic compounds. Inorganic nitrogen compounds are assimilated by algae and phytoplankton and thus transfer into organic forms that serve as food for living organisms. An expenditure of inorganic nitrogen supplies is compensated by atmospheric precipitation, river run-off, and mineralization of organic remains in the process of the living organisms functioning and dying-off. According to Ivanov (1978), nitrogen fluxes in sea water can be schematically shown in Fig. 1.35. Of course, not all nitrogen fluxes available in nature have been taken into account. A diversity of the ways of nitrogen transformation in water has been studied inadequately, though the available information is sufficient for the global model. The processes, such as replenishing the nitrogen supplies in water due to lysis of detritus and functioning of living organisms, the nitrogen exchange between photic and deep layers of the ocean, as well as nitrogen fixation at photosynthesis and denitrification, have been thoroughly studied and described in literature. Also, there are rough estimates of nitrogen supplies in the ocean, according to which one can assume, on the average, that $N_U = N_P = 0.77 \times 10^4$ t km^{-2} and $N_L = N_F = 10^5$ t km^{-2}. More detailed spatial distributions of nitrogen supplies in the hydrosphere can be calculated from the data on biomass, dissolved organic matter, and concentration of dissolved oxygen. The volume relationships of dissolved nitrogen are related to the volume of oxygen as mℓ N_2/l = 1.06 + 1.63 mℓ O_2/l.

The nitrogen supplies in water bodies are replenished due to bacterial decomposition of organic sediments and dissolved organic matter. Consider the component D denoting the content of dead organic matter in the water medium. On its basis, one can write the following relationship: $H_{18}^N = \lambda_D D(\varphi, \lambda, z, t)$, where λ_D is the indicator of the nitrogen content and the rate of detritus lysis. The free nitrogen supplies in the water medium are also replenished in the process of functioning of various organisms. With account of phytoplankton Φ and nekton r, we have:

$$H_4^N = H_r^N + H_\Phi^N; \quad H_r^N = \lambda_r T_r; \quad H_\Phi^N = \lambda_\Phi T_\Phi,$$

where T_r and T_Φ are the characteristics of metabolic processes, respectively, in nekton and phytoplankton, λ_r and λ_Φ are the coefficients. To determine the averages of these coefficients, assume $T_r = 0.194$ t km^{-2} year^{-1}, $T_\Phi = 0.125$ t km^{-2} year^{-1}, $H_r^N = H_\Phi^N = 0.83 \times 10^{-3}$ t km^{-2} year^{-1}. Then $\lambda_r = 0.00428$, $\lambda_\Phi = 0.00664$.

The process of denitrification in the water medium delivers considerable amounts of nitrogen into the atmosphere: $H_{20}^N = \lambda_5(\theta_2)^{\Delta T} N_U(\varphi, \lambda, z, t)$, where λ_5 and θ_2 are the constants.

The biological fixation of nitrogen in the water medium constitutes about 10×10^6 t year^{-1}, reaching 20.7×10^6 t year^{-1} in the photic layer of the ocean, and (36–1,800) $\times 10^4$ t km^{-3} year^{-1} in small lakes. For the World Ocean, $H_{17}^N = 0.0277$ t km^{-2} year^{-1}, on the average. Assuming $H_{17}^N = \lambda_R R_\Phi$, where R_Φ is the phytoplankton produce averaging 168.8 t km^{-2} year^{-1}, we obtain $\lambda_R = 0.164 \times 10^{-3}$.

The characteristic feature of the nitrogen cycle in the water medium is its transport due to gravitational sedimentation, vertical convection, turbulent diffusion, and convergence. The processes of nitrogen transport by migrating animals are almost negligible and can be neglected in the global model. The simplest form of description of the vertical fluxes of nitrogen is reduced to the model $H_{14}^N = \lambda_\kappa \Delta N_\kappa, H_{15}^N = \lambda_\rho \Delta N_\rho$, where $\kappa = (U, P, L)$, $\rho = (P, L, F)$.

The present contribution of human activity to the general biospheric cycle of nitrogen has reached a level when the consequences of the introduced changes become unpredictable and probably rather catastrophic. The epidemiological studies testify to the growth of respiration diseases on the territories with high concentrations of nitrogen and sulfur oxides as well as photochemical oxidizers. The harmful effect of nitrogen oxides on living organisms starts manifesting itself when the level 940 m kg m^{-3} is exceeded. In general, the consequences of the nitrogen pollution of the biosphere are more complicated. For instance, on the one hand, the technogenic accumulation of nitrogen from the atmosphere at fertilizer production plays a positive role by raising the productivity of land and water ecosystems, and on the other hand, it causes an undesirable eutrophication of water basins. Removal of nitrogen from the atmosphere for industrial and agricultural needs is compensated for by a technogenic input of nitrogen into the atmosphere with the burning of solid and liquid fuel. A considerable share is contributed here by the transport emitting nitrogen oxides reaching, for instance, in the USA, 11.7×10^6 t year^{-1}. However, even an observance of this physical equilibrium cannot substitute for the chemico-biological balance. Therefore in this multi-functional hierarchical set of global fluxes of nitrogen, the most vulnerable knots and bonds should be revealed, which is only possible within a well-planned numerical experiment.

Preliminary estimates of an increasing anthropogenic pressure on the nitrogen fluxes between the biospheric elements suggest a hypothesis about a strong correlation existing between the fertilizers production H_9^N and population density G, anthropogenic accumulation of nitrogen at fuel burning H_2^N and mineral resources expenditure R_{MG}, anthropogenic input of nitrogen into the atmosphere H_{19}^N, and intensity of emissions of general pollutions Z_{VG}. The quantitative characteristics of the factors of these dependences can be obtained from the known trends. From some estimates, an amount of nitrogen oxides emitted into the atmosphere is proportional to the weight of the used fuel with a 4 % annual increasing trend. The scales of the industrial fixation of nitrogen for the last 40 years increased by a factor of 5, reaching a value that could have been fixed by every ecosystem on the Earth before the use of the present agricultural technology. In 1968 the global industry gave about 30×10^6 t of fixed nitrogen and in 2000 this value reached 1 billion.

Formalize these correlations as the following models:

$$H_9^N = \min\{U(K)\bar{G}(K,t); N_A\sigma_K/\sigma\},\ H_2^N = \lambda_{AG}R_{MG},\ H_{19}^N = \lambda_{GA}Z_{VG}$$

where K is the number of an economic region, $\bar{G}$ is the average population density of the K region, σ_K is the area of the K region. The coefficients U, λ_{AG}, and λ_{GA} are determined from analysis of available information about the processes in (4.27). If we assume that $H_2^N = 0.154$ t km^{-2} year^{-1}, $H_{19}^N = 0.102$ t km^{-2} year^{-1} and $H_9^N = 0.283$ t km^{-2} year^{-1}, then at $\bar{G} = 24.4$ men km^{-2}, $R_{MG} = 30.5$ oil units km^{-2} year^{-1}, and $Z_{VG} = 3.39$ t km^{-2} year^{-1}, we obtain $U = 0.283$, $\lambda_{AG} = 0.504 \times 10^{-2}$, and $\lambda_{GA} = 0.03$.

1.7.4 Oxygen and Ozone Cycles Modeling

If we denote by $R_\Phi(\varphi, \lambda, t)$ and $R_\kappa(\varphi, \lambda, t)$ the produce of phytoplankton Φ and land surface of the κ type at the Earth surface point (φ, λ) at a time moment t, then the oxygen fluxes to the hydrosphere and from land to the atmosphere can be described by relationships:

$$H_1^O = a_\Phi R_\Phi\quad H_2^O = a_\kappa R_\kappa\quad H_{21}^O = a_S R_\Phi$$

where the coefficients a_Φ, a_κ, and a_S depend on phytoplankton species and the type of vegetation. For their averaging we use the data on the fluxes: $H_1^O = 140$ t O_2 km^{-2} year^{-1}, $H_2^O = 70$ t O_2 km^{-2} year^{-1}, $H_2^O = 600$ t O_2 km^{-2} year^{-1}, $R_\Phi = 401.3$ t km^{-2} year^{-1}, $R_\kappa = 102.4$ t km^{-2} year^{-1}. Then $a_\Phi = 0.35$, $a_\kappa = 0.68$,

$a_S = 1.49$. Of course, these estimates have a considerable spatial and temporal scatter. In particular, using the data on the productivity of some oceans, we obtain the values of the coefficient a_Φ: the Atlantic Ocean—0.53; the Indian Ocean—0.25; the Arctic Ocean—11.1; the Pacific Ocean—0.64.

Apart from photosynthesis, photolysis can be a source of oxygen in the atmosphere, that is, the decomposition of water vapour under the influence of UV radiation in the upper layers of the atmosphere. However, the intensity of this source in the present conditions is negligible. Nevertheless, denote this flux by $H_3^O = a_H W_A$, where W_A is the water vapour content in the atmosphere, a_H is the empirical coefficient. If we assume that in the upper layers of the atmosphere a constant share of W_A can reside, then at $H_3^O = 0.0039$ t O_2 km^{-2} year^{-1} and $W_A = 0.025$ m, we have $a_H = 1.56 \times 10^{-7}$ per year.

Vernadsky (1944) considered rocks metamorphism, basaltic volcanism, and underground radioactive waters as possible sources of oxygen. However, there are no sufficiently reliable estimates of these fluxes and therefore it is impossible to parameterize them.

The oxidation process both on land and in the water medium is the basic consumer of oxygen on the Earth. The ability of oxygen to react with many elements of the Earth crust forms the fluxes of oxygen leaving the biospheric reservoirs. The balance between the income-and-expenditure fluxes of oxygen has been reached in the course of the biospheric evolution.

Oxygen is spent on respiration of plants, animals, humans, and on the dead organic matter decomposition both in the hydrosphere and on land. To parameterize the income parts of the oxygen balance, linear models are used in MGBO unit.

Atmospheric ozone constitutes 0.64×10^{-6} of the atmospheric mass and belongs to optically active gases. It absorbs UV solar radiation in the range 200–300 nm, strongly affecting thereby the thermal regime of the stratosphere. Besides, ozone has a number of vibration-rotation bands of absorption in the IR spectral region (9.57 μm) and partially absorbs visible radiation in the Chappuis band (0.6 μm). The formation and destruction of ozone have been described in detail (Kondratyev and Varotsos 2000).

Ozone forms in the upper stratosphere from molecular oxygen under the influence of UV solar radiation. In the lower stratosphere and troposphere, the source of ozone is the decomposition of nitrogen dioxide under the influence of UV and visible radiation. The formation of the vertical profile of the ozone concentration is connected with its meridional and vertical transports. The general characteristic of this profile is the total amount of ozone measured by the thickness of its layer in the Dobson units (1DU = 0.001 cm).

Ozone was first measured in the mid-19th century. At that time, a maximum of ozone, for instance, over Europe and in the region of Great Lakes varied within 17–23 ppbv. At present, the ozone layer over western regions of North America in April–October is characterized by quantities 30 ± 5 ppbv. Due to a rapid economic growth of many Asiatic regions followed by increased volumes of consumed fossil fuels and respective increase of NO_x and SO_2 emissions (5 % per year, on the average), there is an increasing trend of monthly mean ozone concentration by 2–6 ppbv per year with the rate preserved till 2013. This is despite the attempts undertaken in Europe and North America on reduction of emissions to the atmosphere of chemical compounds by 8–10 %. Therefore local measures on the ozone layer stabilization on a global scale have no prospects.

The ozone destruction is a complex set of photochemical reactions with participation of the compounds of hydrogen, nitrogen, and chlorine. From the available estimates, 50–70 % of ozone are destructed by nitrogen compounds, 20–30 %—by oxygen O, 10–20 %—by water-containing particles of HO_x, and less that 1 %—by chlorine compounds. The prevailing role of nitrogen compounds in ozone destruction has been confirmed for all latitudes. The equation of photochemical equilibrium between concentrations of ozone and nitrogen oxides is $[NO] \cdot [O_3]/[NO_2] = \mu$, where the equilibrium constant μ depends on the solar radiation intensity and can range from 0 to 0.02.

There are various approaches to a parameterization of the process of formation and destruction of the ozone layer. The complexity of derivation of the dynamic models of the ozone cycle in the atmosphere is connected with its participation in more than 75 chemical reactions, a qualitative and quantitative description of which is impossible without deriving detailed models of numerous minor gas components of the atmosphere. Nevertheless, there are empirical models of the ozone layer, which make it possible, in the present climatic situation, to obtain adequate spatial distributions of ozone. For instance, Bekoriukov and Fedorov (1987) derived a simple empirical model of the total ozone content confirmed by the observation data for the Southern Hemisphere:

$$O_3(\varphi, \lambda) = \sum_n \sum_{n \le m} P_n^m(\varphi)\left[a_{n,m}\cos(m\lambda) + a_{n,-m}\sin(m\lambda)\right],$$

where P_n^m are non-normalized spherical functions of the n degree, of the m order, $a_{n,m}$ and $a_{n,-m}$ are the empirical coefficients whose values are given in Bekoryukov and Fedorov (1987) and in Krapivin and Kondratyev (2002).

There are also static models to describe the vertical profile of the ozone density distribution. One of such models is the Kruger formula:

$$O_3(h) = 51.4\, exp[-(h-40)/4.2] \;\; (\text{МкГ}/\text{М}^3).$$

With the combined use of static and prognostic models it is possible to predict the levels of O_3 concentration on a real time scale. But in this case it is necessary to describe the photochemical reactions with other components of the atmosphere and, to a greater extent, NO_2 taken into account (Agirre-Basurko et al. 2006). Other models of ozone have been reviewed by Kondratyev and Varostos (2000).

The simplest dynamic model of the ozone layer can be written in the form of a balance equation that reflects its income-expenditure components. The ozone supplies are replenished due to the impact of UV radiation on oxygen $\left(H_{11}^O = e_3 O_A\right)$ and nitrogen dioxide $\left(H_{12}^O = e_2 N_A\right)$. The ozone layer is destructed at a rate $H_{14}^{O} = O_3/T_3$, where T_3 is the lifetime of ozone molecules depending on atmospheric pollution: $T_3 = T_3^O - e_1 B$. The lifetime T_3^O of the ozone molecules in a perfect atmosphere averages 50–60 days. Nitrogen oxides participating in the contrary to H_{12}^{O} cycle of ozone destruction contribute much to the magnitude of B.

Studies of the history of biospheric evolution reveal a close correlation between the oxygen production intensity and the development of life on Earth. And although the expected relative oscillations of the oxygen concentration in the near future do not exceed 10 %, the considered impacts on the biosphere do not cover all potential anthropogenic trends, and therefore cannot be considered reliable. Therefore let us analyze the constituents of possible mechanisms for violation of the natural balance of oxygen. Naturally, our concern is for not only an increase but also a decrease of the oxygen content in the atmosphere.

The oxygen cycle is complicated by its ability to take part in a lot of chemical reactions giving a multitude of epicycles. This fact makes the oxygen cycle sufficiently stable but hinders an assessment of its stability.

Anthropogenic forcing on numerous epicycles of oxygen manifests itself both directly through its involvement in other cycles of substances at fuel burning and production of various materials, and indirectly through environmental pollution and biospheric destruction. Therefore a parameterization of the anthropogenic impact on the oxygen balance is realized within other units of the global model. The flux H_{18}^{O} taken into account in the scheme in Fig. 1.38 completely covers direst consumption of oxygen both in industry and in agriculture. Assume $H_{18}^{O} = y_1 R_{MG}$, where R_{MG} is the rate of the natural resourcesexpenditure, and y_1 is the coefficient (≈0.084).

The fluxes H_{15}^{O} and H_{16}^{O} are strongly affected by anthropogenic forcings. Their variations are caused by discharges of high-temperature industrial sewage containing considerable amounts of oxidizers as well as by the oil-polluted water bodies. The quantitative characteristics of the change of oxygen dissolved in water as a function of temperature have been studied comprehensively. The empirical formula to calculate the concentration of the seawater dissolved oxygen has the form (Ramad 1981): [O_2 dissolved] = 80/(0.2T_O − 7.1), where [O_2] is expressed in mg/ℓ, T_O in °C. The estimates of oxygen solubility in water are well known (Krapivin and Kondratyev 2002).

The fluxes H_9^O and H_{19}^{O}, balancing in natural conditions the oxygen fluxes into the water medium, at an anthropogenic forcing increase, as a rule, due to more active aerobic bacteria and increasing metabolic needs of animals. For instance, a 10 °C increase of water temperatureincreases the oxygen expenditure on respiration of marine animals by a factor of 2.2.

One of the negative manifestations of the anthropogenic impact on the oxygen cycle is a depletion of the ozone layer, especially marked in the polar regions. There are various hypotheses on the causes of sharply changing concentrations of ozone, as well as discussions on the so-called "ozone hole" over the Antarctic. The main cause of all violations is connected with the progressing human activity accompanied by the growing volumes of long-lived components emitted to the atmosphere (e.g., freons). The consequences of these violations are very serious, and the real scales of danger threatening the life on the Earth can be estimated only with the use of the global model of the nature-society system.

A diversity of the anthropogenic impacts on the global biogeochemical cycle of oxygen is determined by direct and indirect causes of breaking the natural balance of oxygen. According to the equation of photosynthesis, the gram-molecular amounts of assimilated CO_2 and emitted O_2 are equal. Also equal are the gram-molecular amounts of assimilated O_2 and emitted CO_2 for the dead organic matter decomposition and fuel burning. Hence, for time periods of tens and hundreds of years, a change of the CO_2 amount in the atmosphere is accompanied by the same change of O_2, but in the opposite direction. For instance, a CO_2 doubling in the atmosphere leads to a decrease of the amount of O_2. But, since the volume concentration of CO_2 in the atmosphere is now estimated at 0.031 %, and that of O_2 at 20.946 %, in this case a decrease of O_2 will constitute only 0.15 % of the total O_2 content in the atmosphere.

Imagine the following situation. Let the total biomass of the biosphere ($\sim 9.6 \times 10^{11}$ t C), all organic matter of soil ($\sim 14 \times 10^{11}$ t C), and all fossil chemical fuel, the explored supplies of which constitute 128×10^{11} t of conditional fuel (64×10^{11} t C) be burnt. Then the amount of CO_2 in the atmosphere increases by a factor of 12.5, and that of O_2, respectively, decreases but only by 1.75 %. Hence, the amount of oxygen during hundreds of years has to be practically constant.

However, it should be borne in mind that the region of excess anthropogenic emissions of CO_2 and, hence, O_2 assimilation is concentrated on a small area of cities and forest fires. Since the concentrations in the atmosphere do not equalize instantly, a gradient of O_2 concentrations can appear around these sites, when the oxygen provision will be insufficient for animals and humans. Therefore the model of the global oxygen cycle (MGBO unit) reflecting the spatial heterogeneities in the distributions of O_2 concentrations, enables one to reveal such dangerous territories.

An interaction between the cycles of oxygen, nitrogen, sulfur, phosphorus, and carbon manifests itself through the processes of oxidation and decomposition. The level of detailing the global model units does not permit one to reflect all the diversity of these processes. Therefore in the simplest case, when only averaged characteristics of the oxygen cycle elements are taken into account, the scheme in Fig. 1.38 of the global O_2 fluxes can be presented as the schemes in Figs. 1.36 and 1.37. The indicated stability of the O_2 concentration in the atmosphere makes it possible to simplify a description of the MGBO unit, using a single balance equation:

$$\frac{\partial O}{\partial t} + V_\varphi \frac{\partial O}{\partial \varphi} + V_\lambda \frac{\partial O}{\partial \lambda} = k_0 R_F + k_L R_L - v_L T_L - b_G G - v_F T_F - v_G T_G - \mu_Q R_Q$$

where k_O and k_L are the indicators of the rate of O_2 emission due to photosynthesis in the ocean and on land, respectively, v_s is the indicator of the role of respiration of land vegetation ($s = L$), animals ($s = F$), and humans ($s = G$) in the removal of oxygen from the atmosphere, μ_Q is the rate of O_2 consumption at the decomposition of the soil dead organic matter.

1.7.5 Global Methane Budget Modeling

Methane is produced and released from a range of natural and human sources main of which are shown in Figs. 1.39 and 1.40. Understanding the Earth's methane cycle is a necessary step for evaluating the effectiveness of emission reduction schemes, detecting changes in biological sources and sinks that are influenced by climate feedbacks, and predicting, and perhaps mitigating, future methane emissions.

According to Bazhin (2000), the flux $F^1_{CH_4}$ (Fig. 1.39) in every water basin with a vertical stratiform structure forms in an active layer beneath the water layer. Practically, all aqua-geo-systems have such a structure. The layer, where methane forms, has two areas. In the bottom area located at a depth h, methane has the form of bubbles. Above this layer, due to diffusion, the concentration of methane decreases, and the bubbles disappear. Let us denote as $D_{CH_4}(z)$ the coefficient of methane diffusion at a depth z, then the stationary behaviour of the whole system shown in Fig. 1.39 is described by the equation:

$$\frac{d}{dz}\left[D_{CH_4}(z)\frac{d}{dz}M(z)\right] - F_{CH_4}(z) + F^1_{CH_4} + F^2_{CH_4} = 0$$

Model calculations and field measurements performed by Bazhin (2000), show, for instance, that on the rice fields h_b = 1.3 m, $F_{CH_4} = (1.3-1.7) \times 10^{-12}$ mol cm^{-3} s^{-1}. According to Khalil et al. (2000), the rice fields play a significant role in the gas balance of the atmosphere due to emissions of CH_4, CO, N_2O, H_2, and $CHCl_3$. For instance, in China the rice fields deliver these gases to the atmosphere at the following rates (mg/m^2 h): CH_4—900 ÷ 50,000; CO—80–100; H_2—5–30; N_2O—50–1,000; CHCl—1–8. A wide scatter of these estimates is explained by highly unstable fluxes of these gases due to the rice growing technology. For instance, the use of sulphates on the rice fields increases emissions of methane by 12.0–58.9 % depending on other characteristics of these fields (Liping et al. 2000).

Thus an estimation of the flux $F^1_{CH_4}$ as a function of a given territory with account of natural and anthropogenic processes taking place there requires first of all a detailed inventory of methane sources as well as natural and anthropogenic systems functioning on this territory. Examples of such an inventory given above serve the basis for development of studies in this direction.

The dynamics of the CH_4 content H_A in the atmosphere can be parameterized by a simple balance relationship:

$$\frac{\partial H_A}{\partial t} + V_\varphi\frac{\partial H_A}{\partial \varphi} + V_\lambda\frac{\partial H_A}{\partial \lambda} = F^1_{CH_4}(t, \varphi, \lambda, \Xi) - \frac{H_A(t, \varphi, \lambda)}{\tau_H}$$

where Ξ is the identifier of the type of the natural or technogenic system.

On the whole, the fluxes of methane in the environment are rather diverse. The scheme in Fig. 1.39 and data in Fig. 1.40 reflect, to some extent, this diversity.

1.7.6 Global Sulfur Cycle Modeling

Modeling of global sulfur cycle is particularly needed in connection with acid rains. There are numerous models that are parameterized global sulfur cycle taking into consideration of different hypotheses and assumptions concerning the chemistry and transport components. One of the known global models for the sulfur cycle is STOCHEM-Ed model (Collins et al. 1998). It is 3-D Lagrangian chemistry-transport model. Stevenson et al. (2003) studied modified STOCHEM-Ed model taking into consideration of the volcanic component. It was shown that:

- volcanic emissions constitute 10 % of the global SO_2 source to the atmosphere, but from 26 % of the SO_2 burden, and 14 % of the sulfate aerosol burden; and
- volcanic SO_2 dominates large regions of the free troposphere, particularly in the tropics.

The version of the sulfur unit proposed here, in contrast to the known hydrodynamic models of the long-distance transport, takes into account the fluxes of sulfur compounds between the hydrosphere, atmosphere, soil, and biota. The model does not consider the vertical stratification of the atmosphere. The characteristics of sulfur fluxes averaged vertically are calculated for both the land and ocean (Fasham 2003; Zaikov et al. 1991; Zlatev et al. 1992). The spatial digitization of the biosphere and the World Ocean corresponds to the criterion inherent to the global model. The elements of the block-scheme of the model of the biogeochemical cycle of sulfur are described in Tables 1.16 and 1.17. This scheme is realized in every cell Ω_{ij} of the Earth's surface and in every compartment Ω_{ijk} of the World Ocean. Interaction between the cells and compartments is organized through the climate unit of the global model. Therefore the equations of the sulfur unit lack the terms reflecting the dynamic patter of the spatial transformation of the sulfur reservoirs. With notations assumed in Tables 1.16 and 1.17 taken into account, the equations describing the balance relationships between the reservoirs of sulfur compounds will be written in the form:

$$\frac{dAH2SL}{dt} = C_1 + C_2 + C_3 - C_4 + C_{21},$$

$$\frac{dASO2L}{dt} = C_4 + C_5 + C_6 - C_7 - C_8 - C_9,$$

$$\frac{dASO4L}{dt} = C_9 + C_3 + C_{20} - C_{11} - C_{12},$$

$$\frac{dS}{dt} = C_{17} - C_{16} - C_{19},$$

$$\frac{dSO4L}{dt} = C_{10} + C_{11} + C_{12} + C_{16} - C_3 - C_{13} - C_{14},$$

Table 1.16 Characteristics of land and hydrospheric fluxes of sulphur in the structure of Fig. 1.44

Sulfur flux	Land		Hydrosphere	
	Identifier	Estimate	Identifier	Estimate
Volcanic invasions				
H_2S	C_1	0.018	H_3	0.0068
SO_2	C_5	0.036	H_5	0.0073
SO_4^{2-}	C_{20}	0.035	H_9	0.0074
Anthropogenic emissions				
H_2S	C_2	0.072	H_1	0.00076
SO_2	C_6	0.92	H_6	0.038
SO_4^{2-}	C_{10}	0.47		
Oxidation of H_2S to SO_2	C_4	1.13	H_2	0.3
Oxidation of SO_2 to SO_4^{2-}	C_9	1.35	H_8	0.16
Dry sedimentation of SO_4^{2-}	C_{12}	0.37	H_{11}	0.11
Fall-out of SO_4^{2-} with rain	C_{11}	1.26	H_{10}	0.38
Biological decomposition and emission of H_2S to the atmosphere	C_3	1.03	H_4	0.31
Assimilation of SO_4^{2-} by biota	C_{15}	0.41	H_{13}	1.09
Biological decomposition and formation of SO_4^{2-}	C_{16}	1.13	H_{17}	0.43
			H_{23}	0.12
Sedimentation and deposits	C_{18}	0.22	H_{15}	0.98
	C_{19}	0.11	H_{16}	0.55
			H_{19}	0.0076
			H_{25}	0.036
Wind-driven return to the atmosphere	C_{13}	0.25	H_{12}	0.33
Replenishing sulphur supplies due to dead biomass	C_{17}	0.86	H_{14}	1.1
Assimilation of atmospheric SO_2	C_7	0.46	H_7	0.18
Washing-out of SO_2 from the atmosphere	C_8	0.27	H_{24}	0.061
River run-off of SO_4^{2-} to the ocean	C_{14}	1.17		
Transition of gas-phase H_2SO_4 to H_2S	C_{21}	0.018	H_{26}	0.0076
Assimilation of the washed-out part of atmospheric SO_2 by biota	C_{22}	0.036	H_{27}	0.015
Oxidation of H_2S to SO_2 in water medium			H_{18}	0.045
			H_{22}	0.19
Advection of SO_2			H_{20}	0.38
Advection of H_2S			H_{21}	0.37

Numerical estimates of fluxes (mg/m^2 day) are obtained by averaging over the respective territories

Table 1.17 Initial data taken into account under simulation experiments

Reservoir	Identifier of the GSM	Preliminary estimation of reservoir (mg m^{-2})
Atmosphere above the ocean		
H_2S	*AH2SO*	10
SO_2	*ASO2O*	5.3
SO_4^{2-}	*ASO4O*	2
Atmosphere above land		
H_2S	*AH2SL*	36.9
SO_2	*ASO2L*	17.9
SO_4^{2-}	*ASO4L*	12.9
Land		
SO_4^{2-}	*SO4L*	11.2
Biomass	*FIX*	600
Soil	*S*	5,000
Ocean photic layer		
H_2S	*H2SOU*	1.9
SO_4^{2-}	*SO4OU*	19×10^7
Phytomass	*FI*	66.5
DOM	*DU*	730
Deep ocean layers		
H_2S	*H2SOD*	2×10^6
SO_4^{2-}	*SO4OD*	3.4×10^9
DOM	*DD*	13,120

$$\frac{dFIX}{dt} = C_7 + C_{15} - C_{17} + C_{22},$$

$$\frac{dH2SO4L}{dt} = C_8 - C_{18} - C_{21} - C_{22},$$

$$\frac{dAH2SO}{dt} = H_1 + H_3 + H_4 + H_{26} - H_2,$$

$$\frac{dASO2O}{dt} = H_2 + H_5 + H_6 - H_7 - H_8 - H_{24},$$

$$\frac{dASO4O}{dt} = H_8 + H_9 + H_{12} - H_{10} - H_{11},$$

$$\frac{\partial SO4OU}{\partial t} + v_z \frac{\partial SO4OU}{\partial z} + k_z \frac{\partial^2 SO4OU}{\partial z^2} = H_7 + H_{10} + H_{11} + H_{20} + H_{22} + H_{27} + C_{14} - H_{12} - H_{13},$$

$$\frac{\partial H2SOU}{\partial t}+v_z\frac{\partial H2SOU}{\partial z}+k_z\frac{\partial^2 H2SOU}{\partial z^2}=H_{21}+H_{23}-H_4-H_{22},$$

$$\frac{\partial H2SOD}{\partial t}+v_z\frac{\partial H2SOD}{\partial z}+k_z\frac{\partial^2 H2SOD}{\partial z^2}=H_{17}-H_{18}-H_{21},$$

$$\frac{\partial SO4OD}{\partial t}+v_z\frac{\partial SO4OD}{\partial z}+k_z\frac{\partial^2 SO4OD}{\partial z^2}=H_{18}-H_{19}-H_{20},$$

$$\frac{\partial DU}{\partial t}+v_z\frac{\partial DU}{\partial z}+k_z\frac{\partial^2 DU}{\partial z^2}=H_{14}-H_{15}-H_{23},$$

$$\frac{\partial DD}{\partial t}+v_z\frac{\partial DD}{\partial z}+k_z\frac{\partial^2 DD}{\partial z^2}=H_{15}-H_{16}-H_{17},$$

$$\frac{\partial FI}{\partial t}+v_z\frac{\partial FI}{\partial z}+k_z\frac{\partial^2 FI}{\partial z^2}=H_{13}-H_{14},$$

$$\frac{\partial BOT}{\partial t}=H_{16}+H_{19},$$

where v_z is the velocity of the vertical water motion in the ocean, m day^{-1}; k_z is the coefficient of the turbulent mixing, m^2 day$^{-1.}$

The equations of the MGBS item in each cell of the spatial division of the ocean surface are supplemented with initial conditions (Table 1.17). The calculation procedure to estimate the sulfur concentration consists of two stages. At each time moment t_i, first for all cells Ξ_{ij}, the equations of the MGBS item are solved by the method of quasi-linearization, and all reservoirs of sulfur are estimated for $t_{i+1} = t_i + \Delta t$, where a time step Δt is chosen from the condition of convergence of the calculation procedure. Then at the moment t_{i+1} with the use of the climate unit of the global model these estimates are specified with account of the atmospheric transport and ocean currents over the time Δt.

The sulfur supplies in the reservoirs are measured in mg S m^{-3}, the sulfur fluxes have the dimensionality mg S m^{-3} day^{-1}. The sulfur supplies in the water medium are calculated with the volumes of the compartments Ω_{ijk} taken into account. To estimate the sulfur supplies in the atmosphere, it is assumed that an effective thickness of the atmosphere h is an input parameter either introduced into the model by the user or prescribed as the constants from Table 1.17, or received from the climate unit of the global model. The quantitative estimates of the fluxes in the right-hand parts of the equations are obtained in different units of the global model. The anthropogenic fluxes of sulfur H_1, H_6, C_2, C_6, and C_{10} are simulated in the unit of scenarios. The fluxes H_3, H_5, H_9, C_1, C_5, and C_{20} are prescribed either by the climate unit or formed in the unit of scenarios. The accuracy of different functional presentations of the fluxes in the equations of the MGBS item corresponds to the accuracy of similar fluxes of the biogeochemical cycles of hydrogen, phosphorus, and nitrogen. The rate of emission of H_2S into the atmosphere at humus decomposition is described by the linear function $C_3 = \mu_1(pH) \cdot SO4L \cdot T_L$, where μ_1 is the proportion coefficient depending on soil acidity, day^{-1} K^{-1}; T_L is the soil temperature, °K. The initial value of SO4L is estimated proceeding from the humus supply considering that the content of sulfur in humus is prescribed by the parameter a_g, %. According to the available observations of the input of H_2S into the atmosphere from the ocean, the flux H_4 varies widely from low values to high values at the transition from the stagnant waters to zones of upwellings. The flux H_4 is assumed to be the function of the ratio of the rates of H_2S oxidation in the photic layer to the rate of the vertical uplifting of water. Therefore to describe the H_4 flux, use the parameter t_{H2SU}, which reflects the life-time of H_2S in the water medium: $H_4 = H_2SU/t_{H2SU}$. Determine the value of t_{H2SU} as a function of the rate of the vertical advection v_z and concentration of oxygen O_2 in the upper layer Z_{H2S} thick: $t_{H2SU} = H2SOU \cdot v_z(\theta_2 + O_2)/[O_2(\theta_1 + v_z)]$, where the constants θ_1 and θ_2 are determined empirically, the value of O_2 is either calculated by the oxygen unit of the global model or prescribed from the global database.

The reaction of oxidation of H_2S to SO_2 in the atmosphere, on land, and over the water surface is characterized by a rapid process of the reaction of hydrogen sulphide with atomic and molecular oxygen. At the same time, the reaction of H_2S with O_3 in the gas phase is slow. It is impossible to describe within the global model the diversity of the situations appearing here, however, an inclusion of the fluxes H_2 and C_4 into the unit of sulfur enabled one to take into account the correlation between the cycles of sulfur and oxygen. These fluxes are parameterized with the use of the indicator t_{H2SA} of the life-time of H_2S in the atmosphere: $C_4 = AH_2\ SL/t_{H2SA}$, $H_2 = AH_2SO/t_{H2SA}$. The mechanism to remove SO_2 from the atmosphere is described by the fluxes H_7, H_8, H_{27}, C_7, C_8, and C_9. Schematically this mechanism consists of a set of interconnected reactions of SO_2 with atomic oxygen under the influence of various catalysts. A study of the succession of reaction enables one to estimate the life-time of SO_2 for oxidation over land t_{SO2L} and water surface t_{SO2A1}. This makes it possible to assume the following parameterizations of the fluxes H_8 and C_9: $H_8 = ASO_2O/t_{SO2A1}$, $C_9 = ASO_2L/t_{SO2L}$.

Sulfur dioxide is assimilated from the atmosphere by rocks, vegetation, and other Earth's covers. Over the water surface this assimilation is connected with the intensity of turbulent gas fluxes and surface roughness. We describe a dry deposition of SO_2 over the vegetation by the model $C_7 = q_2RX$, where $q_2 = q_2' \cdot ASO_2L/(r_{tl} + r_s)$, r_{tl} is the atmospheric resistance to the SO_2 transport over the vegetation of the l type, day m^{-1}; r_s is the surface resistance of the s type to the SO_2 transport, day m^{-1}; RX is the produce of vegetation of the X type, mg m^{-2} day^{-1} (calculated by the biogeocenotic unit of the global model); q'_2 is the proportion coefficient. The parameters r_{tl} and r_s are functions of the types of the soil-vegetation formations and estimated, respectively, at 0.05 and 4.5 for the forests, 0.9 and 3 for grass cover, 0.5 and 2 for bushes, 0.8 and 1 for bare soils, 1.9 and 0 for water surfaces, 2 and 10 for snow cover.

The process of washing-out of SO_2 from the atmosphere with changing phase to H_2SO_4 and a subsequent neutralization on the surface of the l type is described by the function: $C_8 = q_{1l}W \cdot ASO_2L$ with the Langmuir coefficient (Mon et al. 2006) q_{1l} and precipitation intensity $W(\varphi, \lambda, t)$. An interaction of the acid rain with the Earth surface elements is reflected by the fluxes C_{18}, C_{21}, and C_{22} for land and H_{25}, H_{26}, and H_{27} for water surface. To parameterize these fluxes, assume the hypothesis that the reservoirs of H_2SO_4L and H_2SO_4O are spent in proportion to the out-fluxes, and the coefficients of this proportion are the controlling parameters of the numerical experiments: $C_{18} = h_1 \cdot H_2SO_4L$, $C_{22} = h_2 \cdot RX \cdot H_2SO_4L$, $C_{21} = h_3T_a \cdot H_2SO_4L$, $H_{25} = h_6\, H_2SO_4O$, $H_{26} = h_4T_a\, H_2SO_4O$, $H_{27} = h_5 \cdot RFI \cdot H_2SO_4O$, $h_1 + h_2 \cdot RX + h_3T_a = 1$, $h_4T_a + h_5 \cdot RFI + h_6 = 1$, where $T_a(\varphi, \lambda, t)$ is the air surface temperature. Parameterize the fluxes H_7 and H_{24} by the relationships: $H_7 = ASO_2O/t_{SO2A2}$, $H_{24} = q_{1l}W \cdot ASO_2O$, where t_{SO2A2} is the life-time of SO_2 over the water surface.

Sulphates interacting with the ecosystems and establishing an interaction of the sulfur cycle with other biogeochemical processes are one of the most important elements in the global cycle of sulfur. Numerous complicated transformations of sulphates in the environment are described by a set of fluxes H_7, H_8, H_{10}, H_{11}, H_{12}, C_9, C_{11}, C_{12}, C_{13} for the atmospheric reservoir and the fluxes H_{13}, H_{18}, H_{19}, H_{20}, H_{22}, C_3, C_{14}, C_{15}, C_{16} for land and the World Ocean.

Physical mechanisms for the transport of sulphates from the atmosphere to the soil and water medium are connected with dry and wet sedimentation. An efficient model of the wet removal of particles and gases from the atmosphere was proposed by Langmann (2000): a substitution of the mechanism of the aerosols and gases by a simplified binary model enables one to match it with other units of the global model: $H_{10} = \mu W \cdot ASO_4O$, $H_{11} = \rho v_O \cdot ASO_4O$, $C_{11} = b_3W \cdot ASO_4L$, $C_{12} = d_1v_a \cdot ASO_4L$, where v_O and v_a are the rates of the aerosols dry deposition over the water surface and land, respectively; μ, b_3, ρ, and d_1 are the constants.

The return of sulphates from the soil and water medium to the atmosphere is connected with rock weathering and spray above the rough water surface: $C_{13} = d_2 \cdot RATE \cdot SO_4L$, $H_{12} = \theta \cdot RATE \cdot SO_4U$, where RATE (φ, λ, t) is the wind speed over the surface, m/s; d_2 and θ are the empirical coefficients.

The flux C_{14} relates the surface and water reservoirs of sulfur. Let σ be the share of the river system area on land and d_3—the proportion coefficient, then $C_{14} = d_3W \cdot SO_4L + (C_{11} + C_{12})\sigma$. The surface part of the sulfur cycle is connected with the functioning of the system atmosphere-vegetation-soil. Plants adsorb sulfur from the atmosphere in the form of SO_2 (fluxes C_7 and C_{22}) and assimilate sulfur from the soil in the form of SO_4^{2-} (flux C_{15}). In the hierarchy of the soil processes two levels can be selected defining the sulfur reservoirs as "dead organics" and "SO_4^{2-} in soil". The transitions between them are described by the flux $C_{16} = b_2ST_L$, where the coefficient $b_2 = b_{2,1}b_{2,2}$ reflects the rate $b_{2,1}$ of transition of sulfur contained in dead organics into the form assimilated by vegetation The coefficient $b_{2,2}$ indicated the content of sulfur in dead plants.

The fluxes of sulfur in the water medium according to studies by Bodenbender et al. (1999), depend on the biological processes in the water bodies and constitute an isolated part of the global cycle of sulfur that contains only the fluxes that connect it with the atmospheric and surface cycles. Rough estimates show that the rates of the sulfur cycle in the water of the seas and oceans do not play a substantial role for the remaining parts of its global cycle. And though this specific feature does exist, for the purity of the numerical experiment, in the proposed model the internal hydrospheric fluxes of sulfur compounds are separated in space and parameterized with the same details like other fluxes of sulfur in the atmosphere and on land. This excessiveness is important for other units of the global model as well. In particular, it is important for the parameterization of photosynthesis whose rate *RFI* affects the closure of other bio-geochemical cycles. Finally, assume: $H_{13} = \gamma \cdot RFI$, $H_{14} = b \cdot MFI$, $H_{15} = f \cdot DU$, $H_{16} = p \cdot DD$, $H_{17} = q \cdot DD$, $H_{18} = H_2SOD/t_{H2SOD}$, $H_{19} = u \cdot SO_4D$, $H_{20} = a_1 v_D \cdot SO_4D$, $H_{21} = b_1 v_D \cdot H_2SOD$, $H_{22} = H_2SOU/t_{H2SOU}$, $H_{23} = g \cdot DU$, where *MFI* is the mass of dead phytoplankton, t_{H2SOU} and t_{H2SOD} is the time of complete oxidation of H_2S in the sea water in the photic and deep layers, respectively; γ, b, f, p, q, u, a_1, b_1, and g are the constants.

The anthropogenic input to the sulfur unit is realized through the prescribed fluxes C_2, C_6, C_{10}, H_1, and H_6 in the form of the functions of the spatial coordinates and time.

1.7.7 Global Phosphorus Cycle Modeling

The flux diagram of the model of the global phosphorus cycle (MGBP) is presented in the scheme in Fig. 1.45, according to which the balance system of equations will be:

$$\frac{\partial P_A}{\partial t} + V_\varphi \frac{\partial P_A}{\partial \varphi} + V_\lambda \frac{\partial P_A}{\partial \lambda} = H_1^P + H_{19}^P + \begin{cases} -H_{16}^P, & (\varphi,\lambda) \in \Xi_O; \\ H_7^P - H_8^P, & (\varphi,\lambda) \in \Xi/\Xi_O; \end{cases}$$

$$\frac{\partial P_U}{\partial t} + v_\varphi \frac{\partial P_U}{\partial \varphi} + v_\lambda \frac{\partial P_U}{\partial \lambda} + v_z \frac{\partial P_U}{\partial z} = H_{11}^P + H_{15}^P + H_{16}^P - H_9^P - H_{10}^P;$$

$$\frac{\partial P_L}{\partial t} + v_\varphi \frac{\partial P_L}{\partial \varphi} + v_\lambda \frac{\partial P_L}{\partial \lambda} + v_z \frac{\partial P_L}{\partial z} = H_{12}^P + H_{14}^P - H_{13}^P - H_{15}^P;$$

$$\frac{\partial P_S}{\partial t} = H_2^P + H_8^P + H_9^P + H_{10}^P - H_6^P - H_7^P - H_{11}^P;$$

where $P_U = P_{U1} + P_{U2} + P_{U3}$, $P_S = P_{S1} + P_{S2}$. With this detailing we have:

$$\frac{\partial P_{U1}}{\partial t} + v_\varphi \frac{\partial P_{U1}}{\partial \varphi} + v_\lambda \frac{\partial P_{U1}}{\partial \lambda} + v_z \frac{\partial P_{U1}}{\partial z} = H_{17}^P - H_9^P - H_{10}^P - H_{20}^P;$$

$$\frac{\partial P_{U2}}{\partial t} + v_\varphi \frac{\partial P_{U2}}{\partial \varphi} + v_\lambda \frac{\partial P_{U2}}{\partial \lambda} + v_z \frac{\partial P_{U2}}{\partial z} = H_{20}^P - H_{18}^P;$$

$$\frac{\partial P_{U3}}{\partial t}+v_\varphi\frac{\partial P_{U3}}{\partial \varphi}+v_\lambda\frac{\partial P_{U3}}{\partial \lambda}+v_z\frac{\partial P_{U3}}{\partial z}=H_{11}^P+H_{18}^P-H_{17}^P;$$

$$\frac{\partial P_{S1}}{\partial t}=H_3^P-H_4^P-H_5^P;$$

$$\frac{\partial P_{S2}}{\partial t}=H_2^P+H_4^P+H_5^P+H_8^P-H_3^P-H_6^P-H_7^P;$$

Determine now the functional and dynamic characteristics of the fluxes of phosphorus (Table 1.18) based on analysis of the existing ideas about their nature. The atmospheric cycle is governed by rock weathering, volcanic eruptions, and by washing-out of phosphorus by precipitation. From the available estimates, the phosphorus content in the lithosphere constitutes 0.093 %, and the processes of weathering deliver annually into the atmosphere about 0.2×10^6 t P, on the average. Since these processes are of a complicated stochastic nature and their models are absent, in the first approximation the fluxes H_1^P and H_{19}^P will be considered constant.

The continental cycle of phosphorus is determined by ten fluxes (Fig. 1.45) closed by a single component P_S indicating the phosphorus supplies on land in the soil-vegetation formations and in animals. The supplies of phosphorus in soils are replenished due to the fluxes H_l^P $(l = 2, 4, 5, 8, 9, 10)$. The loss of phosphorus from the soil is determined by the fluxes H_j^P $(j = 3, 6, 7, 11)$. With a more complicated detailing of the surface reservoirs of phosphorus and consideration of more ingenious effects in the interaction of these reservoirs, the classification of the surface fluxes of phosphorus becomes more complicated. Within the assumed detailing of the surface reservoirs of phosphorus the following functional presentations of the fluxes $\{H_i^P\}$ can be considered:

$$H_2^P = p_1 GMG_0^{-1}M_0^{-1};\quad H_3^P = p_V R_V;\quad H_4^P = p_4 M_L;$$
$$H_5^P = p_5 H_F;\quad H_6^P = p_6 P_{S2}/P_{S2,0};\quad H_9^P = p_3 I;\quad H_7^P = p_7 \theta^{\Delta T} P_{S2}/P_{S2,0};$$
$$H_{11}^P = p_2 P_{S2}[1-exp(-k_{su}W_{so})]/P_{S2,0};$$

where p_V is the content of phosphorus in the living biomass of plants, p_4 and p_5 is the content of phosphorus in organic matter of the vegetable and animal origin, respectively, I is the amount of produce extracted from the ocean, G is the population, M is mineral resources, R_V is the produce of the vegetation photosynthesis, H_F is the biomass of the unassimilated food of animals, $M_L = \mu_V L$, μ_V is the rate of vegetation dying-off, L is the vegetation biomass, θ is the temperature coefficient of the rate of the dead organic matter decomposition on land, ΔT is the SAT variation with respect to the control value, W_{so} is the volume of the river run-off into the oceans, p_i (i = 1, …, 7) are the constants. The index “0” in G_0, M_0, and $P_{S2,0}$ attributes these parameters to some control time moment t_0, when all parameters of MGBP unit are known.

Table 1.18 The characteristics of fluxes (10^6 t/year) and reservoirs (10^6 t) of phosphorus in the biosphere/climate/society system

Reservoirs and fluxes of phosphorus	Identifier	Estimate
Phosphorus supplies		
In the atmosphere	P_A	3
On land	P_S	1,546
In the photic layer of the World Ocean	P_U	2×10^4
In deep layers of the World Ocean	P_L	12×10^4
Volcanic emissions	H_1^P	0–2
Fertilization	H_2^P	19
Assimilation by plants	H_3^P	45.34
Input with dead plants	H_4^P	39.34
Input due to organisms functioning		
On land	H_5^P	5
In the World Ocean	H_{20}^P	81.5
Transition into the unassimilable form	H_6^P	2.9
Weathering	H_7^P	5
Falling-out with precipitation		
On land	H_8^P	1.8
On the oceans	H_{16}^P	2
Removal with fish catch	H_9^P	0.06
Removal by birds	H_{10}^P	0.04
Washing-out and sink into the World Ocean	H_{11}^P	4–14
Input due to detritus lysis in the oceans		
Photic layer	H_{18}^P	550
Deep layers	H_{12}^P	159
Exchange between photic and deep layers of the ocean		
Lifting	H_{15}^P	96.1
Descending	H_{14}^P	22
Precipitation	H_{13}^P	13–83.9
Rock weathering	H_{19}^P	1
Photosynthesis	H_{17}^P	630–1,300

Describe the hydro-chemical cycle of phosphorus by a totality of fluxes $H_k^P(k = 9{-}18, 20)$: $H_{12}^P = p_{14}R_{DL}$, $H_{13}^P = p_8 P_L/P_{L,0}$, $H_{14}^P = p_9 P_U/P_{U,0}$, $H_{15}^P = p_{10} P_L/P_{L,0}$, $H_{16}^P = p_{12} R_{WO} P_A/P_{A,0}$, $H_{17}^P = p_{13} R_\Phi$, $H_{18}^P = p_{15} R_{DU}$, $H_{20}^P = p_{16} M_\Phi$, where R_D is the rate of the dead organic matter decomposition, R_{WO} is precipitation over the ocean, R_Φ is the produce of phytoplankton and other living organisms in the ocean, M_Φ is the rate of the living biomass dying-off, p_i (i = 8–16) are the proportion constants.

1.8 Climate Unit of the Global Model

The climatic component of the GIMS is the greatest difficulty in the synthesis of the global model, because it is characterized by a lot of feedbacks, most of which being unstable. Among them are such as ice-albedo, water vapour-radiation, cloudiness-radiation, aerosol-radiation, and many others. The functioning of the Earth climatic system is determined by the state of the atmosphere, oceans, cryosphere, surface of the continents with land biota, lakes, rivers, ground waters, and various anthropogenic structures. Therefore the construction of the model of climate requires a consideration of numerous factors, whose role in its formation in many cases has been studied inadequately. So far, attempts to derive a numerical model of the Earth climatic system have not given the results that could be used in the GIMS. Also, it is not safe to say that a realization of the GCP plans will clarify the problem. But without a parameterization of climate the questions concerning the problem of the greenhouse effect will remain unanswered.

There are two approaches to the synthesis of the climatic system. One approach is based on inclusion of biospheric components into the existing climate models or in those under development. The other approach consists in the development of a unit within the numerical model of the biosphere, which would simulate dependences of biospheric components on climate parameters. In the former case, problems appear of unstable solutions of the respective systems of differential equations, which make it difficult to predict global environmental changes. In the latter case there is a possibility to obtain steady prognostic estimates of the environmental changes, but their reliability depends on the accuracy of parameterization of correlations between the elements of climate and the biosphere. The second approach is preferable, because it permits one to combine the climate models, which can be described on the level of scenario, with the numerical model of the biosphere. A detailed analysis of the problems of climate modeling and assessment of the present state of the problems can be found in Kondratyev et al. (2004a). Some models of individual components of the BCSS, which correspond to the second approach, are discussed here. Among them are models of the atmospheric general circulation, atmosphere-ocean interaction, sensitivity of climate parameters to boundary conditions on the Earth's surface, interaction of biogeochemical and climatic processes, and others.

The climatic system is a physical-chemical-biological system with unlimited degrees of freedom. Therefore, any attempts to model this complex system are connected with serious difficulties. This explains a diversity of parametric descriptions of some processes in this system. For the global model with a time step of digitization up to 1 year, the use of two versions can be acceptable. The first version consists in a combined use of correlations between particular processes of the formation of the climatic situation on a given territory together with the climate scenarios. The second version is based on the use of the global monitoring data, which are the basis for the formation of data series about climatic parameters with their territorial and temporal attachment and used to retrieve a complete pattern of their spatial distribution. One of the widespread correlation functions is the dependence of variations of average temperature ΔT of the atmosphere on the CO_2 content in it:

$$\Delta T = \begin{cases} 2.5[1 - \exp\{-0.82(\xi - 1)\}], & \xi \geq 1; \\ -5.25\xi^2 + 12.55\xi - 7.3, & \xi < 1 \end{cases} \tag{1.8}$$

where ξ is the ratio between the present content of CO_2 in the atmosphere $C_a(t)$ and its pre-industrial level $C_a(1850)$.

As seen from (1.8), T is an increasing function of the amount of atmospheric CO_2. A 20 % increase of CO_2 in the atmosphere raises the temperature by 0.3 °C. A doubling of atmospheric CO_2 causes an increase of T by 1.3 °C. A detailed analysis of the function (1.8) and a comparison of the observed combined variations of ΔT and ξ show that the use of the model (1.8) makes it possible to simplify the climate unit of the global model. In particular, if $(\Delta T)_{2[CO_2]}$ is calculated by (1.8) with a doubled concentration of atmospheric CO_2, then to assess the present trends in changes of T_g the following formula can be used:

$$\Delta T = (\Delta T)_{2[CO_2]} ln\xi / ln2 \,, \tag{1.9}$$

where, from the assumed estimates, the pre-industrial value of $C_a(1850) = 270$ ppm. The formula (1.9) satisfactorily approximates the known data with an error of about 50 %. It follows from (1.9) with $C_a(1980) = 338$ ppm that $\Delta T = 1.3$ °K, whereas a real warming is estimated by many authors at 0.6 °K.

Of course, the results of the recent discussions on the greenhouse effect in connection with the growth of the CO_2 partial pressure in the terrestrial atmosphere should be reflected in GIMS. The formula (1.9) takes into account the impact of CO_2. According to Mintzer (1987), there is a possibility to broaden a consideration of the temperature effect of other greenhouse gases:

$$\Delta T_\Sigma = \Delta T_{CO_2} + \Delta T_{N_2O} + \Delta T_{CH_4} + \Delta T_{O_3} + \Delta T_{CFC11} + \Delta T_{CFC12},$$

where

$$\Delta T_{CO_2} = -0.677 + 3.019 \cdot ln[C_a(t)/C_a(t_o)],$$
$$\Delta T_{N_2O} = 0.057\left\{[N_2O(t)]^{1/2} - [N_2O(t_o)]^{1/2}\right\},$$

$$\Delta T_{CH_4} = 0.19\left\{[CH_4(t)]^{1/2} - [CH_4(t_o)]^{1/2}\right\}, \quad \Delta T_{O_3} = 0.7[O_3(t) - O_3(t_o)]/15,$$

$$\Delta T_{CFC11} = 0.14[CFC11(t) - CFC11(t_o)],$$
$$\Delta T_{CFC12} = 0.16[CFC12(t) - CFC12(t_o)].$$

The value of t_o is identified with the year 1980, when the GHG concentrations were known. To calculate ΔT_{CO_2}, one can use a more accurate formula by Krapivin (1993).

Among simple formulas of calculations of the latitudinal distribution of average temperature over the globe one can indicate the schemes proposed by Sergin (1974):

$$T(\varphi) = T_g + \gamma(\sin^2\varphi_T - \sin^2\varphi), \tag{1.10}$$

where φ is the latitude in radians, γ is the pole-equator temperature difference, φ_T is the latitude at which $T(\varphi) = T_g$. The latitudinal temperature variations can be described:

$$T_a(\varphi,t) = T_e - 2\varphi(T_e - T_N)/\pi \quad \text{for the Northern hemisphere,}$$
$$T_a(\varphi,t) = T_e - 2\varphi(T_e - T_S)/\pi \quad \text{for the Southern hemisphere,}$$

where

$$T_N = \begin{cases} T_{N,\min} + 2t(T_{N,\max} - T_{N,\min})/t_\Delta, & t \in [0, 0, 5t_\Delta]; \\ T_{N,\min} + 2(t_\Delta - t)(T_{N,\max} - T_{N,\min})/t_\Delta, & t \in (0, 5t_\Delta, t_\Delta]; \end{cases}$$

$$T_S = \begin{cases} T_{S,\max} + 2t(T_{S,\min} - T_{S,\max})/t_\Delta, & t \in [0, 0, 5t_\Delta]; \\ T_{S,\max} + 2(t_\Delta - t)(T_{S,\min} - T_{S,\max})/t_\Delta, & t \in (0, 5t_\Delta, t_\Delta]; \end{cases}$$

$T_{N,min}(T_{S,min})$ and $T_{N,max}(T_{S,max})$ are minimum and maximum temperatures at the North (South) Pole, respectively, °C; t_Δ is the duration of the year in units of measurements Δ, T_e is the equatorial temperature of the atmosphere, °C. Many authors use the estimates $T_{N,min} = -30$ °C, $T_{N,max} = 0$ °C, $T_{S,min} = -50$ °C, $T_{S,max} = -10$ °C, $T_e = 28$ °C.

Of course, these zonally averaged temperatures scatter widely, which leads to considerable errors. To more accurate reflect the role of various factors in changes of the basic climatic parameter—temperature, it is necessary to calculate the contribution of each factor separately. This can be done with an assumed additivity of the role of feedbacks: $\Delta T_{a,final} = \Delta T_a + \Delta T_{a,feedback}$. In some models authors introduce the feedback coefficient β: $\Delta T_{a,final} = \beta \cdot \Delta T_a$. The β parameter is expressed through the amplification index g: $\beta = 1/(1 - g)$. The value of the g index is equivalent to albedo α, which on a global scale is a function of T_a. This dependence can be roughly approximated by the formula

$$\alpha(T_a) = \begin{cases} \alpha_{ice} & \text{for } T_a \le T_{ice}, \\ \alpha_{free} & \text{for } T_a \ge T_{free}, \\ \alpha_{free} + b(T_{free} - T) & \text{for } T_{ice} < T_a < T_{free} \end{cases}$$

Here T_{ice} and T_{free} are average planetary temperatures at which the whole surface of the Earth is ice-covered or ice-free, respectively; b is the coefficient of transition between critical conditions of the Earth's albedo. Usually it is assumed that $T_{ice} \in [263, 283]$ °K.

Simple and rough climate models can be specified through account of the characteristic times of feedbacks functioning. Some estimates of the time of establishing the equilibrium states in the interaction of climate sub-systems are given in Table 1.19. The lag time of reactions within BCSS varies widely, and its consideration is necessary when assessing the consequences of changes within one or several climate sub-systems. In particular, the supplies of cold in the Antarctic ice sheet are so huge that to raise its temperature to 0 °C, the average temperature of the World Ocean should be lowered by 2 °C, that is, to obtain $T_o = 3.7$ °C instead of $T_o = 5.7$ °C. With account of the data in Table 1.19, it will take hundreds of years to reach it. The observed rate of the anthropogenic climate warming does not dispose of such energy supplies.

The mechanism of the anthropogenic impact on the climate system is manifested through GHG emissions, albedo change due to reconstruction of the Earth's covers, interference into water cycle, and atmospheric pollution. Aerosol particles with the radius $10^{-7} \pm 10^{-2}$ cm are observed almost at all altitudes of the atmosphere. Particles of non-anthropogenic origin get to the atmosphere from land or ocean surface and result from in-situ chemical reactions. Particles of anthropogenic origin result mainly from fuel burning. Table 1.20 gives an idea about relationships between these fluxes of particles into the atmosphere.

Table 1.19 Time periods of establishing an equilibrium state for some subsystems of the earth climate system

Domain of influence of the climate system	Period of restoration of the equilibrium state
Atmosphere	
Free	11 days
Boundary layer	24 h
World Ocean	
Mixed layer	7–8 years
Deep ocean	300 years
Sea ice	From 24 h to 100 years
Continents	
Lakes and rivers	11 days
Soil-plant formations	11 days
Snow cover and surface ice	24 h
Mountain glaciers	300 years
Ice sheets	3,000 years
Earth's mantle	30 million years

Table 1.20 Estimates of the fluxes of particles with the radius <29 μm either emitted to the atmosphere or formed in-situ

Type of particles	Amount of particles (10^6 t/year)
Particles of natural origin (weathering, erosion, etc.)	100–500
Particles from forest fires and forestry wastes burning	3–150
Sea salt	300
Volcanic dust	25–150
Particles from gas emissions	
Natural processes	
Sulphates from H_2S	130–200
Ammonium salts from HN_3	80–270
Nitrates from NO_x	60–430
Hydrocarbons from vegetative compounds	75–200
Anthropogenic processes	
Sulphates from SO_2	130–200
Nitrates from NO_x	30–35
Hydrocarbons	15–90
Amounts of particles emitted to the atmosphere	
Natural	773–2,200
Anthropogenic	185–415
Total flux of particles into the atmosphere	958–2,615

The mechanism of the effect of particles on atmospheric temperature is explained by that solar radiation coming to the Earth, mainly within the range 0.4–4 μm, is partially reflected and absorbed by them. The global albedo of the surface-atmosphere system varies, too. Besides, the particles affect the processes of moisture condensation in the atmosphere, since clouds, rain, and snow form with their participation. Take the heat balance equation of the system "surface-atmosphere":

$$(1-\alpha)\,E_o^* + E_a - \sigma\,T_s^4 = 0, \qquad (1.11)$$

where T_s is the average efficient temperature of the system's emission, close to the temperature of the average energy level near the surface 400 mb, $E_o^* =$ 0.487 cal/cm^2/min is the mean-hemispherical intensity of the incoming solar radiation; α is albedo; $\sigma = 8.14 \times 10^{-11}$ cal/cm^2/min is the Stefan-Boltzmann constant, E_a is total intensity of the anthropogenic sources of energy per unit surface.

Let albedo $\alpha = \alpha_o - \Delta\alpha$, where $\alpha_o = 0.35$ is the albedo in present conditions, Δα is a small part of albedo determined by the impact of anthropogenic aerosols. From the Eq. (1.11) we obtain an expression for temperature:

$$T_s = \left[E_o^*(1-\alpha)/\sigma\right]^{1/4}\left[1+\Delta\alpha(1-\alpha_o)^{-1}+\left(E_a/E_o^*\right)(1-\alpha_o)^{-1}\right]^{1/4} \quad (1.12)$$

Assuming $\Delta\alpha \ll 1$ and $E_a/E_o^* \ll 1$, expand the function of the right-hand part of the Eq. (1.12) into Taylor series by degrees $\Delta\alpha$ and E_a/E_o^* and write first terms of the series:

$$T_s \approx \left[E_o^*(1-\alpha_o)/\sigma\right]^{1/4}\left\{1+0.25\Delta\alpha(1-\alpha_o)^{-1}\left[1+E_a/E_o^*\right]\right\} \quad (1.13)$$

It follows from Eq. (1.13) that the temperature at not so strong anthropogenic impacts is the sum of the terms describing the bonds in the system "surface-atmosphere", without account of anthropogenic factors, and the terms T_1 and T_2, representing the contribution of heat and aerosols emissions, respectively:

$$T_1 = 0.25(1-\alpha_o)^{-1}\left[E_o^*(1-\alpha_o)/\sigma\right]^{1/4}E_a/E_o^* \approx 96.046E_a/E_o^*,$$

$$T_2 = 0.25(1-\alpha_o)^{-1}\left[E_o^*(1-\alpha_o)/\sigma\right]^{1/4}\Delta\alpha \approx 96.046\Delta\alpha.$$

Note that the contribution of T_1 in present conditions is very small. Assuming that $E_a = 4 \times 10^{-5}$ cal/cm^2 min and, hence, $E_a/E_o^* = 8.21 \times 10^{-4}$, then $T_1 = 0.0079$ °C. Thus the direct impact of the global energy on the atmospheric average temperature is now negligible. It follows from the expression for T_1 that to raise the atmospheric temperature by 0.5 °C due to thermal emissions, the condition $E_a/E_o^* = 0.0052$ should be satisfied, and this means an increase of anthropogenic heat fluxes to the environment by a factor of 63.4. This is equivalent to energy release at the annual burning of 570 × 10^9 t of conventional fuel.

If we assume that the energy production is proportional to the size of population, then $T_1 = 96.046k_{TG}G\sigma_S/E_o^*$, where G is the population density, men/km^2; σ_S is the land area, km^2; k_{TG} is the per capita amount of produced energy, cal/min.

If we neglect the impact of aerosol on the atmospheric thermal regime, then the direct radiation E, its change dE and the change of atmospheric turbidity dB will be related as: $dE/E = -k_B dB$, where $k_B = 0.1154$ km^2/t is the proportion coefficient, B is the amount of anthropogenic aerosols, t/km^2. After integrating this equation, we obtain: $E = E_o^*(1-\alpha_o)exp(-k_B B)$. On the other hand, according to the definition of albedo, $E = E_o^*(1-\alpha) = E_o^*(1-\alpha_o+\Delta\alpha)$. Equating these expressions for E, we obtain $\Delta\alpha = -(1-\alpha_o)[1-exp(-k_B B)]$. Hence, a temperature change due to pollution of the atmosphere by anthropogenic aerosols is equal to:

$$T_2 = -0.25\left[E_o^*(1-\alpha_o)/\sigma\right]^{1/4}[1-exp(-k_B B)] = -62.43[1-exp(-k_B B)].$$

Since the average emission of anthropogenic aerosols is estimated by many authors at 300 × 10^6 t/year, and the average period of the aerosols residence in the atmosphere is estimated at 3 weeks, 17.262 × 10^6 tons of particles, on the average, reside in the atmosphere. In this case it follows from the formula for T_2 that the atmospheric temperature should decrease by 0.84 °C/year.

Some experts, instead of the index β, consider the factor of atmospheric turbidity B_T, defining it as a ratio of the coefficient of attenuation of solar energy in the real atmosphere α_r to the coefficient of attenuation in an ideal atmosphere α_I: $B_T = \alpha_r/\alpha_I = (\alpha_I + \alpha_W - \alpha_A)/\alpha_I$, where α_W and α_A are the coefficients of attenuation by water vapor and by aerosols, respectively. The following estimates are assumed in GIMS:

$$B_T = \begin{cases} 3.0 \text{ in mid-latitudes,} \\ 3.5 \text{ in tropical latitudes,} \\ 2.0 \text{ at reduced contents of dust and water vapour.} \end{cases}$$

An experience of the global climate modeling testifies to the fact that the desire of many experts to accurately and completely take into account all possible feedbacks and compounds of the climate system leads to complicated mathematical problems, the solution of which a huge amount of data is needed, and in most cases the solutions of the respective equations turn out to be unstable. Therefore, the use of these complicated models as a unit of the global model of the GIMS, leads to a negative result, that is, to impossibility to construct an efficient model. The most encouraging approach is to combine the climate models with the data of global monitoring. The scheme of this combination is very simple. The existing surface and satellite systems of monitoring the climate-forming processes cover some part of the cells $\{\Xi_{ij}\}$ of the Earth surface. Over these cells, measurements are made of temperature, cloudiness, water vapor content, aerosols and gases, albedo, and various parameters of energy fluxes. Use of simple climate models and of methods of spatial and temporal interpolations enables one to reconstruct, based on these measurements, a complete pattern of distribution of climatic parameters over the whole territory Ξ.

1.9 Conclusion

Main idea of global information-modeling technology was proposed by Krapivin et al. (1982). First versions of the GIMS were successfully tested in Former USSR (Russia, Moldova, Uzbekistan, Turkmenistan), where different applied tasks were solved (Krapivin and Phillips 2001a; Chukhlantsev et al. 2004; Krapivin et al. 2007c; Krapivin and Shutko 2012; Shutko et al. 2010):

- earlier detection of the forest fire isolated areas;
- monitoring of surface soil moisture in the interests of agriculture;
- determination of depth to a shallow water table (down to 2 m in humid areas and down to 3–5 m in arid/dry areas); and
- assessment of underground moistening level.

During last years the GIMS was successfully tested and applied in many countries in the World including Bulgaria, Poland, Hungary, Vietnam, Cuba, USA and Holland (Krapivin and Shutko 2012). It has been proven that the advantages of GIMS application consist first of all in the following capabilities:

- contours of water seepage through hydrotechnical constructions (levees, dams, destroyed drainage systems, different kinds of leaks);
- biomass of vegetation above a water surface or wet ground;
- increase in temperature in land, forested and volcano areas;
- changes in salinity/mineralization and temperature of a water surface;
- water surface pollution, oil slicks on a water surface;
- on-ground snow melting; and
- ice on a water surface and on the roads, runways.

Thus, the GIMS provides remote sensing, on-ground (in situ) samplings, using GIS-information and mathematical modeling of physics-chemical processes in selected areas. GIMS-technology can effectively be applied for solving many agricultural, hydrological, environmental and many other Earth related problems. In common, GIMS-technology solves numerous environmental tasks that GIS technology struggles with, such as:

- The study of environmental processes and systems that change quickly and are unstable.
- Assessing the ecosystems in regions where monitoring systems do not exist.
- Optimization of the environmental monitoring systems.
- Reconstruction of spatial image of the system basing on the restricted information.
- Revealing the characteristic features of the environmental system evolution.

Chapter 2
Remote-Sensing Technologies and Data Processing Algorithms

2.1 Remote Sensing Methods

Remote sensing of the land covers, atmosphere and World Ocean aquatories is based on the registration of background or reflected and scattered electromagnetic radiation. Such knowledge is necessary to get data about the properties of environmental elements as it explains the dependence of thermal emission on its own physical and geothermal parameters. Also the scattering mechanisms and active radiation reflection are functions of these parameters (Borodin 2008; Bui Ta Long 1998; Shutko et al. 1994, 2010; Krapivin and Shutko 2002, 2012; Petty 1995; Schimel 1995; Golovachev et al. 2004; Cheremisin 2013; Savinikh et al. 2007).

The waves (or frequencies) of electromagnetic emission used for remote sensing in environmental monitoring systems occupy a wide spectrum from 0.3 μm to 1.3 m and cover the subranges: near ultraviolet (0.3–0.4 μm), visible (0.4–0.76 μm), near infrared (IR: 0.76–1.5 μm), middle and far IR (1.5 μm–1 mm), super high frequency (SHF: 1 mm–1.3 m). The SHF-range is divided into three basic subranges: millimeters (1–10 mm), centimeters (1–10 cm), and decimeters (10–130 dm). Two subranges called usually as L-band (15.8–63 cm) and P-band (63–100 cm) are used by many authors.

The real range choosed for objective study of the environment has to take on board many conditions such as the absorption and scattering of electromagnetic waves by the Earth's atmosphere and its interactions with the Earth's land covers and water surfaces. The atmosphere is an extremely limiting factor when choosing a working range for remote monitoring. For instance, the P-band can used for abyssal sensing from the flying laboratories, but its application in satellite monitoring systems is beset with problems because the ionosphere acts as a screen. A similar situation is found with ultraviolet radiation which is intensively absorbed by atmospheric gases. In order words, there is a question about the atmosphere clarity for specific wavelengths. For example, some wavelengths in the IR and submillimeter ranges are intensively absorbed by water vapor which practically precludes

V.F. Krapivin et al., *New Ecoinformatics Tools in Environmental Science*,
Environmental Earth Sciences, DOI 10.1007/978-3-319-13978-4_2

them from consideration as remote-sensing tools. Clouds have a powerful effect on light radiation and create obstacles to land surface observation from space over many territories (Klyuev 2000; Justice et al. 1995; Grankov 1994; Asrar and Dozier 1994).

Knowledge of the atmosphere admitting function in the given synoptical and geographical conditions is the key task under the synthesis of the remote monitoring system. As a rule this task is solved during the realization of measurements. For this, one or several channels use waves that are relatively powerfully absorbed or are dispersed by different atmosphere components; these are added to the basic channels. The content of relative component is determined by the effect of these waves relaxation on the various latitudes (or the integrated content, depending on the task). The measurement results correction is introduced by the basic channels using these additional data.

The specified physical special features of the environment remote sensing are inherent to the super high frequence and ultra short waves ranges which are the foundation of the radiometric geoinformation monitoring systems. These systems allowed the many tasks involved in operative identification of natural phenomena to be solved and new methods and technologies for remote diagnosis of natural and natural/technogenic systems to be developed.

Microwave methods can be conditionally divided into two classes: active and passive. The methods that study the character of the reflection, scattering and absorption of the waves emitted by the source with known spectral density are defined as active. The optical range includes such sources: the Sun, lasers and other light emitters. Active methods used widely in radiolocation systems are in the radio range. The power of reflected and dispersed radiation, its spectral composition and polarization characteristics, phase and propagation time are the study objects in this case.

Passive methods are based on the heat emission analysis of natural formations. The fact that thermal emission is determined by substance temperature and its physical parameters is taken into account here. That is why passive methods are used for the temperature measurement and for the determination of different environmental parameters. These measurements are effective particularly under multi-channel sensing co-ordinated with the data processing algorithms (Sarkisyan 2000; Sellers et al. 1996; Krapivin and Mkrtchyan 1991; Bondur et al. 2010; Krapivin et al. 2010a, b).

Let us now move on and consider the specific features of radiophysical methods (i.e., those in which the radio waves are used). Basic attention is here concentrated on the SHF-range since namely this range is applied in the remote sensing systems located on fly-laboratories and satellites (Kondratyev 2000a; Kondratyev et al. 1990a; Krapivin and Phillips 2001; Shutko et al. 2010). Microwave radiometer systems are used intensively on numerous satellites. These systems secure the operative control of more then twenty basic geophysical parameters forming weather and climatic processes (Bondur et al. 2009).

Main specific features of radiometric methods are connected with high radio-clearness of the atmosphere. It is one of its advantage in comparison with optical and infrared methods. The possibility of the letters is restricted mainly by absorbing and dispersing qualities of the atmosphere. The main problem here is cloud which is opaque in these wavelengths and, hence, the receipt of necessary data about the environment state is prevented. This makes it difficult to obtain operative data when such tasks as the diagnosis of extraordinary and disastrous natural or technogenic anomalies need to be solved. Certainly, the use of devices of optical and infrared ranges is possible only in the regions with small cloudiness. Hence, main advantage of radiophysical methods for the remote monitoring is its all-weatherness. Nevertheless, there are still problems here, which scientists from many countries are addressing. For example, the absorption line of water vapors exists for $\lambda = 1.35$ cm, and the oxygen absorption zone effects for $\lambda = 0.5$ cm.

Radio waves can penetrate under the vegetation cover canopy and into the soil layer depth. That is why the use of radiophysical methods allow to assess the vegetation and soil state and to define many its properties (Krapivin 2000b; Krapivin et al. 1998b; Metternicht 1998; Burkov and Krapivin 2009; Krapivin and Vilkova 1990).

The remote sensing topics cover many theoretical and experimental investigations of the most current specializations within the fields of geoscience, ecology, radiophysics, biochemistry, oceanology, applied mathematics, cybernetics, climatology, and other sciences (Klyuev 2000; Grankov and Milshin 2004, 2010; Haarbrink et al. 2011). This chapter mainly considers the role played by the microwave range in the measurement of environmental parameters. It consists in development and experimental validation of the high-precision remote sensing methods with aim:

(1) to study natural and anthropogenic processes,
(2) to establish relationships between atmospheric, ionospheric and surface phenomena,
(3) to find the long-scale changes connected with climatic, meteorological, seismological and human activity,
(4) to investigate the field of microwave and optical remote sensing for detection of areas of water seepage through irrigation constructions, levees and dykes and for revealing of areas with dangerously high groundwater level, and
(5) to develop a synergetic remote sensing technology for raised groundwater and seepage detection by using jointly spectral microwave and optical approach.

This chapter describes an experience in the remote sensing of the Earth from space. This experience allows obtaining the radio holographic data base for establishing regular and statistical relationships describing influence of the natural and anthropogenic factors on the ionosphere, atmosphere, and developing the adaptive models for simulation of the greenhouse effect.

Existing experience in remote sensing is characterized by different applications. For example, remote sensing methods for the land surface monitoring given possibility to introduce series of indices that help to solve numerous tasks:

(1) NDVI—Normalized Difference Vegetation Index is an index of plant "greenness" or photosynthetic activity, and is one of the most commonly used vegetation indices.
(2) EVI—Enhanced Vegetation Index was specifically developed to:

   - be more sensitive to changes in areas having high biomass (a serious shortcoming of NDVI),
   - reduce the influence of atmospheric conditions on vegetation index values, and
   - correct for canopy background signals.

   EVI tends to be more sensitive to plant canopy differences like leaf area index (LAI), canopy structure, and plant phenology and stress than does NDVI which generally responds just to the amount of chlorophyll present.
(3) SAVI—Soil-adjusted Vegetation Index was developed as a modification of the NDVI to correct for the influence of soil brightness when vegetative cover is low
(4) MSAVI—Modified Soil-adjusted Vegetation Index is soil adjusted vegetation indices that seeks to address some of the limitation of NDVI when applied to areas with a high degree of exposed soil surface.
(5) SATVI—Soil-adjusted Total Vegetation Index is a modification of several earlier vegetation indices that correlates with the amount of green and senescent vegetation present on the ground.
(6) Microwave Vegetation Index (MVI) uses microwave readings from two different frequencies to determine the cover of vegetation. Each of these different frequencies has the ability to penetrate vegetation canopyto a different degree, and the relative different in reflectance between these two can be interepreted to give an index of vegetation cover.
(7) Normalized Burn Ratio (NBR) was defined to highlight areas that have burned and to index the severity of a burn using Landsat TM imagery.
(8) Biological Crust Index uses a unique feature of the phycobilin pigment found in cyanobacterial soil crusts, resulting in a relatively higher reflectance in the blue spectral region compared to soil without cyanobacteria.
(9) Burned Area Reflectance Classification (BARC) is a method using remotely sensed data to assess the severity of fire damage to vegetated areas developed by the US Forest Service and US Geological Survey.

Remote sensing methods widely penetrate to other areas. The fundamental task that is subject of many studies consists in development and experimental validation of the high-precision methods and algorithms for restoration of the time-spatial distribution of the physical parameters of the atmosphere, ionosphere and the Earth's surface using the radio fields measured on some interface or curve in space (radio hologram). Radio hologram can be registered using small satellite in space or an airplane in the atmosphere. Application of the satellite radio holography for global geophysical monitoring is justified by existence of the space radio systems designed for radio navigation, radio communication and other civilian tasks. Many of these radio systems are emitting permanently radio signals having high coherence and stability properties. This allows one to develop radio holographic methods that can be realized with help of the high stability and coherence of the radio signals of the space radio systems. The radio holographic analysis of the satellite radio holograms can be used as a tool for elaborating the data base containing important geophysical information on the global scale with aim to reveal natural and anthropogenic processes, to establish relationships between atmospheric, ionospheric and surface phenomena and to find the long-scale changes connected with climatic, meteorological, seismological and human activity. The database can be elaborated by means of radio holographic processing of the radio holograms registered by exiting and future satellite missions (e.g. Mir, GPS/MET, CHAMP, SAC-C radio occultation experiments). The next fundamental and practical problems are needed to solve:

1. Elaborating the radio holographic methods for high-precision determination of the spatial distribution of the physical parameters of the atmosphere, ionosphere and the Earth's surface.
2. Development of the database for global geophysical monitoring of the physical parameters of the atmosphere, ionosphere and the Earth's surface using radio holograms registered by low orbital satellites.
3. Analysis of the elaborated data base with aim to find the radio-physical parameters that are sensitive to long-scale changes in the state of the atmosphere, ionosphere and the Earth's surface.
4. Elaboration of the radio holographic technology for bistatic radiolocation of the atmospheric boundary layer and the terrestrial surfaces. The aims of this technology are:
   - to determine the humidity and the vertical gradients of the refractivity in the boundary layer of the atmosphere; and
   - to measure parameters of the terrestrial surfaces (wind roughness and long-scale waves in the oceans, ice distribution in the polar regions).

Elaboration of the polarimetric and interferometer methods for studying physical parameters of the land surfaces: the electric permittivity, conductivity, soil moisture, foliage parameters and the long-scale relief changes connected with seismic and anthropogenic activity. Elaboration of the database of the global geophysical monitoring of the Earth's surface and the atmospheric boundary layer by means of the processing of the radio holograms registered by MIR, GPS/MET, CHAMP, SAC-C missions.

5. Investigation of the ionospheric effects of the natural and anthropogenic processes by means of analysis of the radio holograms registered on the communication line satellite-satellite and Earth-satellite.
6. Modeling the greenhouse effect by means of the analysis of the developed global geophysical database.

The experience of elaborating the theory of the radio holographic investigation of Earth and its gas-plasma environment from space was described in publications by Liou et al. (2003, 2010), Pavelyev et al. (2002, 2003, 2010).

Positive scientific results are achieved owing to the combined use of remote sensing and ecoinformatics tools. With the global effect of climate change, ecoinformatics is an emerging drench of research where new techniques, tools and infrastructure are being developed to enable more in-depth study of processes in the nature/society system with effective employment of scientific data collection instruments such as sensors, satellites, scientific experiments and simulations. As Myers and Atkinson (2008) remark, ecoinformatics is the combination of multiple environmental datasets and modeling tools used to test ecological hypotheses and derive information.

2.2 Remote Sensing Techniques

Remote sensing is the observation and measurement of objects from a distance, i.e. instruments or recorders are not in direct contact with objects under investigation. Remote sensing depends upon measuring some kind of energy that is emitted, transmitted, or reflected from an object in order to determine certain physical properties of the object. One of the most common types of remote sensing is microwave radiometry, which along with many other techniques is used for the equipment of flying laboratory and satellites.

There are several types of instrumentation that can be used for environmental investigations. For examination of soil type, or surface moisture, the ones that come to mind are thermal infrared radiometers, active radar systems, and the passive microwave radiometers. A system of non-scanning single-beam microwave radiometers was designed in Kotelnikov Institute of Radioengineering and Electronics (Shutko et al. 2010). The basic technical parameters of this system are given in Tables 2.1 and 2.2. Common view of portable radiometers is given in Figs. 2.1 and 2.2.

Table 2.1 Set of portable microwave radiometers

Parameter	Characteristic
Frequency (GHz)	1.41; 1.67; 5.0
Polarization	V or H
Antennas	
Slot array	19.7″ × 19.7″; 16.3″ × 16.3″
Dipole array	7.9″ × 6.5″
Incidence angle	Fixed angles 0–30° off nadir
Sensitivity (K)	0.5
Beamwidth (3-dB)	30°
Platform	Boom truck, land-rover, small aircraft
Weight (kg)	12
Power required (VA)	30

Table 2.2 Characteristics of microwave radiometers designed in Kotelnikov Institute of Radioengineering and Electronics, Russian Academy of Sciences (Shutko and Krapivin 2011; Krapivin and Shutko 2012)

Frequency (GHz)	Wavelength (cm)	Range	Pixels/scan	Spatial resolution (H)	Mode
37	0.8	Ka	32	0.04	Scanning
15.2	2	X	16	0.08	Scanning
13.3	2.25	X	12	0.09	Scanning
5.5	5.5	C	6	0.13	Scanning
1.4	18/21	L	2	0.65	Twin-beam

H is height above ground

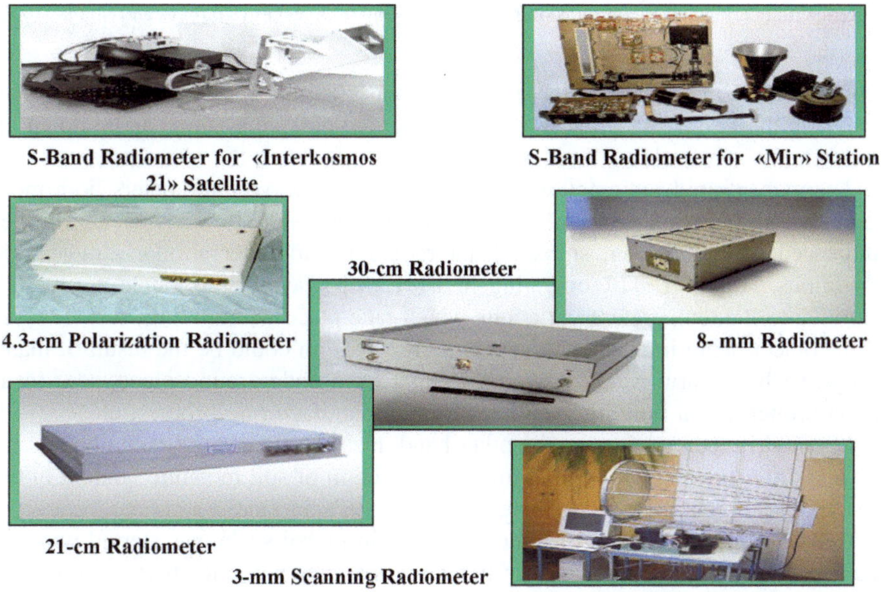

Fig. 2.1 Set of portable radiometers produced by Special Design Office of Kotelnikov Institute of Radioengineering and Electronics of Russian Academy of Sciences (Krapivin and Shutko 2012; Verba et al. 2014)

The radar systems, including Interferometric Synthetic Aperture Radar (IFSAR), share the same frequency band as the microwave radiometers. Therefore they have similar wave propagation characteristics, although the radar systems are actively transmitting and measuring the return values. The radar/IFSAR systems rely on surface texture to reduce surface boundary conditions.

Passive microwave systems only record the naturally emitted radiation, which is a function of the soil moisture content within the first several centimeters/decimeters and is less influenced by surface roughness. From a given height above the

Fig. 2.2 21 cm microwave radiometer produced by Special Design Office of Kotelnikov Institute of Radioengineering and Electronics of Russian Academy of Sciences (Krapivin and Shutko 2012)

ground surface radar systems have a much smaller ground sample distance, or pixel size, than the passive systems. To compensate, the passive system can be flown on light aircraft at very low altitudes, as low as 500 ft above the mean terrain.

Thermal infrared radiometers (TIR) and the microwave radiometers both measure natural radiation from a thermodynamic process. TIR sensors are further influenced by cloud, fog, smoke, solar radiation and other similar atmospheric conditions. However, TIR operating in 10–12.5 μ wavelengths range are mainly sensitive to the temperature variations which take place within a thin surface layer of the object being imaged. This temperature variation could be the result of many things, such as warm water, or steam leaking from underground pipes, poor insulation properties on buildings, effluent from a manufacturing process, etc.

Microwave radiometers, on the other hand, measure the natural radiation at mm, cm and dm wavelengths that is primarily a function of soil moisture content in the first couple of centimeters/decimeters of the Earth's surface.

Considerable experience of remote sensing in global scale was started with the launch of orbital space station "Mir" which was complex science-research complex (Fig. 2.3). This station was launched at February 20, 1986 and was submerged at March 23, 2001 in the Pacific Ocean.

Instrument complex of Mir space station for remote sensing functions was equipped in specialized modules that were named as Spectr and Priroda (Nature). Module Spectr has optical-visual and photometric instruments as well as lidar and photographic cameras for the monitoring of the atmosphere, oceans, land surface, as well as for medical-biological investigations. Priroda (Nature) conceived for remote-sensing research from orbit, was also sitting on the ground partially completed until Russian-US cooperation revived the project. It became the last element, which completed the construction of Mir. Module Mir had specialized complex of sensors:

- trace, panoramic and scanning microwave radiometers (radiometric system Ikar, Tables 2.3 and 2.4);

Fig. 2.3 Russia's Mir space station (http://spaceflight.nasa.gov/gallery/images/shuttle/sts-91/html/91727051.html)

- 64-channels IR-spectrometric system Istok;
- UV-spectrometric systems of visual and IR-range; and
- active radiophysical and optical systems.

Solution of specific environmental tasks requires the optimization of instrumental and algorithmic means. It can be realized by means of the GIMS-technology methods. The ratio between experimental and theoretical methods in this case is defined by the complexity of environmental system that is to be studied. Cumulative experience of the GIMS application is based on the solution of national problems in many countries where different remote sensing platform were created. Historically this experience was started in 1980 with equipping of flying laboratories based on bi-plane AN-2, twin-engine plane IL-14, 4-engine plane IL-18, helicopters MI-2, MI-8 and Ka-26 (Krapivin and Shutko 2011, 2012; Shutko et al. 2010; Haarbrink et al. 2011). In the ensuing years, this set of carriers was widened.

Table 2.3 Trace and panoramic radiometers of radiometric system Ikar

Radiometer characteristic	Panoramic	Trace Ikar-N
Wavelength (cm)	2.25; 6.0	0.3; 0.8; 1.35; 2.25; 6.0.
Fluctuating sensitivity (time for integration is 1 s) (K)	0.15	0.15
Measuring range of radiobrightness temperature (K)	50–350	50–350
Antenna beam width (3 dB level) (°)	12	9
Spatial resolution (for altitude of 400 km) (km)	80	60
Observation angle (°)	40	0, 0
Swath (for altitude of 400 km) (km)	750	60
Required power (W)	30	100
Mass (kg)	40	50
Informativity (Kb/s)	0.5	1.0
Sampling digit capacity (bits)	10	10

Table 2.4 Scanning radiometers of radiometric system Ikar

Radiometer characteristic	P-400	Delta
Wavelength (cm)	4.0	0.8; 1.35; 2.25
Fluctuating sensitivity (time for integration is 1 s) (K)	0.15	0.15–0.4
Measuring range of radio brightness temperature (K)	50–350	50–350
Antenna beam width (3 dB level) (°)	5.0	1.1; 1.7; 3.0
Beam shift angle at channels (°)	0.0	0.0; −1.1; 2.4
Spatial resolution (for altitude of 400 km) (km)	50	10.0; 15.5; 28.0
Observation angle (from nadir) (°)	40	40.0; 37.9; 42.4
Scan time (s)	3.4	2.56
Swath (for altitude of 400 km) (km)	400	660
Угол сканирования, град	72	165
Потребляемая мощность (Вт)	30	50
Масса (кг)	46	100
Информативность (кбит/с)	1.5	10
Разрядность квантования (бит)	10	10

One of examples when multi-purpose flying laboratory was used successfully is IL-18-based (hull No 75423) regional monitoring system (Borodin et al. 1982). Il-18 on-board complex is represented in Fig. 2.4.

Flying laboratory IL-18 has given a possibility to solve numerous environmental tasks practically in real time. On-board system represented in Fig. 2.4 registered environmental characteristics and at the same time processed existing and delivered by sensors information. Figure 2.5 explains a structure of on-board information-modeling system that interrogated all sensors with time period equaled to 0.01 s and taking into consideration of navigation information all results were referred to the geographical coordinates.

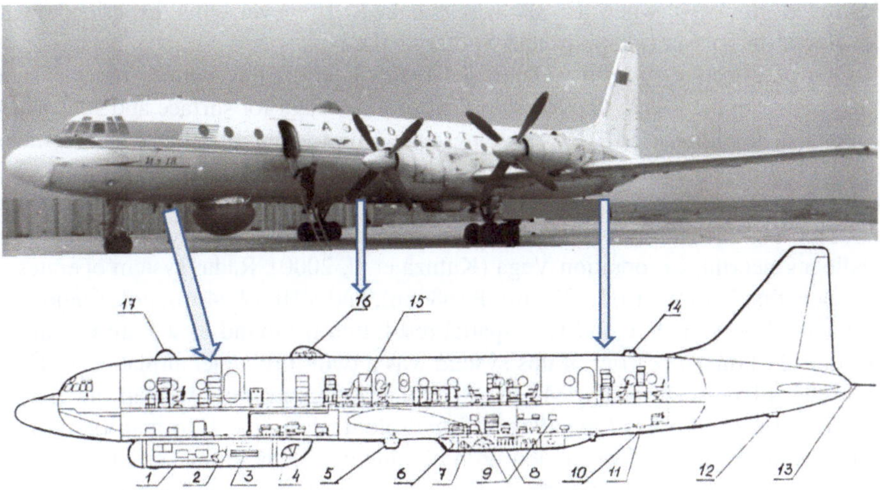

Fig. 2.4 Scheme of positioning the antenna systems and photo-hatches on-board of flying laboratory IL-18. Notation: *Antennas*: *1*, *3*—radio locators with synthetic aperture, wavelengths—2.0 and 10.0 cm; *2*, *6*—trace polarimeters, wavelength—0.8 and 2.25 cm; *4*—six-channel scanning polarimeter, wavelength—0.8, 1.35, 2.25, 10, 20 and 27 cm; *7*, *9*—precision altimeter and interferometer of side looking, wavelength—2.2 cm; *13*—sub-surface sounding station of decimetric range; *Photo-hatches*: *5*, *10*, *12*—large-format and frame TV, aerocamera; *11*, *14*—radiometers of mm range; *16*—trace radiometers, wavelength—0.8, 1.35, and 2.25 cm; *15*—gravimetric and inertial devices; *17*—astro-hatch

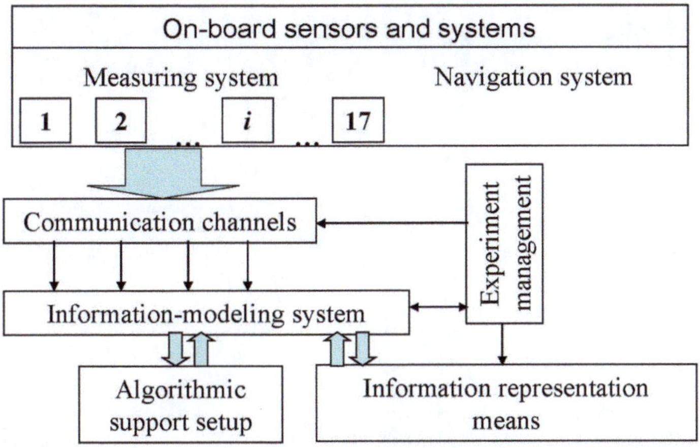

Fig. 2.5 Typical structure of on-board information-modeling system realized in flying laboratory IL-18

Long-term exploitation of flying laboratory IL-18 in Former USSR has showed the availability of using the multi-purpose remote sensing systems under solution of environmental problems on large territories. Unfortunately, this laboratory was destroyed in 1990 when USSR was breaked. However, this experience was developed in Russia (Krapivin and Shutko 2012).

One of effective multi-frequency polarimetric synthetic aperture radar system was created in the 90s by the Russian Corporation Vega for surface and subsurface sensing. It is airborne IMARC SAR system. This Corporation created radiometer system Radius operating at the wavelengths of 0.8, 2, 5.5, 21 and 43 cm (Fig. 2.6, Table 2.5). Main features of the IMARC SAR complex are given in Table 2.6.

The IMARC is a four-wavelength polarimetric airborne SAR system designed at Radioengineering Corporation Vega (Kutuza et al. 2000). Radar system operates at wavelengths: X (3.9 cm), L (23 cm), P (68 cm), and VHF (2.54 m); polarizations in all bands: VV, HH, VH, and HV; spatial resolution is around 12 ± 8 m; maximum swath is 24 km. The carrier of this system was a twin-turbine jet airplane TU-134A and other planes. The main IMARC SAR mission goals were to map the characteristics of Earth covers (including soil hydrological regimes), to map the ground terrain in a presence of vegetation with eliminating of the influence of vegetation and, to produce elevation models, to detect areas with on-ground and underground irregularities, etc. (Shutko et al. 2010; Krapivin and Shutko 2012).

Airborne SAR system IMARC can use for solution the following tasks:

- surface sensing of ocean, ice and vegetation;
- subsurface sensing of scattering objects and deep layers.

Fig. 2.6 General view of the scanning SHF—radiometric system "Radius" (2 and 5 cm) (Shutko et al. 2006a, 2007, 2010; Krapivin and Shutko 2012; Verba et al. 2014)

Table 2.5 Scanning radiometer "Radius" characteristics

Frequency (GHz)	Wavelength (cm)	Band	Pixels/scan	Resolution (H)	Mode
37	0.8	Ka	32	0.04	Scanning
15.2	2	X	16	0.08	Scanning
5.5	5.5	C	6	0.13	Scanning
1.4	21	L	2	0.5	Scanning
0.7	43	S	1	–	Non scanning

Power consumption is 200–300 W, power supply is 27 VDC; weight is 100–120 kg
H is height above ground

Table 2.6 The IMARC SAR complex parameters

Parameter	Value			
Frequency	X	L	P	VHF
Wavelength (cm)	3.9	23	68	254
Polarization	VV, HH, VH, HV			
Resolution (m)	4–6	8–10	10–15	15–20
Antennas: gain (dB)	30	14–17	14–17	9–11
Width in azimuth (°)	18	24	24	40
Width in elevation (°)	24	24	24	60

Multi-frequency polarimetric SAR gives new possibilities for operative remote sensing of sea, soil, vegetation, ice cover and other Earth surfaces. The algorithms used for deep-layer remote sensing, which is undertaken layer-by-layer, make it possible to solve a narrow set of environmental problems.

The larger the wavelength, the higher the influence of deeper soil layers—this fact allows the development of methods of thick-layer deep sensing using multi-frequency radar systems. For subsurface sensing the use of long waves of P and VHF bands is required.

Information about soil properties (soil moisture) profiles can be received from the analysis of measurements of scattering at different wavelengths. To calculate the influence of soil moisture profile into backscattering cross section it is necessary to develop the models of reflection from the layers situated at different depth. Solution of the inverse problem can be obtained by setting the measurement of backscatter at several wavelengths and at different polarization modes. For complete information about soil moisture profile it is necessary to solve the problem of images interpretation in broad band of wavelengths including meter band where attenuation in soil and vegetation is comparably low.

The results of multi-band radar survey obtained with the help of 4-bands airborne SAR IMARC (Radioengineering Corporation Vega) illustrate the possibility of measuring of hydrological soil regimes and water lenses allocation in Kara-Kum desert. Lenses of underground water at the depth of 50–70 m were detected. Results were validated by control well-boring. We can see at radar images:

(1) dry river-Uzboy bed;
(2) sand dunes of 6–15 m height;
(3) underground water lenses;
(4) transmission facilities.

Application of GIMS technology depends on just how complex the solution to the environmental tasks under consideration is in a given time period at a given precision. An effective instrument using GIMS technology was synthesized by Miramap (a private-sector Dutch\Company) in response to a European Space Incubator initiative from the ESA Technology Transfer & Promotion (TTP) Office. The TTP office is contributing to the capitalization of space-based technology and know-how for the benefit of Europe's economy and science. The innovation of microwave radiometer (MR) mapping company (Miramap) was nominated for the Holland Innovation Price in 2005 and was quoted in several newspapers and magazines such as the Dutch Financial Times.

The Miramap instrument consists of three microwave sensors in X-band, C-band and L-band which are all Global Navigation Satellite System (GNSS) integrated (Fig. 2.7, Tables 2.7, 2.8, 2.9 and 2.10). The X-band and C-band sensor makes a conical scan at constant incidence angle over a wide swath, while the L-band sensor makes a twin-beam oscillating scan. The small instrument sizes and weights enable use of a low-cost light aircraft as the observing platform, providing decision makers with a new affordable tool. The platform on which these instruments are flown is a

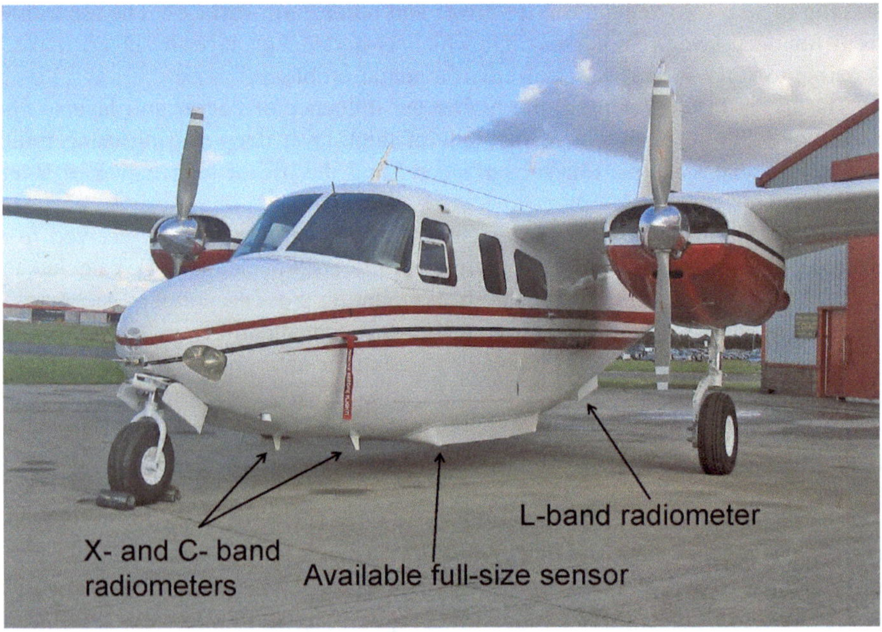

Fig. 2.7 Miramap sensor aircraft (Shutko et al. 2010; Krapivin and Shutko 2012)

Table 2.7 Miramap microwave sensor specifications

Parameter/band	X-band	C-band	L-band
Frequency (GHz)	15.2	5.5	1.4
Wavelength (cm)	2.0	5.5	21
Pixels/Scan	16	6	2
Incidence angle (°)	30	30	15
Beam width (°)	3.5	5	25
Polarization	H	H	H
Sensitivity (K/s)	0.15	0.2	1
Absolute accuracy (K)	±5	±5	±5

Table 2.8 Microwave Radiometer Mapping Company (Miramap, Noordwijk, the Netherlands): Sensor specification

Sensor	Type	Wavelength	Project specs	Use
Digital photo camera	Rollei AIC 50 mm lens	Visible 0.4–0.7 μ	10 cm GSD sub-pixel precision	Detailed visible interpretation
Lidar scanner	Optech altimeter	SW infrared 1,064 nm	2 cm GSD 0.1 m precision	Elevation model
Passive microwave scanner	Radius (IREE-Vega design)	Microwave 2, 5, 21 cm	5 m GSD 0.15 K	(Sub) surface detection of wet and dry areas
Thermal camera	Flir systems	LW infrared 7.5–13 μ	3 m GSD 0.1 °C	Surface temperature

Table 2.9 Miramap company platforms and remote sensing instrumentation

Platforms/ instrumentation	Aircraft laboratory	Unmanned plane/helicopter	Specific car/rover	In situ data collecting instruments
Microwave radiometers	Exists	Exis	Exists	Exists
Infrared radiometers	Exists	Exists	Exists	Exists
Optical color digital cameras	Exists	Exists	Exists	Exists
Lidar (3-D) land surface relief easurer	Exists	No exists	No exists	No exists
Georadar	No exists	No exists	Exists	Exists

Table 2.10 Miramap product specifications

Parameters	Operating range	Max abs error
Soil moisture (g/cm^3)	0.02–0.5	0.07
Depth to water Table (m)	0.05–5	0.3–0.6
Plant biomass (kg/m^2)	0–3	0.2
Pollutant concentrations (ppt)	1–30	1–5

reliable and safe twin Aero Commander (Krapivin and Shutko 2012; Krapivin et al. 2012a). The aircraft is specially modified to simultaneously carry a range of other instruments, such as (digital) photogrammetric cameras, lidar scanners, thermal infrared and multi-spectral sensors. The capability to measure such a comprehensive range of remotely sensed parameters from a single low-cost airborne platform is unique worldwide.

With whole set of sensor shown in Tables 2.7, 2.8 and 2.9, company Miramap provides the customers exactly with the parameters and environmental conditions given in Table 1.18 including:

- surface soil moisture,
- underground moistening,
- depth to a shallow water table (down to 2 m in humid areas and down to 3–5 m in arid/dry areas),
- located on the surface and shallowly buried metal objects of a reasonable size under the conditions of dry ground,
- contours of water seepage through hydrotechnical constructions (levees, dams, destroyed drainage systems, different kinds of leaks),
- biomass of vegetation above a water surface or wet ground,
- increase in temperature in land, forested and volcano areas,
- changes in salinity/mineralization and temperature of a water surface,
- water surface pollution, oil slicks on a water surface,
- on-ground snow melting,
- ice on a water surface and on the roads, runways.

Some indices of the effectiveness of the GIMS-technology realized in framework of Miramap's flying research laboratory are the following:

(1) Soil moisture content

- operating range is 0.02–0.5 g/cm^3
- maximum absolute error is:
  - when vegetation biomass is less than 2 kg/m^2–0.05 g/cm^3;
  - when vegetation biomass is greater than 2 kg/m^2–0.07 g/cm^3.

(2) Depth to a shallow water table

- operating range is:
  - for humid or swampy areas—0.2–2 m;
  - for dry arid areas, deserts—0.2–5 m;
- maximum absolute error is 0.3–0.6 m.

(3) Plant biomass (above wet soil or water surface):

- operating range is 0–3 kg/m^2;
- maximum absolute error is 0.2 kg/m^2.

(4) Salt and pollutant concentration of water areas (off-shore zones, lakes):

- operating range is 1–300 ppt;
- maximum absolute error is 1–5 ppt;
- relative error is 0.5 ppt.

Large-scale investigations on the creation of microwave research carrying platforms were realized by Microwave Remote Sensing Division (MRSD) in 2002 through 2005 within the NASA Center for Hydrology, Soil Climatology and Remote Sensing (HSCaRS) at Alabama Agricultural and Mechanical University (AAMU). This Division was capable of performing microwave radiometric data interpretation and conducting studies in field conditions, from mobile platform and unmanned helicopter.

The antennas, radiometers, data collection system and an embedded Global Positioning System (GPS) receiver were mounted on the manned Rover type mobile platform and unmanned helicopter platform Microwave Autonomous Copter System (MACS) for measuring the soil-plant system radiation. All radiometers were mounted on a folding mounting panel to observe horizontally polarized radiation when folding the panel is situated between the nadir through zenith looking angle. The GPS information was used to register the microwave reading to a common coordinate system of the study area. The data capture rate was set to 1 measurement per second in each of the radiometric channels and GPS readings.

The manned "Rover" type mobile platform is a modified Gator utility vehicle. This two-seater vehicle has a 286 cc, air-cooled, 4-cycle gasoline engine. Its towing capacity is 500 lb (226 kg) with a top speed of 20 mph (32 km/h). The instrument platform (or mounting frame) for the radiometers and other instruments was assembled at AAMU research station. The aluminum folding panel of 1.5 m × 1.5 m connecting all system components was designed so that the incidence angle from 0° (nadir) to 180° (sky) could be easily obtainable (Figs. 2.8 and 2.9 and Table 2.11). The radiometers were mounted with the antennas viewing off to the right hand side of the platform at an incidence angle of 10°. The radiometer "Rover" shuttled back and forth in a north-south direction at a speed of 2–5 mph. using the developed remote sensing system data obtained from a height of 2 m provided the spatial resolution of 1.4 m of land area.

The unmanned helicopter platform MACS was equipped with a 6 cm radiometer (incidence angle 5°) mounted on the nose of the AutoCopter^{TM} onto a stabilized gimbal with pan/tilt interface which attenuates vibrations (Fig. 2.10, and Table 2.12). The MACS is a modified AutoCopter^{TM}, a small unmanned helicopter platform that can fly autonomously (fly a pre-programmed flight path) or semi-autonomously (with an operator directing the maneuvers). This is a product of Neural-Robotics, Inc. (NRI) of Huntsville, AL.

The unmanned helicopter advantage is its patented flight control system consisting of multiple neural network modules working together. The result is an autonomous helicopter that adapts to changing conditions and provides an extremely stable platform for hundreds of applications. The AutoCopter^{TM} is 2.18 m in length (from tip of tail rotor to tip of main rotor) and weighs approximately 13.6 kg.

Fig. 2.8 General view of the manned "Rover" type mobile platform equipped with three portable microwave radiometers, operating at the wavelengths 6, 18, and 21 cm. Also equipped with a folding panel of 1.5 m × 1.5 m for instrumentation installation along with GPS receiver, data acquisition system and power supply battery (Shutko et al. 2010; Krapivin and Shutko 2012; Verba et al. 2014)

It carries a payload of up to 6.8 kg. Basic avionics consist of a PC/104 computer, altitude and heading reference system (AHRS), GPS receiver (WAAS-compatible), downward pointing range finder (ultrasonic sensor), barometric pressure sensor, and heading-hold gyro. The standard transmitter is used as the "ground station."

The Ground Control Station (GCS) with a flight planning program "Way Planner" was used with the AutoCopter^{TM}. This is a self-contained Windows-based application that unlocks the power of fully autonomous flight. The program enables mission planning in 2D using stored satellite images. Flight plans were uploaded to the AutoCopter^{TM} via data link enabling the aircraft to takeoff, climb, fly its programmed route, and land fully autonomously. The programmed flight consisted of an autonomous launch with 16 waypoints, a climb to 30 m, followed by a transition to forward flight at a velocity of 2 m/s and auto-landing.

During flight the aircraft has the ability to state data (aircraft altitude, speed, and other parameters in real time on the GCS screen in 2D and 3D. While maintaining airspeed, altitude and heading, 8 flight lines were flown at a distance of ~500 m at 30 m intervals. The time it took the helicopter to fly from waypoint to waypoint (north-south direction) was ~4.5 min totaling ~40 min of flight time. Because the

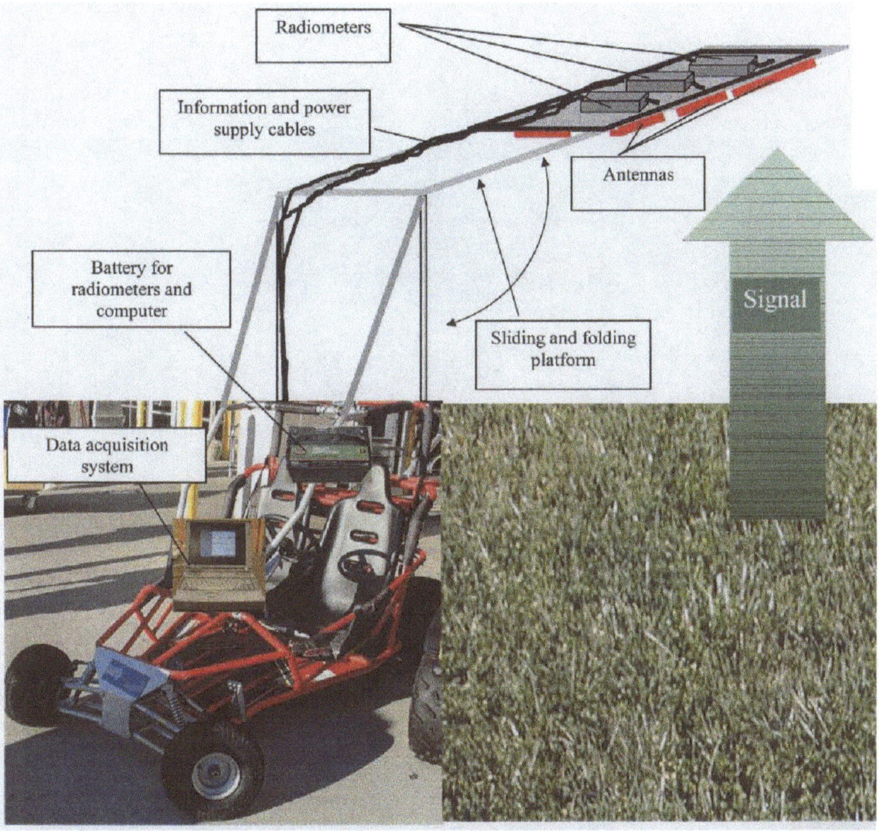

Fig. 2.9 Subsystems and microwave radiometers on mobile platform "Rover" (Shutko et al. 2010; Krapivin and Shutko 2012)

Table 2.11 Characteristics of the microwave platform "Rover"

Parameter	Value
Frequencies	1.41, 1.67 and 5 GHz
Polarization	H or V
Antennas	19.7″ × 19.7″; 16.3″ × 16.3″; slot array; 7.9″ × 6.5″; dipole array
Incidence angle	Fixed angles 0–30° off nadir
Sensitivity	0.5 K
Beam width	(3-dB); 30°
Weight	12 kg
Power required	30 VA

Fig. 2.10 *Photograph* of the unmanned aerial vehicle (UAV) helicopter "Microwave Autonomous Copter System" (MACS) equipped with a 6 cm radiometer, data acquisition system, GPS receiver and power supply battery (Shutko et al. 2010)

Table 2.12 Characteristics of the microwave autonomous copter system (Krapivin ans Shutko 2012)

Parameter	Value
Physical characteristics	
Length	3.58 m
Weight	22.68 kg
Main blades	1.03 m
Engine	120 cc
Range	161 km
Payload capacity	20.41 kg
Fuel tank	32 oz
Passive sensor packages	
Wavelengths	3, 6, or 21 cm
Sensitivity	0.3–0.5 K
3-db beam width	30°
Power supply	27 V DC
Power consumption	15 VA
Weight	2–6 kg
Others	
Frequencies	1.4 or/and 5 GHz
Polarization	H or V
Antennas	Flat panels w/30° 3-dB
Incidence angle range	Fixed angles 0–30°
Sensitivity	0.5 K
Flight time	2 h

aircraft flew below 500 ft it was exempt from FAA regulations. Using the developed remote sensing system, data obtained from an altitude of 30 m provided the spatial resolution of 20 m of land area.

2.3 Microwave Radiometry and Remote Sensing of the Environment

Microwave radiometry, or passive microwave remote sensing, is one of the radiophysical methods used for remote observations of the environment. It is based on measurements of the natural electromagnetic radiation of objects in the millimeter to decimeter range of wavelengths. The theoretical aspects of radiophysical monitoring are determined by organizing a measurement procedure in the active or passive regimes of electromagnetic radiation propagation. An active sensing technology is based on the radiolocation methods, and a passive one is based on the background radiation registration. In both cases, theoretical tasks of microwave monitoring are connected with the study of electromagnetic waves propagation in the environment and, of course in the first place, in the atmosphere and near ionosphere (Kondratyev and Fedchenko 1992; Yakovlev 2001; Yakovlev et al. 2009; Chukhlantsev 2006; Gulyaev et al. 2009; Pavelyev et al. 2010).

In the active regime the talk is about the propagation of electromagnetic waves of various ranges from the emitter to the receiver. Two cases are possible here. In the first case, the transmitter and receiver have different position in the environment and the decision about the assessment of the environment parameters is accepted on the base of registered distortion of transmitted signal. In the second case, the transmitter and receiver are combined so that estimation of the environment parameters is made on the base of analysis of the emitted signal absorption and distortion of received its reflection. Clearly, the time it takes the signal to propagate in the environment is the information parameter. An active form of microwave monitoring is typical for the atmosphere radiotranslusense method.

Basic advantage of microwave remote sensing lies in its largely immune to clouds, precipitation, smoke, etc. and a capacity to penetrate sand, soil, rock, vegetation, dry snow, ice, etc. Knowledge of the physical processes associated with energy emission at microwave range gives a possibility to calculate a variety of surface and atmospheric parameters, including air temperature, sea surface temperature, salinity, soil moisture, sea ice, precipitation, the total amount of water vapor and the total amount of liquid water in the atmospheric column directly above or below the instrument.

The passive methods of microwave monitoring are based on SHF-radiometry. The existence of resonance absorption fields in SHF range allows remote reconstruction of meteorological atmosphere parameters such as the vertical profiles of temperature, humidity, total water vapor mass, water storage in the clouds, and precipitation intensity, etc. The possibility to receive data not only about the properties of water and land surfaces but also about deep characteristics depends on the choice of electromagnetic range. In the infrared range, total emission is formed into very thin surface layer. Electromagnetic waves of microwave range are absorbed powerfully by the land and water surfaces. The depth of its spread into the water environment is measured in hundredth of millimetre; whereas dry soils, continental ices, and dry snow this value can reach several tens of wavelengths. It allows to realize the remote investigations of soil, ice and snow covers up to considerable depths.

Penetrating ability of radio-waves gives the advantage particularly under the sensing of land covers. Vegetation without a dense canopy (grass, cereals, etc.) absorbs radio waves weakly and this is the reason radio-observation of soil covers through this vegetation is possible. Radio-waves can spread across the soil up to the depth reaching one metre.

Main defect of microwave radiometric observations consists in the comparative low spatial resolution in comparison with the optical range. In the radio-range the high spatial resolution is achieved by the multichannel application and specific data processing methods. It demands large economic investments.

Let us consider some aspects of active location. Let us supposed that irradiation at the nadir of a smooth plot is realized. In this case the signal power reflected from the land surface equals $W = pGA|\kappa_0|^2(16\pi H^2)^{-1}$, where p is the emitted power, G is the coefficient of directed action of antenna, A is the efficient antenna area, H is the antenna height above the ground surface, κ_0 is the surface reflection coefficient. If all instrumental parameters and antenna height are known the ratio of power accepted to that of the one received determines the value of the reflection coefficient. In the case when the soil is uniform by depth the reflection coefficient equals $\kappa_0 = (\varepsilon^{1/2} - 1)/(\varepsilon^{1/2} + 1)$, where ε is the dielectric ground permittivity. For the estimation of ε the following approximate formula exists

$$\varepsilon = \left(\varepsilon_w^{1/2} p_w + \varepsilon_s^{1/2}[1 - p_w]\right)^2, \tag{2.1}$$

where ε_w is the dielectric water permittivity, p_w is the relative volume concentration of free moisture in the soil, $\varepsilon_s \approx (1 + 0.5\rho_s)^2$ is the dielectric permittivity of dry ground, and ρ_s is the dry ground density (1–2 g cm^{-3}).

From Eq. (2.1) it is seen that reflected signal power give a possibility to determine the reflection coefficient and, consequently, dielectric ground permeability. The value of ε defines the ground density when it is dry or the water content if it is wet.

Dielectric properties of land covers are basic to use the radiometric methods for its diagnosis under different situations. Specifically, the knowledge of spectral coefficient for the emission in the millimetre range allows to estimate the properties of snow or ice layer.

The reflection coefficient is determined as well as by passive methods using the measurements of radiothermal emission intensity. The intensity of background radiation at microwave range according to the Kirhgof law is characterized by the brightness temperature

$$T_j = \kappa T_e \tag{2.2}$$

where κ is the emission coefficient (or absorbing surface ability, or the blackness level), T_e is the effective surface temperature.

Expression (2.2) characterizes the thermal emission of the surface but does not take into account the emission falling on the surface or reflected by it. The coefficient κ is described by Frenel reflective formula. The thermodynamic and brightness temperatures are measured in Kelvins: T(K) = t(°C) + 273. The emissivity is a function of dielectric permittivity of the object/surface of observation. For a land surface, the dielectric permittivity is first of all a function of soil moisture. The higher a soil moisture content, the higher the permittivity of soil, the lower the emissivity/intensity of radiation/brightness temperature of this piece of land.

For a water surface, the dielectric permittivity is first of all a function of electric conductivity of a water solution that is dependent on a concentration of salts, acids, on a presence of oil films and many other chemical substances. For example, the higher salinity of water, the higher the dielectric permittivity of water solution, the lower the emissivity/intensity of radiation/brightness temperature of this water body. Within the 2–30 cm band, for t = 10–30 °C, the radiation characteristics of several surface types are shown in Table 2.13. Table 2.14 shows the sensitivity of radiation in the X-band (2–3 cm) and L-band (18–30 cm) to the changes in free water content in bare soil, soil density, salinity and temperature of the soil surface. These data show that the main parameter affecting the intensity of a bare soil radiation, practically independent of spectral band, is the soil moisture. Based on this sensitivity it is feasible to estimate the value of soil moisture without a priori data on the soil parameters.

The power registered by completely concerted receiver equals $W = \kappa T_j \Delta f$ where Δf is the received emission branch. In the case when the model of soil with the flat

Table 2.13 Basic microwave radiation characteristics of some typical surface types

Surface	T_j(K)	κ
Metal	0	0
Water surface	90–110	0.3–0.4
Very wet soil	160–180	0.55–0.65
Very dry soil	250–270	0.85–0.93

Table 2.14 Sensitivity of a bare soil microwave radiation to variations in soil moisture (W), soil density (D), salinity (S), and surface temperature (T)

Wavelength (cm)	Spectral band	$\Delta T_j/\Delta W$ (K/g/cm^3)	$\Delta T_j/\Delta D$ (K/g/cm^3)	$\Delta T_j/\Delta S$ (K/ppt)	$\Delta T_j/\Delta T$ (K/°C)
2–3	X	−200	−15	0.05	0.5
18–30	L	−(200 to 300)	−10	−0.5	0.1

surface is considered the emission and reflection coefficients are correlated by the formula $\kappa = |\kappa_0|^2$. Consequently the knowledge of emission coefficient allows to estimate the reflection coefficient and electrophysical properties of ground. Under this it is necessary to know the soil temperature T or to calibrate the radiometer using the land covers with emission coefficients which are known. Theoretical bases of remote methods, as a rule, have approximate character. This has to do with the need for simplified prepositions. For example, among the simplifications often used are uniformity of environmental parameters distribution by the depths or height, absolute transparency of the area, the land covers smoothness. The models synthesized under these simplifications reflect restricted spectrum of the studied phenomenon properties. That is why the estimation of model adequation is necessary.

The problem of wavelengths choice and the combination of its ranges with the classes of solved tasks is now within sight of many investigators. Theoretical foundation of this choice lies in the area of the thermal radiation transmission theory. Specifically, for the monochromatic case the transmission equation has

$$dI(z)/dz = J(z) - \alpha I(z), \tag{2.3}$$

where α is the absorption coefficient, $J(z)$ is the emission source distributed by z, $I(z)$ is the emission intensity in the point z. Under this for the unscattering atmosphere the following equation is valid: $J(z) = \varepsilon_1(z)B(z)$, where ε is the atmosphere emissivity,

$$B(z) = 2hf^3c^{-2}\left[\exp\{hf(kT)^{-1}\}\right]^{-1};$$

h, c and k are fundamental constants, $f = \nu c$ is the frequency. For the microwave branch where the observed frequency is in the region of a few hundred gigahertz or less it takes place a condition $hf \ll kT$ and the Rayleigh-Jeans approximation is valid: $B(z) = 2kT\lambda^{-2}$. Here $k = 1.38 \times 10^{-20}$ mW Hz^{-1}K^{-1} is the Boltzmann's constant, λ is the wavelength.

The Eq. (2.3) has solution

$$T_j = \kappa T_e \tau(0,\infty) + \int_0^\infty T(u)Q(u)du, \tag{2.4}$$

where

$$\tau(a,b) = \exp\left[-\int_a^b \alpha(z)dz\right].$$

Table 2.15 The value of brightness temperatures which are typical for the some land covers

Land cover	Wavelength, λ (cm)			
	0.8		1.35	
	$T_{j,\min}$	$T_{j,\max}$	$T_{j,\min}$	$T_{j,\max}$
Desert	251	261	245	270
Cultivated steppe	255	268	257	277
Steppe covered by snow	246	261	241	260
The ice covered by snow	247	263	231	253
The ice	239	247	235	254
Sea surface	136	171	110	168

The weight function $Q(u) = \partial\, \tau(u, \infty)/\partial u$ describes an influence of various atmospheric layers to the receiving signal. The formula (2.4) makes the expression (2.2) into a whole. The examples of specific values T_j are given in Table 2.15.

Microwave measurement calibration is important in remote monitoring. Calibration procedure can be defined as the process of relating signals that can be extracted from a measurement radiometric system to the physical characteristics of a target such as its brightness temperature. The main goal of a calibration process is in getting all measurements to account with a unique scale. In the passive regime for this aim the objects having known brightness temperature are used (see Table 2.16). Usually it is selected the forest or the freshwater reservoir. In this case the radiobrightness temperature of forest is equaled to the thermodynamical temperature of the trees canopy, and the radiobrightness temperature of reservoir is calculated with the known algorithms (Yakimov 1996; Ferm and Hultberg 1999; Burkov and Krapivin 2009; Krapivin and Shutko 2012).

2.3.1 Remote-Sensing Technologies in the Infrared and Optical Bands

There exist no less than 10 technologies for remote sensing of the environment. Between them are those not widely known and not widely applicable for our case, namely: ultra-sound methods, radar sensing beyond the horizon, solar radiation sensing and a few others.

Table 2.16 Emission coefficient for the some natural and anthropogenic microwave standards (Shutko 1987)

Land cover	Microwave range	
	Centimetres	Decimetres
Dry sand	0.93 ± 0.03	0.93 ± 0.03
Concrete	0.84 ± 0.02	0.82 ± 0.02
Asphalt	0.85 ± 0.03	0.83 ± 0.02
Dense forest	0.95–0.98	0.92–0.96
Fresh water	0.38 ± 0.01	0.36 ± 0.01
Watering soils	0.58 ± 0.03	0.56 ± 0.03

For our case—that is, using the GIMS as the key instrument to solve environmental problems, for decision-making procedures in the GIMS, for geoscience and remote sensing, and for oceanic engineering—and for the very typical example of diagnosing and preventing at an earlier stage the buildup of extensive moisture in the soil, water barrier failure, and flood development the most acceptable, useful and effective, instruments, and technologies that have been proven in a number of domestic and international campaigns are the following:

(1) Optical (digital photo and/or video cameras, spectrometers).
(2) Lidar (Laser 3-D scanner).
(3) Thermal IR-radiometers.
(4) Microwave radiometers.
(5) Radar (Synthetic Aperture Radar—SAR).
(6) Short pulse nadir viewing subsurface dielectric irregularities finder.
(7) Georadar (subsurface dielectric irregularities finder of on-ground location, also nadir viewing).

Table 2.17 presents the effectiveness estimates of first 5 of total 7 rather widely used instruments with regard to their application for remote sensing of water and land areas. Crosses in the Table 2.7 indicate the degree of capability (and thus of usefulness) of different sensors to registration of different parameter variations or changes in different situations in typical/characteristic environments. Three crosses indicate the highest possible rank of effectiveness, one cross—minimum one.

Why are five of the seven, rather than seven of the seven, presented types of sensors discussed? The answers are as follows:

(a) sensors 6 and 7 are still not used so widely as the first 5 sensors,
(b) these two sensors were not used in the development of the GIMS, and
(c) special conditions are needed to organize a flight campaign and to include in it these sensors and Radar, including Synthetic Aperture Radar, which are effective and useful, but heavy, of a big size and expensive tools for aircraft and spacecraft remote sensing.

These technologies are under the development primarily in JPL and Vega Radio Corporation in Russia and available for potential users.

Table 2.17 gives rough but effective assessments of how basic remote sensors, operating at both passive and active microwaves in the infrared and optical bands, behave under the basic parameters and conditions inherent in the following scenarios:

- Oceanic, sea or fresh/water surface.
- Plane and rough water surface.
- Foam-covered and oil spill-covered water surface.
- Water-vegetative canopy system.
- Bare land surface.
- Vegetation-covered soil.
- Atmosphere-tree crown system (when sensing upward through the canopy).

Table 2.17 Typical remote sensing technologies, their information content and effectiveness

Sensors and parameters they determine	O	L	T	M	R
Sea/water surface temperature			++ or +++	++ or +++	
Sea/water surface natural salinity				++ or +++	
Oil pollution on water surface	+		+	++	+++
Water pollution with dissoluble chemicals	+			++	+
Water pollution with waste outflow	+			++	+
Soil moisture	+		+ or ++	+++	+ or ++
Sub-surface moisture (depth to water table)				+++	+ or ++
Zones of water seepage through water barriers	+	+	+	+++	+
Flooding	+ or ++	+ or ++	+	+++	+++
General: water regime of terrain	+ or ++	+	+	+++	++
Terrain relief	+	++ or +++	+	+ or ++	++ or +++
Water areas and/or wet lands over-growing with vegetation	+	+	+	++	+++
Vegetation classification	++	+		+	+ or ++
Vegetation biomass, dynamics, seasonal changes	+	+		+	++
Monitoring crowns of separately standing trees				++ or +++	
Monitoring vegetation sick rate	+ or ++		+		+
Deforestation	+ or ++		+		+
Monitoring timber/forest clearance	+ or ++		+		+
Risk of forest or peat fire	+ or ++		+ or ++	++ or +++	+
Revealing epicenter and/or border of fire in foggy and smoky condition	+ or ++		++ or +++	++ or +++	
Monitoring terrain traffic ability	+ or ++	+ or ++	++ or +++	++ or +++	++ or +++

O Optical (digital photo and/or video cameras, spectrometers), *L* Lidar (Laser 3-D scanner), *T* Thermal IR-radiometers, *M* Microwave radiometers, and *R* Radar (SAR) (Krapivin and Shutko 2012)

As we mentioned above, the crosses in Table 2.17 characterize the information content (usefulness) of different sensors, beginning from a single cross (the lowest rate of usefulness) and moving to three crosses (highest usefulness). Thus this table allows one to obtain preliminary assessments of the effectiveness of the expected results of some experimental campaign under the development.

Of course, Table 2.17 reflects the expert-level opinion regarding the usefulness of different sensors. This opinion was created by them based on multi year studies of the spectral properties of different environments. This personal experience was used to develop, test, and put into practice truly multi-spectral remote-sensing technologies providing state-of-the art information by means of an optimum set of sensors, such as those installed on board of the Dutch/U.S./Russia Miramap aircraft equipped with the above mentioned multi-spectral passive/radiometric microwave, thermal infrared, color digital orthophoto cameras, and lidar (laser 3-D scanner working in the near-infrared band).

2.4 Monitoring of the Soil-Plant Formations

Land cover change is an important environmental process. Studying and monitoring this change are significant when assessing the variations and dynamics of the global biogeochemical cycles of greenhouse gases. One of effective satellite instruments that gives a possibility to detect the disturbances of the globe's land surface is the Moderate Resolution Imaging Spectroradiometer (MODIS) of NASA's Earth Observing System (EOS).

MODIS is a key instrument aboard the Terra (EOS AM) and Aqua (EOS PM) satellites. Terra's orbit around the Earth is timed so that it passes from north to south across the equator in the morning, while Aqua passes south to north over the equator in the afternoon. Terra MODIS and Aqua MODIS are viewing the entire Earth's surface every 1–2 days, acquiring data in 36 spectral bands, or groups of wavelengths. These data will improve our understanding of global dynamics and processes occurring on the land, in the oceans, and in the lower atmosphere. MODIS is playing a vital role in the development of validated, global, interactive Earth system models able to predict global change accurately enough to assist policy makers in making sound decisions concerning the protection of our environment.

Understanding Earth system processes and their interdependencies through observation, theory, and modeling, is the objective of many international programs such as the International Geosphere-Biosphere Programme (IGBP), the International Human Dimensions Programme (IHDP), the World Climate Research Programme (WCRP), and the Global Carbon Project (GCP). However, this is only possible if reliable and operative information about the land covers is available. Undoubtedly, the use of this information demands the creation of algorithms and models to improve the quality of environmental monitoring.

A remote monitoring of the soil-plant formations is aimed at assessing biological productivity, understanding interactions inside the *soil-vegetation-atmosphere* system, evaluating biome dynamics, modeling biogeochemical cycles paying particular attention to the role played by vegetation, and controlling vegetation resources.

Land covers are characterized by high landscape diversity having various soil and plant types, water-salt and temperature regimes, the cultivating of surface layer, and macrorelief. The soil includes solid particles, water (soil mixture) and air. Solid particles consist of mineral and organic matter. Basic mass of majority soils forms mineral substances (mineral soils): particles of size more 3 mm form the stony soil part, particles from 0.01 to 1 mm are called as "physical sand", and particles less 0.01 mm are "physical" clay. The minerals that make up the soil consist basically of two oxides: SiO_2 and Al_2O_3. The content of organic substances (mainly humus) fluctuates in such soils between 1 and 10 % by the weight. All these circumstances, when the monitoring system of land covers is designed, lead to the necessity of microwave channels choice to be informative. The many lengthy investigations in this field facilitate choosing the most effective wave range for monitoring many types of vegetation and soils. Studies of agricultural systems, forest, and arid zones are inevitable very detailed. Such areas as the control of hydrological soil regime under the forest canopy and the study of reforestation dynamics in the regions of clearing and forest fires are perspective scientific fields to be studied at the nearest time in the microwave monitoring. This can be done by using the decimeter range in which the forest canopy is transparent in the microwave range and, consequently, soil up to the ground waters level can be sensed. Experimental characteristics of the natural microwave L-band emission of a forest are described by Chukhlantsev and Shutko (1988), Phillips et al. (2004), and Chukhlantsev (2006).

A possibility of wood-bushes vegetation inventory by means of radiolocation sensing is based on the correlation between the coefficients of SHF waves back scattering and the structure and dielectric properties of vegetation. Using radiolocation systems to control forests is stipulated by the possibilities of additional data being received together with remote measurements in the visible and IR ranges (Savin and Bondur 2000).

Global mapping of vegetation in the scale determined by economic needs of given region is main task of the satellite based radiolocators under the forest observations. Specifically, the forest protection from the fires is one of directions in the development of remote monitoring to assess the vegetation state. Present satellite- and aviation- based monitoring systems solve many significant problems:

- Operative assessment of the forest's fire risk;
- The forest watering mapping under the putting out fire in the forest;
- The mapping of the forest fires contours through the smoke layer and trees canopy;
- The determination of the energetic fire parameters;
- The forest state determination after fire;

Investigations of radiothermal emission from natural layers of a forest that are combustible such as the lichen, mosses, grass rags, died pine needles, and scrub have brought about methodologies to identify forest fire parameters based on the measurements of radiothermal emission in microwave range of electromagnetic spectrum. The examples of radiobrightness temperature in the forest burning zone are given in Figs. 2.11 and 2.12. The leading and rear edges of fire and the burnt forest in between can clearly be distinguished.

Comparing radiobrightness temperatures registered in different wavelengths shows that forest fires emissivity increases with the wavelength decrease and

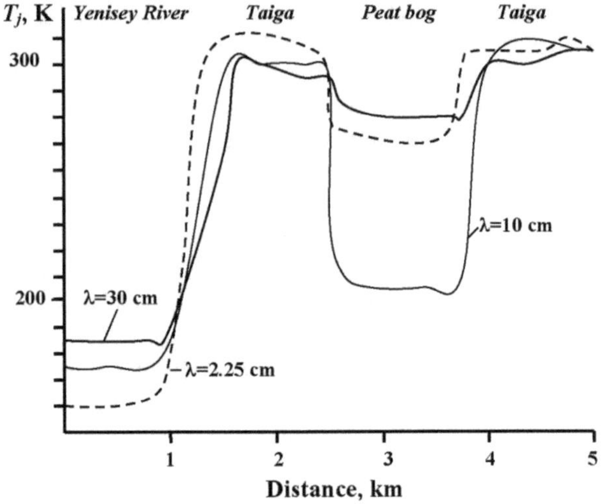

Fig. 2.11 An example of the registerogrammes received by means of flying laboratory IL-18 near Eniseysk city using radiometers of 2.25 cm (1), 10 cm (2), and 30 cm (3) wavelengths

depends on the fire type. Given in Fig. 2.12 averaged results for the dependencies of radiobrightness temperatures spectrum from various components of forest fire under different wavelengths allow to classify these components with high precision.

The GIMS synthesis with the functions of forest fire control on the large areas demands the creation of applied models describing the forest fire-ripening processes. The foundation of such models is, certainly, the knowledge how the intensity of radiothermal emission of forest landscape element depends on its water-content. Many observations show that statistical characteristics of the forest

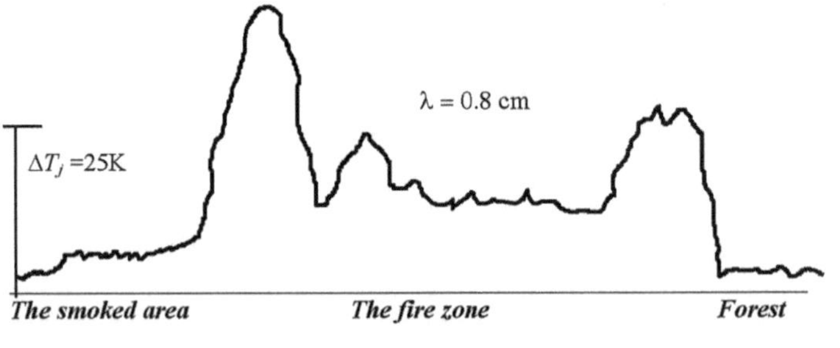

Fig. 2.12 Fragment of the registerogrammes in the wildfire zone near Krasnoyarsk received by means of flying laboratory IL-18

microwave emissions are changed with the fire-ripening growth for the layers of combustible timbers. Specifically, such a dependence can be represented in the following form (Yakimov 1996):

$$T_j = \kappa(\lambda, w)T_S(1 - \exp\{-\alpha(\lambda, w)h\}) + T_{jS}\exp\{-\alpha(\lambda, w)h\},$$

where κ and α are emission and absorption coefficient, respectively; w is the water content in the timber, h is the thickness of emitting layer, T_S is the soil temperature, T_{jS} is the soil brightness temperature. The following deciding factor, based on the calculating the first-order (M) and second-order (σ) moment for the set of T_j corresponding to the forest territory and registered on the wavelength $\lambda = 2.25$ cm were proposed by Yakimov (1996). The value M (mean) is compared with the threshold equaled to $d = 7.251\sigma + 249.876$. If $M \geq d$ then forest plot has average or higher fire-dangerous. If $M < d$ then fire-dangerous of given forest plot is low. This deciding rule is to be extended by the forest fire model. An example of such a model is one describing forest fire as a running wave or as self-supporting process of local energy release into the active environment.

The millimeter range (1–30 mm and 30–300 GHz bands) is best for land cover diagnosis. As numerous experimental investigations show, addicting millimeter range channels to the monitoring system enhances the accuracy of soil-plant formations classification by separating out roughnesses and small formations. It is especially important for anthropogenic landscape monitoring.

McCloy and Lucht (2004) developed a method for comparing the complex spatio-temporal patterns present in two long-time series of data of the seasonal cycles of vegetation for a large part of the global land surface. It is based on the use of two datasets derived from global satellite observations (AVHRR) and Lund-Potsdam-Jena dynamic global vegetation model (LPJ-DGVM). McCloy and Lucht (2004) showed that the LPJ model generally corresponds well with natural vegetation whereas the former is primarily dependent on climate. Due to a lack of representation of agriculture in the model, the correspondence with actual vegetative status was poor.

2.5 Microwave Monitoring of the Soil Moisture

2.5.1 Introduction

A series of world wide large scale Soil Moisture Experiments (SMEX) have been conducted from 1997 to 2005 by the National Aeronautics and Space Administration (NASA) and the United States Department of Agriculture (USDA), Hydrology and Remote Sensing Laboratory, with the involvement of institutions in the USA, Japan and Brazil (Jackson et al. 2002). The primary objective of those experiments was to evaluate the capabilities of satellite microwave sensors for soil moisture mapping on a planetary scale. The experiments to date have all utilized a combination of aircraft microwave measurements and in situ measurements to achieve the goal.

Soil moisture was recognized by the NASA Post 2002 Program as a critical measurement. Operational determination of soil moisture spatial-temporal variability plays a leading role in regional and global studies of the water, biogeochemical and energy cycles. The efforts are targeted on the improvement of spacecraft microwave radiometric assessments of soil moisture to the level of data acceptability when reliable predictions of regional to global hydrologic regimes can be provided. Soil moisture is a key state variable in hydrology: it is the switch that controls the proportion of rainfall that percolates, runs off, or evaporated from the land. It is the life-giving substance for vegetation. Soil moisture integrates precipitation and evaporation over periods of days to weeks and introduces a significant element of memory in the atmosphere-land system. There is strong climatological and modeling evidence that the fast recycling of water through evapotranspiration and precipitation is the primary factor in the persistence of dry or wet anomalies over large continental regions during summer. As a result, soil moisture is possibly one of the most significant land surface boundary conditions controlling the exchange of water and energy with the atmosphere.

Remote sensing of the environment by means of passive microwaves or microwave radiometry is based on retrieval algorithms that relate the environmental parameters (such as soil moisture, vegetation type and biomass, forest stem volume, water equivalent, physical temperature, and others) to the remotely sensed radiation characteristics (spectral brightness temperature, polarization). Microwave emission from a natural object is governed by the interaction of electromagnetic waves with a natural medium that is determined by its geometrical structure and dielectric and electric properties of the medium constituents. In turn, these properties are connected with geophysical and biophysical parameters of the medium that enables the retrieval of these parameters from remotely measured characteristics of electromagnetic waves emitted or scattered by the object. Therefore, dielectric and electric properties of natural media constituents and their relationships to geophysical and biophysical parameters play a crucial role in microwave remote sensing of the environment. Besides, the dielectric permittivity of a medium or its real part and its imaginary part determines the penetration depth of electromagnetic waves in the medium that is very important for subsurface remote sensing of soil moisture and for soil moisture sensing under vegetative canopies.

The impact of microwave radiometer pixel size on the reliability of soil moisture assessments for different landscapes and minimization of the number of in situ locations for sampling soil moisture, temperature and vegetation parameters are key elements of the study. As a result, a combined approach for soil moisture mapping from a regional to global scale using microwave radiometric observations, laboratory tests, in situ measurements and mathematical modeling of the soil-vegetation system inherent in different landscapes is to be developed as a highly calibration and validation input to the series of the NASA and USDA Soil Moisture Experiments (SMEX) in the interests of hydrological science, technology and applications. In this aspect, the following problems are arising in the microwave monitoring of soil moisture:

- development of models parameterizing biometrical, soil properties and radiometric data for typical sites;
- comparative evaluation of seasonal patterns in series of satellite, ground-truth (and/or aerial) and laboratory data and simulations of soil-vegetation model;
- conduct intercomparisons of radiation data and mean values of soil moisture data at different depths with regard for the type of vegetation cover and amount of biomass;
- choice of constructive scheme for the data calibration basing on the available radiative transfer model and results of measurements;
- spectral radiometric data processing related to soil moisture and vegetation cover parameters and results of modeling;
- development of mathematical model for soil-vegetation system dynamics using satellite, field and laboratory data;
- combining approach to parameterization of data needed for description of water balance in soil-vegetation system and of data delivered from microwave satellite (AMSR-E/Aqua, etc.) spacecraft observations, ground based microwave observations, laboratory tests, in situ measurements (sampling) and mathematical modeling; and
- using data of attenuation in vegetation and dielectric properties of soil and vegetation samples as calibration and validation input to the GIMS.

2.5.2 Microwave Technology

The fundamental principles of soil moisture retrieval by means of the use of passive microwave sensors and microwave radiometric technique for measuring soil moisture properties have been established by Shutko (1986) and his disciples. During a series of experiments, microwave radiometers were put on satellites, aircraft, and ground-based vehicles. L-band passive microwave measurements were the most informative in these experiments. Numerous investigations of sensitivity of microwave measurements to vegetation biomass and soil moisture have shown that remote sensing data give a possibility to solve important tasks from hydrology and agriculture and to realize operational diagnostics of damage of stressful natural processes. For instance, Ferrazzoli et al. (1992) carid out a comparative evaluation of the potential of active and passive radiometers in estimating vegetation biomass and soil moisture content. It was shown that low frequency data (L band) at a steep incidence angle (10°) confirm that both the backscattering coefficient and the normalized temperature are correlated and sensitive to soil moisture content.

Interrelations between the characteristics of the microwave emission field of moisted soil and of the soil liquid water content, soil density, temperature and mineralization level of liquid water are the properties studied both theoretically and by field measurements. Both research and field studies show that microwave radiometric measurements permit estimates of 7–10 moisture levels in the top 0.1–1.0 wavelengths of soil and 3–7 grades of the subsurface water level between 0 and 1.5–3.5 m.

The soil moisture is divided into solidly united, loosely united, and free. United moisture is the water adsorbed by ground particles at the surface and takes the form of a film of thickness equal to no more than six to eight molecular layers. The volume of united moisture in the soil layer is determined by the soil type and is fluctuated in the wide interval from 2–3 % for sandy soil to 30–40 % for clay and loess soils. United water is unattainable by plants and does not influence the salt regime of soil. That is why the monitoring system is to realize that kind of moisture classification in the soil.

The soil moisture is expressed in percent of dry soil weight. Radiation models of different types of soil moisture that consider soil density, temperature, and salinity have been studied in great detail. The data required to ascertain moisture levels can be obtained by means of brightness temperature contrasts, degree of polarization, and spectral characteristics at centimeter and decimeter wavelengths. Wavelengths 2.25, 18 and 30 cm proved to be the most informative to solve this task. The microwave radiation model describing the land cover emittively under the conditions of heterogeneously moistened layer is based on the existence of vertical heterogeneity in the dielectric permeability coefficient (see Formula (2.1)): $\varepsilon \cong (1 + 0.5\rho_s)^2$. The value of ε is informative indicator of the moisture soil variation. As soil moisture begin to grows the ε increases at first slowly, if soil was dry, then it increases more rapidly. The weak dependence of ε on the initial moisture stage, explained by moisture tied up in the soil, can be characterized as having small dielectric permeability. The variety of experimental dependencies of ε from the soil types and moisturizing levels was given by many authors (Shutko 1986; Engman and Chauhan 1995; Borodin and Krapivin 1998; Schmugge 1990; Tsegaye et al. 2005). Such a knowledge base combined with application of GIMS technology allows ascertaining water content, $W_s(z)$, diagnostics in the soil layer. Basic types of moisture profiles $W_s(z)$ of real soils and corresponding profiles, $\varepsilon(z)$, of dielectric constant were described by Reutov and Shutko (1986). Four major types of vertical moisture profiles have been identified from an analysis of the moistening characteristics of the various climatic zones in the Former Soviet Union. In other climatic zones it is necessary to have such identification. As a matter of fact the task is reduced to the function $W_s(z)$ reconstruction when the values $W_{si} = W_s(z_i)$ are known and the following conditions are satisfied:

$$\partial W_s/\partial z|_{z=0} \le C, \quad 0 \le W_s(z) \le K_s(z), \tag{2.5}$$

where $K_s(z)$ is the given function.

Practice has shown that, irrespective of the climatic zone, microwave radiometers need to be used to achieve acceptable precision when reconstructing the vertical profile of moisture in a soil layer to a depth of one meter. There is *a priori* information about average soil moisture for depths 50 cm (W_{s2}) and 100 cm (W_{s3}). The value W_{s1} is estimated by means of microwave radiometers with wavelengths $\lambda = 10$ and 30 cm. Then the following approximation $W_s^*(z) = az^3 + bz^2 + cz + d$ is considered, where unknown coefficients are calculated from the condition of minimal deviation between the $W_s^*(z_i)$ and W_{si} ($i = 1, 2, 3$). For this to be done conditions (2.5) have to be realized. This methodology allows soil moisture to be determined by remote monitoring over large areas with an error no greater than 0.3 g cm^{-3} for a biomass of vegetable cover no smaller than 2 kg m^{-2} and with an error up to 0.07 g cm^{-3} for a biomass greater than 2 kg m^{-2}.

The knowledge of function $W_s(z)$ allows to use the model of water balance of territory to reconstruct the dynamical soil properties and other water balance elements as the functions of geophysical and ecological parameters. Experimental valuation of thickness of upper soil layer on the lower boundary of which every 6 days after rain or watering the capillary connections breaking is taken place (Fig. 2.13). Typical thickness of this dried up layer oscillates between 3 and 5 cm.

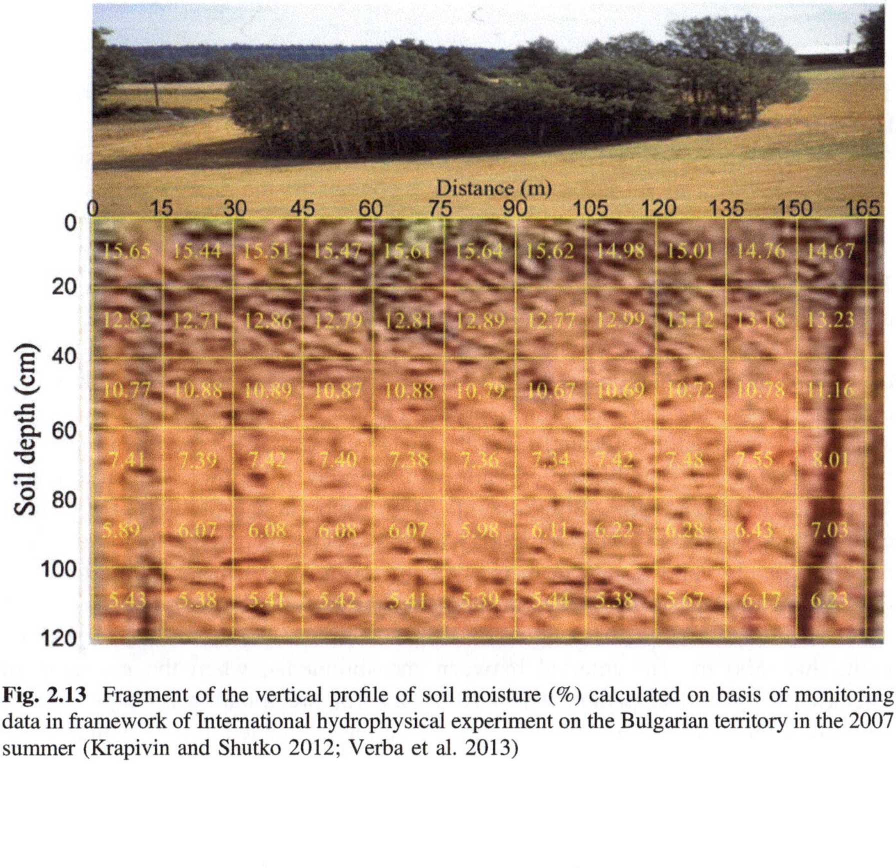

Fig. 2.13 Fragment of the vertical profile of soil moisture (%) calculated on basis of monitoring data in framework of International hydrophysical experiment on the Bulgarian territory in the 2007 summer (Krapivin and Shutko 2012; Verba et al. 2013)

Table 2.18 The reflection coefficients for the microwaves under different soil moistening

Moisture content in the soil (g cm^{-3})	Thickness of the soil layer (cm)				
	0.5	1.0	2.0	3.0	5.0
λ = 18 cm					
0.001	0.98	0.94	0.95	0.96	0.97
0.02	0.91	0.92	0.93	0.95	0.97
0.05	0.88	0.89	0.92	0.95	0.96
0.1	0.82	0.84	0.89	0.95	0.96
0.2	0.73	0.76	0.85	0.94	0.95
0.3	0.65	0.69	0.83	0.94	0.93
0.4	0.59	0.64	0.82	0.94	0.93
0.5	0.53	0.59	0.82	0.92	0.92
λ = 30 cm					
0.001	0.93	0.93	0.94	0.95	0.96
0.02	0.91	0.91	0.92	0.93	0.96
0.05	0.88	0.88	0.89	0.92	0.96
0.1	0.82	0.83	0.85	0.89	0.96
0.2	0.72	0.73	0.78	0.85	0.96
0.3	0.64	0.66	0.73	0.83	0.95
0.4	0.57	0.59	0.69	0.82	0.94
0.5	0.52	0.55	0.66	0.81	0.92

The reflection coefficient estimations for the microwave range reflected from this layer are given in Table 2.18.

Optimal choice of the electromagnetic radiation range and model synthesis based on methodologies to measure the moisture content in the soil demands further investigation of the conditions that influence emission fields in the *atmosphere-soil-vegetation* system (Grankov and Mil'shin 1994; Engman and Chauhan 1995). These investigations are to be correlated with application of remote sensing to agricultural crop forecasting and in selected water resources management.

Figure 2.14 presents the sample of microwave radiometers application and the use of spatial interpolation methods to solve the task of agriculture fields monitoring providing the soil moisture mapping. This can be done radiometers with 18 and 27 cm wavelengths located on a flying laboratory. The experiments of such type are realized in the climatic conditions of Moldavia, Ukraine, Bulgaria, Cuba, Vietnam, Russia, and USA (Shutko et al. 2010; Krapivin and Shutko 2012). These experiments showned GIMS technology allows specific geophysical conditions to have practically valid results from monitoring. To reconstruct the spatial distribution of soil moisture with precision of 20 %, the flying laboratory in most cases needs to fly parallel routes at height $H \approx 200$ m with a distance between them of no more than 500 m. The interval between measurements, when the precision of forecast for synoptical parameters is 10 % and for precipitation is 15 %, can be monthly if the precision of soil moisture estimation is planned to be less that 50 %.

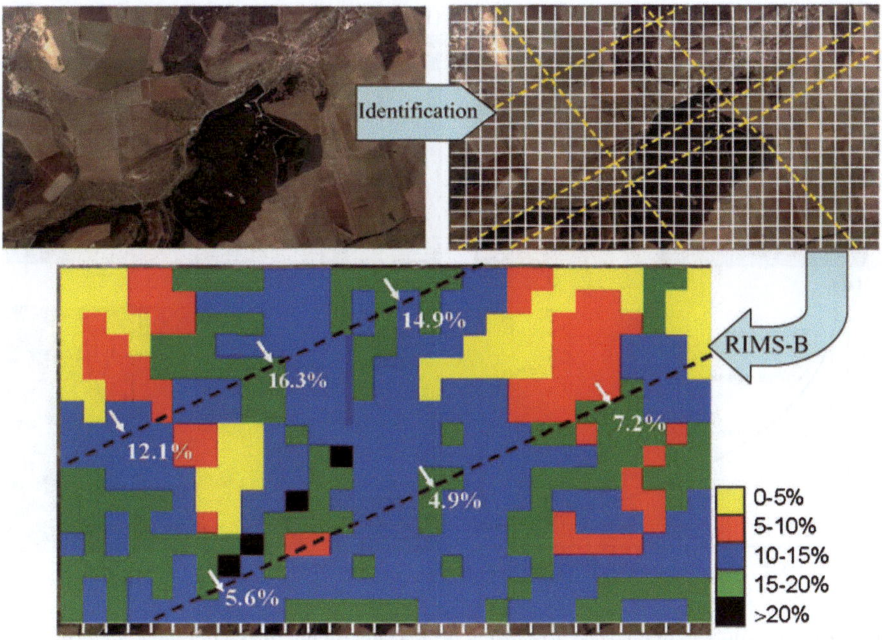

Fig. 2.14 Stages of the monitoring procedure realization to assess the soil moisture distribution in the Nikolovo village environs using the land, remote and computer calculations during the International Experiment GIMS-Bulgaria-2007 (Shutko et al. 2010; Shutko and Krapivin 2011). Spatial resolution is 250 m. *Arrows* show the places where the soil moisture was measured in situ. *Dashed lines* correspond to the traces of flight-laboratory (Soldatov 2011)

Guha et al. (2003) considered a series of experiments using the Electronically Scanned Thin Array Radiometer (ESTAR). The Southern Great Plains Experiment (SGPE-99), within the framework of which ESTAR-derived soil moisture fields were studied, took place in a region of the United States that is characterized by a relatively homogeneous geography, variable surface flux properties, and large seasonal variation in temperature and specific humidity. In addition to ESTAR, 4 different microwave radiometers operating in the L-(1-2 GHz), S-(2-4 GHz), and C-(4-8 GHz) bands were flown on aircraft. The instruments included the Polarimetric Scanning Radiometer (PSR/C) the Step Frequency Microwave Radiometer (Step-C), and the Airborne C-band Microwave Radiometer (ACMR) operating in the C-band and the Passive and Active L- and S- Band System (PALS) in the L- and S- bands. Passive microwave data were also collected by the Tropical Rainfall Measuring Mission (TRMM) satellite and the Special Sensor Microwave Imager (SSM/I) sensor.

Within these measurements, the problem of distinguishing between the water contents in soil and vegetation cover was solved. An empirically derived piecewise function was used to solve this problem. The vegetation water content (VWC) was calculated with the following correlation:

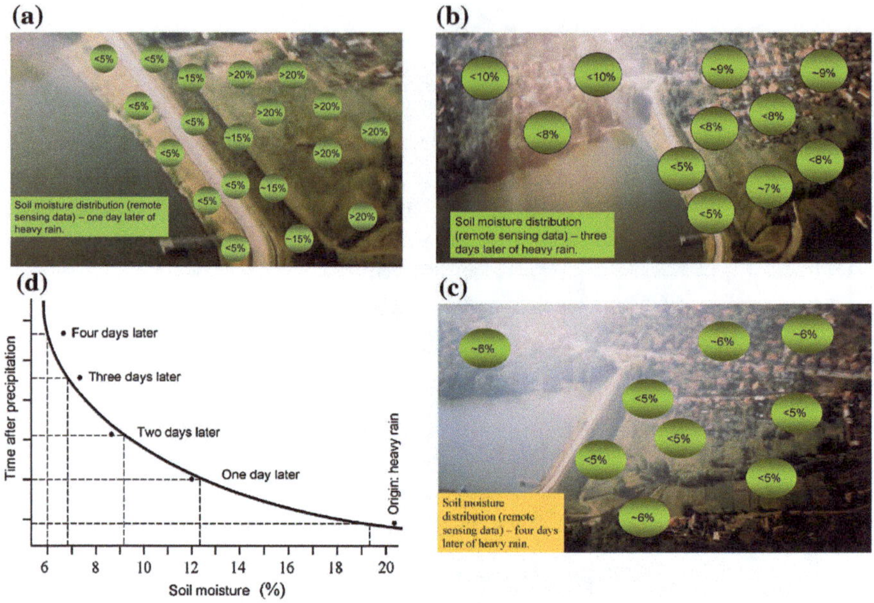

Fig. 2.15 Results of soil moisture retrieval in its dynamics in the Rusenski Lom region (Bulgaria) as results of combined use of remote sensing measurements and the RIMS-B application

$$VWC = \begin{cases} 1.9134(NDVI)^2 - 0.3215(NDVI) \text{ when } NDVI \leq 0.5; \\ 4.2857(NDVI)^2 - 1.5429 \text{ when } NDVI > 0.5; \end{cases}$$

where VWC is measured in kg m^{-2}.

Figure 2.15 demonstrates some results received by means of the GIMS using for remote sensing data processing. Detailed and geo-referenced maps identifying the locations of saturated and dry levees can be produced using microwave radiometric measurements from a light aircraft or helicopter, and integrated with GPS for positioning and orientation. The development of synergetic remote sensing technology for raised groundwater and seepage detection by the joint use of microwave and optical data along with GIS databases is an effective and most contemporary way of supporting risk assessment and facilitating disaster prevention and management.

2.5.3 Geoinformation System to Monitor Agriculture

The GIMS technology was set up to solve the wide range of problems arising in agriculture (Reutov and Shutko 1986; Shutko 1986; Chukhlantsev 1986; Chukhlantsev and Shutko 1988; Shutko et al. 2006a, b, 2010; Krapivin and Shutko 2012). The properties of the interaction of electromagnetic energy with the soil-plant formations have been of considerable interest to scientists and engineers who designed the Geo-Information Monitoring System of Agricultural Function (GIMSAF).

Microwave radiometric technique for measuring soil moisture characteristics have been developed and widely used in many regions by Shutko et al. (2010). This experience was used in the GIMSAF to find a procedure to determine the water content and moisture profiles in soil layers a meter thick and without vegetation cover by means of remote sensing. The procedure is based on the dependence of microwave emissivity of soil on its moisture. The procedure takes into account the variations of temperature, density and roughness of natural soils.

The GIMSAF performs:

- collection of information pertaining to the current parameters of the *soil-vegetation* system (moisture, shallow water table, soil salinity, biomass of the vegetation) over regions of various area;
- processing of the information above within the framework of a model simulating the biogeocenotic, physical and chemical processes in the *soil-vegetation* system with the aim of evaluating its current state and predicting its development and potential productivity;
- represents current and predicted information as a schematic maps with spatio-temporal detailing within a controlled structure;
- making technological decisions that are optimum from the point of view of providing the maximal biological productivity of agriculture vegetation;

The GIMSAF structure includes units of:

- collection of information;
- initial information processing and accumulation of data for simulating the agricultural vegetation productivity and water conditions;
- predicting the geosystem state;
- evaluation of the disparity between the measured and predicted trends in the geosystem processes;
- making decision pertaining to the measurement scheduling and geosystem controlling;
- service support in operation with input and output data.

The data collection unit is based on remotely sensed data and prior knowledge of the state of an object and meteorological parameters. Remote pickups switch on a microwave radiometric complex provided with computer programs that permit to obtain operational area information at the desired degree of detailing concerning the soil moisture content to a depth of 0–100 cm, the subsoil water level to a depth of 0–3 m, the salt concentration of the soil in a range of 3–90 g L^{-1}, the vegetation biomass, the degree of mineralization and contamination in reservoirs. Moisture measurement operating range is 0.0–0.6 g cm^{-3} with maximal absolute error equaled to the vegetation biomass less and more 200 t ha^{-1} to 0.05 and 0.05–0.08 g cm^{-3}, respectively.

The operation of the information processing units is based on a computer program package providing spatio-temporal interpolation of remote trace measurements and point surface measurements in addition to information flow stabilization in the event of the information collection system failing. The program package also realizes the description of the dynamics of the moisture and biological, physical, chemical and energetic processes in the interaction between the soil, vegetation, atmosphere and energy sources including factors of anthropogenic effects.

The adaptation of the information processing units to the actual conditions in the region studied is accomplished by introducing sets of coefficients and characteristic dependencies.

The trend-correcting and decision making units are based on evaluating the state of the object under consideration, predicting its development, and ascertaining the sensitivity of the system to potential production and technological measures put forward by experts.

Let us now list some of the features of the GIMSAF and some of its differences from similarly designated systems:

- use of remote radiophysical sensors pickups providing area, operative and quantitative information on the state of objects;
- ability to function at various levels of information support (remote, surface and a priory data set or part of them or without such data);
- fully open to further development regarding technological and software support and full compatibility with any sensor, information, consulting, and expert systems;
- algorithms that have the ability to adapt make the system self-regulating and, in particular, noise immune;
- compatible means of bringing about production and technological decisions.

Application results are expected to include:

- obtaining reliable operative, area and quantitative information on the parameters of the object and its general state with prediction;
- evaluation of the expediency and efficiency of anthropogenic effect on the agriculture and melioration objects;
- optimum productional and technological decisions made with the help of a simulational experiment and providing an increase in the object control efficiency.

Developing GIMSAF based on a joint use of remote measurements, in situ data, GIS information and mathematical modeling of temporal-spatial variations in soil-vegetation systems in local, regional and global scales needs the following sub-systems and phases of whole system accomplishment:

- Dual-channel multi-polarized SAR measurements from spacecraft with a spatial resolution of 5–10 m for obtaining basic/strategic/global information about the condition of corn crop fields.
- Selective dual-channel multi-polarized SAR measurements from aircraft with a spatial resolution of 1.5–5 m for obtaining synchronous with spacecraft information about the condition of agriculture vegetation in a regional scale.

- Selective four-channel, dual-channel and dual-polarized microwave radiometric measurements from aircraft for determination of corn crop germination and corn crop biomass in the areas with big and medium size fields; comparison of this data with SAR estimates of vegetation biomass and yield assessments; SAR estimates calibration and validation.
- Technical calibration of SAR and microwave radiometric signals with corresponding technical means.
- The real-time or near-real-time data processing in form of maps and interpretation in terms of crop germination at early phase of growth, crop biomass before harvesting, vegetation homogeneity inside the corn fields and finally of the expected yield.
- Additional information: Disaster management, revealing flooding, drought, hurricanes, others, observing annual dynamics inside and around the crop fields.

2.6 Microwave Radiometric Observations of Temperature Anomalies

The microwave range is an informative component of the multi-channel systems that monitor temperature anomalies on the Earth surface such as forest fires, peat bog fires, areas of geothermal activity, etc. Measurements of radiation contrasts make it possible to localize on a land surface the areas with anomalously high and low brightness temperature values due to the temperature variations with respect to a mean value of effective background radiation. Numerous experimental investigations shown that radiobrightness contrasts registered by the radiometers with wavelengths 0.8 and 3.4 cm in the areas of the burnings in the forests, peat bogs, and peateries can achieve values $\Delta T_j \in [200, 300]$K. The most important result is that the microwave range is sensitive to the surface temperature variations, and hence, the detection of possible burning hotbeds can be preventive. Specifically, microwave monitoring of the peat bogs allows to fix interior zones of the temperature anomalies when external signs are absent.

Wildfires and peat burning are extreme events for many regions of Russia, Australia, USA, Canada and other countries. Wildfires occur on every continent except Antarctica. They are a force that we cannot really control, and thus understanding, appreciating, and learning to live with wildfire is ultimately our wisest public policy. These events cause principal changes to forest ecosystem dynamics and influence on regional climate. Population considers these events as disasters bringing economical losses and victims. Therefore, earlier detection of these events with timely warnings is main task of environmental monitoring systems. Solution of tasks arising here is subject of numerous publications (Alvarado et al. 1998; Johnson and Miyanishi 2001; Wuerthner 2006; Johnson 1996; Fuller 1991; Perona and Brebbia 2010; Brebbia 2012).

Johnson (1996) assembles the relevant studies of fire intensity, rate of spread, fuel consumption, fire frequency, and fire weather in the North American boreal forest. The central thesis is that the North American boreal forest has at least four wildfire characteristics that are important in understanding the dynamics of its plant populations:

- the large size of the burns with respect to dispersal distances;
- the short recurrence time of fire with respect to tree lifespans;
- the high mortality of plants due to the predominance of crown fires; and
- a good germination surface due to the large area of the forest floor that is covered by ash.

Perona and Brebbia (2010) note that as in the past, future forest fire scenarios are impacted by climatic trends and changes in climatic extremes, as well as by anthropogenic pressure. It is to be expected that future trends, especially in the Mediterranean regions, will certainly lead to an increasing impact of human pressure on the natural environment. Increases in tourism and enlargement of urban residential areas invading the countryside are basic causes of anthropogenic changes in the environment. Forecasting the effects of both factors (climatic and anthropogenic) and separating their effects on forest fires frequencies may be particularly difficult, but is essential to improve our knowledge of forest fire occurrence probability and to better organize prevention and fighting activities. At the same time, estimation of the possible increase of fire risk over coming years is important, taking into account also the diverse fire prone environments present in the Mediterranean as well as many other areas (mountain slopes, coastal zones, large islands, etc.). Book written by Perona and Brebbia (2010) contains peer-reviewed papers presented at the Second International Conference on Modeling, Monitoring and Management of Forest Fires held in Kos, Greece, in 2010. The papers covered important topics in the field of prevention and fighting of forest fires, including: Computational Methods and Experiments; Air Quality and Health Risk Models; Detection, Monitoring and Response Systems; Decision Support Systems; Resource Optimization; Risk Assessment and Vulnerability Assessment.

Brebbia (2012) includes selected papers presented at the Third International Conference on Modeling, Monitoring and Management of Forest Fires, held in May 2012. The Conference is the third in a series of biennial conferences organized by the Wessex Institute of Technology on the topic of applying computer techniques to the difficult field of forest fire prevention, management, and mitigation. In order to prevent or reduce major loss of life and property, and damage to the environment, it is important to be able to analyze the behavior of forest fires. To do so requires the development of computer codes that can take into consideration a large number of different parameters.

Main problem of wildfire control is earlier detection of fire point. The problems arising here are solved by many authors (Artemov et al. 1974; Borodin et al. 1976, 1978; Borodin and Mitnik 1977; Solimini 1995; Bonafoni et al. 2011). One of efficient methods is microwave radiometry technology which gives possibility to detect temperature anomalies in forested areas. Numerous experimental investigations have shown that wavelengths 0.8 and 3.4 cm register radio-brightness contrasts $\Delta T_b \in [200, 300]$ °K. The most important fact is that microwave range is sensitive to the surface temperature change what allows the detection of heat sources on the peat bogs and in forests. It is indicated that microwave monitoring of peat bogs and forests can detect the fire point with warning.

In common case, passive radiation of the surface registered at the wavelengths λ_1 and λ_2 gives $\Delta T_b \approx \kappa \Delta z dT_s/dz$ where $\Delta z = z_2 - z_1$ $(z_2 > z_1)$ is the difference of the depth between sub-surface layers, $dT_s/dz = -Q/k_T$, Q is the heat flux intensity, k_T is the thermal-conductivity coefficient for layer Δz. SHF-radiometer sensitivity usually equals $dT_s/dz \approx 0.1$–1.0 °/m. It allows measurement of water content in vegetation biomass what gives the possibility to assess fire dangerousness using corresponding models. Effectiveness of microwave monitoring of forested areas depends on the model structure of vegetation environment. Ground fire is the most problematic object for its detection because of presence of radiation attenuation in the vegetation cover. Multi-channel sensing promotes the solution of many tasks, arising here, by means of cluster analysis application.

Use of three-layers model of the flame-smoke-tree crown gives $\alpha_P = (1-\exp\{-\kappa_1\})\exp(-\kappa_2)\exp(-\kappa_3)$ where α_P is the emissivity of fireplace of the lower forest fire, κ_I is the attenuation coefficient for flame radiation ($i = 1$), smoke ($i = 2$) and the whole of crown ($i = 3$). Attenuation coefficient of tree crown is function of tree types. It is determined that absorption in crown grows strongly with decrease of wavelengths. Figure 2.16 shows a dependence of radio-brightness contrasts from wavelengths.

Observations show that statistical characteristics of SHF-radiation for forested areas are changed with growth of fire readiness for layers of combustible forest materials. As an example, such dependence can be written in the form (Yakimov 1996):

$$T_b = \kappa(\lambda, w)T_s(1 - \exp\{-\alpha(\lambda, w)h\} + T_{bs}\exp\{-\alpha(\lambda, w)h\},$$

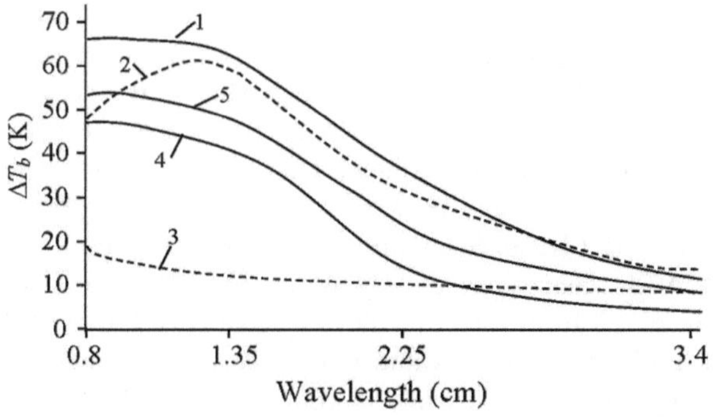

Fig. 2.16 Averaged spectr of radio-brightness contrasts for separate elements of the forest fire: *1*—smoked zone, *2*—fire front, *3*—burned-out zone, *4*—fire back, *5*—boundary between fire front and burned-out zone

where κ and α are the radiation and attenuation coefficients, respectively; w is the medium moisture content; h is the radiative layer depth; T_s is the soil temperature; T_{bs} is the soil brightness temperature. Yakimov (1996) proposed the following decision rule based on the calculation of mathematical average M^* and standard deviation σ^* for series of T_b for forested area using wavelength 2.25 cm. Value M^* is compared with threshold $d = 7.2514\sigma^* + 249.876$. If $M^* \geq d$ then considered forested area has average or higher forest fire dangerous. If $M^* < d$ then fire dangerous of considered area is low.

The forest fire scale depends on the series of factors such as water content in vegetations, topography, and wind direction. Water content in the plants is the most important factor regulating the speed of forest fire evolution. Maki et al. (2004) and Rahimzadeh-Bajgiran et el. (2012) studied relations between different indexes of forest fires and determined that the most perspective indexes are:

- vegetation dryness index (VDI);
- temperature vegetation dryness index (TVDI);
- water deficit index (WDI); and
- normalized difference water index (NDWI).

To assess the water content in vegetation it is needed to use indicators FMC (fuel moisture content) and EWT (equivalent water thickness):

$$\mathrm{FMC}/100\,\% = (\mathrm{FW} - \mathrm{DW})/\mathrm{FW},\;\; \mathrm{EWT}\left(\mathrm{g/cm^2}\right) = (\mathrm{FW} - \mathrm{DW})/\sigma,$$

where σ is the area of leaves in crown (cm^2), FW is the wet weight of vegetations (g) and DW is the dry weight of vegetation (g).

Indicators NDWI, WDI and VDI are calculated by means of the following formulae:

$$\mathrm{NDWI} = \left(T_{b,NIR} - T_{b,SNIR}\right)/\left(T_{b,NIR} + T_{b,SNIR}\right),$$
$$\mathrm{WDI} = 1 - \Delta_1/\Delta_2,\;\; \mathrm{VDI} = 1 - \Delta_3/\Delta_4,$$

where $T_{b,NIR}$ is the brightness temperature registered in near IR range, $T_{b,SNIR}$ is the brightness temperature registered in shot-wave IR range, Δ_1 is the distance on the phase plane (NDWI, NDVI) between conditions when water deficit (D) and excess (C) take place under low-density vegetation, Δ_2 is the distance between conditions of dense (A) and low-density (B) vegetation, Δ_3 and Δ_4 are distances between measured NDVI and NDWI and between sides of parallelogram with corners in points A, B, C and D (dense vegetation and moisture excess) on plane (NDWI, NDVI) under NDVI = const, respectively. Index NDWI is calculated basing on the measurements received from satellite SPOT in channels 430–470, 610–680, 780–890 and 1,580–1,750 nm. NDVI value is usually assessed with use of measurements received by means of sensors in red and near IR ranges from satellite Landsat.

Table 2.19 Values of coefficients in (2.6)

Vegetation	a_{EWT}	b_{EWT}	R_{EWT}	a_{FMC}	b_{FMC}	R_{FMC}
Nerium oleander var.*indicum*	−0.10697	0.62268	0.82	155.89	−34.087	0.06
Betula platyphylla var.*japonica*	−0.03136	0.3339	0.88	135.59	212.72	0.26
Liriodendron tulipifera	−0.00951	0.17521	0.87	68.312	539.65	0.56
Others	−0.02442	0.27963	0.79	119.42	204.69	0.28

Maki et al. (2004) proposed linear approximations:

$$\mathrm{EWT} = a_{\mathrm{EWT}} + b_{\mathrm{EWT}}\mathrm{NDWI}, \quad \mathrm{FMC} = a_{\mathrm{FMC}} + b_{\mathrm{FMC}}\mathrm{NDWI} \tag{2.6}$$

coefficients of which are shown in Table 2.19.

Experimental results of the fire dangerous assessment are represented in Figs. 2.17, 2.18, 2.19, 2.20 and 2.21. Figure 2.16 shows profile of radio-brightness temperature fixed by means of radiometer 0.8 cm under monitoring of burned peat piles, peat bog and forest fire. Increase of radio-brightness temperature for case of burned peat piles equals to 100–200 K. Increase of radio-brightness temperature for case of peat bog monitoring is 100–300 K. Finally, increase of radio-brightness temperature for forest fire case is 30–70 K. These variations exceed levels of phone radiation variations (5–20 K) and equipment noise (1–5 K).

Figures 2.18 and 2.19 show radio-thermal specters of peat bogs from which it is followed that emissivity coefficient of peat bog has anomalous decrease for wavelengths 10–20 cm and has considerable increase of absorption coefficient. These features follow from capillary-porous peat structure with abrupt differences of volumetric and dielectric characteristics of peat particles in space between of which there exist large quantity of coupled water and air.

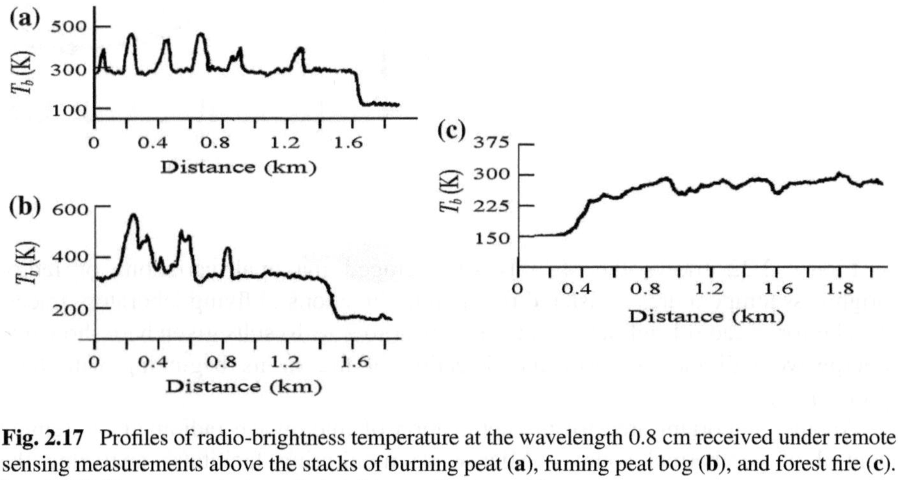

Fig. 2.17 Profiles of radio-brightness temperature at the wavelength 0.8 cm received under remote sensing measurements above the stacks of burning peat (**a**), fuming peat bog (**b**), and forest fire (**c**). Level 150 K corresponds to radio-brightness of water surface

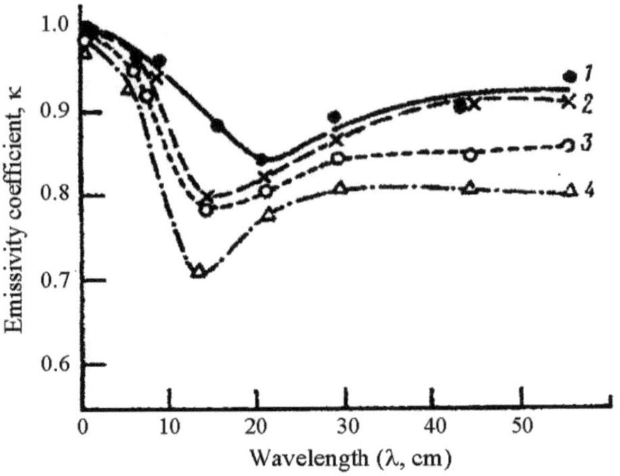

Fig. 2.18 Spectral characteristics of peat emissivity coefficient under weighed humidity: *1*—45 %; *2*—54 %; *3*—67 %; *4*—78 %

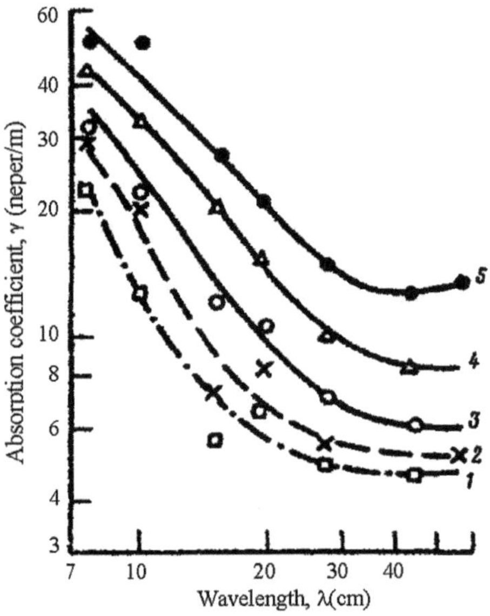

Fig. 2.19 Spectral characteristics of absorption coefficient for peat under weighed humidity: *1*—40 %; *2*—53 %; *3*—63 %; *4*—70 %; *5*—73 %

Figure 2.22 shows the stability of averaged maximal variations of radio-brightness temperatures registered in different directions of flying laboratory traces. Total error of radio-brightness contrasts is near to 2 K. Results given here show that microwave radiometers allow the detection of fire in its beginning with high probability.

Another important area for the application of microwave radiometry is remote control of geothermal sources. Microwave monitoring facilitates analyzing the

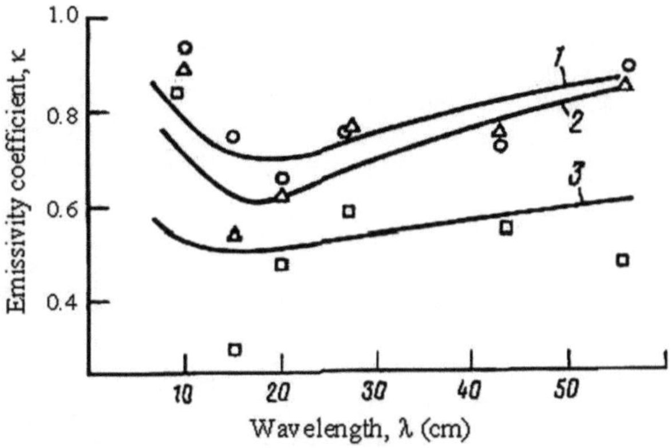

Fig. 2.20 Spectral dependence of peat emissivity coefficient under weighed moisture: *1*—45 % (marker of measurement data is *circle*); *2*—53 % (marker of measurement data is *triangle*); *3*—78 % (marker of measurement data is *square*)

Fig. 2.21 Spectral characteristics of slash fire (*chain line*), peat bog (*solid line*), and sphagnous bog (*dotted line*) in dynamics of drying and moistening: *1*—during rainfall; *2*—after heavy rainfall; and *3*—2 days later rainfall

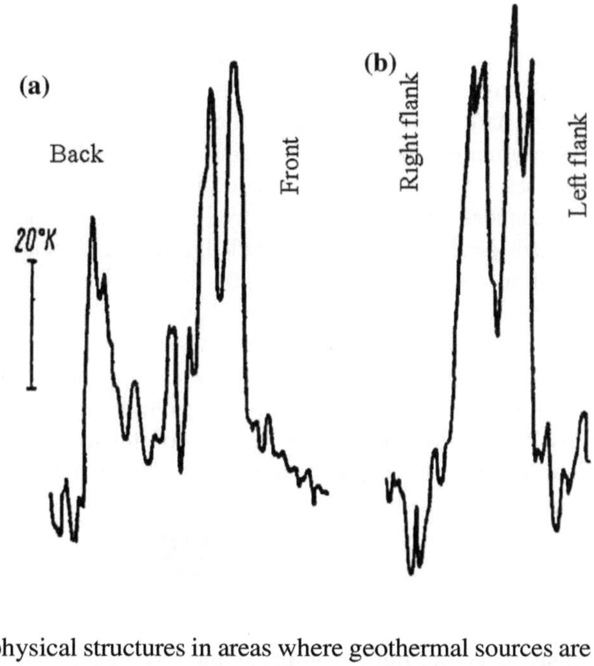

activity and classifying geophysical structures in areas where geothermal sources are active. In such areas the water content of rocks is insignificant and, hence, the depth of electromagnetic wave spread is increased to equal the inverse value of the absorption coefficient per unit of depth z. Multi-channel measurements based on use of the microwave range use and application of GIMS technology allow soils to be reliably classified and thermal situations mapped. Figure 2.23 demonstrates the map of Tolbachik volkano zone (Ostry Tolbachik—55° 57′N, 160° 25′E;

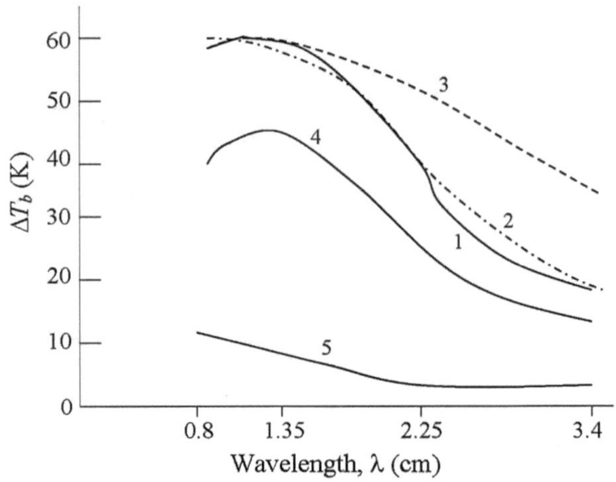

Fig. 2.22 Averaged spectral characteristics of different fire zones: *1*—front, *2*—left flank, *3*—right flank, *4*—back, and *5*—faded zone

Plosky Tolbachik—55° 49′N, 160° 24′E). This map was formed on the base of radiobrightness temperature spectres represented in the Fig. 2.24.

In connection with the change climate problem the forest fires monitoring gains specific role in the global ecoinformatics. The forest state and its influence on the greenhouse gases concentration is impossible for the reliable assessment of the present and future tendencies in the environment change. That is the reason a global geoinformation monitoring system has to have built into it a function to effect such an assessment. For such a function to be brought about there needs to be a model that can identify the forest state from radiothermal radiation. For such a model to be created there needs to be knowledge of the dependencies of forest radiothermal radiation on different wavelengths to reconstruct the moisture content of forest timber at various layers and its combustibility. Investigations of many authors shown that application of microwave sensing in the range 0.8 ÷ 20 cm allows to estimate the fire-dangerous for the waterlogged forests.

Multi-channel sensing allows using cluster analysis algorithms to distinguish and classify fire risk classes. The accuracy of this methodology depends on the detail of description by the model of the forest structure reflecting the canopy state and trees density. The most difficult to identify are forest edge fires. In this case the three-layers model of the *flame-smoke-canopy* system is effective:

$$\alpha_P = (1 - \exp\{-\kappa_1\})\exp(-\kappa_2)\exp(-\kappa_3),$$

where α_P is the emittive ability of the forest near-land fire hotbed, κ_i is the emission relaxation coefficient by the flame ($i = 1$), by the smoke ($i = 2$), and by the canopy ($i = 3$). The emission relaxation coefficients are functions of the trees type. It was established that absorption in the canopy increases with high speed as the wavelength is decreasing. Consequently, the receipt of emissions and the discovery of forest near-land fires become difficult when trees have a compact structure. Nevertheless, wavelengths 0.8–1.5 cm enable forest fires to be estimated with high

Fig. 2.23 A map of radiobrightness temperature T_j distribution in the area of Greet Tolbatchik Volcano on the Kamchatka by means of flying laboratory IL-18 at wavelength λ = 27 cm. This map was reconstructed by means of the GIMS-technology basing on the trass data

reliability in most cases. It is connected with that main contribution to the microwave emission by the fire hotbed gives the flame particles having the sizes up to 100–300 μm. The burning zone gives the radiobrightness temperature increase on the wavelength 0.8 cm equaled about 200 K. It exceeds the radiometers sensitivity limit. Therefore, microwave monitoring of forest fires in contradiction to optical methods does not depend on the weather conditions.

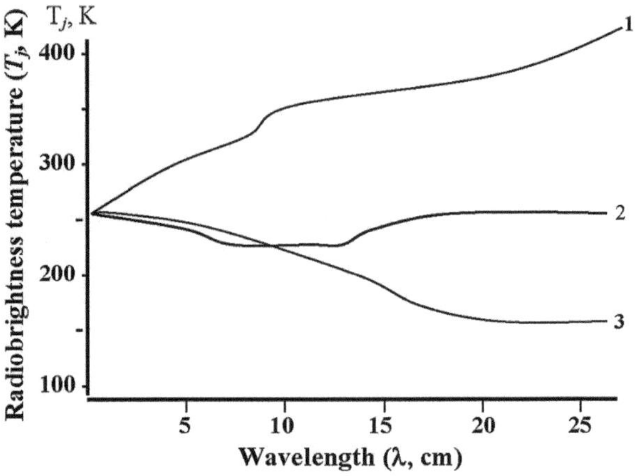

Fig. 2.24 The radiobrightness temperature spectra for Greet Tolbachik Volcano. Notations: *1*—fresh lava flow, *2*—isothermic cold lavas, *3*—slag fields with the presence of subsurface moisture (Borodin and Krapivin 1998)

The problem of temperature anomalies diagnosis demands the development of specific algorithms for the data processing. Joint use of microwave radiometers and model of *atmosphere-soil-vegetation* system allow to have the efficient technology to solve this problem. Microwave observations provide estimates of the vertical profiles of temperature in the ground as a result of the dependence of effectively emitted layer on wavelength. Two wavelengths are normally used in practice. Applied model describing the temperature profile is based on the assumption about the profile form. Thus, the effectiveness of microwave radiometry for thermal source detection and intensity measurement depends on just how adequate this assumption is. The minimum temperature gradient determined from the radiation data at two wavelengths is approximately 0.3–1 $°cm^{-1}$. Note that effectiveness of microwave range is much higher then infrared radiometry due to the ability of electromagnetic microwaves for penetration into a soil.

2.7 The Atmosphere Microwave Monitoring

The present state of industrial technologies, energy supply, and transportation worldwide coupled with the, growing number of massive industrial complexes are in continuous conflict with ecological standards for natural environment quality, leading to greater risks of accidents, anthropogenic and natural catastrophes, and natural disasters in many parts of the world. All of which highlights the necessity for uninterrupted control of the most vulnerable environment subsystem—the atmosphere. A role of microwave methods to solve this problem consists in the expansion of the atmosphere monitoring system functions basing on the optical control principles. Since the atmosphere is almost limpid for many radio waves its effectiveness under the atmosphere parameters measurement turn out to be insignificant especially in the local scales. Nevertheless, numerous experimental investigations using a flying laboratory and satellites show that microwave sensing of the atmosphere is not as good as optical methods on the short traces but is indispensable for atmospheric control over large areas. A choice of wave range for the atmosphere sensing is determined both by the special features of given task and by spatial scales.

A concept of assumed ecological loading on the atmosphere essentially depends on the level of spatial-temporal consideration of the proceeding processes in the interaction between natural environment and anthropogenic factors. Four levels corresponding to the ecological monitoring systems can be shown.

The global level (planetary) predicts the impact of anthropogenic influences on the atmosphere during intervals from several months to tens of years. Macrolevel (continental) considers the processes connected with the trans-boundary moving of the atmosphere pollutants between the countries or continents. A functioning of monitoring systems is regulated by legislative statements. Spatial and temporal scales spread thousands km and several months to year, respectively. Mesolevel (regional) considers the atmosphere pollution process over restricted territory with the selection of pollution sources and its chemical components. The spatial structure of considered processes is detailed by 100 m–100 km. Temporal scales is changed from several hours to units days. The monitoring systems of this significance level control the atmosphere state in the large industrial cities. Microlevel (local) considers the processes of atmosphere pollution over the areas having linear sizes from several to hundreds meters with the timescales from units to 10 min. The monitoring systems of this level registers single phenomenon with detail study of spatial distributions of atmosphere parameters in the local area. Such systems usually are used to control the atmosphere contamination in the zone of dangerous industries.

The atmosphere pollution sources are divided on stationary and unstationary one. Each source is characterized by the quantity of emitted contaminants with the specific properties per unit time, by the temperature of emitted aerial mixture, by the altitude above the earth surface, by the velocity of emitted mixture, and by the geographical coordinates. Under this the source types are distinguished by the geometric form (point, linera, areal) and by the emission regime (continuous action, periodical, emitting a volley). Moreover, atmospheric pollution sources can be classified by their hazard levels. In the first place large chemical industries, ground and underground oil- and gas- reservoirs, gas- and product- pipes, industrial reservoirs of liquid ammonia, chlorine, concentrated sulphuric acid, and also industries connected with the emission of sulphur gas, nitrogen oxides, and hydrogen sulfide are selected (Straub 1989; Liou et al. 2010; Krapivin and Nazaryan 1997).

Classificational characteristics of scaleness and of physical-chemical state of atmosphere pollution sources simplify the monitoring system design. Although pollution source categories can be precisely classified, adequate monitoring systems often do not exist. Nevertheless, in each concrete case such correlations are inevitably established when an atmospheric pollutant dynamics model is created.

Remote sensing methods of the atmosphere are based on the study of electromagnetic and acoustic waves propagation. In the laser sensing optical radiation scattering on the aerosol and atmosphere gases molecules is used. In the radiolocation sensing the effects of turbulent pulsation is taken into account. Laser sensing is certainly effective at determining the chemical composition of the atmosphere, wind speed, temperature, cloud parameters, and dust formations (Klyuev 2000). The diagnostics of water content in the clouds, compound of atmosphere gases and aerosols by its thermal radiation in the microwave range is based on the spectral measurements (Strelkov and Soldatova 1989).

Many authors use the millimeter wavelength band to diagnose atmospheric gases. Strelkov (1995, 1996) implemented the estimates of atmospheric track optical depths between two satellites on admixture gases CO and N_2O resonance absorption frequency in millimetre and submillimetre bands with taking into account the data on their atmospheric concentrations. It was shown that optical depths are completely defined by admixture gases from the sighting distances $\approx$15 km and can be measured, which indicates the use of the active remote sensing method of the atmosphere for CO and N_2O contents definition. Intrusion of these gases to the atmosphere is caused by anthropogenic factors to a considerable extent. But it is problem for many biogeochemical processes such as ozone cycle. That is why possibility of the measurement of these gases concentration with satellite systems is significant achievement.

Strelkov (1996) developed new method for measuring the concentration of water vapor in the mesopause by active sensing at resonance wavelengths of the rotational spectrum. He showed that for two-frequency sensing, simultaneous measurement of the height profiles of the concentration of water vapor and of the temperature within the mesopause is possible. This method is based on the fact that within the mesopause water vapor completely determines the optical depth of the sensing path in the centres of its spectral lines with wavelengths $\lambda \geq 0.5$ mm. Moreover the absorption coefficient in the centre of the line with resonance frequency f_i is described by the expression

$$\alpha_i = \frac{8\pi^2 N}{3hcG(T)} f_i |\mu_i| \left\{ \exp\left[-\frac{hcf_{i1}}{kT} \right] - \exp\left[\frac{hcf_{i2}}{kT} \right] \right\} \frac{c}{\Delta f_i(P, T)},$$

where N is the concentration of water vapor molecules, h is the Planck's constant (=6.55×10^{-34} J), c is the speed of light (=3×10^8 m s^{-1}), k is the Boltzmann's constant (=1.3×10^{-23} J K^{-1}), $G(T)$ is the rotational statistical sum, f_{i1} and f_{i2} are the lower and upper energy levels of transition with frequency $f_i = f_{i2} - f_{i1}$, μ_i is the dipole moment matrix element for transition with frequency f_i, Δf_i is the half-width of the line which depends on the pressure P and the air temperature T, z is the height.

Scientific and technical achievements in the satellite monitoring of atmosphere gases are important in connection with the greenhouse effect and ozone problems (Kondratyev and Varotsos 2000; Zuev 2000; Krapivin and Nitu 2011). The active and passive sensing of some atmospheric gases from satellite board with millimetre and submillimetre wavelengths allow to measure the content of O_2, O_3, N_2O, H_2O, CO, BrO, ClO and HCl. For example, the possibility of active sensing of the added gas is equivalent to the possibility of reliable measurement of the inter satellites trace optical depth, τ, stipulated by the gas:

$$\tau(f, z_m) = 2\int_{z_m}^{H} \frac{\gamma(t, z)(R + z)dz}{\sqrt{(R + z)^2 - (R + z_m)^2}},$$

where γ is the gas absorption coefficient, f is the sensing frequency, z is the height, z_m is the minimal distance of trace from Earth's surface, R is the Earth's radius, H is the height of satellites orbits. A possibility of passive sensing of ClO, for instance, is defined by the expression for the limb brightness temperature:

$$T_j(f, z_m) = 2\int_{z_m}^{H} \frac{T(z)\gamma(f, z)(R + z)}{\sqrt{(R + z)^2 - (R + z_m)^2}} \times [Q_1(z) + Q_2(z)]dz,$$

where

$$Q_1(z) = 0.5\exp\left(-\int_{z}^{H} \frac{\gamma(f, \xi)(R + \xi)d\xi}{\sqrt{(R + \xi)^2 - (R + z_m)^2}}\right),$$

$$Q_2(z) = 0.5\exp\left(-\tau(f, z_m) + \int_{z}^{H} \frac{\gamma(f, \xi)d\xi}{\sqrt{(R + \xi)^2 - (R + z_m)^2}}\right)$$

The radio-trauslusense method with using two satellites at frequencies of strong gas absorption lines proves to be effective approach to the ozone layer characteristic measuring (Strelkov 1995; Yakovlev 2001; Yakovlev et al. 2009). Enhancement of this method is given by Yakovlev et al. (2009). Radio-eclipsing method of the Earth sensing is proposed. Principal scheme of this method is characterized in Fig. 2.25.

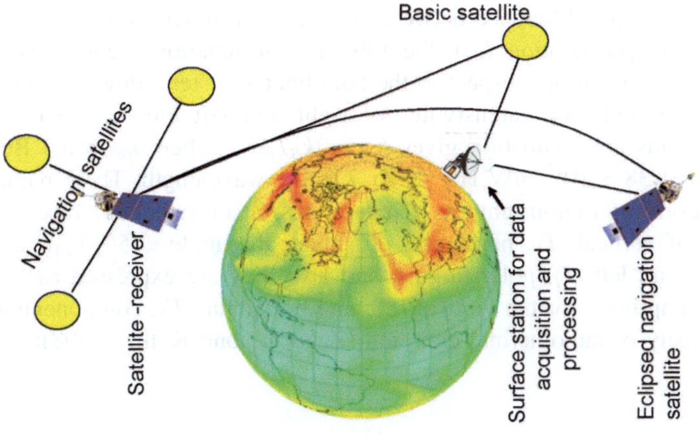

Fig. 2.25 Schematic diagram of global monitoring system based on the radio translusense method

This method is based on the effect of variations in the intensity of electromagnetic field, as well as in the phase and frequency of radiowaves under the transmission of atmosphere and ionosphere. Eclipse monitoring of the atmosphere and ionosphere is dependent on the inverse problem being solved so that it can define the vertical profiles of electron concentrations, refraction coefficients and atmospheric temperatures. Subsequent development of this method would involve creation of algorithms that might make it possible to identify atmospheric and ionospheric characteristics based on regular measurements.

According to the estimations by many authors the greatest role of microwave radiometry is displayed under the precipitation and clouds monitoring. For the first time these possibilities were demonstrated by the experiments with satellite "Cosmos-243" in 1968, and then many space microwave measurements were realized by means of SMMR (Scanning Multichannel Microwave Radiometer), SSM/I, TRMM, and multichannel on-board systems placed on the cosmic station "Mir" what was sunk in 2001.

Local measurements of water vapor concentration in the atmosphere reliable are registered with wavelength 1.35 cm. Under this the dependence of atmosphere radiobrightness temperature on the integrated content of water vapor can be well described by a linear function with a slope equal to ≈ 19 K g^{-1} cm^{-1}.

The application of radio waves from the centimeter and millimeter ranges to diagnose atmospheric properties has been pretty successful because of weak, unlike the optical and IR ranges, interaction with cloud particles, dust, and atmospheric gas molecules. Radio wave interaction with rain drops, on the contrary, is resonant and expressed by the intensive absorption and scattering of wave energy. Therefore, microwave devices of the active and passive sensing at present time are unique means giving a possibility to receive from space the data about the precipitation parameters in global scale.

Microwave observations of precipitation are based on the measurements and calculations of four Stock; s parameters (Zagorin 1999; Zagorin and Kutuza 1998; Kutuza et al. 1998, 2000): $S = \{S_1, S_2, S_3, S_4)$, where S_1 is the total intensivity of radiation, S_2 is the difference between the radiation intensivities of horizontal and vertical signal polarization, S_3 is the difference of radiation intensivities with linear orthogonal polarizations respect to the coordinates system turned by 45°, S_4 is the difference of radiation intensivities of right and left circular polarization. The Reileigh-Jeans approximation gives $S_1 = 2k_B T_j \lambda^{-2}$, where k_B is the Boltzmann's constant (=1.38 × 10^{20} mW $Hz^{-1}K^{-1}$), λ is the wavelength. These parameters are the functions of radiobrightness temperatures T_j of radiation, receiving by the channels of vertical, T_V, horizontal, T_h, under the angle ±45°, $T_{\pm 45}$, and circular (right, T_r, and left, T_l) polarizations, and also they are expressed as functions of complex amplitude for the vertical, E_V, and horizontal, E_h, components of electric field intensity of radiothermal emission (Zagorin abnd Kutuza 1998):

$$S = \begin{pmatrix} S_1 \\ S_2 \\ S_3 \\ S_4 \end{pmatrix} = \begin{pmatrix} T_V + T_h \\ T_V - T_h \\ T_{+45} - T_{-45} \\ T_l - T_r \end{pmatrix} = \frac{\lambda^2}{2k_g} \begin{pmatrix} |E_V|^2 + |E_h|^2 \\ |E_V|^2 - |E_h|^2 \\ 2\mathrm{Re}(E_V E_h^*) \\ 2\mathrm{Im}(E_V E_h^*) \end{pmatrix}$$

The Stock's parameters, measured directly in field experiments, together with other measurements and using cluster analysis allow clouds and rains types to be realized. Figures 2.26 and 2.27 give data about the correlation of the Stock's parameters with the rain intensity calculated by Zagorin (1999). The maximum

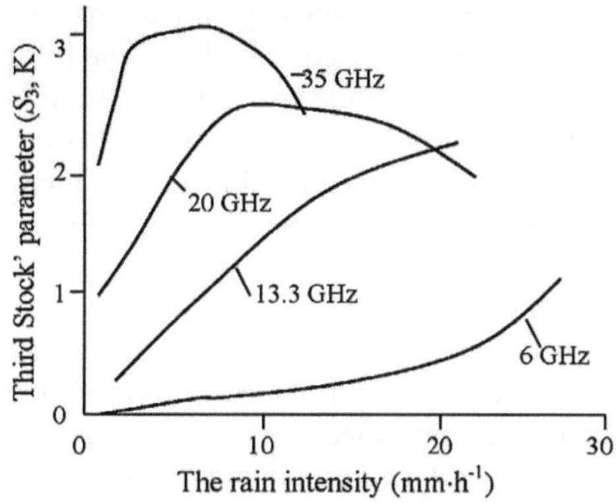

Fig. 2.26 A dependence of third Stock's parameter from the rainfall intensity. A thickness of the rain layer is 3 km. A descending microwave emission with the angle of 50° to the zenith is considered. The drops slope angle is equaled to 5° (Zagorin 1999)

Fig. 2.27 A spectral dependence of fourth Stock's parameter for the microwave Sun radiation dispersed by the rainfall drops in the layer of 4 km. (Zagorin 1999; Zagorin and Kutuza 1998)

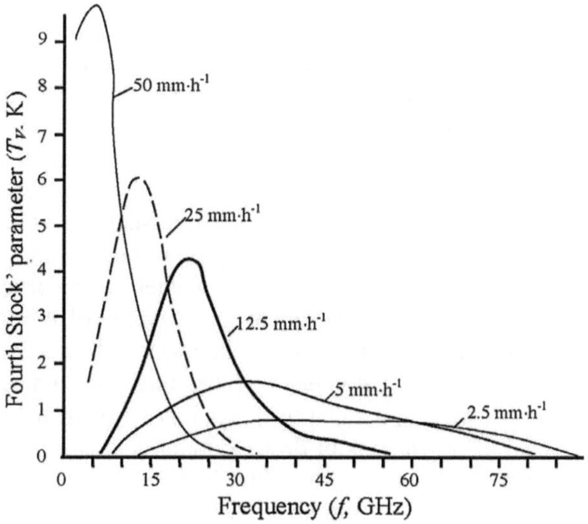

spectral dependencies of the second, third, and fourth Stock's parameters on microwave Sun radiation dispersed as a result of rain intensity increasing are displaced to the low-frequency region (up to ≈6 GHz). So, for the rain layer having the thickness in 4 km under the Sun observation under corner in 20° regarding the horizont the maximal values of S_2, S_3 and S_4 are achieved under the rain intensity $r = 12.5$ mm h^{-1} and for the frequency ≈13.5 GHz. They are equaled: $S_2 = 220$ K, $S_3 = 44.6$ K and $S_4 = 5.95$ K. When $r = 50$ mm h^{-1} and frequency ≈6 GHz these parameters are the following: $S_2 = 483$ K, $S_3 = 101.5$ K, and $S_4 = 28.5$ K.

Polarization radiometry of precipitation is based on two effects: the nonspherical form of rain drops and the existence of a distinct direction in which the symmetrical axes of falling drops are oriented. This means that the emission and absorption of radio waves having vertical and horizontal polarizations that differ in directions from the orientation of symmetrical axes of rain drops can be distinguished. For this reason the azimuth simmetry of radiothermal field of the rain emission is absent. And this means that microwave emission of the rain is partly polarized. Therefore, the task of precipitation diagnostics is reduced to finding the degree of dependence of the emission polarization parameters (polarization degree, slop angle of polarization plane, polarization ellipse form, etc.) on characteristics of the state of the environment in which rain is dispersing.

The first polarization radiometric measurements of the rain descending microwave emission were realized by Kutuza (1977), and then were motivated and developed by the experiments and theoretically in the subsequent investigations. The results of measurements of T_j in general and difference channels of microwave radiometer at $\lambda = 2.25$ cm (13.3 GHz) given the estimations of degree of linear polarization of rain radio-emission. When intensity of steady downpour was changed between 0.5 and 2.5 mm h^{-1} it was equaled to 5.5 %. The rains having the convective type with the intensity more 5 mm h^{-1} had the degree of linear polarization which was increased until 8.5 %. These estimations guarantee the distinct delimitation of the rain zones.

The methodology set up to estimate the parameters involved in the distribution of rain drops by size developed by Gasiewski and Kunkee (1994) is based on the measurements of two Stock's parameters, S_1 and S_2. The gist of methodology consists in solving an inverse task under the preposition that non-spherical drops have an exponential distribution with two unknown parameters determined by the data about S_1 and S_2. The contribution of values of third and fourth Stock's parameters to the solution of rain diagnostics task is determined by the expansion of estimated parameters. So, S_3 is proportional to the differential relaxation, Δ_1, of radiowave. The fourth Stock's parameter is proportional to the product of Δ_1 and differential phase displacement, Δ_2. A value of $\Delta = \Delta_1 + i\Delta_2$ is called as the rain anisotropy parameter and it can be measured directly. Theoretical dependencies of Δ from different parameters of the rain are studied by Gasiewski and Kunkee (1994). It was shown that the spreading in the distribution by the slop corner to the vertical of projections of drops symmetry axis in the plane which is orthogonal to the direction of wave spreading exercise largest influence on the precision of Δ estimation. Statistical models describing the orientation of drops allow the average value and dispersion to be connected with the Gaussian distribution law of drops according to the orientations of wind field parameters (Zagorin 1999). It was discovered that when dispersion of drops orientation less 30 % theoretical and experimental estimations of Δ are similar in the value. It allows to use the developed models of rain in practice.

The atmosphere microwave control supposes the existence of radiation source in the microwave range. Such a source can be Sun, the radio-emission of which in its calm state (when the flashes are absent) within the frequency region $\lambda \in [2,100]$ GHz can be characterized by $T_j \in [5 \times 10^3, 1.5 \times 10^4]$ K. Non-polarized microwave radiation of Sun is relaxed and polarized by the rain, and average relation $S_2/S_1 \approx 0.0528$. The radiometers dislocation and choice of its ranges for such microwave source is actual task of present radiophysics. This task is combined with the formation of T_j level on the radiometer output as the function of fields composition on its antenna. This composition depends on the character of litter surface the repeated reflections of radiowaves from which are to take into account under the solution of inverse task arising in the atmosphere radiometry. In the conditions when the atmosphere-litter surface system has a symmetry by the azimuth, and when emitting and scattering atmosphere particles are oriented chatically, and the litter surface is mirrory- or diffusely- reflecting, then $S_1 \neq 0$, $S_2 \neq 0$, $S_3 = S_4 = 0$. When the symmetry by the azimuth is absent, and when atmosphere particles have the orientation direction differing from vertical, and when spreading surface roughness is characterized by the asymmetrical distribution of slopes we receive: $S_1 \neq 0$, $S_2 \neq 0$, $S_3 \neq 0$, $S_4 \neq 0$. The fourth Stock's parameter S_4 which characterizes the polarization ellipticity, becomes unequal to zero when the radiothermal emission undergoes no less two-fold scattering into the hydrometers volume or on the surface roughness (Zagorin 1999).

An important stage in precipitation monitoring is clearly optimizing registered radiation frequencies. This was partly done by Sasaki et al. (1989), who based on measurements showed that the frequencies near 1 GHz are the most informative. This result is corroborated by the curves given in Figs. 2.28 and 2.29.

The perspectivism of the radiophysical methods used for the design of an atmospheric control system is substantiated by the many theoretical and empirical investigations. The basic problem consists in forming a knowledge base to maintain correlations between atmospheric characteristics (properties) and the behavior of electromagnetic waves inside the atmosphere. The basis on which atmospheric process models have been built reflecting the combinations of spatiotemporal scales is again an important element in the study of these correlations.

As it follows from the investigations by Yakovlev (2001) effective methodics of the atmosphere monitoring system design is connected with the radio occultation studies of the atmosphere and ionosphere. The setup necessary to bring about such measurements is based on placing the transmitter on a satellite orbiting Earth and placing the receiver on a geostationary satellite. As result such structure secures continuous regime of atmosphere vertical transection with the obtaining of its parameters distribution by the altitude. The task is reduced to identifying the values

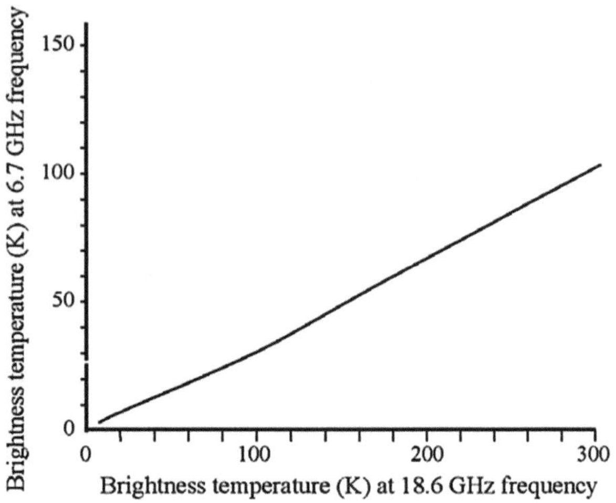

Fig. 2.28 The state image of informational space of brightness temperatures in the section of the most representative frequencies

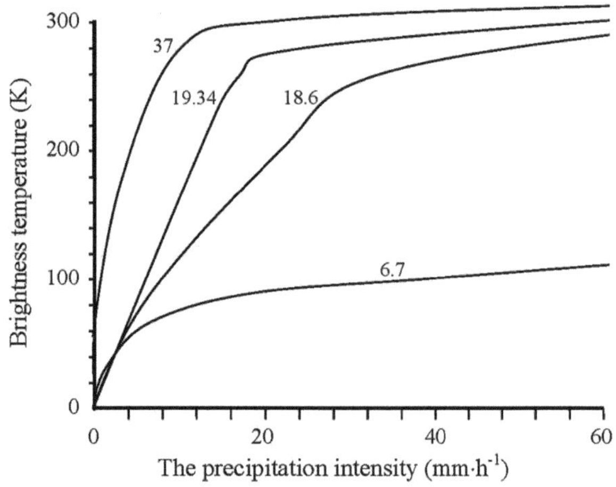

Fig. 2.29 The comparative rainfall rate dependence of the sky brightness temperature for different frequencies (they are shown on the curves, GHz) (Sasaki et al. 1989)

of these parameters using data about signal absorption and fluctuation as a result of propagating along the changing tracks between satellites.

The influence of the atmosphere and ionosphere on the parameters of refraction, fluctuation, relaxation, absorption, decay, and change in signal frequency has been studied in detail. Depending on the orbit position of the satellite this influence makes it possible to receive data about the atmosphere and ionosphere in concrete regions of the planet. A total spatial image for part or of the atmosphere can be had by data processing using GIMS technology.

It is possible to distinguish between the direct and inverse tasks of radio-translusense. The direct task arises when the atmosphere model is used to determine the above mentioned changes in electromagnetic waves for a given satellite trajectory. The inverse task consists in the determination of the atmosphere and ionosphere parameters through experimental data describing the signal variations.

The method of atmosphere radio-translusense can be used when the microwave emission receiver is sited on the Earth's surface. Using the Doppler effect on wavelength 1.35 cm (22.235 GHz) in this case allows the vertical distribution of water vapor in the uppermost atmosphere to be received with acceptable precision.

Atmospheric microwave sensing can be expanded to consider the possibility of estimating a wider set of its parameters. To do so would mean solving a complex mathematical task connected with the prognosis of the microwaves relaxation according to sensing routes in the atmosphere. Millimetre range causes special interest in this respect. According to the investigations by Meriakri (1992) and Strelkov (1995) many gases have discrete absorption spectre in the waves of millimetre and submillimetre ranges. This allows the average concentration of the gas component along the sensing route to be calculated based on estimating electromagnetic wave absorption. An example of such a calculation is given in Table 2.20.

A special feature of the millimeter range is the possibility to form narrowly directed radio-bundles because of the relatively small apertures of the emitters. The task of atmosphere parameter estimation can be transformed to registering variations in signal relaxation on the sensing route between the receiver and transmitter followed by solving a suitable integral equation that takes into account the refraction of electromagnetic waves in the troposphere, cross-diffraction of signal bundles, and possible variations in the vertical distribution of refraction indicators along the route. Monitoring systems of low atmosphere also call for the influence of land covers on the propagation of millimeter waves to be taken into consideration.

Millimetre and submillimetre ranges under the using of active and passive sensing regimes ensure the reliable estimation of content in the atmosphere of certain chlorine compounds, carbon oxide, nitrogen dioxide, ozone, and water vapors, as well allow to determine the wind spead and temperature (Strelkov 1995, 1996). The prospect of these ranges bringing about bi-static radio-translusense of the atmosphere is down to the moderate price of field measurements and existing theoretical results concerning the methodologies used to reconstruct atmospheric spatial distribution parameters along observational routes.

Table 2.20 Calculation of the gases concentrations basing on the measurements of signal relaxation at the frequence ν_a on the atmosphere trace of 10 km (Meriakri 1992)

Gas	Gas concentration (mg m^{-3})	ν_α (MHz)	Atmosphere relaxation (db km^{-1})
CO	1,500	0.3	115.271
	330	6	235.789
N_2O	1,600	0.2	100.492
	240	6	351.666
SO_2	2,300	0.2	104.029
	560	0.4	135.696
H_2S	9	1.8	168.763

The application of microwave range for the atmosphere diagnostics can not be effective without the creation of measurement data processing systems. Such systems are based on simulation models that make reconstructing atmospheric characteristics possible by making good use of multichannel sensing data. What is more, the obligatory presence of the optical and IR ranges between the channels of such systems results in an essential gain in monitoring system informativeness.

In framework of the GIMS technology application, a simulation system for the atmosphere pollution physics (SSAPP) was designed as the experimental system for the control and identification of sources polluting the atmosphere basin (Krapivin et al. 1996, 1997a, b; Bondur et al. 2009). The SSAPP is intended to reconstruct multi-component non-stationary pollutant fields and identify local pollutant sources. The SSAPP uses the procedure of adaptive synthesis of simulation model describing the three-components of pollution field as basic algorithm. This procedure consists in dynamic development of the simulation model as a result of estimating the divergence between model output and data registered by means of measuring devices in various ranges of electromagnetic spectrum.

Putting current *a priory* information about the coordinates of measuring devices, pollutant sources, and parameters into a simulation model is brought about through coordination with a topographical map of the territory being controlled. The model optimization and ranges set control are realized with consideration of the presence database and available technical resources. The system is adapted automatically to the spatial scales of territory the pollution of which is simulated. The set of tasks solved by the SSAPP includes:

- calculation of influence area for each source included in the system database;
- determination of sensitivity area for each measuring device;
- reconstruction of spatial distribution of pollutants;
- identification of the causes behind dangerous pollution zones arizing;
- search of unknown pollution sources.

The SSAPP foundation is set of models describing the atmosphere dynamics and contaminants spreading over the restricted territory up to whole planet. The meso and micro levels can be imitated by a set of Gaussian models, the choice of which depends on the parameters provided by the monitoring system's database. Global- and macro-levels of the atmospheric contaminants spreading are described with the models of Euler and Lagrange types.

2.8 Microwave Radiometry in Remote Monitoring of the Ocean

The formation of radiothermal field emission of the ocean in the microwave range depends on the water temperature and salinity, surface roughness, concentration of chlorophyll and other components of oceanic environment. Remote registration of this field demands to take into account also such factors as evaporation, precipitation, foam-formation, cloudiness, intensity variations of direct and scattered solar radiation. That is why the design of microwave remote monitoring systems on the base of registration of ocean emission in microwave range comes to the solution of set of complex mathematical tasks. Parameters reflecting the state of the *atmosphere-ocean* system have an important influence on how the spatial structure of the radiothermal field of this system is formed. So, outside zones of rain, snow and hail clouds when for wavelength $\lambda \geq 0.8$ cm total absorption by the atmosphere is insignificant, phenomenological theory of radiation transmission is applied to determine the radiobrightness temperatures, T_j, for the *atmosphere-ocean* system. In zones of drop clouds presence, relative contribution of scattering processes for the emission on wavelength $\lambda \geq 0.8$ cm to the formation of field $\{T_j\}$ consists from 5 to 50 %. This means that in these zones the scattering effects are to be taken into account under the microwave radiometry data processing received from the ocean surface.

In common case, when the registration of T_j is realized by satellite-based system from the height H under the observation angle θ respect to the local vertical for wavelength λ the following functional correlation of radiobrightness temperature with thermodynamic and physical characteristics of *atmosphere-ocean* system is valid (Savorskij 1992):

$$T_j(\lambda,\theta) = Q_1 + \cos^{-1}\theta \int_0^H Q_2(u)du + Q_3 \int_0^\infty Q_4(u)du, \tag{2.7}$$

where

$$\begin{aligned} Q_1 &= \kappa_\lambda T_{SO}\exp[-\tau_\lambda(0,H)/\cos\theta], \\ Q_2 &= T(\mathrm{u})\gamma_\lambda(u)\exp[-\tau_\lambda(u,H)\cos^{-1}\theta], \\ Q_3 &= (1-\kappa_\lambda)\exp[-\tau_\lambda(0,H)\cos^{-1}\theta]\cos^{-1}\theta, \\ Q_4 &= T(\mathrm{u})\gamma_\lambda(u)\exp[-\tau_\lambda(0,u)\cos^{-1}\theta], \end{aligned}$$

$\gamma_\lambda(u)$ is the total relaxation coefficient by the atmosphere on the altitude u, T(u) is the thermodynamic temperature on the altitude u, κ_λ is the surface emittive ability, T_{SO} is the ocean surface temperature, $\tau_\lambda(u_1,u_2)$ is the optical thickness of atmosphere layer situated between the levels u_1 and u_2 on the wavelength λ:

$$\tau_\lambda(u_1,u_2) = \int_{u_1}^{u_2} \gamma_\lambda(z)dz.$$

Formula (2.7) is basic for the calculation of the ocean parameters using the measurements $\{T_j\}$. The variety of modifications and simplifications of expression (2.4) is determined by the radiation models set of the *atmosphere-ocean* system, where the dependencies κ_λ and γ_λ from the parameters of this system have theoretical or empirical character. The coefficients of water surface emission on the horizontal, κ_{λ}^{h}, and vertical, κ_{λ}^{V}, polarizations are calculated with the Frenel formulae:

$$\kappa_{\lambda}^{h} = 1 - \left| \frac{\varepsilon_{\lambda} \cos\theta - (\varepsilon_{\lambda} - \sin^2\theta)^{1/2}}{\varepsilon_{\lambda} \cos\theta + (\varepsilon_{\lambda} - \sin^2\theta)^{1/2}} \right|^2, \tag{2.8}$$

$$\kappa_{\lambda}^{V} = 1 - \left| \frac{\cos\theta - (\varepsilon_{\lambda} - \sin^2\theta)^{1/2}}{\cos\theta + (\varepsilon_{\lambda} - \sin^2\theta)^{1/2}} \right|^2, \tag{2.9}$$

where ε_λ is the water dielectric permittivity at the wavelength λ being the function of salinity, temperature and other parameters of upper ocean layer; $\varepsilon_{\lambda} = \varepsilon'_{\lambda} - i\varepsilon''_{\lambda}(\varepsilon'_{\lambda} = 5.5 + a_1; \varepsilon''_{\lambda} = \lambda_S \lambda^{-1} a_1)$,

$$a_1 = [\varepsilon_S - 5.5]/[1 + (\lambda_S/\lambda)^2], \varepsilon_S = 0.0081t^2 - 0.40885t + 88.2, t = T_{SO} - 273,$$
$$\lambda_S = 1.466\exp\{-0.0634t\} + 0.000136t^2 - 0.0272t + 1.8735.$$

A correction to the value ε_λ with consideration of water salinity variations S can take into account by the approximate formula: $\varepsilon_S(S) = \varepsilon_S(0) - 17.2S$. From expressions (2.8) and (2.9) under vertical observation ($\theta° = 0°$) we receive: $\kappa_{\lambda}^{h} = \kappa_{\lambda}^{v} = \kappa_{\lambda}^{o} = P_{\kappa}$, where

$$P_\kappa = 4|\varepsilon_\lambda|^{1/2}\cos(\delta_\lambda/2)]/[|\varepsilon_\lambda| + 2|\varepsilon_\lambda|^{1/2}\cos(\delta_\lambda/2) + 1,$$

angle δ_λ is determined by the expression $tg\delta_{\lambda} = \varepsilon''_{\lambda}/\varepsilon'_{\lambda}$.

Andreev et al. (1984) analyzed Shepovalov's invariant $\left(\kappa_{\lambda}^{h}\right)^2/\left(2\kappa_{\lambda}^{h} - \kappa_{\lambda}^{v}\right) = 1$ and proposed a constructive algorithm to calculate the surface temperature T_S with the measurements of $T_{j,h}$ and $T_{j,V}$ under the observation angle $\theta = 45°$: $\grave{O}_S = (\grave{O}_{j,h})^2/(2T_{j,h} - T_{j,v})$. This algorithm has the precision less ±3 K. It is valid for the calculation both T_{SO} and land thermodynamic temperature. The T_j practically depends on T_{SO} linearly with the slope from 0.35 to 0.5. An increase of salinity from 0 to 40 ‰ is provided with the increase of T_j under $\lambda < 3$ cm by values no exceeding 3 K and with essential decrease of T_j under $\lambda \in [8, 10]$ cm by the values up to 40 K. The dependence slope of this decreasing as the function of S is function of T_{SO}.

Knowledge of the radiation dependencies of water surface properties on the total spectrum of water surface parameters is the essential basis of a radiometric control system of the ocean surface.

In the case of water object with great conductivity (sea water) for the decimetre range the following correlation is valid: $\kappa \cong 0.5(f/\sigma)^{1/2}$, where f is the radiowave frequency, σ is the water conductivity. This expression allows to estimate the water conductivity if the water thermal emission is measured. If the substance dissolved in the water and its conductivity are known the dissolved substance concentration is calculated by the using of thermal radioemission. Hence the microwave radiometry can be used for the monitoring of reservoirs pollution.

The emissive and reflective abilities of a sea surface in microwave region is very connected with ocean environment parameters such as the water temperature, salinity, dielectric constant, and roughness. Sasaki et al. (1988) shown that 6.7–18.6 GHz band has high informativity to diagnose the *atmosphere-ocean* system. It was established that correlation between radiobrightness temperatures of atmosphere and sea surface has linear character and that reflective ability on horizontal polarization at the frequency 18.6 GHz is higher then on vertical polarization. Moreover, it is observed that there exists an obvious dependence of radiobrightness temperature fluctuations of the sea surface on wind speed. It equals for the horizontal polarization to 0.5 and 1.0 K m^{-1} s at 6.7 and 18.6 GHz, respectively. In the case of vertical polarization these values equal to 0.3–0.8 and 0.6–1.5 K m^{-1}s, respectively.

The sensitivity of the sea surface brightness temperature, T_{BP}(K), for the polarization P-type to the variations in the wind speed, V(m s^{-1}) is determined by value (Figs. 2.30 and 2.31):

$$\omega = \frac{\partial T_{BP}}{\partial V}\Big|_{a_1,\ \ldots,a_k=const},$$

where a_i (i = 1, …, k) are the oceanic environmental parameters contributing to T_{BP} change except for wind speed. The value of ω is almost linearly correlated with wind speed at constant sea surface temperature and salinity, so $\partial^2 T_{BP}/\partial V^2 \approx 0$. Then $T_{BP} = T_{BOP}(\theta) + V\partial T_{BP}(\theta)/\partial V$, where T_{BOP} is the brightness temperature of the calm sea surface (K), θ is the zenith angle of incident radiation. According to the Kirchoff's law we have the following wind speed sensitivity of reflectivity:$\partial\gamma_P(\theta)/\partial V = -\partial\varepsilon_P(\theta)/\partial V$. It gives the following expression for reflectivity at

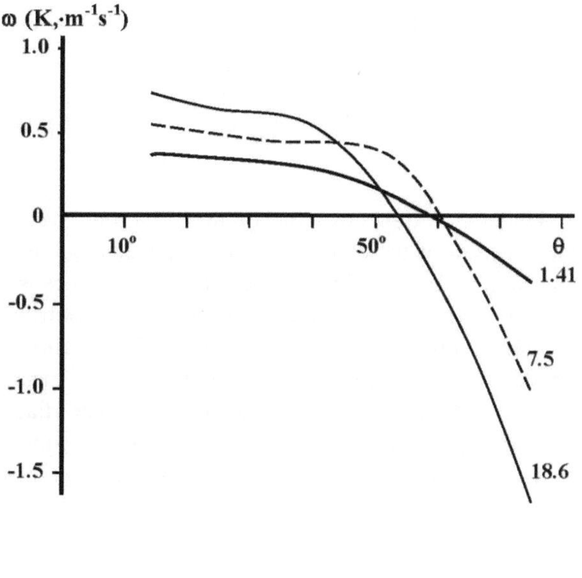

Fig. 2.30 Empirical dependence of brightness temperature of ocean surface on the wind speed at various wavelengths and the observation angles (Sasaki et al. 1988). The frequencies in GHz are given on the curves

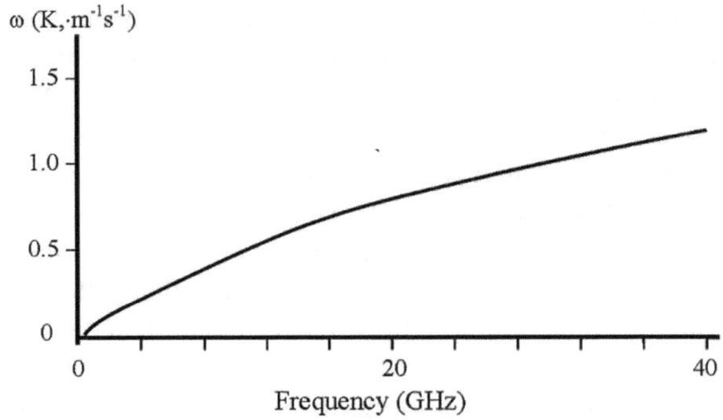

Fig. 2.31 Empirical dependence of brightness temperature sensitivity on the wind speed when the observation is realized by the nadir (Sasaki et al. 1988)

arbitrary wind speed: $\gamma_P(\theta) = \gamma_{OP}(\theta) - (V/T_{S\hat{\imath}}) \times \partial T_{BP}(\theta)/\partial V$, where T_{SO} is the sea surface temperature, γ_P and γ_{OP} are the reflectivity of the roughned and the calm sea surface, respectively.

According to Sasaki et al. (1987a, b, 1988) the wind speed sensitivity of the sea surface brightness temperature at satellite level is

$$\frac{\partial T_{BP}^{sat}(\theta)}{\partial V} = \frac{\partial T_{BP}(\theta)}{\partial V} \cdot \frac{T_{SO} - \left\{T_B^{\downarrow sky}(\theta) + T_B^{space}\right\}}{T_{SO}},$$

where $T_B^{\downarrow sky}$ is the downward sky brightness temperature at sea surface (K), T_B^{space} is the galactic noice (K).

The sea surface brightness temperature at arbitrary wavelength can be represented in the form:

$$T_{BP}^{\lambda}(\theta) = \varepsilon_{1P}(\theta)T_{SO} + \gamma_P(\theta)T_B^{\lambda,sky}(\theta_S, \phi_S),$$

where θ_S and ϕ_S are the zenith and azimuthal angles,

$$\gamma_P(\theta) = \frac{1}{4\pi}\int_0^{2\pi}\int_0^{\pi/2} \gamma_P(\theta, \theta_S, \phi_S)\sin\theta_S d\theta_S d\phi_S.$$

Under conditions of wind-induced sea roughness, when sea wave are so big that radio waves of the centimeter and decimeter ranges cannot be used, scattering and emission characteristics can be estimated by means of Kirchoff's approximation.

In the range of centimetre waves, where electrophysical properties of water are comparatively known, the expression (2.7) allows to determine water surface temperature when its radiothermal emission is measured.

In common case the task of ocean surface diagnosis comes to the reverse of function $T_j = F(a_1, \ldots, a_k)$ in the relation of geophysical parameters set $\{a_i\}$. The function F reflects the type of aquatory radiation model. Linear models allowing to calculate a standard deviation of T_j are widely used: $\delta_i = k_i\delta a_i$, $k_i = \partial F/\partial a_i$, where δa_i is the standard deviation of parameter a_i. The estimations of κ_i are given by Nelepo et al. (1985).

Having synthesized a radiation model, F can then be used to control ocean surface pollution and the *atmosphere-ocean* system is considered as a three-layers environment: *atmosphere-pollutant-water*. For calculation of Frenel's coefficients to estimate the emittive property κ_λ of this system, knowledge of the dielectric permeability of the pollutant layer is needed. Therefore, for oil spills in the microwave range there is weak dependence of the dielectric constant on λ and the values of losses are typically small. For example, the raw oil spill is characterized by value $\mathrm{Re}\varepsilon_\lambda \in [1.6, 3]$. Specifically, for $\lambda = 0.8$ cm we have $\varepsilon_\lambda = \delta_i = k_i\delta a_i$, $k_i = \partial F/\partial a_i$. Certainly ε_λ is changed depending on the oil quality and time interval of its placing in the water.

Any pollutant of water environment forms on the water surface some emulsion layer which exists during certain time interval. Dielectric permeability of this layer can be calculated by means of the mixture formula: $\varepsilon_\lambda = [(1 - W_1)(\varepsilon_{\lambda,1})^{1/3} + W_1 (\varepsilon_{\lambda,2})^{1/3}]^3$, where W_1 is the water content in the emulsion volume, $\varepsilon_{\lambda,1}$ and $\varepsilon_{\lambda,2}$ are the dielectric permeabilities of pollutant and water, respectively. Figure 2.32 demonstrates the frequency of ε_λ for the case of oil pollution.

The methodics of radiation model synthesis for three-layers environment is applied to the microwave diagnostics of the ice cover. Here ε_λ is explained as dielectric permeability of the ice layer (Melentyev et al. 1998).

Variety of theoretical and experimental investigations of the ocean/atmosphere system includes the application of remote sensing technologies based on microwave monitoring. For example, Grankov and Milshin (2009) studied experimentally (in the laboratory) a behavior of characteristics of the water surface being initially under the room temperature and sharply cooled with liquid nitrogen: variations of an intensity of natural microwave radiation at the wavelength 2.25 cm and infrared heat radiation in the window 10.5 μm, as well as a value of the heat content

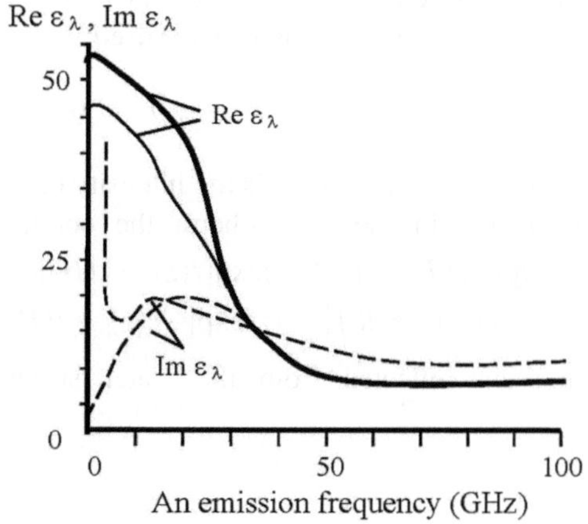

Fig. 2.32 Dielectric properties of oil mixture with the fresh (*solid line*) and saliny (*dashed line*) water

(enthalpy) in the upper water surface layer. A comparison of evolutions of the heat contents computed from the difference between temperatures of upper and bottom boundaries of the subsurface layer with evolutions of microwave and infrared radiation intensity is conducted.

Here, it is again necessary to remark that difference between microwave and infrared ranges consists in their response on the atmosphere state:

- intensity of super high frequency is defined by means of the product of emissivity coefficient (κ) and thermodynamic temperature (T) within the limits of efficient radiation layer (l_{ef}) where $dT_b/dT \leq 0.5$ and l_{ef} equals to the part of centimeter; and
- IR-range methods can be used only in the cases when the atmosphere is free from aerosols, hydrometeors, and clouds as well as l_{ef} equals to the part of millimeter.

Usually, the values of sensible q_{h} and latent q_{e} heat at the air-sea boundary are calculated by means of well known in the meteorology formulae:

$$q_{\mathrm{h}} = c_{\mathrm{p}}\, \rho\, c_{\mathrm{t}}(t_{\mathrm{s}} - t_{\mathrm{a}})V; \quad q_{\mathrm{e}} = L\, \rho\, (0.622/P)c_{\mathrm{e}}(e - e_{\mathrm{o}})V, \tag{2.10}$$

where t_{a} is the air temperature, P is the atmospheric pressure; e is the atmosphere humidity, V is the wind speed in the near-surface atmosphere, t_{s} is the ocean surface temperature, e_{o} is the the air humidity, c_{t} is the numbers of Schmidt (heat exchange), c_{e} is the number of Dalton (moisture exchange), L is the specific heat of evaporation, c_{p} is the specific air heat under constant pressure, and ρ is the atmosphere density.

Formulae (2.10) can be used for calculation of sensible and latent heat fluxes only in aquatories that are serviced by meteorological stations. Figure 2.33 represents such calculations for NDBC SMKF1 during hurricane Katrina evolution.

Capability of satellite SHF-radiometric methods to evaluate the characteristics of large-scale heat and dynamic interaction of the atmosphere and ocean is based on the use of resonance absorption of radiowaves 1.35 cm by atmospheric moisture. In this case different models are used (Skou 2008; Pampaloni and Paloscia 2000). One of such simple model is usually used:

$$I = I_1 + I_2 + I_3$$

where $I_1 = I_s\exp[-\tau\,(H)]$ is the intensity of natural radiation of the oceanic surface I_s attenuated in the atmosphere; the coefficient $\exp[-\tau(H)]$ is the index of the absorption; $I_2 = \int_0^H I_a(z)\exp[\tau(z) - \tau(H)]dz$ is the intensity of upgoing atmospheric radiation; $I_3 = R\int_0^H I_a(z)\exp[-\gamma(z) - \tau(H)]dz$ is the intensity of atmospheric emission reflected from the water surface; $\tau(z) = \int_0^H \gamma(z')dz'$ is the integral

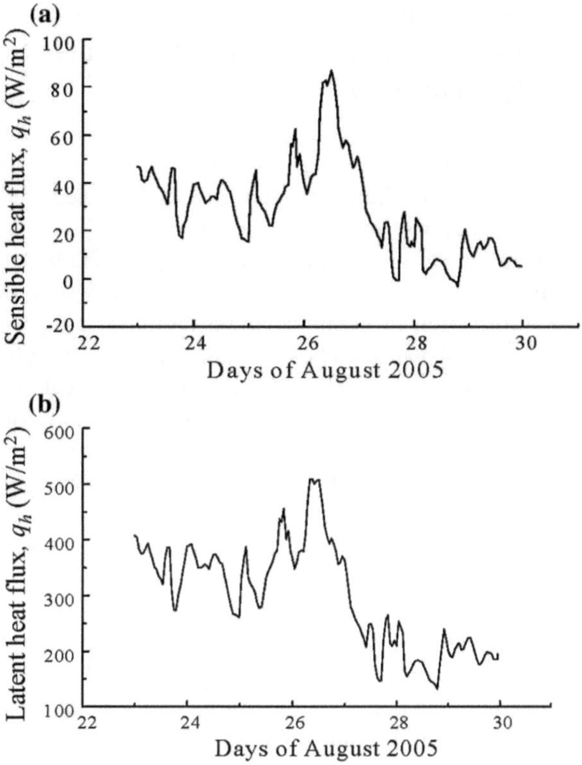

Fig. 2.33 Variations of sensible (**a**) and latent (**b**) heat fluxes at the ocean surface in an area of the station SMKF1 responsibility during the period of tropical hurricane Katrina effect in August 2005

attenuation in the atmosphere; H is the height of the troposphere; R is the reflectivity of water surface.

An intensity of natural radiation of the water surface is determined as:

$$I_s = \begin{cases} \kappa T_s \text{ at microwave range,} \\ \delta B(T_s) \text{ at infrared range.} \end{cases}$$

where T_s is the thermodynamic (kinetic) temperature of the ocean surface, κ and δ are the emissivity coefficient of the ocean surface at microwave and infrared ranges, respectively.

The intensity of atmospheric radiation related to the level z is determined as:

$$I_a(z) = \begin{cases} \gamma(z)T_a(z) \text{ at microwave range,} \\ \gamma(z)B(T_a) \text{ at infrared range.} \end{cases}$$

2.9 An Adaptive Technology to Classify and Interpret Remote-Sensing Data of the Water Surface Qualitatively

Collecting and processing information in a geoinformation monitoring system can only be done by effective monitoring of the object under consideration and involves using simulation modeling, information collecting, and information processing (Armand et al. 1987; Burkov and Krapivin 2009).

From the position of system analysis, the system of collection and processing of the information in geoinformation monitoring represents the structure uniting the computers of various classes, databases and the advanced problem-oriented software. Creation of such system demands the development of formalized description of the information flows and unique methodology of its processing.

Development of geoinformation monitoring systems requires the decision of a set of problems related to the formation of data measurement flows to be solved. The problem of classification of aquatories using the remote sensing measurements is one of important among them. Various algorithms of the theory of image recognition, statistical decisions and cluster analysis are used to solve this problem.

At the present time, there are many image recognition methods, mainly because of the variety of statements about concrete tasks. The problem of recognition consists in the division of some group of objects into the classes at the base of certain requirements. The objects having general properties are related to one class. An initial data for the solution of a recognition problem are results of some observations or the direct measurements that are named initial attributes.

Method of taxonomy (clustering) is one of the important methods of recognition and classification of images.

Let's assume set of M it is required to divide by not crossed subsets (clusters), and the elements included in the same clusters should be close to each other enough from the point of view of the chosen criterion of nearness, and elements from different clusters should be far enough from each other. In one of many possible statements of this task two numbers, a and b ($0 < a < b$), are given. It is considered, that two elements x and y are close to each other enough, if $p(x, y) < a$, and are far enough from each other, if $p(x, y) > b$.

King's method is well known in taxonomy and gives good results when the quantity of available information in assumed clusters is moderate. According to this method the distance between groups of points in space of attributes is defined as distance between centers of masses of these groups. Clustering in this case is based on the assumption that sites of the increased density in the space of attributes correspond to similar situations.

The feature of remote measurements is information acquisition, when the data of measurements, acquired during tracing of flying system along routes of survey, are directed to input of the processing system. As result the two dimensional image of investigated object is registered. Statistical model of spottiness for investigated space is one of models for this image.

In real conditions, the study of spots, the acquiring of their statistical characteristics and their using in a problem of detection is enough a complex problem. Criteria have to be developed to distinguish spots from other phenomena. For example, it is necessary to determine such threshold the exceeding of which is the spot indicator. Also it is necessary to develop model presentation of processes of spots detection.

The method of the thresholds determination is the most obvious and simple way for spots definition. In this case that part of space belongs to area of spots, on which the parameter of environment measured within the chosen channel exceeds value (l^+) or, on the contrary, does not exceed value (l^-) a threshold. Let $y = y\ (x_1, x_2)$ is function of coordinates (x_1, x_2) of points within considered region. If "level surface" y = const is outlined at the region surface, then the closed curves of level y that bound the spots are projected on it.

Algorithms for simulation of spottiness are based on the numerical solution of the algebraic inequalities determining coordinates of internal points of spots. It is impossible to write the equation of spots contours in a general.

Therefore contours of spots are described by system of the simple algebraic equations connected among themselves by equation $\Sigma\varphi_i(x, y) = 0$, where $\varphi_i(x, y)$ is the equation of an elementary curve. For simplification of software realization of simulation of spottiness image as the equations $\varphi_i(x, y)$ the equation of a circle with varied coordinates of the centre and radius is accepted. Complex forms of spots are formed by overlapping on a plane of the drawing of several circles with different parameters that is defined by system of inequalities of a kind:

$$\sum\left\{(x - a_i)^2 + (y - b_i)^2 - r_i\right\} \leq 0$$

where x, y are the cartesian coordinates of internal points of spots, a_i, b_i, and r_i are coordinates of the centre and radius ith circle, respectively, n is quantity of the circles composing the modeled image.

To simulate the randomness of background distribution for spottiness the spottiness model parameters a_i, b_i, and r_i are set by means of random-number generators. By changing laws of distribution of random numbers and their statistical parameters, it is possible to receive statistically different spottiness images.

The list of software items of the simulation system for classification of the phenomena on a terrestrial surface is given in Table 2.21. An important point about the system's algorithms and the software is the possibility of spatial interpolation and data restoration using remote and in situ measurements.

One of main aspects of the practical importance of developed system is qualitative interpretation and visualization of results of remote measurements.

For primary processing of remote measurements it is useful to apply an owerage-connecting method of cluster analysis to detect the specific informational zones. That method is effective under small volumes of sampling. Two variants of this approach can be brought about by the way in which algorithms and index spaces are organized.

Table 2.21 Software modules of the simulation system for classification of the phenomena on a terrestrial surface

Software	The functional characteristic of a software
REICM	Reduction of the experimental information in a computer memory
DRHIM	Data reconstruction by means of the harmonics interpolation method
SIDSM	Spatial interpolation of the data by means of a spline method
DRMOT	A method of optimal interpolation
CASRS	Cluster analysis focused on sings space of remote sensors
CALRS	Cluster analysis focused on the account of local reading of sensors
RSC	Research of spottiness characteristics
RRAMDA	Realization of recognition algorithm by a method of the discriminant analysis
CSAIRM	Sorting and accumulation of the in-situ and remote measurements
CMS	Computer mapping of the spots

In the system designed to classify phenomena on a terrestrial surface automatically the two variants of this approach are program modules CASRS and CALRS, distinguished by the way in which attribute spaces and comparison algorithms are organized.

Software item CASRS is focused on the attribute space of remote sensors. The CALRS module allocates areas of equal instability based on local variations in sensor data. Distinction between the algorithms of comparison in these items consists in the taking into account or neglecting of interrelations of the neighbouring counts of sensors. Item CASRS forms clusters without taking into account a geographical generality of radiometers indications. Item CALRS forms continuous spatial clusters.

CASRS and CALRS automatically exclude from consideration the calibrations information. Item CASRS consists from six groups of operators. The first group carries out the organization of data files. Averaging of the information using final set of radiometer readings is made by the second group of operators. The third group of operators finds minimal cluster distances then two points on which this distance is realized are allocated. The fourth group of operators investigates changes of minimal cluster distances on the given step of procedure and in case of occurrence of sharp changes carries out visualization of clusters structure. The fifth group of operators unites the nearest clusters and recalculates their characteristics. The sixth group of operators develops a criterion for the algorithm to stop.

Analysis of statistical characteristics of “spottiness” for three types of areas of Atlantic and Pacific oceans was conducted. These statistical characteristics were determined for the most informative thresholds. At that time statistical characteristics of “spottiness” for the same areas, selected using criteria of minimal value of coefficient of correlation for joint sample of positive and negative spots. Analysis of these characteristics showed, that the statistical characteristics of “spottiness” coincide for areas with temperate sea roughness and storm zones. Minimum for the coefficient of correlation ρ_{min} is run down for a case of most informative thresholds. But for quiet area the situation is different.

From the aforesaid follows that statistical characteristics for "spottiness" of brightness temperatures in microwaves can be used for detection and classification of the phenomena on a surface of the ocean that was caused by a degree of sea roughness.

Analysis of empirical histograms for spottiness of brightness temperatures in microwaves shows that in most cases (l^+, l^-) are characteristics that can be coordinated with the exponential distribution while their amplitude can be coordinated with the normal distribution. Therefore, to detect and classify phenomena on ocean surfaces it is necessary to apply the best possible algorithms when programming/teaching the computer to make statistical decisions about distributions.

2.10 A Device to Measure Geophysical and Hydrophysical Parameters

According to algorithms of the multi-channel microwave monitoring discussed by Krapivin and Shutko (2012), it is possible to promptly measure the wind speed, sea surface roughness and water temperature as the base characteristics to calculate other parameters of the environment. From these measurements, a mutual control of the accuracy of correction of the estimates of the atmosphere-ocean gas exchange is realized. Solution of the correction equations in a general case requires the use of the criterion of discrepancy between theoretical and empirical estimates of the parameters or correlations. Usually for this purpose, a criterion of mean square deviation is used. The reliability and efficiency of determination of solutions for the system of equations relating the theoretical and empirical estimates can be increased if their dispersive characteristics are taken as a criterion of discrepancy of these solutions (Nitu et al. 2013a).

Let the proper emission of the sea be measured simultaneously with the help of n radiometers, each operating at a fixed wavelength λ_j ($j = 1, \ldots, n$). As a result at the moment t_i at the output of each radiometer the values Z_{ij} ($i = 1, \ldots, M$) are fixed, so that $Z_{ij} = T_j + \xi_{ij}$, where T_j is the brightness temperature of a sea site at the wavelength λ_j, and ξ_{ij} is the random value (noise) with the zero mean and dispersion σ_j^2.

Let the correlation between the brightness temperature and m parameters of the sea be linear:

$$\begin{aligned} A_{11}x_1 + \cdots + A_{1m}x_m &= T_1 + \xi_1 \\ &\cdots \\ A_{n11}x_1 + \cdots + A_{nm}x_m &= T_n + \xi_n \end{aligned} \tag{2.11}$$

where A_{ij} are the coefficients determined in the regime of monitoring the sea site with the known parameters $\{x_i\}$. The x_i^* parameter should be estimated for the x_i^0 parameters so that, first, the average x_i^* coincided with the accurate solution of the system (2.11), that is, $\bar{x}_i^* = x_i^0$ ($i = 1, \ldots, m$), and second, a dispersion of the x_i^*

estimate was at a minimum, that is, $D\left[x_i^*\right] \leq D[\tilde{x}_i]$, where $\tilde{x}_i$ is the estimate of the x_i^0 parameter obtained by any other method.

And finally, third, let the procedure of estimation of the x_i^* parameters be as simple as possible and permit to judge about the best discrete set of radiometric channels needed to realize the two conditions above.

Let the totality of the x_i^* estimates meeting these conditions be σ-solution of the system of linear equations with the disturbed right-hand part. Multiply step by step the ith solution of the system (2.11) by $c_{1i}, \ldots, c_{mi}$, and let

$$\sum_{i=1}^{n} c_{ji} A_{il} = \delta_{jl}, \tag{2.12}$$

where

$$\delta_{jl} = \begin{cases} 1 & for\, j = l \\ 0 & for\, j \neq l \end{cases}$$

As a result, we have

$$x_1^0 = \sum_{i=1}^{n} c_{1i} T_i, \quad \tilde{x}_1 = \sum_{i=1}^{n} c_{1i} T_i + \sum_{i=1}^{n} c_{1i} \xi_i, \quad D[\tilde{x}_1] = \sum_{i=1}^{n} c_{1i}^2 \sigma_i^2$$

For $\tilde{x}_i$ ($i \geq 2$) we write similar relationships. Derive an auxiliary expression:

$$\varphi(c_{11}, \ldots, c_{1n}) = \sum_{i=1}^{n} c_{1i}^2 \sigma_i^2 + \mu_1 \left(\sum_{i=1}^{n} c_{1i} A_{i1} - 1 \right) + \sum_{j=2}^{m} \mu_j \sum_{i=1}^{n} c_{1i} A_{ij},$$

where μ_j are the non-estimated Lagrange multipliers. Putting first derivative functions φ to zero, we obtain:

$$2c_{1k}\sigma_k^2 + \sum_{j=1}^{m} \mu_j A_{kj} = 0 \quad (k = 1, \ldots, n)$$

These equalities together with conditions (2.12) form the system (m + n) of equations whose solution enables one to find desired optimal values of c_{ij}^*. As a result, $D[x_1] = -\mu_1/2$, and the remaining μ_j ($j \geq 2$) values meet a system $m \times n$ of equations:

$$\sum_{j=1}^{m} \mu_j \sum_{i=1}^{n} A_{ij} A_{i1} \sigma_i^{-2} = -2, \quad \sum_{j=1}^{m} \mu_j \sum_{i=1}^{n} A_{ij} A_{i1=} \sigma_i^{-2} = 0, \quad (l = 2, \ldots, m)$$

For m = 2 we have:

$$c^*_{1,k} = \left(A_{k1} \sum A^2_{i2}/\sigma^2_i - A_{k2} \sum_{i=1}^{n} A_{i1}A_{i2}/\sigma^2_i \right) \Delta^{-1}\sigma_k^{-2},$$

$$c^*_{2,k} = \left(A_{k2} \sum A^2_{i1}/\sigma^2_i - A_{k1} \sum_{i=1}^{n} A_{i1}A_{i2}/\sigma^2_i \right) \Delta^{-1}\sigma_k^{-2},$$

$$\Delta = \sum_{i=1}^{n} A^2_{i1}\sigma_i^{-2} \sum_{i=1}^{n} A^2_{i2}\sigma_i^{-2} - \left(\sum_{i=1}^{n} A_{i1}A_{i2}\sigma_i^{-2} \right)^2,$$

$$x^*_j = \sum_{i=1}^{n} c^*_{ji} Z_i, \ (j = 1, 2); \quad D[x^*_1] = \Delta^{-1} \sum_{i=1}^{n} A^2_{i2}\sigma_i^{-2};$$

$$D[x^*_2] = \Delta^{-1} \sum_{i=1}^{n} A^2_{i1}\sigma_i^{-2}$$

For the real use of this algorithm in the radiometric monitoring system the following arrangement can be suggested. Put the n-channel commutation unit and n units of dispersion calculation between the outputs of radiometers and respective inputs of the microprocessor in series. At the radiometers' inputs we plug in switches, which are connected through a control unit with the first group of outputs of the microprocessor, and to the second group of the microprocessor's outputs are linked up the *resolver*, whose second group of outputs is connected, respectively, with the radiometers' outputs. Each unit to calculate dispersion consists of the in-series connected quantizers of the mean value calculator, divisor, summator and second divisor, and the quantizers output is connected with the second input of the summator. The *resolver* consists of identical channels, each containing a summator, n outputs of which are connected with n multipliers. Some inputs of the multipliers form the first group of the *resolver's* inputs, others—the second group of its inputs, the number of channels being equal to the number of the parameters to be measured.

Figure 2.34 shows the structural scheme of an arrangement to measure the parameters of geophysical and hydrophysical objects. The scheme in Fig. 2.35 explains the structure of the unit to calculate dispersion. The *resolver* is characterized in Fig. 2.36. The apparatus operates as follows. When radiometers 1 are calibrated, the signals Z_{ij} from their outputs through the multi-channel switching unit 2 enter the unit 3 of the dispersion calculator. The antennas are switched off from radiometers. In unit 3 the dispersion is calculated using the algorithm.

$$\sigma_j^2 = \frac{1}{M-1} \sum_{i=1}^{M} (Z_{ij} - M_j), \quad M_j = \frac{1}{M} \sum_{i=1}^{M} Z_{ij}$$

The M_j parameters are calculated in the units of the mean calculations. The summator calculates:

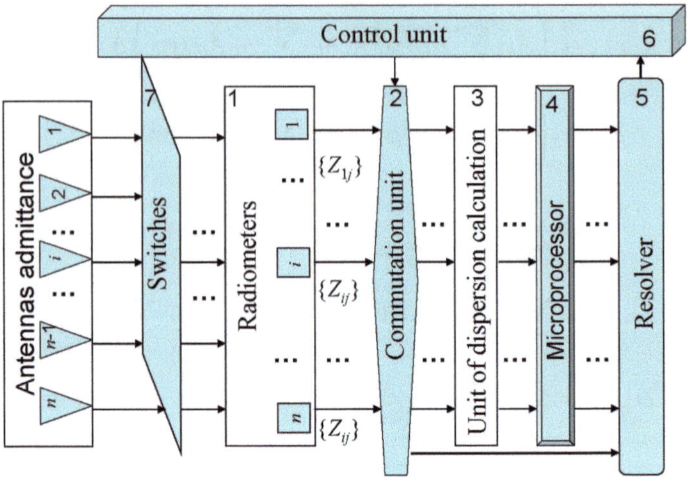

Fig. 2.34 The structural scheme of the arrangement to measure the geophysical and hydrophysical parameters in the regime of the microwave monitoring

Fig. 2.35 Scheme of operations in the unit of dispersion calculation

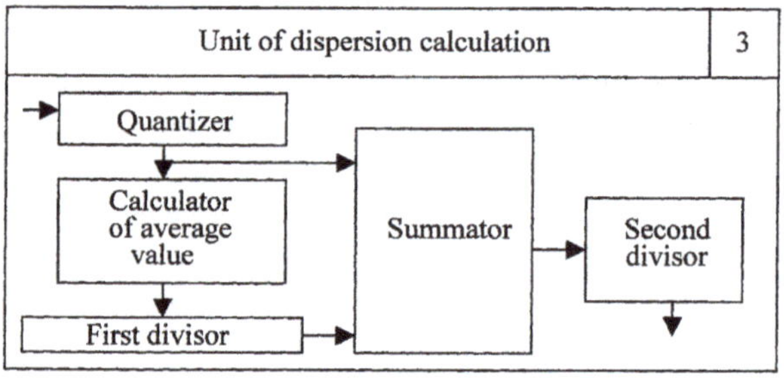

$$\sum_{i=1}^{M} (Z_{ij} - M_j)$$

The calculated values σ_j^2 enter the microprocessor 4 to calculate partial coefficients c_{jl}. From the first group of its outputs they enter the *resolver* 5 where they are used to solve a system of equations, and from the second group of outputs they enter the control unit 6 which forms the signal that switches off radiometers 1, for which the partial coefficients are equal or close to zero, that is, the information channels automatically get excluded. At the same time, the systems of equations to be solved get compressed. This step makes it possible to contract not only the volume of calculations but also to increase the accuracy of estimating the parameters of the environmental objects being sounded.

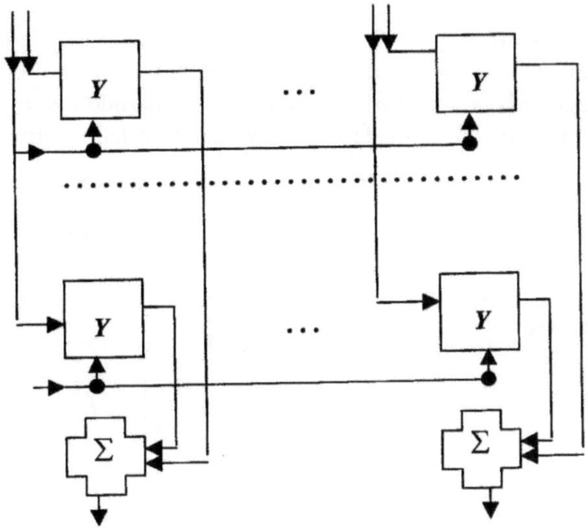

Fig. 2.36 Resolver. *Note* Y = multiplier, Σ = accumulator

Once dispersions have been determined, the commutator 2 blocks the input of data from non-information channels. The data from information channels enter the *resolver* 5. The signal at the multiplier's output corresponds to the value of one of the parameters to be estimated.

2.11 Direct and Inverse Problems of Microwave Monitoring

2.11.1 Typical Inverse Task of the Microwave Radiometry

An approach to the inverse problem of microwave monitoring is based on assuming a linear functional correlation between radiobrightness contrasts and the characteristics of the monitored object. The search of analytical relationships relating the radiobrightness temperature of the environment to its thermodynamic and electrophysical characteristics covers more general classes of the functional models. The most general form of such dependence is as follows:

$$T_{bi} = \kappa_i \int_0^{\infty} T(h)\gamma_i(h) \exp\left[-\int_0^{h} \gamma_i(x)dx\right] dh, \tag{2.13}$$

where $T(h)$ is the thermodynamic profile of the environment, T_{bi} is the radiobrightness temperature at the frequency ν_i, $\gamma_i(h)$ is the profile of the absorption coefficient of the environment, κ_i is the emissivity of the environment.

Relationship (2.13) makes it possible using $T(h)$ and $\gamma_i(h)$ to find T_{bi} (direct problem) and by T_{bi} to calculate $T(h)$ and $\gamma_i(h)$ (inverse problem). The proposed method to solve these problems is based on the use of $T(h)$ and $\gamma_i(h)$ and their derivatives at the point $h = 0$ with some additional limitations resulting from the transformations made below. Write the relationship (2.13) in the form:

$$T_{b\lambda} = -\kappa_\lambda \int\limits_0^\infty T(h) dF_\lambda(h), \tag{2.14}$$

where

$$F_\lambda(h) = \exp\left[-\int\limits_0^h \gamma_\lambda(x)dx\right]$$

λ is the wavelength.

Integrating (2.14) by parts, we obtain:

$$T_{b\lambda} = \kappa_\lambda\left[T_0 + \int\limits_0^\infty T'(h)F_\lambda(h)dh\right] = \kappa_\lambda[T_0 + J_1] \tag{2.15}$$

Let $J_1 = 0$, then $T_{b\lambda} = \kappa_\lambda T_0$. If $dT(h)/dh \equiv 0$, then $J_1 = 0$. Therefore find conditions under which $J_1 = 0$. Let the following relationships be valid:

$$\gamma_\lambda(h) \equiv c_\lambda > 0, \quad T(h) = \sum_{k=0}^n B_k h^k \tag{2.16}$$

It follows from (2.16) that if the coefficients B_k meet the condition:

$$\sum_{k=1}^n k! \frac{B_k}{c_\lambda^k} = 0, \tag{2.17}$$

then $J_1 = 0$.

Integrating (2.15) by parts, we obtain:

$$T_{b\lambda} = \kappa_\lambda\left[T_0 + \frac{T_0'}{\gamma_\lambda(0)} + \int\limits_0^\infty \frac{T''(h)\gamma_\lambda(h) - \gamma_\lambda'(h)T'(h)}{\gamma_\lambda^2(h)} F_\lambda(h)dh\right] \tag{2.18}$$

Equate the integral in (2.18) to zero. It gives

$$T(h) = c\int \gamma_\lambda(h)dh,$$

where c is an arbitrary constant. The integral in (2.18) will be equal to zero if

$$\sum_{k=2}^{n} B_k k(k-1)k! c_\lambda^{1-k} = 0 \tag{2.19}$$

Further integrating of (2.18) by parts gives

$$T_{я\lambda} = \kappa_\lambda\left[T_0 + \frac{T_0'}{\gamma_\lambda(0)} + \left(\frac{T'}{\gamma_\lambda(h)}\right)'\frac{1}{\gamma_\lambda(h)}\Big|_{h=0} + J_3\right],$$

where

$$J_3 = \int_0^\infty \frac{T'''(h)\gamma_\lambda(h) - 3T''(h)\gamma_\lambda(h) - T'(h)\gamma_\lambda''(h) + 3T'(h)\left[\gamma_\lambda'\right]^2}{\gamma_\lambda^4} F_\lambda(h)dh$$

This integral will be equal to zero at

$$T(h) = c\int \gamma_\lambda(h)\int \gamma_\lambda(u)dudh$$

If the condition (2.16) is not met then $J_3 = 0$ at

$$\sum_{k=3}^{n} B_k k(k-1)(k-2)k! c_\lambda^{2-k} = 0$$

With the process of integration of (2.18) continued N times, we obtain

$$T_{я\lambda} = \kappa_\lambda\left[\sum_{k=1}^{N} P_k(0) + \int_0^\infty P_N'(h)F_\lambda(h)dh\right], \tag{2.20}$$

where

$$P_1(h) = T(h), \quad P_2(h) = T'(h)/\gamma_\lambda(h)\ldots, \quad P_k(h) = P_{k-1}'(h)/\gamma_\lambda(h)$$

but

$$J_N = \int_0^\infty P_N'(h)F_\lambda(h)dh = 0$$

for

$$\sum_{k=N}^{n} B_k k(k-1)\cdots(k-N+1)k!c_\lambda^{N-1-k} = 0, \quad (n > N)$$

Consider the case when

$$\gamma_\lambda(h) = \exp\{c_\lambda h\}$$

From (2.20) we obtain:

$$T(h) = \sum_{k=0}^{N-1} B_k \exp[kc_\lambda h]$$

Now we move on to the inverse problem. Let measurements be made at wavelengths $\lambda_1, \ldots, \lambda_N$. Solve the equation:

$$P'_{N-1}(h) = c\gamma_\lambda(h)$$

and find

$$P_{k-1}(h) = \int \gamma_\lambda(h) P_k(h) dh, \quad (k = 1, \ldots, N)$$

Let $\gamma_\lambda(h) = \varphi(\lambda)\psi(h)$, that is, the dependence of the absorption coefficient on the frequency and depth of the layer be divided by the product of the known function $\varphi(\lambda)$ and an unknown function $\psi(h)$. As a rule, the function $\varphi(\lambda)$ is set from the empirical table and its analytical approximation can be constructed. Then, excluding from (2.20) the second right-hand term (it is equal to zero), we write the relationship:

$$T_{b\lambda} = \kappa_\lambda \sum_{k=1}^{N} T^{(k-1)}(0) F_k \left[\varphi(\lambda), \psi^{(N-k-1)}(0), \ldots, \psi'(0), \psi(0)\right] \quad (2.21)$$

and the solution of the inverse problem is reduced to the solution of a system of algebraic equations with regard to functions:

$$T(0), T'(0), \ldots, T^{(N-1)}(0), \psi(0), \ldots, \psi^{(N-1)}(0)$$

Consider the case $\psi(h) = \text{const}$, that is, $\gamma_\lambda = \varphi(\lambda) > 0$. Equation (2.21) is re-written:

$$T_{b\lambda} = \kappa_\lambda \sum_{k=1}^{N} T^{(k-1)}(0) / \varphi^{(k-1)}(\lambda) \quad (2.22)$$

In the simplest version, if $T_{b\lambda}$ was measured at N different wavelengths $\{\lambda_i; i = 1, \ldots, N\}$, then the determinant of the system (2.22) will be equal to:

$$\Delta = \prod_{i=1}^{N} \kappa_i \prod_{i>k}^{N} [\varphi(\lambda_i) - \varphi(\lambda_k)]/[\varphi(\lambda_i)\varphi(\lambda_k)]$$

If $\varphi(\lambda_i) \neq \varphi(\lambda_\pi)$ with none of combinations $i \neq k$, then $\Delta \neq 0$ and the system (2.22) has a single solution. Now let

$$\psi(h) = h^m$$

Then, it follows from (2.20) that

$$T(h) = \sum_{k=0}^{N-1} B_k h^{k(m+1)}$$

and we obtain a system of algebraic equations

$$P_k(0) = (m[k-1])\underset{(m+1)}{!\ldots!}\, b_{k-1}/\varphi^{k-1}(\lambda)$$

with the following determinant:

$$\Delta = \prod_{i=1}^{N} \kappa_{\lambda_i} \prod_{p=1}^{N} [m(k-1)]^{N+1-p} \prod_{t>k}^{N} [\varphi(\lambda_t) - \varphi(\lambda_k)]/[\varphi(\lambda_t)\varphi(\lambda_k)]$$

The case when the function ψ is approximated by a polynomial is considered in a similar way.

2.11.2 Estimation of Radiobrightness Response Function of the Ocean-Atmosphere System on Variations in Heat Fluxes

The problem of using the satellite passive microwave radiometric methods for the analysis of the roles of the ocean and atmosphere (in respect of their priority) in heat interaction over various space and time scales is one of important for solution in tropical cyclogenesis. This problem is long attracts the specialists engaged in the oceanology, meteorology, atmospheric sciences, etc. Some data of measurements from the Meteor-3M MTVZA device (Module of Temperature and Vlazhnost (humidity) Zonding of the Atmosphere), data of the DMSP (SSM/I) and EOS-Aqua (AMSR-E) radiometers can be used for solving this problem.

The phenomena of a time delay of the ocean/atmosphere system brightness temperature response in the resonant spectral domains 5.9 and 13.5 mm on the variations of surface heat fluxes is described by the Duamel's integral equation:

$$T_b(t,\lambda) = \int_0^t q(\tau)r(t-\tau)d\tau \tag{2.23}$$

were $T_b(t,\lambda)$ is radiobrightness temperature for wavelength λ (K), $q(t)$ is heat flux on the ocean-atmosphere interface (W/m^2/h), τ is time delay (hours), $r(t)$ is response delay function (K/(W/m^2/h)).

Solution of Eq. (2.23) can be theoretically realized by traditional methods which are used for solution of inverse tasks in the form of integral equation of the first type. But in real case, functions $T_b(t,\lambda)$ and $q(t)$ have discrete form as restricted in the time numerical sets. Therefore, traditional methods such as Laplace transformation or iterative procedure (Tricomi 1960) lead to large errors. More effective method for solution of Eq. (2.23) was proposed by Grankov and Soldatov (2005). It consists in the use of exponential approximation:

$$r(t) = \sum_{i=1}^{N} a_i \exp(-b_i t)$$

where coefficients a_i and b_i are defined from condition:

$$M = \min_{\{a_i,b_i\}} (1/n)\sqrt{\sum_{i=1}^{n} [T_Я(t_i) - F(t_i)]^2},$$

where $t_i = i\Delta t$,

$$F(t_i) = \sum_{s=0}^{i-1} q(t_i - t_s) \sum_{k=0}^{N} (a_k/b_k)\exp(-b_k t_s)[1-\exp(-b_k\Delta t)]$$

Numerical calculations in Fig. 2.37 show that such approach to solution of Eq. (2.23) gives errors no more 5–7 %. We see that time delay exists between radiobrightness temperature and heat flux. This is important for calibration of satellite measurements. Calculation of the function $r(t)$ of the brightness temperature T_b response (sensitivity) to the total heat fluxes q the variation is clearly demonstrated. In this case, brightness temperatures were calculated from the oceanographic, meteorological, and aerological data, observed in the experiment with ATLANTEX-90 synchronized with use of the DMSP F-08 heat fluxes data q obtained from the vessel *Victor Bugaev* during the cyclone of 9–13 April 1990.

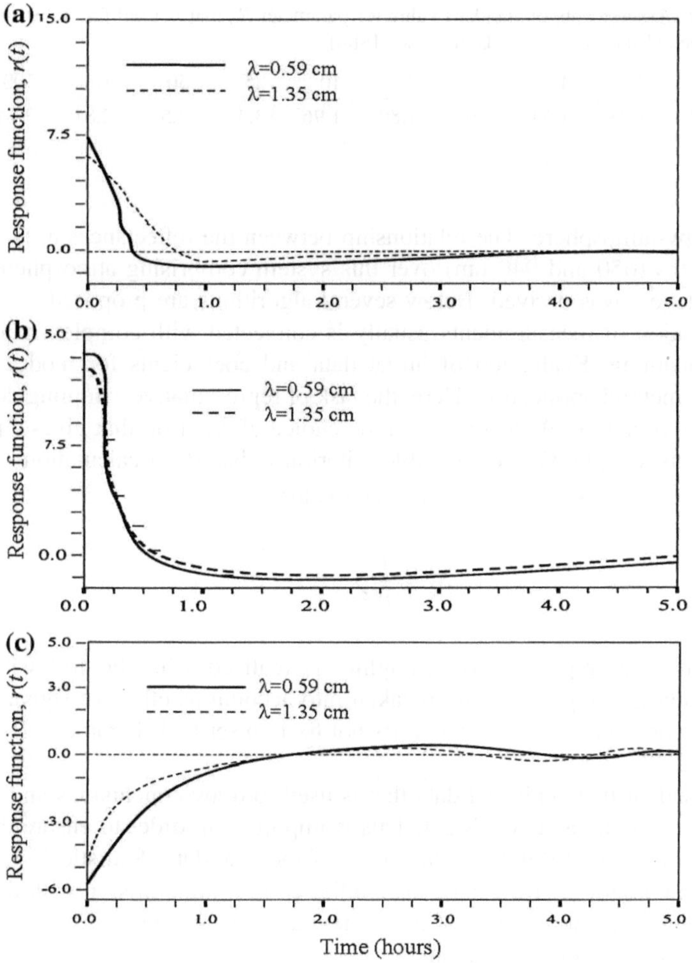

Fig. 2.37 Calculation of response function $r(t)$ (K/(W/m^2/h)) as solution of Eq. (2.66) for measurement data received by ships Bugaev (**a**), Volna (**b**) and Musson (**c**) during ATLANTEX-90

2.12 Algorithms for Remote Data Processing

2.12.1 Introduction

Spatial reconstruction of the vegetation radiometric image, based on fragmentary data, demands the development of new algorithms. There exist many approaches to this task. For example, Yoshioka (2004) discussed the problem of extraction of useful information from satellite imageries by means of algebraic manipulation. The concept of vegetation isolines was underpinned for the three-layer system

Table 2.22 Assessments of standard value for parameter R_0 that is used for statistical regulation of experimental measurements (Dally et al. 1984)

n	2	3	4	5	7	10	25	50	100	300	500
R_0	1.15	1.38	1.54	1.65	1.80	1.96	3.33	2.57	2.81	3.14	3.29

soil-canopy-atmosphere. The relationship between the reflectances at two different wavelengths (650 and 940 nm) over this system comprising atmosphere, canopy, and soil layers was derived. Below several algorithms are proposed.

Realization of measurements usually is connected with complex model choice and its planning. Evaluation of initial data and coefficients for model needs the statistical method application. Here, the task of representative sampling definition is arisen. Important problem consists in the choice of criterion that gives a possibility to form this sample. One of possible criterion is based on calculation of value of $R = \left(\xi_i - \overline{\xi}\right)/S_\xi$, where $\overline{\xi}$ is the average value,

$$S_\xi^2 = \frac{1}{n}\sum_{i=1}^{n}\left(\xi_i - \overline{\xi}\right)^2$$

Formation of representative sampling is realized with the following rule: if $R \le R_0$ then ξ_i is registered and is taken into account as effective signal; if $R > R_0$ then ξ_I is considered as noise and does not fix into set $\{\xi_i\}$. Parameter R_0 is defined in Table 2.22.

It is evident that statistical data that is used to draw conclusions and inferences should be accurate and consistent. This is important in order to ensure the validity of all the inferences drawn on the basis of the raw data. Statistical analysis as a research methodology for ecological studies comes into collision with the problem of raw data non-stationary what needs new approaches and tools for the decision making and scientific conclusions.

2.12.2 Data Reconstruction Using the Harmonic Functions

The process of spreading of heat in the plane homogeneous media G with constant thermophysical characteristics (density ρ, specific heat C and conductivity coefficient K; $\rho, C, K = const > 0$) is described by the equation:

$$\partial T/\partial t = a^2(\partial^2 T/\partial\varphi^2 + \partial^2 T/\partial\lambda^2), \tag{2.24}$$

where $T = T(\varphi,\lambda,t)$ is the temperature of the media at the point with spatial coordinates $(\varphi,\lambda) \in G$ at time t; $a^2 = K/C\rho$ is the temperature-conductivity coefficient of G, φ and λ are the latitude and longitude respectively.

If the thermal transport process is stationary then Eq. (2.24) transforms to the Laplace equation:

$$div \cdot gradT = \partial^2 T/\partial\varphi^2 + \partial^2 T/\partial\lambda^2 \tag{2.25}$$

The solution of Eq. (2.25) is the harmonic function of the spatial coordinates φ and λ. In accordance with the Rayleigh-Jeans law (DeWitt and Nutter 1988) the brightness temperature as a result of remote microwave measurements is $T_a(\varphi,\lambda) = T_a(\varphi,\lambda,\mu,\theta)$, where μ is the wavelength, θ is the observation angle and (μ,θ = *const*). It is proposed that for the any small area V_M at an arbitrary point $M \in G$ the brightness temperature is a linear function of the temperature of the media:

$$T_a(\varphi,\lambda) = A_M + B_M T(\varphi,\lambda); \quad (\varphi,\lambda) \in V_M, \tag{2.26}$$

where A_M and B_M are constants.

Formula (2.26) follows from the theoretical and experimental estimations of T_a (Krapivin and Potapov 2001; Burkov and Krapivin 2009). For media which are homogeneous in depth, the formula $T_a = \kappa T$ is valid. Here $\kappa = \kappa(\mu,\theta,\varepsilon)$ is the media radiation coefficient where ε is the dielectrical permeability. According to the experimental estimations the radiance of fresh water measured by means of radiometers with wavelengths of 5–8 cm is a linear function of *T*. The slope of this dependence is 0.35–0.50 K/°C. An increase in the water salinity *S* from 0 to 16 ‰ is accompanied by a decrease in the sensitivity of the irradiation field to the temperature variations. This effect is observed for wavelengths from 10 to 50 cm. The sensitivity of the irradiation field to variations in *T* is a minimum when the following conditions are realized: $\mu S \cong 700$; $0 \le T \le 30$ °C; $0 \le S \le 180$ ‰; $0 \le \theta \le 25°$.

From Eq. (2.26) it follows that T_a at every point $M \in G$ satisfies the following condition:

$$T_a(\varphi,\lambda) = \frac{1}{2\pi}\int_0^{2\pi} T_a(\varphi + r\cdot\cos a, \lambda + r\cdot\sin a)da,$$

where the integral is over a circle of radius *r* centered at (φ,λ).

This condition is valid for any *r* ($0 < r < r_M$) where r_M is the radius of the area V_M. Therefore T_a is a harmonic function within *G*. A typical task here is the search of the harmonic function $T_a(\varphi, \lambda)$ within *G* when $T_a(\varphi,\lambda) = \tilde{T}_a(u)$ on the boundary Γ of *G* ($u \in \Gamma$, $u = \varphi + i\lambda$). Such a function is the real part of some analytic function *W*(*z*) given by:

$$W(z) = \frac{1}{2\pi}\int_\Gamma \mu(\zeta)/(\zeta - z)d\zeta, \tag{2.27}$$

where μ(ζ) is the real density, $\zeta \in \Gamma$ and $z = \varphi + i\lambda$ is an arbitrary internal point of *G*. We have Re[W(*u*)] and Im $[d\zeta/(\zeta - u)] = -\cos(r,n)d\sigma/r$ where *r* is the distance between ζ and *u*, *d*σ is an element of Γ and *n* is the external normal to Γ. As $z \to u \in \Gamma$, μ(*u*) is approximated by the solution of the integral equation:

$$\mu(u) - \frac{1}{\pi} \int_{\Gamma} \mu(\zeta) r^{-1} \cos(r, n) d\sigma$$

When G is the circle $|z - z_o| < R$ the solution of this task is the Poisson integral:

$$T_a(r, \psi) = \frac{1}{2\pi} \int_0^{2\pi} \tilde{T}_a(a) \frac{R^2 - r^2}{R^2 + r^2 - 2Rr\cos(\psi - a)} da,$$

where $\varphi + i\lambda = z_o + re^{i\psi}(r < R, 0 \le \psi \le 2\pi)$; and

$$\tilde{T}_a(a) = \tilde{T}_a(z_0 + R \cdot e^{ia}), \quad 0 \le a \le 2\pi.$$

The combination of this procedure with other algorithms of spatial-temporal interpolation gives a full representation of the environmental objects on the observation area by means of the parametrical estimations. For example, the GIMS database was formed on the basis of the measurements traced by the flying laboratory using the set of analogous harmonic algorithms. Table 2.23 illustrates the

Table 2.23 An example of the hydrophysical field reconstruction in the lagoon Nuoc Ngot

U (m^3 s^{-1})	M (m^3 s^{-1})	Error (%)	R (m^3 s^{-1})	Error (%)
10.3	11.3	10.1	11.6	12.9
−4	−3.6	9.5	−4.4	9.8
−8.7	−15.3	12.4	−9.9	13.3
−14.2	−12.3	13.1	−16	12.6
−24.1	−22.7	5.7	−22.2	7.7
−22.4	−24.4	8.9	−20.4	9
−16.3	−17.9	9.6	−18.2	11.5
2.3	2.6	11.5	2	11.4
15	13.2	12.3	12.6	15.8
35.1	30.9	12.2	30.1	14.2
39.2	35.1	10.4	43	9.7
38.2	42.6	11.6	33.2	13.1
34.8	39.6	13.8	30	13.7
26.5	25.2	4.8	28.5	7.6
19.2	21.1	9.9	21.2	10.3
Average error (%)	10.4	–	11.5	–
Maximal error (%)	13.8	–	15.8	–
Minimal error (%)	4.8	–	7.6	–

Field measurements was realized during March–April of 2001. Signs correspond to the water inflow ("+") and outflow ("–"). This lagoon is situated in the Central Vietnam, South-China Sea U is the measured value of water flow (m^3 s^{-1}) on the boundary *lagoon-sea*, M and R is the calculated values of this flow with the methods differential approximation and harmonic functions, respectively

precision of this algorithm as compared with the measured values and the differential approximation method.

2.12.3 Method for Parametric Identification of Environmental Objects

Radiometers determine the brightness temperatures Z_{ij} ($i = 1, \ldots, M; j = 1, \ldots, n$) given by $Z_{ij} = T_j + \xi_{ij}$, where M is the number of measurements, n is number of radiometers, T_j is the real value of the brightness temperature for wavelength μ_j and ξ_{ij} is the noise with zero mean and dispersion σ_j. The problem is to determine the correlation function $T_j = f_j(X)$, where $X = \{x_1, \ldots., x_m\}$ are geophysical, ecological, biogeochemical or other parameters. There are many algorithms for the definition of the function f. As a general rule, the mean-square criterion is used for this purpose. But such an approach has one defect: the impossibility of taking the dispersion properties of the noise $E = \{\xi_{ij}\}$ into consideration.

Let the function f be linear. Then we have the following system of equations for parameters A_{ij}:

$$\| A_{ij} \| X = T + E \tag{2.28}$$

It is necessary to solve Eq. (2.28) such that its solution has minimum dispersion. Such a solution is called the σ—solution.

The ith equation of system (2.28) is multiplied by the set of parameters c_{1i}, ..., c_{mi}. An additional condition is given:

$$\sum_{i=1}^{n} c_{ji} A_{il} = \delta_{jl} \tag{2.29}$$

where

$$\delta_{jl} = \begin{cases} 1 & for\ j = l \\ 0 & for\ j \neq l \end{cases} (l,j = 1, \ldots, m) \tag{2.30}$$

Under the conditions (2.29) and (2.30) we have

$$x_1^o = \sum_{i=1}^{n} c_{1i} T_i \tag{2.31}$$

From (2.28) to (2.31) we obtain:

$$\tilde{x}_1 = \sum_{i=1}^{n} c_{1i}T_i + \sum_{i=1}^{n} c_{1i}\xi_i \tag{2.32}$$

The dispersion of solution (2.32) is

$$D[\tilde{x}_1] = \sum_{i=1}^{n} c_{1i}^2\sigma_i^2 \tag{2.33}$$

Dispersions of $\tilde{x}_i$ (i = 2, …, m) are calculated by analogy with (2.33). To calculate the *min* $D[\tilde{x}_1]$ the following additional equation is used

$$\psi(c_{11}, \ldots, c_{1n}) = \sum_{i=1}^{n} c_{1i}^2\sigma_i^2 + \tau_1\left(\sum_{i=1}^{n} c_{1i}A_{i1} - 1\right) + \sum_{j=2}^{m}\tau_j\sum_{i=1}^{n} c_{1i}A_{ij}$$

The first derivatives of ψ are equal to zero, giving the following set of equations:

$$2c_{1k}\sigma_k^2 + \sum_{j=1}^{m}\tau_jA_{kj} = 0, \quad (k = 1, \ldots, n) \tag{2.34}$$

The conditions of (2.29), (2.30) and (2.34) consist of a system of (m + n) equations to be solved.

We have $D[x_j] = \tau_j/2$, where the set of τ_j are defined as solution of the following equations:

$$\sum_{j=1}^{m}\mu_j\sum_{i=1}^{n}\frac{A_{ij}A_{i1}}{\sigma_i^2} = -2; \quad \sum_{j=1}^{m}\mu_j\sum_{i=1}^{n}\frac{A_{ij}A_{il}}{\sigma_i^2} = 0, \; (l = 2, \ldots, m)$$

These algorithms are used as sub-blocks of the Aral-Caspian Expert System (ACES) (Bondur et al. 2009). The forecast of the ACES state is obtained from the GIMS.

2.12.4 Method of Differential Approximation

Databases of the environment monitoring systems not always correspond to the parametrical fullness in framework of the GIMS technology standard. Therefore an algorithm that allows to adapt the database to this standard is considered. Let us suppose that N characteristics, x_i (i = 1, …, N), of environment are measured at the times t_s (s = 1, …, M). Formal dependence between x_i is represented by the system of differential equations with unknown coefficients $\{a_{ijk}, b_{ij}\}$:

$$\frac{d\xi_i}{dt} = \sum_{k,j=1}^{N} \left[a_{ijk}\xi_j(t)\xi_k(t) + b_{ij}\xi_j(t)\right] \tag{2.35}$$

Putting the initial conditions as

$$\xi_i(0)\ (i = 1, \ldots, N) \tag{2.36}$$

the reconstruction task of $\xi_i(t)$ for the arbitrary time $t \in [0,T]$ comes down to the simple task of determining unknown coefficients based on the criterion:

$$E = \sum_{s=1}^{M} \left\{ \sum_{i=1}^{N} [\xi_i(t_s) - \xi_i(t_s)]^2/N \right\}/M \tag{2.37}$$

where $0 \leq t_1 \leq \cdots \leq t_M \leq T$.

Coefficients $\{a_{ijk}, b_{ij})$ can be evaluated as solution of the following optimization task:

$$E_0 = \min_{\{a_{ijk}, b_{ij}, \xi_i(0)\}} E \tag{2.38}$$

There are many methods to solve this task. One of them is based on the Bellman's dynamic programming method (Bellman and Roth 1966; Krapivin and Kondratyev 2002; Krapivin 1969; Ram 2010). In this case search of minimal value of function E comes to the dynamic programming task. Let us supposed that coefficients $\{a_{ijk}, b_{ij})$ are functions of time:

$$Y(t) = \left\| \begin{matrix} \xi_1(t) \\ \vdots \\ \xi_N(t) \\ a_{111}(t) \\ \vdots \\ a_{NNN}(t) \\ b_{11}(t) \\ \vdots \\ b_{NN}(t) \end{matrix} \right\| \tag{2.39}$$

We can consider that $a_{ijk} = a_{ikj}$. Then addicting the system (2.35) Cauchy problem (2.35), (2.36) comes to the following task:

$$dY/dt = G(Y) \tag{2.40}$$

where function G has components:

$$G_i(Y) = 0 \quad \text{for} \quad i = N+1, \ldots, N_c,$$
$$G_i(Y) = \sum_{k,j=1}^{N} [a_{ijk}\ \xi_j(t)\xi_k(t) + b_{ij}\xi_j(t)] \text{ for } i = 1, \ldots, N; \tag{2.41}$$

where $a_{ijk}(0) = \bar{a}_{ijk}, b_{ij}(0) = \bar{b}_{ij},\ N_c = N + N^2 + N^2(N+1)/2$.

The Newton-Raphson method (Ram 2010) gives a possibility to solve system (2.40), (2.41). Let, introduce series of functions $Y^{(1)}(t), \ldots, Y^{(n)}(t)$ such that $Y^{(1)}(t)$ is the first approximation of solution for system equations (2.40). Then, nth approximation is solution of the following linear system:

$$d\,Y_i^{(n)}(t)/dt = G_i\left[Y^{(n-1)}(t)\right] + \sum_{j=1}^{N_c} \left\{dG_i\left[Y^{(n-1)}(t)\right]/dY\right\}\left[Y_j^{(n)} - Y_j^{(n-1)}\right] \tag{2.42}$$

As it was shown by Bellman and Dreyfus (1962), iterative process (2.42) is convergent by the quadratic law. Solution of Eq. (2.42) is described in the following form:

$$Y^{(n)}(t) = P(t) + \sum_{k=1}^{N_c} C_k H^{(k)}(t) \tag{2.43}$$

where $P(t)$ is the particular solution of the Eq. (2.42), $H^{(k)}(t)$ is vector solution of homogeneous system. To define function $P(t)$ it is needed to solve the Eq. (2.42) under initial conditions: $Y_i(0) = 0$ ($i = 1, \ldots, N_c$). Functions $H^{(k)}(t)$ are determined as solutions of the Cauchy problem:

$$\mathrm{d}\,Y_i^{(n)}(t)/dt = \sum_{j=1}^{N_c} \left\{dG_i\left[Y^{(n-1)}(t)\right]/dY\right\}[Y_j^{(n)} - Y_j^{(n-1)}]\ (i = 1, \ldots, N_c) \tag{2.44}$$

$$H^{(1)}(0) = \left\|\begin{matrix}1\\0\\\vdots\\0\end{matrix}\right\|,\ H^{(2)}(0) = \left\|\begin{matrix}0\\1\\\vdots\\0\end{matrix}\right\|,\ \ldots, H^{(N_c)}(0) = \left\|\begin{matrix}0\\0\\\vdots\\1\end{matrix}\right\|, \tag{2.45}$$

As it is followed from Eqs. (2.42)–(2.45), constants C_k are unknown initial conditions of system of equations (2.41). Therefore, constants C_k are defined from condition:

$$E = \sum_{s=1}^{M}\sum_{i=1}^{N}\left[P_i(t_k) + \sum_{k=1}^{N_c} C_k H_i^{(k)}(t_s) - \hat{\xi}_i(t_s)\right]^2 = \min_{\{C_k\}} \quad (2.46)$$

Let demand:

$$\partial E/\partial C_k = 0 \quad \text{for } k = 1, \ldots, N_c \quad (2.47)$$

From Eqs. (2.46) and (2.47) it is followed that

$$\sum_{k=1}^{N_c} A_{km} C_k + B_m = 0, \; m = 1, \ldots, N_c, \quad (2.48)$$

where

$$A_{km} = \sum_{s=1}^{M}\sum_{i=1}^{N} H_i^{(k)}(t_s) H_i^{(m)}(t_s), \quad B = \sum_{s=1}^{M}\sum_{i=1}^{N} [P_i(t_s) - \hat{x}_i(t_s)] H_i^{(m)}(t_s)$$

Finally, every iteration of (2.42) needs to solve the system of equations (2.48). Unfortunately, the convergence of this procedure depends on the successful selection of initial conditions.

2.12.5 Quasi-Linearization Method

The number of problems that crop up in ecoinformatics leads to the necessity of integrating generally non-linear integro-differential equations; but, in a majority of cases, these equations are not integrable by elementary or special functions. To solve them, as a rule, it is necessary to make use of the latest achievements of calculating methods and technics. In many problems the use of well-known numerical methods to solve initial value problems—even by means of modern high-speed electronic computers—does not come up with desired results. The existing approximate methods of solving integro-differential equations are based as rule on replacing derivatives by finite differences and represent a complicated multi-step process, which in practical problems cannot be solved on computers reasonably quickly. Therefore in solving practical problems we have to search other means of approximate solutions for integro-differential equations, without using the finite-difference methods.

In the method considered here, the integro-differential equation is substituted in each subinterval of the independent variable by an easily integrable ordinary differential equation with constant coefficients; this method is not a new theoretical idea for it was known to Euler. However, here the error estimations are obtained for the first time, and methods applicable to various problems are developed in detail.

2.12.5.1 Method of Solution and Estimation of Error

Let us consider the equation

$$L[y] - \lambda W[y] = f(x, y), \tag{2.49}$$

$L[y]$ being the differential operator

$$L[y] = \sum_{i=0}^{n} P_i(x, y, y', \ldots, y^{(m_i)})y^{(n-i)}, \quad (m_i < n) \tag{2.50}$$

and $W[y]$ the generalized Volterra operator

$$W[y] = \int_a^x \sum_{j=0}^{r} K_j(x, \xi)y^{(j)}(\xi)d\xi, \quad (r < n) \tag{2.51}$$

λ—a real number, $P_i(x, y, y', \ldots, y^{(m_i)}$ and $f(x,y)$—continuous functions with respect to their arguments in the finite interval $[a,b]$, $P_0 \neq 0$ and kernels $K_j(x,\xi)$, $j = 0, 1, \ldots, r$ are continuous functions in the region $G\{a \leq \xi \leq x \leq b\}$.

The initial conditions are

$$y^{(s)}(a) = y_0^{(s)}, \quad s = 0, 1, \ldots, n-1 \tag{2.52}$$

Assuming that Eq. (2.49) with the initial conditions (2.52) has a unique continuous solution $y(x)$, let us construct an approximate solution $\tilde{y}(x)$ in $[a,b]$. Let us divide the interval $[a,b]$ by a sequence of points $x_0 = a, x_1, \ldots, x_m = b, h_k = x_{k+1} - x_k$. On each subinterval $[x_k, x_{k+1}]$, $k = 0, 1, \ldots, m-1$ let us replace Eq. (2.49) by the following linear differential equation of the nth order with constant coefficients

$$\tilde{L}_k[y] = \lambda \tilde{W}_k[y] + f(x_k, \tilde{y}_k) \tag{2.53}$$

with the initial conditions:

$$y^{(s)}(x_k) = \tilde{y}_k^{(s)}, \quad s = 0, 1, \ldots, n-1, \tag{2.54}$$

where

$$\tilde{L}_k[y] = \sum_{i=0}^{n} P_i(x_k, \tilde{y}_k, \tilde{y}_k', \ldots, \tilde{y}_k^{(m_i)}\tilde{y}^{(n-i)}, \tag{2.55}$$

$$\tilde{W}_k[y] = \sum_{j=0}^{r} (K_{j,k,0}\tilde{y}_0^{(j)}h_0 + K_{j,k,1}\tilde{y}_1^{(j)}h_1 + \cdots + K_{j,k,k}\tilde{y}_k^{(j)}h_k) \tag{2.56}$$

The general solution of Eq. (2.53) is known:

$$\tilde{y} = \tilde{y}(x, c_1^{(k)}, c_2^{(k)}, \ldots, c_n^{(k)}), \qquad (2.57)$$

where the constants are determined from the initial conditions at the beginning of each interval $[x_k, x_{k+1}]$. The calculations are carried out successively beginning with the interval $(k = 0)$.

Let us estimate the error in solution of Eq. (2.49). Let $y(x)$ and $\tilde{y}(x)$ by the exact and the approximate solutions respectively. Let us denote

$$\begin{aligned} &\tilde{P}_{ik} = P_i(x_k, \tilde{y}_k, \tilde{y}'_k, \ldots, \tilde{y}_k^{(m_i)}), \quad \tilde{f}_k = f(x_k, \tilde{y}_k), \\ &\varepsilon_k = y(x_k) - \tilde{y}(x_k) \end{aligned} \qquad (2.58)$$

Let us integrate times Eqs. (2.49) and (2.53) from x_k to x, and consider the final results for $x = x_{k+1}$. For the sake of convenience and brevity let us denote

$$\underbrace{\int_{x_k}^{x_{k+1}} \int_{x_k}^{x} \cdots \int_{x_k}^{x}}_{n} \varphi(x)\, \underbrace{dx \cdots dx}_{n} = \underset{n}{\int_{x_k}^{x_{k+1}}} \varphi(x) dx,$$

we have

$$\begin{aligned} y_{k+1} = y_k + y_k^{©} h_k + h_k^2 \sum_{s=2}^{n-1} y_k^{(s)} \frac{h_k^{s-2}}{s!} - \sum_{i=1}^{n} \underset{n}{\int_{x_k}^{x_{k+1}}} P_i y^{(n-i)} dx \\ \underset{n}{\int_{x_k}^{x_{k+1}}} f(x, y) dx + \lambda \underset{n}{\int_{x_k}^{x_{k+1}}} W[y] dx \end{aligned} \qquad (2.59)$$

$$\begin{aligned} \tilde{y}_{k+1} = \tilde{y}_k + \tilde{y}'_k h_k + h_k^2 \sum_{s=2}^{n-1} \tilde{y}_k^{(s)} \frac{h_k^{s-2}}{s!} - \sum_{i=1}^{n} \underset{n}{\int_{x_k}^{x_{k+1}}} \tilde{P}_{ik} \tilde{y}^{(n-i)} dx \\ + \underset{n}{\int_{x_k}^{x_{k+1}}} \tilde{f}_k dx + \lambda \underset{n}{\int_{x_k}^{x_{k+1}}} \tilde{W}_k[y] dx \end{aligned} \qquad (2.60)$$

From (2.59) and (2.60) we get:

$$\varepsilon_{k+1} = \varepsilon_k + \varepsilon_k^{'} h_k + h_k^2 \sum_{s=2}^{n-1} \varepsilon_k^{(s)} \frac{h_k^{s-2}}{s!} - \sum_{i=1}^{n} \int_{x_k}^{x_{k+1}} [P_i y^{(n-i)} - \tilde{P}_i \tilde{y}^{(n-i)}] dx$$
$$n$$
$$+ \int_{x_k}^{x_{k+1}} [f - \tilde{f}_k] dx + \lambda \int_{x_k}^{x_{k+1}} \{W[y] - \tilde{W}_k[y]\} dx$$
$$n \qquad n$$

We know that

$$\int_{x_k}^{x_{k+1}} \tilde{f}_k dx = \frac{\tilde{f}_k h_k^n}{n!}, \quad \int_{x_k}^{x_{k+1}} \tilde{W}_k[y] dx = \frac{\tilde{W}_k[y]}{n!}$$
$$n \qquad n$$

Let us denote

$$E_k = \max_j \left| \varepsilon_k^{(j)} \right|, \quad h_{\max} = \max_k h_k, \quad p_i = \max_{[a,b]} |P_i|, \quad M_{n-i} = \max_{[a,b]} \left| y^{(n-i)} \right|,$$
$$L_i = \max_{[a,b]} \left| \tilde{P}_{ik} \right|, \quad N_{n-i} = \max_{[a,b]} \left| \tilde{y}^{(n-i)} \right|, \quad F = \max_{[a,b]} |f|, \quad G_0 = \max_{[a,b]} \left| \tilde{f}_k \right|,$$
$$T = \max_G \left| \tilde{W}_k[y] \right|, \quad s = |b - a| \sum_{j=0}^{r} \max_G \left| K_j(x, \xi) \right| M_j \geq \max_G |W[y]|,$$
$$l = \frac{1}{n!} \left[\sum_{i=1}^{n} (p_i M_{n-i} + L_i N_{n-i}) + F + G_0 + |\lambda| (T + s) \right],$$
$$g = 1 + \sum_{s=2}^{n-1} \frac{h_{ax}^{s-1}}{s!},$$

and $M_{n-i} \approx N_{n-i}$. Then we get the following recurrent error estimation

$$E_{k+1} \leq (1 + g h_{\max}) E_k + l h_{\max}^n \tag{2.61}$$

Hence we get

$$E_k \leq (1 + g h_{\max})^k \varepsilon_0 + l h_{ax}^{n-1} g^{-1} [(1 + g h_{\max})^k - 1] \tag{2.62}$$

where ε_0 is the maximum error in the initial data. Obviously, if $\varepsilon_0 = 0$, then from (2.62) it follows that if $h_{\max} \to 0$ then $E_k \to 0$, i.e. $\tilde{y}(x_k) \to y(x_k)$.

In case that the Eq. (2.49) has the form:

$$L[y] = f(x,y) + \int_a^x F(x,y,y',\ldots,y^{(m_l)})dx, \quad x \in [a,b],$$

then the Eq. (2.62) will read

$$E_k \le (1 + hp^{(0)})^k \varepsilon_0 + \frac{p^{(1)}h^n}{p^{(0)}}[(1 + hp^{(0)})^k - 1],$$

where

$$p^{(0)} = \sum_{s=1}^{n-1} \frac{h^{s-1}}{s!} + \frac{h^{n-1}}{n!}\left[\tilde{b} + \sum_{i=1}^{n}(p_i + (m_l+1)\gamma_i\beta_{n-i}) + (b-a)c(m_l+1)\right],$$

$$p^{(1)} = \frac{1}{(n+1)!}\left(\tilde{a} + \tilde{b}\beta_1 + 2B + \sum_{i=0}^{n} p_{0i}\right) + \frac{(b-a)q}{n!},$$

$$\tilde{a} = \max_{[a,b]}\left|\frac{\partial f}{\partial x}\right|, \quad \tilde{b} = \max_{[a,b]}\left|\frac{\partial f}{\partial y}\right|, \quad \gamma_i = \max_{s,[a,b]}\left|\frac{\partial P_i}{\partial y^{(s)}}\right|, \quad \beta_i = \max_{[a,b]}\left\{\left|y^{(i)}\right|, \left|\tilde{y}^{(i)}\right|\right\},$$

$$c = \max_{s,[a,b]}\left|\frac{\partial F}{\partial y^{(s)}}\right|, \quad q = 0.5\left(A + C\sum_{s=0}^{m_l}\beta_{s+1}\right), \quad p_{oi} = \beta_{n-i}\left(l_i + \gamma_i\sum_{s=0}^{m_l}\beta_{s+1}\right) + 2\alpha_i\beta_{n+1-i},$$

$$A = \max_{[a,b]}\left|\frac{\partial F}{\partial x}\right|, \quad B = \max_{[a,b]}\{|F|, |\tilde{F}|\}$$

2.12.5.2 Solution of Equation $y^{(n)} = f(x,y,y', \ldots, y^{(n-1)})$

Let us apply the approximate method of the solution presented in Sect. 2.12.5.1 to integro-differential equations to solve the initial value problem:

$$y^{(n)} = f(x,y,y', \ldots, y^{(n-1)}),\ (x,y) \in G, \tag{2.63}$$

$$y(x_0) = y_0, \quad y^{(j)}(x_0) = y_0^{(j)},\ j = 1, \ldots, n-1, \quad (x_0, y_0) \in G, \tag{2.64}$$

where the function f satisfies the Lipschitz condition

$$\left|f(x, y+\delta_0, \ldots, y^{(n-1)} + \delta_{n-1}) - f(x,y,y', \ldots, y^{(n-1)})\right| \le K\sum_{i=0}^{n-1}|\delta_i| \tag{2.65}$$

Let us divide the interval $[a,b]$ by a sequence of points $x_0 = a, x_1, \ldots, x_s = b$ into elementary intervals. Let $E = \{x_0, \ldots, x_s\}$. On each interval $[x_\nu, x_{\nu+1}]$, let us solve the initial value problem:

$$y^{(n)} = f(x, \hat{y}_\nu, \hat{y}'_\nu, \ldots, \hat{y}_\nu^{(n-1)}), \quad (x, \hat{y}_\nu) \in G, \quad \nu = 0, 1, \ldots, s-1;$$
$$y(x_\nu) = \hat{y}_\nu, \quad y^{(j)}(x_\nu) = \hat{y}_\nu^{(j)}, \quad j = 1, \ldots, n-1, \quad (x_\nu, \hat{y}_\nu) \in G$$

Then, if the function f satisfies condition (2.65) and $\max_i[x_{i+1} - x_i] = h$, the solution of this problem

$$\hat{y} = \{y_0, \hat{y}_1, \ldots, \hat{y}_s\}, \quad \hat{y}^{(j)} = \{y_0^{(j)}, \hat{y}_1^{(j)}, \ldots, \hat{y}_s^{(j)}\}, \quad j = 1, \ldots, n-1$$

when $h \to 0$ tends to the solution of Eqs. (2.63), (2.64) and the estimation for the rate of convergence is as follows:

$$\max_l \left| y_r^{(n-l)} - \hat{y}_r^{(n-l)} \right| \le \varepsilon_0 (1 + h\alpha_0)^r + \frac{h\alpha_1}{2\alpha_0}[(1 + h\alpha_0)^r - 1], \quad r = 1, 2, \ldots, s$$

where

$$\alpha_0 = \sum_{i=1}^{n-1} \frac{h^{i-1}}{i!} + K\left(n + \frac{h}{2}\sum_{j=0}^{n-1}\sum_{s=0}^{n-j-2}\frac{h^s}{s!}\right),$$
$$\alpha_1 = K\sum_{j=0}^{n-1}\left[M\frac{h^{n-j-1}}{(n-j-1)!} + \sum_{s=0}^{n-j-2} N_{s+j+1}\frac{h^s}{s!}\right],$$
$$M = \max_{[a,b]}\left| f(x, y, y', \ldots, y^{(n-1)}) \right|, \quad N_{s+j+1} = \max_r \left| \hat{y}_r^{(s+j+1)} \right|$$

If the initial conditions are exactly given, the error estimation has the form:

$$\max_{1 \le l \le n-1} \left| y_r^{(n-l)} - \hat{y}_r^{(n-l)} \right| \le Dh^2, \quad r = 1, \ldots, s,$$

where

$$N = \max_{1 \le r \le s} |N_r|, \quad D = K(M + nD)\frac{\left[1 + he^h(1 + 0.5Knh) + Kn\right]^r - 1}{4[1 + Kh(0.5h + e^{-h})]}$$

2.12.5.3 Solution of a System of Ordinary Differential Equations

For the sake of simplicity let us confine ourselves to the important case of equations, having the canonical form

$$y_i^{(m_i)}(t) = f_i(t, y_1, y_1', \ldots, y_1^{(m_1-1)}, \ldots, y_n^{(m_n-1)}), \quad i = 1, \ldots, n \qquad (2.66)$$

The system (2.66) can be replaced by an equivalent system of $m = m_1 + \cdots + m_n$ equations of the first order, relative to the derivatives for all m unknown functions. Then one of the standard software can be used to solve the last system.

Let the functions f_i, $i = 1, \ldots, n$ be continuous and differentiable with respect to all arguments. Let us suppose that the solution of system (2.66) with the initial conditions

$$y_i(t_0) = (y_i)_0, \quad y_i'(t_0) = (y_i')_0, \ldots, y_i^{(m_i-1)}(t_0) = (y_i^{(m_i-1)})_0$$

exists and is unique in $t_0 \leq t \leq T$.

Divide the interval $[t_0,T]$ into elementary intervals $\Delta k = [t_k,t_{k+1}]$ by a sequence of points $t_0 < t_1 < \cdots < t_l = T$. On each such interval, let us search the solution of system (2.66) in the form of a series:

$$\tilde{y}_i(t) = \tilde{y}_i(t_k) + \sum_{j=1}^{m_i-1} \frac{(t-t_k)^j}{j!} \tilde{y}^{(j)}(t_k) + \frac{(t-t_k)^{m_i}}{(m_i)!} (\tilde{f}_i)_k,$$

$$\tilde{y}_i^{(j)}(t_k) = \tilde{y}_i^{(j)}(t_{k-1}) + \sum_{s=1}^{m_i-j-1} \frac{(t_k-t_{k-1})^s}{s!} \tilde{y}_i^{(s+j)}(t_{k-1}) + \frac{(t_k-t_{k-1})^{m_i-j}}{(m_i-j)!} (\tilde{f}_i)_k, \quad (j = 1, \ldots, m_i - 1)$$

The error of such a solution can be easily estimated, considering the exact expansion of the functions $y_i(t)$ and $y_i^{(j)}(t)$ in a Taylor series:

$$|\varepsilon_i(t_{k+1})| \leq |\varepsilon_i(t_k)| + \sum_{j=1}^{m_i-1} \frac{h_k^j}{j!} \left|\varepsilon_i^{(j)}(t_k)\right| + M_i \frac{h_k^{m_i+1}}{(m_i+1)!} + M_i \frac{h_k^{m_i}}{(m_i)!} \sum_{j=1}^{n} \sum_{s=0}^{m_j-1} \left|\varepsilon_j^{(s)}(t_k)\right| \tag{2.67}$$

$$\left|\varepsilon_i^{(j)}(t_k)\right| \leq \left|\varepsilon_i^{(j)}(t_{k-1})\right| + \sum_{s=1}^{m_i-j-1} \frac{h_{k-1}^s}{s!} \left|\varepsilon_i^{(s+j)}(t_{k-1})\right| + M_i \left(\frac{h_k^{m_i-j+1}}{(m_i-j+1)!} + \frac{h_k^{m_i-j}}{(m_i-j)!} \sum_{j=1}^{n} \sum_{s=0}^{m_j-1} \left|\varepsilon_j^{(s)}(t_{k-1})\right| \right) \tag{2.68}$$

where

$$M_i = \max_{[t_0,T]} \left\{ \left|\frac{\partial f_i}{\partial t}\right|, \left|\frac{\partial f_i}{\partial y_1}\right|, \ldots, \left|\frac{\partial f_i}{\partial y_n^{(m_n-1)}}\right| \right\}$$

Formulas (2.67) and (2.68) give a recurrent estimation for error. From them it is possible to obtain an error estimation applicable to the entire interval $[t_0, T]$:

$$E_k \leq \varepsilon_0 (1 + hp_0)^k + \frac{hP_1}{p_0} \left[(1 + hp_0)^k - 1 \right],$$

where the following notations are introduced

$$h = \max_k h_k, \quad E_k = \max_{i,j} \left| \varepsilon_i^{(j)}(t_k) \right|, \quad M = \max_i M_i, \quad v = \min_s m_s,$$

$$P_1 = M \frac{h^{v-1}}{(v+1)!}, \quad p_0 = mM \frac{h^{v-1}}{v!} + \sum_{s=1}^{\mu-1} \frac{h^{s-1}}{s!}, \quad \mu = \max_s m_s$$

2.12.5.4 Solutions of Equations with Known Moments on the Right Side

Let us consider a particular case of Eq. (2.49):

$$L[y] = y^{(n)} + \sum_{i=1}^{n} p_i y^{(n-i)} = f(x), \quad x \geq 0 \tag{2.69}$$

where p_i is the constant coefficients, $f(x)$ is a single-valued and differentiable function, $f(x) \to 0$ as $x \to \infty$ and its moments are known:

$$M_v f(x) = \int_0^{\infty} x^v f(x) dx < \infty, \quad v = 0, 1, \ldots, m \tag{2.70}$$

It is necessary to solve the Eq. (2.69) with the following initial conditions:

$$y^{(s)}(x_0) = y_0^{(s)} \quad (s = 0, 1, \ldots, n-1)$$

We shall approximate $f(x)$ in the following manner:

$$f(x) \approx \exp(-kx) \sum_{i=0}^{m} a_i x^i = P_m(x) \exp(-kx), \tag{2.71}$$

where $m > 0$ is an integer, $k > 0$ and a_i are constants to be determined. Then, from (2.70) and (2.71) we have

$$\tilde{M}_v f(x) = \int_0^{\infty} x^v \left[\exp(-kx) \sum_{i=0}^{m} a_i x^i \right] dx = \sum_{i=0}^{m} a_i \frac{(v+i)!}{k^{v+i+1}}$$

k is fixed from the conditions of best approximation by (2.71). Then the Eq. (2.69) is replaced by the approximate equation:

$$L[\tilde{y}] = \exp(-kx)\sum_{i=0}^{m} a_i x^i, \tag{2.72}$$

which can be easily solved. For the error $\varepsilon(x) = y(x) - \tilde{y}(x)$, we obtain an equation from (2.69) to (2.72):

$$L[\varepsilon(x)] = f(x) - \exp(-kx)P_m(x) \equiv R_m(x) \tag{2.73}$$

Solving Eq. (2.73), we have

$$|\varepsilon(x)| \leq \frac{\eta}{n!}\sum_{k=0}^{\infty}\frac{(b-a)^{n+k}}{k!}M^k,$$

where

$$M = \max_{a \leq s \leq x \leq b}|K(x,s)|, \quad \eta = \max_{[a,b]}|R_m(x)|, \quad K(x,s) = \sum_{i=1}^{n} p_i \frac{(x-s)^{i-1}}{(i-1)!}$$

2.12.5.5 Refinements of Approximate Solutions of Volterra Integral Equations

Let us consider Volterra integral equation of the first and of the second kind arising in the remote monitoring problems:

$$\lambda \int_a^x G(x,y)\varphi(y)dy = g(x), \tag{2.74}$$

$$\varphi(x) - \lambda \int_a^x K(x,y)\varphi(y)dy = f(x), \tag{2.75}$$

where $x \in [a,b]$, the kernel $K(x,y)$ and its derivatives $K'_x(x,y)$ are continuous in the region $R\{a \leq y \leq x \leq b\}$, $f(x)$ is a continuously differentiable function in (a,b), the kernek $G(x,y)$ and $g(x)$ are twice continuously differentiable functions of x, $G(x,x) \neq 0$. Then, as it is known, Eqs. (2.74) and (2.75) have unique solutions $\varphi_1(x)$ and $\varphi_2(x)$ respectively which are continuous and differentiable in $[a,b]$ for any value of λ. The case when $G(x,x) = 0$ for some point in the interval $[a,b]$ or for the entire interval needs special consideration. In our case, Eq. (2.74) is equivalent to the equation of second kind

$$\varphi(x) + \int_a^x \frac{G'_x(x,y)}{G(x,x)} \varphi(y)dy = \frac{g'(x)}{\lambda G(x,x)}$$

Therefore, the argument used to find an approximate solution of (2.75) is valid for Eq. (2.74) as well.

In traditional way the Eq. (2.75) is solved by replacing the integral of the equation by a finite sum of some quadratic formula. Applying this approach let us divide the interval $[a,b]$ by a sequence of points $x_0 = a < x_1 < x_2 < \cdots < x_m = b$ into elementary intervals $\Delta j = [x_j, x_{j+1}]$, and instead of (2.75) let us write the equation:

$$\varphi(x_j) - \lambda \sum_{i=0}^{j-1} \int_{x_i}^{x_{i+1}} K(x_j, y)\varphi(y)dy = f(x_j), \quad (j = 0, 1, \ldots, m) \tag{2.76}$$

Further, because of the assumptions made on $\varphi(x)$ and $K(x,y)$ we can write

$$\varphi(x) = \varphi(x_i) + (x - x_i)\varphi'(x_i) + \frac{(x - x_i)^2}{2!}\varphi''(\xi_i),$$
$$(x_i \le \xi_i \le x \le x_{i+1}) \tag{2.77}$$

and assuming the existence and differentiability of $K'_y(x,y)$ we have

$$K(x_j, y) = K(x_j, x_i) + (y - x_i)K'_y(x_j, x_i) + \frac{(y - x_i)^2}{2} K''_{yy}(x_j, \eta_i),$$
$$(x_i \le \eta_i \le y \le x_{i+1}) \tag{2.78}$$

substituting (2.77) and (2.78) in (2.76) we get:

$$\varphi(x_j) - \lambda \sum_{i=0}^{j-1} \{K(x_j, x_i)\varphi(x_i) + \frac{h_i}{2}[K(x_j, x_i)\varphi'(x_i) + K'_y(x_j, x_i)\varphi(x_i)]$$
$$+ \frac{h_i^2}{3} K'_y(x_j, x_i)\varphi'(x_i)\}h_i + R_j = f(x_j)$$

where

$$R_j = -\lambda \sum_{i=0}^{j-1} \{\frac{h_i^3}{2}\varphi''(\xi_i)[\frac{1}{3}K'_y(x_j, x_i) + \frac{h_i}{4}K'_y(x_j, x_i)]$$
$$+ \frac{h_i^3}{2}K''_{yy}(x_j, \eta_i)[\frac{1}{3}\varphi(x_i) + \frac{h_i}{4}\varphi'(x_i)] + \frac{h_i^5}{20}\varphi''(\xi_i)K''_{yy}(x_j, \eta_i)\}$$

Neglecting the small quantity R_j in this expression, we get a recurrent formula for determining $\varphi(x_j)$ from the values of the function $\varphi(x)$ at $x = x_0, x_1, \ldots, x_{j-1}$. By differentiating (2.75) we get a formula to calculate the values of the derivative $\varphi'(x)$:

$$\varphi'(x) = f'(x) + \lambda K(x,x)\varphi(x) + \lambda \int_a^x K'_x(x,y)dy$$

From this equation, we have at $x = x_i$

$$\varphi'(x_i) = f'(x_i) + \lambda K(x_i,x_i)\varphi(x_i) + \lambda \sum_{s=0}^{i-1} \int_{x_s}^{x_{s+1}} K'_x(x_i,y)\varphi(y)dy \tag{2.79}$$

where

$$\varphi(x_0) = f(a), \varphi'(x_0) = f'(a) + \lambda K(a,a)\varphi(a)$$

Neglecting the quantity

$$r_i = \lambda \sum_{s=0}^{i-1} \{K''_{xy}(x_i,Q_s)\frac{h_s^2}{2}\left[\varphi(x_s) + \frac{2}{3}\varphi'(x_s)h_s + \frac{h_s^2}{4}\varphi''(\xi_s)\right]$$
$$+ K'_x(x_i,x_s)\frac{h_s^2}{6}\varphi''(\xi_s)\}, \quad (x_s \le \xi_s, Q_s \le y \le x_{s+1})$$

Equation (2.79) can be written as:

$$\tilde{\varphi}'(x_i) = f'(x_i) + \lambda K(x_i,x_i)\tilde{\varphi}(x_i) + \lambda \sum_{s=0}^{i-1} h_s K'_x(x_i,x_s)\left[\tilde{\varphi}(x_s) + \frac{h_s}{2}\tilde{\varphi}'(x_s)\right]$$

We get finally the equation for determining $\tilde{\varphi}(x_j)$:

$$\tilde{\varphi}(x_j) - f(x_j) - \lambda \sum_{i=0}^{j-1} \left[\tilde{\varphi}(x_i)K(x_j,x_i)l_{ji} + h_i\gamma_{ji}\delta_i\right]h_i = 0, \tag{2.80}$$

where, for sake of brevity, the following notation is introduced:

$$l_{ji} = 1 + \frac{h_i}{2}\left[\lambda K(x_i, x_i) + \frac{K'_y(x_j, x_i)}{K(x_j, x_i)}\left[1 + \frac{2}{3}\lambda h_i K(x_i, x_i)\right]\right]$$

$$\gamma_{ji} = \frac{1}{2}K(x_j, x_i)\left[1 + \frac{2}{3}h_i \frac{K'_y(x_j, x_i)}{K(x_j, x_i)}\right]$$

$$\delta_i = f'(x_i) + \lambda \sum_{s=0}^{i-1} h_s K'_x(x_i, x_s)\left[\tilde{\varphi}(x_s) + \frac{h_s}{2}\tilde{\varphi}'(x_s)\right]$$

Thus, starting the calculation by formula (2.80) from $j = 1$. We get the values of solution $\tilde{\varphi}_1, \tilde{\varphi}_2, \ldots, \tilde{\varphi}_m$ with the error $\varepsilon(x_i)$. Let us estimate the value of modulus of the error $\varepsilon(x)$ from above. Let us introduce the following notation:

$$M = \max_{[a,b]}|\varphi(x)|, \quad N = \max_{[a,b]}|\varphi'(x)|, \quad L = \max_{[a,b]}|\varphi''(x)|,$$

$$Q = \max_R\left\{\left|K'_y(x,y)\right|, \quad \left|K'_x(x,y)\right|\right\}, \quad G = \max_R|K(x,y)|,$$

$$B = \max_R\left\{\left|K''_{yy}(x,y)\right|, \quad \left|K''_{xy}(x,y)\right|\right\}, \quad h = \max_s|h_s|,$$

$$l = \frac{m}{6}\left[B\left(M + \frac{3h}{4}N\right) + L\left(G + \frac{3}{4}Qh + \frac{3}{10}Bh^2\right)\right],$$

$$n = \frac{|\lambda|Bm}{2}\left(M + \frac{2h}{3}N + \frac{h^2}{4}L\right) + \frac{h}{6}BL$$

Then we have

$$|\varepsilon_j| \le |\lambda|h\left(G + \frac{2}{3}Qh\right)\sum_{i=0}^{j-1}\left(|\varepsilon_i| + \frac{h}{2}|\varepsilon'_i|\right) + |\lambda|lh^3,$$

$$|\varepsilon'_i| \le |\lambda|G|\varepsilon_i| + |\lambda|Qh\sum_{s=0}^{i-1}\left(|\varepsilon_s| + \frac{h}{2}|\varepsilon'_s|\right) + nh^2,$$

where $\varepsilon_j = \varphi(x_j) - \tilde{\varphi}(x_j)$, $\varepsilon'_i = \varphi'(x_i) - \tilde{\varphi}'(x_i)$. From these formulas we get roughly

$$|\varepsilon'_i| \le |\lambda|\left(G|\varepsilon_i| + hQ\sum_{s=0}^{i-1}|\varepsilon_s|\right) + (n + |\lambda|QNm)h^2,$$

$$|\varepsilon_j| \le |\lambda|h\left(G + \frac{2}{3}Qh\right)\sum_{i=0}^{j-1}\left[\left(1 + \frac{|\lambda|Gh}{2}\right)|\varepsilon_i| + \frac{|\lambda|Gh}{2}\sum_{s=0}^{i-1}|\varepsilon_s|\right] + |\lambda|th^3,$$

where

$$t = l + \frac{h}{2} m \left(G + \frac{2}{3} Qh \right) (n + |\lambda QNm|)$$

Let us denote

$$T = |\lambda|(G + 2Qh/3), b = 1 + |\lambda| Gh/2, r = |\lambda| \left(t + QMTm^2 \right)$$

Then we get

$$|\varepsilon_j| \leq hTB \sum_{i=0}^{j-1} |\varepsilon_i| + rh^3$$

or, finally, we have

$$|\varphi_j - \tilde{\varphi}_j| \leq E_j = \varepsilon_0 Z_1^j + rh^3 (1 - Thb_j)^{-1} \leq \varepsilon_0 + rh^3 (1 - Thb_j)^{-1},$$

where $h \leq (Tb)^{-1}$, Z_1 is the real root of the equation

$$Z^{j+1} - TbhZ^{j-1}/(Z - 1) = 0$$

between $Z = 1$ and $Z = hTb$.

Chapter 3
Environmental Decision-Making

3.1 Introduction

Environmental decisions are made based on different kinds of information. In a specific decision-making situation the choice of the appropriate tool depends on the decision context. In fact, there are numerous available tools which provide decision support. Adoption and substantiation of decisions are essential parts of the whole set of tasks related to artificial intelligence. Decision making procedures play a key role as far as classification and recognition tasks are concerned. In suppressing number of cases, decision rules are synthesized under the assumption that characteristics describing the objects for study are independent. In the case of environmental monitoring, the characteristics of nature and man-made systems are strongly correlated. Therefore, in this case, decision-making procedures have to be stable and take into account evolution processes.

In artificial intelligence, an expert system is a computer system which emulates the decision-making ability of a human expert (Jackson 1999; Murphy 2012; Russel and Norvig 2009). Expert systems are designed to solve complex problems by reasoning about knowledge, like an expert, and not by following the procedure of a developer as in the case of conventional programming. Both knowledge base and decision making procedures are crucial in expert system technology. Expert systems operate like a human brain and thus they offer many advantages for users when compared to traditional programs.

Every expert system has a major flaw, which explains its low success despite the principles on which it is based having existed for 70 years: knowledge collection and its interpretation into rules, or knowledge engineering. Most developers have no automated method to perform this task; instead they work manually, increasing the likelihood of errors. Expert knowledge is generally not well understood. For instance, rules may either be contradictory or poorly written and as a result unusable, or they even may not exist. Worse still, most expert systems use a computational engine incapable of reasoning. As a result, an expert system often works

V.F. Krapivin et al., *New Ecoinformatics Tools in Environmental Science*,
Environmental Earth Sciences, DOI 10.1007/978-3-319-13978-4_3

poorly and the project is abandoned (Laudon 2011). However, correct development methodology can mitigate these problems. There exists software capable of interviewing a true expert on a subject and automatically writing the rule or knowledge base, depending on the answers. The expert system can then be simultaneously run before the true expert's eyes, performing a consistency of rules check. Experts and users can check the quality of the software before it is finished.

The main mission of expert systems is to automatize and to computerize the decision making processes in different applied areas. The applications of expert systems find their way into most applications. For example, Karasulu and Korukoglu (2012) applied subtraction method that uses the frame differencing in order to develop an expert system for moving object detection. The basic idea is to subtract the current frame from a background image and to classify each pixel either as foreground or as background by comparing the difference with a threshold. Therefore, the moving object is detected and tracked by using frame differencing and by learning an updated background model. In addition, simulated annealing is an optimization technique for soft computing in the artificial intelligence area. This expert system can be used for tropical cyclone tracking.

There are numerous expert systems in different practical applications of knowledge. Such systems are used for prospecting medical diagnosis or as educational aids. Moreover, they are used in engineering and manufacture in the control of robots, where they interrelate with vision systems. The initial attempts to apply artificial intelligence to generalized problems made limited progress. However, more significant progress could be made in a restricted field of interest.

The period from the end of the 20th century to the beginning of the 21st century is characterized not only by the exacerbation of the environment but also by prognosis of negative consequences in global scale, because of anthropogenic impacts. Particularly, this exacerbation was intensified after the Kyoto Protocol. A settlement of arising disputes concerning the processes of the environment evolution is undoubtedly related to the efficiency of environmental monitoring systems synthesis, on which remote sensing technologies—both passive and active type—depend (Krapivin et al. 2009b, 2010a, 2011).

Methodical support of remote sensing systems of land covers and radio-communication stations by data about the character of screening effect of the vegetation cover at fixed moment of the time and in the concrete geographical point, requires the development of a technology which could reconstruct spatial images using the episodic in the time and fragmentary in the space observations.

The environmental monitoring regime can foresee decision making situations in real time based on the data accumulated up until the moment of the decision making or as a result of the prior data analysis without correlation to the current time. The statistical analysis of several events monitored by the system can be carried out by numerous methods. The applicability of these methods is determined in each case by a set of probabilistic parameters which characterize the phenomenon under study. However, non-steady state and parametrical uncertainty demand a search for new methods capable of making the decision using fragmentary time and space data. Such methods are proposed by a sequential analysis and evolutionary technology (Bukatova et al. 1991; Nitu et al. 2004; Lai 2001; Soldatov 2010, 2011; Soldatov et al. 2010a, b; Moberg 2006).

With the development of alternative methods of making statistical decisions, the problem of searching an objective estimation of the parameters of the processes taking place in the environment has obtained a new substantiation. This problem can be approached by two methods; on the one hand, a classical one based on the procedure of a-priori limitation of the number of observations, and on the other hand a successive analysis based on the procedure of the step-by-step resolve making. The development of computer technologies enables one to realize both approaches in the form of a single system of making statistical decisions.

Making decisions on detection of some effect in the process of continuous environmental monitoring depends on the scheme of measurements organization. The classical approach directs the observation system to a collection of a fixed volume of data, with their subsequent processing revealing certain effects or properties in the area under study. The method of successive analysis does not divide these stages but it alternates them. In other words, the monitoring data processing is carried out after each measurement. Hence, the algorithmic load in the successive procedure changes dynamically, whereas in a classical case the data are processed at a final stage. From the viewpoint of the formation of the structure of an adequate automated *resolver*, these approaches should be realized in the form of individual units, the choice between which can vary in the monitoring dynamics. However, for this purpose high professionalism is required. Therefore, in the proposed structure of the *resolver* the operator only controls the parameters of the choice between the alternative hypotheses.

The decision theory makes us consider the above-formulated problem of using sequential analysis in the procedure of the choice between the competing hypotheses in a more general form and with regard to all aspects appearing in such a problem. The list of the problems which should be solved before making a resolve covers the following subjects:

- selection of an efficient criterion to estimate the parameters;
- determination of the character of probabilistic characteristics of the process under study;
- a-priori assessment of possible losses on the accuracy of decisions.

One of the features of the automated system of making statistical resolves is the presence of a wide spectrum of functions, which should—if possible—provide a solution for most of intermediate and final problems of assessment of the situation in the monitoring regime. Therefore, the system should include the following units:

(1) visualization of the measured data in the form of direct images, histograms, sum and frequencies by time intervals;
(2) calculation of statistical characteristics of the observation results (mean, dispersion, moments of the third and higher order, excess, coefficient of asymmetry, entropy, etc.);
(3) construction of empirical density and function of probabilities distribution;

(4) construction of theoretical distribution of probabilities with its assessment by one of the statistical criteria;
(5) calculation of characteristics for the procedure of adopting hypotheses by the Neyman-Pearson classical method with the resulting assessment of the procedure by the current volume of samples;
(6) calculation of characteristics of the procedure of successive analysis and visualization of its state;
(7) realization of the functions of the operator's access to the system's units at any step of the system's functioning with possibilities to make resolves on changing the parameters of the procedure or ceasing measurements; and
(8) visualization of the resolves.

This set of units ensures the formation of the model of the measurement procedure depending on the a-priori information about the parameters and the character of the process under study. At the same time, the operator is able to test the accuracy of the input information and promptly change the strategy of monitoring. On the whole, the set of the indicated units is an automated system for making statistical decisions.

Moberg (2006) considered environmental systems analysis (ESA) tools as a quantitative and multidisciplinary research field aimed at combining, interpreting and communicating knowledge from the natural and social sciences and technology. In other words, ESA delivers methods and tools for the environmental assessment of human-made systems using a systems perspective. Between the tools that are most frequently used, Moberg (2006) indicated on the following tools:

- SEA (Strategic Environmental Assessment) that is a procedural tool for handling environmental (and sustainability) aspects in strategic decision-making (policies, programmes and plans); and
- EIA (Environmental Impact Assessment) that is a procedural tool required by law in some situations and that describes the environmental impact of a suggested project and its alternatives (e.g., the construction and localization of a waste incineration plant).

In general, not only these but also other tools play a crucial role in managing and minimizing the risk and the uncertainty arising because of the environmental decision making procedures. Uncertainty, which may be both scientifically and socially based, is a typical feature of decision-making processes. Specifically, environmental decisions are complex, since environment comprises many components, numerous processes, complex interconnections and feedback mechanisms. Uncertainty concept has different forms and kinds. The main problem of environmental decision-making is how the uncertainty can be reduced. This chapter describes constructive algorithms and methods as a solution to this problem.

3.2 Procedures for Classical and Sequential Decision-Making

3.2.1 Classical Neyman-Pearson Decision-Making Procedure

The classical decision-making procedure is based on the given volume, n, of the measurements. The value of n is determined by a priori information about the probability density function $f_a(x_1, \ldots, x_n)$ where the random variables $\{x_i\}$ are the observation data. The hypotheses H_0 and H_1 are that a 0 or a 1 is prescribed, respectively. The distinction between these hypotheses is based on the synthesis of the boundary for the optimal critical area, E_1, in the hyper-surface of the form:

$$L_n = L_n(x_1, \ldots, x_n) = f_{a1}(x_1, \ldots, x_n)/f_{a0}(x_1, \ldots, x_n) = C \tag{3.1}$$

where

$$f_a(x_1, \ldots, x_n) = \prod_{i=1}^{n} f_a(x_i),$$

$f_a(x)$ is the probability density function of the variable x with the unknown parameter a and C is a constant determined under the condition that E_1 has a given level of the error of the first kind α.

The ratio of conditional probabilities in Eq. (3.1), which is called likelihood ratio, provides the final choice between the above hypotheses:

(1) if $L_n \le C$ then hypothesis H_0 is accepted,
(2) if $L_n > C$ then hypothesis H_1 is chosen.

In general, there are many criteria such as Bayesian minimum error, minimax, etc. Dalton and Dougherty (2011) derived a closed-form analytic representation of the Bayesian minimum mean-square error estimator for linear classification assuming Gaussian models. This is presented in a general framework permitting a structure on the covariance matrices and a very flexible class of prior parameter distributions with four free parameters. Closed-form solutions are provided for known scaled identity and arbitrary covariance matrices. Minimax theory is developed in the framework of game theory (Krapivin 1972; Krapivin and Klimov 1995, 1997; Myerson 1997).

The content of hypotheses H_0 and H_1 depends on the specific conditions of the task. In fact, there are two steps to use for these criteria. The first step is the synthesis of the empirical function of the probability distribution for the observed variable x. The second step is its transformation to $f(x)$ using, for example, the Neyman-Pearson criterion (Fig. 3.1).

Let us consider the case of a uniform data set where the selected values x_i ($i = 1, \ldots, n$) are independent realizations of the same casual value ξ having the density $f_{a0}(x)$ for the hypothesis H_0 and the density $f_{a1}(x)$ when the hypothesis H_1 is

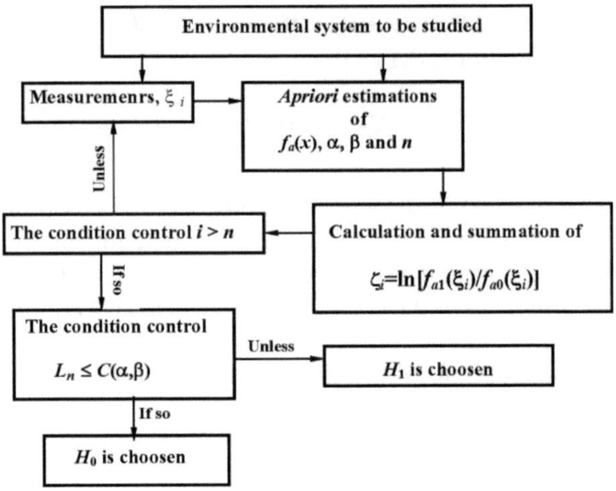

Fig. 3.1 Classical Neyman-Pearson decision-making procedure to choose between the two hypotheses H_0 and H_1

true. In this case, the parameter a of the real density f_a can be equal to neither a_0 or a_1.

Errors of the first kind, α, and the second kind, β, satisfy the following formulas respectively:

$$\alpha \approx \exp[-0.5\{(E_{a1}\xi - E_{a0}\xi)(D_{a0}\xi)^{-1/2}\}^2 n], \tag{3.2}$$

$$\beta \approx \exp[-0.5\{(E_{a1}\xi - E_{a0}\xi)(D_{a1}\xi)^{-1/2}\}^2 n], \tag{3.3}$$

where

$$E_a\xi = \int_{-\infty}^{\infty} \ln[f_{a1}(x)/f_{a0}(x)] f_a(x)\, dx \tag{3.4}$$

$$D_a\xi = \int_{-\infty}^{\infty} \{\ln[f_{a1}(x)/f_{a0}(x)]\}^2 f_a(x)\, dx - (E_a\xi)^2 \tag{3.5}$$

Neyman-Pearson theory focuses on testing a statistical hypothesis provided that the volume of observations allows the calculation of basic statistical characteristics including the empirical distribution function. In this case, theoretical probability density function $f_a(x)$ is assessed on the base of χ^2-criterion. Unfortunately, empirical results of environmental monitoring usually do not satisfy this condition. In this case a sequential procedure is used where the sample size is not fixed in advance. Instead, data are evaluated as they are collected, and further sampling is stopped in accordance with a pre-defined stopping rule as soon as significant results are observed. Thus, a conclusion may sometimes be reached at a much earlier stage than would be possible with more classical hypothesis testing or estimation and consequently at lower financial and/or human cost.

3.2.2 Sequential Decision-Making Procedure

The sequential procedure of the decision making process allows the minimization of the detection time when the existing phenomenon is taking place (Fig. 3.2). The sequential procedure has the following basic characteristic (Wald 2004):

$$L(a) \approx \left[A^{h(a)} - 1\right] / \left[A^{h(a)} - B^{h(a)}\right], \tag{3.6}$$

where A and B are the boundaries for the $L_n(x)$ and $h(a)$ is the solution of the equation:

$$\int_{-\infty}^{\infty} \left[f_{a1}(x)/f_{a0}(x)\right]^{h(a)} f_a(x)\, dx = 1 \tag{3.7}$$

The values of A and B have the estimations:

$$B \approx \beta/(1-\alpha), \quad A \approx (1-\beta)/\alpha \tag{3.8}$$

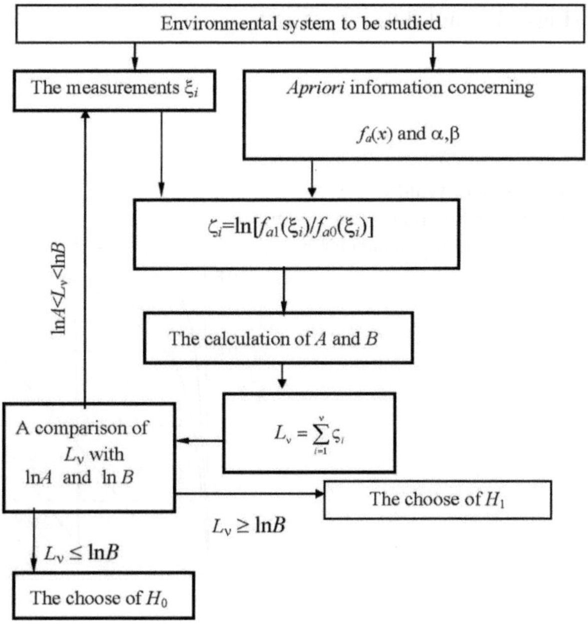

Fig. 3.2 Scheme of the experimental realization using the sequential analysis procedure to decide between hypotheses H_0 and H_1

Accordingly, $L(a_0) = 1 - \alpha$ and $L(a_1) = \beta$, and it thus follows that the average number of observations in a sequential procedure equals to:

$$\begin{aligned} E_a v &= [(1-\alpha)\ln[\beta/(1-\alpha)] + \alpha\ln[(1-\beta)/\alpha]/E_{a0}\xi, && \text{when } a = a_0, \\ E_a v &= [\beta\ln[\beta/(1-\alpha)] + (1-\beta)\ln[(1-\beta)/\alpha]/E_{a1}\xi, && \text{when } a = a_1 \end{aligned} \quad (3.9)$$

For $a = a^*$ and when $E_{a^*}\,\xi = 0$ and $E_{a^*}\,\xi^2 > 0$ we have:

$$E_{a^*} v \approx [-\ln[\beta/(1-\alpha)]\ln[(1-\beta)/\alpha]/E_{a^*}\xi^2 \quad (3.10)$$

According to Eqs. (3.9) and (3.10), the number of observations of a sequential procedure is a random variable v, the average value of which ($E_a v$) can be smaller or larger than n. It is necessary to have the distribution $P(v = n) = P_a(n)$ in order to judge the possible values of v:

$$E_a v P_a(n) = w_c(y) = c^{1/2} y^{-3/2} (2\pi)^{-1/2} \exp\left[-0.5c\left(y + y^{-1} - 2\right)\right], \quad (3.11)$$

where

$$\begin{aligned} &0 \le y \le n|E_a\xi| < \infty, \quad c = K|E_a\xi|/D_a\xi = (E_a v)^2/D_a v > 0, \\ &D_a v = K D_a\xi/(E_a\xi)^3, \quad E_a v = K/E_a\xi, \\ &K = \ln A \text{ for } E_a\xi > 0 \text{ and } K = \ln B \text{ for } E_a\xi < 0. \end{aligned}$$

According to (3.11), the Wald's distribution function $W_c(y)$, has the form (Figs. 3.3 and 3.4):

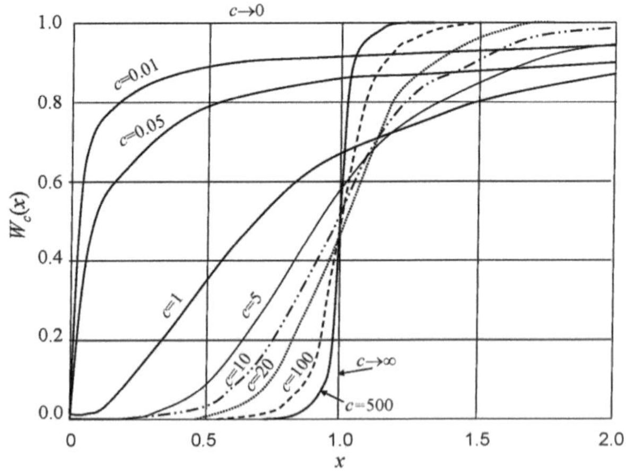

Fig. 3.3 The Wald's distribution function depending on the parameter c

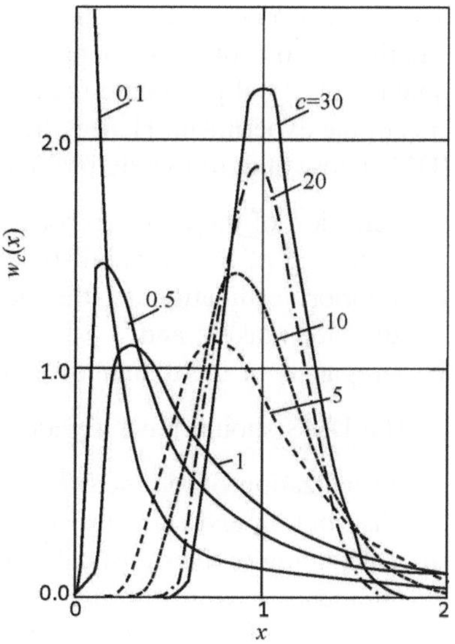

Fig. 3.4 Response of the density of the Wald's distribution function to the variations of parameter *c*

$$W_c(x) = \int_0^x w_c(z)\,dz, \tag{3.12}$$

where $w_c(z) = (c/2\pi)^{1/2} z^{-3/2} \exp[-0.5c(z + z^{-1} - 2)]$ and $c = [E(v)]^2[D(v)]^{-1}$.

The universality of the distribution (3.12) follows from its duality to the Gaussian distribution. As far back as 1960 Wald showed that if $|E_a\xi|$ and $D_a\,\xi$ are sufficiently small in comparison to ln*A* and ln*B*, the distribution of v/E_a *v*defined by the expression (3.11) will be a close approximation to the real one even for ξ not distributed by the Gaussian law.

Theoretical aspects of the universality of the distribution $W_c(x)$ are important for the integrated estimation of the sequential procedure efficiency. However, these are not the principal aspects for experimental applications. For that reason, as a rule, the synthesis of the decision making system is perceived without these considerations. In fact, the volume of the measurements, as a rule, is small and the central limit theorem doesn't work. The statistical difficulties arising from this can be overcome by evolutionary modeling (Bukatova et al. 1991), intelligent technology (Nitu et al. 2000b) and the use of other algorithms.

3.3 Decision-Making Procedure Using the Sequential Analysis

In contradiction to the Neyman-Pearson criterion (Yan and Blum 2001), the sequential procedure doesn't separate the stages of measurement and data processing but it alternates them. This is represented schematically in Fig. 3.2 incomparison to the classical procedure which is shown in Fig. 3.1. According to these figures, the algorithmic load of the sequential procedure changes dynamically, while at the same time the classical procedure realizes the data processing stage only on the finishing step of the experiment. Hence, the synthesis of the efficient decision making system (DMS) poses the following problems:

i. selection of the criterion for the parameters estimation;
ii. revealing of the probabilistic characteristics of the process studied;
iii. a priori estimation of the possible losses concerning the precision of the decisions taken; and
iv. prognosis of the dynamic stability for the experiment results.

The DMS should have a wide spectrum of functions:

i. visualization of the measurement data in the form of direct soft copy, discrete distribution and statistical parameters;
ii. calculation of the statistical characteristics (mean, central second and third order moments, asymmetry and excess coefficients, expression of the entropy, etc.);
iii. synthesis of the empirical and theoretical distribution functions;
iv. valuation of the parameters which are used in the Neyman-Pearson and the sequential procedures of the hypotheses decision; and
v. realization of the user access to all the functions of the decision making system.

According to the scheme represented in Fig. 3.5 the decision-making system should have an expert control level. The unit DMSP-I controls the decision making procedure through its inputs and outputs. According to the functions of the sub-units described in Table 3.1 the user can promptly interfere in any arbitrary stage of the computer experiment, correcting the parameters of the decision making procedure or even ceasing it.

The sub-unit CTT manages the calculation process taking into account the character of the task. It forms the variants that correspond to the concrete combination of the errors of the first and second kind, α and β. Based on this combination, the sub-unit CTT produces a set of parameters to manage the other sub-units. Depending on α and β, the simplified procedures are possible. For example, two variants for the asymmetric thresholds A and B are:

(1) $B = \beta/(1-\alpha) \to 0$, $A = (1-\beta)/\alpha \to \text{const}$; and
(2) $B = \beta/(1-\alpha) = \text{const}$, $A = (1-\beta)/\alpha \to \infty$.

In other words, the errors α and β are unequal in value. Particularly:

(1) $\beta \to 0$, $\alpha = \text{const}$; or
(2) $\beta = \text{const}$, $\alpha \to 0$.

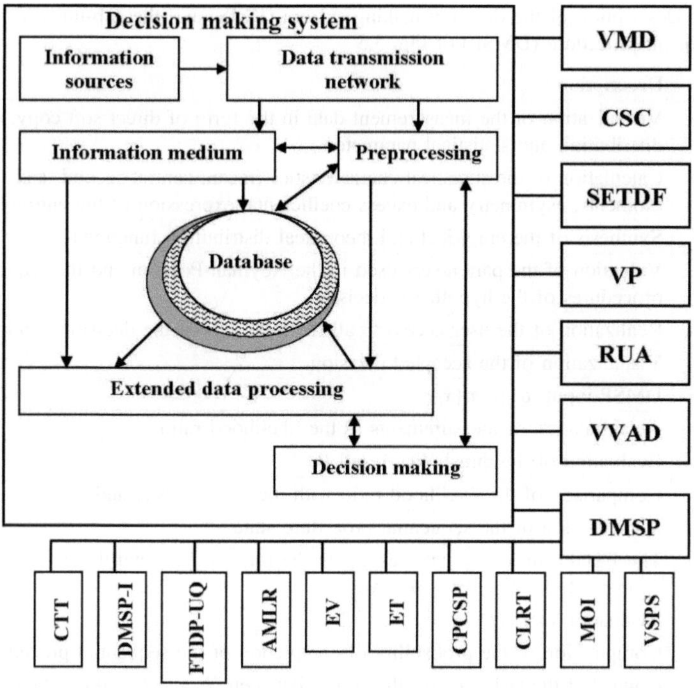

Fig. 3.5 Principal scheme of the decision-making system based on the sequential analysis procedure. A description of the sub-units is given in Table 3.1

In these cases, the sequential procedure will finish with the probability equal to 1 if the following conditions are satisfied:

(1) $E_a\,\zeta > 0$, $\zeta = \ln[f_{a1}(\xi)/f_a(\xi)]$; or
(2) $E_a\,\zeta_0 < 0$.

The probability of completion of the procedure is small when one of the following conditions is not satisfied:

(1) $\beta >> \alpha$, $E_a\,\zeta > 0$; or
(2) $\alpha >> \beta$, $E_a\zeta < 0$.

The user can visualize the state of the procedure as shown in Fig. 3.6.

Unit VMD coordinates system input with the separate channels of measured data. As each random process $\{x_i\}$ is executed, analyses are carried out to eliminate errors and to represent input signals in a form that is acceptable for the other units. The unit CSC calculates signal characteristics:

Table 3.1 Description of the decision making system (DMS) and the sub-units of the decision making sequent procedure (DMSP) of Fig. 3.5

Sub-unit	Description
VMD	Visualization of the measurement data in the form of direct soft copy, discrete distribution and statistical parameters
CSC	Calculation of the statistical characteristics (mean, central second- and third-order moments, asymmetry and excess coefficients, expression of the entropy, etc.)
SETDF	Synthesis of the empirical and theoretical distribution functions
VP	Valuation of the parameters used in the Neyman-Pearson and the sequential procedures of the hypotheses decision
RUA	Realization of the user access to all the functions of the decision-making system
VVAD	Visualization of the accepted decision
DMSP-I	DMSP input: α, β, $a, \{x_i\}$
AMLR	Accumulation of measurements as the likelihood ratio
ET	Evaluation of the thresholds A and B
CLRT	Comparison of the likelihood ratio with the thresholds A and B
VSPS	Visualization of the sequential procedure state
FTDP-UQ	The FTDP unit query with respect to the form of $f_a(x)$ and the activation of the appropriate knowledge base level
EV	Evaluation of E_v, D_v, c
CPCSP	Computation of the probability of completion of the sequential procedure
CTT	Control of the task type for the choice between H_0 and H_1 taking the errors of the first and second kind into consideration
MOI	Management of the operative intervention to the functioning of the DMSP unit

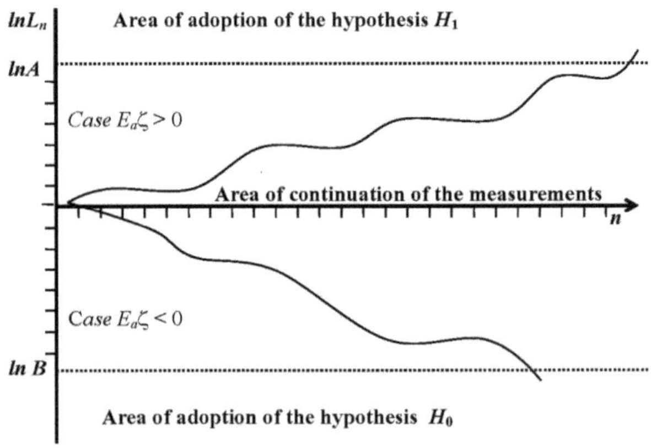

Fig. 3.6 An example of the dynamics of the accumulated sum of the likelihood function logarithm visualized by the sub-unit VSPS

$$M_1 = \frac{1}{n}\sum_{i=1}^{n} x_i, \quad M_2 = \frac{1}{n}\sum_{i=1}^{n}(x_i - M_1)^2, \quad r_1 = M_3 M_2^{-3/2}, \quad V = \frac{\sqrt{M_2}}{M_1}\cdot 100\,\%,$$

$$M_3 = \frac{n}{(n-1)(n-2)}\sum_{i=1}^{n}(x_i - M_1)^3, \quad r_2 = \frac{M_4}{M_2^4} - \frac{3(n-1)^2}{(n-2)(n-3)},$$

$$M_4 = \frac{n(n+1)}{(n-1)(n-2)(n-3)}\sum_{i=1}^{n}(x_i - M_1)^4, \quad H = -\int_{-\infty}^{\infty} f_a(x)\ln[f_a(x)]\,dx, \quad R_i = x_{i,\max} - x_{i,\min},$$

$$M_6 = \frac{M_4}{M_2^2} - 3.$$

These characteristics and other standard parameters are used to reconstruct $f_a(x)$. For instance, if $M_3 \approx 0$ and $r_1 \approx 0$, f_a should be searched for in symmetric distributions. Unit RUA performs this search by selecting continuous distributions from the knowledge base and estimating their distance from the empirical distribution using the criterion:

$$\chi^2 = \sum_{i=1}^{L} \frac{[m_i - f_a(x_i)\Delta x_i]^2}{f_a(x_i)\Delta x_i}$$

The continuous distribution f^* is chosen to satisfy the following condition:

$$\min_{f_a} \chi^2(f_a) = \chi^2(f^*).$$

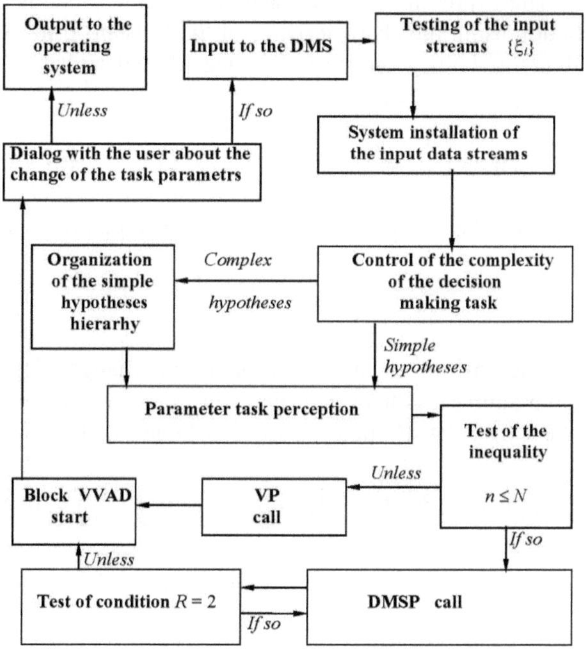

Fig. 3.7 Block scheme of the data processing in the sequential analysis procedure

The decision-making system is synthesized according to the principal scheme of Fig. 3.5. Its functioning scheme is shown in Fig. 3.7.

3.4 Important Parameters of the Sequential Analysis Procedure

An investigation of the sum of x_n of independent random values having the same distribution imposes a double task on the distribution function, $P(x_n < x) = F_1(x)$, both for fixed n and for the variable case. The situation of comparison between x_n and some level C arises in both cases. However, in the second case this task is transformed into the study of the distribution function, $P(\nu < n) = F_2(n)$, of the chance number ν of components by which x_ν first exceeds the level $C = C(\alpha, \beta)$: $x_i < C$ $(i = 1, \ldots, \nu - 1)$, $x_\nu \geq C$.

In conformity with the central limit theorem, in the first case the distribution $F_1(x)$ approaches the normal distribution when $n \rightarrow \infty$. In the second case, we have the distribution represented by expression (3.12). The following correlation between these distributions exists:

$$W_c(x) = \Phi\left[(x-1)(c/x)^{1/2}\right] + \Phi\left[-(x+1)(c/x)^{1/2}\right]\exp\{2c\}, \tag{3.13}$$

where Φ is the normal distribution function.

The expression (3.13) makes it possible to study the sequential analysis distribution using the characteristics of Φ. As seen from (3.11), the distribution function $W_c(x)$ is defined on the half-space $[0, \infty]$ and it has one maximum at the point $x = m_c$. In fact, we have:

$$dW_c(y)/dy = (d/dy)\{c^{1/2}/[y^{3/2}(2\pi)^{1/2}]\exp[-0.5c(y+y^{-1}-2)]\} = 0$$

After the set of transformations, this equation is solved to give (Fig. 3.8):

$$y = m_c = \left[(9+4c^2)^{1/2} - 3\right]/(2c) \tag{3.14}$$

The position of the $W_c(y)$ maximum changes depending only on the parameter c remaining less than one ($m_c < 1$). Moreover, $m_c \rightarrow 0$ when $c \rightarrow 0$ and $m_c \rightarrow 1$ when $c \rightarrow \infty$. Comparing Eqs. (3.1) and (3.14), we find:

$$W_c(m_c) = (3/(2\pi)^{1/2})\left(1/\left[m_c(1-m_c^2)^{1/2}\right]\right) \times \exp\{-3(1-m_c)/[2(1+m_c)]\} \tag{3.15}$$

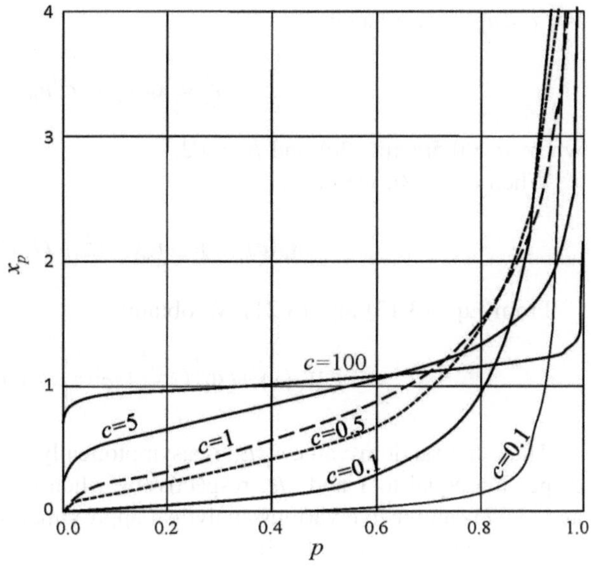

Fig. 3.8 The p-quantil of the Wald's distribution function depending on parameter c

Hence, the function $W_c(y)$ degenerates under $c \to 0$ or $c \to \infty$ into the delta-functions $\delta(y)$ or $\delta(y-1)$, respectively.

From Eq. (3.12) and taking into account that $W_c(x) \to 1$ when $x \to \infty$, we have:

$$W_c(x) = \int_0^x W_c(y)dy = \int_{1/x}^{\infty} zW_c(z)dz = 1 - \int_0^{1/x} zW_c(z)dz \tag{3.16}$$

The important parameters of the distributions Φ and W_c are solutions of the following equations (Fig. 3.8):

$$\Phi(u_p) = p, \quad W_c(x_p) = p \tag{3.17}$$

It is easy to see that when $c > u_p^2/4$ we find:

$$\begin{aligned} &1 + (u_p - \varepsilon)c^{-1/2} + (u_p^2 - \varepsilon)c^{-1}/2 + (u_p^3 - \varepsilon)c^{-3/2}/8 \\ &\quad - (u_p^5 - \varepsilon)c^{-5/2}/128 < x_p(c) < 1 + u_p c^{-1/2} + u_p^2 c^{-1}/2 + u_p^3 c^{-3/2}/8, \end{aligned} \tag{3.18}$$

where

$$\varepsilon = \Phi\left(-2c^{1/2}\right)\exp\{2c\} < 0.5(2\pi c)^{-1/2} \tag{3.19}$$

For $c < u_p^2/16$ we have:

$$c/u_a^2 < x_p(c) < c/u_b^2, \tag{3.20}$$

where $a = 0.5p\exp(-2c)$ and $b = p/2$.

When $c \to \infty$, we have:

$$x_p(c) = 1 + u_p c^{-1/2} + O(1/c) \tag{3.21}$$

From Eqs. (3.13) and (3.21) we obtain:

$$W_c(x) = \Phi\left[(x-1)c^{1/2}\right] + O(1/c) \tag{3.22}$$

Thus, the random value v/E_v is asymptotically normal with average value and dispersion equal to 1 and $1/c$, respectively, when $c \to \infty$. The expressions (3.13) and (3.22) can receive various analytical approximations of W_c. For example, let us represent $\Phi(y)$ by the following:

$$\Phi(y) = \Phi(y_0) + \varphi(y_0)(y - y_0)[1 - 0.5y_0(y - y_0)], \tag{3.23}$$

where $y = (x-1)(c/x)^{1/2}$, $y_0 = (x_0 - 1)(c/x_0)^{1/2}$ and x_0 is some point where the value of Φ was estimated.

The following formula can be easily derived using Eqs. (3.12), (3.13) and (3.23):

$$\begin{aligned} W_c(x) = W_c(x_0) &+ \varphi(y_0)(y - y_0)[1 - 0.5y_0(y - y_0)] \\ &+ \varphi(v_0)(v - v_0)\exp(2c)[1 - 0.5v_0(v - v_0)], \end{aligned} \tag{3.24}$$

where

$$v = -(x+1)(c/x)^{1/2}, v_0 = -(x_0 + 1)(c/x_0)^{1/2} \text{ and}$$
$$\varphi(z) = \varphi(z_0)\left[1 - z_0(z - z_0) + 0.5\left(z_0^2 - 1\right)(z - z_0)^2\right].$$

Let us designate

$$H(x) = \frac{2}{\sqrt{\pi}} \int_0^x \exp(-t^2)dt \tag{3.25}$$

We have

$$H(x) = \begin{cases} \frac{2}{\sqrt{\pi}} \sum_{k=1}^{\infty} (-1)^{k+1} \frac{x^{2k-1}}{(2k-1)(k-1)!} & \text{when} \quad |x| \le x_1, \\ \frac{2}{\sqrt{\pi}} \exp(-x^2) \sum_{k=0}^{\infty} \frac{2^k x^{2k+1}}{(2k+1)!!} & \text{when} \quad x_1 < |x| < x_2, \\ 1 - \frac{1}{\pi} \exp(-x) \sum_{k=0}^{\infty} (-1)^k x^{-(k+1/2)} \Gamma(k+1/2) & \text{when} \quad |x| \ge x_2. \end{cases}$$

The functions Φ and H are related through the obvious correlation:

$$\Phi(x) = [1 + \sin(x) H(2^{-1/2}|x|)]/2, \tag{3.26}$$

from which we have:

$$\Phi(x) = \begin{cases} g(x) & \textit{for} \quad x \le 0, \\ 1 - g(x) & \textit{for} \quad x > 0, \end{cases}$$

where

$$g(x) = \begin{cases} 1/2 - \frac{1}{\sqrt{2\pi}} \sum_{k=1}^{\infty} (-1)^{k+1} \frac{|x|^{2k-1}}{2^{k-1}(2k-1)(k-1)!} & \textit{for} \quad |x| \le x_1, \\ 1/2 - \frac{1}{\sqrt{2\pi}} \exp(-x^2/2) \sum_{k=0}^{\infty} \frac{|x|^{2k+1}}{(2k+1)!!} & \textit{for} \quad x_1 < |x| < x_2, \\ \frac{1}{\sqrt{2\pi}} \exp(-x^2/2) \sum_{k=0}^{\infty} (-1)^k \frac{(2k-1)!!}{|x|^{2k+1}} & \textit{for} \quad |x| \ge x_2. \end{cases}$$

Thus, formula (3.26) allows us to calculate $W_c(x)$ for various values of c and x. The free parameters x_1 and x_2 influence the calculation error. Practically, the values $x_1 = 2.2$ and $x_2 = 7.5$ are acceptable. Moreover, a problem of convergence of the above rows appears when parameter c is increasing. It is easy to see that $W_c(x) \approx \Phi[(x - 1)c^{1/2}]$ for $c \gg 1$ and $W_c(x) \approx 1 - \exp(-cx/2)$ when $x \to \infty$.

The formula (3.13) is the basic expression for the calculation of $W_c(x)$. The large factor, $\exp(2c)$, can be normalized by the following expression:

$$\exp(2c)\Phi\left[-(x+1)\sqrt{c/x}\right] = \frac{1}{\sqrt{2\pi}} \exp(2c - x^2/2) \sum_{k=0}^{\infty} (-1)^k \frac{(2k-1)!!}{x^{2k}}$$

The calculation of $W_c(x)$ can also be realized by using the Bessel (*J*) and Whittaker (Ψ) functions:

$$W_c(x) = \sqrt{\frac{c}{2\pi}} \exp\left[c\left(1 - \frac{1}{2x}\right)\right] \sum_{k=-\infty}^{\infty} \frac{x^l}{c^m} J_k(-c) \Psi_{-l,-s}(-c/x),$$

where $l = (2k + 3)/4$, $m = (6k + 9)/4$ and $s = (1 + 2k)/4$. The following correlations are very useful as well:

$$\exp[(1/z - z)c/2] = \sum_{k=-\infty}^{\infty} J_k(-c)z^k, \quad \int_u^{\infty} e^{-x}x^{-v}dx = u^{-v/2}e^{-u/2}\Psi_{-v/2,(1-v)/2}(u),$$

Finally, the following algorithm is considered to evaluate the distribution function $W_c(x)$:

$$w_c(x) = A_0\varphi_0(x) + A_1\varphi_1(x) + \cdots + A_m\varphi_m(x) + \cdots,$$

where

$$A_i = \frac{(-1)^i}{i!} \int_{-\infty}^{\infty} w_c(z)R_i(z)dz$$

and $R_i(z)$ is the Chebyshev-Hermite polynomial approximation: $R_i(z) = (-1)^i\varphi_i(z)/\varphi_0(z)$, where $\varphi_0(z) = (2\pi)^{-1}\exp(-t^2/2)$, $\varphi_i(z) = d^i\varphi_0(z)/dz^i$, $(i = 1, 2, \ldots)$.

Additionally, the normalization requirement should be satisfied:

$$\int_{-\infty}^{\infty} \varphi_0(z)R_i(z)R_j(z)dz = \begin{cases} 0, & i \neq j, \\ 1, & i = j \end{cases}$$

Thus, the sequential analysis distribution should be written in the form:

$$W_c(x) = \varphi_0\left[\sqrt{c}(x-1)\right] + \sum_{k=1}^{\infty} (-1)^k c^{-k/2} \frac{(2k+1)!!}{(k+2)!} \varphi_{k+1}\left[\sqrt{c}(x-1)\right]$$

As mentioned above, the sequential method helps to tackle some unsolved problems related to minimizing the detection delay time, and to detect a point where safe environmental process becomes dangerous. As seen from Eq. (3.12) and Figs. 3.3 and 3.8, there are environmental processes, for which parameter c tends to large values. In these cases, the approximations of function $W_c(x)$ given above can help to overcome this situation.

3.5 Processing the Multichannel Information

3.5.1 Introduction

The schematic diagram of a monitoring system for detecting anomalies on the Earth's surface involves many levels. The search organization structure may contain a more profound hierarchy including processing of information from satellites, airborne and floating laboratories, and stationary systems. In spite of the possibility of definite realization of this structure and the applicable computer technique, there exists, nevertheless, one general problem whose solution determines the detection efficiency. This task involves the organization of in-line data processing at all detection system levels under real-time conditions, ensuring an uninterrupted matching of the operation of all the cybernetic devices of the system and also taking into account the limitation on their functional characteristics.

Statistical analysis of the information at each level of the monitoring system is completed by making the following two subsequent decisions:

(1) storage of features of the landscape elements for the accumulation of additional information about them,
(2) completion on the presence of an anomaly and transmission of its characteristics to the next data processing level.

These two stages determine the efficacy of the detection system. At the first stage selection and storage of parameters of the elements of landscapes suspected of being anomalous are dependent on the character of the algorithm used for singling out the two-dimensional signal against the background noise. They are also associated with the gradual filling of the main storage. The second stage determines the magnitude of the detection probability, depending on the criteria of registration of the landscape elements selected at the first stage for further analysis. As a consequence, the problem of matching the flow of information between the processing stages under monitoring conditions and at minimal losses arises. The solution to this problem involves two procedures of parallel component processing, forming the intermediate information delay either as a constant computer memory buffer size or for a fixed storage time.

3.5.2 Multi-channel Statistical Analyzer

Real monitoring systems have series of informational channels the efficiency of whom depends on the solved task. Usually, the principal scheme of monitoring system has the form of some transformation F. Its structure and functions are defined by the items realizing separate operations on the analysis and transformation of measurement data. This is given schematically in Fig. 3.9.

The general scheme of sensing the environmental elements assumes that the flux of data $\{T_b\}$ from the j-th measuring device can be analyzed both separately and together with the data from other channels. In fact, there is a digital processor of the type of a personal computer with a speed equal to V operations per second. The task is set to identify an unknown vector parameter $T_b = \{T_{b1}, \ldots, T_{bn}\}$ characterizing the state of the controlled object (e.g., in the problems of microwave monitoring` these are brightness temperatures). To solve this problem, an algorithm of the parallel-in-time analysis of the vector T_b components is proposed, which makes it

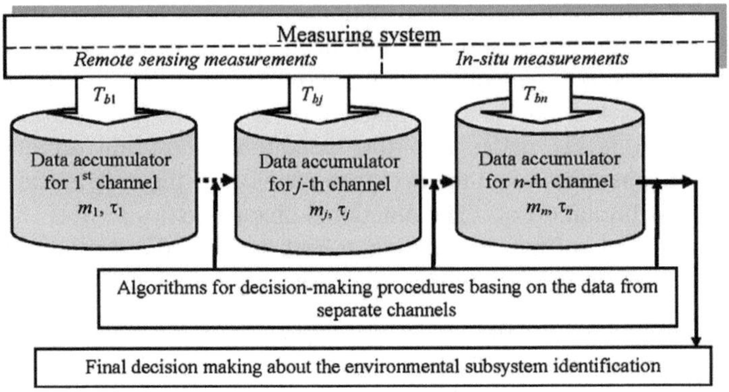

Fig. 3.9 Scheme for realization of parallel-sequential procedure to make a statistical decision

possible to reduce the time of data processing, but in this case the problem of the data flow agreement exists.

Two situations are possible:

(1) the system has time to analyze the whole chain of vectors without delay in individual channels, and
(2) at some moments, a delay occurs in some channels. In a general case, two types of delay are possible: in time and in volume.

Assume constant delays in time $\tau_1, \ldots, \tau_k$, and delays in volume $m_1, m_2, \ldots, m_k$. In the first case, the process of choosing the vector T_b is realized by a channel-to-channel procedure when the vector components T_{bj} are accumulated with the time delay intervals τ_i and then they are transmitted to the chain $\{T_{bi}\}$. The number of delayed estimates of T_{bi} in each channel is random and therefore the buffer memory of each channel should be reserved. In the second case, a random value is the time of delay τ_i ($i = 1, \ldots, k$), and the volumes of the buffers m_i are fixed. In both cases, it is necessary to assess the probability not to miss a real value of the vector T_b and to find optimal values of τ_i and m_i corresponding to the maximum level of this probability.

Let the on-board computer memory has M cells. The statistical analyzer realizes with time the simultaneous parallel selection of components of the "suspected" variant of the vector parameter $b = b^*$. This is achieved by work of the K consecutive devices or by algorithms of selection of component values b_i ($i = 1, \ldots, K$) corresponding to the space of anomalous features. To "stabilize" the process of selecting the anomaly variant, it is possible to create two kinds of constant delays between the selection devices of the anomalous components: constant delay with number of "candidates" and constant delay with time.

The two delay modes allow one to have a more flexible organization of the distribution of M memory elements for storing the intermediate values of the "suspected" components. Because of the random character of the whole process, the statistical analyzer is apt to make transient errors due to the overflow of the computer memory capacity intended for the delays. Therefore it is necessary to obtain an assessment for the probability of appearance of errors and to find memory capacities for delays $M_1 + \cdots + M_k < M$ and for the distribution of time delay intervals $t_1, \ldots, t_K$.

Let selection b_i^* be realized in the volume sample n_i, basing each selection on N_i reasonable values b_i. In this case, the non-anomalous value b_i is taken as the anomalous value with probability α_i and it is rejected with probability $1 - \alpha_i$. The probability of appearance of "candidates" during time $t_i = r_i\, \alpha_i\, S_i$ is given by:

$$P\{\mu_i = S_i\} = C_{r_i}^{S_i} \alpha_i^{S_i} (1 - \alpha_i)^{r_i - S_i} = v_i(S_i).$$

When $\mu_i \leq r_i\alpha_i$, the "candidates" arrive rarely and have time to be processed without delay. If $\mu_i > r_i\alpha_i$, the "candidates" $\{b_i^*\}$ arrive often and do not have time to be analyzed before the arrival of $\{b_{i+1}^*\}$. Therefore, the variants are delayed at F_i. The probability that the number of "candidates" will not exceed the mean values $r_i\, \alpha_i$ by more than ε_i, is given by:

$$P\{\mu_i \leq r_i\alpha_i + \varepsilon_i\} = \sum_{s=0}^{m(i)} v_i(s),$$

where $m(i) = r\, \alpha_i + \varepsilon_i$. Considering that r_i is sufficiently large and according to the Laplace's limit theorem we obtain:

$$P\{\mu_i = s_i\} \approx \phi\left[(s_i - r_i\alpha_i)\{r_i\alpha_i(1 - \alpha_i)\}^{-1/2}\right],$$

where $\phi(u) = (2\pi)^{-1/2}\exp(-u^2/2)$.

Denoting the memory capacity intended for delay of the ith component of b by M_i and fixing the condition emerging from the limitation, we obtain: $r_i\alpha_i + \varepsilon_i \leq M_i$ $(i = 1, \ldots, K)$. Then the probability of non-overflow of memory M_i will be: $P\{\mu_i \leq M_i\} = \Phi(u_i)$, $(i = 1, \ldots, K)$, where

$$u_i = \varepsilon_i\{r_i\alpha_i(1 - \alpha_i)\}^{-1/2} = (M_i - r_i\alpha_i)\{r_i\alpha_i(1 - \alpha_i)\}^{-1/2} \qquad (3.27)$$

Utilizing the Boolean formula, let us calculate the probability of non-overflow of memory M_i on the i'th component during delay with time t_i and the uninterrupted transfer of the "candidates" from F_i to F_{i+1} without delay. Let us denote this probability by $P(1, 2, \ldots, N_i)$:

$$P(1, 2, \ldots, N_i) \geq 1 - N_i[1 - \Phi(u_i)], \quad i = 1, \ldots, K;$$

where N_i is the number of possible variants of values b_i.

Let this probability differ from unity by no more than δ_i. Then we obtain an equation for determining the delay value t_i $(i = 1, \ldots, K)$:

$$\Phi(u_i) = 1 - \delta_i/N_i \tag{3.28}$$

From Eqs. (3.27) and (3.28) we obtain:

$$\Phi\left[(M_i - r_i\alpha_i)\{r_i\alpha_i(1 - \alpha_i)\}^{-1/2}\right] = 1 - \delta_i/N_i \tag{3.29}$$

For $u \gg 1$ we may write approximately: $\Phi(u) \approx 1 - \exp(-u^2/2)$. Then Eq. (3.29) assumes the form:

$$r_i^2 - 2r_i[\alpha_i/M_i + (1 - \alpha_i)\ln(N_i/\delta_i)] + (M_i/\alpha_i)^2 = 0, \quad i = 1, \ldots, K \tag{3.30}$$

For an unambiguous determination of the numerical value of r_i let us find the probability of "non-emptying" of the delay, i.e. the probability that during the exhaustive search for N_i values of the i'th component, at least one "candidate" will be under delay, and then demand that this probability is within the permissible limits. Thus,

$$P\left\{\mu_i > r_i\alpha_i - v_i[r_i\alpha_i(1 - \alpha_i)]^{1/2} = 0\right\} = 1 - \Phi(-v_i) = \Phi(v_i) \tag{3.31}$$

The probability $Q(1, 2, \ldots, N_i)$ of fulfilling the inequality in Eq. (3.31) during the procedure of an exhaustive search for b_i according to the Boolean formula, is estimated by the inequality:

$$Q(1, 2, \ldots, N_i) \geq 1 - N_i[1 - \Phi(v_i)] \tag{3.32}$$

Thus, by solving Eq. (3.30) and inserting its root into Eq. (3.32), we obtain two probable values of emptying the delay: $q_i = N_i[1 - \Phi(v_i)]$. Hence, that root of Eq. (3.30) is chosen to be reasonable with respect to the quality of the numerical delay value.

If the binomial distribution $v_i(s)$ cannot be approximated precisely enough by the normal distribution, we can use the Poisson distribution: $v(s) \approx p(s) = (r\alpha)^s \exp(-r\alpha)/s!$. We obtain:

$$p_i(s_i) \approx (2\pi s_i)^{-1/2} \exp\{-r_i\alpha_i - s_i\{1 - \ln(r_i\alpha_i s_i)\}$$

Furthermore, all the arguments regarding the inference from this equation are similar to Eq. (3.29); for determining t_i they remain the same.

Generally, the calculation procedure for t_i can be realized by the computer for any expression $v_i(s)$. But since $N_i\alpha_i \gg 1$ is practically always the same, the agreements presented are extremely real and simple.

Thus, at a given time delay the probability of error of the whole anomaly detection system is given by:

$$P_t = 1 - \prod_{i=1}^{K} Q(1, 2, \ldots, N_i)$$

Let us consider another variant of constant delay:the delay of a constant number of "candidates". The expectation time for a complete filling of this delay is a random value with a Pascal distribution:

$$P\{\tau_i = t_i^*\} = C_{R_i-1}^{m_i-1} \alpha_i^{m_i} (1 - \alpha_i)^{R_i - m_i},$$

where $R_i = t_i^*/n_i$ is the number of b_i variants surveyed for time t_i^* and n_i is the sample capacity.

In the case of $\tau_i \geq m_i n_i/\alpha_i$, the variants from F_i enter F_{i+1} more rarely than on average. According to the matching condition of the whole flow, they have time to pass by without delay. At $\tau_i < m_i n_i/\alpha_i$, the b_i^* variants do not have time to be processed at F_{i+1} and are delayed at F_i. It is necessary to determine the numerical values of m_i ($i = 1, \ldots, K$) so that $m_i \leq M_i$ and in the process of an exhaustive search for the ith component there should be a definite probability of no overflow and no delay emptying.

Let us take advantage of the approximate expression of Pascal's distribution via distribution (3.12). When $R_i\alpha_i \gg 1$, the following expression is known

$$P\left\{\tau_i = t_i^*\right\} = \alpha_i m_i^{-1} w_{c_i}(y_i)$$

To ensure the predetermined probability of non-overflow of M_i memory, it is necessary to choose m_i in such a way that the time between the arrivals of the "candidates" should be close to average. Let us calculate the probability that τ_i exceeds the value $m_i n_i/\alpha_i$ by more than some constant d_i

$$P\{\tau_i \geq m_i n_i/\alpha_i - d_i\} = 1 - W_{c_i}\lfloor (m_i n_i - d_i \alpha_i) m_i^{-1} n_i^{-1} \rfloor \tag{3.33}$$

The Q(1, 2, …, N_i) probability of fulfilling inequality (6.33) during the whole procedure of an exhaustive search for values of the ith component has the evaluation:

$$Q(1, 2, \ldots, N_i) \geq 1 - N_i W_{c_i}\lfloor (m_i n_i - d_i \alpha_i) m_i^{-1} n_i^{-1} \rfloor$$

Let us try to make this probability differ from unity by not less than the value of ε_i. Then the equation for determining the delay value m_i will take the following form:

$$W_{c_i}\left[(m_i n_i - d_i \alpha_i) m_i^{-1} n_i^{-1}\right] = \varepsilon_i/N_i \tag{3.34}$$

At $c_i > 0$, involving the designated approximation of the normal distribution function, we have $m_i \approx d_i\alpha_i[2c_i \ln(N_i/\varepsilon_i)]^{-1/2}/n_i$ ($i = 1, \ldots, K$). Furthermore, let us demand that the probability of appearance during time $m_i n_i/\alpha_i$—d_i being greater than M_i "candidates" be anywhere near zero. Then we have:

$$P\{\mu_i \leq M_i\} \approx \Phi\lfloor (M_i - R_i\alpha_i)/\sqrt{R_i\alpha_i(1-\alpha_i)} \rfloor \tag{3.35}$$

The probability of realizing $\mu_i \leq M_i$ during the whole procedure of an exhaustive search for values of the *i*th component according to the Boolean formula is estimated by the inequality:

$$\Omega(1, 2, \ldots, N_i) \geq 1 - N_i\left[1 - \Phi\left\{(M_i - R_i\alpha_i)/\sqrt{R_i\alpha_i(1-\alpha_i)}\right\}\right] = 1 - \Delta_i$$

Given that a certain small value is expressed by Δ_t, we obtain an extra condition—except from (3.34) to determine m_i:

$$\Phi\lfloor (M_i - q_i\alpha_i)/\sqrt{q_i\alpha_i(1-\alpha_i)} \rfloor = 1 - \Delta_i/N_i, \tag{3.36}$$

where $q_i = m_i/\alpha_i - d_i/n_i$.

Thus, if the delay capacities are calculated from Eq. (3.36), the probability of error for the whole system will equal to:

$$P_m = 1 - \prod_{i=1}^{K} \Omega(1, 2, \ldots, N_i)$$

because of the overflow of the memory.

The evaluation of probability of the on-board computer memory overflowing under monitoring conditions of processing information through the *K* channels as considered above enables one—in definite situations—to calculate the parameters of the system, to estimate its efficacy, and to choose one of the two processing methods which has already been indicated above. Both delay variants discussed above are equivalent in the requisite additional memory capacity and operation time. Indeed, if $\Delta_i = \sigma_i$, then $r_i = R_i$, i.e. the delay in the first variant equals the number of variants considered in the second. Similarly, assuming $q_i = \varepsilon_i$, we obtain the equality $r_i = R_i$. Therefore, the choice of the kind of delay should be determined by technical considerations of its realization.

3.6 Applications of the Sequential Decision-Making Procedure

Sequential analysis applications are mainly occurred when the under study stochastic processes are time-dependent and the researcher is not able to form representative samples. Carvalho and Lopes (2007) developed and implemented a simulation-based sequential algorithm to estimate a univariate Markov switching stochastic volatility model and showed that the sequential algorithm can perform accurate sequential inferences. It was shown that a useful tool for modeling time varying variables and for measure of risk is stochastic volatility models.

Sequential method helps to understand how the model's output depends on its input, in some region of the input parameter space. In other words, sequential screening allows the focusing only on the important parameters what fasters the study process that doing the analysis over all the parameters. For example, the problem of earlier detection of tropical cyclone beginning is solved taking into account numerous meteorological and geophysical characteristics of the ocean/atmosphere system delivered by the GOOS including TAO/TRITON/PIRATA/RAMA system (Krapivin and Soldatov 2009b; Sukov et al. 2008a). It is evident that restricted number of between these parameters is important.

The sequential decision making procedure is used to estimate the parameters of the land surface based on the remote data received by the flying laboratory Il-18 (Fig. 2.4) which was used for a long time in Former Soviet Union and which was unique remote-sensing system. The measurements were obtained by radiometers of 0.4, 0.8, 2.25, 11, 21, and 27 cm wavelength. The sets of radiobrightness temperatures $\{T_j\}$ are registered and $T_{j,\min}$, $T_{j,\max}$ and $T_{j,\mathrm{mean}}$ are calculated. As a result, they are transformed to the normalized sets $\{T_j/T_{j,\max}\}$ or $\{\Delta T_j = T_j - T_{j,\mathrm{mean}}\}$. Figure 3.10 gives examples of such transformations. The registered radiobrightness temperatures are typical for the land covers of the eastern Caspian Sea coast. This is a region of saline lands, dry sands, saline waters, dense bushes, and populated landscapes. As it is seen in Fig. 3.11, multi-channel measurements provide the data sets, making it possible to distinguish the land surface cover. Moreover, an application of the sequential analysis procedure guarantees the solution to this task in real-time on-board the flying laboratory.

The radiobrightness temperature sets, made discrete by the quasi-uniform subsets, are transformed to combinations of distribution characteristics which make possible the formation of the multi-dimensional image state of the land surface. An example of the cluster analysis application is given in Fig. 3.12. Real brightness

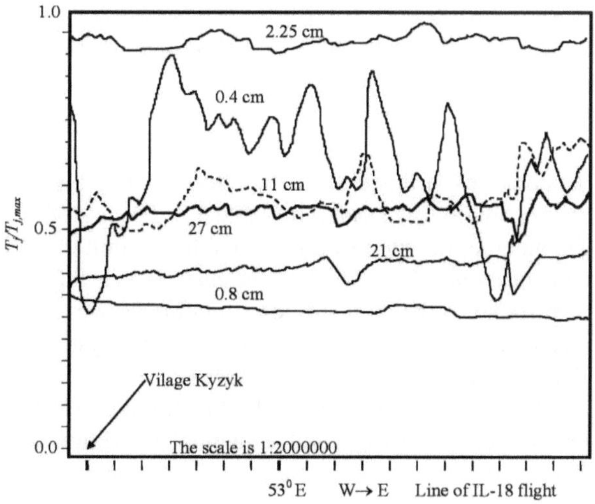

Fig. 3.10 Sample of the radiobrightness temperature registration on board the airborne laboratory IL-18 flying east of the village Kyzyk on the eastern shore of the Caspian Sea

Fig. 3.11 Radiobrightness contrasts in the area of the Kara-Bogaz-Gol Gulf on the eastern shore of the Caspian Sea as registered on board the flying laboratory IL-18

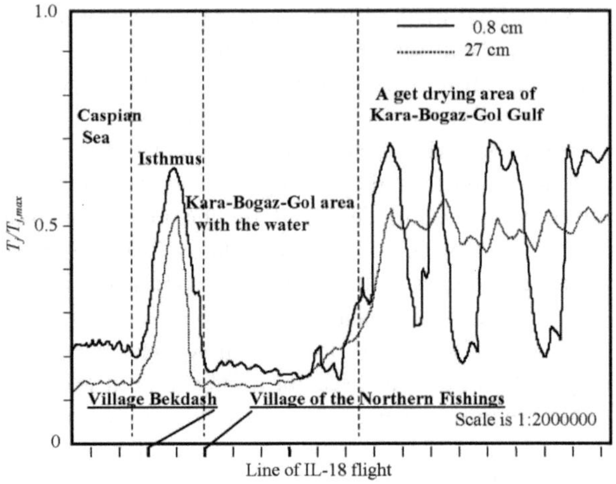

Fig. 3.12 Profile of the state system image for the '*Sor Barsa Kelmes*' saline land—'*Ustyurt*' plateau—formed on the basis of the IL-18 in-flight laboratory measurements. *Symbols designate* ♦—saline land, *—plateau, +—the boundary between them

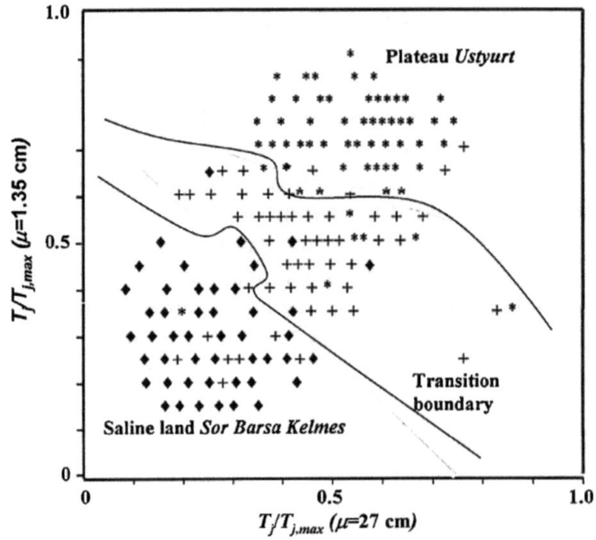

temperatures of this example are given in Fig. 3.13 and the statistical parameters of the distributions are given in Table 3.2. The use of brightness temperatures of multi-channel measurements allows reliable recognition of land surfaces regardless of the climatic zone (Fig. 3.14). It is easy to see that the varied distributions do not correspond to the land surface structure. However, the parametric space expansion increases the efficiency of the decision system by including the distribution characteristics.

The optimal procedure of the functioning of the decision system consists in the joint use of the classical and sequential algorithms for data processing (see Fig. 3.5). This procedure saves time and gives the operative parameters of the distributions

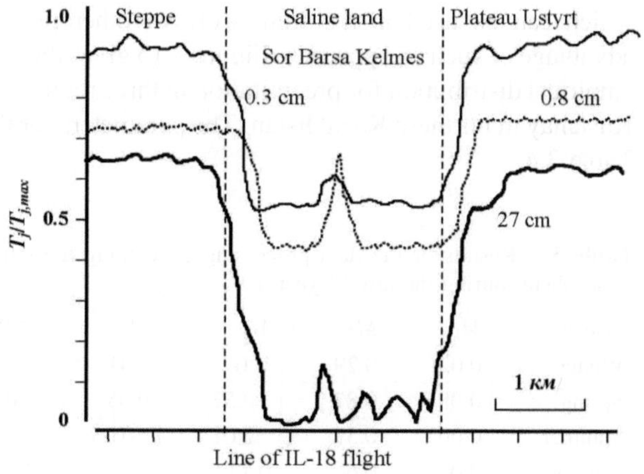

Fig. 3.13 Fragment of the record from the IL-18 flying laboratory on the boundary of the Ustyurt Plateau

Table 3.2 Data processing results of multi-channel radiometric measurements by the flying laboratory IL-18 in the area of the saline land Sor Barsa Kelmes bordering the Aral Sea

Channel (cm)	Distribution			Procedure used for the decision-making		
	Steppe	Saline land	Boundary	Steppe	Saline land	Boundary
0.8	*R*	*G*	*E*	*W*	*W*	*W*
1.35	*R*	*L*	*G*	*W*	*N–P*	*W*
2.25	*R*	*L*	*G*	*W*	*N–P*	*W*
21	*R*	*L*	*E*	*W*	*W*	*W*

Designations: *R* is the Rayleigh distribution, *E* is the exponential distribution, *L* is the lognormal distribution, *G* is the Gaussian distribution, *V* is the Weibull distribution, *W* is the sequential procedure and *N–P* is the Neyman-Pearson procedure

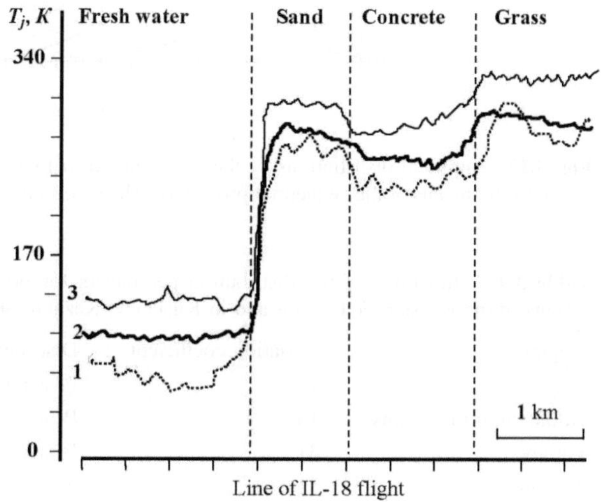

Fig. 3.14 An example of the registerogrammes received by the flying laboratory IL-18 near Yeniseysk city airport using radiometers of 2.25 cm (*1*), 10 cm (*2*), and 30 cm (*3*) wavelengths

which can be used in the framework of other procedures. Table 3.3 shows the advantage of such an approach. Figure 3.15 gives the results of reconstructions of the empirical distribution for precipitation in three regions of the Russian Federation and Kustanay in northern Kazakhstan. The parameters for these distributions are given in Table 3.4.

Table 3.3 Results of the data processing concerning the temperature anomalies for the arctic atmosphere (during the last 10 years)

Season	M_1	M_2	M_5	M_6	NW	Δ, %	Distribution
Winter	0.02	0.29	0.16	0.41	44	56	*K*
Spring	0.08	0.87	−0.52	0.05	108	0	*St*
Summer	0.00	0.39	−0.01	−0.65	51	49	*Si*
Autumn	0.09	0.64	0.11	4.74	73	27	*G*
Year	0.05	0.55	0.26	−0.75	59	41	*G*

Notation: M_1 is the first order moment (mean), M_2 is the central second order moment (dispersion), M_5 is the asymmetry coefficient, and M_6 is the excess coefficient, *NW* is the number of the measurements in the framework of the sequential procedure, Δ is the advantage of the sequential procedure in comparison with the fixed level of 100 measurements in the classical procedure, *G* is the Gaussian distribution, *K* is the Cauchy distribution, *Si* is the Simpson distribution and *St* is the Student distribution

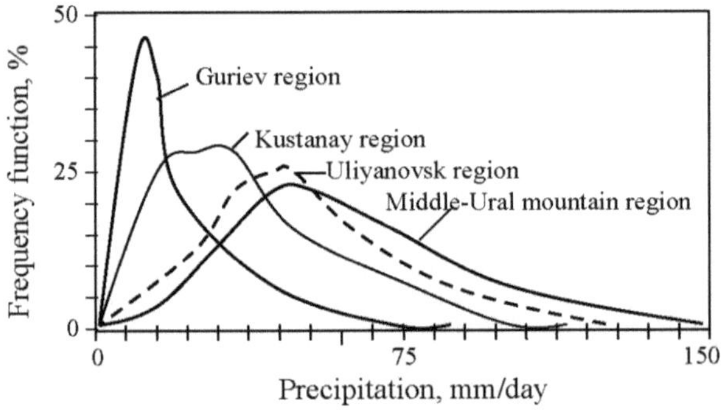

Fig. 3.15 Empirical distributions of the precipitation in four regions of Russia and Kazakhstan estimated by means of the sequential procedure. The distribution parameters are given in Table 3.4

Table 3.4 Estimations of the distribution parameters for precipitation (mm month^{-1}) in three regions of the Russian Federation and in Kustanay, Kazakhstan

Region	Variation coefficient	Distribution	Mean	Dispersion
Uliyanovsk	16	Pearson	310	51
Middle-ural mountains	15	Pearson	435	67
Guriev	31	Pearson	109	33
Kustanay	31	Pearson	211	61

3.7 Disaster Decision-Making

It is evident that methods of local diagnostics of the environment do not permit a complex assessment of the state of a natural object or process, especially if this element of the environment covers a vast area. In the problems of geoinformation monitoring, situations occur when it is necessary to make a decision in real time, with limited capabilities of the applied means of data collection and processing. In this connection, the scheme of Fig. 3.16 proposes a monitoring procedure that formalizes the decision making on the operative diagnose of the studied environmental subsystem. According to this scheme, the regular control of the environmental elements is realized with the use of technical measuring means and algorithms for observational data processing (Kondratyev and Krapivin 2005a).

With the development of the civilization, the problem of the future environment changes forecasting and bound with them changes of the living conditions for people becomes all more current importance. The origin and spread of dangerous natural phenomena do not only ruin the lifes of living beings, but they also cause serious economic damage (Kondratyev and Krapivin 2005; Milne 2004; Gardner 2002). Such phenomena are known as natural catastrophes. Historically, natural anomalies on different spatio-temporal scales have played an important role in the evolution of nature, causing and actuating the mechanisms of natural systems regulation. Wildfires can serve a special example of the similar sort of anomalies (Kondratyev et al.

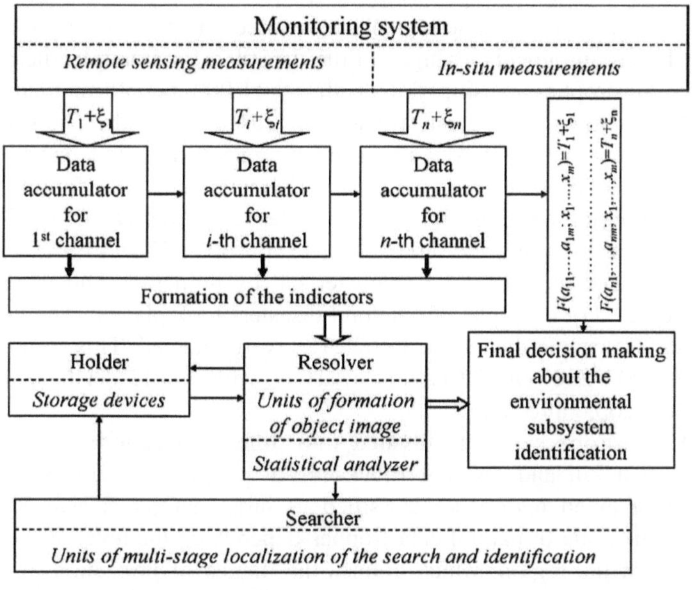

Fig. 3.16 Block-scheme of the monitoring system for the detection and identification of the natural subsystem in the environment in real time regime, with combined use of measuring devices and data processing algorithms

2006b). With the development of industry and the growth of the population density, these mechanisms have suffered significant change and the threat to life is thereby intensified (Nazaretyan 2004; Podlazov 2001; Shahramantyan 2003). This is mainly connected with growth and spreading of the amplitude of anthropogenic changes in the environment. Numerous studies of the problems that have cropped up in recent years showed that the frequency and the power of disastrous natural phenomena is growing and is posing increased risk of greater losses of human lifes, economic losses, and breaches of the social infrastructure. For instance, only in 2001 about 650 natural catastrophes occurred in the world. These natural catastrophes took away the lifes of more than 25 thousands of people and caused an economic damage of more than US$ 35 bn. The overall picture of the last years is dominated by the accumulation of severe earthquakes to an extent seldom experienced in recent decades. Several major catastrophes in 2010 resulted in substantial losses and in an exceptionally high number of fatalities. Altogether, 950 natural catastrophes were recorded, nine-tenths of which were weather-related events like storms and floods. According to this total, 2010 has been the year with the second-highest number of natural catastrophes since 1980, markedly exceeding the annual average for the last 10 years (785 events per year). The overall losses amounted to around US$ 130 bn, of which approximately US$ 37 bn was insured. This puts 2010 among the six most loss-intensive years for the insurance industry since 1980. The level of overall losses was slightly above the high average of the past 10 years. Year 2011 continues this tendency. An exceptional accumulation of very severe natural catastrophes makes 2011 the highest-ever loss year on record, even after the first half-year. Unfortunately, there is no reliable prognosis of the future level of global natural disasters.

Such factors are mainly functions of the readiness of the territory to reduce the risk of the losses and greatly change at time. In 2002, for example, there were 700 significant catastrophes causing the death of 11,000 people and an economic damage of US$ 55 bn. The greatest loss was caused by floods. The spatial distribution of the catastrophes showed a typical strong spottiness (i.e., unevenness). For example, the list of human victims in 2002 is characterized by the following factors: Africa—a number of the catastrophes 51 (661 deaths), America—181 (825), Asia—261 (8,570), Australia and Oceania—69 (61), Europe—136 (459) (Bondur et al. 2009). Subsequent years are also characterized by numerous human victims and economic losses. For example, 354 natural disasters took place during 2008 and as a result 236,000 people died and 211,628,186 were affected. This compares to the average for 2000–2007 of 397 natural disasters, with 66,812 deaths and a total of 231,588,104 people affected.

Natural catastrophes can be divided into different categories. Environmental phenomena that kill and displace thousands of people who are essential to the economy of a given region are classified as large natural catastrophes. It thus follows that the scale of natural catastrophes depends on the level of the economic development of the region, which defines the degree of protection of population from disastrous natural phenomena. So, the study of a phenomenon, in accordance with the natural catastrophes, must be accompanied by the analysis of a poverty level for the population of given region. The results of studies during the last

25 years show that losses from natural catastrophes are much greater in developing countries than in economically developed regions. Taking into account that in recent decades the number and power of natural catastrophes has increased by about five times and the danger they pose by about nine times, the threat to the population of these countries in the near future becomes abundantly clear. For that reason, forecasting and warning of crisis phenomena must be a subject of worry for all countries, regardless of their economic development.

At present, the theory of catastrophes and the analysis of risk are well developed (Brebbia 2004). Their application to events and processes in the real environment requires to synthesize a global model of the NSS that employs the technical facilities of satellite monitoring. The solution of the problems that arise here is the subject of ecoinformatics, which provides the joining of analytical simple, semi-empirical and complex nonlinear models of ecosystems with updated global database. Many international and national programs of environmental study, directed at problem solving and spatial orientation, have recently raised the level of thematic coordination to reach the necessary level of efficiency. An example of such coordination is the Global Carbon Project and Earth Observing System, which between them contain information and technical facilities of the highest quality for estimation and forecasting of NSS dynamic features.

The development of the constructive ways to forecast natural catastrophes requires decisions regarding a set of problems:

- Adaptation of ecoinformatics methods to the problem of diagnosing and forecasting natural catastrophes in all their variety.
- Formation of the statistical characteristics of the natural catastrophes in their history aspect, selecting categories and defining spatial and temporary scales of the disastrous change of the human habitation. The analysis of the natural catastrophes history is important for understanding the modern dependencies between crisis in nature and society. The statistical characteristics of the natural catastrophes in their dynamics allow the formulation of the basic positions to mathematical theory of the catastrophes and define the priority directions of the studies.
- Developing a concept and synthesizing a survivability model to estimate the influence of natural catastrophes on human habitation.
- The study of the objective laws of the interaction between different components and processes in global NSS in interconnection with such capacious notion, as biological complexity of the ecosystem (biocomplexity), considering it as a function of biological, physical, chemical, social, and behavioral interactions of the environmental subsystems, including living organisms and their community. The notion of biocomplexity is linked to objective laws that govern the way the biosphere functions by uniting its ecosystems and natural/economic systems on different scales, from local to global. A joint formalized description of the biological, geochemical, geophysical, and anthropogenic factors and processes occurring at a given level of the spatiotemporal hierarchy of scales is called for

here. It is also important to evaluate the possibility of using different indicators, such as biocomplexity, to show the approach of a natural catastrophe.

- Investigating correlations between survivability, biocomplexity, and NSS evolution using global-modeling technologies. Creating global model blocks, describing regularities and trends in the environment, to ascertain stressful situations initiated by economic or political activity.
- Consideration of the demographic premises of the arising of the natural catastrophes and discovery of mechanisms regulating the nature environment and preventing the realization of these premises.
- Synthesis of the NSS model and the development of a computer method to carry out calculations in order to assess the consequences of a realization of various anthropogenic activity scenarios.

Walker (2003) fairly noticed that the notion of the natural catastrophe is vague and its determination depends on many factors. Grigoryev and Kondratyev (2001) defined the natural catastrophe as "exceeding and dangerous situation in vital activity of the population, caused by essential disadvantage changes in the environment", or "as uneven changes in the system, appearing in the manner of its sudden answer to fluent changes of the external environment". The amount of such critical situations in the environment is increasing. At present, the following phenomenon pertains to natural catastrophes: floods, droughts, hurricanes, storms, tornado, tsunami, earthquakes, landslides, collapses, coldness, volcanic eruptions, wildfires, windstorms, sandstorms, avalanches, strong frosts, heat, epidemics, plagues of locust, and many other natural phenomena. In the future, this list may enlarge to account of the origin of new type of natural catastrophes such as collisions with cosmic body and generated person—bioterrorism, nuclear catastrophes, magnetic field reversal, plague, robot aggression, alien aggression, cultural decline, etc. For this reason, it is important to develop efficient quantitative technologies and criteria so that warnings about the appearance of dangerous disastrous natural phenomena can be issued with high reliability.

Many authors consider the notion of the natural catastropheto be associated with the notion of ecological safety, which first appeared when the need to estimate the risk facing the population in a given area to get insurance to cover health, buildings, or other property as a result of changing environmental parameters arose. These changes can be triggered by both natural and anthropogenic causes. In the first case, the dangerous situation appears at the expense of fluctuations in natural processes connected with change of the ecological situation, origin of epidemics or by natural disaster. In the second case, danger appears as nature reaction to anthropogenic activity. For instance, Gardner (2002) investigated the change in the natural environment of the Himalayas in India and came to the conclusion that factors such as reforestation and change in vegetable cover induce and amplify instability in this region. This results in land resource degradation and increased destruction of the natural environment because of floods. Field and Raupach (2004) and Abrahamson (1989) interfaced the change of regularities of the appearance of natural catastrophes with growth of instability in the carbon-climate-society system. According to

Field et al. (2002), like instability can increase in nearest two decades by a considerable extent at the expense of the change of the many features of the World Ocean ecosystem. Milne (2004), analyzed the history of different large catastrophes and made pessimistic forecasts for the fate of humankind, using terms like "doomsday".

In general case, the rise of ecological danger on given territory is due to deflections of environmental parameters for limits, where living organism under long stay begins to change on direction not corresponding to natural process to evolution. The notion "ecological danger" or "ecological safety" are connected with notions of stability, vitality, and wholeness of biosphere and its elements. Moreover, the NSS, being self-organizing and self-structuring system that is developed according to the laws of evolution, creates collections of ecological niches, the acceptability of which for the population of given territory is defined, as a rule, in accordance with national criteria (a set of environmental quality criteria, religious dogmas, national traditions, etc.).

When considering the viability conditions for life on Earth, it is necessary to move away from human-oriented standards to assess the level of environmental degradation, since local and regional changes in the environment in the course of time become global. The amplitudes of these changes are defined by mechanisms of the NSS functioning, which provide the optimality of changes in its elements. Humankind has increasingly deviated from this optimality in its interaction with environmental abiotic and biotic formations. However, human society—as just another NSS element—will in time understand its large-scale relationship with nature, as a result of many sciences and studying the causal relationship in this system (Kondratyev et al. 2003b).

The ecological environment in which people are living is a complex dynamic system. Its stability in time depends on the regularity of its structure, material composition, energy balance, and stability of its reactions on the same external influences. The system stability can be disturbed as a result of influences from both passive and active external impacts. In other words, in modern conditions nature N and human society H, forming a united planetary system and having hierarchical structures $(|N|, |H|)$, interact pursuing their own goals ($\underline{N}$, $\underline{H}$). From a formal standpoint, this interaction can be considered a casual process $\eta(x, t)$ with an unknown distribution law, presenting the tension level between subsystems N and H or evaluating the state of one of them. Here $x = \{x_1, \ldots, x_n\}$ is a set of the identification features of the subsystems N and H, being components of the possible indicator of the arising natural catastrophe i.e. deflections $\eta(x, t)$ for limits, where state of the subsystem N becomes threatening for H. It thus follows that the purposes and behaviors of subsystems N and H are functions of the factor η, and their behaviors can be antagonistic, indifferent, or cooperative depending on η. The main goal of the subsystem H consists in the achievement of the high living standard with warranty of the permanent survival in the comfort conditions. The goal and behavior of the subsystem N are defined by the objective laws of co-evolution (it follows to call the attention to this context on concept of the biotic regulation of environment (Kondratyev et al. 2003b, 2004b). In this sense, the division between

N and H is conditional and can be interpreted as the division of all natural processes into controlled and non-controlled. As world population increases, natural catastrophes will intensify the sensation of unconformity reflected in the social and cultural conditions of many regions.

From the viewpoint of the theory of systems, H and N are open systems. Without going into philosophical and methodical aspects of this division, we assume that systems are symmetrical from the viewpoint of their simulation. In addition, the system H disposes technology, science, economic potential, industry and agriculture, sociological structure, size of the population, etc. The process of the interaction between the systems N and H brings about a change of η, the level of which influences upon the structures of $\underline{H}$ and H. Really, there exists the threshold η_{max}, outside of which mankind ceases to exist, but nature survives. Asymmetry of the subsystems N and H in this sense causes change of the goal and strategy of system H. Probably, in modern conditions the value of η approaches to η_{max} with high velocity, but therefore separate components of the $\underline{H}$ possible to refer to the cooperative class. Since modern social-economic structure of the world is presented by the countries, it is reasonable to consider the country as functional system element of H. The function $\eta(x, t)$ reflects the result of the interaction between countries or between countries and nature. The collection of the results of these interactions can be described by matrix $B = \|b_{ij}\|$, each element of which carries a symbolic semantic load:

$$b_{ij} = \begin{cases} + \text{ for the cooperative behavior,} \\ - \text{ under the antoganistic interaction,} \\ 0 \text{ for the indifferent behavior.} \end{cases}$$

Many scientific theories are dedicated to the study of complex systems interaction of the different nature. In the asymmetrical case considered here, the question is about the survivability of system H and about finding a way to estimate the future dynamics of system H. According to Podlazov (2001), reflexive behavior of H will eventually help mankind to find technology of the behavior that will be capable to compare advantages and danger, to understand the principal restrictions of our possibilities and to realize the new threats in good time.

As a result of the many aspects of the environmental change in recent decades, many authors have come up with multiple concepts to describe the NSS on a global basis and models of different complexity to parameterize the dynamics of biospheric characteristics. The presence of the global database containing various information characteristics allows to consider and to evaluate the consequences of a possible realization of the different scenarios of the subsystem H development. Traditional approaches to synthesizing global models are founded on considering collections of balance equations, in which parameters $\{x_i\}$ take the form of functions, arguments, factors, and conditions of the transition between parametric descriptions of environmental processes. Moreover, other approaches based on evolutionary and neural network algorithms have been used (Nitu et al. 2004). The organization of the NUH global model functioning can be presented in the manner of a conceptual scheme described by Kondratyev et al. (2004a). The realization of this scheme is performed by the introduction of the geographical cell $\{\varphi_i, \lambda_j\}$ with spatial steps $\Delta\varphi_i$ and $\Delta\lambda_j$ for the land surface and World Ocean by the latitude and longitude, respectively. So, all processes and NSS elements are considered as uniform and are parameterized by point models within the pixel $\Omega_{ij} = \{(\varphi, \lambda): \varphi_i \leq \varphi \leq \varphi_i + \Delta\varphi_i, \lambda_j \leq \lambda \leq \lambda_j + \Delta\lambda_j\}$. The choice of the pixel size is defined by a set of conditions that depend on the spatial resolution of satellite measurements and the availability of a necessary global database. In the case of water surface, the water body of pixel Ω_{ij} is divided by depth z to layers, i.e. three-dimensional volumes $\Omega_{ijk} = \{(\varphi, \lambda, z): (\varphi, \lambda) \in \Omega_{ij}, z_k \leq z \leq z_k + \Delta z_k\}$ are formed. All elements of Ω_{ijk} are considered as uniform. Finally, atmosphere above the pixel Ω_{ij} is digitized by the height in accordance with the atmospheric pressure levels, or on typical layers by height.

It is clear that the creation of a global model is possible only with attraction of the knowledge and data on given multidisciplinary level. Among the ensemble of global models we consider the best to be that described by Kondratyev et al. (2004a). A block diagram of this model is submitted in Fig. 1.21. Kondratyev et al. (2002) describe an adaptive procedure for a global model into a geoinformation monitoring system.

The approach of the moment of the arising of the natural catastrophe is characterized by hit of the vector $\{x_i\}$ in a certain cluster of multi-dimensional space X_c. In other words, going from purely verbal discourses to quantitative determination of this process, we shall enter the generalized feature $I(t)$ of the natural catastrophe and shall identify it with graduated scale Ξ, for which the presence of the relations of the type $\Xi_1 < \Xi_2$, $\Xi_1 > \Xi_2$ or $\Xi_1 \equiv \Xi_2$ is postulated. This means that there always exists such value of $I(t) = \rho$, which defines the level of vicinity of the arising of the natural catastrophe of a given type: $\Xi \rightarrow \rho = f(\Xi)$, where f is certain transformation of the notion "natural catastrophe" in number. As a result, the value $\theta = |I(t) - \rho|$ defines the expected interval of time before the catastrophe comes.

Let us try to find a satisfactory model to portray a natural catastrophe by means of notions and signs that comply with a formalized description and transformation. For this purpose, we shall select m element-subsystems of undermost level in system NUH. The interaction between these element-subsystems is defined by the matrix function $A = ||a_{ij}||$, where a_{ij} is a factor of level of dependencies of the relations between subsystem i and j. Then, the characteristic $I(t)$ can be defined as the following sum:

$$I(t) = \sum_{i=1}^{m} \sum_{j>i}^{m} a_{ij} \tag{3.37}$$

It is clear that in general case we have $I = I(\varphi, \lambda, t)$. For limited territory Ω with the area σ the indicator I can be defined as an average value:

$$I_\Omega(t) = (1/\sigma) \int_{(\varphi,\lambda)\in\Omega} I(\varphi, \lambda, t) d\varphi d\lambda \tag{3.38}$$

Entering indicator I brings a scheme to monitor and forecast natural catastrophes much closer. We now present the possible structure of a monitoring system that has functions to search for, forecasts, and plots the course of a natural catastrophe. Three levels are selected in the system: holder, resolver and searcher, the blocks of which have the following function:

(1) periodic examination of the Earth's surface elements;
(2) memorizing suspicious elements;
(3) forming of traces of the moving anomaly from suspect elements;
(4) accumulation in time of data on fixed elements of the Earth's surface to carry out a statistical analysis in order to decide whether the fixed suspicious elements are of noise or signal in character; and
(5) stage-by-stage localization of the procedure of the search for an anomaly.

The efficiency of such a monitoring system depends on the parameters used by technical facilities to take measurements and on the algorithms for data processing. The environmental model, used in parallel with formation and statistical test of the row $\{I_\Omega(t)\}$ and adapted to mode of the monitoring in accordance with GIMS-based adaptation scheme, plays an important role.

Employing these criteria when studying natural catastrophes reveals that both form and behavior $I_\Omega(t)$ have a typical type for each process in the environment. One of the difficult problems consists in determination of these forms and corresponding to their categorizations. For instance, a such often appearing of dangerous natural phenomena, as landslides and mudflows, has typical signs of the preliminary change in the relief and landscape, which are successfully registered with satellite in optical range. Moreover, aerial images and in situ measurements of relief slopes, hillsides, and hydrologic networks permit forecasting such events several days before. However, the limited possibility of the optical range under the cloudiness or vegetable cover has to be overcome by introducing the systems of the remote sensing in microwave range of the electromagnetic spectrum. Then, in addition to the above-mentioned indicators of landslides and mudflows, it is possible to add such informative parameters as soil moisture and biomass, since increase in soil moistureleads to the appearance of landslides, while increase in the biomass is indicative of strengthening the restraining role played by vegetation cover in preventing soil rocks moving. Particularly, this is important when supervision is snowed-stone or simply snow avalanches. Creating a catalog of such indicators for all possible natural catastrophes and adding them to the knowledge base of a monitoring system is a necessary stage in increasing its efficiency.

The knowledge of the informative indicators set $\{x_i^j\}$ characterizing the natural catastrophe of j-th type and a priori determination of its cluster X^j in space of these indicators allows to calculate the velocity v_j of the approach of the point $\{x_i^j\}$ to the center of X^j during the satellite observation and, thereby, to evaluate the time of the coming of the catastrophe. Other algorithms can be used to forecast natural catastrophes. For instance, forecasting of the forest fire is possible by means of dependencies of forest radiothermal radiation on different length of the waves from the combustible timber moistening, as a rule, being situated by the layers. Knowledge of such dependencies facilitates assessing the risk of forest fire taking into account the water content in plant cover and in the upper layer of soil.

Many studies have shown that there is real possibility to assess the fire dangerous of the waterlogged and marsh forest taking into consideration the water content in vegetable cover and upper soil layer, using the microwave measurements within the range of 0.8–30 cm. Multi-channel observations allow to solve the task of the forests classification by the fire-mature categories by means of algorithms of the cluster analysis. The efficiency of such methods depends on the detailed descriptions of forest structure in model, reflecting the state of the canopy and density of trees. Undergrowth fires are the most dangerous and difficult to detect. The three-layer model of the soil-stem-canopy system is efficient in this case when it is coordinated with the fire risk indicator $I(\lambda_1, \lambda_2) = [T_b(\lambda_1) - T_b(\lambda_2)]/[T_b(\lambda_1) + T_b(\lambda_2)]$. For instance, under $\lambda_1 = 0.8$ cm and $\lambda_2 = 3.2$ cm indicator I changes approximately from -0.25 in zone of the absence of the forest burning to 0.54 in zone of the fire. In the field of appearance of first signs of the litter burning $I \approx 0.23$. In addition, the value of I weakly depends on the distribution of combustible layers in the forest such as lichens, mosses, grassy rags, died pine-needles or leaves.

Realization of such a three-level mode for decision making about the imminence of natural catastrophedepends on the spatiotemporal scales of the monitoring system corresponding with characteristics of the natural phenomena. The most complex for decision making are delayed action natural catastrophes, the realization of which can take decades. Ozone holes, global warming, desertification, biodiversity reduction, overpopulation of the Earth, etc. belong to expected catastrophes. The solution to the basic task of reliable forecasting such undesirable natural phenomena can be realized via the GIMS using global databases, current satellites, and in situ measurements.

Chapter 4
Ecoinformatics Problems of the World Ocean

4.1 The Present Day of the World Ocean

A more detailed description of the production processes in the ocean ecosystems is needed in order to increase the reliability of assessing the role of the World Ocean in the global carbon cycle. The biological processes play an important role along with the physical and chemical processes of transformation and motion of carbon in the ocean medium. In particular, phytoplankton, like the nutrient elements, assimilates dissolved CO_2 from the salt water. As a result, an organic substance, that partially goes to the nutrient chains of the trophic pyramid of a given basin of the World Ocean and partially descends to bottom sediments, is formed. A totality of all processes of carbon motion in the ocean medium creates a gradient of the CO_2 concentration between surface and deep waters. Therefore, the study of the structure and functioning of the ocean ecosystems becomes one of the most important and rapidly developing directions of marine biology. Its various aspects are being developed in many countries within the framework of the International Biological Programme. In particular, the international programme JGOFS (Joint Global Ocean Flux Study) is dedicated to the study of biochemical processes in the ocean in order to obtain a deeper knowledge of the response of the ocean to external forcings. One of the goals of such studies is to obtain a method to predict the system's behaviour as a result of changes of some of its parameters. However, due to the unique nature and broad spatial extent of the World Ocean, it is difficult to quantitatively estimate all the elements of the system at different moments of its development and in different regions of the World Ocean and, moreover, to assess the effect of their change on the functioning of the system on the whole. With this aim in view, regular expeditions are organized, databases of the satellite monitoring of the World Ocean basins are accumulated, and methods of using the model approach for these purposes are developed. Of course, an accuracy of assessing the role of water ecosystems on global changes of GHGs cycles in general strongly depends on a

V.F. Krapivin et al., *New Ecoinformatics Tools in Environmental Science*,
Environmental Earth Sciences, DOI 10.1007/978-3-319-13978-4_4

thorough and accurate description of photosynthesis as a function of the environmental parameters and especially considering its seasonal variations.

The ocean covering approximately 71 % of the planet's surface gives 1 % of a substantial amount of food products consumed by humans, the remaining 99 % they obtain from the cultivated land. The ocean contains 97 % of the Earth's water, and oceanographers have stated that only 5 % of the World Ocean has been explored. The total volume is approximately 1.3 billion cubic kilometers with an average depth of 3,682 m (Matthew and Walter 2010). It provides food in the form of fish and shellfish—about 91 million tons are caught each year. At the same time, the total amount of organic matter produced in the ocean is approximately equal to that produced by land vegetation. By rough estimates, the total biomass of nekton constitutes 5.3×10^9 t. The catch of fish and other species from the World Ocean is estimated at 70×10^6 t/year, which constitutes 20 % of protein consumed by humans. The catch of traditional species is close to a limit ($\sim$90–100 $\times 10^6$ t/year). However, it is not a limit to the industrial ability of the ocean ecosystems in general, since the supplies of krill and other biological objects are still used little.

This disagreement between the role of the land and ocean ecosystems in food production is explained, first of all, by the fact that an intensive cultural economy is realized on land, whereas this cultural economy is weakly developed in the seas and oceans. Therefore, possible ways of increasing the ocean bioproductivity have not been applied yet. First, humankind uses the final trophic chains of natural communities of the World Ocean: fish and whales. Each successive trophic level obtained about 0.1 share of energy accumulated at a previous level. On land, two levels of organisms (vegetation, herbivorous animals) are used, while in the ocean and in the seas up to 5 levels. The direct use of non-fish species will make it possible to sharply increase an amount of proteins obtained from the ocean.

Second, a question about the transition from free fishing to cultural methods of economy in the World Ocean is posed; the question of an artificial increase of productivity of the biological communities of the ocean. To do this, it is necessary, first of all, to study the methods of controlling the output of the final produce in the biological systems of the World Ocean. To determine the rational ways of affecting the ocean communities, it is necessary to study their structure and functioning, in order to understand the production processes, transformations of matter and energy flux at different trophic levels of the ocean ecosystems. It is also necessary to develop a theory of control in the biological systems of the coastal waters and the open ocean, which differ both in natural hydrophysical and biogeochemical parameters as well as in the extent of anthropogenic impacts.

Marine communities are complicated biological systems of populations of individual species. As a result of their interaction, communities are in dynamic development. Their spatial structure is mostly determined by the composition of numerous biotic and abiotic factors, which depend on a totality of the oceanic parameters. The latter are determined by the laws of general circulation of the ocean waters, including tides and ebbs, zones of convergence and divergence, wind and thermohaline currents, etc.

In the late 20th century an urgent problem of predicting the dynamics of the ocean systems in conditions of increasing anthropogenic impacts (chemical poisoning, mechanical liquidation of living organisms, environmental changes) as well as assessing their role in the dynamics of the whole biosphere arose. So, recent studies of the climatic impact of greenhouse gases have shown that the role of the World Ocean in this process has been underestimated. In particular, according to Kondratyev and Johannessen (1993) and which concern the role of the arctic basins of the World Ocean in the formation of the global CO_2 cycle, previous assessments of this role were incorrect. This is connected with the fact that an account of biological and gravitational processes playing the combined role of a pump that pumps carbon dioxide from the atmosphere to deep layers of the ocean was inadequate in the earlier models of the global biogeochemical carbon cycle. Therefore, specified models of the working regime for this pump with climatic zones taken into account may play an important role in predicting estimates of the greenhouse effect.

The impact of the ocean ecosystems on the biogeochemical cycles is manifested through the atmosphere-water border and is usually parameterized based on the observational data. However, in this impact the vertical structure of the processes taking place in the ocean medium is very important. The nature of these processes depends much on external factors. For instance, according to Legendre and Legendre (1998), in the arctic zones of the World Ocean the patchy structure of the springtime blossoming of phytoplankton is determined by the winter conditions of ice formation and the subsequent ice melting. In other zones, these external factors are pollution of the atmosphere and ocean surface, changes in the phytoplankton living conditions and functioning of the carbonate system.

Phytoplankton is at one of the initial stages of the trophic hierarchy of the ocean system. As field observations have shown, the World Ocean has a patchy structure formed by a combination of non-uniform spatial distributions of insolation level, temperature, salinity, concentration of nutrient elements, hydrodynamic characteristics, etc. The vertical structure of the phytoplankton distribution is less diverse and possesses rather universal properties. These properties are manifested through the existence of one-to-four vertical maxima of phytoplankton biomass.

The variability of the patchy topology and vertical structure is connected with seasonal cycles and has been well studied experimentally in many climatic zones of the World Ocean. The typical qualitative and quantitative indicators of this variability have been found. The combined distributions of abiotic, hydrological, and biotic components of the ocean ecosystems have been studied.

The complexity and mutual dependence of all the processes in the ocean substantially hinder the uncovering of the laws of formation of phytoplankton spots and establishing the correlations between various factors of regulating the trophic relationships intensity in the ocean ecosystems. For instance, many studies revealed a close relationship between primary production and phytoplankton amount. At the same time, this relationship gets broken depending on a combination of the synoptic situation and insolation. It turns out that an extent of breaking depends much on a combination of the groups of phytoplankton (Legendre and Legendre 1998).

An analysis of the accumulated observation data on assessments of the produce of seas and oceans and an attempt of many authors to uncover the laws of the produce formation characteristic of various water basins have revealed various laws in local relationships between productivity and environmental parameters.

An efficient method of studying the vertical structure of the ocean ecosystems is their numerical modeling based on measurements of their characteristics (Kuck et al. 2000). To derive the model, it is necessary to know the structure of trophic relationships in the ecosystems, specific features of hydrological conditions and information about other characteristics of the environment. An experience of the modeling has shown a possibility of an efficient prediction of the dynamics of the World Ocean communities. An example of such models is a 3-D model of the ecosystem of the Peruvian current (Krapivin 1996), of the Okhotsk Sea (Aota et al. 1993; Mkrtchyan and Krapivin 2011; Kaevitser et al. 2013; Krapivin and Thu 2001) and others. In all these models the unit of parameterization of the vertical structure of the ecosystem is central.

4.2 World Ocean and Inland Reservoirs

Acuteness of problems arising because of global climate change is mainly explained by the uncertainty of many environmental problems connected with the World Ocean and inland seas. As a result, many countries develop scientific programs oriented to the study of global and regional hydrophysical processes both for understanding environmental processes and for the progress in geopolitical, social-economic and technical interests (Krapivin et al. 1990, 1991, 1994a, b, 1998c, 2005a, b, 2006a, 2007c). It is clear that the progress in this direction depends on the complex approach that should be based on new information technologies. An example of such a program is the Federal Purposeful Program "World Ocean" of the Russian Federation.

The Basic Program objective is the solution to the following tasks:

- investigation of the World Ocean environment and key processes occurring in it;
- fundamental investigations of interaction processes between the ocean and the atmosphere including processes of global scale (greenhouse effect, energy and mass exchange, biogeochemical cycles of carbon and oxygen, etc.);
- investigation of natural environment and resource potential of continental shelf;
- study and monitoring of the World Ocean conditions and hydrometeorological situation in bordering seas;
- study of ecosystem dynamics and sea biological resources to discover new aquatories for fishing;
- investigation of structure and evolution of the earth's crust on the bottom of oceans and seas;
- prognosis and assessment of the World Ocean mineral resources;
- navigation-hydrographic and hydrometeorological maintenance of national tasks;

- research of natural and anthropogenic extreme processes in sea aquatories and coastal zones (earthquakes, tsunami, wildfires, floods, volcano eruptions, oil leakages, etc.); and
- monitoring of the World Ocean to detect polluted zones and to assess ecosystem state.

The solution to these tasks demands the development and the application of new information technologies. Specific interest is stimulated by the growing pollution of coastal seas. An example of such a sea is Baltic Sea, where the influence of anthropogenic factors plays a significant role. North-Western International Cleaner Production and Environment Management Centre was organized in Sankt-Petersburg (Russian Federation) to control these factors. One principal and very important problem of Baltic Sea environment is the oil hydrocarbon pollution. In connection with this, Ship Analytics International Company (USA) created the system CRISIS™ oriented to the warning and the elimination of oil hydrocarbon flood. This system can also model the crisis situation and realize education of disaster overcoming. The main objective of this system is realized by means of the creation of a base station for liquidation of oil floods. This station is located in the zone of Baltic shipyard and its function is to collect oil products and to promote them for utilization.

Ships are a basic source of Baltic Sea pollution in the Russian zone, because they make unapproved emissions of processed waters including oil products as well as Neva River outflow. Kondratyev et al. (2002) proposed the solution of the problems arising here by means of the synthesis of the GIMS, which is described in Chap. 1. The application of the GIMS to solve the tasks mentioned above can support:

- training information-measuring optical and microwave system of expert level for operative diagnostics of water quality both in situ and laboratory conditions;
- intellectual simulation model of regional water balance;
- database renewed in on-line mode including the data about pollutant sources of given territory with characteristics of sewage (volumes, pollutant compositions, exceeding of limits);
- dislocation and operational status of measuring systems controlling the water chemistry of dangerous sites; and
- new architecture of antirecessionary situation-analytic centre which collects information from measuring systems and co-ordinates the GIMS functions to decision making in operative mode.

Sea oil and gas condensation deposits are usually situated in zones of sufficiently intensive anthropogenic influences on the environment. Consequently, the task of designing and creating the ecological monitoring system to control the sea aquatory must consider the state of the atmospheric and water environments for the adjoining aquatories.

For the sea aquatory of the oil deposit zone, the practical realization of the GIMS—technology concept demands a thorough analysis of hydrophysical and synoptic characteristics and the choice of technical systems to measure the environment parameters under the local climate (low temperatures, icing, rough sea state, etc.) taking into consideration the distance from land. The variations of the interannual climatic regimes that are connected with changes of the water surface state and of the insolation are also important. Climate conditions of the functioning of the ecological monitoring system in the zone of oil extraction put definite restrictions on the structure of the measurement system. These restrictions are usually connected with the absence of possibilities for the free arrangement of the measuring and transmitting devices in any arbitrary point of the aquatory; and for this reason the tasks of measurements planning and the choice of effective algorithms for data processing are of great concern.

An investigation of such a dynamic environment as the water systems demands the creation of complex algorithms for the data processing of the field observations including a set of various models giving a possibility to reconstruct the spatial image of the studied object basing on the incomplete information. The water systems (oceans, seas, rivers, etc.) were investigated during last decades by means of flying and floating laboratories. This gave the opportunity to develop the effective methods for the description of the dynamics of the water systems having different spatial scales. In most cases, the field and satellite observations are connected with the measurements of temperature, salinity, dissolved oxygen, current directions and velocities. The set of hydrophysical observations is more detailed in the anthropogenic zones.

Let us designate a point on the water surface by φ latitude and by λ longitude. A rectangular coordinate system (φ, λ, z), where the z axis has the value $z = 0$ on the water surface and value $z > 0$ with depth, is selected to describe the sets of the observation data. In these coordinates any measurement is represented as a function $\xi(\varphi, \lambda, z, t)$ which is a time-dependent random variable. The processing of the sets of such values demands the application of special methods. As a rule, these methods include non-stationarity reduction procedures using spatio-temporal discretization. The choice of spatio-temporal scales is defined by the dynamic characteristics of the water space and by the specific tasks that must be reached. Table 4.1 describes the scheme of Fig. 4.1 and explains a possible structure of the interconnections between spatial and temporal scales. It is observed that the accumulation of data sets by the monitoring system may be realized taking into consideration discrete situations distinguished in the framework of a problem that should be solved. A certain level of reliability in the assessment of the system state under study can be achieved only when a mathematical model is used. It combines space and time into a single interwoven continuum.

The data sets containing the experimental estimations of the circulation characteristics for the sea environment always include periodical, unperiodical, stationary and non-stationary fluctuations. That explains why the sets of measurements should be corrected, taking into account the scalability of the processes under study. Hence, follows the planning of the special features of the measurements.

The study of the oil pollution of the World Ocean is one of the most important problems of environmental monitoring. The global scale of this process demands the application of monitoring systems, giving the possibility of controlling the water environment over enormous areas. Satellite-based systems are an example of such

Table 4.1 Scheme for the coordination of spatial and temporal scales of the process that should be taken into consideration under the sea environment modeling

Time scales	Spatial scales		
	Micro	Meso	Macro
Centuries			Inter-century variability. Wavelengths: 1,000 km
Decades			Inter-century variability. Wavelengths: 1,000 km
Years			Inter-year variability. Wavelengths: 1,000 km
Months		Synoptical changeability, oceanic meanders and whirlwinds. Waves range: 100 km	Fluctuations of the currents velocity and direction. Wavelengths: 1,000 km
Days		Inertia currents. Waves range: 100 km	
Hours		Rising tides. Waves range: dozens km	
Minutes	Detail vertical kinematic structure. Waves range: m	Interior gravitation waves. Waves range: km	
Seconds	Turbulence, vertical kinematic microstructure. Waves range: cm and mm		

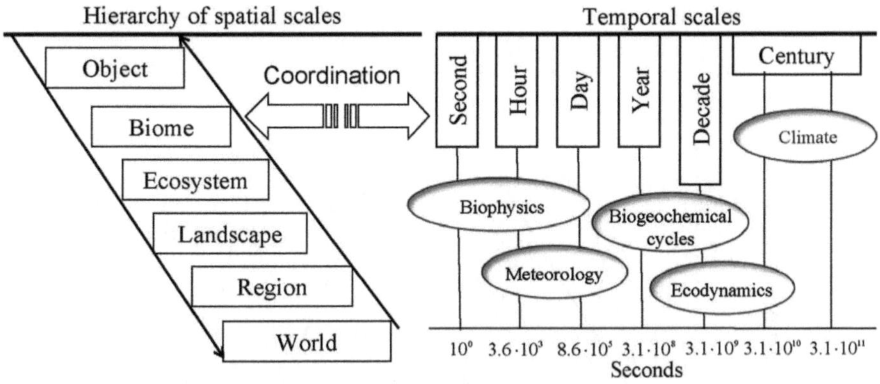

Fig. 4.1 Principal scheme for coordination of spatial and temporal scales under the choice of a model to parameterize the environmental processes

approaches. An analysis of measurement data shows that remote sensing of the water surface by means of devices using various wavelengths makes possible to detect oil spills on the water surface in order to determine the type of oil and to estimate the oil spill parameters (area, thickness, volume). Remote sensing techniques based on microwave radiometry allow to determine the oil pollution of the water surface under arbitrary weather conditions.

An extensive application of remote sensing techniques depends on the knowledge of the processes of interaction between oil and sea water, its optical and electric characteristics, the impact of the atmosphere and other factors on the propagation of electromagnetic waves. The combination of microwave and IR (infrared) ranges with mathematical modeling techniques is an effective method to discover oil spills on the water surface. The distinction between the emittance and the temperature of polluted and fresh water surface areas is the physical base for the remote sensing of oil spills by means of microwave—and IR—radiometers. The emittance κ of the three-layer system atmosphere-oil-water having smooth boundaries of the sections can be calculated with α formulae suggested by Kondratyev et al. (1996). Figure 4.2 gives an example of the dependence of the emittance variations $\Delta\kappa$ for the atmosphere-oil-water system as a function of the oil spill thickness α.

The oil spill dielectric properties occupy an intermediate position between free space (atmosphere) and water. As a result, the arising of a film leads to the effect of medium co-ordination (to the decrease in the reflection coefficient) and to the growth of the surface radiobrightness temperature. As the oil film thickness increases, the value $T_Y = \kappa T_0$ (T_0 is the surface temperature) at first increases, but then an interchange of maxima and minima is observed. From Fig. 4.2 it is obvious that it is necessary to simultaneously measure the radiobrightness temperature by means of radiometers with different wavelengths in order to remove the uncertainty in the film thickness determination.

Thin films, arising under small volumes of spilled oil or in the near-boundary area of the oil spills, do not change the emittive ability of a smooth water surface in

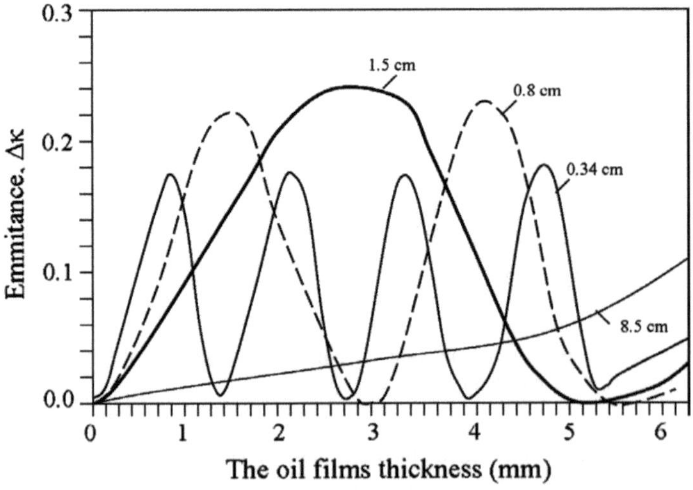

Fig. 4.2 The emittive ability of oil spills having different thickness (Mitnik 1977). Wavelengths (in cm) given on the *curves*. The water temperature is 10 °C and the oil dielectric permeability ε is 2.2

microwave-range. However, disturbed surface areas covered with thin oil films are characterized by lowered values of T_Y, due to the suppression of high-frequency components in the rough sea spectrum. The value and sign of the radiation contrast of the spills on the clean water background depend on the thickness and optical properties of the oil films, the hydrometeorogical conditions, the time of the day, etc.

Fundamental experimental investigations of water oil pollution by means of microwave- and IR-radiometers have been described by many authors (Mitnik 1977; Bogorodsky et al. 1976). Field experiments have shown that satisfactory results were attained when radiometers with wavelengths of 8–12 μm, and 0.34, 0.8, 1.5 and 8.5 cm were used. The sensitivity of the radiometer relative to its antenna inputs equaled 0.1–0.3 K under the time constant of 1 s. The experiments were performed with a flying laboratory at a height from 100 to 200 m. The radiometers were calibrated using blackbody radiobrightness temperatures or by means of calculations for a calm water surface under a cloudy sky. Thin films are recognized with high accuracy by means of IR-radiometers. Most thick films are detected with high reliability by means of microwave radiometers.

The oil film thickness can be estimated by the dependence of the radiobrightness temperature variation ΔT_Y, on the emittance ability, $\Delta\kappa$:

$$\Delta T_Y = \Delta\kappa\left[T_0 - \left(1 - T_{Y,\mathrm{atm}}/T_0\right)\right],$$

where T_0 is the surface temperature, $T_{Y,\mathrm{atm}}$ is the atmosphere radiobrightness temperature calculated by radiosensing data and the value $(1 - T_{Y,\mathrm{atm}}/T_0)$ characterizes the influence of the atmosphere.

The geometric parameters of the oil films are defined by means of photogrammetric methods the base of which is the spectrozonal photo-picture on the various wavelengths. The ranges 0.4–0.5 and 0.7–0.8 μm are the most informative for solving this task. The oil products registered with wavelengths 0.4–0.5 μm are a light spot upon the dark background of the water image. The image registered by wavelengths 0.7–0.8 μm helps to decipher the water surface.

The registration of the oil spills can be realized by means of active sensing methods. For example, the oil spill exposed by the near ultraviolet radiation begins fluorescence in the visible range (0.6–0.7 μm). This fluorescence can be registered by the adaptive identifier in the real time regime (Krapivin et al. 2001).

The above methods allow us to consider two versions of the monitoring system for controlling the gas extraction zone. The first version corresponds to the oil extraction system situated completely below the water surface where the stationary position of remote sensing systems is impossible. In this case, the monitoring system structure has submerged measuring subsystems fixed by anchors and emerged subsystems placed on flying or floating laboratories. The estimation of the concentration of pollutants emitted to the atmosphere is realized by modeling calculations. For this aim, the gradient of the gas components and the advection speed are measured in the surface layer. Moreover, it is possible to use the emerged measuring sub-systems.

Specific attention is given to the seas and inland reservoirs. A typical example is the Baltic Sea (Knuuttila et al. 2011). This region is controlled by the Helsinki Commission which is concerned about reliable data on inputs into the Baltic Sea from land-based sources, as well as information about the significance of different pollution sources. Data on waterborne inputs to the Baltic Sea and the development in these inputs have been even more important after the agreement on the HELCOM Baltic Sea Action Plan (BSAP) in 2007. The BSAP recognizes the need to reduce nutrient (nitrogen and phosphorus) inputs to a maximum allowable level. BSAP is an ambitious programme to restore the good ecological status of the Baltic marine environment by 2021.

The Baltic Sea is a relatively young and ecologically vulnerable water body. The state of the pelagic and coastal waters has suffered from human activities. Rheinheimer (1998) has given a detailed analysis of the pollution problems related to the Baltic Sea at the end of the 20th century. The health of the Baltic Sea has been seriously damaged since the 1960s due to the excessive pollution from the countries in its catchment area. The Baltic Sea is almost totally surrounded by land and, therefore, more endangered by pollution than other marine areas. The sources of marine pollution are municipal and industrial waste that outputs directly into the sea or via rivers, and atmospheric inputs mainly from traffic and agriculture. The increase of inorganic plant nutrients (NH_3, NO_x, PO_4) caused eutrophication and, consequently, oxygen depletion in coastal bottom waters as well as in the depths of the open sea. In the anoxic sediments, hydrogen sulfide can be produced by protein-decomposing and sulfate-reducing bacteria. The bottom fauna will be destroyed and only H_2S tolerant microorganisms can survive. Originating from cellulose manufacturing and from paper mills, large amounts of poisonous chlorinated compounds contaminated the coastal waters of Sweden and Finland until the 1980s. Most of this material is still present in sediments of the central Baltic Sea and can be resuspended by near bottom currents. To reduce pollution and improve the situation in the Baltic Sea, the surrounding countries organized the Helsinki Convention, which came into force on 3.5.1980. The Helsinki Commission (HELCOM), founded in 1974, acts as coordinator and is responsible for the enforcement of the Baltic monitoring program and international research projects. The activities of HELCOM have led to the reduction of dangerous pollutants which in turn has caused the regeneration of flora and fauna in some areas. Rheinheimer (1998) expressed that further improvements can be expected.

The beginning of 21st century is characterized by attempts to harmonize the policies and actions in the coastal countries through various conventions and programs in order to improve the state of the Baltic Sea. Knuuttila et al. (2011) present the main results from the PLC-5 Project. These include quantified annual waterborne total loads of nitrogen and phosphorus from 1994 to 2008 to provide a basis for evaluating any trends in the total flow normalized waterborne inputs to the Baltic Sea. The contributions from different point and diffuse sources, including the natural background load, are also estimated and described. An estimate of the total inputs of some heavy metals is also included. Furthermore, the report includes assessments of the major remaining gaps to reach the HELCOM Baltic Sea Action Plan (BSAP) nutrient reduction target as well the need for further measures.

Russia's Baltic Sea pollution problems are mainly connected with St-Petersburg's flow of industrial waste into the sea. Neva River and ports are basic sources of pollutants carried out to sea from Russian territory. The Neva is the only river flowing from Lake Ladoga. It flows through the city Saint Petersburg, three smaller towns of Shlisselburg, Kirovsk and Otradnoye, and dozens of settlements. The river is navigable throughout and is part of the Volga–Baltic Waterway and White Sea—Baltic Canal. This means that Baltic Sea pollution problems have to be solved in the framework of Russian national ecological strategy.

Marine pollution occurs when harmful or potentially harmful effects result from the entry of chemicals, particles, industrial, agricultural and residential waste, noise, or the spread of invasive organisms into the ocean. Most sources of marine pollution are land based. The pollution often comes from nonpoint sources such as agricultural runoff, wind-blown debris and dust. Nutrient pollution, a form of water pollution, refers to the contamination by excessive inputs of nutrients. It is a primary cause of eutrophication of surface waters, in which excess nutrients, usually nitrogen or phosphorus, stimulate algal growth. There are many different ways to categorize, and examine the inputs of pollution into marine ecosystems:

- Pollutants enter rivers and the sea directly from urban sewerage and industrial waste discharges, sometimes in the form of hazardous toxic wastes.
- Surface runoff from farming, as well as urban runoff and runoff from the construction of roads, buildings, ports, channels, and harbours, can carry soil and particles laden with carbon, nitrogen, phosphorus, and minerals.
- Ships can pollute waterways and oceans in many ways.
- Pathway of marine pollution through the atmosphere is realized by means of wind transport of dust and debris.

The solution to many problems connected with the marine pollution can be realized by means of the GIMS technology application.

4.3 Interactions Between the Atmosphere and the Ocean

The knowledge of exchange processes between the atmosphere and the ocean is very important in assessing the greenhouse effect of CO_2. In most of the known models of the global CO_2 cycle the relationship between partial pressure of CO_2 dissolved in water (p_c) and general concentration C_U of inorganic carbon in the surface waters is calculated based on the buffer coefficient:

$$\xi = \left(p_c - p_{c,o}\right) C_{U,o} \left(C_U - C_{U,o}\right)^{-1} / p_{c,o},$$

where the "*o*" index refers to the pre-industrial period. However, numerous observations show that the use of only this coefficient when calculating the fluxes H_2^C and H_3^C (Fig. 1.30 and Table 1.10) gives rough estimates, without taking into account the effect of many factors.

One of the first attempts to consider the spatial heterogeneity of the World Ocean and to simulate the impact of temperature gradients on the CO_2 exchange between the upper layer of the ocean and the atmosphere has been made by Bjorkstrom (1979). An idea to divide the World Ocean basins into two parts corresponding to warm and cold waters has been later developed by numerous authors (Nefedova and Tarko 1993). Pervaniuk (2001) developed this scheme dividing the World Ocean surface into 211 homogeneous basins. This discretization was based on combining the basins $4^o \times 5^o$ in size. The adjacent cells of this grid with similar directions of the vertical components of the water flow velocities are combined into one water basin. The average annual field of the water flow velocities was considered at a depth of 75 m. The size of water basins was determined taking into account the dependence of CO_2 fluxes on the climatic factors. A maximum of the latitudinal size of the basin was assumed to be 8°. Above 80^oN the World Ocean is presented as one site.

A most complete study of the physical mechanisms of the CO_2 transport under different conditions of the water-air interface has been carried out in Alexeev et al. (1992). Here for the first time, detailed measurements of CO_2 fluxes have been made, and their dependence on the parameters of the state of the atmosphere-ocean interface has been analyzed, taking into account the wind-driven mixing, the appearance of foam on the water surface, waves breaking, and pollution. Parametric descriptions of the process of the ocean-atmosphere gas exchange have been proposed for each type of the water-air interface conditions (McGillis et al. 2001). For instance, it was shown that the intensity of gas exchange grows substantially as the sea roughness and foam layer increase. For a foam layer 5 cm thick, the rate of gas exchange exceeds the gas exchange across a free surface by a factor of 2.4. This fact is important for the evaluation of the gas exchange in the tidal zones and strong-storm regions, where the foam stretches may be several cm thick. The result connected with the availability of the surface active substances (SASs) in the upper layer of the ocean is also very interesting. When SASs reach about 7.8×10^{-4} % (volume), other conditions being equal, the rate of gas exchange reduces to 60 %. However, where there is foam formation this effect decreases substantially.

On the whole, over the World Ocean basins the estimates of the CO_2 flux between the atmosphere and the upper layer of the ocean vary between 16 and 1,250 mol/m^2/year). This variability means that the global model of the CO_2 biogeochemical cycle should accurately consider every feature mentioned.

In a stationary state the hydrosphere and atmosphere are in a certain equilibrium with respect to CO_2 exchange, being broken with fluctuations of temperature, ocean surface level, vertical circulation regime, etc. An amount of CO_2 assimilated and emitted in the process of exchange between the ocean and the atmosphere constitutes 55.6×10^9 tC/year. The algae assimilate 16.7×10^9 tC/year from the atmosphere. These values are non-uniformly distributed over the water surface and strongly vary in time. The characteristic place in the CO_2 exchange is occupied by the Arctic Ocean, whose low temperatures determine a high absolute content of CO_2 dissolved in the surface layer in any season. On the average, in the summer, the arctic waters mainly assimilate CO_2 and emit O_2 to the atmosphere, whereas in winter they emit CO_2 and assimilate O_2. This clearly expressed seasonal change is smoothed down in the basins of the North Atlantic seas where a decrease of the ice cover, an increase of the period of photosynthetic activity and the existing fall-winter vertical convection lead to an intensification of the gas exchange with the atmosphere towards prevailing of the H_3^C flux. Carbon constitutes 0.18×10^9 tC/year of the amount of CO_2 assimilated from the atmosphere. The contribution of the arctic water basins into the removal of excess carbonic acid from the atmosphere is small compared to the total assimilation of CO_2 by the World Ocean, which is estimated at 5.2–6.6 × 10^9 tC/year, (Table 4.2). However, this assessment is rather underestimated, considering the results obtained by Kelley (1987). The average estimates show that the partial pressure of CO_2 in the atmosphere exceeds that in the arctic seas by 110 ppm. The CO_2 deficit in the marine medium takes place mainly during springtime blossoming and is estimated at ~450 gC/m^2. This means that the CO_2 flux from the atmosphere into the marine medium can vary from 1.5 to 4.0 gC/m^2/day. This assessment changes considerably as a function of longitude. For instance, the Norwegian Sea has a deficit of the CO_2 partial pressure between 20 and 50 ppm, and the Bering Sea $\Delta p_{CO2} \cong 70$ ppm. A clear linear correlation is observed with the proportion coefficient 10 ppm CO_2 per °C between the CO_2 content in the sea water and its temperature.

Table 4.2 Annual budget of CO_2 exchange with the atmosphere for water bodies of the Arctic Basin and northern seas (10^6 tC/year) (McGuire et al. 2010)

Region of the ocean	Summer			Winter			Year
	Assimilation	Emission	Balance	Assimilation	Emission	Balance	
Arctic Basin	4.4	0.0	+4.4	–	–	–	+4.4
Arctic Seas							
East Sibrian Sea	1.0	2.1	−1.1	0.0	0.7	−0.7	−1.8
Kara Sea	11.4	2.4	+9.0	0.0	0.2	−0.2	+8.8
Laptev Sea	2.4	3.6	−1.2	0.0	2.9	−2.9	−4.1
Chuckchee Sea	11.2	0.0	+11.2	0.0	0.6	−0.6	+10.6
Total	26.0	8.1	+17.9	0.0	4.4	−4.4	+13.5
North Atlantic Seas							
Barents Sea	71.9	0.0	+71.9	0.0	0.0	0.0	+71.9
Greenland Sea	19.1	0.0	+19.1	8.7	0.0	+8.7	+27.8
Norwegian Sea	64.8	0.0	+64.8	0.0	6.0	+6.0	+5.8
Total	155.8	0.0	+155.8	8.7	6.0	+2.7	+158.5
Arctic Ocean on the whole	186.2	8.1	+178.1	8.7	10.4	−1.7	+176.4

In conclusion, the arctic waters play the role of a pump that pumps away carbon dioxide from the atmosphere. Therefore, a detailed study of this process is needed to specify the global model of carbon cycle and to raise the accuracy of the estimation of the greenhouse effect due to emissions of anthropogenic CO_2.

On the whole, in the synthesis of the global model of the CO_2 biogeochemical cycle, the unit that simulates the ocean part of this cycle must describe the functioning of the ocean carbonate system. Alexeev et al. (1992), analyzed the system CO_2-HCO_3-CO_3^{2-} and the distribution of pH values of the ocean waters and found out that more than 80 % of dissolved carbon dioxide are in the form of hydrocarbonate ion of HCO_3. This means that only a first stage of dissociation of carbonic acid can be reliably considered in the synthesis of the model of the ocean carbonate system.

On global scales, in the World Ocean regions with cold waters (northern latitudes, upwelling zones) the flux H_3^C prevails while in the warm waters the flux H_2^C. The dynamics of this regime is maintained by reactions that take place in the hydrosphere between CO_2 and water resulting in the formation of carbonic acid:

$$HCO_3 + H_2O \Leftrightarrow H_2CO_3 + OH^-, \quad CO_3^{2-} + 2H_2O \Leftrightarrow H_2CO_3 + 2OH^-,$$

In the salt water, there are dissolved carbon dioxide, non-dissociated molecules of H_2CO_3, anions of HCO_3, and $CO_3{}^{2-}$. They all are in equilibrium:

$$CO_2 + H_2O \Leftrightarrow H_2CO_3 \Leftrightarrow HCO_3^- + H^+ \Leftrightarrow CO_3^{2-} + 2H^+$$

An equilibrium state between carbonic acid bicarbonate and carbonate ions establishes rapidly. The solution of CO_2 and formation of H_2CO_3 take place much slower. However, since many authors dispute the availability of H_2CO_3, the summarized concentration of CO_2 and H_2CO_3 should be considered, denoting it as $[CO_2]$. Let us introduce notations to characterize the total concentration of intermediate components

$$\sum C = [CO_2] + [H_2CO_3] + \left[HCO_3^-\right] + \left[CO_3^{2-}\right] \quad (4.1)$$

Another characteristic of the hydrosphere is alkalinity:

$$A = \left[HCO_3^-\right] + 2\left[CO_3^{2-}\right] + [H_2CO_3] + [OH^-] - [H^+] \quad (4.2)$$

where $[Ca^{2+}]$ and $[H^+]$ are the concentrations of ions of calcium and hydrogen respectively, and $\left[HCO_3^-\right]$ and $[CO_3^{2-}]$ are the concentrations of bicarbonate and carbonate ions correspondingly.

Using these notations, the equilibrium conditions are formulated:

$$K_0 p_c = [CO_2], \tag{4.3}$$

$$[HCO_3^-] = K_1[CO_2]/[H^+], \tag{4.4}$$

$$[CO_3^{2-}] = K_2[HCO_3^-]/[H^+], \tag{4.5}$$

$$[Ca^{2+}] = L_p[H^+]^2 / (K_1 K_2[CO_2]), \tag{4.6}$$

where K_0 is the indicator of CO_2 solubility (=250 mmol/ℓ atm at 30 °C and 640 mmol/ℓ atm at °C), L_p is the solubility indicator for $CaCO_3$, K_1 and K_2 are the first and second apparent coefficients of dissociation of carbonic acid (depend on temperature and pressure).

From Eqs. (4.1)–(4.6) we obtain:

$$\sum C = [CO_2]\left(1 + K_1/[H^+] + K_1K_2/[H^+]^2\right), A = [CO_3^{2-}]\left(K_1/[H^+] + 2K_1K_2/[H^+]^2\right)$$

Eriksson (1963) used a differentiation operator δ and calculated the relative changes of $[CO_2]$, $[Ca^{2+}]$, p_c, and A. Any of these components are denoted as U and taking $\Delta U = \delta U/U$, we obtain:

$$\Delta p_c = \Delta[CO_2] - \Delta K_0 \tag{4.7}$$

$$\Delta[Ca^{2+}] = \Delta L_p + 2\Delta[H^+] - \Delta K_1 - \Delta K_2 - \Delta[CO_2] \tag{4.8}$$

$$\Delta\Sigma C = \Delta[CO_2] + F_1\Delta K_1/F_0 + a_2\Delta K_2/F_0 - F_2\Delta[H^+]/F_0 \tag{4.9}$$

$$\Delta A = \Delta[CO_2] + \Delta K_1 + 2a_2\Delta K_2/F_2 - F_3\Delta[H^+]/F_2 \tag{4.10}$$

where

$$F_0 = 1 + a_1 + a_2, \quad F_1 = a_1 + a_2, \quad F_2 = a_1 + 2a_2, \quad F_3 = a_1 + 4a_2,$$
$$a_1 = K_1/[H^+], \quad a_2 = K_1K_2/[H^+]^2.$$

Based on (4.7) through (4.10), and considering the experimental data, Eriksson (1963) showed that a 1 °C increase of the temperature of the upper layer of the marine medium causes an increase of the partial pressure of CO_2 in the atmosphere by 4.2–5.8 %, a 1 % decrease of the water volume implies an increase of the partial pressure of CO_2 in the atmosphere by 3 %, with 1 % of carbon dioxide in the water medium precipitating as $CaCO_3$. With variations of the equilibrium state in deep layers of the hydrosphere a 1 % output of CO_2 in the form of gas is followed by a 1 % sediment of CO_2 in the form of $CaCO_3$. Moreover, variations in the CO_2 partial pressure correlate with changes of the concentration of phosphorus P, so that a decrease of P in deep layers by 1 % leads to an increase of p_c by 5.6 %.

Eriksson (1963) found out that a 1 % increase of alkalinity causes a 2.26 % decrease of the CO_2 partial pressure in the atmosphere and a 98 % reduction of total CO_2 supply in the hydrosphere. When there is no external input of Ca into the hydrosphere, a 1 % increase of alkalinity causes an increase of the rate of $CaCO_3$ deposition and the total alkalinity decreases by 0.92 %. Thus, a 1 % increase of the water alkalinity is equivalent to an increase of *pH* by 0.5.

An equilibrium between various components of the hydrospheric carbonate system depends on temperature and pressure, a combination of which correlates with *pH* so that at a given temperature and pressure the equilibrium is a function of only $pH = -lg[H^+]$. The effect of temperature on *pH* in the first approximation can be described by the Merrey law (Ivanov 1978): $\Delta pH = -0.0111\Delta T$, valid at $pH \in [7.5, 8.4]$, $T \in [1–30]$ °C, and salinity from 10 to 40‰. The dependence of *pH* on pressure p_c follows the Buch-Grippenberg law: $pH = d\Delta p_c$, where, on the average, $d = -0.0254$. A more accurate presentation of this law is given in Table 4.3.

The connection between the equilibrium condition of CO_2 exchange and *pH* on the atmosphere-ocean border is such that when the CO_2 pressure in the atmosphere reaches 330×10^{-6} atm, equilibrium occurs at 20 °C for $pH = 8.16$ and at 0 °C for $pH = 8.11$. For a lower pH value, the ocean assimilates CO_2, while for a higher *pH* value the ocean emits CO_2. Hence, the structure of the ocean carbonate system should be thoroughly studied in order to describe the functions of the fluxes H_2^C and H_3^C on the atmosphere-ocean border. A simplified description of these fluxes is usually based on the comparison of the partial pressures of CO_2 in the atmosphere and in the ocean. According to the data in Alexeev et al. (1992), the fluxes H_2^C and H_3^C are well approximated by the function $H_i^C = k_i(p_a - p_c)^{1/2}$, where p_a and p_c are the partial pressures of CO_2 in the atmosphere and in the ocean, respectively. The partial pressure of CO_2 in the atmosphere at the level of the ocean can be calculated using the formula:

$$p_a = 0.421542 \cdot 10^{-18}\, M_C(273.15 + T),$$

where M_C is the CO_2 mass in tons and T is the air temperature in °C.

According to Bjorkstrom (1979), the functional dependence of p_c on the parameters of the ocean carbonate system can be presented in the form of $p_c = [CO_2]/K_0$. From the condition of chemical equilibrium, according to the Eqs. (4.4) and (4.5), it follows:

$$[CO_2] = [H^+]\sum C/a,$$

where $a = [H^+]^2 + [H^+]K_1 + K_1K_2$, $\Sigma C = C_U/W_U$, W_U is the volume of an elementary reservoir.

Table 4.3 Empirical dependence of pH on atmospheric pressure

pH at atmospheric pressure	7.5	7.7	7.9	8.1	8.3
ΔpH with pressure increasing by 1,000 dbar	−0.035	−0.028	−0.023	−0.021	−0.02

Denoting

$$K_W = [H^+][OH^-], \quad K_B = [H^+]\left[B(OH)_4^-\right]/\left[B(OH)_3\right],$$
$$B_T = \left[B(OH)_3\right] + \left[B(OH)_4^-\right]$$

and using Eq. (4.2) we obtain:

$$A = K_W[H^+]^{-1} + K_B B_T (K_B + [H^+])^{-1} + C_U([H^+]K_1 + 2K_1K_2)/(aW_U) - [H^+] \text{ or}$$

$$A = [CO_2]\left(\frac{K_1}{[H^+]} + \frac{2K_1K_2}{[H^+]^2}\right) + \frac{B_T}{1 + [H^+]/K_B} + \frac{K_W}{[H^+]} - [H^+] \tag{4.11}$$

The solution of this equation relative to $[H^+]$ enables one to determine p_c as a function of C_U. According to Bjorkstrom (1979), $K_B = 2 \times 10^{-9}$ and $K_W = 10^{-14}$.

The right-hand part of Eqs. (4.2) and (4.11) implies an account of all weak acids in the salt water. Ions of other compounds are taken into account through the dependence of the equilibrium constants on salinity or chlorine content. The characteristic parameter of the equilibrium of the carbonate system is a variable:

$$X = \frac{(K_1K_2)^{1/2}}{[H^+]} = \left[\frac{[CO_3^{2-}]}{[CJ_2]}\right]^{1/2}$$

In the terms of this parameter we have:

$$\Sigma C = \left(1 + KX + X^2\right)[CO_2] \tag{4.12}$$

$$A = \left(KX + 2X^2\right)[CO_2] + W(X), \tag{4.13}$$

where $K = \sqrt{K_1/K_2}$,

$$W(X) = B_T/\left\{1 + [K_1K_2]^{1/2}X^{-1}\right\}/K_B + K_W X [K_1K_2]^{-1/2} - [K_1K_2]^{1/2}X^{-1}$$

Excluding $[CO_2]$ from the Eqs. (4.12) and (4.13):

$$(2C - A')X^2 - K(A' - C)X - A' = 0, \tag{4.14}$$

where $A' = A - W(X)$.

Since $W(X) << A$, the following iterative procedure can be used beginning with an estimation of $X = X_1$ ($\cong 1$) in order to solve the Eq. (4.14):

$$A'_i = A - W(X_i), X_i = 0.5\left[-b + \left(b^2 - 4ac\right)^{1/2}\right]/a(i \geq 2),$$

where $a = 2C - A'_{i-1}$, $b = K(A'_{i-1} - C)$.

The exchange processes taking place on the atmosphere-ocean border have been experimentally studied by Kiseleva (1990), Zaitsev (1988), and others. In their works it has been shown that at high wind speeds the rate of gas exchange sharply increases. This is connected with the mechanism of the foam formation on wave crests as well as an intensive catching of air bubbles and their penetration beneath the water surface. The dependence of the amount of spray Q on the height over the water surface is well approximated by a linear function. For instance, for the wind speed $V = 11.1$ m/s this approximation is (Kiseleva 1990):

$$Q = \begin{cases} Q' \exp\{-14.5z + 0.684\} & \text{without wave breaking,} \\ Q' \exp\{-7.48z + 0.842\} & \text{with wave breaking,} \end{cases}$$

where $Q'[\mathrm{cm}^{-2}\mathrm{s}^{-1}]$ is the dimensional multiplier.

Experimental estimates of the net gas transport due to the droplet mechanism have shown that the CO_2 flux between water and air can vary from 1.44 mg CO_2/m^2/h in the absence of wave breaking to 26.6 mg CO_2/m^2/h. The rate of the gas exchange on the atmosphere-ocean border is markedly affected by the foam formation on the ocean surface. In this case, the rate of the gas exchange can increase up to 28 %. All this suggests the conclusion that a detailed modeling of the mechanisms of the atmosphere-water CO_2 transport is needed. The satellite World Ocean monitoring allows one to identify the zones of foam formation and, hence, along with the modeling of the processes of foam formation, to perform direct measurements of the areas of the foam-covered basins.

The mechanism of the air bubbles behaviour in the upper layer of the ocean has been poorly studied. The available theoretical results are based, as a rule, on a number of suppositions, which in many cases can drastically distort the ideas about real processes of the gas exchange between the atmosphere and the ocean. Among these suppositions the following have been used most often:

- air bubbles in the upper layer of the ocean practically instantly acquire the temperature of the environment, since the molecular coefficient of heat conductivity exceeds that of gases by 2–3 orders of magnitude;
- air bubbles in water do not affect its dynamics; and
- air bubbles do not merge in water.

The size of air bubbles in water varies constantly because of their gradual solution. The gas flux through the bubbles is described by the relationship:

$$Q = 4\pi RD|\Delta p|\mathrm{S} \cdot Nu,$$

where $|\Delta p|$ is the difference of the partial pressures of gas in the water and in the atmosphere, R is the bubble's radius,

$$Nu = \begin{cases} 1 + 0.5Pe + 0.5Pe^2 \ln(2Pe) & \text{for} \quad R \le 20\,\mu\mathrm{m}, \\ \sqrt{2/3\pi Pe^{1/3}} & \text{for} \quad 20\,\mu\mathrm{m} < R \le 200\,\mu\mathrm{m}, \\ 0.45Pe^{1/3}\mathrm{Re}^{1/3} & \text{for} \quad 200\,\mu\mathrm{m} < R \le 400\,\mu\mathrm{m}, \end{cases}$$

Re is the Reynolds number ($Re = 2vR/D$), *Pe*—is the Pekle number ($Pe = vR/\nu$), ν is water viscosity, *v* is the velocity of the bubble's motion, *D* is the coefficient of the gas molecular diffusion.

Of course, the ocean surface condition substantially affects its gas exchange with the atmosphere. The size of the basins covered with either foam or white caps directly depends on a combination of the parameters such as wind speed, water temperature, and sea currents. An analysis of the statistical characteristics of the patchy pattern of the ocean surface made by many experts makes it possible to describe the percentage distribution of the areas covered with foam (S_f) and white caps (S_l) with the following binary functions of wind speed *V* (at a height of 10 m):

$$S_f = \begin{cases} 0 & \text{for} \quad V < 5 \text{ m/s}, \\ 0.65\{1 + 4.76 \times 10^{-2}(V - 5^2)\} & \text{for} \quad V \geq 5 \text{ m/s}, \end{cases}$$

$$S_l = \begin{cases} 0 & \text{for} \quad V < 5 \text{ m/s}, \\ 0.015\{1 + 2.2 \times 10^{-2}(V - 5^3)\} & \text{for} \quad V \geq 5 \text{ m/s}. \end{cases}$$

According to Kiseleva (1990), the relationship between S_f and S_l at $V \geq 5$ m/s obeys the following rule:

$$S_f/S_l = 50 - 3.4(V - 5)$$

It is clear that the structure of the atmosphere-ocean border can affect the gas exchange within widely varying values of fluxes H_2^C and H_3^C (see Fig. 1.30). Unfortunately, the level of knowledge of the laws of changes of the atmosphere-ocean border structure as a function of synoptic situations does not permit a clear estimation of the limits of detailing the processes taking place on this border, to obtain the values of the fluxes H_2^C and H_3^C close to real ones. This means that at this stage of the synthesis of the global model of the CO_2 cycle some uncertainty still exists.

4.4 The Zonal Model of the Global Carbon Cycle in the Atmosphere-Ocean System

Consider one of the versions of parameterization of the carbon dioxide global dynamics in the atmosphere-ocean system proposed and studied by Nefedova (1994). The spatial heterogeneity of the World Ocean is approximated by the zonal scheme in accordance with the latitudinal dependence of the climatic processes as well as processes of mixing in the atmosphere and in the oceans. There are 14 latitudinal zones 10° in size. In the vertical, there are three layers in the World Ocean: the upper quasi-homogeneous layer (UQHL) with a time-dependent thickness, thermocline, and deep ocean. As a result, the World Ocean is divided into 42 volume parts. An upwelling is assumed to exist between 40°N and 40°S along with a downwelling in high latitudes. In the UQL the water flows poleward from the equator, and in deep layers the water flows in the opposite direction (Fig. 1.27). Over each water basin, the atmosphere is simulated by a point model. The carbon exchange between the zones of the atmosphere takes place due to advection H_i^a and turbulent diffusion H_i^d:

$$H_i = H_i^a + H_i^d, \quad (i = \overline{1.14}); \; H_i^a = 2\pi R_0 h_a V_i^* \left(C_i^a - C_{i+1}^a\right) \cos \phi_i,$$
$$H_i^d = [2\pi R_0 A_h h_a / \Delta\phi_i]\left(C_i^a - C_{i+1}^a\right) \cos \phi_i,$$

where $C_i^a = M_i^a / V_i^a$ is the concentration of carbon in the i-th zone of the atmosphere, M_i^a is the carbon mass in the i-th zone of the atmosphere, V_i^a is the volume of the i-th zone of the atmosphere, R_0 is the Earth's radius, ϕ_i is the latitude of the southern boundary of the i-th zone, h_a is the altitude of the atmosphere (10 km), V_i^* is the average velocity of the meridian transport of air masses in the atmosphere (0.2–1.0 m/s), A_h is the coefficient (10^5 m^2/s).

The CO_2 exchange on the atmosphere-ocean border is described by the traditional law:

$$H_i^{a0} = k(u_i)\left(P_i^o - P_i^o\right),$$

where

$$P_i^a = k_a M_i^a R T_i^a S_i^{-1} \mu^{-1},$$

$k(u_i)$ is the proportion coefficient depending on wind speed, P_i^o and P_i^o are partial pressures of CO_2 in the i-th zone of the atmosphere and the ocean respectively, k_a is the share of the 100 m air column mass in the mass of a 10 km column (≈0.01602), R is the universal gas constant (8.31451 J/mol/K), T_i^a is the air temperature at the level of the ocean in the i-th zone $\left(T_i^a = T_i^{as} + \Delta T_i^a\right)$, T_i^{as} is the seasonal temperature component, ΔT_i^a is the average annual change of air temperature caused by the increased CO_2 content in the atmosphere.

A SST change is assumed to take place in phase with a change of air temperature by the same value:

$$T_i^o = T_i^{os} + \Delta T_i^a$$

The partial pressure of CO_2 dissolved in the surface waters is proportional to its concentration in the water and inversely proportional to its solubility. This dependence is established by solving the system of Eqs. (4.12) and (4.13) that describe the functioning of the ocean carbonate system. For the quantitative solution of this system one can use, for instance, the methods of secants. As a result, we obtain [CO_2] and P_i^o. Based on the data on the temperature dependence of the equilibrium constants for the respective chemical reactions, we find:

$$P_i^o = P_i^o\left(C_i^1, T_i^o\right), \; \left(i = \overline{1.14}\right)$$

The turbulent fluxes of carbon at the lower boundary of the UQL (H_{UPL}) and the upper boundary of thermocline (H_T) are described by the functions:

$$H_{UPL} = \begin{cases} (C_i^1 - C_i^2)(V_i + dh_i/dt) & for \quad dh_i/dt + V_i > 0, \\ 0 & for \quad dh_i/dt + V_i \leq 0 \end{cases}$$

$$H_T = \begin{cases} (C_i^1 - C_i^2)(V_i + dh_i/dt) & for \quad dh_i/dt + V_i \leq 0, \\ 0 & for \quad dh_i/dt + V_i > 0 \end{cases}$$

where C_i^1 and C_i^2 are the concentrations of inorganic carbon in the UQHL and thermocline, respectively, h_i is the UQL depth of the i-th zone of the ocean (depends on the season), V_i is the water lifting (in upwelling) or lowering (in downwelling) (0.001–0.1 cm/s).

The turbulent fluxes of carbon on the thermocline—deep ocean border is considered to be proportional to the coefficient k_T of the difference of carbon concentrations in the bordering layers:

$$H_{TF} = k_T\left(C_i^3 - C_i^2\right),\ \left(i = \overline{1.14}\right)$$

where C_i^3 is the concentration of inorganic carbon in the deep ocean of the i-th zone.

Production and circulation of organic carbon in the ocean have been studied in detail and parameterized by many authors. In the model by Nefedova and Tarko (1993) the following approximations have been assumed:

$$B_i^p = B_i^o\{1 + (\delta_i - 1)/(\delta_i + 1) \cdot \sin(wt - \pi/2)\} \text{for}$$
$$\varphi \in [60°\text{N}, 90°\text{N}] \cup [60°\text{S}, 90°\text{S}],\ (i = 1, 14);$$
$$B_i^p = B_i^o\{1 + (\delta_i - 1)/(\delta_i + 1) \cdot \sin(2wt - \pi/2)\} \text{ for}$$
$$\varphi \in [10°\text{N}, 60°\text{N}] \cup [10°\text{S}, 60°\text{S}],\ (i = 2 - 6, 9 - 13);$$
$$B_i^p = B_i^o, \text{ for } \varphi \in [0°, 10°\text{N}] \cup [0°, 10°\text{S}], \quad (i = 7, 8),$$

where B_i^o is the average annual production of phytoplankton (150–550 × 10^9 t/year), δ_I is the ratio of maximum and minimum values of the rate of organic matter production.

Organic matter produced in the UQHL decomposes and descends to deeper layers. Let B_i^{1d}, B_i^{2d} and B_i^{3d} be the rates of the organic matter decomposition in the UQHL, thermocline, and deep ocean, respectively; γ—the ratio of the average annual amount of organic matter decomposing in the UQHL to the average annual production of phytoplankton, τ—the lag time of the organic matter decomposition, D—an average depth of the ocean (3,653–3,795 m), H—the depth of the thermocline bottom (30–200 m). Then

$$B_i^{1d} = \gamma B_i^p(t - \tau),\ \left(i = \overline{1.14}\right);\ B_i^{2d} = (1 - \gamma)(H - h_i)/(D - h_i)\left\langle B_i^{1p}\right\rangle;$$

$$B_i^{3d} = (1 - \gamma)(D - H)/(D - h_i)\left\langle B_i^{1p}\right\rangle$$

Here the angular brackets <> denote the calculation of the average value for the annual cycle.

So, the seasonal model of the global carbon cycle developed by Nefedova and Tarko (1993) and Nefedova (1994) is simulated by a system of 56 ordinary differential equations with periodic coefficients ($i = 1, \ldots, 14$):

$$\frac{d}{dt}C_i^1 h_i - \left(\frac{dh_i}{dt} + V_i\right)C_i^1 = H_i^{ao} - H_{UPL} + L_i^1 - B_i^p + B_i^{1d},$$

$$\frac{d}{dt}C_i^2(H - h_i) + \left(\frac{dh_i}{dt} + V_i\right)C_i^2 = H_T + H_{TF} + B_i^{2d};$$

$$\frac{d}{dt}C_i^3(D - H) + V_i C_i^3 = L_i^3 - H_{TF} + B_i^{3d};$$

$$\frac{dM_i^a}{dt} = \gamma_i^a + V_i^a(H_i - H_{i-1}) - 0.012 S_i H_i^{ao};$$

where γ_i^a are the anthropogenic carbon emissions in the i-th zone of the atmosphere.

4.5 Modelling the Carbon Cycle in the World Ocean

Along with the physical-chemical processes of transformation of carbon compounds in the hydrosphere mentioned above, the general circulation in the World Ocean plays an important role in their long-range transport. According to the available estimates, about 80 % ($\sim$1.7 Gt) of organic matter formed in the process of photosynthesis descends to deeper layers and oxidizes giving CO_2. This process is balanced by a slow lifting of water, and thereby a situation for a stable CO_2 exchange is formed. However, in local situations there are strong deviations from the stable conditions, which can be described only by the scheme of the spatial-temporal structure of the ocean waters motion. The block-scheme of this structure identifies the surface, intermediate, deep, and bottom waters as well as the currents topography on every horizon. The simulation models of the World Ocean circulation widely vary in the scientific literature, therefore the development of this part of the global model unit is not difficult.

Following Nitu et al. (2000b), by the depth z the ocean can be divided into four basic layers: photic to well-heated depths ($\Omega_U = \mathrm{U}[z_i, z_{i+1}]$, $z_0 = 0$; $i = 0, 1, \ldots, m - 1$); intermediate ($\Omega_P = \mathrm{U}[z_i, z_{i+1}]$, $i = m, \ldots, m + n - 1$); deep ($\Omega_L = \mathrm{U}[z_i, z_{i+1}]$, $i = m + n, \ldots, m + n + l - 1$), and bottom Ω_F. By the hydro-physical-ecological characteristics the layer Ω_U as a function of latitude φ, longitude λ, and season t can be attributed to warm or cold waters, the layer Ω_P is photic but always with low water temperatures, in the layer Ω_L the phytoplankton is not produced. Finally, the layer Ω_F plays the role of a boundary layer.

On the average, the layer Ω_U has the area 360×10^{12} m^2 and depth $z_m \cong 75$ m, the warm waters covering area 240×10^{12} m^2. The layer Ω_P is located at depths from z_m to $z_{m+n} \cong 200\ m$. In concrete model realizations given in Krapivin and Kondratyev (2002) the following assumptions have been made: $m = 10$, $n = 2$, $l = 8$.

The vertical CO_2 transport in the ocean is determined by advective fluxes $H_{19,ij}^C$ and gravitational sedimentation of dead organic matter (the flux $H_{20,ij}^C$). An advective transport from the i-th into the j-th reservoir of the ocean is considered proportional to the concentration of carbon in the respective reservoirs: $H_{19,ij}^C = \lambda_{2,ij} C_{a,i}$ ($a = U,P,L$), where $\lambda_{2,ij} = V_{ij}/V_i$, V_{ij} is the water volume transported per unit time from the i-th reservoir into the j-th reservoir; V_i is the volume of the i-th reservoir.

The following algorithm is widely used for the parameterization of the process of carbon sedimentation. The flux under unit area of the ocean is supposed to decrease exponentially with depth. If we denote the inflow of organic matter in the i-th reservoir as g_i and the net outflow of organic matter from the water surface as H_{20T}, we obtain:

$$H_{20,1}^C = H_{20,T};\ H_{20,i}^C = g_{i-1}(\sigma_i/\sigma_{i-1})exp[-(z_i - z_{i-1})/D_s],\ (i = 2, \ldots, m+n+l),$$

where σ_i is the area of the i-th reservoir, D_s is the characteristic time of the organic matter particles sedimentation before their decomposition. The rate of decomposition in each reservoir is equal to:

$$R_{D,i} = H_{20,i}^C - H_{20,i+1}^C,\ (i = 1, \ldots, m+n+l); R_{D,F} = H_{20,m+n+l}^C - H_{16}^C.$$

However, if the time of the transition of the organic matter particles from one layer to another is short compared to D_s, then it is better to take $H_{20,i}^C = \lambda_1 C_{a,i}$, $H_{16,i}^C = \lambda_4 C_{F,i}$. In addition to these fluxes, one should take into account the fluxes of detritus decomposition, solution of bottom sediments, and carbon consumption in the process of photosynthesis:

$$H_{17,i}^C = Const.,\ H_{18,i}^C = \lambda_3 D_{L,i},\ H_{22,i}^C = \lambda_3 D_{U,i},\ H_{21,i}^C = C_{31} R_{\Phi,i}.$$

The estimates for the modeled parameters of the particular oceanic processes of carbon cycle scatter widely. For instance, from the data of various authors the estimates of assimilation of carbon from the hydrosphere in the process of photosynthesis range from 10 to 155 Gt C/year. The value 127.8 Gt C/year is the most widely used. However, because of large dispersions of these estimates the calculation of the C_{31} coefficient is fraught with a great uncertainty and therefore numerical experiments with the use of other, more accurate data are required in order to be specified.

Finally, let us suppose the surface layers of the ocean are filled with carbon due to its run-off from the land $H_{24,i}^C = C_7 W_{sio}$, $H_{23,i}^C = C_8 W_{Hio}$, where W_{sio} and W_{Hio} are the river and underground run-offs into the World Ocean, respectively.

4.6 Organic Carbon and Oceanic Ecosystems

4.6.1 Models of the Oceanic Ecosystems

Each element of the ocean ecosystem A can be described by α number of parameters, and the connection between the elements can be presented as the connection between the respective parameters. Then, on the whole, the ecosystem A can be described by N time-dependent parameters $x^*(t) = \{x_j(t), j = 1,\ldots,N\}$. The structure $|A(t)|$ and behaviour $\bar{A}(t)$ of the ecosystem A which can be observed in more or less detail, are the functions of these parameters. Therefore the ecosystem itself $A(t) = \{|A(t)|, \bar{A}(t)\}$, as a collection of the structure and behaviour, is the function of these parameters:

$$A(t) = F(x^*(t)) \tag{4.15}$$

Hence, according to (4.15) with a change of the time t the ecosystem A will be characterized by the trajectory in the $(N + 1)$-dimensional Euclidean space. The following abstract formation will be considered a model of the ecosystem $A(t)$:

$$A_M(t) = F_M\left(x^{*,M}\right), \tag{4.16}$$

that depends on $M \leq N$ components of the vector $x^*(t)$ $(\{x^{*,M}(t)\} \in \{x^*(t)\})$ and considers all the connections existing between them.

The nearer M is to N and the more completely the connections between the vector $x^{*,M}(t)$ components are taken into account, the less is a disagreement of the trajectories of the ecosystem $A(t)$ and its model $A_M(t)$. The latter can be measured by any natural measure, for instance, by a maximum of absolute difference of all the respective coordinates of trajectories or by the integral of absolute difference of all respective coordinates for a final time period. In other words, introduce some goal functional

$$V = Q(\{x_i(t)\}) \tag{4.17}$$

on the trajectory of the ecosystem $A(t)$. The form of the Q function is determined by the requirements made by the system A to the environment. The natural evolutionary process is considered to lead to an optimal system and therefore the system $A_{opt,M}(t)$, which ensures an extremum of the functional (4.17), can be considered an optimal model of the ecosystem A.

The value of disagreement between the trajectories of the ecosystem A and an optimal model $A_{opt,M}(t)$ is affected by the degree of correspondence of the goal functional V chosen by the relationship (4.17) to a real goal $\underline{A}$ of the ecosystem A. The set $\underline{G} = \{\underline{A}_1, \ldots, \underline{A}_r, \ldots, \underline{A}_m\}$ of possible reliable goals $\{\underline{A}_r\}$ of the ecosystem can be derived on the basis of the experience accumulated by oceanologists. Then, determining $A_{opt,M,r} = g_r(\underline{A}_r)$, $\underline{A}_r \in \underline{G}$, we obtain a limited assembly of possible optimal systems $A_{opt,M,r}$ $(r = 1, \ldots, m)$, whose trajectories together with the trajectory of the ecosystem A, are within the space of possible trajectories. If we find

$A_{opt,M,r0}$, whose trajectory has a minimum disagreement with $A(t)$, then, based on this, we can find the most reliable goal of the ecosystem

$$\underline{A}_{r0} = (g_r)^{-1}\left(A_{opt,M,r0}(t)\right).$$

According to the principles mentioned above, the derivation of a numerical model of the ocean ecosystem A (Fig. 4.3) requires either a detailed description of its states or derivation of an adequate complex of numerical models of energy exchange between the trophic levels, taking place in A, as well as interactions of the biotic, abiotic, and hydrophysical factors. Of course, in this case an availability of a certain set of hypothesis is assumed concerning the character of the balance relationships in the ecosystem A.

The basic hypothesis is that in the ecosystem A the only original source of energy and matter for all forms of life is the solar radiation energy (E). According to numerous theoretical and experimental studies, the sun light penetration into deep layers of the ocean follows an exponential law

$$\begin{aligned} E(\varphi,\lambda,z,t) &= uE_0 \exp\left[-\int_0^z \{\delta p(\varphi,\lambda,x,t) + \beta d(\varphi,\lambda,x,t) + vZ(\varphi,\lambda,x,t)\}dx - \alpha z\right] \\ &\quad + (1-u)E_0 \exp(-\zeta z), \end{aligned} \tag{4.18}$$

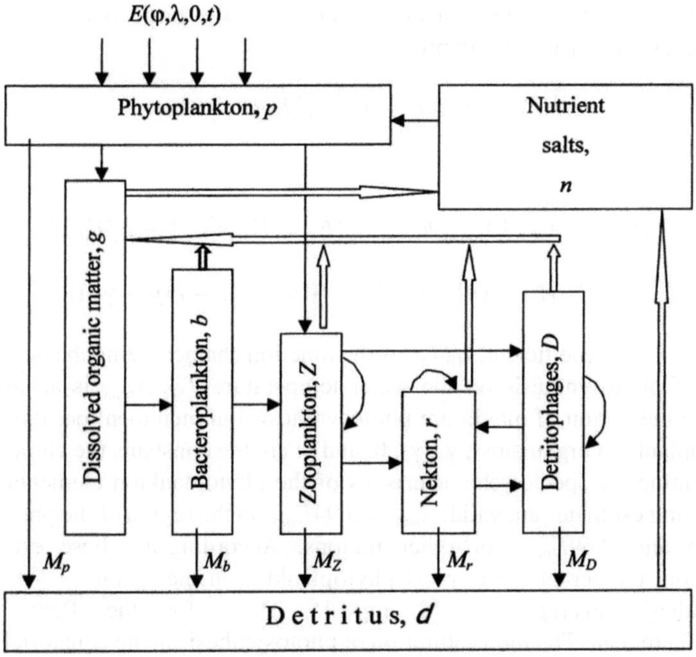

Fig. 4.3 The standard block-scheme of the model of the ocean ecosystem. The *arrows* show the energy flux directions

where $E_0 = E(\varphi, \lambda, 0, t)$ is the illumination of the ocean surface, α is the coefficient of light absorption by the filtered sea water; δ, β, and ν are the coefficients of light attenuation due to phytoplankton (*p*), detritus (*d*), and zooplankton (*Z*), respectively; *u* and ζ are the parameters chosen in a concrete situation to bring $E(\varphi, \lambda, z, t)$ closer to a real pattern of illumination changing with depth. It must be noted that the impact of the biomass of other trophic levels on water transparency is considered to be negligibly small.

Illumination affects the rate of photosynthesis R_p. The R_p parameter as a function of *E* has a maximum at an optimal value of E_{max}, which drifts from this critical value when the illumination increases or decreases. A maximum of R_p at various latitudes φ is located at depths varying as a function of season (i.e. sun elevation). Thus, this variability with depth is most pronounced in the tropical zones. On the average, the photosynthesis maximum is located at depths 10–30 m, and in the open water bodies it can be observed at depths below 30 m. Here E_{max} = 65–85 cal/cm^2/day. Beginning from the depths where E = 20–25 cal/cm^2/day, photosynthesis decreases in proportion to *E*. An apparent suppression of phytoplankton by light is observed at $E > 100$ cal/cm^2/day. These estimates are quite different in the northern latitudes, where the photosynthesis maximum is located—as a rule—on the surface.

The rate of photosynthesis at a depth *z* depends on the water temperature T_W, the concentration of biogenic elements *n*, and the phytoplankton biomass *p*, as well as on other factors, which are not considered here. To express this dependence, various equations are used, which in either form reflect the limiting role of the elements *E, n,* and *p*. Considering that $\partial p/p\partial z \rightarrow 0$ at $n \rightarrow 0$ and $\partial p/p\partial z \rightarrow const$ with increasing *n*, the following function is taken as the basic one to describe the photosynthesis intensity at a depth *z*:

$$R_p(\varphi, \lambda, z, t) = k_0(T_W)K_T f_2(p) f_3(n), \qquad (4.19)$$

where

$$K_T = Af_1(E),\ A = kA_{max}/E_{max},\ f_1(E) = Eexp[m(1 - E/E_{max})], \qquad (4.20)$$

$$f_2(p) = [1 - exp\{-\gamma_1 p\}],\quad f_3(n) = [1 - exp\{-\gamma_2 n\}]^{\theta},$$

k is the proportion coefficient; $k_0(T_W)$ is the function characterizing the dependence of the rate of photosynthesis on the water temperature T_W; A_{max} is an assimilation number in the region of maximum photosynthesis (an increment per unit weight of the phytoplankton organisms); γ_1, γ_2, θ, and *m* are the constants the choice of which can determine the species characteristics of the phytoplankton elements. For A_{max} the following estimates are valid: $A_{max} = 5.94E_{max}$ in the region of the photosynthesis minimum and $2.69E_{max}$ for other regions. According to these estimates, an assimilation number of the tropical phytoplankton in the region of the maximum photosynthesis averages 11–12 mgC/h. Thus, for the Peru upwelling A_{max} = 6.25 mgC/h. The light saturation of photosynthesis in the equatorial regions is reached at 9 cal/cm^2/day.

As for the dependence of $k_O(T_W)$, the specific intensity of the phytoplankton photosynthesis first increases with the temperature changing from low to high values, reaches in some interval of temperatures a maximum value optimal for p, and then, with the further increase of the temperature it begins to decrease. Near the maximum the following approximation is often used:

$$k_O(T_W) = exp\{(T_W - T_{W,opt})ln(\theta_0)\},\ 0 < \theta_0 \leq 2.$$

The dependence of the rate of photosynthesis on the concentration of nutrient elements $n(\varphi, \lambda, z, t)$ (phosphorus, silicon, nitrogen, and others), expressed in the formula (4.19) by the exponential term, is, of course, more complicated. The nutrient elements are one of the most important parts of the ecosystem, since they regulate the energy flux in the ecosystem. The nutrient elements supplies are spent in the process of photosynthesis at a rate R_n, usually approximated by the expression $R_n = \delta R_p$, where δ is the proportion coefficient. The nutrient elements supplies are replenished due to their sweeping out with the lifting of deep waters, where their supplies are formed as a result of chemical processes of the dead organic matter decomposition. This process is controlled by several abiotic conditions which are characteristic of various climatic zones of the World Ocean. The vertical flux of nutrient elements is determined by conditions of water mixing. In the tropical zones, where the vertical structure of the water has a clear 3-layer configuration with a layer where the temperature leaps suddenly (the layer of thermocline), the vertical motion of nutrient elements is confined to this layer. In the water bodies where the thermocline is located at depths 40–100 m, the upper layer is usually poor in nutrient elements, and their input to this layer takes place only in the zones of upwelling. In this case, the average rate of the vertical water lifting beneath the thermocline varies within 10^{-3}–10^{-2} cm/s, and in the zones of upwelling (break-through of the thermocline) it can reach 0.1 cm/s.

4.6.2 Equations of the World Ocean Ecosystem Dynamics

The entire volume of ocean water is considered as a single biocenosis, in which the flux of organic matter produced in the surface layers and then descending to maximum depths of the ocean is the main connecting factor. All model parameters are assumed to be able to change as functions of place and time, and their parametric description is made by average characteristics (i.e., deterministic models).

Let us suppose the food bonds between trophic levels are adequately described by the Ivlev model (i.e., the consumption of various kinds of food by the i-th trophic level is proportional to the efficiency of their biomasses). Taking into account the diagram of food bonds shown in Fig. 4.3 and the structure of the trophic pyramid of a typical ocean ecosystem, we can consider each trophic level in detail.

Bacterioplankton b plays an important role in the trophic chains of the ocean. By the available estimates, not less than 30 % of the bacterioplankton mass is in natural aggregates exceeding 3–5 μ in size, therefore the filtrates can feed on this biomass. This fact must be taken into account when deriving a model of the ecosystem, since in many regions of the World Ocean the bacteria production is comparable with the production of phytoplankton (Kolber et al. 2001). Bacteria taking a special place in the trophic pyramid, differ by a variable exchange, strongly decreasing with the deficit of food, which is followed by the respective decrease of the rate of their growth. The food for bacteria is mainly detritus d and dissolved organic matter g emitted by phytoplankton (Hansell and Carlson 2002). As a result, the ration for bacteria can be described by the Ivlev formula

$$R_b = k_b b[1 - exp(-k_{1,d}d - k_{1,g}g)], \tag{4.21}$$

where k_b, $k_{1,d}$, and $k_{1,g}$ are the coefficients determined experimentally.

The equation describing the dynamics of the bacterioplankton biomass is written in the form:

$$\begin{aligned} \partial b/\partial t + V_\varphi \partial b/\partial \varphi + V_\lambda \partial b/\partial \lambda + V_z \partial b/\partial z = R_b - T_b - M_b \\ - \sum_{s \in \Gamma_b} C_{bs} R_s + k_{2,\varphi} \partial^2 b/\partial \varphi^2 + k_{2,\lambda} \partial^2 b/\partial \lambda^2 + k_{2,z} \partial^2 b/\partial z^2, \end{aligned} \tag{4.22}$$

where T_b and M_b are losses of the bacterioplankton biomass due to energy exchange with the environment and the dying-off; Γ_b is a set of the trophic subordination of bacterioplankton (in a typical case Γ_b is one element Z); $C_{b,s}$ is the share of bacterioplankton in the food ration of the s-th element of the ecosystem. The parameters T_b and M_b are described by the following relationships:

$$T_b = t_b b \tag{4.23}$$

$$M_b = max\{0, \mu_b(\mathrm{b} - \underline{B}_\mathrm{b})^\xi\}, \tag{4.24}$$

where t_b is the specific expenditure on the exchange with the environment; μ_b is the rate of the bacteria's dying-off; $\underline{B}_b$ and ξ are the constants that determine the dependence of the intensity of the bacteria's dying-off on their concentration. The coefficient $k_2 = (k_{2,\varphi}, k_{2,\lambda}, k_{2,z})$ determines the process of the turbulent mixing of the ocean waters.

The dynamic equation for the phytoplankton biomass is:

$$\begin{aligned} \partial p/\partial t + V_\varphi \partial p/\partial \varphi + V_\lambda \partial p/\partial \lambda + V_z \partial p/\partial z = R_p - T_p - M_p \\ - \sum_{s \in \Gamma_p} C_{ps} R_s + k_{2,\varphi} \partial^2 p/\partial \varphi^2 + k_{2,\lambda} \partial^2 p/\partial \lambda^2 + k_{2,z} \partial^2 p/\partial z^2 \end{aligned} \tag{4.25}$$

where T_p is the set of the trophic subordination of phytoplankton, C_{ps} is the share of the phytoplankton biomass in the food ration of the elements of the s-th trophic level of the ecosystem, T_p is the expenditure on the energy exchange with the environment, M_p is the dying-off of the phytoplankton cells. The latter parameters are determined by the following relationships:

$$M_p = max\{0, \mu_p(p - \underline{p})^{\theta}\} \tag{4.26}$$

$$T_p = t_p p, \tag{4.27}$$

where t_p is the specific expenditure on respiration of the phytoplankton cells, μ_p is the coefficient of the phytoplankton dying-off, $\underline{p}$ and θ are the coefficients characterizing the dependence of the rate of the phytoplankton cells' dying-off on their concentration.

Zooplankton is an important trophic element in the ocean ecosystem. In the scheme of Fig. 4.3 it is presented by an integral level *Z*, which implies the presence of a large number of sub-levels with the intersecting trophic bonds. Zooplankton consumes phyto- and bacterioplankton. Zooplankton itself feeds a lot of animals, in the scheme of Fig. 4.3 denoted as nekton *r* and detritofugs *D*.

Describe the zooplankton ration by the Ivlev formula:

$$R_Z = k_Z\left(1 - \exp\left(-\nu\overline{B}\right)\right), \tag{4.28}$$

where $\overline{B}$ is the biomass of the accessible food $\left(\overline{B} = \max\{0, B - B_{\min}\}\right)$, k_Z is a maximum of the ration with an excess of food, ν is the coefficient characterizing the level of starvation. The ration maximum is assumed to be equal to the needs of food, which—in turn—are determined by the exchange intensity T_1 and a maximum possible increment P_1 at a given intensity of exchange. Both latter parameters are related to the coefficient $q_2 = P_1/(P_1 + T_1)$, so that we obtain:

$$k_Z = T_1 u\left(1 - q_{2,max}\right),$$

where $1/u$ is the food assimilation, $q_{2,max} = max\ q_2$.

The formula (4.28) implies that with a small amount of food the ration of zooplankton grows in proportion to the amount of food, but as the ration approaches to a maximum of k_Z, its dependence on *B* diminishes. Since in fact one trophic level never eats out the other, there is a limitation in (4.28) that establishes $R_Z = 0$ at $B \leq B_{min}$, where B_{min} is a minimum of the unconsumed food biomass. In (4.26) the $\underline{p}$ parameter plays the same role, but in the process of the phytoplankton cells' mortality.

Thus, a change of the zooplankton biomass follows the law described by the following differential equation:

$$\begin{aligned} \partial Z/\partial t + V_{\varphi}\partial Z/\partial\varphi + V_{\lambda}\partial Z/\partial\lambda + V_z\partial Z/\partial z &= R_Z - T_Z - M_Z - H_Z \\ \sum_{s\in\Gamma_Z} C_{Zs}R_s + k_{2,\varphi}\partial^2 Z/\partial\varphi^2 &+ k_{2,\lambda}\partial^2 Z/\partial\lambda^2 + k_{2,z}\partial^2 Z/\partial z^2 \end{aligned} \tag{4.29}$$

where T_Z is a set of the trophic subordination of zooplankton, C_{Zs} is the share of the zooplankton biomass in the food ration of the *s*-th trophic level, H_Z, T_Z, and M_Z are the losses of the zooplankton biomass due to unconsumed food, expenditures on respiration and mortality, respectively. Let us describe the latter three parameters by relationships

$$H_Z = h_Z R_Z,\ T_Z = t_Z Z,\ M_Z = (\mu_Z + \mu_{Z,1} Z)Z, \tag{4.30}$$

where the coefficients h_Z, t_Z, μ_Z, and $\mu_{Z,1}$ are determined empirically for a concrete species of zooplankton.

As seen from (4.29), zooplankton is considered as a passive element of the ecosystem subject to physical processes of transference in space with the motion of water masses. However, zooplankton is known to migrate mainly in the vertical direction. In the given model a simple mechanism can be used to simulate the process of the vertical migration of zooplankton. For this purpose, let us divide the whole water thickness into two layers: $0 \le z \le z_0$ and $z_0 < z \le H$. The zooplankton migration between these layers is supposed to depend on food availability. Hence, some part of zooplankton from the layer $[z_0, H]$ can satisfy its needs of food in the layer $[0, z_0]$. So, the whole vertical profile $B(\varphi, \lambda, z, t)$ is considered, having taken into account B_{min}.

The coefficients C_{as} $(a = p, Z)$ in the formulas (4.25) and (4.29) are determined on the supposition that the consumption of various kinds of food in the s-th trophic level is proportional to their efficient biomasses:

$$C_{as} = k_{sa}\overline{B}_a \left[\sum_{a \in S_s} k_{sa}\overline{B}_a\right]^{-1}, \tag{4.31}$$

where $\overline{B}_a$ is an efficient biomass of the a-th food, S_s is the set of trophic subordination of the s-th component, k_{sa} is the proportion coefficient that determines the significance of the s-th constituent in the food ration of the a-th element.

According to the scheme in Fig. 4.3, the equations which describe the dynamics of the biomass of nekton, detritus feeders, detritus, dissolved organic matter, and nutrient salts will be

$$\partial r/\partial t = R_r - H_r - T_r - M_r - \sum_{s \in \Gamma_r} C_{rs} R_s \tag{4.32}$$

$$\partial D/\partial t = R_D - H_D - T_D - M_D - \sum_{s \in \Gamma_D} C_{Ds} R_s \tag{4.33}$$

$$\begin{aligned} \partial d/\partial t + V_\varphi/\partial\varphi + V_\lambda \partial d/\partial\lambda + V_z \partial d/\partial z = M_b + M_D + M_r + M_p + M_Z + H_Z \\ + H_r + H_D - \mu_d d - C_{dD} R_D + k_{2,\varphi}\partial^2 d/\partial\varphi^2 + k_{2,\lambda}\partial^2 d/\partial\lambda^2 + k_{2,z}\partial^2 d/\partial z^2 \end{aligned} \tag{4.34}$$

$$\partial n/\partial t + V_\varphi \partial n/\partial \varphi + V_\lambda \partial n/\partial \lambda + V_z \partial n/\partial z$$
$$= \mu_a d - \delta R_p + k_{2,\varphi}\partial^2 n/\partial \varphi^2 + k_{2,\lambda}\partial^2 n/\partial \lambda^2 + k_{2,z}\partial^2 n/\partial z^2 \quad (4.35)$$

$$\partial g/\partial t + V_\varphi \partial g/\partial \varphi + V_\lambda \partial g/\partial \lambda + V_z \partial g/\partial z = T_p + T_b + T_r + T_D + T_Z - C_{gb}R_b$$
$$+ k_{2,\varphi}\partial^2 g/\partial \varphi^2 + k_{2,\varphi}\partial^2 g/\partial \varphi^2 + k_{2,\lambda}\partial^2 g/\partial \lambda^2 + k_{2,z}\partial^2 g/\partial z^2$$
$$(4.36)$$

where $H_a = (1 - h_a)R_a$ is the unassimilated food of the a-th element ($a = r, D$), $T_a = t_a a$ is the expenditure on energy exchange, $M_a = (\mu_a + \mu_{a,1}a)a$ is the mortality, ρ_g is the indicator of the rate of replenishing the supplies of nutrient elements due to decomposition of the dissolved organic matter, δ is the coefficient of consumption of nutrient elements in the process of photosynthesis.

As follows from Eqs. (4.32) and (4.33), a supposition that the elements of nekton and detritus consumers do not move in space with the motion of the water masses is made in the model. These elements are assumed to migrate independent of the hydrophysical conditions of their environment. Let us consider two possible versions of modeling the processes of migration. The first version is connected with an addition to the right-hand parts of the Eqs. (4.32) and (4.33) of the terms describing the turbulent mixing with the coefficients $k_2^* > k_2$. In other words, the process of migration is identified with the process of an intensified turbulent mixing, that means it is accidental. However, the process of fish migration manifests some expediency in the choice of the direction of their movement. According to the biological principle of adaptation, the migration of fish is subject to the principle of the complex maximization of the efficient food ration, with the environmental parameters remaining within the conditions of their habitat. Hence, the motion of fish in space with characteristic velocities ensures their location in the regions where at this time the food and other abiotic conditions (temperature, salinity, dissolved oxygen, chemicals concentration) are most favourable. This means that fish migrates in the direction of a maximum gradient of the efficient food with the preserved limitations of the environmental parameters.

4.6.3 The Vertical Structure of the Ocean

To qualitatively assess the role of the World Ocean ecosystems in the global carbon cycle, it is useful to consider some aspects of the laws of formation of the vertical distributions of organic matter. According to the Eq. (4.25), the number s of maxima in its solution, $p(\varphi, \lambda, z, t)$ at a fixed time at any point (φ, λ) is determined by the number l of changes of the sign of the derivative $\partial p/\partial z$, with $s \leq 0.5(1 + l)$. It is important to know the location of the phytoplankton biomass maxima as a function of the depth in order to understand the whole range of the biogeocenotic processes in the ocean ecosystem. Of course, the amount and the location of the

p function maxima are determined by many characteristics of the water medium, among which the hydrological heterogeneity between various layers of the water space is one of the important ones. The results of numerous field studies show that there can be four maxima of phytoplankton and by the depth these maxima can locate from the surface down to 500 m. Of course, in most cases there are one-two maxima down to 50–100 m.

A theoretical analysis of the Eq. (4.25) is possible after its simplification:

$$dp/dz = F(z)p - \nu N(z) - w(z), \tag{4.37}$$

where F, N, and w are the rate of photosynthesis, the zooplankton density, and the rate of sedimentation, respectively, at a depth z. The concentration of nutrient elements n(z) is supposed to be taken into account in the form of the function $F(z)$.

The Eq. (4.37) is linearly related to p, and therefore it will not change if it is multiplied on both sides by the value

$$\exp\left\{-\int_0^z [F(\xi) - \nu N(\xi)]\right\}.$$

We have

$$\exp\left\{-\int_0^z [F(\xi) - \nu N(\xi)]\right\}\{dp/dz - [F(z) - \nu N(z)]p\} = w(z)\exp\left\{-\int_0^z [F(\xi) - \nu N(\xi)]\right\} \tag{4.38}$$

and

$$\exp\left\{-\int_0^z [F(\xi) - \nu N(\xi)]\right\}\{dp/dz - [F(z) - \nu N(z)]p\} = (d/dz)\left[p\exp\left\{-\int_0^z [F(\xi) - \nu N(\xi)]\right\}\right] \tag{4.39}$$

Substituting (4.39) into (4.38) and integrating both parts with respect to z, we have upon simplifications:

$$p(z)\exp\left\{-\int_0^z [F(\xi)-\nu N(\xi)]\right\} = -\int_0^z w(y)\left\{\exp\left[-\int_0^y [F(x)-\nu N(x)]dx\right]\right\}dy + C \tag{4.40}$$

Substituting $z = 0$ into (4.40), we obtain: $C = p(0)$. Finally, from (4.40) we obtain the formula of the vertical distribution of the phytoplankton biomass:

$$p(z) = \left\{p(0) - \int_0^z w(y)\exp\left\{-\int_0^y [F(x)-\nu N(x)]dx\right\}dy\right\} \times \exp\left\{\int_0^z [F(\xi)-\nu N(\xi)]d\xi\right\} \tag{4.41}$$

The character of the $p(z)$ distribution from the viewpoint of the number of maxima and their location is important for the specification of the role of various depths in the functioning of the ocean carbonic system. So, it will be investigated based on (4.41). For:

$$g(h) = \int_0^h N(z)dz, \quad \psi(h) = \int_0^h F()dz,$$

then the expression (4.41) is re-written as:

$$p(h) = \{p(0) - \int_0^h w(y)\exp[\nu g(y) - \psi(y)]dy\} \cdot \exp[\psi(h) - \nu g(h)] \tag{4.42}$$

Thus, $p(h)$ is the functional of $g(h)$ and $\psi(h)$. The g function has a simple interpretation if $h_2 > h_1$, then the difference $g(h_2) - g(h_1)$ represents the zooplankton biomass located in the layer $h_1 \le z \le h_2$. The functions $g(h)$ and $\psi(h)$ are monotonically non-decreasing, since $T(h)$ and $F(h)$ are not negative.

Before proceeding to a qualitative analysis of the function $p(h)$, let us obtain some auxiliary results. The following sets are introduced:

$$H^+ = \{h \ge 0 : F(h)/N(h) = \nu/F'(h)/N'(h) > \nu\},$$
$$H^- = \{h \ge 0 : F(h)/N(h) = \nu/F'(h)/N'(h) < \nu\}$$

and the function

$$\varphi(h) = vg(h) - \psi(h).$$

Lemma 4.1 *The* $\varphi(h)$ *function is nondecreasing at* $\forall h \in \Omega_1^+ \cup \Omega_2^+$, *where*

$$\Omega_1^+ = \{h \geq 0 : h \in [0, h_1^+]/h_1^+ = \inf H^+, F(0)/N(0) < v\},$$
$$\Omega_2^+ = \left\{h \geq 0 : h \in \left[h_i^-, h_j^+\right]/h_j^- \in H^-, h_j^+ \in H^+, \left(h_i^-, h_j^+\right) \cap H^- = 0\right\}$$

Lemma 4.2 *The* $\varphi(h)$ *function is nonincreasing at* $\forall h \in \Omega_1^- \cup \Omega_2^-$ *where*

$$\Omega_1^- = \{h \geq 0 : h \in [0, h_1^-]/h_1^- = \inf H^-, F(0)/N(0) < v\},$$
$$\Omega_2^- = \left\{h \geq 0 : h \in \left[h_i^+, h_j^-\right]/h_i^+ \in H^+, h_j^- \in H^-, \left(h_i^+, h_j^-\right) \cap H^+ = 0\right\}$$

The validity of the Lemmas 4.1 and 4.2 follows from the elementary properties of the function $\varphi(h)$. In particular, we have: $\varphi(0) = 0, \varphi'(0) = vN(0) - F(0), \ldots,$ $\varphi^{(i)}(0) = vN^{(i-1)}(h) - F^{(i-1)}(h)$.

Note 1. The conclusion of the lemma 4.1 remains also valid at $F(0)/N(0) = v$, provided

$$(\exists s)\left[F^{(s)}(0)/N^{(s)}(0) < v,\ F^{(j)}(0)/N^{(j)}(0) = v,\ j = 0, 1, \ldots, s1\right].$$

Note 2. The conclusion of the lemma 4.2 remains also valid at $F(0)/N(0) = v$, provided

$$(\exists s)\left[F^{(s)}(0)/N^{(s)}(0) > v,\ F^{(j)}(0)/N^{(j)}(0) = v,\ j = 0, 1, \ldots, s1\right].$$

According to Eq. (4.42) the function $p(h)$ is the function of $\varphi(h)$:

$$p(h) = \left\{p(0) - \int_0^h w(y) \exp[\varphi(y)]dy\right\} \exp[-\varphi(h)] \tag{4.43}$$

Based on the results of the lemmas 4.1 and 4.2, from (4.43) one can easily formulate the monotonicity properties of the function $p(h)$. Since $\varphi(h)$ does not decrease in the set $\Omega^+ = \Omega_1^+ \cup \Omega_2^+$, the function $exp[-\varphi(h)]$ will be nonincreasing at $\forall h \in \Omega^+$. At $w(h) \geq 0$ the function $p(0) - \int w(y)exp[\varphi(y)]dy$ is nonincreasing at any values of $h \geq 0$. Therefore $p(h)$ is the nonincreasing function of $\forall h \in \Omega^+$.

In the set $\Omega^- = \Omega_1^- \cup \Omega_2^-$, the function $p(h)$ is the product of the nondecreasing and nonincreasing functions. Hence, in Ω^- $p(h)$ is either nondecreasing, or nonincreasing or has a maximum. Since for $\forall h \in \Omega^- F(h)/N(h) > v$, then $p(h)$ will be the nonincreasing function at $\forall h \in H_0 = \{h : h \geq h_0\}$, where

$$p(0) = \int_0^{h_0} w(y) \exp[\varphi(y)]dy$$

The multitude of values $h \geq 0$, for which the following condition is valid, are denoted by H_1:

$$w(h)[F(h) - vN(h)]^{-1} exp\left[\varphi(h) + \int w(y)exp[\varphi(y)]dy > p(0)\right]$$

Then $p(h)$ will be the nonincreasing function at $\forall h \in H_1 \cap \Omega^-$.

The obtained results are formulated as the following theorem.

Theorem *The function $p(h)$ is nonincreasing at $\forall h \in H^+ \cup (H_0 \cap H^-) \cup (H_1 \cap H^-)$ and nondecreasing at $\forall h \in H^- \cap (E \backslash H_0) \cap (E \backslash H_1)$, where $E = \{h\!: h \geq 0\}$.*

In a more general case $\partial p/\partial$ can be presented in the form of the sum:

$$\partial p/\partial z = (\partial p/\partial z)_{EB} + (\partial p/\partial z)_O + (\partial p/\partial z)_\tau + (\partial p/\partial z)_M + (\partial p/\partial z)_Z + (\partial p/\partial z)_T, \tag{4.44}$$

where each of the terms on the right-hand side reflects a change of the phytoplankton biomass due to changes in illumination and concentration of nutrient elements (*EB*), sedimentation (*O*), temperature changes (τ), mortality (*M*), consumption by zooplankton (*Z*) and turbulent mixing (*T*). According to Eq. (4.43), it can be supposed that

$$(\partial p/\partial z)_{EB} \approx F(t, p, z) = Af_1(E)f_2(n)[1 - 10^{-\gamma p}].$$

At $\gamma p << 1$ we obtain

$$p^{-1}(\partial p/\partial z)_{EB} \approx Af_1(E)f_2(n).$$

With an increasing depth in other stationary conditions the water temperature is known to monotonically decrease, dropping sharply in the layer of the thermocline (i.e., $(\partial p/\partial z)_\tau \leq 0$). The similar term $(\partial p/\partial z)_M$ has the form of the sectionally continuous non-positive function. However, due to the physical processes of mixing a situation can appear when this monotonicity gets broken.

The values of the constituent $(\partial p/\partial z)_Z$ depend on the distribution of $N(z)$. The term $(\partial p/\partial z)_T$ is totally determined by the hydrological conditions through the water thickness, and its behaviour can change arbitrarily. Apparently, in the layer above the thermocline $(\partial p/\partial z)_O \leq 0$.

Studies of all components in totality will make it possible to determine the number of changes of the sign of the derivative $(\partial p/\partial z)$, which uniquely determines the number of the phytoplankton maxima. In particular, if $\partial p/\partial z$ changes the sign seven times in the layer under study, and $(\partial p/\partial z)_{z=0} > 0$ and $(\partial p/\partial z)_{z\to\infty} < 0$, then the function $p(z)$ has four maxima.

Finally, assuming that zooplankton uniformly eat out phytoplankton through the depth of the layer considered, and neglecting in Eq. (4.43) all constituents apart from the first one, based on Eq. (4.41) we obtain the following expression for the specific rate of the change of the phytoplankton biomass:

$$p^{-1}(\partial p/\partial t) \approx b - ap, \tag{4.45}$$

where

$$a = (k\gamma^2/2)E \cdot 10^J,\ b = k\gamma E \cdot 10^J,\ m = (E_{\max}/E_0) \cdot 10^Y, J = m(1 - E/E_{\max}),\ Y = \alpha H_{\max}$$

The Eq. (4.45) has the solution:

$$p(t, h) = [\gamma/2 + \exp\{-tk\gamma E \cdot 10^S\}],$$

where

$$S = (E_{\max}/E_0) \cdot 10^c,\ c = \alpha H_{\max}(1 - E/E_{\max})$$

Since $E \to 0$, at $h \to \infty$ then $p(t, h) \approx [\gamma/2 + Kexp\{-tk\gamma E\}]^{-1}$, where $K = \exp\{-10^{-m}\}$.

4.6.4 Model of the Peruvian Current Ecosystem

Each region of the World Ocean has characteristic hydrophysical and ecological properties. Therefore, with these properties taken into account, a specification of the model of the carbon cycle in the World Ocean will raise the reliability of the model assessment of the role of the ocean systems in the assimilation of excess CO_2 from the atmosphere. With this aim in view, let us consider the region of the Peruvian Current characterized by high concentrations of nutrient elements (mkg-at/ℓ): nitrites (0.6–0.8); nitrates (5.5–14.9); phosphates (0.6–1.31); silicic acids (3–45), and large values of primary production (cal/m^2/day): 17.9–69.3 for phytoplankton; 3.26 for bacteria; 0.12–0.73 for zooplankton, and 0.7–3 for nekton (mainly, anchovy). Near the Peruvian coastline there is the most powerful on the Earth upwelling of deep waters that feeds the photic layer of the bordering region of the ocean with nutrient elements. Here the main object of the industrial fishing is anchovy whose average production is estimated at 100×10^8 kg/year. From the available data, anchovy constitutes above 95 % of the catch, and therefore the derivation of the model of the functioning of the Peruvian ecosystem is also important for the economy of the bordering states. Of course, the real limits of the interaction of the natural-anthropogenic system of the concrete state of South America with the global environment are impossible to assess without taking into account the role of the ecosystem of the Peruvian Current.

The Peruvian Current refers to the regions of the World Ocean where complicated interconnected and interdependent physical, biological, and chemical processes take place on a relatively small area. Near the coastlines of Peru and Chili the production of organic substance is so large that the available oxygen is not enough for it to be oxidized, and as a result, a layer with an oxygen deficit appears at depths 100–800 m. This is accompanied by underfeeding of fauna.

In the hierarchic structure of the trophic relationships of the Peruvian Current community, the following components can be selected (Fig. 4.4): B_1—bacterioplankton; B_2—phytoplankton; B_3—phytophag; B_4—predatory zooplankton; B_5—industrial anchovy; B_6—anchovy grubs; B_7—predatory fish; B_9—anchovy youngsters; B_{10}—detritus; B_{13}—dissolved organic matter (cal/m^3); B_8—birds (birds/m^2); B_{12}—biogenic elements (mg/m^3); B_{14}—dissolved oxygen (mℓ/ℓ); B_{11}—volume of anchovy catch (t/km^2). The characteristic feature of the trophic graph of the system is the availability of connections between the components of the ocean biocenoses (B_5, B_7, B_9) and the land biocenoses (B_8). An introduction of this connection into the model has turned out to be essential, since the maritime birds (cormorants, pelicans, etc.) in the regions of the Peruvian current markedly affect the dynamics of the whole system. From the available estimates, anchovy constitutes from 80 to 96 % of the food spectrum for birds. Therefore the characteristic feature of the developed model is an account of the element of interaction of the land and ocean

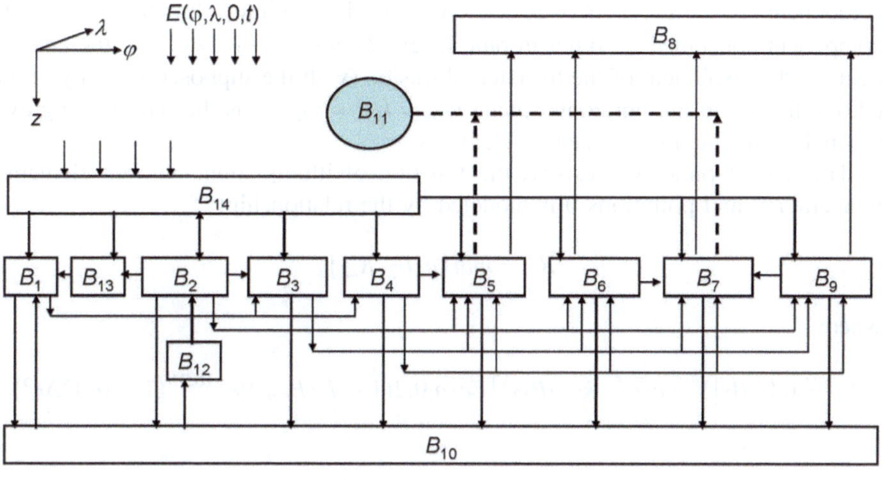

Fig. 4.4 The block-scheme of the Peruvian Current ecosystem model (PCEM)

biocenoses expressed by the introduction of trophic bonds between birds and anchovy. The trophic pressure on anchovy from other fish is expressed by a single connection with the component B_7—which is mainly Chili pelamyd (*Sarda chiliensis*), though anchovy is in the food spectrum for other animals, too.

The block-scheme of the PCEM reflects the elements of the atmosphere-ocean interaction reduced to a mutual exchange with oxygen, with the given rates of invasion and evasion.

The whole system receives the solar radiation energy E whose intensity is estimated by the formula (4.18) at $\alpha = 0.013$, $\delta = \beta = \nu = 5 \times 10^{-5}$. A change of the biomass of the i-th component (i = 1–7, 9) is described by the balance equation similar to (4.29):

$$\partial B_i/\partial t + V_\varphi \partial B_i/\partial \varphi + V_\lambda \partial B_i/\partial \lambda + V_z \partial B_i/\partial z = R_i - T_i - M_i$$
$$- \sum_{j \in \Gamma_i} C_{ij} R_j + k_{2,\varphi} \partial^2 B_i/\partial \varphi^2 + k_{2,\lambda} \partial^2 B_i/\partial \lambda^2 + k_{2,z} \partial^2 B_i/\partial z^2 - v_i \partial B_i/\partial z + Q_i - H_i \tag{4.46}$$

where Q_i is the characteristic of the age transitions between the stages of the anchovy development (i = 5, 6, 9), Q_i = 0 for i = 1–4, 7; R_i is the produce, T_i is the expenditure on energy exchange; M_i is an amount of the died-off biomass; H_i is the unassimilated food (cal/m^3/day);

$$C_{ij} = k_{ij} \overline{B}_i / \sum_{m \in S_j} k_{jm} \overline{B}_m;$$

T_i is the set of the trophic subordination of the i-th component; S_j is the food spectrum of the j-th component; k_{ab} is the indicator of the degree of satisfaction of the needs of food of the a-th component at the expense of the biomass of the b-th component; $\overline{B}_i = \max\{0, B_i - B_{i,\min}\}$, $B_{i,min}$ is the minimum biomass of the i-th component consumed by other trophic levels; k_2 = ($k_{2,\varphi}$, $k_{2,\lambda}$, $k_{2,z}$) are the constituents of the coefficient of the turbulent diffusion (with the supposed isotropy of the eddy diffusion in the horizontal plane $k_{2,\varphi} = k_{2,\lambda} = \nu_H$); ν_i is the rate of the gravitational sedimentation of inactive elements.

The rate of photosynthesis as the function of illumination, nutrient elements, temperature, and pollutions θ is modeled by the relationship:

$$R_2 = min\{\rho_2 B_2, R_{22}\},$$

where

$$R_{22} = 0.07 B_2 \left[1 - exp\left(-k_{2,12} B_{12}\right)\right] E exp[0.2(1 - E/E_{max})] e^{-10.7\,\theta} (1 + 0.02 \Delta t^{o}).$$

The produce R_i on the remaining trophic levels is calculated by the formula

$$R_i = R_{i,\max}\left[1 - \exp\left(-\sum_{j\in S_i} k_{ij}\overline{B}_j\right)\right](1 + 0.02\Delta t^o)B_i$$

The function T_i (i = 2–9) is postulated as the dependence $T_i = \tau_i B_i^{\omega_i}$, where ω_I = 0.7–0.95. The coefficients τ_i are functions of the water temperature t^0 and oxygen concentration B_{14}:

$$\tau_i(t^o, B_{14}) = \begin{cases} t_i^*(t^o) & for \quad B_{14} > B_{14,i}^*; \\ t_i^*(t^o)B_{14}/B_{14,i}^* & for \quad 0 \le B_{14} \le B_{14,i}^*; \end{cases}$$

$$t_i^*(t_o) = \begin{cases} \tau_i^* & for \quad t^o \in \left[t_{1,i}^o, t_{2,i}^o\right]; \\ \tau_i^*[a_i \exp(-b_i t^o)]^s & for \quad t^o \notin \left[t_{1,i}^o, t_{2,i}^o\right]; \end{cases}$$

where $s = (t^{\rm o} - t_i^{\rm oo})/m_i$, $t_{1,i}^o$, and $t_{2,i}^o$ are the lower and upper boundaries of the temperature zone of adaptation of the i-th trophic level; $t_i^{\rm oo} = 0.5\left(t_{1,i}^{\rm o} + t_{2,i}^{\rm o}\right)$; $a_i, b_{i,}$, and m_i are constant values, selected from the condition of the measurement data approximation.

In contrast to other hydrobionts, bacteria are characterized by a substantial dependence of gas exchange on the food concentration. Therefore, the following relationship is assumed in order to describe T_1:

$$T_1 = t_1 B^{\omega_1}\left[1 - \exp\left(-t_1'\overline{B}_{10} - t_1''\overline{B}_{13}\right)\right],$$

where t_1 is the expenditure of bacteria on energy exchange per unit biomass; ω_1 is the indicator of the dependence of expenditures on energy exchange on the amount of the bacterioplankton biomass; t_1' and t_1'' are indicators of the dependence of expenditures on energy exchange on the efficient concentrations of detritus and dissolved organic matter, respectively.

For phytoplankton, the losses on respiration are characterized by reduced assimilation of carbon in photosynthesis. For the Peruvian current these losses average 74.6 % of assimilated oxygen, which constitutes 1.45 g/cm^2/day. Determine the dynamic indicator of mortality M_i based on the assumptions made above in the form of the dependence $M_i = (\mu_i + \mu_i' B_i)B_i$, where $\mu_i = 0.01-0.02$; $\mu_i' = 0.1\mu_i$. The H_i parameter is calculated by the formula $H_i = h_i R_i$.

Among living components, anchovy takes a special place in the model, and for its description the following three stages have been chosen: grubs, youngsters, and industrial anchovy. It is assumed that the rate of the transition from the i-th stage to the j-th one is constant and it takes 1 year for these three stages.

The inert components in the model are described by the following equations:

$$
\begin{aligned}
\partial B_{10}/\partial t + (V \cdot \nabla)B_{10} &= \sum_{s=1}^{9} (M_s + H_s) - \rho_{10}B_{10} - v_{10}\partial B_{10}/\partial z \\
&\quad - \overline{B}_{10}\left(k_{1,10}R_1/P_1 + k_{3,10}R_3/P_3 + k_{4,10}R_4/P_4\right) + \left(k_2\nabla^2\right)B_{10}; \\
\partial B_{12}/\partial t + (V \cdot \nabla)B_{12} &= \rho_{10}B_{10} + \rho_1\left(T_9 + \sum_{s=1}^{7} T_s\right) - \delta_{12}R_2 + \left(k_2\nabla^2\right)B_{12} + \beta_{12}\partial B_{12}/\partial z; \\
\partial B_{13}/\partial t + (V \cdot \nabla)B_{13} &= k_{13,0}R_2 - \overline{B}_{13}k_{1,13}R_1/P_1; \\
\partial B_{14}/\partial t + (V \cdot \nabla)B_{14} &= \delta_{14}R_2 - \sum_{s=1}^{7} \xi_s T_s - \xi_9 T_9 + \left(k_2\nabla^2\right)B_{14} \\
&\quad - \delta_{10}\rho_{10}B_{10} + \max\{0, (z_{14} - z)/|z - z_{14}|\}w_{14}
\end{aligned}
\tag{4.47}
$$

where

$$P_i = \sum_{j \in S_i} k_{i,j}\overline{B}_j,$$

ρ_{10} is the velocity of detritus decomposition per unit of biomass; ν_{10} is the velocity of gravity settling; k_2 is the turbulent diffusion coefficient; ρ_1 is the part of exchange biomass losses transforming to nutrients; δ_{12} is the nutrient assimilation velocity by the photosynthetic process per unit of the phytoplankton production; δ_{14} is the oxygen production per unit of phytoplankton production; ξ_i (i = 1–7, 9) is the oxygen losses at the expense of transpiration; z_{14} is the maximal depth of oxygen exchange of sea water with the atmosphere; w_{14} is the difference between the invasion and evasion processes.

The equation describing the changes in the size of birds is written as

$$\partial B_8/\partial t = (1 - h_8)R_8 - M_8,$$

where

$$R_8 = max\{0, (\Delta^* z - z)/|z - \Delta^* z|\}k_{8,0}B_8(1 - exp[-V_8]),$$

$$V_8 = \sum_{s=5}^{7} k_{8,s}\overline{B}_s + k_{8,9}\overline{B}_9 + \gamma'_8, \qquad M_8 = \mu_8 B_8$$

γ'_8 is the share of the non-fish food in the ration of birds.

The Q_i function of the age transitions is presented only for the anchovy elements and written as a linear approximation:

$$Q_6 = \lambda_{56}B_5 - \lambda_{69}B_6;\ Q_5 = \lambda_{95}B_9;\ Q_9 = \lambda_{69}B_6 - \lambda_{95}B_9.$$

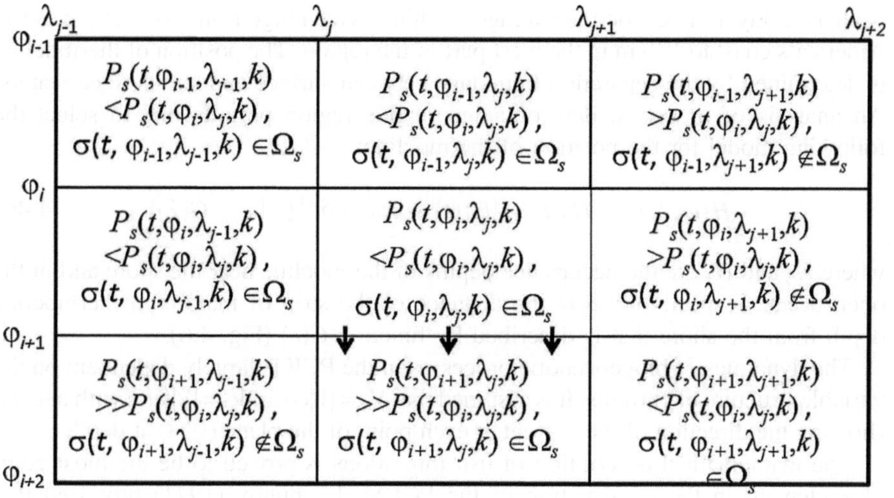

Fig. 4.5 The example of simulation scheme for horizontal fish migration realized in the PCEM. Notation: Ω_s is the interval of favourable temperature for s-th element of the PCE ($s = 5, 7, 9$); $P_s(t, \varphi_i, \lambda_{j-1}, k)$ is the effective level of food for s-th element of the PCE ($s = 5, 7, 9$); $\sigma(t, \varphi_i, \lambda_j, k)$ is the water temperature. The same principle is realized for vertical fish migration

The block of climatic and anthropogenic impacts on the PCE is described according to the procedure of GIMS. It was assumed that the scope of pollution of the PCE area is determined by the mean pollution level of the Pacific and is qualitatively characterized by the quantity $0.13 \le \theta \le 0.27$ mg/m^2 of conventional pollutant.

Temperature variations were incorporated into the model by using a database in the form of the charts of its distribution in space and allowing for seasonal variations according to the diagram in Fig. 4.5. In this case, the discretization of data in time and space agrees with discretization steps adopted in the computer realization of the PCEM as the GIMS item. Temperature anomalies that take place in the area of the PCE are known as El Niño years by the magnitude of quantity $4 \le \Delta\sigma_\varepsilon \le 6$ K. We would like to note that a more accurate simulation of El Niño is an independent task.

The temperature regime of the PCE is affected by strong variations in solar radiation which are due to varying cloudiness near the coasts and in the open ocean. In the narrow off-shore zone, where cloudiness is practically non-existent, radiation is 1.5 times as high as in the adjoining ocean areas and attains magnitudes of the order of $E(\varphi, \lambda, 0, t) = 2.64$ W/m^2. The mean energy of solar radiation is 1.85 W/m^2. In subsequent computations the following dependence was used in a scenario with free parameter R ($R = 1$ for standard conditions):

$$E(\varphi, \lambda, 0, t) = R(230 - 7/3\lambda)(1 + 0.1 \sin[2\pi t/365]) \tag{4.48}$$

The temperature regime of the PCE along the z-coordinate is determined by an intricate complex of hydrological factors. The depth of the lower boundary of the isothermal layer varies, on the average, within a wide range from 10 m at the South America's coast to 130 m in the west part of the region. The position of thermocline is determined by the separation boundary between surface and sub-surface waters. An analysis of measured data obtained in this region permits one to select the following model for the position of thermocline:

$$H(\varphi, \lambda, t) = H_0(t) + [H_1(t) - H_0(t)]\delta^{-1}(t)[\varphi - G(\lambda)] \tag{4.49}$$

where H_0 and H_1 are the occurrence depths of thermocline near the shore and in the open ocean, respectively; δ is the distance of the strip of maximum thermocline depth from the shore that is described by function $G(\lambda)$ (Fig. 4.6).

The dynamics of biogeocenotic processes in the PCE is largely dependent on the variable structure of currents. It is assumed that $V_\varphi = |V|\cos\alpha$, $V_\lambda = |V|\sin\alpha$ with angle α showing the direction of the current at each point of the plane (φ, λ) at depth z.

The mathematical description of fish migrations is proved to be the most complex element in the construction of the PCEM. Lyapunov (1971) proposed that the migration process should be identified with the intense turbulent diffusion with the coefficient $|v^*| >> |v|$. Krapivin (1978) realized this algorithm and has shown that the estimation of v^* to be a very incorrect task. The investigations by Radakov (1972) revealed that the process of fish migration is accompanied by an external appearance of purposeful behavior. Therefore, the law of migration which follows from the general biological principle of adaptation and long-term adaptability is formulated: migrations of anchovy, predatory fishes and birds are subordinated to the principle of complex maximization of effective nutritive ration P_i ($i = 5, 7, 8, 9$),

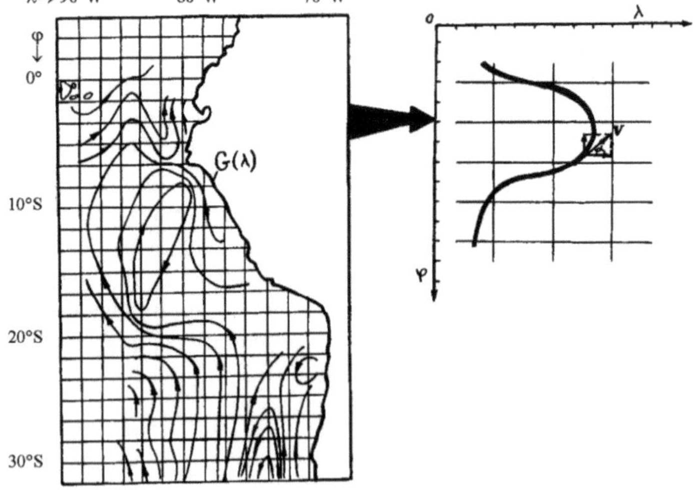

Fig. 4.6 A scheme for the superposition of spatial model structure with the current map

given the preservation of favourable temperature conditions. In other words, traveling of migrating species takes place at velocities characteristic for them in the direction of the maximum gradient of effective food, given the adherence to temperature restrictions.

The PCEM makes it possible to carry out computer experiments for an estimation of the PCE response to anthropogenic impacts on the exterior conditions or to climate change. The initial data and model coefficients are given in Table 4.4. For the young and commercial anchovy the temperatures $\sigma \leq 289$ K and for predatory fishes $\sigma \geq 288$ K were deemed to be acceptable. The rate of fish migration was determined by the value of 0.03 m/s, and for birds 0.14 m/s. The thermocline position was described by the binary function (4.49) with $\delta = \delta_s$ in spring, summer

Table 4.4 The PCEM coefficients used for simulation experiments

Coefficient	Value	Coefficient	Value	Coefficient	Value
$k_{1,10}$	0.3	ω_8	0.7	ν_2	0.1
$k_{1,13}$	0.3	ω_9	0.9	λ_{95}	0.02
$k_{1,14}$	5.0	$k_{5,6}$	0.5×10^{-2}	λ_{56}	0.02
$k_{2,12}$	100.0	$k_{5,7}$	0	λ_{69}	0.02
$k_{2,14}$	5.0	$k_{5,9}$	0.5×10^{-2}	ν_{10}	50.0
$k_{3,1}$	0.2	$k_{5,10}$	0.5×10^{-2}	ρ	3.0
$k_{3,2}$	0.05	$k_{5,14}$	5.0	ρ_1	10^{-4}
$k_{3,10}$	0.005	$k_{6,1}$	0.05	δ_{12}	0.012
$k_{3,13}$	0.5×10^{-3}	$k_{6,2}$	0.3	$k_{13,0}$	0.9
$k_{3,14}$	5.0	$k_{6,3}$	0.1	δ_{14}	0.01
$k_{4,1}$	0.1	$k_{6,4}$	0.1	W_{14}	0.03
$k_{4,2}$	0.05	$k_{6,6}$	0.05	$k_{9,2}$	0.1
$k_{4,3}$	0.3	$k_{6,10}$	0.5×10^{-3}	$k_{9,3}$	0.1
$k_{4,4}$	0.05	$k_{6,14}$	5.0	$k_{9,4}$	0.1
$k_{4,6}$	0.04	$k_{7,2}$	0.5×10^{-2}	$k_{9,5}$	0.01
$k_{4,10}$	0.5×10^{-3}	$k_{7,3}$	0.05	$k_{9,10}$	0.001
$k_{4,13}$	0.5×10^{-3}	$k_{7,4}$	0.1	$k_{9,14}$	5.0
$k_{4,14}$	5.0	$k_{7,5}$	0.3	$k_{9,9}$	0.1
$k_{5,2}$	0.2	$k_{7,6}$	0.1	γ'_8	0.1
$k_{5,3}$	0.1	$k_{7,7}$	0.1	α	0.013
$k_{5,4}$	0.05	$k_{7,9}$	0.2	t_1'	0.2
$k_{5,5}$	0.5×10^{-2}	$k_{7,10}$	0.1	t_1''	0.2
h_3	0.265	$k_{7,14}$	5.0	ω_2	1.0
h_4	0.2293	$k_{8,5}$	0.1	ω_3	1.0
h_5	0.14	$k_{8,6}$	0.1	ω_4	1.0
h_6	0.15	$k_{8,7}$	0.1	ω_5	0.9
h_7	0.13	$k_{8,9}$	0.1	ω_6	0.9
h_8	0.1	$k_{8,10}$	0.01	ω_7	0.9
h_9	0.14	ω_1	1.0	–	–

and autumn and $\delta = \delta_w$ in winter. In further results, it was taken $\delta_w = 0.5°$, $\delta_s = 1°$ (Krapivin 1996).

Seasonal and latitudinal temperature variations are known to occur in the PCE area. The natural cycles of such variations have been incorporated in the model by consulting the available real data and it appears interesting to consider herein the departures of temperature from these model values. Figure 4.7 shows the biomass of the anchovy and predatory fishes versus temperature fluctuations. As seen from Fig. 4.7, if the system is functioning under normal temperature conditions, a clear-cut division between the habitation areas of the anchovy and predatory fishes is observed. In this case, the maximum concentration of the anchovy biomass is observed to occur in the offshore zone about 1.5°–2° wide in longitude, whereas the biomass of predatory fishes attains its maximum concentration in the open ocean. This pattern is observed throughout the area and appears to be stable 30–40 days later for the model distributions of temperature. This pattern falls apart as soon as the overall temperature background is subject to change. A reduction in the general temperature level by 1 K accounts for the fact that the habitation areas of anchovy and predatory fishes increasingly intersect. Moreover, in this case their total biomass rises by about 10 %. In the El Niño periods the effect of diffusion of the ecosystem into the offshore and open-ocean parts disappears completely, being

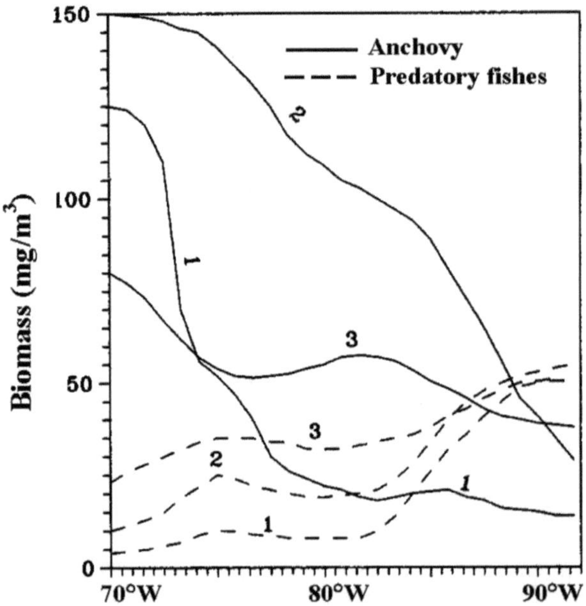

Fig. 4.7 Effect of the water temperature variations on predatory fishes and anchovy biomass in PCE at 20°S referred to the moment of time at = 100 days. Notation: *1* normal temperature regime; *2* the temperature is decreased by 1 K; *3* El Niño regime. The initial data for t = 0 were assumed to be uniform throughout the area: $B_1 = 200$, $B_2 = 400$, $B_3 = 80$, $B_6 = B_7 = B_9 = 8$, $B_4 = 35$, $B_{10} = 700$, $B_{13} = 3.5 \cdot 10^4$, $B_5 = 16$, $B_{12} = 200$, $B_{14} = 65$ (mg/m^3), $B_8 = 2$ specimens/km^2

accompanied by a decline in the total biomass of anchovy and predatory fishes. These effects were observed by Marti (1971) and Parin (1971).

To estimate the survivability of the PCE, Krapivin (1996) has introduced the survivability function:

$$J(t) = \frac{\sum_{i=1}^{9} \iiint_{(\varphi,\lambda,z)\in\Omega} B_i(\varphi,\lambda,z,t)\,d\varphi\,d\lambda\,dz}{\sum_{i=1}^{9} \iiint_{(\varphi,\lambda,z)\in\Omega} B_i(\varphi,\lambda,z,t_0)\,d\lambda\,dz} \quad (4.50)$$

Figure 4.8 shows the survivability function (4.50) for various departures of temperature from its model value. We will consider the system to be in a living state if the condition $J(t) > \kappa\, J(t_0)$ is carried out for $t \geq t_0$, where $\kappa < 1$ is the level of survivability. It is obvious that temperature fluctuations by ±7 K bring the system to a 'dead state' after 70 days for +7 K and after 190 days for −7 K. Water temperature fluctuations within ±5 K turn out to be non-dangerous to the system, but they may be the cause of its conversion into a different quasi-stationary state. This follows from the comparison of the phase patterns of behaviour of the system trajectories similar to these given in Fig. 4.9 where the PCE dynamics in the plane of $B_5 \times B_7$ is given under different initial states.

It is supposed that fluctuations of solar radiation energy cause proportional variations of the water temperature. A climate block of GIMS is used for the calculation of atmospheric temperature σ_A in Ω. A temperature regime of PCE is described by the

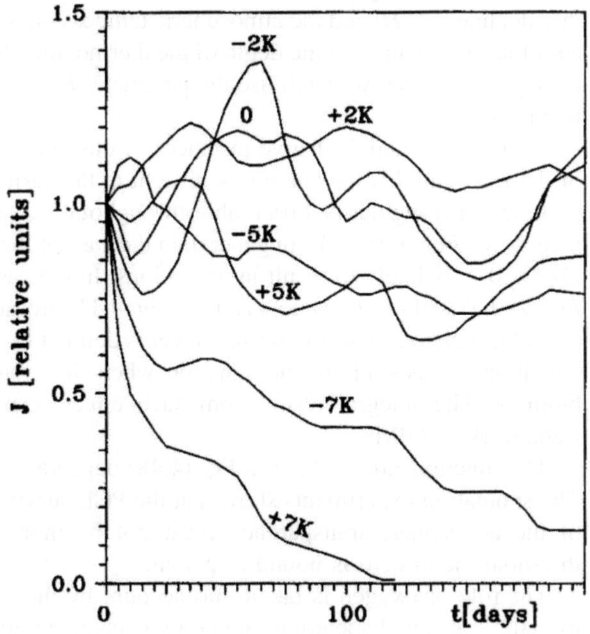

Fig. 4.8 The estimation of the PCE survivability under the temperature variations. The figures on the *curves* show the water temperature variations from the normal conditions

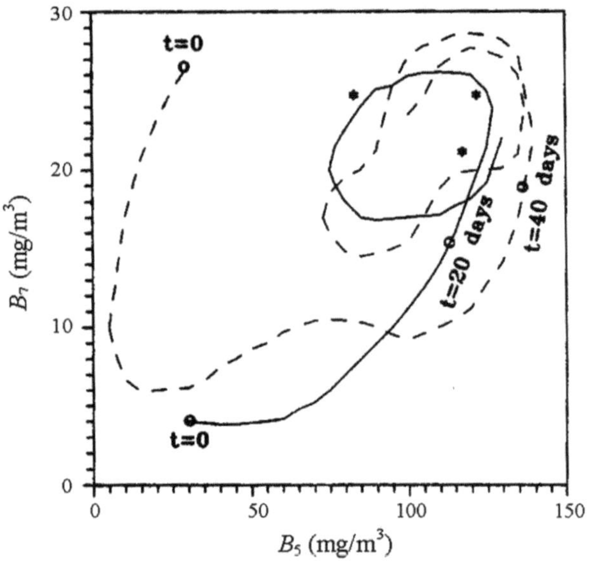

Fig. 4.9 Trajectories of the PCE on the plane of $B_5 \times B_7$ under different initial states. The *asterisks* correspond with the data of Marti (1971)

model of Vinogradov et al. (1975), Vinogradov (1975). Additional conditions are also taken into account as follows. Suppose that the difference $\Delta\sigma(\varphi, \lambda, z, t)$ between the water temperature above the thermocline ($z < H$) and σ_A linearly increases with higher latitudes. This difference is $\Delta\sigma(0, \lambda, z, t) = 0$ and $\Delta\sigma(30°\text{S}, \lambda, z, t) \leq 0.2\sigma_A$. $\Delta\sigma$ is independent of longitude. The difference between the water temperature under the thermocline ($z > H$) and the atmospheric temperature is independent of geographical coordinates. Changes in the depth of the thermocline due to light fluctuations should be neglected. Then we shall use the parameter R as a free parameter to assign light fluctuations.

Figures 4.10 and 4.11 give instances of the calculations of the vertical structure of the PCE and $J(t)$ for variations of light at the surface. These results point to the presence of a regularity observable throughout the area and are indicative of the increase of $E(\varphi, \lambda, 0, t)$. If, in general, in the case of the biosphere a 2 % reduction in $E(\varphi, \lambda, 0, t)$ is likely to result in glaciations. In the case of the PCE, fluctuations of $E(\varphi, \lambda, 0, t)$ within −60 % to +700 % (Fig. 4.11) are safe for this system. This effect has a local character when the safe level is defined by the condition $J(t) \geq 0.5$. This condition corresponds to the situation when the system loses more that 70 % of biomass. The integral result from such effects can be understood only in the framework of GIMS.

The function $E(\varphi, \lambda, 0, t)$ in Eq. (4.48) depends on the state of the atmosphere. The simulation experiments show that the PCE survivability permissible reductions of the atmosphere transparency must not be more than 2.5 times. Above this threshold the system is bound to perish.

The role of oxygen is taken into account by the model in terms of production, mortality rates and spending on energy exchange within the environment. Simulation experiments point to great inhomogeneities of the saturation of waters with

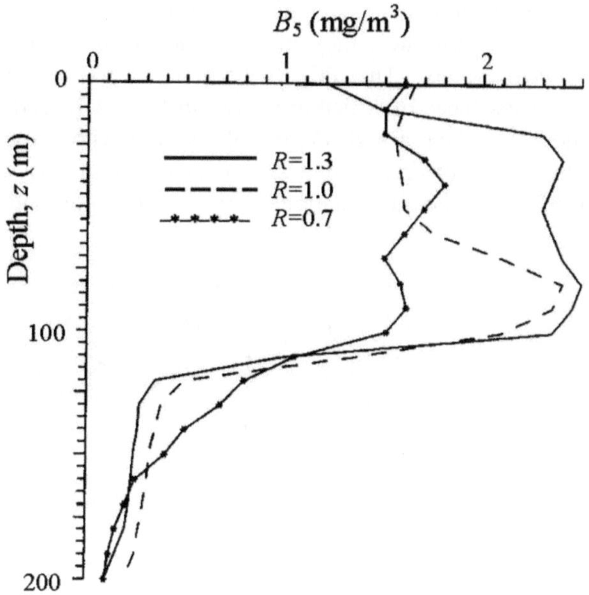

Fig. 4.10 The PCE vertical structure as a function of the variations in $E(\varphi, \lambda, 0, t)$ under $t = 50$ days, $\varphi = 12°S$, $\lambda = 82°W$. Initial conditions are the same as in Fig. 4.8. R is the scenario parameter reflecting the level of variations in incoming solar radiation

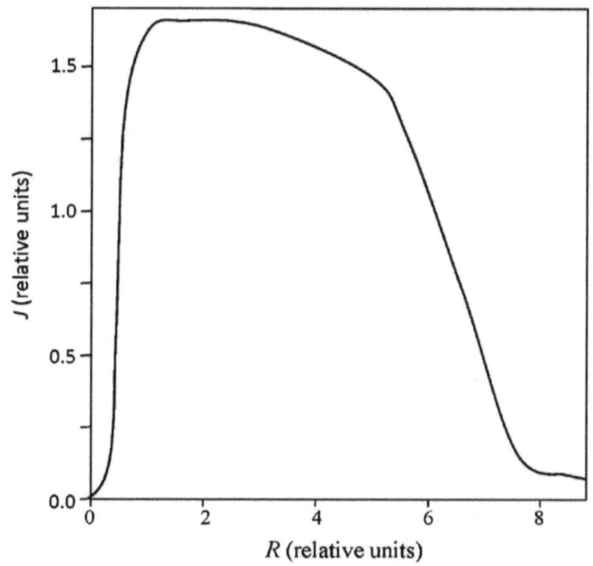

Fig. 4.11 Relationship between $J(t)$ and the variations in $E(\varphi, \lambda, 0, t)$ in the framework of scenario: $E(\varphi, \lambda, 0, t) = R(3.04 - 0.03\varphi)(1 + 0.1 \sin[2\pi t/365])$, where $R = 1$ under normal conditions of system functioning

oxygen both in horizontal sections and in depth. At the places of oxygen deficit (<6 mg/m^3) a drastic reduction in the biomass of the higher trophic levels is observed, but no stable pattern is noted either with respect to the waters little saturated or those oversaturated. Accordingly, for the region as a whole these zones are observed to be travelling apparently due to hydrodynamic processes on a non-stable pattern without causing any perceptible damage to the entire community.

Fluctuations in the saturation of waters with oxygen may have an anthropogenic origin, for instance, they may be due to oil pollutions or to large discharges of sewage waters. The PCEM permits us to set some hypothetic experiments, some instances of which are given in Figs. 4.12 and 4.13. It is noticeable that the model displays a high degree of stability with respect to the initial saturation of water with oxygen and from the behavior of $J(t)$ we see how rapidly the PCE proceeds to the quasi-stationary regime of functioning at $B_{14}(\varphi, \lambda, z, t_0) \geq 60$ mg/m^3

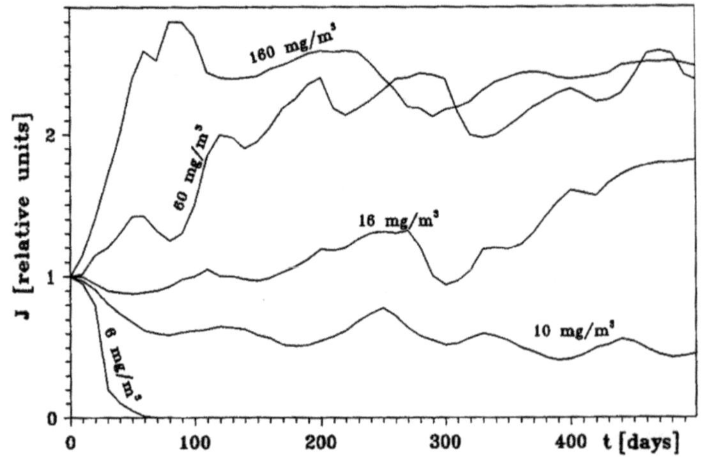

Fig. 4.12 The dynamics of the survivability function $J(t)$ under different initial conditions of water saturation with oxygen (it is shown on *curves*)

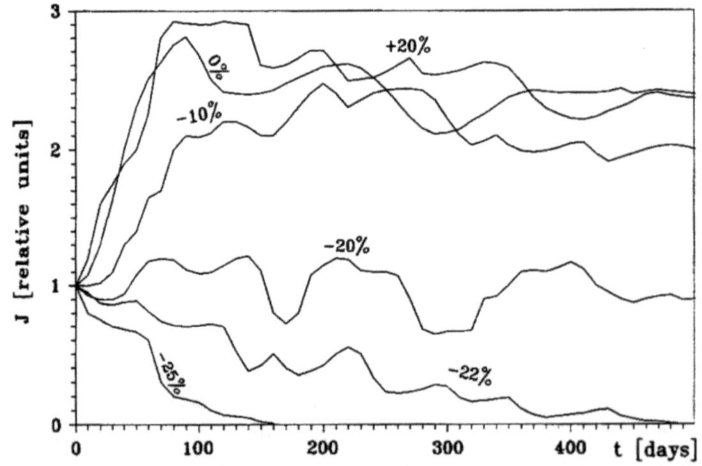

Fig. 4.13 The dynamics of the survivability function $J(t)$ under variations of the rate of photosynthetic oxygen production in percents of the model value ($\delta_{14} = 0.01$)

and how long it takes to overcome the initial shortage of oxygen in the case of $B_{14}(\varphi, \lambda, z, t_0) = 16$ mg/m^3. For $B_{14}(\varphi, \lambda, z, t_0) < 16$ mg/m^3 the PCE is unable to proceed to the stationary regime of functioning.

The PCEM is observed to be even more sensitive to a dynamic effect on the process of water saturation with oxygen. This clearly follows from the behaviour of $J(t)$ in Fig. 4.13. A reduction of the oxygen production by 10 % does not influence the PCE dynamics. A subsequent decrease of δ_{14} in Eq. (4.47) leads up to the displacement in the stable state of the PCE. The PCE does not survive when the parameter δ_{14} is decreased by 22 %. Figure 4.11 illustrates the important qualitative result when Eq. (4.50) gives a possibility for the search of the dangerous zones in Ω.

Table 4.5 gives the results obtained in computing the PCE dynamics in the case of a variation of the concentration of nutrients at the initial moment of time t_0 and at the depth $z > 150$ m. From the comparison of the results contained herein it follows that a variation of the nutrient concentration within a wide range at the moment $t = t_0$ does not practically affect the behaviour of the system at the moments $t >> t_0$. The system 'health' with time from the 'blows' it has suffered and proceeds to the same functioning level.

It is different when at depths $z > 150$ m B_{12} becomes close or inferior to a certain value $B_{12,\min}$. In this case, the system is unable to make up for the introduced fluctuations and at $B_{12} >> 0.1$ mg/m^3 the system begins to experience the effect of strong limitation throughout the area. Such a reduction in the nutrient concentration is, for instance, possible in the case of contamination of bottom sediments with oil products.

The effect of variations in the velocity of vertical advection plays an important role in the PCE evolution. To estimate turbulent escape of nutrients into layers overlying the thermocline, it is assumed that the velocity of water uprise is equal to 10^{-3} cm/s in the PCEM. The obtained data point to the fact that, on the average, the integrated pattern of the distribution of community elements is not subject to any significant variations within the velocity range from 3×10^{-4} to 10^{-2} and even 10^{-1} cm/s, but is observed that it is drastically distorted under a higher. Slow water lifting ($<10^{-4}$ cm/s) results the ecosystem extinction.

The simulation experiments enable us to answer the question about securing an increase in the productivity of the PCE through the creation of artificial upwellings. For the PCE this can be achieved within up to 40 %. In the open ocean the higher velocity of vertical advection improves, on the one hand, nutrients supply for surface-adjoining waters and, on the other it brings down the water temperature, thereby drastically restricting the productivity of the system. Moreover, on account of the lengthy trophic chain in the open ocean, the ultimate effect of increased nutrients saturation proves to be insignificant. In the shelf zone the nutrients limitation is absent.

In terms of the GIMS, the PCEM can be considered as one step to the hierarchical biosphere model. The PCEM proposed herein is based on the experience of preceding investigations. The choice of the composition of the system mainly reflects the concepts of the trophic and physiochemical nature of links between the components of the PCE, as are maintained by marine ecologists. In contrast to the earlier proposed models of marine and oceanic ecosystems, the PCEM is

Table 4.5 Simulation results in which various nutrient concentrations are used (φ = 12°S, λ = 82°W)

Depth (m)	Biomass of phytoplankton (kg/m^2)					
	$B_{12}(\varphi, \lambda, z, t_0)$ = 0.05 mg/m^3			$B_{12}(\varphi, \lambda, z, t_0)$ = 2 mg/m^3		
	$t - t_0$ (days)			$t - t_0$ (days)		
	10	50	100	10	50	100
Initial nutrient concentrations are held in $z \le 150$ m						
0–10	0.08	3.49	5.73	3.92	2.14	28.06
10–20	8.43	19.48	12.05	22.99	5.19	56.31
20–30	9.67	25.28	33.54	25.45	38.89	75.49
30–40	9.64	27.98	33.27	26.59	66.41	86.36
40–50	9.26	28.27	33.26	30.88	40.52	27.93
50–60	9.39	28.23	29.18	28.31	24.39	21.04
60–70	11.97	27.77	26.59	27.81	8.63	84.69
70–80	6.16	15.23	13.98	21.84	2.29	63.49
80–90	2.79	7.52	6.56	10.74	0.29	22.75
90–100	0.78	3.09	3.52	5.91	0.08	6.16
100–150	1.45	3.93	4.23	3.99	0.43	8.63
150–200	1.15	3.81	2.63	4.09	0.46	11.63
Initial nutrient concentrations are held in $z > 150$ m						
0–10	0.08	0	0	3.07	4.19	19.61
10–20	0.09	0	0	4.13	4.35	35.59
20–30	0.15	0	0	31.05	19.65	40.57
30–40	0.21	0	0	28.02	42.65	72.46
40–50	1.52	0	0	35.64	66.36	69.56
50–60	3.53	0.03	0	19.64	55.78	44.79
60–70	3.22	0.17	0	13.32	21.79	12.84
70–80	2.49	0.13	0	11.69	13.22	6.09
80–90	1.41	0.13	0	4.51	5.96	2.92
90–100	1.28	0.05	0.01	2.09	2.67	3.07
100–150	0.93	0.18	0.01	2.19	2.96	2.54
150–200	0.75	0.12	0.01	2.91	1.86	3.53

characterized by a wider coverage of processes taking place in the system. It allows for their variations in time and space.

The structure of the PCEM software provides an opportunity for an early correction of functional dependencies in the model equations, for introducing and exclusion of components, reorientation of trophic graph and introduction of other hypothetic situations.

A further improvement of the PCEM involves consideration of new data on ecosystem elements and processes taking place in the system, allowance for water heterogeneity and salinity and specification of the trophic graph on its higher levels.

The model of El Niño should be developed in the form of an additional block. Progress towards a more detailed parameterization of the PCE depends on the database development and the use of satellite observations.

4.6.5 Model of the Upwelling Ecosystem

The zones of the deep waters lifting in the World Ocean called upwellings are known to exhibit a high productivity. An upwelling results from the surface waters wind-driven from the shore, dispersing currents, or water retreat from the shore for other reasons. The rates of water lifting and the upwelling stability are determined by several synoptic parameters. A most characteristic value of the vertical velocity of water lifting in the zone of upwelling is 0.77×10^{-3} m/s. The depths from which the water starts lifting vary widely, but the depths around 200 m prevail.

High oceanic productivity occurs in areas of upwelling in the ocean, particularly along continental shelves (red areas on map in Fig. 4.14). The coastal upwelling in these regions is the result of deep oceanic currents (Fig. 4.15) colliding with sharp coastal shelves, forcing nutrient-rich cool water to the surface. Over 90 % of the world's living biomass is contained in the oceans, yet only about 0.2 % of marine production is harvested. High oceanic productivity occurs in areas of upwelling in the ocean, particularly along continental shelves (red areas on map). The coastal upwelling in these regions is the result of deep oceanic currents colliding with sharp coastal shelves, forcing nutrient-rich cool water to the surface. Over 90 % of the world's living biomass is contained in the oceans, yet only about 0.2 % of marine

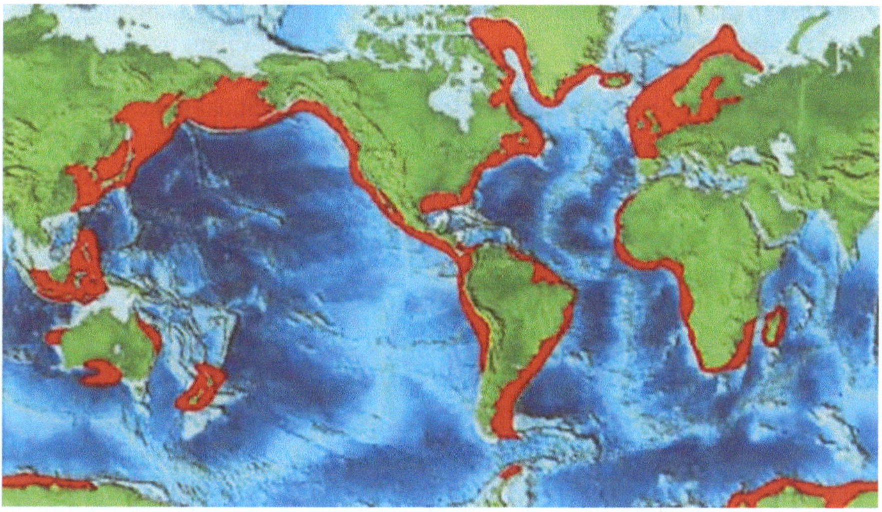

Fig. 4.14 Coastal upwelling zones (http://en.wikipedia.org)

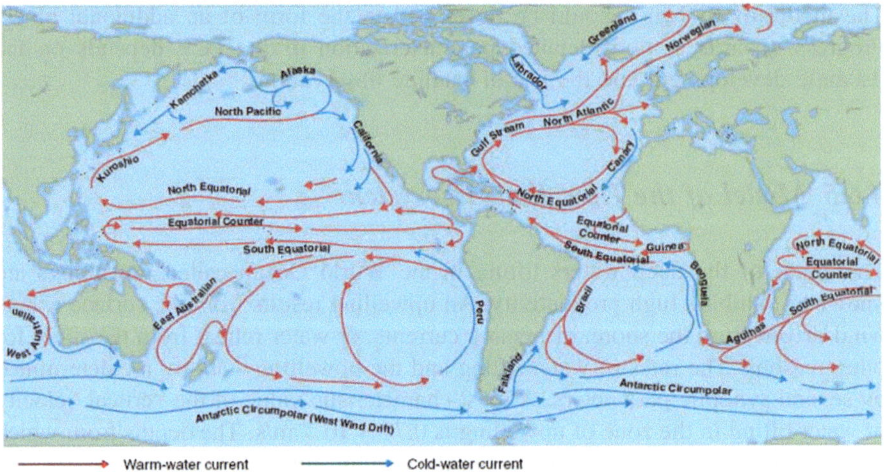

Fig. 4.15 The map of global currents in the World Ocean (http://en.wikipedia.org)

production is harvested. The Peruvian upwelling is a 300 × 300 mile area adjacent to the coast and is the most biologically productive coastal upwelling system on Earth. Carbon levels (an indicator of production) are tens of times higher than those of the next most productive upwelling region, the California current.

High productivity is observed in ecosystems located in zones of coral reefs (Fig. 4.16). Coral reef ecosystems contain distinct zones that represent different kinds of habitats. Usually, three major zones are recognized: the fore reef, reef crest, and the back reef (frequently referred to as the reef lagoon). All three zones are physically and ecologically interconnected. Reef life and oceanic processes create opportunities for exchange of seawater, sediments, nutrients, and marine life among one another.

Let us consider the ocean ecosystem in the zone of upwelling proceeding from the concept of the successive development of community from the moment of its formation in the region of the deep waters lifting to the climax state in the oligotrophic region of convergence, and derive a model of the upwelling ecosystem

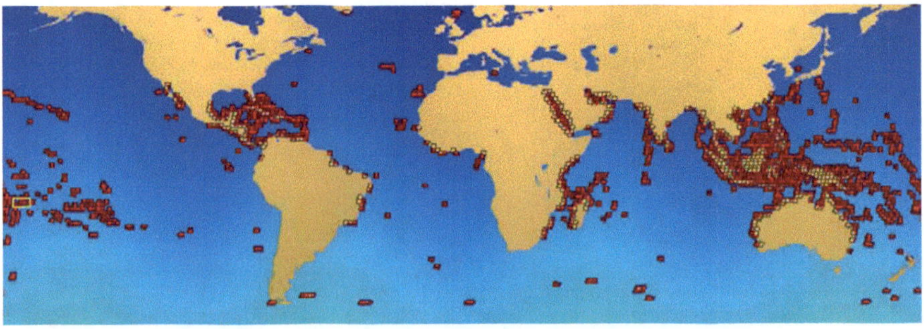

Fig. 4.16 Global distribution of coral reefs (http://en.wikipedia.org)

(MUE). Between these moments the system developed in time and, respectively, moves together with the water flow. In this case, changes take place both in the total energy supply in the community and in its structure (spatial, trophic, specific). The basic moments of these changes have been generally studied in numerous field investigations. Therefore, the correspondence of the model calculations to these ideas can be one of the criteria of the model adequacy.

Supposing that the ecosystem develops uniformly horizontally from the zone of upwelling, changes are observed both through the depth z with a step $\Delta z = 10\,м$ and in time with a step Δt = 1 day. The horizontal velocity of water flow from the zone of upwelling is assumed to be constant and equal to $V = V_\varphi = V_\lambda$, so that the distance of a unit volume of water from the zone of upwelling is equal to $\Delta r = (\Delta\varphi^2 + \Delta\lambda^2) = V\Delta t$.

The state of the ecosystem on each horizon $z = const$ is determined by illumination $E(z, t)$, concentration of nutrient elements $n(z,t)$ and detritus $d(z, t)$, biomass of phytoplankton, $p(z, t)$, bacteria $b(z, t)$, protozoa $Z_1(z, t)$, microzooplankton $Z_2(z, t)$, small filtraters $Z_3(z, t)$, large filtraters $Z_4(z, t)$, small predatory cyclopoids $Z_5(z, t)$, predatory calanoids $Z_6(z, t)$, large predatory chaetognaths and polychaete $Z_7(z, t)$. Protozoa include infusorian and radiolarian, microzooplankton includes naupliuses of all copepod. Based on studies of the food for tropical plankton animals, small filtraters (<1 mm in size) are plankton animals such as *Oikopleura*, copepods *Clausocalanus, Paracalanus, Acartia, Lucicutia*, small ostracode *Conchoecia*. Large filtraters include copepod larger than 1 mm, such as *Undinula, Pleuromamma, Centropages, Temora, Scolecythrix,* and others, baby euphasiids, pteropod, etc. In the trophic pyramid the impact of nekton elements is considered negligibly small. The food bonds between the enumerated elements are shown in the scheme in Fig. 4.17.

It is assumed that 30 % of the mass of bacteria is in natural aggregates >3–5 μm in size, which can be eaten and assimilated by filtraters (Z_3, Z_4); nauplii (Z_2), protozoa (Z_1), and small filtraters (Z_3) can also feed on non-aggregated bacterioplankton.

The trophic bonds between the elements are described based on the energy principle. Biomass, rates of producing and exchange, and rations are expressed in energy units (cal/m^3 or cal/m^2).

Like in the general model of the ocean ecosystem, the original source of energy and substance in the community is the primary production of phytoplankton (R_p). The solar radiation energy (E) and nutrient elements (n) come from outside. The vertical structure of the water medium is described by a 3-layer model with one layer above the thermocline, one layer of sharp gradient, and one layer of lower gradients beneath the thermocline. The upper (z_b) and lower (z_l) boundaries of the layer with sharp gradients are assumed to be located (depth in metres):

$$z_b = \begin{cases} 10 + 2,2t & \text{for} \quad 0 \le t \le 50 \quad \text{days}, \\ 120 + 0,6(t - 50) & \text{for} \quad t > 50 \quad \text{days}, \end{cases}$$

$$z_l = \begin{cases} 30 + 2,4t & \text{for} \quad 0 \le t \le 50 \quad \text{days}, \\ 150 + 1,4(t - 50) & \text{for} \quad t > 50 \quad \text{days}. \end{cases}$$

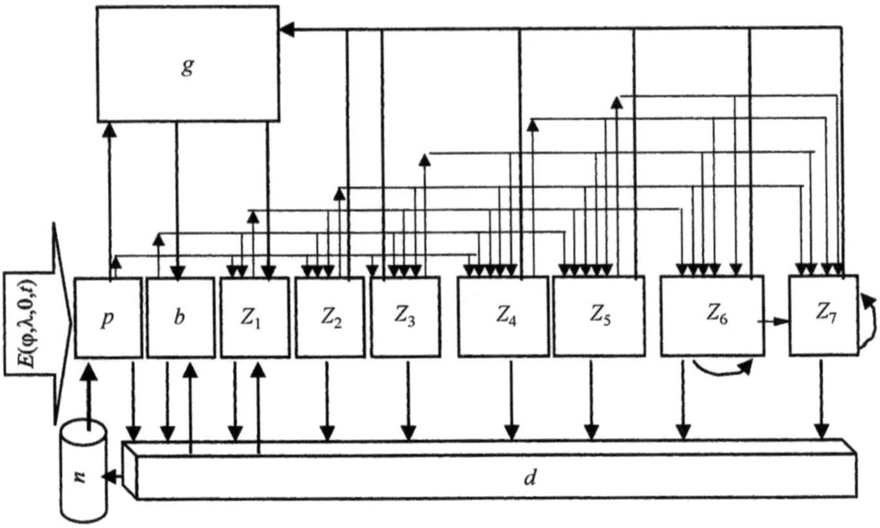

Fig. 4.17 The scheme of energy fluxes (cal/m^3/day) through the community of pelagic organisms residing in the 200 m surface layer in the oligotrophic tropical region of the World Ocean

Thus, the thermocline at an initial time moment $t = 0$ is located in the layer 10–30 m, descending gradually down to 120–150 m on the 50th day and to 150–190 m on the 100th day.

The supply of nutrient elements in the layer 0–200 m is replenished at the expense of detritus decomposition and emission of unassimilated food by living elements of the community, as well as at the expense of the input from deep layers ($z > 200$ m) due to turbulent mixing (coefficient k_2) and vertical advection (velocity V_Z). The content of nutrients in organic matter is constant and constitutes 10 %.

Taking into account the assumptions mentioned, the Eq. (4.19) becomes

$$R_p = k_T\left[1 - 10^{-0.25p\gamma(t)}\right]\left[1 - 10^{-0.1\,n}\right]^{0.6},$$

where the function $\gamma(t)$ characterizes the time dependence of the production/biomass ratio (p/b-coefficient) for phytoplankton. If a maximum of p/b for phytoplankton at the point of upwelling ($t = 0$) is assumed to be 5, then the p/b-coefficient decreases and on the 15th day it reaches 1 and remains constant. Hence, the Eq. (4.25) is rewritten:

$$\partial p/\partial t = R_p - t_p p - \mu_p p - \sum_{j \in \omega_0 \backslash p} C_{pj} R_j + k_2 \partial^2 p/\partial z^2 + \left(V_z - w_p\right)\partial p/\partial z,$$

where the coefficient w_p (cm/day) describes the process of gravitational sedimentation and is estimated at

$$w_p = \begin{cases} 50 & \text{for} \quad 0 \le z \le z_b, \\ 10 & \text{for} \quad z_b < z < z_l, \\ 30 & \text{for} \quad z \ge z_l \end{cases}$$

Bacteria feed on detritus and dissolved organic matter emitted by phytoplankton. The allochthonous dissolved organic matter is neglected. In the Eq. (4.36) an approximation is considered:

$$T_p + T_b + T_Z + T_r + T_D = 0.3R_p.$$

It is also assumed that $\mu_b = 0.01$, $t_b = 0.75\left[1 - 10^{\zeta}\right]$, $R_p = 3b\left[1 - 10^{\zeta}\right]$, $\zeta = -0.2d - 0.3R_p$.

A limitation that during a 24 h bacterioplankton can consume not more than 10 % of all detritus located in the same water layer is established.

On these assumptions, the Eq. (4.35) is written as:

$$\partial n/\partial t = -0.1R_p + 0.1d + 0.05\sum_{i\in\omega_0} t_i B_i + k_2 \partial^2 n/\partial z^2 + V_z \partial n/\partial z,$$

where $\omega_0 = \{p, b, Z_j\ (j = 1 - 7)\}$, B_i is the biomass of the i-th element.

The coefficients k_2 (cm²/day) and V_z (cm/day) are assumed to be:

$$V_z = \begin{cases} 5 & \text{for} \quad 0 \le z \le z_b, \\ 0.1 & \text{for} \quad z_b < z < z_l, \\ 1 & \text{for} \quad z \ge z_l, \end{cases} \qquad k_2 = \begin{cases} 200 & \text{for} \quad 0 \le z \le z_b, \\ 50 & \text{for} \quad z_b < z < z_l, \\ 150 & \text{for} \quad z \ge z_l. \end{cases}$$

The formula (4.18) for the case considered is simplified:

$$E(\varphi, \lambda, z_n, t) = E_0 \cdot 10^{-\alpha z_n},$$

where z_n is the depth of the nth layer (m);

$$\alpha = 0.01 + 0.001\sum_{k=0}^{n} [p(z_k, t) + d(z_k, t)]/z_n$$

A marked suppression of phytoplankton is observed at $E > 100$ cal/cm^2/day. Hence, the formula (4.20) is re-written:

$$K_T = 0.041E \cdot 10^{0.25[1 - E/E_{\max}]},$$

where $E_{max} \approx 70$ cal/cm^2/day.

For this case, the formula (4.19) is reduced to the form:

$$\partial b/\partial t = R_b - \mu_b b - t_b b - \sum_{j\in\omega_0\backslash(p,b)} C_{bj}R_j + k_2\partial^2 b/\partial z^2 + (V_z - w_p)\partial b/\partial z$$

The rate of the change of the biomass of each element of zooplankton $\partial Z_i/\partial t$ (i = 1–7) like in the Eq. (4.29) is determined by the level of consumption (ration) $R_{Z,i}$, food assimilation $1/u_{Z,i}$, expenditures on energy exchange $T_{Z,i} = t_{Z,i}Z_i$, the rate of dying-off $\mu_{Z,i} = 0.01Z_i$, eating-out of the i-th element by the j-th element with the coefficient C_{ij} and the age transition of nauplii into copepod of filtraters and predators. Thus

$$\partial Z_i/\partial t = \Omega_{m,i} + (1 - h_{Z,i})R_{Z,i} - (t_{Z,i} + \mu_{Z,i})Z_i - \sum_{j\geq i}^{7} C_{ij}R_{Z,j},\ (i = \overline{1.7}),$$

where $h_{Z,i} = 1 - 1/u_{Z,i}$, $\Omega_{m,i}$ characterizes the age transition of nauplii to another category of zooplankton. The transition of nauplii Z_2 to copepod of filtraters (Z_3, Z_4) takes place with an intensity

$$\Omega_{m,3} = Z_2Z_3\{15(Z_3 + Z_4 + Z_5 + Z_6)\}^{-1},$$
$$\Omega_{m,4} = Z_2Z_4\{20(Z_3 + Z_4 + Z_5 + Z_6)\}^{-1},$$

and for predatory copepod (Z_5, Z_6) with an intensity

$$\Omega_{m,4} = Z_2Z_5\{15(Z_3 + Z_4 + Z_5 + Z_6)\}^{-1},$$
$$\Omega_{m,6} = Z_2Z_6\{20(Z_3 + Z_4 + Z_5 + Z_6)\}^{-1}.$$

To calculate the ration for the elements of zooplankton by the formula (4.28), the coefficient ν is determined from the condition $\nu = 0.01/B_{min}$. When calculating the coefficients C_{ij}, it is assumed that the food needs refer to various food objects in proportion to their biomasses taking into account food selectivity. The coefficients C_{ij} for all elements of the ecosystem are given in Fig. 4.18.

The daily migrations of zooplankton are simulated by an addition to the daily ration of the elements Z_j (j = 4–7), located in the layer 0–50 m, of some share (k_Z) of total food needs of the same elements of the community but in the layer 50–200 m. The coefficient k_Z is considered to be time-dependent:

$$k_z = \begin{cases} 0.02 + 0.0016t & \text{for} \quad t \leq 50 \quad \text{days,} \\ 0.1 & \text{for} \quad t > 50 \quad \text{days.} \end{cases}$$

Various scenarios both of migration and nutrition of the elements of zooplankton are possible. Only for certain it is known that some share of the zooplankton elements from deep layers satisfy their needs for food in the upper layers of the ocean.

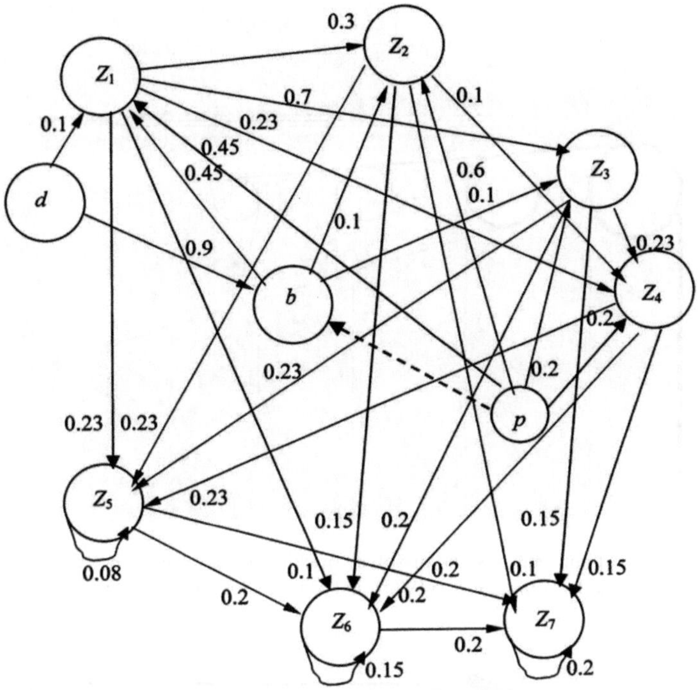

Fig. 4.18 The scheme of trophic loads between the elements of the upwelling ecosystem. The *letter* identifiers correspond to Fig. 4.17. The *arrows* indicate the values of the coefficients C_{ij}

In this case, the Eq. (4.34) is written as:

$$\partial d/\partial t = \sum_{i\in\omega_0} (H_{Z,i} + M_{Z,i}) - \sum_{j\in\omega_0\backslash p} C_{dj}R_j + k_2\partial^2 d/\partial z^2 - \mu_d d + (V_z - w_d)\partial d/\partial z,$$

where the coefficient of the rate of the gravitational sedimentation of detritus is taken in the form (cm/day):

$$w_d = \begin{cases} 25 & \text{for} \quad z_b \le z \le z_l, \\ 50 & \text{for} \quad z < z_b, z > z_l. \end{cases}$$

Figures 4.19 and 4.20 schematically represent the vertical structure of the typical upwelling with extraction of specific zones:

1. The rising of nutrient elements and dissolved organic matter from deep layers.
2. Multiple use of nutrient elements within the production/destruction cycles of ecosystem functioning in upper water layers.
3. Loss of nutrient elements due to their capture by the falling organic debris and with migrating organisms.

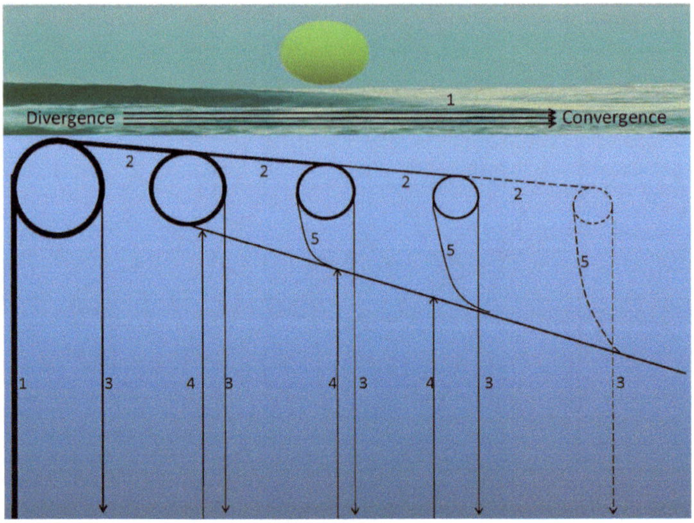

Fig. 4.19 Scheme of nutrient salts and dissolved organic matter cycle in the ecosystem succession in the upwelling zone by Vinogradov (1975) and Vinogradov et al. (1996)

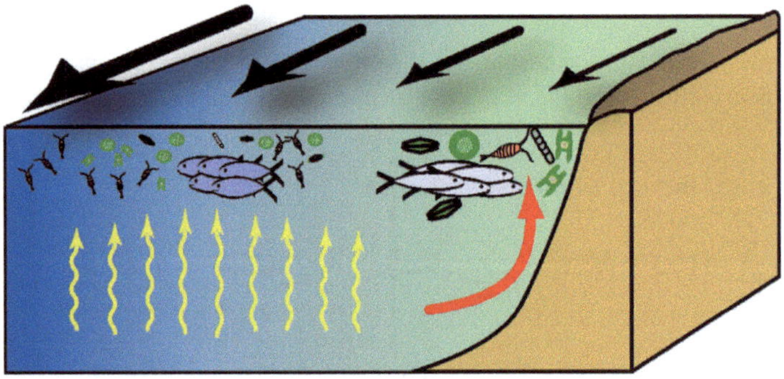

Fig. 4.20 Upwelling diagram that reflects the influence of ocean winds on the pelagic ecosystem in upwelling regions (http://www.pnas.org/content/105/6/1965/F1.large.jpg)

4. Turbulent lifting of nutrient elements and dissolved organic matter from deep layers and their delay in the zone of lower maximum.
5. Inclusion of nutrient salts lifted from lower maximum with migrating elements to the production cycles of surface community.

It is supposed that all ecosystem elements are uniformly distributed by depth z and $n(\varphi_O, \lambda_O, z, t_0) = 250$ mg/m^3, $d(\varphi_O, \lambda_O, z, t_O) = 0$, $b(\varphi_O, \lambda_O, z, t_O) = 1$, $p(\varphi_O, \lambda_O, z, t_0) = 0.5$, $Z_i(\varphi_O, \lambda_O, z, t_O) = 0.01–0.5$ $(i = 2–7)$, $Z_1(\varphi_O, \lambda_O, z, t_O) = 0.0001$ cal/m^3 where φ_O and λ_O are the geographical coordinates of upwelling zone center. Simulation experiments have shown that under variations of these initial values by 50 times ecosystem during 50–70 days arrives at stationary regime. It means that the ecosystem of the upwelling zone 'forgets' these fluctuations in initial concentrations.

Upper (z = 0) and lower (z = 200 m) boundaries of the ecosystem are characterized by zero gradients for all its elements except for n and d:

$$\partial n/|_{z=0} = 0, \partial d/\partial z|_{z=0} = 0, \partial d/\partial z|_{z=220} = -0.5, n(\varphi, l, 200, t) = 250\ \mathrm{mg/m^3}$$

It is supposed that water covers the distance from the upwelling zone to the oligotrophy zone in more that 60 days. Figure 4.21 characterizes this evolution process. It is seen that biomasses of phytoplankton and bacterioplankton reach maximal levels at 5–10 and 10–15 days, respectively. Small filtraters are slow in their evolution, but large filtrators are evolved more slowly. Biomass of large filtrators reaches maximal level only in the 30 day. Existing feedbacks control the evolution processes for all living ecosystem elements and their biomasses slowly decrease as a system is moving from the upwelling zone. Flows of nutrient salts that lift from deep waters through thermocline and vertical turbulence play main role in this evolution. Ecosystem results quasi-stationary level at 50–60 days characterized low biomasses of all living elements. Under this situation, stable balance is achieved between photosynthesis intensity and nutrient salts inflow from deep layers.

Numerous experimental observations for the ecosystems of the upwelling zones have verified these model results concerning a dissociation of maximal biomasses of living elements during ecosystem evolution. Table 4.6 characterizes the precision of the modeling results. Bacterioplankton biomass proves to be larger in real data. It is evident that this follows from the failure to take account of allochtonous organic matter as element of the bacterioplankton food. In general, modeling and observational results are similar. This conclusion is confirmed by results represented in Fig. 4.22 where the vertical structure of the ecosystem is given. One can see that phytoplankton biomass is uniformly big during the first 5 days in the layer 0–50 m.

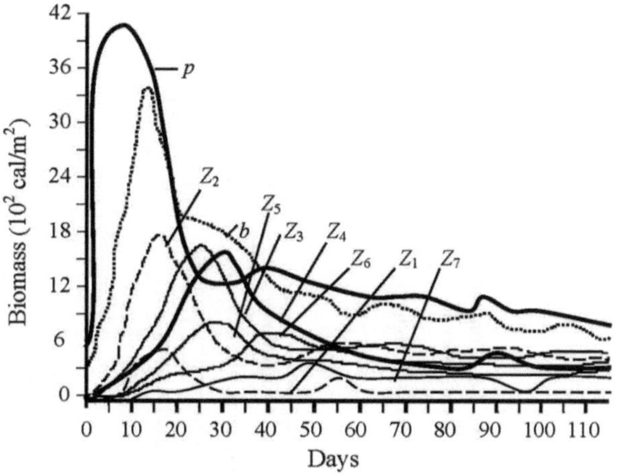

Fig. 4.21 Integral biomasses dynamics of living elements in the water layer 0–200 m

Table 4.6 Comparison of the modeling and observational results concerning the upwelling ecosystem evolution under typical conditions of tropical zone of the Pacific Ocean

Ecosystem element	Community is situated on middle stage of the maturity (30–49 days)			The maturity community (60–80 days)	
	Modeling results		Measurements (Station 6,429)	Modeling results	Measurements (Station 6,493)
	$\Delta t = 30$ days	$\Delta t = 40$ days		$\Delta t = 70$ days	
Phytoplankton, p	1,319	1,092	2,000	827	900
Bacterioplankton, b	1,673	864	4,100	564	2,180
Microzooplankton, Z_2	394	303	321	300	–
Small filtraters Z_3	1,338	612	525	290	74
Large filtraters, Z_4	1,416	726	420	252	165
$Z_3 + Z_4$	2,754	1,338	945	542	238
Small predatory cyclopoids, Z_5	624	491	495	203	236
Predatory calamoids, Z_6	288	600	610	191	175
Large predatory chaetognaths and					
Polychaete, Z_7	184	183	15	102	51
$Z_5 + Z_6 + Z_7$	796	1,274	1,110	496	462

Measurements were realized during 50th expedition (cruise 1972) of research vessel "Vityaz"

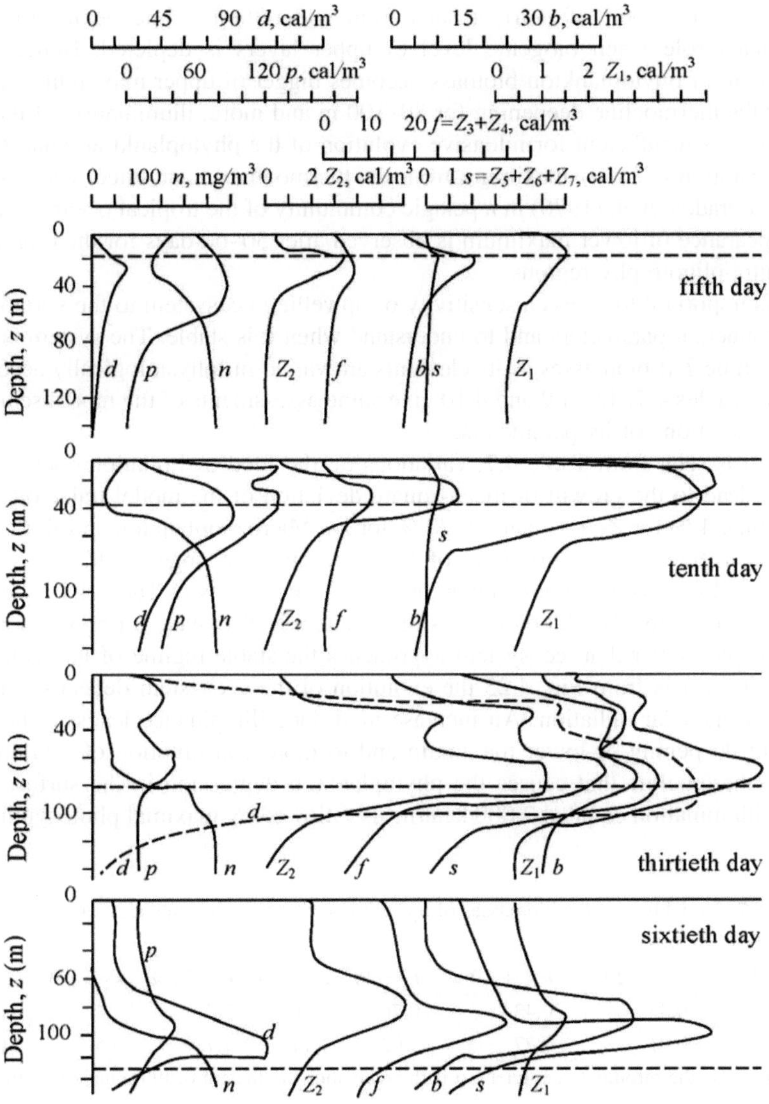

Fig. 4.22 Dynamics of vertical structure of the upwelling ecosystem elements

All of the other living elements have more or less sharply defined and enough narrow maximum by the depth. However, reserve of nutrient salts is almost fully used up in upper layer during 10 days, but nevertheless maximum of phytoplankton biomass is kept at the depth of 10–20 m. Two-maxima structure is formed beginning from 20–30 days of the system evolution when nutrient salts enter across thermocline to upper water layers. In this case, the thermocline depth is 55–75 m.

Vertical transport of nutrient salts from under thermocline begins to play a significant role when biogenic level of upper layers is depleted. Hence, lower maximum of phytoplankton biomass becomes bigger of upper maximum. Later on under the thermocline deepening for 80–100 m and more, illumination of its upper boundary is insufficient for intensive evolution of the phytoplankton what leads to the separation of lower maximum from the thermocline. Such effect was observed by Vinogradov et al. (1970) in a pelagic community of the tropical ocean. Complete disappearance of lower maximum is observed after 50–60 days for the oligotrophy and ultra-oligotrophy regions.

It is important to assess a sensitivity of upwelling ecosystem to the variations of environmental parameters and to understand when it is stable. The system is stable during time *T* if biomasses of its elements are varied in 'physiologically acceptable limits'. Tables 4.7, 4.8, 4.9 and 4.10 give some assessments of the model sensitivity to the variations of its parameters.

As it is seen from Table 4.7, variations on the food assimilation coefficient by ±20 % lead to the growth of mean-square deviation of the model trajectory by no more than 4 % for $Z_3 + Z_4$ and by 70 % for Z_2. Microzooplankton and filtraters are more sensitive to the variations of the energetic exchange coefficient (*T*) and undigested food coefficient (*H*) than predators. The results of Table 4.10 show that repetitive change of initial biomasses of predators is displayed during no more than 20–30 days. After that, ecosystem approaches the stable regime of its evolution.

As it follows from Fig. 4.23 the evolution of the ecosystem depends considerably on the solar radiation. An increase in surface illuminance leads to the insignificant deepening of lower maximum and to more consumption of nutrient salts above thermocline that causes the phytoplankton depression in the surface layer. Daily illumination equaled 2,000 kcal/m^2 is followed by maximal photosynthesis in

Table 4.7 Variability of the biomasses of upwelling ecosystem elements under the change of its parameters

Δ	Model	$u^{-1} - 20$ %	$u^{-1} + 20$ %	$t_z - 2\sigma$	$t_z + 2\sigma$	$\alpha + 50$ %	$k_2 = 0.3$	$k_2 = 0.2$
Δ_1	0.29	0.35	0.43	0.76	3.36	7.9	1.44	2.6
Δ_2	0.75	0.91	1.27	1.02	1.71	1.78	3.5	6.0

Notation k_2 is the turbulent coefficient, α is the relaxation coefficient of illumination by the depth,

$$\Delta_1 = \left[(1/200) \int_0^{200} \left(Z_3(\varphi, \lambda, z, 40) + Z_4(\varphi, \lambda, z, 40) - \overline{Z_3} - \overline{Z_4} \right) dz \right]^{1/2},$$

$$\Delta_2 = \left[(1/200) \int_0^{200} \left(Z_2(\varphi, \lambda, z, 40) - \overline{Z_2} \right) dz \right]^{1/2},$$

$\overline{Z_2}$, $\overline{Z_3}$, and $\overline{Z_4}$ are the observed values of Z_2, Z_3, and Z_4, respectively (Vinogradov et al. 1970); u^{-1} is the food assimilation coefficient; t_z is the energy expenses coefficient.

Table 4.8 Mean-square deviations of biomasses of tropical levels of upwelling ecosystem under variations in the energetic exchange (*T*) and undigested food (*H*)

Δ_i	$T - 20$ %	$T - 10$ %	$T + 10$ %	$T + 20$ %	$H - 20$ %	$H - 10$ %	$H + 10$ %	$H + 20$ %
Δ_2	1.7	1.2	0.6	1.9	2.4	0.7	1.9	2.6
Δ_3	0.3	0.2	0.3	1.0	0.5	0.5	0.9	1.3
Δ_4	0.5	0.4	0.6	1.3	0.9	0.4	0.6	1.5
$\Delta \sum_{i=2}^{4} Z_i$	0.6	0.5	0.6	1.2	0.9	0.6	1.0	1.8
Δ_5	1.4	0.9	0.7	1.2	0.9	0.6	1.2	1.1
Δ_6	0.1	0.1	0.1	0.1	0.1	0.1	0.1	0.2
$\Delta 7$	0.8	0.4	0.1	0.2	0.5	0.2	0.1	0.2
$\Delta \sum_{i=5}^{7} Z_i$	1.8	1.2	1.2	1.3	1.2	0.6	1.1	1.3

Notation $Z_{i,\exp}$ is the modeling result when ecosystem parameter was changed. $\Delta_i = \left\{(1/200) \int_1^{200} \left[Z_i(\phi, \lambda, z, 40) - Z_{i,\exp}(\phi, \lambda, z, 40)\right]\right\}^{1/2}$

Table 4.9 Mean-square deviations of the vertical distributions of ecosystem elements after 30 days under the change of initial concentration of nutrient salts

Δ_i	Change of the nutrient salts, $n(\varphi_0, \lambda_0, z, t_0)$		
	$0.5n(\varphi_0, \lambda_0, z, t_0)$	$0.8n(\varphi_0, \lambda_0, z, t_0)$	$1.2n(\varphi_0, \lambda_0, z, t_0)$
Δ_2	5.0	0.6	0.9
Δ_3	0.4	0.4	0.3
Δ_4	1.3	0.1	0.1
$\Delta\sum_{i=2}^{4} Z_i$	1.5	0.4	0.4
Δ_5	1.1	0.5	0.4
Δ_6	1.1	0.04	0.03
$\Delta 7$	0.1	0.03	0.02
$\Delta\sum_{i=5}^{7} Z_i$	1.2	0.5	0.4

Notation is given in Table 4.8

Table 4.10 Ecosystem structure dynamics (cal/m^3) under the change of initial biomass of small predatory cyclopoids (Z_5), predatory calanoids (Z_6) and large predatory chaetognats and polychaete (Z_7)

Ecosystem element	Fifth day			Thirtieth day		
	Initial biomass, $Z_\Sigma(\varphi_0, \lambda_0, z, t_0)$			Initial biomass, $Z_\Sigma(\varphi_0, \lambda_0, z, t_0)$		
	0.3 cal/m^3	30 cal/m^3	300 cal/m^3	0.3 cal/m^3	30 cal/m^3	300 cal/m^3
Bacterioplankton, b	1,789	2,263	2,263	2,269	2,159	1,995
Phytoplankton, p	3,264	3,462	3,794	3,971	2,086	2,614
Protozoa, Z_1	31	12	3	18	3	2
Microzooplankton, Z_2	178	106	101	109	265	465
Small filtraters, Z_3	70	85	80	80	201	415
Large filtraters, Z_4	67	14	74	145	156	191
Small predatory cyclopoids, Z_5	18	14	9	114	119	53
Predatory calanoids, Z_6	1	25	41	14	25	22
Large predatory chaetognaths and polychaete, Z_7	36	234	183	213	437	821

Notation $Z_\Sigma = Z_5 + Z_6 + Z_7$

upper layer. Variations of surface illumination in range of 2,000–7,000 kcal/m^2 under sufficient concentration of nutrient salts influence the vertical structure of the ecosystem elements with the same forms.

The effects of the nutrient salt concentrations at the initial time of the upwelling ecosystem formation are shown in Table 4.11. Variations of nutrient salts from 100 to 400 mg/m^3 and more, practically are displayed only in the beginning of the ecosystem evolution. The MUE gives a possibility to study different hypothetical scenarios taking into consideration many situations. For example, Fig. 4.24 characterizes the ecosystem dynamics for absence of predatory elements. In this case,

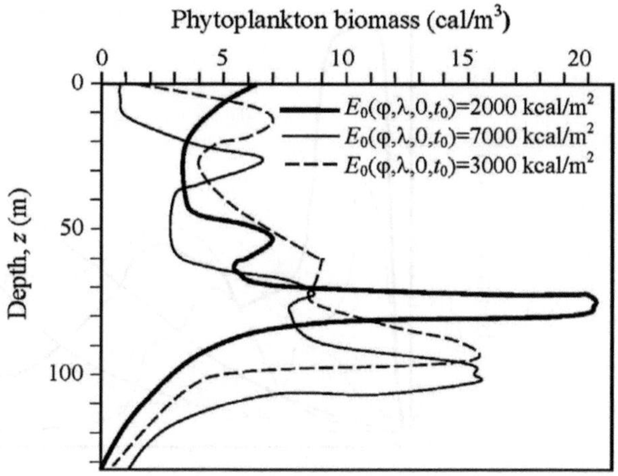

Fig. 4.23 Influence of solar radiation intensity on the vertical structure of upwelling ecosystem after the 30th day of its evolution

Table 4.11 Phytoplankton biomass dynamics as function of initial concentration of nutrients salts

Layer, m	$n(\varphi_0, \lambda_0, z, t_0)$ = 100 mg/m³			$n(\varphi_0, \lambda_0, z, t_0)$ = 400 mg/m³		
	The ecosystem lifetime (days)					
	10	50	100	10	50	100
0–10	0.3	0.1	0	0.3	0.1	0
10–20	31.7	8.7	10.9	37.4	16.1	15.4
20–30	39.6	9.4	10.7	289.7	7.8	10.4
30–40	39.4	9.2	10.6	207.7	7.8	10.4
40–50	36.6	9.7	10.9	295.1	7.9	10.7
50–60	36.7	9.5	10.9	153.7	7.1	10.5
60–70	41.7	11.4	10.9	62.1	18.9	8.8
70–80	13.4	15.7	11.2	15.6	10.3	17.9
80–90	1.9	31.4	17.0	2.1	39.0	17.9
90–100	0.6	3.6	12.8	0.6	4.2	1.4
100–150	3.1	2.1	2.1	3.1	2.2	3.4
150–200	0.9	0.4	0.5	0.8	0.7	0.7

upwelling ecosystem does not achieve the stability and this means that predatory elements are the regulating factor.

Ratio of phytoplankton production to its biomass (*p/b*—coefficient) is, generally, one of the important characteristics of oceanic ecosystem and it defines the ecosystem stability character. For example, Fig. 4.25 shows that upwelling ecosystem evolution for its *p/b*—coefficient always equals to 3. In fact, *p/b*—coefficient changes depends on the time, beginning from 15 in upwelling zone and finishing by 1 in the oligotrophy zone (Vinogradov et al. (1970, 1975).

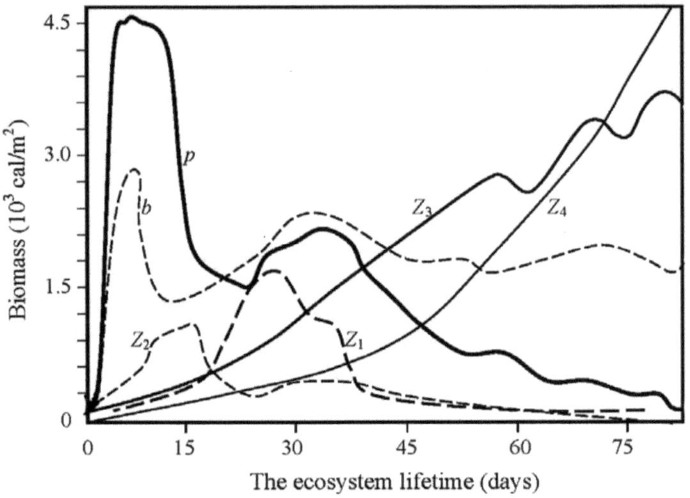

Fig. 4.24 Hypothetical dynamics of the upwelling ecosystem elements under the absence of the predatory elements

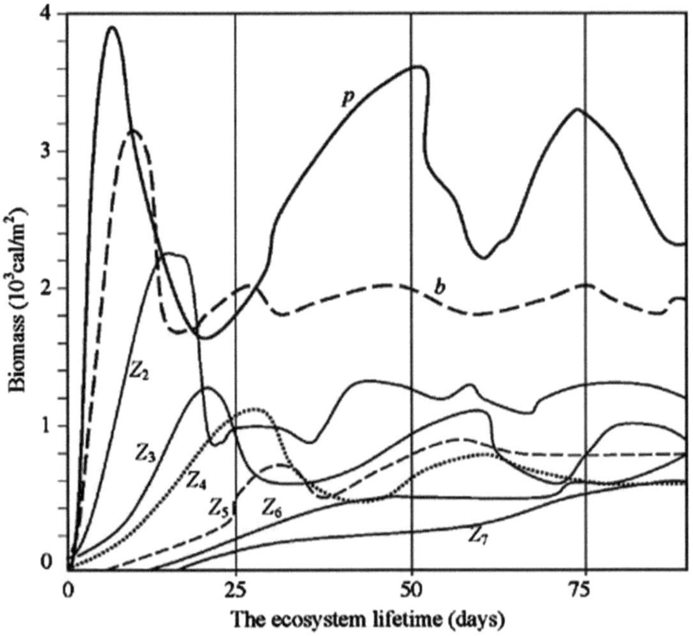

Fig. 4.25 Evolution of the upwelling ecosystem when its *p/b*–coefficient is constant equaled to 3

Simulation experiments analogous to these given above, help to assess the ecosystem characteristics, the measurement of which in real conditions is impossible or inaccurate. One of such characteristic is the vertical water lifting velocity V_z, the existing evaluations of which vary in a range from 10^{-4} to 0.1 cm/s. The MUE experiments show that when $V_z < 10^{-4}$ cm/s upwelling ecosystem evolves during no more than several days, but when $V_z > 10^{-1}$ cm/s it rapidly collapses without stable state.

4.7 Biocomplexity Indicator as a Predictor of the Ecosystem State

The ecosystem complexity is formed in transition of its relations with environment. Almost every ecological system is complex. Classical examples of complex ecosystem include the behaviour of upwelling ecosystems discussed above. For instance, tropical cyclones bringing heavy destructions on the land generate the zones of upwelling in the ocean and create conditions for high productivity of phytoplankton due to the lifting of biogenic elements from depths to surface waters. As a result, the trophic pyramid of the zone of the tropical cyclone's impact becomes complex leading to the growth of biocomplexity of the ecosystem of this zone. In general, humans influence on oceanic ecosystems is growing and as a result, it is necessary to search the indicators for the control of oceanic ecosystem productivity and survivability. Biocomplexity is one of such indicators that help to resolve much of the controversy about the impact that humans are having on health and stability of oceanic ecosystems relative to the effects of natural changes in the ocean. Biocomplexity indicator can help to investigate any environmental system taking into consideration the existing nonlinearities and feedback processes (Krapivin 2008a).

According to Szathmáry and Griesemer (2003), the ratio between living and non-living substances determines the vitality of the natural system and answers the question whether it is alive. The natural system accomplishes the transition between its extreme states due to its changing complexity, whose indicator can serve a forerunner of a critical state. In particular, such transitions can be caused by climate change. The successful search of such indicators depends on our knowledge of the laws of the living world and its evolution. The basic concepts of ecology developed by Beeby and Brennan (2003) suggest the conclusion that living systems of any level respond to an approaching natural catastrophe.

The problem of the interaction of various elements and processes in the global nature-society system (NSS) in common, and in the ocean-atmosphere system (OAS) in particular, has recently attracted the attention of many scientists. Attempts to assess and forecast the dynamics of this interaction have been made in different scientific directions. One of these attempts is the Biocomplexity Program set up in the U.S.A. by the National Science Foundation, which plans to study and understand relationships between the dynamics of complexity of biological, physical, and social systems and trends in changes of the present habitat. Within the framework of this program, the complexity of the system interacting with the environment is connected with phenomena appearing in case of the contact of a living system with the environment on a global scale.

Biocomplexity is a derivative of biological, physical, chemical, social and behavioural interactions of the environmental subsystems, including living organisms and global population. As a matter of fact, the notion of biocomplexity is closely connected with the laws of biospheric functioning as a unity of the forming ecosystems and natural-economic systems of various scales, from local to global. Therefore, a combined formalized description of biological, geochemical, geophysical, and anthropogenic factors and processes taking place at a given level of the spatial-temporal hierarchy of scales is needed in order to define biocomplexity and assess it.

Manifestation of biocomplexity is a characteristic indicator of all systems of the environment connected with life. The elements of this manifestation are studied within the theory of stability and survivability of ecosystems. It should be noted that the formation of biocomplexity includes indicators of the degree of mutual modification of the interacting systems, and it means that biocomplexity should be studied with due regard to both spatial and biological levels of organization. The difficulty lies in the complicated behaviour of the object under study, especially if the human factor is considered as a result of which the number of stress situations in the environment is constantly growing.

Humankind has accumulated knowledge about the environmental systems. The use of this knowledge to study biocomplexity is possible within the framework of synthesizing a global model that reflects the laws of interactions between environmental elements and permits assessing just how efficient it is at constructing different scenarios in the development of human society, based on the actual data of ground and satellite measurements. Consequently, this problem serves the basis of all questions set forth by the Biocomplexity Program.

Studies of the interaction process are aimed—as a rule—at understanding and assessing the consequences of a given interaction. The reliability and accuracy of such assessments depend on criteria taken as the basis in conclusions, examination, and recommendations. At present, there is no agreement on what technique to choose such criteria for lack of a single scientifically substantiated approach to the ecological criteria of economic impacts on the environment. The accuracy of the ecological examination of the existing and planned production systems as well as the representativeness of the data of global geoinformation monitoring depends on the choice of such criteria.

Processes taking place in the environment can be represented as a totality of interactions between its subsystems. Since a human is one of its elements, it is impossible, for example, to simply divide the environment into interacting biosphere and society: everything on Earth is correlated. The problem is to find mechanisms to describe relevant correlations and interdependences which would reliably reflect the environmental dynamics and answer the questions formulated in "Biocomplexity" programme:

1. How does the complexity of biological, physical, and social systems in the environment occur and change?
2. What are the mechanisms for spontaneous development of many events in the environment?
3. How do the environmental systems with the living components including those created by man respond and adapt themselves to stress situations?
4. How do information, energy and matter move within the environmental systems and through their levels of organization?
5. Is it possible to forecast the adaptability of the system and to give prognostic estimates of its changes?
6. How does humankind affect and respond to biocomplexity in natural systems?

Many other, also important and significant questions can be added to this enumeration. For instance, up to what level of complexity should space-borne observation systems be improved so that their information to be of a sufficiently high standard to reliably estimate the state of the environment, even only at the moment of receiving this information? The question of an optimal location of the geoinformation monitoring means at different levels of its established organization is also important. It is also crucial to understand, which forerunners of natural disasters can be measured using the existing space-borne systems?

Environmental biocomplexity is, to some extent, an indicator of interaction between its systems. In this way, one can introduce the unit Ξ of biocomplexity varying with conditions when all interactions in the environment are reduced to the level corresponding to the natural process of evolution. Thus, we obtain an integral indicator of the environmental state, on the whole, with bioavailability, biodiversity and survivability taken into account. This indicator characterizes all kinds of interaction of the environmental components. So, for instance, in biological interactions connected with relations of the type "predator-prey" or "competition for energy resource", there is some minimal level of available food when it becomes practically inaccessible, and the consumer-producer interaction stops. The chemical and physical processes of interaction of environmental elements also depend on the sets of certain critical parameters.

All this emphasizes that biocomplexity refers to categories difficult to measure and express quantitatively. However, we shall try to give formalized quantitative evaluations. Proceeding to quantitative numerical scale Ξ, we postulate that there are relations of the type $\Xi_1 < \Xi_2$, $\Xi_1 > \Xi_2$ or $\Xi_1 \equiv \Xi_2$. In other words, there is always a value of this scale ρ, which determines the level of biocomplexity $\Xi \rightarrow \rho = f(\Xi)$, where f is a transformation of the notion of biocomplexity into a number.

Let us try to find a satisfactory model which will transform the verbal definition of biocomplexity into notions and features that can be formalized and converted. With this aim in view, we select in NSS or OAS m elements—subsystems of the lowest level, whose interaction is determined by binary matrix function $A = \|a_{ij}\|$, where $a_{ij} = 0$ if elements i and j do not interact; $a_{ij} \neq 0$ if elements i and j interact.

a_{ij} can be interpreted as a level of interaction of elements i and j. Then, any point $\xi \in \Xi$ is determined as the sum

$$\xi = \sum_{i=1}^{m} k_i \sum_{j=1}^{n} \beta_j a_{ij}$$

where weight coefficients k_i ($i = 1,\ldots, m$) and β_j ($j = 1,\ldots, n$) determine the meaning of appropriate elements from interacting systems. Nature of these elements depends on their role within the systems.

It is clear that $\xi = \xi(\varphi, \lambda, t)$ where φ and λ are geographic latitude and longitude, respectively, t is the current time. For some territory Ω the indicator of biocomplexity is determined as an average value:

$$\xi_\Omega(t) = (1/\sigma) \int_{(\varphi,\lambda)\in\Omega} \xi(\varphi, \lambda, t) d\varphi d\lambda, \tag{4.51}$$

where σ is the area of the territory Ω.

Thus, the indicator $\xi_\Omega(t)$ plays the role of the integral index of NSS complexity, reflecting the individuality of its structure and behaviour at each time moment t in a space Ω. In accordance with the laws of natural evolution, a decrease (increase) of the parameter ξ_Ω will follow an increase (reduction) of biodiversity and ability of natural-anthropogenic systems to survive. Since a decrease of biodiversity breaks the completeness of biogeochemical cycles and leads to an increase of load on irreversible resources, the binary structure of the matrix A favors resource-depleting technologies, and the vector of energy exchange between NSS subsystems shifts to the state when the level of its survivability lowers.

The NSS consists of elements-subsystems B_i ($i = 1,\ldots, m$), whose interaction is formed in time depending on many factors. The NSS biocomplexity is the sum of structural and dynamic complexities of its elements. In other words, the NSS biocomplexity is formed in the process of interaction of its parts $\{B_i\}$. In due course, the subsystems B_i can change their state and, consequently, the topology of their bonds. The evolutionary mechanism for adapting the subsystems B_i to the environment suggests the hypothesis that each subsystem B_i, independent of its structure, possesses the structure $B_{i,S}$, behavior $B_{i,B}$ and goal $B_{i,G}$. Thus, $B_i = \{B_{i,S}, B_{i,B}, B_{i,G}\}$. The goal $B_{i,G}$ of the subsystem B_i is to achieve a certain preferable state. The expedience of the structure $B_{i,S}$ and the purposefulness of behaviour $B_{i,B}$ of the subsystem B_i is estimated by the efficiency of reaching the goal $B_{i,G}$.

This can be exemplified by the process of migration of nekton elements. Fish migrates towards a maximum gradient of nutritional ration with due regard to possible limitations of the water medium parameters (temperature, salinity, oxygen concentration, pollution, etc.). Hence, the goal $B_{i,G}$ of the elements of nekton is to enlarge their ration, and their behaviour $B_{i,B}$ consists in calculation of the trajectory of shifting in the process of formation of shoals, which can be presented in terms of $Bi,_S$ for each kind of the nekton elements.

Since the interaction of subsystems $\{B_i\}$ is connected with chemical and energy cycles, it is natural to suppose that each subsystem B_i organizes the geochemical and geophysical transformations of matter and energy in order to preserve its stability. The formalized approach to this process rests on the supposition that in the NSS structure an exchange takes place between subsystems B_i of some amount V of spent resources for some amount W of consumed resources ((V,W)-exchange). Natural disasters prevent a group of subsystems B_i of a certain region from accomplishing a profitable (V,W)-exchange during some time. In a general case, $W = W(V, B_i, \{B_k, k \in K\})$, where K is a multitude of the numbers of subsystems contacting with the subsystem B_i. Denoting that $B_K = \{B_k, k \in K\}$, the interaction of the subsystem B_i with its environment B_K will result in the following (V,W)-exchanges:

$$W_{i,0} = \max_{B_i} \min_{B_K} W_i(V_i, B_i, B_K) = W_i(V_i, B_{i,opt}, B_{K,opt});$$

$$W_{K,0} = \max_{B_K} \min_{B_i} W_K(V_K, B_i, B_K) = W_K(V_K, B_{i,opt}, B_{K,opt}).$$

Hence, when determining the levels V_i and V_K, the goal of the subsystem B_i becomes slightly vague. Since there are limiting factors in nature, in this case it is natural to assume the presence of some threshold $V_{i,min}$, at the reaching of which the energy resource of the subsystem stops being spent on getting an external resource. This means that at $V_i \leq V_{i,min}$ the subsystem B_i moves on to the regime of regeneration of the internal resource. In other words, at $V_i \leq V_{i,min}$ the biocomplexity indicator $\xi_\Omega(t)$ decreases due to broken connections of the subsystem B_i with other subsystems. In a general case, V_{min} is the structural step function, so the transition a_{ij} from the state $a_{ij} \neq 0$ to the state $a_{ij} = 0$ does not take place for all j simultaneously. In other cases the interaction of subsystems $\{B_i\}$ can stop depending on various combinations of their parameters. The formalized description of possible situations of the interaction of subsystems $\{B_i\}$ can be carried out using a simulation model of NSS functioning.

The study of the structure and functioning of the ocean's ecosystems becomes one of the most important and rapidly developing directions of marine biology. Its various aspects are being developed in many countries within the International Biological Program (IBP) (Mourant 1964). One of the goals of this study is to obtain a possibility to forecast the system's behavior as a result of changing some of its parameters. However, due to uniqueness and great spatial extent of the World Ocean's ecosystems, it is difficult to quantitatively evaluate all the elements of the system at different moments of its development and in different regions of the ocean, and the more so, to assess the impact of their change on the functioning of the system, on the whole. Therefore, the use of a model approach is one of the ways to solve these problems.

4.8 Oil and Gas Extraction in Seas

The gas extraction zone according to the GIMS technology should have a structure with subsystems enumerated in Table 4.12. The design of the gas extraction zone is connected with the interpretation and detailing of the structural and functional properties of these subsystems. The experience of such investigations shows that the first stage demands the solution of the following tasks in order to design the system of environmental monitoring:

- an elaboration of a science-based and economically covered criterion to assess the atmosphere and water environment state in the area of the gas extraction zone influence;
- an analysis of the hydrophysical structures and aquatory formation to be taken into account by the gas extraction zone monitoring system;
- the preparation of propositions on the structure of the software and hardware for the gas extraction zone monitoring system taking into consideration the real conditions of the initial informational field, the presence of developed technologies for monitoring the data processing, and environment quality standards;
- an elaboration of a version of the project describing in detail the hierarchy of the subsystems and their elements with instructions for the algorithmical and technical infrastructure of the environmental monitoring system;
- preparation of the technical project for the environmental monitoring system with the indication of the final variant chosen, and with the substantiation of the set of devices, the software, the informational network components, the control structure and recommendations for the operation of the system; and
- preparation of the technical documentation for the full set of monitoring system components with detailed calculation of the reliability, functional effectivity and stability, and with recommendations of possible modernization stages reflecting the consideration of the system operational experience.

Table 4.12 List and characteristics of the environmental monitoring system subsystems

Subsystem	Functions of the subsystem
Measuring subsystem	This subsystem includes contact and remote devices to measure synoptic, hydrophysical, atmospheric, ecological, and hydrochemical characteristics of the environment and to register the oil leakage from the means of its transportation
Subsystem of accumulation, analysis and interpretation of measurement data	This subsystem includes technical means for the data transmission to a single informational centre which is equipped with means for data processing, including software for realizing algorithms of environmental state assessment, identification of pollution sources and visualization of measurement results

The monitoring system design depends on the conditions of its functioning. These conditions are connected basically with the hydrophysical properties of the environment. The synoptical situation influences the speed of the degradation processes and of the oil product transport through the food chains. All aquatories can be placed by the hydrophysical parameters into two classes: freezing and nonfreezing. It is evident that from the view point of modeling, the second class is a special case of the first class when the snow and ice cover thickness equals zero. Therefore, the first class is considered in detail. The Shtockman's gas extraction zone situated in the Barents Sea with high variability of the synoptic situation is accompanied during the year by periods of sharp change in the hydrophysical state of the sea environment. In this zone, the air masses brought from the Atlantic Ocean and the Central Arctic aquatories collide and mix up. The monsoon character of the Barents Sea climate is shown by the presence of winds in the low atmospheric layers which blow from the ocean to the land in summer and in the opposite direction during the whole winter. The winter season (October–March) is characterized by strong cyclonic activity (the largest number of days with storms and maximal repetition of strong winds). The ice cover achieves its maximal size in April. During the June–August the period of cyclonic activity is minimal.

The properties mentioned above allow us to make the obvious conclusion that the technical realization of the system to measure the oil pollutants of Shtockman's gas extraction zone demands the realization of devices protected from great physical loads and considerable temperature variations. At the same time, powerful turbulence of the lower atmosphere layer near the water surface in the winter reduces the quantity of measurements points in space which are required to give the necessary information. If the wind speed is more than 5 m s^{-1} and the atmosphere turbulence is high then the atmosphere can be considered as well mixed with a sufficient precision on the territory of 50 × 50 km. In this case, the measurements can be realized at two or three sites of the aquatory. On the contrary, in the summer (June–August), when the intensity of the atmosphere turbulence is decreased, the measurements need to be realized in the area of influence of each oil well. The assessment of dispersion of the pollutant and the calculation of its spatial distribution are realized by means of corresponding models of atmosphere dynamics.

The most important stage of the analysis and design of the measuring system to determine the oil pollution level consists in the description of the sea environment dynamics. It is known that an oil spill spreading on the sea surface is subordinated to the superposition of two processes. The first is the spill drift due to sea current, wind and surface waves. The second is the spill spreading over a calm surface.

For the Barents Sea the second process can be not be taken into account under the synthesis of the gas extraction zone measurement network. This process should be taken into account only in the hydrophysical model to reflect the total range of the hydrophysical processes.

An analysis of many models describing the oil spreading over the sea surface gives the following results. The velocity of oil spill motion equals 60 % of the current speed and 2–4 % of the wind speed. When ice cover exists the wind component is absent.

The character of the gas condensation deposit measuring subsystem securing the assessment of oil concentration into the sea water thickness essentially depends on the possibility of putting algorithms into the information processing unit and of reconstructing the forms of the oil spatial distribution based on data fragmentary in space and episodical in time. The synthesis of an adequate model to describe the oil product kinetics in the sea environment under the arctic conditions is an important procedure.

Calculations show that oil pollution in the arctic climate can persist for several years. Therefore, the danger of its accumulation is great. The transformation processes of oil pollution to other forms under the arctic climate are essentially slowed down in comparison with the analogous processes in warmer waters.

It is known that contributions to the process of sea water self-cleaning equal 0–70 % for evaporation, 15–30 % for photo-oxidation, and 2–7 % for biological utilization. In summer conditions the transformation of 0–60 % of the raw oil mass is realized during 40 days. In winter, these estimations are decreased by approximately three times. In this period, the process of oil products accumulation by the ice enters into the action. The intensity of this process depends on the ice porosity. A perceptible contribution to the aquatory cleaning is introduced by the drifting ice which secures oil removal in quantities estimated at 25 % of its own weight. However, the participation of ice in the process of sea surface cleaning from oil pollution has had negative results. The oil captured by the ice is moved to other aquatories and practically the same volume is returned to the water environment when the ice melts. This exchange between aquatories should be taken into consideration under the monitoring system design, since the oil pollutants can reach a controlled aquatory by such a way. That is why in winter the control of the ice pollution level in adjoining aquatories is needed.

Shtockman's gas condensation deposit zone is characterized by such generalized synoptic and hydrophysical situations which provide the conditions when this zone is covered by ice during 5 months (from February to June). From February to March, the ice drifts to the North or to the North–West. Under these conditions, the drift speed is estimated at 100 km month^{-1}. As a result, the polluted ice reaches the Central Arctic Basin where it thaws and the oil comes into the water.

The Shtockman's gas condensation deposit zone is under the influence of the West Novaya Zemlya current branch that forms the east boundary of the cyclonic cycle in the region of the central depression. The configuration of the formation of the moving water masses suggests that it is necessary to control its southern and eastern boundaries in order to discover the pollutants brought to the Shtockman's gas condensation deposit aquatory from other aquatories.

The oil products are characterized by a multi-component process that expands the situations set of its behaviour in the sea water. Among the most important processes of transformation of the oil products in the sea water it is necessary to indicate the following: dissolving, evaporation, spreading to the deep layers in the form of drops, oxidation, absorption by suspended organic matter, biosedimentation, and bacterial decomposition. The objective laws of formation of all these processes were studied and, therefore, it is necessary to take this into account under the monitoring system design. This will allow the reduction of the standards to the composition of the measuring equipment and to the regime of the measurements realization.

The biodegradation of the oil hydrocarbons is an important process of oil elimination in the sea environment. This process is connected with bacterioplankton, phytoplankton and other sea animals functioning. Taking this process into consideration in the monitoring system is possible at the expense of including the ecosystem model of Shtockman's gas condensation deposit aquatory in the monitoring system software. The conceptual model base is represented in Fig. 4.26.

An elaboration of the Shtockman's gas field ecosystem model demands the consideration of the physical structure of the environment. This structure has a seasonal character. The parametrization of its vertical structure is possible under the consideration of results received earlier by many authors (Legendre and Krapivin 1992; Legendre and Legendre 1998; Emery and Thomson 2001; Vinogradov et al. 2001).

The scheme of Fig. 4.26 represents the typical structure of vertical division of the sea environment. This permits the realization of the scheme of Fig. 4.27.

An ecological monitoring system of practically any anthropogenic-loading object demands the consideration of all components foreseen by the GIMS technology. However, natural conditions of the monitoring system functioning demand the revision of the GIMS formal structure. It is quite evident that in the case considered, the following standards have to be provided:

- the measuring devices should function reliably under low temperatures;
- the measuring network infrastructure should not demand any additional construction;
- the informational network should secure the data concept in the framework of international standards;

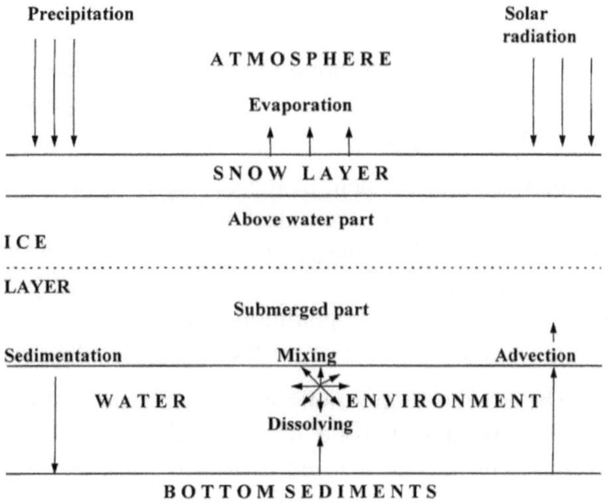

Fig. 4.26 Vertical structure of the ice-covered aquatory

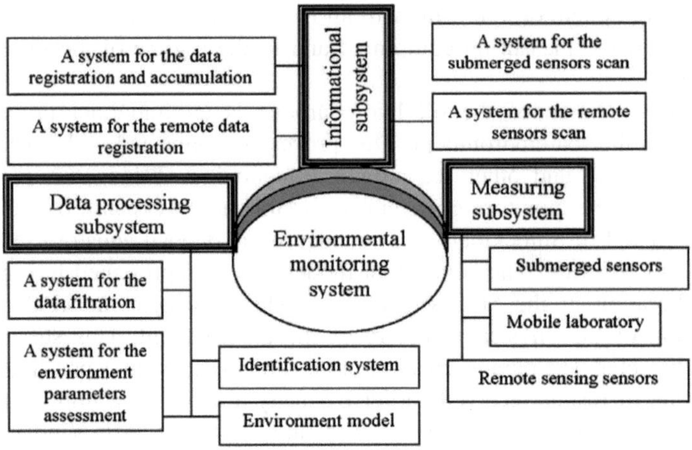

Fig. 4.27 Principal structure of the system of environmental monitoring in reference to the gas condensation deposits zone

- the data processing subsystem should reduce the conditions made by the demands on the measuring subsystem;
- the data presentation subsystem should allow to the gas condensation deposit administration the possibility of multi-shape assessment of the environment state;
- the monitoring system should be combined with other informational systems (regional, national, and international).

The experience of synthesis of GIS and expert systems having the function of nature-protection calls for the necessity of observing the conception and subsequent stages under the realization of the natural environmental monitoring systems (Krapivin and Shutko 2012) to propose the SEM infrastructure in the form represented by Fig. 4.27.

The environmental monitoring system software package must be oriented to the realization of an algorithmical set giving the possibility to realize the following minimal set of functions:

- integrated assessment of the ecological state of the aquatory;
- assessment of the ecological state of the aquatory for each of its components;
- assessment of the integrated ecological state of the aquatory in the spot, on the square area and for the total space;
- identification of the pollution source; and
- predictional assessment of the ecological state of the aquatory.

The answers to questions arising under the realization of these functions and the decision making demand transformation of the measurement data in acceptable format. This can be realized by means of the software package listed below:

- measurement data calibration;
- measurement data filtration;
- technical tools for scanning and blending functions of the measuring devices with the informational unit;
- decision making about the existence of an external situation;
- spatial-temporal coordination of multiple data types;
- spatial interpolation of the measurement data and formation of spatial images;
- reconstruction of the spatial distribution of environmental characteristics on the basis of measurements which are fragmentary over the aquatory and the depth;
- calculation of kinetic characteristics of the pollutants in the sea water under climatic conditions of the sea region;
- calculation of gas contaminants and solid particles content in the atmosphere over the gas condensat deposit aquatory;
- assessment of the ecological situation of the controlling zone in accordance with the given criterion;
- synchronization of informational fluxes and securing of its dumping to the data processing centre in a volley;
- realization of physical-chemical processes in the system atmosphere-sea-gas field;
- formation of short- and long- term forecastings of common ecological situations in the gas field zone;
- identification of sources causing the disturbance of ecological standards on all controlled parameters taking into account the accepted criteria set;
- realization of computer cartography algorithms;
- choice of the program system for database control, the elaboration of data maintenance in the monitoring system informational network and its accumulation to the database;
- software for the user interface;
- software for database maintenance;
- reconstruction of omitted functions under incomplete and inexact information;
- accumulation of knowledge about different specific and typical situations; and
- formation of the operative data under emergency situations.

Algorithmical support of the monitoring system database has a double load. On the one hand, the accumulation of data about the functioning of the ecological system of the gas field zone allows us to increase the reliability of receiving assessments of the ecological situations and to decrease the demands to the measuring subsystem. On the other hand, the monitoring system database can be used as a fragment of the regional or national database.

The monitoring system functioning is secured by the correlation between the measuring subsystem, the regenerated database and the model. According to Aota et al. (1993), continuous monitoring can be realized with sufficient stability only under adaptive algorithmical support. The dynamic correlation of the functioning regimes for all subsystems is represented schematically in Fig. 4.28. According to this scheme, the measuring subsystem can work in discrete regimes together with the model. Detectors of emergency situations work only in the continuous regime.

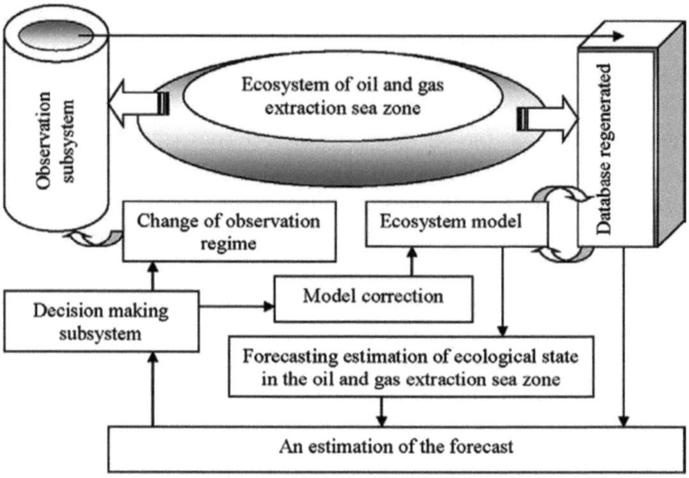

Fig. 4.28 A mutual adaptive scheme using the ecosystem model and measurements for realizing the GIMS technology to thc monitoring system functioning in the sea zone of oil and gas extraction

4.9 Monitoring of the Oil and Gas Extraction Zone in the South-China Sea

Optical sensors and their effectiveness in the framework of the expert system for ecological control of estuary zone (ESECEZ)—as described in detail in Chap. 9—were tested more than once during several Russian/American and Vietnam/Russian hydrophysical expeditions (Hong et al. 1992, 1994; Rochon et al. 1996; Phillips et al. 1997; Mkrtchyan et al. 2004, 2011, 2013; Man et al. 2010). The results of these expeditions spread over many contaminants of anthropogenic origin. Here, only one example of oil pollution is considered. Namely it is a matter of investigations by means of Adaptive Identifier of some water bodies situated on the territory of South Vietnam. The most important aquatory of South Vietnam is the area Ω of South-China Sea bordering on Vung Tau city where there exists the exploration and extraction of oil from the sea bottom. In 1975, the Vietnam Oil and Gas General Directorate was set up. The majority of the oil companies (Vietsov-petro, Petrolimex, PetroVietnam) realize national and international projects concerning the exploration and exploitation of the oil and gas extraction zone.

The quality of sea water in this area mainly depends on the oil and gas extraction zone and the flows of the Mekong and Saigon rivers. The aquatory Ω is restricted on the north by the coast line running in a north-easterly direction, on the west and east by the meridians 106°E and 109°E, respectively, and on the south by the parallel 8.5° north latitude. The water current scheme of this aquatory is considered as given. According to the existing data it can be approximated by a seasonal scheme with the attachment to winter (τ_w), spring (τ_s), summer (τ_u) and fall (τ_a).

On average during the fall and winter the currents are directed along the coast line in a north-easterly direction ($\alpha = 45^o$). During the remaining time the currents have the opposite direction ($\alpha = 225^o$). The speed value is varied insignificantly both in space and in time. Near the coastline this speed equals to 2–6 km h^{-1}. The further one gets from coastline the speed variations are 0.7–4.8 km h^{-1}. It is supposed that depths the h_{ij} are constant on the whole aquatory Ω. The vertical gradient of the water temperature is considered as negligibly small. The synoptic regime during the year is approximated by a binary situation: the rainy season exists during May–October, while from November to April the dry season takes its place.

Table 4.13 consists of some results of oil pollution measurements by means of Adaptive Identifier. Simulation results are given in Figs. 4.29 and 4.30. As follows from Fig. 4.29 under the above suppositions, the average annual distributions of oil hydrocarbons on the aquatory Ω are essentially nonuniform. Taking an average by the seasons eliminated the influence of pollution drift by the currents and revealed the places where oil hydrocarbons always come to the upper sea layer. The degree of oil hydrocarbon accumulation in the surface-water layer varies between the limits of 25–130 mg m^{-2}. This is sometimes higher than the oil hydrocarbon concentration in the open ocean and significantly lower than in the very polluted waters.

Figure 4.30 forecasts the oil hydrocarbon distribution on the whole aquatory Ω. It was supposed that the oil hydrocarbon concentration (O) on the boundary Ω_B is a constant equal to 0.001 mg L^{-1} and is brought into Ω in conformity with the scheme of the currents stored in the ESECEZ database.

Table 4.13 Results of measurement data processing during the field study of oil products content in the water (mg L^{-1}) on the South Vietnam territory

Depth (m)	Sai Gon River			Dong Nai River		
	Day hours			Day hours		
	9	15	18	9	15	18
0.1	0.052	0.101	0.047	0.019	0.019	0.011
0.5	0.061	0.094	0.06	0.012	0.013	0.015
1.0	0.063	0.085	0.053	0.012	0.012	0.013
1.5	0.066	0.103	0.061	0.010	0.010	0.011
2.0	0.054	0.077	0.088	0.009	0.010	0.007
2.5	0.051	0.049	0.064	0.008	0.007	0.006
3.0	0.062	0.068	0.048	0.006	0.008	0.005
4.0	0.043	0.094	0.034	0.004	0.009	0.004
5.0	0.048	0.083	0.026	0.001	0.002	0.006
6.0	0.059	0.074	0.055	0	0	0
7.0	0.062	0.052	0.084	0	0	0
8.0	0.041	0.043	0.054	0	0	0
9.0	0.042	0.049	0.077	0	0	0
10.0	0.037	0.055	0.038	0	0	0

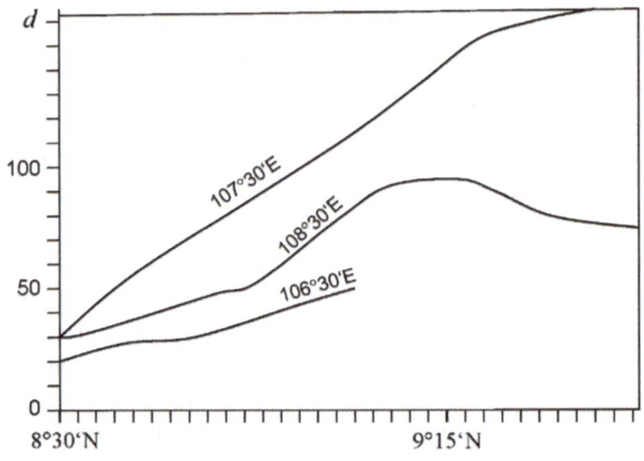

Fig. 4.29 The ratio *d*, of oil hydrocarbon content in the upper sea layer of 1 m depth to the content in the thick water layer as a function of geographical coordinates. The modeling results are averaged by all seasons. It is supposed that the oil hydrocarbon concentration in $\Omega_R U \Omega_P$ and in oil and gas extraction zone is constant and is equal to 0.06 mg L^{-1}

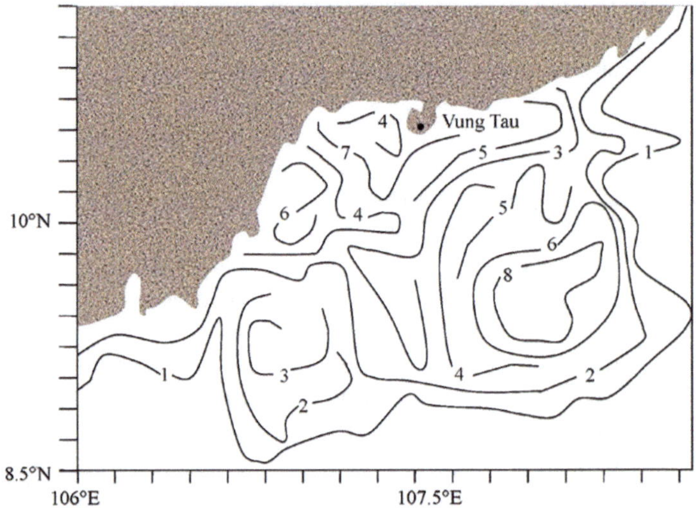

Fig. 4.30 The map-scheme of yearly-averaged distribution of the oil hydrocarbons in the oil and gas extraction zone of Vung Tau city in 2001. These estimations are calculated by means of the model based on the initial data of 1994. The scale step is 0.01 mg L^{-1}

As seen from Fig. 4.30, the stabilized distribution $\int O(\varphi, \lambda, z, t)dz$ is sufficiently well co-ordinated with the direction of the currents. The formation of a field with increased oil hydrocarbon concentration situated near the Vung Tau shelf zone is explained by the supposition that in the oil and gas extraction site the following conditions occur: $O(\varphi, \lambda, z, t) = 0.09\ \mathrm{mg\,L^{-1}}$ in the upper layer, and $O(\varphi, \lambda, z, t) = 0.05\ \mathrm{mg\,L^{-1}}$ in the remaining water depth. Under other suppositions, the structure of the distribution $O(\varphi, \lambda, z, t)$ is does not change. Only while the pollution scales change.

The creation of the monitoring system for the oil and gas extraction zone situated in the South-China Sea foresees the distinguishing of oil pollution sources. That is why Adaptive Identifier was used to estimate the oil pollution of the main rivers of South Vietnam (Saigon and Dong Nai). Apart from this, the task of distinguishing between oil and salt spills on the water surface of reservoirs, where industrial biological production is realized, was solved. Moreover, the control of water quality in the Dongnai river is important for the ecological service of HoChiMinh City.

Table 4.14 and Fig. 4.31 represent some experimental data. It is obvious that the Saigon river is characterized by unstable formations of oil clots moving randomly by depth. The predictability of the Saigon river pollution level equals 60–70 %

Table 4.14 Results of model-based calculations of pollution levels of the Saigon river based on the measurements by Adaptive Identifier

	Water quality factors					
x	NO_3 (mg L^{-1})	P_2O_5 (mg L^{-1})	TDS (mg L^{-1})	TSS (mg L^{-1})	BOD_5 (mgO_2 L^{-1})	Oil (mkg L^{-1})
0	0.80	0.10	50.1	15.5	2.2	0.40
1	0.81	0.12	50.1	15.5	2.4	0.40
2	0.83	0.15	49.9	15.4	2.5	0.41
3	0.92	0.21	50.1	15.5	2.8	0.42
4	0.96	0.24	52.3	15.9	2.9	0.44
5	1.11	0.35	54.2	16.7	3.3	0.45
6	1.26	0.42	60.9	17.2	4.0	0.46
7	1.35	0.44	65.8	19.8	5.1	0.47
8	2.09	0.91	70.3	25.1	6.2	0.53
9	3.41	1.23	77.7	30.2	6.8	0.56
10	4.18	2.42	80.1	37.3	7.7	0.58
11	6.01	3.09	90.4	41.8	8.3	0.60
12	6.12	3.17	91.3	42.7	8.4	0.61
13	5.99	3.18	91.2	41.9	8.3	0.63
14	6.03	3.21	91.1	41.8	8.5	0.65
15M	6.07	3.19	91.1	41.9	8.6	0.65
15E	6.8	7.32	4.01	130.5	58.3	0.92
Δ	17.08	20.45	30.19	28.13	30.08	29.34

The river velocity is equal to 3 km h^{-1}. Measurements are realized during 15 November 1994 from 10:00 to 15:00. *Notation* x is the distance from HoChiNinh river port along the Saigon river downstream (km), Δ is the model error (%), *TDS* (Total Dissolved Solids), *TSS* (Total Suspended Solids), *BOD* (Biological Oxygen Demands), the symbols M and E correspond to the model and experiment, respectively

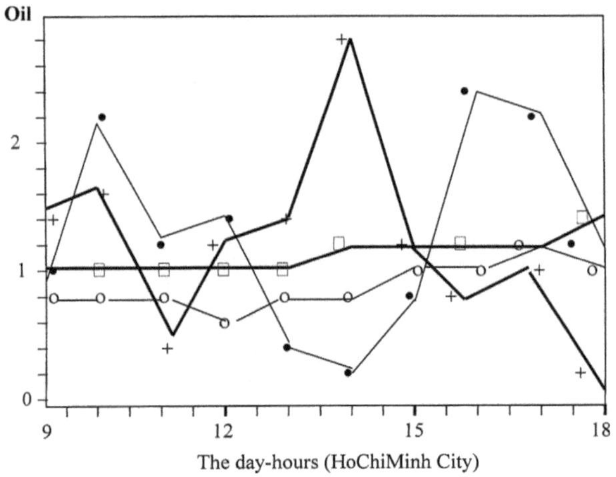

Fig. 4.31 The results of field experiments during Vietnamese/Russian hydrophysical expedition of 1994 when the oil pollution of the Sai Gon and Dong Nai rivers is studied by means of Adaptive Identifier. The results of the measurements are marked by the symbols "•" and «o» for the Saigon river and by the symbols "+" and " " for Dong Nai River at 14 and 15 November, respectively. The scale step is 0.05 mg L^{-1}

during 1 day. It is clear that including in the monitoring system functions for the control of Saigon river pollution demands additional information about the dynamics of the rise and fall of the tides. This information has to be parameterized and stored in the monitoring system knowledge base, or come from the measuring subsystem.

At present, the Dong Nai River has a low pollution level. Measurements data received by Adaptive Identifier shows a high stability of oil pollution distribution in this river. Therefore, the realization of the monitoring system function to control oil pollution of the Dong Nai River can be considered as a future stage.

Finally, Fig. 4.31 represents hourly dynamics of the oil pollution level for both rivers. The measurements by Adaptive Identifier are realized for the Saigon River at its middle in the HoChiMinh city center and for the Dong Nai River in the drinking water station at the distance of 30 m from the waterfront. Comparison of the curves of Fig. 4.31 shows that control of the oil pollution near this station should be realized by means of Adaptive Identifier in the continuous regime and has the possibility to control the water standard. It is more difficult to control the water standard in the Saigon River, where hourly variations of the pollution level oscillate between the limits of 100–300 %. Such oscillations are caused by complex hydrodynamic conditions of the spreading of the pollutants. A combination of high current velocity, rise and fall of the tides and high turbulence causes the formation of pollutant clots which move as closed water volumes.

4.10 Estimation of Oil Hydrocarbon Pollution Parameters in Sea Water

The problem of identification and assessment of oil pollution parameters in the water environment has been studied sufficiently. The existing methods of oil discovery in the water may be divided into two groups, surface film and suspension, corresponding to the state of the oil hydrocarbons. The oil and oil products, by their physical-chemical properties, can form water films, clots, emulsions, and solutions in the sea. The oil films have thicknesses from fractions of a millimetre to several centimetres. All this determines the methods of oil pollution monitoring.

The choice of monitoring method is determined by the oil pollution level. It is known that oil pollution exercises influence on the physical-chemical processes taking place in the sea water. Specifically, the surface strength for oil and oil products is two-four times less than for non-polluted water. The thermal conductivities and thermal capacities of water and oil are distinct and equal respectively to 0.599 W $m^{-1}K^{-1}$ and 4.187 kJ $kg^{-1}K^{-1}$ for water and to 0.15 W $m^{-1}K^{-1}$ and 1.7–2.1 kJ $kg^{-1}K^{-1}$ for oil. These distinctions influence the many processes in the atmosphere-water system. Oil pollution films decrease thermal conductivity and thermal capacity of the upper sea layer. They change the evaporation process decreasing it by 1.5 times or more, they disturb the gas exchange between the atmosphere and sea water, and they change the water temperature. All these effects are used for the design of measuring devices to estimate the oil pollution level (Klyuev 2000; Marhaba et al. 2000). Microwave methods are described by Krapivin (2000). Inhere we consider optical methods.

On a level with the above mentioned effects of sea water oil pollution there exists an optical effect which is determined by the change in sea surface albedo and by the variation of the interior optical properties of sea water. Theoretical approaches to the problem of light spreading in the sea water environment or its reflection from the sea surface are connected with the consideration of various tasks. For example, calculation of the reflection coefficient under different sea environment states is given by Kabanov et al. (2000). Theoretical and experimental studies show the existence of contrasts in the reflecting properties of oil films and a non-polluted water surface (Fig. 4.32). Certainly, these contrasts are functions of many parameters: wavelength, oil film thickness, vision angle, salinity, water roughness, light intensity, and content of other contaminants. As shown by Hong et al. (1994), the use of spectral measurements at a range of 380–700 nm allows one to have a reliable technique for the detection of oil pollution on the water surface.

The Adaptive Identifier is designed on the principle shown in Figs. 9.11 and 9.12 . Adaptive Identifier consists of a multi-channel device registering the light scattered in the water or reflected from the surface in the visible range, of an informational interface and of software. The principle of the functioning of the system consists in the registration in numerical code $n \in [7, 128]$ of spectra $\{T_n(\lambda): \lambda \in [\lambda_i, \lambda_{i+1}], i = 0, 1, \ldots, n - 1\}$ on the basis of which the solution about the existence of oil pollution is taken and the assessment of its parameters is realized. Specifically, for

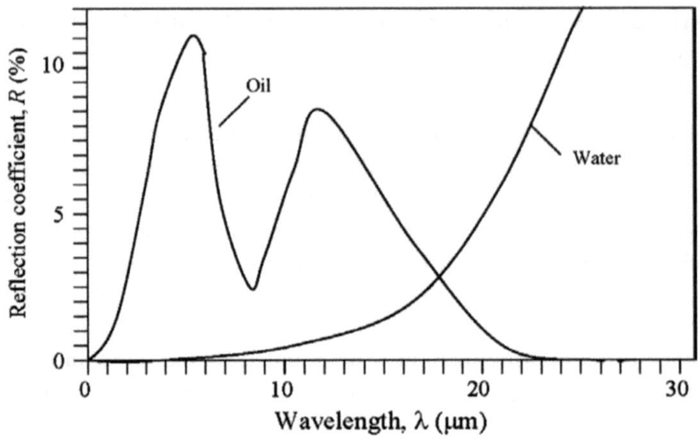

Fig. 4.32 Spectral reflection coefficients for oil and water

$n = 7$ the basic spectral characteristics of the channels are given in Fig. 4.33. It is easy to see that the spectral characteristics of the Adaptive Identifier interference filters have pronounced maxima and insignificant intersections. The maximal conductance of the channels is observed under $\lambda_1 = 398$ nm, $\lambda_2 = 439$ nm, $\lambda_3 = 480$ nm, $\lambda_4 = 510$ nm, $\lambda_5 = 583$ nm, $\lambda_6 = 631$ nm, and $\lambda_7 = 680$ nm. A typical correlation matrix with the addition of the solar radiation spectrum $[\lambda_0, \lambda_n]$ is represented in Table 4.15. The stability of this matrix equals to 5 %.

An increase in the size of n certainly expands the possibilities for oil pollution classification. Adaptive Identifier gives such possibilities. A combined use of real-time spectrometry measurements and data processing methods has been implemented for the first time in an Adaptative Identifier. Its creation becomes possible due to the new approach in the field of polarization optics. An effective elemental basis for polarization optics and a method of discrete modulation of the polarization mode have been developed (Kovalev et al. 1999). The application of simple and highly effective polarization switches and sets of silicon photodiodes with arbitrary access to them has essentially simplified the problem of creating compact and cheap polarization-optical devices: spectro-photometers, spectro-polarimeters, spectro-ellipsometers etc.

The technology of combined use of spectrometry and algorithms of identification and recognition has—for the first time—allowed the creation of a standard integral complex of instrumental, algorithmical, modular and software tools for the collection and processing of data on the aquatic environment with forecasting and decision-making functions. The adaptive identifier of a second version has a set of modifications intended on application under different conditions. A stationary version covers the complete set of technical and algorithmic tools providing for real-time measurements. This version can be used if a power supply line with a voltage of 220 V is available. A handhold field version of the Adaptive Identifier provides two applications. If a notebook computer is available under field

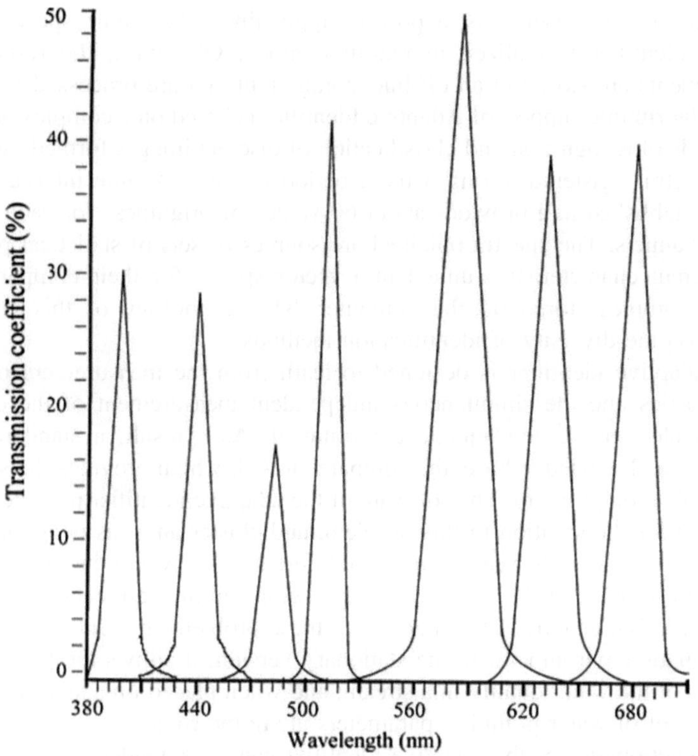

Fig. 4.33 Basic spectral characteristics of the Adaptive Identifier light filters (Aleshin and Klimov 1992)

Table 4.15 Averaged correlation matrix of Adaptive Identifier reflecting the sky emission in December in HoChiMinh City (Hong et al. 1994; Krapivin et al. 2001)

N	*Channel*						
	1	2	3	4	5	6	7
	(380 nm)	(440 nm)	(500 nm)	(560 nm)	(620 nm)	(680 nm)	(740 nm)
1	1.00	−0.11	−0.24	0.03	0.29	−0.06	0.02
2	−0.11	1.00	−0.01	0.13	0.09	0.10	0.06
3	−0.24	−0.01	1.00	0.24	−0.27	−0.03	0.06
4	0.03	0.13	0.24	1.00	0.05	0.09	-10^{-3}
5	0.29	0.09	−0.27	0.05	1.00	−0.07	0.04
6	−0.06	0.10	−0.03	0.09	−0.07	1.00	0.01
7	0.02	0.06	0.06	-10^{-3}	0.04	0.01	1.00

Channels N, are characterized in Fig. 4.33

conditions (in the absence of a power supply line), the whole spectrum of the adaptive identifier is realized in real-time mode. Otherwise, the results of the measurements are stored in an off-line storage unit and are processed later.

The algorithmic support of Adaptive Identifier is based on a complex application of methods of recognition and classification of discrete images formed on the basis of 35 spectra registered during a fixed period of time. A time interval of 1 s is usually established and provides about 60 values of brightness for each of the 35 optical channels. The spectra obtained are sources of sets of statistical parameters and different characteristics united into vector spaces for their comparison with standard samples stored on the computer. The technology of this comparison depends on the diversity of identification methods.

The adaptive identifier is designed to learn from the measurement of spectral characteristics and the simultaneous independent measurement of the content of chemical elements in the aquatic environment. As a result, a standard bank is formed in the knowledge base, the comparison with which provides the solution to the identification problem. The software of the adaptive identifier provides different algorithms for the solution to this problem, and cluster analysis is among them.

The adaptive identifier can be used in different fields where the quality of water should be estimated or the presence of a particular set of chemical elements should be revealed. The adaptive identifier solves these problems by real-time monitoring of the aquatic environment. In the stationary version, it allows the tracking of the dynamics of the water quality in a stream, and when placed on a ship, it allows the measurement of water pollution parameters along the route.

The functionality of the adaptive identifier can be extended by increasing the volume of standards in the knowledge base. The use of a natural light source allows the examination of soils, the identification of oil products on a water surface, the determination of the degree of pollution of atmospheric air and the estimation of the conditions of other objects of the environment, whose spectral images may change. That is why the adaptive identifier is a universal measuring device the use of which in the monitoring system allows one to have operative information, not only on the arising of oil pollution, but also about other pollutants which may appear in the oil and gas extraction zone.

4.11 Adaptive Technologies and Sea Navigation

The GIMS technology is instrument for the increase in effectiveness of decision making under management complex processes and systems subjected to the influence of external factors and being in steady development or moving. One of the GIMS application areas is sea navigation where management of physical system ship-sea is connected with solution of many tasks in operative regime of measurement and identification of environmental characteristics. Remote hydrophysical experiment is more reliable method for the control and management of ship-sea system (Krylov et al. 1981).

The ship moving on the dynamical sea surface is described by the theory of automatic control equations. These equations are extended by hydrophysical correlations reflecting functional dependencies between the ship moving parameters and characteristics of sea currents and turbulent flows. Complete series of equations describing the ship-sea system dynamics has a set of parameters which determine the state of this system. Methods for assessment of these parameters developed in the experimental hydrophysics give in most cases averaged by time and space values that does not correspond to the demand of efficacy and do not give the possibility to have in full measure the complex database for the ship management in real time regime, for example, by means of their realization in the autopilot mode. Specifically, this is important in zones of ports where calculation of safety conditions for ship has to be operative.

Remote monitoring methods of sea surface in the aggregate with data processing algorithms are characterized by responsiveness. The development of these methods with reference to the sea navigation problems is one of the ecoinformatics tasks (Krapivin and Thu 2001). The reception of initial information for the calculation of ship velocity and course basing on the whole data received from navigation systems and on remote sensing data is considered as experimental ecoinformatics task the solution of which is realized by the choice of the most reliable information sources and rational methods for its processing.

Moving or stopping ship is subjected by action of forces formed by complex of processes taking place into the sea arising of which connects with the wind, temperature gradients, water density variations, gravitation waves etc. Trajectory tasks of the ship handling are divided into two classes. The first class has local character when the operative control of physical processes in the ship-sea system is needed. These tasks are arisen in the connection with the necessity to prevent the ship intense heaving or to decrease the yaw angle. In this situation, the measuring remote sensing system is installed on the ship board. The second class of tasks is connected with an assessment of sea state on long distances of ship course under information security for its management systems. Generation of such information is possible by using satellites, flying laboratories or coastal systems.

The task is in the regulator design which is located on the ship provides optimal control of the ship course basing on the receiving information about the sea state and meteorological parameters. Since environmental parameters during the ship moving are changing, the regulator must have adaptability function with time interval that is less time between changes of environmental parameters.

The principal scheme for adaptive regulation of the ship control parameters, using GIMS, consists in the determination of the ship location and assessment of discrepancy in forecasting ship course basing on the data from on-board navigation sensors and radio-navigation satellite systems. Correcting controlling parameters of the ship moving course are calculated on the base of this discrepancy value. The ship course warning interval is selected by the navigator depending on the real situation or is optimized automatically if this decision is not taken.

The synthesis of the ship-sea system model is the base of the GIMS technology use in sea navigation. This model parameterizes the correlations between physical factors of the ship-sea system reflecting cycles and frequencies in the impacts of ship and waves. As a result, the prognosis of the ship course depends on the relative spatial distribution of the ship course and rough sea as well as on different restrictions for the ship moving (other ships, waterway etc.). Adaptive regulator of the ship course provided with this model gives in real time recommendations for the ship course change minimizing the risk of emergency conditions (strike with barrier, beaching etc.).

Figure 4.34 explains the interrelation between physical factors in the ship-sea system. According to this scheme, period and frequency of collision between ship and waves is calculated using the following formulas:

$$T_e = L_\omega/[V_\omega - V\cos\chi], \quad \omega_l = 2\pi[V_\omega - V\cos\chi]/L_\omega, \tag{4.52}$$

where T_e is the period of the ship collision with wave components, ω_l is the angle frequency of collision ($=2\pi/T_l$), L_ω is the wavelength, V is the ship speed, V_ω is the wave velocity, χ is the corner between the ship speed vector and the direction of waves.

From Fig. 4.34 and Eq. (4.52) it is easy to see that if sea wave is moving stern of ship ($90° < \chi < 270°$, $\cos\chi < 0$), the frequency of collisions is more than in other cases ($0° \le \chi \le 90°$, $270° \le \chi \le 360°$). Ship moving before the wave can be associated with small values of $V_\omega - V\cos\chi$ and, therefore, frequency of collisions will be also low in contrast to other situations. Hence, physical factors of the ship-sea system dynamics are essentially diverse into two zones of the angle χ changes. The knowledge of these factors gives the possibility to control a ship management.

In the case of an astern wave, according to (4.52) there are three situations:

- If $V_\omega > V\cos\chi$ there is an overtaking wave. In this case, waves overhaul the ship and the frequency of collisions between the ship and waves is low.
- If $V_\omega = V\cos\chi$ there is a semi-static situation. In this case, the frequency of collisions between the ship and waves is near zero and if at the time $t = 0$ the ship is situated in stable position regarding to waves, this state is retained at any time $t = t_1$.

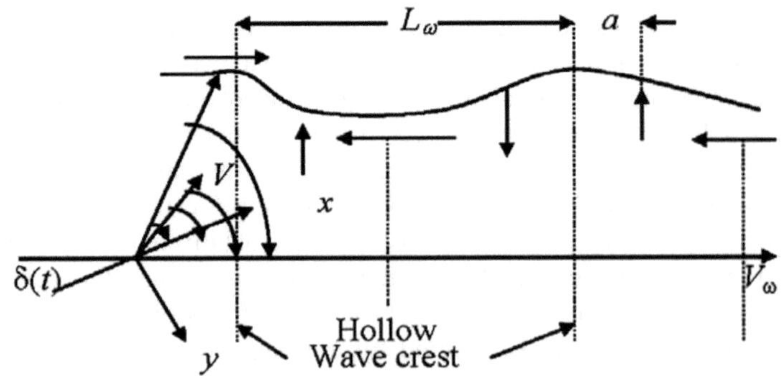

Fig. 4.34 Diagram for the interaction between physical factors in the sea-ship system

- If $V_\omega < V\cos\chi$ there is an escaping wave. This situation corresponds to the case when the frequency of collisions as in the case $V_\omega > V\cos\chi$ can be lower but opposite by sign, and ship will overhaul the wave.

For example, for a ship having length 152.4 m ($0.5 < L_\omega/L < 1.5$) with ratio height/length = 1/48 under $\chi = 30°$ and = 60°, $V = 9.3$ m/s when $V_\omega > V\cos\chi$. In this case, maximal lateral movement amplitudes can reach about 2.5°. Amplitude will be growing linearly with an increase of relation wave height/wavelength, while frequency will be changing intensively as $L^{1/2}$ and collision amplitude will be increasing with growth of the ship velocity but will sharply decrease if the wavelength becomes lower.

Practical navigation takes into account the physical forces which stimulate the ship drift using an approximate model (Matusevich 1956)

$$\alpha = K_\alpha \left(\frac{W}{V}\right)^2 \sin q,$$

where K_α is the shift coefficient, q is the course angle of apparent wind, W is the wind speed. The coefficient K_α is determined from an experimental formula:

$$K_\alpha = \sum p_i K_{\alpha i} / \sum p_i,$$

where $K_{\alpha i}$ is a single value of shift coefficient defined for shift angle α_i in concrete conditions

$$K_{\alpha i} = \alpha_i \left(\frac{W}{V}\right)_i^{-2} \sin^{-1} q,$$

where p_i is weight of single shift angle.

Problems of the ship management under its approach to the port and the calculation of powers affecting it under berthing in the port have specific place in navigation. The solution of these problems demands the study of wave energy formation causes that is connected with different sources of water mass fluctuation, due to the variety in physical influence of sea environment on the ship. Methods based on the wave energy balance equations are used to solve the tasks arising here.

Elevation $\zeta(x, y, t)$ of free sea surface is described by the following function (Krylov 1966)

$$\zeta(x, y, t) = a_0 u(x, y) \mathrm{ch}(kH) \exp(-i\omega t), \tag{4.53}$$

where t is the time, a_0 is the wave amplitude, $k = 2\pi/\lambda$ is the wave number, H is the sea depth, x and y are the rectangular coordinates, λ and ω is the length and frequency of incoming waves, respectively.

The following approximation is valid for plain progressive waves spreading in direction x without barriers

$$u(x, y) = A(x, y)\exp(ikx) \tag{4.54}$$

The function (4.54) is solution of equations

$$\Delta u + k^2 u = 0, \tag{4.55}$$

$$\frac{\partial A}{\partial x} = \frac{i}{2k}\left(\frac{\partial^2 A}{\partial x^2} + \frac{\partial^2 A}{\partial y^2}\right)$$

The diffraction coefficient is

$$K_d = |A(x, y)|$$

when $\partial^2 A/\partial x^2 \gg \partial_2 A/\partial y^2$ the following relation takes place

$$\frac{\partial A}{\partial r} = \frac{i}{2k}\frac{\partial^2 A}{\partial l^2} - \frac{A}{2r},$$

where $l = r\theta$ and relation $A/(2r)$ reflects wave front crookedness, θ and r are polar coordinates.

Seiche oscillation takes place in the port zone, due to the penetration of low-frequency oscillations from the open sea to the port zone. This oscillation stimulates a very dangerous phenomenon called as harbor oscillation that leads to disagreeable consequences for the ship. Cycles of harbor oscillations range from 0.5 s to 4 min with an amplitude of horizontal oscillations equal to 4 m. According to the theory of formation of low-frequency oscillations two systems of wind-induced waves exist in the sea. One of them is a resonance system with phase speed equal to the wind speed. The second system is under-resonance with smaller phase speed that steadily increases. Two maxima correspond to these systems in the energetic spectrum of wind-induced sea roughness. The origin of such a situation is impossible under long strong stable wind above vast aquatory (linear range about 100 km). The calculation of low-frequency oscillations in coastal zone can be realized using the following formula (Krylov 1966):

$$h_{lf} = 2\pi^{3/2}\frac{h^2}{g\tau^2}f_2\left(\frac{H}{g\tau^2}\right), \tag{4.56}$$

where H is the aquatory depth; g is the acceleration of gravity; h is the height of wind-induced wave; τ is the time period of wind-induces wave. Function f_2 is defined empirically and its view is given in Fig. 4.35 (Krylov 1981).

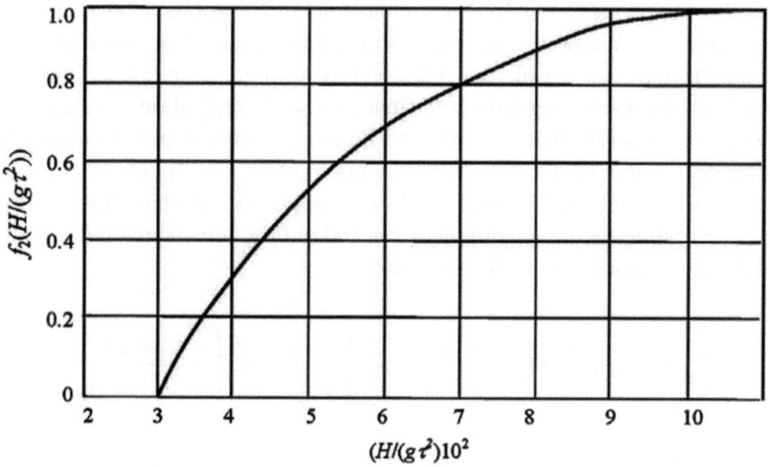

Fig. 4.35 Function f_2 in (4.56)

A physical model of functional correlation between elements of low-frequency oscillations and wind-generated waves was developed by Krylov (2001):

$$h_{lf} = \alpha \frac{h^2}{g\tau^2} \Psi\left(\frac{H}{g\tau^2}\right), \; \tau_{lf} = \beta\tau, \tag{4.57}$$

where Ψ is an empirical function which satisfies the following conditions: $\Psi \to 1$ when $H/(g\tau^2) \to \infty$ and $\tau = 6.2$ s for $\tau_{lf} = 62$ s and $\beta = 10°$ then $\Psi = 0.16\,\text{s}^{-1}$ if $\alpha = 11.5$. Characteristic Ψ is defined for every basin. As example, Table 4.16 gives the estimates of parameters from (4.57) for South-China Sea.

The physical effect of sea oscillation on the ship is shown in the form of its transverse-horizontal rocking with oscillation amplitude of the ship mass center:

$$A_\eta = \text{k}_\eta a = 0.5h\text{k}_\eta \text{ch}[2\pi(H - z)/\lambda]/\text{sh}[2\pi H/\lambda],$$

where a is the amplitude of transverse-horizontal oscillations of water particles on the level of the ship mass centre; H is the sea depth; z is the elevation of the ship

Table 4.16 Elements of wind-generated waves and low-frequency oscillations for Vietnamese coastal waters of South-China Sea

Aquatory	h (m)	h_{lf} (m)	τ (s)	τ_{lf} (s)	F	α	β
Tam Zhang	1.89	0.12	5.9	61	0.44	11.3	9.9
Nai	1.71	0.15	5.2	63	0.65	12.1	11.0
Cam Ranh Bay	1.44	0.13	4.6	62	0.76	11.4	13.0
Van Phong Port	1.87	0.14	5.1	60	0.64	12.3	12.0
Thi Nai Port	1.66	0.12	4.7	65	0.75	11.5	14.0
Nuoc Ngot Lagoon	1.57	0.13	4.8	66	0.73	11.2	13.8

Notation $F = 2\pi H g^{-1}\tau^{-2}$

mass centre above the water-line; κ_η is the reducing coefficient of transverse-horizontal oscillation; h and λ is the height and length of harmonic wave, respectively.

The ship moving on sea surface is defined by numerous factors which were listed above as well by Earth magnetic field, inaccuracy of navigation devices and errors in the measurements of distance up to the lightships. This implies that ship moving has stochastic character. Let the ship start his moving from point O of rectangular system of coordinates $\varphi O\lambda$ at time $t = 0$. It is supposed that ship shift on the axis φ is forced by regular factor shifting it by $\Delta\varphi \geq 0$ at time unit. Ship shift $\Delta\eta$ on axis λ has stochastic character with distribution

$$P(\Delta\eta < \Delta\lambda) = \Phi\left(\frac{a - \Delta\lambda}{\sigma}\right) = \frac{1}{\sqrt{2\pi}} \int_{-\infty}^{\Delta\lambda} \exp\left\{-\frac{(a - z)^2}{\sqrt{2\sigma}}\right\} dz,$$

where a and σ^2 are the average of distribution and dispersion of $\Delta\eta$, respectively.

As a result, the ship will be in coordinates (φ_T, η) at time $t = T$ under the final goal to achieve a point $(\varphi_T, 0)$. It is supposed that axis φ and axis t are identical. Time T for the ship moving from point O to point (φ_T, η) equals to $T = \varphi_T/\Delta\varphi$.

Let us divide the ship path into intervals from which it passes during time periods $t_1, \ldots, t_{n+1}$ $(t_1 + \cdots + t_{n+1} = T)$ and it is supposed that $a = 0$ when $t = 0$. Then at the moment t_1 the ship will be to the position (φ_{t1}, η_1) where $(|\eta_1| \leq u_p \sigma \sqrt{t_1}$, u_p is the p-fractile of normal distribution). At this moment, the law of the ship coordinate change is

$$P(\Delta\eta < \Delta\lambda) = \Phi\left(\frac{a_1 - \Delta\lambda}{\sigma}\right),$$

where $a_1 = \eta_1/(t_1 - T)$.

The ship has inertia factor and its mobility is restricted by $|a| \leq h$ where h is some constant particularly equal to the tangent of the turn corner that is permissible for this ship type. We have

$$P\{|\eta_1| < u_p \sigma \sqrt{t_1}\} = p$$

and as a result

$$u_p \sigma \sqrt{t_1} \leq h t_2^*,$$

where $t_2^* = T - t_1$.

Then dispersion minimum of the ship deviation from position $(\varphi_T, 0)$ at the moment T under $n = 1$ is achieved when the management is realized at the moment

$$t_1 = T + 0.5y^2 - y\sqrt{T + 0.25y^2}, \quad y = u_p \sigma / h$$

According to this, the ship will have position η with average $E\eta = \eta_1 + E[(T - t_1)\Delta\eta] = \eta_1 - \eta_1(T - t_1)/(T - t_1) = 0$ and dispersion $D\eta = (T - t_1)\sigma^2$. The case of arbitrary of management actions demands the fragmentation of controlling values of $|a|$ by partial intervals $[h_i, \quad h_{i+1}]$. Then $u_p\sigma\sqrt{t_i}/[T - t_1 - \cdots - t_i] \leq h_i - h_{i-1}(i = 1,\ldots,n)$, $h_0 = 0, h_n = h$. As a result, we have:

$$D\eta = \sigma_n^2 = \sigma^2 t_{n+1} = \sigma^2\left(T - \sum_{j=1}^{n} t_j\right)$$

To determine the optimal management moments $\{t_i\}$ as functions of T, it is necessary to form relevant equations. It is evident that value t_j does not depend on the number of subsequent control, but is fully defined by the interval [0, h] segmentation. So, supposing that moments $\{t_i\}$ are defined and fixed, we have

$$t_1 = T + 0.5y_1^2 - y_1\sqrt{T + 0.25y_1^2},$$
$$t_{i+1} = T - T_i + r_i\left[r_i - 2\sqrt{r_i^2 + T - T_i}\right],$$

where $r_i = y_i/[2(h_{i+1}/h_i - 1)]$, $y_i = u_p\sigma/h_i(i = 1,\ldots,n-1)$; $t_{n+1} = T - T_n, T_i = t_1 + \cdots + t_i$.

Above considered the case of the random deviations of the ship course as considered above, supports the choice of the parameter a which provides the ship arrival to an in advance defined position after each control. But another case is possible when a is defined on every management stage. In this case, the ship arrivals to the following management moment with null ordinate. If i.e. $h_i = h$, the ship hunting is not controlled, and

$$u_p\sigma\sqrt{t_i} \leq ht_{i+1}\ (i = 1,\ldots,n-1)$$

From this follows

$$t_{n-j} = \mu^s t_{n+1}^l, \quad s = 2(2^j - 1), \quad l = 2^j, \quad \mu = 1/y,$$
$$T = t_{n+1} + y^2\sum_{j=1}^{n}\left(\mu^2 t_{n+1}\right)^l$$

These equations fully and definitely determine the decomposition of interval (0, T).

Finally, there is a third controlling procedure when the ship management is realized only if it goes out of some ε-stripe. In this case, the ship starting at moment $t = 0$ from point O with parameters $(0, \sigma^2)$ reaches some position (T,η) with $E\eta = 0$ and $D\eta = T\sigma^2$ at the moment T. The management of the ship course by changing the parameter a when the ship goes out of ε-stripe is realized at the moments distributed with Wald law

$$P(\tau < y) = W_c(y) = \int_0^y w_c(z)dz$$

where $\tau = \nu/E\nu$, ν is number of intersections of ε-stripe,

$$w_c(z) = \sqrt{c/2\pi z}^{-3/2} \exp\left[-c(z + z^{-1} - 2)/2\right], \quad c = (E\nu)^2/D\nu$$

In this case, $D\eta = \sigma^2(T - t_\mu) > \varepsilon\sigma^2/h$, where $t_\mu \leq T - \varepsilon/h$.

This reasoning gives the possibility to propose adaptive regulator of physical factors in real time (Fig. 4.36). There exists an evident restriction on the adaptation time τ_a: $\tau_a \leq \max_i |t_i - t_{i-1}|$

Adaptive regulator improves navigation parameters realizing the following operations:

- measurement of navigation parameters to determine the ship position;
- processing and analysis of navigation information to determine the ship moving elements;
- calculation of forecasting drifts and measured elements of the ship course;
- decision making based on the criterion of the ship drift assessment
- control of the ship course elements corresponding to the on-board devices and observations; and
- prognosis of the ship position for the next time interval.

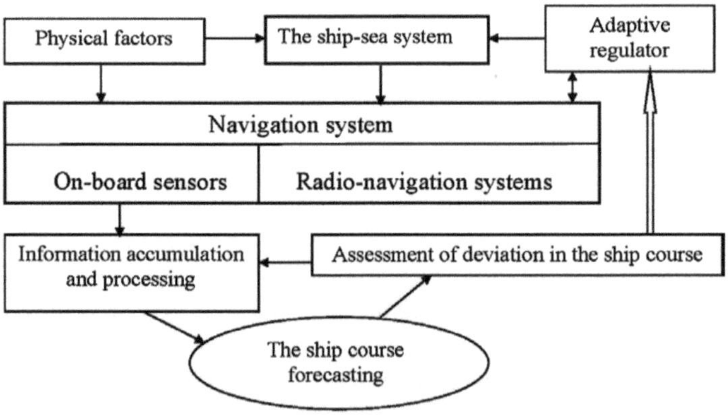

Fig. 4.36 Principal scheme for adaptive management of the ship course parameters

4.12 Conclusion

The World Ocean is among the biggest reservoirs for life on the Earth. Therefore, it is important to learn about the issues—pollution, overfishing, and global warming chief among them—and the possible solutions for protecting the ocean. There are many problems a solution to which should be the subject of present society:

- Overexploitation of the World Ocean resources reaches unacceptable levels. For example, the biomass of global fish stocks has been allowed to run down to the point where it is no longer possible to catch the amount of fish that could be caught.
- The oceans are rising and getting warmer faster than predicted. Global warming is creating a climate time bomb by storing enormous amounts of heat in the waters of the World Ocean. Warmer waters change trophic pyramids and global circulation processes in the atmosphere-land-ocean system. The destruction of the ocean's most important predators has significant consequences that ripple down the basic food chains.
- Ocean acidification is destined to be one of the biggest issues humanity has ever over faced. Ocean acidification refers to the reduction in the pH of the ocean over an extended period time, caused preliminary by the uptake of carbon dioxide from the atmosphere. Research and understanding of this progressive process is important for fishing and biological processes. The average global surface ocean pH has already fallen from a pre-industrial value of 8.2–8.1, corresponding to an increase in acidity of about 30 %. Values of 7.8–7.9 are expected by 2100, representing a doubling of acidity.
- One of the critical ocean issues is coral reefs. Live coral are colonies of small animals embedded in calcium carbonate shells. Coral reefs are some of the world's most productive ecosystems, providing complex and varied marine habitats that support a wide range of other organisms. Coral reefs are dying around the world. In particular, coral mining, agricultural and urban runoff, pollution (organic and inorganic), overfishing, blast fishing, disease, and the digging of canals and assess into islands and bays are localized threats to coral ecosystems. Protection and restoration of coral reef environments is very prominent for the World Ocean management.
- One of the exacerbated dilemmas of the oceans is offshore drilling for the oil production. When oil is extracted from the ocean floor, other chemicals like mercury, arsenic, and lead come up with it. Furthermore, the infrastructure projects to transport the oil often create worse problems, eroding the coastline. The mercury is absorbed by organisms on the bottom of the food chain. Bigger fishes eat smaller fishes. The consumption of fish is by far the most significant source of ingestion-related mercury exposure in humans and animals.
- Ocean dead zones are widened. Dead zones are hypoxic (low-oxygen) areas in the world's oceans caused by extensive nutrient pollution (nitrogen and phosphorous). Low oxygen levels recorded along numerous coastal waters where intensive economics is developed. Today, more than 500 dead zones are growing around the globe.

Ocean health protection has not alternative. A focus on how to protect the World Ocean is important in order to solve the climate warming problem. Ecoinformatics methods can help to solve the tasks arising here. It is impermissible for humanity to realize large-scale anthropogenic projects without their preliminary consideration by means of simulation experiments. The GIMS technology helps to realize such simulation experiments taking into consideration different hypothetical scenarios.

Chapter 5
Ecoinformatics Problems of Global Climate Change

5.1 Common Questions of Global Climate Change

Climate is the average pattern of weather over the long term. The earth's climate has warmed and cooled for millions of years, since long before people appeared on the scene. The unprecedented growth of interest in the problems of climate observed during the last decades has stimulated both scientific and applied developments which have considerably advanced our understanding of the causes of present climate changes, the laws of paleoclimate. They have also advanced the development of scenarios for possible future changes of climate, though it must be stressed that scenarios are not predictions and that their potential for developing predictions must be assessed as doubtful. Unfortunately the growing interest in climate is, in part, explained by the important role played by various speculative exaggerations and apocalyptic 'predictions' (e.g., complete melting of the arctic sea ice in the first half of this century).

The problems of climate change, formulated as anthropogenic global warming, have therefore, become a focus of geopolitics and global environmental policy. Presidents and prime ministers of several countries now discuss whether the Kyoto Protocol (KP) should be considered as a scientifically justified document. Confusion is caused, in particular, by the lack of sufficiently clear and agreed terminology. Ignoring the very complicated notion of climate itself (which needs a separate discussion), one should remember that climate change was defined as being anthropogenically induced. One of the main unresolved problems is the absence of convincing quantitative estimates of the contribution of anthropogenic factors to the formation of global climate, though there can be no doubt that anthropogenic forcing of climate does exist.

Some international documents containing analyses of current ideas of climate refer to a consensus with respect to scientific conclusions contained in these documents. This wrongly assumes that the development of science is determined not by different views and relevant discussions, but by a general agreement and even

V.F. Krapivin et al., *New Ecoinformatics Tools in Environmental Science*,
Environmental Earth Sciences, DOI 10.1007/978-3-319-13978-4_5

voting. Apart from the question of definitions, the issue of uncertain conceptual estimates concerning various aspects of the climate problem remains of importance. In particular, this refers to the main conclusion of the summary of the IPCC reports (IPCC 2001, 2005, 2007) which claims that: "... An increasing body of observations gives a collective picture of a warming world and most of the observed warming over the last 50 years is likely to have been due to human activities".

It is a pity that in the recent (2003) article in the British newspaper *The Guardian* the former chairman of the IPCC Working Group 1 Professor J. Houghton (Houghton 2003; Houghton et al. 2001) compared the threat of anthropogenic climate changes with weapons of mass destruction, and accused the USA of their refusal to support the concept of global warming and KP as being the basic cause of this threat. No matter how paradoxical it may be, such a statement has been made against a background of an increasing understanding of imperfections in present global climate models and the absence of their adequate verification, which makes the predictions on the basis of numerical modeling no more than conditional scenarios. As for the USA, their huge efforts to support climate studies should be encouraged. These efforts have a special emphasis on the improvement of the observing system and on developments in the field of climate problems in general (Kondratyev and Krapivin 2001a; McBrean 1998; Randall et al. 2007).

The statement of the Intergovernmental Group G-8 published on 2 July 2003 (G8 2003) has justly emphasized that in the years to come efforts will be concentrated on three directions:

(1) co-ordination of global observations strategies;
(2) provision of a pure, more stable and efficient application of energy; and
(3) provision of sustainable agricultural production and preservation of biodiversity.

The Earth's climate system has indeed changed markedly since the industrial revolution, with some changes being of anthropogenic origin. The consequences of climate change do present a serious challenge to the policy-makers responsible for the environmental policy. This alone makes the acquisition of objective information on climate change, of its impact and possible responses, most urgent. With this aim in mind, the World Meteorological Organization (WMO) and the UN Environmental Programme in 1988 organized the Intergovernmental Panel on Climate Change (IPCC) divided into three working groups (WG) with spheres of responsibility for the:

(1) scientific aspects of climate and its change (WG-I);
(2) effects on and adaptation to climate (WG-II);
(3) analysis of possibilities to limit (mitigate) climate changes (WG-III).

The IPCC has so far prepared series of detailed reports as well as several special reports and technical papers (Pachauri and Reisinger 2007; Griggs and Noguer 2002; Alcamo et al. 1994; WMO 2000; Woodcock 1999a, b). The fourth synthesis IPCC-2007 Report provides integrated view of climate change paying attention to the following six themes:

- role of information on observed changes in climate and the effects of past climate change in natural the NSS;
- causes of change considering both natural and anthropogenic drivers of climate change;
- future climate change and its impacts;
- adaptation and mitigation options and responses related with sustainable development;
- the long-term perspectives and analyses scientific, technical and socio-economic aspects relevant to adaptation and mitigation; and
- robust findings and key uncertainties.

Analysis of the observational data as contained in TAR led to the conclusion that global climate change is taking place. The Reports IPCC (2001, 2005, 2007) give a detailed review of the observational data of the spatial-temporal variability of the concentrations of various GHGs and aerosols in the atmosphere. The adequacy of numerical models was discussed from the viewpoint of the climate-forming factors and the usefulness of models to predict climate change in the future. The main conclusion about anthropogenic impacts on climate was that "there is new and stronger evidence that most of the warming observed during the last 50 years has been determined by human activity". According to all prognostic estimates considered in TAR, both surface air temperature (SAT) increase and sea level rise should take place during the 21st century.

When characterizing the IPCC data for the empirical diagnostics of climate, Folland et al. (2002) drew attention to the uncertainty of the definitions of some basic concepts. According to IPCC terminology, climate changes are statistically substantial variations of an average state or its variability, whose stability is preserved for long time periods (for decades and longer). Climate changes can be natural in origin (connected both with internal processes and external impacts) and/or may be determined by anthropogenic factors, such as changes in the atmospheric composition or land use. This definition differs from that suggested in the Framework Climate Change Convention (FCCC), where climate changes are only of anthropogenic origin, in contrast to natural climate change. In accordance with the IPCC terminology, climatic variability means variations of the average state and other statistical characteristics (MSD, repeatability of extreme events, etc.) of climate on every temporal and spatial scale, beyond individual weather phenomena. Hence, climate variability can be both of natural (due to internal processes and external forcings) and anthropogenic origin: possess both of internal and external variability. As Folland et al. (2002) have noted, seven key questions are most important for the diagnostics of observed changes and the climate variability:

(1) How significant is climate warming?
(2) Is currently observed warming significant?
(3) How rapidly has the climate changed in the distant past?
(4) Have precipitation and atmospheric water content changed?
(5) Do changes in the general circulation of the atmosphere and ocean take place?

(6) Have climate variability and climate extremes changed?
(7) Are observed trends internally coordinated?

In order to answer the above questions, the reliability of observational data is fundamental. Adequate empirical diagnostics of climate remain impossible without such observational data. Yet the information concerning numerous meteorological parameters, so very important for documentation, detection and attribution of climate change, remains inadequate for the drawing of reliable conclusions. This is especially true for the global trends of those parameters (e.g., precipitation), which are characterized by a great regional variability.

Folland et al. (2002) have answered some of the questions above. A comparison of the secular change of global average annual sea surface temperature (SST), land surface air temperature (LSAT), and nocturnal marine air temperature (NMAT) over the ocean for the period 1861–2000 on the whole revealed some similarity, though the warming in the 1980s from LSAT data turned out to be stronger, and the NMAT data showed a moderate cooling at the end of the 19th century not demonstrated by SST data. The global temperature trend can be interpreted cautiously as equivalent linear warming over 140 years, constituting 0.61 °C at a 95 % confidence level with an uncertainty range of ±0.16 °C. Later on, in 1901 a warming by 0.57 °C took place with an uncertainty range of ±0.17 °C. These estimates suggest that at the beginning from the end of the 19th century, an average global warming by 0.6 °C took place, with the interval of estimates corresponding to a 95 % confidence level equal to 0.4–0.8 °C.

The spatial structure of the temperature field in the 20th century was characterized by a comparatively uniform warming in the tropics and by a considerable variability in the extratropical latitudes. The warming between 1910 and 1945 was initially concentrated in the Northern Atlantic and the adjacent regions. The Northern Hemisphere was characterized by cooling between 1946 and 1975, while in the Southern Hemisphere some warming was observed during this period. The temperature rise observed during the last decades (1970–2000) turns out, on the whole, to have been globally synchronous and clearly manifested across Northern Hemisphere continents in winter and spring. In some Southern Hemisphere regions and in the Atlantic, however, a small all-year-round cooling was observed. A temperature decrease in the Northern Atlantic between 1960 and 1985 was later followed by an opposite trend. On the whole, the climate warming over the period of measurements was more uniform in the Southern Hemisphere than in the Northern Hemisphere. In many continental regions between 1950 and 1993, the temperature increased more rapidly at night than during daytime (this does not refer, however, to coastal regions). The rate of temperature increase varied from 0.1 to 0.2 °C/10 years.

According to the data of aerological observations, the air temperature in the lower and middle troposphere was increasing after 1958 at a rate of 0.1 °C/10 years, but in the upper troposphere (after 1960) it remained more or less constant. A combined analysis of the aerological and satellite information has shown that in the period 1979–2000 the temperature trend in the lower troposphere was weak, whereas near the land surface it turned out to be statistically significant and reached

0.16 ± 0.06 °C/10 years. The statistically substantial trend of the difference between the Earth's surface and the lower troposphere constituted 0.13 ± 0.06 °C/10 years, which differs from the data for the period 1958–1978, when the average global temperature in the lower troposphere increased more rapidly (by 0.03 °C/10 years) than near the surface. The considerable differences between the temperature trends in the lower troposphere and near the surface are most likely to be real. So far, these differences cannot be convincingly explained. The climate warming in the Northern Hemisphere observed in the 20th century was the most substantial over the last 1,000 years according to the observational warming (Mikolajewicz et al. 2004; Islam et al. 2005; Jacobson 2000).

Special attention has been paid in the IPCC Reports (IPCC 2001, 2005, 2007) to the possibility for predicting future climatic changes. The chaotic character of the atmospheric dynamics limits the long-term weather forecasts to 1 or 2 weeks and prevents the prediction of a detailed climate change (e.g., it is impossible to predict precipitation in Great Britain for the winter of 2050). However, it is possible to consider climate projections, that is, to develop scenarios of probable climate changes, due to the continuing growth of GHGs concentrations in the atmosphere. Such scenarios, if credible, may be useful for decision-makers in the field of ecological policy. The basic method to make such scenarios tangible involves the use of numerical climate models Climate that simulate interactive processes in the atmosphere-ocean-land surface-cryosphere-biosphere climatic system (Jönsson et al. 2004). As Collins and Senior (2002) noted, because there are so many such models, the serious difficulty arises as to which is the best model to choose. As this problem of choice is insoluble, there remains the possibility of comparing the climate scenarios obtained by using various models.

According to the IPCC recommendations, four levels of projection reliability are considered:

(1) from reliable to very probable (in this case there is an agreement between the results for most of the models);
(2) very probably (an agreement of new projections obtained with the latest models);
(3) probable (new projections with an agreement for a small number of models); and
(4) restrictedly probable (model results are not certain but changes are physically possible).

A principal difficulty in giving substance to the projections is the impossibility of determining agreed predictions on how GHGs concentrations will evolve in future, which makes it necessary to take into account a totality of various scenarios. The huge thermal inertia of the World Ocean dictates a possibility of delayed climatic impacts of the GHGs concentrations, which has already increased.

Calculations of annual average global SAT using the energy-balance climate model with various scenarios of the temporal variations of CO_2 concentrations have led to SAT intervals in 2020, 2050, and 2100 to be 0.3–0.9, 0.7–2.6, and 1.4–5.8 °C respectively. Due to the ocean thermal inertia, a delayed warming should manifest itself within 0.1–0.2 °C/10 years (such a delay can take place over several decades).

The following conclusions can be attributed to the category of projections with the highest reliability (Collins and Senior 2002):

(1) surface air warming should be accompanied by a tropospheric warming and stratospheric cooling (the latter is due to a decrease of the upward longwave radiation flux from the troposphere);
(2) faster warming on land compared to oceanic regions (as a result of the great thermal inertia of the ocean); a faster warming in the high-mountain regions (due to albedo feedbacks);
(3) aerosol-induced atmospheric cooling holds a SAT increase (new estimates suggest the conclusion about a weaker manifestation of the aerosol impact);
(4) presence of the warming minima in the North Atlantic and in the circumpolar regions of the oceans in the Southern Hemisphere due to mixing in the oceanic thickness;
(5) decrease of the snow and sea ice cover extent in the Northern Hemisphere;
(6) increase of the average global content of water vapor in the atmosphere, enhancement of precipitation and evaporation, as well as intensification of the global water cycle;
(7) intensification (on average) of precipitation in the tropical and high latitudes, but its attenuation in the sub-tropical latitudes;
(8) increase of precipitation intensity (more substantial than expected as a result of precipitation enhancement, on average);
(9) summertime decrease of soil moisture in the middle regions of the continents due to intensified evaporation;
(10) intensification of the El Niño regime in the tropical Pacific with a stronger warming in the eastern regions than in the western ones, which is accompanied by an eastward shift of the precipitation zones;
(11) intensification of the interannual variability of the summer monsoon in the Northern Hemisphere;
(12) more frequent appearance of high-temperature extremes but infrequent occurrence of temperature minima (with an increasing amplitude of the diurnal temperature course in many regions and with a greater enhancement of nocturnal temperature minima compared to the daytime maxima);
(13) higher reliability of conclusions about temperature changes compared to those about precipitation;
(14) attenuation of the thermohaline circulation (THC) that causes a decrease of the warming in the North Atlantic (the effect of the THC dynamics cannot however compensate for the warming in West Europe due to the growing concentration of GHGs); and
(15) most intensive penetration of the warming into the ocean depth in high latitude where the vertical mixing is most intensive.

As for the estimates characterized by a lower level of reliability, the conclusion (at level 4) about the lack of an agreed view on the changing frequency of storms in middle latitudes, is of special interest here, as is a similar lack of agreement about the changing frequency of occurrence of tropical cyclones under global warming. An important future task is to improve climate models, aimed at reaching eventually a level of reliability that would enable the prediction of climatic changes.

Allen (2002) has discussed the basic conclusions contained in the "Summary for policy-makers" (SPM) of the Third IPCC Report and especially of its main conclusion that "There is new and stronger evidence that most of the warming observed during the last 50 years should be attributed to human activity". This conclusion supplements the statement according to which "as follows from the present climate models, it is very unlikely that the warming taking place during the last 100 years was determined only by the internal variability" ("very unlikely" means that there is less than one chance of ten for an opposite statement to be well-founded).

Clearly, the reality of such a statement depends on an adequate modeling of the observed climatic variability. Analysis of the results of the relevant calculations using six different models has shown that three of six models reproduce climate variability on time scales from 10 to 50 years which agrees with the observational data.

Another conclusion in SPM (TAR) is that "reconstruction of data on climate for the last 1,000 years shows that the present warming is unusual and it is unlikely that it can be only of natural origin" ("unlikely" means that there is less than one chance of three for an opposite conclusion).

This conclusion is supplemented with the following: "Numerical modeling of the response to only natural disturbing forces does not explain the warming that took place in the second half of the 20th century". This view is based on the analysis of the results from the numerical modeling of changes in the average global SAT during the last 50 years. It follows from this that a consideration of natural forcings (solar activity, volcanic eruptions) has demonstrated a climatic cooling (mainly due to large-scale eruptions in 1982 and 1991) which has allowed the conclusion that the impact of only natural climatic factors is unlikely. However, there is only one chance in three that it was so: such carefulness is due to insufficient reliability based on indirect information concerning natural forcings in the past.

Results of numerical modeling cannot explain the pre-1940 climate warming with only anthropogenic factors taken into account, but are quite adequate considering both natural and anthropogenic impacts (GHGs and sulphate aerosol). As it was mentioned in SPM of TAR, "these results… do not exclude possibilities of contributions of other forcings". It is possible therefore that good agreement of the calculated and observed secular trends of SAT may in part be determined by a random mutual compensation of uncertainties. Another important illustration of the inadequacy of the numerical modeling results is their difference with observations concerning temperature changes near the Earth's surface and in the free troposphere. If, as according to models, the tropospheric temperature increases more rapidly than near the surface, then the analysis of observational data between 1979 and 2000 reveals that the temperature increase in the free troposphere is slower and probably is absent at all (Chou et al. 2002; Collins et al. 1998; Conant et al. 2002).

When assessing the content of the IPCC-2001 Report, Griggs and Noguer (2002) argued that this report:

(1) contains a most complete description of the current ideas about the known and unknown aspects of the climate system and the associated factors;
(2) is based on the knowledge of an international group of experts;
(3) is prepared and based on open and professional reviewing;
(4) is based on scientific publications.

Sadly, none of these statements can be convincingly substantiated. The IPCC-2001 Report has therefore been strongly criticized in the scientific literature (Cracknell et al. 2009a, b). In fourth report since 1990, the IPCC offered its strongest language yet that Earth's climate is warming and humans are largely responsible:

> Warming of the climate system is unequivocal, as is now evident from observations of increases in global average air and ocean temperatures, widespread melting of snow and ice, and rising global mean sea level.
>
> Most of the observed increase in globally averaged temperatures since the mid-20th century is very likely due to the observed increase in anthropogenic greenhouse gas concentrations. Discernible human influences now extend to other aspects of climate, including ocean warming, continental-average temperatures, temperature extremes, and wind patterns.

Earth has warmed, sea levels have begun to rise at an accelerated rate, and Northern Hemisphere snow cover has decreased substantially over the past 150 years (Fig. 5.1). These facts are not controversial. The big change from the IPCC's last report, in 2001, is the level of confidence on if humans are to blame. In that report, human-emitted (anthropogenic) greenhouse gases were estimated to be likely responsible for Earth's temperature increase (67–90 % chance), while the new report says it is very likely (greater than 90 % chance).

On 24th December 2009 the UN General Assembly adopted a Resolution (A/RES/64/236) agreeing to hold the United Nations Conference on Sustainable Development (UNCSD) in 2012—also referred to as 'Rio+20'. The Conference seeks three objectives:

- securing renewed political commitment to sustainable development,
- assessing the progress and implementation gaps in meeting already agreed commitments, and
- addressing new and emerging challenges.

The Member States have agreed on the following two themes for the Conference: green economy within the context of sustainable development and poverty eradication, and institutional framework for sustainable development. The U.N. Conference on Sustainable Development (Rio+20), a once-a-decade meeting aimed at reconciling economic and environmental aspirations, has produced a nonbinding declaration, committing the world's politicians to modest goals. The proposals set out at the beginning, such as providing universal energy access and doubling renewable by 2030.

Discussions about global climate change and sustain development problems which dialog politics, is useful only in restricted sense. Real solution of these problems is possible in the framework of international working groups used to complete the mathematical models of the nature-society system.

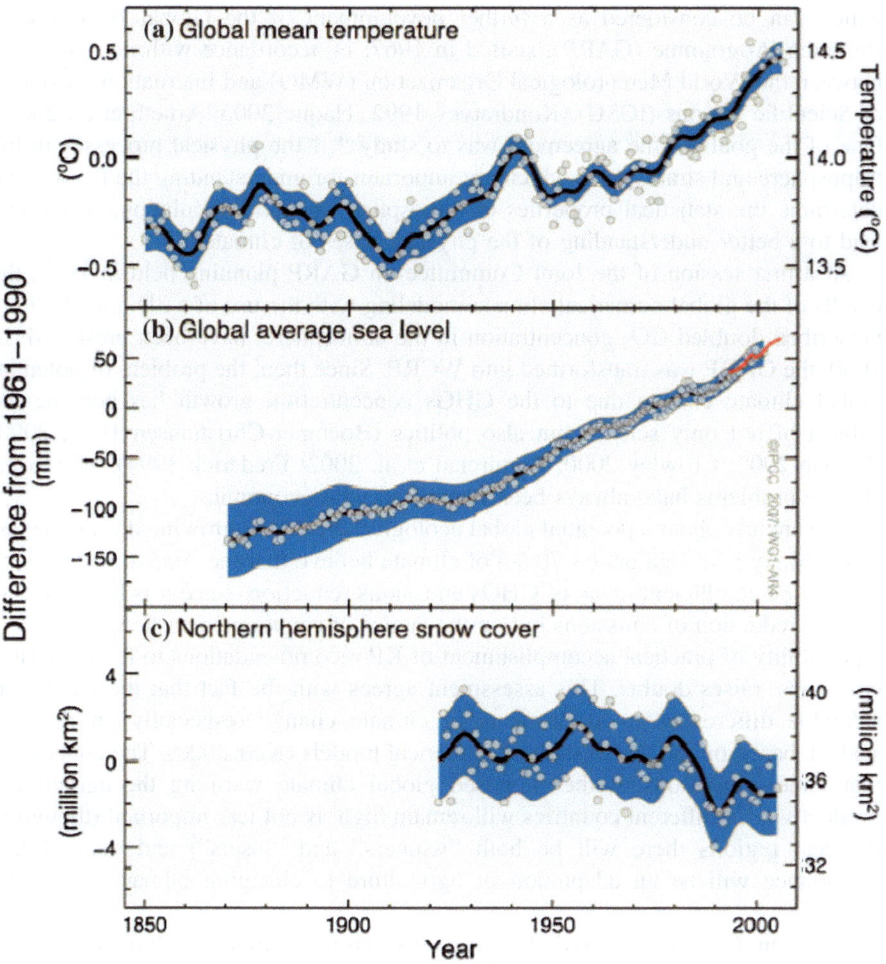

Fig. 5.1 Observed changes in **a** global average surface temperature; **b** global average sea level rise from tide gauge (*blue*) and satellite (*red*) data and **c** Northern Hemisphere snow cover for March–April. All changes are relative to corresponding averages for the period 1961–1990. *Smoothed curves* represent decadal averaged values while *circles* show yearly values. The *shaded areas* are the uncertainty intervals estimated from a comprehensive analysis of known uncertainties (**a** and **b**) and from the time series (**c**) (IPCC 2007)

5.2 General Problems of Global Climate Dynamics

A growing interest in the problems of global climate change determined by its great practical value and contradictory character of assessments of anthropogenic contribution to climate change, dictates, first of all, a necessity of analysis of the available observational data. Studies of global climate change are connected, first of all, with an accomplishment of the World Climate Research Programme (WCRP) which can be considered as a further development of the Global Atmospheric Research Programme (GARP), started in 1967, in accordance with an agreement between the World Meteorological Organization (WMO) and International Council of Scientific Unions (ICSU) (Kondratyev 1992; Haque 2005; Arneth et al. 2002). One of the goals of the agreement was to study "… the physical processes in the troposphere and stratosphere which are important for understanding the factors that determine the statistical properties of atmospheric general circulation, which can lead to a better understanding of the physical bases of climate".

At a first session of the Joint Committee on GARP planning held in 1967, the results of the global numerical climate modeling with the use of a 3D model with a prescribed doubled CO_2 concentration in the atmosphere, have been presented. In 1980, the GARP was transformed into WCRP. Since then, the problem of potential global climate change due to the GHGs concentration growth has become the subject of not only science but also politics (Boehmer-Christiansen 1997, 2000; Zillman 2000; Crowley 2000; Demircian et al. 2002; Frederick 1994). Moreover, climate problems have always been connected with economics.

The anxiety about a potential global ecological crisis was growing during the last years. Many investigators ($\sim$70 %) of climate believe that the "emissions trading" will not be an efficient mean of GHGs emissions reduction, since it is based not on the real reduction of emissions but on the tactics of economic measures. Therefore, a possibility of practical accomplishment of KP recommendations to reduce GHGs emissions, raises doubts. This assessment agrees with the fact that there are considerable differences in assessments of climate change (especially on regional scales), based on the use of various numerical models (Kerr 2000). The conclusion that under conditions of the supposed global climate warming the agricultural productivity in different countries will remain high, is not less important (though in different regions there will be both "winners" and "losers") and thus of key importance will be an adaptation of agriculture to changing climate (Sirotenko 2000).

Crichton (2005) discussed the problems arisen in modern climatology concerning the survivability of people when global temperature will increase. General consensus about future trends in global temperature does not exist among experts. Some say temperatures will increase by 1.5 °C over the next century, some say 5 °C. Majority of experts tries to solve this problem (Kondratyev 2001, 2002). But there exist experts who propose to see what happens. As Diamond (2005) points Crichton's (2005) strategy of "just it out" is a sure path to dirty air, famine, water shortages, war, and death. Diamond (2005) examines why certain ancient societies, such as Easter Islanders and Norse Greenlanders, disappeared, and how many modern societies are on the same path of destruction due to bad choice made in response to environmental circumstances.

In accordance with the 1987 UN Resolution on the Environment and Development, the ICSU and WMO decided in November 1988 to organize the IPCC. Representatives of about 30 countries have come to the conclusion about a necessity not only to analyze the existing ideas of possible anthropogenic climate change but also to consider its probable impacts on the environment and society as well as to discuss measures needed to prevent undesirable consequences of such impacts. An important result of IPCC efforts was the development of the FCCC.

The third Conference of representatives of FCCC signatory powers in Kyoto in December 1997 ended with signing the Kyoto Protocol, the most important part of which consists in recommendations of the levels of CO_2 emissions reduction by the countries listed in Appendix 1 of the KP (Table 5.1). Relevant discussions were focused mainly on discussion of economic consequences of emissions reduction and ignored the problem of substantiation of recommended levels of reduction,

Table 5.1 The KP recommendations on CO_2 emissions reduction by different countries by the years 2008–2012 (compared to 1990 levels) (Najam and Sagar 1998)

Level of emissions (%)	Countries	Annual emissions of CO_2 (10^3 t)	Per-capita GDP (US$)	Per-capita CO_2 emissions (tons)	CO_2 emissions per unit GDP (t/10^3 US$)
	Some countries from the list of Appendix 1				
108	Australia	288,965	16,516	16.91	0.98
94	Canada	462,643	19,705	17.44	0.81
92	Czechia	165,792	3,049	16.00	5.25
92	France	366,536	20,966	6.49	0.31
92	Germany	1,014,155	21,861	12.76	0.59
92	Ireland	30,719	11,349	8.77	0.68
92	Japan	1,155,000	24,205	9.35	0.39
92	Luxemburg	11,343	34,614	30.41	1.10
99	The Netherlands	174,000	18,939	11.22	0.61
100	New Zealand	25,476	12,192	7.61	0.59
101	Norway	35,514	26,387	8.37	0.31
94	Poland	414,930	1,459	10.87	7.04
92	Portugal	42,148	6,774	4.00	0.63
100	Russia	2,388,720	3,897	16.11	4.13
92	England	577,012	16,723	10.08	0.59
93	USA	4,957,022	22,046	19.83	0.90
	Some countries not recorded				
	Brazil	197,905	2,877	1.34	0.41
	China	2,340,635	313	2.06	6.60
	India	681,248	347	0.80	2.28
	Indonesia	213,422	613	1.20	1.87
	Iran	187,986	2,139	3.37	1.56
	Mexico	313,826	2,856	3.76	1.27
	South Korea	243,434	5,875	5.68	0.96
	South Africa	294,107	2,758	7.93	2.76

Notation GDP gross domestic production

which causes a serious concern, especially from the viewpoint of further prospects of development of global biospheric processes.

The 2012 United Nations Climate Change Conference was the 18th yearly session of the Conference of the Parties (COP) to the 1992 United Nations Framework Convention on Climate Change (UNFCCC) and the 8th session of the Meeting of the Parties (CMP) to the 1997 Kyoto Protocol (the protocol having been developed under the UNFCCC's charter). The conference took place from Monday 26 November to Saturday 8 December 2012, at the Qatar National Convention Centre in Doha.

At the 2012 Doha climate change talks, Parties to the Kyoto Protocol agreed to a second commitment period of emissions reductions from 1 January 2013 to 31 December 2020, which takes the form of an amendment to the Protocol. The 37 countries with binding targets in the second commitment period are Australia, all members of the European Union, Belarus, Croatia, Iceland, Kazakhstan, Norway, Switzerland, and Ukraine. However, a last minute objection at the conference by Russia, Ukraine, Belarus and Kazakhstan indicates that they will likely withdraw from the Protocol or not ratify the Protocol amendment. Collectively, these countries will reduce their emissions 18 % below their 1990 level between 2013 and 2020. The targets may be strengthened by 2014. The emissions targets specified in second commitment period will apply to about 15 % of the world's greenhouse gas emissions. Several Annex I Parties who participated in Kyoto's first-round have not taken on new targets in the second commitment period, and are Japan, New Zealand, and Russia. Other Annex I Parties without second-round targets are the United States (which never became a member to Protocol) and Canada (which withdrew from the Kyoto Protocol effective 2012).

As for the situation with developing countries, in this connection, the main fact consists in inability of developed countries to perceive the key significance of the principle of equality (this refers, in particular, to the problem of per-capita emissions). It is easy to predict that the discussed arbitrariness in resolving the problem will remain in the future, though it is apparent that the principles of equality, responsibility, and efficiency formulated in the Climate Convention should serve the basis. The KP recommendations are in apparent contradiction with these principles.

The key aspect of the FCCC consists, in the series of recommendations for reduction, of GHGs emissions to the atmosphere, in order to prevent undesirable anthropogenic climate changes. As Hulme and Parry (1998) noted, an adoption of the Protocol on GHGs emissions reduction at the Third FCCC Conference held in Kyoto in December 1997 can serve a manifestation of the anxiety of the world community about potential danger for sustainable development of anthropogenic impacts on climate. The logic of GHGs emissions reduction proceeds from two assumptions:

(1) a reduction of emissions will make it possible to avoid in the future a dangerous climate change; and
(2) it is cheaper to reduce emissions and provide softer climate changes than to adapt to the changed climate.

However, the problem is that neither scientists nor politicians have adequately studied and estimated the destructions, which can be avoided in accomplishing any scenario of GHGs emissions reduction, and relative expenditures on adaptation or softening. This is a very complicated problem and therefore it is most difficult to substantiate an adequate strategy.

In this connection, Hulme and Parry (1998) obtained and discussed estimates of climate change for Great Britain based on the use of four scenarios of GHGs emissions reduction (only CO_2 has been considered) for the period 1990–2100: IS92 (uncontrolled increase of emissions without their reduction); KYOTO (a 5.2 % reduction of emissions by the year 2010 compared to 1990 by the "first list" countries, that is, industrial countries, with subsequently stabilized emissions (other countries follow the "trajectory" of IS92); KYOTO+ (after 2010, the "first list" countries continue to reduce emissions by 1 % per year; for other countries—IS92); KYOTO++ (the same as KYOTO+ for the "first list" countries; after 2020, other countries reduce emissions by 1 % per year). The MINUS-60 scenario of global CO_2 reduction by 60 % by the year 2010 has been considered, too.

Data in Table 5.2 illustrate the estimates of respective changes (increase) of the annual mean SAT in Great Britain (in these calculations the impact of atmospheric aerosol was left out of account). As it is seen, all discussed measures on CO_2 emissions reduction provide only a negligible restraining of climate warming. This means a necessity to pay a much more serious attention to development of the strategy of adaptation to climate change.

McBean (1998) emphasized that though in the context of FCCC the emphasis has been placed on the problem of GHGs emissions reduction, clause 4 of FCCC foresees also the necessity:

(1) to support efforts (scientific, technological, technical, socio-economic) in the interests of further scientific developments, systematic observations and creation of the climate data archive, bearing in mind to get a better understanding of the causes of climate change and its socio-economic consequences;

Table 5.2 Estimates of the annual mean temperature increase in Great Britain with various scenarios of GHGs emission reductions (Spedicato 1991)

Scenario	Annual mean temperature increase (°C)	
	By 2050	By 2100
IS92a	1.39	2.54
KYOTO	1.33	2.39
KYOTO+	1.29	2.26
KYOTO++	1.24	1.79
MINUS-60	0.78	0.11

(2) to provide an exchange of the respective scientific, technological, technical, socio-economic, and juridical information to substantiate the strategies of responding to climate change; and
(3) to intensify measures on development of education, improvement of qualification, and informing the population about climate change.

At the Intergovernmental meeting on WCP held in April 1993, a decision was made on the expedience of complex development on climate problems which was called Climate Agenda (CA) aimed at providing a more efficient coordination of respective efforts. The CA supported by WMO, UNEP, ICSU, UNESCO, FAO, WHO was concentrated on four directions:

(1) Prospects of development of science of climate, and weather forecast.
(2) Climate services needed to ensure sustainable development.
(3) Study of climatic forcings and development of strategies of responding to the forcings to minimize their consequences.
(4) Adequately planned observations of the climate system.

For thc organized realization of the CA, the Interdisciplinary Committee on Climate Agenda (IACCA) was formed in March 1999 in Paris as a result of a substantial reorganization of the Coordination Committee on CA.

According to the adopted agreement, scientific developments are the sphere of responsibility of WCRP, whereas IPCC answers for negotiations and recommendations of the needed environmental policy. Efforts have been performed on the basis of a close cooperation between WCRP and IPCC, though not without some problems consisting, first of all, in underestimation by politicians of the significance of the problem of improving the observation means and in insufficient understanding by scientific community of specific character of information needed for politicians.

As Bolin (1998, 1999) has noted, in the context of the latter, the most important conclusion of the Second IPCC-1995 Report consisted in that "the balance of evidence supposes a marked impact of man on global climate", though on the other hand, it was noted that "our ability to quantitatively estimate the impact of man on global climate is now limited, since the expected signal appears just now at the background of natural variability as well as in view of available uncertainties (in estimates) of key factors". Such conclusions have been mainly supported by scientific community and stimulated further developments, the results of which have been generalized in the Third IPCC-2001 Report.

In this connection, Bolin (1998, 1999) emphasized that "negotiations between countries that have ratified the FCCC, have shown that substantial political actions are scarcely probable, if the impact of humans' activity on global climate is not proved". The absence of the needed understanding of the impact of climate on humans and ecosystems has determined the resistance of some countries to measures on limitation of the anthropogenic impact on climate. An important aspect of the problem is groundless statements of the "green" about the growth of the frequency of extreme natural phenomena such as storms, tropical cyclones, floods, and others. This question requires, however, further thorough analysis (Grigoryev and Kondratyev 2001).

One of the key unresolved problems is the forecast of development of energy production, increase of CO_2 concentration in the atmosphere, and, respectively, radiative forcing (RF). According to the Special IPCC-1994 Report, various scenarios determine a possible increase of CO_2 emissions by 2,100 within 2–60 GtC (compared to the current value 7.5 GtC), that is, the obtained estimates differ by a factor of 30 (with the use of all coal and natural gas resources known, the concentration of CO_2 can increase to 450 ppm).

Data in Table 5.3 illustrate possible changes of RF at different levels of CO_2 concentration stabilization in the atmosphere (the key unresolved problems consist, however, in that what level of stabilization can be considered acceptable). In this connection, special attention should be fixed on further development of global carbon cycle studies, especially bearing in mind the unresolved problem of the "lost sink", which is partially determined, apparently, by the functioning of land ecosystems as CO_2 sinks (on the whole, the role of the biospheric dynamics in the formation of carbon cycle is far from being clear). Quite urgent (but also very complicated) is the modeling of the nature-society interaction with long-term nonlinear changes of climate system taken into account. The concept of the risk as applied to the problem of climate change moves up to the foreground.

As follows from the above-said and a huge amount of publications on the key aspects of climate, one of the sufficiently problematic tasks is a regulation at scientific, economic, and political levels of the problem brought forth by the Kyoto Protocol. Perhaps, Barenbaum (2002, 2004) and Syvorotkin (2002) were first who formulated its reconciliatory solution, practically explained the mechanisms of closing global carbon cycle, and demonstrated its correlation with the cycles of other GHGs and water. Semenov (2004) pointed to the important role of the phase transitions of water, especially between liquid and solid ones. Masses of ice and permafrost are buffers of temperature changes, playing the role of the climate-forming function. Based on the model calculations, Semenov (2004) analyzed the impact of climate change on the vertical distribution of temperature under conditions of permafrost and determined the laws of the deep seasonal melting of the continental permafrost with the growing greenhouse effect. The main conclusion is that the vertical heat transfer and water phase transitions in the surface layer of the northern-latitude lithosphere are climatic stabilizers.

Table 5.3 The level of CO_2 concentration stabilization, radiative forcing, and change of global mean SAT with prescribed SAT sensitivity to a doubled CO_2 concentration within 1.5–4.5 °C

Level of CO_2 concentration stabilization (ppm)	Radiative forcing (W/m^2)	Change of global mean SAT (°C)		
		Minimum	Middle	Maximum
450	3.6	1.2	2.1	3.6
550	5.0	1.7	2.9	5.2
650	6.2	2.1	3.5	6.4
750	7.1	2.5	4.1	7.4

As has been mentioned above, the prevailing "anthropogenic" concept of global climate change is based on information on the secular change of the annual mean global mean SAT that was calculated from the data of observations at the land meteorological networks and ship observations of water temperature in the upper layer of the World Ocean, as well as numerical modeling of SAT change during the last one and a half century. Meanwhile, there is no doubt that the information content of the notion of global mean SAT needs further analysis, and being calculated from the data of observations, the SAT values are not only aggravated with errors which are difficult to estimate, but also are insufficiently representative. On the other hand, it is obviously that reliable results concerning climatic variability can be obtained only by considering the 3D air temperature fields and other climatic parameters, that is, by analysis of climate variability on regional scales. In this connection, substantiation of strategies of national and regional studies becomes especially urgent.

The World Climate Conference (Moscow, 29 September–3 October 2003) has stimulated the publication of a number of summarizing papers on the current studies of global climate dynamics (Demirchian and Kondratyev 2004; Kondratyev and Krapivin 2004a, b) with due regard to the earlier results (Kondratyev 1982; Kondratyev and Moskalenko 1984; Kondratyev and Johannessen 1993). The related key conclusion consists in establishing the presence of serious uncertainties in results in the fields of both empirical diagnostics and numerical climate modeling. Of exclusive importance becomes adequate substantiation and efficient realization of further developments in order to reduce the level of existing uncertainties and to assess the reliability of global climate forecasts.

Planning of further efforts in the field of global climate studies should be based on the two international programmes: WCRP and IGBP. A new programme COPES (Coordinated Observations and Prediction of the Earth System) approved by the WCRP Joint Scientific Committee is aimed at substantiation of a new strategy and goals of WCRP for the period 2005–2015 (COPES 2004), the beginning of which coincides with the 25th anniversary of WCRP. As it is known, two main goals of WCRP consist in assessing of the climate predictability and the degree of anthropogenic impact on climate. Achievement of these goals should be based on accomplishing the interdisciplinary studies of the climatic system "atmosphere-hydrosphere-lithosphere-cryosphere-biosphere" influenced by various external factors (including anthropogenic). Such studies should foresee, in particular:

- observations of changes taking place in the climate system;
- better understanding of the climate variability (mechanisms that determine this variability) on regional scales;
- assessment of substantial trends of regional and global climates;
- development of numerical modeling methods able to simulate the spatial-temporal variability in a wide range of scales and useful for operational predictions;
- study of the climate system's sensitivity to natural and anthropogenic forcings, quantitative assessment of contributions of various forcings;

- analysis of the nature and predictability of the annual change and interannual variability of the climate system on regional and global scales to create the scientific basis for operational predictions of climatic variability in the interests of sustainable development; and
- detection of climate change and recognition of its causes (especially of anthropogenic origin) in the context of the UN Framework Convention on Climate Change (FCCC) and developments of IPCC.

The 25th anniversary of WCRP is one of the stimuli to summarize the main results and substantiate the perspectives of further developments in connection with an appearance of new possibilities and advancements which include the solution of complex interdisciplinary problems such as:

(a) Forecasting the climate system's dynamics on time scales of weeks—seasons—years—decades—centuries considering possibilities of climate forecast for centuries with the use of initial data for observed numerous parameters characterizing the state of the climate system. Solution of the respective problems requires not only radical improvement of the existing means of numerical climate modeling but, first of all, adequate substantiation of requirements to observational data and substantial development of the global observational system (these questions have been discussed in detail elsewhere (Kondratyev 1998a, b; Kondratyev and Cracknell 1998).

(b) Development of methods for numerical modeling of global ecodynamics and—in this context—climate dynamics with due regard to interactive chemical processes, formation of global carbon cycle, land and ocean ecosystem dynamics (Kondratyev and Krapivin 2003a, b; Kondratyev et al. 2003a, b, c, d). In other words, the problem of global climate change should be considered as an interactive component of global changes. One should also emphasize the significance of the problems of risk connected with natural and anthropogenic catastrophes (Kondratyev et al. 2006a, b), as well as unexpected "turns" in the evolution of the global biosphere. The latter are illustrated by new (and demanding further thorough investigation) supposition about an important role of the fine structure of the spectrum of solar radiation (determined by the presence of Fraunhofer lines) in evolution of the biosphere and, in particular, in development of immunodeficiency (Kondratyev et al. 2004a, b).

(c) Analysis of significance of an adequate understanding and reliable climate forecasts from the viewpoint of provision of sustainable development (the urgency of this problem has become especially acute in connection with the ratification of the KP by Russia and, thus, its coming into effect).

These general goals can be achieved only by means of an efficient coordination of efforts within international programmes such as WCRP, IGBP, IHDP (socio-economic aspects of global change), DIVERSITAS (International Biodiversity Programme), and others.

The main goal of the present stage of the WCRP development is an adequate (well coordinated) solution of the problems of obtaining needed observational data, a deeper understanding of the laws of the climate system, and provision of reliable forecast of its dynamics. This strategy planned for the period 2005–2015 should be realized within the COPES programme, based on accomplishment of the projects included into WCRP and directed at the following problems of development (COPES 2004):

(1) Determination of possibility and assessment of expected reliability of the seasonal forecasts of climate for all regions of the globe with the use of available climate models and observation data, as well as more adequate information with further development of models and systems of observations.
(2) Further development and test of adequacy of the ensemble forecasts of natural and anthropogenic variability of climate.
(3) Scientific substantiation and analysis of adequacy of methods of regional climate forecasts on different temporal scales.
(4) Development of reliably verified models of climate dynamics with chemical processes taken into account.
(5) Realistic numerical modeling of the glacier-interglacier cycles with the use of general circulation models.
(6) Shortening of the intervals of uncertainties of the estimates of sensitivity of numerical climate modeling results to external forcings.
(7) Improvement of numerical modeling of arid climates and analysis of their reliability.
(8) Improvement of numerical modeling of monsoon climates and analysis of their reliability.
(9) Cooperation with IPCC WG-1 in the interests of preparation of AR4 Fourth IPCC Report on climate problems.
(10) Analysis of reasons why, how, and where the modes of climate variability change in response to anthropogenic forcings, and affects of such processes on the long-term climatic variability.
(11) Assessment of predictability of intraseasonal climate changes based on the use of interactive models.
(12) Determination of spatial-temporal scales of predictability of temperature, salinity, and general circulation of the ocean.
(13) Specification of forecasts of anthropogenic rise of the World Ocean level.
(14) Support to efforts on re-analysis of data on variability of the climate system's components.

This enumeration is far from being complete, and this clearly follows from information discussed in recent overviews (Kondratyev 2004a, b; Kondratyev et al. 2003a, b, c, d). As it has been mentioned, one of the important problems of COPES consists in analysis of the information content of the data of observations of various climatic parameters from the viewpoint of evaluation of contribution of these data, to the increase of reliability of climate forecasts on different spatial-temporal scales. Resolving of problems connected with this aspect of COPES, will require coordinated efforts on collection, accumulation, and re-analysis of the climatological information in order to obtain an internally coordinated (dynamically balanced) structure ensuring an adequate simulation of variability of the climate system in the interests of climate forecast. Naturally, first steps in this direction should consist in substantiation of an adequate global observational system and detailed requirements to observation data (these problems are far from being resolved). One of the related steps was an accomplishment of the COPES project of coordinated frequented observations in 2002–2004 within the sub-programme of the experiment on study of global energy and water cycles (GEWEX) as part of WCRP, the key questions of which were (COPES 2004):

- Do variations of the rate of water cycle take place due to climate change?
- To what degree do local weather changes result from anthropogenic (or natural) global climate change?
- Is it possible to forecast precipitation on spatial-temporal scales determined by requirements connected with solving of applied problems?
- How can the total impact of processes determining water cycle be taken into account (parameterized) in weather forecasts and climate models?

A new stage of the GEWEX programme is connected with accomplishment of a complete global description of water and energy cycles, as well as the development of improved forecasts of precipitation and water cycle variability (with due regard to natural and anthropogenic climate changes), bearing in mind resolution of various practical problems, especially those concerning water resources. As Lawford (2004) pointed out, the following items are playing an increased role: the use of space-borne observational means such as an experimental satellite CloudSat for radar measurements of the vertical structure of clouds and their properties; the system of polar-orbiting environmental satellites (NPOESS); means of remote sensing of soil moisture and ocean water salinity (SMOS); programme to study the state of the hydrosphere (Hydros) with the global survey of changes of soil moisture and conditions of surface permafrost melting and global precipitation measurement (GPM). The Integrated Global Observing Strategy (IGOS) planned for 10 years and combined systems of global Earth observations (GEOSS) favour the coordination of such developments. Within the programme of partnership IGOS (IGOS-P) plans are being made to create observational systems with the use of routine (in-situ) and satellite observational means. In this connection, the problem-oriented special complex field experiments such as CEOP (Koike 2004) aimed mainly at studies of global water cycle should play a special role.

The main goal of CEOP project started in October 2002 and completed by the end of 2004 was to assess the impact of the sources and sinks of heat and moisture on land on the global climate formation. The CEOP was considered as a pilot project in accomplishing global observations of water cycle within the partnership in strategy of global observations (IGOS-P) (Wanders et al. 2012).

Another important observational component of GEWEX and CLIVAR (programmes of climate studies) is the project of multi-disciplinary analysis of the African Monsoon Multidisciplinary Analysis (AMMA), which is a continuation of the project CATCH to study the tropical atmosphere and water cycle interaction.

AMMA has two phases: (1) 2002–2010 and (2) 2010–2020 (http://badc.nerc.ac.uk/view/badc.nerc.ac.uk__ATOM__dataent_amma). AMMA is an international project to improve our knowledge and understanding of the West African Monsoon (WAM) and its variability. AMMA is motivated by an interest in fundamental scientific issues and by the societal need for improved prediction of the West African Monsoon and its impacts on West African nations. The societal need to develop strategies that reduce the socioeconomic impacts of the variability of the West African Monsoon forces the AMMA will facilitate the multidisciplinary research required to provide improved predictions of the WAM and its impacts. AMMA promotes international coordination of ongoing activities, basic research and a multi-year campaign over West Africa and the tropical Atlantic. AMMA develops close partnerships between those involved in basic research of the West African Monsoon, operational forecasting and decision making, and it will establish blended training and education activities for African technical institutions and schools.

The observational programs of special importance include the Third International Polar Year (IPY) planned for 2007–2009 with coordination carried out by the scientific enterprising group CliC on the problems of climate and cryosphere. First IPY was realized in 1932–1933 and second IPY occurred 75 years after the first IPY (1957–1958). The latest IPY brought about the most ambitious Arctic climate change research project ever undertaken in Canada, a CAD research program called the Circumpolar Flaw Lead (CFL) System Study. Led by University of Manitoba Professor David Barber, a Canada Research Chair, the project involved more than 300 scientists from 16 countries, including over 40 faculty members, research associates, graduate students, technicians and support staff from the University of Manitoba.

The THORPEX program should be more informative and longer-term. This international programme of atmospheric studies planned for 10 years is carried out under aegis of the WMO Commission on Atmospheric Sciences and is a part of the World Weather Research Programme (WWRP). The main goal of the programme is to work out methods of weather prediction (especially its manifestations which especially strongly affect humans' life and economic activity) with an adequate use of the data of conventional and satellite observations. The THORPEX should culminate in a 1-year global meteorological experiment during the period 2010–2015. The fundamental problem of exclusive importance is the development of a full-scale system of global climate observations (GCOS) to obtain long and homogeneous series of high-quality data on the climate system parameters, bearing in mind the solution of the following set of problems (Steffen et al. 2004):

- Climate monitoring for quantitative estimates of natural climate change in a wide range of spatial-temporal scales and for recognition of anthropogenic climate change.
- Detection and quantitative estimates of various (especially anthropogenic) factors of climate change.

- Diagnostic analysis to get a deeper understanding of the dynamics of climate system and its components, including evaluation of contributions of natural climate change.
- Development and analysis of reliability of various hypotheses on the factors of climatic variability on different spatial-temporal scales, as well as assessment of climate predictability.
- Studies of various climate-forming dynamic, physical, chemical, and biological processes with their interactivity taken into account.
- Further improvement of climate models for climate forecasts on time scales from inter-seasonal to inter-annual variability.

The latter problem is especially urgent: despite serious achievements in development of methods of numerical climate modeling, the current climate models are still inadequate. This concerns, first of all, consideration (parameterization) of chemical and biological processes in the climate system. Still unresolved fundamental problems of climate theory include:

- interactive numerical modeling of climate and carbon cycle (Cracknell et al. 2009; Kondratyev and Krapivin 2004a; Kondratyev et al. 2003c);
- consideration of biospheric dynamics including continental and marine ecosystems in this context of importance, is the concept of biotic regulation of the environment (Kondratyev et al. 2003d, e); and
- parameterization of chemical processes in the troposphere and stratosphere, simulation of paleoclimate dynamics (Lohmann and Sirocko 2004; Lohmann et al. 2004; Mikolajewicz et al. 2004; Schneider et al. 2004; Sirocko et al. 2004), etc.

The analysis of adequacy of the numerical modeling results by comparing with observational data is of decisive importance. The problem of primary importance is the development of approaches to NSS numerical modeling. This will require an accomplishment of an efficient coordination of efforts within four main international programmes: WCRP, IGBP, IHDP, and DIVERSITAS. In this connection, one should mention a timely appearance in IGBP of the programme AIMES (analysis, integration, and numerical modeling of the Earth system). The challenge for AIMES is to achieve a deeper and more quantitative understanding of the role of human perturbations to the Earth's biogeochemical cycles and their interactions with the coupled physical climate system. AIMES will focus on:

- The functioning of global biogeochemical cycles, including interactions and feedbacks with the physical climate system;
- The interplay between human activities and biogeochemical cycles, both in the past and into the future;
- How the biogeochemical cycles function on different time scales.

The program of future climate studies is connected, first of all, with the WCRP perspectives based mainly on results obtained in the course of completed programmes, such as TOGA (study of Tropical Ocean and Global Atmosphere:

1985–1994); WOCE (World Ocean Circulation Experiment: 1982–2002); ACSYS (study of the Arctic climate: 1994–2003). The scientific goal of the WCRP's Arctic Climate System Study (ACSYS) project is to ascertain the role of the Arctic in global climate by attempting to find answers to the following related questions (Lemke and Jacobi 2012):

- What are the global consequences of natural or human-induced change in the Arctic climate system?
- Is the Arctic climate system as sensitive to increased greenhouse gas concentrations as climate models suggest?

Problems of TOGA and WOCE will be continued within the project CLIVAR to study the climate variability on time scales from seasons (with emphasis on the problem of monsoons) to centuries (in this case, detection and assessment of anthropogenic impacts on climate will play a key role), the main goals of which are:

- Description and understanding of physical processes responsible for variability and predictability of climate on seasonal, inter-annual, decadal, and centennial scales based on the use of available observations data and interactive models of climate system.
- Use of paleoclimatic information for a better understanding of present (and possibly future) climate change.
- Development of methods to forecast climate change on scales from seasonal to inter-annual based on the use of global interactive prognostic models.
- Understanding and prediction of the climate system's response to increasing concentrations of MGCs and aerosol, as well as subsequent comparison of numerical modeling results with observational data with emphasis on separation of anthropogenic changes from natural climate "signals".

An accomplishment of the CLIVAR project should answer the following questions, in particular:

- Will El Niño take place next year?
- What climate will be in Asia next summer and will the next monsoon bring droughts or floods?
- What will be the next winter in Northern Europe: "warm and wet" or "cold and dry"?
- What is the global climate warming due to anthropogenic impacts?
- How high can be the World Ocean level in the 21st century?
- Will the global climate warming cause an increase of the frequency of extreme weather phenomena and will these phenomena intensify and become widespread?
- Are sudden climate changes possible?

As has been mentioned, studies of global systems of monsoons are of primary importance for CLIVAR. This has been favoured by the successful accomplishment of the field experiment SALLJEX on studies of the south-American jet stream in the troposphere, which was part of a wider project VAMOS (Wood et al. 2011) to analyze the variability of the system of American monsoons. Also of great importance are studies of the role of processes in the ocean in climate formation, including problems such as variability of thermohaline circulation in the Atlantic Ocean; dynamics and predictability of the intra-tropical convergence zone in the Atlantic Ocean and its impact on regional climate; atmospheric forcings, their connection with processes in the upper layer of the ocean and feedbacks with SAT, including an assessment of the role of Kuroshio Current; and upwelling in the Pacific Ocean. In this connection, close attention is devoted to improvement of the observational network in cooperation with GCOS (Global Climate Observation System) and GOOS (Global Ocean Observation System).

An important place in the WCRP programme is occupied, as before, by the project SPARC, the main goal of which is to study the climatic impact of the stratosphere, taking into account interactive chemical, dynamic, and radiative processes that control the atmospheric circulation in the stratosphere and its chemical composition, including the ozone content whose changes determine variations of the UV solar radiation that reaches the Earth's surface (Kondratyev and Varotsos 1996, 2000; Wauben et al. 1997). A substantial task of SPARC is recognition and quantitative description of the long-term trends of the water vapour and ozone content in the stratosphere, as well as temperature. The beginning of the second decade of the SPARC functioning has been marked by close attention to such problems as stratosphere-troposphere interaction (parameterization of the stratosphere in climate models), interactive impact of chemical processes on climate, stratospheric indicators of climate change, various aspects of the problems of assimilation of the data of observations of stratospheric parameters. Creation of new models of the stratosphere has opened up possibilities to reliably simulate the dynamics of the "ozone hole" in the polar stratosphere. By the end of 2004, the estimation of the stratospheric aerosol dynamics was completed, and a new initiative is to study the processes of formation and evolution of polar stratospheric clouds.

General SPARC themes and activities address areas of societal concern such as:

- *Climate variability and change*. A central issue within SPARC is the impact of changing atmospheric composition on circulation and weather, from increasing surface temperatures to severe weather events and rising sea-levels. The stratosphere represents a significant source of variability and internal memory of the climate system.
- *Ozone*. Ozone and the ozone layer are of major concern to life on Earth. SPARC provides key scientific support to the WMO/UNEP Ozone Assessments that are produced every 4 years in accordance with the mandate of the signatories of the Montreal Protocol.
- *Atmospheric chemistry and aerosols*. Atmospheric chemistry is a cornerstone for understanding the evolution of the climate system. The effect of greenhouse gases (GHGs) and ozone depleting substances (ODSs) has been to warm the Earth's surface, while cooling the stratosphere. SPARC has been actively involved in advancing our understanding of the chemical processes that affect climate.

- *Polar processes*. Melting ice sheets and shrinking sea-ice extent and thickness greatly affect not only the polar areas, but coastal regions all over the globe. SPARC is teaming up with other interested research institutes to bring its expertise to bear on scientific issues specific to the polar areas.

The project CliC, being a continuation and development of the programme CCSS, is directed at further studies of the impact of processes in the Arctic on the global climate formation. The related problems include: determination of the role of the Arctic Ocean waters and ice cover dynamics as factors of strong changes of the arctic climate; study of sensitivity of the arctic climate system to the growing concentration of GHGs in the atmosphere. The urgency of these and other problems is determined, in particular, by obtaining new data of observations, demonstrating substantial changes of the global cryosphere, including:

- a small extent of multi-year arctic ice cover (with minimum values in September 2002 and in 2003);
- intensive melting of Greenland glaciers from the beginning of satellite observations in 1980 (note that according to the latest radio-altimeter data, the coastal parts of glaciers have been melting, whereas in the centre of ice cover there was an accumulation of ice); and
- destruction of the ice shelf Larsen B in the region of Western Antarctic in 2002; accelerated melting of mountain glaciers on all continents.

One of the important tasks of CliC is to study possible additional GHGs emissions to the atmosphere in permafrost melting (Fig. 5.2). It is known that permafrost, as a critical part of the Earth's climate system, covers 24 % of the land in the northern hemisphere. It also stores approximately 1.5 trillion tons of carbon—twice the amount of carbon currently in the atmosphere. The permafrost carbon feedback is not present in the global climate change models used to estimate how warm the earth could get over the next century. As a result of climate change, permafrost is at risk of melting, releasing the stored carbon in the form of carbon dioxide and methane. As it follows from Fig. 5.2, permafrost structure is not uniform and depends on the land covers. Therefore, a creation of permafrost database with spatial distribution of its characteristics is principal stage of climate modeling processes.

An important circumstance is the impossibility of reliable short-range and long-range climate forecasts without an adequate consideration of cryospheric processes (Kondratyev and Johannessen 1993; Marshall 2011; Vallis and Thoumi 2012; Kerr 2000). This concerns, in particular, the solution of an important problem, such as forecast of possible rise of the World Ocean level under conditions of global climate warming. Some other global aspects of CliC include studies of: thermohaline circulation (fresh water inflow to the North Atlantic); snow and ice covers; reliable methods of taking into account the permafrost and ice cover dynamics in climate models; changes of solid precipitation with the use of new methods of observations.

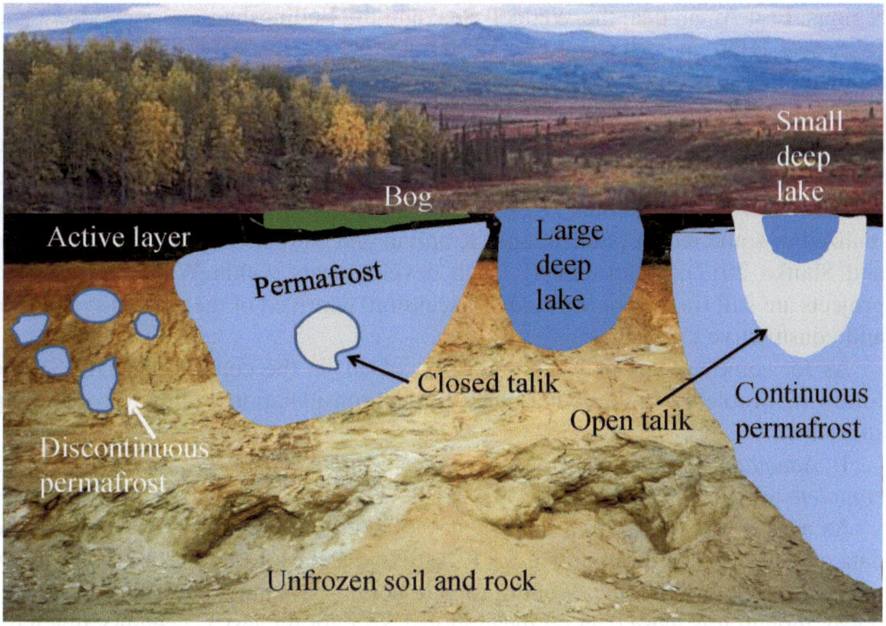

Fig. 5.2 Idealized permafrost cross section

In addition to above, permafrost is structurally important, and its melting has been known to cause erosion, disappearance of lakes, landslides, and ground subsidence. It will also cause changes in plant species composition at high latitudes. Thawing permafrost has significant effects on surface and subsurface regimes, including those governing hydrology and energy and moisture balance. Ecosystem diversity, composition, and productivity are not only impacted by increasing air temperatures, but by the associated effects of increasing ground temperatures as well. Because of this, thawing permafrost has significant impacts on infrastructure and ecosystems. Where ground ice contents are comparatively high, permafrost degradation can have significant impacts (Turetsky et al. 2007).

The international IPAB programme of observations using the instruments mounted on sea buoys in the SH oceans started really in 1995 under aegis of WCRP, and deserves further support. IPAB is a self-sustaining project of the WCRP, and provides a link between institutions with Antarctic and Southern Ocean interests. IPAB was formally established, following a 1 year pilot phase, at a meeting in Helsinki, Finland in June 1994. IPAB aims to establish and maintain a network of drifting buoys in the Antarctic sea-ice zone, which monitor ice motion, pressure and temperature. The IPAB Antarctic Drifting Buoy Data archive, presently spanning the years 1995–1998, includes measurements of buoy position, atmospheric pressure, air temperature, and sea surface temperature. Data are organized by daily and 3-h averages and the raw, instantaneous, non-interpolated data values. Data were collected from buoys initially deployed in three study regions: East Antarctica; the Weddell Sea; and the Bellingshausen, Amundsen, and Ross Seas (Heil and Allison 1999; WMO 2000).

First of all, it is important to perceive the fact that the discussed climate problems can be solved only on the basis of a complex study of interactions of all major climate system components. Moreover, the climate problem should be considered as a part of a much more general and extremely difficult problem of study and simulation of NSS dynamics. This approach is based and has been constructively realized in some recent publications (Krapivin and Varotsos 2007, 2008; Krapivin and Shutko 2012). Unfortunately, such developments within WCRP and related projects are still (in the context under discussion) deprived of the needed systematic and constructive character.

As for some aspects of IGBP (and first of all, in the context of COPES), the following problems, in particular, need further consideration and more convincing substantiation:

1. *Adequacy of available observational database from the viewpoint of its completeness and reliability.*

As an example illustrating the urgency of such problems, one can refer to the data on mean annual mean global SAT. As Essex and McKitrick (2002) have justly noted, even definition and meaning of this notion have not been clearly substantiated, though it is these data on mean global SAT for the last 1.5 century that have formed the basis of the most important conclusions about the nature of present climate change and various practical recommendations (including the KP). Of no less importance is the fact that the global mean SAT is the product of averaging with the use of rather fragmentary information (especially it refers to the late 19th to early 20th century), consisting of a combination of the data of air temperature observations on land (the problem of filtering out the contribution of the urban "heat islands" from observations at meteorological stations has not been adequately solved as yet) and results of ship observations of water temperature in the upper layer of the ocean, methodically complicated and far from being representative. One of the convincing illustrations of discrepancy (and insufficient reliability) of available meteorological information is a heated discussion on principally important differences between the data on SAT trends and results of tropospheric temperature retrieval from the data of satellite microwave observations (Kondratyev 2004a, b; Christy and Spencer 2003). One more illustration is unreliable manifestation of the so-called "hockey stick" (a sharp increase of mean global SAT for the last decades at the background of a weak variability during the previous millennium, which excludes, in particular, phenomena such as "Little Ice Age" which is of principal importance for characteristic of the global warming (McIntyre and McKitrick 2003). On the one hand, it means a necessity to continue analysis of reliability of information on global mean SAT values (there is no doubt that the "orthodox" estimation of global warming 0.6 ± 0.2 °C is conditional). On the other hand, there is an acute need of substantiation of a single global system of climate observations in the context of global change.

As for the latter circumstance, at first sight, the situation is quite safe: during the last years, some programs have been developed, such as GCOS, GOOS, GTOS (global system of land surface observations), IGOS (Integral Global Observational System), and others. In fact, a single global observational system is needed in the interests of NSS dynamics monitoring, containing the problem-oriented units based on substantiation of priorities. The most important aspect of substantiation of such a system is an optimization of relationship between the use of conventional and satellite observational means. In this connection, there is no doubt that multi-billion expenditures on creation of the EOS system consisting of three satellites equipped with complexes of the most up-to-date instrumentation have turned out to be not properly justified (Goody 2002).

2. *The role of biospheric dynamics as an interactive component of climate system.*

The fact of biotic regulation of the environment still has not been acknowledged (Kondratyev et al. 2003d, 2004b), though the Amsterdam Declaration reflects a substantial progress achieved in this direction. The first step in this direction should be an interactive consideration of global carbon cycle instead of a-priori prescription of arbitrary scenarios of changes of CO_2 concentration in the atmosphere (Krapivin and Varotsos 2008).

3. *Greenhouse effect.*

With tremendous importance attached to anthropogenic changes of the atmospheric greenhouse effect as the most important climate-forming factors, the absence in the COPES programme of a section specially dedicated to these problems seems a paradox. Therefore, some contradictory opinions appear about the contributions of various components of the greenhouse effect to the formation of global climate. So, for instance, Hansen et al. (2002) believe (though one should note that opinions of this group of specialists sometimes suffer sudden changes), that since there is an approximate mutual compensation of RF of different signs due to carbon dioxide and aerosol, the consideration of contribution to the atmospheric greenhouse effect enhancement, determined by the growth of methane concentration, is of major importance for the prognostic RF values. On the whole, it is necessary to take into consideration all GHGs effects and to take into account the specific characteristics of water vapors, methane, ozone, carbon dioxide, nitrous oxide, etc.

4. *Atmospheric aerosol and clouds.*

In view of a more important role of atmospheric aerosol and clouds in the RF formation (and, hence, climate change) compared to MGCs, the respective problems should attract much more serious attention (Kondratyev 2005). It means, in particular, a necessity of the aerosol-cloud sector in the COPES programme. The same refers to the problems of “atmospheric ozone (both stratospheric and tropospheric) and climate” (Braesicke and Pyle 2004; Kondratyev and Varotsos 2000; Friedrich 2001).

The basic conclusion made by Kondratyev et al. (2006a) about the mechanisms of formation and properties of atmospheric aerosols is that existing climate models are still far from being adequate. This situation is especially substantial in the context of consideration of aerosol as one of the most important climate-forming factors. On the one hand, it is obvious that an interactive consideration of aerosol, such as a climate-forming atmospheric component, is needed, and on the other hand, there is no doubt that the still highly uncertain data on the global spatial-temporal variability of aerosol properties and mechanisms of formation do not permit an adequate parameterization of the aerosol dynamics in climate models, where the most them predict only a few general trends.

5. *Natural disasters.*

The problems of risk and natural disasters are of special importance in the context of analysis of the role of non-linear dynamics of climate system, as a factor of sudden and strong climate change in reaching certain threshold levels of forcing.

Knutson et al. (2010) discuss the problem concerning the existence of correlation between climate change and tropical cyclogenesis processes. Unfortunately, the historical data do not confirm the presence of such correlation with high reliability. However, the existing climate models show the presence of globally averaged increase in the tropical cyclone intensities to 2100 by 2–11 %, under the reduction of their frequency by 6–34 %.

6. *Solar radiation.*

Still neglected in "orthodox" climate studies, solar radiation is an important problem of possible impacts of solar activity on climate, via various mechanisms of formation of positive feedbacks with solar activity-induced changes of the content of some MGCs in the atmosphere (in particular, nitrogen dioxide) and cloud cover properties (Douglass et al. 2004). In this connection, the work of Solanki et al. (2004) deserves attention. They detected a strong anomaly of solar activity during the last decades, compared to its quieter level during the preceding 11,000 years.

Thus, the main conclusion is that further serious efforts are needed to substantiate a new strategy of global climate studies. However, it is important to remark that this strategy cannot be realized adequately without more precise account of the problem of nature-society interaction (Krapivin and Varotsos 2007, 2008).

5.3 Empirical Diagnostics of the Global Climate

5.3.1 Introduction

The main cause of contradictions in studies of the present climate and its changes, represents the inadequacy of the available observational databases. They remain incomplete and of poor quality. In this connection, Mohr and Bridge (2003) have carried out a thorough analysis of evolution of the global observing system. As it is well known, climate is characterized by many parameters, such as air temperature and humidity near the Earth surface and in the free atmosphere; precipitation (liquid or solid); amount of cloud cover and the height of its lower and upper boundaries, microphysical and optical properties of clouds; radiation budget and its components; microphysical and optical parameters of atmospheric aerosols; atmospheric chemical composition, and more. However, the empirical analysis of climatic data is usually limited by the results of SAT observations, with data series available for no more than 100–150 years. Even these data series are heterogeneous, especially with regard to the global database as the main source of information for proving evidence for global warming idea. Also, it should be borne in mind that the globally averaged secular trend of SAT values is based, to a large extent, on the use of imperfect observed data of sea surface temperature (SST).

The most important (and controversial) conclusion concerning the anthropogenic nature of present day climate change is based on the analysis of the SAT and SST combined data, that is on the secular trend of mean global average annual surface temperature (GST). There exist both positive and negative anomalies of the GST that indicate the observed temperature deviation from the reference value. In this connection, two questions arise:

(1) about the information content of the notion of GST (this problem was formulated by Essex and McKitrick (2002); and
(2) about the reliability of GST values determined, in particular, by fragmentary data for the Southern Hemisphere, as well as the still unresolved problem of urban "heat islands".

An urban heat island (UHI) is a metropolitan area that is significantly warmer than its surrounding rural areas due to human activities. The phenomenon was first investigated and described by Luke Howard in the 1810s, although he was not the one to name the phenomenon (Howard 1820). The temperature difference is usually larger at night than during the day, and is most apparent when winds are weak. Seasonally, UHI is seen during both summer and winter. The main cause of the urban heat island is the modification of the land surface by urban development, which uses materials which effectively retain heat. Waste heat generated by energy usage is a secondary contributor (Li and Zhao 2012). As a population center grows, it tends to expand its area, and increase in its average temperature. The less-used term heat island refers to any area, populated or not, which is consistently hotter than the surrounding area.

Studies on the reliability of the SAT observations are continuing from the perspective of observational techniques. For more than 100 years, SAT was measured with the glass thermometers, but now arrangements to protect the thermometers from direct solar radiation and wind have been repeatedly changed. This dictates a necessity for filtering out SAT data to provide homogeneous data series. In the period from April till August 2000 at the station of the Nebraska State University, USA (40° 83′N; 96° 67′W), Hubbard et al. (2004) carried out comparative SAT observations over smooth grass cover with the use of various protections of thermometers. At the same time, direct solar radiation and wind speed were measured. Analysis of observations has shown that differences of observed data can reach several tenths of a degree. Therefore, a technique has been proposed to increase the homogeneity of observation series which substantially increases the homogeneity of the series. However, it does not permit the exclusion of the effect of calibration errors and drift of the temperature sensor's sensitivity.

For the diagnostics of the observational data, emphasis should be put on the analysis of climate variability in which a consideration not of averages but moments of higher orders is important. Unfortunately, there have been no attempts to use this approach. The same approach refers to estimates of the internal correlation of observation series. McKitrick (2010), having analyzed the secular trend of SAT, showed that with the filtered-out contribution to temperature variations during the last several decades at the expense of internal correlations (i.e., determined by the climatic system's inertia), it turns out that practically the temperature has not changed. There is a paradox: an increase of the global average SAT during the last 20–30 years is the principal basis for the conclusion concerning the anthropogenic contribution to the present-day climate changes.

5.3.2 Air Temperature

According to SAT observations discussed in IPCC (2001), during the period from 1860 to the present its annual and global averages increased by 0.76 ± 0.19 °C Figs. 5.3 and 5.4 show the time series of the combined global land and marine surface temperature record from 1880 to 2013. This is approximately 0.15 °C higher than the value given in the IPCC-1996 report, which was explained by a high SAT level between 1995 and 2000. The observed data revealed a strong spatial-temporal variability of the mean annual SAT on the globe. This manifested itself, for instance, in that the climate warming in the 20th century took place during two

Fig. 5.3 Average global air temperature anomalies, 1880–2012 (www.cru.uea.ac.uk/cru/info/warming)

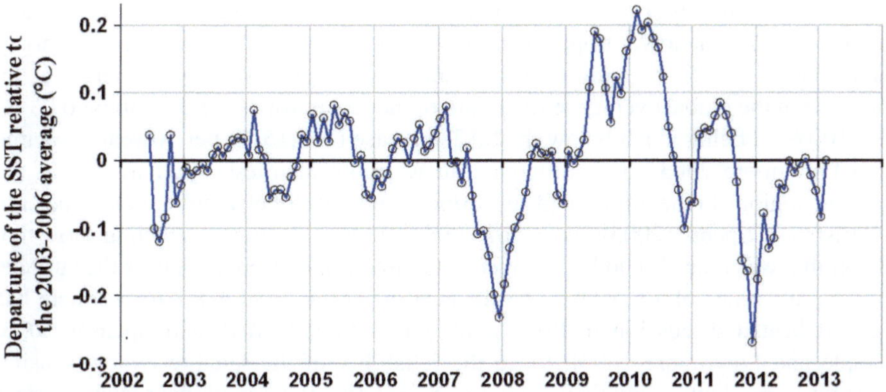

Fig. 5.4 The global average sea surface temperature (SST) update for Feb. 2013 is −0.01 °C, relative to the 2003–2006 average (http://www.drroyspencer.com/global-microwave-sea-surface-temperature-update-for-feb-2013)

time periods: 1919–1945 and from 1976 until now. It follows from global climate diagnostics that the warming in the Northern Hemisphere in the 20th century was, apparently, the strongest during the last 1,000 years, the 1990s being the warmest decade, and the year 1998, the warmest year (Hansen and Sato 2001; Hansen et al. 1997, 2010).

The period 2001–2010 (0.477 °C above 1961–1990 mean) was 0.217 °C warmer than the 1991–2000 decade (0.270 °C above the 1961–1990 average). The warmest year of the entire series was 2010, with a temperature of 0.540 °C above the 1961–1990 mean. The value for this year, given the uncertainties discussed in Morice et al. (2012) is not distinguishable from the years 1998 (0.523 °C) and 2005 (0.534 °C). The coldest year of the 21st century (2008 with a value of 0.383 °C) was warmer than all years in the 20th century with the exception of 1998 and 1997 (0.390 °C). The first 2 years of the present decade (2011 and 2012) are cooler than the average for 2001–2010, but warmer than all years before 2000 except for 1998.

An important feature of the climate dynamics was that, on the average, the rate of increase of nocturnal (minimum) SAT values on land was almost twice as high compared to that of diurnal (maximum) SAT values, starting from 1950 (0.2 °C against 0.1 °C per 10 years). This favored an increase in the duration of the frost-free periods in many regions of mid- and high latitudes.

The IPCC-2001 Report does not mention a previously assumed enhanced increase of climate warming in the NH high latitudes as a characteristic indicator of anthropogenic global warming. However, an analysis of direct SAT measurements at the "North Pole" stations during 30 years (Adamenko and Kondratyev 1990), as well as of dendroclimatic indirect data for the last 2–3 centuries, shows that there had been no homogeneous enhancement of such warming. Climatic changes during the last century and the last decade were characterized by a strong spatial-temporal heterogeneity: in the Arctic, the regions of both warming and cooling of climate were formed simultaneously (see also Moritz et al. 2002).

From satellite observations (beginning from 1979), the trend of global average temperature for the lower troposphere (0–8 km) was +0.07 °C per 10 years (Christy and Spenser 2003). According to the data of aerological soundings there was an increase of the global average temperature of the lower troposphere by about 0.03 °C per 10 years, being much below the SAT increase ($\sim$0.15 °C per 10 years) (Waple and Lawrimore 2003). This difference in warming manifests itself mainly in the oceanic regions in the tropics and sub-tropics, and it is not clear why such is the case (Christy and Spenser 2003). The results of the numerical climate modeling show that the global warming should be stronger in the free troposphere than near the surface.

The difference of temperature trends near the surface and in the troposphere has caused heated discussion in the scientific literature (Christy and Spenser 2003; Waple and Lawrimore 2003). Since the reliability of the satellite remote sensing data raises no doubts, and their spatial representativeness (on global scales) is more reliable than that of the data of surface measurements, this difference should be interpreted as necessitating further analysis of the SAT and SST data adequacy.

Data on changes in the height of the tropopause have recently attracted rapt attention. The height of the tropopause depends on the location, notably the latitude, as shown in the Fig. 5.5. As Santer et al. (2003) noted, starting from 1979, the height of the tropopause increased by several hundred meters, agreeing with the results of numerical climate modeling, taking into account the growth of GHGs concentrations, whose contribution prevails, again, in "enigmatic" agreement of the observed and calculated data. As it follows from Fig. 5.4 the tropopause height does not gradually drop from low to high latitudes. Rather, it drops rapidly in the area of the subtropical and polar front jets (Linacre and Geerts 1997).

Studies of the dynamics of the tropical tropopause layer are of great interest for quantitative estimates of climate change and an understanding of mechanisms for the troposphere-stratosphere interactions. These circumstances have stimulated recent serious attention to studies of the climatic structure and variability of the tropical tropopause, as well mechanisms responsible for the formation of this

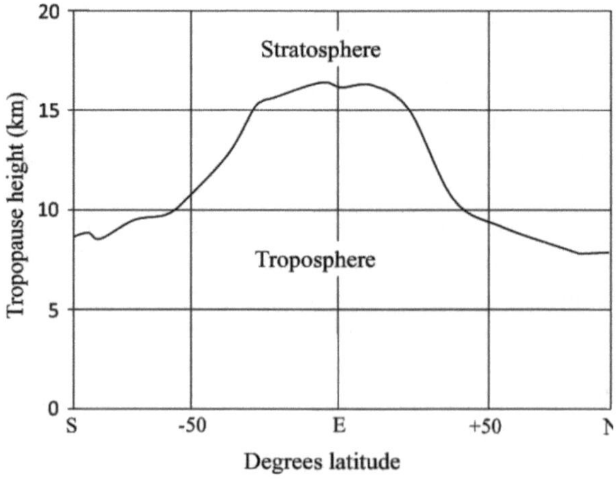

Fig. 5.5 The height of the tropopause

structure. Serious attention has also been paid to analysis of data on the content of water vapor in the stratosphere and mechanisms for the formation of thin cirrus clouds in the tropics. Randel et al. (2003) undertook studies of the structure and variability of the temperature field in the upper troposphere and lower stratosphere of the tropics (at altitudes about 10–30 km) from the data of radio-occultation observations for the period from April 1995 till February 1997, using the satellite system designed for geodetic measurements (GPS). A comparison with a large number (several hundreds) of synchronous aerological soundings has shown that a retrieval of the vertical temperature profiles from GPS/MET data provides reliable enough information.

Analysis of the obtained results suggested that the spatial structure and variability of the tropopause altitude determined by a "cool point" (minimum temperature) of the vertical temperature profile are governed mainly by wave-like fluctuations like Kelvin waves. A strong correlation was observed between temperature from GPS/MET data and outgoing longwave radiation, which can serve as an indirect indicator of penetrating convection in the tropics. This correlation confirms a reality of temperature fluctuation revealed from GPS/MET data and opens possibilities of quantitative assessments of the response of large-scale temperature field in the tropics to time-varying conditions of convection, revealing coherent wavelike variations at altitudes between 12 and 18 km.

5.3.3 Snow and Ice Cover

Snow and ice cover estimation is important for climate change studies and successful water resource management. Sea ice is important component in the global climate system. Sea ice regulates of heat, moisture and salinity in the polar oceans. Sea ice area is considerable varied in both hemispheres. In the Northern Hemisphere, the total extent of sea ice varies from a minimum of about 7.8×10^6 km^2 in September to a maximum of about 14.8×10^6 km^2 in March. In the Southern Hemisphere, the extent varies from about 4.0×10^6 km^2 in February to about 20.0×10^6 km^2 in September (Parkinson 1994; Hall et al. 1995; Parkinson et al. 1987; Orheim 1999). Since the end of the 1960s, a 10 % decrease of the snow cover extent has been observed, as well as a 2-week reduction of the annual duration of lakes and river ice-cover in NH middle and high latitudes, while in the non-polar regions mountain glaciers are retreating. In 2002, NH snow cover extent constituted 25.4 million km^2, on the average by 0.2 million km^2, less than during the preceding 30 years. The annual trend of snow cover changes from 2.7 (August) to 46.9 (January) million km^2 (Waple and Lawrimore 2002). Also, starting from the 1950s, the extent of NH ice cover in spring and summer has been decreasing by 10–15 %. During the last decades (in the periods "late summer—early fall") the Arctic sea ice cover thickness has probably decreased by about 40 %, with winter decrease being less substantial. From regular satellite observations (starting from the 1970s) no marked trend in the extent of ice cover in the Antarctic has been observed.

In recent years, satellite data have indicated an even more evident reduction in regional ice cover. Table 5.4, Figs. 5.6 and 5.7 contain some characteristics of sea ice extent and its variations. All estimated values in Table 5.4, were based on the NSIDC passive microwave satellite data, which represent the best method to monitor sea ice, because of the ability to snow data through most clouds and during darkness. Observations of polar oceans derived from microwave instruments are essential for tracking the ice edge, estimating sea ice concentration, and classifying sea ice types.

Figure 5.8 characterizes yearly distribution of the arctic sea ice area. This and analogous data regularly delivered by different satellite instruments mainly of the U. S. National Snow and Ice Data Center (NSIDC). Kelley et al. (2003) developed an algorithm to estimate global snow cover volume from spaceborne passive microwave remote sensing observations. This algorithm is a simple dynamic approach to retrieve global snow depth estimation. This approach, in contrast to statistic approach, assumes that snowpack properties are spatially and temporally dynamic and requires two simple empirical models of density and snowpack grain radius evolution, plus a dense media radiative transfer model based on the quasicrystalline approximation and sticky particle theory.

Ananicheva (2005) notes that the main parameter, which determines the extent and evolution of much of the terrestrial cryosphere, is the position of the bottom boundary of equal snow line altitudes, where potential snow accumulation and melting are of the same value. There exist and other indicators that characterize climate zones:

- climatic snow line (CSL) and
- glacier equilibrium line altitude (ELA).

Table 5.4 Characteristics of the yearly dynamics of the ice extent

Year of september average extent	Extent (million km^2)	Anomaly relative to 1981–2010 average (million km^2)	Anomaly relative to 1981–2010 average (%)	Anomaly relative to previous record (million km^2)	Anomaly relative to previous record (%)	Linear trend since 1979 (km^2 per year)	Linear trend since 1979 relative to 1981–2010 average (% per decade)
2002	5.96	−0.56	−8.6	−0.17	−2.8	−51,000	−7.8
2003	6.15	−0.37	−5.7	0.19	3.2	−52,800	−8.1
2004	6.05	−0.47	−7.2	0.09	1.5	−54,600	−8.4
2005	5.57	−0.95	−14.6	−0.39	−6.5	−59,400	−9.1
2006	5.92	−0.6	−9.2	0.35	6.3	−60,200	−9.2
2007	4.3	−2.22	−34	−1.27	−22.8	−71,600	−11
2008	4.73	−1.79	−27.5	0.43	10	−77,700	−11.9
2009	5.39	−1.13	−17.3	1.09	25.3	−78,200	−12
2010	4.93	−1.59	−24.4	0.63	14.7	−80,700	−12.4
2011	4.63	−1.89	−29	0.33	7.7	−84,000	−12.9
2012	3.63	−2.89	−44.3	−0.67	−15.6	−91,400	−14.0
2013	5.35	−1.16	−17.9	1.74	48.2	−89,530	−13.7

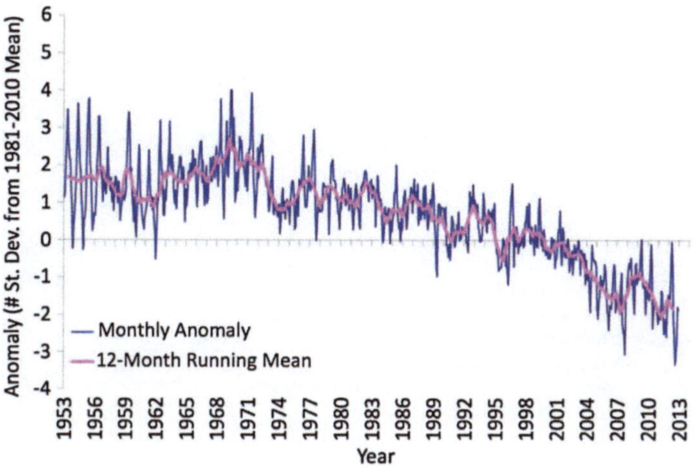

Fig. 5.6 Arctic sea ice extent standardized anomalies, January 1953–December 2012. (http://nsidc.org/cryosphere/sotc/sea_ice.html)

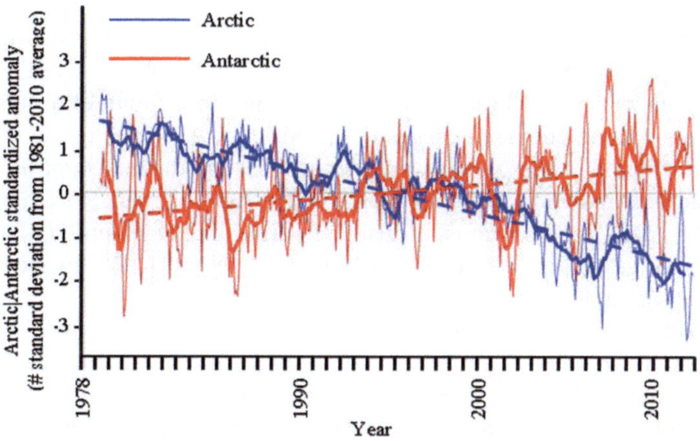

Fig. 5.7 Arctic and Antarctic standardized anomaly and trend, November 1978–December 2012. (http://nsidc.org/cryosphere/sotc/sea_ice.html)

The CSL is the point above which snow and ice cover the ground throughout the year. The actual snow line may seasonally be significantly lower. Typical CSLs are given in Table 5.5. The ELA marks the area or zone on a glacier where annual accumulation and ablation are equal.

Numerical modeling using global climate models has shown (from considering the growing concentration of GHGs and aerosols) that climate warming should increase in the Arctic because of a feedback determined by the melting of the sea ice and snow cover, causing a decrease in surface albedo. On the other hand, from

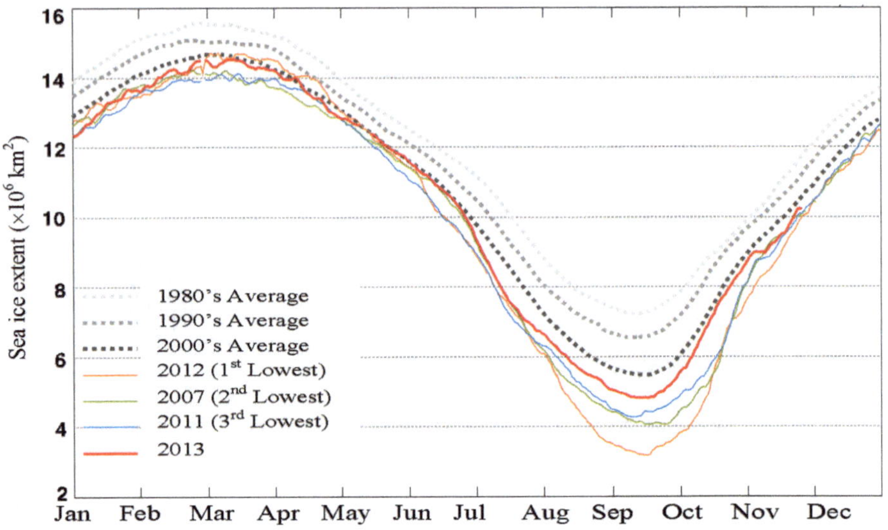

Fig. 5.8 Arctic sea ice extent during year (http://www.ijis.iarc.uaf.edu/en/home/seaice_extent.htm)

the observed data, SAT has increased during the last decades over most of the Arctic. One of the regions where a warming has taken place is northern Alaska (especially in winter and in spring). In this connection, Stone et al. (2002) have analyzed the data on climatic changes in the North of Alaska to reveal their impact on the annual trend of the snow cover extent (SCE) and the impact of SCE changes on the surface radiation budget (SRB) and SAT. The NASA/GEWEX SRB Release-3.0 data sets contains global 3-hourly, daily, monthly/3-hourly, and monthly averages of surface and top-of atmosphere (TOA) longwave and shortwave radiative parameters on a $1° \times 1°$ grid.

The Northern Hemisphere EASE-Grid 2.0 Weekly Snow Cover and Sea Ice Extent Version 4 product combine snow cover and sea ice extent at weekly intervals from 23 October 1978 through 31 December 2010, and snow cover alone from 03 October 1966 through 31 December 2010. Snow cover extent for this data set is based on the NOAA/NCDC Climate Data Record (CDR) of Northern Hemisphere (NH) Snow Cover Extent (SCE) regridded to the EASE-Grid. The NOAA/NCDC CDR of Northern Hemisphere Snow Cover Extent data were derived from the manual interpretation of AVHRR, GOES, and other visible-band satellite data (Helfrich et al. 2007; NOAA 2012). Sea ice extent is regridded to EASE-Grid from Sea Ice Concentrations from Nimbus-7 SMMR and DMSP SSM/I-SSMIS Passive Microwave Data. Designed to facilitate the study of Northern Hemisphere seasonal fluctuations of snow cover and sea ice extent, the data set also includes monthly climatologies, describing the average extent, probability of occurrence, and variance. Data are provided in flat, unsigned binary files and are available via FTP.

Table 5.5 Levels of the climatic snow line in different regions (http://en.wikipedia.org/wiki/Snow_line)

Region	Latitude	Altitude range (m)
Svalbard	78°N	0300–0600
Scandinavia at the polar circle	67°N	1,000–1,500
Iceland	65°N	0700–1,100
Eastern Siberia	63°N	2,300–2,800
southern Scandinavia	62°N	1,200–2,200
Alaska Panhandle	58°N	1,000–1,500
Kamchatka (coastal)	55°N	700–1,500
Kamchatka (interior)	55°N	2,000–2,800
Alps (northern slopes)	48°N	2,500–2,800
Central Alps	47°N	2,900–3,200
Alps (southern slopes)	46°N	2,700–2,800
Pyrenees	43°N	2,600–2,900
Caucasus	43°N	2,700–3,800
Pontic Mountains	42°N	3,800–4,300
Karakoram	36°N	5,400–5,800
Transhimalaya	32°N	6,300–6,500
Himalaya	30°N	4,800–6,000
Mount Kenya	0°	4,600–4,700
New Guinea	2°S	4,600–4,700
Andes in Ecuador	2°S	4,800–5,000
Kilimanjaro	3°S	5,500–5,600
Andes in Bolivia	18°S	6,000–6,500
Andes in Chile	30°S	5,800–6,500
North Island, New Zealand	37°S	2,500–2,700
South Island, New Zealand	43°S	1,600–2,700
Tierra del Fuego	54°S	0800–1,300
Antarctica	70°S	0000–0400

5.3.4 Sea Surface Level and Heat Content of the Ocean Upper Layer

Sea surface level is usually presented as mean sea level, that is the average height of the sea's surface. Detection of long-time variations of mean sea level can provide information on climate change. During the 20th century, the World Ocean surface level rose by 0.1–0.2 m. Apparently, this was caused by the thermal expansion of sea water and ice melting on land due to the global warming. The rate of the World Ocean level rise in the 20th century apparently exceeded that observed during the last 3,000 years by a factor of ten. Beginning from the end of the 1950s (when SST changes became large-scale), the heat content of the ocean upper layer has also been increasing. Between 1870 and 2004, global average sea levels rose 195 mm (7.7 in) (Church and White 2006). From 1950 to 2009, measurements show an average annual rise in sea level of 1.7 ± 0.3 mm per year, with satellite data showing a rise of 3.3 ± 0.4 mm per year from 1993 to 2009 (Nicholls et al. 2010), a faster rate of increase than previously estimated (IPCC 2007). For at least the last 100 years, sea level has been rising at an average rate of about 1.8 mm per year. Most of this rise can be attributed to the increase in temperature of the sea and the resulting slight thermal expansion of the upper 500 m of sea water.

Operative data about sea level is delivered by the Global Sea Level Observing System (GLOSS) that aims at the establishment of high quality global and regional sea level networks for application to climate, oceanographic and coastal sea level research. The main component of GLOSS is the Global Core Network (GCN) of 290 sea level stations around the world for long term climate change and oceanographic sea level monitoring (Fig. 5.9). Figures 5.10 and 5.11 characterize data about sea level dynamics. Figure 5.10 shows the sea level change as a function of temperature. Hourly, daily and monthly sea level data are provided and hosted by Joint Archive for Sea level (JASL)-NODC and the University of Hawaii Sea Level Center (UHSLC) from stations in tropical and subtropical areas of the Pacific, Atlantic, and Indian Oceans. Most data are from tide gauges but some data are delivered from bottom-mounted pressure gauges. Global sea level data are also delivered from satellite altimeter data (http://data.nodc.noaa.gov/accession/0053056).

Levitus et al. (2001) analyzed data on the warming of some components of the climatic system during the second half of the 20th century. These data were derived from the growth of the heat content of the atmosphere and ocean as well as from estimates of the heat losses due to melting of some components of the cryosphere. These findings have led to the conclusion that the heat content of the atmosphere and ocean is rising. The growth of the heat content in the 3-km ocean layer between

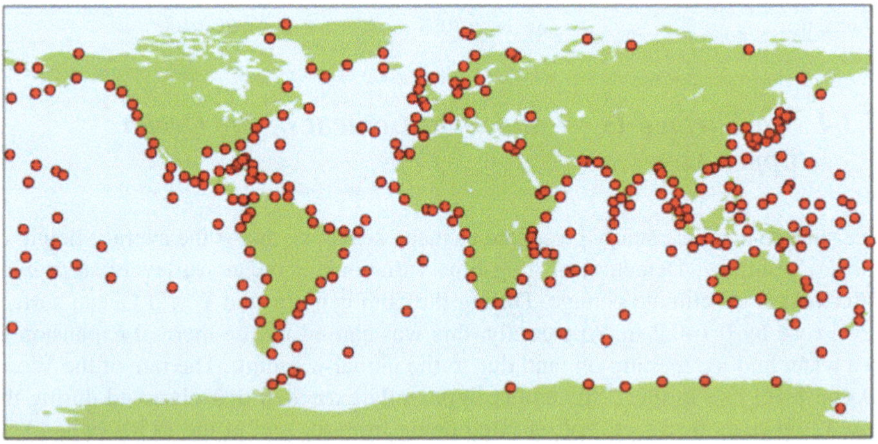

Fig. 5.9 The Global Sea Level Observing System core network (http://www.gloss-sealevel.org)

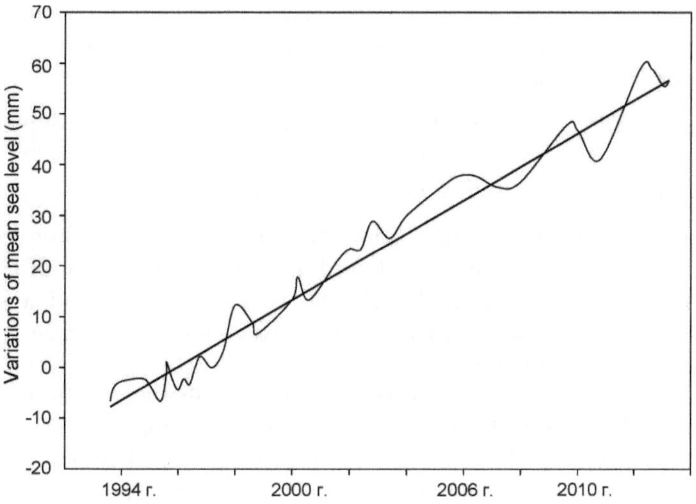

Fig. 5.10 Changes of the global mean sea level in time

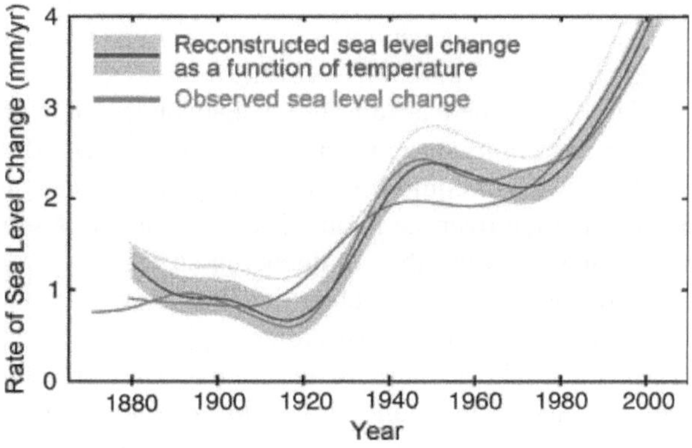

Fig. 5.11 Observed rate of sea-level rise (*red*) compared with reconstructed sea level calculated from global temperature (*dark blue* with *light blue* uncertainty range). *Grey line* is reconstructed sea level from an earlier, simpler relationship between sea level and temperature (Vermeer and Rahmstor 2009)

1950 and 1990 exceeded, at least, by an order of magnitude, the increase of the heat content in other components of the climate system. While the observed increase of the ocean heat content between 1955 and 1996 reached 18.2×10^{22} J, then for the atmosphere it constituted only 6.6×10^{21} J. As for the values of latent heat due to water phase transformations, they were: 8.1×10^{21} J (a decrease of the mass of glaciers on land); 3.2×10^{21} J (a decrease of the sea ice cover extent in the Antarctic); 1.1×10^{21} J (melting of mountain glaciers); 4.6×10^{19} J (a decrease of the snow cover extent in the Northern Hemisphere); 2.4×10^{19} J (melting of permafrost in the Arctic).

The observed data were compared with results of numerical modeling using the GFDL interactive model of the "atmosphere-ocean" system taking into account the radiative forcings due to:

(1) the observed growth of GHGs concentrations, changes of the sulphate aerosol content in the atmosphere and extra-atmospheric insolation; and
(2) volcanic aerosol; as well as GHGs and sulphate aerosol only.

The resulting comparisons have led to the conclusion that the observed changes of the ocean heat content can be explained, mainly, by the growth of GHGs concentrations in the atmosphere, though one should bear in mind a substantial uncertainty concerning the estimate of radiative forcings (RF) due to sulphate aerosol and volcanic eruptions.

Cai and Whetton (2002) drew attention to the fact that the ocean dynamics can considerably affect future global-scale precipitation. Developments in these difficult problems are based on the use of both observed data and results of numerical modeling, and have led to quite different conclusions. The climatic warming of the last decades was characterized by the spatial structure similar to that of the El Niño/Southern Oscillation (ENSO) event. But since there are no data on such a structure for the whole century, the observed structure of warming is assumed to be a manifestation of the multi-decadal natural variability of climate, not the result of the greenhouse forcing.

Moritz et al. (2002) revealed a substantial inadequacy of climate models as applied to the Arctic conditions. In most cases, calculated AO (Arctic oscillations) trends turned out to be weaker compared to observed ones. The calculated climate warming is greater in the fall over the Arctic Ocean, while the observed warming is at a maximum in winter and over the continents in spring.

Prestrud et al. (2007) analyzes data about ice and snow in polar and mountain regions of the world covering all parts of the cryosphere (the world of ice): snow, land ice, sea ice, river and lake ice, and frozen ground. They present information on the trends in ice and snow, the outlook for this century and beyond, and the consequences to ecosystems and human well-being of changes in ice and snow. But this book is also about sea-level rise and the islands in the South Pacific, and about shrinking glaciers and the great rivers of Asia. It is about the way the Earth's climate is changing and causing ice and snow to melt—and how melting ice and snow are changing the Earth's climate.

Singh et al. (2011) published encyclopedia that mainly focuses on the various aspects of snow, ice and glaciers, but also covers other cryospheric branches, and provides an up-to-date information and basic concepts on relevant topics. It includes alphabetically arranged and professionally written, comprehensive and authoritative academic articles by well-known international experts in individual fields. The encyclopedia contains a broad spectrum of topics, ranging from the atmospheric processes responsible for snow formation; transformation of snow to ice and changes in their properties; classification of ice and glaciers and their worldwide distribution; glaciation and ice ages; glacier dynamics; glacier surface and subsurface characteristics; geomorphic processes and landscape formation;

hydrology and sedimentary systems; permafrost degradation; hazards caused by cryospheric changes; and trends of glacier retreat on the global scale along with the impact of climate change.

Singh et al. (2011) note that the earth's cryosphere, which includes snow, glaciers and ice caps, ice sheets, ice shelves, sea ice, river and lake ice, and frozen ground, contains about 75 % of the earth's fresh water. It exists at almost all latitudes, from the tropics to the poles, and plays a vital role in controlling the global climate system. It also provides direct visible evidence of the effect of climate change, and, therefore, requires proper understanding of the complex dynamics.

5.3.5 Other Climatic Parameters

Data on global soil temperature (GST) are important for climate diagnostics. As Majorovicz et al. (2002) have noted, an analysis of the GST data obtained in different regions of Canada by measuring the ground temperature in bore-holes, revealed considerable spatial differentiation both in GST increase observed in the 20th century, and in the onset of the warming. For instance, from measurements in 21 bore-holes covering a period of the last 1,000 year, warming was detected (within 1–3 °C) during the last 200 years. The warming was preceded by a long cooling trend in the region 80°–96°W, 46°–50°N, which continued until the beginning of the 19th century. According to data for ten bore-holes in central Canada, the temperature had reached a minimum about the year 1820 with a subsequent warming by about 1.5 °C. In western Canada, during the last 100 years the warming reached 2 °C.

An analysis has been made by Majorovicz et al. (2002) of more adequate information on GST from data of measurements in 141 bore-holes at a depth of several hundred meters. The holes were drilled in 1970–1990. The results obtained revealed an intensive warming that started in the 18–19th centuries, which followed a long period of cooling (especially during the Little Ice Age), continuing during the rest of the millennium. The time of the onset of the present warming differed between regions. An analysis of the spatial distribution of the GST changes over the territory of Canada revealed a substantial delay in the onset of the present warming in the east-to-west direction, with a higher level of the GST increase in the 20th century in western Canada. This conclusion is confirmed by the data of SAT observations. It should be noted that the GST increase in eastern Canada had begun about 100 years before the industrial era.

Characteristics of the atmospheric general circulation are important components of climate diagnostics. Wallace and Thompson (2002) pointed out that the west-eastern zonal wind component averaged over the 55°N latitudinal belt can be a representative indicator of the primary mode of the surface air pressure anomalies—the Northern Hemisphere annual mode (NAM) that is hemispheric scale patterns of climate variability. Both NAM and a similar index SAM for the Southern Hemisphere are typical signatures of symbiotic relationships between the meridional profiles of the west-eastern transport in the respective hemisphere and wave disturbances superimposed on this transport. Their index determined (using a respective normalization) as a coefficient for the first term of NAM expansion in empirical orthogonal functions, can serve as the quantitative characteristic of the modes. The presence of the positive NAM (or SAM) index denoted the existence of a relatively strong west-eastern transport.

During the last years, it was recognized that dynamic factors contribute much to observed temperature trends. For instance, in 1995 a marked similarity was observed between the spatial distributions of the SAT field and NAM fluctuations for the last 30 years, with a clear increase of the NAM index. The increasing trend of the index was accompanied by mild winters, changes of the spatial distribution of precipitation in Europe, and the ozone layer depletion in the latitudinal belt >40°N. Similar data are available for the Southern Hemisphere. The main conclusion is that along with the ENSO event, both NAM and SAM are the leading factors of the global atmospheric variability. In this regard, attention should be focused on the problem of the 30-year trend of NAM towards an IM increase (after 1995 the index lowered). It is still not clear whether this trend is a part of long-term oscillations.

Figure 5.12 represents NAM for the Northern Hemisphere. When there's a deficit of air mass in the red areas, there tends to be a surplus in the blue areas, and vice versa. At the same time, the jet stream winds tend to speed up or slow down in the direction of the green arrows. A similar pattern exists in the Southern Hemisphere.

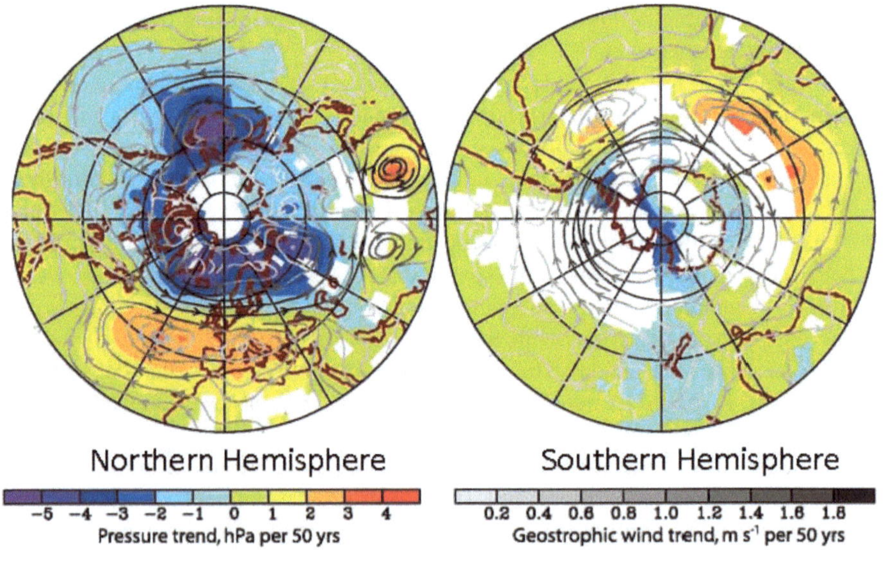

Fig. 5.12 The North Annular Mode pattern, which describes a large fraction of the monthly to decadal changes in northern hemisphere climate (IPCC 2007)

Feldstein and Franzke (2006) examined composites of the sea level pressure, 40-h Pa, and 300-h Pa streamfunction fields during NAO and NAM persistent events to test null hypothesis. Standard statistical techniques were used, such as *t* tests for the differences of means. This study addresses the question of whether persistent events of the NAO and the NAM teleconnection patterns are distinguishable from each other. Standard daily index time series are used to specify the amplitude of the NAO and NAM patterns. It is found that the null hypothesis cannot be rejected even at the 80 % confidence level. The wave-breaking characteristics during the NAM life cycle are also examined. Both the positive and negative NAM phases yield the same wave-breaking properties as those for the NAO. The results suggest that not only are the NAO and NAM persistent events indistinguishable, but that the NAO/NAM events are neither confined to the North Atlantic, nor are they annular.

The observational data show that during the 20th century an increase of precipitation constituted 0.5–1 % per 10 years over most of land surface in the middle and high latitudes of the Northern Hemisphere, but a decrease (by about 0.3 % per 10 years) took place over most of land surface in sub-tropical latitudes, which has recently weakened, however. As for the World Ocean, the lack of adequate observational data has not permitted the identification of any reliable trends of precipitation. In recent decades, intensive and extreme precipitation in the middle and high latitudes of the Northern Hemisphere has probably become more frequent. Beginning from the mid-1970s the ENSO events have been frequent, stable and intensive. This ENSO dynamics was reflected in specific regional variations of precipitation and SAT in most of the zones of the tropics and sub-tropics. Data on the intensity and frequency of occurrence of the tropical and extratropical cyclones as well as local storms still remain fragmentary and inadequate and do not permit to draw conclusions on any trends.

Changes in the biosphere are also important indicators of climate. One of them is the bleaching of corals. It is important to recognize that enhanced atmospheric forcings on coral reefs do not lead to their disappearance but to their transformation into more resistant species (Chin 2011; Hughes et al. 2003). Changes of sea water properties are one more indicator (Broecker 2003). The properties of sea water are a function of temperature, salinity (i.e. total dissolved solids in g kg^{-1}) and pressure (Table 5.6).

5.3.6 Concentrations of Greenhouse Gases and Anthropogenic Aerosol in the Atmosphere

A greenhouse gas (GHG) is a gas in an atmosphere that absorbs and emits radiation within the thermal infrared range. This process is the fundamental cause of the greenhouse effect. The primary greenhouse gases in the Earth's atmosphere are water vapor, carbon dioxide, methane, nitrous oxide, and ozone. Table 5.7 contents list of greenhouse gases and their characteristics related to the climate. Aside from water vapor, which has a residence time of about 9 days, major greenhouse gases

Table 5.6 Mechanical and thermal properties of sea water at salinity 35 g kg^{-1} and atmospheric pressure

Property	0 °C	20 °C
Dynamic viscosity	1.88×10^{-3} Pa s	1.08×10^{-3} Pa s
Kinematic viscosity, ν	1.83×10^{-6} m^2 s^{-1}	1.05×10^{-6} m^2 s^{-1}
Thermal conductivity	0.563 W m^{-1} K^{-1}	0.596 W m^{-1} K^{-1}
Thermal diffusivity, κ	1.37×10^{-7} m^2 s^{-1}	1.46×10^{-7} m^2 s^{-1}
Prandtl number, ν/κ	13.4	7.2
Specific heat capacity, C_p	3,985 J kg^{-1} K^{-1}	3,993 J kg^{-1} K^{-1}
Thermal expansion coefficient		
Pressure = 0.1 MN m^{-2}	52×10^{-6} K^{-1}	250×10^{-6} K^{-1}
Pressure = 100 MN m^{-2}	244×10^{-6} K^{-1}	325×10^{-6} K^{-1}
Ratio of specific heat capacities, C_p/C_v	1.0004	1.0106
Velocity of sound	1,449 m s^{-1}	1,522 m s^{-1}
Compressibility	4.65×10^{-10} Pa^{-1}	4.28×10^{-10} Pa^{-1}
Freezing point		−1.910 °C
Boiling point		100.56 °C

http://www.kayelaby.npl.co.uk/general_physics/2_7/2_7_9.html

Table 5.7 Gases relevant to radiative forcing only (IPCC 2007)

	Mole fractions and their changes		Radiative forcing	
Species	2005	Change since 1998	2005 (W m^{-2})	1998 (%)
CO_2	379 ± 0.65 μmol/mol	+13 μmol/mol	1.66	+13
CH_4	1,774 ± 1.8 nmol/mol	+11 nmol/mol	0.48	–
N_2O	319 ± 0.12 nmol/mol	+5 nmol/mol	0.16	+11
CFC-11	251 ± 0.36 pmol/mol	−13	0.063	−5
CFC-12	538 ± 0.18 pmol/mol	+4	0.17	+1
CFC-113	79 ± 0.064 pmol/mol	−4	0.024	−5
HCFC-22	169 ± 1.0 pmol/mol	+38	0.033	+29
HCFC-141b	18 ± 0.068 pmol/mol	+9	0.0025	+93
HCFC-142b	15 ± 0.13 pmol/mol	+6	0.0031	+57
CH_3CCl_3	19 ± 0.47 pmol/mol	−47	0.0011	−72
CCl_4	93 ± 0.17 pmol/mol	−7	0.012	−7
HFC-125	3.7 ± 0.10 pmol/mol	+2.6	0.0009	+234
HFC-134a	35 ± 0.73 pmol/mol	+27	0.0055	+349
HFC-152a	3.9 ± 0.11 pmol/mol	+2.4	0.0004	+151
HFC-23	18 ± 0.12 pmol/mol	+4	0.0033	+29
SF_6	5.6 ± 0.038 pmol/mol	+1.5	0.0029	+36
CF_4 (PFC-14)	74 ± 1.6 pmol/mol	–	0.0034	–
C_2F_6 (PFC-116)	2.9 ± 0.025 pmol/mol	+0.5	0.0008	+22

Units μmol/mol = ppm = parts per million (10^6); nmol/mol = ppb = parts per billion (10^9); pmol/mol = ppt = parts per trillion (10^{12})

are well-mixed, and take many years to leave the atmosphere. Greenhouse gases in the atmosphere absorb and re-emit some of the outgoing energy radiated from the Earth's surface, causing that heat to be retained in the lower atmosphere. The global warming potential (GWP) depends on both the efficiency of the molecule as a greenhouse gas and its atmospheric lifetime (Table 5.8).

From 1750 till now the CO_2 concentration in the atmosphere has increased by about one third, reaching the highest level for the last 420 thousand years (and, probably, during the last 20 million years), which is illustrated by the data from ice cores (IPCC 2001, 2007). The growth of CO_2 concentration by about two thirds during the last 20 years is explained by emissions to the atmosphere from fossil fuel burning (contributions of deforestation and, to a lesser extent, cement industry constitute one third). Figure 5.13 characterizes this data. It is of interest that by the end of 1999, CO_2 emissions in the USA exceeded their 1990 level by 12 %, and by 2013 their further increase should raise this value by another 10 %. Nevertheless, according to the Kyoto Protocol, the emissions should be reduced by 7 % by 2013 compared to the level of 1990, which requires a total reduction in emissions of by about 25 %.

According to available observational data, both the World Ocean and land are currently global sinks for CO_2. In the ocean, both chemical and biological processes are responsible; on land we observe an enhanced "fertilization" of vegetation due to increased concentrations of CO_2 and nitrogen, as well as with changes in land use. Yet much remains unclear about the global carbon cycle (Kondratyev et al. 2003c). In particular, contradictions in the estimates of the role of the biosphere and ocean in the global carbon cycle remain to be resolved.

There is no doubt that fossil fuel burning will remain the main factor in the growth of CO_2 concentration in the 21st century. The role of the biosphere (both the ocean and land) as a barrier to the growth of CO_2 concentration will be reduced in time. According to the IPCC-2001 Report, the probable interval of CO_2

Table 5.8 Atmospheric lifetime and GWP relative to CO_2 at different time horizon for various greenhouse gases (http://en.wikipedia.org/wiki/Global-warming_potential)

Gas name	Chemical formula	Lifetime (years)	Global warming potential (GWP) for given time horizon		
			20-year	100-year	500-year
Carbon dioxide	CO_2	30–95	1	1	1
Methane	CH_4	12	72	25	7.6
Nitrous oxide	N_2O	114	289	298	153
CFC-12	CCl_2F_2	100	11,000	10,900	5,200
HCFC-22	$CHClF_2$	12	5,160	1,810	549
Tetrafluoromethane	CF_4	50,000	5,210	7,390	11,200
Hexafluoroethane	C_2F_6	10,000	8,630	12,200	18,200
Sulfur hexafluoride	SF_6	3,200	16,300	22,800	32,600
Nitrogen trifluoride	NF_3	740	12,300	17,200	20,700

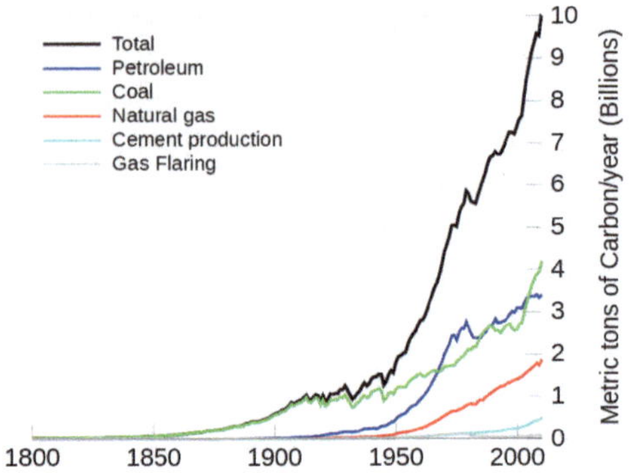

Fig. 5.13 Global annual fossil fuel carbon dioxide emissions through year 2007, in million metric tons of carbon (http://www.upload.wikimedia.org/wiki/File:Global_Carbon_Emission_by_Type_to_Y2004.png and Global_Carbon_Emission_by_Type_to_Y2004.png)

concentration values by the end of the century will constitute 540–970 ppm (pre-industrial and present values are, respectively, 280 and 367 ppm). Changes of land use are also an important factor of the global carbon cycle, but all carbon emitted to the atmosphere due to land use will be assimilated by the land biosphere. This could only lead to a decrease of CO_2 concentration within 40–70 ppm. As for the prognostic estimates of the concentrations of other GHGs, they vary widely. So, for instance, it is concluded from some estimates that the role of tropospheric ozone as a greenhouse gas can be equal to the contribution of methane and will be substantial, as being the factor of deterioration of air quality over most of the northern hemisphere.

The concentration of methane in the atmosphere increased by a factor of 2.5, compared with that observed in 1750, and continues to grow. The annual rate of CH_4 increase was reduced, however, and became more variable in the 1990s compared with the 1980s. From the beginning of 1750, nitrous oxide concentration has increased by 16 %. With the implementation of the Montreal Protocol, concentrations of several halocarbon compounds functioning as GHG and ozone-destructing gases either, have increased more slowly or started decreasing. However, concentrations of their substitutes and some other synthetic compounds started to grow rapidly (e.g., perfluorocarbons, PFC, and sulphur hexafluoride, SF_6).

As for the properties of atmospheric aerosol and its climatic impact, respective current information has been reviewed in detail in (Kondratyev et al. 2003c; Krapivin and Varotsos 2008). In this connection, it is pointed out again that the supposed anthropogenic nature of the present global climate warming was explained by the warming caused by the growth in GHGs concentrations (primarily CO_2 and CH_4), as well as cooling due to anthropogenic aerosols. However, if the estimates of the "greenhouse" warming can be considered as sufficiently reliable, then the respective calculations of radiative forcing (RF) due to aerosol are very uncertain. Of no less importance is the fact that, while the global distribution of the "greenhouse" RF is comparatively uniform, the case of the "aerosol" RF is characterized by a strong spatial-temporal variability (including changes of the sign of RF).

5.3.7 Paleoclimatic Information

Paleoclimatic information is an important source of data for the comparative analysis of the present and past climates. Analysis of the data of paleoclimatic observations reveals large-scale abrupt climate changes in the past, when the climate system had exceeded certain threshold levels. Though some mechanisms for such changes have been identified and the existing methods of numerical climate modeling are being gradually improved, the existing models still do not permit a reliable reconstruction of past climatic changes. With emphasis on the climatic implications of the growth of GHGs concentrations in the atmosphere, less effort has been made to study possible sudden climate changes that may be of natural origin, though possibly intensified by anthropogenic forcings.

Since such changes lie beyond the problems addressed in the UN FCCC, Alley et al. (2003) undertook a conceptual evaluation of the problem of large-scale abrupt climate changes. Though the available long-term stabilizing feedbacks have determined the existence on the Earth of comparatively persistent global climate for about 4 billion years, with characteristic time scales from 1 year to 1 million years, feedbacks prevailing in the climate system had favored an enhancement of forcings on climate. So, for instance, changes of global average SAT within 5–6 °C during the glaciation cycles apparently resulted from very weak forcings due to variations of the orbital parameters.

It is still more surprising that during several decades and in the absence of external forcings, regional changes have taken place, reaching 30–50 % of those, that had taken place in the epochs of glaciations. Data from the period of instrumental observations have revealed abrupt climatic changes, quite often accompanied by serious socio-economic consequences. So, for instance, the warming in many northern regions in the 20th century took place in two rapid “steps”, which enables one to suppose that in this case there was a superposition of the anthropogenic trend on inter-annual natural variability. Special attention was paid to the role of the ENSO event. The latter also refers to a sharp change of the climate system in the Pacific region in 1976–1977.

Considerable abrupt changes of regional climate in the period of Paleocene were detected from paleoclimatic reconstructions. They had been manifested as changes of the frequency of occurrence of hurricanes, floods, and especially droughts. Regional SAT changes reaching 8–16 °C had happened in the periods of 10 years and shorter. Dansgaard-Oeschger (DO) oscillations can serve as an example of large-scale sudden changes. DO events are rapid climate fluctuations that occurred 25 times during the last glacial period.

The climatic system involves numerous factors that intensify climatic changes with minimum forcings. The withering or death of plants, for example, may cause a decrease of evapotranspiration and hence lead to precipitation attenuation, which may further increase drought conditions. In the cold-climate regions the snow cover formation is accompanied by a strong increase of albedo, which favours further cooling (the so-called “albedo effect”). Substantial climatic feedbacks are associated with the dynamics of the thermohaline circulation.

While the factors of enhancement of either changes or stability of climate are comparatively well known, the understanding of the factors in the spatial distribution of anomalies over large regions, including the globe, is of great importance. In this connection, further studies of various modes of the general circulation of the atmosphere and the ocean (ENSO, DO oscillations, etc.) are important, as it is the respective improvement of general circulation models. Most important here are the potential effects of abrupt climatic changes on ecology and economy, as current estimates are generally based on the assumption of slow and gradual change.

Abrupt climate changes were especially substantial in periods of the transition of one climatic state to another. Therefore, if anthropogenic forcings of climate can favor the drifting of the climate system towards a threshold level, the possibility of raising the probability of abrupt climate changes also increases. Of great importance has not only the amount but also the rate of anthropogenic forcings on the climate system. So, for instance, a faster climate warming should favor a stronger attenuation of the thermohaline circulation, as this may promote an acceleration of the shift to the threshold of climatic changes (it is important that under these conditions, thermohaline circulation dynamics becomes less predictable). To accept adequate solutions in the field of ecological policy, a deeper understanding of the whole spectrum of possible sudden climate changes is extremely important. Difficulties in the identification and quantitative estimation of all possible causes of sudden climate change and low predictability near threshold levels, testify the fact that the problem of abrupt climate changes will always be aggravated by more serious uncertainties, than the problem of slow change. Under these conditions the development of ways to provide the stability and high adaptability of economics and ecosystems is of great importance.

5.4 Interactive Components of Climate System

5.4.1 Anomalous Situations and Climate

Kosaka and Xie (2013) considered the hiatus in global warming mentioned that despite the continued increase in atmospheric GHG concentrations, the annual-mean global temperature has not risen in the twenty-first century. Figure 5.14 represent variations of temperature for 2002–2012. Compared to seasonal norms, in September 2013 the coolest area on the globe was south of South Africa in the southern ocean, where temperatures in the troposphere were about 2.49 °C cooler than normal. The warmest area was in the Wilkes Land area of the east Antarctic, where tropospheric temperatures were 5.20 °C warmer than seasonal norms.

Climate change manifests itself both on global and regional scales. One of the important features of climate formation not only on regional but also on global scales consists in considerable variability determined by internal dynamics of climate system. One of the most substantial factors of internal dynamics is the El Niño/Southern Oscillation (ENSO) event. One of the recent climate warming due to

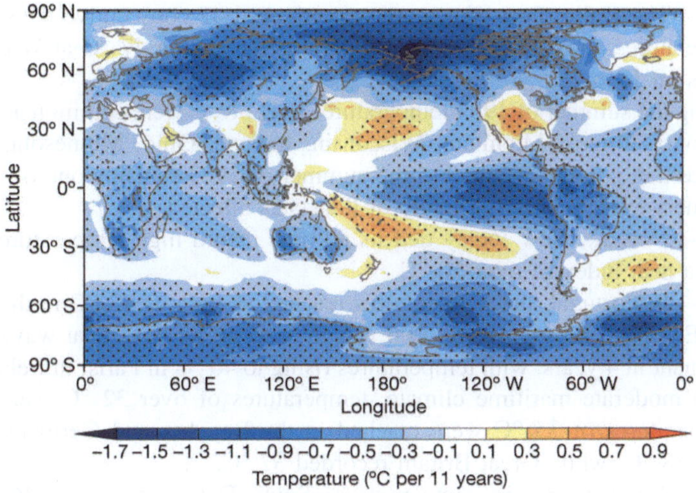

Fig. 5.14 11-year trends of annual-mean SAT composited for 34 hiatus events in history (http://www.nature.com/nature/journal/vaop/ncurrent/full/nature12534.html?WT.ec_id=NATURE-2)

ENSO started in October–November 2002 and ended in March–April 2003. However, despite the end of ENSO in the period of boreal spring, the ENSO-induced warming has led to regional anomalous rains in a wide range of the Pacific Ocean, including the formation of the zone of increased moisture content along the western coast of South America and the region of moisture deficiency in the eastern part of Australia as well as in the south-western sector of the Pacific Ocean.

The global mean SAT in 2003 turned out to be close to three maximum values observed during the period from 1880, but below the record level of SAT in 1998. An increase of global mean SAT in 2003 compared to an average value for 1961–1990 constituted 0.46 °C. According to the data of satellite thermal sounding, the global mean temperature of the middle troposphere in 2003 was third by the level of warming, compared to the average value for 1979–1998 period. The annual global average surface temperature anomaly for 2011 was +0.07 °C, with the 1981–2010 average as a baseline. 2012 was the ninth warmest year with global average temperature 14.45 °C. This is 0.45 ± 0.10 °C above the 1961–1990 average. The trend of global average temperature is 0.68 °C/Century. August 2013 was the 342nd consecutive month (more than 28 years) with a global temperature above the 20th-century average. These data show that SAT is changed with some dispersion relative to some climatic trend that is represented in Fig. 5.9.

In general, history represents the temperature anomalies regularly:

- 1923/1924—During a period of 160 days from 31 October 1923 to 7 April 1924, the Western Australian town of Marble Bar set a world record for the most consecutive days above 38 °C.
- On July 14 1954, the thermometer reached 47 °C at East St. Louis, Illinois, which remains the record highest temperature for that state.

- 1980—The Dallas/Fort Worth area experienced 42 consecutive days with high temperatures above 38 °C, with temperatures reaching 47 °C at Wichita Falls, Texas on June 28.
- During the summer of 1983 temperatures over 38 °C were common across Iowa, Missouri, Illinois, Michigan, Wisconsin, Indiana, Ohio, Minnesota, Nebraska and certain parts of Kentucky; the summer of 1983 remains one of the hottest summers ever recorded in many of the states affected.
- 2001—Newark, New Jersey tied its all-time record high temperature of 41 °C with a heat index of over 50 °C.
- 2003—In Portugal, the temperatures reached as high as 47 °C in the south.
- The European heat wave of 2006 was the second massive heat wave to hit the continent in 4 years, with temperatures rising to 40 °C in Paris; in Ireland, which has a moderate maritime climate, temperatures of over 32 °C were reported. Temperatures of 35 °C were reached in the Benelux and Germany (in some areas 38 °C, while Great Britain recorded 37 °C.
- 2006—Temperatures in some parts of South Dakota exceeded 46 °C. Also, California experienced temperatures that were extraordinarily high, with records ranging from 40 to 54.4 °C. On July 22, the County of Los Angeles recorded its highest temperature ever at 48 °C.
- 2007—Bulgaria experienced its hottest year on record, with previously unrecorded temperatures above 45 °C. The Indian city of Datia experienced temperatures of 48 °C.
- 2008—Alice Springs in Australia's Northern Territory recorded ten consecutive days of temperatures above 40 °C with the average temperature for that month being 39.8 °C. In March, Adelaide, South Australia experienced maximum temperatures of above 35 °C for 15 consecutive days, 7 days more than the previous longest stretch of 35 °C days. The March heat wave also included eleven consecutive days above 38 °C. The heat wave was especially notable because it occurred in March, an autumn month, in which Adelaide averages only 2.3 days above 35 °C.
- 2009—Adelaide, South Australia was hit by a heat wave with temperatures reaching 40 °C for 6 days in a row, while many rural areas experienced temperatures hovering around about mid ~40 °C. Kyancutta on the Eyre Peninsula endured at least 1 day at 48 °C. In neighbouring Victoria recorded 3 consecutive days over 43 °C, and also recorded its highest ever temperature 8 days later in a secondary heatwave, with the mercury peaking at 46.4 °C.
- 2010—On 26 May, at Mohenjo-daro, Sindh province in Pakistan, a national record of high temperature of 53.5 °C occurred.
- In late July and early August 2011, temperatures in Iraq were exceeding 49 °C (Baghdad—52 °C).

The summer of 2003, in some regions of Western Europe, was one of the warmest summer seasons with heat waves affecting mainly Central and Western Europe. Two anomalous heat waves taking place in June and July–August (especially the second wave) were especially powerful. Droughts accompanying them have caused forest fires, which covered a considerable part of the territory in the south of France and in Portugal in July and August. The summer of 2003 in Western Europe was apparently the hottest over the period after the year 1540. The heat wave in France killed 11,000 people. In Germany that summer was the hottest in the 20th century and (except some regions of northern and north-western Germany) the hottest over the whole period of instrumental observations.

The most substantial anomalous situations that happened in March 2003 included:

(1) extremely intensive precipitation in the middle part of the Atlantic Ocean, in the south-east and eastern coast of the USA;
(2) extremely low SAT values and unusual snowfalls over the European territory of Russia;
(3) 546 tornados in May in the USA, which was unprecedented;
(4) a long-term drought in the west of the USA, where in some regions it was the fourth and 50 year of a considerable deficiency of precipitation;
(5) heavy brush fires in the eastern part of Australia in January and powerful forest fires in the south of California in October;
(6) anomalously intensive precipitation in Western Africa and in Sahel;
(7) return to the normal level of precipitation on the Indian sub-continent in the period of summer monsoon; and
(8) close-to-record extent of the "ozone hole" in the Antarctic reaching a maximum of 28.2 million km^2 in September 2003.

The globally averaged temperature for 2012 (14.6 °C) marked the 10th warmest year since record keeping began in 1880. It also marked the 36th consecutive year with a global temperature above the 20th century average. The last below-average annual temperature was 1976. Including 2012, all 12 years to date in the 21st century (2001–2012) rank among the 14 warmest in the 133-year period of record. Only 1 year during the 20th century—1998—was warmer than 2012.

Last years have been marked by an increased interest in study of the present climate change in high latitudes of the Northern and Southern Hemispheres, which is mostly determined by decision to conduct in 2007–2008 the Third International Polar Year. The major conclusions concerning the Arctic climate diagnostics are concentrated on analysis of the spatial-temporal variability of polar climate instead of exaggerated attention to unfounded simplification of the situation as manifesting itself through a homogeneous anthropogenic enhancement of climate warming in high latitudes. In this context, of great interest are new results of the paleoclimatic analysis of an ice core from the station "Vostok" (Vakulenko et al. 2004), which demonstrated a negative correlation between changes of CO_2 concentration in the atmosphere and air temperature. Paleoclimatic developments become a more and more actual way to study the laws of present climate dynamics (Widmann et al. 2004).

From the data on the Antarctic discussed in the report (Levinson and Waple 2004), it follows that the last decade in this region was anomalously cold. From the late 1970s to the mid-winter of 1990, the sea ice cover extent round the Antarctic continent was growing.

The monograph of Filatov (2004) dedicated to studies of climate of Karelia can serve an example of the informational analysis of the regional features of climate. New developments dedicated to the climate of cities (Mayers 2004) and analysis of individual long series of meteorological observations (Alessio et al. 2004; Garcia-Barrón and Pita 2004) have made an important contribution to studies of regional climate change. A new important stage in comprehending the data of empirical diagnostics of climate was the development and application of interactive models of climate system and an ensemble approach to numerical climate modeling.

There exist many climate models which are differed between them by spatial resolution, input information, conditions of use, and by algorithms. For example, in Java Climate Model climate change is influenced by complex interlinked processes. This interactive model lets you explore the system and how we can change it, simply by adjusting with your mouse parameters and observing the effect instantly on diverse plots ranging from socioeconomic drivers to climate impacts. The core science calculation methods are calibrated to be consistent with results from the Intergovernmental Panel on Climate Change, implemented efficiently in the java language, to enable anybody to access this tool via the internet and explore diverse scenarios and the sensitivity of projections to risk/value assessment and scientific uncertainties.

As has been repeatedly emphasized, the interactive components of the present climate system include a broad spectrum of natural and natural-anthropogenic subsystems and processes, without a complex study of which, it is impossible to reliably select prevailing trends in climate change. In this connection, one should enumerate the most important ones:

- Global water cycle. Effect of "cloud" feedbacks.
- Global carbon cycle. Interaction of water and carbon cycles.
- Land use and land surface changes.
- Present trends of the GHGs content in the atmosphere and mechanisms of their control.
- Interaction of climate and land ecosystems' productivity.
- Effect of the climate regime shifts on marine ecosystems.
- Control of natural resourcesNatural to neutralize negative consequences of humans' activity.
- Socio-economic aspects of ecodynamics and climate and their analysis for optimization of land use strategy.
- Interactions between processes in the geosphere and biosphere and their dependence of cosmic impacts.

5.4.2 Climate Change, Forests, and Agriculture

Forest and agriculture ecosystems are the environmental components most sensitive to climate change. The former, determine many characteristics of biogeochemical cycles of GHGs, and the latter form the man-environment interaction. Problems appearing here are thoroughly studied within many international programmes on the environment and climate. Especially, they are emphasized in the national programmes of the USA. In particular, analysis of consequences with the use of various scenarios of possible changes of global climate has led to the conclusion that in the case of several scenarios, the impact on forestry and agriculture in the USA will be economically favourable. Partially, it is connected with the growth of forests productivity (due to the growth of CO_2 concentration) and determined by the ability of forests to adapt to climate change. As for agriculture, according to available prognostic estimates for the period up to 2060, the positive impact of the process of global warming on agriculture in the USA will be less economically favourable than it follows from the earlier estimates.

Unfortunately, in view of the global and poorly studied character of correlation between climate change and behaviour of vegetation cover (forest ecosystems, in particular) at present there are no reliable estimates of the consequences of climate change for their productivity. The study of the problems appearing here has been just started.

One of the principal aspects of the anthropogenic impact on the environment is the evaluation of the consequences of CO_2 emissions into the atmosphere. The published results estimating the greenhouse effect and excess CO_2 distribution in the biosphere, which bear on this problem, are widely varying and sometimes contradictory or else they are too flatly stated. This is a natural consequence of all kinds of simplifications adopted in modeling the global CO_2 cycle. The GIMS makes it possible to create an effective monitoring system allowing the estimation of the spatial distribution of the carbon sinks and sources in real time.

Before this, some problems should be solved to assess the role of the anthropogenic use of the Earth's surface. In particular, among these problems there is the problem of the formalized description of the processes of change of the structure of the Earth land covers, such as afforestation, forest reconstruction, deforestation and the associated carbon supplies. Understanding of the meteorological processes as functions of the greenhouse gases is one of the key problems of humankind in the first decade of the third Millennium. Only an adequate knowledge of the meteorological phenomena on various spatial-temporal scales under conditions of varying supplies of CO_2 and other greenhouse gases will enable one to make correct and constructive decisions in the field of global environmental protection.

The dynamics of the surface ecosystems depends on interactions between biogeochemical cycles, which during the last decade of the 20th century suffered significant anthropogenic modification, especially to the cycles of carbon, nitrogen, and water. The surface ecosystems, in which carbon remains in the living biomass, decomposing organic matter, and the soil, play an important role in the global CO_2 cycle. Carbon exchanges between these reservoirs and the atmosphere take place through photosynthesis, respiration, decomposition, and burning. Human interference into this process takes place through changing the structure of the vegetation covers, pollution of the water basin surfaces and of the soil areas, as well as through direct emissions of CO_2 into the atmosphere.

The role of various ecosystems in the formation of carbon supplies to the biospheric reservoirs determines the rate and direction of changes to the regional meteorological situations and to the global climate. The accuracy of assessment of the level of these changes depends on the reliability of the data on the inventory of the surface ecosystems.

Existing environmental data show that knowledge of the rates and trends of carbon accumulation in the surface ecosystems are rather uncertain. However, it is clear that the surface ecosystems are important assimilators of excess CO_2. Understanding the details of such assimilation, is only possible through modeling the process of plant growth; that is, considering the effect of the nutrient elements of the soil and other biophysical factors on plant photosynthesis. Therefore, the forest ecosystems and associated processes of natural afforestation, forest reconstruction, and deforestation should be studied in detail. In a forested area, the volume of the reservoir of CO_2 from the atmosphere is a function of the density of the forest canopy, and in a given period of time a change of this volume is determined by the level and character of the dynamic processes of transition of one type of forest into another. The causes of this transition can be natural, anthropogenic, or mixed. Biocenology tries to create a universal theory of such transitions, but so far there is only a qualitative description of the observed transitions.

Let us return briefly to the problem of anthropogenic changes of the structure of forest ranges. In the Kyoto Protocol its solution is connected with the problem of definition of such notations as forest, afforestation, tree planting, deforestation, biome, surface cover, land use, degradation, etc. Anyhow, a question arises here: to what extent is the use of either definition justified at making a decision and how does the result of its realization depend on their quality? The Kyoto Protocol not only does not answer this question, it even does not put it. In this respect, Table 5.9 is demonstrative and serves the basis for drawing various conclusions. In fact, a huge range of uncertainty exists here due to which the processes of afforestation and

Table 5.9 Directions of carbon supplies changes with different forms of soil cover reconstructions (Watson et al. 2000)

Type of land cover reconstruction	Direction of changes in C supplies					
	Biomass		Forest cover/ wood		Q	M
	A	U	S	L		
Cultivated land → forest	↑	↑	–	↑	↑	↑
Uncultivated land → forest	↑	↑	–	↑	?	↑
Forest → cultivated land	↓↓	↓	↓	↓	↓	–
Forest → pasture	↓↓	↓	↑	↓	?	–

Notations A above ground, *U* under ground, *S* short-lived, *L* long-lived, *Q* organic matter, *M* wood production

deforestation cannot be estimated uniquely. Clearly, a single correct way to solve the principal problem and the respective partial problems is to make a detailed inventory of the forest ranges and to use the obtained data in the global model. The specific features of this inventory should be determined in an adaptive regime of using the model, gradually changing the spatial-temporal scales, starting from the known model with a mean-annual time step and the one of geographical grids 4° × 5°, 1° × 1° or 0.5° × 0.5°.

5.4.3 Observational Data

Analysis of observational data is reduced, as a rule, to the consideration of two categories of information:

(1) SAT changes for the last 1.5 century (and especially during the last 20–30 years, when an increase of the global mean annual mean SAT was at a maximum).
(2) Paleoclimatic changes. They attract attention from the viewpoint of their comparison with present climatic trends and, to some extent, as an analog to possible climate change in the future (such attempts are being continued, though inadequacy of paleoanalogs for future climate forecast has been repeatedly and convincingly argumented).

By definition, climate is characterized by values of meteorological parameters averaged over 30 years (so, for instance, climate anomalies in 2010 are determined as deviations from averages over the period 1981–2010). An analysis of the spatial-temporal climate variability for individual years is also widely practiced. In particular, the World Meteorological Organization (WMO) publishes annual surveys of global climate. In these reports attempts are being made to answer the most important questions:

- Does a climate warming take place?
- Is the moisture cycle intensity changing?
- Is the general circulation of the atmosphere and ocean changing?
- Do extreme climate changes (storms, droughts, floods) intensify?
- Is a reliable estimate of the anthropogenic contribution to climate change possible?

The decade of the 1990s, on the whole, was the warmest over the whole period of meteorological observations (beginning from 1860), and the year 1999 was the fifth year by the level of anomalies of the global mean annual mean SAT (+0.33 °C) for the period 1860–1999 (also the fifth was the average SAT anomaly (+0.45 °C) in the Northern Hemisphere, but in the Southern Hemisphere it was only the tenth (+0.20 °C)).

The band of maximum annual mean SAT extended from the continent of North America eastward across the Atlantic Ocean and the Eurasian continent to the equatorial band of the western sector of the Pacific Ocean. Minimum SAT anomalies were observed in a broad band of the central and north-eastern regions of the Pacific Ocean (including a decrease of SAT). An analysis of observational data revealed the prevailing of positive temperature anomalies in 1999 in many regions of the globe. The most apparent anomalous situations include both warming and cooling, including:

(1) The cold wave observed in January has brought a SAT decrease in Norway, Sweden, and in some regions of Russia to the levels not observed since the late 19th century.
(2) A temperature decrease in February in Western Europe was followed, in particular, by heavy snowfalls in the Alps.
(3) In Western Australia the SAT decreased to values below the norm, though an extreme warming in early January has led to intensive brush fires.
(4) The March temperature in Iceland was at a minimum for the last 20 years.
(5) In April, powerful heat waves formed in the northern and central regions of India, and in July and August—in the north-eastern and middle-western regions of the USA.
(6) Unusually hot and dry weather was observed in the western part of Russia (SAT anomalies in the central and north-western regions of the European territory of Russia exceeded 5 °C).
7) On the Australian continent, maximum average SAT in November—December turned out to be the lowest after 1950.
(8) The second half of the year was colder (than usual) in central and southern Africa; the Sahel region was colder, wetter, and with more clouds than during the preceding years.
(9) The warming in the USA during the last 50 years was weaker than over the rest of the globe, with a weak cooling in the eastern part of the USA.

The land and ocean surface temperature decrease in the tropics in 1999 was determined by the year-long La Niña event. This year was characterized by a great number of destructive meteorological catastrophes, especially floods. In Australia, USA, and Asia there was a multitude of tropical storms, in Europe—heavy snowfalls, avalanches, and storms; in the USA—the drought and tornado.

The global mean annual SAT value in the late 20th century exceeded by more than 0.6 °C the value recorded in the late 19th century (the error of this estimate ±0.2 °C corresponds to a 95 % confidence level). Analysis of the SAT observational data suggested the conclusion that from the beginning of 1850, there has been an irregular but substantial trend of climate warming on a global scale. This trend was very weak in the period from the mid-19th century till 1910, and then it increased to 0.1 °C during 10 years (from combined data on SAT and SST in the period 1910–1940 and during the last two decades). Two positive episodes of cooling were separated by the interval of a weak cooling, especially in the Northern Hemisphere. In the periods from 1951–1960 to 1981–1990 the sign of the inter-hemispherical difference of temperatures changed: the Northern Hemisphere became colder than the Southern Hemisphere.

As Moron et al. (1998) noted, the present global warming was considered by some specialists as connected with sudden changes in the region of the Pacific Ocean about 1976 or with a gradual warming of the tropical band of the Pacific Ocean as well as with other regional-scale phenomena. The irregular trend mentioned above and an attempt to recognize the external forcings on it (anthropogenic or natural), complicated by the presence of the internally determined variability of the climate system, have been traditionally based on the interpretation of the trend as red noise (or, later, on the idea that the trend's variability is determined by that it is regularly "immersed" in noise).

The existence of such regularities has been well established and ascribed mainly (if not completely) to instability of the interactive "atmosphere-ocean" system in the tropical Pacific Ocean. Periodicities of about 4–6 years and 2–3 years connected with the ENSO event have been detected. Such regularities on scales of decadal and inter-decadal variability were more difficult to detect in view of insufficient length of the observation series (Krapivin and Nitu 2009).

In this connection, Moron et al. (1998) undertook a detailed analysis of all available data on the spatial-temporal variability of the SAT fields of the World Ocean and for its individual regions with the use of a multi-channel singular spectral analysis (MSSA). The main goal of analysis was to detect the laws of variability and inter-basin relationships between SST on time scales from inter-annual to inter-decadal. The length of observational data series was sufficient for a reliable analysis of SST variability on time scales 2–15 years, though the statistical reliability of results for longer periods is more difficult to guarantee.

In view of a great interest in SST variability in the Atlantic Ocean, the major attention was given to this region. The strongest climatic signal was an irregular long-term SST trend. The use of the MSSA method for data processing for the 20th century revealed the well-known regularities mentioned above: a gradual increase of SST in both hemispheres in 1910-1940, with the subsequent increase of SST in the Northern Hemisphere till the mid-1950s; but a lower SST in the Southern Hemisphere; northern hemisphere ocean cooling in the 1960s until the end of the 1970s; and an initial stability and then increase of SST in both hemispheres in the 1980s with a small weakening of this trend during the last years.

An insufficient length of the series of instrumental observations makes impossible to interpret the enumerated global laws as a manifestation of more or less monotonic increase of SST or as part of long-tern centennial oscillations (according to indirect data, oscillations were observed with periods from 65 to 500 years). Possible external factors of variability include: the growth of CO_2 concentration, change of extra-atmospheric insolation, volcanic eruptions. A new and surprising result was a detection of the fact that a large-scale warming and cooling was preceded by the same SST variability near the southern edge of Greenland and (soon after that) in the central part of the Pacific Ocean in the Northern Hemisphere. This reflects an important role of high-latitude processes in the North Atlantic and possible interaction (via the atmosphere) with the Pacific Ocean.

On time scales of about decadal variability (7–12 years) there have not been observed any regular oscillations, which are coherent on global scales. In the

Northern Atlantic, there were 13–15-year and 90-year oscillations. Near Cape Hatteras there were inter-decadal oscillations which propagated along Gulf-stream to the zone of the North Atlantic, where their phases change (similar results obtained earlier were rather contradictory). In the context of the search of inter-decadal oscillations of ENSO the data considered did not reveal any substantial maximum of SST variability with periods longer than 10 years in either the Pacific Ocean or the whole World Ocean, but in the Indian Ocean there were observed 20-year SST oscillations, especially regular during the first half of the 20th century (oscillations of this kind have been observed earlier).

Analysis of 7–8-year oscillations revealed the contradictory character of their phase in the sub-tropical and sub-polar cycles of the North Atlantic. As for the inter-annual variability (2–6 years), three dominating periods were recorded: 24–30, 40, and 60–65 months. The first period is a well known quasi-biannual ENSO component which most strongly manifests itself in the tropics of the eastern sector of the Pacific Ocean, with anomalies of the constant sign propagating along the western coastline of North and South America (in other oceans such a variability is negligibly small).

An important new result consists in detection of two clearly differing low-frequency modes of oscillation, combined by the common physical nature of the quasi-four-year mode and characterized by a drastic change of periodicity in 1960—from $\sim$5 years to almost 4 years. This set of observational data suggests the conclusion that the ENSO irregularity occurs due to interaction of the internal instability of the atmosphere-ocean system in the tropical Pacific Ocean with the annual change. Since the mode of quasi-biannual oscillations exists also in other oceans, but does not correlate practically with the index of the southern oscillations, one should assume that a stronger quasi-four-year signal forming here can remotely propagate beyond the Pacific Ocean, whereas the opposite process is practically impossible.

A weak oscillation with periods about 28–30 months is observed in the SST field in the SH Atlantic Ocean and agrees with the Hermanito event observed earlier, which is, probably, an ENSO analog. The results discussed are a stage of developments, aimed at comparison of numerical modeling results with the data of observations using the 2- and 3-D models of the atmosphere-ocean system.

An important contribution into the idea of SST changes in the past has been made by analysis of the data of temperature observations in boreholes. So, for instance, Bodri and Čermák (1999) noted that while the amplitude of long-term SAT changes in transitions from glaciations to inter-glacial periods reached 10–15 K, during the Holocene (the last 10–14 thousand years) changes of the order of several degrees K took place on time scales from decades to several centuries. In this connection, an analysis has been made of the data on the vertical profiles of temperature, measured at different depths in bore-holes on the territory of the Czechia and maps were drawn of SAT changes on the territory of Czechia taking place during 1,100–1,300 years to nowadays (little climatic optimum), 1,400–1,500 years, and 1,600–1,700 years (the main phases of the Little Ice Age).

Huang et al. (2000) discussed the results of processing the data on temperature measured at different depths in 616 bore-holes in the Southern Hemisphere, which has opened up possibilities to retrieve the change of global mean temperature for five centuries. The data for 479 holes revealed the global warming by about 1.0 K taking place during the last five centuries. Only during the 20th century, the warmest one, an increase of the continents' surface temperature reached 0.5 K (about 80 % of climate warming fell on the 19th–20th centuries). The warming during the five centuries had been stronger in the Northern Hemisphere (1.1 K) than in the Southern Hemisphere (0.8 K). On the whole, the results obtained agree with conclusions drawn on the basis of data on tree rings, though the latter demonstrate a weaker centennial SAT trend, which can be explained by special features of dendroclimatic methods.

Analysis of paleo-information on SAT obtained from the data on oxygen isotopes in Greenland ice cores for the Quaternary period has shown that long-term temperature changes are superimposed by faster changes on time scales from millennium to 10 years (Bowen 2000). An analysis of the Antarctic ice cores revealed similar changes. In particular, in both polar regions, substantial changes of temperature had taken place in the period of Holocene.

The data on plankton foraminifera *Neogloboquadrina pachyderma* from the north-eastern region of the Atlantic (west of Ireland) made it possible to retrieve the SST and trace the Heinrich events connected with icebergs' outbrakes.

Information on the content of carbon dioxide and methane in air bubbles contained in ice cores reflects an important role of those MGCs in climate formation. For instance, it was shown that during four interglacial periods temperature in the Antarctic had changed about 4,000 years earlier than changes of CO_2 concentration.

New results of numerical modeling of the dynamics of the El Niño event caused by variations of the orbital parameters, have satisfactorily followed "Milankowich" frequencies during the last 150,000 years as well as variations on a time scale of about a millennium (Morén and Påsse 2001). However, most surprising was a detection of climate variations with the period of 1,450 years from different data for different regions of the globe, which had regularly repeated, in particular, in Greenland during the last 110,000 years, including the last glaciation and Holocene (an increase of the amplitude of such changes was observed in the periods of glaciation). These results reflect a radical reorganization of the climate system taking place during comparatively short time periods. Holocene looked (compared with these changes) like a period of comparatively stable climate. There is no doubt that in the absence of climatic feedbacks, the growth of GHGs concentration in the atmosphere should bring forth a climate warming. However, a real situation turns out to be much more complicated, and to understand it, a reliable detection and quantitative estimates of the role of feedbacks are needed. Otherwise, a reliable forecast of climate change in the future is impossible. Since one of the very important sources of respective information is peat-bogs, they should be thoroughly protected.

Having analyzed the data of satellite observations of SST started from 1982, Strong et al. (2000) noted a warming over most of the tropics and in mid-latitudes of the Northern Hemisphere (with the global mean trend +0.005 °C per year not exceeding the limits of observation errors). Less representative SH SST data reflect the existence of the opposite cooling trend (the problem of SST data reliability needs serious attention).

According to the data of Levitus et al. (2000), during the last 50 years (1948–1998) the World Ocean has substantially warmed. The upper 300-m layer has warmed most (by 0.31 °C, on the average), whereas the temperature of the 3-km layer increased by 0.06 °C. This increase of the temperature of the upper layer of the ocean had preceded the SAT increase started in 1970.

Satellite data on the sea ice cover extent are important as an indicator of the global climate dynamics. Gloersen et al. (1999) detected a statistically substantial decrease of the global area of sea ice, constituting (−0.01 ± 0.003) 10^6 km^2 per 10 years.

Microwave remote sensing data is quite special, analysis of which has not revealed any substantial changes of average temperature of the lower troposphere during the last decades. This is confirmed also by results of aerologic soundings. From the data of Woodcock (1999a, b), the global mean SAT in October 1999 was 0.2 °C below the average value for the period 1979–1999. The earth's average temperature rose by 0.8 °C during the last century. That change is in the mean global temperature that has been calculated for each year since 1880. The history of average global temperature is based on data gathered around the world at weather stations, on ships and by satellites. Despite the overall increase, global temperatures have not climbed steadily during the decades. The average global annual temperature hovered around 13.7 °C from the 1880s through the 1910s. During the 1920s–1940s, temperatures climbed about 0.1 °C each decade. The mean global temperatures were then stabilized at roughly 14.0 °C until the 1980s. The world has mainly grown hotter since 1980, at a rate of nearly 0.2 °C per decade. The annual global temperature from 2000 to 2009 was 0.61 °C higher than the average temperature for 1951–1980. If the current rate of increase continues, the world will warm by 2 °C in the next century. The years 2010–2013 were warmest in the GISS analysis:

- The global land and ocean surface temperature (land and ocean) for 2010 tied (with 2005) as the warmest of the 131-year record.
- The year 2010 was also the wettest year on record, globally, in terms of precipitation measured. Precipitation in 2010 was much more variable than temperature on smaller scales (region-to-region and even within regions). This is typical of any year.
- The year 2010 tied with 2005 in all three global-scale components: the global land temperature, the global ocean temperature, and the global land and ocean surface temperature.
- The year 2010 was the 34th consecutive year with a global temperature above the 20th century average. The last year with a global temperature below the average was 1976.
- December was the 310th consecutive month with a global temperature above the 20th century average. The last month with below average temperatures was February 1985.

- March, April and May of 2010 were each the warmest on record (for the month). This ties 2010 with 1998, as the years with the most such "warmest months" on record, with three each.
- The current 2012 global average temperature is 14.45 °C. This is 0.45 ± 0.10 °C above the 1961–1990 average.

Santer et al. (2000) have discussed the causes of different trends of SAT and the lower troposphere temperature. Having analyzed the SAT data for the periods 1925–1944 and 1978–1999, Delworth and Knutson (2000) came to the conclusion that the main cause of SAT changes was a combined impact of anthropogenic RF and unusually substantial many-decadal internal variability of the climate system.

An important indicator of climate dynamics can be satellite data on changes of the balance of the mass of Greenland glaciers. Results of laser altimetry in northern Greenland, for the period 1994–1999, show that, on the whole, at altitudes above 2 km the ice sheet was balanced, with local changes of different signs. A decrease of the glacier's thickness dominated at low altitudes, exceeding 1 m per year, which is enough to raise the World Ocean level by 0.13 mm per year (this is equivalent to about 7 % of the observed rise of the ocean's level).

The data of observations of the moisture cycle parameters still remain fragmentary. An exception are publications such as Russo et al. (2000), in which an analysis has been made at the Observatory of the Genoa University of the change of the diurnal sum of precipitation for 1833–1985, and a decrease of the quantity of rainy days for the whole period of observations has been revealed, as well as a considerable growth of rain rate starting from 1950. During the last 30 years there was a considerable increase of the number of days with intensive precipitation.

Yu et al. (1999) performed an analysis of available climate data on the heat balance of the atmosphere with the use of results of both observations and calculations. The atmospheric radiation budget was found from the data of satellite observations of the fluxes of outgoing shortwave and longwave radiation and radiation fluxes at surface level retrieved from satellite data. Quantities of turbulent heat fluxes at surface level were taken from the data of observations within the COADS programme, and the horizontal heat transport was calculated with the use of the respective meteorological information. To minimize random errors, the spatial-temporal averaging was made: the zonally averaged components of the atmospheric heat balance components for the latitudinal band 50°N—50°S, as well as values for this latitudinal band have been considered.

An analysis of the data discussed has shown that it is impossible to close the atmospheric heat balance: additional 20 W m^{-2} are needed. Attempts to use different versions of the input volumes of information did not help to remove this "disbalance". Since the closing of water vapour balance with the use of the same data was successful, one can assume that the cause of this "disbalance" is an inadequacy of estimates of the atmospheric radiation budget, manifesting though underestimated solar radiation absorbed by the atmosphere.

Having analyzed the completeness and reliability of the available data of climatic observations, Folland et al. (2000) have come to the conclusion that the existing volume and quality of data make it impossible to give adequate answers to the questions enumerated above. In this connection, anxiety is aroused because of the degradation of the systems of conventional meteorological observations, taking place during the last decades, which are of importance also for calibration of the satellite remote sensing results. Therefore, even calculations of decadal mean values of climate parameters are difficult for some regions of Africa and vast regions of the World Ocean.

Discussing the observed regularities of global climate change with time and its causes, Wallace (1998) noted the first-priority to consider the following problems:

(1) periodic climate change due to variations of extra-atmospheric solar radiation;
(2) quasi-periodic climate variability (its most vivid manifestation—quasi-biennial oscillations in the equatorial stratosphere);
(3) ENSO event (in view of a wide range of frequencies, this event cannot be considered quasi-periodic);
(4) inter-decadal climate variations, which are to a great extent determined by the internal intra-seasonal and intra-annual variability of the climate system;
(5) climate variability on time scales from inter-decadal to centennial;
(6) analysis of statistical significance of estimates of unprecedented events and "shifts of regimes" in the light of time-dependent series of many climatic parameters; and
(7) revealing the phase relationships between climate change on tine scales from inter-annual to inter-decadal.

An important fact is that most of climate variability can be described by a separate consideration of dependencies on time and space coordinates. In view of an exceptional complexity of the climate system, with its numerous degrees of freedom and a multitude of feedbacks, highly regular structures and modes of climate evolution can be rather an exception than a rule. "Rough" schemes of parametrization for reconstructing the structure and evolution of climate anomalies without superfluous detail should have a higher degree of stability. Important climatic "signals" considered in solving the problems of detection and forecast of global climate change should be seen "with the naked eye". A much more complicated problem than usually supposed is an assessment of statistical significance of some quantitative characteristics of climate variability, especially unprecedented events and "shifts of regime" from the data on time series of limited duration (as a rule, in view of time-dependence of such series).

Apparently, one of the perspective directions of studies of climatic time-dependence and related catastrophic events of the type of the tropical storm Katrina, which 13 years after Hurricane Andrew was the most powerful in the history of Miami (Florida), caused the USA a huge economic damage, completely submerging New Orleans and ruining a huge amount of constructions in late August 2005, is a search of connections between temperature variations of different scales in different water bodies of the World Ocean. For instance, Chang et al. (2000) and Yamagata et al. (2004) have shown that there is a stable correlation between changes of water surface temperature in the Indian and Pacific Oceans, which especially strongly manifest themselves in the season of monsoons in the Indian

Ocean basin. The ENSO event favouring the propagation of the sub-tropical anti-cyclone over the Western Pacific plays a marked role in stirring up of feedback mechanisms. Studies of appearing correlations are successfully carried out with the use of an interactive ensemble CGCM developed at the Centre for Ocean-Land-Atmosphere Studies (COLA).

A new method to simulate and diagnose the observed climate variability has been developed by Schneider (2006). A method, which can be used to retrospectively simulate and understand the evolution of surface temperature using reanalysis surface fluxes and a coupled GCM, has been developed. The method has two parts:

(1) weather noise surface fluxes are extracted from the reanalysis surface fluxes by removing the ensemble mean response of an AGCM ensemble to the observed surface temperature evolution; and
(2) an interactive ensemble (IE) CGCM (an ensemble of atmospheric models coupled to a single OGCM) is then forced by this noise.

The IE CGCM will reproduce the observed surface temperature evolution up to errors in the model, the reanalysis surface fluxes, and the ocean initial state given certain conditions. These conditions are that there is no internal ocean "weather noise" or coupled instabilities. External forcing must also be taken into account properly (Schneider et al. 2009; Despres et al. 1986).

In the Pacific Ocean there are two key regions which play an important role in variability of the upper water layer temperature. These are western and central sectors of the northern Pacific Ocean. Changes taking place here affect the climatic situation in many regions of Asia and in more remote areas (Nakamura and Yamagata 1997). Therefore, study of complicated climatic situations in the Pacific region is important for detection of latent dependences between stimulators of global climate change in the future (Timmermann et al. 1999).

The Secretariat of the Pacific Regional Environment Programme (SPREP) has been designated by Pacific heads of government as the lead agency to coordinate the region's response to this challenge. Climate change is already disproportionately affecting the islands of the Pacific. Although islanders have done little to contribute to the cause—less than 0.03 % of current global greenhouse gas emissions—they are among the first to be affected. Most islands are experiencing climate change impacts on communities, infrastructure, water supply, coastal and forest ecosystems, fisheries, agriculture, and human health. The consequences of sea level rise, sea temperature increases, ocean acidification, altered rainfall patterns, and overall temperature rise will be increasingly felt (Faletau et al. 2012).

5.4.4 Climate-Forming Factors

Discussing the prospects of developments within the CLIVAR program of study of climate variability, Bolin (1999) emphasized that "IPCC was very careful in its assessments in order to stick to conclusions known from scientific literature, which serve the basis for such assessments. The key fact is that it is necessary to distinguish between something that can be considered real and which remains uncertain. As for the future climate forecasts, many uncertainties still remain. This approach has determined the confidence in scientific community when making concrete decisions and should be preserved in future".

Bolin (1999) emphasized that though the IPCC-1995 assessment report contains the statement that global climate change taking place in the 20th century is partially determined by humans' activity, this conclusion was formulated very carefully. An evaluation of the probable contribution of random climate variations independent of humans' impact in this context was of principal importance. Results of recent studies have clarified this question, showing that random variations of global mean SAT on time scales from decades to centuries for the last 600 years were within ±0.2 °C or less (this conclusion, as has been mentioned above, disagrees with data of observations from which it follows that SAT changed in the past in wider limits). In this connection (as Bolin (1999) believes), the skeptics about estimates of the contribution of anthropogenic warming for the last 50–75 years should be asked how one can explain a much stronger global warming observed during the last decades.

In reports of official representatives of several countries at Conferences in Kyoto and Buenos Aires and in mass media, weather and climate anomalies like tropical hurricanes and unusual El Niño were ascribed to the impacts of global warming. These opinions should, however, be thoroughly tested scientifically, though a possibility of more frequent anomalous events under conditions of global warming is not excluded. Therefore, especially urgent becomes the development of methods of climate forecasts on regional scales, bearing in mind, first of all, the period of 2008–2012.

In Bolin's (1999) opinion, to assess the socio-economic consequences of the accomplishment of measures foreseen by the Kyoto Protocol on GHGs emissions reduction, it is very important to develop integral models—a combination of models of climate, carbon cycle, as well as power engineering and socio-economic development, which will require much more time and much effort. In this respect, a difficult problem is to validate such models in order to analyze their reliability. The absence of an adequate validation means that the results of numerical modeling with the use of integral models can be considered only as possible scenarios but not forecasts.

Characterizing the climatic forcings, Hansen et al. (1998, 1999) pointed out that they are still not determined with an accuracy that is sufficient for reliable climate forecasts. There is reliable information about the GHGs content in the atmosphere, which determines a positive RF, but serious difficulties are connected with assessments of the impacts caused by such factors as atmospheric aerosol, clouds, land use change, causing a negative RF, which determines a partial compensation of the "greenhouse" climate warming. One of the consequences of this compensation consists in a much more significant role of changes of extra-atmospheric insolation (solar constant—SC) as a climate-forming factor than it was supposed earlier, based on numerical modeling with only GHGs contribution taken into account (the "greenhouse" RF due to the growth of CO_2 concentration in the period from the beginning of the industrial revolution to nowadays was estimated at about 1.5 W m^{-2}).

In connection with these circumstances, Hansen et al. (1998, 1999) obtained new estimates of global mean RF. The data in Table 5.10 characterize the results of analytical approximation (with an error of about 10 %) of various components of the "greenhouse" RF (the recent detection of SF_5CF_3 as a substantial GHG shows that the problem of substantiation of GHG priority cannot be considered resolved completely). The GHG concentrations are expressed in ppm (CO_2, c); ppb (CH_4, m); ppb (CFC-11, x; CFC-12, y); CFC—chlorofluoroorganic compounds (freons). The total RF value is 2.3 ± 0.25 W m^{-2}. The fact that the rate of RF increased from 0.01 to 0.04 W m^{-2} per year over the period 1950–1970, during subsequent 20 years decreased down to 0.03 W m^{-2} per year, in connection with the decreasing rate of the growth of CO_2 concentration (despite the continuing growth of CO_2 emissions), is of great interest, although the reasons remaining unclear. A certain contribution was made also by the reduction of the growth of CO_2 concentration, but also for unknown reasons.

The RF due to the growing concentration of tropospheric ozone was estimated at 0.4 ± 0.15 W m^{-2}. A drop of stratospheric ozone could result in the RF equal to −0.2 ± 0.1 W m^{-2}. Although these changes of the sign are mutually partly compensated, it does not mean their insignificance, since variations of the ozone content in the troposphere and stratosphere affect substantially and in a different way the formation of the vertical profile of temperature.

As for the RF due to aerosol, its determination is still unreliable for lack of adequate information on the real atmospheric aerosol. A numerical modeling, with anthropogenic sulphate, organic, and soil aerosol taken into account (with a latent consideration of soot aerosol by prescribed realistic aerosol absorption), for prescribed global distributions of aerosol optical thickness has made it possible to evaluate the global distributions of RF and balanced surface temperature, and then to obtain respective global mean values of changes of RF (ΔF) and surface temperature (ΔT_s) for purely scattering aerosol (single scattering albedo ω = 1) and more realistic aerosol (Table 5.11).

Table 5.10 Greenhouse radiative forcing *F* over the period after the beginning of industrial revolution (Kondratyev et al. 2006b)

Gas	Radiative forcing
CO_2	$F = f(c) - f(c_0)$, where $f(c) = 5.04\ \lg(c + 0.0005c^2)$
CH_4	$0.04(m^{1/2} - m_0^{1/2}) - [g(m, n_0) - g(m_0, n_0)]$; $g(m, n) = 0.5\ \lg[1 + 0.00002(mn)^{0.75}]$
N_2O	$0.04(n^{1/2} - n_0^{1/2}) - [g(m_0, n) - g(m_0, n_0)]$;
CFC-11	$0.25(x - x_0)$
FC-12	$0.30(y - y_0)$

Table 5.11 Global mean RF for three types of anthropogenic aerosol

Type of aerosol	$\varpi = 1$		"More realistic" ϖ	
	ΔF (W/m^2)	ΔT_s (°C)	ΔF (W/m^2)	ΔT_s (°C)
Sulphate	−0.28	−0.19	−0.20	−0.11
Organic	−0.41	−0.25	−0.22	−0.08
Dust	−0.53	−0.28	−0.12	−0.09
Total	−1.22	−0.72	−0.54	−0.28

Hansen et al. (1998) noted that most reliable value of total RF due to aerosol constitutes -0.4 ± 0.3 W m^{-2} instead of the value −0.54 given in Table 5.6, though in any case this estimate remains very uncertain due to unreliable input data on aerosol properties.

Anthropogenic RF changes due to clouds are undouptedly more substantial than those due to aerosol, but they are even more uncertain. Such changes (including the impact of aircraft contrails) mainly result from the indirect impact of anthropogenic acrosol, which causes (functioning as condensation nuclei) variations of cloud droplets size distribution and optical properties. Rough estimates give the "cloud" RF from −1 to −1.5 W m^{-2}, but it can change by an order of magnitude (depending on prescribed input parameters). Conditional value $-1^{+0.5}_{-1}$ W m^{-2} can be assumed. Some increase of cloud amount observed in the 20th century may be attributed to indirect impact of aerosol. Specification of the obtained estimates requires an accomplishment of complex observational programmes in different regions of the globe.

The contribution of land-use changes into RF variations is connected with processes of deforestation, desertification, and biomass burning, which affect the surface albedo and roughness as well as evapotranspiration. It is also important that the bare surface albedo changes more strongly being covered with snow than vegetation. Approximate estimates of the ERB change due to land-use evolution were -0.2 ± 0.2 W m^{-2}.

The natural RF due to SC changes during the last century (including an indirect impact on the ozone layer) can be assumed to equal 0.4 ± 0.2 W m^{-2}. Since the total RF constitutes only about 1 W m^{-2}, the contribution of extra-atmospheric insolation variability could play a substantial role. Volcanic eruptions cause RF changes from 0.2 to −0.5 W m^{-2} (these estimates are, however, conditional). For analysis of possible anthropogenic impact on global climate, extremely important are estimates of sensitivity of the climate system to external forcings. Hansen et al. (1998) assumed that the change of global mean SAT at a doubled CO_2 concentration should constitute 3 ± 1 °C. Since the RF estimates are not reliable enough, it is expedient to use different scenarios of RF change. One of the developments in this sphere is a study of Tett et al. (1999) and IPCC (2007).

The simulations of the 19th and 20th centuries are based upon changes in long-lived greenhouse gases (LLGHGs) that are reasonably constrained by the observational record. Therefore, the models have qualitatively similar temporal evolutions of their radiative forcing time histories for LLGHGs (e.g., see Fig. 5.15). However,

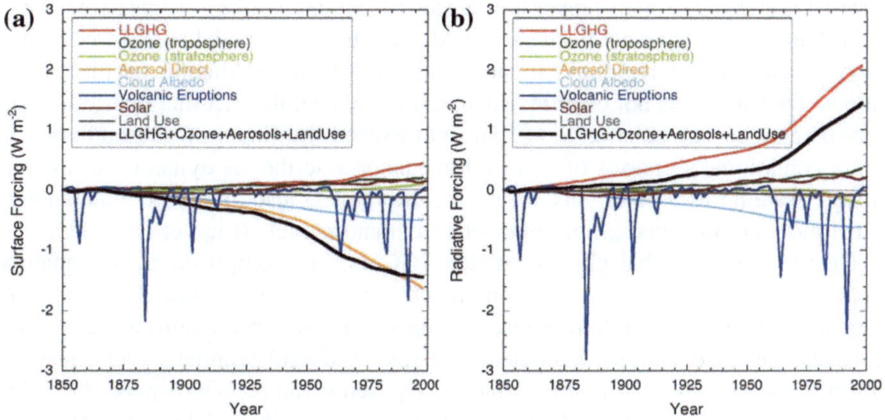

Fig. 5.15 Globally and annually averaged temporal evolution of the instantaneous all-sky RF (**a**) and surface forcing (**b**) due to various agents, as simulated in the IROC+SPRINTARS model (Watanabe et al. 2011; Takemura et al. 2005). This is an illustrative example of the forcings as implemented and computed in one of the climate models participating in the AR4. Note that there could be differences in the RFs among models. Most models simulate roughly similar evolution of the LLGHGs' RF

estimates of future concentrations of LLGHGs and other radiatively active species are clearly subject to significant uncertainties. The evolution of these species is governed by a variety of factors that are difficult to predict, including changes in population, energy use, energy sources and emissions. For these reasons, a range of projections of future climate change has been conducted using coupled AOGCMs. The future concentrations of LLGHGs and the anthropogenic emissions of sulphur dioxide (SO_2), a chemical precursor of sulphate aerosol, are obtained from several scenarios considered representative of low, medium and high emission trajectories. These basic scenarios and other forcing agents incorporated in the AOGCM projections, including several types of natural and anthropogenic aerosols, are discussed in IPCC (2007).

Crowley (2000) estimated the contribution of various factors to climate formation (SAT change) for the last 1,000 years using the energy-balance climate model. According to the results obtained:

(1) Changes of global mean SAT during the last 1,000 years may be explained as a result of the combined impact of the known RFs (in the pre-industrial epoch 41–64 % of SAT changes had taken place due to the extra-atmospheric insolation and volcanic activity).
(2) Global warming observed in the 20th century was mainly of anthropogenic ("greenhouse") origin, substantially exceeding the internally caused variability of the climate system.

Unfortunately, the argumentation contained in the study of Crowley (2000) is unconvincing even from the viewpoint of explanation of the centennial change of global mean SAT. For instance, the causes of climate cooling in the late 19th—early 20th century have not been explained. Of course, the model considered cannot simulate changes of regional climate. The role of North-Atlantic oscillations in climate formation has not been revealed. On the whole, the fact that possibilities of approximate energy-balance models do not exceed the limits of substantiation of rather conditional scenarios of climate but cannot describe the dynamics of the real climate system raises no doubts. This conclusion refers also to results obtained with much more complicated global interactive climate models (Liao et al. 2011).

On the whole, global climate models (GCMs) are comprised of fundamental concepts (laws) and parameterisations of physical, biological, and chemical components of the climate system. These concepts and parameterisations are expressed as mathematical equations, averaged over time and grid volumes. The equations describe the evolution of many variables (e.g. temperature, wind speed, humidity and pressure) and together define the state of the atmosphere. These equations are then converted to a programming language, defining among other things their possible interacting with other formulations, so that they can be solved on a computer and integrated forward in discrete time steps.

A very simple model of the radiative equilibrium of the Earth is:

$$(1 - \alpha)S\pi r^2 = 4\sigma\pi r^2 \varepsilon T^4 \tag{5.1}$$

where the left hand side represents the incoming energy from the Sun and the right hand side represents the outgoing energy from the Earth, calculated from the Stefan–Boltzmann law assuming a constant radiative temperature; T is global mean temperature that is to be found; S ($\approx$1,367 W m^{-2}) is the solar constant—the incoming solar radiationper unit area; α ($\approx$0.3) is the Earth's average albedo; r ($\approx 6.371 \times 10^6$ m) is Earth's radius; σ ($\approx 5.67 \times 10^{-8}$ J K^{-4} m^{-2} s^{-1}) is the Stefan-Boltzmann constant; $\in$ ($\approx$0.612) is the effective emissivity of earth.

The constant πr^2 can be factored out, giving the temperature

$$T = \sqrt{4}(1 - \alpha)S/(4\varepsilon\sigma) \tag{5.2}$$

Noting that climate models with a low spatial resolution ($\sim$3°–6° lat.) cannot reliably simulate (and, moreover, forecast) climate changes on regional scales, Mearns et al. (1999) showed that to resolve such problems, two approaches can be used:

(1) statistical (with regard to observational data) scaling (reducing to a higher spatial resolution) of the numerical modeling results obtained with the help of low-resolution models;
(2) expanding such models by including the "nested" regional models with a higher resolution.

In this connection, Mearns et al. (1999) undertook a comparison of scenarios of anthropogenic climate change (with a doubled CO_2 concentration) calculated with the use of the NCAR "nested" model RegCM2 and semi-empirical method of scaling (SDS). In both cases large-scale numerical modeling was carried out with the use of the GCM model developed in the Australian Scientific-Industrial Centre (CSIRO).

The results obtained show that the RegCM2 model reveals a stronger spatial variability of the fields of temperature and precipitation than the SDS model, which leads, however, to a greater amplitude of the annual change of temperature, than the models RegCM2 or GCM. The diurnal change of temperature turned out to be weaker in the cases of SDS and GCM than RegCM2, and the amplitude of the diurnal change of precipitation varied in the interval between those corresponding to SDS and RegCM2. Calculations with RegCM2 reproduce both an increase and decrease of probability of precipitation with a doubled CO_2 concentration, whereas SDS gave only an increase of precipitation.

One of the causes of differences mentioned above could be the fact that the semi-empirical model SFS is based on the data that refer only to surface level 700 gPa, whereas in two other models the vertical structure of the atmosphere is taken into account. This comparison does not permit, however, to draw a conclusion, which of the results obtained reflects "correctly" the impact of forcing. To answer this question and establish the cause of these differences, the numerical modeling should be further improved.

One of the important illustrations of highly uncertain theoretical estimates of the causes of climate change is a re-assessment of the role of the "Milankowich mechanism", as the main factor of paleodynamics of climate observed during the last years.

According to the Milankowich theory (Morén and Påsse 2001), changes of paleoclimate had been determined by latitudinal re-distribution of extra-atmospheric insolation and in the annual change as a result of variations of the parameters of the Earth's orbit (especially it refers to glacial-interglacial cycles during the Quaternary period) which include:

(1) inclination of the rotation axis with respect to the orbital plane (fluctuating within 22–24.5° with the current estimate 23.4°; average periodicity of variations constitutes 41,000 years and affects mainly the high-latitude insolation);
(2) precession of equinoctial points affecting the time of the onset of equinoxes and solstices, which tells mainly on low-latitude insolation (precession is characterized by dual periodicities 19,000 and 23,000 years); and
(3) eccentricity of the Earth's orbit which changes from almost circular to strongly elliptical with a periodicity of about 95,800 years (these changes cause a modulation of precession.

Milankowich supposed, in particular, that the summertime low-level insolation in high latitudes is the cause of the onset of glaciations and formation of ice sheets. The resulting increase of surface albedo determines the functioning of a positive feedback which intensifies the impact of insolation decrease. The summertime low-level insolation was observed at a minimum angle of the orbit's inclination, high eccentricity, and at an apogee in the NH summer. According to Milankowich's calculations, this configuration took place 185, 115, and 70 thousand years ago.

Although during the last years the concept of Milankowich has been acknowledged (but not completely), its individual aspects have been critically analyzed. This had two reasons:

(1) many new geological data of analysis of the cores of sea bottom rocks and ice cores have appeared; and
(2) a considerable progress has been achieved in numerical climate modeling.

An analysis of both these sources of information has shown that an adequate explanation of paleoclimate changes is only possible with regard for not only variations of the orbital parameters, but also other climate-forming factors—in particular, variations of the GHGs content in the atmosphere which include carbon dioxide as most important.

In this connection, Palutikof et al. (1999) performed an analysis of new geological data on paleoclimate changes, with a higher temporal resolution, in the context of present ideas of the global climate dynamics. The data of analysis of ice cores and pollen obtained in 1990 have led to the following two important general conclusions:

(1) part of observational data does not confirm the glaciation cycles having been determined by the Milankowich mechanism (this refers especially to data on $\delta^{18}O$ in calcite veins Devils Hole in the U.S.S. Nevada, which testify to opposite phases of the cycles of glaciation and theory of Milankowich); and
(2) it follows from other data that this mechanism can explain only slow quasi-periodic variations but not the short-term variability (on time scales from decades to millennia), with respect to which it has turned out that it had happened much more often than it was supposed.

This variability of the global mean temperature could reach several degrees during several decades. In particular, a large-scale sudden climate cooling had happened in the Ames interglacial period ($\sim$122,000 years ago), when climatic conditions had been very close to the present ones. A typical example of a short-term climatic variability is the Heinrich and Dansgaard/Oeshger events (Keigwin and Boyle 2000). Such events can be repeated in the future.

Many uncertainties remain also, concerning the impacts of present changes of extra-atmospheric insolation on climate. Soon et al. (2000) have demonstrated, for instance, the presence of hyper-sensitivity of the climate system to changes in UV insolation, whose effect is intensified by feedback due to statistical stability of clouds, effect of tropical cirrus clouds, and stratospheric ozone (the "ozone-climate" problem needs a special analysis).

An interactive consideration of the biospheric dynamics as a climate system's component is of particular interest. The significance of this problem can be exemplified by estimates of the climatic impact of deforestation in the tropical Amazon basin, obtained by Bunyard (1999). Nature in the Amazon basin (first of all, wet tropical forests WTF) performs a number of important functions, still not completely taken into account of including the energy input from the tropics to higher latitudes, which, however, is under threat in view of a high rate of WTF destruction.

According to present assessment, every year up to 17 million ha of tropical forests are removed, with ~6 million ha falling on the basin of Brazil Amazon. By the end of 1988, 21 million have been deforested, and 10 years later this area reached 27.5 million ha, which exceeds the territory of Great Britain.

The WTF destruction is important from the viewpoint of the impact on the global carbon cycle, since there is a danger of transformation of the WTF zone from the sink to the source of carbon for the atmosphere. Not less substantial are the ecological aspects of WTF elimination, in view of ecological uniqueness of the tropical forests of Central and South America. According to available estimates, deforestation only in the basin of Brazil Amazon (over the area 360 million ha) will result in the annual sink of carbon up to 0.56 billion t, and on global scales the level of this source can reach 4 billion t C per year. Bearing in mind that in 1988, as a result of forest fires, the tropical forests were burned over an area of about 9 million ha, then due to only this source the atmosphere could get 1–2 billion t C.

Over the virgin tropical forests, about 75 % of the incoming solar radiation are spent on evapotranspiration. Therefore, removal of WTF will result in radical changes of energy exchange and global atmospheric circulation. Changes of local climate should be even more substantial, especially from the viewpoint of precipitation, which can be reduced by 65 %. Of key importance is the fact that the threshold level of WTF elimination that determines the loss by the ecosystem of its self-support ability, remains unclear. For instance, if it is 20 %, then this threshold has been exceeded already.

Bengtsson (1999) drew attention to the fact that since non-linear processes exhibit the prevailing impact on climate system's variability, it is impossible to establish any simple connection between external forcings (e.g., the growth of the GHGs content or variability of extra-atmospheric insolation) and response of the climate system to such forcings. With the unpredictability of some factors of climate taken into account, the difficulty of distinguishing between anthropogenic and natural variability of climate becomes apparent and even increases due to the fact that both internally and externally forced modes of climate variability are determined by the same mechanisms and feedbacks.

Although a considerable progress has been recently achieved in numerical modeling of the climate system, it refers mainly to the atmosphere, which is testified by correspondence of the results of numerical modeling of atmospheric circulation to observational data. Results of the "ensemble" numerical experiments indicate that the 3D atmospheric circulation in the tropics is mainly determined by the impact of boundary conditions, whereas in high latitudes the impact of atmospheric dynamics prevails. The simulation of water cycle in the atmospheric turned out to be rather realistic.

A considerable progress in modeling of the interactive atmosphere-ocean system has made it possible to successfully predict the seasonal and inter-annual variability and, in particular, El Niño events. An adequate consideration of the processes on land surface has ensured a substantial increase of reliability of hydrological forecasts (including the river run-off).

In this context, Bengtsson (1999) discussed the progress in numerical climate modeling in three directions. A successful accomplishment of the TOGA program has made an important contribution to provision of the transition to operational forecasts of the seasonal and inter-annual variability considering the prescribed SAT changes in the tropics, which has determined the critical importance of reliable SAT data. The second of these directions is connected with numerical modeling of climate change on scales of decades and longer and especially with explanation of the centennial change of global mean annual mean SAT. Apparently, in the case of long-term climatic variability the stochastic forcing can be considered as a zero hypothesis.

A consideration of the impact of low-frequency climatic fluctuations on the level of the Caspian Sea has shown that the long-term variability of the level is connected first of all with SST anomalies in the eastern sector of the tropical Pacific Ocean. It turned out that positive SST anomalies correlate with enhanced precipitation in the basin of the Volga watershed and vice versa. The main cause of variations of the Caspian Sea level is the long-term dynamics of ENSO events, which should be considered chaotic.

An important part of the discussed problem is the study of anthropogenic climate change. Calculations have shown that a doubling of CO_2 concentration should result in the outgoing longwave radiation at the level of tropopause decreasing by 3.1 W m^{-2} and the downward longwave radiation flux in the stratosphere growing by about 1.3 W m^{-2}. Thus the total RF at the tropospheric top level will constitute 4.4 W m^{-2}. Calculations of the resulting SAT change with the use of 11 climate models revealed a warming within 2.1–4.8 °C, as well as intensification of global mean precipitation within 1–10 %.

According to Bengtsson (1999), before 1980, the centennial change of the global mean SAT was characterized by the prevailing contribution of natural variability with the subsequent increase of the anthropogenic contribution. An important task of subsequent developments is to improve the numerical modeling (first of all, from the viewpoint of a more adequate consideration of various mechanisms of feedbacks) in order to provide reliable forecasts on regional and local scales. Quite an urgent problem of global modeling consists in the interactive consideration of biogeochemical cycles.

One of the most important aspects of numerical climate modeling is an assessment of the contribution of anthropogenic climate-forming factors. In this connection, Allen et al. (1999) discussed the possibilities of recognition, evaluation, and forecast of the contribution of anthropogenic global climate change, characterized by the SAT, with regard for available data of observations and numerical modeling. The latter is carried out bearing in mind the internal variability of the climate system, as well as the impact of the greenhouse effect (and respective climate warming) and sulphate aerosol (the effect of climate cooling).

The four global 3D models of interactive atmosphere-ocean system predicted an increase of the global mean SAT for the decade 2036–2046 (compared to pre-industrial level) within 1.1–2.3 K. Calculations of the climate system sensitivity to a doubled CO_2 concentration gave values within 2.5–3.5 K. According to

HadCM2-G5 climate model developed at the Hadley Centre (Great Britain), the global mean "greenhouse" warming in the period 1996–2004 should constitute 1.35 K, and the "sulphate" cooling will constitute 0.35 K, and thus the resulting warming will reach 1 K.

With the anthropogenic SAT increase in the 20th century equal to 0.25–0.5 K per 100 years, the calculations made with a simple climate model, predict the 1–2 K uncertainty of the balanced SAT forecast by the year 2040. Such estimates can, however, be sufficiently reliable only with an adequate consideration of the characteristic time of the ocean's adaptation. A critically important aspect of prognostic estimates is a necessity to take into account possible sudden non-linear climate changes, which seriously limits the lead time of forecasts.

The most attractive perspective for assessment and forecast of anthropogenic SAT changes is connected with analysis of the spatial-temporal variability of the SAT fields that takes into account the impacts of the greenhouse effect and aerosol. Realization of this approach is seriously complicated, however, by impossibility of reliable prescription of aerosol forcing on the SAT field. A serious problem is also a necessity to take into account the climatic impact of changes in the content of stratospheric and tropospheric ozone.

5.4.5 Contradictoriness of the Climate Study Results

The problem of anthropogenic global warming is now in the centre of attention of not only specialists but also wide circles of population. With a deeper understanding of this problem, there appears a feeling of inconsistency of the results obtained in this field, especially during the last decade. In this connection, Mahlman (1998) carried out an overview of such results aimed at analyzing of the fundamental scientific aspects of the problem under discussion, in which the emphasis is placed on the role of numerical modeling and analysis of observational data in understanding of the present climate change. The monograph by Weber (1992) has been dedicated to the same theme.

The main difficulty in understanding the causes of climate change is connected with a still preserved impossibility of an adequate consideration of climatic feedbacks. First of all, this refers to the cloud-radiation feedback, direct and indirect (through the impact on radiative properties of clouds) climatic impact of atmospheric aerosol as well as the impact of the atmosphere-ocean interaction on climate formation.

The often ignored specific manifestation of the "greenhouse" climate warming is the system's large inertia (the thermal "memory" of deep layers of the ocean spans centuries and even thousands of years). Schlesinger et al. (2000) pointed out, for instance, the existence of the global SAT oscillations with a period of about 65–70 years from the observational data for the period 1858–1992. It is important that principal differences between numerical modeling (and forecasts) of weather and climate should be taken into account. In the case of numerical climate modeling, it is important to use "tuning" and adjustment because of the difficulty of an adequate consideration of a complicated totality of interactive processes and spatial-temporal scales. In this context, the use of paleoclimatic data plays a substantial role, though they cannot be analogs to a possible climate change in the future.

A serious anxiety has been caused by inadequacy of global observation systems and degradation of ground observations, especially manifesting themselves in some cases. Mahlman (1998) emphasized that the controversy of the problem of anthropogenic climate change consists in the absence of reliable quantitative estimates of relationship between the contributions of natural and anthropogenic factors of changes. This circumstance creates serious difficulties in practical realization of recommendations contained in the Kyoto Protocol.

Such conclusions, discussed in detail before, have received universal recognition, illustrated by the recent review of Grassl (2000). In this context, quite surprising is the wide use of the term "climate change" as determining only the anthropogenic change. The substitution of the notion of "climate change" (in its true meaning) with the term "global warming" is also incorrect, since both observational data and the results of numerical modeling indicate a highly inhomogeneous present climate change, far from being reduced to only SAT increase.

Such terminological misunderstanding is not accidental, however. It is aimed at disinformation for the sake of establishing a false conception of anthropogenic ("greenhouse") global warming, which has been convincingly explained by Boehmer-Christiansen (1997, 2000), who analyzed the political motivation of this concept. Cracknell (2009) continues of discussion about global warming.

In August 1997, the U.S. Minister of geology V. Babbitt, addressing about 3,000 participants of the annual Congress of the U.S. Ecological Society, said that they should implement their civil obligation—help to convince skeptical American public that global warming is both real and dangerous: "We have a scientific consensus but we have not a public consensus". In this connection, Morris (1997) carried out an overview of available scientific information in order to analyze the grounds for this opinion, since many specialists do not share the apocalyptic predictions of anthropogenic global warming. The emphasis in the overview has been placed on the problem of distinguishing between natural and anthropogenic climate change.

In the mid-1970s the forecasts of global cooling due to sulphate aerosol predicted, for instance, that this impact will limit an increase of global mean temperature due to enhanced greenhouse effect of the atmosphere by less than 2 °C even with a 8-fold increase of CO_2 concentration. The warming trend observed in the 1980s has attracted attention to the problem of climate warming. At the height of the summer of 1988, (Hansen et al. 1988; Hansen 1998) declared at the U.S. Congress a 99 % probability of anthropogenic global warming and its destructive consequences for the ecosystems in the future, as well as a consensus reached among specialists on this problem, though many meteorologists and climatologists did not share those views: respective developments did not permit to establish reliable cause-and-effect relationships between anthropogenic GHGs emissions and observed the climate change.

The political motivation of the support to the global "greenhouse" warming conception has determined a gigantic growth of governmental financing of such developments in the USA in the period 1990–1995 from 600 million dollars to 1.8 billion dollars.

In the context of contributions of different factors to the formation of global climate, Morris (1997) emphasized an importance of the combined consideration of contributions of the "greenhouse" warming and solar activity, which during the period of instrumental observations constituted 0.31 and 0.41 °C, respectively. According to these estimates, a doubling of CO_2 concentration will lead to a global warming within 1.26–1.33 °C. With such climate changes, there are no grounds to expect often predicted catastrophic consequences, such as the rise of the World Ocean level and an increase of epidemic diseases. However, there is no doubt that one should expect an increase of the agricultural yield. On the whole, the climate warming consequences should be positive. Although reliable forecasts of ecological consequences of GHGs emissions are still impossible, it is apparent that their reduction will strongly affect the economy. Therefore, the main conclusion which may be made, is that one should not try to prevent still unreliably predicted climate change but adapt to it.

In view of the climate problem complexity, relevant scientific publications usually emphasize a serious uncertainty of the available estimates, which favours the possibility to continue studies, but on the other hand, is fraught with serious consequences in the context of ecological policy. As Boehmer-Christiansen (1997) noted, "ecological" bureaucrats and experts as well as well-organized representatives of the sphere of fuel-energy industry are interested in the proof of the existence of anthropogenic warming and want to obtain the needed scientific support. These three groups with common interests support non-governmental ecological organizations and thoroughly consolidate them after the Conference in Rio.

The main target of the "green" pressure is the governmental hierarchy which, in view of a wide range of its responsibility, as a rule, either does not want or cannot realize the needed measures including the ecological taxes, subsidizing development of the renewable energy sources, nuclear energy, public transport, energy efficiency, and others. The World Bank (WB) and Global Ecological Fund (GEF) are also lobbied for, to support numerous and expensive developments which are controlled by numerous experts and specialists in the sphere of finances. These experts are the most vigorous advocates of measures to prevent the supposed undesirable climate change, and the WB and respective UN organizations more and more use their recommendations, in particular, those connected with the so-called "Joint Implementation".

This system is based on such fundamental documents as the International Framework Convention on Climate Change, and the scientific bases are the IPCC reports. To achieve these goals, the following components should be provided: science that substantiates the threats; the "green movement" that expresses emotions; rhetoric and "principle of precaution"; developments of new technologies aimed at reducing the GHGs emissions; bureaucratic bodies that prepare the needed plans and develop strategies.

And only after that can politicians, who should ensure the financial support in order to transform plans into concrete actions, start acting. In this context, Boehmer-Christiansen (1997) discussed various circumstances concerning "scientific provision". Clearly, the IPCC Reports may be the fundamental documents, if anthropogenic impact on climate is really dangerous. However, the difficult problem is how to determine the criteria of "danger" and who should be responsible for it (considering also the socio-economic and political factors).

Still there are many uncertainties in this problem. In particular, in connection with the discussion of necessity to reduce GHGs emission, the emphasis is placed on decreasing the scale of the use of coal, especially in electric power production (while for natural gas the conditional "factor of CO_2 emissions" constitutes 15, in the case of coal it reaches 25 and more, being very variable). An "attack" on coal was not, however, connected with ecological motives. The question of ecology did not rise at all till the mid-1980s, when the price of oil and gas dropped (especially in 1986). As to resolve the appearing economic problems, some ecological allies were needed to provide the competitive ability of "pure" energy for the sake of "sustainable development".

It is no mere chance that the IPCC was forethought in 1995, really planned in 1987, and started functioning in 1988. Governments of various countries support the plans of GHGs emissions reduction not for ecological but for other reasons: enhancement of national nuclear energy (Germany); an increase of the export potential for electric power produced by nuclear power stations and gas (France, Norway); an increase of the size of financial support, etc. Naturally, the countries where electric power depends on coal, are most skeptical about the "greenhouse" warming.

Having analyzed the role of various international organizations and programs in the problem of global climate change, Boehmer-Christiansen (1997) emphasized: "Climate policy cannot be understood without a deeper analysis of the role of science and scientific understanding of the coalition of non-ecological interests (both commercial and bureaucratic) which serve as a driving force of development of events on international scales. Where this coalition will lead us, remains unclear, so far".

Constructive prospects of resolving this problem are connected with the development and application of complex models to assess possible changes of climate and socio-economic development. In the detailed overview of methods and results of numerical modeling of global climate change with regard for dynamics of socio-economic processes, prepared by Parson and Fisher-Vanden (1997), the main aspects of the so-called integral assessments (IA) have been discussed. The main goal of such developments is to substantiate recommendations for people making respective decisions concerning the ecological policy.

Four concrete goals include the following problems:

(1) assessment of possible response to climate change;
(2) analysis of the structure of scientific bases of modeling and characteristic uncertainties of the results obtained;

(3) comparative estimates of possible risks; and
(4) analysis of the achieved scientific progress.

During the last several decades, two approaches prevailed in developments of IA models:

(1) use of assessments obtained by interdisciplinary groups of experts; and
(2) formal numerical modeling.

The former approach is characteristic of IPCC efforts and developments within the Montreal Protocol, whereas the latter approach was realized by individual specialists. The emphasis was placed on the use of IA models to analyze possible impacts of climate change on the development of energy and economy in the context of the problem of CO_2 emissions to the atmosphere.

The general assessment of the results obtained consists in that the IA models cannot be used to substantiate the highly specialized measures in view of the insufficiently detailed character of such models. However, the use of such models is important for assessment of possible uncertainties and, hence, expediency of making any decisions. In this case, it is important that considerable uncertainties concerning the system on the whole or recommendations on the needed ecological policy, can turn out to be not those uncertainties that are most important for understanding the processes responsible for understanding changes taking place in the environment or characterized by most substantial variability.

Considering the data for concrete models, it is possible to range uncertainties by their significance, but this ranging depends on specific character of the models. Although the use of IA models has made it possible to substantiate some correlations of dynamics of the socio-economic development and variability of the environment, the relevant results should be considered only preliminary. So far, the use of IA models has made only a small contribution to estimation of comparative risks and to obtaining the answer to the fundamental question, to what degree and in which respects a possible climate change is most substantial.

In view of these circumstances, the results of the use of IA models to substantiate an adequate ecological policy have been, so far, rather limited, and this refers especially to purely didactic estimates based on the use of simple models. The main problems of the IA models' improvement include: overcoming insufficient understanding of probable impacts and possibilities of adaptation; poorly substantiated or absent at all descriptions of social and behavioural processes in developing countries; very limited ideas of a scarcely probable, but radical climate change. Despite these and other unresolved problems, the urgency of further efforts in development of IA models is obvious.

In conclusion, again, it is emphasized that an improvement of the global observation system is of great importance. The urgency of this problem has been illustrated in Demirchian et al. (2002) where the complicated spatial-temporal SAT variability in high latitudes of the Northern Hemisphere has been demonstrated, as well as the groundlessness of the "greenhouse" warming conception, according to which under conditions of the Arctic the effect of the growth of GHGs concentration should manifest itself especially strongly.

5.5 Climate, Energetics and Global Urbanization

5.5.1 Climate and Civilization

Cities are one of the most substantial and evident phenomena of the present civilization development. Despite an apparent controversy of their development, for most of the planetary population they are a symbol of progress. But nevertheless, "cities do not deserve their epoch, they are not worthy of us". This opinion of the outstanding French architect and town-builder Le Corbusie (1977), voiced as far back as 1925, is valid even now. One of the major causes of this assessment is an unfavourable ecological situation, and another one is a continuous growth of the urban population. During the next 20 years most of the population in developing countries will live in cities. This fact, like the problems of large cities, attracts serious attention of many investigators (Grigoryev and Kondratyev 2004; Bulkeley and Betsill 2003; Kondratyev et al. 2006a).

The main feature at the turn of the 20th century is the continuing growth of megalopoliscs. If in 1970 the urban population constituted 35 % of the planetary population, in 2000 it was already almost 50 %. It was supposed that by the year 2023 more than 3.3 billion people will live in cities. The prevailing contribution (90 %) is made by the cities of developing countries (Grigoryev and Kondratyev 2004).

Forecasts of urbanization development for the nearest 30 years show that from the viewpoint of population location, cities head the list. In 2030, the size of the urban population can reach 60 % of the global population.

Serious progress in many spheres of vital activity, which is connected with the growth of cities, is accompanied by heavy negative consequences. They are rather diverse and connected, in particular, with the poverty of many city-dwellers, low level of health, high mortality, environmental degradation, as well as shortcomings in the infrastructure and development of the municipal services. Of these problems, the most important is the environmental situation and lack of readiness of infrastructures to protect population against natural disasters. That's the initial cause of other numerous negative aspects of the urban life (first of all, city-dwellers' health and poverty). This hinders the progressive positive development of cities. Of course, specific features of urbanization in the regions of different continents differ drastically. The highest level of urbanization is observed in the countries of North America (77.2 %), Latin America (75.3 %); and Europe (74.8 %). In other regions these indicators are as follows: Oceania and Australia—70.2 %, Asia—36.7 %, and Africa—37.9 %. This statistics is quite natural and reflects the difference of the countries by the level of industrial development.

For individual countries, the urbanization phenomenon is characterized by the following indicators: Belgium—97.3 %, Iceland—92.5 %, Uruguay—91.3 %, Great Britain—89.5 %, the Netherlands—89.4 %, Germany—87.5 %, Venezuela—86.6 %, Argentina—85.9 %, Chili—85.7 %, Denmark—85.3 %, Sweden—83.3 %, Brazil—81.3 %. Analysis of these and other indicators of urbanization shows that there is a dependence of population concentration on the frequency of unfavourable and dangerous natural phenomena on a given territory. In particular, in Iceland the settling is affected by volcanic activity, in the Netherlands—by the floods of lowlands. In Chili, vast territories are covered with deserts and mountains, in Uruguay floods happen frequently. In Brazil and Venezuela large areas are covered with tropical forests. Apparently, in other regions, natural disasters also affect the settling of population. However, this dependence remains poorly studied.

In North America, despite the general high level of population concentration in the cities, in the two greatest countries of the continent—the USA and Canada—in 2000 the level of urbanization did not reach 80 % (77.2 and 77.1 %, respectively). By 2015, only in the USA the level of urbanization will exceed this threshold and reach 81 %. This is explained, at least, by two reasons. One of them is the presence in these countries of vast flat territories within which there is plenty of safe (in natural respect) and comfortable places for settling. Of course, the holding factor of the growth of large cities is also a high standard of life in both countries, both in cities (large and small) and at rural localities.

At present, the least urbanized are the countries in Africa. At present time, the level of urbanization in Africa is characterized by the following data: Libya (88 %), Gaboon (81.4 %), Djibouti (83.3 %), Western Sahara (95.4 %), Tunisia (66.2 %), Algeria (57.7 %), South Africa (57.7 %) and Equatorial Guinea (49.3 %). A low level of the urban population will remain here by 2015, not exceeding 50 % in most countries. The urbanization of most of Africa is moving fast forward, especially south of the Sahara. During 2000–2005 urbanization growth was 4.7 % in Eastern Africa and 2.09 % in South Africa under average 3.76 % for all Africa.

In Asia, in the year 2000, among 50 countries 80 % threshold of the urban population was exceeded only in 11 countries. Large countries on this list is only Korea (81.9 %). In most densely populated countries the level of urbanization is: China (32.1 %), India (28.4 %), Indonesia (40.9 %), and Pakistan (37.0 %). A high level of urbanization is observed in Japan (78.8 %), which is explained by the impact of unfavourable natural factors, such as tsunami, volcanic eruptions, earthquakes, and typhoons. Asian cities are growing rapidly. Another 1.1 billion people will live in the region's cities in the next 20 years. By 2030, more than 55 % of the population of Asia will be urban.

One of the features of the current urbanization is the rise of cities and urban agglomerations. In North America, by the year 2000 such cities numbered 41 (37 in the USA and 4 in Canada). Among them were 3 cities with population exceeding 5 millions: New York (16,649 thousands), Los Angeles (13,140 thousands), and Chicago (6,951 thousands). Among the European countries, Russia and Germany head the list (11 and 13 cities-millionaires, respectively). In Europe there are only four cities with population more than 5 millions: Paris (9,264 thousands), Moscow (9,321 thousands), Essen (6,541 thousands), and London (7,640 thousands). In Africa, there are three such cities: Kinshasa (5,064 thousands), Cairo (10,552 thousands), and Lagos (13,427 thousands). By the year 2015, about 500 mega-cities with population more than 1 million should rise.

Thus, an analysis of the general trend in settling the planetary population carried out by Grigoryev and Kondratyev (2004) has shown that on the whole, the number of cities-millionaires is growing, but the rate of the growth differs in different regions. The global characteristics of the present population dynamics in the cities are as follows:

1. On the whole, at present the urbanization is characteristic of all countries in the world. Cities are growing in number, the percentage of the urban population increases. However, the rate of urbanization is slowing down.
2. The cities-millionaires are growing in number. But in the countries with a high level of development the growth of population in the cities-millionaires is at a minimum. At present it is slowing down, and in many cities it stopped altogether. In developing, poor countries the growth of population in the cities-millionaires is slow but to a lesser extent.
3. The characteristic feature of the development of urbanization now and in future is the formation of cities with population above 5 and even 10 millions, which are called mega-cities. It is especially interesting to study these cities, since they exhibit features of the future process of urbanization, and, as a matter of fact, the features of the future of humankind. They concentrate a multitude of positive aspects from the viewpoint of humans' activity. At the same time, in such cities every negative aspect of the phenomenon discussed clearly shows itself, especially in the countries with developing economy.

The 21st century is an epoch of the transition of the planetary population to new living conditions within the globalization of many processes. Of course, the elements of globalization of various kinds of humans' activity have been observed before, even many centuries ago, due to trade, navigation, and military expansions. However, its present features have greatly changed. Globalization of any phenomena has accelerated and is covering larger areas. The range of phenomena incorporated in planetary processes has drastically broadened. The main thing is that not only chains of events have appeared and propagated on a global scale. The whole world and various kinds of human activity—economy, politics, science, technology, engineering—have turned out to be involved in the process of globalization. Moreover, the whole world has turned out to be not simply in globalization chains, but in a web of chains of events which are propagating across the globe.

Cities, especially large ones, have become the key point of the process of globalization. Globalization in cities is accompanied by many positive phenomena. Among them are, for instance, prompt information on possible acts of terrorism, approaching tropical cyclones, which can threaten the human life, information on global propagation of dangerous infectious diseases, information about new medicines, and many others. Negative aspects of globalization are also known, which vividly and painfully manifest themselves in the cities. Among them are: propagation of drugs, epidemics, criminality, etc. Ecological problems in the cities also contain negative components. These are pollution of the atmosphere and hydrosphere, high risk of propagation of epidemics resulting from destructions after earthquakes, hurricanes, and floods.

Contrasts in the ecological situation in the cities connected with the socio-economic factors, are clearly seen with an access of population to objects of sanitation as an example. People's security in case of natural disasters depends substantially on this indicator. The best access to such objects is provided in the cities of North America (100 % access), Europe (99 %), Australia and Oceania (99 %). A much lower level of sanitary-hygiene provision is founded in the cities of Africa (84 %) and Latin America (87 %). A low level of access of city-dwellers to normal services is in the cities of Asia (78 %), and the worst sanitary conditions are in India (46 %), Nigeria (61 %), Ghana and Pakistan (53 %).

The cities of the world differ substantially in their waste. For instance, every year, on average, the cities with populations of 1 million or more in Russia dump about 3.5 million t of solid and concentrated waste. Its main constituents are (thousands of tons): wood waste (400), paper (9), textile (8), and glass breakage (3). Most of waste is in the cities of highly developed industrial countries, such as the USA and Canada, where per-capita waste is estimated at 2 and 1.5 kg day^{-1}. In the cities of Latin America it is 1.0 kg day^{-1}. Of course, of greater importance is not the amount of waste but possibilities and practice of its processing. In contrast to the cities of the USA and Canada, where the waste is processed practically completely, in most of the cities of South America, from 20 to 50 % of garbage are not removed at all (Gurberlet 2003).

Large cities concentrate a lot of anthropogenic processes on small territories, the state of which is determined by the ecological, material, sanitary-epidemiological, social, cultural, political, religious, and other specific features of society. All these features define input variables when determining the danger threatening the city-dwellers in case of a natural disaster. Clearly, for the optimal growth of cities in harmony with the environment, will be the development of criteria of the ecological risk in cities and their indication; and development of an efficient system of indicators of the ecological situations in the cities. At present, there are a lot of such indicators. As a rule, they concern individual elements of the environment. Examples are information systems of assessment of the quality and condition of the water medium, sanitary-epidemiological situation, quality of atmospheric air, etc. Alimov et al. (1999) undertook a multi-criterial assessment of the medium quality and the state of the urban ecosystem of St.-Petersburg. Flood (1995) described the system of indicators used by the U.N. having studied more than 100 cities of the world.

Indicators of various phenomena in the cities are efficient means to analyze and monitor the interaction between the cities and the environment. Among the indicators of the ecological situation in the cities, two large groups can be selected. The first group includes indicators of the environment, directly affecting the humans' activity and the ecological situation. First of all, it concerns the indicators of air and water quality. Such indicators as the area of green plantations are widely used and to a lesser extent the characteristics of their state. During the last decades emphasis has been placed on several geoecological characteristics of the urban environment of technogenic origin. Among them are: the level of noise, radioactive pollution, impact of local electromagnetic fields. The second group includes indicators characterizing the level of poverty, health, special features of infrastructure, transport, the state of available housing, and quality of the environmental control. A detailed analysis of some of them has been made by Grigoryev and Kondratyev (2004).

In Europe, in 1999, within the framework of conception "Towards sustainable development at a local level: European indicators of sustainability", ten indicators have been proposed (Grigoryev and Kondratyev 2004):

- satisfaction of citizens with the standard of living at a settlement;
- contribution of the cities to the global climate change;
- population mobility and passenger traffic;
- availability of public accommodations and services;
- air quality at a settlement;
- transportation of children to school and from school;
- control according to principles of sustainable development;
- noise pollution;
- stable land use; and
- output of products that favours the sustainable development.

The main apparent shortcoming of these indicators is their inadequacy and some vagueness. Here concrete indicators are absent, such as:

- quality of drinking water;
- rate of mortality (including infant mortality);
- per-capita gross production;
- quantity of paupers and waifs; and
- crime rate.

From the viewpoint of the urban population security during natural disasters, the use of such indicators would make it possible to resolve many problems of averting great human losses and reducing the economic damage. Of course, all of these problems can be resolved by using a global model that describes maximum amount of processes and sub-systems of the environment. Such a model has been proposed in Kondratyev et al. (2004a). It can be realized only within the framework of the agreement signed by 60 countries on unification of national systems of observation of the environment into a global system.

Major disparities in the level of urbanization exist among development groups. Thus, whereas the proportion of urban population in the more developed regions was already nearly 54 % in 1950, it will still take another decade for half of the population of the less developed regions to live in urban areas (Fig. 5.16).

The world urban population is expected to increase by 72 % by 2050, from 3.6 billion in 2011 to 6.3 billion in 2050. By mid-century the world urban population will likely be the same size as the world's total population was in 2002. Virtually all of the expected growth in the world population will be concentrated in the urban areas of the less developed regions, whose population is projected to increase from 2.7 billion in 2011 to 5.1 billion in 2050. Over the same period, the rural population of the less developed regions is expected to decline from 3.1 billion to 2.9 billion. In the more developed regions, the urban population is projected to increase modestly, from 1 billion in 2011 to 1.1 billion in 2050 (UN 2012).

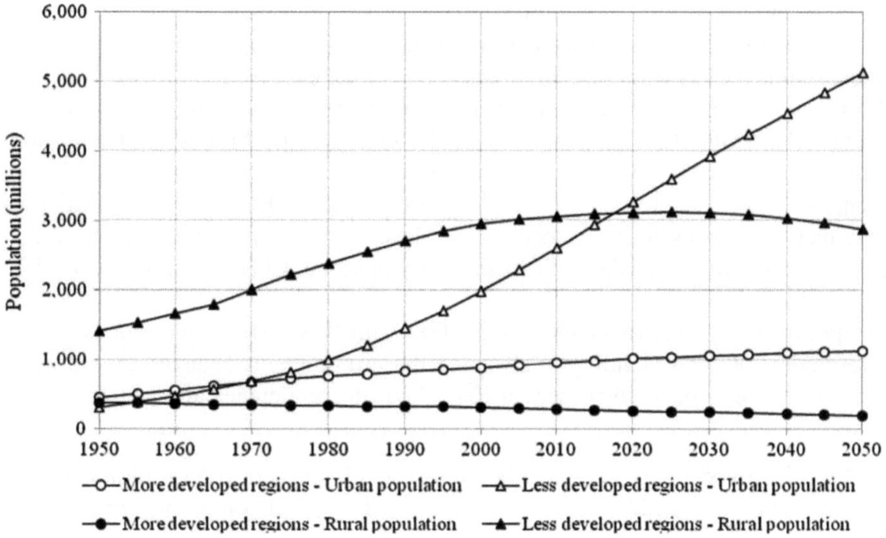

Fig. 5.16 Urban and rural population by development group, 1950–2050 (UN 2012)

The world rural population is expected to reach a maximum of 3.4 billion in 2021 and to decline slowly thereafter, to reach 3.05 billion in 2050. These global trends are driven mostly by the dynamics of rural population growth in the less developed regions, which house today almost 92 % of the world rural population. Whereas the rural population of the more developed regions has been declining steadily during the second half of the twentieth century and will continue to do so for the foreseeable future, the rural population of the less developed regions more than doubled since 1950 and will likely continue to grow until 2021 before a long-term decline sets in (UN 2012).

5.5.2 Climate and Energetics

Globalization processes and their present tendencies put many problems connected with the search of strategy for the control by the waste products of different sectors of the economics. Regions with unstable structure of the NSS dynamics cause special anxiety. Particularly, the presence of the uncertainty in future development of oil and gas import from reservoirs of Middle East and North Africa for the reason of the extension of China and India role leads to the uncertainty in the prediction of future fluxes of anthropogenic CO_2. It is expected that under the conservation of present tendency in the economic development in these regions CO_2 emission can increase by 52 % to 2030 in comparison with present level (WEO 2002, 2004, 2005, 2013).

The need for energy in Middle East and North Africa is regulated by the requirements of growing population and by growth of the economics. The need for the primary energy in this region to 2030 can be more than double. At the same time oil extraction can increase approximately by 75 % and gas by three times. A quota of this region within the oil extraction will increase from 35 to 44 % in 2030. All this will be possible under the investments of 56 billion dollars per year to the energetic infrastructure development in the countries of Middle East and North Africa. It is connected with defined levels of the risk for the countries G8. Different scenarios of the development for the regions of Middle East and North Africa are here naturally arisen to be connected with levels of investments and prices a consideration of which introduces the uncertainty in global CO_2 emission (Cozzi 2003; IEA 2013c).

Special IPCC Reports (IPCC 2005, 2007) and IEA (2013a, b, c) analyzing the information of long standing analysis of the environment changes give a set of assessments and make some conclusions about the tendencies and causes of these changes:

- World proved oil reserves at the end of 2012 reached 1668.9 billion barrels, sufficient to meet 52.9 years of global production. World oil production increased by 1.9 million b/d in 2012, more than double the growth of global consumption. World proved natural gas reserves at end-2012 stood at 187.3 trillion m^3, sufficient to meet 55.7 years of global production. World natural gas production increased by 1.9 % in 2012. World proved reserves of coal in 2012 were sufficient to meet 109 years of global production. Global coal production grew by 2 %.
- Global energy consumption and related CO_2 entering to the atmosphere in 21st century continues to increase. Fossil fuels deliver 86 % of world energy and are the source for 75 % of anthropogenic emission of CO_2 to the atmosphere. World energy consumption increases from 524 quadrillion British thermal units (Btu) in 2010 to 630 quadrillion Btu in 2020 and 820 quadrillion in 2040, a 30-year increase of 56 %.
- Global consumption of primary energy increased in the average with speed equaled to 1.4 % during 1990–1995 and with speed 1.6 % from 1995 to 2001. Under this, the rates of energy consumption increase are distributed by the economics sectors with the following shares: 0.3–0.9 % in industry, 2.1–2.2 % in the transport, 2.1–2.7 % in the building sector and −0.8 to −2.4 % in the agriculture. According to IEA (2013a, b, c) data from 1990 to 2012, the average energy use per person increased 10 % while world population increased 27 %. Regional energy use also grew from 1990 to 2008: the Middle East increased by 170 %, China by 146 %, India by 91 %, Africa by 70 %, Latin America by 66 %, the USA by 20 %, the EU-27 block by 7 %, and world overall grew by 39 %. In 2008, total worldwide energy consumption was 474 EJ (132,000 TWh). This is equivalent to an average power use of 15 TW (2.0×10^{10} hp). The annual potential for renewable energy is: solar energy 1,575 EJ (438,000 TWh), wind power 640 EJ (180,000 TWh), geothermal energy 5,000 EJ (1,400,000 TWh), biomass 276 EJ (77,000 TWh), hydropower 50 EJ (14,000 TWh) and ocean energy 1 EJ (280 TWh).

- The production of electricity in 2009 was 20,053 TWh. Sources of electricity were fossil fuels 67 %, renewable energy 16 % (mainly hydroelectric, wind, solar and biomass), and nuclear power 13 %, and other sources were 3 %. The majority of fossil fuel usage for the generation of electricity was coal and gas. Oil was 5.5 %, as it is the most expensive common commodity used to produce electrical energy. Ninety-two percent of renewable energy was hydroelectric followed by wind at 6 % and geothermal at 1.8 %. Solar photovoltaic was 0.06 %, and solar thermal was 0.004 %. Greenhouse gas emissions from power plants and other industrial facilities declined by 4.5 % from 2011 to 2012 as utilities continued to switch from coal to natural gas to generate electricity and produced slightly less power overall. Taking into consideration of emission factor equaled 7.0555×10^{-4} metric tons CO_2/kWh, 2,250 MtC was emitted to the atmosphere in 2011 out of the electric power production. It is expected that coal will be occupy the energetic niche 36 % to the 2020.
- An average CO_2 emission was increased by 1 % during 1990–1995 and by 1.4 % from 1995 to 2001. The most input to the CO_2 emission gives the electric power generation. Under this, the contribution of separate economic sectors was: 1.7–2.0 %—industry, 2.0–2.3 %—building, and −1.0 to −2.8 %—agriculture.
- Many experts will expect that the level of 550 ppm CO_2 stabilization on can be reached to 2100 due to the reduction of its emission by 7–70 %. Such uncertainty is determined by the variety between the estimations of CO_2 sources and sinks both natural and anthropogenic.
- Total emission of CO_2 under the fossil fuel combustion for 2001 was assessed as 24 $GtCO_2$ (6.6 GtC) per year. Under this 47 % of this emission was realized by developed countries, 13 % by countries with transient economics, and 25 % by developing countries from Pacific sector of Asia. Input of separate sources was: 263 MtC/year—automobile industry, 1,173 MtC/year—building and manufactories, 1,150 MtC/year—traveling transport, and 520 MtC/year—residential areas.
- Input of greenhouse gases to the variation of the Earth's radiation balance is 20 % for CO_2 and 6 % for nitrogen oxides.

Tables 5.12, 5.13 and 5.14 characterizes a distribution of CO_2 emission by the economic sectors and regions. It is expected that growth of CO_2 emissions by total volume is changed from 24,409 $MtCO_2$ in 2002 to 33,284 $MtCO_2$ in 2015, and it achieved 38,790 $MtCO_2$ to 2025 exceeded the level of 1990 by 81 %.

Despite the fact that Kyoto Protocol after its ratification by Russia 16 February 2005 came into force for more than 55 countries its recommendations on the CO_2 emission reduction no simplify the realization of real prognosis of the energetics development during nearest decades. Nevertheless,supposing that all countries signed the Kyoto Protocol will follow to its recommendations, we receive that average rate of CO_2 emission increase due to fossil fuels combustion during 2002–2025 will be to equal to 2 %. The prognosis can be improved to a certain extent, taking into consideration of data given in Table 5.15 where a comparison between

Table 5.12 List of basic stationary CO_2 sources emitting annually more 0.1 $MtCO_2$ (WEO 2002)

Source	Volumetric concentration of CO_2 in emitted gas (%)	Volume of the emitted CO_2 (MtC)	Part from all emissions of CO_2 (%)	An average value of CO_2 emission per source, MtC per source
Coal	12–15	2181.4	59.69	1.08
Natural gas	3–10	205.5–207.4	5.65	0.21–0.28
Petroleum	3–8	89.1–178.7	2.43–4.89	0.15–0.35
Other fossils	–	16.7	0.45	0.21
Hydrogen	–	0.8	0.02	0.35
Cement industry	3–20	218.0–254.6	5.97–6.97	0.22–0.34

Table 5.13 A distribution of CO_2 emissions due to the energy production by economic sectors and regions (MtC) (IEA 2013a, b, c)

Region	Indicator of the economic sector								
	PEHP	VP	OES	MAC	T	CPS	R	OS	Σ
Countries with the transit economic	307.3	107.5	29.3	143.3	87.1	15.9	85.9	35.1	811.4
The west sector of OECD	298.7	36.1	60.9	197.3	284.4	47.8	135.1	26.3	1086.6
USA	622.3	36.9	74.4	179.8	470.0	61.6	101.5	11.7	1558.2
Pacific sector of OECD	139.9	23.8	17.0	82.3	94.1	26.0	20.7	9.8	413.6
Central regions of South-East Asia	254.3	28.4	37.7	145.7	123.4	13.9	50.7	10.8	664.9
Asian countries with planned economics	366.0	10.3	37.8	267.3	67.0	19.8	60.5	32.4	861.1
Middle East	77.1	1.8	32.4	52.7	46.9	4.5	24.8	30.7	270.9
Africa	75.9	4.3	11.0	37.6	39.2	1.4	12.2	9.5	191.1
Latin America	61.1	10.1	36.7	76.3	108.2	4.9	22.1	11.3	330.7
Total	2202.6	259.2	337.2	1182.3	1320.3	195.8	513.5	177.6	6188.5

PEHP is the production of electric power and heat for the population, *VP* is the vehicle production, *OES* is other energy sources, *MAC* is the manufactures and construction, *T* is the transport, *CPS* is the commerce and population service, *R* is the residence, *OS* is other sectors, *Σ* is total energy

all economics and the most representative regions is given by the index of their development effectiveness. It is necessary to take into account the presence of the effect in the reduction of GDP efficiency, along with the population growth and GDP increase for the enhancement of the human development index (HDI). In fact,

Table 5.14 Distribution of CO_2 emissions by the economic sectors and regions with the prognosis to 2025 (IEA 2005b)

Region/country	CO_2 emissions in 2002 under the energy production (MtC year^{-1})				Prognosis of total volume for CO_2 emission (MtC year^{-1})		σ
	Due to oil combustion	Due to gas combustion	Due to the coal combustion	Total emission	2010	2025	
USA	671.3	328.7	565.6	1565.6	1853.3	2339.1	1.7
Canada	74.9	43.7	42.1	160.7	191.3	236.6	1.7
Mexico	69.9	22.7	6.3	98.9	120.8	173.5	2.5
West Europe	522.1	221.9	225.4	969.4	1055.2	1134.4	0.7
Japan	179.5	40.2	102.5	322.2	337.7	360.4	0.5
Australia/New Zeeland	38.0	16.9	67.5	122.4	145.4	181.1	1.7
Russia	97.3	212.0	106.6	415.9	500.0	629.5	1.8
Other CIS countries	61.5	101.4	76.8	239.7	298.9	397.3	2.2
East Europe	52.5	37.2	108.5	198.2	237.2	305.7	1.9
China	182.8	19.4	705.5	907.7	1583.6	2524.0	4.5
India	76.2	13.1	190.7	280.0	389.1	603.6	3.4
South Korea	65.8	13.1	44.0	122.9	157.4	231.1	2.8
Other Asian countries	205.2	73.0	106.3	384.5	535.0	845.6	3.5
Middle East	219.7	124.9	27.3	371.9	504.4	730.6	3.0
Africa	103.0	38.0	92.3	233.3	317.8	464.8	3.0
Brasil	75.1	7.1	11.2	93.4	125.4	207.1	3.5
Other countries of Central and South America	119.4	48.4	8.7	176.5	248.6	350.8	3.0
Countries with well-developed market	1556.0	674.0	1009.3	3239.3	3703.6	4425.4	1.4
Countries with transit economics	211.2	350.5	291.8	853.5	1036.3	1332.2	2.0
Countries with developing economics	1047.5	336.6	1186.3	2570.4	3861.2	5957.9	3.7

Designation σ is the average annual percent of CO_2 emission change for the period 2002–2025

the ideal model of globalization is to provide for the HDI alignment by all countries out of the surplus resources redistribution. However, such model is far from reality and therefore existing predictions of increase in CO_2 emissions have considerable varieties by the estimations of rates and volumes. Data of Table 5.16 indicate

Table 5.15 1973 and 2010 regional shares of CO_2 emissions (KWES 2012)

Source of CO_2 emission	1973 (% to global)	2010 (% to global)
Non-OECD Americas	2.6	3.5
Asia excluding China	3.0	11.0
China	5.9	24.1
Non-OECD Europe and Euroasia	16.2	8.6
Midle East	0.8	5.1
OESD	66.1	41.0
Africa	1.8	3.1
Bunkers	3.6	3.6

Table 5.16 Current indicators for the state of global consumption society

Region/ country	Part of population (%)	Part of resources (%)	Resources consumption (%)
USA	5	6	40
OPEC	15	10	40
Russia	3	25	5
Third world	77	59	15

eloquently a current heterogeneity in the world distribution of resources, and consequently the role of separate regions in the environment improvement.

Rates of changes of the anthropogenic carbon fluxes intensity in the future could vary depending on the technologic achievements, weighed economic policy, and replacement of priorities of energetics by the laws satisfied to the stabilization conditions of environment quality indicators. An expected increase of science-technical progress efficiency is with consideration of such progress in future to decrease continuously relative rates of anthropogenic CO_2 fluxes to the atmosphere. And this means that achievement of high HDI level in the countries with transition and developing economics can realized without violent changes of climate due to anthropogenic causes. As it is seen from Table 5.16 USA that did not sign the Kyoto Protocol occupy special place. USA projected with own initiative (RGGI) according to which the elaboration of the reduction strategy of greenhouse gases is planned. This strategy will be satisfied only to the USA national interests. This initiative assumes a development of the methods for the management of greenhouse gases emission to be the regulator of marketable interrelations between states concerning their exchange by the carbon quotes.

Important globalization aspect is an impossibility to assess with reliability the development of separate processes within present conditions of the NSS functioning and without the analysis of all correlations that exist in real world. In this sense, numerous International Programs can have significance only as means of data and knowledge accumulation, but no as technology for the environment study. Specific dependence is observed between the energetics and biogeochemical cycles.

Development of present energetics and its nearest future is closely connected with the use of natural gas, oil and coal. The processing of these fossils in the industry and other sectors of economics leads inevitably to the emission of many other gases to the atmosphere with more high potential of the atmosphere heating (Fig. 5.17). Prognosis of the energetics development in the future depends on the reliable parameterization of socio-economic processes in the GMNSS. According to EIA (2005),world has several clusters that differ by their economic development:

- Countries with developed market economics covering 15 % of the world population. It is countries of North America, West Europe, Australia and New Zealand.
- Countries with transition economics having 6 % of world population and including East Europe and countries of Former Soviet Union.
- Countries with developing economics (78 % of world population):

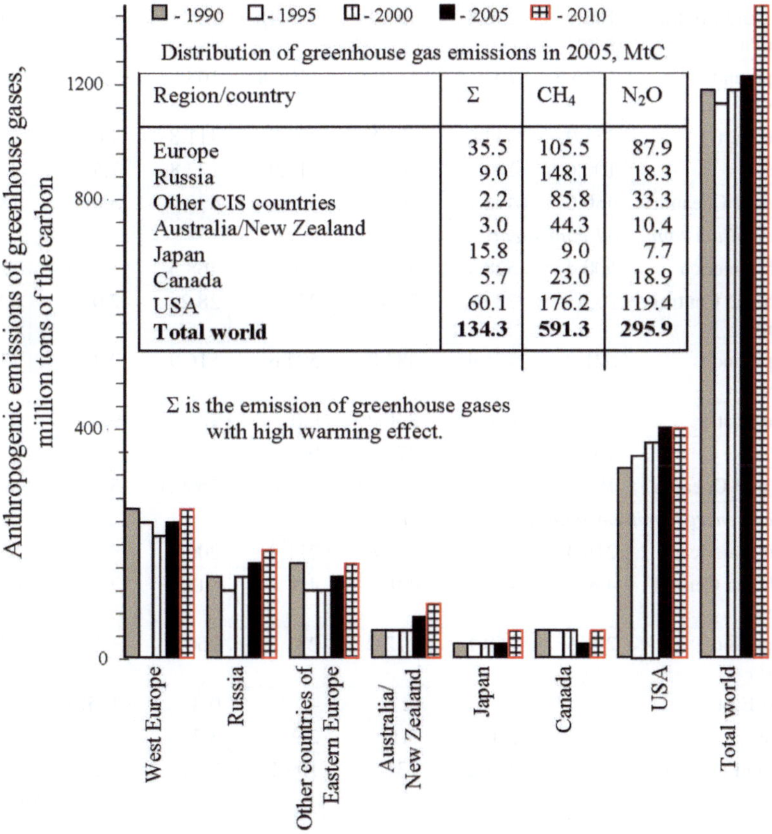

Region/country	Σ	CH_4	N_2O
Europe	35.5	105.5	87.9
Russia	9.0	148.1	18.3
Other CIS countries	2.2	85.8	33.3
Australia/New Zealand	3.0	44.3	10.4
Japan	15.8	9.0	7.7
Canada	5.7	23.0	18.9
USA	60.1	176.2	119.4
Total world	**134.3**	**591.3**	**295.9**

Fig. 5.17 Regional distribution of non-CO_2 greenhouse gases emission in the developed countries with the prognosis to 2010 (EPA 2001, 2005, 2006)

Table 5.17 Energy consumption activities for individual regions in the beginning of 21 century (BP 2005, 2013)

Region	Energy consumption					Δ_1 (%)	Δ_2 (%)
	2008	2009	2010	2011	2012		
Oil (thousand barrels daily)							
North America	23,860	22,969	23,464	23,397	23,040	−1.8	24.6
South and Central America	5,892	5,921	6,222	6,406	6,533	2.0	7.3
Europe and Eurasia	20,017	19,149	19,057	18,974	18,543	−2.5	21.3
Midle East	7,185	7,526	7,861	7,992	8,354	4.5	9.1
Africa	3,218	3,302	3,483	3,359	3,523	5.1	4.0
Asia and Oceania	2,5881	26,206	27,766	28,754	29,781	3.7	33.6
Natural gas (billion cubic metres)							
North America	821.5	816.1	849.6	868.0	906.5	4.0	27.5
South and Central America	140.9	136.8	152.2	156.4	165.1	5.3	5.0
Europe and Eurasia	1136.3	1049.5	1129.6	1105.8	1083.3	−2.3	32.6
Midle East	331.9	344.6	376.8	394.7	411.8	4.0	12.4
Africa	100.6	100.1	107.8	114.0	122.8	7.5	3.7
Asia and Oceania	480.0	496.7	560.4	593.6	625.0	5.0	18.8
Coal (million tonnes oil equivalent)							
North America	600.4	528.4	557.5	526.7	468.5	−11.3	12.6
South and Central America	23.5	22.0	25.4	27.6	28.2	2.0	0.8
Europe and Eurasia	521.0	471.4	484.8	504.6	516.9	2.2	13.9
Midle East	9.0	8.9	8.8	9.0	9.9	9.2	0.3
Africa	105.0	99.7	97.4	96.7	97.5	0.6	2.6
Asia and Oceania	1997.4	2108.4	2290.2	2464.2	2609.1	5.6	69.9
Nuclear energy (million tonnes oil equivalent)							
North America	216.4	213.0	213.8	211.9	206.9	−2.6	36.9
South and Central America	4.8	4.7	4.9	4.9	5.0	3.0	0.9
Europe and Eurasia	276.5	268.1	272.9	271.5	266.9	−2.0	47.6
Midle East	–	–	–	–	0.3	1252.0	0.1
Africa	2.7	3.1	3.1	2.9	3.2	11.3	0.6
Asia and Oceania	119.7	128.2	131.7	109.1	78.1	−28.7	13.9

(continued)

Table 5.17 (continued)

Region	Energy consumption					Δ_1 (%)	Δ_2 (%)
	2008	2009	2010	2011	2012		
Hydroelectricity (million tonnes oil equivalent)							
North America	146.3	151.4	147.2	166.3	156.3	−6.3	18.8
South and Central America	153.2	157.8	158.6	167.9	165.7	−1.6	19.9
Europe and Eurasia	183.0	184.9	197.9	179.0	190.8	6.3	23.0
Midle East	3.2	2.8	4.0	4.3	5.1	18.5	0.6
Africa	21.4	22.0	23.4	22.5	24.1	7.2	2.9
Asia and Oceania	214.6	218.9	251.0	254.7	289.0	13.2	34.8
Renewable energy (million tonnes oil equivalent)							
North America	33.9	38.9	45.0	50.9	57.0	11.6	24.0
South and Central America	9.0	9.9	11.0	123.0	15.6	20.1	6.6
Europe and Eurasia	54.7	61.4	71.1	85.9	99.1	15.1	41.7
Midle East	–	–	0.1	0.1	0.1	27.6	0.1
Africa	0.9	1.0	1.2	1.3	1.4	4.8	0.6
Asia and Oceania	24.7	30.7	40.2	54.4	64.1	17.7	27.0

Notation Δ_1 is the charge 2012 over 2011, Δ_2 is the share of total

- developing Asia (53 %);
- Africa (14 %);
- Middle East (4 %);
- South America (7 %).

In other aspects, countries can be selected by groupings (Table 5.17):

- Countries that ratified, accepted, supported or approved of the Kyoto Protocol.
- Countries of EU.
- Countries of G8.
- Three countries of NAFTA.
- Thirty countries of OECD.
- Eleven countries of OPEC.
- Eight countries of developing Pacific basin.
- Seven countries of the Persian Gulf.

Absolutely, it is obvious that coordinated strategies of the energetics development exist within each of these clusters, but there exist the contradictions. An optimality of such structure can be found by means of the (V, W)-exchange model described by Kondratyev et al. (2006b). A methodic developed in EIA (2005) for the prognosis of energetics development up to 2025 gave the possibility to make weighed estimations for the rates of needs in the energy sources, the realizing of which leads to the achievement of CO_2 atmospheric concentration equaled to

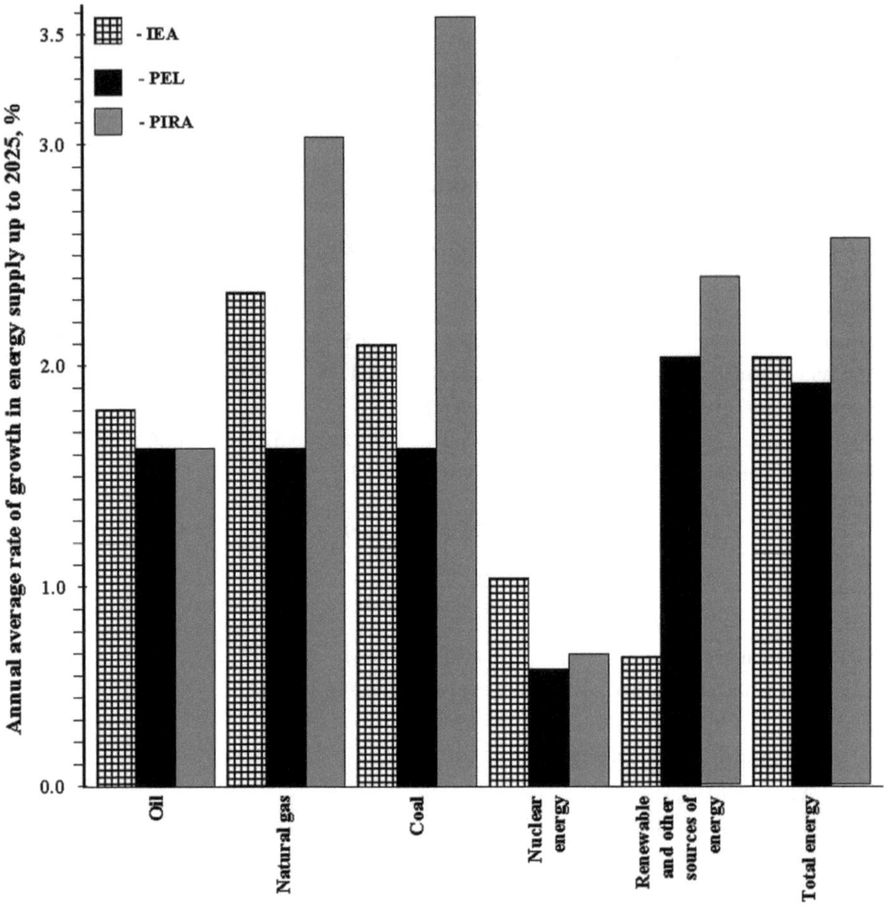

Fig. 5.18 Forecasts of rates for average annual increase of the energy supply made by IEA, PEL and PIRA (IEA 2005a, b, 2013c)

550 ppm in the end of 21 century (Fig. 5.18). In some degree, this will be connected with the priorities redistribution between sectors of the economics. It is expected that significantly increase the role of transport loads on the environment. If in 1990 the transport part was equaled to 20 % of consumed energy, but to 2005 this level approach to 40 %. But at the same time, the progress in the industry will bring to the reduction of consumed energy from 38 % in 1990 to 17 % in 2095. It is expected that role of such sectors of the economics, as electric power production and building is significantly not changed. The forecasts given in Fig. 5.18 and realized by different experts, characterize the presence of high uncertainty in the initial data of biogeochemical block of GMNSS. From this, it is seen that synthesize of global model for economic development has special attention for the models of biogeochemical cycles and for the assessment of their role in the formation of the environment state.

5.6 Thunderstorms as a Component of the Global Ecodynamics

Thunderstorms are the most powerful storms in the Earth's atmosphere. Annually, an estimated of 16 million thunderstorms happen across the world. Thunderstorms are capable of causing extensive damage to both life and property. Thunderstorms affect small areas when compared with hurricanes and winter storms. The typical thunderstorm is 15 miles in diameter and lasts an average of 30 min. Nevertheless, thunderstorm activity is closely connected with the atmosphere instability in global scale, which impacts on global biogeochemical cycles (Kurz and Grewe 2002; Mo et al. 2002).

The level of understanding the photochemical processes in the troposphere that affect its composition, depends substantially on adequacy of ideas about the budget and distribution of nitrogen oxides NO_x ($NO + NO_2$). Due to reactions of the catalytic cycle with the participation of peroxides radicals, NO_x play the decisive role in the processes of tropospheric ozone formation and affect the cycle of HO_x ($OH + HO_2$). One of the most important unclear aspects of the problem under discussion is the insufficient reliability of data on the contribution of lightning strikes to the global budget of NO_x. According to the earlier estimates, this contribution varied within 2–20 Tg N year^{-1}, with most probable values of about 2–6 Tg N year^{-1} (Kondratyev et al. 2006a).

Crawford et al. (2000) carried out an analysis of results of aircraft measurements within the program of the field experiment SONEX on studies of ozone and nitrogen oxides in the troposphere. Results of the analysis revealed the presence in the upper troposphere of episodes of NO_x concentration increase determined, apparently, by the impact of lightning strikes. The correlation with specific periods of lightning activity could be found out from analysis of "reverse" trajectories of the motion of air masses as well as from the data of observations at the national network of lightning record. The Lagrangian numerical modeling of "reverse" trajectories has been carried out to trace the evolution of NO_x plumes during a 1–2-day interval between their occurrence and obtaining of aircraft measurement data. Calculations were made for pre-calculation of expected changes of concentrations of HNO_3, H_2O_2, CH3COOH, HO_2, and OH. Depending on conditions, the initial concentration of NO_x varied within 1–7 ppb. Since the calculated estimates of HNO_3 concentration turned out to be overestimated compared to those observed, the reasons of such differences were analyzed and it was shown that the observed values of H_2O_2 concentration agree with the supposed removal of H_2O_2 from the atmosphere due to convection. Although it is possible that in the upper troposphere the concentration of CH_3COOH grows due to convection, the numerical modeling has led to the conclusion that this increase can be only of short duration (less than 2 h), which excludes the possibility to detect this component during aircraft measurement. The possibility of CH_3COOH concentration increase is excluded too, in view of a high level of NO concentration.

The reaction between NO and HO_2 in all the cases considered, favors a decrease of HO_2 concentration. In some cases the calculated values of OH concentration doubled, but at a maximum level of NO_x concentration the loss of hydroxyl due to the reaction OH + NO_2 compensates the formation of hydroxyl due to reactions NO_2 + NO. An additional increase of OH by 30–60 % can, however, result from convective transfer of CH_3COOH.

To form an estimate of the contribution of lightning and anthropogenic sources to the formation of NO_x concentration on global scales, Kurz and Grewe (2002) undertook a numerical modeling of the frequency and global distribution of lightning strikes and the resulting formation of NO_x using the complex model of climate and chemical processes taking place in the atmosphere ECHAM4.L39 (DRL)/CHEM(E39/C), whose important feature is a realistic simulation of processes responsible for the formation of penetrating convection.

Calculated estimates of the height of convective clouds, on the whole, agree well with those observed. But in mid-latitudes of Western Europe the calculated values were underestimated. The calculated spatial distribution of lightning strikes agrees well with the observed distribution, though the calculated ratio of the density of strikes over land and over the oceans turned out to be underestimated. In the mid-latitudes of Western Europe the density of strikes is underestimated compared to that observed, which can be explained by inadequate scheme of convection parameterization.

According to numerical modeling results, the NO_x formation is most intensive in the tropics and in mid-latitudes, and is clearly separated (in space) from NO_x emissions due to aviation. The lightning-induced maximum level of NO_x emissions is located at altitudes approximately 5 km below the level of the tropopause, with the level of emissions exceeding 3 times the respective anthropogenic level. Besides this, it is important that maximum aircraft emissions of NO_x tend to be at much higher altitudes and farther north (being concentrated mainly within the North-Atlantic corridor 30–60°N at a level of about 200 h Pa, that is, 12 km) compared to NO_x formed due to lightning strikes.

Zhang et al. (2003a, b) analyzed the possibilities of numerical modeling of nitrogen oxides formed due to lightning strikes using the 2D version of the model that simulates an electrization of the storm. It is supposed that the formation of NO is determined by dissipation of energy, whose value is calculated from the value of the electric field before and after a lightning strike. The rate of formation of energy responsible for NO formation is 9.2×10^{16} molecules J^{-1}. Considering a limited set of chemical reactions in which NO, NO_2, and O_3 participate, a numerical modeling has been carried out of processes taking place in a small storm with 10 intra-cloud strikes during 2 min. The level of dissipation of energy varied between 0.024 and 0.28 GJ. After cessation of lightning strikes the integration continued during 18 min more. Analysis of the modeling results has shown that the mixing ratio of NO formed within a cloud (by order of magnitude) is 10 ppb after most powerful strikes and 1–2 ppb on the windward side of the thundercloud anvil at the end of the integration interval. These estimates agree with the data of observations. A comparison with results of the earlier numerical modeling with the use of different 2D model revealed, on the whole, an agreement of estimates, but the results discussed are characterized by greater volumes of energy at high altitudes, which can partially result from a longer integration time and lacking of consideration of the strikes onto the Earth in the model in question, as well as from some assumptions made earlier. On the whole, the numerical modeling of lightning formation carried out recently and their impact on NO formation should be considered more adequate than in previous studies. The necessity of further improvement of the model has determined the development of a new 3D model of the processes of formation of electric charges in thunderclouds with a more complete than earlier consideration of chemical processes of nitrogen oxides formation under the influence of lightning strikes (Zhang et al. 2003b).

A totality of considered GHGs includes in this case NO, NO_2, O_3, CO, CH_4, OH, and HNO_3. The numerical modeling has been carried out for concrete conditions of thunderclouds (storm) observed on 19 July 1981 with 18 intra-cloud strikes during 3 min. The numerical modeling has been carried out for a period of 38 min, before the thundercloud dissipated. The level of energy dissipated at lightning strikes varied within 0.91–2.28 GJ. The maximum level of the NO_x mixing ratio due to lightning strikes reached 35.8 ppb. At the cloud's dissipation, after cessation of lightning strikes, maximum concentrations of NO and NO_2 (in both cases) constituted about 6.3 ppb and were observed at an altitude of about 4 km. The NO mixing ratio in an anvil reached a maximum of about 2 ppb at an altitude of about 10.5 km. These results agree well with observational data.

Quite surprising was the formation of the NO_2 plume at a concentration of about 0.5 ppb, which reached the surface. In the case of NO there was no plume. On the other hand, NO was transported from the cloud's centre to the anvil absent in the case of NO_2, which was determined, probably, by the impact of photolysis. The ratio of concentrations NO_2/NO decreased with altitude in accordance with observational data. The formation of NO calculated per unit length averaged 2.03×10^{22} molecules m^{-1}. The results obtained show that the short-lived storms determine the formation of the vertical profile of NO_x concentration, that differs from the earlier observed C-shape profile.

Mansell et al. (2002) carried out a numerical modeling of lightning strikes with the use of a stochastic model of dielectric break and parameterization of electrification mechanisms. This model enables one to simulate a 2D development of a strike as a stochastic "step-by-step" process. The strike channels propagate over a homogeneous spatial grid, and the direction of propagation (including diagonals) for each step is considered random with the probability of the choice of a certain direction depending on the total electric field. After each step, the electric fields are calculated anew with the use of the Poisson equation in order to take into account the impact of the channel's conductivity. The applied parameterization of the process of lightning strikes formation, provides a realistic 3D simulation of ramified strikes. The model is able to simulate the formation of different kinds of lightning strikes, including intra-cloud strikes, negative and positive strikes "cloud-ground" (CG).

According to numerical modeling results, the hypothesis that negative strikes appear only when the region of the positive strike is located beneath the centre of the negative strike can be considered substantiated. The calculated positive CG strikes were observed only in the parts of the thundercloud, where two charged layers located near the Earth had an approximately similar "normal dipole" structure (i.e., positive charges were above negative ones).

Brown et al. (2002) studied the relationships between CG lightning strikes and the stages of the vertical motions development in the region of thunder centre, taking as an example the data of observations of a multi-cell thundercloud centre formed on 11 July 1989 in the region of the city of Elgin (North Dakota). Radar observations of reflectance and vertical velocity enabled one to identify some cells within the thunder centre and trace the evolution of each cell. The evolution process took place in accordance with the model of thunder centre developed in the late 1940s and in the following succession (Bréon 2004):

- maturing of upward motions and increase of the cells vertical extent at the stage of cumulus clouds development;
- maximum development of vertical motions in the upper part of a cell and downward fluxes with rains from the middle part of the cell; and
- light rains at the stage of the cell's dissipation.

On 11 July there was observed a trend in the formation of the cell clusters, each cluster consisting of both growing (at the stage of maturing) and disintegrating cells. There were no CG strikes when the zone of storm's convection contained only one cluster of cells. The strikes took place only in the presence of two or more clusters. Except from two cases, lightning strikes were observed in the zone of the storm's convection, as a rule, closer to the growing (mature) than to dissipating cells. The discussed observational data favor the hypothesis according to which rains falling at the sites of downward fluxes determine the conditions favoring the formation of lightning strikes that reach the Earth. However, it follows from observations that the complicated structure of the electrical field caused by superimposed fields of several cells especially favors an appearance of CG strikes. This situation can promote an earlier formation of CG strikes in the process of new storm cells formation.

Although studies of lightning were started in the late 1700s, only in the early 20th century observations of the electric field were carried out to retrieve the spatial distribution of charges in thunderclouds responsible for the formation of lightning. Then intensive studies began of the distribution of low-frequency electromagnetic waves generated by lightning strikes (atmospherics). Random natural fields received on the Earth's surface in the frequency range from several Hz to MHz are created mainly by thunderstorm sources. Their study for a formalized description in the interests of radar observations has been carried out by Remizov (1985). Radio-noise caused by thunderstorms is disastrous for communication systems just as thunderstorms themselves are disastrous for population as a factor that can lead to fires and destructions.

Chronis and Anagnostou (2003) discussed the preliminary results of the functioning of the experimental network recording the lightning strikes (ZEUS), which consists of six receivers located in Western Europe and functions from July 2001.

The receivers make it possible to record the lightning-generated low-frequency electromagnetic waves (5–15 kHz) which propagate via the waveguide "Earth—ionosphere" over distances of several thousands kilometers. Estimates have been obtained of the errors of observations at the network ZEUS (from the viewpoint of reliability of detection, localization, and characteristic of lightning strikes) by comparing with the data of independent observations. Such comparisons have been made for three regions: the eastern coast of the USA/north-western sector of the Atlantic Ocean, Africa, and Spain. Data for comparison came from the results of observations made with the use of the lightning sensor mounted on the TRMM satellite to study precipitation in the tropics, as well as data of the Spanish network for lightning observations. Results of the comparisons have shown that the errors of localization (determination of coordinates) of lightning vary within 40–400 km at distances up to 5,000 km and farther. Within the territory on which the network is located the errors do not exceed 40 km.

The formation of positive charges in the lower parts of clouds revealed by calculations with the use of the so-called tripole model of thunderclouds, was verified by results of direct balloon soundings and remote sensing from the data of ground measurements of the electric field. However, since the sources of these charges have not been understood clearly, Mo et al. (2002) undertook direct aircraft measurements of positive charges in the lower parts of clouds with the use of two aircraft flying close by.

Data of observations obtained near New Mexico (USA) on 10 August 1997 demonstrated (at least, in some cases) that near the bottoms of clouds (at an altitude of about 3.4 km) there were located positive charges formed, apparently, under the influence of lightning strikes. The charge recorded in one of the cases at an altitude of about 4 km was ~ 1.25 C, which agrees with the data of balloon and ground measurements of electric fields. Numerical modeling with electric fields prescribed from the data of aircraft observations near the idealized charge dipole with the instant introduction of the positive charge gave results which agree well with observations. The observational data indicate that at a ripe stage of formation of the horizontal distribution of the charge near the cloud bottom, this distribution can be very complicated and is characterized by a combination of the contacting regions with opposite charges.

Rakov and Tuni (2003) studied an adequacy of numerical modeling of the lightning electric field at a great distance using the model of the transmission line (TL) and modified model MTLE taking into account an exponential decrease of current with altitude depending on polar angle (elevation) and the rate of propagation of the opposite charge. The shape of the wave of the latter was approximated by a step function. The same presentation was used for the TL model, whereas in the case of the MTLE model it was supposed that the electric field increases instantly to the level corresponding to TL and then decreases exponentially. The exponential decrease with altitude (in the case of MTLE) results in a considerable decrease of the electric field intensity during about 1 microsecond after reaching a maximum, especially at low values of the polar angle and high rate of propagation.

Calculations made by Marshall and Stolzenburg (2002) showed that in the case of positive CG strikes at which Q-flashes occur, the level of energy constitutes about 1×10^{10} J, and the area of strike is $\sim 40 \times 40$ km^2. An estimation of total electrostatic energy stored in two stratus clouds of mesoscale convective cloud system gave values 5×10^{11} and 2×10^{12} J. These levels of energy are sufficient to provide hundreds or thousands of typical lightning strikes but only 10–100 positive CG strikes with the accompanying Q-flashes.

Using the Pockels sensors (Miki et al. 2002) in the International Center for Lightning Research and Testing (ICLRT) in the State of Florida, Miki et al. (2002) carried out measurements of the shape of the electric field wave at horizontal distances from 0.1 to 1.5 m from the lightning channel. The dynamic range of the measuring system varied between 20 kW m^{-2} and 5 MW m^{-2}, and the band width in the interval from 50 GHz to 1 MHz. Also, electric fields were measured near the bottom of the channel and at distances 5, 15, and 30 m from the lightning channel. Using the Pockels sensors, measurements of electric fields were made for 36 strikes in 9 "trigger" flashes. 8 of 36 strikes measurements were also made of horizontal electric fields. According to the results obtained, the shape of the electric field wave looks like an impulse with its front edge determined by the strike leader, and the rear edge determined by the inverse strike. Six of 36 studied shapes of the electric field wave were close the V-shape, whereas the other 30 were characterized by much slower variations in the phase of reverse strike than at the leader stage. The vertical electric field reached a maximum in the interval from 176 kW m^{-1} to 1.5 MW m^{-1} (an average of 577 kW m^{-1}), and the horizontal electric field—in the range between 495 kW m^{-1} and 1.2 MW m^{-1} (an average of 821 kW m^{-1}). These values are characterized by a 40 % underestimation.

5.7 The Numerical Modeling of the 3-D Distribution of Aerosol and Climate

Developments concerning the 3-D field of concentration and properties of aerosol in the context of substantiation of air quality models have contributed much to numerical modeling of the role of aerosol in the formation of climate.

The aerosol component of CMAQ model described by Binkowski and Roselle (2003) for multiscale assessments of air quality, is aimed at efficient and time-saving simulation of the atmospheric aerosol dynamics. The aerosol size distribution is represented as a superposition of three log-normal modes of aerosol which include the small-sized mode PM-2.5 ($D < 2.5$ μm) consisting of two sub-modes—Aitken nuclei ($D < 0.1$ μm) and accumulation sub-mode ($D = 0.1$–2.5 μm), as well as large-sized mode PM-10 ($D = 2.5$–10 μm). The process of the aerosol properties evolution is described with coagulation, growth and formation of new particles.

In consideration of the aerosol components, the PM-2.5 and PM-10 modes of primary emissions of elemental and organic carbon, as well as dust and other particles have been taken into account. Secondary components of aerosol were sulphate, nitrate, ammonium, water, and secondary organic compounds, of natural and anthropogenic origin. A parameterization of aerosol light extinction in the visible spectral region has been made with the use of Mie formulae and empirical relationships verified by observational data. Algorithms have been described by Binkowski and Roselle (2003) which simulate interactions between aerosol and cloudiness. The obtained results have been illustrated by calculated data with the use of the box and 3-D models.

Mebust et al. (2003) performed a preliminary analysis of the adequacy of Model-3 developed by a group of scientists for a multi-scale assessment of air quality (CMAQ) by comparing calculated and observed values of visibility indices and concentration of various components of aerosol. Comparisons with the data of meteorological visibility observations at 130 airports of the USA for the period 11–15 July 1995 have shown that calculated values of CMAQ parameters, on the whole, reliably reflect the main laws of the spatial-temporal variability of visibility, including the spatial gradients and extreme levels of visibility.

However, an application of both calculation techniques used in CMAQ to calculate light extinction (with Mie formulae and empirical methods) has led to an underestimation of visibility reduction (i.e., to its overestimation). In the case of calculations with Mie formulae, the normalized mean difference (NMD) and normalized mean error (NME) are −21.7 and 25.41 %, respectively, and in case of empirical method −35.5 and 36.2 %. In most cases the accuracy of the calculated values agree with the observed ones only with the accuracy of the coefficient 2, though the correlation coefficient is 0.25 and 0.24.

A special comparison has been made in Mebust et al. (2003) with the use of observed concentrations of sulphate, nitrate, PM-2.5, PM-10, and organic carbon at 18 stations in June 1995. In this case, comparisons have led to the conclusion that except good results for sulphate (the mean systematic difference of the mass concentration constitutes 0.15 $\mu g\ m^{-3}$, and NMD = 3.1 %), the model gave a systematic underestimation of concentration in aerosol of nitrate (−0.10 $\mu g\ m^{-3}$, −33.1 %), PM-2.5 (−3.9 $\mu g\ m^{-3}$, −30.1 %), PM-10 (−5.66 $\mu g\ m^{-3}$, −29.2 %), and organic carbon (−0.78 $\mu g\ m^{-3}$, −33.7 %). The adequacy of simulation on concentrations of various components is as follows: sulphate (r^2 = 0.63, average error 1.75 $\mu g\ m^{-3}$, NME = 36.2 %); PM-2.5 (0.55, 5.00 $\mu g\ m^{-3}$, 38.5 %); organic carbon (0.25, 0.94 $\mu g\ m^{-3}$, 40.6 %); PM-10 (0.13, 9.85 $\mu g\ m^{-3}$, 50.8 %), and nitrate (0.01, 0.33 $\mu g\ m^{-3}$, 104.3 %). Except for nitrate, in 75–80 % of cases the calculated values of concentration agree with those observed within the coefficient 2.

Solar radiation absorption in the atmosphere by BC-containing aerosol can cause a change of the radiative warming of the atmosphere and surface, which, in turn, affects the dynamics and hydrological processes responsible for the cloud cover formation. In this connection, Conant et al. (2002) considered a new microphysical mechanism of the BC impact on climate, consisting of the fact that the radiative warming due to solar radiation absorption in the presence of BC-containing CCN slows down or even prevents CCN from functioning as centers of cloud droplets formation. The temperature of BC-containing droplets turns out to be higher (as a result of their warming due to solar radiation absorption) by shares of degree with respect to the environment, which leads to an increase of water vapor pressure on droplets and limits an activation caused by most strongly absorbing CCN.

In connection with this, it has been proposed to generalize the Köhler theory which determines the dependence of the equilibrium water vapor pressure on the level of water assimilation, with the size of CCN and the share of BC taken into account. This dependence manifests itself most clearly in case of CCN with the volume of BC greater than a sphere with D = 500 nm. In the presence of aerosol with a relative content of BC by mass less than 10 % (per each particle), a 10 % decrease of CCN number density can occur due to a solar warming, with a 0.01 % level of critical oversaturation. On the other hand, the effect of warming due to absorption by BC on activation with a −0.1 % level of critical oversaturation is negligibly small.

Analysis of observational data has led to the conclusion about a global stratosphere cooling during the last decades, though the rate of cooling depended on duration of observation series and was specific for various regions of the globe. While in high latitudes of the Southern Hemisphere the cooling trend has been observed beginning from 1980, in the Northern Hemisphere most substantial changes took place after 1990. The common opinion about the nature of stratospheric cooling in the Southern Hemisphere is based on the fact that it is connected mainly with chemically induced ozone depletion. Here a mechanism of radiative-chemical feedback forms, since a cooling leads to ozone depletion. A quite different situation is observed in the Northern Hemisphere, where the internally induced climate system's dynamics should play an important role.

Manzini et al. (2003) undertook a numerical modeling of the atmospheric circulation sensitivity to ozone depletion and to the growth of GHGs concentration, with chemical reactions in the atmosphere taken into account. With this aim in view, three series of numerical modeling have been carried out under fixed boundary conditions for the recent past (1960) and present conditions (1990 and 2000), considering changes of GHGs concentrations, total organic chlorine and average SST. Changes of the ozone content were calculated with simulation of interactivity of the processes considered.

The numerical modeling results indicate a decrease of the ozone content in the stratosphere. Under conditions of the Antarctic, in 1990 and 2000 an "ozone hole" formed, which was absent, however in 1960, as observations showed. In the stratosphere and mesosphere, the temperature was decreasing during the whole period under consideration, which was most clearly manifested at the level of stratopause and in the lower stratosphere in the region of the South Pole, which agrees with observational data. In the lower stratosphere of the Arctic, a cooling in March (compared to 1960 conditions) was detected only from calculations for 2000.

The activity of long waves propagating in winter from the troposphere in 1960 and 2000 turned out to be comparable. This suggests the conclusion about the contribution of ozone depletion and an increase of GHGs concentration to the cooling taking place in the lower stratosphere of the Arctic in 1960 and 2000. The obtained results suggest the conclusion that the extremely low temperatures observed in March during the last decade, could result from radiative and chemical processes, though a possible impact of other factors cannot be excluded.

A comparison of numerical modeling results for 1960 and 2000 revealed an enhancement of downward fluxes in the mesosphere in the period of cooling of the lower stratosphere (in March in the Arctic and in October in the Antarctic). An enhancement of downward fluxed (downwelling) in the mesosphere can be explained as a response of gravity waves to wind intensification connected with a cooling of the lower stratosphere. The downward shift of the enhanced downwelling with a time shift of about 1 month can be partially explained by the impact of planetary waves. An enhancement of the dynamically induced warming connected with an enhanced downwelling favors a limitation of stratospheric cooling and intensification of the circumpolar vortex in the lower stratosphere and, thus, favors an ozone layer reconstruction due to such a feedback. Both in the Arctic and in the Antarctic a cooling due to ozone depletion covers a region where polar stratospheric clouds (PSC) form in spring, whose extent has increased between 1960 and 2000. The increase of PSC amount could lead to ozone depletion in 2000 as revealed by calculations.

Based on the use of the NCAR CCM-3 (version 3) model, Kristjansson (2002) obtained new estimates of the indirect climatic impact of sulphate and BC aerosol due to the aerosol impact on cloud cover dynamics. Two versions of the aerosol impact on clouds have been considered. One of them (the first indirect effect or the effect of radius) is connected with that the appearance of additional aerosol particles as CCN leads to a decrease of the size of cloud droplets. The second indirect effect manifests itself through a suppressed coalescence of droplets due to a decrease of the size of droplets and hence, an increase of clouds lifetime (the effect of lifetime). Both these effects raise the cloud albedo.

The global fields of aerosol concentration have been simulated in Kristjansson (2002) with the use of submodels of aerosol formation built in the climate model, with the respective "life cycles" taken into account. Besides, characteristics of background aerosol and the dynamics of the background aerosol size distribution have been described. The droplets number density in liquid-water clouds was calculated with the prescribed levels of oversaturation. The obtained size distribution of cloud droplets and the outgoing SWR fluxes agree well with the results of satellite observations.

With the use of the data on aerosol properties contained in IPCC-2001 Report, it has been shown that in the case of global averaging a 5.3 % decrease of the cloud droplets radius (by 0.58 μm for the average radius of droplets 10.31 μm) and a 4.9 % increase of cloud water content due to the impact of anthropogenic aerosol take place. Maximum changes of these two parameters take place (by the order of their significance) in the regions of south-eastern Asia (here the content of sulphate aerosol is at a maximum and the solar zenith angle is at a minimum), Northern Atlantic, Europe, Siberia, and the eastern USA. Similar changes are also observed in the values of indirect RF, whose global average value is $-1.8\ \mathrm{W\ m^{-2}}$, with contributions of changes of droplets' radius and lifetime being, respectively, -1.3 and $-0.46\ \mathrm{W\ m^{-2}}$.

A repetition of the numerical experiment with the use of calculated data for 2100, according to the IPCC A2 scenario, has not changed the global average estimate of RF, but revealed a shift of maximum indirect effect of aerosol to the tropical zone. Experiments on sensitivity have led to the conclusion that the impact of changes of cloud droplets radius is about 3 times stronger than that of changes of clouds lifetime, as well as that the contribution of carbon aerosol is small compared to total indirect climatic impact of aerosol. An approximate character of the obtained results is determined, first of all, by neglecting some types of aerosol—for instance, organic carbon aerosol (this is connected with the absence of needed information about aerosol), as well as an exclusion from consideration of ice clouds and the impact of the processes discussed on the LWR transfer. The climate model is planned to be further developed by taking into account the ocean-atmosphere interaction with stronger emphasis on regional effects.

Cook and Highwood (2004) have been the first who undertook a numerical modeling of the climatic impact of aerosol absorbing solar radiation within the "interim" model of atmospheric general circulation, IGCM, developed at Reading University (USA). The 22-layer global model IGCM of the system "atmosphere—2-m mixed layer of the ocean" has been realized over the grid 5° lat. × 5° long. and ensures a simulation of equilibrium climate after integration for a period of 5 years (in fact, calculations have been made for 30 years). The results obtained in Cook and Highwood (2004) indicate that the sign of direct ARF cannot be considered as a representative characteristic of the sign of real changes of global average annual average SAT. The related important fact is that the climatic sensitivity to radiation absorbing aerosol is much stronger than the climatic sensitivity to the "greenhouse" warming due to the growth of CO_2 concentration. The situation is determined by a change of cloud amount in the process of climate formation (SAT change).

When single-scattering albedo of the aerosol is <0.95, the low-level cloud amount decreases, which determines the functioning of the respective positive feedback. On the other hand, changes of the upper-level cloud amount manifest themselves as the process of negative feedback formation. The total impact of clouds is determined by the balance between two effects of the opposite signs and depends strongly on the choice of the parameterization scheme for the cloud cover dynamics. These results of numerical modeling revealed a distinct "semi-direct" impact of aerosol on climate. In this connection, a supposition has been made in Cook and Highwood (2004) that in all GCM models, the aerosol-cloud feedback due to absorbing aerosols is implicitly present. This situation suggests the conclusion about critical importance of an adequate parameterization of cloud cover dynamics (in particular, in the context of hypothesis of "semi-direct" ARF proposed by Hansen et al. (1997)). According to IGCM model, clouds play an extremely important role in the formation of the climatic system's response to the atmospheric aerosol dynamics.

Takemura et al. (2005) applied a coupled model of the long-range transport and atmospheric general circulation to estimate changes of cloud cover characteristics, precipitation, and temperature due to direct and indirect ARF. The obtained estimates of the efficient radius of cloud droplets, cloud-induced RF (CRF) are in satisfactory agreement with the data of satellite observations. The numerical modeling revealed a global-scale decrease of an efficient radius of droplets due to indirect anthropogenic ARF. On the other hand, changes of the cloud water and precipitation are determined by a variation of the dynamical hydrological cycle with a temperature change by the aerosol direct and indirect effect rather than the second indirect effect itself. The global mean direct and indirect anthropogenic ARFs at the tropopause by anthropogenic aerosols are calculated to be −0.1 and −0.9 W m^{-2}, respectively. It has been suggested therefore that aerosol particles reduce the increase in the SAT due to GHGs approximately by 40 %, on the global mean.

Quaas et al. (2004) have used three ensembles of simulations with the LMDZ general circulation model developed at the Laboratoire de Mètèorologie Dynamique to assess the radiative impacts of five GHGs (CO_2, CH_4, N_2O, CFC-11, and CFC-12), as well as sulphate aerosols for the time period 1930–1989 (sea surface temperature was prescribed). It has been shown that the GHGs concentration increase results in a reduction of clouds at all atmospheric levels, thus decreasing the total greenhouse effect in the longwave spectrum and increasing absorption of solar radiation by reduction of cloud albedo. Since different changes in high- (decrease) and low-level (increase) cloudiness have been obtained, the final result was a slightly positive net effect due to cloudiness changes. The total aerosol effect including the aerosol direct and indirect effects remains strongly negative.

Lohmann (2004) has studied the global impact of aerosols on the riming rate of snow crystals of different shapes with cloud droplets in stratiform clouds using the AGCM general circulation model to answer the question: can anthropogenic aerosols decrease the snowfall rate? The final answer was that in accordance with the present-day simulation there had been a slight increase of the snowfall rate over pre-industrial times.

The opinion expressed by Jacobson et al. (2002) that control of emissions of BC aerosol and organic matter to the atmosphere due to fossil fuel burning can become an efficient method of slowing the global warming, has caused a hot discussion. Feichter et al. (2003) noted this connection that according to estimates obtained by Jacobson, emissions to the atmosphere of carbon aerosol (BC and organic carbon, OC) due to fossil fuel burning can cause an increase of global average SAT by 0.35 K. However, there is doubt with respect to grounds for conclusions in this study connected with, that the emissions under discussion contain also sulphur dioxide (which determines the formation of sulphate aerosol). Besides, emissions affect both absorption and scattering of shortwave radiation, which they can cause both warming and cooling. Still, more important is the fact that the climate model GATOR-GCMM used by Jacobson has not been tested for adequacy by comparing with other models and with observational data, which is necessary, especially since calculations for 6 years cannot ensure a simulation of an equilibrium climate. Penner (2003) has made some comments.

Chock et al. (2003) have made remarks on the results of Jacobson (2002) concerning the adequacy of the applied climate model, reliability of input information, and substantiation of the needed ecological policy. As for the input information, it was assumed that annual emissions of BC and OC constituted 6.4 and 10.1 TgC $year^{-1}$. The assumed ratio OM:BC = 3.1:1 is equivalent to global emissions of OM and OM-submicron aerosol reaching 19.9 and 15.9 TgC $year^{-1}$, respectively, that is, much higher than the estimates mentioned above. Besides, drawbacks have been noted such as incomplete consideration of climatic indirect ARF, which leads to a climate cooling, and an unacceptable simplification of the climate model. In his reply to these and other comments, Jacobson (2003 a, b, c) rejected them as incorrect and groundless.

With respect to the discussion mentioned above, it should be emphasized that the most important thing, as follows from the above review, is that all earlier obtained estimates of the impact of aerosol on global climate are highly uncertain.

5.8 An Expert System for the Physics of the Atmospheric Pollution

5.8.1 The Structure of the Expert System

Figure 5.19 gives a list of units of the expert system for physics of atmospheric pollution (ESPAP). The software items for the ESPAP functioning is reproduced in Table 5.18. The input information in the ESPAP is assumed to concentrate in thematic files, which can be formed on the basis of various sources, such as remote

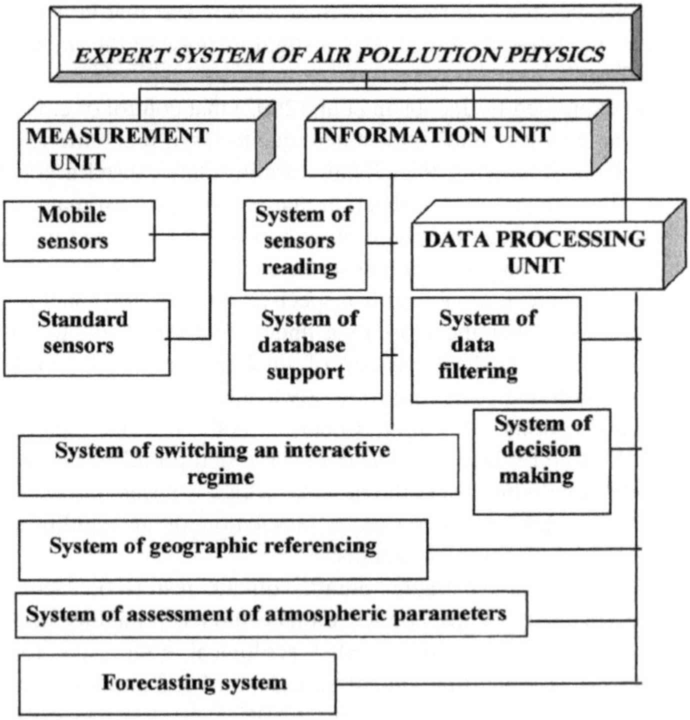

Fig. 5.19 The ESPAP structure with functions of complex assessment of air pollution over a given territory

Table 5.18 The ESPAP software

Identifier of software product	Characteristic of software product
CSII	Calibration and scaling of input information
IDF	Input data filtration
STA	Spatial-temporal agreement of various types of data
SIOD	Spatial interpolation of observational data and formation of spatial images
DR	Data restoration
GTM	Gauss-type model
ETM	Euler-type model
CSDB	Choice of the spatial digitization grid and transfer to block modeling
RFDB	Removal from data base of information needed for modeling
ASI	Modeling the processes of pollutants' transport due to atmosphere-sea interaction
APS	Modeling the processes of pollutants' transport with due regard to functioning of the atmosphere-plant-soil system
MSF	Model of sulphur fluxes in the environment
CPMR	Cartographic presentation of modeling results
AMD	Analysis of modeling data and their visualization with MCL levels taken into account
UI	User interfaceInformation
MNF	Model of nitrogen fluxes in the environment
MOB	Model of oxygen balance in the natural-anthropogenic medium
MCC	Model of CO_2 cycle in the environment
CI	Control and interaction of the ESPAP units
DMP	Decision making procedure based on sequential analysis method
FOI	Formation of output information

MCL is maximum (permissible) concentration level

sensors, lines of information communication, transported standard and non-standard accumulators of observational data, direct input of data from the keyboard or another compatible input unit. The base version of ESPAP is calculated for the latter case. Switching of ESPAP to other regimes requires additional software elements, adjusting the medium and providing its transformation into the file unit. The interaction between all elements of the ESPAP is schematically shown in Fig. 5.20.

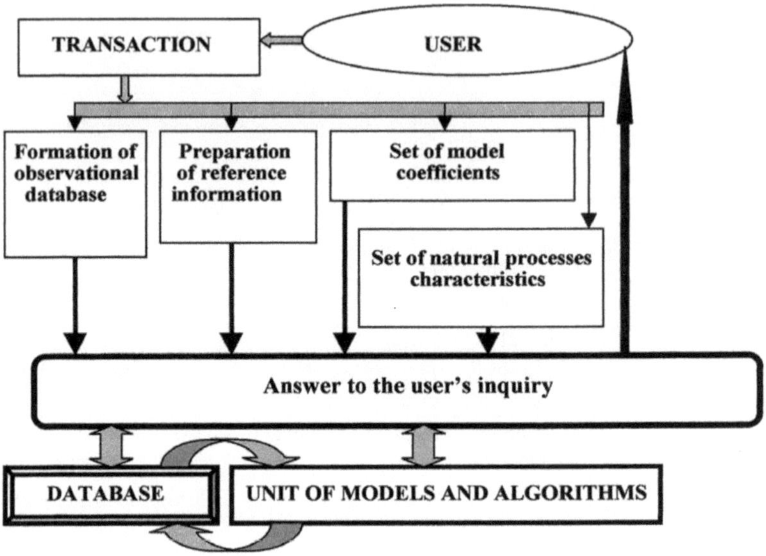

Fig. 5.20 Approximate scheme of ESPAP functioning in the regime of dialogue with the user

5.8.2 Formation of the Database Components

Algorithms of formation and processing of spatial data bases have been reliably verified in the GIS system. To refer the ESPAP to a concrete object, we choose the matrix hierarchical topologic structure (spatial identifier) which reflects the spatial and object structure of the territory Ω so that

$$\Omega = \bigcup_{(i,j)} \Omega_{ij}, \ \Omega_{ij} = \bigcup_{(s,l)} \Omega_{i_s j_l}$$

it is shown schematically in Fig. 5.21. According to this algorithm, the territory Ω in the ESPAP data base is presented by a set of matrices:

$$A_k = \left\| a_{ij}^k \right\|, \quad k = \overline{1, N};$$

where

$$a_{ij}^1 = \begin{cases} 0 & for \quad (\varphi, \lambda) \notin \Omega, \\ a_1 & for \quad (\varphi, \lambda) \in \Omega, \end{cases}$$

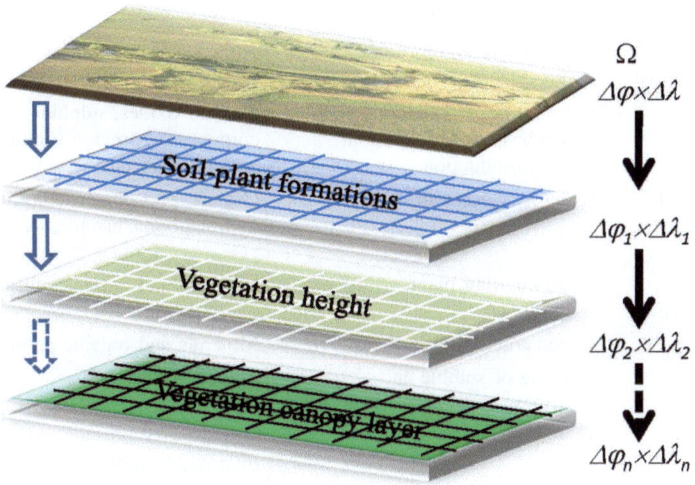

Fig. 5.21 Procedure of topological referencing of the territory to the ESPAP data base components

$$a_{ij}^{2} = \begin{cases} A & - \quad \text{forest,} \\ B & - \quad \text{grass,} \\ C & - \quad \text{agriculture (crops),} \\ D & - \quad \text{plough land,} \\ E & - \quad \text{highway,} \\ F & - \quad \text{boggy locality,} \\ G & - \quad \text{urban territory,} \\ H & - \quad \text{industrial zone,} \\ J & - \quad \text{fruit trees,} \\ I & - \quad \text{agricultural vegetation (others),} \end{cases}$$

$$a_{ij}^{3} = \begin{cases} A & - \quad \text{no pollution sources} \\ B & - \quad \text{pollution sources of the type ``B''} \\ C & - \quad \text{pollution sources of the type ``C''} \\ D & - \quad \text{other type of pollution sources} \end{cases}$$

Matrices A_4 and A_8 contain geographic information about the territory levels with respect to the sea level and data on rain rate, respectively. The A_5 identifier gives the distribution of population density in the region. The A_6 and A_7 structures determine mean statistical data on wind speed and direction over the territory Ω. All information is prescribed for a certain time period and, hence, the ESPAP database stores the sets of matrices $\{A_{ij}\}$, marked with the respective time interval. Each element of these matrix structures is hierarchically decoded by levels but not necessarily in the same way. The A_3 matrix takes into account the types of pollution sources enumerated in Table 5.19. For a better familiarization of ESPAP, the database foresees the possibilities to identify the atmospheric pollution sources with

Table 5.19 Classification of the atmospheric pollution sources

Category of the pollution source	Examples of enterprises of a given category	Atmospheric pollutants
Chemicals plants	Oil refining, woodworking, superphosphate excavation, cement production	Sulphur oxides, sulphides, fluorine compounds, organic vapour, particles, aromatic substances
Spraying and dispersing in agriculture	Control of pesticides and weeds	Organic phosphates, chlorinated hydrocarbons, arsenic, lead
Crushing, grinding, sifting	Road-building plants	Minerals and organic substances
Destruction	Reconstruction of cities	Minerals and organic particles
Burning of fields	Burning of stubble and clearings	Smoke, ashes, soot
Fuel burning	Power stations	Sulphur and nitrogen oxides, carbon monoxide, smoke and aromatic substances
Fuel production	Gas scattering	Fluorine compounds
Printer's ink production	Photography and printing	Fluorine compounds
Metallurgical works	Metal melting, steel and aluminum production	Lead, arsenic, zinc, chlorine compounds, sulphur oxides
Mills, car engines	Elevators, transport means	Sulphur and nitrogen oxides, aromatic substances, carbon monoxide, smoke
Use of nuclear fuel	Nuclear reactors	Argon-41
Extraction of ore	Grinding, crushing, sifting	Uranium and beryllium dust
Burning of garbage and waste	Dust-heaps, garbage-processing plants	Sulphur oxides, smoke, ashes, organic vapour, aromatic substances
Waste processing	Stores of scrap metal, car-heaps, processing of second-hand materials	Smoke, organic vapour, aromatic substances, soot

a set of pollutants. The structure of the elements of the A_9 identifier, identifies each spatial cell Ω_{ij} with the type of the polluting enterprise.

5.8.3 A Subsystem for Statistical Decisions

A totality of units DMP and FOI performs the function of decision making about the signal output for the user in correspondence with the a-priori prescribed elemental structure of maximum concentration level (MCL). To preserve the generality, it is assumed that MCL can be prescribed in each cell Ω_{ij} or in some their totality by an independent set of elements. This information contains the A_{10} identifier.

In the process of measurements or calculations, for each pollutant of the type ξ the function of concentration distribution $P(\varphi, \lambda, h, t)$ is formed. Depending on time, at each point (φ, λ, h) a series of data $P_{\xi 1}, P_{\xi}^2, \ldots, P_{\xi}^n, \ldots$ is formed. Because of measurements and calculations errors due to incorrect data, the choice of situations $P_\xi < \mathrm{MCL}_\xi$ or $P_\xi \geq \mathrm{MCL}_\xi$ is made at each time moment t on the basis of the respective statistical assessment in distinguishing between two versions:

(1) the decision is made on the basis of a totality of indicators, and
(2) the decision is made on the basis of each indicator separately.

In the latter case, the inequality $P_\xi \geq \mathrm{MCL}_\xi$ is tested for stability during a prescribed time interval Δt. In the case of meeting the condition $P_\xi \geq \mathrm{MCL}_\xi$ during this time interval the display gives a signal with the respective information. The time Δt is determined at expert level in the process of the system's exploitation to minimize the amount of false alarms or, from condition of maximum probability not to miss a situation with a violated MCL. The actual choice of Δt depends on the multitude of local anthropogenic and other factors and refers to the user's competence. To make it easier, as to obtain an expert estimate or to make a decision in the real time regime, especially in the case of the use of an integral criterion, the user receives estimates of all indicators on a homogeneous normalized scale.

Let $q(l, i, j, k)$ be the indicator of the atmospheric quality in the compartment $\Omega_{i,j,k}$ by the l-th parameter. To reduce all indicators to a single scale, we normalize them:

$$Q_l = Q(l, i, j, k) = q(l, i, j, k) / \max_{i,j,k} q(l, i, j, k)$$

Then the results of normalization are smoothed by height (k), latitude (i), longitude (j) and the whole territory Ω. The smoothed results are used to assess the air quality of territories by the user's choice. The final decision is made on the basis of comparison of Q_l estimates with a set of threshold values included in data base. The complex assessment of the state of the atmosphere over a given part of the territory Ω is made by the vector parameter $Q = (Q_1, \ldots, Q_n)$. This n-dimensional vector belongs to some indicator space X, each point in which corresponds to a concrete combination of MCL_ξ (prescribed by the user). The whole space X is divided into two parts: X_0—pollution does not exceed acceptable levels; X_1—pollution reaches a dangerous level. A concrete realization of Q is estimated by belonging to one of these sub-spaces.

Thus, the main stages of ESPAP functioning are described by the scheme presented in Fig. 5.22. Of course, the quality of the decisions made depends on the adequacy of information in the data base. Table 5.20 shows a fragment of the data base. These database fragments and various modifications of parametric descriptions of partial processes of aerosol transformation are controlled by a totality of identifiers which the user disposes of through such schemes of interaction with the system as scrolling windows and hierarchical menu. Therefore, a multitude of identifiers is used at the level of the system's manager who sees to the compatibility of its functions.

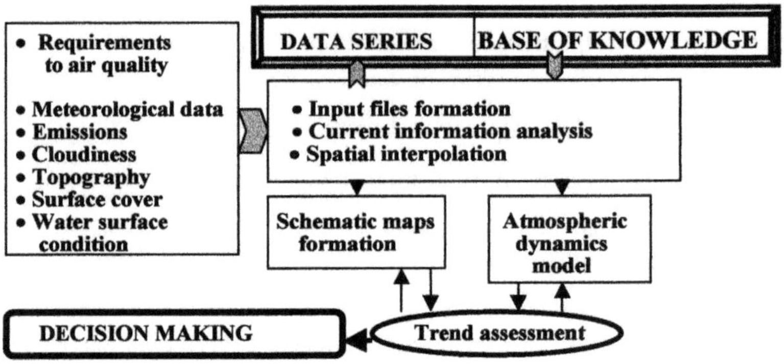

Fig. 5.22 Basic components of ESPAP functioning

Table 5.20 Components of the ESPAP database

Element of Λ_2 identifier	Rate of dry deposition (cm s^{-1})	
	SO_2	NO_2
Water surface	0.2–1.5	0.01–0.04
Lime surface	0.3–1.0	0.03–0.1
Acid dry soil	0.1–0.5	0.01–0.05
Acid wet soil	0.1–0.8	0.01–0.08
Plants <10 cm	0.1–0.8	0.02–0.25
Plants from 10 cm to 1 m	0.2–1.5	0.4–0.7
Forest	0.2–2.0	0.1–0.8

5.8.4 A Subsystem for Control and Visualization

The ESPAP provides a wide hierarchical dialogue for the user with the computer version of the simulation system of aerosol propagation in the atmosphere. An adaptation to the territory in the regime of dialogue is accomplished at the expense of identifiers which ensure the formation of database fragments responsible for the structure of surface covers, meteorological situation, and environmental parameters including information about the sources of aerosols and their characteristics. The operator's interaction is shown schematically in Fig. 5.23.

The ESPAP operates with schematic maps containing conditional images of the Earth surface or distributions of the atmospheric parameters. The schematic presentation of the modeling results is connected with their transformation into one of the following forms of distribution of the parameter to be estimated:

- schematic representation (1-D, 2-D, 3-D);
- symbolic representation of a 2-D image;
- tabulated representation by a set of formats;

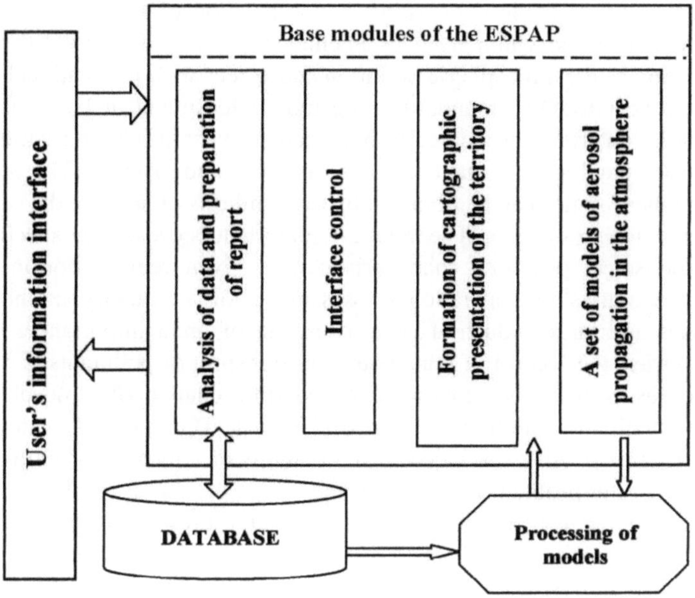

Fig. 5.23 An approximate scheme of the operator—ESPAP interaction in the regime of its adaptation to specific conditions of simulation experiment on assessment of aerosol propagation over a given territory

- false-colour image reflecting the representation of modeling results by a fixed digital scale; and
- cartographic representation by isolines drawn on the basis of GIS-technology.

5.9 Modeling Aerosol Transport in the Atmosphere

5.9.1 Relationships Between the Scales of Atmospheric Mixing Processes and the Choice of Models

An experience gained from many studies on modeling the atmospheric processes of pollutants transport dictates a necessity to classify these processes in accordance with interrelationships between spatial and temporal scales (Aloyan 2004; Dimov et al. 2004; Mehdizadeh and Rifal 2004). The need of this classification has been substantiated in many international programs, such as "Global Changes", "Global Atmospheric Chemistry" (IGAC), "Modeling of Global Biogeochemical Cycles" (MGBC), and others. The need and even necessity of classification of physical processes of atmospheric mixing is dictated by parameters of the systems for measurements (monitoring) of atmospheric characteristics, requirements of simplicity of the models of the aerosol and gases transport in the atmosphere, as well as by limitations of available data bases. The interaction of these causes leads to a range of spatial scales, which provides an efficient parameterization of the processes of propagation of the atmospheric pollution and agrees with international standards. At present, the GEODAS standard is the most widespread, it has nine levels of spatial resolution from one degree to half-second in longitude and latitude. Seven scales of cartographic presentation of data are provided (Table 5.21).

Data on these scales can be obtained by the synthesis of satellite data and data of national monitoring systems. The latter are important because they specify the data domain and select priorities characteristic of a given region. For instance, for developed countries the operational assessments of air quality in the zones of megapolises and large industrial enterprises, are of great importance. For developing countries, the control of trans-boundary transport of pollutants is of principal importance as well as an assessment of possible change of atmospheric air in connection with the building of industrial plants. The ratio of scales given in Table 5.21 corresponds to a majority of situations of atmospheric monitoring.

Along with the problem of choosing the scales of spatial resolution, there is a problem of their agreement with time scales. This problem is important for the formation of the structure of numerical models which describe the pollutants' dynamics in the atmosphere. According to preliminary estimates of the International Geosphere-Biosphere Program, there is a scale of transitions in temporal and spatial measurements between complexity and depth of the hierarchic structure of connections considered in the model. So far, the developed models practically ignored this fact, and therefore it was often impossible to apply them to the natural object under study. An agreed discretization scale for natural phenomena to be used in models, proposed by Nitu et al. (2000a, b, 2004), which enables one to classify the natural phenomena with due account of their hierarchic subordination on temporal and spatial scales. This classification is based on the fundamental understanding of the hierarchy in the general theory of systems. According to this theory, the behavior of any complicated system is determined by the triad of frequencies of its variability, which provides an agreement between coherence and stability of the system. This makes it possible to exclude from consideration unnecessary details in the model's structure, by the prescribed time scale or to establish a minimum time step from data on spatial scale. For instance, if in the model the time step is chosen to be equal to 1 year, there is no sense in taking into account such processes as atmospheric turbulence. In other words, in this case the atmosphere can be

Table 5.21 List of scales of cartographic information presentation characteristic of the developed monitoring systems

Spatial resolution (km)	Scale	Spatial resolution	Scale
0.5 × 0.5	1:1,250	40 × 40	1:50,000
1 × 1	1:2,500	250 × 250	1:250,000
5 × 5	1:10,000	500 × 500	1:625,000
10 × 20	1:25,000		

described with a point model, and all attempts to develop complicated constructions to describe the processes of atmospheric motion, cannot improve the reliability of the model but will increase its complexity.

A more strict theoretical substantiation of this approach is a combined description of dynamic processes with different characteristic time scales, which results in parameters of mathematical description referring to different processes and strongly (by orders of magnitude) differing from each other. This enables one to divide these processes into three groups:

- processes referring to a chosen time scale;
- processes which, with respect to time scale, can be considered as being in dynamic equilibrium (quasi-stationary) and for which some parameterization can be introduced (fast processes); and
- processes which, with respect to the chosen time scale, can be considered constant, that is, static (slow processes).

As an example, we take a numerical model of the biogeochemical cycle of carbon which includes both the geological-time-scale processes and fast processes (photosynthesis, respiration). If the aim of the model is to study the dynamics of CO_2 content in the atmosphere over several decades, the geological processes with characteristic timescales of millions of years, should be excluded from consideration and the processes should be parameterized with characteristic timescales of days, for instance, processes of living biomass formation with photosynthesis. In general, the problems mentioned have not been solved yet, and there is no constructive mechanism for matching the spatial and temporal scales. Each scientist follows a principle of his own in choosing the model elements for their subsequent realization. Unfortunately, at this stage there appear unavoidable deviations of the model from reality.

Characteristic time intervals of variability for most natural processes are well known. Here are some of them:

- processes of deposition—minutes, hours;
- plants' transpiration—hours, days;
- formation of plants' biomass—days, months;
- changes in communities of plants and animals—months, years;
- soil formation—years, centuries; and
- geomorphological processes—centuries, millennia.

Therefore, a systematization of timescales for the processes considered should follow a synthesis of the system of models. As a result, priorities for models and their units as well as software structure can be determined.

Because of difficulties to parameterize atmospheric processes, there are many models of atmospheric dynamics. The type of the model correlates strongly with the spatial scale. A consideration in the models of the processes of physical mixing together with chemical processes taking place in the atmosphere is determined by the accuracy level. The developed models, depending on the spatial scale, take into account processes of physical transformation of pollutants from micro-processes in

Table 5.22 A fragment of the scale for development of the structure of the atmospheric pollution dynamics model

Spatial resolution of the model	Processes recommended for consideration in the model
Industrial region, landscape, megapolis, city (up to 50 km)	Use of the Gauss-type models. Burning of wastes, deforestation and reconstruction of surface covers, contamination of drinking water and water basins, industrial emissions of aerosol, soil contamination, washing-out of pollutants with rains, production processes, medico-biological assessment of the territory, division of the atmosphere into many levels
Large region, district, country (up to 1,000 km)	Use of Lagrangian and Eulerian models. Large-scale atmospheric circulation with selection of the upper and lower atmosphere, irrigation and other aquatic systems, integral areal sources of biospheric pollution, biogeochemical cycles, erosion, large-scale fires, desertification and swamping, succession of surface covers, river runoff, interactions on shelfs
Continent, globe (>1,000 km)	Use of block models. Averaged characteristics of the atmosphere and climate, oceanic circulation, interactions in the system "atmosphere-land-ocean", biogeochemical cycles, succession of large tracts of forests

clouds to large-scale atmospheric motions. Depending on this, models are divided into dispersive, Gaussian, Eulerian and Lagrangian (Rasmussen and Williams 2006; Rodean 1996; Laudon and Traver 2011). Within this system, there is anhierarchy of models taking into account or neglecting the vertical structure of the atmosphere, atmosphere-surface (land, water) interaction, exchange processes between clouds, and vertical air fluxes depending on relationships between synoptic and physical parameters of the atmosphere. One of the examples of such studies is a series of versions of the ICLIPS model with the spatial resolution 500 × 500 km and a time step of 1 year (Integrierte Abschatzung von Klimaschutzstrategien) (Toth et al. 2000). A more accurate model ECMWF (European Centre for Medium Range Weather Forecast) has the spatial resolution 150 × 150 km and a time step of 6 h (Gregory et al. 2000).

It is impossible to establish an unambiguous connection between the scale of the model and its internal infrastructure without taking into account various characteristics of the model. Therefore, the estimates given in Table 5.22 should be considered as recommendations, directed at the modeler's aim giving him a possibility to assess an expedience of including certain components into the model.

5.9.2 Interrelationship Between the Types of Models and Aerosol Characteristics

The formation of the fields of atmospheric pollution from natural and anthropogenic sources depends strongly on the physical characteristics of pollutants. Clearly, for a complete understanding of the processes of formation and growth of clouds and precipitation, it is necessary to consider all dynamic microphysical interactive processes. These processes are determined by a combination of physico-chemical parameters of the atmosphere itself and pollutant components, which are characterized by strongly variable characteristics such as weight, concentration, size, shape, phase state, and electric charge. For instance, the classification of atmospheric aerosol pollution adopted by the US National Oceanic and Atmospheric Administration (NOAA) includes three basic classes and eight sub-classes. This classification is sufficient to be used in models which do not take into account the size of aerosol particles and do not include the ion-level processes.

The existing classification of particles by the size covers the particles' diameters from 0.0001 μm to 1 cm. Within this range, solid particles with D = 0.0001–1 μm are considered as smoke components, while liquid particles are elements of fog. Particles exceeding 1 μm in diameter are interpreted as dust or spray. Depending on the size, the role of particles in dynamic processes of atmospheric pollution changes, too. Particles with diameter less than 1 μm form smog, tens of μm—clouds and fog, hundreds of μm—haze and drizzle, thousands of μm—rain. This classification simplifies the choice of the model's structure, if the nature of processes of atmospheric pollution is known. In a more complicated situation, when the size spectrum of aerosol particles is sufficiently broad, the division of pollutants by their physical and chemical characteristics enables one to synthesize the complicated model as a totality of hierarchically submitted partial models and to simplify thereby the procedure of calculation of dynamic characteristics of the polluted atmosphere.

The physical characteristics of atmospheric pollution include also the rate of gravitational deposition, residence in the atmosphere, and phase state. Some of the atmospheric gas components such as N_2, O_2, He, Ne, Ar, Kr, Xe, and H_2 have a very long lifetime. The lifetime of CO_2, O_3, N_2O, and CH_4 is from several years to decades. Such gases as H_2O, NO_2, NO, NH_3, SO_2, H_2S, CO, HCl, and I_2 live in the atmosphere only several days or weeks. Depending only on this characteristic to describe the dynamics of various gases in the atmosphere, one can choose an adequate model with minimum requirements for database (Kondratyev et al. 2006a; Bornstein 1999).

For the choice of the type of the polluted atmosphere dynamics model, the size of an aerosol particle is of great importance. The mentioned intervals only partly cover a possible classification of aerosols. Additional information is needed about the source of pollutant which further specifies the parametric space of the model. The knowledge of the cause of pollution simplifies the choice of the model type. Of course, the classification and typification of aerosols and gases can be more detailed. For instance, there are tens of the types of smoke. The size of smoke particles can be 1–0.01 μm for resin smoke, 0.15–0.01 μm for tobacco smoke, etc. Here, in the model, it is necessary to consider micro-processes connected with the motion of these particles. For instance, the run of a particle of carbonaceous smoke during *t* seconds averages 0.00068 t/D cm. Filling the base of knowledge of the monitoring system with such dependences is one of the first-priority problems of ecoinformatics.

In modeling, the scattering of gases and particles in the atmosphere is important to know the difference between the polluted and clear atmosphere properties. Also, one should always bear in mind the vertical heterogeneity of the atmosphere. On a global scale, the formation of air quality depends on processes at all levels of the atmosphere: the troposphere, stratosphere, hemosphere, and ionosphere. For instance, in the problem of the impact of aviation on the atmosphere, it is necessary to take into account the interaction between the troposphere and stratosphere. In studying the fluxes of pollutants from the surface sources, the motion of the lower atmosphere is considered first of all. Of course, here of importance is the spatial scale and, hence, the time interval of the pollutants' residence in the atmosphere. Data known for the clear atmosphere should be used in the control simulation experiments.

Pollutants emitted to the atmosphere are subject to gravitational sedimentation, turbulent mixing, wind-driven transport, and washing out by rain. A totality of these impacts determines the character of behavior of the polluting cloud, shape and type of the flux of pollutants, as well as the spatial structure of the distribution of aerosol number density over a given territory.

Smoke and other atmospheric aerosols are gravitationally influenced and interact with solar radiation, gases and ions. In the surface layer, this interaction is supplemented with different effects of the surface (vegetation cover, soil, land surface roughness, and sea surface roughness).

The role of sedimentation is more substantial in case of large particles, larger in diameter than sub-micron particles. Small particles sediment are much slower compared to their transport by the moving atmosphere, and therefore in many models this vertical constituent is neglected. Note that for the process of sedimentation the diameter of particles is less important than their density. For instance, soot structures with a low efficient density and large aerodynamic cross-section are easily wind-driven and sediment much slower than the compact particles of the same mass. The rate of sedimentation of particles with $D = 0.1–1\ \mu m$ averages $0.001\ m\ s^{-1}$, which is negligibly small compared to the rate of atmospheric transport.

In case of heavily polluted formations one can observe the process of photophores consisting in particles' lifting due to non-uniform solar heating. However, the possibility of this phenomenon and its characteristics have been poorly studied, therefore in a first approximation, many experts leave it out of account, especially because during time intervals longer than a day, due to Brownian motion, an irregular heating of particles decreases. Finally, such physical process should be pointed out as particles coagulation consisting in capture of one particle by another due to different rates of motion. In this case, particles can either stick together or be repelled, so that diverse situations of their interactions arise which determine the shape of the cloud of pollutants and can prolong their lifetime in the atmosphere.

The washing-out is a very important process of removal of pollutants from the atmosphere. Here two situations are possible. One is connected with a simple capture of particles by rain droplets, the other—with the so-called nucleation process. This process is connected with oversaturated water vapor condensation on the surface of aerosol particles, which leads to the formation of water droplets or ice crystals with subsequent deposition onto land surface. Therefore, one of the ways to specify the model of aerosol dynamics in the atmosphere is to include them in a unit of parameterization of water cycle in its different phase states.

The final estimation of the residence time for a given pollutant in the atmosphere is realized with the use of the respective models. This estimation has been given in detail in Picket (1987), Baltrenas et al. (2003). Here the meteorological features of pollutants propagation in the atmosphere are described, the scales of transport and scattering of pollutants are analyzed, models are constructed which predict the concentration of pollutants, algorithms of parameterization of the processes of clouds and pollutants' jets formation are simulated and the ratios are given to describe the vertical structure of the atmosphere. The Earth's radiation budget components are analyzed and the simplest characteristics of relationships between pressure, wind, temperature, and humidity are given. The state of the atmosphere is classified as neutral, unstable and stable by the scale of the vertical temperature lapse rate, which considerably simplifies the process of parameterization of the vertical gradients and rates. The scale of atmospheric phenomena is estimated over the time interval from 1 s to 1 month, and with spatial scale varying from 20 to 1,000 km. Within this scale, processes of transport and scattering of atmospheric pollutants are analyzed from point sources, as well as moving and covering the final territory.

In general, a change of the concentration of any pollutant C is described with the following equation:

$$\partial C(t, \varphi, \lambda, h)/\partial t + \nabla \cdot \vec{V}C = \nabla D \cdot \nabla C + R \tag{5.3}$$

where $\vec{V}\left(V_{\varphi}, V_{\lambda}, V_{h}\right)$ is the wind speed, φ is the latitude, λ is the longitude, h is the height, t is the time, D is the coefficient of molecular diffusion, R is the change due to atmospheric turbidity, emission, and mixing.

Detailed description of the terms of Eq. (5.3) needs an analysis of specific processes of atmospheric propagation of pollutants and construction of respective units of a general model (dynamic, correlative, probabilistic, system, evolutionary, etc.). As examples of such units, we shall consider the parameterizations used successfully in models ICLIPS, ECMWF, and others.

Problems of chemical interaction of atmospheric pollutants are also important, and their consideration in modeling further complicates the study. Therefore, most of the models of pollutants propagation in the atmosphere assume a-priori that all components are mutually neutral. However, in some cases a parameterization of the processes of chemical conversion of pollutants is possible due to the use of statistical characteristics of chemical reactions or by describing the laws of phase transitions. In particular, a simple model of SO_2 conversion into H_2SO_4 turned out to be sufficiently efficient (Kondratyev et al. 2004a).

$$d[H_2SO_4]/dt - d[SO_4]/dt = W[SO_2]$$

where W = 0.1 % day^{-1} in the daytime and 0.01 % day^{-1} at night.

Numerous models have been created to simulate the process of sedimentation of pollutants. So, for instance, Sinik et al. (1985) proposed several parameterizations for the coefficient of aerosol washing out from the atmosphere:

$$r = 10^{-4} I^{1/2};\ r = \theta \cdot I^{a};\ r = -C^{-1} dC/dt;\ r = 3.3 \times 10^{-4} I^{0.9};$$

where $I = RR/(24\ N)$ is the rain rate (mm/h), RR is the precipitation amount per month (mm), N is the number of days precipitation, θ and a are parameters.

The following diffusion equation is widely used

$$\frac{\partial C}{\partial t} + V_{\varphi}\frac{\partial C}{\partial \varphi} + V_{\lambda}\frac{\partial C}{\partial \lambda} + V_{h}\frac{\partial C}{\partial h} = -\frac{\partial}{\partial \varphi}\left(\overline{V_{\varphi}C}\right) - \frac{\partial}{\partial \lambda}\left(\overline{V_{\lambda}C}\right) - \frac{\partial}{\partial h}\left(\overline{V_{h}C}\right) \quad (5.4)$$

If we suppose that in Eq. (5.4) an advection prevails over diffusion in the direction h, that is, $\partial\left(\overline{V_h C}/\partial h\right) \ll V_h \partial C/\partial h$, then Eq. (5.4) with respect to λ gives:

$$\frac{\partial C_{\varphi}}{\partial t} + V_{h}\frac{\partial C_{\varphi}}{\partial h} + V_{\lambda}\frac{\partial C_{\varphi}}{\partial \lambda} = -\frac{\partial}{\partial \lambda}\int_{-\infty}^{\infty} \overline{V_{\lambda}C}\, d\varphi,$$

where

$$C_{\varphi} = \int_{-\infty}^{\infty} C\, d\varphi.$$

As a result of this transformation, the problem becomes two-dimensional.

Chobadian et al. (1985) proposed two formulas to estimate the depth of the mixed atmospheric layer, which is important in determination of the model's vertical structure:

$$H = 8.8 x_1 U_a^{-1} \Delta\theta,\ H = \frac{U_*}{U_a}\left\{\frac{x_1|\theta_1 - \theta_2|}{|\Delta\theta/\Delta x_3|}\right\}^{1/2},$$

where x_1 is the rate of shifting with respect to land surface (m s^{-1}), Δθ is the vertical gradient of potential temperature in the inverse layer (°C), U_* is the rate of friction over the leeward surface, θ_1 is the lower level of potential temperature over the source of the pollution (°K), $|\Delta\theta/\Delta x_3|$ is the absolute value of the rate of motion over the source.

The desire to simplify the parameterization of individual sub-processes of the atmospheric pollution dynamics, led in many cases to the development of sufficiently simple and efficient models, requiring a small database. Numerous studies have been dedicated to the classification of situations taking place in the real atmosphere with pollutants emitted from the point sources (stacks, etc.). Sufficiently complete classification can be found in Scorer (1997), where ten types are given of the behavior of the pollutants' jet (smoke) in the vicinity of an isolated point source (stack): folded, spiral, flagwise; sedimenting, broadening, expanding, breaking, bifurcational, fumigating, and rising. In each of these cases a Gauss-type model can be used with a minimum of input information. The monitoring system should only be able to distinguish between these situations based on measurements of meteorological parameters. For instance, the folded shape of the jet can be formed due to a rough surface (high buildings, hilly locality) in clear sunny weather. Smoke particles are wind-driven in zig-zag inside the broadening and gradually sedimenting jet. The breaking jet is typical of the second half of day in summer, when the atmosphere is well heated, and convective motions of air masses prevail.

One of the units of the atmospheric pollution model describes the process of emission of aerosol to the atmosphere. In most cases the Gauss-type model is used:

$$C(r,h) = f(h)\frac{M}{2\pi\sigma_r^2}\exp\left(-r^2/\left(2\sigma_r^2\right)\right) \tag{5.5}$$

where r is the distance to the center of the pollutants' jet, σ_r is the horizontal size of the jet, $f(h)$ is the vertical distribution of the mass of pollutants, $M = Q\Delta t$ is the mass of pollutants in the jet, Q is the rate of emission from the source, Δt is the time step. The f function can be approximated by the following dependence: $f(h) = 1/H$ for $0.5H < h < 1.5H$ and 0 in the opposite case (H is the effective height of the jet). For instance, to calculate H, the following formula has been proposed Despres et al. (1986):

$$H = \begin{cases} Z + 0.29V^{-1}Q_h^{1/2} & \text{normal conditions of stability} \\ Z + 0.49V^{-1}Q_h^{0.29} & \text{Weak conditions of stability} \end{cases}$$

where Q_h is the rate of the heat flux from the source (stack), cal s^{-1}; Z is the height of the source, m.

To simplify the process of simulation of the pollutants field in the zone of the source, the flux can be quantized into individual formations (small clouds, dust masses), each being considered separately as a homogeneous cloud. In particular, for quantization of the flux into n parts over the time Δt, the following formula can be used: $n = 2V\Delta t/\mathrm{H}$. For instance, at the wind speed $V = 5$ m/s and $Z = 100$ m we obtain that some clusters can be emitted from the stack every 10 s. As a result, the process of simulating the dynamics of the whole emitted jet is much simplified. So, for instance, in modeling the propagation of sulphur one can additionally distinguish between sulphur-forming components.

5.9.3 Types of Aerosol Models

Models of the atmospheric pollution dynamics are divided into Gaussian, Eulerian, and Lagrangian methods of parameterization which describe processes of pollutant scattering. Within each type there are statistical, box, correlative, and determinate sub-models. The most important division of the models' space is the division by the methods applied, since in this case the models are strictly divided by their information bases.

The Gauss-type models need information about the height of the source of pollutants and dispersion characteristics inside the pollutants' cloud. Different approximations of the Gauss jet in the vicinity of a point source depend on parameters of atmospheric surface layer stability. The model's configuration is affected by the geophysical characteristics. With the source present in the zone of transition from water surface to land, clusters can be selected with stable and unstable behavior of the pollutants jet. With respect to the horizontal coordinate x on land, with wind direction along this coordinate, three zones can be selected with characteristic features: the zone of undisturbed dispersion, the zone of deposition; and the zone of delay. Detailed description of the related models is given in Eqs. (1.2)–(1.5). These models describe well the fields of pollutant over territories on scales up to 100 km. For large territories, other types of models are used.

The information base of the Euler- and Lagrange-type models is formed according to the Cauchy problem for Eq. (1.5). Depending on a variety of real situations the data base composition ensuring a realization of the model of the pollutants transport becomes hierarchical. This hierarchy determines a relationship between spatial and temporal scales, it determines an accuracy of the model and its other characteristics. Numerous global data bases consist, as a rule, of information levels with an object orientation.

Knowledge of the wind field components is necessary for modeling the processes of atmospheric pollutants propagation over the territory of the region independent of a spatial scale. In the presence of many sources of atmospheric pollution in the region, information is needed about the 3-D wind field, and this is only possible with combined measurements and modeling. Usually, for large regions or in studies of conditions of pollutants propagation at the land-water boundary, the grid Euler-type model is used:

$$\partial u/\partial x + \partial v/\partial y + \partial w/\partial z = 0 \tag{5.6}$$

$$\partial u/\partial t + u\partial u/\partial x + v\partial u/\partial y + w\partial u/\partial z = f(v - v_g) + \partial(K_M\partial u/\partial z)/\partial z, \tag{5.7}$$

$$\partial v/\partial t + u\partial v/\partial x + v\partial v/\partial y + w\partial v/\partial z = f(u_g - u) + \partial(K_M\partial v/\partial z)/\partial z, \tag{5.8}$$

$$\partial T/\partial t + u\partial T/\partial x + v\partial T/\partial y + w\partial u/\partial z = \partial(K_H(\Gamma + \partial T/\partial z))/\partial z, \tag{5.9}$$

where f is the Coriolis parameter, Γ is the rate of dry adiabatic deposition, T is the air temperature, K_M and K_H are coefficients of diffusion, u_g and v_g are the thermal wind components:

$$u_g = -\frac{gT}{f}\int_0^Z \frac{1}{T^2}\left(\frac{\partial T(t,x,y,\xi)}{\partial y}\right)d\xi,\ v_g = \frac{gT}{f}\int_0^Z \frac{1}{T^2}\left(\frac{\partial T(t,x,y,\eta)}{\partial x}\right)d\eta. \quad (5.10)$$

For practical application of Eqs. (5.6)–(5.10), the following should be taken into account: deviation and rising of the flux over the hills, thermal friction, tunneling effect in valleys, thermal impacts of islands, lakes, and mountains. Additions to the given system of equations are made proceeding from an actual configuration and topology of the region.

The Gauss-type models are used, as a rule, to parameterize the processes of pollutants' propagation near high sources. The best studied modifications of the PGT dispersion model (Pasquill–Gifford–Turner) adequately describes the surface fluxes of pollutants and less accurately the fluxes formed from the highly located sources. Raaschou-Nielsen et al. (2000) have modified the PGT model providing it with more universal functions for an arbitrary point source. The modification is based on the base Gauss-type model to describe the concentration of pollutants in the surface layer of the atmosphere:

$$C(\varphi,\lambda,0) = \frac{Q}{\pi\bar{u}\sigma_z\sigma_\lambda}\exp\left[-0.5\left(\frac{h_{ef}}{\sigma_z}\right)^2\right]\exp\left[-0.5\left(\frac{\lambda}{\sigma_\lambda}\right)^2\right] + \text{reflection terms} \quad (5.11)$$

where Q is the source power, h_{ef} is the average height of the pollutants cloud, $\bar{u}$ is the effective rate of transport.

Important parameters of the model (5.11) are parameters of the pollutant jet scattering. They are complex functions of the meteorological situations. Theoretical and observational studies of many authors have shown that the turbulence and diffusion in the convective boundary layer are controlled by two important parameters: the height of mixing z_i and the scale of convection rate w_*:

$$w_* = \left(\frac{g}{T\rho c_p}Yz_i\right)^{1/3} \quad (5.12)$$

where Y is the surface sensible heat flux, g is the acceleration of the Earth's gravity, T is the air temperature, c_p is the air specific heat at a constant pressure.

The vertical σ_w and horizontal σ_v constituents of the rate of deviation of the turbulent flux from the centre of the convective cloud are proportional to w_* and, hence, depend on z_i. The total energy of turbulence is a composition of two energies generated by mechanical (σ_{wm}) and convective (σ_{wc}) forces:

$$\sigma_w^2 = \sigma_{wm}^2 + \sigma_{wc}^2. \tag{5.13}$$

It is assumed here that mechanical and convective forces do not correlate. Then, by analogy, we can write:

$$\sigma_z^2 = \sigma_{zm}^2 + \sigma_{zc}^2. \tag{5.14}$$

The σ_{zc}^2 value is calculated on the assumption that

$$\frac{d}{dt}\sigma_{zc} = \sigma_{wc}(z'), \text{ where } \quad \mathrm{t} = \varphi/\overline{u}. \tag{5.15}$$

Here z' is the effective height at which the vertical turbulence is calculated. Dependences on the height for σ_{wc} are parameterized with relationships:

$$\sigma_{wc}^2 = \begin{cases} 1.54 w_*^2 (z/z_i)^{2/3} & \text{for} \quad z < 0.1 z_i; \\ 0.33 w_*^2 & \text{for} \quad z \geq 0.1 z_i; \end{cases} \tag{5.16}$$

where the level $z = 0.1z_i$ corresponds to similar values of the vertical constituent of turbulence.

Let h_s be the height of the pollutants' jet and t have the scale $x/\overline{u}$. Then using Eqs. (5.15) and (5.16) we obtain the relationship between h_s and z_i. For $h_s \geq 0.1z_i$ the following approximation is valid:

$$\sigma_{zc}^2 = 0.33 w_*^2 t^2 \tag{5.17}$$

With $h_s < 0.1z_i$ we have:

$$\sigma_{zc}^2 = \begin{cases} 1.54 w_*^2 (h_s/z_i)^{2/3} t^2 & \mathit{for} \quad \sigma_{zc} < h_s, \\ \left(0.83 w_* z_i^{-1/3} t + 0.33 h_s^{2/3}\right)^3 & \mathit{for} \quad h_s \leq \sigma_{zc} < 0.1 z_i, \\ \left(0.581 w_* t + 0.231 h_s^{2/3} z_i^{1/3} - 0.05 z_i\right)^2 & \mathit{for} \quad \sigma_{zc} \geq 0.1 z_i \end{cases}$$

The mechanical constituent in Eq. (5.14) is calculated on the assumption that variations of the vertical gradient due to mechanical mixing are constant in the boundary layer and determined with the relationship:

$$\sigma_{wm}^2 = 1.2 u_*^2, \tag{5.18}$$

where u_* is the friction rate.

For the parameter of mechanical scattering σ_{zm} the ratio to the rate of mechanical turbulence σ_{wm} is not so simple as in the case of convective turbulence. This is connected with the fact that the size of the scale characterizing the mechanical scattering, in contrast to the case of convective changes with altitude, is less than the convection scale. Therefore, for the unstable state the following dependences are used:

$$\sigma_{zm}^2 = \begin{cases} \sigma_{zmu}^2 = 1.2u_*^2t^2 \exp(-0.6tu_*/h_s) & for \quad tu_*/h_s < 1, \\ \sigma_{zmu}^2 = 1.2u_*^2t^2 \exp(-0.6) & for \quad tu_*/h_s \geq 1 \end{cases}$$

For stable conditions the following approximation is valid:

$$\sigma_{zm}^2 = \sigma_{zmu}^2/(1 + 1.11tu_*/L)$$

where L is the Monin-Obukhov length.

Thus, both terms on the right in formula (5.14) are estimated. Now determine the σ_λ parameter in Eq. (5.11). The characteristic of horizontal scattering of the pollutant is well approximated with the formula:

$$\sigma_\lambda = \left(\frac{0.25w_*^2}{1 + 0.9\varphi w_*/uz_i} + u_*^2\right)^{1/2} \frac{\varphi}{\overline{u}} \tag{5.19}$$

In this expression the first and second terms in brackets represent the contribution of convective and mechanical turbulence, respectively. For nocturnal conditions, the first term is equal to zero. For the stable atmosphere and at weak winds, the ratio $u_*/\overline{u}$ decreases rapidly with the growing stability. Nevertheless, numerous observations have shown that the horizontal scattering with an hourly averaging can exceed the values for the unstable atmosphere. The horizontal rate fluctuates with the constant amplitude about 0.5 m s^{-1}. As the horizontal scattering remains proportional to fluctuations of horizontal wind, the u_* parameter in Eq. (5.19) for the case of stable stratification can be changed for 0.5 m s^{-1}. It means that σ_λ will never be less than $0.5\varphi/\overline{u}$.

Estimate the parameter of horizontal scattering:

$$\sigma_{\lambda f} = \left(\Delta d\big/\sqrt{2\pi}\right)\varphi,$$

where Δd is the change of wind direction in radians.

According to the Eq. (5.11), the accuracy of calculation of the aerosol concentration at a point with coordinates (φ, λ) depends substantially on wind speed. An estimate of the aerosol density in a given space volume depends on the accuracy of its estimation at a given height. For instance, in the case of unstable stratification the wind speed is approximated with its value at a height of the source, and under stable conditions the wind speed is calculated by the vertical averaging of its values. According to Monin and Obukhov (1954), we have

$$u(z) = \frac{u_*}{k}\left[\text{In}\left(\frac{z+z_0}{z_0}\right) - \psi_m\left(\frac{z}{L}\right) + \psi_m\left(\frac{z_0}{L}\right)\right], \tag{5.20}$$

where z_0 is the linear size of surface roughness, Ψ_m are universal functions of similarity calculated, for instance, in Paulsen (1970). Based on these calculations, the wind speed is estimated with the formula (5.20) at $z \le z_B$ and is assumed to be equal to $u(z) = u(z_B)$ for all heights $z > z_B$, where $z_B = \max\{0.1z_i, |L|$.

Practical application of the Gauss-type models has many aspects appearing, depending on the multitude of the factors of the natural-anthropogenic environment. This includes an effect of buoyancy of aerosol particles in the atmosphere, intermittent atmospheric layers, and many others. Raaschou-Nielsen et al. (2000) proposed to calculate the h_{ef} parameters using the following formulas:

$$H_{ef} = h_s + \Delta h,$$

where

$$\Delta h_f = \min\{\Delta h_{in}, \Delta h_f\},\ \Delta h_{in} = 1.6\left(\frac{F}{u}\right)^{1/3}\varphi^{2/3},\ F = \frac{V}{\pi}\frac{g}{T_e}(T_e - T_a),$$

Here F is the buoyancy flux, h_s is the stack height, V is the flux volume, T_e is the temperature of the flux emitted from the stack, T_a is the air temperature.

The height of the aerosol jet due to the effect of its buoyancy is calculated with the formula:

$$\Delta h_f = \begin{cases} 1.3\frac{F}{uu_*^2}\left(1+\frac{h_s}{\Delta h_f}\right), & \text{neutral conditions,} \\ 4.3\left(\frac{F}{u}\right)^{3/5}\left(\frac{Yg}{T_a\rho c_p}\right)^{-2/5}, & \text{convection,} \\ 2.6\left(\frac{F}{us}\right)^{1/3}, & \text{stability,} \end{cases}$$

where $s = (g/T_a)\partial\theta_a/\partial z$. Here $\partial\theta_a/\partial z$ is the potential air temperature gradient at the stack's level.

A multitude of possible situations with the use of the Gauss-type model is determined by locality topology, the presence of the boundaries of the sea-land type, mountains, and others. A consideration of a concrete situation enables one to simplify the formulas and raise their reliability. In particular, Wang and Christiansen (1986) considered such situations at the lake-land boundary. At the water-land boundary the air flux is continuously adjusting itself to the wind field configuration forming into jets, corresponding to mechanical and temperature conditions. Under conditions of a persistent coastal heat layer the gradient of aerosol concentration can be stable, which depends on the distance of the source from the shore. Here the following formula is valid:

$$\frac{3X}{\beta'^2 U s z_b^3} = \frac{3}{2}\left[\frac{z_{eq}}{z_b}\right]^2 \left[\frac{z_{eq}}{z_b} - 1\right] + \frac{2}{9}, \tag{5.21}$$

where z_{eq} is the stable height of the jet over the stack, z_b is the height of the bottom of the stable atmospheric layer over the stack, U is the wind speed at the height of the stack, β' is the effective coefficient of aerosol capture ($\beta' \cong 0.4$), s is the parameter of stability, X is the jet buoyancy parameter. The s parameter depends on the potential temperature gradient θ and air temperature $T_a : s = (g/T_a)\partial\theta/\partial z$ (g is the gravity coefficient). The F parameter can be calculated as the function of velocity v of the jet emitted from the stack, internal radius R of the stack, and temperature T of the emitted jet:

$$F = gvR_2(T - T_a)/T$$

With the z_d value known from (5.21) one can easily find z_{eq}, and the effective height Y of the jet is found as the sum: $Y = z_{eq} + h_s$, where h_s is the height of the stack. The whole zone of the aerosol cloud propagation with wind s blowing from the water towards land is divided into zones of stability, instability, and the aerosol cloud lowering. Such a digitization of space simplifies the model, reducing the needed data base and increases its reliability. In each zone, the use of the Gauss-type model leads to a simplified parameterization of the process of aerosol scattering and facilitates an interpretation of the modeling results.

In the undisturbed dispersion zone, the distribution of the aerosol jet in the homogeneous stable atmospheric layer is described with the base Gaussian equation of scattering:

$$C(\varphi,\lambda,z) = \frac{Q}{2\pi U\sigma_\lambda\sigma_z}\exp\left[-\frac{1}{2}\left(\frac{\lambda}{\sigma_\lambda}\right)^2\right] \times\left\{\exp\left[-\frac{1}{2}\left(\frac{z-Y}{\sigma_z}\right)^2\right] + \exp\left[-\frac{1}{2}\left(\frac{z+Y}{\sigma_z}\right)^2\right]\right\}$$

where U is the wind speed, Y is the effective height of the jet, σ_λ and σ_z are parameters of the horizontal and vertical dispersion.

In the zone of fumigation, the forces get activated which cause a distortion of the jet due to instability of heat fluxes in the surface layer. In this zone the following approximation is valid:

$$C(\varphi,\lambda,z) = \frac{Q}{(2\pi)^{3/2}}\int_0^{\varphi}\frac{1}{\sigma'\sigma_z(u,\varphi,\varphi\prime)U}\exp\left[-\frac{1}{2}(p'^2 + \left(\frac{\lambda}{\sigma'}\right)^2)\right] \times\left\{\exp\left[-\frac{(z-Y)^2}{2\sigma_z^2(u,\varphi,\varphi')}\right] + \exp\left[-\frac{(z+Y)^2}{2\sigma_z^2(u,\varphi,\varphi')}\right]\right\}\frac{dp'}{d\varphi'}d\varphi',$$

where φ' is the leeward distance of intersection of the aerosol jet with the upper boundary of the internal thermal boundary layer (ITBL), $p' = [L(\varphi) - Y]/\sigma_z\ (s, \varphi), L(\varphi)$ is the height of ITBL at the distance φ, $\sigma_\lambda(s, \varphi)$ is the σ_λ value for the stable atmospheric layer at the distance $\sigma_z(s, \varphi)$ is the σ_z value for the unstable atmospheric layer at the distance $\varphi, \sigma_\lambda(u, \varphi, \varphi')$ is the σ_λ value for the unstable atmospheric layer φ after passing the leeward distance $\varphi', \sigma_z(u, \varphi, \varphi')$ is the σ_z value for the unstable atmospheric layer at the distance φ after passing the leeward distance φ',

$$\sigma' = \sqrt{\sigma_\lambda^2(s, \varphi) + \sigma_\lambda^2(u, \varphi, \varphi')}$$

In the zone where the aerosol cloud top is close to $L(\varphi)$ and it is totally within the ITBL, the following approximations are valid:

$$C(\varphi, \lambda, z) = \begin{cases} xEG/\sigma_z(u, \varphi) & for \quad \sigma_z < 0.47L(\varphi), \\ xE/L(Q) & for \quad \sigma_z > 0.8L(\varphi), \end{cases} \tag{5.22}$$

where

$$x = Q/[\pi U \sigma_\lambda(u, \varphi)], \quad E = \exp\left[-\frac{1}{2}\left(\frac{\lambda}{\sigma_\lambda(u, \varphi)}\right)^2\right],$$

$$G = \exp\left[-\frac{1}{2}\left(\frac{z - Y}{\sigma_z(u, \varphi)}\right)^2\right] + \exp\left[-\frac{1}{2}\left(\frac{z + Y}{\sigma_z(u, \varphi)}\right)\right].$$

For the leeward distance φ, where $0.47L(\varphi) \leq \sigma_z(\varphi) \leq 0.8L(\varphi)$, the concentration of aerosol is calculated by interpolation between its estimates in the boundary zones according to formula (5.22).

Modeling the processes of aerosol propagation in the surface layer is connected with the simulation of dependences on wind speed, air temperature and humidity as 3-D functions of time and space coordinates. One of such models of the planetary boundary layer (PBL) has been described in Bornstein (1999). The model contains two units. The first unit analytically simulates the PBL processes. The field of aerosol is averaged and presented by a 1-D structure. Another transitional layer of the atmosphere is described by hydrodynamic and thermodynamic equations solved numerically with the use of the finite difference schemes. Equations for the atmospheric transition layer are derived on the assumption that the atmosphere is incompressible and hydrostatic, water vapour does not change its properties, the potential temperature is calculated with the formula $\theta = T + \Gamma z$, where T is the air temperature, Γ is the rate of dry adiabatic cooling, z is the vertical coordinate. Equations for the atmospheric transition layer are as follows:

$$\frac{\partial u}{\partial t}+\frac{\partial(uu)}{\partial x}+\frac{\partial(vu)}{\partial y}+\frac{\partial(wu)}{\partial z}=-\frac{1}{\rho_a}\frac{\partial p_M}{\partial x}+f(v-v_g)+\frac{\partial}{\partial z}\left(K_m\frac{\partial u}{\partial z}\right)$$
$$+K_H\left(\frac{\partial^2 u}{\partial x^2}+\frac{\partial^2 u}{\partial y}\right),$$

$$\frac{\partial v}{\partial t}+\frac{\partial(uv)}{\partial x}+\frac{\partial(vv)}{\partial y}+\frac{\partial(wv)}{\partial z}=-\frac{1}{\rho_a}\frac{\partial p_M}{\partial y}+f(u-u_g)+\frac{\partial}{\partial z}\left(K_m\frac{\partial v}{\partial z}\right)$$
$$+K_H\left(\frac{\partial^2 v}{\partial x^2}+\frac{\partial^2 v}{\partial y^2}\right),$$

$$\frac{\partial u}{\partial x}+\frac{\partial v}{\partial y}+\frac{\partial w}{\partial z}=0,$$

$$g\frac{(T_M+T_n)}{\theta_a}-\frac{1}{\rho_a}\frac{\partial p_M}{\partial z}=0,$$

$$\frac{\partial\theta}{\partial t}+\frac{\partial(u\theta)}{\partial x}+\frac{\partial(v\theta)}{\partial y}+\frac{\partial(w\theta)}{\partial z}=\frac{\partial}{\partial z}\left(K_h\frac{\partial\theta}{\partial x^2}\right)+K_H\left(\frac{\partial^2\theta}{\partial x^2}+\frac{\partial^2\theta}{\partial y^2}\right)+\frac{1}{\rho_a c_p}\frac{\partial QN}{\partial z}$$
$$\frac{\partial q}{\partial t}+\frac{\partial(uq)}{\partial x}+\frac{\partial(vq)}{\partial y}+\frac{\partial(wq)}{\partial z}=\frac{\partial}{\partial z}\left(K_q\frac{\partial q}{\partial z}\right)+K_H\left(\frac{\partial^2 q}{\partial x^2}+\frac{\partial^2 q}{\partial y^2}\right),$$

where

$$u_g=-\frac{1}{f\rho_a}\frac{\partial p_n}{\partial y},\ v_g=\frac{1}{f\rho_a}\frac{\partial p_n}{\partial x},$$

(u,v,w) are projections of wind speed onto rectangular coordinates x, y and z, respectively, K_H is the coefficient of horizontal turbulent mixing, ρ_a is the air density, q is the specific humidity. A strict solution of these equations is impossible in general, and therefore it is necessary to choose a suitable discrete grid on which these equations are substituted with the finite-difference equations. The step of time quantization and the size of discrete space grid should meet the condition: $\Delta t\leq\min\{\Delta x/u,\ \Delta y/v,\ \Delta z/w\}$.

In practice, instead of this sufficiently cumbersome scheme, simulation schemes are often used based on the acceptable quantization of space. In this case the territory of a region or the globe is divided, as a rule, into homogeneous sites with linear sizes $\Delta\varphi$ in latitude and $\Delta\lambda$ in longitude: $\Omega=\{\Omega_{ij};\ i=1,\ldots,n; j=1,\ldots,m\}$. Setting the vertical structure of the atmosphere and taking into account the physical and chemical processes of aerosol transformation, relations recurrent with respect to the time step Δt are derived to re-calculate their concentrations. With available control measurements at several points in space $\Xi=\{(\varphi,\lambda,z),\ (\varphi,\lambda)\in\Omega,\ 0\leq z\leq H\}$, the steps of quantization of the geographic grid and time are optimized on the basis of one or several criteria of accuracy. As a rule, such an approach provides a needed stability of the modeling results.

Denote as $C_s(t, \varphi, \lambda, h)$ the concentration of aerosol of the s-th type at the height h over the point with coordinates (φ, λ) at a time moment t. For the Euler-type model, general equations of aerosol transport in the environment according to (5.3) have the following form:

$$\frac{\partial C_s}{\partial t} + V_\varphi \frac{\partial C_s}{\partial \varphi} + V_\lambda \frac{\partial C_s}{\partial y} + V_h \frac{\partial C_s}{\partial h} = \frac{\partial}{\partial \varphi}\left(K_\varphi \frac{\partial C_s}{\partial \varphi}\right) + \frac{\partial}{\partial y}\left(K_\lambda \frac{\partial C_s}{\partial \lambda}\right) + \frac{\partial}{\partial h}\left(K_h \frac{\partial C_s}{\partial h}\right) + E_s(t, \varphi, \lambda, h) + P_s(C_1, \ldots, C_q) - \nu_{1s} - \nu_{2s}$$

where K_φ, K_λ, and K_h are coefficients of the turbulent diffusion, E_s is the characteristic function of the sources of emission of aerosols of the s-th type, P_s is the operator describing physical and chemical transformations of aerosol, ν_{1s} is the rate of aerosol washing out with precipitation, ν_{2s} is the rate of dry deposition, $V = \{V_\varphi, V_\lambda, V_h\}$ are the wind speed components.

The model of this type is used to calculate the aerosol concentration in the atmosphere, as a rule, for the scales of territory exceeding 50 km. Simplified schemes of calculation of the C function are drawn by dividing the space into units $\Delta\varphi \times \Delta\lambda \times \Delta h$, and at each height h_k a step-by-step calculation of concentration $C_s(t, \varphi_i, \lambda_j, h_k)$ is made. The calculation scheme can be further simplified by dividing the procedure into two stages. First, for each level of the vertical digitization of space the distribution of $C_s(t, \varphi_i, \lambda_j, h_k)$ and then the processes of the vertical transition of aerosols is taken into account. This scheme enables one to easily move on to vertically averaged levels depending on available information about the parameters of the atmospheric vertical stratification. Convergence of such a procedure depends on relationships between the parameters Δt, $\Delta\varphi$, $\Delta\lambda$, Δh.

The Euler-type model contains many degrees of freedom including the consideration of various scenarios. The base of knowledge of the simulation system (the description of ESPAP is given below) contains sets of parameterizations of partial processes of aerosol transformation, and a concrete choice is made by the user. Information in data and knowledge bases are structured according to a multitude of spatial and object identifiers—matrix structures $\{A_m\}$. In particular, in the “default mode” regime the following parameterizations are used.

In accordance with the structure of the identifier describing the sources of pollution, in each compartment $\Delta\varphi \times \Delta\lambda$ the passport information about the rates of emission of the s-th pollutant is put in: minimum and maximum rates $E_{s,min}$ and $E_{s,max}$, respectively. In the absence of additional information about some source, the E_s value is calculated according to the procedure of the uniform distribution over the interval $[E_{s,min}, E_{s,max}]$ or another law of distribution is assumed (e.g., the Gauss law).

The functional description for ν_{1s} and ν_{2s} is important, and the model's adequacy depends on it. Therefore, many studies have been dedicated to this problem. It has been proven that the following linear approximation of the functions ν_{1s} and ν_{2s} is most acceptable:

$$V_{is} = K_{is}C_s,\ i = 1, 2$$

The coefficient of proportionality K_{is} can be function of time and space coordinates and depends on the physical state of the surface. With the use of the identifier $\|\mu_{ij}\|$ this dependence can be broadened. For instance, the version $\mu_{ij} = 0$ can mean that the dry deposition in the model is neglected at all. For $\mu_{ij} = 1$ the scheme proposed in [128] is taken into account]:

$$K_{2s} = K_{2s}^* v_{ds} d / H_{mix}(t, \varphi, \lambda),$$

where K_{2s}^* is the coefficient of proportionality, v_{ds} is the rate of deposition of the s-th type of aerosol (e.g., $v_{ds} = 0.008$ m s^{-1} for SO_2, 0.002 m s^{-1} for SO_4, 0.001 m s^{-1} for NO_2 and HNO_3), H_{mix} is the height of the atmospheric mixed layer (m), d is the parameter considering the physical state of the surface ($d = 1$ for dry surface, $d = 2$ for wet surface, etc.).

With the use of the μ_{ij} parameter the system of modeling can govern the choice of various dependences of the rate of dry deposition in a wide range of synoptic parameters and other environmental characteristics contained in the base of knowledge. For instance, for $\mu_{ij} = 2$ the following model is chosen:

$$v_{2s} = \begin{cases} 0.002U_* & \textit{for} \quad L \geq 0, \\ 0.002U_*\left[1 + (-300/L)^{2/3}\right] & \textit{for} \quad L < 0, \end{cases}$$

where L is the (Monin-Obukhov length), U_* is the rate of friction.

By analogy with the previous case, in modeling the rate of wet deposition of aerosols from the atmosphere, it is expedient to consider the problem from different positions, since in general, the v_{1s} value is the function of many factors, some of them not always being estimated reliably. For convenience of computer modeling, the identifier $\|\text{v}_{ij}\|$ is introduced and, for instance, for $\text{v}_{ij} = 0$ we assume automatically $v_{1s} = 0$ (no washing out). For $\text{v}_{ij} = 1$ there is a version when

$$v_{1s} = K_{2s}^* C_s r,$$

where r is the rain rate.

With $\text{v}_{ij} = 2$ we have the model of washing out:

$$\partial C_s / \partial t = -\phi r(t, \varphi, \lambda) C_s(t, \varphi, \lambda, h)$$

At ϕ = const we obtain a simple explicit scheme of C_s recalculation in each cell:

$$C_s(t + \Delta t, \varphi_i, \lambda_j, h_k) = C_s(t, \varphi_i, \lambda_j, h_k) \exp\{-\phi r(t, \varphi_{,i}, \lambda_j,) \Delta t\}$$

5.10 Ordinary Climate Parameterizations

The GIMS climate item is the most problematic element for the realization. The use of existing global climate models as the GIMS item leads to high uncertainty within the modeling results. In this case, precision and prognosis depth are characterized by instability and low reliability. The use of small climate models gives the possibility to reduce these negative circumstances, but leads to other restrictions. Actual climate predicting procedures as a background process in GIMS are usually changed by climate scenarios. Different international and national institutes try to elaborate such scenarios. The Climate Scenario section of the Data Distribution Centre (DDC) of the Intergovernmental Panel on Climate Change (IPCC) is one of them. The DDC distributes a number of datasets, derived from various climate modeling experiments, using general circulation models (GCMs) that are commonly used in the construction and application of climate change scenarios for climate change impacts assessments. The intention is to provide a common set of scenario data for use in climate impacts research that may get reported in the Fourth Assessment Report of the IPCC and beyond (IPCC 2007; Pachauri and Reisinger 2007).

Mu et al. (2002) investigated the uncertainties caused by the errors of the initial states and the parameters in the numerical model. Three problems of predictability in numerical weather and climate prediction are proposed, which are related to the maximum predictable time, the maximum prediction error, and the maximum admissible errors of the initial values and the parameters in the model respectively. The three problems are then formulated into nonlinear optimization problems. Effective approaches to deal with these nonlinear optimization problems are provided. The Lorenz' model is employed to demonstrate how to use these ideas in dealing with these three problems.

Therefore, one of the most important problem within the global and regional studies of the NSS is the generation of climate forecast systems of various seasonal-to-decadal time scales. These systems are to have input information which can be delivered by existing or monitoring systems planned. Energy use statistics has principal importance in this aspect. Figure 5.24 shows a huge discrepancy between the highest energy users and the lowest. Knowledge this statistics can help to form the more realistic climate scenarios.

Authors of NRC (2013) note that more accurate forecasts of climate conditions over time periods of weeks to a few years could help people plan agricultural activities, mitigate drought, and manage energy resources, amongst other activities. However, current forecast systems have limited ability on these time-scales. Models for such climate forecasts must take into account complex interactions among the ocean, atmosphere, and land surface. Such processes can be difficult to represent realistically. To improve the quality of forecasts, NRC (2013) makes recommendations about the development of the tools used in forecasting and about specific research goals for improving.

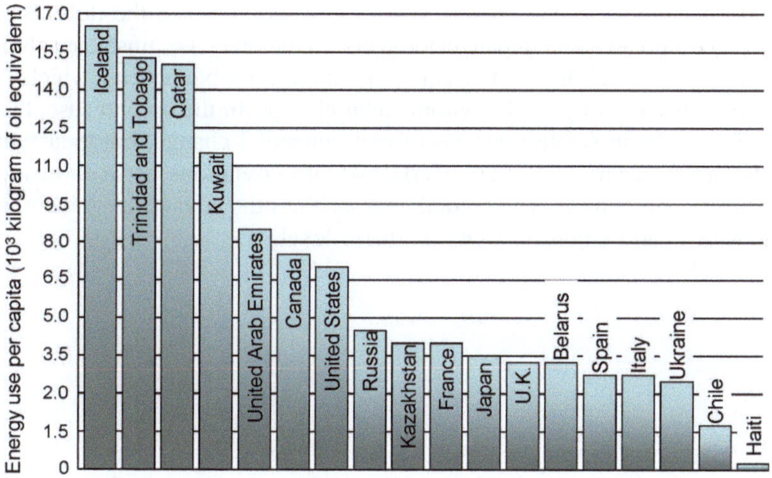

Fig. 5.24 Comparative analysis of energy use by separate countries in 2013 (IEA 2013c; BP 2013)

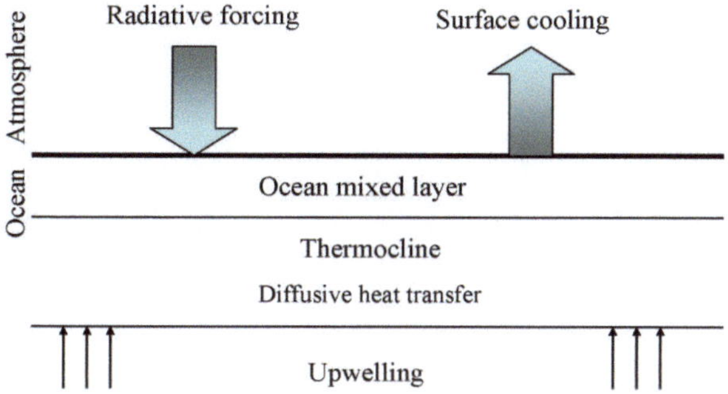

Fig. 5.25 Global climate model item related to the ocean-atmosphere heat interaction

Development of global model for the assessing nature-society system dynamics is impossible without adequate climate model. Existing climate models practically have not biospheric components. Complexity of these models results to the instability of prognosis. Therefore, really climatic blocks of future global model are to be based on simple climate models such as it is shown in Fig. 5.25.

It is evident that the climatic component of global biospheric model is most difficult for the synthesis, since it is characterized by numerous feedbacks, most of which are unstable. As it was above mentioned these feedbacks include ice-albedo, watervapor-radiation, cloud-radiation, and aerosol-radiation. That is why there are two main approaches to climate model synthesis. One approach is based on inclusion of biospheric components in climate models that have already been created or developed. The other approach consists in development of a unit within the numerical model of the biosphere which would simulate the dependences of biospheric components on climate parameters. In the first case, unstable solutions of respective systems of differential equations appear problematical, which hinders prognostic estimates of global environmental change. In the second case, there is a probability to obtain stable forecasts of environmental change, but their reliability depends on the accuracy of parameterization of correlations between elements of the biosphere and climate. The second approach has the advantage of incorporating models which can be described at a scenario level.

Chapter 6
The Arctic Environmental Problems

6.1 Introduction

Arctic region is currently undergoing unprecedented warming and drying, as well as dramatic changes in human use. Global climate change science includes important themes related to the understanding of annual and inter-annual balances of carbon, methane, water and other greenhouse gases and their stability. In cases when the concentration of the main greenhouse gas is continuing to increase at an alarming rate, the Arctic Basin obtains important role as possible global climate regulator. Specific importance in this aspect belongs to arctic soils in the tundra and permafrost. Due to microorganism activity, thawing tundra releases the CO_2 with increasing rate when atmospheric temperature is growing.

Arctic- wide inter-governmental cooperation protecting the Arctic environment formally began in 1991 with the establishment of the Arctic Environmental Protection Strategy (AEPS), which was later transformed into the Arctic Council in 1996. Recent growing attention to the Arctic environmental problems is motivated by a number of circumstances, including (AMAP 2010; Walker et al. 1998, 2008; Callaghan et al. 2011; Girnis et al. 2003; Krapivin and Mkrtchyan 2009b, c, 2010, 2012; Nitu et al. 2010; Peterson and Krantz 2008; Prestrud et al. 2007; Schneider 2006; Shapovalov 2010; Shultze et al. 2001):

(1) a stronger sensitivity of high latitude environment to various external forcings;
(2) better understanding of the importance of numerous interactions and dependencies between components of the Earth's system; and
(3) increasing necessity to use natural resources located at high latitudes (Arctic shelf especially).

It is fair to say (Vörösmarty et al. 2001) that "the Arctic system constitutes a unique and important environment with a central role in the dynamics and evolution of the Earth system". Some of the recent scientific results have been highlighted in the ACIA Implementation Plan (ACIA 2000):

V.F. Krapivin et al., *New Ecoinformatics Tools in Environmental Science*,
Environmental Earth Sciences, DOI 10.1007/978-3-319-13978-4_6

- There has been increased coastal erosion in the Bering Sea from storm surges, resulting from reduced sea ice.
- Sea ice extent in the Arctic has decreased overall by 0.35 % per year since 1979. During summer of 1998, the record reduction of sea ice coverage was observed in the Beaufort and Chukchi Seas. Sea ice area is assessed by 11,827,773 km^2 on December 2013.
- Sea ice thickness has also been reduced by one to two meters in most parts of the Arctic Ocean and the sub-Arctic seas. On October 2013, the multiyear ice had been on average at about 20 %, or around 30 cm, thicker than last year.
- Stream flow discharge of major Siberian Rivers into the Arctic Ocean was increased in recent years and is associated with a warmer climate and enhanced precipitation in the river basins.
- Since 1970, the Arctic Oscillation, which is a measure of the strength of the circumpolar vortex, has strengthened. This has been found to be consistent with temperature change in the Arctic.
- There has been a raise in temperatures of the water in the Arctic Ocean's Atlantic layer and an approximate 20 % greater coverage of Atlantic waters.
- Record of low levels of ozone was measured in 2000 in the Arctic, with increasing evidence that these levels are likely to continue for at least the next 20 years.
- Ongoing studies indicate that the current UV levels can have a significant effect on the fish larvae survival rates.
- General warming of the soils in regions with permafrost, specifically from Alaskan regions, has been observed over recent years.

It has been emphasized in ACIA (2000) that past assessments indicated that the Arctic is important to global-scale processes in at least four important ways:

- The thermohaline circulation dominated by the Arctic Ocean and Nordic Seas is responsible for a considerable part of the Earth's pole ward heat transport and may also serve as a sink for CO_2. Alterations of this circulation, as have been observed during climatic changes of the past, can affect global climate and in particular the climate of Europe and North America.
- The melting of the Arctic land ice sheets can cause sea level rise around the world. A compilation of studies suggests that a global warming of 1 °C will lead to ~1 mm raise per year of the sea-level from small ice caps and glaciers. The Arctic will supply over half of this total, with an additional 0.3–0.4 mm per year contribution from Greenland, although uncertainties remain about the mass balance of the Greenland ice sheet.
- Arctic soils can act as either sinks or sources of greenhouse gases, depending on temperature and moisture changes within the Arctic. Moisture has contradictory effects on the concentrations of the two major trace gases: CH flux declines with soil drying while CO_2 flux initially increases. These changes can influence greenhouse gas warming globally.

- Our current understanding of the Arctic climate system suggests that positive feedbacks in high-latitude systems, including the snow and ice albedo effect, amplify anthropogenic ally-induced atmospheric changes. These disturbances in the circumpolar Arctic climate may substantially influence global climate.

In the context of the health of the Arctic marine environment (Fig. 6.1), from the viewpoint of proper functioning of economically important ecosystems, Drewry and Orheim (2000) have formulated a number of key questions:

- How was the polar basin-formed, where are the plate boundaries?
- What has been the detailed paleo-climatic history of the high Arctic Ocean during the last 1 million years?
- Do decreases in the ice extent and upper stratification of the ocean signal a different sea ice regime?

Fig. 6.1 The Arctic environment (http://arctic.cbl.umces.edu)

- What is the stability of the sea ice cover, what are the effects of radiative feedback in the Arctic and how do they modulate global ocean circulation?
- What is the role of continental shelves in the cycling of C, N, Si and other chemicals?
- What is the productivity of the Arctic Ocean, and what is the structure and diversity of higher trophic levels?
- What are the effects of environmental changes, both of climate and of pollutants and contaminants such as the introduction of POPs into the food chain?

There are numerous other problems related to the Arctic Basin. For example, glaciers and ice sheets modulate global sea level by storing water deposited as snow on the surface and discharging water back into the ocean through melting and via icebergs. At present time, the study of these processes is made easier by means of satellite observations. Existing satellites make possible monitoring of Polar Regions, location of the ice layer and measurements of the topography and properties of the glacier bed as well as ice thickness.

Of particular interest is the high-latitude climate dynamics. According to Weller and Lange (1999), "While considerable uncertainty still exists about the exact nature of the future impacts of global climate change, there can no longer be any doubt that major changes in the climate have occurred in recent decades in the Arctic, with visible and measurable impacts following the climatic changes. Greater impacts are likely in the future and while some of them will be positive, others will be detrimental to human activities".

Recent analysis of ice cores from the Arctic (Everett and Fitzharris 2001) has revealed large-scale and rapid paleo-climate changes. Rapid warming took place ~11,500 years ago, at the end of the last glacial period. The coldest parts of ice cores had been as much as 21 °C colder than the present temperature in Central Greenland; and the temperatures increased by more then 10 °C in a few decades. There is evidence of even more rapid change in the precipitation pattern, rapid changes of atmospheric circulation, and periods of rapid warming during the past 20,000 years. Rapid temperatures increase of ~10 °C in a few decades during the last glacial period in Central Greenland had been followed by periods of slower cooling over a few centuries and then a generally rapid return to glacial conditions. About 20 such intervals, each lasting between 500 and 2,000 years, occurred during the last glacial period.

It has been emphasized (Everett and Fitzharris 2001; Singer 2008; Vakulenko et al. 2004) that the polar systems are extremely sensitive to the variations in temperature, and several aspects of these systems will be affected by any further climate change. The primary impact will be on the physical environment, including ice, permafrost, and hydrology; on biota and Ecosystems, including fisheries and terrestrial systems; and on human activities, including social and economic impacts on settlements, on resource extraction and transportation, and on existing infrastructure. Scenario predictions of potential future global warming indicate a necessity to particularly take into account various phenomena such as thermocarst erosion in lowland areas, thawing of permafrost accompanied by hydrological and climatic changes. Climate change will affect terrestrial ecological systems through changes in permafrost as well as direct climatic changes, including changes in precipitation, snow cover, and temperature. Terrestrial ecosystems are likely to change from tundra to boreal forests, although vegetative changes are likely to lag behind climatic change. Major shifts in biomass will be associated with changes in microbiological (bacteria, algae, etc.) and insect communities (some of them may diminish while others prosper).

It has been pointed out (Everett and Fitzharris 2001) that in the recent geologic past, the tundra was a carbon sink, but recent climate warming in the Arctic, coupled with the concomitant drying of the active layer and the lowering of the water table, has shifted areas of the Arctic from sinks to sources of CO_2 (this problem is, however, far from being solved). An important potential consequence of permafrost thawing is emissions of methane—a greenhouse gas. As for tropospheric ozone, another greenhouse gas, changes might happen due to warming of the troposphere (Kondratyev and Varotsos 2000; Varotsos 2005).

An interesting illustration of potential future surprises due to interactions and feedbacks has been discussed by Stevenson et al. (2000) who obtained future estimates of tropospheric ozone radiative forcing and methane turnover in the context of the impact of climate change (it should be pointed out that studies of the contribution of tropospheric ozone, O_{3T}, as a greenhouse gas as well as assessments of potential impact of global warming on permafrost melting and methane emissions are still at the preliminary stage of development). Interactive simulations of climate dynamics and O_{3T} changes during the period of 1990–2100, for scenarios of "high" (A2) or "central" (B2) cases of CO_2 emissions, resulted in tropospheric ozone radiative forcing (RF) equal to +0.27(A2) or +0.09(B2) Wm^{-2}, if climate-ozone coupling was neglected, then relevant RF values were equal to +0.43 (+0.22) Wm^{-2}. When climate change was included, CH_4 lifetime fell by 0–5 %. Hence, climate warming exerts a negative feedback on itself by enhancing O_{3T} and CH_4 destruction.

The three principal achievements have stimulated during the recent years the progress in studying the Arctic environment (Dickson 1999; Baker 2011; Vallis and Thoumi 2012):

(1) further development of observation programs with the usage of various observation means (including satellites and submarines);
(2) declassification of the military Soviet-American archive of ocean "climatology" data; and
(3) discovery of the fact that the climatic forcing in the Arctic region and northern seas in the 1990s has increased in comparison with that observed during the previous century [similar situation also took place with respect to climate dynamics indicators (such as Arctic Oscillation (AO) and North-American Oscillation (NAO))].

Overland and Adams (2001) have pointed out that "decadal differences between the 1990s and 1980s in winter sea-level pressure and 300 hPa zonal winds have an

Arctic-centered character with nearly equal contributions from the Atlantic and Pacific sectors. In contrast, the differences between positive and negative AO composites defined from monthly values of Principal Components from the same period have similar magnitudes in the Pacific and Arctic, but have additional large NAO signature in the Atlantic sector. Thus Arctic changes of decadal scales are more symmetric with the pole than suggested by the standard AO index definition. Change point analysis of the AO shows that a shift in value near 1989 is an alternative hypothesis to a linear trend. Analysis of zonal and meridional winds by longitudinal sectors shows the importance of the standing wave pattern in interpreting the AO, which supplements the view of the AO as a simple zonal average (annular) mode". Thus the Arctic Oscillation should be considered as a physical phenomenon connected with the enhancement of circumpolar vortex and relevant mass and temperature changes in the stratosphere.

By the end of the 1980s—the beginning of the 1990s a very strong NAO enhancement resulted in powerful transport of warmer and fresher Norwegian Atlantic waters to the north of the Fram Strait and the Barents Sea. Entering the Arctic, the sub-layer of Atlantic waters was becoming thinner, warming (by about 2 °C) and increasing its horizontal extent ($\sim$20 %). At smaller depths, the cold haloclyne (which thermally isolates the sea ice cover from the warm Atlantic layer located below) shifted towards the euro-asiatic basin, which resulted in substantial changes of mass and energy balances of the ice cover surface. This and other phenomena have been studied within a number of recent programs such as ASP (Arctic Studies Program), ISAC (International Study of Arctic Change), SEATCH (NOAA Study of Environmental Arctic Change), etc. The climatic impact of polynyas is of particular interest (Schrank 2007).

Alekseev (1998) has emphasized that the Arctic is in many respects a key part of the global climatic system, where the strongest natural fluctuations of climatic characteristics develop. The global impact of the Arctic is primarily accomplished through the Arctic Ocean, which is capable of changing its structure circulation regime under the influence of changes in fresh water, salt and heat exchange with the non-polar parts of the global system. The freshened upper layer and sea ice located above it turn out to be the most active components, with fresh water, heat and salt transport being the major processes responsible for coupling between the high-latitude environment and its lower-latitude parts.

Specific features of the arctic atmosphere such as phenomena of Arctic Haze as well as extended cloudiness and radiation, have been studied during the period of the First GARP (Global Atmospheric Research Programme) Global Experiment—FGGE (Kondratyev et al. 1992; Kondratyev 1999).

Important progress has been achieved in the field of arctic climate diagnostics. Basic features of arctic climate dynamics have been demonstrated, such as strong space and time variability of various scales. For example, it was shown, that as far as inter-annual lower troposphere variations are concerned, before the 1980s negative anomalies had prevailed, while later on, for the whole troposphere, positive temperature anomalies were typical. However, the total polar atmosphere energy (potential plus internal) during the previous 40 years has not changed.

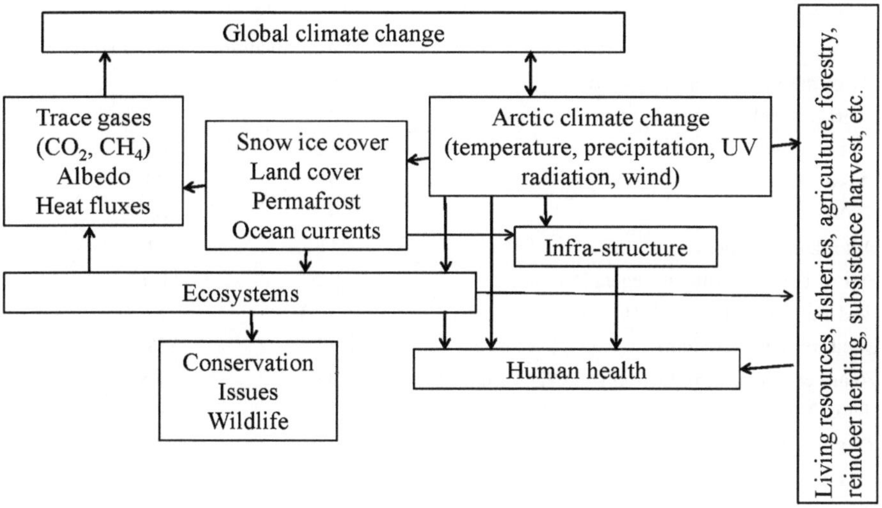

Fig. 6.2 Key interactions among major components of the Arctic climate system and the impact on human activities in the North

Much more difficult situation exists in the field of numerical modeling of high-latitude climate change. It has been mentioned in the Report (2001) that current estimates of future changes to the Arctic vary significantly. The model results disagree as to both the magnitude of changes and the regional aspects of these changes.

Climate variability and change have become important issues in the Arctic region over the recent decades. These issues have also prevailed in numerous international and national programs of scientific research. The most essential results of a penetrating analysis of these researches consist in the improvement of arctic science about key interactions among major components of the Arctic climate system in conditions of human activity growth (Fig. 6.2).

6.2 Climate and Cryosphere

6.2.1 Cryosphere Interactions with the Global Climate System

Walker et al. (2004) mention that the atmosphere-permafrost system plays important role in the global gas exchange processes which is comparable with the role of tropical ecosystems in the regulation of biogeochemical cycles. After all, knowledge of organic matter decomposition processes and vital functions of poor vegetation covered in arctic latitudes, without doubt, allows the significant reduction of uncertainty in the assessment of sources and sinks of greenhouse gases and, most of

Table 6.1 Comparison of the ecosystem characteristics for acid and no acid tundra

The tundra ecosystem characteristics	Acid medium	No acid medium
PH of the upper mineral horizon	6.9	5.2
Average height of the vegetation canopy (cm)	3.9	6.5
Leaf area index (m^2/m^2)	0.5	0.84
Mossy cover (%)	65	79
NDVI	0.28	0.41
O-horizon thickness	11	21
Part of the area without vegetation (%)	7.5	0.8
Heat flux from the soil ($MJ/m^2/day$)	3.13	1.83
Depth of soil thaw (cm)	52	39
Total net primary production ($mgCO_2$-$C/m^2/day$)	940	1,820
Net CO_2 absorption	670	950
Respiration losses	270	870
CH_4 extraction ($mgCH_4/m^2/day$)	69	449
Organic carbon content in the metric layer (kg C/m^3)	40	88
Calcium storage in active soil layer (mole(+)/m^3)	400	35

all, carbon dioxide. In particular, Table 6.1 shows that division of the tundra soil into acid and no acid, influences on this assessment. Study of arctic ecosystems and their role in biogeochemical cycles is realized in framework of series scientific programs between of which LAII (Land-Atmosphere-Ice Interactions) and ATLAS (Arctic Transitions in the Land-Atmosphere System) are notable for investigation of fluxes of water, energy and GHGs in the Arctic as well as for mapping the spectral reflection and zonal boundaries between ecosystems.

An important step forward in studying the Arctic environment is the Climate and Cryosphere (CliC) Project (Allison et al. 2001). The term "cryosphere" describes those portions of the Earth's surface, where water is in solid form. This includes all kinds of ice and snow and frozen ground, such as permafrost. The main components of the cryosphere are snow, river and lake ice, sea ice, glaciers and ice caps, ice shelves, ice sheets, and frozen ground. In terms of the ice mass and its heat capacity, the cryosphere is the second largest component of the climate system (after the ocean). Its influence for climate variability and change is based on physical properties, such as its high surface reflectivity (albedo) and the latent heat associated with phase changes, which have a strong impact on the surface energy balance.

The cryosphere is an important part of the global climate system. It is strongly influenced by temperature, solar radiation and precipitation, and, in turn, influences each of these properties. It also has an effect on the exchange of heat and moisture between the Earth's surface (land or sea) and the atmosphere, on clouds, on river flow (hydrology), and on atmospheric and oceanic circulation. Parts of the cryosphere are strongly influenced by changes in climate. The cryosphere may therefore act as an early indicator of both natural and human-induced climate changes.

The cryosphere on land stores about 75 % of the world's freshwater. The volumes of the Greenland and Antarctic Ice Sheets are equivalent to approximately

7 and 57 m of sea level rise, respectively (Heil and Allison 1999). Changes in the ice mass on land have contributed to recent changes in sea level. On a regional scale, many glaciers and ice caps play a crucial role in freshwater availability. Presently, ice permanently covers 10 % of the land surface, of which only a tiny fraction lies in ice caps and glaciers outside Antarctica and Greenland. Ice also covers approximately 7 % of the oceans in the annual mean. In midwinter, snow covers approximately 49 % of the land surface in the Northern Hemisphere (NH). Frozen ground has the largest area of any component of the cryosphere. All parts of the cryosphere contribute to short-term climate changes, with permafrost, ice shelves and ice sheets also contributing to longer-term changes including the ice age cycles.

As a core project of the World Climate Research Programme, the CliC project encourages and promotes research into the cryosphere and its interactions as part of the global climate system. It seeks to focus attention on the most important issues, encourage communication between researchers with common interests in cryospheric and climate science, promote international co-operation, and highlight the importance of this field of science to policy makers, funding agencies, and the general public. CliC also publicizes significant findings regarding the role of the cryosphere in climate, and recommends directions for future study.

CliC aims to improve understanding of the cryosphere and its interactions with the global climate system, and to enhance the ability to use parts of the cryosphere for detection of climate change. The scientific goals of CliC are to:

- improve understanding of the physical processes through which the cryosphere interacts within the climate system;
- improve the representation of cryospheric processes in climate models;
- assess and quantify the impacts and consequences of past and future climatic variability on components of the cryosphere; and
- enhance the observation and monitoring of the cryosphere.

To attain these goals, CliC seeks to develop and coordinate national and international activities aimed at increasing the understanding of four main scientific themes:

1. Interactions between the atmosphere and snow and ice on the land surface.
2. Interactions between glaciers and ice sheets and sea level.
3. Interactions between sea ice, oceans, and the atmosphere.
4. Interactions of the cryosphere with the atmosphere and oceans on a global scale.

CliC encourages the use of observations, process studies and numerical modeling within each of the above topic areas. In addition, CliC promotes the establishment of new cryospheric monitoring programs. The cryosphere is also considered as an indicator of climate variability and change. It has been pointed out in (Allison et al. 2001): "Atmosphere-snow/ice-land interactions are concerned with the role of the terrestrial cryosphere within the climate system and with improved understanding of the processes, and of observational and predictive capabilities applicable over a range of time and space scales. Better understanding of the interactions and feedback of the land/cryosphere system and their adequate parameterization within climate and hydrological models are still needed. Specific issues include the interactions and feedback of terrestrial snow and ice in the current climate and their variability; in land surface processes; and in the hydrological cycle. Improved knowledge is required of the amount, distribution, and variability of solid precipitation on a regional and global scale, and its response to a changing climate. Seasonally-frozen ground and permafrost modulate water and energy fluxes, and the exchange of carbon, between the land and the atmosphere. How do changes of the seasonal thaw depth alter the land-atmosphere interaction, and what will be the response and feedback of permafrost to changes in the climate system? These issues require improved understanding of the processes and improved observational and modeling capabilities that describe the terrestrial cryosphere in the entire coupled atmosphere-land-ice-ocean climate system.

Over a considerable fraction of the high-latitude global ocean, sea ice forms a boundary between the atmosphere and the ocean, and considerably influences their interaction. The details and consequences of the role of sea ice in the global climate system are still poorly known. Improved knowledge is needed for the broad-scale time-varying distributions of the physical characteristics of sea ice, particularly ice thickness and the overlying snow-cover thickness, in both hemispheres, and the dominant processes of ice formation, modification, decay, and transport which influence and determine ice thickness, composition and distribution. We do not know how are accurate the present model predictions of the sea ice responses to climate change, since the representation of much of the physics is incomplete in many models. It will be necessary to improve coupled models considerably to provide this predictive capability.

Key issues on the global scale are: understanding the direct interactions between the cryosphere and atmosphere, correctly parameterizing the processes involved in models, and providing improved data sets to support these activities. In particular, it is needed an improved interactive modeling of the atmosphere-cryosphere surface energy budget and surface hydrology, including fresh-water runoff.

The scientific strategy for CliC project is similar in each of the areas of interaction: a combination of measurement, observation, monitoring and analysis, field process studies and modeling at a range of time and space scales. A CliC modeling strategy must address improved parameterization in models of the direct interactions between all components of the cryosphere, the atmosphere, and the ocean. It will need to do this at a variety of scales from the regional to global; and with an hierarchy of models ranging from those of individual processes to fully coupled climate models. It will also be essential to provide the improved data sets needed for validation of models and parameterization schemes."

Table 6.2 data characterizes major components of the cryosphere. It has been mentioned in (Allison et al. 2001) that the processes in the coupled cryosphere-climate system involve three time scales—intraseasonal-interannual, decadal-centennial, and millennial or longer. The longest time scale is addressed through the IGBP PAGES program, although abrupt climate shifts evidenced in ice core and ocean sediment records (Heinrich events, involving extensive deposition of ice-rafted detritus in the

Table 6.2 Areal and volumetric extent of major components of the cryosphere (Allison et al. 2001)

Component	S	I	H
Land snow cover[b]			
Northern Hemisphere Late January	46.5	0.002	
Late August	3.9		
Southern Hemisphere Late July	0.85		
Early May	0.07		
Sea ice			
Northern Hemisphere Late March	14.0[c]	0.05	
Early September	6.0[c]	0.02	
Southern Hemisphere Late September	15.0[d]	0.02	
Late February	2.0[d]	0.002	
Permafrost (underlying the exposed land surface, excluding Antarctica and S. Hemisphere high mountains)			
Continuous[e]	10.69	0.0097–0.0250	0.024–0.063
Discontinuous and Sporadic	12.10	0.0017–0.0115	0.004–0.028
Continental ice and ice shelves			
East Antarctica[f]	10.1	22.7	56.8
West Antarctica and Antarctic Peninsula[f]	2.3	3.0	7.5
Greenland	1.8	2.6	6.6
Small Ice Caps and Mountain Glaciers	0.68	0.18	0.5
Ice Shelves[f]	1.5	0.66	–

S is the area (10^6 km^2), I is the ice volume (10^6 km^3), H is the sea level equivalent (m)[a]

[a] Sea level equivalent does not equate directly with potential sea-level rise, as a correction is required for the volume of the Antarctic and Greenland ice sheets that are presently below sea level. 400,000 km^3 of ice is equivalent to 1 m of global sea level

[b] Snow cover includes that on land ice, but excludes snow-covered sea ice

[c] Actual ice areas, excluding open water. Ice extent ranges between approximately 7.0 and 15.4×10^6 km^2

[d] Actual ice area excluding open water. Ice extent ranges between approximately 3.8 and 18.8×10^6 km^2. Southern Hemisphere sea ice is mostly seasonal and generally much thinner than Arctic sea ice

[e] Data calculated using the digital circum-arctic map of permafrost and ground-ice conditions and the GLOBE-1 km elevation data set

[f] Ice sheet data include only grounded ice. Floating ice shelves, which do not affect sea level, are considered separately

North Atlantic) are also highly relevant to CliC. The other two time scales are commensurate with WCRP interests, as manifest in ACSYS, GEWEX and CLIVAR. In the space domain, cryospheric processes and phenomena need to be investigated over a wide range of scales from meters to thousands of kilometers.

It is evident that changes in the Arctic's climate are a result of complex interactions between the cryosphere, atmosphere, ocean, and biosphere (Callaghan et al. 2011). Studying the cryosphere dynamics is of importance for many applications. In particular, changes in cryosphere play significant role in the socio-economic sectors worldwide. Main cryosphere factor such as ice volume can lead to catastrophic consequences. Upon the whole, cryosphere factors include (Allison et al. 2001):

- Land/ice melt contribution to sea level.
- Snow/glacier melt runoff.
- Thaw of clathrates.
- Loss of glaciers.

Tables 6.2 and 6.3 data illustrate some of possible effects depending on the climate-cryosphere system dynamics. Discrepancy between some data of these tables says about existing uncertainties in this area. It is evident that different parts of the cryosphere play various roles in the climate change. The amount of snow and ice that melts each summer is increasing because of global warming. Main focus is made on the potential role of the Greenland and Antarctic ice sheets a total melting of which can raise global sea level by 7.3 and 56.6 m, respectively. Forecasting this process is main task of present science. Solution of this task is possible by means of the GMNSS taking into consideration of information about time scales of the cryosphere components.

Four overarching goals that address major concerns for the WCRP can be identified (Allison et al. 2001). These are:

1. To improve understanding of the physical processes and feedbacks through which the cryosphere interacts within the climate system.
2. To improve the representation of cryospheric processes in models to reduce uncertainties in simulations of climate and predictions of climate change.
3. To assess and quantify the impacts of past and future climatic variability and change on components of the cryosphere and their consequences, particularly for global energy and water budgets, frozen ground conditions, sea level change, and the maintenance of polar sea-ice covers.

Table 6.3 The cryosphere components and their characteristics (http://www.unep.org/geo/geo_ice/graphics.asp)

Components of the cryosphere	Area covered (million km^2)	Ice volume (million km^3)	Potential sea level rise (cm)
Snow on land in the Northern Hemisphere (annual minimum–maximum)	1.9–45.2	0.0005–0.005	0.1–1.0
Sea ice, Arctic and Antarctic (annual minimum–maximum)	19–27	0.019–0.025	0
Ice shelves	1.5	0.7	0
Ice sheets (total) Greenland Antarctica	14.0 1.7 12.3	27.6 2.9 24.7	6,390 730 5,660
Glaciers and ice caps (lowest and [highest] estimates)	0.51 [0.54]	0.05 [0.13]	15 [37]
Permafrost (Northern Hemisphere)	22.8	4.5	~7

4. To enhance the observation and monitoring of the cryosphere in support of process studies, model evaluation, and change detection.

Specific questions that help define the primary tasks of CliC are:

(i) How stable is the global cryosphere?

- How well do we understand and model the key processes involved in each cryospheric component of the climate system?
- How do we best determine the rates of change in the cryospheric components?

(ii) What is the contribution of glaciers, ice caps and ice sheets to changes in global sea level on decadal-to-century time scales?

- How can we reduce the current uncertainties in these estimates?

(iii) What changes in frozen ground regimes can be anticipated on decadal-to-century time scales that would have major socio-economic consequences, either directly or through feedback on the climate system?

(iv) What will be the annual magnitudes, rates of change, and patterns of seasonal redistribution in water supplies from snow- and ice-fed rivers under climate changes?

(v) What will be the nature of changes in sea-ice mass balance in both polar regions in response to climate change?

(vi) What is the likelihood of abrupt climate changes resulting from regime changes in ice shelf—ocean and sea ice—ocean interactions that impact the ocean thermohaline circulation?

(vii) How do we monitor cryospheric components as indicators of change in the climate system?

Monitoring cryosphere dynamics is a key aspect of high-latitude environmental studies, especially because of a rather controversial information concerning ice cover dynamics. It is true, for instance, for ice thickness observations. Holloway and Sou (2001) have pointed out that "while submarine records had indicated a stunningly rapid thinning, model results show that the position of submarine observations was exceptionally biased towards regions of thinning. A conclusion is that observation to date, along as with much physics as models represent, imply little or no overall thinning".

6.2.2 Some Problems of the Permafrost Melting Feedback to Global Climate Change

Northern ecosystems play significant role in global climate change. This fact is studied by many experts. Particularly, Eastern Siberia refers to these ecosystems. The territory of Eastern Siberia (7.3 million km^2) occupies a significant part of the Northern Hemisphere and plays an important role for the climate system of the Earth.

Cryosphere and atmosphere are two (of the five) sub-systems of the climate system, which in turn is a sub-system of the global system. Nowadays, a crucial problem is that of climate change and its impacts. Frequently, with the term "climate change" we mean the occurrence of abnormal meteorological phenomena (e.g. heat-waves, tornadoes, floods, etc.), considering sometimes that it is a result of the uncontrolled human activities, ignoring completely the natural variability. Thus, we reach to conclusions about the climate system states, taking into account one or a few parameters of one of its sub-systems only, assuming that the rest parameters and the other sub-systems remain constant. It is obviously a flaw that leads to a vicious circle. We should not ignore that the climate subsystems are in a continuous interaction that is taking place, in general, between non-linear interactive processes. A typical example is the interaction between cryosphere and atmosphere, which is not only limited at their interface. Many authors are aimed in current research to find new links connecting these interactive processes which could shed light to various climate mechanisms that still remain under investigation.

Nowadays, it is generally recognized that under the global warming conditions, the high latitude regions will be more sensitive to the induced impacts. Specifically, the increase in the surface air temperature (SAT) will be more significant at the above-said geographical area and the response of the cryosphere to this atmospheric effect will be more considerable over Arctic and Sub-Arctic, than elsewhere. Since the late of 19th century, observational data revealed that, SAT in Eastern Siberia has increased by 10 °C during wintertime, and by almost 2.0–3.5 °C on an annual basis. In particular, the SAT increase was most pronounced in the second half of the 20th century (especially in the 1970–1980s).

The environmental changes at the Arctic and Sub-Arctic regions are of crucial importance for the additional reason that their ecosystems are considerably more vulnerable to anthropogenic forcings, than the corresponding ecosystems of other latitudes, mainly because the processes in the Arctic and Sub-Arctic environments are relatively slower. For example, the living organisms in the permafrost ecosystems live in the polar climate, where the vegetation period is seriously restricted. There, the feedback mechanisms operate with significant time delays and the capacity to neutralize the effects of human activity appears to be weak. In addition, the Arctic and Sub-Arctic ecosystems have specific boundary conditions that are closely associated with the permafrost boundary that reduce its ability to recover.

A quantitative forecast of changes in the balance of water, heat and carbon in the cryogenic ecosystems of the Eastern Siberia (e.g. forest and tundra) under global climate change conditions, requires:

- long time-series of ground-based and satellite observations in order to regularly monitor the accumulating ability of cryolithozone ecosystems, as well as
- development of adequate models providing the regional balance of carbon.

Another serious impact of the SAT increase is the ice melting, which results in the release into the troposphere and sometimes in the stratosphere of substantial amounts of the heat-trapping carbon dioxide and methane emitted from the cryosphere, thus accelerating the climate change (e.g. by enhancing the formation of clouds) and the stratospheric ozone loss. The latter would allow substantial enhancement to the biologically effective ultraviolet radiation to reach Earth, with potential adverse effects on ecosystems.

In conclusion, a multi-component and large-scale field campaign over Arctic and Sub-Arctic regions will address questions about the major adverse effects of the global warming to the local ecosystems, improving the reliability of relevant models and the current understanding of the above-mentioned key processes.

6.3 Arctic Basin Pollution Problems

The Arctic region exploration strategy in a broad context of biospheric studies has been discussed in details by Morgan (2008). A necessity of ecosystem's approach to studying land and marine biota has been particularly emphasized, as well as, conditions of socio-economic development in high-latitude regions. Aibulatov (2000) have discussed general problems of high-latitude environmental dynamics with special emphasis on radioactive pollution as an echo of the cold war. Aibulatov (2000) has analyzed principal sources of artificial radioisotopes in the Russian Arctic seas such as: atomic explosions at Novaya Zemlya, the global radionuclide background as a result of the overall nuclear tests conducted on the planet, Russian chemical and mining plants, Chernobyl accident, West-European radiochemical plants, solid and liquid radioactive waste dumping in the Barents and Kara Seas, the Northern Military Marine and its bases, atomic submarine construction and maintenance facilities and Atomflot (atomic fleet) of the Murmansk Shipping Company.

Studying of the distribution of ^{137}Cs, ^{90}Sr and $^{239,240}Pu$ in the water masses of the Northern, Norwegian, Barents, Kara, White and Laptev Seas has resulted in the following conclusions (Aibulatov 2000):

1. The general level of radioactive contamination of the waters of Arctic seas, except for several local areas, is characterized at the present time by little difference in comparison with background level ($\sim$6 Bq/kg).
2. The radioactive pollution of the waters of the Northern and Norwegian Seas is entirely due to the emissions from radiochemical plants located in Western Europe.
3. The contamination of waters of the Barents, White, Kara, and Laptev Seas is due to both local (Russian) sources and West-European plants.
4. The field observations in the Kara Sea in 1992–1995 have resulted in the conclusion that there have been no substantial radioactive emissions from the burial sites in the area.
5. The contribution of the Ob and Yenisei river runoff to the overall radioactive transport is not significant at the present time, except cases of extremely heavy floods which happen very rarely.
6. Compared to open waters of the Arctic Ocean, shelf seas of the Russian Arctic are more heavily contaminated.

Aibulatov (2000) has pointed out that judging from the ^{137}Cs-distribution patterns in the Kara Sea, it becomes evident that the Yenisey and Ob rivers (less evident, however, in the latter case) should be considered as transport channels for inputs of technogenic radionuclides to the Arctic Ocean waters. There are radioactive sources in the ocean as well. The ^{137}Cs activity level reached its maximum in 1984 and was equal to 245 Bq/kg (in open sea); during the 1990s (1993) this level was found to be equal to 100 Bq/kg (in the Yenisei estuary).

Arctic fjords have been classified into categories of comparatively clean, contaminated, heavily contaminated, and potentially contaminated. Contaminated areas include, for instance, Kola Gulf and, probably, all the fjords of the northern Kola Peninsula west of Murmansk. The content of radionuclides in phytobenthos, in the coastal zone east of Murmansk, is low. Evidently, there has not been recently any serious radionuclide penetration into this area. A rather low gamma-nuclide level (1–3 Bq/kg) is typical for the zoobenthos of the Barents Sea. This is also true for the Kara Sea.

An impact of all the sources of radioactivity in the zone of the Arctic coast on the local population has not been assessed reliably enough. It was particularly difficult to separately identify natural and anthropogenic components of such an impact. Aibulatov (2000) has discussed future research of the Russian Arctic radioactive pollution, including:

- Development of a coordinated program Russian Arctic Sea Radioactivity Ecological Monitoring.
- Assessments of impacts of different radioactive sources on the contamination of the Arctic marine environment including water basins, land, and atmosphere.
- Studies of detailed space and time variability of various long-lived technogenic radionuclides in bottom sediments.
- A detailed examination of all Novaya Zemlya fjords in connection with dumping radioactive waste.
- Research of the impact of radioactive pollution on the Arctic marine ecosystem dynamics.
- Studying medical aspects of environmental pollution in the Arctic.

The fundamental study of radioactivity of the Arctic and sub-Arctic marine ecosystems have been undertaken by Matishov and Matishov (2001), which resulted in the substantiation of a new branch of science—radiational ecological oceanology. Investigations have been conducted of the level of artificial radionuclide concentration in both environment and biota of the bays and inlets (the Kola, the Chernaya, the West Litsa), where radioactively dangerous objects are located. In this context, a classification has been suggested for coastal areas (bays, gulfs, fjords) in accordance with contamination levels for bottom sediments. The discovery of a bio filter of pelagic and coastal zone during the processes of self-purification of water reservoirs and transport of radionuclides from water to bottom sediments, is of great importance. For the first time the levels of ^{137}Cs, ^{90}Sr, $^{239,240}Pu$ concentrations for different types and populations of sea organisms were measured. Migrations of radioisotopes along the trophic chains (from macrophytes and plankton to zoobenthos, fish, birds, seals, and whales) were studied, as well. The assessments of comparative contributions of global, regional, and local sources of radioactive environmental contamination during the time-period from nuclear tests till recent time have been analyzed and used as a source of information for environmental predictions. An important optimistic conclusion concerning consequences of potential accidents is that for all prescribed scenarios of radioactive emissions, it is highly improbable that large-scale contamination of the Arctic Ocean will take place with ruinous impacts on marine bio resources. High biological assimilation capacity in combination with specific features of hydrodynamic and other processes is supposed to serve a barrier against dangerous pollution of the Arctic Ocean.

Kalabin (2000) has accomplished a study of environmental dynamics and industrial potential of the Murmansk region, the most urbanized and industrially developed trans-polar region of the planet. Under these conditions, specific features of environmental dynamics result in the enhancement of anthropogenic impacts. In this context, Kalabin (2000) has analyzed critical environmental loads for some of the northern ecosystems and emphasized a necessity to investigate their assimilation (buffer) capacity, as a principal aspect of sustainable functioning natural systems. The solution of regional problems of sustainable development requires a careful analysis of the interaction between eco-dynamics and socio-economic development.

The progress achieved in studying the Arctic environment variability is due to the accomplishment of a number of international research programs. The Arctic Climate System Study (ACSYS) project developed in 1991, as well as the WCRP initiative is of particular importance and practicable program for the next decade, to assess the role of the Arctic in global climate. Five areas are emphasized:

(1) Ocean circulation;
(2) Sea ice climatology;
(3) The Arctic atmosphere;
(4) The hydrological cycle; and
(5) Modeling.

The scientific goal of ACSYS, which started its main observational phase on January 1994 and will continue for a 10-year period, includes the three main objectives:

(1) understanding the interaction between the Arctic Ocean circulation, ice cover, and the hydrological cycle;
(2) initiating long-term climate research and monitoring programs for the Arctic;
(3) providing a scientific basis for an accurate representation of Arctic processes in global climate models.

The Arctic Ocean Circulation Program of ACSYS consists of the four components:

(1) the Arctic Ocean Hydrographic Survey to collect a high-quality hydrographic data-base representative of the Arctic Ocean;
(2) the Arctic Ocean Shelf Studies which are aimed at understanding how the shelf processes partitions salt- and fresh-water components; at defining the dynamics and thermodynamics of the shelf waters as well as other processes;

(3) the Arctic Ocean Variability Project designed to assess the variability of the circulation and density structure of the Arctic Ocean; and
(4) the Historical Arctic Ocean Climate Database Project aimed to establish a universally available digital hydrographic database for the Arctic Ocean for analysis of climate-related processes and variability, and to provide a data set suitable for initialization and verification of arctic climate and circulation models.

The ACSYS sea ice program includes three main components:

(1) establishing an Arctic Basin-wide sea ice climatology database;
(2) monitoring the export of sea ice through the Fram Strait; and
(3) arctic sea ice processes studies.

One of the main tasks of the ACSYS arctic sea ice program is to establish a climatology of ice thickness and ice velocity. Such data will be supplied by the WCRP Arctic Ice Thickness Project, the International Arctic Buoy Program, sonar profiling from naval submarines and unmanned vehicles, airborne oceanographic lidars, and polar satellites carrying appropriate instruments.

The arctic atmosphere provides the dynamic and thermodynamic forcing of the Arctic Ocean circulation and sea ice. Key directions of research include problems such as: cloud-radiation interaction, air-sea interaction in the presence of ice cover (impacts of polynyas and leads are of special interest), arctic haze, etc.

Primary ACSYS efforts within the project of the Hydrological Cycles in the arctic region are aimed at:

(1) the documentation and inter-comparison of solid precipitation measurement procedures used in high latitudes; and
(2) the development of methodologies for determining areal (regional) distributions of precipitation from station data.

There are two relevant data-archiving efforts: Arctic Precipitation Data Archive (APDA) and Arctic Run-off Data Base (ARDB).

The principal purpose of the ACSYS modeling program is simulations of climate variations in polar regions which arise from the interaction between atmosphere, sea ice and ocean.

Apart from the ACSYS project described above, a number of new research programs have been developed, such as the Study of Environmental Arctic Change (SEARCH), which is an interdisciplinary, multi-scale program dedicated to understanding the complex interrelated changes that have been observed in the Arctic environment in the past few decades (SEARCH 2001a, b). SEARCH is envisioned as a long-term effort of observations, modeling, process studies, and applications with emphasis on five major thematic areas:

- human society;
- marine/terrestrial biosphere;
- atmosphere and cryosphere;
- ocean, and
- integrated projects/models/assessment.

The Arctic System Science (ARCSS) Program is an interdisciplinary program with the principal goals to

(1) understand the physical, geological, chemical, biological, and socio-cultural processes of the arctic system that interact with the total Earth's system and thus contribute to or are influenced by global change, in order to
(2) advance the scientific basis for predicting environmental change on a seasonal-to-centuries time scale, and for formulating policy options in response to the anticipated impacts of global change on humans and societal support systems. The following four scientific thrusts include central aims of ARCSS:

- to understand the global and regional impacts of the arctic climate system and its variability;
- to determine the role of the Arctic in global biogeochemical cycling;
- to identify global change impacts on the structure and stability of arctic ecosystems;
- to establish the links between environmental change and human activity.

ARCSS has four linked ongoing components: Ocean/Atmosphere/Ice Interactions (OAII); Land/Atmosphere/Ice Interactions (LAII); Paleoenvironmental Studies (including GISP2, Greenland Ice Sheet Project Two), and Paleoclimates of Arctic Lakes and Estuaries (PALE); Synthesis, Integration, and Modeling Studies (SIMS), and Human Dimensions of the Arctic System (HARC).

Aagard (1998) discussed basic problems with a multidisciplinary look at the Arctic Ocean, including: physical and chemical studies; biological studies; contaminant studies; measurements of the properties and variability of the ice cover and of the surface radiation budget; studies of atmospheric chemistry; geological observations.

LAII research has three main goals:

(1) to estimate important fluxes in the region, including the amount of carbon dioxide, and methane reaching the atmosphere, the amount of river water reaching the Arctic Ocean, and the radiative flux back to the atmosphere;
(2) to predict how possible changes in the arctic energy balance, temperature and precipitation will lead to feedback affecting large areas; this incorporates changes in water budget, duration of snow cover, extent of permafrost, and soil warming, wetting, and drying; and
(3) to predict how the land and fresh-water biotic communities of the Arctic will change, and how this change will affect future ecosystem structure and function.

A major LAII research project is the Flux Study; its principal purpose is a regional estimate of the present and future movement of materials between the land, atmosphere, and ocean in the Kuparuk river basin in northern Alaska.

Of the nineteen LAII projects, three are part of the International Tundra Experiment (ITEX), which looks at the response of plant communities to climate change. Three others are concerned with atmosphere processes, including weather pattern affecting snowmelt, Arctic-wide temperature trends, and water vapour over the Arctic and its relationship to the atmospheric circulation and surface conditions. One project deals with response of large birds to climate and sea-level change at river deltas, and one studies the balance and recent volume changes of the McCall Glacier in the Brooks Range.

Synthesis, integration and modeling studies are intended to foster linkages and system-level understanding. Research on the past contemporary relationship to humans to global climate change is thought to be critical to understanding the consequences of global change in the Arctic.

There are a number of ARCSS data projects, including: LAII Flux Study Alaska North Slope data sampler CD-ROM; OAII Northeast Water (NEW) polynya project CD-ROM; Arctic solar and terrestrial radiation CD-ROM, etc.

A list of the OAII components includes the joint US/Japan cruise, the Western Arctic Mooring project, and the Northeast Water Polynya project. Among other OAII projects the most notable are the US/Canada Arctic Ocean Section and the Surface Heat Budget of the Arctic Ocean (SHEBA) project.

An outstanding effort has been accomplished in 1994 within the Canada/US 1994 Arctic Ocean Section, when the two icebreakers entered the ice in the northern Chukchi Sea on 26 July 1994, reached the North Pole on 22 August, and left the ice northwest of Spitsbergen on 30 August, completing thereby the first crossing of the Arctic Ocean by surface vessels. This voyage altered greatly our understanding of biological productivity, the food web, ocean circulation and thermal structure, and the role of clouds in the summer radiation balance, as well as the extent of contamination and spreading pathways (especially related to radionuclides and chlorinated organics), and the extent and effects of sediment transport by sea ice.

In connection with the SHEBA project, the U.S. Department of Energy's Atmosphere Radiation Measurement (ARM) program indicated its intention to develop a Cloud and Radiation Testbed (CART) facility on the North Slope of Alaska. The principal focus of this program will be on atmospheric radiative transport, especially as modified by clouds (which impacts the growth and decay of sea ice), as well as testing, validation, and comparison of radiation transfer models in both the ice pack and arctic coastal environment.

A special place is occupied by the Russian-American Initiative on Shelf-Land Environments in the Arctic (RAISE) with the principal goal of facilitating ship-based research in the Russian Arctic (RAISE 2001). Earlier relevant land-based research projects under the RAISE umbrella included studies of:

- organic material and nutrient fluxes from Russian rivers;
- seasonal flooding dynamics along rivers, and
- reconstruction of late Pleistocene glacial and sea-level history on Wrangel Island.

New scientific topics in the near-shore waters of the Russian continental shelf will include a broad range of studies: from the biogeochemical fate of organic materials contributed to the Arctic Ocean by shoreline erosion and river runoff to the social and biological impacts of changes in sea-ice distributions.

The Western Arctic Shelf-Basin Interactions (SBI) project, sponsored by ARCSS Program and the U.S. Office of Naval Research, is investigating the Arctic marine ecosystem, to improve our capacity to predict environmental change. The SBI Phase II Field Implementation Plan (2002–2006) focuses on three research topics in the core study area:

- northward fluxes of water and bioactive elements through the Bering Strait input region;
- seasonal and spatial variability in the prediction and recycling of biogenic matter on the shelf-slope area; and
- temporal and spatial variability of exchanges across shelf-slope region into the Canada Basin.

The meeting of the International Arctic Science Committee (IASC) has identified the following four science priorities:

(1) arctic processes relevant to global systems;
(2) effects of global change on the Arctic and its peoples;
(3) natural processes within the Arctic; and
(4) sustainable development in the Arctic.

The following areas in Arctic global change research have been considered the most significant:

(1) terrestrial ecosystem;
(2) mass balance of glaciers and ice sheets;
(3) regional cumulative impacts; and
(4) human dimensions.

An important aspect of studying high-latitude environmental dynamics is an assessment of the impact of potential anthropogenic climate warming. In this context Frederick (1994) has formulated key issues of integrated assessments of the impact of climate change on natural resources. Specific project objectives include:

(1) characterizing the current state of natural science and socioeconomic modeling of the impacts of climate change and current climate variability on forests, grasslands, and water;
(2) identifying what can be done currently with impact assessments and how to undertake such assessments;
(3) identifying impediments to linking biophysical and socioeconomic models into integrated assessments for policy purposes; and
(4) recommending research activities that will improve the state of the art and remove impediments to model integration.

The following questions are supposed to be answered:

- How will the overall system (physical-biological-economic) respond to various imposed stresses?
- How do the uncertainties in the component models add up to give an overall system response uncertainty?

- Is society made more vulnerable to extreme natural events either by changing those events or by reducing human ability to respond with corrective action?
- How likely is it that the consequences of climate change will be severe or catastrophic?
- What is at risk and when is it a risk?
- What are the likely impacts on the landscape and the hydrological system?
- How might the boundary conditions and the overall productivity of the forests, grasslands, and other rangelands be affected?
- How might increasing carbon dioxide levels affect crops and food supplies for humans, livestock, and wildlife?
- What are the socio-economic consequences of these physical and biological changes?
- What are the likely consequences for ecosystems of mitigation actions?
- Can the costs associated with climate change be reduced through natural adaptation of ecosystems or policy-initiated adaptation?

Frederick (1994) has emphasized that the accumulated results of many regional and local climate impact assessments may help provide informed answers to these questions. Nevertheless, the uncertainties surrounding both the nature and the impacts of any future climate change are likely to remain very large, precluding precise estimates of the net benefits associated with alternative policy responses. Even if the range of uncertainty were diminished, it might still be difficult to justify specific measures on narrow economic grounds, because (as noted above) the impacts on natural resource systems are apt to be poorly reflected in standard benefit-cost analysis.

Mendelsohn and Rosenberg (1994) have formulated the following questions relevant to global warming effects in the area of ecological and water resources:

- Do changes in ecosystems provide important feedbacks to the natural carbon, nitrogen, and methane cycles? For example, will the natural sinks or emitters be affected by changing precipitation, temperature, and CO_2 levels?
- What are the appropriate output measures of ecosystem component models? What are the ecological effects of climate change that policy analysts use to determinate the importance of an ecosystem change?
- What climate change-driven shifts in ecosystem boundaries can be predicted?
- Will these effects be subtle and small or large and dramatic and over what time frame and spatial dimensions?
- Will climate change cause a change in the productivity of valuable market or non-market species? For example, to what extent will some forests grow more quickly or more slowly. Will desired non-market species, such as bear, elk, and bald eagles, be more or less plentiful?
- What species could be lost with rapid climate changes? How do the vulnerable species break down by type and geographic distribution? How should conservation policies adapt to a world requiring change?
- How are the ecosystems likely to change as the climate evolves over time: will there be a large increase in early succession species and where?

- How will average flows in rivers change with greenhouse warming? How will these flows change over seasons? Will the probabilities of catastrophic events change?
- What values do people assign to the changes in ecosystems by climate change? Which changes are important and which are minor? Can a value be assigned to non-use?
- How much should society be willing to pay to reduce the probability of losing specific species? If different scenarios favor different species, how should society trade between these outcomes?
- What impact do ecosystem changes have upon the economy? For example, how will climate change affect grazing, commercial fishing, timber, or commercial tourism?

It has been suggested that priority program areas and relevant projects are as follows:

(a) Impacts of global changes on the arctic region and its people.

- Regional cumulative impacts.
- Effects of increased UV radiation.

(b) Arctic processes of relevance to global systems.

- Mass balance of glaciers and ice sheets.
- Terrestrial ecosystems and feedback on climate change.

(c) Natural processes within the Arctic.

- Arctic marine/coastal/riverine systems.
- Disturbance and recovery of terrestrial ecosystems.

(d) Sustainable development in the Arctic.

- Sustainable use of living resources
- Dynamics of arctic populations and ecosystems
- Environmental and social impacts of industrial development.

Future priorities of the ARCSS include the following research questions (ARCUS 1998):

- How will the Arctic climate change over the next 50–100 years?
- How will human activities interact with future global change to affect the sustainability of natural ecosystems and human societies?
- How will changes in arctic biogeochemical cycles and feedbacks affect arctic and global systems?
- How will changes in arctic hydrological cycles and feedbacks affect arctic and global systems? Are predicted changes in the Arctic detectable?

Important perspectives are connected with paleo-environmental studies by the Paleo-environmental Arctic Sciences (PARCS) community (PARCS 2000) which have a principal aim of answering the question: how much do recent observations of climate change in the Arctic reflect natural climate cycles? Relevant major topics include:

- the medieval warm period (apparently, AD 1000–1400) and Little Ice Age (approximately AD 1400–1850);
- high-amplitude Holocene climate cycles, and
- the possible connection of the onset of neo-glacial (a mid-Holocene cooling, particularly evident at high northern latitudes) with shifts in the frequency and amplitude of such climate cycles.

According to the PARCS, there are very warm past scenarios that can serve as analogues for future climate warming:

- the early Holocene, when the Arctic had experienced high summer insolation anomalies, and
- the last interglacial period (marine isotope stage 5), which appears as a very strong warming in the paleo-record approximately 125,000 years ago.

Key topics investigated in relation to these periods are:

- feedbacks and nonlinear changes (surprises) as consequences of strong warming—particularly the role of sea ice, ice sheets, and land surface cover;
- implications of strong warming for arctic and global carbon budgets.

To summarize what has been mentioned in connection with recent Arctic environmental programs, it must be emphasized that relevant information was not at all exhaustive (IASC 2001). An obvious conclusion is that the number of programs is too great. A clear necessity exists of better coordination of all on-going efforts and their "regularization".

Vörösmarty et al. (2001) are right in their conclusion that "understanding the full dimension of arctic change is a fundamental challenge to the scientific community over the coming decades and will require a major new effort at interdisciplinary synthesis. It also requires an unprecedented degree of international cooperation".

As it was noted earlier, the Arctic region plays a unique role in the global environmental processes, forming numerous adverse correlations with the Earth's climatic system. Low temperatures, a high level of atmospheric circulation and the availability of large ice-covered aquatories are notable for the high latitudes, quite unlike the other areas of the globe.

The intensive industrial development of the northern territories has led to significant environmental change in these regions, especially in the Russian territories. Oil and gas extraction on the Yamal and Taimir peninsulas in northwestern Siberia, coal and gold extraction in Yakutia and Chukot as well as the mining industry on the Kola Peninsula have led to a strong anthropogenic intervention into the natural environment. Plant cover has been violated over great territories, the area and productivity of reindeer pastures diminished and the hydrologic regime of rivers disturbed. The greatest quantities of pollution substances are brought to the northern coast of Russia by the rivers, violating the ecosystems of the northern seas.

Further adverse influence on the Arctic ecosystems can lead to rising regional or even global scale negative consequences. Thus the problem of the development of the northern territories of Russia, the United States of America, Canada and the Scandinavian countries needs to be solved by careful analysis of all types of ecosystem dynamics, by real time data collection and formation of national *databases* and by defining effective ways to coordinate development of the natural and anthropogenic processes.

6.4 The Spatial Simulation Model of the Arctic Ecosystem

6.4.1 Simulation Model as a Tool for the Study of the Arctic Ecosystem

Environmental investigations of the Arctic ecosystem encounter with many problems arising by reason of technical realization of monitoring systems. Therefore, this restriction is overcome by the use of modeling technology (Shaver et al. 2013; Luus et al. 2013; Krapivin and Varotsos 2008). Norwegian Polar Institute in January 2013 organized scientific seminar "Arctic Ecosystem modeling: State of the art, challenges, and a possible way forward" that given an overview of background and current research in marine ecosystem modeling. Several models were realized and used:

- Integrated Ecosystem Model (IEM) for Alaska and Northwest Canada. The IEM is an effort to try and forecast landscape change within the Alaska and Northwest Canada region. The IEM brings together climate driven changes to disturbance, vegetation succession, permafrost, and hydrology to illustrate how arctic and boreal landscapes may be altered in the future.
- The Terrestrial Ecosystem Model (TEM). The TEM includes four main modules: (1) the environmental module, (2) the dynamic soil layer module (DSL), (3) the ecological module, and (4) the fire disturbance module. The dynamic soil layer module includes explicit simulation of soil carbon vertical distribution and change of the thicknesses of organic layers based on the soil carbon content. The thickness of the soil organic layer may change due to ecological processes or fire disturbance. Note that this module includes the explicit representation of permafrost dynamics. The ecological module of TEM includes a dynamic vegetation model (TEM-DVM) with multiple vegetation pools, including the leaf, wood and root pools.
- Single regression model predicts net ecosystem exchange (NEE) as a function of leaf area index (LAI), air temperature and photosynthetically active radiation (PAR). The model was developed in concert with a survey of the light response of NEE in Arctic and subarctic tundras in Alaska, Greenland, Svalbard and Sweden. Model parameterizations based on data collected in one part of the Arctic can be used to predict NEE in other parts of the Arctic with accuracy

similar to that of predictions based on data collected in the same site where NEE is predicted. The principal requirement for the dataset is that it should contain a sufficiently wide range of measurements of NEE at both high and low values of LAI, air temperature and PAR, to properly constrain the estimates of model parameters. Canopy N content can also be substituted for leaf area in predicting NEE, with equal or greater accuracy, but substitution of soil temperature for air temperature does not improve predictions.

Riseborough et al. (2008) provides a review of permafrost modelling advances, primarily since the 2003 permafrost conference in Zürich, Switzerland, with an emphasis on spatial permafrost models, in both arctic and high mountain environments. Models are categorized according to temporal, thermal and spatial criteria, and their approach to defining the relationship between climate, site surface conditions and permafrost status. The most significant recent advances include the expanding application of permafrost thermal models within spatial models, application of transient numerical thermal models within spatial models and incorporation of permafrost directly within global circulation model (GCM) land surface schemes. Future challenges for permafrost modeling will include establishing the appropriate level of integration required for accurate simulation of permafrost-climate interaction within GCMs, the integration of environmental changes such as tree line migration into permafrost response to climate change projections, and parameterizing the effects of sub-grid scale variability in surface processes and properties on small-scale (large area) spatial models.

The purpose of this chapter is to develop and to investigate a simulation model of the pollution dynamics in the Arctic Basin. There are many experimental and theoretical results giving estimates of the growing dependencies between the pollution dynamics in the World Ocean and the state of the continental environment. The problem of the Arctic Basin pollution causes the most anxiety to investigators (Krapivin and Phillips 2001b; Bobylev et al. 2003). It is known that the ecosystems of the Arctic seas are vulnerable to a considerable extent in comparison with the ecosystems of other seas. Processes that clean the Arctic Ocean are slower and marine organisms of the Arctic ecosystem live in the polar climate, where the vegetation period is restricted. Some feedback mechanisms operate with significant time delays and the capacity to neutralize the effects of human activity is feeble. Apart from these reasons, the Arctic ecosystem has specific boundary conditions connected with the sea-ice ergocline, which reduce its survivability level.

In connection with this circumstances, the Arctic Basin is the object for investigations in the frameworks of many national and international environmental programs, such as the International Geosphere-Biosphere Program, U.S. Global Change Research, the International Arctic System Science program (ARCSS), (McCauley and Meier 1991; ARCSS 2003), the U.S. Arctic Nuclear Waste Assessment Program (ANWAP) and the International Arctic Monitoring and Assessment Program (AMAP). The research strategies of these programs include the theoretical and experimental study of the tundra ecosystems, Siberian rivers and near-shore and open arctic waters. The main problems arising here consist of determining a set of the principle matters that are necessary to study in the first place. These include the following:

(1) *Transport modeling of pollutants in the arctic ecosystems.* Based on the experimental database, it is necessary to prepare a complete set of models and their corresponding computer realizations for describing the processes of transfer and transformation of pollution substances in the Arctic natural ecosystems. This set includes the following models:

- a model of the transformation of organic pollution in the ecosystems of the fresh water basins and streams;
- models of the self-cleaning processes for oil, radionuclides, heavy metals and other pollutants;
- a model of radionuclide and heavy metal accumulation in the river ecosystems of the Far North;
- a model of the transport of radionuclide, heavy metal and organic pollution with river flow into the near-coast Arctic waters;
- a model of pollution wash-out during the spring season in the tundra and forest-tundra zones;
- a model of the kinetics and transformation mechanisms for the biosphere elements in the water systems;
- a kinetics model of radionuclides and heavy metals in the food chains of the land ecosystems for boreal zones;
- a model of the surface flow of chemical elements and compounds from territories in mining zones with open-cast mines under the climatic conditions of the Far North; and
- a model of the seasonal influence of pollution on the phytoplankton and primary production in the Northern seas.

(2) *Modeling the exchange processes of carbon dioxide and methane between the tundra ecosystems and the atmosphere.* The global interaction of the Arctic ecosystems with the biosphere and with the Earth's climatic system is carried out particularly through the influence upon the biogeochemical cycles of carbon dioxide and methane. Existing models of the global circulation of these greenhouse gases are incomplete, considering that they do not take into account this interaction. Present estimations of the gas exchange between the Arctic ecosystem and the atmosphere, confirm however, the necessity of making such an account. To create a model set related to the gas exchange in the Arctic reservoirs, it is necessary to compile a catalog of soil-plant formations, ice fields, land-based and oceanic reservoirs. It is necessary also to put in the database estimations of evapotranspiration, dead vegetation decomposition rate and productivity of vegetation communities. With the aid of this model set it will be possible to evaluate the role of the tundra ecosystems in forming the greenhouse effect.

(3) *Modeling the hydrological regime and estimation of the pollutant flows in the Arctic Basin.* It is necessary to prepare a set of models to describe the dynamics of the separate aquatories and the whole hydro-system of the Arctic ocean, including:

- a complex model of the water circulation in the Arctic Basin;
- regional models of the water circulation in the Arctic seas;
- a model of the kinetics of radionuclides, heavy metals and organic pollutants in the trophic structures of the Arctic marine ecosystems;
- a model of the spread of a pollutant concentration from a point-like source in the near-coastal zone of the Arctic Basin;
- a model of the transfer of radionuclides, heavy metals and organic pollutants due to vertical mixing of the Arctic waters;
- a model of the conservation and release processes due to freezing and thawing of the ice cover.

(4) *Modeling the Arctic ecosystems under conditions of anthropogenic impacts.* Anthropogenic influence in the Arctic Basin and on adjacent territories is connected with local, regional and global activities. Therefore, it is necessary to work out the following models:

- models for the influence of radionuclides, heavy metals and oil hydrocarbons upon the dynamics of the marine ecosystems under Arctic climate conditions;
- models for the limiting of vegetation cover due to the different types of pollution that are brought to the land ecosystems with precipitation and surface flows;
- models for the dynamics of vegetation covers subjected to a physical influence;
- models/scenarios for the town and settlement structures under development;
- models/scenarios for the changes in area of the traditional seasonal regions of activity of the nomadic population; and
- models/scenarios for the social development of the scattered peoples in the Far North.

(5) *Modeling the biogeochemical carbon cycle in the "Atmosphere/Arctic Ocean" system.* As has been shown by the experience of modeling the carbon dioxide global cycle, the estimations of the role of the World Ocean in redundant carbon absorption are rather rough (Nitu et al. 2004). For the models to be more precise, they need to be reinforced by more reliable parameterizations of the physical processes, related to the interaction between the bordering layers of the and the Arctic aquatories. According to numerous laboratory and natural observations, the directivity of these processes depends considerably on many factors. The most significant of these are the speed of the driving wind, the presence of ice cover and the vertical distribution profile of the water temperature. A complex composition of these factors determines the variety of possible models and their details. One of the significant problems is to explain the powerful flow of seaweeds during the spring season and hence to construct a parameterization system for the dynamics of photosynthetic processes under conditions of snow and ice covers along with their spring thawing.

(6) *Development of a complex model for the functioning of the hydrologic and biogeochemical systems in Arctic.* In addition to the model set intended for local and fragmentary processes in the Arctic ecosystems, as well as for understanding their global role, it is necessary to synthesize a single model for the whole complex

of biogeochemical, biogeocenotic and hydrologic processes that occur in the boreal systems. Creation of such a model will ease the development of the database requirements and will make it possible to obtain means for estimating the consequences of anthropogenic projects. Using this model, the consequences can be estimated for forest cutting and fires, for the broadening of zones with disturbed land cover, for land and basin pollution with oil, hydrogeological changes on the separate territories due to submerging of landscapes, territorial pollution with waste materials from the mining industry, etc.

(7) *Stability estimation of the Arctic systems under variable global climate conditions*. Human activities in the delicate ecosystems of the Far North need to be conducted with great care. But how is the natural balance to be safe from the rising influence of human civilization with its industrial machines?—how can the survival of these ecosystems be evaluated in different situations?—these and other questions need to be answered as the program outlined above is realized.

An understanding of the environmental processes in the Arctic regions, a prerequisite for finding scientific solutions to the problems arising there, can be found only by combining many disciplines, including ecology, oceanography, mathematical modeling, and system analysis. This chapter synthesizes many data sources and knowledge from various scientific fields in the form of a Spatial Simulation Model of the Arctic Ecosystem (SSMAE). Separate blocks of the SSMAE were created earlier by many authors (Riedlinger and Preller 1991; Muller and Peter 1992; Legendre and Krapivin 1992; Krapivin 1995). The sequence of these blocks in the SSMAE structure and the adaptation of it to the Spatial Global Model (SGM) provide a technology for computer experiments (Krapivin 1993).

The goal of the NSF ARCSS Program is to answer the question: What do changes in the arctic system imply for the future? To address this question, researchers must:

- Advance from a component understanding to a system understanding of the Arctic.
- Understand the behavior of the arctic system—past, present and future.
- Understand the role of the Arctic as a component of the global system.
- Include society as an integral part of the arctic system.

AMAP is an international organization established in 1991 to implement components of the Arctic Environmental Protection Strategy (AEPS). Now a program group of the Arctic Council, AMAP's current objective is "providing reliable and sufficient information on the status and threats of the Arctic environment, and providing scientific advice on actions to be taken in order to support Arctic governments in their efforts to take remedial and preventive actions relating to contaminants" (http://www.amap.no; AMAP 2009).

The present chapter describes a simulation system based on sets of computer algorithms for processing data from monitoring of the Arctic regions and for applying mathematical models of natural and anthropogenic processes.

The basic blocks of the SSMAE are oriented on the description of the dynamics of any given pollutant. For consideration of a specific pollutant it is necessary to include in the SSMAE an additional block with the description of its physical and chemical characteristics. This procedure is demonstrated by examples of blocks that simulate the characteristics of radionuclides, heavy metals and oil hydrocarbons. The consideration of these pollutants is restricted to elements with averaged properties.

The Arctic Basin aquatory (Ω) which is studied in this chapter has boundaries which include the peripheral Arctic seas as well as the coast line and southern boundaries of the Norwegian and Bering Seas.

6.4.2 Structure of the Simulation Model

A conceptual diagram and the block contents of the SSMAE are shown in Fig. 6.3 and Table 6.4. The functioning of the SSMAE is supported by the SGM and by the climate model (CM). The SSMAE inputs are the data about the pollutant sources of the near-shore Arctic Basin, the ice areas and the current maps. The SSMAE contains three types of blocks: mathematical models of the natural ecological and hydrophysical processes, service software and the scenarios generator. The marine biota block (MBB) describes the dynamics of the energy flows in the trophical chains of the Arctic Basin ecosystem. The hydrological block (HB) describes the spatial discretization of Ω on the water circulation in the Arctic. The pollution simulation model (PSM) contains the anthropogenic scenarios and the service control block (SCB) provides the control of the simulation experiment.

The Arctic ecosystem has a unique, complex food web that is fashioned by its distinctive plankton, animal species, and environmental factors. Carbon also cycles through the web from atmosphere to seawater and back. Phytoplankton and algae take up carbon dioxide from seawater and transform it into the organic carbon of their tissue. Then it flows through successive levels of eating animals that convert

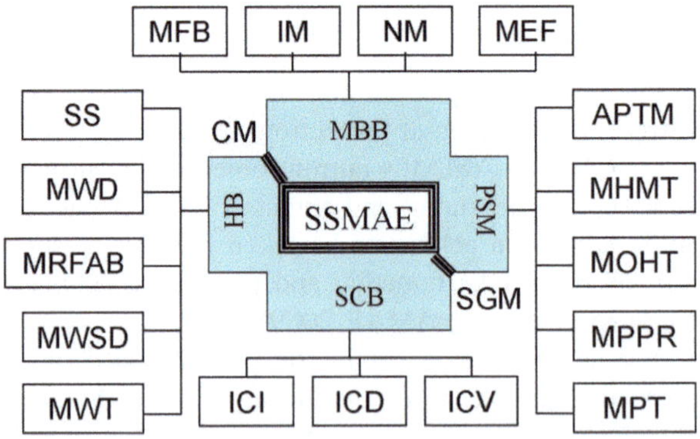

Fig. 6.3 Block-diagram of the SSMAE. Descriptions of the blocks are given in Table 6.4. SGM is the spatial global model (Krapivin 1993). CM is the climate model or the climate scenario

Table 6.4 Description of the SSMAE blocks (Fig. 6.3)

Block	Block description
MBB	Marine Biota Block containing the set of models for energy flows in the trophical chains of the Arctic Basin ecosystem (Nitu et al. 2000a; Kondratyev et al. 2003a)
HB	Hydrological Block describing the water circulation in the Arctic seas and the movement of ecological elements (Krapivin 1995, 1996)
PSM	Pollution Simulation Model of the Arctic Basin including the set of anthropogenic scenarios (Krapivin 1993, 1995)
SCB	Service Control Block to manage the simulation experiment
APTM	Air Pollution Transport Model (Kondratyev et al. 2006a)
MFB	Model of the Functioning of the Biota under the conditions of energy exchange in the trophical chains of the Arctic Basin ecosystem (Legendre and Legendre 1998; Legendre and Krapivin 1992)
SS	Simulator of Scenarios describing the ice fields, the synoptical situations and changes in the hydrological regimes
MWD	Model for the Water Dynamics of the Arctic Basin (Riedlinger and Preller 1991)
MHMT	Model for Heavy Metals Transport through the food chains (Krapivin et al. 1998a)
IM	The Illumination Model (Nitu et al. 2000b)
NM	The Nutrients Model (Legendre and Krapivin 1992; Legendre and Legendre 1998; Krapivin 1996)
MPT	Model for Pollution Transport through water-exchange between the Arctic Basin and the Atlantic and Pacific Oceans (Bourke et al. 1992)
MOHT	Model of Oil Hydrocarbon Transport to the food chains (Payne et al. 1991)
MPRT	Model for the Process of Radionuclide Transport to the food chains (Thiessen et al. 1999; Krapivin 1995)
MRFAB	Model of River Flow to the Arctic Basin (Krapivin et al. 1998a)
MWSD	Model of Water Salinity Dynamics (Nitu et al. 2000b)
MEF	Model for Energy Flow transport in the Arctic Basin ecosystem (Hrol 1993)
MWT	Model for the Water Temperature calculation (Nitu et al. 2000b)
ICI	Interface for Control of the Identifiers
ICD	Interface for Control of the Database
ICV	Interface for Control of the Visualization

their prey's carbon into their own tissues or into sinking fecal pellets. Along the way, some carbon dioxide escapes back to the atmosphere through the organisms' respiration. General representation of the Arctic ecosystem is shown in Figs. 6.4 and 6.5.

Let us designate the Arctic Basin aquatory as $\Omega = \{(\varphi, \lambda)\}$, where φ and λ are latitude and longitude, respectively. Spatial inhomogeneity of the Arctic Basin model is provided for by the set of cells ΔT_{CO_2} with latitude and longitude steps of $\Delta\varphi$ and $\Delta\lambda$, respectively. These cells are the basic spatial structure of Ω for the realization of the computer algorithms. The cells Ω_{ij} are heterogeneous as to their parameters and functioning. There are a set of cells which are adjacent to the river mouths (Ω_R) and to the ports (Ω_P), bordering on the land (Ω_Γ), in the Bering Strait

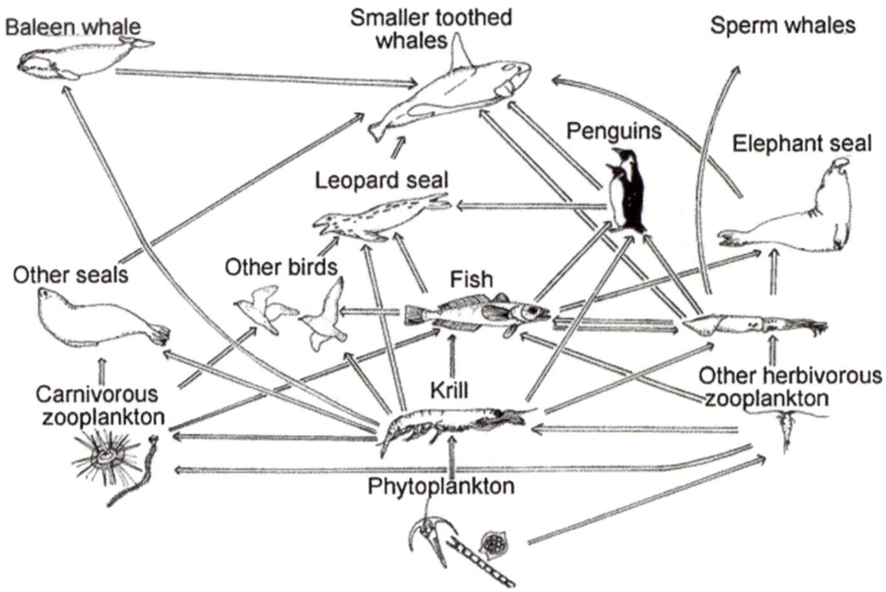

Fig. 6.4 Arctic marine food web

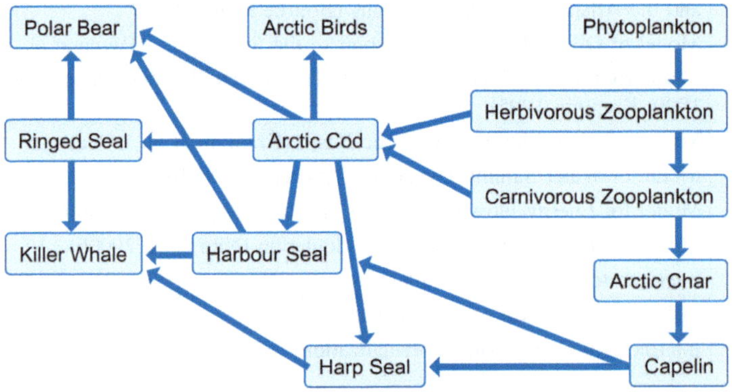

Fig. 6.5 The scheme of trophic loads between the elements of the Arctic ecosystem

(Ω_B) and on the south boundary of the Norwegian Sea (Ω_N). The aquatory Ω is divided in depth z by step Δz. The distribution of depths is given as the matrix $H = \|h_{ij}\|$ where $h_{ij} = H(\varphi_i, \lambda_j)$, $(\varphi_i, \lambda_j) \in \Omega_{ij}$. As a result, the full water volume of Ω is divided into volumetric compartments

$$\Xi_{ijk} = \{(\varphi, \lambda, z) | \varphi_i \leq \varphi \leq \varphi_{i+1};\ \lambda_j \leq \lambda \leq \lambda_{j+1};\ z_k \leq z \leq z_{k+1}\}$$

with volumes $\sigma_{ijk} = \Delta\varphi_i \Delta\lambda_j \Delta z_k$. Within Ξ_{ijk} the water body is considered as a homogeneous structure. The water temperature, salinity, density and biomass of the compartments Ξ_{ijk} are described by box models. The anthropogenic processes on the aquatory Ω are described for the four seasons: τ_w—winter, τ_s—spring, τ_u—summer, τ_a—autumn.

The procedure of spatial discretization is provided for via the ICI block of the SSMAE database, including the set of identifiers $A_k = \left\| a_{ij}^k \right\|$, where a_{ij}^k is a specific symbol to identify a real element of Ω_{ij} in the computer memory. Identifier A_1 reflects the spatial structure of the Arctic Basin and adjoining territories ($a_{ij}^1 = 0$ for $(\varphi_i, \lambda_j) \notin \Omega$; $a_{ij}^1 = 1$ for $(\varphi_i, \lambda_j) \in \Omega$ when (φ_i, λ_j) belongs to the land (islands), and a_{ij}^1 equals the aquatory identifier symbol from the second column of Table 6.5 when (φ_i, λ_j) belongs to a given sea.

Identifier A_2 shows the position of the cells Ω_R, Ω_P, Ω_N, Ω_S, Ω_Γ and describes the spatial distribution of the pollutant sources. Other identifiers A_k are used to describe ice fields ($k = 3$), the spatial distribution of solar radiation ($k = 4$) and the dislocation of upwelling zones ($k = 5$).

The user of the SSMAE in free-running mode may choose different ways to describe the many input parameters. Blocks ICI and ICD realize the on-line entry to A_k and to the database. For example, if the user has data about the spatial distribution of ice fields in Ω, he can form identifier A_3 with $a_{ij}^3 = 0$ for an ice-free water surface, $a_{ij}^3 = 1$ for new ice and $a_{ij}^3 = 2$ for old ice. In this case block *SS* enables the input of data from the climate model concerning the ice fields.

The block structure of the SSMAE is provided for with a C++ program. Each of the blocks from Table 6.4 is a C++ function. The main function provides the interactions between the SSMAE, SGM and CM. This functional specification supports overlapping output and input streams of the SSMAE blocks. With the conversational mode the user can toggle the data streams between the slave blocks.

Table 6.5 Initial data for the distribution of pollutants in the Arctic aquatories at the moment t_0

Aquatory	Identifier	Concentration			
		Radionuclides (Bq/l)		Heavy metals (mg/l)	Oil hydrocarbons (mg/l)
		^{137}Cs	^{60}Co		
Greenland Sea	Γ	0.05	0.05	0.5	0.2
Norwegian Sea	*N*	0.05	0.05	0.7	0.4
Barents Sea	*B*	0.07	0.07	0.8	0.6
Kara Sea	*K*	0.1	0.1	1.0	0.4
White Sea	∇	0.1	0.1	1.1	0.4
Laptev Sea	Λ	0.05	0.05	0.9	0.5
E. Siberian Sea	*E*	0.01	0.01	0.9	0.5
Bering Sea	*S*	0.02	0.02	0.8	0.7
Chukchi Sea	*X*	0.01	0.01	0.8	0.6
Beaufort Sea	Φ	0.05	0.05	0.7	0.2
Central Basin	*U*	0	0	0.1	0.1

The calculating procedure is based on the subdivision of the Arctic Basin into boxes $\{\Xi_{ijk}\}$. This is realized by means of a quasi-linearization method (Nitu et al. 2000a). All differential equations of the SSMAE are substituted for in each box Ξ_{ijk} by easily integrable ordinary differential equations with constant coefficients. The water motion and turbulent mixing are realized in conformity with current velocity fields which are defined on the same coordinate grid as the $\{\Xi_{ijk}\}$ (Krapivin et al. 1998a).

6.4.3 Description of Ecological Processes

The ice-water ergocline plays an important role in the biological productivity of the north seas. According to the hypothesis of Legendre and Legendre (1998), energetic ergoclines are the preferential sites for biological production in the Arctic Ocean. The primary production in the food chains of the Arctic Basin ecosystems is determined by the phytoplankton productivity. This is connected with complex variations in the meteorological, hydrodynamic, geochemical and energetic parameters of the sea environment. The problem of parameterization of the phytoplankton production in the north seas was studied by Legendre and Legendre (1998). Table 6.6 gives the structure for the seasonal composition of conditions affecting the primary production in Ω. This scheme is applied for each of the Ξ_{ijk}.

The block MWT calculates the water temperature T_w by means of the averaging of temperatures for the mixed water volumes. In addition the following correlations are applied:

$$T_g = T_r = T_f = \begin{cases} -0.024b + 0.76T_0 + 8.38 & \text{when} \quad b \leq 50\,\text{cm} \\ -0.042\,b + 0.391T_0 - 0.549 & \text{when} \quad b < 50\,\text{cm} \end{cases} \tag{6.1}$$

where $b = g + r + f$, T_0 is the surface temperature, g is the snow depth, r is the thickness of the floating ice and f is the depth of the submerged ice below the water surface. If we designate by ρ_g, ρ_r and ζ the density of snow, ice and seawater respectively, we obtain for the depth of ice beneath the surface:

Table 6.6 The vertical structure of the Arctic Basin's water bodies

Layer (A)	Δz	Parameters of the layer				
		T_A	E_A	k_A	α_A	β_A
Surface	0	T_0	E_0			
Snow	g	T_g	E_g		α_g	β_g
Floating ice	r	T_r	E_r		α_r	β_r
Drowned ice	f	T_f	E_f		α_f	β_f
Water	$z - f$	T_W	E_W	k_W	α_W	β_W

Δz layer thickness, T_A temperature, E_A illumination, k_A turbulence coefficient, α_A coefficient of illumination attenuation, β_A coefficient of light reflection

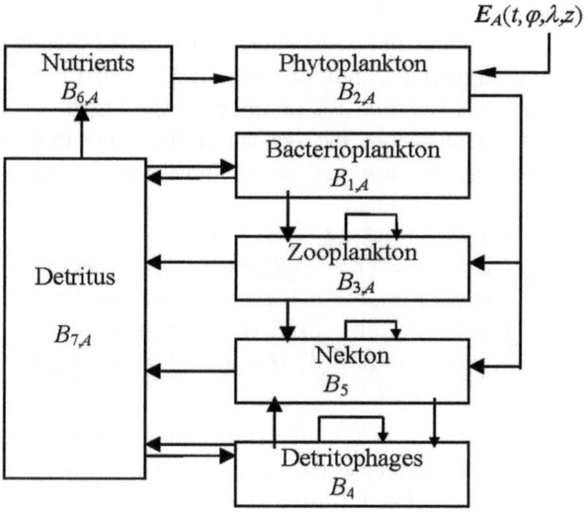

Fig. 6.6 Block-diagram of the energy flows (cal m^{-3} day^{-1}) in the trophic pyramid of the Arctic Basin ecosystem

$$f = (g\rho_g + r\rho_r)/(\zeta - \rho_r)$$

Figures 6.6 and 6.7 show a conceptual flowchart of the energy in an ecological system. The energy input during time interval t is provided by solar radiation $E_A(t, \varphi, \lambda, z)$ and the upward transport of nutrients from the deep-sea layers.

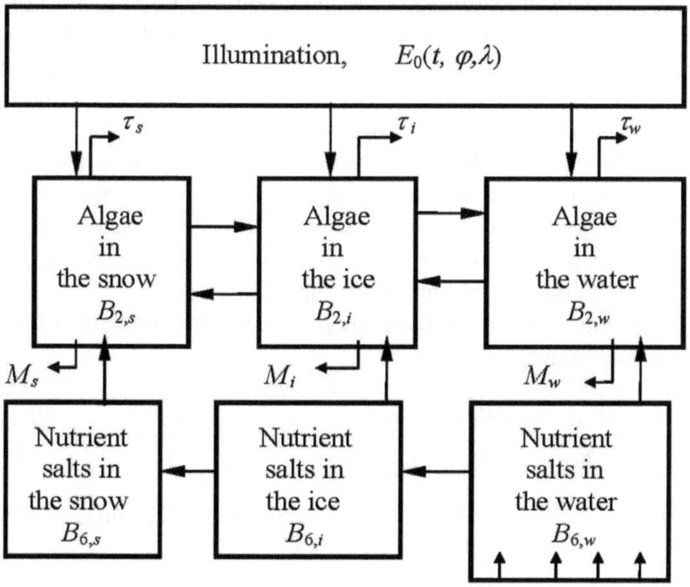

Fig. 6.7 Block-diagram of the energy flows (cal m^{-3} day^{-1}) in the snow-ice-water interface

The concentration of nutrients $B_{6,A}$ $(t,\ \varphi,\ \lambda,\ z)$ at the depth z is determined by photosynthesis R_{pA}, advection and destruction of suspended dead organic matter B_7. The role of the hydrodynamic conditions manifests itself in maintenance of the concentration of nutrients required for photosynthesis which occurs via transport from other layers or aquatories of the sea where the concentration is sufficiently high. Taking into account the designations of Table 6.5 we have:

$$E(t,\varphi,\lambda,z) = \begin{cases} E_0(t,\varphi,\lambda) \quad \text{when } z \leq -(g+r), \\ E_g(t,\varphi,\lambda,z) = (1-\beta_g)E_0\exp(\alpha_g z) \quad \text{when } z\in[-(g+r),-r], \\ E_r(t,\varphi,\lambda,z) = (1-\beta_r)E_g(t,\varphi,\lambda,-r)\exp(\alpha_r z) \quad \text{when } z\in[-r,0], \\ E_f(t,\varphi,\lambda,z) = (1-\beta_f)E_r(t,\varphi,\lambda,0)\exp(-\alpha_f z) \quad \text{when } z\in[0,f], \\ E_w(t,\varphi,\lambda,z) = (1-\beta_w)E(t,\varphi,\lambda,f)\exp(-\alpha_w z) \quad \text{when } z>f. \end{cases}$$

where the values of α_A (A = g, r, f, w) depend on the optical properties of the Ath medium. The irradiance E_0 arrives at the surface Ω. The estimate of E_0 is obtained from monitoring or is calculated from the climatic model. The flow of E_0 is attenuated by snow, ice and water according to the scheme of Table 6.5. In each cell Ω_{ij} the structure of these layers is changed corresponding to the time of year. Within each layer, the attenuation of the irradiance with depth is described by an exponential model. The parameters α_A and β_A are functions of the salinity, turbidity, temperature, and biomass. The form of this dependence is given as a scenario or else the standard functions are used (block IM).

Block-diagram of Fig. 6.6 is realized as block MEF. The boxes with elements denote the generalized trophic levels of the arctic ecosystem. All of the elements are described by means of averaged parameters for the biological community of the north seas. It is supposed that this trophic pyramid takes place in each of the Arctic Basin seas. The trophic relations between the elements of the model are described on the basis of the energetic principle (Nitu et al. 2000b). Biomasses, the rates of production and exchange (respiration), and food rations are expressed in energy units. Total nitrogen serves as a "nutrient salts" prototype in the model.

As a basic scheme for the flow of nutrients in the water, the scheme proposed by Krapivin (1996) is accepted as corrected for the conditions of the Arctic Basin by Legendre and Legendre (1998). It is supposed that the spatial distribution of upwelling zones is given with seasonal variations. Block NM realizes this scheme in reference to the current structure of the upwelling regions.

The dynamic equation for the nutrients $B_{6,A}$ in the environment, where A = {S—snow, I—ice, W—water}, is given by

$$\frac{\partial B_{6,A}}{\partial t} + v_\varphi^A\frac{\partial B_{6,A}}{\partial \varphi} + v_\lambda^A\frac{\partial B_{6,A}}{\partial \lambda} + v_z^A\frac{\partial B_{6,A}}{\partial z} = Q_A + k_2^W\frac{\partial^2 B_{6,A}}{\partial^2 z} + \beta_V\frac{\partial B_{6,A}}{\partial z} + \rho_1\sum_{i=1}^{5}T_i - \delta R_{pA} + \varepsilon_1^A H_1^\varepsilon \tag{6.2}$$

where $v_\varphi^A, v_\lambda^A, v_z^A$ are velocity projections of motion in the environment; Q_A is the input of biogenic elements to A resulting from the decomposition of detritus $(Q_A = \delta_n R_D^A)$ with $R_D^A = \mu_A B_7$; δ_n is the content of nutrients in the dead organic matter; μ_A is the rate of decomposition of detritus into the environment A; k_2^W is the kinematical coefficient of vertical diffusion; δ_1 is the velocity of nutrient assimilation by the photosynthetic process per unit of the phytoplankton production; ε_1^A is the proportional part of the ε-th radionuclide which is chemically analogous to $B_{6,A}$ on substrate A; H_1^ε is the rate of input flow of the ε-th radionuclide; T_i is the rate of exchange with the environment; ρ_1 is the part of the biomass losses due to exchange which transform into nutrients (Legendre and Legendre 1998); and β_V is the upwelling velocity. Equation (6.1) is the basic element of block NM.

As a basic scheme for the flow of nutrients in the water, the scheme proposed by Krapivin (1996) is accepted as corrected for the conditions of the Arctic Basin by Legendre and Legendre (1998). It is supposed that the spatial distribution of upwelling zones is given with seasonal variations. Block *NUM* realizes this scheme in reference to the current structure of the upwelling regions.

The irradiance E_0 arrives at the surface of Ω. The estimate of E_0 is obtained from monitoring or is calculated from the climatic model. The flow of E_0 is attenuated by snow, ice and water according to the scheme of Table 6.4. In each cell Ω_{ij} the structure of these layers is changed corresponding to the time of year. Within each layer, the attenuation of the irradiance with the depth is described by an exponential model (Legendre and Krapivin 1992). The parameters α_A and β_A are functions of the salinity, turbidity, temperature, and biomass. The form of this dependence is given as a scenario or else the standard functions are used (block IM).

The phytoplankton production R_{pA} in the environment A is a function of the solar radiation E_A, the concentration of nutrients n_A, the temperature T_A, the phytoplankton biomass p_A, and the concentration of pollutants ξ_A. There are many models for the description of the photosynthesis process (Legendre and Legendre 1998; Legendre and Krapivin 1992). For the description of this function in the present study, the equation of the Michaelis-Menten type is used (block MFB):

$$R_{pA} = a_A k_I^A p_{A,max} / \left(E_A + K_I^A\right) \tag{6.3}$$

where K_I^A is the irradiance level at which $R_{pA} = 0.5 R_{pA,\max}$, and $p_{A,\max}$ is the maximum quantum yield (Legendre and Legendre 1998). The coefficient a_A reflects the dependence of the phytoplankton production on the environment temperature T and the concentration of nutrients $B_{6,A}$. The block MFB realizes the following equation for the calculation of a_A:

$$a_A = a_1 K_0(T,t) / \left[1 + B_{2,A} / (a_2 B_{6,A})\right] \tag{6.4}$$

Here a_1 is the maximal rate of the absorption of nutrients by the phytoplankton (day^{-1}), a_2 is the index of the rate of saturation of photosynthesis, and

$$K_0(T,t) = a_3 \max\left\{0, \frac{T_c - T}{T_c - T_{opt}} \exp\left[1 - \frac{T_c - T}{T_c - T_{opt}}\right]\right\} \tag{6.5}$$

where a_3 is the weight coefficient, and T_c and T_{opt} are the critical and optimal temperatures for photosynthesis, respectively (°C).

Equation (6.3) adequately fits the laboratory data. The relationships (6.4) and (6.5) make the description of the phytoplankton production more accurate for critical environmental conditions when the concentration of nutrients and the temperature have high fluctuations. The coefficients of these relationships are defined on the basis of estimates given by Legendre and Legendre (1998).

The dynamic equation for the phytoplankton biomass in the environment A has the following form:

$$\begin{aligned} &\partial B_{2,A}\Big/\partial t + v_\varphi^A \partial B_{2,A}/\partial\varphi + v_\lambda^A \partial B_{2,A}/\partial\lambda + v_z^A \partial B_{2,A}/\partial z \\ &= R_{pA} - \theta_p^A - M_p^A + k_2^A \partial^2 B_{2,A}/-\partial z - \left[k_{Zp}^A R_{ZA}\Big/\xi_Z^A + k_{Fp}^A R_{FA}\Big/\xi_F^A\right] B_{2,A} \end{aligned} \tag{6.6}$$

where $R_{ZA}\left(\xi_Z^A\right)$ and $R_{ZA}\left(\xi_F^A\right)$ are the production (the food spectrum) of zooplankton B_3 and nekton B_5, respectively; M_p^A is the mortality; and θ_p^A is the rate of exchange. The balance equations for the other ecological elements of Fig. 6.6 are given by Krapivin (1995, 1996).

The energy source for the entire system issolar radiation energy E_A (t, φ, λ, z), the intensity of which depends on time t, latitude φ, longitude λ, and depth z. The equations that describe the biomass dynamics of the living elements are:

$$\begin{aligned} \frac{\partial B_i}{\partial t} + \xi_i\left[V_\varphi \frac{\partial B_i}{\partial\varphi} + V_\lambda \frac{\partial B_i}{\partial\lambda} + V_z \frac{\partial B_i}{\partial z}\right] &= R_i - T_i - M_i - H_i - \sum_{j\in\Gamma_i} C_{ij} R_j + \xi_i\left[\frac{\partial}{\partial\varphi}\left(\Delta_\varphi \frac{\partial B_i}{\partial\varphi}\right)\right. \\ &\left. + \frac{\partial}{\partial\lambda}\left(\Delta_\lambda \frac{\partial B_i}{\partial\lambda}\right) + \frac{\partial}{\partial z}\left(\Delta_z \frac{\partial B_i}{\partial z}\right) + \beta_V \frac{\partial B_i}{\partial z}\right], (i = 1,3,4,5); \end{aligned} \tag{6.7}$$

where $V(V_\varphi, V_\lambda, V_z)$ are the components of the water current velocity $\left(V_\varphi = v_\varphi^W,\right.$ $\left.V_\lambda = v_\lambda^W, V_z = v_z^W\right)$; R_i is the production; M_i is the mortality; H_i is non-assimilated food; and Γ_i is the set of the trophic dependence of the i-th component: $C_{ji} = k_{ji}F_i/\sum_{m\in S_i} k_{jm}F_m$; S_i is the food spectrum of the j-th component; k_{jm} is the index of satisfaction of the nutritive requirements of the j-th component at the expense of the m-th component biomass; $F_i = \max\{0, B_i - B_{i,min}\}$; $B_{i,min}$ is the minimal biomass of the i-th component consumed by other trophic levels; $\Delta(\Delta_\varphi, \Delta_\lambda, \Delta_z)$ are components of the turbulent mixing coefficient (on the assumption of isotropism of vertical mixing in the horizontal plane $\Delta_\varphi = \Delta_\lambda = \Delta_H$); β_V is the upwelling velocity. Functions R_i, M_i, H_i and T_i are parametrized according to the models by Krapivin (1996) and Legendre and Krapivin (1992). The equations describing the dynamics of the abiotic elements are represented in conformity with Kondratyev and Krapivin (2001b). Functions M_4 and M_5 include the biomass loses at the expense of fishing. Parameter ξ_i characterizes the subjection of the i-th component relative to the current. It is supposed that $\xi_i = 1$ for $i = 1, 2, 3$ and $\xi_i = 0$ for $i = 4, 5$.

The inert components are described by the following equations (Krapivin 1996):

$$\begin{aligned}\frac{\partial B_7}{\partial t} + V_\varphi \frac{\partial B_7}{\partial \varphi} + V_\lambda \frac{\partial B_7}{\partial \lambda} + V_z \frac{\partial B_7}{\partial z} &= \sum_{i=1}^{5} (M_i + H_i) - \mu_W B_7 - (v* - \beta_V)\partial B_7/\partial z \\ &\quad - \left(k_{1,7}R_1/{+P_1} + k_{3,7}R_3/P_3 + k_{4,7}R_4/P_4\right) + k_{5,7}R_5/P_5\big)B_{7,min} \\ &\quad + \frac{\partial}{\partial \varphi}\left(\Delta_\varphi \frac{\partial B_7}{\partial \varphi}\right) + \frac{\partial}{\partial \lambda}\left(\Delta_\lambda \frac{\partial B_7}{\partial \lambda}\right) + \frac{\partial}{\partial z}\left(\Delta_z \frac{\partial B_7}{\partial z}\right)\end{aligned} \tag{6.8}$$

where $P_i = \sum_{j \in S_i} k_{i,j} B_{j,\min}$; μ_W is the velocity of decomposition of detritus per unit of biomass; v_* is the velocity of settling due to gravity; and k_{ij} is a coefficient of the relation of the i-th element with respect to the j-th element of the ecosystem.

The Eqs. (6.2)–(6.8) are used in the complete volume only when $(\varphi, \lambda, z) \in W$. In the other cases (in the layers of ice or snow) these equations are automatically reduced in accordance with the scheme represented in Fig. 6.7.

6.4.4 Description of the Hydrologic Cycle

The circulation of the waters in the Arctic Basin is a complex system of cycles and currents with different scales. Block HB simulates the dynamics of the Arctic Basin waters by a system of sub-blocks presented in Fig. 6.3. The water dynamics in Ω is presented by flows between the compartments Ξ_{ijk}. The directions of water exchanges are represented on every level $z_k = z_0 + (k-1)\Delta z_k$ according to Aota et al. (1992) in conformity with the current maps assigned as the SSMAE input. The external boundary of Ω is determined by the coast line, the bottom, the Bering strait, the south boundary of the Norwegian Sea, and the water-atmosphere boundary.

The hydrological data are synthesized via a four-level structure according to the seasons (block MWD). The velocity of current in the Bering Strait is estimated by the following binary function:

$$V(t) = \begin{cases} V_1 \text{ for } t \in \tau_u \cup \tau_a \\ V_2 \text{ for } t \in \tau_w \cup \tau_s \end{cases}$$

The water exchange through the south boundary of the Norwegian Sea is V_3. The water temperature T^W_{ijk} in Ξ_{ijk} the (block MWT) is a function of evaporation, precipitation, river flows and inflows of water from the Atlantic and Pacific Oceans. Its change with time in Ξ_{ijk} is described by the equation of heat balance:

$$\zeta C\sigma_{ijk}\frac{\partial T_{ijk}^{W}}{\partial t} = \sum_{s,l,m}\left(W_{slm}^{ijk} + f_{slm}^{ijk}\right) - W_{ijk} \qquad (6.9)$$

where ζ is the seawater density (g cm^{-3}); C is the water thermal capacity, (cal g^{-1} grad^{-1}); σ_{ijk} is the volume of Ξ_{ijk}; W_{slm}^{ijk} is the heat inflow to Ξ_{ijk} from Ξ_{slm}; f_{slm}^{ijk} is the heat exchange between Ξ_{slm} and Ξ_{ijk} caused by turbulent mixing; and W_{ijk} is the total heat outflow from Ξ_{ijk} to the bordering boxes. Heat exchange with the atmosphere is calculated in accordance with empirical Eq. (6.1).

It is considered that the dissipation of moving kinetic energy, geothermic flow on the ocean bed, heat effects of chemical processes in the ocean ecosystem, and freezing and melting of the ice are not global determinants of the water temperature fields. The SSMAE does not consider these effects.

The dynamics of the water salinity $S(t, \varphi, \lambda, z)$ during the time interval t are described by the balance equation as block MWSD. The ice salinity is defined by a two-step scale: s_1—old, s_2—new. It is supposed that $S(t, \varphi, \lambda, z) = s_0$ for $z > 100\,\text{m}$, $s_2 = k_s S(t, \varphi, \lambda, f)$ for $r + f > H_{\max}$ and $s_1 = k_r s_2 H_{\max}/(r + f)$ for $r + f < H_{\max}$ where coefficients k_s and k_r are determined empirically and $H_{\max}$ is the maximal thickness of new ice. In accordance with the estimations by Krapivin (1995), the simulation experiments are realized for $H_{\max} = 50$ cm, $k_s = k_r = 1$. The river flows, ice fields and synoptic situations are described by scenarios given in the MRF and SS blocks and formed by the user of the SSMAE.

The snow-layer thickness $g(t, \varphi, \lambda)$ may be described via statistical data with given dispersion characteristics: $g = \bar{g} + g_0$ where the value $\bar{g}$ is defined as the mean characteristic for the chosen time interval and the function $g_0(t, \varphi, \lambda)$ gives the variation of g for the given time interval.

An alternative description is the parameterization of the snow-layer dynamics process in the framework of the atmospheric process simulation algorithm (block APTM) relating the thickness of the growth and melting of the snow-layer to the temperature and precipitation:

$$g(t + \Delta t, \varphi, \lambda) = g(t, \varphi, \lambda) + S_F - S_M,$$

where S_F is the part of the snow precipitated at temperatures close to freezing (265 K $\leq T_0 \leq$ 275 K) and S_M is the snow ablation (i.e. evaporation + melting).

Block SS gives to the user the possibility to select between these algorithms. When statistical data on the snow-layer thickness exists, the function $g(t, \varphi, \lambda)$ is reconstructed for $(\varphi, \lambda) \in \Omega$ by means of the approximation algorithm at the time and polynomial interpolation in the space (Krapivin 2000c, d; Nitu et al. 2000b).

6.4.5 Simulation of Dispersion Processes for the Water Pollutants

The block PSM simulates the pollution processes over the territory Ω as a result of the atmosphere transport, river and surface coastal outflow, navigation, and other human activity. The variety of pollutants is described by three components: oil hydrocarbons O, heavy metals (e—solid particles, ψ—dissolved fraction) and radionuclides ε. It is supposed the pollutants get into the living organisms only through the food chains.

The rivers bring a considerable contribution to the level of pollution of the arctic waters. The concentration of pollutant κ in the river γ is γ_κ. The pollutant κ enters compartment $\Omega_{ijk} \in \Omega_R$ with the velocity c_γ reflecting the mass flow per unit time. Subsequently, the spreading of the pollutant in Ω is described by other sub-blocks.

The influence of water exchanges between the Arctic Basin and the Pacific and Atlantic Oceans on the pollution level in Ω is described by block MPT. It is supposed that the watersheds of the Norwegian Sea Ω_N and the Bering Strait Ω_B are characterized by currents with varying directions given as a scenario.

The atmospheric transport of heavy metals, oil hydrocarbons and radionuclides is described by many models (Phillips et al. 1997; Payne et al. 1991; Sportisse 2000). The application of these models to the reconstruction of the pollution distribution over Ω makes it possible to estimate optimal values of $\Delta\varphi$, $\Delta\lambda$ and the step in time Δt. The present level of the database for the Arctic Basin provides for use of a one-level Euler model with Δt = 10 days, $\Delta\varphi = \Delta\lambda = 1°$ (block APTM). It is supposed that pollution sources can be located on the Arctic basin boundary. Their detailed distributions are given as the SSMAE input. The transport of pollutants to the Arctic Basin and the formation of their spatial distribution are realized in conformity with the wind velocity field, which is considered as given (Krapivin and Phillips 2001b).

It is postulated that the oil hydrocarbons $O(t, \varphi, \lambda, z)$ are transformed by the following processes (Payne et al. 1991): dissolving H_O^1, evaporation H_O^2, sedimentation H_O^3, oxidation H_O^4, biological adsorption H_O^5, bio-sedimentation H_O^6, and bacterial decomposition H_O^7. The kinetics equation for the description of the dynamics of the oil hydrocarbons in the Arctic Basin is given by

$$\partial O/\partial t + v_\varphi^w \partial O/\partial\varphi + v_\lambda^w \partial O/\partial\lambda + v_z^w \partial O/\partial z = Q_O + k_2^w \frac{\partial^2 O}{\partial z_2} - \sum_{i=1}^{7} H_O^i, \quad (6.10)$$

where Q_O is the anthropogenic source of the oil hydrocarbons.

The process of the diffusion of the heavy metals in the seawater depends on their state. The dissolved fraction of the heavy metals (ψ) takes part in the biogeochemical processes more intensively than the suspended particles (e). But as suspended particles, the heavy metals fall out more rapidly to the sediment. A description of the entire spectrum of these processes in the framework of this study is impossible. Therefore, block MHMT describes the processes which have estimates. The transport of heavy metals in the sea water includes absorption of the dissolved fraction ψ by plankton $\left(H_Z^{\psi}\right)$ and by nekton $\left(H_F^{\psi}\right)$, sedimentation of the solid fraction $\left(H_1^e\right)$, deposition with the detritus $\left(H_D^{\psi}\right)$, adsorption by detritophages from bottom sediments $\left(H_L^{e\psi}\right)$, and release from bottom sediments owing to diffusion $\left(H_a^{e\psi}\right)$. As a result, the dynamic equations for the heavy metals become:

$$\frac{\partial e_W}{\partial t} + v_{\varphi}^W \frac{\partial e_W}{\partial \varphi,} + v_{\lambda}^W \frac{\partial e_W}{\lambda} + v_z^W \frac{\partial e_W}{\partial z} = \sum_{i=1}^{3} \alpha_2^i Q_{e\Psi}^i - H_1^e + \alpha_1 H_a^{e\Psi} \tag{6.11}$$

$$\begin{aligned}\frac{\partial \psi_W}{\partial t} + v_{\varphi}^W \frac{\partial \psi_W}{\partial \varphi} + v_{\lambda}^W + \frac{\partial \psi_W}{\partial \lambda} + v_z^W \frac{\partial \psi_W}{\partial z} + v_{\varphi}^W &= (1 - \alpha_1) H_a^{e\psi} + k_2^W \frac{\partial^2 \psi_W}{\partial z^2} \\ &\quad - H_Z^{\psi} - H_F^{\psi} - H_D^{\psi} - H_a^{\psi}\end{aligned} \tag{6.12}$$

$$\frac{\partial e^*}{\partial t} = H_1^e - \alpha_1 \left(H_L^{e\psi} + H_a^{e\psi}\right) \tag{6.13}$$

$$\frac{\partial \psi^*}{\partial t} = H_D^{\psi} - (1 - \alpha_1)(H_L^{e\psi} + H_a^{e\psi}) \tag{6.14}$$

where e_w, ψ_w and e^*, ψ^* are the concentrations of heavy metals in the water and in the bottom sediments as solid and dissolved phases, respectively; H_a^{ψ} is the output of heavy metals from the sea to the atmosphere by evaporation and spray; $Q_{e\psi}^i$ is the input of heavy metals to the sea with river waters (i = 1), atmospheric deposition (i = 2) and ship's wastes (i = 3); α_2^i is the part of the suspended particles in the i-th flow of heavy metals; and α_1 is the part of the solid fraction of heavy metals in the bottom sediments.

Each radionuclide of ε-th type is characterized by the half-life τ^{ε}, the rate H_1^{ε} of input flow to the water area Ω, the accumulation rate H_{α}^{ε} in the living organisms $\alpha(p_A, B_A, Z, F, L)$ and the removal rate H_D^{ε} with the dead elements of the ecosystem. As a result, the concentration Q^{ε} of the radionuclide ε in Ω_{ijk} is described by the following system of equations:

$$\begin{aligned}\frac{\partial Q^{\varepsilon}}{\partial t} + v_{\varphi}^W \frac{\partial Q^{\varepsilon}}{\partial \varphi} + v_{\lambda}^W \frac{\partial Q^{\varepsilon}}{\partial \lambda} + v_z^W \frac{\partial Q^{\varepsilon}}{\partial z} &= \frac{\sigma_{ijk}}{\sigma} H_1^{\varepsilon} + k_2^W \frac{\partial^2 Q^{\varepsilon}}{\partial z^2} - H_{\alpha}^{\varepsilon} \\ &\quad - H_D^{\varepsilon} - \frac{\text{In}2}{\tau^{\varepsilon}} Q^{\varepsilon} + H_*^{\varepsilon},\end{aligned} \tag{6.15}$$

$$\frac{\partial Q_*^{\varepsilon}}{\partial t} = H_D^{\varepsilon} - H_*^{\varepsilon} - \frac{\ln 2}{\tau^{\varepsilon}} Q_*^{\varepsilon} \tag{6.16}$$

where Q_*^ε is the concentration of ε-th radionuclide in the bottom sediments and H_*^ϕ is the rate of the output flow of the ε-th radionuclide from the bottom sediments via desorption. The exchange of radionuclides between the water layers by migration of living elements is ignored as it has a small value in comparison with the flow H_D^ε.

6.4.6 Simulation Results

6.4.6.1 The Assumptions

The SSMAE allows for the estimation of the pollution dynamics of the Arctic Basin under various a priori suppositions about the intensities of the flows of pollutants and under other anthropogenic impacts on the ecosystems of this region. Here, some of the possible situations are considered. The thermal regime of the Arctic Basin is given by a normal distribution with average temperatures and with dispersions on the aquatories as given by the SEDAAR (Strategic Environmental Distributed Active Archive Resource). The scheme of transport of pollutants in the atmosphere is adopted from Christensen (1997). The estimates of parameters for the blocks of Table 6.7 are given by literature sources or personal recommendations.

The vertical distribution of pollutants in the initial moment t_0 is taken as homogeneous. The average diameters of the solid particles are estimated to be in the range from 0.12 to 1,000 μm and the vertical velocity of sedimentation is 0.003 m/s. The concentration of nutrients in the ice and snow equals 0. Also it is supposed that the deep water temperature is defined as $Y(t, \varphi, \lambda) = 0$ °C and the surface ice temperature as $f_1(t, \varphi, \lambda) = -3$ °C for $(\varphi, \lambda) \in \Omega$. It is supposed that $\varepsilon_1^A = 0$ and the phytoplankton productivity in the ice layer is 2.5 % of the primary production in the water column $[(R_{p,r} + R_{pf})/R_{p,w} = 0.025]$.

Let the ratio between solid and dissolved phases of heavy metals at the moment $t = t_0$ equal 1:2, i.e. $e(t_0, \varphi, \lambda, z)/\psi(t_0, \varphi, \lambda, z) = 0.5$. The flows of heavy metals, $H_Z^\psi, H_F^\psi, H_D^\psi$ and H_L^ψ, are described by linear models, $H_1^e = 0.01\, e_w$, $H_a^{e\psi} = 0$. The boundaries of the Norwegian and Bering Seas are approximated by lines with $\varphi_N = 62$°N and $\varphi_B = 51$°N, respectively. Values for the other parameters are defined by Wielgolaski (1997), Wania et al. (1998), Valette-Silver (1999), Preller and Cheng (1999), Bard (1999) and Rudels et al. (1991). The initial data are defined in Tables 6.5 and 6.8.

6.4.6.2 The Dynamics of Arctic Basin Radio Nuclear Pollution

The intensity of external flows through the boundaries of the Arctic Basin and the internal flows due to dead organisms H_D^ε, sediment H_*^ε and living organisms H_α^ε are described by linear models in accordance with Krapivin and Phillips (2001b). Some results of the simulation experiment are given in Figs. 6.8, 6.9 and 6.10. Figure 6.8

Table 6.7 The values of some parameters in the framework of the simulation experiments using the SSMAE

Parameter	Estimate of the parameter
Step of space digitizing by	
Latitude, $\Delta\varphi$	1°
Longitude, $\Delta\lambda$	1°
Depth, Δz	
$z \leq 100$ m	1 m
$z > 100$ m	h-100 m
Coefficient of the ice heat conductivity, λ_1	2.21 W/m/°
Coefficient of the water heat conductivity, λ_2	0.551 W/m/°
Coefficient of turbulent mixing, k_2^W	
For open waters	0.5×10^{-4} m^2/s
For ice-covered water bodies	5×10^{-6} m^2/s
Characteristic heat of ice melting, q	334 kJ/kg
Content of biogenic elements in dead organic matter, Δ_n	0.1
Intensity of detritus decomposition, μ_A	
$A = g, r, f$	0
$A = W$	0.01
Velocity of current in the Bering Strait, V_i	
$i = 1$	0.2 m/s
$i = 2$	0.05 m/s
Water heat capacity, C	4.18 kJ/kg/K
Ice salinity, s_i	
$i = 1$	5 ‰
$i = 2$	1 ‰
Water salinity at $z > 100$ m, s_0	34.95 ‰
Area of the Arctic Basin, σ	
Half-life period of radionuclides, τ_g	16,795,000 km^2
$\varepsilon = {}^{60}$C	5.271 years
$\varepsilon = {}^{137}$Cs	30.17 years
Critical temperature for photosynthesis, T_c	−0.5 °C

shows the tendency versus time of the average content of the radio nuclear pollution on the whole Arctic water area. The distribution with depth is represented by a three layer model, upper waters ($z < 1$ km), deep waters ($z > 1$ km) and sediments. The bottom depth is taken as 1.5 km. A more realistic depth representation of the shallow seas and the deeper Arctic Basin will be considered in a future refinement of the model. The curves describe the vertical distribution with time of the radionuclide content in the two water layers and in the sediments. The transfer of radionuclides from the upper waters to the deep waters occurs with a speed which results in the reduction of radio nuclear pollution in the upper waters by 43.3 % over 20 years. These distributions for each of arctic seas are given in Table 6.9.

Table 6.8 The input flows of radionuclides, heavy metals (suspended particles *e* and dissolved fraction ψ) and oil hydrocarbons *O* to Ω by water flows taken into account in the SSMAE

Source	Flow into the basin (km³/year)	Concentration of pollutant				
		^{137}Cs (Bq/l)	^{60}Co (Bq/l)	*e* (mg/l)	Ψ (mg/l)	*O* (mg/l)
Rivers						
Yenisey	600	0.5	0.5	0.3	5.1	2.3
Ob'	400	0.1	0.1	0.4	6.9	4.7
Lena	500	0.0	0.0	1.1	8.8	6.9
Pechora	130	0.1	0.1	0.3	1.5	3.0
Northern Dvina	100	0.0	0.0	0.2	1.1	4.0
Other Siberian rivers	200	0.1	0.1	0.1	0.5	2.3
Rivers of North America	600	0.2	0.2	0.1	1.0	1.0
Evaporation	3,500	0.0	0.0	0.0	0.0	0.0
Precipitation	5,300	0.0	0.0	0.1	0.1	0.0
Southern boundary of the						
Norwegian Sea	12,000	0.1	0.1	0.6	2.2	2.4
Bering Strait	10,560	0.0	0.0	0.5	1.9	1.9

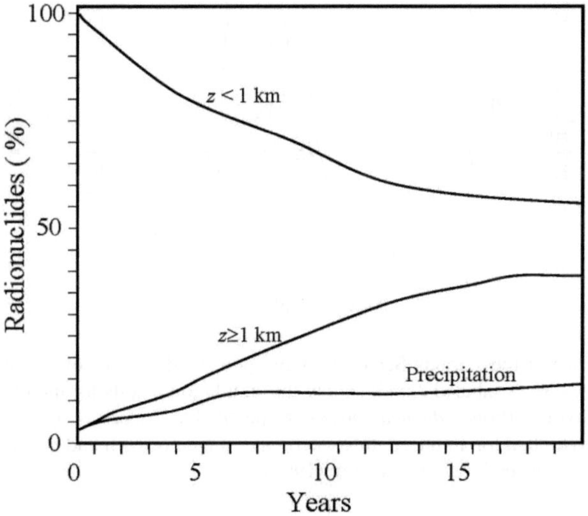

Fig. 6.8 Dynamics of the radionuclide distribution in the Arctic Basin. It is assumed that at the moment $t_0 = 0$ radionuclear pollutants (^{137}Cs, ^{60}Co; see Tables 6.4 and 6.7) are in the upper water layer $z \leq 1$ km only. The *curves* show the radionuclear pollutant distribution with time in the two water layers and in the sediments obtained by the averaging of the simulation results for all of the north seas

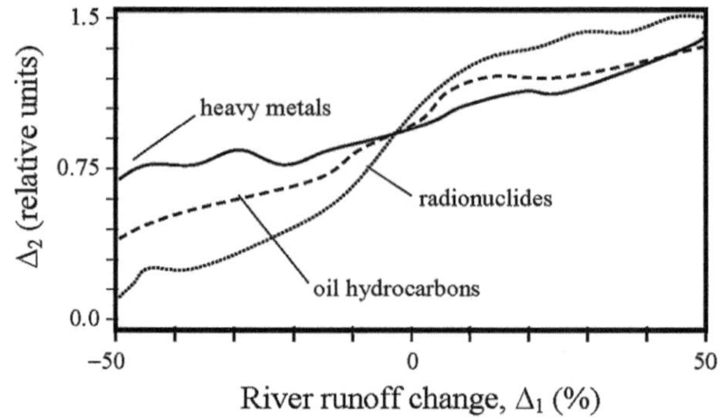

Fig. 6.9 Influence of variations in the river flows on the Arctic Basin pollution level. Here Δ_1 is the percentage variation in the river flow to Ω with respect to the value averaged on Ω_R in the last 3 years and Δ_2 is the content of the pollutant averaged on all rivers in Ω_R and normalized to the initial data (such that $\Delta_2 = 1$ for $\Delta_1 = 0$)

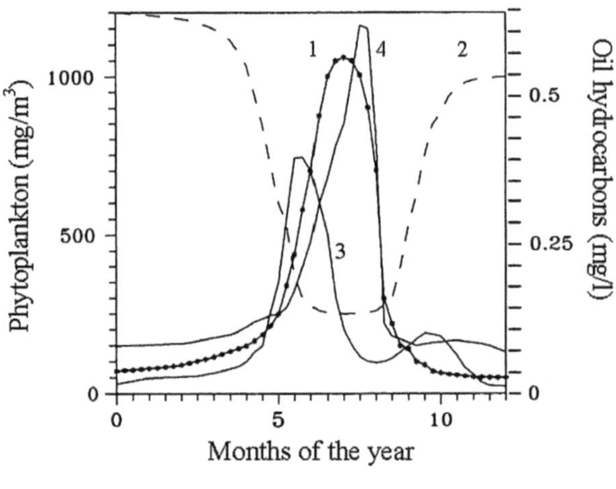

Fig. 6.10 Influence of the Barents Sea ecosystem upon the dynamics of the oil hydrocarbons in the sea water. The curves labeled *1* and *2* show the simulation results for the phytoplankton (*solid curve*) and oil hydrocarbons (*dashed curve*), respectively. Curves *3* and *4* show the yearly distribution of phytoplankton in the south-western and in the northern and north-eastern aquatories of the Barents Sea, respectively, by Terziev (1992)

Local variations in the vertical distribution of radionuclides are determined by both hydrological and ecological conditions. The correlation between these conditions is a function of the season. Table 6.10 gives the estimates of the role of the ecological processes in the formation of the vertical distribution of the radio nuclear pollution of the arctic seas. These estimates show that the biological community plays a minor role in the radionuclide transport from the upper layers to the deep ocean.

Table 6.9 Distribution of radionuclear pollution in the arctic aquatories 30 and 50 years after t_0 (in percent)

Aquatory (see Table 6.5)	Γ	N	B	K	∇	∧	E	S	X	Φ	U
30 *years after* t_0											
$z \leq 1$ km	50	60	69	46	73	44	43	57	58	61	68
$z > 1$ km	49	39	29	52	26	54	54	39	39	34	29
Bottom	1	1	2	2	1	2	3	4	3	3	3
50 *years after* t_0											
$z \leq 1$ km	65	57	70	66	70	50	49	62	59	58	70
$z > 1$ km	30	38	24	27	21	47	46	32	37	34	26
Bottom	5	5	6	7	9	3	5	6	4	8	4

Table 6.10 A portion of the simulation experiment results using the SSMAE for the estimation of the vertical distribution of radionuclides in the Arctic Basin

Aquatory	Season							
	Winter		Spring		Summer		Fall	
	τ_w		τ_s		τ_u		τ_a	
	p_w	ξ	p_w	ξ	p_w	ξ	p_w	ξ
Greenland Sea	3.2	2	8.4	10	5.7	5	6.3	5
Norwegian Sea	2.9	2	7.8	9	5.9	5	6.7	6
Barents Sea	2.1	1	8.9	11	6.8	6	7.1	6
Kara Sea	2.4	1	9.2	12	5.3	5	6	5
White Sea	2.2	1	7.6	9	6.3	6	6.4	5
Laptev Sea	0.9	1	2.4	4	1.3	2	1.4	2
E. Siberian Sea	1.3	1	2.7	4	1.9	3	2.1	3
Bering Sea	2.5	2	7.1	9	3.9	4	5.3	4.0
Chukchi Sea	2.3	2	6.9	8	4.1	4	5.1	4
Beaufort Sea	1.9	2	5.7	7	4.8	4	4.9	4
Central Basin	1	1	1.7	2	1.5	2	1.6	2
Average value	2.1	1.5	6.2	7.7	4.3	4.2	4.8	4

The contribution of ecological processes to the formation of the vertical distribution in the radionuclide content of the water is represented by the parameter $\xi(\%)$. The average content of the phytoplankton biomass is represented by the parameter $p_w(g/m^2)$

The aquatories of the White, Laptev, East-Siberian, and Chukchi Seas are subject to visible variations in the radio nuclear pollution. An accumulation of radionuclides is observed in the Central aquatory of the Arctic Basin. The aqua-geosystems of the Greenland and Kara Seas have some conservative character, leading to buildup of radio nuclear pollution, but in the Norwegian Sea there is even a decrease in the pollution level.

A somewhat stable situation is observed in the vertical distribution of the radionuclides. It is generally achieved during 5–7 years following the initial moment t_0 with the exception of the East-Siberian, Laptev and Kara Seas where the stabilization processes of the vertical distribution are delayed by 10–12 years compared to the other water areas of the Arctic Basin.

The results of the simulation experiments show that variations of the initial data by ±100 %, change the stabilization time by no more than 30 %, so that the distributions take shape in 4–8 years. One unstable parameter is the river flow into the Arctic Basin. Figure 6.9 shows the variations in the simulation results under a change in the river flow to the Arctic Basin. The radio nuclear pollution is reduced by 80 % when the river flow decreases by 50 %. While the river flow increases by 50 %, the radio nuclear pollution of the Arctic Basin increases by only 58 %. Hence, a ±50 % error in the river flow estimate can cause a <100 % deviation of the simulation results for radio nuclear pollutants. As follows from the other curves of Fig. 6.6, such deviations are still less for heavy metals and oil hydrocarbons.

The SSMAE allows for the estimation of a wide spectrum of radio nuclear pollution parameters. Thus, this study shows a dependence of the biological transformation mechanism on the initial prerequisites. The biological transport of radionuclides downward out of the mixed layer, varies in a wide interval from months to scores of years. The vertical transport by biological elements is divided between 11 % for migration by living organisms and 89 % for transport by dead organisms. There is a stable result in that the lower trophic levels of the arctic ecosystem have a greater concentration of radionuclides than the higher trophic levels. But it is evident that for a higher precision of block MPPR the model of the biological processes must consider separately each radioactive element and its interaction with the plankton. The variant of MPPR realized here considers the physical processes as the major forces.

6.4.6.3 The Dynamics of Arctic Basin Pollution by Heavy Metals

The results of the simulation experiment are given in Table 6.11. We see that the stabilization of the average content of heavy metals on the full water area of the Arctic Basin is achieved in 3–5 years. Under this stable regime, the concentration of heavy metals in the compartments $\Omega_R \cup \Omega_P$ (river mouths and ports) is six times higher than in the Central aquatory and two times higher than in $\Omega_\gamma \cup \Omega_B \cup \Omega_N$ (the near-shore waters, the Bering Strait and the south boundary of the Norwegian Sea). The concentration of heavy metals in the phytoplankton is 18 % lower than in the zooplankton and 29 % lower than in the nekton. There is observed a process of accumulation of heavy metals in the upper trophic levels; moreover the relation of the concentration of heavy metals in the phytoplankton to their concentration in the other trophic levels varies from 0.3 in $\Omega_R \cup \Omega_P$ to 0.5 in the open part of the Arctic Basin.

The spatial features of the distribution of heavy metals in the sea water is characterized by more rapid accumulation into the water areas adjoining the west coastline of Novaya Zemlya and situated at the boundary between the Jan-Mayen and East-Iceland currents. The central water area of the Arctic Basin has a quasi-uniform distribution of heavy metals concentration. The vertical gradients of heavy metals vary in the interval from 0.11 to 1.54 mg m^{-2} day^{-1}. The forms of the vertical

Table 6.11 Results of the simulation experiment on the estimates of the dynamical characteristic parameters for pollution of the arctic waters by heavy metals (Kondratyev et al. 2002, 2003a)

Parameter	Estimate of the parameter after Δt years					
	$\Delta t = 1$	$\Delta t = 3$	$\Delta t = 5$	$\Delta t = 10$	$\Delta t = 15$	$\Delta t = 20$
Average concentration of heavy metals in the biomass (ppm)						
Phytoplankton	0.011	0.012	0.016	0.024	0.036	0.037
Zooplankton	0.013	0.014	0.019	0.028	0.041	0.043
Nekton	0.015	0.017	0.022	0.04	0.07	0.07
Detritophages	0.033	0.037	0.048	0.088	0.15	0.16
Average content of heavy metals in the arctic waters (ppm)	0.022	0.027	0.036	0.037	0.038	0.038
Flow of heavy metals from the upper layer to the deep waters (mg m^{-2} day^{-1})						
Norwegian Sea	0.71	1.07	1.14	1.17	1.19	1.22
Barents Sea	0.72	1.08	1.25	1.19	1.24	1.16
Greenland Sea	0.26	0.62	0.71	0.82	0.76	0.89
White Sea	0.11	0.23	0.24	0.21	0.19	0.2
Kara Sea	0.34	0.47	0.57	0.61	0.63	0.64
Laptev Sea	0.55	0.78	0.81	0.89	0.74	0.77
East-Siberian Sea	0.59	0.79	0.95	0.97	1.02	1.07
Chukchi Sea	0.88	0.83	1.54	1.49	1.31	1.44
Beaufort Sea	0.34	0.67	0.66	0.81	0.74	0.69

distributions of the dissolved fraction (ψ_W) and the suspended particles (e_W) of the heavy metals are not obviously expressed. The average relation of the concentration of heavy metals in the sediments and the water ($[\psi^* + e^*]/[\psi_W + e_W]$) varies over the area of the Arctic Basin from 1.9 to 5.7. For example, this relation for the Bering Sea is 3.3. The contribution of the bio-sedimentation process in the vertical distribution of the heavy metals is defined by values, which vary from 0.23 to 1.24 mg m^{-2} day^{-1}. The SSMAE provides the possibility of estimating the characteristics for separate types of heavy metals. An example of such calculations is given in Table 6.12.

6.4.6.4 The Dynamics of Arctic Basin Pollution by Oil Hydrocarbons

Spatial distributions of the forecast of oil hydrocarbons in the arctic water areas for 5 years after t_0 were obtained under the following restrictions: $O(t_0, \varphi, \lambda, z) = 0$, $H_O^1 = 0.1$ mg m^{-3} day^{-1}, $H_O^2 = 0$ for $g(t) > 0$ and $H_O^2 = 0.01$ mg m^{-3} day^{-1} for $g(t) = 0$ and $z \geq \Delta z_1$, $H_O^3 = 0.01$ mg m^{-3} day^{-1}, $H_O^4 = 0.01$ mg m^{-3} day^{-1}, $H_O^5 = k_D D_A^{1/3}$ (the adsorption coefficient $k_D = 0$ for $A = g, r, f$ and $k_D = 0.005$ day^{-1} when $A = W$), $H_O^6 = k_Z Z_A^{1/4}$ (the biosedimentation coefficient $k_Z = 0$ when $A = g, r, f$ and $k_Z = 0.004$ day^{-1} for $A = W$), $H_O^7 = k_B B_A$ (the bacterial destruction coefficient $k_B = 0.01$ day^{-1} for $A = g, r, f$ and $k_B = 0.05$ when $A = W$). The simulation experiments

Table 6.12 Estimates of heavy metal flows to and from the atmosphere (Kondratyev et al. 2002)

Heavy metal	Parameter	
	Atmospheric deposition, $Q^2_{e\psi}$ (mg/m^2/h)	Evaporation and spray, H^{ψ}_{α} (t/year)
Ag	0.7	7
Cd	1.1	58
Co	0.3	5
Cr	1.8	188
Cu	15	169
Fe	599	894
Hg	0.6	3
Mn	4.2	283
Ni	5.5	60
Pb	48	5
Sb	0.05	123
Zn	109	4,471

show that the intensity of the anthropogenic sources of the oil hydrocarbons (Q_O) estimated by McIntyre (1999) is transformed to other forms by 56 % in the environments of the surface snow, floating ice and submerged ice and by 72 % in the water. The stabilization of the distribution of the oil hydrocarbons is realized in 3 years after t_0.

The average level of oil pollution in the Arctic Basin reaches the value of 0.005 mg/l. This is lower than the natural level of World Ocean pollution. In the influence zone of the Gulf-stream current and in the Pacific waters an insignificant excess of this level is observed. The Barents and Kara Seas are the most polluted. Here the concentration of oil hydrocarbons reaches the value 0.03 mg/l. An average summary content of oil hydrocarbons in the Arctic Basin is 65,331 t (metric tons) with a dispersion of 32 %. The hierarchy of flows H^i_O $(i = 1, \ldots, 7)$ (see Eq. 6.10) is estimated by the set of $H^2_O > H^4_O > H^1_O > H^5_O > H^6_O > H^7_O$. This set is changed for each of the Arctic Seas The order of preponderance of the destruction processes H^i_O of the oil hydrocarbons is defined by the seasonal conditions. The oxidation process at the expense of evaporation of the oil hydrocarbons (H^2_O) prevails over the other processes in the summer.

In reality, the oil hydrocarbons evaporated from the surface of the Arctic Seas return to the Arctic Basin with the atmospheric precipitation. These processes are simulated in the blocks APM and PSM. The maximal destruction of oil hydrocarbons is 0.0028 g m^{-2} day^{-1}. The flow (H^7_O) due to bacterial decomposition averages 27 t year^{-1}. It has unequal values for different seas (t year^{-1}): Bering—3.7, Greenland—11, Norwegian—2.2, Barents—3.4, Kara—2.3, White—2.3, Laptev—2.8, East Siberian—2.8, Chukchi—3.4, Beaufort—2.5, Central Basin—0.5.

The total estimate of the role of the Arctic Basin ecosystem in the dynamics of the oil hydrocarbons is traced for each of Arctic Seas. As an example, Fig. 6.9 gives such results for the Barents Sea. The discrepancy between the simulation results (solid curve 1) and the data of curves 2 and 3 (Terziev 1992) is explained by the assumption that the trophic structures of the different Arctic Seas are described by means of the general scheme represented in Fig. 6.3 and the discrimination between the ecological elements in each of the Arctic Seas is not taken into consideration in the SSMAE.

Figure 6.10 indicates that the vegetative period for the phytoplankton in the Barents Sea equals 4.9 months as shown by the effects on the ecosystem contribution to the self-clearing of the oil hydrocarbons (dashed curve.) In the case considered, the Barents Sea ecosystem neutralizes about 25 % of oil hydrocarbons during the vegetative period. The rest of the time, this value oscillates near 3 %. The dispersion of these estimates with latitude reaches 53 %. For example, in the northern part of the Barents Sea the vegetative period varies from 2.6 to 3.1 months, while in the southern latitudes the variation is 5.3–5.8 months. Consequently, the ecosystem role in the sea cleaning of the oil hydrocarbons is 8 and 36 % for the northern and southern water areas, respectively. Such estimates can be calculated for each cell $\Delta\varphi \times \Delta\lambda \times \Delta z$ of the Arctic Basin.

6.4.6.5 The Dynamics of Pollutants in the Arctic Basin

An important problem is that of determining and understanding the role of the various pollutant sources in forming the pollution levels for the different Arctic Basin areas: McCauley and Meier (1991). One major function of the SSMAE is to estimate the pollution dynamics in each of the cells $\Omega_{ij} \subseteq \Omega$ as a function of time. The influence of the pollutant sources on the Arctic ecosystem occurs through the boundary area $\Omega_R \cup \Omega_P \cup \Omega_B \cup \Omega_N \cup \Omega_\Gamma$ and through the atmosphere. The total picture of the spatial distribution of pollutants is formed from the local dynamic processes. The incompleteness of the Arctic database leads to the need for assumption to be made for some scenarios in the framework: Krapivin et al. (1998a) estimated the flow of pollutants to the Kara Sea based on the experimental measurements of radionuclear pollution and heavy metals in the Angara-Yenisey river system. Therefore, the SSMAE is used to estimate the flow of pollutants from the Kara Sea to the other water areas of the Arctic Basin.

The Ob and Yenisey rivers are considered to be the main sources of radio nuclear pollution, heavy metals and oil hydrocarbons for the Kara Sea (Table 6.10). Figure 6.9 shows the influence of the river flow on the volume of pollutants transported by the Kara Sea aquageosystem to the Central Basin. As shown by curves 1 and 2, the transference of heavy metals and radionuclides from the Ob and Yenisey rivers through the Kara Sea aquageosystem to the Central water areas of the Arctic Basin amounts to 2.1 % when the river flow varies from 500 to 1,000 km^3/year and after that it begins to grow linearly up to 7.6 % for 2,000 km^3/year. Hence, there is a critical level of pollution for the Kara Sea ecosystem beyond which it does not have time to dilute the flow of heavy metals and radio nuclear pollutants. Similar estimations can be obtained for the other elements of set $\Omega_{R,}$ using the rivers referred to in Table 6.8.

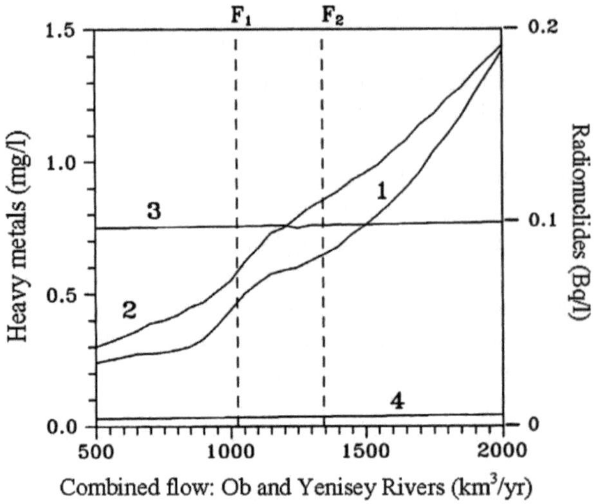

Fig. 6.11 Dependence of the concentrations of heavy metals ($\psi + e$) and radionuclides ($\varepsilon = {}^{137}Cs + {}^{60}Co$) at different geographical points as a function of the flow (F) of the Ob and Yenisey rivers to the Kara Sea. The interval $[F_1, F_2]$ between the *dashed lines* corresponds to the range of variations of F in the real world. Curves *1* and *2* show concentrations of the heavy metals and radionuclides, respectively, at the point with φ, λ = 75°N, 65°E in the northwestern part of the Kara Sea. Curves *3* and *4* show the concentrations of heavy metals and radionuclides, respectively, at the point with φ, λ = 72°N, 150°W above the Alaskan North Coast

The SSMAE allows for the estimation of the flow of pollutants between the different water areas of the Arctic Basin. For example, the transport of heavy metals and oil hydrocarbons from the Barents Sea to the Kara Sea is 631 and 473 kg year^{-1}, respectively. The total flow of pollutants from the Russian coast line to Alaska varies in the framework of Table 6.8 between 0.3 and 0.9 % of the initial flow. As is evident from curves 3 and 4 of Fig. 6.11, the flow of the Ob and Yenisey rivers have practically no influence on the pollution level of the arctic waters near Alaska. This effect does not change over time.

6.4.7 Summary and Conclusions

We will discuss here three aspects of the SSMAE which are very important. The first concerns the incorporation of ecological, hydrophysical, climatic and biogeochemical relationships in a model simulating the dynamics of the Arctic Basin pollution. The main problem here is how to parameterize these relationships to achieve the satisfactory precision. The second, concerns the key problem of database conformity to the model. In this case, the task is in the adaptation of the spatial-time scale to the database. The third, concerns the user's ability to run the SSMAE in the scenarios space.

This investigation shows that the simulation experiment provides the possibility of studying both the common and spatial distributions of the pollutant dynamics in the Arctic Basin. The precision of the results is a function both of the scenarios and the forms of the parameterization of the elements in Eqs. (6.1)–(6.16). It is obvious that the SSMAE is not effective when the climate conditions are varied to a critical state or when the anthropogenic impacts are increased to a critical value. But in the SSMAE the Arctic Basin acts as a stabilizing subsystem of the biosphere. When the atmospheric temperature is reduced by −1 °C, there occurs an inverse connection in the water balance of the atmosphere-land-sea system, which acts to stabilize the estimates. The parameterization of such variations in the framework of the SSMAE is not convenient. Nevertheless, the connection of the SSMAE to a global database having estimates of such parameters as ice area, temperature and albedo distributions, will allow the use of the SSMAE in the present form. The effectiveness of the SSMAE will increase with the use of models such as the coupled ice-ocean model described in the paper of Riedlinger and Preller (1991). In general, many different modifications of the SSMAE are possible. But, it is obvious that movement to the optimal SSMAE structure demands greater accuracy of pollutant types, ecosystem structure, water cycle, ice movement and climate model. The main difficulty is to realize the optimal modification at the same time for each of the blocks of the SSMAE.

The pollution level of the Arctic Basin is formed mainly by the river flows. Because of this, the block MRFAB plays a very important role in the SSMAE. Regular monitoring of the water flows and pollutant inputs by the rivers to the SSMAE is impossible. Consequently, the study and measurement of these flows during scientific expeditions and the modeling of the results are significant steps in the investigation of the Arctic Basin pollution.

One example of such a step is the U.S./Russian expedition of 1995 to obtain on-site measurements of the pollution levels in the Yenisey and its tributary the Angara, in order to investigate the likely origins of land-based sources contributing to the pollution levels in the Yenisey estuary.

The problem of verification of the SSMAE is important. However, it will be possible to realize this after essential modifications to the SSMAE, using models of greater precision to account for the hydrological, biogeochemical, ecological and climatic processes. The present SSMAE structure leads to a new technology of Arctic Basin pollution monitoring. Greater, or at least better, accuracy in the SSMAE may be realized by means of simulation experiments where the model parameters are varied over wide intervals. That is beyond the scope of this chapter.

We draw attention to several results of this investigation. In the framework of the nuclear war scenario, for example, the SSMAE shows that the Arctic environmental stability is disturbed 3 months after the impact. From other scenarios, it follows that variations in the velocity of vertical advection from 0.004 to 0.05 cm/s does not affect the Arctic environmental state. An error by 32 % in the ice area estimate leads to a variation of the simulation results by 36 %. When this error is more than 32 %, the simulation results become less stable and can vary by several times. The problem lies in finding the proper criterion to estimate the SSMAE sensitivity to variations in model parameters. As is shown by Krapivin (1996), a survivability function $J(t)$ reflecting the dynamics of the total biomass of living elements enables one to estimate this sensitivity. In this instance

$$J(t) = \frac{\sum_{i=1}^{5} \int_{(\varphi,\lambda)\in\Omega} \int_{0}^{H(\varphi,\lambda)} B_i(t, \varphi, \lambda, z) d\varphi d\lambda dz}{\sum_{i=1}^{5} \int_{(\varphi,\lambda)\in\Omega} \int_{0}^{H(\varphi,\lambda)} B_i(t_0, \varphi, \lambda, z) d\varphi d\lambda dz}$$

The index $J(t)$ provides an estimation of the uncertainty associated with the SSMAE parameters. Although a complete investigation of the influence of the SSMAE parameter variations on model results is an independent task, various estimations are given here. Preliminary simulation results show that the SSMAE permits variations of the initial data in the interval −70 to 150 %. In this case the model "forgets" these variations during ≈40 days. Also a large uncertainty (±50 %) is permitted in the value of such parameters as μ_A, δ_n, V_i, T_c, T_{opt}, ρ_1, k_{ij}. The correlation between variations of these parameters and the model results is linear. However, high model sensitivity is observed under variations of β_A, α_A, E_0, T_0. In general terms, the acceptable variation of these parameters is ±20 %. Moreover the deviation in the model results due to variations of these parameters is nonlinear. For example, fluctuations of the surface temperature T_0 within ±5 K turn out to be not hazardous to the system, causing small variations of $J(t)$ by ±10 %, but fluctuations of T_0 by ±7 K cause much larger variations in the value of $J(t)$ by ±30 %. Under this the temporal dependence of the system dynamics to variations in the parameters is diverse.

The SSMAE structure and its realization do not completely describe the processes taking place in Arctic Basin. The optimal extension of the SSMAE functions is possible by the use of environmental monitoring data to control parametric and functional model inputs. In this framework the prognosis of the Arctic aquageosystem state is realized on the basis of the SSMAE and by processing of the observed data.

6.5 The Angara-Yenisey River System Simulation Model

6.5.1 Introduction

The Arctic region is a mosaic of freshwater, terrestrial and marine ecosystems, intimately interactive with the factors of the Nature/Society system. Interactions include many components such as *Ocean/Atmosphere/Ice*, *Land/Atmosphere/Ice* and *Land/Ocean/Freshwater*. An interpretation and prediction of correlations between the processes occurring in the Arctic environment, is possible only in the framework of the complex approach to the study of these processes. This approach is based on the GIMS—technology (Krapivin and Shutko 2012). The interaction between the atmosphere, land and sea ecosystems under the arctic climate is characterized by a series of spatial-temporal scales. An understanding of the interior correlations at every level, is the key problem of environmental monitoring. Each scale has specific landscape, vegetation cover, topography, character of both hydrophysical and weather structures, and living elements. The discovery of the internal and external correlations between these elements and other elements of the NSS makes it possible to form the knowledge base for the Arctic Basin monitoring system.

Interactions between the arctic components of the global NSS were studied by many authors (McCauley and Meier 1991; Kondratyev and Johannessen 1993; Riedlinger and Preller 1991; Preller and Cheng 1999; Krapivin and Varotsos 2008). The Angara/Yenisey river system of Siberia is considered as an example of these interactions. The intensive industrial development of the Northern Russian territories has led to significant environmental changes in these regions (Morgan and Codispoti 1995). The major quantities of pollution substances are brought to the northern coast of Russia by the rivers, thus violating the ecosystems of the northern seas. One such river system is Angara-Yenisey river system (AYRS).

The Yenisey river flows northward to the Kara Sea along the boundary between the western Siberian flood plain and the central uplands, draining an area of about 2.6 million km^2 during its 4,100 km length. The flow rate of the Yenisey into the Kara Sea has large seasonal variations, averaging 19,800 m^3/s, and as high as 130,000 m^3/s during the spring run-off. The Angara, a major tributary, accounts for about one fourth of the total flow. It flows swiftly northward from Lake Baikal for about one third of its 1,850 km length, before turning westward, toward its confluence with the Yenisey. Recognizing that major sources of the radionuclides found in the Kara Sea, as well as other environmental pollutants, might lie in the Siberian watersheds of the Yenisey and the Angara, a joint Russian-American expedition was undertaken in July and August of 1995. The region where the expedition was conducted includes the five hydroelectric dams at Krasnoyarsk and Sayano-Shushenskoye on the Yenisey and at Irkutsk, Bratsk and Ust-Ilimsk on the Angara. The power output from these facilities has fostered rapid industrial growth in this region. Krasnoyarsk is the major industrial city located on the upper reaches of the Yenisey. Nearby is the nuclear production and processing facility, Krasnoyarsk-26, which is situated approximately 270 km upstream of the Angara-Yenisey junction. Along the Angara River, there are five cities with major industrial activities: Irkutsk, Angarsk, Usolye-Sibirskoye, Svirsk and Bratsk. These cities have facilities producing both radionuclides and chemical pollutants that can contribute to the source terms in the two rivers.

Samples of Angara River water and sediments were taken, beginning from near its source at Lake Baikal and continuing at selected sites of interest, downstream to its junction with the Yenisey river, near the lumber processing village of Strelka. Along the Yenisey sampling was conducted on both sides of the junction, beginning upstream from just below the village of Kazachinskoye and continuing downstream to the town of Lesosibirsk.

As is shown in Krapivin (1995), a complex evaluation of the pollution level in the Arctic Basin as a whole, is possible by synthesis of a mathematical model of pollutant transport by the rivers from adjacent territories.

6.5.2 Model Description

Following the work of Krapivin (1995), the block-diagram of the AYRSSM (AYRS simulation model) is represented in Fig. 6.12. The AYRSSM has three levels of blocks. A description of the AYRSSM blocks is given in Table 6.13. The two blocks of the first level realize the control for the models of different processes and ensure various regimes of the computer experiments. The block HYD simulates the river's flow with consideration of the correlation between the water regime and ecological, topographical and synoptical parameters of the territory studied. The block CON realizes the functions of the informational interface between the user and the other AYRSSM blocks. The main function of CON is forming the database and the identification of the AYRS elements. The four blocks of the second level analyze the input information and synthesize the AYRSSM structure. The twenty six blocks of the third level realize the concrete models and processes during the computer experiment.

It is supposed that the AYRS watershed has an area Ω. The spatial structure of Ω is determined by the spatial discretization of the AYRS surface with a uniform

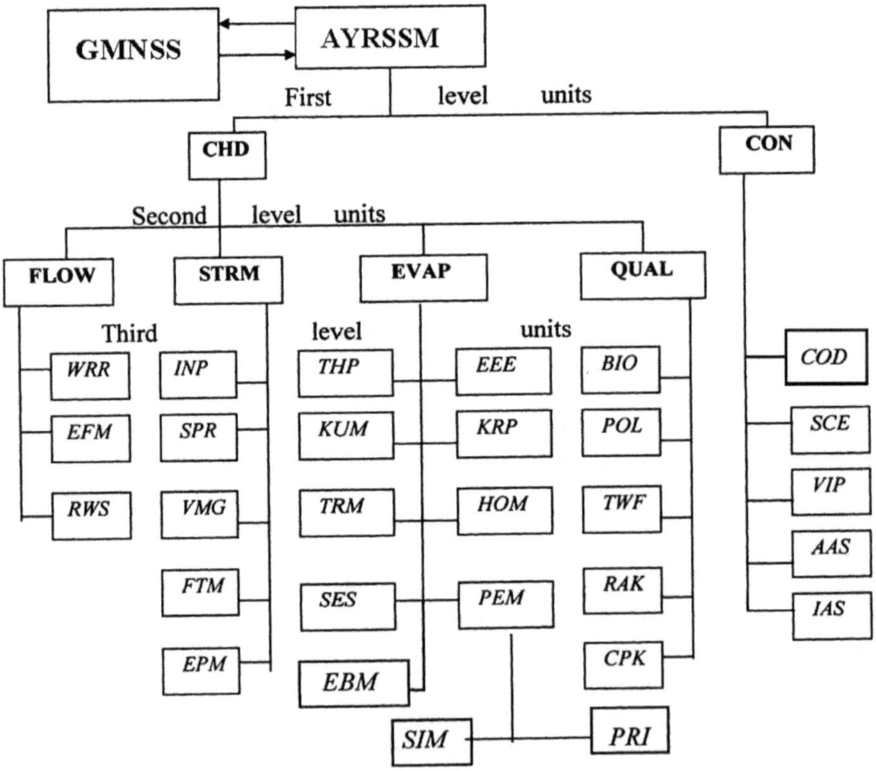

Fig. 6.12 Structure of the AYRSSM. Description of the blocks are given in Table 6.13

Table 6.13 List of units of the AYRSSM whose scheme is presented in Fig. 6.12

Unit	Description of the unit
CHD	Control of parameterization of hydrophysical and hydrological processes in the AY-system
CON	Controlling interface with functions of service of the man-machine dialogue when realizing simulation experiments
FLOW	Model of the river run-off
STRM	Simulation procedure that reflects the water masses motion in the AY-system
EVAP	The choice of the model to parameterize the process of evaporation
QUAL	The choice of the criterion to assess the water quality
WRR	Model of the water regime in a water body
EFM	An empirical model of the river runoff
RWS	Model of spreading the river run-off beyond the river-bed
INP	Model of infiltration
SPR	Model of the sink taking into account the effect of vegetation and soil covers
VMG	Model of the vertical uplifting of ground waters in evaporation, feeding and exfiltration
FTM	Model of filtration
EPM	An empirical model of precipitation
THP	A specified model of transpiration
RUM	The Kuzmin model (Kuzmin 1957)
TRM	A simple model of transpiration
SES	Model of snow melting and evaporation from snow surface
EBM	Parameterization of evaporation using the method of energy balance
PEM	The Penman model (Penman 1948)
PRI	The Penman model as specified by Priestley (Bras 1990)
KRP	The Penman model as specified by Kohler and Richards (1962)
SIM	A simplified version of the Penman model
EEE	Empirical models of evaporation
KOH	The Kohler model (Bras 1990)
HOM	The Horton model (Bras 1990)
BIO	A simulation model of sedimentation and biological assimilation of pollutants
POL	A simulation model of the formation of anthropogenic fluxes of pollutants
TWT	Model of the process of the water temperature formation
RAK	Model of kinetics of the radionuclide pollution of water
CPK	Model of kinetics of the chemical pollution of water
COD	Database correction
SCE	The choice and formation of scenario for simulation experiment
VIP	Visualization of the simulation experiment results
AAS	Adaptation of the AYRSSM to the scenario of the simulation experiment
IAS	Referencing the AYRSSM to the space and time scales

geographic grid with latitude φ and longitude λ, divided into steps of $\Delta\varphi$ and $\Delta\lambda$, respectively. In this study, it is supposed $\Delta\varphi = \Delta\lambda = 0.1°$. As result, the area $\Omega = \cup_{i=1}^{N} \Omega_k$ where Ω_k is part of Ω with area $\sigma_k = \Delta\varphi\, \Delta\lambda$, $\left(\sum_{k=1}^{N} \sigma_k = \sigma\right)$. The cells Ω_k are situated along the AYRS beginning with Ω_1 at the Angara river source up to Ω_N in the Yenisey River mouth. The procedure of spatial discretization is provided by the IAS block via, including in the AYRSSM database the set of identifiers $A_k = \left\| a_{ij}^k \right\|$, $k = 1, \ldots, 5$. The hydrology regime of the AYRS is described by the schematic diagram of Fig. 6.14. The equations for this scheme can now be written in the form of balance correlations on each of the Ω_k ($k = 1, \ldots, N$):

$$\sigma_k \left(\frac{\partial W}{\partial t} + \xi_\varphi \frac{\partial W}{\partial \varphi} + \xi_\lambda \frac{\partial W}{\partial \lambda} \right) = V - B\sigma_k + D + T + L \qquad (6.17)$$

$$\rho_k \sigma_k \left(\frac{\partial C}{\partial t} + \mu \frac{\partial C}{\partial x} \right) = \rho_k \sigma_k B + J + K - V - U - F - M - R \qquad (6.18)$$

$$(1 - \rho_k)\sigma_k \frac{d\Phi}{dt} = U + F + M + N + (1 - \rho_k)\sigma_k B - T - L - K - P \qquad (6.19)$$

$$\sigma_k \left(\frac{\partial G}{\partial t} + v_\varphi \frac{\partial G}{\partial \varphi} + v_\lambda \frac{\partial G}{\partial \lambda} \right) = R + P - J - N - D \qquad (6.20)$$

where ξ_φ and ξ_λ are the projections of the wind speed, ρ_k is the part of the area Ω_k occupied by the river, μ is the river speed, v_φ and v_λ are the speed projections of the ground water motion, x is the direction of river flow, and t is time.

The functions on the right side of Eqs. (6.17)–(6.20) are described by mathematical expressions in accordance with the papers by Krapivin et al. (1996) and Bras (1990). Appropriate models are given in Table 6.13. There are many realizations for some of these functions. This provides the user of the AYRSSM with the possibility of forming scenarios for the computer experiments. Values of ξ, μ and v were estimated on the basis of the Irkutsk Scientific Center database. It is possible for the user to vary these parameters during the calculation process. In this study, average values of these parameters are estimated by $\xi = 3.3$ m/s, $\mu = 1.7$ m/s and $v = 0$. Variations of the parameter μ are realized by adaptation of the left part of Eq. (6.18) to the empirical data illustrated in Fig. 6.13. Boundary conditions for Eqs. (6.17)–(6.20) are formed by the global model (Krapivin 1993). Soil moisture transport between the cells Ω_k is neglected. Synoptic situations are described by a discrete scheme with temporal parameters t_i ($i = 1, \ldots, 4$), where t_1 is the beginning of the summer period, t_2 is the start of winter, t_3 is the end of winter, and t_4 is the time of the spring thaw when the snow and ice are melting. Between these times the synoptical situation does not change.

In the common case, the vertical structure of the river water area in Ω_k ($k = 1, \ldots, N$) is described by block SES (Krapivin 1995). A snow layer of thickness g_k is formed at the expense of flow B_k according to:

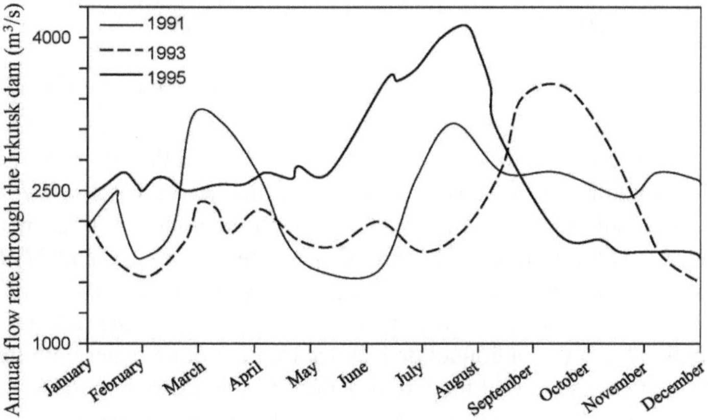

Fig. 6.13 Annual flow rate through the Irkutsk dam for the years 1991–1995

$$\frac{dg_k}{dt} = \begin{cases} 0 & when \ t \notin [t_2, t_3] \\ B_k & when \ t \in [t_2, t_3] \end{cases}. \tag{6.21}$$

For $t \in [t_3, t_4]$ the value of $g_k(t, \varphi, \lambda)$ is decreased linearly from $g_k(t_3, \varphi, \lambda)$ to $g_k(t_4, \varphi, \lambda) = 0$. The functional representation of the other blocks from Table 6.13 is realized at times $t_i (i = 1, \ldots, 4)$ by values which are given by the user under the scenario realization. The dynamics of pollutants in the AYRS is determined by the structure of its hydrological regime, which takes into account the transport of pollutants by water motion and their accumulation in sediments, ice, snow and the living biomass. The pollutant types are divided into radio nuclear elements and heavy metals. The set of radio nuclear pollutants is described by the index v. The set of heavy metals is divided into particles (index e) and the dissolved fraction (index ψ).

The v-th radionuclide is characterized by its half-life t_v, by the rates H_{kv} of input to and output from the area $\Omega_k (k = 1, \ldots, N)$ and by the concentrations Q_v, E_v, Ξ_v and S_v in the water, soil, sediments and ground water, respectively. As a result, the main balance equations of the RAK block are written in the form:

$$\Delta_k \left(\frac{\partial Q_{k,v}}{\partial t} + \mu_k \frac{\partial Q_{k,v}}{\partial x} \right) = H_{1,v} + H_{2,v} + H_{3,v} - H_{4,v} - H_{5,v} + H_{6,v} - H_{7,v} + \rho_k H_{12,v} - \frac{\ln 2}{t_v} \Delta_k Q_{k,v} \tag{6.22}$$

$$\delta_k \frac{\partial E_{k,v}}{\partial t} = H_{8,v} + H_{4,v} + H_{9,v} - H_{10,v} + (1 - \rho_k) H_{12,v} - \delta_k E_{k,v} \frac{\ln 2}{t_v} \tag{6.23}$$

$$\psi_k \frac{\partial \Xi_{k,v}}{\partial t} = H_{5,v} - H_{6,v} - \psi_k \Xi_{k,v} \frac{\ln 2}{t_v} \tag{6.24}$$

$$y_k\left(\partial S_{k,v}/\partial t + v_\varphi \partial S_{k,v}/\partial \varphi + v_\lambda \partial S_{k,v}/\partial \lambda\right) = H_{7,v} + H_{10,v} - H_{2,v} - H_{9,v} - H_{11,v} - y_k S_{k,v}\frac{\ln 2}{t_v} \quad (6.25)$$

where $H_{1,v}$ is the rate of radionuclide washing away from the soil with flow K, $H_{2,v}$ is the radionuclide input to the river from the ground water with flow J, $H_{3,v}$ is the transport of the radionuclide to the area Ω_k through the boundary of Ω by the AYRS tributaries, $H_{4,v}$ is the radionuclide carried out with flows U, F and M; $H_{5,v}$ is the radionuclide sedimentation to the river bottom by gravitation and with the dead biomass of the river ecosystem; $H_{6,v}$ is the washing away of the radionuclide from the sediments; $H_{7,v}$ is the radionuclide leakage to the ground water with flow R; $H_{8,v}$ is the anthropogenic source of the radionuclide; $H_{9,v}$ is the radionuclide input to the soil from the ground water by flow N; $H_{10,v}$ is the radionuclide washing out form the soil to the ground water by P flow; $H_{11,v}$ is the radionuclide losses by sedimentation in the ground water; and $H_{12,v}$ is the radionuclide input by rain. The quantities $\Delta_k = \rho_k\,\sigma_k\,C_k$, $y_k = \sigma_k\,G_k$, $\Delta_k = l_k(1-\rho_k)\,\sigma_k$ and $\psi_k = r_k\,\rho_k\,\sigma_k$, where l_k is the thickness of the efficient soil layer on the area Ω_k, and r_k is the thickness of the sediment layer.

The flows $H_{i,v}$ ($i = 1, \ldots, 12$) are parameterized by linear models according to the papers of Krapivin (1995) and Nitu et al. (2000a). The BIO and CPK blocks are described by similar balance models in analogy with the models described by Somes (1999) and Kram (1999). The AYRS biology is given in the form of a scenario or is described by the model of Legendre and Krapivin (1992). The flows of heavy metals include an assimilation of dissolved fractions by plankton $\left(H_\psi^Z\right)$ and by nekton $\left(H_\psi^F\right)$, the sedimentation of solid fractions $\left(H_e^1\right)$, the absorption from sediments by living elements $H_{e,\psi}^L$, sedimentation with the dead organic matter $\left(H_\psi^D\right)$ and the discharge from sediments $H_{e,\psi}^a$ by erosion, digenesis, turbulence and anthropogenic impacts. The balance equations taking these flows into account have the same form as the Eqs. (6.22)–(6.25):

$$\Delta_k\left(\frac{\partial e_w}{\partial t} + \mu_k\frac{\partial e_w}{\partial x}\right) = \sum_{i=1}^{3}\alpha_2^i Q_{e,\psi}^i - H_e^1 + \alpha_1 H_{e,\psi}^a \quad (6.26)$$

$$\Delta_k\left(\frac{\partial \psi_w}{\partial t} + \mu_k\frac{\partial \psi_k}{\partial x}\right) = (1-\alpha_1)H_{e,\psi}^a - H_\psi^Z - H_\psi^D \quad (6.27)$$

$$\psi_k\frac{\partial e^*}{\partial t} = H_e^1 - \alpha_1\left(H_{e,\psi}^L + H_{e,\psi}^a\right) \quad (6.28)$$

$$\psi_k\frac{\partial \psi^*}{\partial t} = H_\psi^D - (1-\alpha_1)\left(H_{e,\psi}^L + H_{e,\psi}^a\right) \quad (6.29)$$

where (e_w, e^*) and (ψ_w, ψ^*) are the heavy metal concentrations in (water, sediments) as solid and dissolved phases, respectively; $Q^i_{e,\psi}$ is the heavy metal input with the AYRS tributaries (i = 1), by atmosphere precipitation (i = 2) and with industrial wastes (i = 3); α_1 is the solid fraction part in the bottom sediments and α_2^i is the solid particles part in the i-th flow of heavy metals. The removal of heavy metals from water by evaporation and sprays is neglected.

Approximate solutions of the initial value problem for the Eqs. (6.17)–(6.29) are realized by means of the quasi-linearization method (Nitu et al. 2000b).

6.5.3 In-Situ Measurements

To obtain the data for the AYRSSM database, the joint US/Russian expedition to Siberia's Angara and Yenisey rivers was conducted in the summer of 1995 (Krapivin and Phillips 2001b; Phillips et al. 1997; Krapivin et al. 1997a, 1998a). Sampling began at the town of Bolshaya Rechka on the Angara near Lake Baikal and continued past the cities of Irkutsk, Angarsk and Bratsk, with a side trip on the Kitoy River, near Angarsk. The expedition then proceeded to the Yenisey-Angara junction. Sampling began near the village of Kulakovo, upstream on the Angara and continued downstream to the junction with the Yenisey at the logging town of Strelka. On the Yenisey, sampling began near the town of Kazachinskoye above the junction and continued downstream past the industrial complex at Lesosibirsk below the junction. Figure 6.14 shows points of measurement during this expedition.

The water balance of the AYRS was calculated with the data given in Fig. 6.13. Some results from this expedition are given in Tables 6.14, 6.15 and 6.16. A transportable radionuclide analysis system, consisting of a mechanically-cooled germanium gamma-ray spectrometer in a lead shield, was set up on-site in Irkutsk. Results of this analysis are given in Table 6.14. An X-ray-fluorescent spectrometer (XRF) was also set up on-site and used for screening of the samples for heavy metals. After the expedition, samples were sent to a commercial laboratory for elemental analysis. Results of this analysis are given in Tables 6.15 and 6.16. Results for each of these two categories of pollutants are discussed below. Table 6.16 represents comparison of results (ppm) of the laboratory analysis of the 1996 expedition materials on the Angara water quality (Analytical Services Center of Ecology and Environment, Inc., New York).

The manmade radioisotope ^{137}Cs (30-year half-life) was detected in all samples analyzed from above the Irkutsk dam, with concentrations ranging from 2 to 12 Bq kg^{-1} (dry weight). These values are consistent with background levels that can be expected due to global fallout as a residual from atmospheric nuclear testing. Below the Irkutsk dam in the vicinity of the cities of Irkutsk and Angarsk, the measured ^{137}Cs concentrations ranged from <4 to 30 Bq kg^{-1}, indicating that some of the samples contained ^{137}Cs concentrations were significantly higher than the

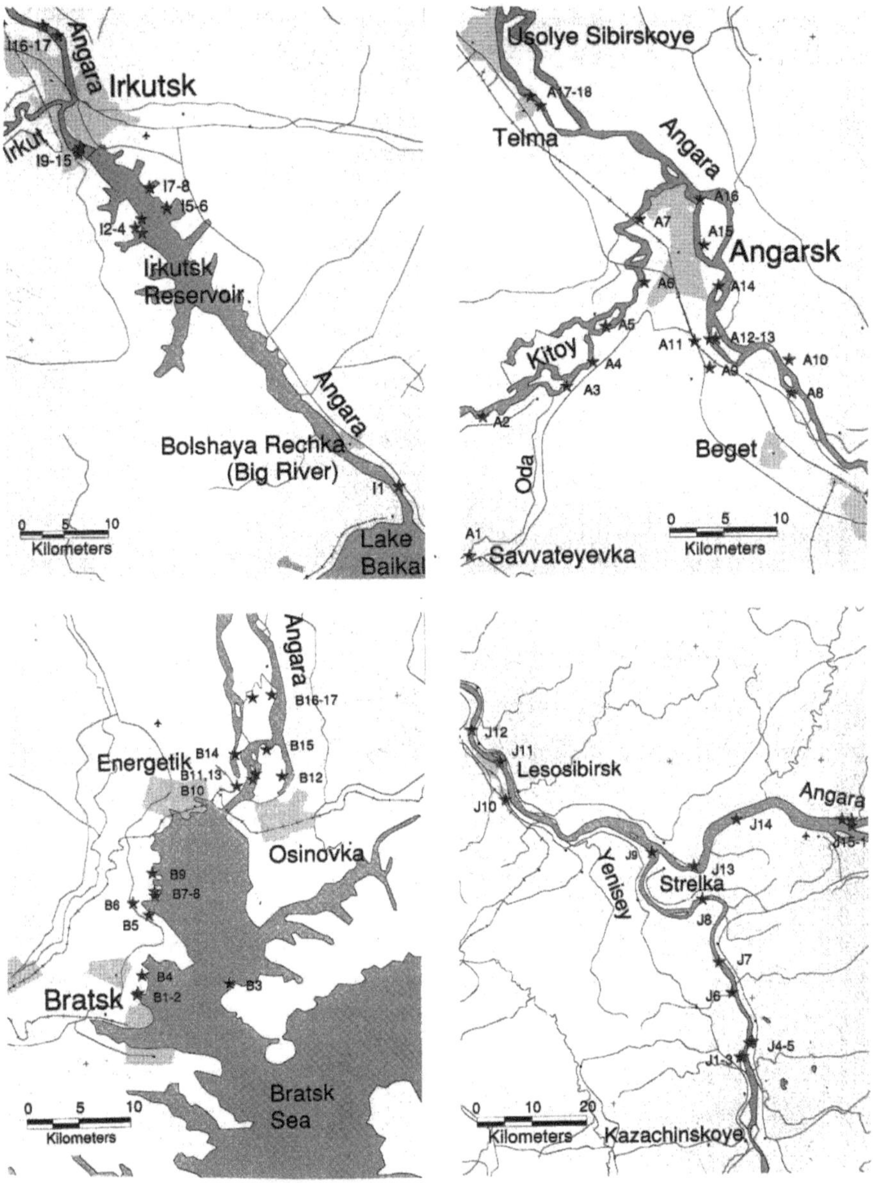

Fig. 6.14 Maps of sample location (marked by *stars*) during the American-Russian ecological expedition of 1996 (Krapivin et al. 1998a)

Table 6.14 Results of measurements of the content of radionuclides in river bottom sediments made in July 1996 (averaged data of isotopic activity, Bq/kg of dry weight)

Place of sampling	^{60}Co	^{137}Cs	^{152}Eu	^{235}U	^{238}U
Angara River					
Up the Irkutsk dam (I1)	<2.0	2.2 ± 1.0	<3.1	<1.9	<280
Down the Irkutsk dam (I2)	<3.5	<3.6	<5.3	27.2 ± 4.6	720 ± 300
Below Irkutsk (I3)	<3.7	30.6 ± 2.6	<5.1	<3.7	<580
Angarsk (technical canal) (A1)	<2.0	<1.8	<3.1	<1.7	<320
Up Angarsk (A2)	<4.1	25.2 ± 3.1	<6.0	<4.2	<630
Up the Bratsk dam (B1)	<2.3	<2.3	<3.9	<2.2	<420
Down the Bratsk dam (B2)	<2.6	3.4 ± 1.2	<4.0	<2.5	<440
Before the flow of Angara into Yenisey River (J1)	<1.9	2.2 ± 1.0	<2.8	<1.7	<290
Yenisey River					
Up Kazachinskoye (J2)	8.6 ± 1.9	22.9 ± 2.3	6.7 ± 3.0	<2.7	<500
Kazachinskoye (J3)	241 ± 11	392 ± 12	151 ± 27	<8.2	<1,340
Down Kazachinskoye (J7)	30.1 ± 3.0	203 ± 5	42.3 ± 9.5	<3.8	<680
Below Lesosibirsk (J6)	<5.7	27.1 ± 3.8	<8.2	<6.4	<860
Strelka, at the junction (J4)	96.9 ± 5.1	211 ± 6	55.7 ± 13.5	<4.5	<840

Places of sampling are marked in the map in Fig. 6.14 and are listed in Table 6.15. Errors given are ±2 standard deviations. The sign "<" indicates less than the minimum detectable concentration, which is given at the 90 % confidence level. Sample J7 was taken in the flood plane of the Yenisey River above the junction. The locations of the rest of the samples are given in Table 6.15

Table 6.15 Laboratory analytical results of heavy metal concentration in sediments and river-water measured in July 1996 during the American-Russian hydrophysical expedition (Krapivin et al. 1998a)

Sample ID#	Location	As	Cd	Cr	Cu	Ni	Pb	Zn
River sediment samples (ppm)								
I4	Bolshaya Rechka	7.2	<0.52	26	19	25	15	<64
I1	Irkutsk, above the dam	5.0	<0.51	41	22	44	11	55
I2	Irkutsk, below the dam	4.5	<0.51	20	11	21	9	17
I3	Below Irkutsk	8.1	<0.52	54	42	60	20	88
A1	Angarsk, technical ditch	0.9	<0.50	7.0	11	9.3	7.9	32
A2	Above Angarsk	4.3	<0.51	48	38	40	14	86
A3	Kitoy River	2.4	<0.50	27	20	37	7.8	40
B1	Bratsk, above the dam	0.5	<0.50	6.4	9	12	2.3	22
B2	Bratsk, below the dam	3.4	<0.50	31	210	39	6.1	50
J1	Angara, below Kulakovo	3.2	<0.50	14	13	18	4.5	32
J3	Below Kazachinskoye	5.8	<0.51	54	43	390	18	150
J4	Strelka, at the junction	5.1	<0.51	47	35		14	100
J5	Below Kazachinskoye	2.2	<0.50	18	8.	37	3.9	47
J6	Below Lesosibirsk	6.9	<0.51	34	31	20	17	100
I2	Irkutsk, below the dam	0.512	<0.50	<10	<20	36	<5.7	120
I3	Below Irkutsk	10	<0.50	<10	<20	<20	<5.7	92
A1	Angarsk	12	<0.50	<10	<20	<20	<5.7	240
B1	Bratsk, above the dam	16	<0.50	<10	<20	<20	<5.7	87
J1	Angara, below Kulakovo	12	<0.50	<10	<20	<20	<5.7	31
J2	Kazachinskoye	8.6	<0.50	<10	<20	<20	<5.7	140
J6	Below Lesosibirsk	13	<0.50	<10	<20	<20	<5.7	92
River water samples (ppb)								
I2	Irkutsk, below the dam	12	<0.50	<10	<20	<20	<5.7	120
I3	Below Irkutsk	10	<0.50	<10	<20	<20	<5.7	92
A	Angarsk	12	<0.50	<10	<20	<20	<5.7	240
B1	Bratsk, above the dam	16	<0.50	<10	<20	<20	<5.7	87
J1	Angara, below Kulakovo	12	<0.50	<10	<20	<20	<5.7	31
J2	Kazachinskoye	8.6	<0.50	<10	<20	<20	<5.7	140
J6	Below Lesosibirsk	13	<0.50	<10	<20	<20	<5.7	92

Places of sampling are marked in the map in Fig. 6.14. The measurement error did not exceed one sigma. Concentration values are given within 1 sigma uncertainty. The sign "<" indicates less than the minimum detectable concentration at the 90 % confidence level

average global background. For samples taken near Bratsk, only background levels of ^{137}Cs activity were detected.

The ^{137}Cs concentrations in samples taken on the undeveloped reaches of the Angara River upstream of the AYRS junction at Strelka, were determined to be at low background levels, about 2 Bq/kg of dry river sediment. Downstream from the

Table 6.16 Results (ppm) of the laboratory analysis of the 1996 expedition materials

Data type	Data of the Irkutsk ecological service on the Angara water quality control						Results of water samples analysis	
Year	1978		1979		1980		1996	
Season	T1	T2	T1	T2	T1	T2	August	
The Ershov water collector (up the Irkutsk dam)							Map ID#-I15	Map ID#-I11
Mg	3.2	3.0	3.3	3.1	3.4	3.3	3.5	4.9
Fe	na	na	0.01	nm	na	na	<0.05	<0.05
As	na	na	na	na	na	na	0.0063	0.012
Zn	na	na	na	na	na	na	0.15	0.12
Cu	0.002	0.002	0.003	0.002	0.001	0.003	<0.02	<0.02
Mn	0.001	0.001	0.003	0.005	0.003	0.006	<0.01	<0.01
Al	0.005	0.005	0.018	0.008	0.004	0.007	<0.1	<0.1
Co	nm	na	nm	na	na	na	<0.02	<0.02
Ni	na	na	nm	0.003	0.001	0.002	<0.02	<0.02
V	na	na	na	na	na	na	<0.02	<0.02
Pb	na	na	na	na	na	na	<0.0057	<0.057
The water collector near the Sukhov station up Angarsk							Map ID#-A14	
Mg	3.4	3.4	3.7	3.5	3.3	3.4	3.2	
Fe	0.02	0.3	0.07	0.001	na	na	<0.05	
As	na	na	na	na	na	na	0.012	
Zn	na	na	0.0053	na	na	na	0.24	
Cu	0.001	0.002	0.004	0.007	0.004	0.008	<0.02	
Mn	0.011	nm	0.018	0.007	0.012	0.011	0.028	
Al	na	0.015	0.02	0.028	0.013	0.012	<0.01	
Co	nm	na	0.004	0.002	na	na	<0.02	
Ni	nm	0.001	0.003	0.002	0.002	0.0005	<0.02	
V	na	na	na	na	0.001	na	<0.02	
Pb	na	na	na	na	na	na	<0.0057	
The water collector 0.5 km down the Bratsk dam							Map ID#-B2	
Mn	4.4	3.8	4.4	3.9	4.8	5.2	4.9	
Fe	na	na	0.02	0.13	na	na	<0.05	
As	na	na	na	na	na	na	0.016	
Zn	na	na	na	na	na	na	0.087	
Cu	na	na	na	na	na	na	<0.02	
Mn	na	na	0.004	0.001	0.002	0.001	<0.01	
Al	nm	0.002	na	0.002	0.003	nm	<0.1	
Co	na	na	na	na	na	na	<0.02	
Ni	na	na	na	na	na	na	<0.02	
V	nm	na	na	0.001	0.001	0.15	<0.02	
Pb	na	na	na	na	na	na	<0.0057	

nm no measurements; *na* beyond measurement accuracy (sensitivity of devices); T1 = $[t_1, t_2]$; T2 = $[t_2, t_4]$

junction on the Yenisey, the ^{137}Cs concentration in the samples ranged from 3 to 27 Bq/kg. The latter value is significantly above the activity levels detected on the Angara River upstream of the junction.

In samples drawn from the Yenisey river at and upstream of the junction, were much higher than background levels were detected, of the neutron activation product ^{60}Co (5.27-year half-life) and of the fission products ^{137}Cs and ^{152}Eu (13.3-year half-life): ranging from 9 to 240 Bq kg^{-1} for ^{60}Co; from 14 to 400 Bq kg^{-1} for ^{137}Cs; and from 7 to 150 Bq kg^{-1} for ^{152}Eu. These are unmistakable indications of reactor products and are consistent with releases from the plutonium production reactors at Krasnoyarsk-26, which is approximately 270 km upstream on the Yenisey River. The large variations in the measured activities were due to sampling location: the higher values came from a dead-end backwater channel with thick sediment (J3) and from a flood-plain area with thick mud deposits (J7), while the lower values came from sandy sediment (J2) where the current was swift and deposits were more likely to be washed away.

Above background levels of ^{235}U and ^{238}U were observed in several samples taken just below the dam at Irkutsk. However, the measured concentrations for ^{238}U have large uncertainties due to the relatively weak gamma-ray emission from this radionuclide. The ratios observed for ^{235}U/^{238}U are larger than the natural abundance of 0.7 % and are outside the statistical uncertainties. However, these are counting errors only, and do not include uncertainties in the interference between the overlying 186 keV gamma rays of ^{235}U and ^{226}Ra and possible errors from self-attenuation due to no uniformity in the samples. These samples were taken in an area undergoing active landfill. As a result, uranium-containing fills could have been brought in from elsewhere.

Using the XRF unit as a screening instrument, a set of heavy metals were identified in all riverbank and river sediment samples, collected during the expedition. After the expedition, twenty-one river sediment and eight river water samples were sent for trace element analysis to EPA, a certified commercial laboratory. Part of the analytical results is shown in Tables 6.15 and 6.16. Five heavy metals (Cd, Cr, Cu, Ni and Pb) are identified in all the samples at concentrations which are within the usual ranges of worldwide background levels. Two heavy metals, As and Zn, were identified in the river water samples at concentrations much higher than the median for natural fresh water (Lal and Stewart 1994). In fact, the Zn concentration in four of the samples exceeded the normal range for natural fresh water. These water samples were collected near to or downstream from industrial complexes on both the Angara and Yenisey rivers.

Table 6.16 gives a comparison of the 1995 US/Russian expedition results to archived 1978–1980 water-quality data taken from the Angara at Irkutsk, Angarsk and Bratsk, performed for the Irkutsk Medical Inspection Service (private communication, Svender G.M. and Krechetov A.A., Chemical Department, Irkutsk State University, Russia). The measurements indicate a significantly higher Zn concentration in the water collected above Angarsk, than was measured previously.

6.5.4 Experiments Using the Angara-Yenisey River System Simulation Model

The AYRSSM database includes estimates of the model coefficients, initial information for the climatic and anthropogenic scenarios and a set of identifiers $\{A_i\}$ describing the Ω area by its boundaries, thus giving the structure of the land-water surface. The CON block provides an interface with the database and allows the user to modify its elements.

It is assumed that the vertical distribution of pollutants in the river water is homogeneous, the pollution of the soil and plants is negligible and the cleaning process of the atmosphere has an exponential character with half-life τ_a ($\tau_a = 10$ days when $t \in [t_1, t_2]$, $\tau_a = 20$ days when $t \in [t_2, t_4]$). It is believed that the pollution sources located on the Ω territory support a constant level of distribution of the pollutants. The distribution function is uniform on the intervals shown in Table 6.17. The values of ξ, μ and ν were determined by average estimates for the last 3 years on the base of published data (Rovinsky et al. 1995) and data of the Irkutsk Scientific Center of Russian Academy of Sciences. The AYRS slope is taken equal to 0.21 m/km. Ecological elements of the AYRS are considered as having unique levels with biomass $\beta(t)$, a constant value at each of the intervals $[t_i, t_{i+1}]$ ($i = 1, 2, 3$). The equation for H_{5v} is written in the form: $H_{5v} = [g_1 + m_1 m_2 \beta] Q_{kv} \Delta$ where g_1 is the gravitation coefficient (0.05 day^{-1}), m_1 is the mortality coefficient (0.01 day^{-1}) and m_2 is the pollutant capture coefficient (0.03 kg^{-1}).

Table 6.17 gives an example of using the AYRSSM for the estimation of ^{137}Cs distribution in the Ω area where the cells Ω_i are situated along the AYRS with the step of discretization at $\Delta x = 10$ km. The ^{137}Cs distribution along the river system has a single maximum which changes in value and coordinate over time. A similar picture is observed for other radionuclides from Table 6.16. Such a result is not

Table 6.17 Relative concentrations of ^{137}Cs in water (γ_w) and in bottom sediments (γ_d) normalized for values at $x = 0$

Distance from Lake Baikal (km)	Time from the beginning of the simulation experiment (days)							
	30		60		90		120	
x	γ_w	γ_d	γ_w	γ_d	γ_w	γ_d	γ_w	γ_d
0	1.0	1.0	1.0	1.0	1.0	1.0	1.0	1.0
250	8.2	20.4	6.7	18.7	7.4	17.8	8.3	17.1
500	9.5	14.4	9.4	12.3	9.3	15.3	8.4	14.9
1,000	10.3	4.1	8.8	9.8	7.8	11.6	7.9	10.3
1,500	9.7	3.0	10.2	5.2	6.3	8.5	4.8	7.6
2,000	8.9	2.9	6.9	2.9	4.9	4.1	4.5	2.8
2,500	6.5	2.3	5.9	2.2	3.1	2.4	3.1	1.8
3,000	3.2	1.9	2.7	1.7	2.9	1.9	3.0	1.4
3,500	2.5	1.6	1.8	1.1	1.8	1.2	2.4	1.1
4,000	1.7	1.1	1.6	0.9	1.8	1.2	1.4	1.1

explained by the location of sources of radio-nuclear pollutants only. It is possible that the variations in the value and position of maximal ^{137}Cs concentration are caused by the high turbulence in the river system and by the existence of reservoirs and eddies. The AYRSSM gives only averaged results. The fourth level blocks in the AYRSSM structure are necessary for the description of hydrological processes with $\Delta x < 100$ m.

The computer experiments show that the input of radio-nuclear pollution to the Kara Sea from the Yenisey River has a stable value with dispersion equal to ±32 %. The role of the AYRS ecosystem in the process of transformation of radio-nuclear pollution is neglected as small (<3 %). The vertical transport by organisms varies from 0.1 to 0.7 %. Such calculations can be accomplished in the framework of different scenarios.

Figure 6.15 shows a distribution of heavy metals concentration along the AYRS constructed by means of computer experiment. We see that there are three maxima of heavy metals concentrations located at distances from Lake Baikal of 200, 1200 and 2000 km. This is the result of the distribution of pollution sources along the river system. The AYRS neutralizes the pollutants over a distance of 600–1,000 km from the source. The locations of the maxima vary as a function of the river flow rate. For the flow data of Fig. 6.14 this variation is 150 km. The river system transforms the pollutant flow such that input into the Kara Sea is estimated at a pollution level which is less than 2 % of the maximal concentration of pollutant in the AYRS.

Figure 6.16 shows how spatial distribution of oil hydrocarbons is changed 30 and 50 years after t_0 when its concentration is zero and processes of transformation of oil products are postulated. Results of simulation experiments show that the functional stabilization over the whole Arctic Basin takes 32 months. The concentration of heavy metals in the compartments $\Omega_R \cup \Omega_P$ (river mouths and ports) exceeds 6 times their level in the Central Basin and twice the level in $\Omega_B \cup \Omega_\Gamma \cup \Omega_N$

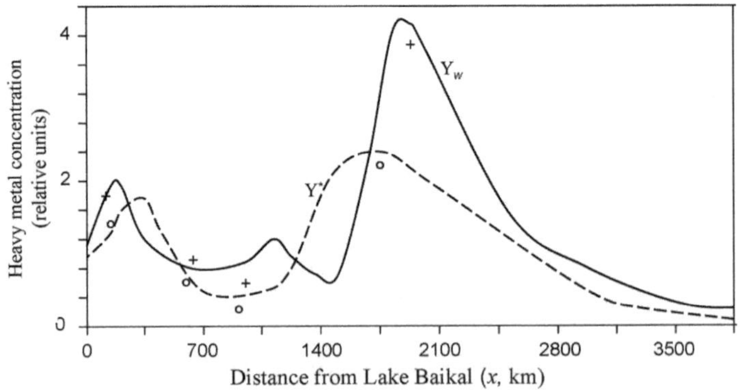

Fig. 6.15 Distribution of heavy metals concentration in water (*dashed line*) and in sediments (*solid line*) as function of distance *x* from Lake Baikal. The signs "0" and "+" correspond to the measured concentrations of metals in the water and sediments, respectively. The quantities $Y_w = [e_w(x) + \Psi_w(x)]/[e_w(0) + \Psi_w(0)]$ and $Y^* = [e^*(x) + \Psi^*(x)]/[e^*(0) + \Psi^*(0)]$

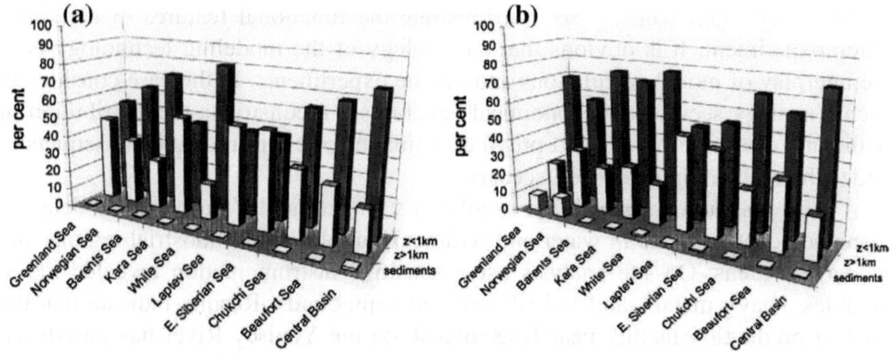

Fig. 6.16 Spatial distribution of oil hydrocarbons in the Arctic Basin 30 years (**a**) and 50 years (**b**) after the beginning of the experiment. It is supposed that transformation of oil products is realized by dissolution = 0.1 mg/m^3/day, evaporation = 0.01 mg/m^3/day, penetration into deep water layers in the form of droplets = 0.01 mg/m^3/day, oxidation = 0.02 mg/m^3/day, adsorption by a weighed organic substance = $k_D(D_A)^{1/3}$ (k_D is the coefficient of adsorption equal to zero for media $A = s, i_1, i_2$ and 0.005 day^{-1} in the medium $A = w$), biosidementation = $k_Z Z^{1/4}$ (k_Z is the coefficient of biosedimentation equal to zero for media A = s, i_1, i_2 and 0.004 day^{-1} in the medium $A = w$), and bacterial decomposition = $k_B B_A$ (k_B is the coefficient of bacterial decomposition equal to 0.01 for media $A = s, i_1, i_2$ and 0.05 day^{-1} in the medium $A = w$). D_A is the detritus biomass, B_A is bacterioplankton biomass, Z is zooplankton biomass

(near the coastline, in the Bering Strait and at the southern boundary of the Norwegian Sea). The concentration of heavy metals in phytoplankton is less by 18 % than in zooplankton and is higher by 26 % than in nekton. There is a process of accumulation of heavy metals at the upper trophic levels. Besides, the ratio of the concentration of heavy metals in phytoplankton to that at other trophic levels varies from 0.3 in $\Omega_R \cup \Omega_P$ to 0.5 in open waters.

The AYRSSM allows one to estimate the dependence of pollution level in the AYRS estuary as a function of anthropogenic activity. Suppose that the intensity of the sources of heavy metals is such that their concentration in the water near Angarsk, Irkutsk, Krasnoyarsk, Bratsk and Ust-Ilimsk is described by a stationary function, supporting the heavy metals concentration at the level h at each city mentioned above. The computer experiments show that there is stable correlation between h, the heavy metals concentration in the AYRS estuary, and the water flow rate μ. An increase in h of 10 % leads to a rise of pollution input to Kara Sea by 2.5 %. An increase in μ of 1 % leads to a rise of pollution input to Kara Sea by 0.7 %. These results are correct only when values μ and h are varied near their average estimates. Near their critical values the estimates are unstable and more detailed models are required.

In conclusion, it is necessary to note that this study demonstrates the possibility of using the modeling technology in the solution of complex environmental problems, demanding the combination of knowledge from different scientific fields. The AYRSSM is one example of such a combination. It gives typical elements for the synthesis of a simulation system for investigation of Arctic Basin pollution.

The results given in this section illustrate the functional features of the simulation experiment. It is obvious that the strategy of the modeling technology is in the interplay of model calculations and on-site experiments. In the case considered, such a strategy secures the economical benefit as it reconstructs the distribution of pollutants along the AYRS and provides estimates of environmental consequences under the realization of chosen scenarios.

A joint US/Russian expedition to Siberia's Angara and Yenisey rivers detected manmade contaminants in water and sediment samples from industrial regions and wilderness areas. On-site analysis using sensitive instrumentation revealed radionuclides, heavy metals, and volatile organic compounds. Results indicate that the nuclear production facility near Krasnoyarsk on the Yenisey River has introduced radioactive contamination far downstream and is a probable source of previously detected radioactivity in the Yenisey estuary at its outlet into the Kara Sea.

The AYRSSM is a complex system having an hierarchical structure with natural and anthropogenic elements. This study of the AYRS is the first step where simulation results are based upon large scale on-site measurements. The method proposed in this paper can be used for investigation of other Siberian river systems. But to expand the experimental base, remote sensing technology must be used (Krapivin and Shutko 2012). Remote monitoring can give more precise data on the structure of the identifiers A_2 and A_3, which describe the spatial distribution in the Arctic of the pollutants and the ice fields, respectively. The authors hope for the continuation of this study. Based on the database created, it is planned to prepare a complete set of models and their corresponding software for description of the process of transfer and transformation of pollution substances in the Arctic natural ecosystems. This set will demand the synthesis of models for the kinetics of radionuclides and chemical compounds in the food chains of the water and land ecosystems for the boreal zones, modeling of the hydrological regime and estimation of pollutant flows in the Arctic Basin.

A final aim of the investigations is the development of an environmental technology which will be the main result of co-operative scientific work directed toward the evaluation of the state of the Arctic ecosystem. The modeling system should be provided with a sufficiently full database to have at its disposal a ramified informational measuring network and a complete set of computer models for the main biogeochemical, climatic and biogeocenotic processes.

6.6 Arctic Biocomplexity

6.6.1 Introduction

Biocomplexity refers to phenomena that result from dynamic interactions between the physical, biological and social components of the Biosphere/Society System (BSS). The investigations of the processes of interaction between the Society and Biosphere are, as a rule, targeted at understanding and estimating the consequences of such interactions. The reliability and precision of these estimations depend on criteria founded on conclusions, expertise and recommendations. At present, there is no unified methodology for selection between the set of criteria due to the absence of a common science-based approach to the ecological standardization of anthropogenic impacts on the natural environment. After all, the precision of the ecological expertise for the functioning and planning of anthropogenic systems, as well as the representativeness of the global geoinformation monitoring data, depend on these criteria.

Biocomplexity is the study of complex structures and behaviors that arise from nonlinear interactions of active biological agents, which may range in scale from molecules to cells to organisms. Almost every biological system is complex. It is characterized by emergent properties, where the ensemble possesses capabilities that its individual agents lack. While physical and chemical processes give rise to a great variety of spatial and temporal structures, the complexity of even the simplest biological phenomena is infinitely richer. The goal of many authors is to meld physical, mathematical, and computational approaches with those of modern biology to understand this complexity in a quantitative and predictive way (Feller et al. 2010).

What is bio-complexity? This is a question which many researchers grapple with, every day. The world of bio-complexity research encompasses the efforts of many different types of scientists all working towards a common goal, how living things interact with and affect the environment and each other. It is very hard to find a good definition for bio-complexity. It includes areas of biology, chemistry, engineering, oceanography, geology, microbiology, genetics, and ecology. The U. S. National Science Foundation explains that it "refers to phenomena that arise from dynamic interactions that take place within biological systems and between these systems and the physical environment" (http://www.nsf.gov).

Arctic biocomplexity project is examining how biological and physical processes interact to control carbon uptake, storage and release in Arctic tundra ecosystems and how the self-organizing nature of these interactions varies across multiple spatial and temporal scales. Approximately 25 % of the world's soil organic carbon reservoir is stored at high northern latitudes in permafrost and seasonally-thawed soils in the Arctic, a region that is currently undergoing unprecedented warming and drying, as well as dramatic changes in human land use. Understanding how changes in annual and inter-annual ecosystem productivity interact and potentially offset the balance and stability of the Arctic soil carbon reservoir is of utmost importance to global climate change science. If there is a net loss of soil carbon to the atmosphere in the form of greenhouse gases (namely CO_2 and CH_4), greenhouse warming could be enhanced. This non-linear, potentially positive feedback response could very quickly cause Arctic terrestrial ecosystems to function in an unprecedented manner and with globally significant implications.

Many international and national projects aim at biocomplexity study for the arctic regions. Project of Alaska Geobotany Center "Biocomplexity of Frost Boil Ecosystems" (http://www.geobotany.uaf.edu/cryoturbation) seeks to understand the complex linkages between biogeochemical cycles, vegetation, disturbance, and climate across the full summer temperature gradient in the Arctic in order to better predict ecosystem responses to changing climate. This project focuses on frost-boils because:

- The processes that are involved in the self-organization of these landforms drive biogeochemical cycling and vegetation succession of extensive arctic ecosystem.
- Arctic ecosystems contain perhaps the most diverse and ecologically important zonal ecosystems in the Arctic and are important to global carbon budgets.
- The complex ecological relationships between patterned-ground formation, biogeochemical cycles, and vegetation and the significance of these relationships at multiple scales have not been studied.
- The responses of the system to changes in temperature are likely to be nonlinear, but can be understood and modeled by examining the relative strengths of feedbacks between the components of the system at several sites along the natural arctic temperature gradient.

The project aims to answer six questions related to the biocomplexity of frost-boil ecosystems:

- How does the self-organization associated with frost boils occur?
- How does the frost heave affect the soil biogeochemical processes within and between the frost boils?
- How do frost heave and biogeochemical processes affect plant communities?
- How do the biological processes in turn feed back to control frost heave?
- How do interactions between biogeochemistry, cryoturbation and vegetation change along the existing arctic climate gradient?
- How do the complex patterns associated with frost boils affect the tundra systems in a hierarchy of spatial-temporal scales?

6.6.2 Biocomplexity of the Arctic Tundra

Global climate change is expected to have a great impact on arctic areas including tundra. Snowbed habitats develop in areas that accumulate large amounts of winter snow and these habitats make up a notable component of the tundra biome. In connection with the arctic environment role in global warming there exist some aspects that are to be investigated:

- the arctic ecosystem ability to adapt to current and predicted global change;
- the dynamics of plant diversity and community structure along the internal snow-gradient within the snowbeds and how these are affected by increased nitrogen deposition;
- seasonal shifts in soil microbial composition;
- plant communities as separate entities with respect to nitrogen turnover and partitioning;

- importance of links between soil biota and plant community structure in determining nitrogen turnover in tundra ecosystems;
- activities of nitrifying and denitrifying microbes; and
- sensitivity of the snowbed plant communities to global change in the tundra landscape.

The biodiversity of the tundra as one of biocomplexity component is low: 1,700 species of vascular plants and only 48 species of land mammals can be found, although millions of birds migrate there each year for the marshes. There are also a few fish species such as the flatfish. There are few species with large populations. Notable animals in the Arctic tundra include caribou (reindeer), musk ox, arctic hare, arctic fox, snowy owl, lemmings, and polar bears (only near ocean-fed bodies of water). Tundra is largely devoid of poikilotherms such as frogs or lizards.

The tundra is a bleak and treeless place. Due to the harsh climate of the Arctic tundra, regions of this kind have seen little human activity, even though they are sometimes rich in natural resources such as oil and uranium. In recent times this has begun to change in Alaska, Russia, and some other parts of the world.

The arctic tundra biome is the treeless fringe of land that surrounds the Arctic Ocean. The terrestrial area of the arctic tundra biome, excluding glaciers and alpine areas, cover about 5.15 million km^2, an area about 3.4 % of the total land surface of the world. Knowledge and understanding of the complex interactions between climate, permafrost hydrology, and biological systems can help to assess the nature of a rapidly changing climate, as well as to understand a character of the nonlinear response of arctic systems to perturbations and to evaluate the magnitude and extent of positive and negative feedback processes involved in such fey processes as heat, CO_2 and water vapor exchange between the biosphere and the atmosphere. In this aspect, important problem is the parameterization of the dynamics of self organization in processes taking place in the arctic environment (Walker et al. 2008).

Arctic tundra region is characterized by the ubiquitous non-sorted circles that are formed and persist as an ecosystem due to complex soil-water-energy-ice-plant relationships and dynamics in the Arctic. Daanen et al. (2008) present model that captures the dynamics of the physical and biological components of the non-sorted circle ecosystem. This model helps to understand formation and persistence of this ecosystem type, especially in a changing climatic environment. Principal structure of the model is presented in Fig. 6.17.

The assessment of arctic tundra ecosystem biocomplexity puts numerous problems that are solved in framework of different international and national arctic programs (Corliss 2002; Walker et al. 2008; Bobylev et al. 2003). For example, the ACCSV project as part of the Land-Atmosphere –Ice Interactions (LAII) and Arctic Transitions in the Land-Atmosphere System (ATLAS) puts the following questions:

1. How does the self-organization associated with frost boils occur?
2. How do the biological processes and climate change feed back to control the formation of frost boils?
3. How does frost heave effect biogeochemical processes within and between the frost boils?

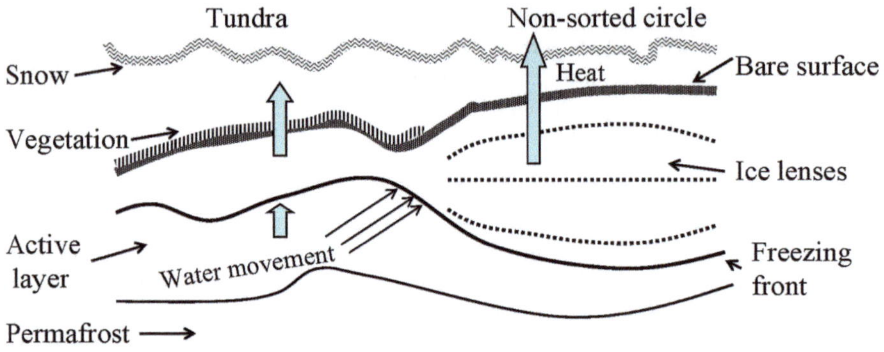

Fig. 6.17 A conceptual diagram of the heat and water fluxes in the active layer of the non-sorted circle ecosystem (Daanen et al. 2008)

4. How do biogeochemistry and cryoturbation affect plant communities?
5. How do the complex patterns associated with frost boils affect the tundra systems in a hierarchy of spatial-temporal scales?
6. How do these patterns and processes change along the existing climate gradient as the pool of available species and plant functional types changes?

Walker et al. (2008) and Peterson and Krantz (2008) studied differences in many key ecosystem processes between moist nonacidic tundra (MNT) and moist acidic tundra (MAT) and represented received results as Table 6.18. More detailed description of ecological trends is given in Table 6.19. We see that surface moisture

Table 6.18 Comparison of ecosystem properties of MNT and MAT (Walker et al. 2004, 2008, 2012)

Ecosystem property	MNT	MAT
pH of top mineral horizon	6.9	5.2
Number of vascular plant species	26	14
Average height of plant canopy (cm)	3.9	6.5
Leaf area index (m^2 m^{-2})	0.50	0.84
Moss cover (%)	65	79
NDVI	0.28	0.41
O-horizon thickness (cm)	11	21
Bare soil (% cover)	7.5	0.8
Soil heat flux (M J m^{-2} day^{-1})	3.13	1.83
Thaw depth (cm)	52	39
Gross primary production (mgCO_2-C m^{-2} day^{-1})	0.40	1,820
Net CO_2 uptake	670	950
Respiration loss	270	870
Methane emission (mgCH_4 C m^{-2} $year^{-1}$)	69	440
Soil organic carbon to 1 m depth (kg C m^{-2})	40	88
Soil Ca in active layer (mol(+1) m^{-2})	400	35

Table 6.19 Potential long-term ecological trends due to climate warming (UNIT TWO 2008)

Characteristic	Phytoplankton	Zooplankton	Fish	Mammals and seabirds
Distribution	Increased spatial extent of areas of high primary production in the central Arctic Ocean	Southern limit of distribution for colder water species to move northward. Distribution of more southerly species to move northward	Southern limit of distribution for colder water species to move northward. Distribution of more southerly species to move northward. Timing and location of spawning and feeding migrations to alter	Species distributions shift northwards towards the pole
Production	Increased production in central Arctic Ocean, and Barents and Bering Sea shelves	Difficult to predict, will depend on the timing of phytoplankton production and sea temperatures	Wind-driven movement patterns of larvae may be critical as well as a match/mismatch in the timing of zooplankton production and fish larval production	Dramatic declines in production by ice-associated marine mammals and increases by more temperate species. Seabird production likely to be primarily affected through food availability, which is unpredictable
Species composition/ diversity	Dependent on mixing depth. Different depths favor different types of phytoplankton	Adaptable arctic copepods, such as *Calanus glacialis*, may be favored	Cod, herring, walleye pollock, and some flatfish are likely to move northward and become more abundant, while capelin, polar cod, and Greenland halibut will have a restricted range and decline in abundance	Declines in polar bear, and in ringed, harp, hooded, spotted, ribbon, and possibly bearded seals. Increased distribution of harbor seals and grey seals. Possible declines in bowhead, narwhal, grey, and beluga whales. Ivory gulls and several small auk species are likely to decline; other changes in bird populations are unpredictable

in the arctic tundra is an integrator of biological, geochemical, physical, and anthropogenic processes and interactions. These processes correlate with regional and global climate system that mainly depends on biogeochemical cycles of carbon and methane. Approximately 25 % of the world's soil organic carbon reservoir is stored at high northern latitudes in permafrost and seasonally-thawed soils in the Arctic. Knowledge of carbon cycle in the arctic tundra and understanding of its dependence on the biological and ecological processes has globally significant implication for the study of non-linear potentially positive feedback response of arctic climate on the very quickly change of arctic terrestrial ecosystems.

The nature of the interactions among climate, permafrost, hydrology, and arctic ecosystems was the topic of a "Biocomplexity, Hydrology and Frozen Ground in Cold Regions" session at the 2006 Fall Meeting of the American Geophysical Union (AGU) where problems of arctic tundra investigations refer in part to understanding:

- the nonlinear response of arctic systems to perturbations;
- the magnitude and extent of positive and negative feedback processes involved in such key processes as heat, CO_2, and water vapor exchange between the biosphere and the atmosphere;
- the dynamics of self organization in processes governing patterned-ground formation; and
- the high degree of interdisciplinary research required to understand the interactions of biological, physical, and chemical processes in these systems.

Biocomplexity of the arctic tundra ecosystem is determined by coopled consideration of interactions between plants, animals and environment. Arctic plants include: arctic moss, arctic willow, bearberry, carbon moss, diamond-leaf willow, labrador tea, pasque flower, tufted saxifrage, reindeer moss, dwarf birch, woodland geranium, arctic fowl-grass, kelp, lichen centraria. Arctic animals include: arctic fox, caribou, ermine, grizzly bear, harlequin duck, musk ox, polar bear, walrus, habitual seal, lemming, northern deer, pink gull, guillemot. Arctic animals migrate during year depending on the seasonal dynamics of living conditions. Total dynamic assessment of the biocomplexity of arctic tundra ecosystem can be realized by the realization of modelling technology for the parameterization of energetic relations between its components. Figures 6.18 and 6.19 gives the principal schemes for possible model realizations. Model that realizes the diagram of Fig. 6.18, demonstrates dual nutrient controls across trophic levels and its biogeochemical significance. Figure 6.19 characterizes trophic interactions between the tundra plants and some living components of the arctic ecosystem.

Biocomplexity $\xi(t)$ of the tundra ecosystem has a seasonal character, changing, depending on the migration regime and climate that is characterized by low temperatures during practically all year. Average temperature of the most warm month (July or August) in the tundra is 10–15 °C. The vegetation reanimation is observed when the summer comes and solar radiation is available. Life processes in the tundra is continued during winter period only in the lakes and marshes, as well as partly in soils. Figure 6.20 shows typical distribution of the biocomplexity in the arctic tundra calculated for meteorological conditions averaged for 2000–2013.

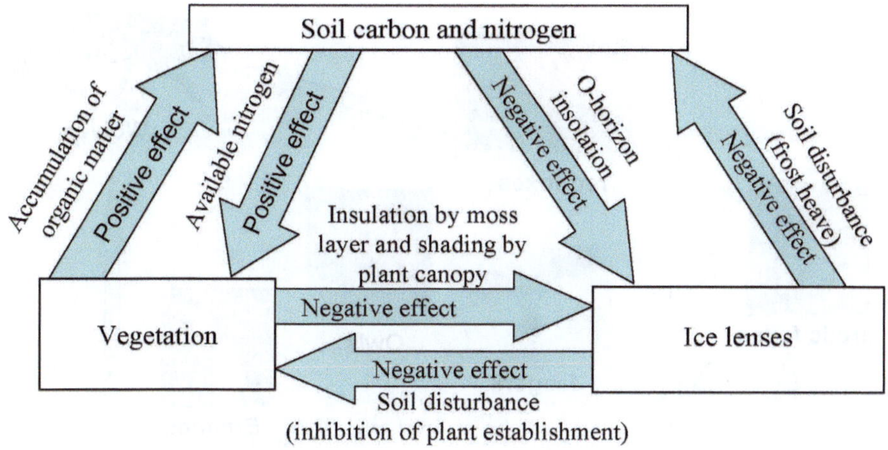

Fig. 6.18 Interactions between key components of the frost boil ecosystem

6.6.3 Biocomplexity Problem Related to Fisheries in the Okhotsk Sea

6.6.3.1 Introduction and Overview

Okhotsk Sea is one of the most unique object of the northern latitudes, where dynamics of ecological processes and their correlation with global changes were studied with poor accuracy, what did not allow to realize reliable forecasting of the ecosystem productivity. Therefore, principal concern is given to the study of the Okhotsk Sea ecosystem. Results of theoretical and experimental investigations that were received earlier, allow to understand the basic schemes for energy flows in the Okhotsk Sea ecosystem and to address to the more detail consideration of their structure, depending on the climate parameters changes and anthropogenic factor impacts, including pollution and fishery.

Certainly, the basic interest to the Okhotsk Sea ecosystem state was appeared and was developed from the direction of Russian Federation and Japan, as countries realizing the use of its ecological resources. There, exist institutes both in Russian Federation and in Japan, where the basic goal of them is the study of conditions of functioning and using the Okhotsk Sea resources. The Moscow Institute of Oceanology and Far-Eastern Institute of Oceanology of Russian Academy of Sciences and others, have collected for the realization of this goal, enough complete series of data about characteristics of various trophic levels of the Okhotsk Sea ecosystem and have studied in detail the geophysical and meteorological conditions of its functioning (Leonov et al. 2007; Dulepova 2002). All Russian Scientific-Research Institute for Fishery and Oceanography and set of other Institutes have published many scientific works with the description of dynamics of Okhotsk Sea aquageosystem parameters, including biological, chemical, physical, and geological

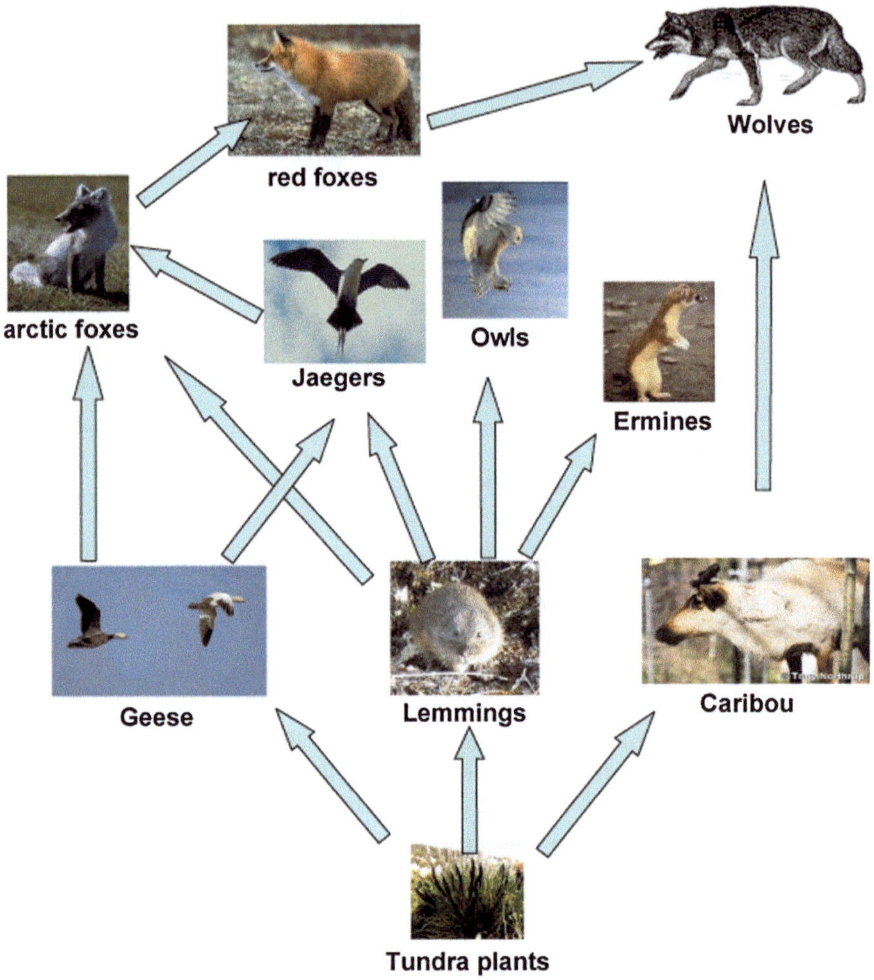

Fig. 6.19 Trophic relations in the arctic tundra ecosystem (http://www.arctic-predators.uit.no/project.html)

components. However, all the results of these investigations have mainly a descriptive character and are oriented on the explanation of particular effects and processes, explaining only partially the functional features of all ecosystem of the Okhotsk Sea and, especially, doing no assessment of such important characteristics as the dependence of ecosystem complexity on the external factors of natural and anthropogenic origin.

Special attention to the study of the Okhotsk Sea is given by many research centers of Japan. In particular, the Annual International Symposium on Okhotsk Sea and Sea Ice were organized during many years in Mombetsu (Hokkaido) by the Okhotsk Sea and Cold Ocean Research Association and Hokkaido University.

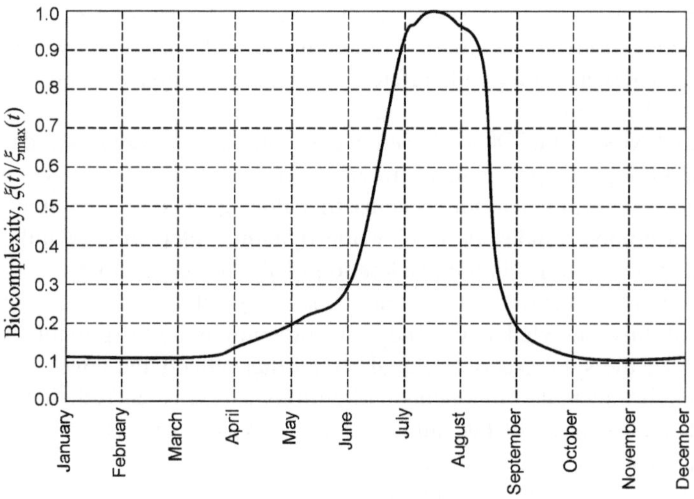

Fig. 6.20 Typical yearly distribution of the biocomplexity for the arctic tundra

The Meteorological Research Institute (Hokkaido) has produced a database of long-term meteorological data, the use of which gives a possibility to exclude the application of inaccurate climatic models for the creation of Okhotsk Sea ecosystem model.

An interest to the Okhotsk Sea ecosystem is caused by the need to assess the consequences of global changes for the ecosystems of northern latitudes and by the requirement of elaboration of effective informational technologies, that guarantee reliable monitoring of these vulnerable natural systems. Such informational technologies have been created in the Kotelnikov's Institute of Radioengineering and Electronics of Russian Academy of Science (Krapivin and Varotsos 2008; Krapivin and Shutko 2012) with reference to the synthesis of monitoring systems, aiming to detect earlier tropical hurricanes, warming on the underflooding and floods, as well as forecasting of fire dangerous in the forests.

This section describes ideas and some results of the application of a new information technology for the diagnostics of complex and many-sided dynamics of the Okhotsk Sea ecosystem. This technology is based on the generalization of experience that was now accumulated under the study of the Okhotsk Sea ecosystem and on the transfer of existing data to narrow range of spatial-temporal scales. This range is defined by the need to receive the assessments for the changes of energy balance in the Okhotsk Sea ecosystem. First attempt to solve this task was undertaken in Rostov-Don State University by means of creation of integral model of the Okhotsk Sea ecosystem (Berdnikov et al. 1989). Next versions of this model were proposed by Krapivin and Varotsos (2008). Integral indicator of the sea ecosystem state based on the calculation of biocomplexity indices was introduced by Krapivin and Mkrtchyan (2009a). These indices summarize such ecosystem characteristics as bioavailability, biodiversity, and survivability. An introduction of the biocomplexity indicator was realized by means of the use of methods of stability and ecosystem survivability theory. Into the framework of this study, it is supposed to adopt and develop this with reference to the conditions of the Okhotsk Sea ecosystem functioning.

Biocomplexity of the Okhotsk Sea ecosystem is considered as a derivative of totality of biological, physical, chemical, social, and anthropogenic interactions between the environmental sub-systems that impact on this ecosystem. Biocomplexity is a complex index describing the dynamic interactions within the environmental system, taking into consideration of the human factors.

Humankind has accumulated sufficient knowledge about environmental systems. The use of this knowledge to study the biocomplexity is possible within the framework of the synthesis of a global model that reflects the laws of interactions between environmental components and permits assess to an "efficiently" of realization of the scenarios of the human society development, based on the data of ground and satellite measurements.

Into the framework of this study, a biocomplexity conception is connected with laws of Okhotsk Sea ecosystem, functioning as unity of its elements that have different temporal and spatial scales. To define and assess it will develop a joint formal description of biological, geochemical, geophysical and anthropogenic factors and processes, taking place on a given level of spatial-temporal hierarchy scales and standards, including the dependencies of these factors and processes on the global changes in the biosphere and climate. The formation of the Okhotsk Sea ecosystem biocomplexity is studied at the expense of existing data about laws of its sub-system functioning, taking into consideration of the spatial and biological levels of their formation. The simulation model of the ecosystem functioning within the arctic climate conditions will be created. The stress full situations that can arise by natural and anthropogenic reasons will be studied, basing on this model. The characteristics of the ecosystem and components survivability will be defined, as well as allowable levels for the anthropogenic impacts will be established. Taking into account of the human factor and of influence of global changes on the Okhotsk Sea ecosystem, will be realized by means of global model of "biosphere-society" system that was developed by Krapivin and Kelley (2009). An integral index for the ecosystem state was introduced as an indicator of biocomplexity for the formalized description of biocomplexity. It allows to characterize the level of correlation between the Okhotsk Sea sub-system and to introduce numerical scale of biocomplexity, which gives a new level for the analysis of biological*, chemical and physical processes of interactions between the ecosystem elements* (Kelley et al. 1992a, b, 1999a, b).

As a result of this study, the oceanologic conditions in all scales of the Sea of Okhotsk, on the basis of satellite supervision is investigated. Also, the oceanologic, hydrochemical and zooplankton conditions are studied in the gulf of Aniva, during various seasons of year on the basis of ship shootings; the analysis of materials of tool measurements of currents (east shelf of island Sakhalin) (Shevchenko et al. 2004).

The results of this study were mainly the realization developed of the new informational technologies for the diagnostics of the Okhotsk Sea ecosystem state, which produces the conditions for the development of such decisions on regional and international levels, giving the possibility to defense the unique region from the unwanted changes. These technologies include the methods for measurement of sea water characteristics together with the use of sensors of spectroellipsometry and microwave radiometry.

In order to achieve the basic goal of this study, it is necessary to solve a set of particular tasks, between of them most important are:

1. Search of correlation between complexity of biological, physical, chemical and social systems in the Okhotsk Sea region.
2. Determination of laws in cycles of energy and substrate within the Okhotsk Sea ecosystem taking into account of levels of their organization.
3. Investigation of oceanological conditions in the whole Sea of Okhotsk on the base of satellite observations.
4. Investigation of oceanological and hydrochemical conditions and zooplankton status in the Aniva Gulf in different seasons from ship expeditions data; mooring current measurements data analysis (eastern Sakhalin shelf).
5. Description of adaptation mechanisms for the Okhotsk Sea ecosystem elements to the changes of environmental characteristics.
6. Search of constructive methods for the parameterization of structure of energy and substance flows in the Okhotsk Sea ecosystem, giving a possibility to construct the effective small-scale simulation model of its functioning.
7. Adaptation of technology of geoecological information—modeling system (GIMS-technology) to the conditions of Okhotsk Sea region.

6.6.3.2 Biocomplexity Model of the Okhotsk Sea

The Okhotsk Sea belongs to the type of seas with high ecosystem productivity (150–300 g C m^{-2} $year^{-1}$) which functions under a rigorous climate. The structure of the spatial-temporal fields of the basic hydrological and ecological characteristics of Okhotsk Sea is heterogeneous. The chemical, physical and biological processes occuring in sea waters, have been studied by many authors to assess its bioproductivity. According to the investigations by Terziev et al. (1993), the following structural discretization of the Okhotsk Sea can be realized. Five ecological layers exist. Layer 1, of maximal photosynthesis, is situated above the thermocline and has a depth of about 20–30 m. In fact, it corresponds to the wind-mixed layer. Layer 2 occupies the water space from 30 to 150 m in depth. It has low temperatures and oxygen saturation of about 80–90 %. Layer 3, characterized by low oxygen saturation (15–20 %), lies between the depth of 150–750 m. Layer 4 extends from 750 m down to a depth of 1,500 m. This layer has minimal oxygen saturation (10–15 %). Lastly, layer 5 is situated deeper than 1,500 m. It is characterized by oxygen saturation of 25–30 %.

The Okhotsk Sea ecosystem includes the following groups: phytoplankton, bascterioplankton, protozoa, herbivorous zooplankton, predatory zooplankton, nonpredatory zoobenthos, predatory zoobenthos, pelagic fishes, demersal fishes, demersal invertebrates, squids, seabirds, marine mammals. Fish community includes the following species: pollock, herring, capelin, deep sea smelt, cyclopteriids, sakhalin plaice. Total fish biomass during 1985–2013 was reduced from 8.9 mollion tonnes to 4.9 million tonnes. Fish biomass reducing process has different anthropogenic forcings: water pollution and overexpluitation of biological resources.

The Okhotsk Sea aquatory is divided into zones having specific ecological features (Suzuki 1992). The spatial distribution of the fish biomass depends on seasonal conditions and to a great extent correlates with the layers outlined above. The use of the sea biological resources is a function of this distribution. The fishing intensity essentially depends on the knowledge of the biomass distribution in the zones with specific environmental conditions. Many authors (Plotnikov 1996; Aota et al. 1992; Krapivin and Soldatov 2009a) have tried to solve this task by means of models simulating the ecosystem dynamics. However, the modeling results do not always turn out to be sufficiently representative and reflect the classification of sea zones by their productivity scale. The biocomplexity indicator is one such simple form, which can identify these zones. In fact, it is shown by many investigators that the Okhotsk Sea zones with high productivity, are characterized by complex many-level trophic charts (Terziev et al. 1993). This effect is not universal to other seas. For instance, the Peruvian current ecosystem has high productivity in zones where the trophic chain is short (Krapivin 1996). These situations are distinguished by the migration processes. That is why the biocomplexity of these ecosystems is formed in various ways.

The following components of the Okhotsk Sea ecosystem are mentioned in Table 6.20: The trophical pyramid $X = \|x_{ij}\|$, where x_{ij} is a binary value equal to "1" or "0" under the existence or absence of a nutritive correlation between the i-th and j-th components, respectively. Define the biocomplexity as function:

$$\xi(\varphi, \lambda, z, t) = \sum_{i=1}^{20} \sum_{j=1}^{19} x_{ij} C_{ij} \tag{6.30}$$

where φ and λ are geographical latitude and longitude; t is the current time; z is the depth; $B_{m,min}$ is the minimal biomass of the m-th component consumed by other trophic levels; $C_{ij} = k_{ij} B_{i,*} / \Sigma_{j+}$ is the nutritive pressure of the j-th component on the i-th component;

$$\sum\nolimits_{i+} = \sum_{m \in S_i} k_{im} B_{m,*}$$

is the actual food storage which is available to the i-th component; $k_{im} = k_{im}(t, T_W, S_W)$ $(i = 1, \ldots, 17)$ is the index of satisfaction of the nutritive requirements of the i-th

Table 6.20 Trophic pyramid of the Okhotsk Sea ecosystem considered in calculations of biocomplexity indicator

Consumers of energy and matter	Sources of energy and matter
	B_1 B_2 B_3 B_4 B_5 B_6 B_7 B_8 B_9 B_{10} B_{11} B_{12} B_{13} B_{14} B_{15} B_{16} B_{17} B_{18} B_{19}
Phytoplankton, B_1	0 0 0 0 0 0 0 0 0 0 0 0 0 0 0 0 0 1 0
Bacterioplankton, B_2	0 0 0 0 0 0 0 0 0 0 0 0 0 0 0 0 0 0 1
Microzoa, B_3	1 1 0 0 0 0 0 0 0 0 0 0 0 0 0 0 0 0 1
Microzoa, B_4	1 1 0 0 0 0 0 0 0 0 0 0 0 0 0 0 0 0 0
Carnivores, B_5	0 1 1 1 1 0 0 0 0 0 0 0 0 0 0 0 0 0 1
Zoobenthic animals, B_6	1 1 1 1 1 0 0 0 0 0 0 0 0 0 0 0 0 0 1
Flat-fish, B_7	0 0 0 0 0 1 1 1 1 1 1 1 1 1 1 1 1 0 0
Coffidae, B_8	0 0 0 0 1 1 1 1 1 1 1 1 1 1 1 1 1 0 0
Ammodytes hexapterus, B_9	0 0 0 0 0 1 1 0 0 1 1 1 0 0 0 1 0 0 0
Mallotus, B_{10}	0 0 0 0 1 1 1 0 0 0 1 0 0 0 0 0 0 0 0
Theragra chalcogramma, B_{11}	0 0 0 0 1 0 0 0 0 1 1 1 0 0 0 0 0 0 0
Salmonidae, B_{12}	0 0 0 0 0 1 1 0 0 1 1 1 1 0 0 1 0 0 0
Coryphaenoides, B_{13}	0 0 0 0 1 1 1 1 1 1 1 1 1 1 1 1 1 0 0
B_{14}	0 0 0 0 1 1 1 1 1 1 1 1 1 1 1 1 1 0 0
B_{15}	0 0 0 0 1 1 0 0 0 0 0 0 0 0 1 0 0 0 0
Crabs, B_{16}	0 0 0 0 0 1 0 1 1 1 0 1 0 0 0 1 1 0 0
Laemonema longipes, B_{17}	0 0 0 0 1 1 0 0 0 0 0 0 0 0 0 0 1 0 0
Biogenic salts, B_{18}	0 0 0 0 0 0 0 0 0 0 0 0 0 0 0 0 0 0 1
Detritus, B_{19}	1 1 1 1 1 1 1 1 1 1 1 1 1 1 1 1 1 0 0
People, B_{20}	1 1 1 1 1 1 1 1 1 1 1 1 1 1 1 1 1 1 1

Notations B_{14} *Reinchardti ushippoglossoi des matsuurae*, B_{15} *Clupeapallasi pallasi Val*

component at the expense of the m-th component biomass; k_{im} (i = 18, 19) is the transformation coefficient from the m-th coponent to the i-th component; k_{i20} is the characteristic of anthropogenic influence on the i-th component; $S_i = \{i\text{: } x_{ij} = 1, j = 1, \ldots, 19\}$ is the food spectrum of the i-th component, T_W is water temperature; S_W is water salinity; $B_{m,*} = max\{0, B_m - B_{m,\min}\}$;

$$x_{ij} = \begin{cases} 1, & if\ B_m \geq B_{m,\min} \\ 0, & if\ B_m < B_{m,\min} \end{cases}$$

Designate the water area of the Okhotsk Sea by $\Omega = \{(\varphi, \lambda)\}$. The value of the biocomplexity indicator for any area $\omega \in \Omega$ is determined by $\xi_\omega(z_1, z_2, t)$, where $[z_1, z_2]$ is the water layer located between depths z_1 and z_2. The maximal value of $\xi = \xi_{max}(\approx 20)$ is reached during the spring-summer period, when nutritive relations into the Okhotsk Sea ecosystem are extended, the intensity of energetic exchanges is increased and horizontal and vertical migration processes are stimulated. In the winter time the value of ξ is reduced to near ξ_{min} ($\approx$8). The spatial distribution of reflects the local variability of the food spectrum for its components. Table 6.21 and

Table 6.21 Estimations of the biocomplexity indicator ξ^* for the different layers in the spring-summer season and in winter

Season	Layers				
	1	2	3	4	5
Spring-Summer	0.89	0.93	0.62	0.34	0.21
Winter	0.31	0.49	0.71	0.39	0.22

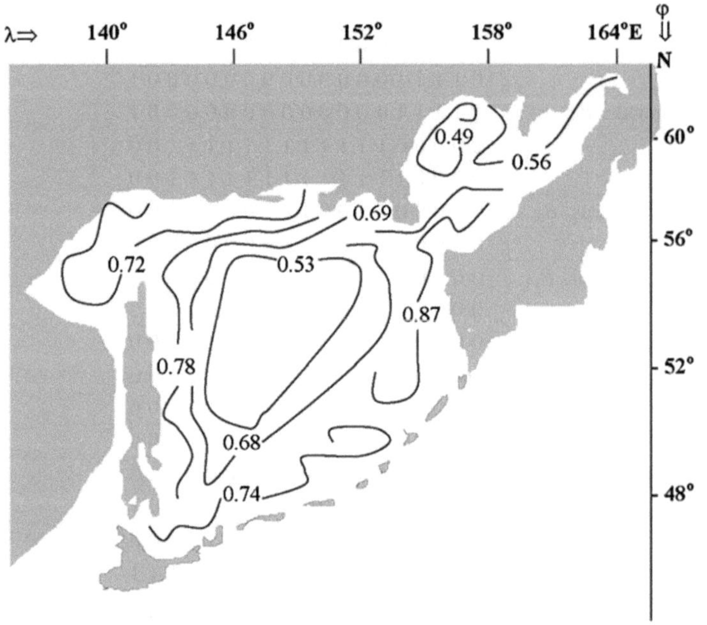

Fig. 6.21 Spatial distribution of biocoplexity indicator $\xi^* = \xi/\xi_{max}$ for the spring-summer time

Fig. 6.21 show an example of such a distribution. Comparison of this distribution with the distribution of zones with industrial fish accumulations (Terziev et al. 1993), shows that there is a correlation between these distributions.

In the common case, the indicator ξ reflects the level of complexity of the Okhotsk Sea ecosystem. A change ξ in is realized as a consequence of migration processes and of the variability of nutritive interactions. The subsystem B_{20} plays a role in these processes as an external source of change in the other components. These changes are interpreted in terms of fishing and other impacts causing variations in the biomass of the components.

Calculations show that basic variability in $\xi^* = \xi/\xi_{\max}$ is caused by migration processes, under which quick redistribution of the interior structure of the matrices X and $\|C_{ij}\|$ occurs. For instance, according to Terziev et al. (1993) many fish species during spring time migrate to the shelf zone, and during winter time they move to the central aquatories of the sea. Therefore, the value of $\xi^* \rightarrow 1$ during spring and $\xi^* \rightarrow 0.6$ during winter for the shelf zone. This means that the biocomplexity of the Okhotsk sea ecosystem in the shelf decreases by 40 % in winter in comparison with spring. For the central aquatories, the value ξ^* changes little during the year. Such stability of the biocomplexity indicator is explained by the balance between nutritive correlations and productivity during the spring, summer and winter seasons.

It can be established that the variability in ξ^* reflects the changes of fish concentrations which are controlled by environmental conditions. Specifically, during the spring time *Clupeapallasi escapes* occupy the area with $T_W < 5$ °C. Other fishes have a preferred depth for their feeding and spawning (Terziev et al. 1993). All these processes exert an influence on the variability of ξ^*. A more detailed investigation of the correlations between the value of ξ^* and the structural and behavioral dynamics of the Okhotsk Sea ecosystem requires additional studies.

This section introduces the main idea of how to move from the verbal description of biocomplexity to its numerical representation. In future studies it is necessary to take into consideration bottom relief, climate trends, ice fields dynamics, detailed components of the trophic pyramid, bottom sediments and the structure of the currents. Also, it is necessary to add to the formula (6.30) the members describing the anthropogenic impacts on the ecosystem considered in a socio-economic sense.

6.6.4 Expert System for Monitoring of the Okhotsk Sea Environment

There are many examples of effective marine models making it possible to study the sea ecosystem dynamics. However, the sea ecosystem modeling runs into difficulties, which requires the development of new modeling methods and monitoring data processing algorithms. Namely, the functioning of the sea ecosystem is complex under unstable conditions of the environment. These conditions are defined for both measured and unmeasured parameters. Therefore, the sea ecosystems are studied with the use of observation data, mathematical models, and satellite observations. A combination of these approaches gives the most useful results. The main difficulties of this method arise because of incompleteness of information. These difficulties are negotiated by new information-modeling technology developed by Krapivin and Kelley (2009).

Aota et al. (1991a, b) and Kaevitser et al. (2013) proposed new view on the Okhotsk Sea monitoring, basing on the Global Information-Modeling System that entails synthesizing a simulation model of the Okhotsk Sea environment (OSE) to describe the associated ecosystem dynamics. These help to describe the spatial dynamics of OSE and to optimize the monitoring regime. Additional application of biocomplexity indicator gives possibility to have spatial distribution of OSE elements with their characteristics.

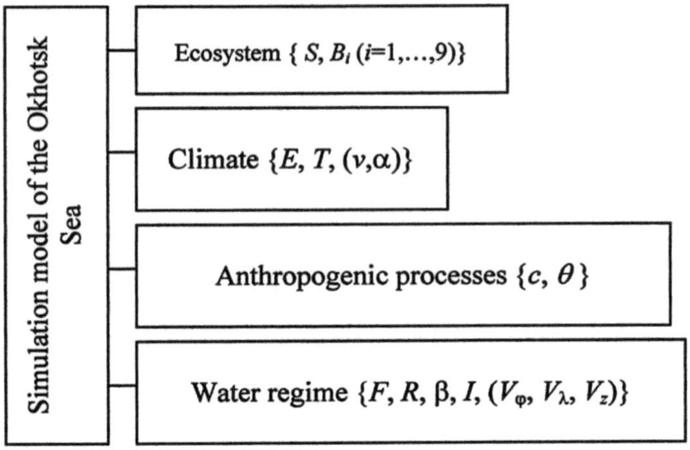

Fig. 6.22 The OSE model structure

The OSE model structure and the scheme of energy fluxes in the ecosystem of the Okhotsk Sea are shown in Figs. 6.22 and 6.23, respectively. A complete system of equations of ecosystem dynamics is formed from the set of traditional balance, hydrodynamic, and biogeochemical equations (Krapivin et al. 2007a).

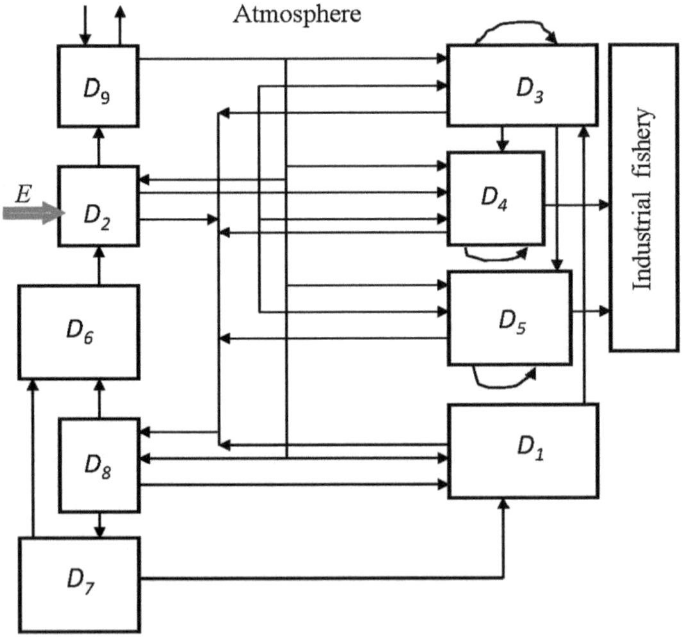

Fig. 6.23 Energy fluxes in the OSE model

The ecosystem block includes the following components: bacterioplankton (D_1), phytoplankton (D_2), zooplankton (D_3), benthos (D_4), nekton (D_5), nutrients (D_6), dissolved organic matter (D_7), detritus (D_8), oxygen (D_9), and salts (S). Climate block simulates the temperature (T), illumination (E), and wind (v , α). The block describing a water regime takes into account the following processes: precipitation (F), evaporation (I), current $\left(V_\varphi, V_\lambda, V_z\right)$, river run-off, and water exchange with Pacific ocean (β). The block considering anthropogenic processes includes the water exchange with Pacific Ocean (β) and pollution (θ).

The scheme to calculate the temperature regime is in fact an approximate simulation. This semi-empirical approach is acceptable for the climatic zone of the Okhotsk Sea and does not require complicated calculations. The exchange processes with water masses between the segments V_i ($i = 1, \ldots, N$) are simulated with account of information on directions and velocities of the currents contained in the data base.

The functioning of the Okhotsk Sea ecosystem is determined by severe climatic conditions. During 6–7 months in a year the sea is almost completely covered with ice 80–100 cm thick. Therefore, in the OSE model the vertical structure of the Okhotsk Sea is presented as a vertical section shown in Fig. 6.24. With account of this scheme, in the zones of the frozen sea the energy fluxes in Fig. 6.23 get complicated due to partition of the component "phytoplankton" into three possible media. Schematically it is shown in Fig. 6.25.

The equations of OSE dynamics have a set of coefficients $X = X_1 \cup X_2 \cup X_3$, where sub-set X_1 contains the coefficients values which are determined with the high precision, sub-set X_2 consists of the coefficients determined by inaccuracy, and sub-set $X_3 = X \backslash (X_2 \cup X_1)$ has the coefficients which are inestimated. Simulation procedure foresees two-tier process of model adaptation. At the first tier the model adaptation process is realized at the expense of sub-set X_3 coefficients determination. This

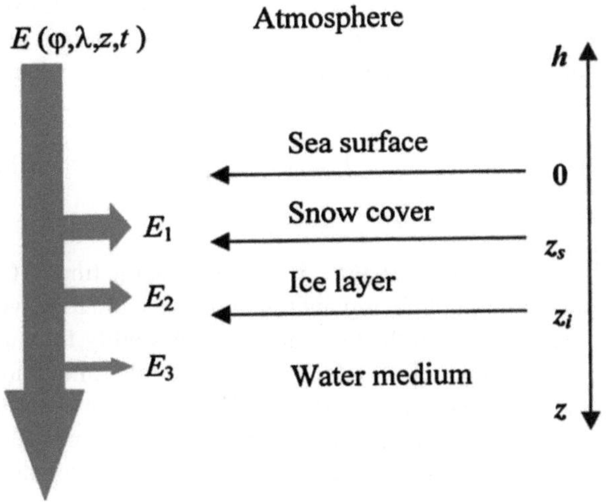

Fig. 6.24 The vertical section of the model medium of an arbitrary sea basin in the freezing conditions

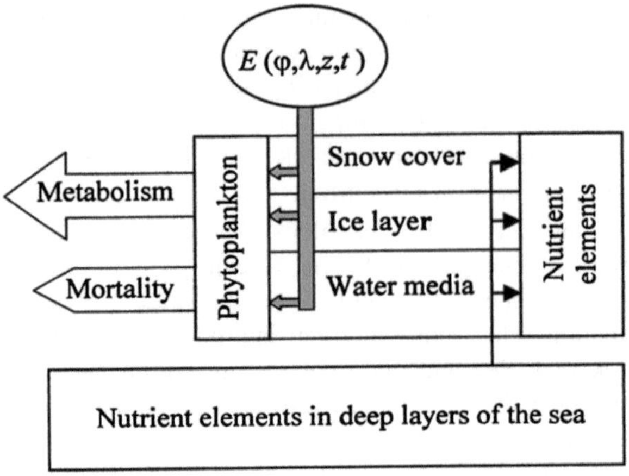

Fig. 6.25 The conceptual structure of the model of the organic matter cycle in the ice-covered sea

process consists in the variation of coefficient values in the ranges fixed by the OSE model user. The quality criterion is formulated by the user proceeding also from the existing experimental data or from the other consideration. Service block of the OSE model has a set of such possibilities. The user can demand to minimize the divergence between his data about some characteristic of OSE and its model estimation. Spectrum of such parameters spreads to main individual and integral elements of OSE.

At the second tier, the model adaptation process is continued by increasing the precision of sub-set X_2. The quality criterion of this process is based on the improvement of model quality reached during the first stage of adaptation process. Thus, the simulation procedure of the SMOSE is a continuous process of model adaptation via the coefficient changes or trophical graph modification. Really there is a set of models $\{A_i\}$. This set is formed with the sorting out of model parameters. Every model A_i is characterized by the quality level Q_i. Complete model A^* is formed as the limit of above models set. The finishing of adaptation procedure depends on the user's definition of quality criterion Q^* and on the values $|Q^* - Q_i|$. Model A^* ensures the $\min_i |Q^* - Q_i|$. It is believed that

$$Q_i = \frac{1}{h_1 \sigma} \int\limits_{(\varphi,\lambda,z)\in\Omega} D_2(t, \varphi, \lambda, z) d\varphi d\lambda dz.$$

Value of Q^*(= 20 g/m^3) is estimated at the time t_* (middle of summer) as average concentration of phytoplankton biomass at the top layer with depth of h_1 (= 100 m).

Simulations of the OSE offer the possibility for studying various aspects of how the OSE functions. Below there are some simulation results received under the values of OSE parameters. Specifically, the present status of the OSE is of great interest in order to understand the role of anthropogenic influence on it. According to the common theory of complex systems, a dynamic system is in the 'living state' at the time interval $[t_0, t_1]$ if the biomass of its elements is within limits $D_i \geq D_{i,min}$. Correlations between the trophic levels and nonliving elements give summary conditions for survivability criteria: $D_{min} \leq \Sigma D_i$. Finally, survivability criteria for the OSE can be written in the form: $J(t) \geq \alpha J(t_0)$, where $J = U(t)/U(t_0)$,

$U(t) = \sum_{k=1}^{9} \int_{(\varphi,\lambda,z)} B_k(t, \varphi, \lambda, x)dx$; t_0 is a moment of time when value of function $A(t)$ is considered as known; $\alpha < 1$ is the level of survivability. Really, we consider the OSE to be in a living state if the condition $J(t) > \alpha J(t_0)$ is carried out for $t > t_0$. Calculations of $J(t)$ for $t > t_0$ demonstrate how the OSE reaction depends on the variability of various environmental conditions. For example, fluctuations in the oxygen saturation of water may be of anthropogenic origin (oil pollution, large discharge of sewage waters, temperature increase, etc.). The $J(t)$ shows that the OSE displays a high degree of stability with respect to the initial saturation of water with oxygen, then how the OSE proceeds rapidly to a quasi-stationary regime of functioning at $D_9(t_0, \varphi, \lambda, z) \geq 1.8$ ml/l and how long it takes to overcome the initial shortage of oxygen in the case of $D_9(t_0, \varphi, \lambda, z) = 1.1$ ml/l. For $D_9(t_0, \varphi, \lambda, z) \leq 0.8$ ml/l, the OSE is unable to proceed to the stationary regime of functioning. The OSE is observed to be more sensitive to dynamic effects when the water is saturated with oxygen. A reduction of the oxygen production by 12 % does not influence the OSE dynamics. However, the OSE does not survive when the oxygen production is decreased as much as 20 %.

Calculations of J(t) offer the possibility for detecting the capability in searching zones at risk for survivability in Ω. For instance, one of particularly dangerous anthropogenic influences is the change in nutrient concentration. Simulation experiments enable us to determine the variations in the vertical uprising velocity of water within the same range.

To estimate the turbulent escape of nutrients into layers, overlying the maximal depth of photosynthetic layer ($z \leq h_*$), it is assumed in the SMOSE that the velocity of water uprise is equal to 10^{-3} cm/s. The obtained data point to the fact that, on the average, the integrated pattern of the distribution of community elements is not a subject to any significant variations within the velocity range from 3.5×10^{-4} to 10^{-2} and even 10^{-1} cm/s, but it is observed to be drastically distorted under a higher, which is the most important under a lower ($<10^{-4}$ cm/s) vertical advection of water.

Table 6.22 gives some of the results obtained in the case of a variation of the concentration of nutrients at the initial moment of time t_0. From the comparison of the results contained herein, it seems that a variation of the nutrients concentration within a wide range at the moment $t = t_0$ does not practically affect the behaviour of OSE at the moments $t \gg t_0$. The system, so to speak, '*healts*' with time from the '*blows*', has suffered and proceeds to the same functioning level. When initial nutrient conditions is limiting, the OSE teaches the stable state during 50 days only. If the nutrient concentration at the time t_0 has unlimited level the OSE had reached the stable state on the 30th day after beginning of simulation experiment.

The OSE model allows to realize other various simulation experiments. Clearly, there exist some problems with parametrical descriptions of OSE functions.

Table 6.22 Simulation results in which various nutrient concentrations are used. Space-time-mean results are given. Initial data for the phytoplankton biomass is $D_2(t_0, \varphi, \lambda, z) = 2$ mg m^{-3}

Depth, m	Phytoplankton biomass (g m^{-3})					
	$D_6(t_0, \varphi, \lambda, z) = 0.05$ mg m^{-3}			$D_6(t_0, \varphi, \lambda, z) = 3$ mg m^{-3}		
	$t - t_0$ (days)					
	10	30	50	10	30	50
0	1.1	5.4	10.7	6.3	11.3	10.9
10	1.9	5.3	12.3	5.4	11.9	15.8
20	2.4	10.2	21.9	7.1	19.2	21.1
30	1.3	6.6	18.4	7.2	14.6	14.9
40	0.6	3.1	9.2	5.5	10.1	9.7
50	0.2	1.2	8.7	4.2	7.7	8.4
70	0.1	0.9	5.3	1.7	4.8	5.1
100	0	0.4	2.1	1.1	1.9	2.2
150	0	0.1	0.9	0.8	1.1	1.3
200	0	0.1	0.1	0.2	0.3	0.2

The first problem for the future modernization of the OSE model is the extension of parametrical description of energy and heat exchange between OSE and the atmosphere to increase the model validity and to establish the dependence between model parameters and satellite measurements. The second problem of the OSE model to be greater in accuracy, consists in the detailed elaboration of OSE elements to extend its applied significance to the fishing. The description in more details of OSE biological balance with the formation of industrial fishes has a priority for future investigations. Finally, the third problem to improve the OSE model by synthesizing an expert system at the center of which the OSE model will be the main unit.

6.7 Carbon Cycle Dynamics in the Arctic System

Studies of the climate-forming processes in the Arctic, whose principal feature is determined by the isolation of the ocean from the atmosphere by the ice cover, have long been of particular concern. The most important impacts of sea ice on the climate, as revealed through numerical modeling, are the following:

(1) Maximum climate warming with increasing CO_2 concentration in the wintertime Arctic due to increased heat input from the ocean through the thinner ice, as a result of the warming;
(2) The effect of albedo on the atmospheric temperature of the more extended sea ice 18,000 years ago, which is compatible with the impact of continental glaciers; and
(3) Possible reversal of the conventional relationship between the amplitude of the annual change of temperature and the depth of the oceanic mixed layer when

the sea ice dynamics are taken into account. A thinner mixed layer favors the strengthening of the wintertime sea ice, which causes a delay of the springtime melting and produces a colder summer. The latest data of numerical modeling has confirmed, on the whole, by Kondratyev and Johannessen (1993), Kondratyev (1990).

When discussing the problem of the role of the polar regions in the formation of global changes, one should have in mind that two aspects of the global ecology are of paramount importance:

(1) the anthropogenically induced redistribution of the components of the heat balance of the Earth as a planet (with emphasis on the problem of the atmospheric greenhouse effect and its climatic impact); and
(2) anthropogenically induced breaking of the global biogeochemical cycles (primarily, this refers to carbon, nitrogen and sulfur).

That is why polar regions are very specific as a component of the globaleco-system. These aspects are not taken into account completely by the existing global models of the Nature/Society system. A new approach to the synthesis of geo-information monitoring systems, proposed by (Krapivin and Shutko 2012), overcomes this shortfall. In the framework of this approach the interchange of CO_2 between high latitude vegetation, the North Ocean and the atmosphere is considered as part of a global biogeochemical cycle, described through the framework of a Global Simulation Model (GSM) (Krapivin 1993). The GSM is comprised of blocks that parameterize the following:

(1) the global hydrological balance;
(2) productivity of soil-plant formations, with 30 types defined;
(3) photosynthesis in the ocean ecosystems taking into account depth and surface inhomogeneity;
(4) demographic processes; and
(5) anthropogenic changes.

The GSM makes it possible to compute the dynamics of industrial CO_2 distribution between the oceans, terrestrial biota and the atmosphere. The GSM describes the World Ocean by a spatial four-layer model with due regard for water chemistry. Spatial heterogeneity of the World Ocean is represented by the structural distribution of surface temperature applied to the upwelling and convergence zones. Sea ice in the polar region is considered through use of CIESIN data (Yohe et al. 2006). The biogeochemical cycle of CO_2 is described by balance equations in accordance with Fig. 1.30. The GSM carbon dioxide block takes into consideration the dependence of flows H_2 and H_3 from water surface processes (wind-wave mixing, rough seas, foaming waves) (Alekseev et al. 1992; Weis et al. 1982). Simulation experiments have shown that these flows have variations from 16 to 1,250 mol m^{-2} year^{-1}. In the Arctic Ocean during the June-September period the partial pressure p_a of CO_2 in the atmosphere exceeds the partial pressure of CO_2 in the sea waters by 20–110 ppm. These variations in the partial pressure of CO_2 have specific distributions for the arctic seas. For instance, the Norwegian Sea and Bering Sea have CO_2 deficits of 18–54 and 33–69 ppm, respectively. The average CO_2 deficit reaches 450 gC m^{-2}. The flow H_3 changes between 1.5 and 4.1 gC m^{-2}day^{-1}. In addition, linear correlations between the partial pressure of CO_2 and the water temperature T are observed, with a proportionality coefficient equal to 9.8 ppm/°C.

The problem of the greenhouse effect is discussed by many authors within the framework of different anthropogenic scenarios. The main conclusion is that global climate change brought about through CO_2 dynamics, will be insignificant during next century if:

(1) the World Ocean pollution, especially by oil products, does not exceed the level of 1990 by 10 %;
(2) agricultural land areas do not expand at the expense of forests;
(3) the rate of fossil fuel consumption remains at the level of 1990 with a dispersion of 15 %; and
(4) alternative energy sources (atomic, wind, etc.) are developed at a rate which does not hinder the food production.

The role of soil-plant formation for the absorption of excess atmospheric carbon dioxide under the above scenario is estimated with the data displayed in a geographic grid of 4° × 5° lat/long (Fig. 6.9 and Table 6.2). The role of the World Ocean is considered by taking into account the water temperature in the surface layer. It has been shown that the atmospheric CO_2 concentration can reach a mixing level of 556.7 ppm during the 21st century. The dynamics of industrial CO_2 distribution between the atmosphere, oceans and vegetation will fluctuate with an amplitude less than 25 %. With an increase in industrial CO_2 emission over the period from 1990 to 2090, the atmospheric portion will rise and the contribution of the oceans to its absorption will increase; whereas that of the biota will stabilize after going through a small maximum. At the end of the 21st century and beginning of the 22nd century, during the highest level of human economic activity, the contribution of the oceans (especially the Arctic Basin) to industrial CO_2 absorption will be considerably higher than that of vegetation. This is due to the fact that, with an increase in CO_2 concentration in the atmosphere, the ability of the upper ocean layers to absorb industrial CO_2 will be supported by the transformation of biogeochemical processes in the deep ocean layers. The restoration of plant cover and ocean pollution are the main short-time problems (Watson et al. 2000). For example, it was shown that if the natural/disturbed land relation will change from 2/3 in 1990 to 3/2 in 2050 then the atmospheric CO_2 concentration would reach no more 497.3 ppm during the 21st century. This illustrates that the role of the biospheric system in global change needs to be investigated more thoroughly (Table 6.23).

Simulation results show that there is an overall exaggeration of the importance of the problem of global climate change caused by anthropogenic CO_2 emissions. Thus, the existing arguments for climate change are not reliable. It is necessary to develop the GSM by inserting new correlations between the elements of the NSS and taking into account the biotic regulation processes.

Table 6.23 A model estimation of the surplus CO_2 absorption by vegetation on the Russian territory

Soil-plant formation	Carbon flow absorbed by vegetation (10^6 t C/year)
Arctic deserts and tundra, sub-arctic grasslands and marshes	3.2
Tundra, mountain- and forest-tundra	10.7
North-taiga forests	11.8
Mid-taiga forests	33.2
South-taiga forests	24.9

The anthropogenicemission of carbon is assumed to be 6.5 GtC/year

This study showed that the role of the Arctic Ocean in the global CO_2 balance was estimated with low precision. Therefore, the Kyoto Protocol reflecting the problems of climate warming via emission of greenhouse gases suffers from a lack of objective reasons. In fact, existing climate models do not reliably reflect the correlation of global temperature with anthropogenic emissions of greenhouse gases. These models do not take into consideration the role of biospheric processes in the land and oceans (biotic regulation of the environment and the functioning of the ocean chemical system). The Millennium Ecosystems Assessment Program can solve the problem of global model synthesis in order to obtain forecasts of the dynamics of the NSS.

A summary of existing simulations of the global carbon dioxide cycle shows the existence of long-term studies concerning this problem. The main conclusion is that the exchange of carbon between the atmosphere and biosphere has spotty spatial characteristics. This spottiness, taken into account in the models of the carbon dioxide cycle, produces high errors. That is why the Kyoto Protocol conclusions were based on incomplete estimations. There are many carbon flows (natural and anthropogenic), mentioned in Fig. 1.30, which have not been satisfactorily described parametrically. The principal key questions relating to the exchange of carbon between the atmosphere and the terrestrial pool of the above-ground biomass, below-ground biomass, soils and hydrospheric systems are discussed by Watson et al. (2000), Wigley and Schimel (2005), Emerson and Hedges (2008). The Arctic Basin is the most weakly studied part of the biosphere in order to understand its global role in the CO_2 absorption processes. Figure 6.26 demonstrates the high variability in the forecasting of the carbon dioxide dynamics under different anthropogenic scenarios:

(1) the pessimistic scenario of Keeling and Bacastow (1977) describes a situation where the ocean's role in the carbon exchange with the atmosphere is restricted only to physical processes;
(2) the optimistic scenario of Bjorkstrom (1979) takes into account the ocean-carbonate system under the parameterization of the H_2 and H_3 flows in Fig. 1.30;
(3) the scenario of the IPCC, Intergovernmental Panel on Climate Change (Watson et al. 2000), is based on specific activity requirements concerning land-use strategy (e.g. planting versus regeneration through silvicultural activities); and

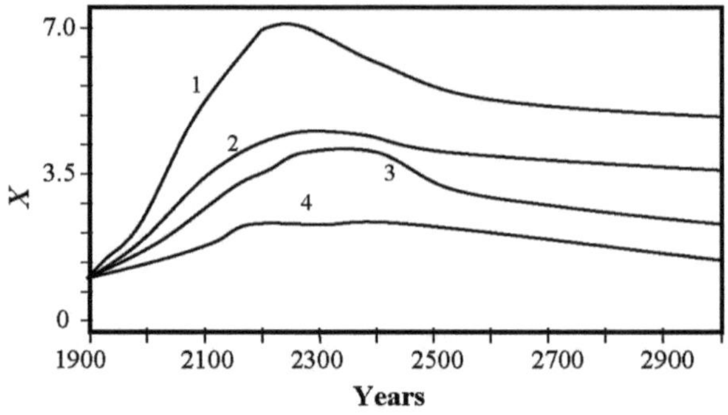

Fig. 6.26 Forecasting of the carbon dioxide content in the atmosphere obtained under different anthropogenic scenarios: *1* Keeling and Bacastow pessimistic scenario; *2* Bjorkstrom optimistic scenario; *3* IPCC scenario; *4* Kondratyev realistic scenario. The ordinate $X = C_a(t)/C_a(1900)$

(4) the realistic scenario of Kondratyev (1999) foresees the existing tendencies in the world's energetic, demographic and urbanization processes. This scenario is realized by the GSM (Krapivin 1993; Kondratyev and Krapivin 2001c, d).

It is obvious that in the framework of the third scenario a greenhouse effect via CO_2 is problematic. To estimate the dynamic of the atmospheric temperature dependence on carbon dioxide it is necessary to use a correlation describing the greenhouse effect. There are many various empirical functional representations of the warming effects. Simple correlations (see (5.3) and (5.4)) approximate the existing empirical dependence of the atmospheric temperature deviation ΔT_{CO_2} upon the variations in the atmospheric carbon dioxide parameter $X = C_a(t)/C_a(1900)$. One of them is:

$$\Delta T_{CO_2} = \begin{cases} 0.911 + 1.509 \ln X - 1.25 \exp\{-0.82(X-1)\} \text{ when } X \geq 1; \\ = 2.63X^2 + 6.27X + 1.509 \ln X - 3.988 \text{ when } X < 1 \end{cases}$$

Finally, according to the GSM calculations, future emissions of CO_2 will commit Earth to a warm-up by 0.2–1.3 °C. These variations in ΔT_{CO_2} depend upon the assumptions about the intensity of the urbanization process and about land-use strategy. The calculation results show that the dynamics of industrial CO_2 distribution between the atmosphere and oceans changes with the increasing preponderance of flows H_3 in the northern water areas. At the end of the 21st century and beginning of the 22nd century, during the highest pressure of human economic activity, the contribution of the oceans to industrial CO_2 absorption will be considerably lower than that of vegetation. This is due to the fact that, with an increase in CO_2 concentration in the atmosphere, the partial pressure in the oceans rises and its ability to absorb industrial CO_2 decreases, whereas the productivity of vegetation does not fall. By the end of the 22nd century, with a projected decrease in human impact, the contribution of the oceans increases due to the growing role of its deep layers. The atmospheric CO_2 will decrease at the same rate as the contribution of the oceans increases.

The entire area of the Arctic Ocean is only 3.8 % of the world's ocean surface but its role in the carbon dioxide absorption varies from 23 to 38 % (16.7–28.9 GtC year^{-1}). This role is greatly influenced by seasonal variations in the ice cover and in the ecosystem productivity. That is why the results of this study have a preliminary character. It is necessary to continue the closer definition of the parametrical descriptions of flows H_i^C from Fig. 1.30. The authors of this book understand that policy-oriented computer tools aimed at supporting decision making processes related to global change, require the design of a new methodology in global modeling, based on simulation models with detailed spatial descriptions of the biosphere systems and with operative informational service defining the parameters.

It is clear that the Arctic Basin role in climate change is more complex than was represented. There exist many informational uncertainties related to the permafrost, methane emissions to the atmosphere. Further development of global information-modeling system lies in the modernization of the GSM by means of detailed elaboration of its blocks especially that parameterize the cycles of greenhouse gases. As example, Figs. 6.27 and 6.28 represent more detailed schemes for biogeochemical cycles of CO_2 and CH_4.

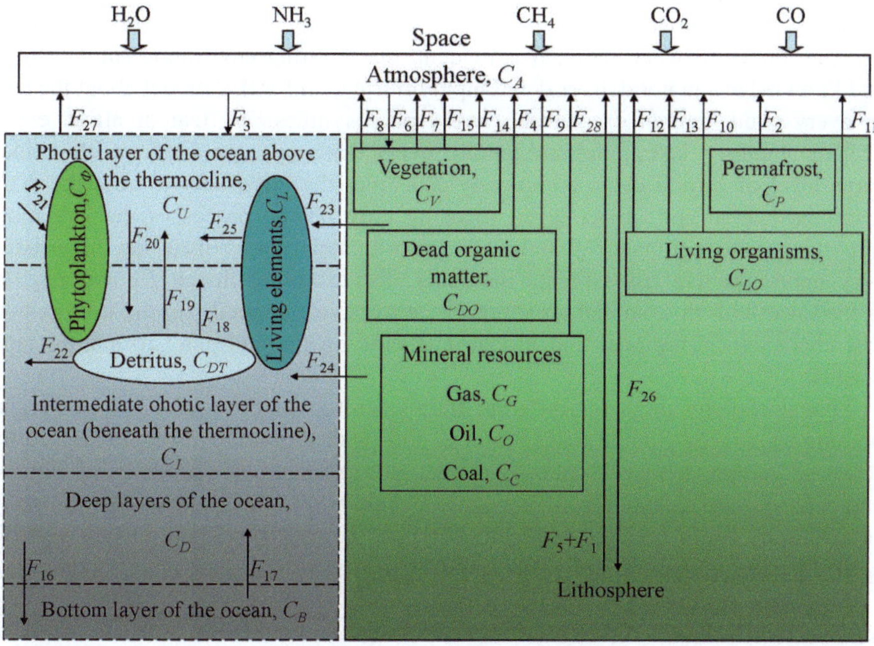

Fig. 6.27 A block-scheme of the global carboncycle model (GCCM) in the atmosphere-ocean-land system. The CO_2 reservoirs and fluxes are listed in Table 6.24

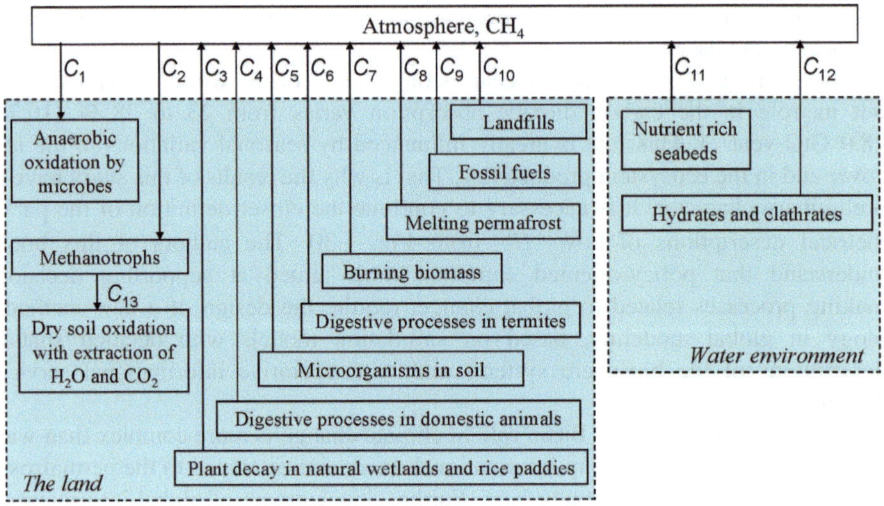

Fig. 6.28 Global methane cycle in the atmosphere-ocean-land system

As it is concluded from Table 6.24, where existing evaluations of carbon reservoirs and fluxes are given, all carbon pools and fluxes are divided by categories. Model of global biogeochemical carbon cycle is to account these reservoirs and to describe fluxes between them. It is evident, that existing environmental databases and knowledge can not deliver detail input information for this model about the role of every plant, animal or microbe, every photosynthesizing leaf or algae, every oceanic aquatory, lake, pond and puddle, every landscape, etc. Figure 6.28 defines some level of carbon cycle complexity and gives the possibility to describe it on restricted spatial and object level, focusing only on the processes that are most important at the regional or global scale. Unfortunately, series of carbon fluxes have very approximative assessments including, for example, fluxes F_3 and F_{26}. In accordance with this, spatial structure of carbon cycle depends on the information that characterizes regional and global databases on the soil-plant formations and oceans.

Spatial ecology proposes different fragmentations of the land ecosystems and oceanic aquatories. Spatial distribution of soil-plant formations is characterized by series of geographical grids. For example, soil geographical databases of Europe and Canada are characterized by scale 1:1.000.000. Such databases of many countries are formed on the principle of typical landscapes distinguishing agricultural, forested, economic and other types. New harmonized global soil database of FAO covers over 15,000 different units of soil mapping that combine existing regional and national updates of soil information worldwide in framework of 1:5,000,000 scale. Existing global GIS datasets are the most acceptable for the synthesis of global carbon cycle model.

Table 6.24 Reservoirs and fluxes of carbon as CO_2 in the biosphere in a simulation model of the global biogeochemical cycle of carbon dioxide shown in Fig. 6.28

CO_2 reservoirs and fluxes	Identifier	Tthe reservoir (GtC) or flux (GtC/year)
Atmosphere	C_A	650–750
Ocean		
Photic layer of the ocean above the thermocline	C_U	380–520
Intermediate photic layer of the ocean	C_I	280–610
Ocean plants (algae)	C_Φ	3
Living elements	C_L	1.5
Detritus	C_{DT}	29
Deep layers of the ocean	C_D	28,500–33,890
Bottom layer of the ocean	C_B	5,000
wLand		
Dead organic matter (in soil and peat)	C_S	1,500–3,000
Vegetation	C_V	550–610
Permafrost	C_P	1,672
Living organisms	C_{LO}	0.5
Fossil fuels		
Coal	C_C	4,500
Oil	C_O	500
Gas	C_G	5,000
Emission due to burning of the vegetation	F_8	6.9
Fossil fuel and cement production	F_{28}	4,000
Rock weathering	F_4	0.04
Volcanic emanations	F_5	2.7
Assimilation by land vegetation	F_6	224.4
Respiration		
Plants	F_7	50–59.3
Human	F_{10}	0.7
Animals	F_{11}	4.1
Emissions		
Decomposed soil humus	F_9	139.5
Plants roots	F_{15}	56.1
Permafrost melting	F_2	190
Vital functions		
Human population	F_{12}	0.3
Animals	F_{13}	3.1
Vegetation decay	F_{14}	31.5–50
Sedimentation to bottom deposits	F_{16}	0.1–0.2
Solution of marine sediments	F_{17}	0.1

(continued)

Table 6.24 (continued)

CO_2 reservoirs and fluxes	Identifier	Tthe reservoir (GtC) or flux (GtC/year)
Decomposition of detritus		
Photic layer	F_{22}	35
Deep layers of the ocean	F_{18}	5
Upwelling with deep waters	F_{19}	45
Sinking with surface waters and due to gravitational sedimentation	F_{20}	40
Photosynthesis	F_{21}	69
Underground sink	F_{23}	0.5
Surface sink	F_{24}	0.5–0.6
Breathing of living organisms in the ocean	F_{25}	25
Degazation processes	F_1	21.16
Sink to the Earth's bowels	F_{26}	1.3
Desorption	F_{27}	97.08

Carbon fluxes represented in Fig. 6.28 and Table 6.24 are used for the balance equations components of which are functions of environmental parameters and time:

$$\frac{\partial C_{As}(\varphi,\lambda,t)}{\partial t}+V_{\varphi s}\frac{\partial C_{As}(\varphi,\lambda,t)}{\partial\varphi}+V_{\lambda s}\frac{\partial C_{As}(\varphi,\lambda,t)}{\partial\lambda}=\sum_{i\in I_s}F_i-\sum_{j\in J_s}F_j$$

where C_{As} is the carbon reservoir in the s-th pixel of spatial structure of the land and ocean; I_s and J_s are spaces of the carbon sources and sinks, respectively; φ is latitude; λ is longitude; t is time; $V_s(V_{\varphi s}, V_{\lambda s})$ is the wind field in the s-th pixel.

The interaction between carbon reservoirs—the atmosphere, land pixels Ξ_{ij} and ocean aquatories Ω_{sk}—is expressed by carbon fluxes formed through ecological, geophysical, and biogeochemical processes, including photosynthesis, respiration, decomposition, burning, volcanic emanations, rock weathering, etc. Detailed description of these processes and their parameterization are given by Kondratyev (2000b), Kondratyev et al. (2003a), Krapivin and Varotsos (2008), Wigley and Schimel (2005), Hanson et al. (2000) and Archer (2010). Some of them are to be corrected and defined more precisely.

Methane fluxes C_i (i = 1–13) are functions of temperature, microbial population, geographical coordinates and other environmental characteristics that are specific for pixels Ξ_{ij}. For example, Panikov and Dedysh (2000) developed the model for methane emission to the atmosphere from snow-covered bogs in West Siberia. This approach can be used for the modeling of anaerobic formation of CH_4 in pixels with the permafrost. In this case flux C_8 can be described by the equation:

$$dC_8/dt = \mu X(1 - Y)/Y,$$

where μ is specific growth rate of respective microbial population with biomass *X*, *Y* is biomass yield per unit of substrate consumed. Microbial biomass *X* is formed with the following law:

$$dX/dt = \mu X,$$

where $\mu = \mu_{max} SR/(S + K_S)$, *S* is catabolic substrate concentration, *R* is the function of physiological state, K_S is saturation constant, numerically equal to that substrate concentration at which microbial specific growth rate attains the half of maximal value ($\mu = 0.5\mu_{max}$).

Natural wetlands and rice paddies deliver to the atmosphere more than 30 % of global CH_4 emission. Flux C_3 can be parameterized by the following equation:

$$C_3 = H_r f_1(T_s) f_2(h) f_3(\text{pH}) f_4(r_p),$$

where H_r is heterotrophic respiration, T_s is soil temperature, *h* is water table position, r_p is redox potential, functions f_i (i = 1–4) parameterized the CH_4 emission rates. Fluxes C_1 and C_2 that characterize major atmospheric CH_4 sinks are mainly parameterized by its reaction with hydroxyl (OH) radical. These fluxes depend on the OH levels and reaction rate. Under this, it is known that increase in methane leads to positive feedback (Xu et al. 2007; Krapivin and Varotsos 2008).

The scheme of Fig. 6.27 is realized by global carbon cycle model (GCCM) blocks of which are characterized in Fig. 6.29 and Table 6.25. Many of the GCCM blocks have form of coupled set of time-dependent advection, diffusion, and reaction equations. Set of theoretical and empirical models for the carbon fluxes of Fig. 6.27 are realized by specific blocks, represented in Fig. 1.30. (Kondratyev et al. 2003b; Degermendzhi et al. 2009; Williams and Follows 2011).

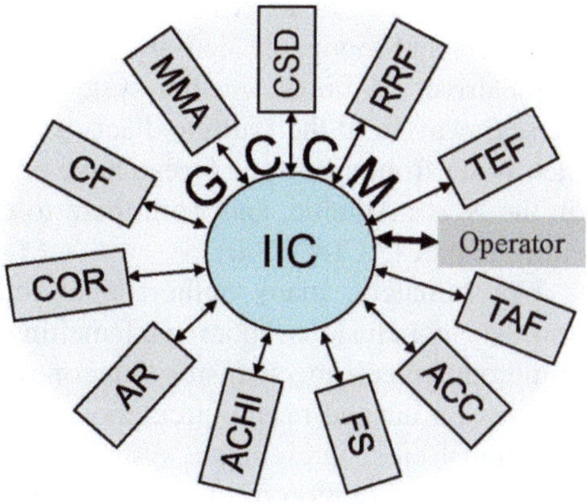

Fig. 6.29 Structure of the GCCM. Notation is given in Table 6.25

Table 6.25 List of blocks of the GCCM, whose schemes are presented in Figs. 6.28 and 6.29

Identifier of the block	Description of the unit
CSD	Coordination of spatial distribution of pixels Ξ_{ij} and aquatories Ω_{sk} with the available database and knowledge base
MMA	Management of models and algorithms for the parameterization of the fluxes F_i ($i = 1, \ldots, 28$) and C_j ($j = 1, \ldots, 13$)
CF	Calculation of the fluxes F_i ($i = 1, \ldots, 28$) and C_j ($j = 1, \ldots, 13$) taking into account of spatial structure of the atmosphere-land-ocean system
COR	Correction of the fluxes F_i ($i = 1, \ldots, 28$) and C_j ($j = 1, \ldots, 13$) taking into consideration of the atmosphere circulation, hydrological processes and oceanic currents
AR	Algorithm for the database and knowledge base renovation
ACHI	Algorithm for the coordination of heterogeneous information
FS	Formation of scenarios that change spatial structure of soil-plant formations, fluxes F_3, F_{27} and F_{28} due to the ocean surface pollution and changes in the technologies of the use of mineral resources
ACC	An assessment of climate change indicators as the consequence of the specific scenario realization
TAF	Taking into account of the fluxes F_i ($i = 1, \ldots, 28$) and C_j ($j = 1, \ldots, 13$) changes under the variations of climatic parameters
TEF	Tracing the environmental feedbacks in their dynamics and under variations of their parameters
RRF	Re-calculation of radiative forcing indices depending on the F_i/C_j dynamics in the appropriate pixels and aquatories
IIC	Information interface and control of information connections between the GCCM blocks

6.8 Conclusion

As it was mentioned above, there exist many implementation plans for an assessment of the consequences of climate variability and change and the effects of increased anthropogenic activity in the Arctic Region. The Arctic's sea ice is a major driver of global weather systems. The light surface of the ice reflects solar energy away from the Earth and acts as a natural refrigerator for the planet. Ice and melt water from the Arctic Ocean have profound affects on ocean circulation pattern on the North Atlantic, and from there to ocean and other climate systems over the entire planet (ACIA 2000).

Unfortunately, many authors evolving the most sophisticated analysis make untimely conclusions about total melting the Arctic's sea ice. Really there exist numerous inverse mechanisms that control complex effects of different natural subsystems on air and ocean circulation, clouds and precipitation stabilizing regional and global biosphere-climate system. Therefore, only coupled models can help to evaluate the evolution of this system. The TOPAZ model system is one of them. It is an operational real time ocean modeling and forecasting system for the North

Atlantic and the Arctic Ocean. This model provides 10 days sea ice forecasts for the Arctic and adjacent seas in addition to 3-D oceanographic fields. The TOPAZ system produces maps with 20–25 km resolution of sea surface temperature, ice concentration and ice thickness on regular bases.

Arctic Basin study is the subject of many national and international scientific programs. For example, the IBCAO project has received wide recognition and acceptance on account of the excellence of its output, and it has been gratifying to see increasing references to IBCAO as an important adjunct to scientific research in the Arctic. The IBCAO continues to be the most popular of the International Bathymetric Charts projects hosted at NGDC.

The NASA Cold Land Processes Working Group (CLPWG) proposed to coordinate implementation of a tiered system for monitoring the Pan Arctic dynamics of the terrestrial snowpack, particularly in regards to its effects on fresh water supply to the Arctic Basin and its significant coupling to the atmosphere. The system of monitoring Arctic terrestrial snow includes three categories of ground measurements: long-term research areas, field campaigns associated with the IPY, and cooperative measurements by educational institutions.

Key aspects of the Arctic Basin study in its evolution are climate predictions. Unfortunately, considerable uncertainties exist in this area. Many authors try to develop numerical models based on the past climate reconstructions. It is clear that main goal is to be the development of a realistic Earth System model that includes all the relevant components of the climate system (atmosphere, ocean, land-biosphere, oceanic carbon cycle, cryosphere). In this aspect, because of limited computer resources, conventional studies often neglect or do not formulate sufficiently certain climate components (Rawlins et al. 2005). Solution of this problem lies in the use of coupled small-scale models (Bartsev et al. 2003; Degermendzhi et al. 2009). Luus et al. (2013) shown that snow outset and melt processes play important role in arctic climate change through the Arctic net ecosystem exchange of CO_2 between the land and the atmosphere. The model developed by Luus et al. (2013) can considered as small-scale model for estimation of the net ecosystem exchange. This model is based on satellite data delivered by Landsat and MODIS.

The key role of the soil-plants and the World Ocean in the formation of the carbon cycle can be assessed by means of the GCCM that allows the realization of different scenarios with the reconstruction of soil-plant formations or the World Ocean pollution. Spatial resolution of distribution of soil-plant formations can be change without the model reconstruction.

Unfortunately, geographical grid $4° \times 5°$ is very rough for the consideration of scenarios with more detailed changes in the structure of vegetation cover, including agriculture fields. Therefore, the GCCM modernization is needed, taking into consideration of spatial grids $1° \times 1°$, $0.5° \times 0.5°$ and $1\ km \times 1\ km$. In this case, additional blocks are formed for the parameterization of many carbon fluxes in Fig. 6.28 basing on the satellite data concerning the vegetation indices such as NDVI. But there exist regions where specific information is needed. One of these regions is cryotic soils (permafrost) where the carbon and methane exchange processes strongly depend on the climate change. Variety of geophysical, geochemical and ecological processes in permafrost complicates an adequate choice model for these processes to be parameterized. It is known that in a periglacial environment the effects of freezing and thawing modify drastically the ground surface and change the gas exchange between the atmosphere and land. All these aspects suggest the necessity to develop detailed GIS maps for regions with permafrost (Polischuk and Polischuk 2013).

There, exist an other unsolved task connected with a link between cosmic dust and climate change. It is known that 100–300 tons of cosmic dust enters the atmosphere each day. It is needed to understand how this dust impacts on global biogeochemical cycles (Ehrenfreund and Spaans 2008).

Chapter 7
Tropical Cyclogenesis and Ecoinformatics Methods

7.1 Introduction

Undoubtedly, the state of the ocean-atmosphere system (OAS) plays an important role in the life of many countries. For instance, the tropical cyclones as hazardous state of the OAS occupy a specific position on the potential damages caused by hurricanes, where the risks for economic losses and victims are very high. So, hurricanes, as the final phase of the tropical cyclone, pose a serious threat to many countries in both economic and human terms. That is why the problem of the prediction of the start, the motion and the intensity of a tropical cyclone is important for the solution of the problem of tropical cyclones.

It has been considered that the tropical cyclogenesis is linked to specific solar processes. For instance, the fluctuations of the flow of galactic cosmic rays were found to be in phase with fluctuations in the vertical temperature profile of the atmosphere initiating tropical cyclones (Callaghan and Power 2011; Bondur et al. 2009; Bondur and Krapivin 2014; Enz 2006; Stauffer and Aharony 1994). Very recently, an out of phase relationship between hurricane activity and the 11-year solar cycle have been noticed (Elsner et al. 2010). The latter revealed that this opposite correlation is not caused by the solar and cosmic variations, but instead, by the fluctuations of the solar ultraviolet (UV) radiation (Kondratyev and Varotsos 1996; Varotsos 2005).

It should be therefore of great importance to investigate the phenomenon of the tropical hurricanes from its initial stages of its formation, in order to go further to its prediction. The development of satellite and other observational methods resulted in the routine observational analysis of the parameters that describe the start of the tropical hurricanes and their evolution in the early stages, thus allowing the reduction of risks of this dangerous natural phenomenon (Cracknell and Varotsos 2007; Soldatov et al. 2011; Nitu et al. 2013a, b; Krapivin et al. 2007b; Topics Geo 2003–2008). In this context, the satellite passive microwave radiometric methods serve as an important tool for the determination of oceanographic and meteorological

V.F. Krapivin et al., *New Ecoinformatics Tools in Environmental Science*,
Environmental Earth Sciences, DOI 10.1007/978-3-319-13978-4_7

parameters, that depict the influence of energy exchange between the sub-systems of the OAS, such as the water surface temperature, wind speed, total water vapor content in the atmosphere, integral content of liquid water in clouds, as well as the vertical turbulent fluxes of heat, which are especially important in studies of cyclonic fields of the ocean. The satellite observations can also give indirect information about the processes in the ocean column from its bottom, which are invisible for means of direct remote sensing (Fedorov and Ginsburg 1988).

The goal of many studies is to develop technologies of diagnostics of the tropical hurricane origin in ocean zones, which are regular sources of the hurricane generation on the basis of the passive microwave radiometric observations conducted by satellites, vessels, buoys and mathematical modeling data of the OAS behavior at different stages, notably (Krapivin 2010; Kondratyev et al. 2006a, b; Krapivin and Shutko 2012; Soldatov et al. 2011; Krapivin and Kondratyev 2002):

(a) the stage preceding the tropical hurricane appearance;
(b) the tropical hurricane appearance; and
(c) the stage of the OAS relaxation after the tropical hurricane appearance.

Soldatov et al. (2010) have recently proposed a new technology for the OAS diagnostics based on the calculation of the instability indicator. The solution to the problem related to the investigation of indicator-precursor for the tropical hurricane origin is associated with the assessment of the OAS state and with the peculiar states of the system at the time of the commencement of its phase transition. Basic phases of the OAS are determined by the Saffir-Simpson scale which classifies hurricanes into five categories according to their intensity (http://www.nhc.noaa.gov/aboutsshws.php).

It has long been recognized that the environmental changes that are associated with the OAS evolution, have all stochastic character (Abu-Zaid and Ahmadi 1989; Mees et al. 1997), as numerous other natural catastrophes that are distinguished between them only by the environmental parameters that affect their start (Callaghan and Power 2011). In connection to this, the indicator-precursors may be investigated only by means of the analytical relationships that describes the functioning of the global climate system. An illustrative example is the employment of the conventional Boolean algebra algorithm and baroclinic models for the forecasting of the tropical cyclone motion (Kondratyev and Krapivin 2005a, b, 2006; Tang et al. 2005; Zhou and Chen 2006).

7.2 Ocean-Atmosphere System Phases

Understanding the ocean-atmosphere interaction is critical for predicting changes in global temperature patterns and climate variability with global warming as well as the prediction of beginning the tropical cyclones. Ocean-atmosphere coupling is a concept of climate dynamics essential for understanding a large number of climate phenomena including the El Niño–Southern Oscillation (ENSO) and climate variability on multi-decadal timescales. Heat exchange between the ocean and the atmosphere (heat fluxes), atmospheric circulation and hence the effect of winds on the ocean surface (wind stress) and a large-scale anomaly of sea surface temperature are principal areas of investigations based on the numerous observations and models.

Ocean-atmosphere coupling in different latitudes remains a subject of intense research and even heated debate. Climate variability is determined to a large extent by random atmospheric variability, also known as the atmospheric "noise," and passive oceanic response to the atmospheric forcing. Numerous efforts to parameterize the ocean-atmosphere interactions within different aspects of simulation modeling did not give some successful results. The main difficulties arising in the solution of tasks are caused by instability of the ocean-atmosphere system (OAS). For this reason, Soldatov (2009a, b) proposed a new approach to the study of the ocean-atmosphere interactions considering the state of this system as the assembly of its phases.

There exist scales that classify the OAS states. The most known of them are Saffir-Simpson Hurricane scale, Beaufort scale and Fujita-Pearson scale.

Saffir-Simpson scale was developed in 1971 by civil engineer Herbert Saffir and meteorologist Bob Simpson, who at the time was director of the U.S. National Hurricane Center (NHC). This scale is a classification used for some Western Hemisphere tropical cyclones that exceed the intensities of tropical depressions and tropical storms. The scale divides hurricanes into five categories distinguished by the intensities of their sustained winds. To be classified as a hurricane, a tropical cyclone must have maximum sustained winds of at least 74 mph (33 m/s; 64 kn; 119 km/h). The highest classification in the scale, Category 5, is reserved for storms with winds exceeding 155 mph (69 m/s; 135 kn; 249 km/h). Really there exist two additional categories that correspond to storm and depression (Table 7.1).

The Beaufort scale (Table 7.2) is an empirical measure of describing wind speed based mainly on observed sea conditions. The scale was created in 1806 by Sir Francis Beaufort, an Irish-born British admiral and hydrographer. The Beaufort scale was extended in 1946, when Forces 13–17 were added. However, Forces 13–17 were

Table 7.1 The Saffir-Simpson scale for the measure of the hurricane magnitude (Webster et al. 2005)

Hurricane category	Maximum wind speed		Minimum atmospheric pressure (hPa)	Storm surge height (m)
	m/s	km/h		
Tropical depression	0–17	0–62		
Tropical storm	18–32	63–118		
One	33–43	119–153	≥980	1.0–1.7
Two	43–50	154–177	979–965	1.8–2.6
Three	50–56	178–209	964–945	2.7–3.8
Four	56–67	210–249	944–920	3.9–5.6
Five	>67	>250	<920	≥5.7

Table 7.2 The modern Beaufort scale

Beaufort number	Description	Wind speed (m/s)	Wave height (m)	Sea conditions
0	Calm	<0.3	0	Flat
1	Light air	0.3–1.5	0–0.2	Ripples without crests
2	Light breeze	1.6–3.4	0.2–0.5	Small wavelets
3	Gentle breeze	3.5–5.4	0.5–1	Large wavelets
4	Moderate breeze	5.5–7.9	1–2	Small waves with breaking crests
5	Fresh breeze	8.0–10.7	2–3	Moderate waves of some length
6	Strong breeze	10.8–13.8	3–4	Long waves begin to form
7	High wind, moderate gale, near gale	13.9–17.1	4–5.5	Sea heaps up
8	Gale, fresh gale	17.2–20.7	5.5–7.5	Moderately high waves with breaking crests forming spindrift
9	Strong gale	20.8–24.4	7.5–9	High waves whose crests sometimes roll over
10	Storm, whole gale	24.5–28.4	9–11.5	Very high waves with over-hanging crests
11	Violent storm	28.5–32.6	11.5–14	Exceptionally high waves
12	Hurricane-force	≥32.7	≥14	Huge waves

intended to apply only to special cases, such as tropical cyclones. Nowadays, the extended scale is only used in Taiwan and mainland China, which are often affected by typhoons. Wind speed on the 1946 Beaufort scale is based on the empirical formula:

$$v = 0.837B^{3/2}\ \mathrm{m/s}$$

where v is the equivalent wind speed at 10 m above the sea surface and B is Beaufort scale number. For example, $B = 9.5$ is related to 24.5 m/s which is equal to the lower limit of "10 Beaufort". Using this formula the highest winds in hurricanes would be 23 in the scale.

Today, hurricane force winds are sometimes described as Beaufort scale 12 through 16, very roughly related to the respective category speeds of the Saffir-Simpson Hurricane Scale, by which actual hurricanes are measured, where Category 1 is equivalent to Beaufort 12. However, the extended Beaufort numbers above 13 do not match the Saffir-Simpson Scale. Category 1 tornadoes on the Fujita and TORRO (TORnado and storm Research Organization) scales also begin roughly at the end of level 12 of the Beaufort scale but are indeed independent scales. The TORRO scale (or T-Scale) is a scale measuring tornado intensity between T0 and T11. It was developed by Meaden (1985) of the TORRO a

Table 7.3 TORRO scale parameters

FI	*v*, m/s	Tornado description	Intensity description	Damage description
FC		Funnel cloud aloft (not a tornado)	Weak	No damage to structures, unless on tops of tallest towers, or to radiosondes, balloons, and aircraft. No damage in the country, except possibly agitation to highest tree-tops and effect on birds and smoke. Record FC when not known to have reached ground level. A whistling or rushing sound aloft may be noticed
F0	17–24	Light		• Loose light litter raised from ground level in spirals • Tents, marquees, awnings seriously disturbed • Some exposed tiles, slates on roofs dislodged. Twigs snapped; trail visible through crops • Wheelie bins tipped and rolled • Garden furniture and pots disturbed
F1	25–32	Mild		• Deck chairs, small plants, heavy litter becomes airborne • Minor damage to sheds • More serious dislodging of tiles, slates • Chimney pots dislodged. Wooden fences flattened • Slight damage to hedges and trees • Some windows already ajar blown open breaking latches
F2	33–41	Moderate		• Heavy mobile homes displaced. Light caravans blown over • Garden sheds destroyed. Garage roofs torn away and doors imploded • Much damage to tiled roofs and chimneys. Ridge tiles missing • General damage to trees, some big branches twisted or snapped off, small trees uprooted • Bonnets blown open on cars • Weak or old brick walls toppled • Windows blown open or glazing sucked out of frames
F3	42–51	Strong		• Mobile homes overturned/badly damaged. Light caravans destroyed. Garages and weak outbuildings destroyed • House roof timbers considerably exposed. Some of the bigger trees snapped or uprooted • Some heavier debris becomes airborne causing secondary damage breaking windows and impaling softer objects • Debris carried considerable distances. Garden walls blown over • Eyewitness reports of buildings physically shaking • Mud sprayed up the side of buildings
F4	52–61	Severe	Strong	• Motorcars levitated. Mobile homes airborne/destroyed • Sheds airborne for considerable distances. Entire roofs removed from some houses • Roof timbers of stronger brick or stone houses completely exposed. Gable ends torn away • Numerous trees uprooted or snapped. Traffic Signs folded or twisted

(continued)

Table 7.3 (continued)

FI	v, m/s	Tornado description	Intensity description	Damage description
				• Some large trees uprooted and carried several yards • Debris carried up to 2 km leaving an obvious trail
F5	62–72	Intense		• Heavier motor vehicles (4 × 4, 4 tonne trucks) levitated • Wall plates, entire roofs and several rows of bricks on top floors removed • Items sucked out from inside house including partition walls and furniture • Older, weaker buildings collapse completely • Utility poles snapped
F6	73–83	Moderately-devastating		• Strongly built houses suffer major damage or are demolished completely • Bricks and blocks etc. become dangerous airborne debris • National grid pylons are damaged or twisted • Exceptional or unusual damage found, e.g. objects embedded in walls or small structures elevated and landed with no obvious damage
F7	84–95	Strongly-devastating		• Brick and Wooden-frame houses wholly demolished • Steel-framed warehouse-type constructions destroyed or seriously damaged • Locomotives thrown over • Noticeable de-barking of trees by flying debris
F8	96–107	Severely-devastating	Violent	• Motorcars carried great distances • Some steel framed factory units severely damaged or destroyed • Steel and other heavy debris strewn over a great distances • A high level of damage within the periphery of the damage path
F9	108–120	Intensely-devastating		• Many steel-framed buildings demolished • Locomotives or trains hurled some distances • Complete debarking of any standing tree-trunks •Inhabitants survival reliant on shelter below ground level
F10	121–134	Super		• Entire frame houses and similar buildings lifted bodily from foundations and carried some distances • Destruction of a severe nature, rendering a broad linear track largely devoid of vegetation, trees and man made structures

Notation FI TORRO intensity; v wind speed (m/s)

meteorological organization in the United Kingdom, as an extension of the Beaufort scale. T-scale (Table 7.3) sets T0 as the equivalent of 8 on the Beaufort scale and is related to the Beaufort scale (B) by the formula: $B = 2(TI + 4)$ and conversely: $TI = (B/2 - 4)$. TORRO scale formula for the wind speed is $v = 2.365(T + 4)^{3/2}$ m/s or $v = 8.511\ (T + 4)^{3/2}$ km/h.

Table 7.4 Characteristic of the Fujita-Pearson scale (Marshall 2001)

Level of scale	Intensity	Wind speed (km/h)	Damage scale	Type of danger
F0	Storm	40–72	Light	Some damages to chimneys, broken tree branches, uprooted trees, signs torn off the buildings
F1	Moderate tornado	73–112	Moderate	Lowest wind speed of the early hurricane. Blows off the roofs, pools down houses from foundation or turns then over, moves the cars, can destroy separate or attached garages
F2	Substantial tornado	113–157	Considerable	High level of danger. Blows off the roofs, destroys mobile dwellings, overturns box garages, breaks or uproots big trees, takes away lightweight objects
F3	Strong tornado	158–206	Severe	Destroys roofs and walls of stationary constructions, overturns trains, uproots the trees
F4	Devastating tornado	207–260	Devastating	Destroys solid buildings, takes away houses with poor foundations, cars, and large objects
F5	Incredible tornado	261–318	Incredible	Throws houses off foundations and takes them away destroying, moved cars at distances up to 100 m, breaks trees, heavily damages steel, concrete-fastened strictures
F6	Inconceivable tornado	319–379	Maximal	Such winds are improbable. Their destructions are difficult to distinguish from destructions by tornados of the type F4 and F5. The emergent vortex exhibit a huge power over a small territory, it can take large volumes of water and transport them at great distances

In most cases, tornado intensity is measured by the Fujita-Pearson scale (Table 7.4) that really reflects the risk level. It is important to know when and how the tornado origins. It is known that many tornadoes have their origin as the product of a tropical hurricane. Therefore, the information about conditions of tropical hurricane beginning and its tracing are principal stages to predict the tornado origin.

The solution of the tasks arisen here depends on data that have different origin. The scale allows a tornado's wind speed to be determined by various means:

- visiting the tornado damage site to make non-engineering assessments;
- obtaining an engineering assessment of the damage;
- using Doppler radar;
- applying photogrammetric analysis; and
- directly measuring observed tornadoes.

Whose solution permits us to forecast the hurricane origin and prognosis of its category.

7.3 Tropical Cyclone as Dynamic Category of the Environmental Phenomena

Tropical cyclone refers to the category of natural disasters the origin and evolution of which does not depend on the human. Walker (2003) and Grigoriev and Kondratyev (2001) noted that number of natural catastrophes increased during last time. For example, up to 1990 during all of foregoing 30 years more thousand tornadoes was registered in USA only in 1973 but this level is exceeded during last time regularly.

Unstable atmospheric systems (tornadoes and hurricanes) have large destructive power but they differ on origin, frequency and spatial scales. Only on the USA territory, 10–15 hurricanes and 1,000–1,200 tornadoes are beginning during a year. Ahurricane can cover a space up to 2,000 km whereas a tornado has a moving band no more than 4 km (Figs. 7.1 and 7.2). Some statistics given in Figs. 7.3, 7.4 and Table 7.5 shows how the number of natural catastrophes changes and the consequences arisen from them.

Thereby, many authors associate the natural catastrophenatural notion with the notion of ecological security. This is arisen in connection with the necessity of danger assessment for population of some territory that can lose health, buildings or property because of the environment change. These changes can be caused by natural or anthropogenic reasons. In the first case, dangerous origins due to fluctuations of natural processes connected with synoptic changes, epidemic occurrence, or natural hazard. In the second case, risk originates from a nature reaction on human actions. For example, Gardner (2002) analyzed environmental changes in the Himalayas on Indian territory and came to the conclusion that such facts as deforestation and vegetation cover change were stimulus and intensifiers of instabilities in this region. This is characterized by degradation of soil resources and increase of consequences from floods.

Historic knowledge about consequences of tropical cyclones practically is absent. It is obstacle to determine a correlation between tropical cyclones and other natural processes. There exists fragmentary information about separate episodes of tropical

Fig. 7.1 Tornado (http://www.fanpop.com/clubs/national-geographic/images/6968505/title/tornado-wallpaper)

Fig. 7.2 Tropical hurricane (http://en.wikipedia.org/wiki/Tropical_cyclone)

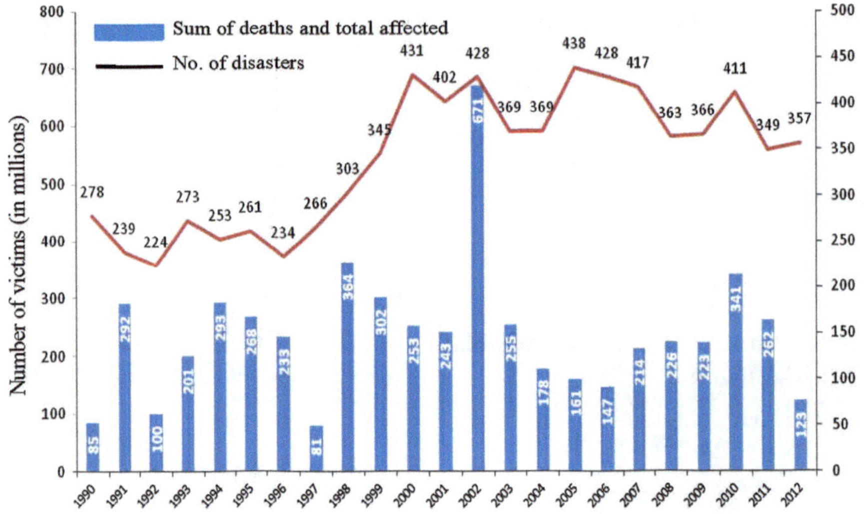

Fig. 7.3 Dynamics of total amount of natural catastrophes and victims beginning from 1990 (Guha-Sapir et al. 2013)

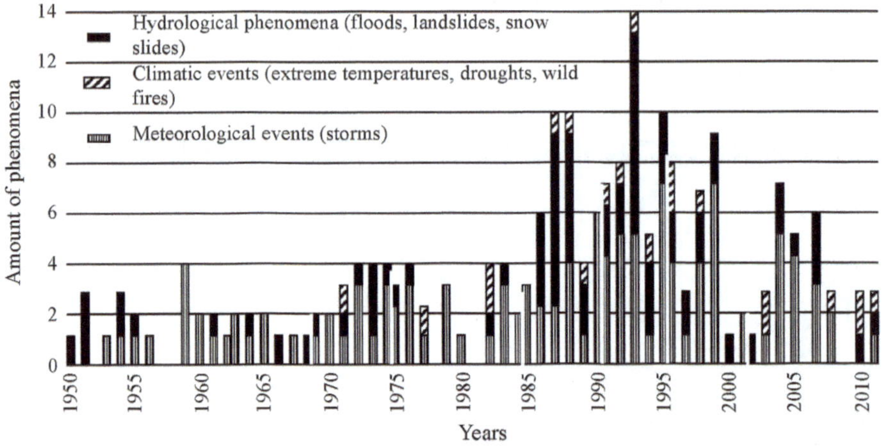

Fig. 7.4 Statistics of large natural catastrophes related with climatic processes

hurricanes. For example, intense tropical hurricane which was in June 1502 between Puerto Rico and Hispanola, as described by Christopher Columbus, destroyed 21 ships and killed about 500 seamen. The presence of such fragmentary data about tropical hurricanes is a subject for discussion of many experts in the field of meteorology about the possibility to reconstruct the correlation between climate change and intensity of hurricanes. Such efforts are made, for instance, by studing sediments in the USA coastal waters which were delivered during the past 1,000 years by hurricanes (Liu and Fearn 1993; Plumlee et al. 2006).

Table 7.5 Relation of damages connected with large catastrophes having climatic character during 1950–2012

Phenomenon	Total	Part of meteorological processes (%)	Part of hydrological processes (%)	Part of climatic processes (%)
Total phenomena	213	56	35	9
Fatalities (person)	1,050,000	76	12	12
Economical losses (billion dollars)	1,560	57	33	10
Insurance costs (billion dollars)	430	86	9	5

Intense destructive hurricanes (4th and 5th categories) historically are infrequent events. But they, certainly, played important role as geomorphic factors in coastal zone change as well as ecological factors in disturbance of ecosystems and in change of their evolution. Many authors (Dunion and Velden 2004; Fink et al. 2010; Wong 2005; DeMaria et al. 2005; Peng et al. 2012) mention clear correlation between the end of the 20-years drought cycle in the Sahara that was observed from 1989, and increase in intensity of hurricanes in the North Atlantic Ocean. In particular, well mixed dry adiabatic air layer with high content of mineral dust is formed above Sahara and Northern Africa and moving in western direction and under the moving off coast is impacted with cold and humid air above ocean surface. This process is observed very well by infrared sensors of GOES which has detected that when air layer above Sahara achieves a zone where tropical cyclone is born, relative decrease on its activity is observed. Certainly, these correlations have to be studied in the future.

Fink et al. (2010) determined the existence of seasonal correlation between activity of tropical cyclones in the Atlantic Ocean and precipitations in Western Africa. Under this, it is marked its instable character and certain variability in years of high water temperature. In common, rate of number of tropical disturbances in different active zones of the World Ocean has irregular distribution (Fig. 7.5). On the whole, more 54 % of tropical disturbances are formed in the Pacific Ocean. In general, the global average temperature fluctuates over recent time dramatically. Figure 7.6 shows trends of the annual atmospheric temperature anomalies for 1880–2012. Present temperature anomaly trend mainly is defined by the oceans that store about 90 % of the additional heat. The oceans are warming up what implies that the Earth must absorb more solar energy than longwave radiation that emits into space. The atmosphere stores only about 2 % because of its small heat capacity. Heat content of the World Ocean during last years has changed from about 1.05×10^{22} J in 1990 to 17.67×10^{22} J in 2013.

Figure 7.7 shows temporal sets for global number of tropical cyclones and storm days. It is known that there exists a ten-day oscillation in storm days. Figure 7.8 represents data about annual frequency and duration of hurricanes during 35 years. Figure 7.9 reflects global distribution of intensity of hurricanes. The number of

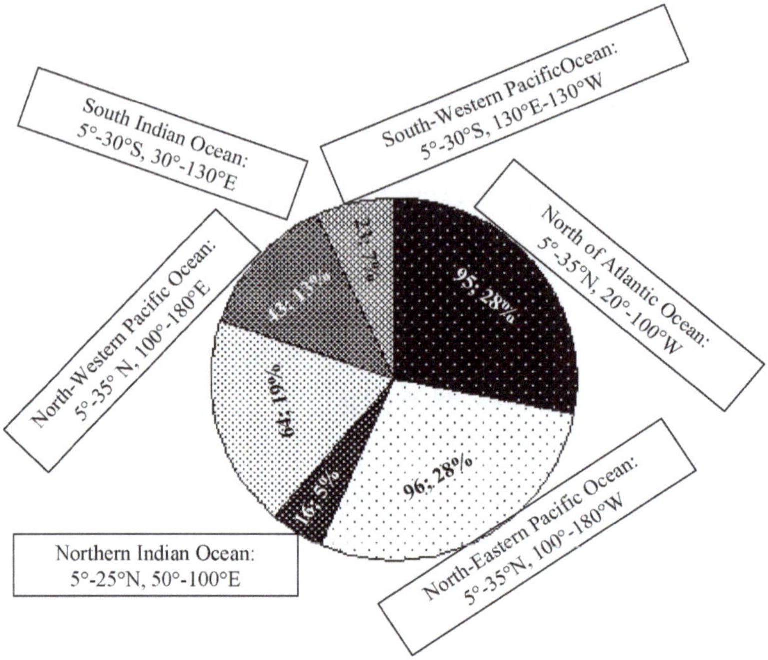

Fig. 7.5 Distribution of number of tropical disturbances in active zones

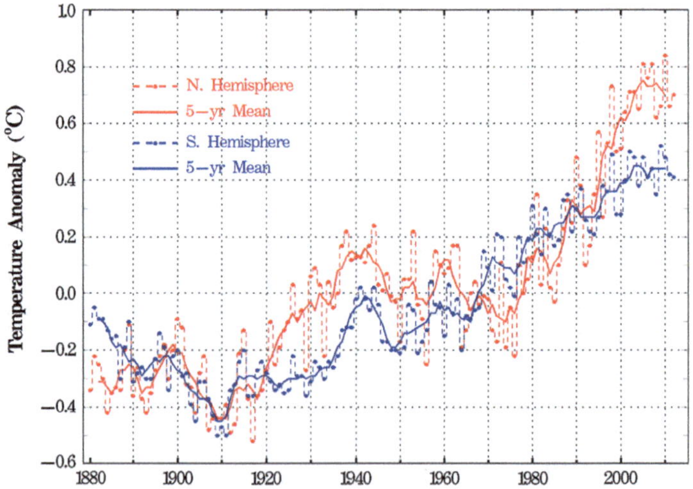

Fig. 7.6 Global temperature anomalies, 1880–2012 (Hansen et al. 2013)

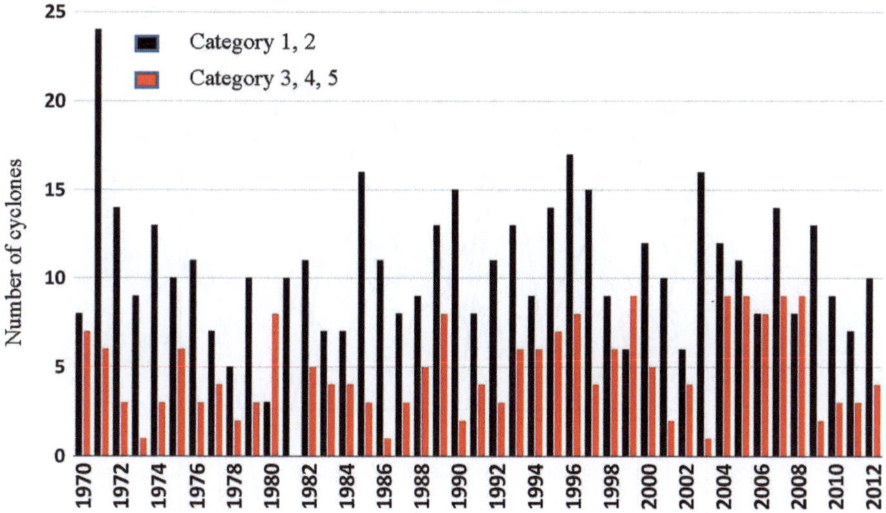

Fig. 7.7 Statistics of global tropical cyclone landfalls (Webster et al. 2005; Weinkle et al. 2012). http://rogerpielkejr.blogspot.com

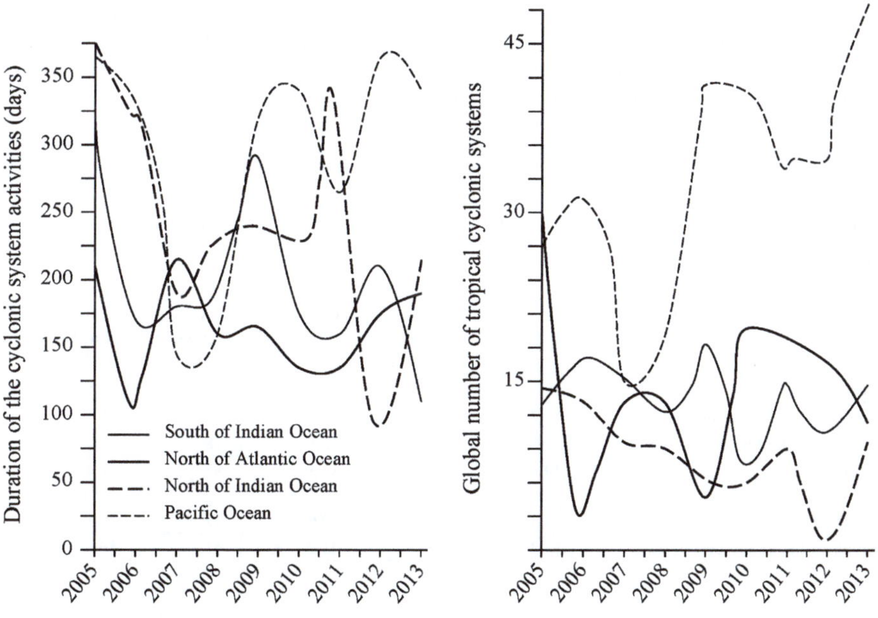

Fig. 7.8 Distributionof numbers of hurricanes and days with storms for aquatories of the World Ocean for 2005–2013

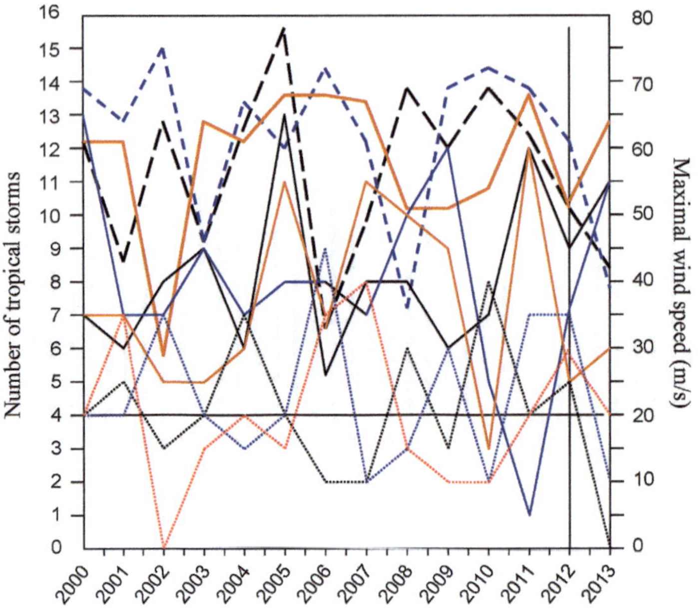

Fig. 7.9 Characteristics of hurricanes of different categories by Saffir-Simpson Scale for 2000–2013. Color lines correspond to different oceans: *dark* Atlantic Ocean, *blue* Pacific Ocean, and *red* Indian Ocean. More *thick dashed lines* correspond to wind speed. More *thick solid lines* correspond to number of storms that no achieve hurricane stage. *Dotty* lives correspond to number of hurricanes of 2nd, 3rd, 4th and 5th categories

Table 7.6 Statistics of hurricanes for 1997–2011 in different zones of the world ocean

Basin	Time interval					
	1997–2001		2002–2006		2007–2011	
	Number of cyclones	Percent of hurricanes (typhoons)	Number of cyclones	Percent of hurricanes (typhoons)	Number of cyclones	Percent of hurricanes (typhoons)
Pacific Ocean	204	38.7	159	62.9	198	45.5
South of Pacific Ocean	96	17.7	70	25.7	94	11.7
East of Pacific Ocean	84	45.2	85	42.4	78	39.7
North of Atlantic Ocean	75	49.3	92	43.5	87	41.4
North of Indian Ocean	44	27.3	51	19.6	47	19.1
South of Indian Ocean	67	35.8	78	41.0	61	50.8

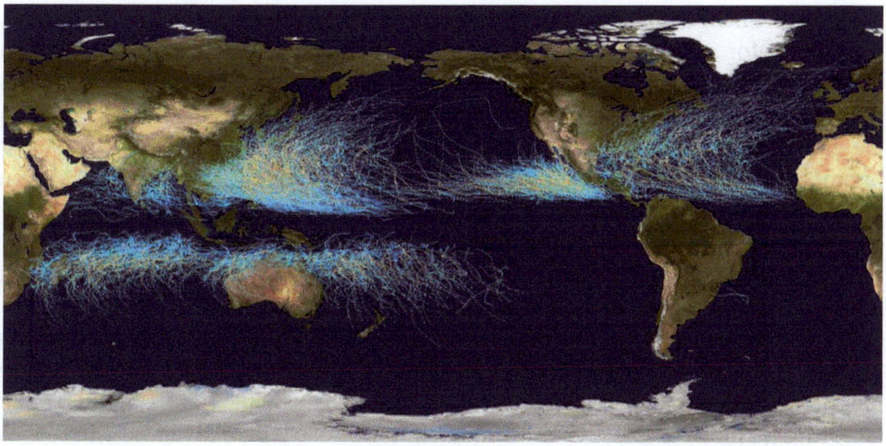

Fig. 7.10 Global tracks of tropical cyclones (http://commons.wikimedia.org/wiki/File:Global_tropical_cyclone_tracks-edit2.jpg)

hurricanes of 1st category remained nearly identical but monotonically decreased as percent of total number of hurricanes. The trend of sum of hurricanes of 2nd and 3th categories is small both by amount and by their duration. As opposed to this, the number of hurricanes 4th and 5th category increased from 31 for 2000–2004 to 36 for 2005–2009. These changes took place in all oceanic basins (Table 7.6), but they have unstable character.

Tropical cyclones take place mainly in latitudes 5°N–25°N and regularly impact on USA (states Florida, Texas, Louisiana), Caribbean Sea, Mexico, Japan, China, Philippines, Korea, Indonesia, Russian East, India, Madagascar, Mauritius, Reunion, Australia, etc. Global zone of hurricanes (typhoons) cover tropical oceans between the parallels 40°S and 40°N (Fig. 4.11). In this vast zone, the frequency of tropical cyclones is distributed non-uniformly. Consequently, losses from tropical cyclones have inhomogeneous distribution (Fig. 7.10).

7.4 Monitoring Data

The existing global Earth-observing system includes several levels of measuring sub-systems. The Global Observation Earth System of Systems (GEOSS) is the base of present monitoring of the ocean-atmosphere system (Charvat et al. 2010). GEOSS is built by the Group on Earth Observations (GEO) on the basis of a 10-Year Implementation Plan running from 2005 to 2015 (Klien 2008). Earth observations include measurements and monitoring of the Earth under water, on the land surface and beneath, air and water quality atmospheric conditions, and measures of the health of humans, plants and animals. Measurements can be made by sensors in contact or remotely, and all these observations support modeling or other tools that create information for environmental decision-making.

Global Change Observation Mission (GCOM) directly relates to retrieval algorithms for geophysical products, product validation, and data application. GCOM seeks to establish and demonstrate a global, long-term satellite observing system. Essential geophysical parameters have to be measured for understanding the global climate change and water cycle mechanism, and eventually contribute to improving future climate projection through a collaborative framework with climate model institutions. Demonstrating capabilities of operational applications providing continuous data to operational agencies is another important purpose. GCOM will take over the Advanced Earth Observing Satellite-II (ADEOS-II) mission and transition into long-term monitoring of the Earth. GCOM will consist of two satellite types and three consecutive generations with a one-year overlap, resulting in over a 13-year observation period in order to achieve a global, comprehensive, long-term, and homogeneous observation. The two satellites are GCOM-W and GCOM-C (Climate). The GCOM-W1 satellite will carry the Advanced Microwave Scanning Radiometer-2 (AMSR2) to contribute to understanding the water and energy cycle. The GCOM-C1 satellite will be equipped with the Second-generation Global Imager (SGLI) to observe the Earth's atmosphere and surface for contributing to the understanding of the carbon cycle and radiation budget (GCOM 2008).

There are numerous data sources about the atmospheric and oceanic processes which can be used for the decision making related to the OSE state. The basic purposes of the most of global Earth observing systems consist in the predicting climate change. For realistic prediction of climate change, these data and oceanic and atmospheric models must be coupled. It is evident that oceanic processes exert a major influence on climate. Moreover, sea-surface temperature determines the heat exchange between the atmosphere and oceans. In order to assess the OSE state, results from different Earth observing systems and general circulation models have to be processed. According to Cracknell et al. (2009), the OSE state is characterized by instable dynamics of its components and therefore it is impossible to use monitoring data because of their temporal incompleteness. In this sense, data of numerous meteorological stations such as The Tropical Atmosphere Ocean/TRIangle Trans-Ocean buoy Network & PIlot Research moored Array in the Tropical Atlantic/Research moored Array for Monsoon Analysis (TAO/TRITON&PIRATA/RAMA) are more acceptable (Fig. 7.11).

The Global Climate Observing System (GCOS) and Global Ocean Observing System (GOOS) (Fig. 7.12 and Table 7.7) deliver large amounts of archive and operative data about different environmental characteristics on a global basis. TAO/TRITON/PIRATA/RAMA is one efficient sub-system of these systems. Namely, the data of this sub-system are the most precise and informative.

Table 7.8 describes the data that are employed in this study. More precisely this data set is based on the archived data of the TAO/TRITON&PIRATA system for the months of hurricane activity. The basic goal of the data processing is to search for a procedure and a criterion to determine with high probability the time at which the hurricane arises and the conditions that precede that event. The tropical moored buoys that deliver many data about different meteorological and geophysical characteristics of the ocean-atmosphere system are located in the Pacific and Atlantic

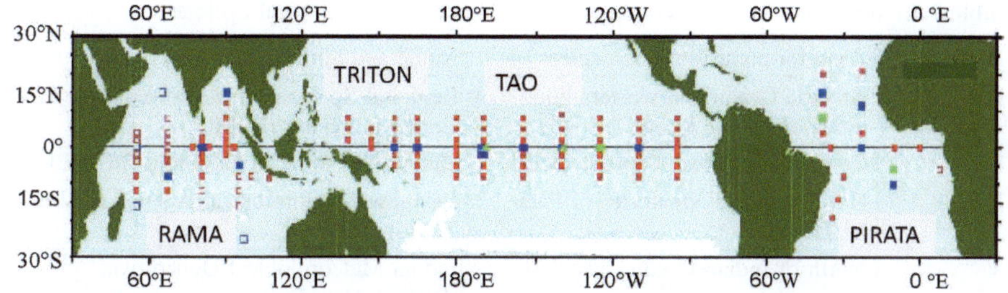

Fig. 7.11 Configuration of the global tropical buoy meteorological stations. *Color* marking of the station types: ■-flux reference site, ■-standard mooring, ■-CO_2 enhanced, ■-flux and CO_2 enhanced, ■-planned. (https://climatedataguide.ucar.edu/climate-data/tropical-moored-buoy-system-tao-triton-pirata-rama-toga)

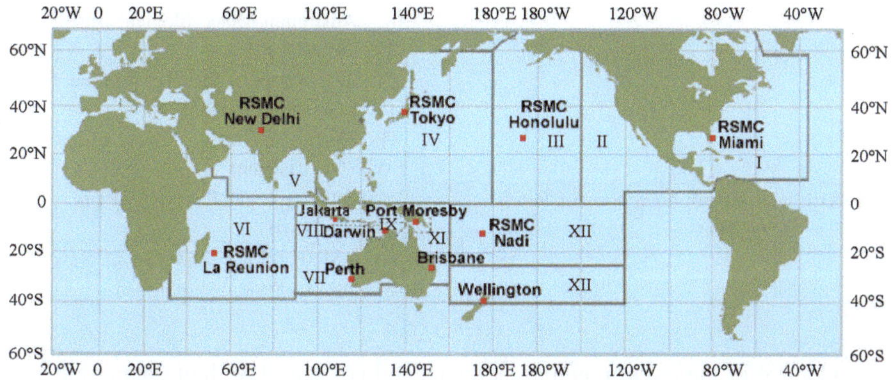

Fig. 7.12 World centers for the observations of tropical cyclones (Table 7.7)

Oceans. The most densely spaced meteorological stations are in the tropical sector of the Atlantic Ocean near the U.S. State of Florida. An assessment of the algorithm that is proposed in this paper is accomplished by means of a choice of such meteorological stations that are located close to the equator, where severe hurricanes are registered by the existing monitoring systems. To this end, the standard meteorological data of meteorological stations of the U.S. National Data Buoy Center were also used.

The introduction of the characteristic I_Ω in (3.37) and (3.38) facilitates the design of the monitoring scheme and the prediction of the OAS phase transition. Figure 7.13 illustrates a possible structure of the monitoring system with functions of the investigation, forecast, and tracing these processes (Krapivin et al. 2012b). Thus, the system has three levels: fixing, decision making and investigation. The corresponding units have the following functions:

Table 7.7 Regional special meteorological warning centers for tropical cyclones

Region	Zone for control by the center	Name and affiliation of the centre
I–II	Atlantic Ocean and western aquatories of Pacific Ocean	Regional Specialized Meteorological Center in Miami (RSMC)
III	Central aquatories of Pacific Ocean	Hurricane Center in Honolulu (RSMC)
IV	North-Western aquatories of Pacific Ocean	Japanese Meteorological Agency (RSMC, Tokio)
V	North of Indian Ocean	Indian Meteorological Department (RSMC, New Deli)
VI	South-West of Indian Ocean	Meteorological Agency of France (RSMC, La Réunion)
VII–XI	South-West of Pacific Ocean and South-East of Indian Ocean	VII—Australian Meteorological Bureau (TCWC, Perth)
		VIII—Indonesian Meteorological Agency (TCWC, Jakarta)
		IX—Australian Meteorological Bureau (TCWC, Darvin)
		X—Meteorological Service of Papua New Guinea (TCWC, Port Moresby)
		XI—Australian Meteorological Bureau (TCWC, Brisbane)
XII–XIII	Southern aquatories of Pacific Ocean	XII—Meteorological Service of Fiji (RSMC, Nadi)
		XIII—Meteorological Service of New Zealand (TCWC, Wellington)

(1) Control of the environmental elements and collection of their data at the state of interest in the regime allowed by the applied technical means.
(2) Identification of the candidate (suspicious) environmental elements and the value of their indicator $R_{\Omega}(t)$ corresponding to the interval of a given type of natural anomaly.
(3) Development of the dynamic series $\{I_{\Omega}(t)\}$ for a suspicious element to make a statistical decision on its noise and signal character. In the latter case, we check the suspicious element by criteria of the next level of accuracy (including the vector $\{x_i\}$ into the cluster, etc.).
(4) Making the final decision about the approach of the moment of the origin of change in the OAS state and delivering of the information to the responsible services for the environmental control.

The efficiency of the above discussed monitoring system depends on the parameters of the measuring means and the algorithms of the observational data processing. The model of the environment used herewith in parallel with the formation and statistical analysis of the series $R_{\Omega}(t)$ and adapted to the monitoring regime in correspondence with the scheme in Fig. 7.13 plays an important role.

Table 7.8 The list of environmental characteristics used for the detection of the beginning of a tropical hurricane

Parameter	Identifier	Parameter characteristic
Ocean surface temperature, °C	x_1	Temperature is measured on the depths 1.0 or 1.5 m depending on the buoy station type with precision ± 0.003 to ± 0.02 °C
Wind speed (m/s) and its direction (in degrees on clockwise from northern direction)	x_2, x_3	Wind parameters are measured on the altitude 3.5 m. Wind speed is measured with precision 3 %. In direction it is determined with precision 5.0–7.8°
Precipitation (mm/h)	x_4	Precipitation are registered every 10 min on the altitude 3.5 m with precision ± 0.4 mm/h.
Sea water density (kg m^3)	x_5	Water density is calculated as function of its salinity
Depth of the isotherm 20 °C	x_6	Depth of isotherm is calculated on the base of vertical temperature distribution by means of linear interpolation with the step 20 m by depth
Water temperature of fixed depths $\{h_i\}$, °C	$x_7(h_i)$	Water temperature is measured up to 750 m on the fixed horizons with the precision ± 0.02 °C
Relative humidity of atmospheric air, %	x_8	Air humidity is measured on the altitude 2.2 m above the sea level with precision ± 2.7 %
Atmospheric pressure, atm.	x_9	Atmospheric pressure is fixed with the precision ± 0.01 % on the altitude 3 m above the sea level
Atmospheric temperature, °C	x_{10}	Atmospheric temperature is measured on the latitude 2.2 m with the precision ± 0.2 °C
Sea water salinity, ‰	x_{11}	Salinity is assessed by the data on the water temperature and its conductivity with the precision ± 0.02 ‰
Significant wave height, m	x_{12}	It is calculated as the average of the highest one-third of all of the wave heights during the 20-min sampling period
Dominant wave period, s	x_{13}	It is the period with the maximum wave energy
Average wave period, s	x_{14}	It is the period (seconds) of all waves during the 20-min period
The direction from which the waves at the dominant period are coming, degrees	x_{15}	The units are degrees from true North, increasing clockwise, with North as 0 (zero) degrees and East as 90°
Dewpoint temperature, °C	x_{16}	Dewpoint temperature taken at the same height as the air temperature measurement
Station visibility, nmi	x_{17}	Buoy stations are limited to reports from 0 to 1.6 nmi

Meteorological parameters are measured by of the TAO/TRITON&PIRATA system

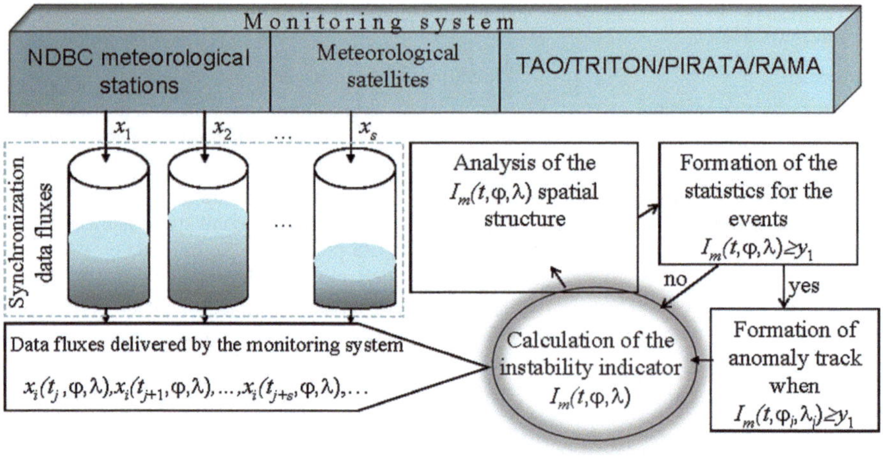

Fig. 7.13 The schematic diagram of the monitoring system for search and detection of the time of the start of the OAS phase transition

From the above-introduced criterion for the approach of the OAS phase change, it is clear that the form and behavior of $I_\Omega(t)$ are characteristic of each type of the process that is taking place in the OAS environment. One of the complex problems consists in determination of these forms and their corresponding classification. For instance, the dangerous natural phenomena, such as landslides and mudflows, have characteristic indicators of preliminary changes of relief and landscape. These are recorded from satellites (in the optical band) and together with the data of aircraft survey and ground measurements of relief slopes, expositions of slopes and state of hydro-network, are used to predict them several days before their occurrence.

However, in the OAS case, such indicators are lacking. Therefore, it is necessary to define the parameter a_{ij} as a function of the OAS phase state. This dependence is the relationship between the OAS indicator instability calculated for i-th and j-th measuring stations and is expressed as the ratio:

$$a_{ij} = I_T(\varphi, \lambda_i, t)/I_T(\varphi_j, \lambda_j, t) \tag{7.1}$$

In this case, the main problem is focused on the determination of the informative space Ω and the formalization of the investigation procedure in the monitoring system (Fig. 7.13). This procedure has the following scheme. The holder registers data from sensors and keeps the "suspicious" elements in the memory accordingly with some algorithm. Decision maker analyses data delivered by holder and decides what character of these data—it is signal about the start of the OAS phase change or it is noise. Then, the finder taking into account the output of the decision maker localizes the anomaly and starts to coordinate the functions of the holder and decision maker on the data specification for the detected anomaly. Finally, if an anomaly (suspicion about the OAS phase transition) with duration τ (that is defined empirically) is detected the decision maker decides about the start of the OAS phase change. Identification of the expectation time τ can be changed by accidental value of τ using the procedure of sequential analysis where I_{Ω} is compared to thresholds.

7.5 Percolation Procedure

Let us consider the cubic grid $\Xi_d = \{\Xi_{i_1,\ldots,i_d}\}$ of dimension d with side h in the phase space of the meteorological parameters enumerated in Table 7.8. Data flow from the system TAO/TRITON&PIRATA is not restricted in time and is characterized by data entry $\bar{x} = \{x_i;\ i = 1,\ldots, 10\}$ regularly at hourly intervals. A limited time interval, T, is chosen with the successive times $\{t_j; j = 1, \ldots, N\}$ when data on the meteorological situation in the previously surveyed site of the ocean-atmosphere system are delivered. In the course of time the cells $\Xi_{i_1,\ldots,i_d}$ in which a point $\{x_1^*(t_j),\ldots, x_d^*(t_j)\}$ is hit are colored. Here the value $x_i^*(t_j)$ is the observed value of the ith parameter in the time $t_j \in T$. Following percolation procedure (Soldatov 2006, 2007, 2009c; Sukov and Soldatov 2007; Sukov et al. 2008b) we form clusters by combining the interfacing of colored cells into a unique chain. Parameter λ denotes the fraction of the total number of cells that are colored and μ is the relative volume of the cluster space; the distribution of the clusters by their size is assessed.

The cluster space structure depends on the interval T, the choice of which is determined by the specific character of the given task. As it follows from the data processing, parameters λ and μ can achieve their asymptotic levels λ^* and μ^* when the observations are gathered over a long time. Under this consideration the character of such a transition between different phase states of the ocean-atmosphere system reflects the objective laws of the start of a hurricane. Consequently, a calculation of percolation cluster parameters for different environmental conditions of the ocean-atmosphere system functioning gives the possibility of constructing a decision making rule to assess the moment at which the phase transition in the ocean-atmosphere system dynamics occurs.

An average cluster size S (number of colored cells) and correlation length ξ (linear size of the cluster) are approximated by the following functions

$$S(\lambda) \approx |\lambda - \lambda^*|^{-\gamma},\ \xi(\lambda) \approx |\lambda - \lambda^*|^{-\nu}, \tag{7.2}$$

where the parameters γ and ν depend on the cluster space dimension d and in a particular case are determined empirically. Functions S and ξ defined in the formulae (7.2) increase sharply at the start of a hurricane. By definition (Essam 1980) an average number N_t of colored sites in the percolation cluster after numerous steps t of the walking of point A and under the absence of a critical transitions equals:

$$<N_t> \ = t^{\frac{-\theta}{2}},$$

where parameter $\theta \in [1.29–2.84]$ is evaluated in each case by means of a modeling procedure.

The use of TAO/TRITON&PIRATA/RAMA data in the monitoring regime with the employment of the percolation model is equivalent to the scheme of the point $\{x_i\}$ walk within the sites of Ξ_d. In the case $d = 2$ we consider the wind speed, x_2, and the atmospheric pressure, x_9, which are two meteorological parameters that can be expected to vary very strongly in the presence of a tropical cyclone. In Fig. 7.14 we plot the hourly values of x_2 and x_9 for August 2005 during which Hurricane Katrina formed, developed and dissipated. The random walk of point $\{x_i\}$ in $\Xi_2(x_9,x_2)$ is a function of a changing meteorological situation that is characterized by the transition from the background to the state when a tropical hurricane occurs. In this diagram the existence of the tropical hurricane within the time period T is characterized by a double percolation transition of the point $\{x_9, x_2\}$ through zone Ξ_2^2 between zones Ξ_2^1 and Ξ_2^3. An analysis of the motion of the point $\{x_9, x_2\}$ is connected with the assessment of probability characteristics for the drift of its projection on the direction η of the zone Ξ_2^2 boundaries. The drift ν of this projection from some initial position A represents the sum Σ_ν of a random number of the components.

The distribution structure of zones in the plane Ξ_2 for the Bahamas in August 2005 (Fig. 7.14) has specific elements bounded by the zone Ξ_2^2 in which the event leading to Hurricane Katrina was arising but it was not detected. When $(x_9, x_2) \in \Xi_2^1$, the meteorological situation was characterized as the background up to 20 August 2005. Hurricane Katrina was recorded in the morning of 23 August 2005 by Landsat and other meteorological satellites above the south-eastern sector of the Bahamas and after that it was classified in the first category by the Saffir-Simpson scale. Moving to the Gulf of Mexico in the direction towards New

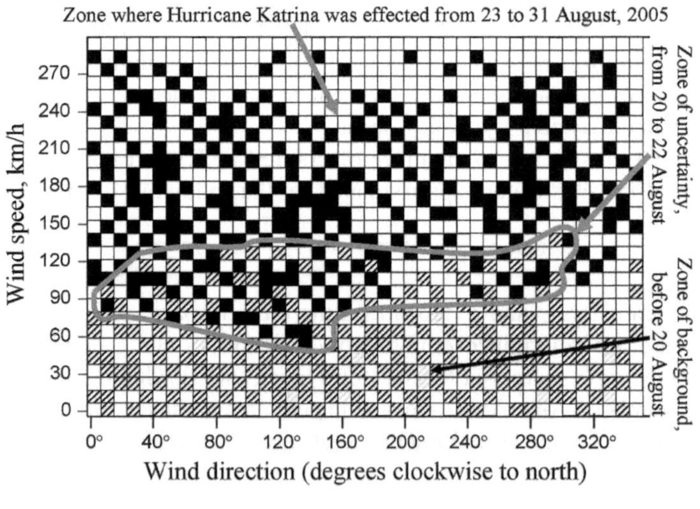

Fig. 7.14 Cluster structure of the meteorological situation that took place during August 2005 in the region of the Bahamas from the data of Meteorological Station No. SPGF1 (26° 42′ 16″N, 78° 59′ 40″W). Hurricane Katrina formed on 23 August, hit New Orleans on 28 August and dissipated on 30 August, 2005

Orleans, Louisiana, it reached category 5 on the morning of 28 August and then its strength decreased slowly until 31 August. Hurricane Katrina passed New Orleans on 29 August 2005 and by 31 August New Orleans was flooded as result of Hurricane Katrina. An analysis of meteorological data shows that the first signs of the environmental instability began to originate on 20 August. Consequently, the interval between 20 and 23 August can be considered as the uncertainty zone (see Fig. 7.14). Thus, it is necessary to search the indicators of environmental instability within this zone. Undoubtedly, the value of Δ, the width of zone Ξ_2^2, and the structure of zone Ξ_2^2 depend on the hurricane category dynamics. Searching such dependency is an important task.

Figure 7.15 represents a percolation cluster that was formed in the region of the Bahamas in August of the 2005 hurricane season. The process of percolation cluster formation shows that there exists a moment when the ocean-atmosphere system changes its phase state. This moment is determined by the distances between λ and λ^*, μ and μ^*. Simulation of this process has shown that the moment at which this transition occurs corresponds to the boundary between a storm and a hurricane of the first category. This is not a very acceptable result. Therefore, it is necessary to introduce an additional indicator.

It can be seen that both percolation cluster sections have the zones of transition to the infinite cluster but are distinguished by the structural characteristics of the percolation cluster. Section (x_2, x_9) has the following parameters: $\lambda^* = 0.2$, $\mu^* = 0.87$ and $S = 7.08$. Section (x_{10}, x_8) is characterized by the parameters $\lambda^* = 0.5$, $\mu^* = 0.94$ and $S = 22.5$. The zone of the phase transition is defined in each case by the geometric reduction of the way between two zones of basic cluster each of which satisfies one state of the environment. It is evident that the moment of this reduction

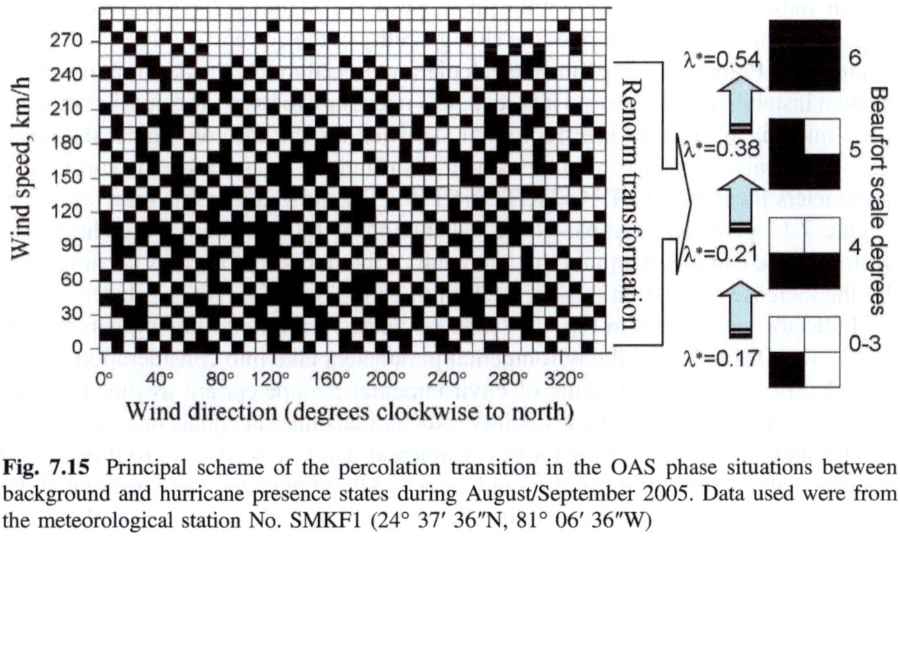

Fig. 7.15 Principal scheme of the percolation transition in the OAS phase situations between background and hurricane presence states during August/September 2005. Data used were from the meteorological station No. SMKF1 (24° 37′ 36″N, 81° 06′ 36″W)

formation corresponds to the hurricane origin. Calculation of the hurricane origin probability depends on the dynamics of cluster space structure.

To formalize the detection procedure for the beginning of a hurricane we introduce the algorithm for the assessment of the ocean-atmosphere system instability $I_m(t)$ (Soldatov et al. 2008; Krapivin et al. 2007a; Sukov et al. 2008a):

$$I_m(k) = \frac{1}{s(m-N)} \sum_{j=1}^{m-N} \sum_{i=1}^{s} \beta_i \alpha_i(t_j) \quad (k = 1, 2, 3, \ldots, K) \tag{7.3}$$

where N is the length of the quasi-stationary period of observations to calculate average values of the ocean-atmosphere system characteristics; m is the length of time interval on which indicator $I_m(k)$ is assessed; s is the number of registered parameters of the ocean-atmosphere system; β_i is the weight coefficient for the i-th characteristic; k is the number of the time interval for the $I_m(k)$ assessment;

$$\alpha_i(j) = \begin{cases} 1 \text{ when } \Delta x_i(j) \cdot \Delta x_i(j+1) \leq 0, \\ 0 \text{ when } \Delta x_i(j) \cdot \Delta x_i(j+1) > 0 \end{cases} \tag{7.4}$$

where

$$\Delta x_i(j) = \frac{1}{N} \sum_{l=m(k-1)+j}^{m(k-1)+j+N-1} x_i(l) - x_i(m(k-1) + j + N),$$

$$\Delta x_i(j+1) = \frac{1}{N} \sum_{l=m(k-1)+j+1}^{m(k-1)+j+N} x_i(l) - x_i(m(k-1) + j + N + 1)$$

An indicator of the ocean-atmosphere system instability $I_m(t)$ defined by the Eqs. (7.3) and (7.4) characterizes the changeability of the current average value of the vector $\{x_i\}$. Figure 7.16 explains the calculation procedure for the ocean-atmosphere system instability indicator. As it is seen, it is necessary to assess the quasi-stationary time interval in order to use traditional statistics for the meteorological data processing. Analysis of the one-dimensional distributions for selected meteorological parameters measured by TAO/TRITON&PIRATA showed that this interval is 7 h. Figure 7.17 gives an example of the calculation of $I_m(t)$ for the case of hurricane Katrina. One can ascertain that the approach of the hurricane origin is accompanied by the increase of $I_m(t)$. In this connection three regions of $I_m(t)$ take place:

Indicator represented in Fig. 7.17 was calculated with the following parameters: $s = 17$ (see Table 7.8). All environmental parameters take into consideration, $\beta_i = 1$ ($i = 1 \ldots 17$; weight coefficients of environmental parameters are assumed equal), $m = 24$ h (time period for the instability indicator calculation equals one day), $N = 7$ (quasi-stationary time interval for environmental data), $K = 31$ days (a time period for the calculation of instability indicator). All parameters were estimated by numerical experiments to have high reliability within the calculation results.

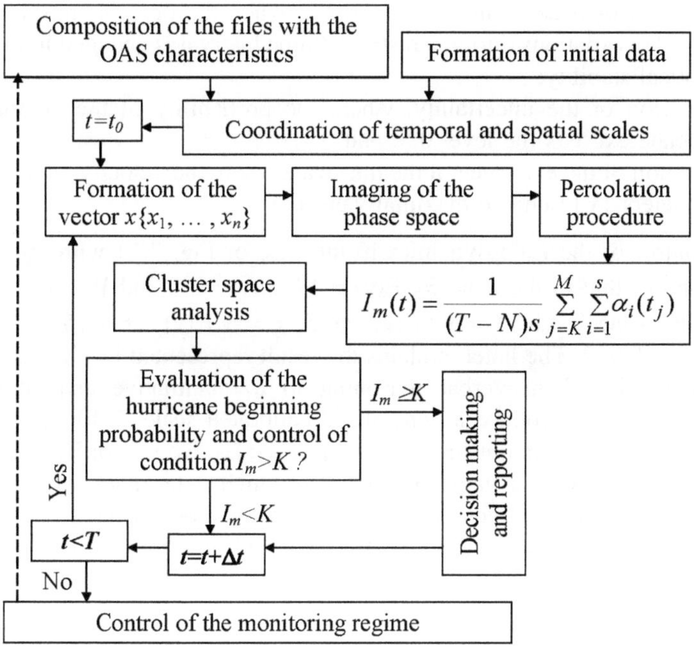

Fig. 7.16 Block diagram of the procedure for calculation of the instability indicator characterizing the phase states of the ocean-atmosphere system

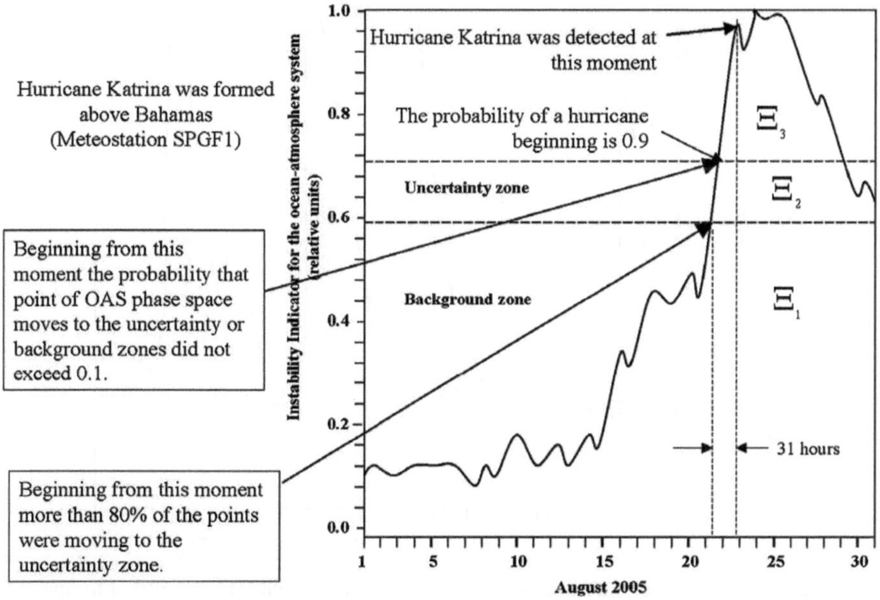

Fig. 7.17 Dynamics of the ocean-atmosphere system indicator instability assessed by the meteorological data from meteorological stations FWYF1 (25° 35′ 25″N, 80° 05′ 48″ W); MLRF1 (25° 00′ 36″N, 80° 22′ 48″W) and SPGF1 (26° 42′ 16″N, 78° 59′ 40″W)

(1) the region of background characteristics when the ocean-atmosphere system parameter variations during a restricted time interval is changed in the limits of statistical stability;
(2) the region of the uncertainty, where the probability of the beginning of a hurricane exceeds the level 0.5; and
(3) the region of the existence of the hurricane where the velocity of changes of all parameters $\{x_i\}$ achieve maximal values.

The control of the $I_m(t)$ dynamics in the case of Fig. 7.14 with the sequential analysis shows that in the zone Ξ_2^1 E(ν) = 17, D(ν) → 0 and $W_c(x) \to 0$. At this time, in the zone Ξ_2^3 E(ν) = 9, D(ν) → 5, x = ν/D(ν) → 1.35, $c \to 16$ and $W_{16}(1.35) \to 0.912$. The latter explains the result represented in Fig. 7.18. In other words, processing from verbal reasoning to a quantitative estimation of the $I_m(t)$ dynamics, we introduce a generalized calibrated scale Ξ, for which we postulate the presence of relationships of the type $\Xi_1 < \Xi_2$, $\Xi_1 > \Xi_2$ or $\Xi_1 = \Xi_2$. It means that there always exists a value of $I_m(t)$ = ρ which determines when the ocean-atmosphere system phase transition of a given type can be expected. As a result, the magnitude $\theta = |I_m(t) - \rho|$ determines the expected time interval before the phase transition occurs.

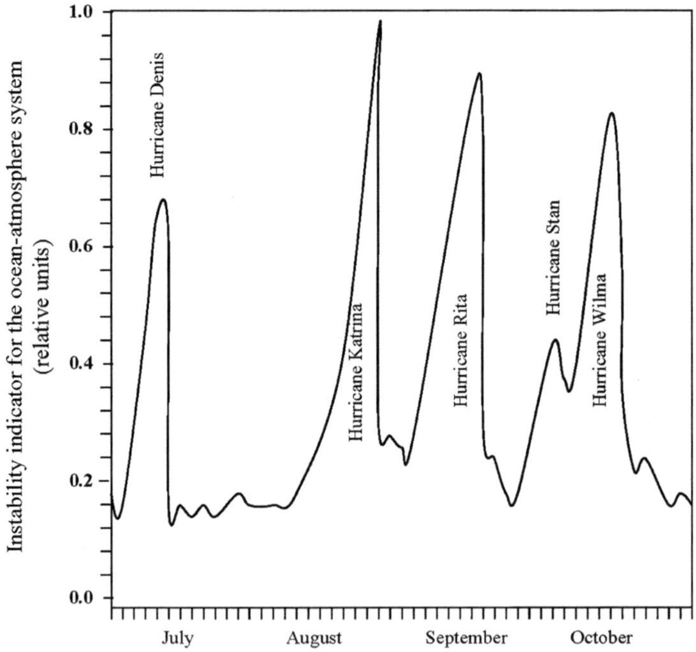

Fig. 7.18 Dynamics of the ocean-atmosphere system instability indicator calculated for the 2005 tropical hurricane season based on data from meteorological station No. SMKF1 (24° 37′ 36″N, 81° 06′ 36″W). Parameters of $I_m(k)$ are s = 17, m = 24 h, N = 7, K = 123 days

The region of the sharp increase in the derivative of $I_m(t)$ corresponds to the ocean-atmosphere system phase transition. Consequently, tracing for the meteorological situation and synchronous calculation of the instability indicator under its comparison with cluster space structure gives a possibility to detect the moment of the ocean-atmosphere system transition from background state to the hurricane origin. Figure 7.17 shows that introduced instability indicator could detect the start of hurricane Katrina at least 31 h before its detection by satellite. Figure 7.18 shows $I_m(t)$ within the 2005 hurricane season. The behavior of the $I_m(t)$ is identical to the registered hurricane series. Such a result is justified by the results given in Fig. 7.19 for the case of tropical hurricane Flossie that formed in the eastern sector of the Pacific Ocean on 8 August and reached Hawaii on 15 August 2007. One can see that the probabilities of the ocean-atmosphere system transition between its phase positions are more than 0.7. Results of Figs. 7.20 and 7.21 again corroborate the indicator information capability. Therefore the indicator $I_m(t)$ reflects the ocean-atmosphere system phase structure with high reliability and can be used to predict the beginning of a hurricane.

Figure 7.22 depicts the results obtained from the calculation of the instability indicator for the two cases of the hurricanes Isaac and Katrina. The data used were derived from two NDBC weather stations during August–September of 2012 and for one NDBC weather station during August of 2005. In Fig. 7.22 the OAS instability level for different distances between station and cyclone is presented.

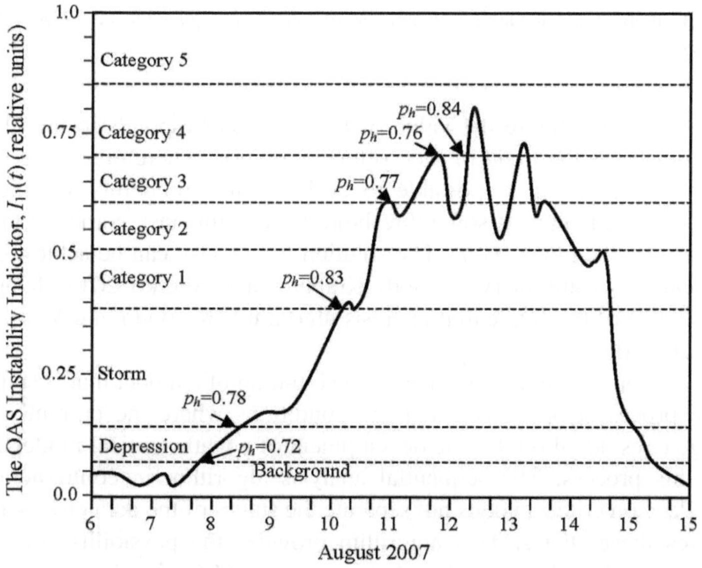

Fig. 7.19 Dynamics of the ocean-atmosphere system instability indicator for the case of tropical hurricane Flossie which occurred on 8–15 August 2007. Data from meteorological station No. 51,003 (19° 22′N, 160° 82′) were used to assess the $I_m(t)$ ($m = 17$) as well as $s = 17$, $m = 24$ h, $N = 7$, $K = 30$ days

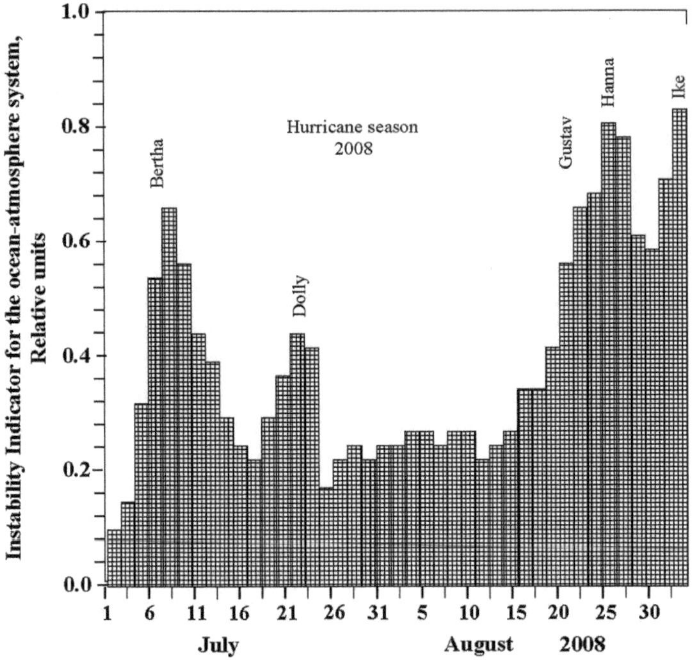

Fig. 7.20 A dynamics of the Instability Indicator for the ocean-atmosphere system during season of tropical hurricanes 2008 calculated by the data of meteorological stations located in the Caribbean Basin and Gulf of Mexico. Parameters of $I_m(k)$ were equaled to $s = 17$, $m = 24$ h, $N = 7$, $K = 62$ days

Furthermore, in this figure the value of $I_T(\varphi, \lambda, t)$ corresponds to the OAS phase state and illustrates the trend of the cyclone evolution. Therefore, using one of the prediction algorithms, it is possible to evaluate the probability when stochastic process $I_T(\varphi, \lambda, t)$ intersects some threshold. In fact, this task comes to the random walk control on the plane (I_T, t). The solution to this task can be achieved with the employment of evolutionary method (Krapivin and Shutko 2012; Krapivin and Varotsos 2007) or the sequential analysis algorithm (Krapivin and Varotsos 2008; Soldatov 2009a).

The evolutionary method consists of the building of a model that is suitable for a stochastic process $I_T(\varphi, \lambda, t)$ under the conditions where the currently available knowledge does not allow for the development of a mathematical model capable to describe this process. The sequential analysis algorithm in contradiction to the Neyman-Pearson criterion does not separate the stages of the data processing, rather it alternates these stages. This algorithm provides the possibility for a decision making after each realization of a random process $\{I_T(\varphi, \lambda, t_i\}$.

Figure 7.23 demonstrates this possibility for the case of the hurricane Irene in 2011. Inspection of this figure shows that the instability indicator responds to the OAS phase transitions with some probability, which depends on the weather station

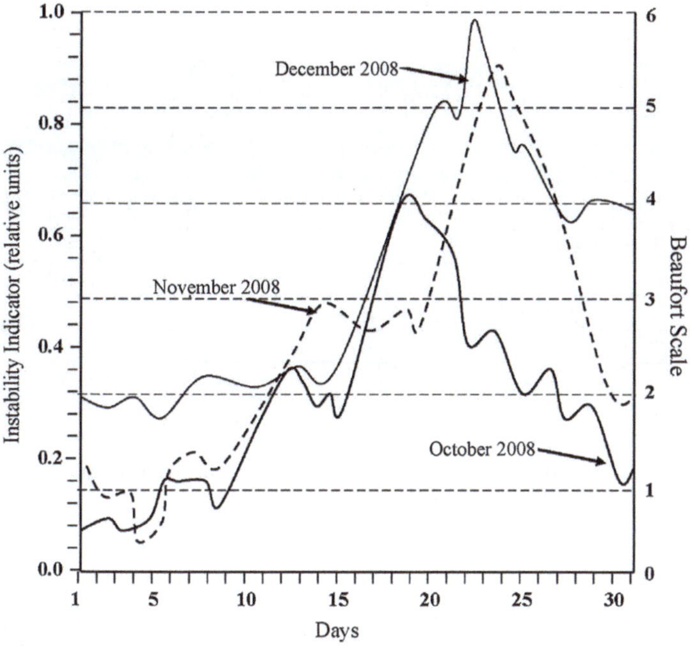

Fig. 7.21 Instability Indicator for the Baltic Sea. Calculations were realized under $s = 2$, $m = 24$ h, $N = 7$, $K = 31$ days. Wind speed and direction are considered only

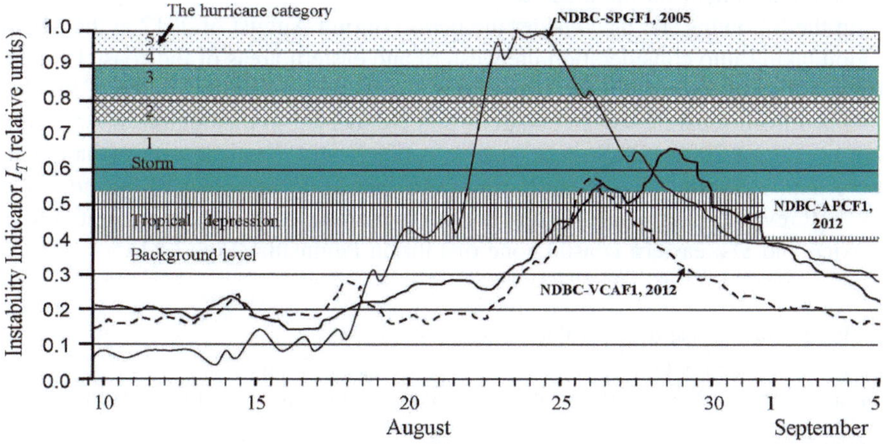

Fig. 7.22 The OAS instability indicator calculated for two cases of tropical cyclones Isaac (2012) and Katrina (2005)

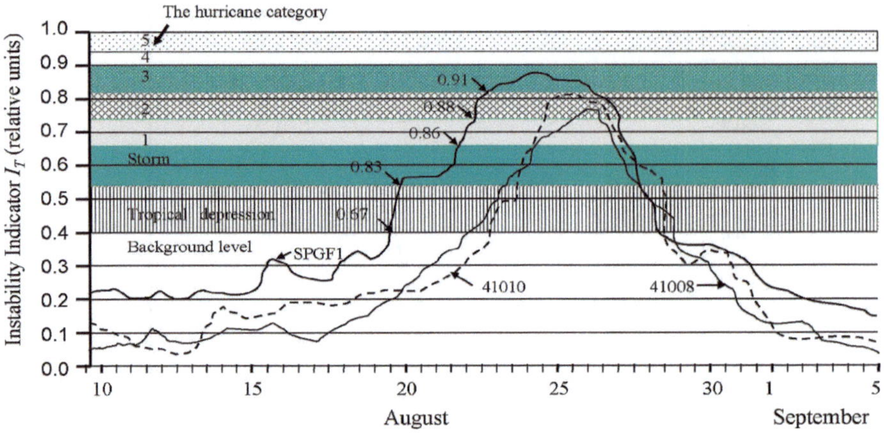

Fig. 7.23 The OAS instability indicator calculated on the data of weather stations NDBC-SPGF1, 41010 and 41008 during August/September, 2011

distance from tropical cyclone location. After the calculation of $I_T(\varphi, \lambda, t)$, the sequential algorithm predicted the OAS phase transition from tropical storm to hurricane of category 1 and category 3 with probabilities 0.86 and 0.91 for 19 and 27 h, respectively. Furthermore, the calculation of $I_T(\varphi, \lambda, t)$ for each weather station of the TAO/TRITON/PIRATA/RAMA system will give the possibility to localize the disturbance process in the OAS of the given region and to start the monitoring procedure with the calculation of matrix A with corresponding control of the environment on the area Ω.

In the following, let us consider the period during August of 2012 at the Miami region taking into consideration the western and eastern areas of the coastal waters. Table 7.9 gives an example of two matrices A that are calculated using of the data obtained from separate weather stations that are located in this region. It was found that in August 20 the OAS phase state was practically identical with variations no more than 9 %. But at August 26, the OAS instability level varied by more than 45 %. It reveals that the area Ω broke up to two areas—Ω_1: eastern zone of Gulf of Mexico and Ω_2: eastern coastal zone of Florida Peninsula (Fig. 7.24).

Table 7.9 An example of calculations using Eq. (7.3)

Time	August 20, 2012					August 26, 2012				
NDBC Station	APCF1	APXF1	41008	SPGF1	42013	APCF1	APXF1	41008	SPGF1	42013
APCF1	1	0.87	1.01	0.76	0.64	1	1.07	0.03	0.18	0.79
APXF1	0.99	1	0.77	0.82	1.18	0.93	1	0.12	0.09	1.24
41008	1.1	1.3	1	1.29	0.93	33.33	8.33	1	0.96	7.68
SPGF1	1.32	1.22	0.78	1	0.89	5.56	11.11	1.04	1	9.14
42013	1.56	0.85	1.08	1.12	1	1.27	0.81	0.13	0.11	1

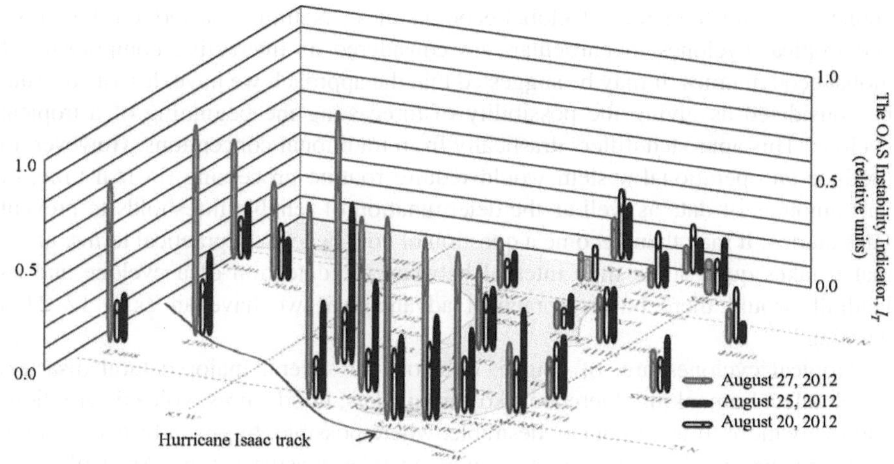

Fig. 7.24 Instability indicator calculated for August 20, 25 and 27 of 2012 using meteorological data from separate NDBC weather stations located in the Florida zone

In addition, the Table 7.9 presents the calculation data for restricted list of weather stations. It is worth to note that the result obtained is the same with the case of all weather stations located in the coastal zone of Florida Peninsula taken into consideration. It is seen that zone Ω_1 is the parameter for the subsequent control of the OAS according to the scheme shown in Fig. 7.14.

7.6 Predicting the Path of a Tropical Cyclone

Storms, hurricanes, and typhoons are examples of natural instability that occurring in the ocean-atmosphere system and cause catastrophic events connected with deaths and economic losses. Many authors try to forecast the extreme processes in the ocean-atmosphere system. To solve this problem, a network of meteorological stations has been built up including the TAO/TRITON/PIRATA system of buoy stations. However, forecasting the development of a tropical cyclone is a complex problem that includes several elements: track forecasting, intensity forecasting, rainfall forecasting, and storm surge forecasting. Another, possibly even more, difficult task is the detection of the origin of a tropical cyclone.

The methods used in forecasting tropical cyclones have changed with the passage of time based on different information sources including satellites. The base of these methods is the tropical cyclone forecast models of which there are mainly three types: statistical, dynamical, or combined statistical-dynamic. Such models utilize meteorological data to calculate paths and intensities only. New forecasting techniques have been developed during recent years by Soldatov (2009b) in connection with the study of global ecodynamics . Natural disasters on the whole and tropical cyclones in particular, are considered as interactive components of global ecodynamics. It may be suggested that the approach we have described could be considered as giving the possibility of forecasting the beginning of a tropical cyclone. This approach differs drastically from traditional conceptions. However, to produce an operational system would require routine processing, in real time, of vast amounts of data as well as the determination of reliable thresholds to prevent false alarms. It may then become a question of cost-benefit. In practical terms, given that it takes quite some time interval between the detection of a cyclone and its landfall, would there be any practical advantage if we have an extra 12–24 h warning?

Tropical cyclones are, of course, only one of several major natural disasters which affect humankind; there are also earthquakes, floods, fires, volcanic eruptions and many more. It is, of course, desirable, where possible, to give advance warning of an impending disaster so that people can take evasive action. An earthquake strikes suddenly and, as we know, often without specific warning and a lot of scientific research has been devoted to attempting to predict earthquakes. In a sense, a tropical cyclone strikes suddenly; it comes in from the ocean and its damage is (mostly) done where and when it reaches the land. But before it makes landfall it has been tracked for quite a while and it has built up as it has travelled across the ocean, gathering strength as it goes. The problem is not so much making an initial prediction but on following the cyclone over a matter of a few days before it makes landfall and predicting exactly where it will strike the land. The issue is not so much detecting the start of the typhoon but trying to predict its path accurately in order to make precise predictions of where and when it will strike the coastline. When it starts, it has not reached its full strength, it is a severe weather condition that is revealed by conventional weather forecasting and observation; navigators, at sea or in the air, can take evasive action. If the study presented is able ultimately to contribute to detecting a cyclone earlier than it would be detected by conventional methods that could be very useful. But we are a long way far from being in that situation yet. Remote sensing has its role to play—and indeed it has a large part in the tracking of a hurricane and in providing input to models for predicting the path of a hurricane. From the point of view of humankind, the issue is not to predict or detect the beginning of a hurricane, but to predict its development and the path that it will travel. Claims for supposed benefits of remote sensing that are not really relevant to benefitting mankind only discredit the subject.

Analysis of numerous calculations of $I_m(t)$ for different tropical cyclones shown that if it is calculated at the same time for a large region Ξ basing on data delivered by meteorological stations and satellites, then $I_m(\varphi, \lambda, t)$ $(\varphi, \lambda) \in \Xi$ is a two-dimensional distribution of instability indicator. In this case, sustained splash of the $I_m(\varphi_0, \lambda_0, t)$ $t \in [t_1, t_2]$ shows that tropical cyclone is starting in coordinates (φ_0, λ_0) and monitoring system is to begin probability control of this zone. Figure 7.25 gives example of such operation.

Consideration of examples that could show the instability indicator characteristics is possible only when cyclone track lies in a zone of meteorological stations

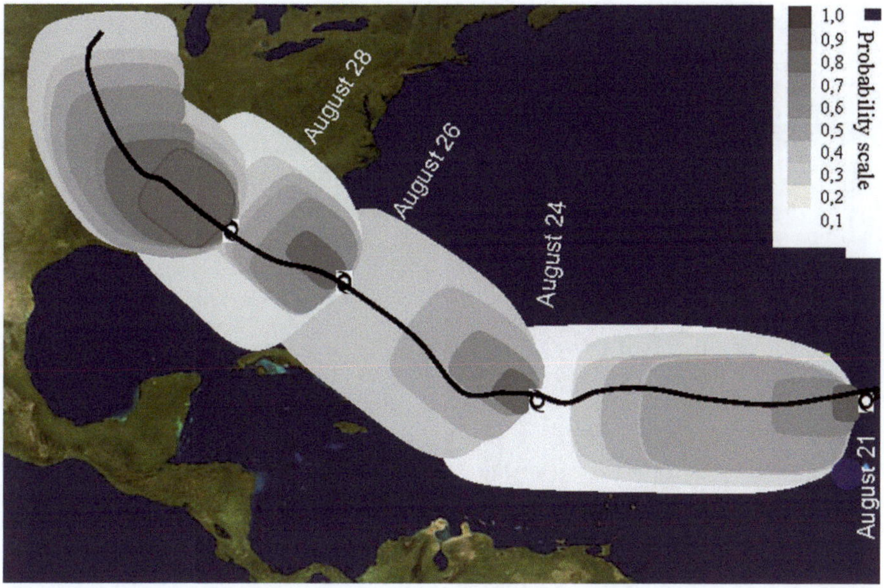

Fig. 7.25 Fragments of probability distributions for phase transitions calculated for tropical cyclone Isaak (2012) basing on the data delivered by NDBC meteorological stations located in Florida (USA) region

where standard meteorological data are registered with small time interval. The zone of meteorological station sensitivity usually varies in the boundary of 100–500 km.

In connection with this, the case of tropical cyclone Sandy is considered. Figure 7.26a shows its track from the beginning as tropical depression in the western Caribbean Sea on October 22, 2012. Sandy moved slowly northward toward the Greater Antilles and gradually intensified. Series of the NDBC meteorological stations that fixed OAS characteristics to calculate Instability Indicators are marked in Fig. 7.26. As it is seen from Fig. 7.26, tropical cyclone Sandy track, as opposed to case of Isaac, was mainly ranged from meteorological stations when it had categories 2 and 3 in Saffir-Simpson Scale.

The distance between NDBC-42057 and NDBC-42058 is about 750 km. These meteorological stations—as it is seen from Fig. 7.26b—had small increased the Instability Indicator but did not show a transition of the OAS in phase of tropical depression. Additional use of satellite data (QuickSCAT) about wind characteristics gives the possibility to fix this phase on October 23 at 20–50 in GMT (Brennan et al. 2009). Figure 7.27 shows the OAS Instability Indicator dynamics for different meteorological stations that are given in Fig. 7.26a. We see that Sandy was a Category 3 storm at its peak intensity on October 25. Then, it became a Category 2 on October 26. Curves in Fig. 7.27 show how Sandy changes its phases moving toward USA and Canada and how its diameter is increased. For example, minimal

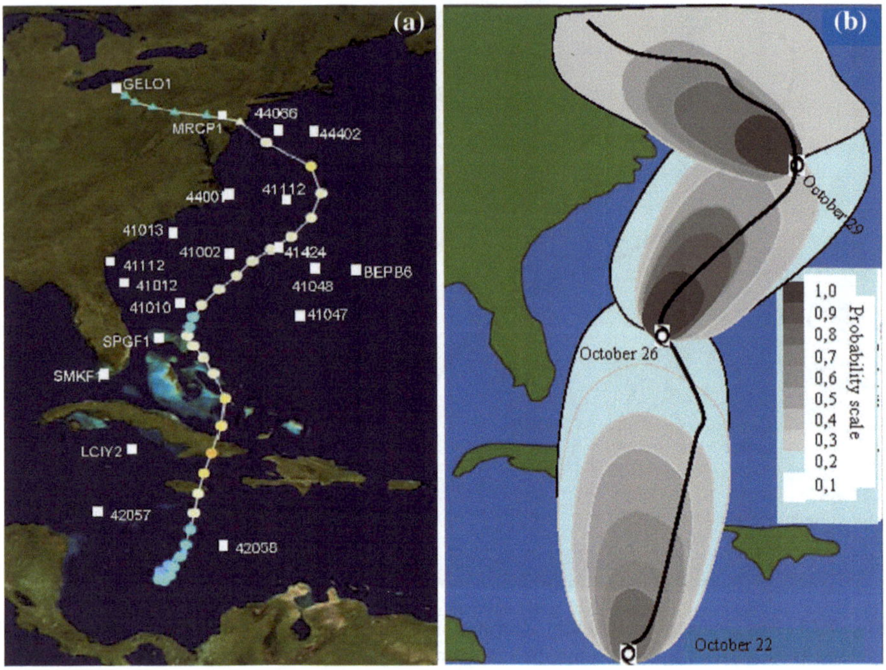

Fig. 7.26 Fragments of probability distribution **b** for instability indicator position in first level of Saffir-Simpson Scale calculated for October 22 and 26 of 2012 using meteorological data from separate NDBC weather stations located in the zone of the Sandy hurricane track **a**

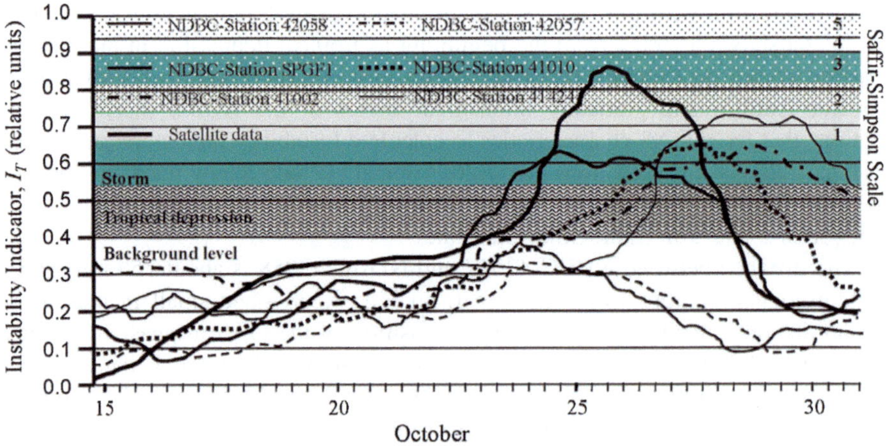

Fig. 7.27 The OAS Instability Indicator dynamics calculated for separate NDBC-meteorological stations located along the tropical cyclone Sandy track

distance NDBC-44001 from Sandy eye was about 600 km but Instability Indicator reacts on the cyclone passing.

In the United States, Hurricane Sandy affected 24 states, including the entire eastern seaboard from Florida to Maine and west across the Appalachian Mountains to Michigan and Wisconsin, with particularly severe damage in New Jersey and New York. Its storm surge hit New York City on October 29. The OAS Instability Indicator showed that at this time Sandy diameter was changed from 700 to 900 km. All weather stations actually located on the southeastern coast given the OAS Instability Indicator jump. Maximal value of the OAS Instability Indicator on October 29 was at NOAA-ACYN4 and it was moved to the direction of NOAA-CBL01 where its maximal level was achieved on October 30. Therefore, calculations of the OAS Instability Indicator for series of weather stations located on given territory and aquatory gives the possibility to synthesize the cyclone track with determination of its impact zones.

7.7 Searching the Indicator-Precursors of Tropical Hurricane Beginning

The solution to the problem related to the search of indicator-precursor for the tropical hurricane origin is connected with the assessment of the ocean-atmosphere system (OAS) state and with the revealing situations when this system begins to change its phase position. It is known that the origin of tropical hurricanes is one of the displays of the cyclone moving. Basic phases of the OAS are (Mazur and Ivanov 2004):

- *tempest*—wind that has velocity 19–23 m/s and corresponds to the 8 balls by the Beaufort scale;
- *storm*—strong wind corresponding to more than 9 balls by the Beaufort scale with wind speed 23–30 m/s; and
- *hurricane*—12 balls and wind speed is more than 35 m/s.

So far, as all environmental changes connected with this process have stochastic character, the solution to the problem of the detection of the time moment when OAS transitions to the origin of storm transformation to the hurricane of first category really needs the chaos modeling, that is the complex problem in his turn (Mees et al. 1997; Abu-Zaid and Ahmadi 1989). So, it is necessary to take into consideration the fact that numerous other natural catastrophes are characterized by the same dynamics, but they differ only by series of environmental parameters that influence their beginning. Therefore, the search of laws describing the tropical hurricane origin needs to analyze other natural phenomena in the atmosphere, and maybe in the more narrow space. For example, as it follows from study by Bondur et al. (2008), the processes on the Sun significantly influence the tropical cyclone beginning, and in general it was determined that there exists a correlation between the flow variation of galactic cosmic rays and variations of vertical temperature profile of the atmosphere. In particular, decrease of the temperature by 9 °C at the altitude 16 km was detected in the case of Hurricane Katrina as a result of a higher high-altitude gradient of the temperature being formed and the intensification of the convention taken place. As a result, the warmer atmospheric layers begin to lift up, causing a temperature rise by 2 °C at the latitudes 5–14 km. This result confirms the theoretical conclusions received by Sharkov (2003).

There exists the opinion that common global warming stimulates such changes in the atmosphere circulation that promote to the intensification of tropical hurricanes. This opinion becomes stronger in connection with 2005 hurricane season when storms and hurricanes raged at the tropical latitudes. In fact, tropical hurricanes of 2005 season functioned as giant heat engines. It was established that intensity and duration of tropical cyclones during the last 30 years in the North Atlantic and Western North Pacific were doubled. In connection with this, we can make the conclusion that the search of the indicator-precursors is realized by means of the analysis of the relationships in the global climate system functioning. Under this complex and unstable climate models can not be used. Searching stable correlations in global climate system can be solution of this problem. There is a demonstrative example of the solution of particular task related to the forecasting of tropical cyclone motion using a baroclinic model (Zhou and Chen 2006). was According to Tang et al. (2005) the Boolean algebra algorithm which is widely used for market basket analysis can be used to analyze the meteorological data sets characterizing OAS in the zones of possible tropical hurricanes beginning. This algorithm consists in the formation of binary set that is based on the wind speed analysis.

Dynamic processes conducting natural phenomena proceed by complex laws that are not studied a lot as a rule. Nevertheless, there exists the ensemble of methods which we can mention: the method of Fourier rapid transformation, the method of maximal entropy, the method of quasi linearization, the method of evolutionary modeling, the method of harmonic functions, etc. An application of these methods to search the relationships in the series of meteorological, geophysical and ecological data that characterize the zones where tropical hurricanes arose selecting the time periods that forewent the hurricane beginning can be productive. In particular, a detection of the specific features in the natural processes within the OAS before the tropical hurricane beginning can give mechanism for the early detection of the moment when OAS will pass from storm state to the phase of hurricane of the first category. Some experience of the solution to problems arising here is given in the publications by Kondratyev and Krapivin (2005b, 2006), Krapivin and Potapov (2002, 2006a, b), Potapov et al. (2006), Sukov et al. (2008a), Grankov et al. (2006), Kondratyev et al. (2006a, b), and Soldatov (2007). This experience is based on the methods for the analysis of unstable processes, and it is undertaken as one of the possible approaches to the creation of constructive algorithms for early detection of the zones where tropical hurricanes could begin.

The history contains rich information about natural disasters which are now called tropical cyclones and which are mainly formed over the oceans in the tropics, regularly falling onto the eastern and near-equatorial regions of the continents. The tropical cyclones are hurricanes and typhoons, taking place in the northern and southern Pacific, in the Bay of Bengal, the Arab Sea, in the southern sector of the Indian Ocean, near the coast of Madagascar, and the north-western coast of Australia. The Atlantic hurricanes form mainly in the zone confined to the coordinates (10°N–20°N, 20°W–6°W) and (7.5°N–17.5°N, 30°W–100°W). The tropical cyclones are usually called by names.

The origin of tropical hurricanes are one of displays of the tropical climate system functioning. Numerous International and National hurricane centers study this functioning. One of them is the US National Hurricane Center that publishes reports with a detailed description of each hurricane case describing all phenomena (speed and direction of the wind, precipitation, etc.) that are going with it. Basic direction of the studies in these centers is associated with detecting and tracking the hurricanes that are arisen. Forecasting of the trace for hurricane moving, certainly, is important to reduce the losses. There exist many models of different complexity and detail by means of which the efforts are undertaken to predict the hurricane trajectory, its intensity and precipitation intensity (Gray et al. 1991; Pielke 1999).

Similar investigations are realized in the Canadian Hurricane Center (CHC) or Atlantic Storm Prediction Centre (ASPC). During the hurricane season forecasters carefully monitor all storms of tropical origin. When a storm begins to threaten Canadian waters or land, specially trained meteorologists in the CHC track the tropical storms and provide guidance to other Canadian storm prediction centers. They work closely with the US hurricane Centre in Miami as well as municipal, provincial and federal emergency measures organizations in order to inform the public of expected conditions.

Forecasters at the ASPC work with many tools and technology to provide forecasts and warnings, including: Doppler weather radars, weather satellites, observations from a network of upper atmosphere and surface monitors, a network of offshore weather buoys, ship reports, volunteer reports and a variety of computer model output. The ASPC operates 24 h a day, 365 days a year, which means forecasters work day and night shifts as well as holidays. Supercomputer Hurricane Prediction is used to solve the arising problems. Of course, basic tendency of these predictions are directed to the assessment of the prospective number of hurricanes and their intensity but not the direct detection of given hurricane beginning. A solution to this problem results from the study by Golitsin and Vasiliev (2001), being possible under the detailed investigation of the conditions when atmospheric convective processes are developed. Under this, regular data flows characterizing the aquatories with dangerous wind conditions are an obligatory condition for the success. It is possible when buoy station and methods of aquatory monitoring by means of oceanographic satellites are jointly used. These data sources give the possibility to have narrow observation of the zones where tropical hurricanes can begin.

Origination of the data about wind fields on the large aquatories is possible with the radio-locators of side observations giving spatial resolution about 1 km and by means of radio-locators with synthetic aperture with spatial resolution of 25 m. Methodical questions of maximal wind speed evaluation basing on the information receiving from radio-locators of side observation on the satellites are given by Mur and Fen (1979).

It is clear that the search of the indicator-predictors of tropical hurricanes is necessary to perform at the same time together with the study of all atmospheric processes that happen above the ocean and offshore. A set of these processes includes tornados, storms, hurricanes, and typhoons. A whirlwind (tornado) is one of the insidious and sudden natural formations in the atmosphere. It is a rotating funnel-shaped cloud which stretches from the bottom of the thundercloud to the Earth's surface. The characteristic tornado wind speeds are 65–120 km h^{-1}, but sometimes this magnitude reaches 320 km h^{-1} and more.

The USA territory is the most characteristic region where tornados are rather frequent phenomena, though tornados happen all over the world. During the period 1961–2004, over the USA territory, tornados took away the lives of 83 people per year, on the average. Most often, tornados take place in the eastern states adjacent to the Gulf of Mexico and in February and March their frequency of occurrence is at a maximum. On the territories of Iowa and Kansas, tornados are most frequent in May-June. The average amount of tornados over the USA territory is estimated at $\sim$800 per year, 50 % of which falls on April–June. The territorial heterogeneity of the frequency of occurrence of tornados in the USA has stable characteristics: in Texas—120 tornados per year, and in the north-eastern and western states—1 tornado/year. For instance, only in April and November 2002 there were more than 100 tornados that caused numerous destructions with more than 600 cases of insurance payments. The element does not spare other countries, too. For instance, the winter hurricane "Janette" in 2002 over Europe caused numerous destructions and insurance payments reaching more than 1 billion dollars. For instance, the hurricane Charlie in the USA in 2004 over the states Florida, Cuba, and Jamaica, causing 19, 2, and 1 human victim, respectively. More than 150 thousand people remained without telephonic communication, and the damage exceeded 11 billion dollars.

Historical chronology of hurricanes happening over the American continent shows that during one year the frequency of occurrence of hurricanes is non-uniform: 92 % of events fell on August-October with an approximately uniform distribution.

Typhoons can cause a local rise of the ocean's level by several meters and, if it takes place near the coastline, a 10-m wave is formed flooding the lowlands at an enormous rate. For example, numerous typhoons took place in 2000 over Taiwan, Japan, South Korea, and other global regions. So, the typhoon "Billis" falling on 20 August at a rate of 360 km h^{-1} onto Taiwan during several hours ruined 25-buildings (banks, schools, offices of companies, and dwelling houses), with the damage of 48 million dollars. In late August, the typhoon "Prapiroon" stormed over the East-China Sea at a wind speed of 43–181 km h^{-1}. The international airport in Shanghai was closed for some time. In the World Ocean there are seven regions of the origin of typhoons located near the equator. The main cause of typhoons' appearance is the heating of several tens of meters of the ocean water above the critical level (26.8 °C). The satellite monitoring of typhoons has made it possible to trace their routes. A trade-wind-caused typhoon moves mainly westward, gradually turning to the north-east and east under the influence of Coriolis force. Up to 47 %

of typhoons move along this parabolic trajectory. The remaining 53 % move along other trajectories and, as a rule, not strictly regularly. But still, a great group of typhoons (21 %) move, as a rule, along half the parabolic territory mentioned above, and then move from south-east to north-west. Seven percent of typhoons move east-to-west and from south-west to northeast, and 4 % of typhoons move from south-to north.

Extreme atmospheric processes that have different external effects, undoubtedly, have general origin. The search of the correlation between them can help to solve the problem of earlier detection of tropical hurricane. However, it is necessary to understand the interactivity character of processes that forego to the tropical hurricane. Especially, this is important when temperature conditions for the hurricane beginning are not realized.

The possibility that formation of hurricanes formed at temperature of an ocean's surface below 27 °C is connected with influence of other factors is not excluded. For example, a constant or frequent-periodic degassing of rift zones of the Earth, connected with magmatic, volcanic and hydrothermal activity on an ocean floor of a tropical zone promotes considerable decrease in the level of "ocean-atmosphere" system stability.

Figures 7.28 and 7.29 explain the role of geo-tectonic factors in tropical cyclogenesis activity. Considering the Fig. 7.29, it is necessary to notice that when the seafloor crust is younger, progress of tectonic processes and processes of degassing is especially probable. According to these schemes, a number of tropical depressions were generated practically over the mid-Atlantic ridge. The mid-Atlantic ridge has the most strongly pronounced features among others mid-oceanic ridges of the World

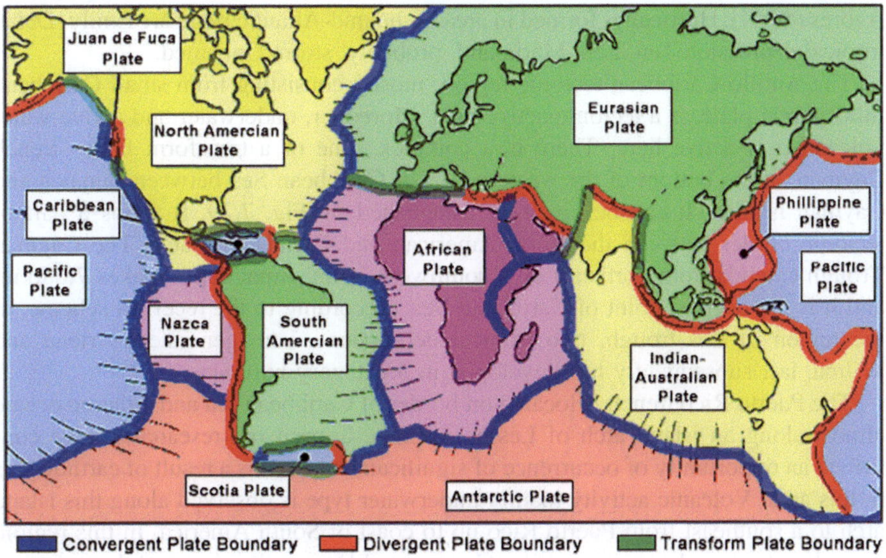

Fig. 7.28 Map of lithospheric plates (http://uccpbank.k12hsn.org)

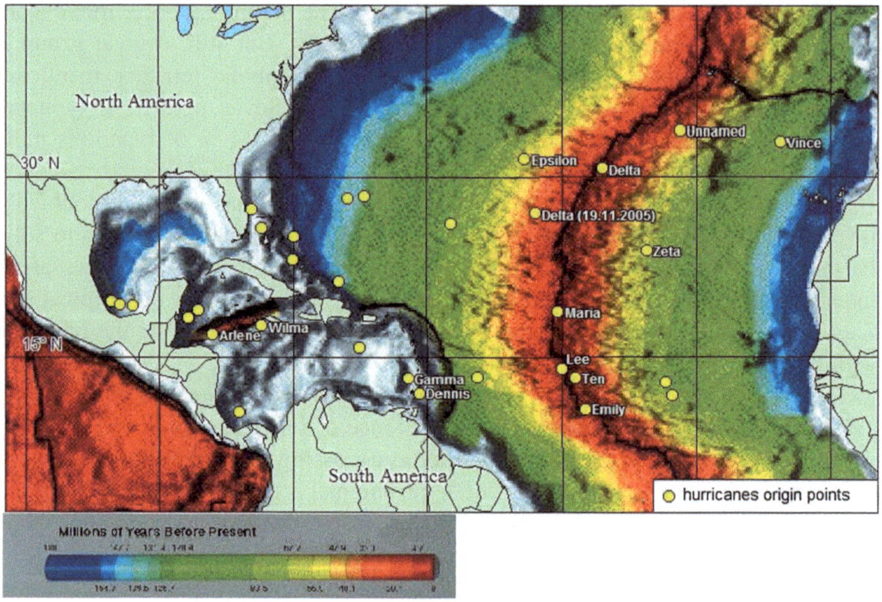

Fig. 7.29 Earth's seafloor crust age (*red color* corresponds to the most young crust) (Grankov 2011)

ocean. Along the ridge arch there is a system of deep longitudinal breaks (rifts), on all extent it is crossed by cross-section (transform) faults. Tectonic activity of a ridge is in a greater degree shown in its northern part. The most active area of display ancient, modern, above-water, underwater and rift volcanism in the mid-Atlantic ridge are the Azores (40°N). Hurricanes formed in area of the mid-Atlantic ridge are Emily, Delta, tropical depression Ten, Lee, Maria and, probably, storm Unnamed.

The Antillo-Caribbean tectonic region, mostly consisting from small Caribbean lithospheric plate, is a tectonic active area. Moreover, underwater and above-water volcanism is active here. There is a complex zone of a transform fault—trench Cayman, at the bottom of the western part of Caribbean Sea between Jamaica and Cayman islands. It can be precisely established on Fig. 7.29. It forms a part of tectonic border between the North-American and Caribbean plates. The Cayman Trough extends from northeast on a southwest, its maximal depth makes 7,686 m, and it is the deepest point of Caribbean Sea. According to the received scheme, in the region of this trough, two tropical depressions were generated: Arlene and Wilma, last subsequently has developed in 5 category hurricane.

The Puerto Rico trench is located on border of Caribbean sea and Atlantic ocean, almost along an island arch of Lesser Antilles. Carried out researches have confirmed an opportunity of occurrence of significant tsunami as a result of earthquakes in this area. Volcanic activity mainly underwater type is observed along this island arch to a southeast from Puerto Rico up to coast of South America. In this region, according to the scheme, hurricanes Gamma and Dennis (hurricane of 4 categories) have been formed.

Thus, from 31 considered tropical depressions, 5 depressions were formed and developed further up to storm and hurricanes at temperatures of an ocean's surface below 27 °C. 10 tropical depressions were formed in areas of high tectonic activity—subduction zones and rift zones of ocean. As it has appeared, tropical storm Delta and Unnamed were formed and at temperature of an ocean's surface below 27 °C, and near to zones of high tectonic activity. In the previous report, assumptions have been formulated and mechanisms of possible active influence of degassing processes of the Earth (especially in tectonic active zones of oceans) on processes initiation and origin of tropical depressions are described. The materials presented by the given report have convincingly shown an opportunity of existence of such processes in the nature on examples of origin of tropical cyclones and depressions in Atlantic Ocean for a season of 2005.

The mechanisms of possible active influence of degassing processes of the Earth (especially in active tectonic zones of oceans) on processes initiation and origin of tropical depressions have been described by Syvorotkin (2002). It was also shown that warm and cold ocean currents and flows of air masses have powerful influence on tropical cyclogenesis. This fact is observed when analyzing the variability of the origin zones of tropical cyclones (TC) and in the study of seasonal and other variations in ocean currents and winds.

We have already carried out a joint analysis of the impact of potential risk factors for the tropical cyclogenesis, depending on the time period for the territory of the Atlantic Ocean. In this study, an attempt was made to analyze a part of the globe for distribution of these TC forming factors. The Western Hemisphere of the Earth was chosen for the analysis. When working, it was decided to use the following method: recalculation of the presence or absence of some TC forming factors on a grid with cell size 1° by 1°.

Coordinate grid was formed containing 32,400 cells (180° × 180°). The numbering of cells is a line from the lower left corner. Cells lying entirely in the territory of land were excluded from further consideration (about 7,000 cells). As an indicator of tectonic activity the following data were collected: (1) data points of recorded earthquakes (for period from 1973 to 2011); (2) the data on the location of major volcanoes; map with boundaries of tectonic plates with different types of faults (subduction zones, rift zones, transform faults, spreading, displacement). It is important to note that the location points of volcanoes are given only to surface volcanoes, the data for submarine volcanoes are absent, so we'll keep in mind that most of the submarine volcanoes are associated with the tectonically active zones.

For earthquakes data it was decided to allocate three gradations of danger (points): 0—no risk, 1—small risk (the number of earthquakes for the entire period of less than or equal to 5), 2—high. Points of volcanoes: 0—no volcanoes nearby, 1—small risk (the number of volcanoes less than or equal to 2), high—2. To determine the proximity of cells to tectonic faults were allocated three grades: 0—no tectonic fault nearby, 1—close to a fault, the proximity was determined by constructing a "buffer zones" all along the faults width of 100 km, 2—the cell is directly at tectonic fault. Data on the tectonic faults were calculated separately for the oceanic and continental faults.

Table 7.10 Database structure

Column name	Content	Data format
Id	Number of cells of coordinate grid	Number, 1–32,400
Land	Indicated that cell belonging to the land area or ocean	Land/ocean
Earthquake	The number of registered earthquakes for the period 1973–2011	0–5310
Earthquake_pt	Points	0/1/2
Volcanoes	The number of volcanoes	0–12
Volcanoes_pt	Points	0/1/2
Geo	Indicates the proximity of grid cell to the faults oceanic crust	0/1/2
Geo_cont	Indicates the proximity of grid cell to the faults continental crust	0/1/2
Streams	Indicates the absence/presence of convergence zones in oceanic currents	0/1
Summa	Total points for columns: Earthquake_sc+Volcanoes_sc +Geo+Geo_cont+Streams	0–6

In addition to these indicators of tectonic activity, data on the possible influence of ocean currents have been included in the database too. The presence of ocean currents is largely caused to constant winds. Effect of Inter-Tropical Convergence Zone (ITCZ—a transition zone between trade winds of the northern and southern hemispheres, or between trades and monsoons, or between trades and equatorial westerly winds) on the formation of tropical cyclones is well known. It is characterized by the convergence of velocity, i.e. the weakening of the wind speed and the convergence of the streamlines, as well as intensive development of convective clouds. The database included data on the presence of convergence zones of oceanic currents: 0—no zones of convergence of oceanic currents, 1—there is such zone.

All data were entered into the database (1° × 1° grid). Further obtained points for all indicators were summed up. Database is shown in the Table 7.10.

7.8 Concluding Remarks and Future Problems

An analysis of observational data characterizing the ocean-atmosphere system dynamics in the tropical sector of the World's oceans shows that differentiation of the ocean-atmosphere system phases by means of transitions in the meteorological percolation cluster allows the division of zones with different levels of the ocean-atmosphere system instability. To achieve a more reliable agreement between the ocean-atmosphere system state and parametrical space of meteorological percolation cluster, it is necessary to establish the algorithm for the calculation of the critical level of the renormalization probability p^* taking into consideration the procedure of renormalization group transformations for more than two dimensions. Only in this case the real-space renormalization process can make more precise the moment of $I_m(t)$ transition from background zone Ξ_1 to the signal zone Ξ_3 (Fig. 7.17).

Thus, the following conclusions are drawn:

(1) An application of percolation procedure gives the possibility to detect the moment when the ocean-atmosphere system switches its phase state from the background position to the state with high level of environmental instability.
(2) An introduced indicator of the ocean-atmosphere system instability allows the detection of the beginning of a hurricane before its detection by existing monitoring systems.
(3) An increase of the precision of the proposed algorithm is possible with the optimization of parametrical structure in the indicator $I_m(t)$.

It is evident that the solution to the problem of reliable prediction of the beginning of a hurricane requires the development of an efficient information technology which must be used in the environmental monitoring systems. This technology should include sections responsible for planning of measurements, development of algorithms for complex processing of data from different sources, and relevant risk assessment. The main difficulty in this lies in the spatial and temporal heterogeneity of existing information obtained from different observational systems. This heterogeneity can be overcome by using the algorithms of the space-temporal interpolation (Kondratayev et al. 2006; Krapivin and Varotsos 2007). Ultimately, however, the proof of the usefulness of a prediction algorithm such as this one can only really be established by attempting to use it "live", i.e. in real time, to make predictions of hurricanes in their initial stages. The serious task consists in the prediction of environmental disaster under minimal false alarm. To a considerable extent it is a matter of the successful definition of thresholds.

Thus, the structural scheme of the multi-channel monitoring system for search and detection of the hurricanes in their beginning stage can be presented by the scheme of Figs. 3.16 and 7.13. One of these systems can be realized on the existing meteorological stations located in the zones where hurricanes or typhoons are beginning. Here T_i and ξ_i is the deterministic and stochastic components of I_m or λ^*, respectively. Meteorological and geophysical data delivered by satellite sensors and NDBC meteorological stations are processing for the calculation of I_m or λ^* that are analyzed in the framework of sequential procedure described in Chap. 3. In fact, the picture of Fig. 7.30 is realized at all times. The calculation of probabilities that characterize the prognosis level for the anomaly moving direction (see Figs. 7.25 and 7.26) gives a possibility to have a cyclone track. This procedure can operate in automatic mode.

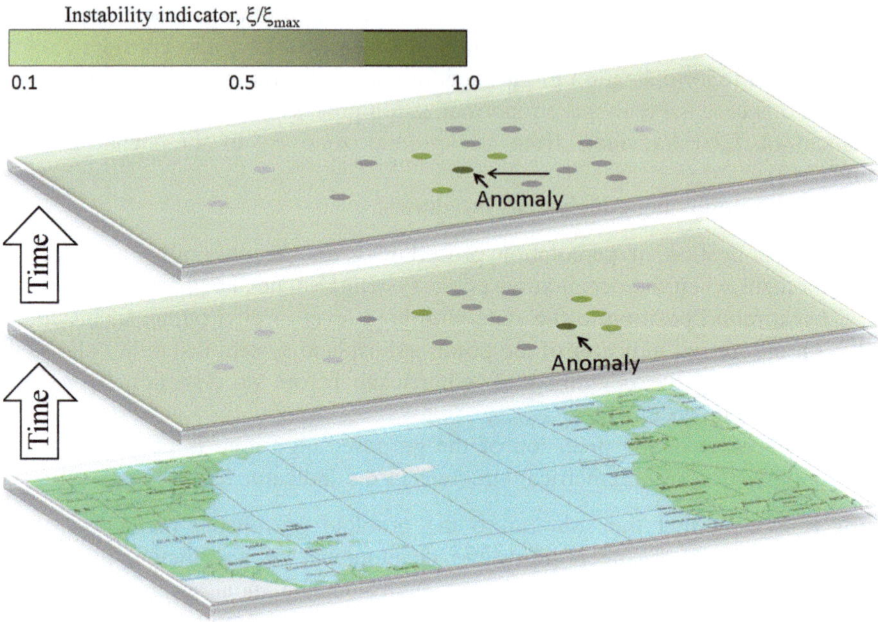

Fig. 7.30 Operational scheme of monitoring regime for the detection of tropical cyclone beginning and its next tracking

The future development of the proposed approach to the hurricane beginning detection, in the first place, requires the solution to the following problems:

- Study of the correlations between hurricane origin and solar activity and understanding the interdependence between state of ocean-atmosphere system and large-scale meteorological processes.
- An assessment of the role that processes of the Earth degassing play in the formation of tropical cyclones and how the ozone layer state correlates with events conducting the hurricanes.
- Search of the technologies to control the hurricanes when they are beginning.
- Perfection of climate models to make their effectiveness acceptable for the detection of hurricane origin.

Chapter 8
Ecoinformatics and Soil-Plant Formations

8.1 Methodology, Theory and Problems

As mentioned above, the field of ecoinformatics provides concepts, methods and standards in order to guide management and analysis of ecological data with particular emphasis on exploration of co-occurrences of organisms and their linkage to environmental conditions and taxon attributes (Boden et al. 1994; Dengler et al. 2011; Fang et al. 2001; Rastetter et al. 2003). This chapter reflects on the development of ecoinformatics and explores its importance for future global change research with special focus on vegetation-plot data. Vegetation covers including forests are the most impacted by the anthropogenic reconstruction and revision. Therefore, the objective of this chapter is threefold:

(1) To present a working methodology for the combined use of modeling technology and microwave remote sensing measurements in the assessment of attenuation of electromagnetic waves by the vegetation cover.
(2) To illustrate this methodology with computer calculations of the attenuation for various soil-plant formations.
(3) To give a perspective of the developed methodology applied to the study of global environmental change, including the radiative forcing problem.

The solution to the majority of applied problems within agro-meteorology, forestry, animal husbandry, and other areas of human activity directed toward the protection of nature is difficult because effective methods of control of the soil-plant formation (SPF) are insufficiently developed. During the last few years, the global carbon cycle problem has acquired a special significance because of the greenhouse effect. The knowledge of the state of the SPF allows to have a real picture of the spatial distribution of the carbon sinks and sources on the Earth's surface.

V.F. Krapivin et al., *New Ecoinformatics Tools in Environmental Science*,
Environmental Earth Sciences, DOI 10.1007/978-3-319-13978-4_8

Specific interest is due to the fact that forest ecosystems play important role in global environmental change. A forest ecosystem is a natural woodland unit consisting of all plants, animals and micro-organisms (biotic) in an area functioning together with all of the non-living physical (abiotic) factors of the environment. The entire assemblage of forest organisms (trees, shrubs, herbs, bacteria, fungi, and animals, including people) together with their environmental substrate (the surrounding air, soil, water, organic debris, and rocks), is forest environment that is studied by many authors (Gyde 2012; Silver 1998; Sudarshana et al. 2012; Krapivin et al. 2013a–d). Forests and woodlands occupy about 31–38 % of the Earth's surface, and they are more productive and have greater biodiversity than other types of terrestrial vegetation. Forests grow in a wide variety of climates, from steamy tropical rainforests to frigid arctic mountain slopes, and from arid interior mountains to windy rain-drenched coastlines. The type of forest in a given place results from a complex of factors, including frequency and type of disturbances, seed sources, soils, slope and aspect, climate, seasonal patterns of rainfall, insects and pathogens, and history of human influence.

The world's total forest area is just over 4 billion ha, which corresponds to an average of 0.6 ha per capita. The five most forest-rich countries (the Russian Federation, Brazil, Canada, the United States of America and China) account for more than the half of the total forest area. Ten countries or areas have no forest at all and an additional 54 have forest on less than 10 % of their total land area (Figs. 8.1, 8.2 and 8.3).

Tropical forests have specific interest. Sudarshana et al. (2012) noted that the astounding richness and biodiversity of tropical forests are rapidly dwindling. This has severely altered the vital biogeochemical cycles of carbon, phosphorus, Global nitrogen etc. and has led to the change in global climate and pristine natural ecosystems. Sudarshana et al. (2012) have defined "Tropical Forests" broadly, into five different themes:

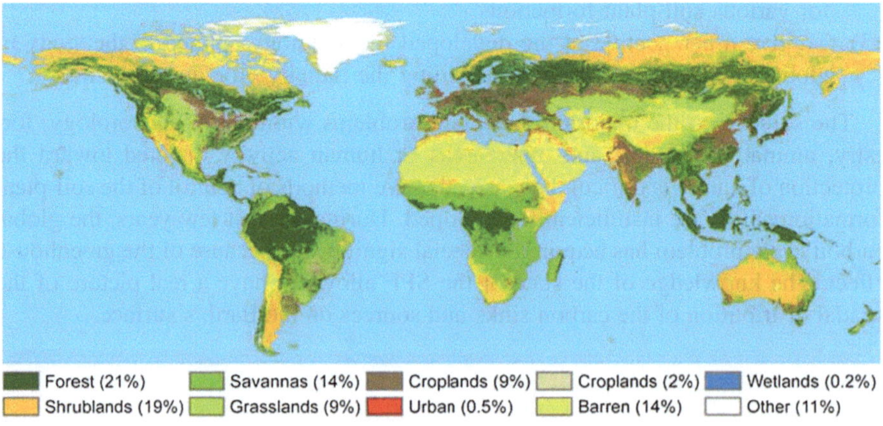

Fig. 8.1 Global distribution of land cover classes (http://edcimswww.cr.usgs.gov/pub/imswelcome/)

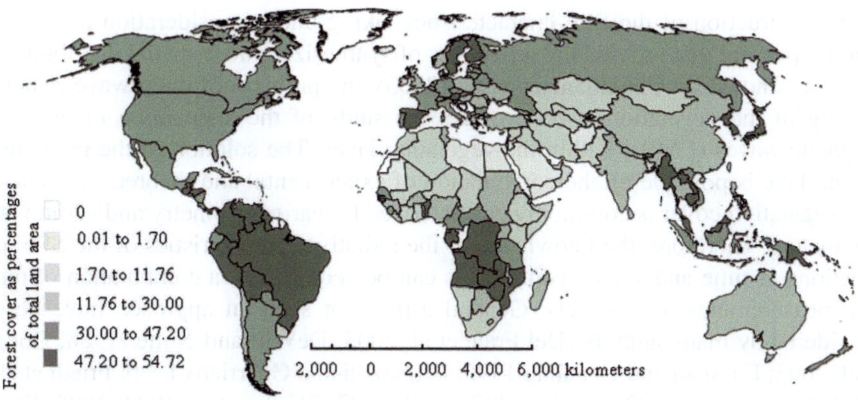

Fig. 8.2 Forest area in percent of land area by country 2010 (FRA 2011)

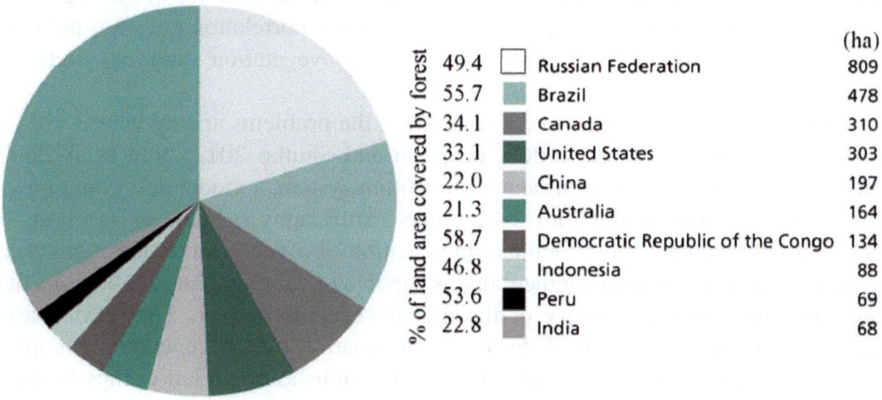

Fig. 8.3 Countries with largest forest area, million hectares (FRA 2011)

(1) tropical forest structure, synergy, synthesis,
(2) tropical forest fragmentation,
(3) impact of anthropogenic pressure,
(4) Geographic Information System and remote sensing, and
(5) tropical forest protection and process.

At present time, many government, state, and private forestry organizations and agencies utilize geospatial technology such as GIS (geographic information systems) and remote sensing for various applications supporting analysis, assessment, and management of forests. It is well known that among the types of remote sensing techniques, microwave radiometry is proved to be effective for observations of SPF environmental parameters. However, these observations are a function of different environmental conditions mainly depending on the SPF type. For this reason, it is necessary to develop data processing methods for microwave monitoring that allow the reconstruction of the SPF characteristics taking under consideration the vegetation types and that provide the possibility of synthesizing their spatial distribution.

As it is noted by Chukhlantsev et al. (2003), the problem of microwave remote sensing of the vegetation cover requires the study of the attenuation of electromagnetic waves (EMW) within the vegetation layer. The solution of the problems arising here is possible by the combination of experimental and theoretical studies. The vegetation cover is commonly characterized by varied geometry and additional parameters. Therefore, the knowledge of the radiative characteristics of the SPF as functions of time and spatial coordinates can be acquired by a combination of on-site measurements and models. General aspects of such an approach have been considered by many authors (Del Frate et al. 2003; DeWitt and Nutter 1988; Dong et al. 2003; Engman and Chauhan 1995; Ferrazzoli and Guerriero 1996; Friedi et al. 2002; Karam et al. 1992; Shutko 1987, 1992, 1997; Shutko et al. 1994, 1995; Font et al. 2002; Kirdiashev et al. 1979; Field 2001; Goody and Jung 1989; Sellers 1985, 1987; Vinogradova et al. 2004). However, these investigations were mainly restricted to the investigation of models describing the dependence of the vegetation medium on environmental properties, as well as the correlation between the morphological and biometrical properties of the vegetation and its radiative characteristics.

One prospective approach to the solution of the problems arising here is GIMS-technology (Krapivin et al. 2006b; Krapivin and Shutko 2012; Nitu et al. 2007, 2013c). Combining an environmental acquisition system, a model describing how a typical geo-ecosystem functions, a computer cartography system, and a means of artificial intelligence should facilitate the creation of a geo-information monitoring system for a typical natural element that is capable of solving many of the tasks that arise in microwave radiometry of the global vegetation cover. The GIMS-based approach, in the framework of the EMW attenuation by the vegetation canopies, allows the synthesis of a knowledge base that establishes the relationships between experiments, algorithms and models. The links between these areas have an adaptive character giving an optimal strategy for experimental design and model structure. The goal of this chapter is to explain and assess the application of the GIMS method to the tasks of reconstructing the spatial and temporal distribution of the SHF radiation characteristics.

An experience of the GIMS-technology application has shown that it opens new prospects of microwave remote sensing monitoring of vegetation covers giving the possibility to combine theoretical and empirical investigations of the vegetation role in the microwave radiation of the land surface. The progress in the applied results depends on the base of the GIMS models. As a result, the GIMS allows the evaluation of the biosphere energy balance with high precision and for this reason the following tasks are currently a subject to intense study:

- Development of the models for description of SPFs under minimal demands to the parametrical environment.
- Analysis of possibilities of different monitoring systems to assess the SPF's characteristics.

- Theoretical and empirical study of microwave radiation and its propagation in the atmosphere-vegetation-soil system.
- Synthesis of the GIMS database including theoretical and empirical assessments for model coefficients, radiation characteristics of vegetation, and spatial distribution of the biomes.

8.2 Global Dynamics of Land Ecosystems

The growing global size of population and the associated growing forcing of human activity on the environment and ecosystems have become not only the main threat to further sustainable development of civilization, in the context of the global ecological safety, but also reflect a dangerous disorder in the normal functioning of various systems of life support (Bolin and Sukumar 2000; Borodin et al. 2008; Krapivin and Varotsos 2007, 2008). In connection with the key role of the ecosystems in the processes of the natural regulation of the environmental properties, of principal importance is an analysis of the available data on the global dynamics of ecosystems and an assessment of possible trends. Important information on these problems can be found, in particular, in recent publications by many authors (Kondratyev and Krapivin 2001c, d; Hackler et al. 1999; Ito and Oikawa 2002; Seppelt and Voinov 2002; Gbondo-Tugbawa and Driscoll 2002; Lischke et al. 2002; Sellers et al. 1997; Bouwman 1990), among which the most informative is the report prepared by the World Resources Institute (WRI 2011) supported by the UN Programme of Development, UNEP, and the World Bank (A Guide… 2000; WRI 2012). This report emphasizes a close relationship between the global ecosystems and the global population—a symbiosis, unique and extremely sensitive to external forcings. These presentations have been thoroughly substantiated by Gorshkov et al. (2000, 2002) in the form of a concept of biotic regulation of the environment. Monographs by Kondratyev (1998a, b, 1999) and new publications (Kondratyev et al. 2000; Kondratyev and Krapivin 2001c; Nitu et al. 2004, 2013a, b; Krapivin and Shutko 2012) contain an analysis of the key aspects of the global changes from the viewpoint of functioning of the global systems of life support and requirements to an adequate ecological monitoring.

Global natural and regulated ecosystems play an important role as a factor of the environmental dynamics ranging from micro-scales (e.g., soil bacteria) to the whole planet and, on the other hand, are vitally important sources of drinking water, food, timber, paper, and other means of life support. As has been mentioned earlier, an urgency of the problem is that the world, on the whole, has already drawn near such limits to the impacts on the ecosystems, an exceeding of which is fraught with irreversible destruction of the global systems of life support. These limits have already been exceeded; though the present enthusiasm for apocalyptic predictions is, so far, unfounded (this especially refers to the so-called "global warming".

An extreme complexity of the discussed problem is that it is necessary to explain (and, as far as possible, to predict) the dynamics of the interactive system "nature–society" (the society should be placed first here since its functioning determines its impact on nature) with its numerous feedbacks, nonlinear nature, and "surprises". Unfortunately, the present stage of studies of the system "nature–society" can be considered not more than initial and preliminary. This refers to even a simple description of the present condition of nature (global ecosystems), which results from the observational data deficit with an apparent abundance of some observational means (especially expensive space-borne means). Therefore the reports of the World Resources Institute (WRI 2011, 2012) are in many respects incomplete, being concentrated only in consideration of five types of ecosystems (the share of land surface is given in brackets, except the Antarctic and Greenland occupied by the respective ecosystem): agricultural ecosystems (28 %), coastal regions (22 % within a 100-km band), forest (22 %), fresh-water (<1 %), and grass (41 %) ecosystems. An abandonment of the World Ocean is, of course, a serious, though justified (in view of information deficit) flaw.

Special attention should be also paid to the soil ecosystems. These ecosystems are very important for the solution of various problems of human life support and regulation of the environmental conditions (Table 8.1). Human use of the land surface includes about 2 % for habitation, roads, and the like; 12 % for crops; 25 % for livestock grazing; and exploitation one way or another of 30 % more in forests or tree farms. The remaining 20 % is mostly deserts, mountaintops, and other places with inhospitable climates. People also mobilize minerals at rates far above those of natural processes and are changing the composition of the atmosphere in potentially dangerous ways. In 1986, a study found that humanity had already destroyed or was co-opting almost the half of the planet's NPP, thus interfering in many ways with the biosphere's present and future capacity to support the human population (Ehrlich and Ehrlich 2009).

The authors of the report of the World Resources Institute (WRI 2011, 2012) have briefly summarized the anthropogenic impacts on the ecosystems during the civilization development, beginning from an intensive application of the irrigation systems during the Shumer civilization, which had led to soil salinization, to the present global processes of atmospheric pollution and the ozone layer depletion (Kondratyev and Varotsos 2000; Sivasakthivel and Reddy 2011). The data of Table 8.2 characterize the most important aspects of anthropogenic impacts on the ecosystems nowadays.

There are numerous examples of destructive (and even catastrophic) impacts on ecosystems and their economic consequences (see for details Grigoryev and Kondratyev 2001). The collapse of cod catch in 1990 in the sea regions of Canada resulted in about 30 thousand fishermen unemployed, and only in the region of Newfoundland brought forth serious economic difficulties in 700 settlements. Material losses in China reaching 11.2×10^9 \$US/year have resulted from a deficit of drinking water due to polluted river and sub-soil waters. In India, the commercial forest cutting and the transformation of the deforested lands into the agricultural

Table 8.1 Ecosystems and their functions of life support

Ecosystem	Human life support	Regulation of the state of the environment and human life support
Agro-ecosystem	• Food (agricultural crops) • Fibre (agricultural crops) • Genetic diversity of agricultural crops	• Maintaining the functions of watersheds (infiltration, regulation of run-offs, partial protection of soils) • Provision of habitat for birds, pollination processes, functioning of soil organisms • Formation of soil organics • Accumulation of carbon from the atmosphere • Provision of population employment
Coastal ecosystems	• Fish and mollusks • Fish as food for animals • Algae (as food and raw material for industry) • Salt • Genetic resources	• Some impact on storms (mangrove thickets, "barrier" islands • Habitat for wild fauna and flora • Biodiversity maintenance • Solution and processing of wastes • Provision of harbours and routes for water transport • Habitation for humans • Provision of population employment • Provision of places for recreation and aesthetic satisfaction
Forest ecosystems	• Wood • Wood fuel • Water for drinking and reclamation • Non-wood products (wine, bamboo, leaves, etc.) • Food (honey, mushrooms, fruit, etc.) • Genetic resources	• Removal of atmospheric pollutants and emission of oxygen • Nutrients cycles • Maintaining the functions of watersheds (infiltration, water cleaning, regulation of run-off, stabilization of soil properties) • Biodiversity maintenance • Accumulation of carbon from the atmosphere • Some weakening of extreme impacts of weather • Soil formation • Provision of habitat for humans and wild animals • Provision of places for recreation and aesthetic satisfaction
Fresh-water ecosystems	• Water for drinking and reclamation • Fish • Hydroelectric energy • Genetic resources	• Regulation of water regime • Solution and removal of wastes • Cycles of nutrients • Biodiversity maintenance • Provision of aquatic habitat • Provision of water transport routes • Provision of population employment • Provision of places for recreation and aesthetic satisfaction

(continued)

Table 8.1 (continued)

Ecosystem	Human life support	Regulation of the state of the environment and human life support
Grass ecosystems	• Domestic animals • Water for drinking and reclamation • Genetic resources	• Maintaining the functions of watersheds (infiltration, water cleaning, regulation of run-offs, stabilization of soil properties) • Cycles of nutrients • Removal of atmospheric pollutants and formation of oxygen • Biodiversity maintenance • Soil formation • Accumulation of carbon from the atmosphere • Provision of habitat for humans and animals • Provision of places for recreation and aesthetic satisfaction

ones have not only changed the traditional way of people's life but also caused a deficit of wood fuel and timber to the detriment of 275 millions of rural population.

As for the estimates of the consequences of the global anthropogenic impacts, the situation with the water resources is an example: about 28 % of the global population have no access to pure drinking water; every year about 5 million people die because of drinking low quality water and anti-sanitary conditions; about 90 % of wastes in the developing countries go to rivers, lakes, and coastal regions of the seas, etc. Intensified emissions of CO_2 into the atmosphere have caused considerable changes in the global carbon cycle (Demirchian and Kondratyev 2004; Kondratyev et al. 2003c; Field et al. 2004; Archer 2010).

The most important fact is that the levels of impact on the ecosystems have become of a global scale. About 75 % of marine fish populations have either decreased due to violation of the permissible amounts of catch or come close to the threshold of their survival. An intensive forest cutting has almost halved the forested areas, and the construction of various economic infrastructures has caused a fragmentation of the forest cover (Kaiser 2000; Keller et al. 2001). About 58 % of coral reefs are seriously affected by fishery, tourism, and pollution. Almost 65 % of arable lands have partially lost their fertility. The scales of economic usage of ground waters exceed the rate of their natural recovery by at least 160×10^9 m^3/year. In most cases the anthropogenic load on the ecosystems has intensified. This especially concerns the pasture ecosystems where the interference of the human's factor has reached the level of direct control (Schulte 2003; Schulte et al. 2003).

It is well known that the main causes of ecosystems' degradation are the growing size of population and, respectively, increased needs for natural resources as well as enhancing loads on the environments (Table 8.2). Concrete detailed data characterizing the present global situation can be found in numerous publications (Kondratyev 1998a, 1999, 2000a; A Guide… 2000; Watson et al. 2000). It is notable that the strongest geographical non-uniformly of the growing scales of

Table 8.2 Anthropogenic impacts on ecosystems

Ecosystem	Anthropogenic impacts	Causes of impacts
Agri-ecosystems	• Use of agricultural lands for the urban constructions or industrial purposes • Pollution of inland waters due to nutrients, run-offs, and silting • Impoverishment of water resources as a result of irrigation • Reduction of soil fertility due to erosion, change of agricultural practice, or decreased input of nutrients • Weather change	• Increasing size of population • Increasing needs for food and industrial products • Urbanization • Governmental subsidies to agriculture • Poverty and unemployment • Climate changes
Coastal ecosystems	• Exhaustion of fish resources (exceeding the permissible levels of catch) • Transformation of wetlands and coastal habitats • Water pollution by agricultural and industrial sources • Fragmentation and destruction of natural barriers by tides and reefs • Intrusion of alien species • Possible rise of sea level	• Increasing size of population • Increasing needs for food and the growth of tourism • Development of urbanization and of recreation places • Governmental subsidies to fishery • Inadequate information about the state of ecosystems, especially from the viewpoint of fishery conditions • Poverty and unemployment • Uncoordinated ecological policy in coastal regions • Climate changes
Forest ecosystems	• Transformation or fragmentation of forest ecosystems as a result of either agricultural use or urbanization • Biodiversity losses due to fires and deforestation, emissions of stored carbon, pollution of the atmosphere and water basins • Acid precipitation due to industrial pollutions • Intrusion of alien species • Over-exploitation of water resources in their economic use	• Increasing size of population • Growing needs for wood • Governmental subsidies connected with deforestation and use of timber • Inadequate estimation of losses caused by industrial pollution of the atmosphere • Poverty and unemployment

(continued)

Table 8.2 (continued)

Ecosystem	Anthropogenic impacts	Causes of impacts
Fresh-water ecosystems	• Over-exploitation of water sources in their economic use • Breaking the norms of fish catch in the inland water basins • Building of dams in the interests of irrigation, creation of hydroelectric power stations, and floods control • Pollution of water basins due to various factors of economic water use • Intrusion of alien species	• Increasing size of population • General deficit of water and natural non-uniform distribution of water resources • Governmental subsidies to maintain the water use • Inadequate estimate of losses due to water basins pollution • Poverty and unemployment • Growing needs for hydroenergetics
Grass ecosystems	• Transformation or fragmentation due to either agricultural use or urbanization • Biodiversity losses due to fires, emissions of stored carbon, and atmospheric pollution • Reduction of soil fertility and water pollution by domestic animals • Unacceptable scales of the use of animals for entertainment	• Increasing size of population • Growing needs for agricultural products, especially for meat • Inadequate information on the state of ecosystems • Poverty and unemployment • Accessibility and easiness of grass ecosystems

consumption is an extremely important feature of the growing scales of consumption reflecting the socio-economic contrasts in the world.

Furthermore, we shall discuss the principal results obtained within the framework of the pilot analysis of the global ecosystems (PAGES) (A Guide… 2000). The main difficulties of this analysis are connected with the deficit of the available information. The PAGES project is focused on the consideration of the three types of the indicators of impacts on the ecosystems dynamics:

(1) anthropogenic loads (increasing size of population, growing level of resources consumption, pollutions, overexploitation of natural and controlled ecosystems);
(2) the spatial extent of ecosystems (size, shapes, localization, geographical distribution);
(3) production of economically important products such as agricultural crops, wood, fish, etc.

A considerable drawback of each of these indicators and all of them is that they do not contain information about the thresholds of ecosystems' ability to perform their functions of life support.

Figure 8.4 illustrates the presentation of the expert assessments of the state of ecosystems and their functioning as the life supporting systems including the respective trends: increasing ability of ecosystems to perform the life supporting

	Agro	Coasts	Forests	Water	Grass
Food/fiber	↘ 2	↘ 3	↗ 2	↑↓ 2	↘ 3
Water quality	↘ 4	↓↑ 3	↘ 3	↘ 4	6
Water amount	↘ 3	6	↘ 3	↘ 3	6
Biodiversity	↘ 4	↘ 3	↘ 4	↘ 5	↘ 3
Carbon supplies	↑↓ 3	6	↘ 3	6	↘ 2
Recreations	6	? 2	6	6	↘ 2
Protection of coasts	6	↘ 4	6	6	6
Wood fuel	6	6	? 3	6	6

Fig. 8.4 Expert estimates of the levels of anthropogenic forcings on ecosystems and their consequences

functions (↗), a reduced ability (↘), mixed trends (↑↓), unknown trends (?). The figures characterize the levels of the ecosystems' functioning as life-supporting compared with the situation 20–30 years ago: 1—excellent, 2—good, 3—satisfactory, 4—weak, 5—unsatisfactory, 6—no estimate. The results of the expert assessments shown in Fig. 8.4 testify, on the whole, to considerable anthropogenic forcings on ecosystems (especially during the 20th century) that have led to considerable changes in their ability to perform the life-supporting functions. Though in some cases (e.g., food production) the ecosystems' productivity has substantially increased, providing thereby the needed level of production, nevertheless, in other cases (drinking water quality, preserving the biomass, etc.) there was a degradation of the ecosystems' functioning.

The biodiversity losses during the last century have acquired the threatening scales and concern practically all ecosystems (mainly—as a result of the loss of habitats). So, for instance, the area of forests has decreased at least by 20 % and probably (on global scales) by 50 %. Some forest ecosystems (like the dry tropical forests of Central America) have disappeared at all. In many countries up to 50 % of mangrove plantations have been lost, the area of wetlands has almost halved, and in some regions the size of territories covered with grass has decreased by 90 %. Only the tundra, arctic, and marine deep-water systems have little changed though in these cases substantial anthropogenic impacts also occur sometimes.

Even when the initial areas of the ecosystems have been preserved, many species are threatened with pollution, over-exploitation, intrusion of "alien" species, and degradation of habitats. From the viewpoint of biodiversity, the fresh-water ecosystems have suffered most. For instance, about 20 % of the fresh-water fish species have been exterminated, and many other species are in a very dangerous condition. The forest, grass, and coastal ecosystems face serious problems of survival. The biodiversity is threatened with frequented diseases of marine organisms, intensive florescence of water bodies, and a considerable reduction of the population of amphibians. Apart from the respective losses for medicine (due to a decreasing volume of raw materials for drugs), banks of genes, and eco-tourism, a reduction of biodiversity means also a threat to the ecosystems' productivity, their integrity and resistance to various external forcings.

And finally, it is clear that there are now numerous signs of anthropogenically caused degradation of ecosystems and their reduced ability to function as systems of life support and control of the environmental properties. The data discussed above show that all the ecosystems considered suffer serious and increasing anthropogenic forcings. Intensive processes of transformation of natural ecosystems take place due to deforestation, pollution (with biogenes, including), dams building and biological intrusions of alien organisms. Anthropogenic forcings result in considerable transformations of the global biogeochemical cycles, whose normal functioning is a guarantee of the ecosystems' safety.

So far, the negative trends of the state of the global ecosystems do not vividly threaten the high levels of production of various products and services. The food and fibre production has never reached such high levels as nowadays, but the dam constructions have ensured an unprecedented regulation of the river run-off to provide the water use in the interests of agriculture. However, the progress in creating the material well-being is fraught with a long-term prospect of reduction of productivity of the global ecosystems and their reduced ability to regulate the environmental quality. The use of new technical means and technologies such as synthetic fertilizers, improved means of fishing, and irrigation systems has masked a decrease of bioproductivity of natural ecosystems, ensuring a needed growth of food and fibre production. As for the long-term prospects (in this context the sustainable development is defined), one should remember a reduced functioning of natural systems of life support, which manifests itself, for instance, in the form of a dangerous decrease of biodiversity, reduced quality of drinking water, intensive emissions of greenhouse gases into the atmosphere, and many other negative phenomena.

The ecosystem approach has the following main features:

1. Complex (systematic) character of analysis of interactions in the "society–nature" system;
2. Not fragmentary but integral analysis of the ecosystems' functioning;
3. Consideration of all the spatial-temporal diversity of the processes with the indispensable priority of long-term prospects;
4. A detailed consideration of interactive processes in the "society–nature" system;
5. Justification and accounting of the limits of the normal functioning of ecosystems, and in this connection, permissible anthropogenic loads.

The Table 8.3 data illustrate the differences between traditional and ecosystem approaches to the use of natural resources as applied to forestry. Of course, advantages of the ecosystem approach raise no doubts. However, the problem is to what extent the conditions of the socio-economic development of either country permit a successive realization of the ecosystem approach. The negative manifestations of the dynamics of various ecosystems reflect vividly great (sometimes insurmountable) difficulties to realize the ecosystem approach. As far as this is concerned, the participants of the PAGES project have formulated the following general recommendations:

Table 8.3 Differences between traditional and ecosystem approaches to studies of natural resources with forestry as an example

Parameter	Traditional approach	Ecosystem approach
Goal	• Maximum possible production of consumer goods • Maximizing the value of produce • Maintaining the state of the forests or the use of forest products at levels equal to or below those ensuring the growth or restoration of the forests	• Maintenance of forest ecosystems as an interconnection component of the natural medium with provided stable production of consumer products • Preserving a possibility of the choice in the future • Providing a persistent productivity of ecosystems with an account of forest aesthetics and social acceptability of the forestry practice
Scale	• Works on the level of forest stand within the boundaries of the states or property	• Works on the levels of ecosystems and landscapes
Role of science	• Consideration of problems of forestry management as the applied science	• Forestry management with scientific data and social factors taken into account
Role of management	• Special attention to final results (production of goods and services) from the viewpoint of provision of needs for wood, solution of recreation problems, maintenance of wild nature, and obtaining the forage • Reaching the management that ensures the needed industrial production • Priority of wood production • Desire to avoid the wood deficit • Consideration of forests as systems of wood production • Assessment of the economic efficiency of forestry	• Emphasis on the safety of forest ecosystems with an account of needed soil properties, biological diversity, and ecology on the whole in the interests of raising the level of production of goods and services • Approach to management that ensures the solution of the nature-protection problems • Complex approach to solution of the production problems considering similar importance of the problems of nature protection • Desire to avoid losses of biodiversity and soil degradation • Approach to forests as safe natural systems • Assessment of economic profits and social acceptability

- to stimulate the development of science and observational means;
- to realize and quantitatively estimate the value of functions performed by the ecosystems;
- to stimulate a broad discussion of the goals, scientific foundations, and practical importance of the ecosystem approach; and
- to provide the participation of all needed specialists in the solution of the problems of the ecosystems management.

The scientific understanding of the laws of ecosystems' functioning and, moreover, a possibility to predict the ecosystems' dynamics are still in the initial phase. To a great extent this is explained by inadequate complex information about the ecosystems and their functioning. So, for instance, the problem of the global carbon cycle attracts serious attention since long ago. Nevertheless, even the system requirements have not been formulated yet for complex observational data, which would make it possible to also obtain adequate information about the laws of the carbon cycle (of course, this requires a combined use of both usual and satellite means of observations).

In this connection, UN Secretary-General Kofi A. Annan justly noted: "It is impossible to justify an efficient ecological policy if it is not based on reliable scientific information. Though in many spheres a considerable progress has been reached in obtaining the observational data, there remain considerable gaps in our knowledge. In particular, so far, no complete global assessment has been obtained for the state of the main ecosystems existing over the globe. A response to this necessity was the planned (Millennium Ecosystems Assessment—MEA), as an important international initiative aimed at mapping the "health" of our planet (Lu and Li 2003). This initiative has been backed by many Governments, as well as by UNEP, UNDP, FAO, and UNESCO. I call upon the Governments—UN members to give the needed financial support in assessing the state of ecosystems for the last Millennium and take an active part in the program".

An implementation of the MEA program started in 2001 is only a first step towards an assessment of the state of the global ecosystems, as well as development and realization of measures needed to preserve the ecosystems on the planet when the need for natural resources grows. On the other hand, there came up an acute problem to preserve not simply the quality of the environment but normal functioning of the life-supporting systems on the Earth, without which further development of civilization is impossible.

8.3 Modeling the Vegetation Dynamics

8.3.1 General Approach to the Modeling in Biocoenology

Numerical biocoenology has become an independent scientific direction due to many authors who combined in their studies the potentials of ecology and numerical modeling (Krapivin and Kelley 2009; Degermendzhy et al. 2009). This combination has resulted in many principles of traditional ecology being re-oriented, creating a kind of information interface between ecology and other sciences and evaluating the role played by complex interdisciplinary studies. The sphere of interests of this new direction includes the problems of complexity and generality in the development of the ecological laws.

The recently developed models cover the problems such as a parameterization of the vegetation covers succession, description of the production process as a function of the environmental factors (Wirtz 2000), modeling the competition for common resources (Alekseev et al. 1992), modeling the natural-anthropogenic systems (Polishchuk 1992; Ugolnitsky 1999), modeling the spatial-temporal dynamics of the ecological systems, and many others.

Many authors solve the problem of modeling a succession of vegetation communities basing on their points of view on succession processes. There are two fundamental interpretations of these processes based on the concepts of endo-ecogenesis and exo-ecogenesis. The most stable idea of the character of the dynamic phenomena in phytocenoses is an idea of a single process of the vegetation development under the influence of both external and internal factors. In this case, the process of succession covers such factors as soil development, layer-by-layer division of the vegetation community, change of the height of plants and their accumulated biomass, microclimate formation, change of the species diversity, and relative stability of the community. The classification of the forms of the vegetation community dynamics and its affecting factors is important for further synthesis of the numerical model. Vasilevich (1983) introduced the following categories of successions: endogenic (autogenic), exogenic (allogenic), continuous, post-disruptive (post-catastrophic). In the case of endogenic succession, the source of changes is located in the vegetation itself and it is affected by changes in the medium. There are no stages in continuous successions. In the course of post-disruptive succession, changes in the community take place from unstable to stable conditions. Exo-ecogenesis is characterized by prevailing external factors.

Population dynamics was a novel development in succession modeling. Thus, Hulst (1979) proposed three types of the models of succession. The first type is a model of succession with response but without competition, that is, there is an effect of reaction of the vegetation community to changes in the environmental conditions. Succession is described by the usual differential equation:

$$dN_i/dt = r_i N_i (k_i - N_i)/k_i$$

where N_i is the number of individuals of the ith species, r_i is the rate of an exponential growth, k_i is a maximum of population density.

The second type of the succession model is based on suppression of one species by another in the process of intrusion. In this case, the coefficient of competition is used:

$$\alpha_{ij} = \partial[dN_i/dt]/\partial N_j/\partial[dN_i/dt]/\partial N_i,$$

and the model itself is written as:

$$dN_i/dt = r_i N_i \left(k_i - \sum \alpha_{ij} N_i\right)/k_i$$

The third type of the model of succession has alternative character. If in the first two models the population density is a function of the size of previous species of vegetation, a maximum of population density k_i depends on time.

The production of organic matter by vegetation community depends on physiological processes in the organisms of the plants, resources of solar radiation, moisture, elements of mineral nutrition, temperature, and many other internal and external factors. The modeling of the production process takes into account photosynthesis, respiration, matter motion, transpiration, division of cells, etc. There are numerous correlations between these factors and production of plants. Lieth's formulas are used most often (Lieth 1985). In order to estimate the annual production of plants R_p (g/cm^2/year) as a function of the mean-annual temperature T (°C) and the mean annual precipitation amount r (mm), the following estimate can be used:

$$R_p = 3{,}000 \min\left\{(1 + \exp(1.315 - 0.119T))^{-1}, (1 - \exp(-0.000664r))\right\}$$

Vegetation biomass density P on a given territory is generally modeled by the balance equation:

$$dP/dt = \min\{\rho_P, R_{PP}\} - M_P - T_P$$

where ρ_P is a maximum *B/D*-coefficient of the considered type of vegetation, R_{PP} is the vegetation productivity for *B/D*-coefficient values less than ρ_P for the present values of *B/D*-coefficient, M_P and T_P are parameters of decay and respiration expenditure per unit time, respectively.

The use of parametric descriptions of productivity in biogeocenotic models is important as an addition to the monitoring of vegetation communities. Unfortunately, the existing database and the knowledge do not give any synthesized information about a huge amount of empirical correlations representing dependences of productivity and other functional components of vegetation communities.

Quite special in the numerical bio-geocenology is the problem of model description of the competition between individual plants. On the whole, the mutual impact of plants is described by the so-called competition indices (Bogatyrev 1988). A freely growing plant is a plant whose functioning is independent of resource and is determined by only species characteristics and external parameters. An interaction between plants starts with their influence on the same resource and with the mutual intrusion on the territories occupied by free growth. Biomes have more complicated and diverse interrelationships and, as a result, they differentiate niches and habitats. Mirkin (1986) introduced five types of strategies for the plants with regard to their functioning in the environment (*K*—violents, *S*—ecotopic patients, S_k—phytocenotic patients, *R*—true explerents, R_k—false or phytocenotic explerents). A systems approach is possible to a model description of competition and survival. All these strategies are inherent in individual species to different degrees and, therefore, synthesizing a model of survival is determined by a concrete set of the plant's requirements to niche parameters. A plant's niche is usually an aggregation of the space it occupies, resources it uses, and additional conditions, for instance, the presence of pollinators, phytofugs, etc.

Each plant or some totality of plants N interacts with the environment H (including other plants) exchanging some amount of resources V_N for an obtained volume of resources W_N. The structure $|N|$ and the behavior $\underline{N}$ of plants or their communities are in an equilibrium state with the structure $|H|$ and behavior $\underline{H}$ of the environment, and this state is described by a minimax criterion:

$$W_{N,opt} = \max_{\{|N|,\underline{N}\}} \min_{\{|H|,\underline{H}\}} W_N(V_N, |N|, \underline{N}, |H|, \underline{H})$$

$$W_{H,opt} = \max_{\{|H|,\underline{H}\}} \min_{\{|N|,\underline{N}\}} W_H(V_N, |N|, \underline{N}, |H|, \underline{H})$$

The goals N_S and H_S of vegetation continuum N and environment H, respectively, can be opposite, partially coinciding, or indifferent. Due to the developed optimal strategy, plants of different kinds adjust to the levels of (V, W)-exchange, which are possible on a given territory. Thus, because of differentiation of niches, there are multi-level vegetation communities, for instance, in the forest the factor of niches differentiation is many-sided. It covers such special features of various kinds of plants as the location of roots at different depths, different periods of blossoming, different response of plants to climate changes, different need of species for mineral nutrition, etc. In general, a plant's niche can vary in time both due to the dynamics of the plant itself and in connection with the drifting of the adjacent niches. The diversity of possible situations brings forth series of models of the functioning of vegetation communities.

One of the characteristic examples of the models of vegetation communities is a discrete model by Galitsky (1985). Let us consider it in detail. The equation that describes a change of the plant's biomass B is written as

$$dB/dt = k(t)(\alpha B^k + (B/B_f)^\gamma B_f - \alpha B^k$$

where $k(t) = \min\{A/A_f(t), 1\}$ is the indicator of the plant's provision with a territory, A is the size of the available territory for the plant's growth, A_f is the size of the territory needed for free growth; B_f is the biomass of a freely growing plant; α is the characteristic of the item of expenditure in the balance (on the biomass maintenance); γ is the index of the deficit of territory.

The gist of the Galitsky model's functioning is that the biomass of each plant located in a cell of its own in a common mosaic, changes independent of other plants of the community until either this plant or any of its neighbors dies for lack of territory. At the moment of dying the adjacent plants rush to distribute the vacant territory, and after changing the parameter A, the equation of the model is valid again. As a result, as individual plants die, the original mosaic changes on a local scale, gradually covering larger areas. The criterion of dying off is simple: a plant is considered dead if its biomass becomes below some threshold level B_{dead} ($B_{\text{dead}}(t) = \varepsilon B_f(t)$, $0 < \varepsilon < 1$).

8.3.2 Biocoenotic Model

The vegetation cover parameters change during the year depending on the weather situation (Table 8.4). The specific biomass Q_i of the ith type of vegetation at time t can be parameterized according to the following equation:

$$\partial Q_i/\partial t = R_i - M_i - E_i, \quad (8.1)$$

where R_i is the biomass productivity and M_i and E_i are the biomass losses at the expense of withdrawal and transpiration, respectively.

The function $M_i(\varphi, \psi, t)$ reflects the set of natural M_{Ni} and anthropogenic M_{Ai} processes leading to vegetation biomass losses ($M_i = M_{Ni} + M_{Ai}$):

$$M_i(\varphi, \psi, t) = \mu_i(t) Q_i(\varphi, \psi, t)$$

where φ, ψ are the latitude and longitude, respectively.

The flux E_i is calculated by the formula:

$$E_i(\varphi, \psi, t) = \frac{\rho c_p \lfloor e^*(T_c) - e_a \rfloor}{\gamma_p (r_c + r_b)},$$

where $e^*(T_c)$ is the saturated vapor pressure inside the canopy foliage (in units of Pa), e_a is the vapor pressure in the canopy air space (Pa), r_c is canopy resistance (sm^{-1}), r_b is the bulk leaf boundary layer resistance of the canopy (sm^{-1}), ρ is air density (kg m^{-3}), c_p is the air specific heat (J kg^{-1} K^{-1}) and γ_p is the psychrometric constant (Pa K^{-1}).

Table 8.4 Dependence of the annual production R (kg $m^{-2}year^{-1}$) on the average annual temperature T_a and the total annual rainfall W, $F(Ta, W)$

W (mm/year)	Atmospheric temperature (°C), T_a											
	−14	−10	−6	−2	2	6	10	14	18	22	26	30
3,125							3.4	3.5	3.7	3.8	3.9	4.0
2,875							3.2	3.3	3.5	3.6	3.7	3.8
2,625							3.0	3.2	3.3	3.4	3.5	3.6
2,375							2.8	2.9	3.0	3.1	3.2	3.3
2,125							2.5	2.6	2.7	2.9	2.9	3.0
1,875						1.6	2.3	2.3	2.4	2.5	2.6	2.7
1,625				0.4	0.6	1.3	2.0	2.1	2.1	2.2	2.3	2.4
1,375		0.2	0.3	0.4	0.7	1.1	1.7	1.9	1.9	2.1	2.1	2.0
1,125	0.2	0.3	0.3	0.4	0.6	1.0	1.6	1.8	1.9	1.8	1.8	1.7
875	0.2	0.3	0.4	0.5	0.8	0.9	1.5	1.4	1.3	1.3	1.2	1.2
625	0.3	0.3	0.5	0.6	0.9	0.9	0.9	0.8	0.8	0.7	0.7	0.7
375	0.4	0.4	0.5	0.7	0.6	0.6	0.6	0.5	0.5	0.5	0.4	0.4
125	0.1	0.3	0.3	0.2	0.2	0.2	0.2	0.2	0.2	0.1	0.1	0.1

As shown in Figs. 1.9 and 1.11, the GIMS comprises 30 or 12 models for the SPF. The list of SPFs given in Table 1.11 was adopted to spatial resolution of 4° × 5° according to Bazilivich and Rodin (1967) classification. The results obtained by Krapivin et al. (1982) were used to synthesize these models. All of these are based on Eq. (8.1) for the balance of the biomass $Q_i(\varphi, \psi, t)$. The actual plant productivity is approximated as follows:

$$R_i = \delta_c^i\left(1 + \alpha_T^i \cdot \Delta T/100\right)\exp(-\beta_i/Q_i)\min\left\{\delta_e^i, \delta_Z^i, \delta_W^i, \delta_B^i\right\},$$

where α_T^i and β_i are indices of dependence of production on the temperature variation ΔT and biomass Q_i, respectively; δ_ζ^i is the index of limitation of production by the factor ζ: e = illumination, Z = pollution, W = soil moisture, B = nutrient salts of the soil and c = atmosphere CO_2 concentration. The δ_ζ^i functions actually used in the framework of real situations are calculated based on existing or preliminary receiving data. Thus, the role of the atmospheric CO_2 concentration C_A in photosynthesis is described by the equation $\delta_c^i = b_i C_A/\left(C_A + C_{0.5}^i\right)$, where $C_{0.5}^i$ is the CO_2 concentration for which $\delta_c^i = b_i/2$. The influence of the solar radiation intensity $e(\varphi, \psi, t)$ on photosynthesis is parameterized by the relation $\delta_e^i = \delta_i^* \exp(1 - \delta_i^*)$, where $\delta_i^* = e/e_i^*$ and e_i^* is the optimal illumination for ith type of plant. A more detailed description of the correlations between the biocenotic processes is given by Kondratyev et al. (2002).

8.3.3 Classification of Soil-Plant Formations and Their Archive

Figure 1.9 gives an example of the ecological subdivision of the SPFs taking into account many abiotic factors. From the standpoint of microwave monitoring, the set of principal vegetation features play a significant role in the solution of the electromagnetic wave attenuation problem. It is known that there are three priority factors for each type of vegetation cover: wet biomass Q, productivity R and water content ρ_P. These factors are functions of the climatic zones. Therefore, the assessment of the impact of climate change on the potential productivity of the vegetation requires a model development to describe the vegetation dynamics. There exist many approaches toward solving this task. Some classifications are based on the reflection of the principle variables describing the structure of the ground ecosystem: land cover and soil types, phenology, biomass, soil moisture and temperature profiles. High precision classifications take into account the vertical structure of the ecosystems (layered ecosystems), their age structure, canopy chemistry, disturbance, decomposition, degree of anaerobiosis of the ecosystem, rate of succession, ecosystem stratification, ratio of C_3–C_4 plants, etc. It is clear that the classification of the vegetation cover is a complex problem of biocenology. The degree of correspondence between published classifications and real distributions of vegetation types influences the final model estimations of EMW attenuation by the vegetation cover. Therefore, GIMS technology stipulates that the spatial structure of the vegetation of the landscape studied is corrected by satellite observations. However, reliable observations are available only for relatively limited areas. Therefore, extrapolations are often used to correct the vegetation structure image.

The effective global spatial structure of SPFs depends on the specific features of problem solved. For instance, Bounoua et al. (2000) using SiB2 model and spatial resolution $7.2° \times 9.0°$ for 12 types of SPFs analyzed seasonal variations of NDVI depending on the climate change. Sud et al. (1996) used horizontal resolution $4° \times 5°$ for 13 types of soil-plant formations to study biogeophysical consequences of a tropical deforestation scenario realization. Certainly, the quality of results is a function of both soil-plant classification and spatial resolution. Unfortunately, at the present time there is no reliable global database with detailed parametrical description of SPFs and this is one of the most important problems of global monitoring.

Biometric and production characteristics of the vegetation covers are important for the solution to many problems when it is necessary to assess the consequences of anthropogenic activity, to decide the economic problems of the nature management, to assess the risk of the land ecosystem reconstruction, and to study the causes of natural disasters. Moreover, there are numerous other problems arising in the NSS with the use of SPFs archives. The base of SPFs archive of long-time storing contains the existing spatial and structural classifications of SPFs. Examples of such classifications are given in Figs. 1.9 and 1.11. The GIMS database and knowledge base is filled with information about the SPFs spatial structure and their characteristics. Consequently, the archive of biometric and production characteristics of vegetation (ABPCV) consists from three basic levels:

1. Scales of spatial geographical discretization of the land surface.
2. Skeleton map containing the vegetation types in each pixel of spatial discretization of the land surface.
3. Biometric and production characteristics of vegetation located in each pixel of the land surface.

Information content of these levels is supported by the ABPCV file structure (Figs. 8.5 and 8.6). Files with the intermediate information include data about illumination, nutrient salts in soil, soil moisture, topographic parameters, etc. Information control system of the ABPCV is presented in Fig. 8.7.

Figure 8.8 explains the ABPCV configuration where file of metadata has documentation giving a possibility to make a correct choice of data that are needed for the user.

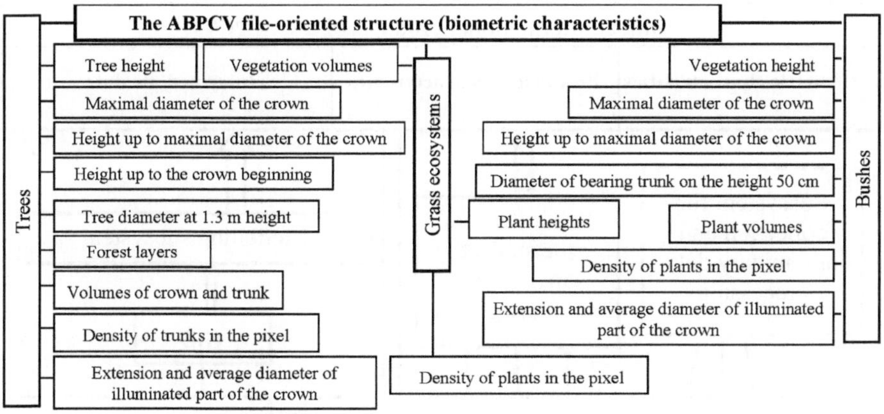

Fig. 8.5 Biometric files of the ABPCV

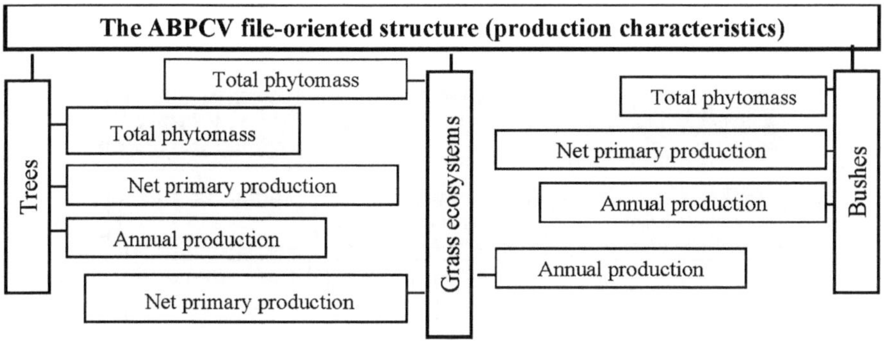

Fig. 8.6 Production files of the ABPCV

8.3.4 Modeling the Forest Ecosystems

The forest ecological system is an important constituent of the global continuum of the soil-plant formations on the Earth (see Table 1.2 and Fig. 1.9). Therefore the development of the models that would describe the forest dynamics is a principal stage of the synthesis of the global model. Of course, the importance of the models of the forest ecosystems is greater, since forest resources and their use affect the economic potentials of the regions.

The variety of forest ecosystem types functioning in different environmental conditions needs the search of acceptable models giving a possibility to assess their dynamics. Hacenauer (2012) discussed modeling technology for these aims displayed on three main areas:

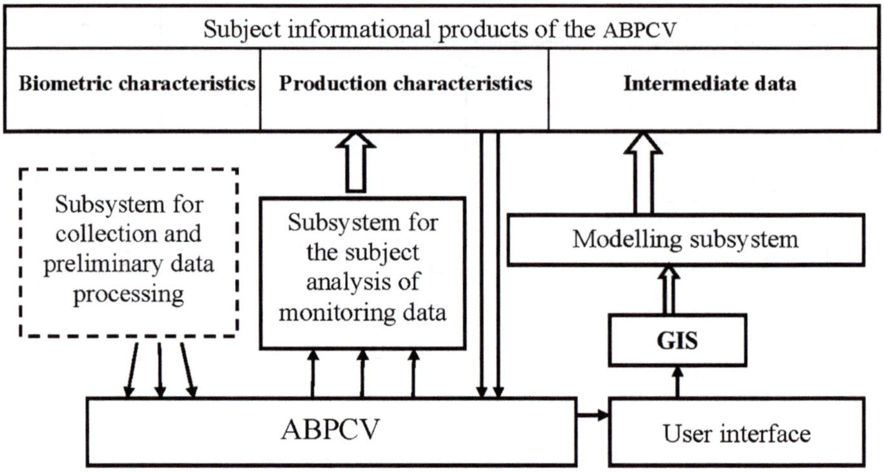

Fig. 8.7 Structure of the ABPCV information control service

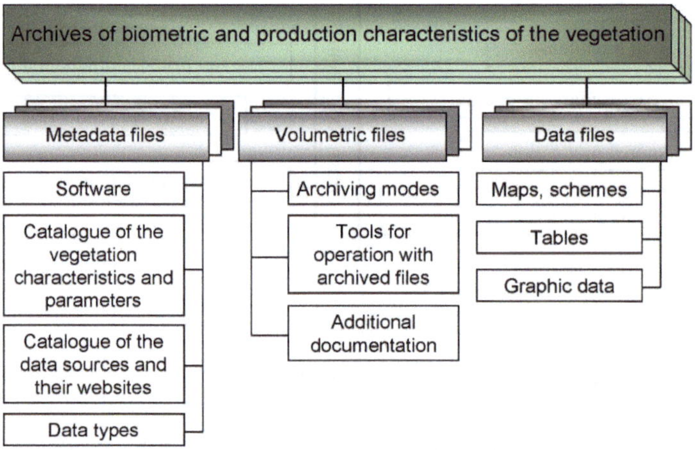

Fig. 8.8 Basic elements of the ABPCV

- models of the growth and productivity;
- succession models; and
- biogeochemical-mechanical models.

The choice of the model type or creation of combined model requests the coordination of spatial and temporal scales as it is shown in Fig. 8.9.

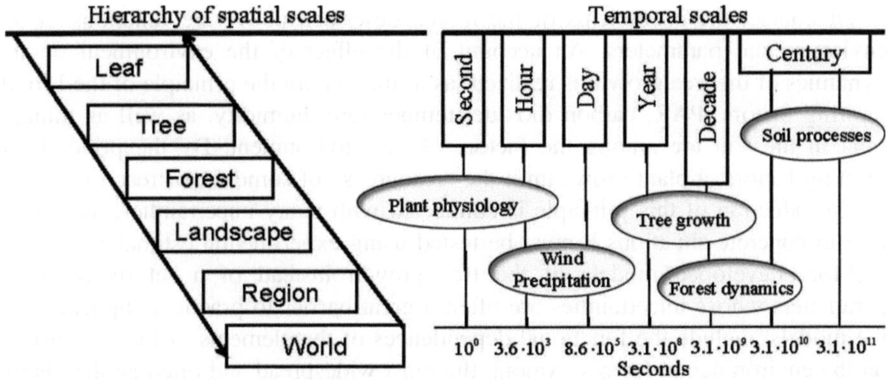

Fig. 8.9 Scheme for the co-ordination of spatial and temporal scales when forest ecosystem model is created

Many authors describe the dynamics of total biomass of an individual plant in the form of a logistic equation:

$$dB/dt = k(B_{\lim} - B)B,$$

where B is the biomass, $B_{\lim}$ is the threshold amount of biomass, k is the coefficient reflecting the effect of the environmental factors and the plant's type on its growth.

To describe the dynamics of the tree growth, the simplest functions are often used, such as Terazaki's equation $B = a \exp(-b/t)$, Koller's function $B = ax^b c^{-ct}$, approximation by Korsun' and Bakman $B = a \exp(b\ln t + c\ln^2 t)$, etc. Bichele et al. (1980) were first to propose a more complicated model of the tree growth described by the system of ordinary differential equations:

$$dB_j = K_\Phi - k_R R_j - V_j + K_M,$$

where $K_\Phi = k_F \sum_{i=1}^{4} \alpha_{ij}\Phi_i$; $K_M = M \sum_{i=1}^{4} \beta_{ij}$; B_j is the biomass of individual organs of a plant (j = 1, 2, 3, 4), M is the dry biomass of the whole plant; Φ_i is an amount of carbon dioxide assimilated during 24 h by the ith organ of the plant ($k_F\Phi_I$ is an equivalent dry biomass, k_F is the coefficient of equivalence); R_i is an amount of carbon dioxide emitted during 24 h by the ith organ of the plant ($k_R R_i$ is an equivalent dry biomass, k_R is the coefficient of equivalency); V_i is the dry biomass of part of the ith organ; α_{ij} is the share of "fresh" assimilates created during 24 h in the ith organ of the plant, moving during the same period of time into the jth organ of the plant; β_{ij} is the exchange of "old" assimilates during 24 h between the ith and jth organs, per gram of dry biomass of the whole plant; K_Φ is an increment of biomass due to photosynthesis and re-distribution of fresh assimilates; K_M is an increment connected with the re-distribution of old assimilates; $k_R R_j$ is the biomass expenditure on respiration.

Of course, the coefficients of the model described above are functions of the environmental parameters. An account of the effect of the environment on the dynamics of the tree growth is realized, as a rule, within the principle of the Libikh limiting factors. PAR, carbon dioxide, temperature, humidity, as well as mineral salts in the soil are among the factors of the environment. By the principle of limiting factors, a plant grows until the "resources" of some factor reach the limit. An introduction of this principle is connected with many uncertainties, and therefore in concrete situations it must be tested using experimental estimates.

More developed models of the tree growth instead of a set of empirical parameters whose uncertainties are often a great barrier to practical application of the models, include the functional dependences of the elements of the tree growth on the environmental factors. Among the most widespread and often applied is the Libikh criterion according which the growth of a plant is confined to an element, whose concentration is at a minimum. By this criterion a plant or its organs grow until the resources of some element, for instance, photosynthetically active radiation (PAR), moisture, temperature, carbon dioxide, nutritious salts are exhausted. There are other criteria testifying to a weighed influence of all environmental factors, these weights are characteristic indicators of the plant type.

The process of tree growing can be presented as a result of interaction of three processes of growth: change of the leaf (needle) mass, growth of the trunk, and development of the root system. This process is schematically shown in Figs. 8.10 and 8.11.

Denote the biomass of the canopy, trunk, and roots of the tree as x_1, x_2, and x_3, respectively. Then, the balance equation of a tree model is written as

$$x_i(t+\Delta t) = x_i(t) + \varepsilon_i(t)\{R_x - T_x\}\Delta t, \ (i = 1,2,3),$$

where Δt is the characteristic time step, R_x is the gross productivity of a tree due to photosynthesis, T_x is the biomass expenditure on respiration, ε_i is the share of new biomass moving into the biomass of the ith organ of the tree.

The ε_i parameters are functions of time as well as other parameters of the environment. Following the principle of maximum survival, a supposition can be made that the principle of maximum primary productivity is valid:

$$\tilde{Y}^*(t+\Delta t) = \max_{\{\varepsilon_i(t)\}} Y\{x_i(t) + \varepsilon_i(t)Y[x_i(t), E(t,\varphi,\lambda), C_A, T_A]\Delta t, E, C_A, T_A\}$$

One of the important limiting factors of the tree growth is water, the movement of which in the soil and the tree body determines the dynamics of the tree biomass change. Following Kirilenko (1990), let us denote the soil as a homogeneous porous layer [0, z_g], where z_g is the depth of ground water. Moreover, the following notations are introduced: W—relative soil moisture (kg H_2O/m^3/[kg soil/m^3], ρ—water density, ρ_s—volume density of soil. A system of coordinates (x, y, z) is also introduced, where the z-axis is directed down from the soil surface (z = 0). Then, the movement of water through the soil can be described by Darsi's 3-D equation:

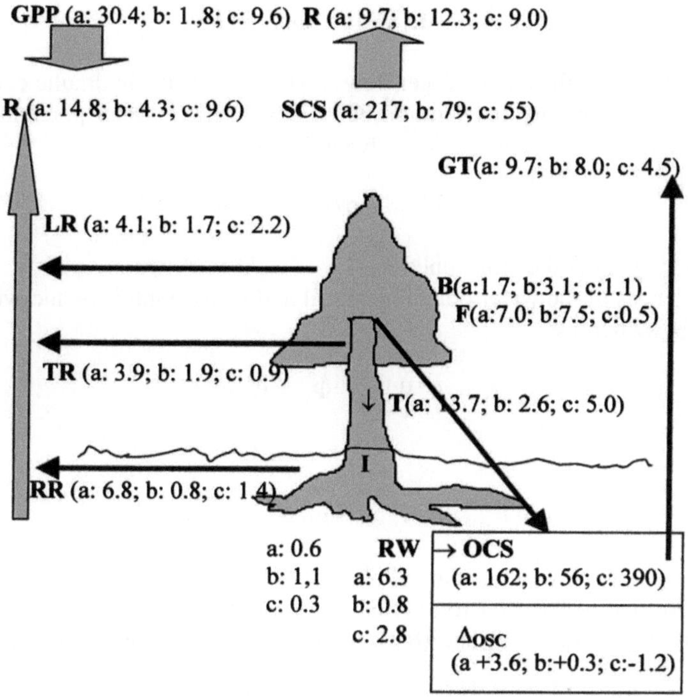

Fig. 8.10 Annual supplies (tC/ha) and fluxes (tC ha^{-1} year^{-1}) of carbon for tropical (*a*), temperate (*b*), and boreal (*c*) forests (Watson et al. 2000). *Note GPP* gross primary production, *RA* autotrophic respiration, *RH* heterotrophic respiration, *RL* foliage respiration, *RW* above-ground wood respiration, *RR* root respiration, *SOC* soil organic carbon stock, Δ_{SOC} net increment in soil organic carbon, *E* below-ground detritus, *K* below-ground net biomass carbon increment, *B* above-ground net biomass carbon increment, *O* above-ground detritus (litter fall and mortality), *T* translocation from above to below ground

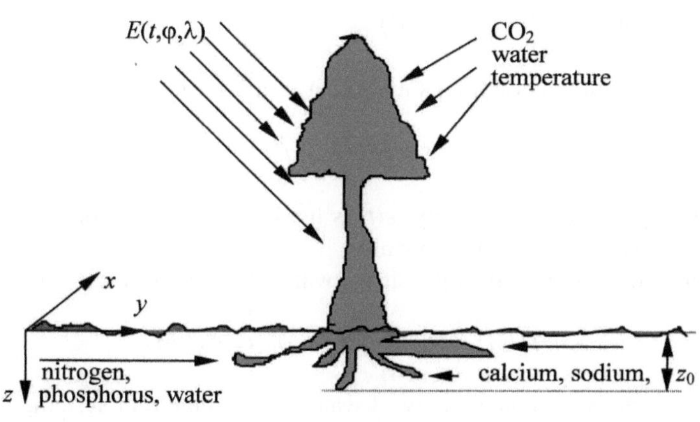

Fig. 8.11 Model of the growth of a tree as an element of the forest biogeocenosis. *Notation* z_0 thickness of the root system layer; *E* solar radiation

$$V = -\mathrm{K}[\nabla(\Phi_s - \rho g z)]/(\rho g), \tag{8.2}$$

where $V(t, x, y, z)$ is the rate of water flow, $K(t, x, y, z)$ is the hydraulic conductivity of water, g is the gravity acceleration, $\Phi(t, x, y, z)$ is the water potential of soil.

The equation of the soil moisture dynamics is

$$\partial W/\partial t = -(\rho/\rho_s)(f + \mathrm{div} V) \tag{8.3}$$

where $f(t, x, y, z)$ is the sink function.

Dependences of the water potential of soil and its hydraulic conductivity on soil moisture is written as

$$K = K_s^* W^m, \quad \Phi_s = \Phi_s^* W^{-n}$$

where

$$K_s^* = K_{FM} W_{FM}^{-m}, \quad \Phi_s^* = \Phi_{MG} W_{MG}^n, \quad m = 2n + 1,$$

W_{FM} is the total water capacity of soil, K_{FM} is the soil conductivity corresponding to total water capacity, m and n are constants.

The system of Eqs. (8.2) and (8.3), with assumed notations, is re-written as:

$$\partial\omega/\partial t = (-\rho/\rho_s)[(K_{FM}/W_{FM})\mathrm{div}\{\mu\,\omega^n\nabla\omega + i\omega^n\} + f/W_{FM},$$
$$V = K_{FM}(\mu\omega^n\nabla\omega + i\omega^n),$$

where $\omega = n\Phi_{FM}/(\rho g)$, i is the unit vector directed along the z-axis.

In the case of homogeneous processes in the plane (x, y) the problem is simplified:

$$\partial\omega/\partial t = (-\rho/\rho_s)[(K_{FM}/W_{FM})\partial/\partial z\{\mu\,\omega^n\partial\omega/\partial z + \omega^n\} + f(t,z)/W_{FM},$$
$$V(t,z) = K_{FM}(\mu\omega^n\partial\omega/\partial z + \omega^n)$$

Boundary and initial conditions are given in the form:

$$\omega(t, z_g) = 1, \;\; V(\mathrm{t}, 0) = P(t) - e(t), \quad \omega(t_0, z_g) = \omega^0(z),$$

where $P(t)$ is precipitation intensity, $e(t)$ is the intensity of evaporation off the soil surface, $\omega^0(z)$ is the initial soil moisture.

The f function describes the intensity of water absorption by a tree's root system:

$$f(t,z) = \xi_0[\Phi_s(t,z) - \psi_0(t)\}ds/dt,$$

where $\xi_0 = \xi_k\sigma_0$ is the conductivity of water entry to the root system, $\xi_k = r_k^{-1}$ is the specific conductivity of water entry to the root, r_k is the specific resistance of water entry to the root, σ_0 is the general area of the sucking roots of the tree per unit soil surface area ($\sigma_0 = \beta_k m_k$), m_k is the mass of roots beneath unit soil area, β_k is an empirical constant, ψ_0 is the water potential within the root system, $s(z)$ is the function of the vertical distribution of the root system ($s(0) = 0$, $s(z_0) = 1$).

To complete the synthesis of the tree's water regime, let us parameterize the processes of water flow in the plant and transpiration. It is assumed that from the root system the water flows by the trunk to canopy (branches, leaves) and then evaporated into the atmosphere (Lai et al. 2000; Lal et al. 1998) and the above ground part of the tree of the height H is divided into n equal layers: $H = n\Delta z$. Across the boundaries of the ith layer the water moves due to the difference between water potentials ψ_i and ψ_{i-1} overcoming the resistance of the xylem vessels:

$$r_{ks}^{*}(z) = \left(S_{ks}^{*}(z)\xi_{ks}\right)^{-1},$$

where ξ_{ks} is the specific conductivity of the xylem, S_{ks}^{*} is the cross-section of the tree at a height z. As a result, with the area of the cross-section supposed to be vertically constant, the following formula describing the rate of water flow in the trunk across the ith layer is obtained:

$$v_{i-1,i}^{*} = -\xi_{ks}S_{ks}^{*}\Delta z^{-1}(\psi_i - \psi_{i-1} + \rho g \Delta z)$$

According to Kirilenko (1990), the intensity of transpiration from the ith layer can be described by the formula:

$$\gamma_i^{*} = \frac{d^i}{r_L^i} S_L^{*i}$$

where d^i is the deficit of saturation of an absolute air humidity in the atmosphere, S_L^{*i} is the leaf area of the ith layer, $r_L^i = r_{st}^i + r_a$, r_{dt}^i is the stomatal resistance, r_a is the air boundary layer resistance.

To complete the description of the model after Kirilenko (1990), a standing wood of the density ρ_F is considered. The model equations are written as

$$d^2\varphi/d\eta = Uv(\eta)\varphi(\eta), \quad \eta \in [0, \alpha]; \tag{8.4}$$

and the boundary conditions are

$$\begin{cases} \varphi_0' = & -[G(\tilde{\varphi}_s - \varphi_0) + F] \\ \varphi_\alpha' = & -F \end{cases}, \; d\varphi/d\eta = -F \text{ for } \eta \in [\alpha, 1]; \tag{8.5}$$

For $\varphi_0 \leq 0$, a system of equations (8.4) and (8.5) becomes:

$$\begin{cases} \dfrac{d\varphi}{d\eta} = & -F \\ \varphi_0 = & \tilde{\varphi}_s \end{cases}$$

with the following notations: $\eta = z/H$, $\varphi(\eta) = 1 + \psi(\eta H)/\psi^*$, ψ^* is the leaf's water potential, $1/D_1$ is a minimum mouth resistance, $U = Hr_{ks}D_1 d\beta m_L / \psi^*$, $F = \rho g H/\psi^*$, $G = Hr_{ks}\,\xi_0$, $\alpha = a/H$, β is the leaf index, $h = -z$,

$$a = \begin{cases} H, & \psi(H) \geq \psi^*, \\ h, \quad \psi(h) = -\psi^*, & \psi(H) < -\psi^* < \psi(0), \\ 0, & \psi(0) \leq -\psi^* \end{cases}$$

8.3.5 Modeling the Energy Fluxes in the Atmosphere-Plant-Soil System

The energy exchange in the system atmosphere-plant-soil (APS) plays an important role in the formation of climatic processes on every spatial and temporal scale, and therefore it should be considered in the models of the greenhouse effect that increases the reliability of the estimates. The appearing problems include the development of methods in order to measure the energy fluxes in the air-surface layer as well as their modeling. A special aspect of these problems concerns the development of technologies of estimation of energy fluxes in the APS system on global scales to take these estimates into account in the global climate systems. There are well-developed methods of the remote sensing of a wide set of climatic parameters including energy fluxes (Krapivin and Kondratyev 2002; Shutko et al. 2010). However, in the remote sounding of the APS system a complicated problem of an assessment of the screening impact of vegetation over the wavelength regions used in the monitoring system arises.

Beginning from the 1980s, studies have been carried out on developing a model of photosynthesis a vegetation bud. Farquhar et al. (1980) showed how a biological model of the CO_2 assimilation by leaves is connected with fermentative kinetics and electronic properties of the chloroplast, which made it possible to rather plausibly describe the process of photosynthesis for the plants of class *C*. Then Ball (1988) and Collatz et al. (1990, 1991, 1992), using a model technology and observation data on the process of the vegetation bud conductivity in photosynthesis, derived a semi-empirical model of the vegetation bud functioning with an account of changes of the environmental temperature, humidity, CO_2 concentration in the atmosphere, and intensity of assimilation of the vegetation species itself.

The problem have been also considered on the consequences of the distribution of the photosynthetic potential of leaves and canopy of plants with respect to illumination and a criterion has been developed to determine a common potential corresponding to maximum photosynthetic abilities of the plant in CO_2 assimilation.

Sellers (1985, 1987) developed methods of the vertical integration of leaf canopies, models of light scattering and absorption, photosynthesis and conductivity of the vegetation bud, known before only for a leaf. These studies revealed an interaction between spectral characteristics of a vegetation species (spectral vegetation index—SVI) and important functional relationships that regulate the processes of photosynthesis and evaporation off leaf canopies. Apart from this, an accomplished analysis revealed a connection between the share of the photosynthetically active radiance (PAR), absorbed by the vegetation canopy with the respective characteristic coefficient of vegetation species (simple ratio vegetation index—SR). Here, we consider SR and relative difference (normalized difference vegetation index—NDVI) as SVI. Spectral characteristics of vegetation species SR and NDVI are based on the remote sounding of the reflected emission in the visible and near-IR regions and determined as follows:

$$SR = a_N / a_V, \quad NDVI = (a_N - a_V)/(a_N + a_V)$$

where a_N and a_N are hemispheric reflections and radiance in the visible and near-IR wavelength regions, respectively.

The parameters a_N and a_N are measured directly above a leaf canopy. Therefore the effects of absorption and scattering of emission by the atmosphere are not considered here. For the space-borne remote sounding a method has been developed to correct the experimental data, FASIR (Fourier-adjusted, solar zenith angle corrected, interpolated and reconstructed data). A theoretical dependence (close to linear) has been established between PAR and SR, confirmed later by experimental measurements in the region of Kansas (Sellers et al. 1995, 1996, 1997). A minimum and maximum of FPAR are determined experimentally, and the final relationship becomes

$$PAR = \frac{\left(SR - SR_{i,\min}\right) \cdot \left(PAR_{\max} - PAR_{\min}\right)}{\left(SR_{i,\max} - SR_{i,\min}\right)} \quad (8.6)$$

where $PAR_{\max} = 0.950$, $PAR_{\min} = 0.001$, but in general, $PAR_{\max}$ and $PAR_{\min}$ depend on the type of vegetation; $SR_{i,\max}$ and $SR_{i,\min}$ can be calculated from NDVI, based on experimental data.

Sellers et al. (1996) use the Π parameter reflecting an amount of vegetation cellulose taking part in the process of photosynthesis. The Π parameter changes from 0 (lack of vegetation) to ≈ 1.5 for bushy species rich in photic vegetation cellulose and can be approximately estimated as

$$\Pi \approx \frac{PAR}{\bar{k}}$$

where $\bar{k}$ is the coefficient of attenuation for PAR averaged in time.

Apart from spectral characteristics, Sellers et al. (1996) considered the photic characteristic (green leaf area index L_g) and the photic share (green part N) for a concrete vegetation species. Many authors derived dependences of L_g on FPAR for species growing separately and in groups. The result can be presented in the form

$$L_g = (1 - F_{cl})\, L_{g,i,\max} \frac{\log(1 - PAR)}{\log(1 - PAR_{\max})} + F_{cl} \frac{L_{g,i,\max}\, PAR}{PAR_{\max}}$$

where F_d is the share of bushy vegetation determined for each of the vegetation classes L_s.

The photic share N of vegetation species can be obtained from the following considerations. The plant canopy consists both of leaves matter (L_g) participating in the process of photosynthesis and of the trunk cellulose L_s (0.076 for forests, 0.05 for meadows and pastures) as well as a dead fraction L_d. The dead cellulose is supposed to remain in the leaf canopy for about a month before leaf falling or another means of extraction. Thus, the photic share of vegetation cellulose is determined as $N = L_g/L_T$, where $L_T = L_s + L_d + L_g$; L_d is the dead fraction of vegetation cellulose, $L_d \geq 0$; L_g is the photic fraction determined from PAR; L_T is the total territorial characteristic of a vegetation species; L_s is the fraction of the trunk cellulose.

8.3.6 A Model of Leaf Canopy Photosynthesis

In the model of the biosphere (SiB2) derived by Sellers et al. (1996) there are equations which determine temperatures, humidity, and evaporation for vegetation canopy and a 3-layer soil cover. Apart from this, a model of radiative transfer (soil–leaf canopy–atmosphere) and a model of photosynthesis were synthesized. The model of the leaf canopy photosynthesis considers a 1-layer vegetation species, surface and root layers, as well as the zone of re-charging. Equations of the biospheric model SiB2 are based on the concept of flux exchange in the system "soil–leaf canopy–atmosphere" (resistance r and conductivity g are associated by: $r = 1/g$). The flux in this case is determined by analogy with the electrostatic form

$$Flux = \frac{Potentials\ difference}{Resistance}$$

Sellers et al. (1996) considered the fluxes of sensible and latent heat transport through evaporation of water vapor from vegetation canopy λE_c and the surface soil layer, as well as the CO_2 flux (A_c and R_{SOIL} for soil). The potentials differences connected with the fluxes are presented in the form of differences of temperatures, pressures, and partial pressure of CO_2 for heat transport, latent heat transport, and CO_2, respectively. In the model of photosynthesis of interest are CO_2 fluxes (A_c and R_{SOIL}) and the flux of latent heat exchange λE_c of leaf canopy and its respiration constituent λE_{ct}.

The model of the leaf canopy photosynthesis derived by Collatz et al. (1990) for vegetation species of classes C_3 and C_4 together with the bud model (Ball 1988) have been used to derive the biospheric model SiB2. In this model, photosynthesis and Greenhouse conductivity are closely connected (Sellers et al. 1996).

According to Collatz et al. (1991), there are three factors limiting the plant photosynthesis intensity. These are an efficiency of the photosynthetic ferment system w_c (the so-called limitation after Rubisco), an amount of PAR absorbed by cellulose chlorophyll w_e, and an ability of a vegetation species to assimilate and emit to the environment the products of photosynthesis w_s. As a result, the value of photosynthesis intensity A is a minimum of the three limitations enumerated above:

$$A \leq \min(w_c, w_e, w_s)$$

where A is the photosynthesis intensity (mol/m^{2}s); w_c is a Rubisco (ferment) limitation of photosynthesis intensity (mol/m^{2}s); w_e is a limitation of assimilation by illumination (mol/m^{2}s); w_s is a limitation of carbon content for a vegetation species (mol/m^{2}s).

The physiological limitation of assimilation w_c is a function of the ferment supplies of species, in other words, it is an ability of cellulose to realize the biochemical reactions. The ferment reserve of the plant is described in the model with the V_{max} parameters, a maximum catalytic supply of ferments (Rubisco) taking part in photosynthesis. The w_c value also depends on the leaf canopy temperature and soil moisture. Collatz et al. (1990) suggested the following expression for w_c for vegetation classes C_3 and C_4:

$$w_c = V_m \left[\frac{c_i - \Gamma^*}{c_i + K_c(1 + O_2/K_o)} \right], \text{ for } C_3,$$
$$w_c = V_m, \text{ for } C_4,$$

where V_m is a maximum catalytic supply of ferments (Rubisco) (mol/m^2/s); c_i is the partial pressure of CO_2 in the leaf (Pa); O_2 is the partial pressure of O_2 in the leaf (Pa); Γ^* is the point of CO_2 compensation (Pa) = O_2/S; S is a special interaction between CO_2 and O_2; K_c is the Michaelis—Menten constant for CO_2, (Pa); K_o is the constant of O_2 suppression (inhibition).

The V_m parameter is expressed through V_{max}, the function of dependence on the soil temperature and moisture. The S, K_c, K_o parameters are functions of temperature.

A limitation of assimilation intensity w_e is described as follows:

$$w_e = (F_\pi \cdot n)\varepsilon_3(1 - \omega_\pi)\left[\frac{c_i - \Gamma^*}{c_i + 2\Gamma^*}\right], \text{ for } C_3,$$
$$w_e = (F_\pi \cdot n)\varepsilon_4(1 - \omega_\pi), \text{ for } C_4,$$

where F_π is a PAR vector (W/m^2); n is the normal to the leaf surface; $\varepsilon_{3,4}$ is quantum efficiency of CO_2 assimilation (mol); ω_π is the dispersion coefficient for PAR.

The third limitation w_s is written as:

$$w_s = \begin{cases} V_m/2 \text{ for } C_3, \\ V_m p/2 \text{ for } C_4 \end{cases}$$

where p is atmospheric pressure (Pa).

The photosynthesis intensity A as a minimum of functions w_c, w_e, and w_s is estimated in a simplest way, but observations show that the transition from one limitation to another is not that spasmodic. Therefore Collatz et al. (1992) proposed a method of combining the contributions of all the three limitations in the form of two equations, which have to be solved with the least values:

$$\beta_{ce} w_p^2 - w_p(w_c + w_e) + w_e w_c = 0,$$
$$\beta_{ps} A^2 - A\left(w_p + w_s\right) + w_p w_s = 0,$$

where β_{ce} and β_{ps} are coefficients of correlation; w_p is a "smoothed" minimum between w_c and w_e, (mol/m^{2}s).

Theoretically β_{ce} and β_{ps} can have values from unity (lack of relationships between limitations) to zero. Real estimates vary from 0.8 to 0.99.

Pure assimilation of CO_2 by a plant A_n is equal to $A_n = A - R_d$, where R_d is CO_2 losses resulting from respiration connected with the content of carboxylase in cellulose: $R_d = f_d V_m$ where $f_d = 0.015$ for C_3, $f_d = 0.025$ for C_4.

The connection of the Collatz model of photosynthesis with the Ball semi-empirical model of vegetation bud conductivity is written as

$$g_s = m\frac{A_n}{c_s} h_s p + b$$

where g_s is the vegetation bud conductivity (mol/m^{2}s); $g_s(\text{m/s}) = 0.0224\,\frac{T}{T_f} \times \frac{p_0}{p} \times g_s\left(\text{mol/m}^2/\text{s}\right)$; m is an empirical coefficient (=9 for vegetation of class C_3, =4 for C_4, =6 for coniferous forests); b is an empirical coefficient (mol/m^2/s or m/s, =0.01 for C_3, =0.04 for C_4); h_s is a relative moisture of the leaf surface; c_s is the CO_2 partial pressure in the stem, (Pa); p is atmospheric pressure (Pa); p_0 is the normal atmospheric pressure $\left(=1.013 \times 10^5\text{Pa}\right)$; T_f is the absolute zero temperature (=273.16 °K).

There is a relationship between partial pressure of CO_2 (c_i, c_s) and relative humidity h_s. This dependence can be obtained by the analysis of the processes taking place on the border "vegetation bud–atmosphere", the respective conductivities g_i and g_s, pure CO_2 flux, and water vapor flux E_{it} as a result of the process of photosynthesis.

This relationship is expressed as

$$E_{lt} = g_l(e_s - e_a)\frac{\rho c_p}{\lambda\gamma} = g_s(e_i - e_s)\frac{\rho c_p}{\lambda\gamma}, h_s = \left[\frac{e_s}{e_i}\right],$$

$$A_n = \frac{(c_a - c_s)}{p}\frac{g_l}{1.4} = \frac{(c_s - c_i)}{p}\frac{g_s}{1.6},$$

where e_a, e_s, and e_i are the partial pressures of water vapor in the open air, on the surface, and in the stem, respectively (Pa); c_p is the specific heat capacity of air (J/kg K); γ is a psychometric constant (Pa/K); g_l is the water vapor conductivity on one side of the leaf (mol/m^{2}s or m/s, = 1/(2*rl*); E_{lt} is respiration intensity (kg/m^2/s); c_a and c_s are partial pressures of carbon dioxide outside and inside the stem, respectively (Pa); $e_i = e^*(T_c)$ (Pa).

Equations of the photosynthesis-conductivity model describe the processes taking place in a vegetation bud, and the next step is to integrate the relationships with respect to a leaf canopy to obtain total values of A_c and g_c. This was made in Sellers et al. (1996) with a supposition that the vertical distribution of vegetation ferments in the canopy is connected with the time-averaged distribution of PAR:

$$V_{\max_L} = V_{\max_0} e^{-\bar{k}L},$$

where $V_{\max_L}$, $V_{\max_0}$ are the $V_{\max}$ values on the level LAI = L and at the top of the canopy (LAI = 0), respectively (mol/m^{2}s); $\bar{k}$ is a time-averaged value of the coefficient of PAR attenuation.

An amount of PAR assimilated by the leaf canopy is:

$$F_\pi \cdot n \approx F_{\pi 0}\left[\frac{G(\mu)}{\mu}\right] e^{-\bar{k}L}$$

where $F_{\pi 0}$ is the value of the PAR flux caught by the leaf canopy (W/m^2); $G(\mu)$ is a projection of the orientation of the canopy leaves towards the direction of the incoming radiation flux (μ). Sellers et al. (1996) established that the use of the time-averaged $\bar{k}$ value instead of the instant k value leads to only negligible uncertainties in calculations.

Since now the *wc, we,* and *ws* parameters are connected (depending on the vertical coordinate) with LAI of the leaf canopy through an exponent, it is possible to use them to obtain an integral estimate of A_c. We have already discussed the difference of bushy and separately growing species ($V < 1$) and the notion of the photic share ($N < 1$). Therefore bearing in mind that in the case of dense thickets the photosynthesis intensity is reduced due to the lack of "green", rich-in-ferment vegetation material, and the integration is realized from 0 to L_T/V:

$$A_c = A_{no} \int_0^{L_T/V} VNe^{-\bar{k}L} dL = A_{n_0}\Pi \tag{8.7}$$

where $A_{n_0} = A_n$ for leaves at the top of the canopy (mol/m^{2}s):

$$A_{no} = f_c(V_{\max_0}, \ldots), f_e(F_{\pi_0}, \ldots), f_s(V_{\max_0}, \ldots);$$
$$\Pi = \frac{VN\left(1 - e^{-\bar{k}L_T/V}\right)}{\bar{k}}; \Pi \approx PAR/\bar{k}.$$

The A_c value is used to determine the respective conductivity of the leaf canopy using the modified expression:

$$g_c = m\frac{A_c}{c_S}h_S p + bL_T$$

where h_S and c_S are the volume analogs to values h_s and c_s relative to the leaf canopy.

The g_c parameter is used to estimate the respiration intensity λE_{ct}:

$$\lambda E_{ct} = \left[\frac{e^*(T_c) - e_a}{1/g_c + 2r_b}\right]\frac{\rho c_p}{\gamma}(1 - W_c),$$

where $e^*(T_c)$ is the saturated water vapor pressure at a temperature T_c (Pa); e_a is the water vapour pressure in an open atmosphere near the leaf canopy (Pa); ρ and c_p are the air density and specific heat capacity, respectively (kg/m^2, J/kg/K); γ is a psychometric constant; W_c is the share of the lead canopy wet area.

To complete a description of equations of the photosynthesis-conductivity model, the following expressions are given that estimate the ferment supply of vegetation cellulose V_m:

$$V_m = V_{\max} f_T(T_c) f_W(W_2); \quad K_c = 30 f_T(T_c); \quad K_o = 30\,000 f_T(T_c);$$
$$S = 2600 f_T(T_c), \quad f_T(T_c) = {}^{2Q_t}/_{\{1 + \exp[s_1(T_c - s_2)]\}}, \text{ for } C_3 \text{ and } V_m;$$
$$f_T(T_c) = 2Q_t/\{1 + \exp[s_1(T_c - s_2)]\} \times \{1 + \exp[s_3(s_4 - T_c)]\}, \text{ for } C_4 \text{ and } V_m;$$
$$f_T(T_c) = {}^{2Q_t}/_{\{1 + \exp[s_5(T_c - s_6)]\}}, \text{ for } R_d \text{ and } V_m;$$
$$f_T(T_c) = 2.1\, Q_t, \text{ for } K_c;\ f_T(T_c) = 1.2\, Q_t, \text{ for } K_o;$$
$$f_T(T_c) = 0.57\, Q_t, \text{ for S};\ Q_t = {}^{(T_c - 298)}/_{10};$$
$$f_W(W_2) = {}^{1}/_{\{1 + \exp[0.02(\psi_c - \psi_r)]\}}.$$

where ψ_c is a critical water potential (m); ψ_s is the value of saturation potential for moisture content in the soil (m); B is an empirical parameter; s_l is the parameter of suppression of high and low temperatures; $\Psi_r = \Psi_s W_2^{-B}$ is the potential of moisture content in the root later of the soil (m).

The physiological processes described by the A_c and g_c parameters do not exhibit an instant response to changes in the control derivatives. It takes about 1 min to stabilize the photosynthesis process after changes in the external conditions, but for conductivity the time step constitutes several minutes. Hence, for time steps longer than one minute the photosynthesis intensity A_c can be calculated using relationship (8.7), and a delay in changes of conductivity can be described reliably by the equation:

$$\frac{\partial g_c}{\partial t} = -k_g(g_c - g_{\inf}) \tag{8.8}$$

where k_g is a time constant for the response of vegetation cellulose (=0.00113 c^{-1}); $g_{c\inf} = g_c$ for $t \to \infty$; $g_{c\inf} = m\frac{A_c}{cs}hsp + b$.

The solution (8.7) is written as:

$$g_c = e^{-\bar{k}_g t} g_{c0} + \left(1 - e^{-\bar{k}_g t}\right) g_{c\inf},\; g_{c0} = g_c \text{ for } t = 0,$$

and a change in the g_c during a time step Δt will be:

$$\Delta g_c = g_c - g_{c0} = \left(1 - e^{-\bar{k}_g \Delta t}\right)\left(g_{c\inf} - g_{c0}\right).$$

These correlations give the opportunity to calculate the energy fluxes in a multi-layer structure of the system “atmosphere-plant-soil”. As a result, one can specify the elements of the climate models, and to assess from the space-borne monitoring data the impact of surface vegetation on the synoptic situation. This is especially important in arid and over-humid zones where the state of vegetation cover strongly depends on small changes of temperature and other meteorological parameters. Another important conclusion from these correlations can be drawn for the study of the effect of vegetation cover on the processes of CO_2 assimilation from the atmosphere. Clearly, the use of such models increases the adequacy of the global carbon cycle modeling.

8.3.7 Modeling the Forest Age Structure

Seppelt and Voinov (2002, 2003) proposed two-stage procedure for the age structure modeling of the forest ecosystem. The first stage uses the simulator and model for the parameterization of carbon balance and heterotrophic ecosystem transpiration only for stand of isolated tree. The second stage synthesizes regional

Table 8.5 List of models simulating the decomposition of basic carbon reservoirs

Carbon reservoir	Decomposition model	Decomposed components (released part)
Surface structure (lignine, C_{1L})	$\Delta C_{1L} = 0.325 f_1(T_s) f_2(M) \exp\{-3L_s\} C_{1L}$	CO_2 (0.3)
		Slow carbon (0.7)
Surface structure (cellulose, C_{1C})	$\Delta C_{1C} = 0.325 f_1(T_s) f_2(M) \exp\{-3L_s\} C_{1C}$	CO_2
		Surface microorganisms (0.4)
Soil structure (lignin, C_{2L})	$\Delta C_{2L} = 0.4 f_1(T_s) f_2(M) \exp\{-3L_s\} C_{2L}$	CO_2 (0.3)
		Slow carbon (0.7)
Soil structure (cellulose, C_{2C})	$\Delta C_{2C} = 0.4 f_1(T_s) f_2(M) \exp\{-3L_s\} C_{2C}$	CO_2 (0.55)
		Soil microorganisms (0.45)
Soil microorganisms (C_3)	$\Delta C_3 = 0.6083 f_1(T_s) f_2(M)(1 - 0,75S) C_3$	CO_2 (0.85 − 0.68S)
		Slow carbon (0.147 + 0.68S − 0.032T_c)
Surface microorganisms (C_4)	$\Delta C_4 = 0.5 f_1(T_s) f_2(M) C_4$	CO_2 (0.6)
		Slow carbon (0.4)
Surface metabolism (C_5)	$\Delta C_5 = 0.2083 f_1(T_s) f_2(M) C_5$	CO_2 (0.6)
		Surface microorganisms (0.4)
Soil metabolism (C_6)	$\Delta C_6 = 1.5417 f_1(T_s) f_2(M) C_5$	CO_2 (0.55)
		Soil microorganisms (0.45)
Slow carbon (C_7)	$\Delta C_7 = 0.01667 f_1(T_s) f_2(M) C_5$	CO_2 (0.55)
		Soil microorganisms (0.447 + 0.009T_c)
Passive carbon (C_8)	$\Delta C_8 = 0.000375 f_1(T_s) f_2(M) C_8$	CO_2 (0.55)
		Soil microorganisms (0.45)

Notation L_s is the carbon part in the lignine form, S is the part of alumina content in the soil

model of carbon balance for the whole of the forest ecosystem taking into account the eight levels of carbon reservoirs (Table 8.5). The model allows the consideration of stratified physical soil structure taking into account the erosion and deposition processes. Two-stage procedure includes sub-blocks that describe:

- circulation of biogenic salts in the plant-soil system (grass ecosystems, agricultural fields, forests, and savannas),
- decomposition of the soil organic matter,
- formation of the forest ecosystem production,
- water balance of the vegetation-soil system, and
- flows of the carbon, nitrogen, phosphorus, and sulphur.

Carbon storage Y_i contained in the ith tree segment (leaves, branches of different thickness, and large roots) is described by the equation:

$$Y_i = D_i \exp(a_i + b_i) \tag{8.9}$$

where D_i is the diameter of the ith tree at the height 1.35 m (cm), a_i and b_i are empirical coefficients. As a result, it is considered that tree biomass contains 50 % of the carbon.

The average annual increase of carbon reserves in ith tree section equals:

$$\Delta Y_i = Y_i(t) - Y_i(t-1)$$

The decomposition process of the litter and dead organic matter of soil depends on the average monthly temperature T_s and soil moisture M:

$$f_1(T_s) = \theta^{(Ts - T_{opt})/10},\ f_2(M) = 0.8(M_{sat})^B + 0.2$$

where T_{opt} is optimal temperature for the decomposition process (≈30 °C), θ is the decomposition constant (≈2.0), and M_{sat} is the saturation level for soil moisture (%),

$$B = \left[\frac{M^d - (M_{opt})^d}{(M_{opt})^d - 100^d}\right]^2,\ T_s = 2.3554 + 0.7042T_a$$

where M_{opt} is the optimal soil moisture under which a restriction of decomposition velocity is no exists, T_a is the average monthly atmospheric temperature, and d is empirical constant.

Soil water regime is described by the following relation:

$$W(t) = W(t-1) + P - E_A,$$

where W is the average monthly water content in the soil, P is the volume of monthly precipitation, E_A is the evapotranspiration:

$$E_A = \begin{cases} \min\{[P + (E_P - P)f_D],\ [P + (W(t-1) - W_m)]\} \text{ when } P < E_P, \\ E_P \text{ when } P \geq E_P \end{cases}$$

where f_D is the factor reflecting limitative effect when precipitation is less potential evapotranspiration E_P, and W_m is the vegetation wilting level.

Limiting indicator f_D is defined by the soil structure and relative water content in the soil (θ_w, $0 < \theta_w < 1$):

$$f_D = \left(\frac{1 + a_s}{1 + a_s \theta_w^{b_s}} \right),$$

where empirical coefficients a_s and b_s determine the soil structure.

The net primary production (NPP) of the forest in the pixel Ω is assessed according to the following formula:

$$NPP = \sum_{j \in \Omega} \sum_{i=1}^{5} \Delta Y_{ij},$$

where index j corresponds to individual tree in the pixel, index I characterizes separate elements of the tree: leafs (i = 1), thick branches (i = 2), thin branches (i = 3), twigs (i = 4), and thick roots (i = 5).

Intensive anthropogenic exploitation of the forests changes their age structure. This effect has to take into consideration under the forest carbon balance modeling. As a whole, net productivity F_Ω of the forest ecosystem can be calculated in the pixel Ω during time interval T:

$$F_\Omega(T) = \sum_{s=1}^{N} F_\Omega(s)\sigma(s),$$

where $\sigma(s)$ is the area occupied by trees of age s, $F_\Omega(s)$ is the net productivity of trees of age s.

Elimination of the wood from the forest biogeocenosis is not connected with natural causes of the carbon losses and is a commercial part of the forest exploitation. Total losses of the carbon are calculated as follows:

$$C_H = \beta \sum_{s=1}^{N} C_{sT} \sigma_H(s),$$

where β is efficient index of the wooding ($0 < \beta < 1$), C_{sT} is the average carbon content in the steam of age s during time interval T, and $\sigma_H(s)$ is the area of withdrawn forest of age s.

Involvement or not-involvement of the carbon withdrawn from pixel Ω within its natural cycle depends on the strategy of the wood usage. As a result, the following parameterization of non-respiratory carbon losses is possible:

$$R_n = C_H \left[\frac{\gamma_L + \lambda(\gamma_S - \gamma_L)}{\gamma_S \gamma_L} \right],$$

where λ is part of carbon in the timber of long time utilization, γ_S and γ_L are the carbon living times under short-time and long-time timber utilization, respectively.

Wooding wastes are multiform and therefore the following equation is used:

$$C_R = (1 - \beta) \sum_{s=1}^{N} C_{sT} \sigma_H(s)$$

This equation completes the Eq. (8.9) taking into account the carbon losses due to the stumps and the top of cut down trees. Thus, net production of the forest biome in the pixel Ω is evaluated:

$$F_B(T) = F_\Omega(T) - R_n$$

8.3.8 Modeling the Production Processes in the Coniferous Forest

It is evident that coniferous forests play an important role in the global carbon cycle. Therefore in this section we shall consider a parameterization of the production process in the forest ecosystem with the pine forest stand of one age proposed and developed by Kirilenko (1990) for practical use in climate models. The model reflects correlations of mean annual and seasonal processes without consideration of the succession of vegetation communities. It describes the dynamics of carbon and water in the ecosystem using a system of five ordinary differential equations, one partial differential equation, and many algebraic equations reflecting the biotic and abiotic processes in the coniferous forest.

The model includes the following values: phase derivatives, input (external) and output (internal) derivatives, as well as parameters (constants, coefficients).

The phase derivatives include (kg/m^2 of soil):

- carbon of the structural mass of young pine-needles m_1 and old ones m_2, $m_L = m_1 + m_2$;
- carbon of the above ground parts of the grown trees (branches, boughs, trunks) and growing roots m_3.
- carbon of thin roots m_4;
- fund of mobile carbon assimilates m_5;

The input variables characterize the environmental conditions:

- the incoming solar radiation intensity E;
- air temperature T and humidity H;
- length of the light period of day τ;
- daily amount of precipitation P; and
- concentration of carbon dioxide in the atmosphere C_o.

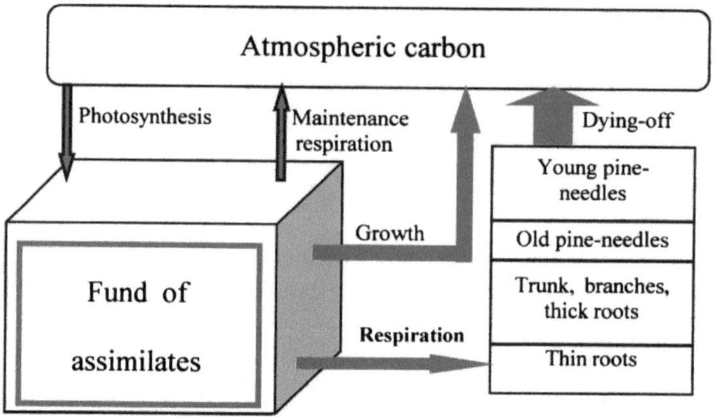

Fig. 8.12 The conceptual scheme of the carbon cycle in the model of the production process in the coniferous forest (Kirilenko 1990)

The output variables are photosynthesis, average height and diameter of trees, transpiration, respiration, increment of the tree's organs, and other characteristics of the coniferous forest ecosystem calculated in the model.

The biomass of the phase variables in the model is measured in carbon units, and the respective scheme of carbon fluxes is shown in Fig. 8.12. The notations for dynamic equations are the following:

- s_3 (m_3) is the area of the trunks per m^2 of soil;
- θ_i (τ_B) is the function of distribution of assimilates, that is, the share of the fund of assimilates spent on the increment of the *i*th organ;
- τ_B is the biological time;
- d_i is the share of the dead biomass of the *i*th organ ($i = 1, \ldots, 4$);
- F is a photosynthesis value;
- r_i is the respiration intensity of the growth of the *i*th organ; and
- p_i is the respiration intensity of the maintenance of the *i*th organ.

The process of photosynthesis can be parameterized using the Chartier formula:

$$F = \frac{\sigma(T_L)\Omega(\tilde{\omega}_S)}{2(r - r_c)}\frac{S_L}{2}\tau\left(C_0 + r\alpha E - \sqrt{(C_0 - r\alpha E)^2 + 4r_c\alpha E C_0}\right),$$

where $\sigma(T_L)$ is a function of the dependence of photosynthesis intensity on the leaf temperature T_L; $S_L = \beta m_L$ is the area of pine-needles; $\tilde{\omega}_S$ is the weighted-mean soil moisture; $\Omega(\tilde{\omega}_S)$ is the function of the impact of the over-moistened root-layer of the soil on photosynthesis; $r = r_a^c + r_{st}^c + r_c$, r_a^c and r_{st}^c are diffuse resistances for CO_2 molecules in the leaf boundary layer and in the leaf stoma, respectively, r_c is an efficient resistance of carboxylizing; τ is the duration of the light period of day; α is an initial angle of inclination of the light curve of photosynthesis:

$$\tilde{\omega}_S = \int_0^{A_0} \omega_S(z)ds(z)$$

where $s(z)$ is the function of distribution of the root system along the vertical cross-section of the soil layer; A_0 is a maximum capacity of the root layer of the soil.

For conifers, $r_a^c \ll r_{st}^c$, and along the height of the tree the value of photosynthesis per unit mass of pine-needles changes negligibly, and therefore we shall consider some point, averaged over the tree height, for which the intensity of the photosynthetically active radiation (PAR) is determined by the Beer-Lambert law:

$$E = \kappa_E E_e \exp\left(0.5 K_e h S_L H^{-1}\right)$$

where E_e is the intensity of the incoming solar radiation over the forest, κ_E is the share of PAR in the total solar radiation; K_e is the coefficient of extinction.

Assuming that the temperature of pine-needles coincides with that of the atmosphere ($T_L = T$), we use the following approximation for the $\sigma(T)$ function:

$$\sigma(T) = \max\left\{0, \frac{T - T_1}{T_0 - T_1}\exp\left[l^{-1} - l^{-1}\left(\frac{T - T_1}{T_0 - T_1}\right)^l\right]\right\},$$

where the parameter $l = 1.8$; T_1 is a minimum temperature at which the photosynthesis still takes place; T_0 is a temperature optimal for photosynthesis.

The Ω function characterizes the dependence of photosynthesis on soil moisture. With a short-term over-moistening of soil the photosynthesis does not decrease, and therefore $\Omega(\tilde{\omega}_S) = 1 \quad \forall \tilde{\omega}_S \in [0, 1]$. In a general case, we use the dependence:

$$\Omega(\tilde{\omega}_S) = \begin{cases} 1, & \tilde{\omega}_S \in (0, \omega_{HB}); \\ (1 - \tilde{\omega}_S)/(1 - \omega_{HB}), & \tilde{\omega}_S \in [\omega_{HB}, 1] \end{cases}$$

The function $\theta_i(\tau_B)$ is introduced in order to describe the strategy of distribution of assimilates, where τ_B is the biological time and to determine the ratio of the sum of efficient temperatures T_l to the mean multi-year sum of efficient temperatures T_Σ accumulated during the vegetation period:

$$\tau_B(j) = \sum_{l=1}^{j} T_l / T_\Sigma,$$

where j is the day number.

The θ_I function is determined from the expression:

$$\theta_i = a_i^\theta \exp\lfloor -b_i^\theta (c_i^\theta - \tau_B)^2 \rfloor \quad (i = 1, \ldots, 4)$$

where the a_i^θ coefficient is equal to a maximum of θ_i, c_i^θ is the biological time of a maximum of assimilates' flux to the *i*th organ, the b_i^θ coefficient determines the steepness of the θ_i function.

The flux of assimilates at a time moment τ_B into the *i*th organ of the plant is equal to $m_5\theta_i(\tau_B)$.

Assume that respiration connected with the growth of the *i*th organ of a tree is proportional to the rate of its increment and equal to $r_i dm_i/dt$, and the respiration for maintenance of leaves and sucking roots is proportional to their mass and equal to $p_i m_i$ (i = 1, 2, 4). The p_i coefficient depends on temperature:

$$p_i = a_i^p \exp[k_i^p (T - T^p)]$$

where a_i^p is the respiration for maintenance at a temperature T^p.

The respiration for maintenance of the trunk and thick roots is in proportion to the trunk area s_3 and is equal to $s_3 p_3 \sqrt{\beta_{st}}$, where β_{st} is the share of the trunk mass in m_3. The dying-off of the organs is considered to be in proportion to their masses with the coefficients d_i (i = 1, …, 4).

The water cycle in a tree is described by a system of equations that parameterize the water motion over the territory, in soil, in a tree and its elements. The water is one of the limiting factors of the tree's growth through a value of the stomatal resistance that affects the photosynthetic activity of plants. The water cycle on the forest territory is the sum of two basic parts—abiotic and biotic, which describe the processes of water motion in soil and its absorption by the root system, its motion in the plant, and transpiration. There are a lot of models of these processes. The difference between them is determined by different suppositions about the character of the soil structure and the location of the root system. Kirilenko (1990) proposed to describe the water motion in the soil with an equation that generalizes the Darsi classical model when the soil is considered a homogeneous porous layer:

$$\frac{\partial \omega}{\partial t} = -\frac{\rho}{\rho_S}\left[\frac{K_{\Pi B}}{\Phi_{\Pi B}}\frac{\partial}{\partial z}\left(\mu\omega^n \frac{\partial \omega}{\partial z} + \omega^n\right) + \frac{f(z,t)}{\Phi_{\Pi B}}\right],$$
$$V(z,t) = K_{\Pi B}\left(\mu\omega^n \frac{\partial \omega}{\partial z} + \omega^m\right)$$

with the boundary conditions:

$$\Phi(z_g, t) = 1, \quad V(0,t) = W(t) - L(t)$$

and the initial condition:

$$\omega(z_g, t_0) = \omega^0(z),$$

where $W(t)$ is precipitation intensity, $L(t)$ is the intensity of evaporation off the soil surface, ρ is water density, ρ_S is the volume density of soil, Φ is the relative soil moisture, $\phi_{\Pi B}$ is the total water capacity of soil; V is the rate of moisture motion in soil, $\mu = n\phi_{\Pi B}/(\rho g)$, $m = 1+2n$, n is a constant, g is a gravity acceleration, $\phi_{\Pi B} = \phi_S \omega^n$, ϕ_S is the water potential of soil, $K_{\Pi B}$ is the conductivity corresponding to soil moisture $\Phi_{\Pi B}$, z is the vertical coordinate. The f function describes the intensity of water absorption by the root system:

$$f(z,t) = \left[\xi_0 \bar{\phi} \varphi(z,t) + v(t)\right] dS/dz,$$

where $\varphi(z,t) = \varphi_S(z,t) - \tilde{\varphi}_S(t)$ is the deviation of the dimensionless water potential from the average value for the root layer of the soil, $\xi_0 = \xi_k \sigma_0$ is the conductivity of the entry of water to the root system, $\xi_k = 1/r_k$ is the specific conductivity of the entry of water to the root, r_k is the specific resistance of the entry of water to the root, σ_0 is the total area of the sucking roots of the tree per unit soil area

$$\varphi_S(z,t) = 1 - \left(\frac{\omega(z,t)}{\bar{\omega}}\right)^{-n}, \quad \tilde{\varphi}_S(t) = 1 + \tilde{\phi}_S / \bar{\phi},$$

where $\tilde{\varphi}_S$ is an average dimensionless water potential for the root layer of the soil, $v = \xi_0 \left(\tilde{\phi}_S - \phi_0\right)$ is the rate of water absorption by roots from the soil.

Water from the root system moves upward along the trunk, through branches into leaves and evaporates into the atmosphere. It is assumed that an amount of the moisture from respiration is equal to that assimilated by the root system, and the rate of motion is equal to v. The integral effect of the impact of the coniferous forest on water cycle in its territory is the sum of the effects of each tree. The practical application of the developed model requires information about spatial heterogeneities and distributions of the characteristics of the forest stand. An application of the model to deciduous and mixed forest systems requires additional units which would describe the processes of the transformation of water potentials as well as carbon cycle in leaves. Intensive studies are being carried out in this direction.

8.3.9 Modeling Succession Processes in the Tundra-Taiga System

Predicting the northern boundary of forests has become problematical as a consequence of worldwide anthropogenic processes and the way in which their effects are inexorably propagating towards northern latitudes. Connections between the observed propagation of this boundary and forest fires, reindeers' grazing, anthropogenic transformations, and climate change have been widely discussed in scientific bibliography (Ji and Stocker 2003; Kondratyev and Grigoryev 2004). Various hypotheses have been expressed about correlations between the natural technogenic processes in the zone of this boundary. For instance, there is the opinion that forest fires can influence the northward propagation of the birch forests. This can lead to considerable changes in the natives' habitation and the reindeers' propagation. In the late 20th century some disagreements were observed between forestry and deer-breeding. Thus, in Finland the areas of the reindeers' propagation moves 400–500 km south from the boundary of the forests, that is, up to the boreal zone. In other words, a change of the forest boundary is followed by a change of the soil layer, and this, due to shifts in biogeocenotic cycles, can direct the succession of the northern boundary of forests in an unpredicted direction. Moreover, this prediction, based on the pre-history of the development of this boundary in natural conditions, cannot be reliable. Therefore the use of the methods of eco-informatics with its models and technologies of monitoring is now a single adequate method for predicted estimates of the dynamics of the northern forests boundary. As far as this is concerned, a lot of ecological problems appear, and thus the following is necessary:

- combining various elements of the forest ecosystems in the zone of the northern boundary of the forests (reasons for the presence or absence of some type of trees);
- the role of various diseases of the northern forests and their sensitivity to climate changes and other external factors;
- the role of fires and their location and power in the change of the structure and configuration of the transfer boundary between forests and tundra;
- the role of insects with an account of the dependence of their propagation and the biomass on climate change in the dynamics of the northern forests boundary;
- patterns of scattering the seeds of various plants in the zone of the forests northern boundary and their effect on its dynamics; and
- the impact of the dynamics of the forests northern boundary on the wild animals' migration and their habitat.

An expected climate warming can result in the substitution of some vegetation ecosystems for others (exogenic succession). It will be especially manifested in high latitudes where taiga transforms into tundra. Bogatyrev (1988) proposed a model, which parameterizes the transformation processes in the tundra-taiga system and makes it possible to reveal some important regularities. The phase variables are the biomass of moss (X_1), vegetation of the grass-bush level (X_2), conifers (X_3), forest cover (X_4), and dead organic matter of soil (X_5). The phase variables are measured in tons of carbon per unit area (t/ha), and their spatial distribution is assumed to be uniform. The carbon flux scheme is given in Fig. 8.13.

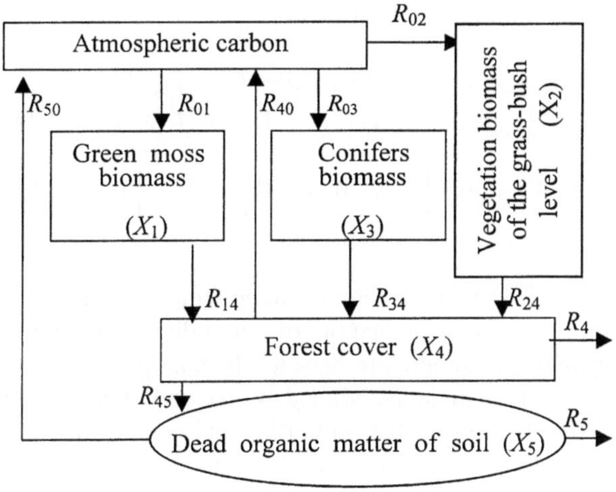

Fig. 8.13 The scheme of the carbon cycle in the tundra-taiga system (Bogatyrev 1988)

The system of balance equations of the model of the transfer processes on the border taiga-tundra according to the scheme in Fig. 8.13 is

$$dX_i/dt = R_{0i} - R_{i4} \qquad (i = 1, 2, 3);$$

$$dX_4/dt = \sum_{i=1}^{3} R_{i4} - R_{40} - R_{45} - R_4;$$

$$dX_5/dt = R_{45} - R_{50} - R_5;$$

where the R_{ij} functions depend on many parameters, and their determination is the principal stage of the model formulation. Bogatyrev (1988) proposed the following formulas for these functions:

$$R_{0i} = F_i\varphi_i(T_M)G_i(X_3)f_i(X_5)H_i(W), R_{i4} = k_iX_i\ (i = 1, 2);$$
$$R_{03} = F_3\varphi_3(T_M)G_3(X_3)f_3(X_5)H_3(W)\chi(X_3); R_4 = m_4X_4; R_5 = m_5X_5; R_{34} = k_3X_3[1 + \kappa(T_M - T_A)]^{-1};$$
$$R_{40} = \sigma k_4X_4\varphi_4(T_M); R_{50} = k_5X_5\varphi_5(T_M); R_{45} = (1 - \sigma)k_4X_4\varphi_4(T_M);$$

where F_i coefficients express the dependence of the biomass increment on its type (the characteristic indicators of vegetation), φ_i functions assign the laws of changes in the biomass increment as a function of atmospheric temperature beneath the canopy T_M, and the G_i, f_i and H_i functions describe deviations of the annual biomass increment from a maximum level at a given temperature T_M due to a deficit of illumination, an amount of mineral resource in the soil, and moisture, respectively. For trees, the annual production also depends on their biomass, and this dependence is represented with the χ function. For vegetation of the grass-bush level and the moss, their annual increment is supposed to be independent of their biomass.

Determine the functions φ_i, G_i, f_i, H_i, and χ:

$$\varphi_i(T_M) = 1 + (\delta_i/100)(T_M - T_Y),\ (i = 1,2);\ \varphi_3(T_M) = 1 + (\delta_3/100)(T_M - T_A);$$
$$G_3(X_3)\ - \exp(-\alpha_i X_3),\ (i = 1,2,3); f_1(X_5) \equiv 1; f_i(X_5) = \exp(-a_i/X_5),\ (i = 2,3);$$
$$H_i(W) = \exp\left\{ -b_i\left(W - \frac{W_A - W_Y}{\sigma_i} \right)^2 \right\} (i = 1,2,3);\ \chi(X_3) = \left(\frac{X_3}{X_{3A}} \right)^\gamma;$$

where T_Y and T_A are atmospheric temperatures in tundra and taiga, respectively, δ_i (i = 1, 2, 3) are indicators of variability of the annual biomass increment with an increasing temperature beneath the canopy by 1 °C; α_i, a_i, b_i, κ, and $\gamma \in (0,1)$ are constant coefficients; W_A and W_Y are soil moisture in taiga and tundra, respectively; k_i (i = 1, 2, 3) are the proportionality coefficients; $\sigma \in (0,1)$ is the share of the forest cover transforming into dead organic matter of soil as a result of decomposition.

To parameterize the atmospheric temperature beneath the tree canopy, it is supposed that it cannot be below the environmental temperature T, increases with the mass of trees up to some level $X_3 = 0.5\beta_2/\beta_1$, and then decreases:

$$T_M(X_3, T) = T \begin{cases} f(X_3), & 1 \le f(X_3) \le T_A/T; \\ T_A/T, & f(X_3) > T_A/T; \\ 1, & f(X_3) < 1; \end{cases}$$

where

$$f(X_3) = 1 - \beta_1 X_3^2 + \beta_2 X_3;$$

β_1 and β_2 are positive coefficients.

The model described above can be used as an independent unit of the global model, providing a transformation of the spatial structure in Fig. 1.23 and specifying the local elements of the model of the water balance on the territory described in Krapivin and Kondratyev (2002). For real application of the model, its parameters should be estimated with a concrete attachment to geographical latitude.

8.4 Wildfires as Components of Global Ecodynamics

8.4.1 Fires and Forest Ecosystem

Fires play a substantial role in the formation of forest ecosystems. Moreover, it can be stated that they are an integral element of their development. Fire in a forest can be caused either by a lightning stroke or anthropogenically. From available estimates, the frequency of lightning strokes in tropical forests reaches 50 per km^2 annually. In the moderate-zone forests this quantity is less and constitutes 2–5 strokes per km^2 annually. Of course, not every stroke leads to a forest fire. Nevertheless, in different regions of land, lightning strokes lead to 2–30 % of forest fires.

During the historical period, geography and quantity of fires had been changing, but their role in the formation of forest landscapes has remained constant. Many species of trees, such as pine, larch, birch, and aspen had a pyrogenic period in their development. Owing to fires these species have preserved their natural habitats and had not been ousted by spruce, cedar, and fir. In the process of evolution, the light-demanding species have become fire resistant due to, for instance, the thick bark and deep root system of the larch and pine, and the birch and aspen's ability to give rich shoots from the roots after a fire. Some species of American pines exhibit protective reaction. For instance, they open their cones only after a fire, when conditions for seeds' sprouting are favorable.

Some investigators believe in an objective idea according to which practically all current taiga forests in Russia had been changing due to fires (Furiayev 1996). Present studies make it possible to confirm the above-said with respect to taiga forests in the historical and pre-historical past. This is confirmed, in particular, by paleoecological developments carried out in the USA and Finland. Studies of fires using the method of pollen analysis and of traces of burning fixed in stratified lake deposits in eastern Finland testify to a large scale of fires. During the last 2,100 years, fires have been an integral factor of development of the taiga forests in this region. The repeatability of large-scale fires in eastern Finland varies from 7–110 to 130–180 years. It was noticed that fires became more frequent after 600 AD, when the impact of humans on forests became stronger.

According to calculations of Kurbatsky (1964), in the early Holocene there had been several thousands of forest fires caused by lightning strokes. Of course, a human had also played a substantial role in the origin of fires in the forest ecosystems (first due to carelessness with fire, and then as a result of loosing control of fire during propagation of the slash-and-burn method used to clear the forests for agriculture). In the 1950s, on all continents, about 200,000 forest fires happened every year, with 3 % of this quantity having been caused by lightning strokes. It is not excluded that all forest ecosystems of the planet in the historical and pre-historical past had suffered from fire. There had long been a version of the pyrogenic origin of the present savannas over a vast territory of the Hindustan Peninsula, considered to have propagated in place of the burned tropical forests.

Thus, the problem of forest fires has two sides—negative and positive. The negative side is connected with apparent damage to forest ecosystems and economic losses for humans. The positive aspect consists in the evolutionary role of forest fires. For instance, the post-fire renewal of pine forests over vast territories is known to take place easily and rapidly compared to clearings during which the seeds get destroyed. At the fire-sites where the seeds have not been destroyed, shoots and underwood appear rapidly and all together. Moreover, in different global regions, an increase of the post-fire productivity of forest ecosystems has been observed. In particular, it was found out that in the regions of Slovenian Karst and Istria at the south-west of the Republic Slovenia, the fires that happened there affected the species composition of plants (Grigoryev and Kondratyev 2005). Immediately after the fire, the diversity of species and living forms grows even more rapidly, but in the course of the age succession to initial forest composition, this diversity decreases.

Restoration of forest communities depends on the frequency and intensity of fires. In the Siberian taiga forests relatively frequent fires of low intensity cause changes in the age and quality of forest stands. Changes of forest planting are connected with intensive but rare fires. Studies of the post-fire dynamics of forest ecosystems at the Far East of Russia show that successions of forest vegetation are connected with fires (Efremov and Sapozhnikov 1997). At the same time, it was shown that fires are a needed element to maintain biodiversity.

The post-fire regeneration of forest ecosystems depends on many factors (climate, soil, rocks composition, relief, etc.). In deciduous forests of Siberia, with its permafrost, the shrub and grass-shrub tiers are restored in 4–5 years after the fire, while moss and lichen grow much slower.

Sometimes the fires affect positively the soil component of the forest ecosystem and, in particular, promote the preservation and provision of soil with nitrogen, which is known to play an important role in forest ecosystems' productivity. This effect has been found out by the scientists from the Montana University in Missoula in the forests of the western part of Montana (Newland and De Luca 2000). Local forests with prevailing yellow pine and Douglas fir develop better in the case of abundant nitrogen-fixing plants and due to this, sufficient content of nitrogen in soil. It was found out that these plants are very important as a factor of compensation of nitrogen, when the forest ecosystems loose it after the fires. In the absence of fires, the quantity of nitrogen-fixing plants decreases, and further on the forest structure changes drastically. On the contrary, periodic fires in the forests of yellow pine and Douglas fir, broaden the propagation of nitrogen-fixing plants and favor the forest ecosystems' productivity, on the whole.

As mentioned above, one of the manifestations of the anthropogenic interference into the forest ecosystems' dynamics is forest burning down for subsequent land use in agriculture. As Krimmer and Lake (2001) have shown, the controlled burning-down of forests maintains their landscape mosaic and promotes the preservation of biodiversity, as well as raises the forest ecodynamics' productivity. From observations of Hoffmann (1998), in savannas serrado in Brazil, the productivity of the ecosystem of 3 species of trees, two species of shrubs and one species of subshrub increased 4–7 times after the burning. Clearly, the burning of the forest sites by natives should be attributed to the means of their harmonious co-existence with nature, and the planned burning of forests should be carried out on a scientific basis and considered as an element of land use optimization (precisely, forest-use). The problem of the balanced use of positive and negative consequences of a forest fire remains unsolved.

One of the main problems of the forestry science is earlier detection of the fire hazard areas. As it was shown in Chap. 1 of this book a solution of this problem can

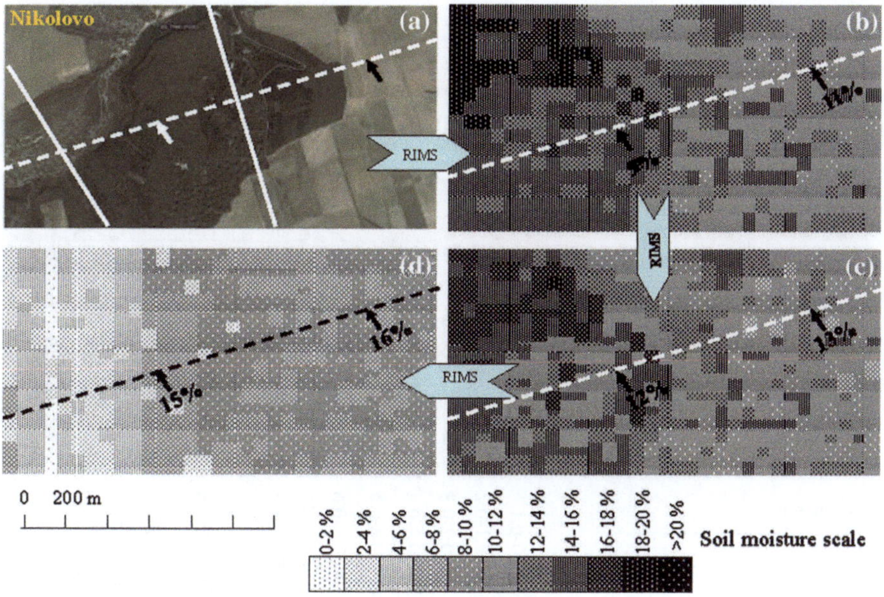

Fig. 8.14 Scheme and results of the RIMS application for the reconstruction of spatial distribution of soil and forest litter moisture parameters in their dynamic in the Bulgarian national park Rusensky Lom. *Solid* and *dotted lines* show the flying laboratory air routes. *Dotted lines* correspond to air routes that were used for the calibration of radiometers. *Arrows* mark the points where in-situ measurements are realized and the RIMS errors are shown. Time dynamic of moisture parameters is characterized by the conditions: **a** heavy precipitation, **b** 1 day later, **c** 2 days later, and **d** 4 days later

be realized by the RIMS with its location on a flying laboratory. An assessment of the moisture soil content allows the evaluation of the forest fire hazard level. Examples of such experiments are given in Figs. 8.14 and 8.15. Bonafoni et al. (2011) deals with passive microwave imaging for fire detection by means of a single-channel ground-based radiometer. The simulation of images sensed in the presence of fire spots under different environmental and operative conditions is presented. A low-cost ground-based radiometer operating at 12.65 GHz was used. Scenarios where fires are un-visible and IR sensors are not useful with respect to a microwave imager were investigated in deep, such as in the presence of vegetation canopy optically masking fire sources and smoke plumes in the early stage. These simulations allow the assessment of limits and capabilities of microwave imaging for the identification of little fires masked by forest areas. With regard to forest areas, the proposed simulations have the aim to establish performances and limits in the identification capability of little fires by means of microwave radiometry. It has been seen that the radiometric contrast can be computed as the difference between two successive scans of the same area.

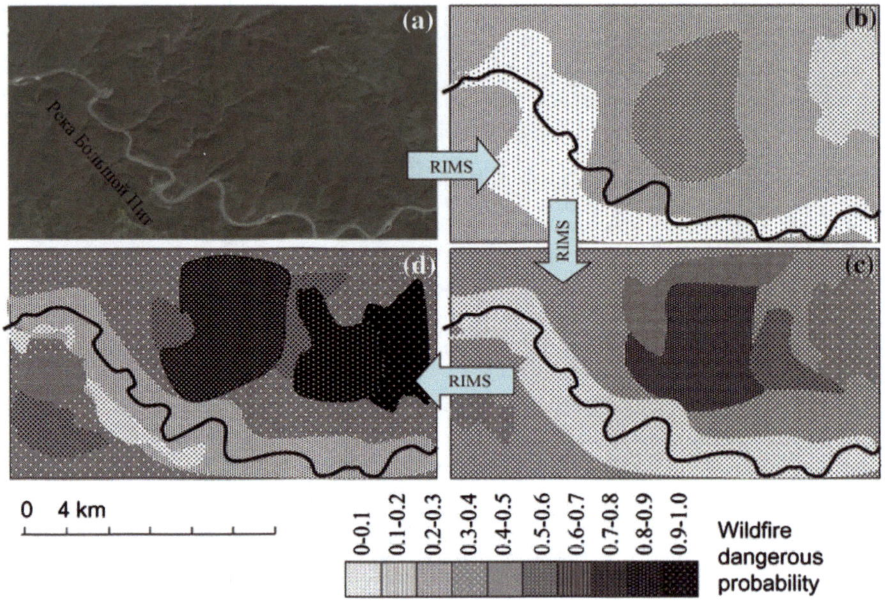

Fig. 8.15 An example of spatial distribution reconstruction for the wildfire dangerous probability basing on the soil and forest litter moisture in their dynamics in village Nikolovo (Bulgaria). Time dynamic of the wildfire dangerous is characterized by the conditions: **a** heavy precipitation, **b** one day later, **c** two days later, and **d** four days later

8.4.2 Wildfires, Dynamics of the Biosphere, and Climate

Poorly studied positive and negative feedbacks of forest fires makes it impossible to draw the final conclusions about their role in the global eco-dynamics. Positive feedbacks mentioned above can be supplemented with facts indicating the opposite post-fire trends in the ecosystems' dynamics. Numerous results of numerical modeling show that the tropical deforestation results in changes of albedo, land surface roughness, leaf area index, and root zone depth, which lead to a decrease of rain rate and relative humidity, but to an increase of SAT and wind speed. Global consequences of such changes will depend on the ratio of scales of reduction of forest areas and measures taken to protect nature. Preliminary results of modeling these consequences obtained by Hoffmann et al. (2003), with the use of the atmospheric general circulation model, CCM 3.2, and data of remote sensing, revealed a decrease of rains and relative humidity during the fires in the Amazon basin. The empirical connection between the scales of desertification in the regions of Amazon, Congo, and Indonesia and the frequency of forest fires, established from the NOAA-12 satellite measurements, revealed a dependence of the index of the fire risk of the territory on their frequency. This means that forest fires directly affect a climate change, which leads to a change of their frequency of occurrence. This interactivity makes it possible to search a regime of forests regulation which would reduce the level of CO_2 concentration in the atmosphere and exclude an undesirable climate change.

It is supposed that boreal forests are a region for carbon sink. However, their fires can transform boreal forests into sources of carbon due to direct emissions of carbon during biomass burning and indirect impacts of fires on the thermal and water regimes, as well as the structure and functioning of the ecosystems. The frequency of fires in boreal forests during the last several decades has increased and can increase further under conditions of continuing global climate warming. Thus, this should lead to a shorter time period of ecosystems' recovery in the periods between the fires and to enhancing emissions of GHGs to the atmosphere. The CO_2 flux from the ground $R_s = H_7^C + H_8^C + H_9^C + H_{15}^C$ (see Fig. 1.30) is a second important flux of carbon in boreal forests which determines its role in the formation of the global carbon cycle. The R_s value changes due to different processes:

- fires lead to a partial removal of vegetation cover and a decrease of soil surface albedo (the latter determines an increase of surface temperature and the rate of decomposition of vegetation remains);
- a fire breaks the process of accumulation of organic matter in soil and changes the balance between the input of detritus and heterotrophic respiration (as a result of a great input of detritus);
- a fire brings forth changes of vegetation succession and composition of its species as well as litter quality.

To better understand the impact of forest fires on the carbon cycle in boreal forests, Wang et al. (2003) performed measurements and numerical modeling of R_s for black spruce (*Picea mariana)* with post-fire chronological succession for seven forest fires in the north of the Manitoba Province (Canada), under conditions of good and poor drainage of soils as an example. Studies have been aimed at:

i. a quantitative characteristic of the dependence of R_s on soil temperature for forests of different age;
ii. study of the post-fire succession dynamics of forest stands; and
iii. estimating the annual CO_2 flux from soil surface.

The CO_2 flux depended strongly on conditions of drainage and age of the forest stand. There was a positive correlation of CO_2 flux with soil temperature (=0.78), with results of numerical modeling of the flux differing strongly depending on a combination of the level of drainage and age of the forest stand. In the period of the vegetation development season the CO_2 flux from a well drained soil was much greater than from a poorly drained one. The annual mean values of the CO_2 flux from the soil in different years constituted 244 (the year 1870), 274 (1930), 350 (1964), 413 (1981), 357 (1989), 412 (1995), and 226 gC $m^{-2}year^{-1}$ (1998). Under conditions of well drained soils the numbers above become 264, 233, 256, 303, 300, 380, respectively, while 146 gC m^{-2} $year^{-1}$ in the case of poorly drained soils. In the winter (from 1 November to 30 April) the values of CO_2 flux varied within 5–19 % with respect to the annual flux. Apparently, a decrease of the flux at the post-fire sites was determined mainly by the decreasing level of respiration due to the trees' roots.

In the course of all chronological succession of fires considered, the CO_2 fluxes from soil changed almost twofold at reaching maximum values before joining of the trees' crowns, when the soil was the warmest and an accumulation took place of both ground and sub-ground biomass. The observed decrease of CO_2 flux in the case of older forest stands can be explained by a lower soil temperature, which is determined by an accumulated heat-isolating organic substance and other factors.

Boreal forests cover the territory of about 14 million km^2 in the circumpolar latitudinal band 50–70°N, which constitutes $\sim$10 % of the global land surface. These forests contain disproportionately large amounts of soil carbon because of climatic conditions which are unfavorable for the processes of organics decomposition. The annual mean temperature in boreal forests is close to 0 °C with a weakly drained soil explained by prevailing lowlands. The characteristic features of the considered latitudinal band are widespread permafrost and isolation of deep ground horizons from summertime heating because of moss and thin roots which burn in forest fires but restore during decades. This heat-insulation favors the preservations of permafrost, which complicates the soil drainage and slows down the processes of decomposition of organic matter stored in deep soil layers beneath the moss layer. Intensification of respiration due to deep layers measured in July and August 1996 reached $\sim$10 kgC ha^{-1} day^{-1}. This enhancement of respiration correlates with an increase of the deep layers temperature. Thus, organic matter of deep layers of the soil is characterized by lability with a low rate of decomposition connected with low temperature.

The data above show that it is necessary to extend the parametric descriptions of the R_s flux components, apparently not only for forested territories but also for other soil-plant biomes in order to raise the accuracy of estimates of sinks and sources of CO_2 on land. Smith et al. (2003) carried out measurements of CO_2 fluxes under conditions of mountain ecosystem of the steppe wormwood in the south-eastern part of Wyoming and found out that the difference between CO_2 fluxes measured at different time of day reached 9 %, and that even under conditions of homogeneous vegetation cover the temporal variability of CO_2 fluxes between the ecosystem and the atmosphere is substantial.

An important consequence of forest fires is the fire-induced changes of atmospheric chemistry due to GHGs emissions. For instance, large-scale forest fires in September–November 1997 in Indonesia, Malaysia and Papua-New Guinea resulted in about 130 TgCO emitted to the atmosphere, which changed the ozone concentration by 10 %. The GHGs emitted to the atmosphere propagated over vast territories, and the total content of CO in December grew by 10–20 % in the latitudinal band 30°N–45°S, by 5–10 %—south of 45°S, and <5 %—north of 45°N. On the whole, global emission of GHGs in biomass burning reached 38–4,300 TgC year^{-1} with a very small contribution due to fires in boreal forests (23 TgC year^{-1}, or 0.6 %). Emissions of GHGs to the atmosphere due to fires in boreal forests in 1998 constituted 290–383 Tg (total carbon), 828–1,105 Tg(CO_2), 88–128 Tg(CO), and 2.9–4.7 Tg(CH_4). The upper level of these estimates corresponds to 8.9 % of total global emissions of carbon due to biomass burning; 13.8 % of global CO emissions due to forest fires, and 12.4 % of global Global emissions of methane due to fires. The contribution of forest fires in Russia constituted 78 % of total emissions (19 % is the share of North America). The contribution of peat-bog fires at Far East to carbon input to the atmosphere in the fall of 1998 reached 40 TgC.

A relative impact of fires on the chemical composition of the troposphere was weaker in the Northern Hemisphere, where the background CO content was greater than in the Southern Hemisphere. The radiative forcing at the surface level reached 10 W m^{-2} over most of the tropical Indian Ocean, and 150 W m^{-2} in the regions of forest fires in Indonesia. Thus, forest fires turned out to be an important factor of the impact on the radiative regime in the tropics. About 700 TgC were emitted to the atmosphere, which constituted 3–4 % of carbon supplies in the peat bogs of Indonesia. All this testifies to a necessity of a detailed analysis of consequences of forest fires and search of strategy of their control. One of the perspective which approaches to solution of this problem is the use of satellite systems with microwave radiometers, which will make it possible to control, on global scales, the moisture content of forest and to determine thereby the level of their fire risk.

8.4.3 Biomass Burning and Atmospheric Chemistry

Biomass burning is one of the most substantial sources of the input of many GHGs to the atmosphere. This process takes place both during a forest fire and in burning the wood remains when logging or in agriculture. Therefore the control of the seasonal and inter-annual variability of biomass burning is important for the estimation of GHGs emissions and the calculation of radiation budget variability. Duncan et al. (2003) proposed a method of control with the use of the satellite scanning radiometer ASTR. The results of measurements in the period 1979–2000 in south-western Asia, Indonesia, Malaysia, Brazil, Mexico, Central America, Canada, Alaska, and the Asian part of Russia did not reveal any long-term trend of CO emissions in forest fires, but revealed a clear inter-annual variability. The annual volume of CO emission in the indicated regions changes from 2 (the Asian part of Russia and China) to 20 Tg $year^{-1}$ (Indonesia and Malaysia) with average global total emissions of 437 TgC $year^{-1}$ (annual values varied from 429 to 565 TgC $year^{-1}$).

In 1987, in the USA the long-term program was launched to study the spatial-temporal variability of CO concentration in the troposphere. Initial air samples taken in the marine atmospheric boundary layer to analyze the concentration of CO_2 and methane were also used to measure the CO concentration. During the last decade, the scientists of the NOAA Climate Modeling and Diagnostics Laboratory carried out an analysis of air samples to estimate the CO concentration from the data of observations at the global network. It follows from the results obtained that the short-term periods of increasing or decreasing CO concentration were superimposed, on the long-term decreasing trend of CO concentration in the troposphere, observed in the 1990s. It turned out that the instability of standard samples used for calibration determined the systematic underestimation of the observed CO concentration. Novelli et al. (2003) showed that the decrease of tropospheric CO is confined mainly to the Northern Hemisphere, where emissions to the atmosphere due to fossil fuel burning were substantially lower, constituting 1.8 ± 0.2 ppb $year^{-1}$. In the Southern Hemisphere, between 1991 and 2001, this trend was not observed.

The globally averaged CO concentration is characterized by a strong interannual variability explained mainly by interannual changes of the scale of biomass burning. During the whole period of observations, the globally averaged CO concentration was decreasing by about 0.5 ppb $year^{-1}$; this decrease was mainly concentrated north of 30°N. A strong increase of CO content in the troposphere observed in 1997–1998 was connected with the impact of unusually widespread and intensive forest fires; more that 300 TgCO were emitted to the troposphere because of fires. In the years of large-scale fires in boreal forests, such emissions can cause changes of the dynamics of redox reactions in the troposphere on global scales.

Wet tropical forests in the Amazon basin play a special role in the global ecodynamics due to huge sizes of their territory (about 5×10^6 km^2) and high biotic activity. Here all-the-year-round emissions to the atmosphere of natural biogenic aerosol take place both of primary (micro-organisms, pollen, detritus of vegetative origin) and secondary aerosol appearing due to gas-to-particles transformation of organic, nitrogen- and sulphur-containing GHGs emitted to the atmosphere by plants. In the dry season period, this background aerosol is "suppressed" by smoke aerosol caused by deforestation and burning of biomass remains.

The processes of intensive convection and advection determine the long-range transport of both types of aerosol to higher latitudes, with far-reaching consequences from the viewpoint of affecting the chemical reactions in the atmosphere, as well as the formation of climate and biogeochemical cycles of nutrients. On the other hand, natural aerosol comes to the Amazon basin from distant regions, including marine aerosol from the Atlantic Ocean and dust aerosol from Sahara, which can serve as significant nutrients for wet tropical forests. In July 2001, near the city of Balbina, in the center of the Amazon region, Graham et al. (2003) carried out measurements of aerosol which showed that the atmosphere contains both coarse-scale and fine-scale fractions, which consisted of organic compounds by 70 and 80 %, respectively. Coarse-scale aerosols also contained a small amount of soil dust and sea-salt particles, while fine-scale aerosols contained some amount of non-marine sulphate. The concentration of coarse-scale aerosol particles averaged 3.9 ± 1.4 μg m^{-3} with the ratio of nocturnal to daytime values 1.9 ± 0.4, whereas the concentration of fine-scale aerosols increased in the daytime (an average of 2.6 ± 0.8 μg m^{-3}, with the ratio 0.7 ± 0.1).

Biomass burning during the last decades has increased with the growing population size. The geographical zones of intensive biomass burning have broadened. Most considerable components of emissions in biomass burning include CO_2, CO,

CH_4, NH_3, HCN, methanol, acidic acid, acetol, and others. The knowledge of the volumes of emitted aerosols is important to study the consequences of their impact on the radiation budget and dynamics of ecosystems of land and the World Ocean. Christian et al. (2003) found out several new components appearing in burning the African fuels; they are chemically active substances which can take an active part in oxidation reactions. These substances include acetaldehyde, phenol, acetol, glycolaldehyde, methylvinilether, furan, acetone, acetonitrile, propylene-nitrile and propanenitrile. Emissions of chemically active oxidizing volatile organic compounds in the fires in African savannas and in Indonesia constitute 70 and 77 % of all emitted aerosols, respectively.

Many uncertainties appear in studies of the impact of biomass burning on the ozone layer, though there is evidence of this impact. In the period 21 January–14 February 1999, when huge amounts of biomass were burned on the western coast of Africa, sonde measurements of the tropospheric ozone content were carried out on the scientific ship along the route from Norfolk (37°N, 76°W) to Cape Town (34°S, 22°E). The earlier conclusion was confirmed: the tropospheric ozone and its precursors connected with biomass burning were concentrated in the lower troposphere mainly because of the absence of an intensive convection on land. In the layer 1,000–799 hPa there was a wind-driven transport of ozone and its precursors towards the equator or west. On the other hand, the effect of the upward adiabatic motions connected with the diurnal change of the processes in the west-African planetary atmospheric boundary layer (ABL), could determine the vertical transport of ozone and its precursors to the free atmosphere over the marine ABL. Besides, lightning strokes in the regions of South Africa, Central Africa, and meso-scale convective systems of the Gulf of Guinea could cause an increase of tropospheric ozone concentration in the middle and upper troposphere of the tropical Atlantic Ocean in the Southern Hemisphere. At the same time, a decrease of the ozone content called the "ozone paradox" was recorded in the tropical Atlantic Ocean in the Southern Hemisphere.

8.4.4 Wildfires and Carbon Cycle

Wood remains (WR) are an important component of all forest ecosystems, since they affect the nutrients' cycle, humus formation, carbon storage, frequency of occurrence of forest fires, water cycle, and are a natural habitat for heterotrophic and autotrophic organisms. Usually the presence of dead trees and WR is left out of account in the estimation of carbon budgets, though the importance of these carbon reservoirs as boreal supplies of carbon has recently attracted attention.

Bond-Lamberty et al. (2003) studied the distribution and respiration dynamics of wood remains under conditions of chronological succession in the fires of boreal forests consisting mainly of black spruce, in the northern part of the Manitoba Province (Canada). The considered chronological succession included seven forest stands that suffered fires in the period 1870–1998. Each of the forest stands was a separate well or poorly drained site. The WR biomass varied from 1.4 to 177.6 mg ha^{-1} decreasing, as a rule, with age. Most decayed wood remains were characterized by a higher humidity, lower density, and higher rate of respiration compared to the less decayed remains. Depending on the wood remains, estimates of the annual emissions of carbon varied between 0.11 and 1.92 mgC $ha^{-1}year^{-1}$. This study has shown that in modeling the carbon exchange in the atmosphere-plant-soil system, it is necessary to select a new component that describes the wood remains (Hoelzemann et al. 2004).

In this connection, the equation of biomass dynamics should contain, at least, three terms that parameterize photosynthesis, respiration, and dying-off. This will make it possible to more adequately reflect the role of the forest ecosystem in the change of CO_2 exchange between the land cover and the atmosphere, by specifying the dependences of the indicated components on climate parameters. In particular, it is known that many boreal ecosystems are sources of carbon in winter and sinks in summer. This process is controlled by air temperature and day duration. In summer, in boreal forests during the season of plants' vegetation, the amount of carbon absorbed from the atmosphere can reach 80–310 gC m^{-2}, and in winter, carbon emissions can reach 60–90 gC m^{-2}. The inter-annual variability of the seasonal exchange of carbon, from estimates of Suni et al. (2003), can constitute from 30 gC m^{-2} in autumn and spring to 80 gC m^{-2} in summer.

Yue et al. (2013) studied the forest age dynamics in connection with the role of fire in the regional forest carbon balance. The vegetation model ORCHIDEE was used to simulate the CO_2 emissions from boreal forest fire and the subsequent recovery after a stand-replacing fire. Simulation results demonstrate that a global vegetation model such as ORCHIDEE is able to capture the essential ecosystem processes in fire-disturbed boreal forests and produces satisfactory results in terms of both carbon fluxes and carbon-stock evolution after fire. This makes the model suitable for regional simulations in boreal regions where fire regimes play a key role in the ecosystem carbon balance. The ORCHIDEE dynamic global vegetation model represents the land surface features of the IPSL coupled atmosphere-ocean-vegetation model. ORCHIDEE has been developed using first order ecophysiological principles to represent both natural ecosystem and managed land carbon, water, and energy dynamics across multiple spatial (site to globe) and temporal (sub-daily to centennial) scales.

One principal question is how does humidity affect forest fire? Humidity has an important impact on forest fires in two major ways. First, long periods of low humidity tend to dry out all forest fuels to a dangerous degree. Even a few days of low humidity can lead to the combustion of steams, branches, foliage, and roots. Post-fire effect can have short-term and long-term consequences for carbon balance on the landscape that was affected by wildfire.

Goto and Suzuki (2013) studied emissions from forest fires that directly affect the global and regional carbon cycles by increasing atmospheric carbon as well as affecting carbon sequestration by forests. It was estimated the release of total carbon, carbon-based trace gases (CO_2, CO, CH_4) and non-methane hydrocarbons (NMHC) emitted from forest fires in Japan during a 30-year period from 1979 through 2008. The area burnt varied widely from year to year but has gradually diminished since the 1980s. The mean annual area burnt during this period was 1878 ha. The mean annual estimate of direct carbon emissions from forest fires in Japan was 15.8 Gg C $year^{-1}$ and ranged between 2.7 and 60.4 Gg C $year^{-1}$. The mean annual trace gas emissions were 49.4 Gg CO_2 $year^{-1}$, 3.4 Gg CO $year^{-1}$, 0.15 Gg CH_4 $year^{-1}$ and 0.18 Gg NMHC $year^{-1}$. Although the carbon emissions varied widely from year to year based on the area burnt, they decreased dramatically from the 1980s onward. The interannual variations in trace gases parallel the total carbon emissions. The direct emissions from forest fires in Japan were substantially lower compared with the mean annual net primary production of Japanese forests or the carbon release in other countries and regions. However, the average annual carbon released per unit area burnt was comparable to that estimated in other regions and rose gradually with the increasing age of plantations.

8.4.5 Wildfires and Biocomplexity

A wildfire is an uncontrolled fire in an area of combustible vegetation. This phenomenon is inseparably linked with biocomplexity of plant community. Wildfire changes a structure of plant community causing the disbalance of ecological processes on burned area. One principal indicator of natural ecosystem is the level of carbon exchange between its elements. Other ecological processes within the ecosystem correlate with the carbon cycle. In other words, carbon cycle characterizes a level of interactions between all living and plant organisms of ecosystem. Wildfire impacts on the carbon cycle. Carbon fluxes a_{ij} between ith and jth ecosystem elements are functions of time t, temperature T, soil moisture W, and solar energy E. Approximate scheme of carbon fluxes in the forest ecosystem can be presented by the chart of Fig. 8.16. Certainly, parameter a_{ij} can be evaluated by the relation of C:N or by other alternative model. Somehow or other, biocomplexity indicator is showing the forest fire consequences. Balance equations for the scheme of Fig. 8.16 have the following form:

$$
\begin{aligned}
x_1/dt &= a_{61} - a_{14} \\
dx_2/dt &= a_{62} - a_{24} \\
dx_3/dt &= a_{63} - a_{34} \\
dx_4/dt &= a_{14} + a_{34} + a_{24} - a_{45} - a_{47} \\
dx_5/dt &= a_{45} - a_{56} - a_{57} \\
dx_6/dt &= a_{56} + a_{46} - a_{62} - a_{61} - a_{63} \\
dx_7/dt &= a_{47} + a_{57}
\end{aligned}
$$

Fluxes a_{6i} ($i = 1, 2, 3$) are parameterized by the photosynthesis equations taking into account a transpiration. Velocities of the litter and soil organic matter

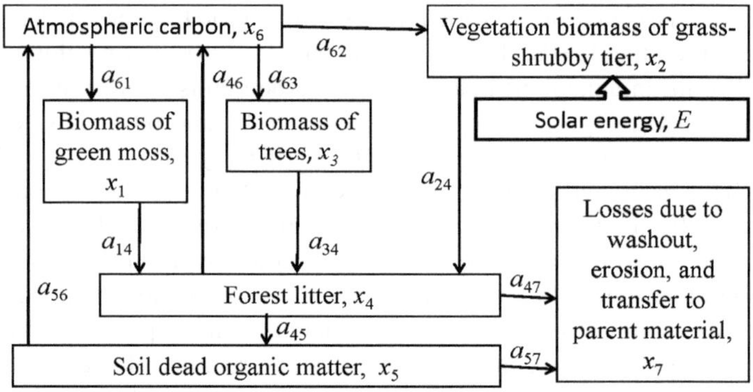

Fig. 8.16 Carbon cycle in the forest ecosystem

decomposition are described by linear or exponential models. Velocities of the biomass dying are modeled by linear models. In general, the choice of model forms to parameterize the carbon cycle depends on the existing database about forest ecosystem parameters and characteristics. Below, the specific models are considered.

Photosynthesis is described by the equations:

$$A_{6i} = \beta_i \varphi_i(T)\Psi_i(E) f_i(x_5) h_i(w),\ (i = 1, 2)$$
$$A_{63} = \beta_3 \varphi_3(T)\Psi_3(E) f_3(x_5) h_3(w)\xi(x_3),$$

where β_i (i = 1, 2, 3) are the coefficients that reflect a dependence of photosynthesis on the vegetation type; and $\varphi_i(T)$ is the dependence of photosynthesis on the temperature T under the forest canopy. Functions Ψ_i, f_i and h_i reflect the photosynthesis dependence on illumination, mineral salts and moisture. Function ξ characterizes a dependence of photosynthesis on the biomass of trees. The other functions have the following view:

$$\varphi_i(T) = 1 + (\delta_i/100)(T - T_{opt})\ (i = 1, 2), \varphi_3(T) = 1 + (\delta_3/100)(T - T_{A,opt}), \Psi_i(E) = \exp(-\alpha_i E)\ (i = 1, 2, 3),$$
$$f_i(x_5) = \exp(-a_i/x_5)\ (i = 2, 3), f_1(x_5) = 1, h_i(w) = \exp\{-b_i[w - (w_{A,opt} - w_{opt})/\sigma_i]\}\ (i = 1, 2, 3), \xi(x_3) = (x_3/x_{3A})^{\gamma},$$

where δ_i, a_i, α_i, b_i, σ_i (i = 1, 2, 3) and γ are constants; T_{opt} and $T_{A,opt}$ are optimal temperatures under the forest canopy and in the free atmosphere for photosynthesis, respectively; w_{opt} and $w_{A,opt}$ are optimal humidity under the forest canopy and in free atmosphere for photosynthesis, respectively.

Carbon fluxes from vegetation tiers of the forest ecosystem are described by the following models:

$$a_{i4} = k_i x_i (i = 1, 2);\ a_{34} = k_3 x_3 / [1 + d(T - T_A)],$$

where k_i (i = 1, 2, 3) are the proportionality coefficients.

The litter and dead organic matter decomposition processes are parameterized by the multiplicative models:

$$a_{47} = k_4x_4(1 - \nu\exp\{-\varepsilon w\}); a_{45} = (1 - \lambda)[x_4 - a_{47}]1 + (\delta_4/100)(T - T_{opt});$$
$$a_{46} = \lambda[x_4 - a_{47}]1 + (\delta_4/100)(T - T_{opt}); a_{57} = (1 - \mu)k_5x_5(1 - \nu\exp\{-\varepsilon w\});$$
$$a_{56} = \mu[x_5 - a_{57}]1 + (\delta_5/100)(T - T_{opt})$$

where k_4, ν, ε, λ, μ, δ_4, k_5 and δ_5 are constants defined depending on the ecosystem type.

Correlation between the temperatures of free atmosphere and under forest canopy is parameterized by the following model:

$$T(x_3, T_A) = T_A \begin{cases} g(x_3)\ \text{при}\ 1 \le g(x_3) \le T_{max}/T_A, \\ T_{max}/T_A\ \text{при}\ g(x_3) \ge T_{max}/T_A, \\ 1\ \text{при}\ g(x_3) \le 1. \end{cases}$$

Here, function $g(x_3)$ is selected under the condition that air temperature under the forest canopy can not be less than free atmosphere temperature:

$$g(x_3) = 1 - c_1x_3^2 + c_2x_3,$$

where coefficients c_1 and c_2 are defined empirically.

Moisture regime of each pixel $\Omega_{ij} \subseteq \Omega$ is described by regional water balance model (see Figs. 1.13 and 1.14) the initial data for which are delivered by microwave monitoring system (Krapivin and Shutko 2012).

In this case, biocomplexity inducator $\xi_\Omega(t)$ (see Eq. 6.30) can be calculated by the expression:

$$\xi_\Omega(t) = (1/\sigma) \int_{(\varphi,\lambda)\in\Omega} \xi(\varphi, \lambda, t)d\varphi d\lambda$$

where $\xi = \sum_{i=1}^{m} \sum_{j>i}^{m} d_j a_{ij}$, weighed coefficient $d_j \in [0,1]$ corresponds to the jth element signification.

Figures 8.17 and 8.18 give two examples of the biocomplexity evaluation for Bulgarian national park Rusenski Lom during the International Hydrophysical Experiment (Krapivin and Shutko 2012; Rashid and Seizov 2012). This park is characterized by large diversity of vegetation communities including oak, lime-tree, hornbeam, maple, ash-tree, pine-tree, etc. This biological diversity is displayed in the wide range of surface variability of the carbon fluxes. Biometric characteristics of vegetation covers were submitted by Bulgarian experts and assessed by numerous

Fig. 8.17 An example of the biocomplexity indicator calculation above the territory of national park Risenski Lom (Bulgaria) realized at July 31, 2007. Park is located in the north-east Bulgaria within 20 km from Russe town. It supposed that $d_i = 1$. *Thick dark arrows* show a direction of flying laboratory track. Spatial resolution is 15 km

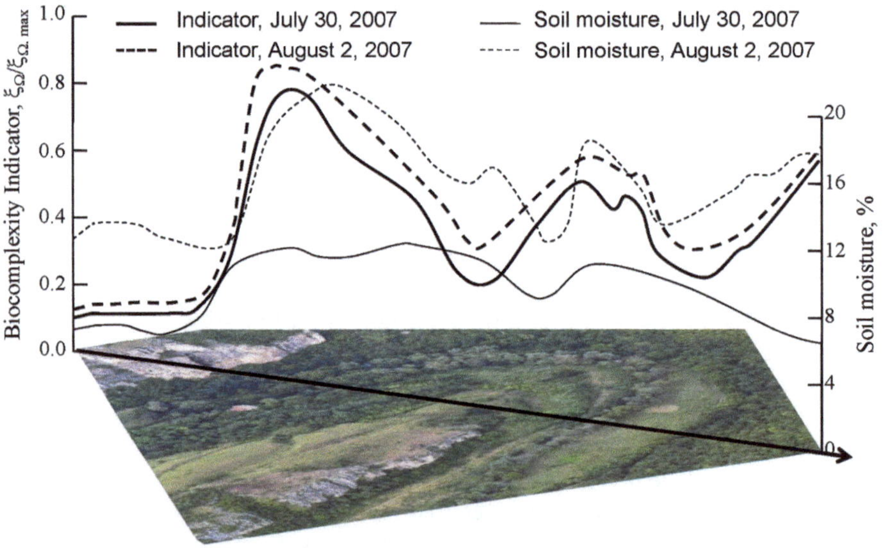

Fig. 8.18 An example of the biocomplexity indicator calculation during International Hydrophysical Experiment in Bulgaria. *Thick dark narrow* shows the trace of flying laboratory above national park Rusenski Lom

publications (Conant et al. 2011; Chen et al. 1999, 2000; Cocknell et al. 2006). The results represented in Fig. 8.18 show that biocomplexity is a nonlinear function of soil moisture.

8.5 The Role of Forests in CO_2 Cycle

One of the principal aspects of the anthropogenic impact on the environment is the evaluation of the consequences of CO_2 emissions into the atmosphere. Published results estimating the greenhouse effect and excess CO_2 distribution in the biosphere widely vary, sometimes they are contradictory, or too blandly stated. This is a natural consequence of all kinds of simplifications adopted in modeling the global CO_2 cycle. The GIMS makes it possible to create an effective monitoring system allowing the estimation of the spatial distribution of the carbon sinks and sources in real time.

First of all, some problems should be solved to assess the role of the anthropogenic use of the Earth's surface. In particular, among these problems there is the problem of the formalized description of the processes of change of the structure of the Earth land covers, such as afforestation, forest reconstruction, deforestation and the associated carbon supplies. Understanding the meteorological processes as functions of the greenhouse gases is one of the key problems of humankind in the first decade of the third Millennium. Only an adequate knowledge of the meteorological phenomena on various spatial-temporal scales under conditions of varying supplies of CO_2 and other greenhouse gases will enable to make correct and constructive decisions in the field of global environmental protection.

The dynamics of the surface ecosystems depend on interactions between the biogeochemical cycles, which during the last decade of the 20th century and first decade of the 21st century have suffered significant anthropogenic modification, especially to the cycles of carbon, nitrogen, and water. The surface ecosystems, in which carbon remains in the living biomass, decomposing organic matter, and the soil, play an important role in the global CO_2 cycle. Carbon exchanges between these reservoirs and the atmosphere take place through photosynthesis, respiration, decomposition, and burning. Human interference into this process takes place through changing the structure of the vegetation covers, pollution of the water basin surfaces and of the soil areas, as well as through direct emissions of CO_2 into the atmosphere.

The role of various ecosystems in the formation of carbon supplies to the biospheric reservoirs determines the rate and direction of changes to the regional meteorological situations and to the global climate. The accuracy of assessment of the level of these changes depends on the reliability of the data on the inventory of the surface ecosystems.

Existing environmental data show that knowledge of the rates and trends of carbon accumulation in the surface ecosystems are rather uncertain. However, it is clear that the surface ecosystems are important assimilators of excess CO_2. Understanding the details of such assimilation is only possible through modeling the process of plant growth; that is, considering the effect of the nutrient elements of the soil and other biophysical factors on plant photosynthesis. Therefore, the forest ecosystems and associated processes of natural afforestation, forest reconstruction, and deforestation should be studied in detail. In a forested area, the volume of the reservoir of CO_2 from the atmosphere is a function of the density of the forest canopy, and in a given period of time a change of this volume is determined by the level and character of the dynamic processes of transition of one type of forest into another. The causes of this transition can be natural, anthropogenic, or mixed. Biocenology tries to create a universal theory of such transitions, but so far there is only a qualitative description of the observed transitions.

The land biota and forests particularly is one of the sinks for atmospheric CO_2. Unfortunately, the available data and knowledge of the processes of plants' respiration suggest only rough integrated estimates of CO_2 flux in the vegetation cover. Actually, the role of plant in assimilation of atmospheric CO_2 changes sharply during 24 h and is a complex function of such factors of the environment as temperature, illumination, and air humidity. Nevertheless, attempts to parameterize the functions of vegetation made in the GIMS makes it possible to assess the role of all types of the soil-plant formations in CO_2 assimilation (mentioned in Tables 1.2 and 1.3).

The land ecosystem dynamics depends on the interaction of biogeochemical cycles that were anthropogenically modified during last decades (e.g., water, carbon, and nitrogen). Land ecosystems, the carbon of which is stored in living biomass, soil and decomposed organic matter, play an important role in global CO_2 cycle. Carbon is exchanged between these reservoirs and the atmosphere at the expense of photosynthesis, respiration, decomposition and burning. Human influence in this process is happened through the change of land cover structure, pollution of surface water reservoirs and soils, as well as through direct CO_2 emission to the atmosphere.

The relation of the role of different ecosystems in the formation of carbon reserves in the biosphere defines the speed and direction of changes in regional meteorological situations and global climate. The precision of the assessment of these changes depends on the reliability of land ecosystem inventories. This is justified by data of Table 8.6 where considerable dispersion of the evaluations of carbon storage in vegetation of various types gives the possibility to speak about the importance of more precise classification of the land ecosystems.

Table 8.7 illustrates the influence of afforestation/deforestation processes on the carbon storage under anthropogenic scenario FAO-2000 where the forest is defined as the land area equaled to not less than 0.5 ha and covered by trees with height of more than 5 m the crone of which occupies more than 10 % of area (FAO 2001). Deforestation is the removal of a forest or stand of trees where the land is thereafter converted to a non-forest use and crone area is less than 10 % of the total area. Afforestation is the establishment of a forest or stand of trees in an area where there was no forest. It is noted that natural expansion of forest is also considered by FAO-2000 scenario as afforestation. At last, direct artificial planting of the trees on the area where forest was earlier moved is considered as afforestation.

Table 8.6 Global carbon dioxide balance and its biospheric exchange

Reservoir	CO_2 reserve (10^9 tonnes C)	Element or biospheric process	Exchange speed (10^9 tonnes C/yr)
Atmosphere	697	Forests	10
Hydrosphere	35,420	Cultivated soils	4
Geosphere Inorganic carbonates Inorganic non-carbonates Coal, gas, oil	 18.3 млн 6.8 млн 7,400	Grass ecosystems Deserts Photosynthesis northern 45°N People transpiration	1 0.2 15 0.14
Litter, detritus In the ocean On the land	 3,220 710	Transpiration of the domestic animals Transpiration of the wild animals Soil transpiration	0.5 0.4 8.4
Biosphere In the ocean On the land	 10 124		

Table 8.7 An assessment of change of average annual carbon storage calculated in framework of afforestation/deforestation scenarios

Region	RF	AF	TR		FR	
			A	B	A	B
Boreal	35	0.4–1.2	0.5	0.1	−18	−185
Moderate	60	1.5–4.5	2.1	1.9	−90	−501
Tropical	120	4–8	13.7	2.6	−1644	−1,352

Notation *A* is the deforestation, *B* is the afforestation, *RF* is the change of average carbon storage under deforestation (tC/ha), *AF* is the average velocity of CO_2 absorption under the afforestation (tC/ha/yr), *TR* is the change of area as result of transition between the forest absence and forest (10^6 ha/yr), *FR* is the forecast of change in the carbon storage during 2008–2012 in framework og the FAO-2000 scenario (10^6 tC/yr)

Figures 8.10 and 8.11 demonstrate the role of the forest vegetation in the CO_2 dynamics. In addition to these results, it must be noted that experiments with global models make it possible to trace the dependence of the atmospheric gas composition on the structure of the planetary forest covers. From the available estimates, the total area of the forests for $t_0 = 1970$ can be estimated at $\sigma_{L0} = \sigma_L(t_0) \approx 40.3\text{–}41.84 \times 10^6$ km^2 (Watson et al. 2000), 1 % constituting national parks and forest reserves. With the formulated scenario, assume $t_L = 2050$, $\sigma_{X0} = \sigma_X(t_0) = 19.5 \times 10^6$ km^2. As seen from Fig. 8.19, the increasing rate of deforestation raises considerably the concentration of CO_2 in the atmosphere. Even with a 10 % reduction of forest areas by 2050 compared to 1970, that is, $\sigma_{L1}/\sigma_{L0} = 0.9$, the atmospheric CO_2 can increase by 44 %

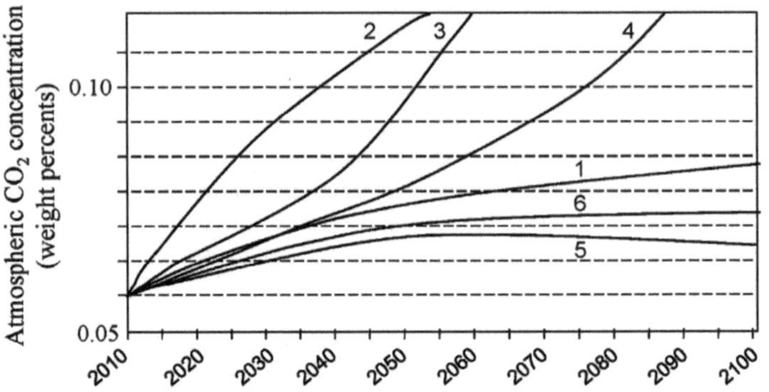

Fig. 8.19 The dynamics of CO_2 concentration for different scenarios of changing forest areas within the framework of the scenario after Demirchian et al. (2002): *1* rates of changes of the forest areas remain as of 1970; *2* by 2050 the forests will be liquidated at all; *3* by 2050 the area of the forests is reduced by 50 %; *4* by 10 %; *5* by 2050 the area of the forests increases by 50 %; *6* by 10 %

by the end of the 21st century. On the contrary, a 10 % increase of the forested areas decreases the concentration of atmospheric CO_2 by 15 %. With a 50 % increase of the forested areas by 2050, a decrease of atmospheric CO_2 by 2100 will constitute 60 % relative to its possible value, with the scale of impacts on the forest ecosystems, observed in the end of the 20th century, preserved. Hence, variations of the forested areas in the biosphere even within ±10 % can substantially change the dynamics of numerous components of the global ecosystem.

Table 8.8 exemplifies the calculation of the CO_2 sinks into the vegetation cover on the territory of Russia. Such calculations with the use of the global model of the N-S system demonstrate the dynamics of the CO_2 fluxes mosaic in the system atmosphere-plant-soil. The knowledge of this mosaic makes it possible to assess the role of concrete types of the soil-plant formations in the regional balance of carbon, and on this basis, to calculate the global fluxes of carbon dioxide on the atmosphere-land border. Similar calculations are also possible for the atmosphere-ocean system.

Table 8.9 demonstrates the consequences of changing the global structure of the soil-plant formations for the dynamics of CO_2 assimilation by vegetation. As seen, the anthropogenic change of the vegetation cover substantially changes the balance components in the global carbon cycle. Clearly, such experiments require a thorough analysis of data on the possibility of transformation of vegetation covers with account of climatic zones and biocenological consistency. Nevertheless, such hypothetical experiments are useful for general assessments of possible ranges of anthropogenic impacts on the global carbon cycle. For instance, natural and anthropogenic cataclysms connected with forest fires introduce annually considerable changes into this cycle, because they change numerous fluxes and supplies of carbon over large territories. Tables 8.10, 8.11 and 8.12 give estimates of deviations in the content of carbon in the basic biospheric reservoirs with forests fires in

Table 8.8 The dynamics of CO_2 assimilation by plants on the territory of Russia

Vegetation formation (Table 1.2)	Rate of CO_2 assimilation, 10^6 tC/year				
	Years				
	1990	2000	2050	2100	2150
A	2.6	2.8	6.7	7.1	6.9
C	3.7	4.6	10.9	12.0	12.1
M	4.0	5.1	12.4	14.5	13.8
L	3.2	3.9	9.2	10.3	10.4
F	11.2	14.8	43.6	47.2	44.2
D	31.6	39.9	110.6	121.9	109.3
G	23.3	29.2	72.2	73.4	70.5
R	5.2	6.2	13.1	13.8	10.7
W	4.7	5.1	8.2	8.8	7.9
V	0.7	0.7	0.9	1.1	0.8
@	2.4	2.6	3.7	3.9	2.9
S	0.6	0.7	1.2	1.4	1.0
Q	1.5	1.6	2.2	2.3	1.8
Total	94.7	206.1	294.9	317.7	292.3

Emission of carbon on this territory in 1990 is assumed to be 1.6 Gt C/year with an annual change after Keeling's scenario (Kondratyev et al. 2003c)

different zones. Large-scale impacts on land biota are damped during 60–100 years. The biosphere turns out to be more resistant to impacts on the forests of southern latitudes and more sensitive to violations of forest areas in temperate latitudes. Hence, the northern hemisphere forests up to 42°N play an important role in stabilizing the carbon cycle in the biosphere.

The scenario of liquidation of forests, as seen from the studies of numerous authors, evokes great interest in studies of the global carbon cycle and the associated climate change. A diversity of possible real situations of the transformation of land covers is so large that it is impossible to assess all the consequences. Note only that, for instance, a liquidation of all northern taiga and mid-taiga forests (types *F*, *D*) in 50 years will lead to a 53 % increase of atmospheric CO_2 with subsequent negative consequences for the flux H_6^C. Similar consequences follow after liquidation of all wet evergreen and deciduous tropical forests (types *Z*, *Y*), but in this case the indicated increase of atmospheric CO_2 will be reached 20 years later.

The GMNSS reflected the interactions of natural and anthropogenic factors taking into consideration that their spatial and temporal heterogeneity and dynamics allow the assessment of expected risks from possible greenhouse effect. An example of such assessments is given in Fig. 8.20. It is evident that GMNNS delivers more optimistic prognosis than climatic model of Hadley Center.

The land cover structure changes not only due to human activity. In some regions of the globe, hurricanes introduce considerable changes in the carbon balance of the forest ecosystems. Thus, on the USA territory every 3 years two

Table 8.9 The dynamics of the ratio of integral rates of (H_6^C) CO_2 assimilation by vegetation covers from the atmosphere with natural distribution of the soil-plant formations (Fig. 1.9) and with its transformation according to the scenario in the second column

Scenario		F_6 (changed Fig. 1.9)/F_6 (Fig. 1.9)			
		Years			
Figure 1.9	Changed Fig. 1.9	2020	2030	2050	2100
A	L	2.54	2.04	1.97	2.15
C	L	0.96	0.94	0.95	1.03
M	L	1.42	1.15	1.01	1.04
F	D	1.63	1.44	1.11	1.12
G	D	1.98	1.66	1.45	1.33
R	D	3.91	3.69	2.86	1.95
P	Z	3.07	2.55	2.44	1.77
U	Z	21.49	19.46	17.95	18.34
W	Z	22.15	18.33	16.32	14.77
E	Z	99.23	77.12	68.54	70.09
H	Z	187.65	153.23	138.39	140.67
Q	Z	790.04	766.42	751.26	767.33
Y	Z	1.42	1.33	1.23	1.25
N	Z	67.56	60.19	56.59	57.34
J	Z	5.98	4.98	4.67	5.08
T	Z	25.99	24.76	23.57	22.65
I	Z	17.11	15.78	14.92	13.99
#	Z	0.95	1.21	0.97	1.09

Table 8.10 Model evaluations of the variations in the carbon reserve under conditions when all coniferous forests in Northern regions (up to 42°N) are burned

Years after influence	Deviation in the carbon reserve (Gt)				Years after influence	Deviation in the carbon reserve (Gt)			
	ΔC_A	ΔC_S	ΔC_U	ΔC_{AL}		ΔC_A	ΔC_S	ΔC_U	ΔC_{AL}
0	150.2	−5.7	15.7	0.1	60	25.1	−30.4	8.5	13.1
10	114.6	−34.1	30.8	3.1	70	16.7	−25.7	6.7	13.5
20	84.1	−44.2	22.5	7.4	80	11.2	−21.5	5.1	14.0
30	64.4	−43.9	19.4	8.5	90	5.9	−17.9	4.3	14.3
40	48.2	−40.7	14.6	10.4	100	2.1	−14.8	2.8	13.9
50	35.2	−33.8	12.1	11.7	200	-9.1	−3.4	−0.9	13.5

hurricanes happen, which accelerate the transition of living biomass of trees into dead organic matter. If on the USA territory, 20 TgC are removed from the forests every year, then 10–15 % of this is the result of one hurricane (McNulty 2002). Hence, the hurricanes accelerate the return of carbon to the atmosphere, and their global inventory is needed to more accurately estimate numerous components of the global carbon cycle.

Table 8.11 Model evaluations of the variations in the carbon reserve when forests of northern regions (up to 42°N) are burned

Years after influence	Deviation in the carbon reserve (Gt)				Years after influence	Deviation in the carbon reserve (Gt)			
	ΔC_A	ΔC_S	ΔC_U	ΔC_{AL}		ΔC_A	ΔC_S	ΔC_U	ΔC_{AL}
0	238.1	−7.9	24.9	0.1	60	44.2	−49.1	14.3	20.7
10	174.2	−31.6	47.9	4.9	70	33.1	−41.6	10.3	21.9
20	138.9	−67.6	39.2	10.0	80	20.7	−35.0	7.5	22.8
30	107.9	−90.3	32.2	13.8	90	12.9	−30.1	5.5	23.3
40	82.0	−64.3	24.1	16.8	100	7.3	−24.7	3.7	23.6
50	60.9	−56.9	18.4	19.1	200	−12.7	−5.9	−1.7	21.7

Table 8.12 Model evaluations of the variations in the carbon reserve when forests of tropical regions are burned

Years after influence	Deviation in the carbon reserve (Gt)				Years after influence	Deviation in the carbon reserve (Gt)			
	ΔC_A	ΔC_S	ΔC_U	ΔC_{AL}		ΔC_A	ΔC_S	ΔC_U	ΔC_{AL}
0	406.2	−20.0	42.2	0.2	60	2.9	−12.4	3.0	22.8
10	264.4	−93.7	74.1	8.0	70	−5.8	−7.5	0.5	22.8
20	162.2	−84.8	48.0	14.9	80	−11.6	−4.2	−0.9	22.6
30	90.6	−38.6	27.9	19.1	90	−13.2	−2.6	−1.7	22.5
40	45.4	−36.5	15.0	21.3	100	−14.5	−1.9	−2.1	21.8
50	18.3	−21.6	7.5	22.4	200	−13.2	−2.3	−1.9	17.7

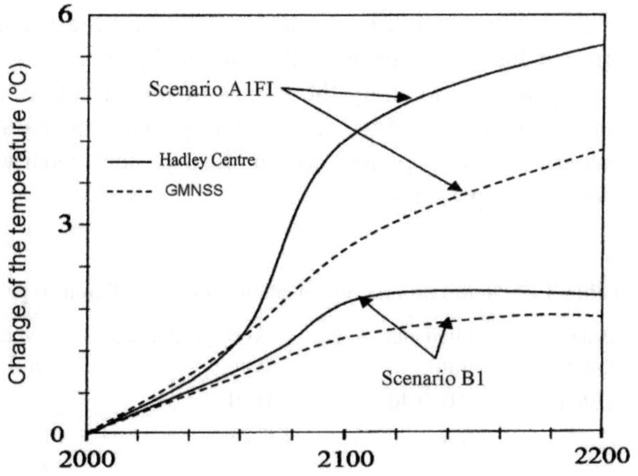

Fig. 8.20 Prognoses of the changes in average global temperature evaluated by means of the Hadley Centre climate model and GMNSS in framework of two scenarios of energy use

8.6 Vegetation Media as the Object Under Study of Attenuation of Electromagnetic Waves

Many investigators (Basharinov et al. 1979; Hansen et al. 2002; Zhan et al. 2002; Burkov and Krapivin 2009; Chukhlantsev 2006) support a constructive technology for microwave monitoring of vegetation covers. However, greater accuracy of this technology is required because various vegetation covers, especially vegetation canopies, are very dynamic and complicated for experimental study or model parameterization. In this respect, there are two aspects. One is connected with the creation of a highly productive technology to estimate the land cover characteristics. The other is connected with the climate change problem through the greenhouse effect (Kondratyev et al. 2003c). In these cases the GIMS-technology recommends a balanced scheme for the solution of problems of land cover monitoring. As follows from Armand et al. (1997), a combination of experimental and theoretical studies of attenuation of microwave radiation by vegetation cover is required. It is actually possible to synthesize the experimental dependence between attenuation and a restricted set of vegetation parameters. An estimation of attenuation of microwave radiation by vegetation cover in real time is possible only with the application of microwave models and interpolation algorithms.

From the point of view of microwave remote sensing, the knowledge of the water content of vegetation is an important element which has to be used for the synthesis of a model of attenuation of microwave radiation by the vegetation. This knowledge can be acquired by making use of the fact that water absorption features dominate the spectral reflectance of vegetation in the near-infrared spectrum (Sims and Gamon 2003). Using such indexes as the normalized difference vegetation index (NDVI), plant water index (PWI), the leaf area index (LAI), the simple ratio vegetation index (SRVI), and the canopy structure index (CSI) it is possible to determine the canopy structure and photosynthetic tissue morphologies. The correlation between the leaf water content and the leaf-level reflectance in the near-infrared has been successfully studied by many authors that linked the leaf and canopy models to study the effects of leaf structure, dry matter content, LAI, and canopy geometry. As a result, it is possible to parameterize the forest structure including the canopy and crown closure, stem density, tree height, crown size, and other forest parameters (Tables 8.13 and 8.14).

Table 8.13 Some estimations of parameters for different types of branches (Karam et al. 1992)

Branch group	Diameter range (cm)	Average diameter (cm)	Average length (cm)	Number density (m^{-3})
1/stems	0.0–0.40	0.10	18	250.00
2	0.5–1.90	1.28	14	11.40
3	2.0–2.90	2.60	32	0.43
4	3.0–6.90	5.00	58	0.33
5/trunk	7.0–17.1	9.00	76	0.14

Table 8.14 Dielectric constant values (Karam et al. 1992)

SPF element	Frequency band	
	L	X
Leaves	19.58 − j5.54	14.9 − j4.9
Branch/stems	27.30 − j8.4	20.0 − j9.7
Soil	5.00 − j0.7	5.0 − j0.4

Quantitative information about SPF properties may be obtained by means of different remote sensing techniques based on passive microwave data. This approach is particularly effective under an estimate of the hydraulic parameters of the soil-plant formations. The knowledge of these parameters allows the water balance of a territory occupied by vegetation of a given type to be parameterized and plant to be modeled. Under this approach a comparison of remote data registered at different EMW ranges may provide a high precision of estimates of the water content in the various layers of the SPF (canopy, trunks, stalks, and soil). Moreover, the combination of data analysis with remote sensing allows one to estimate the parameters of the ground vegetation cover and its role in the fluxes of sensible and latent heat from the surface to the atmosphere.

A problem of wavelength choice and the combination of its ranges with classes of solved tasks is now within the sight of many investigators. The theoretical foundation of this choice lies in the area of thermal radiation transmission theory. Specifically, for the monochromatic case the transmission equation has:

$$dI(z)/dz = J(z) - \alpha I(z) \tag{8.10}$$

where α is the absorption coefficient, $J(z)$ is the emission source distributed by the coordinate z, and $I(z)$ is emission intensity at point z. Under these conditions, the following equation is valid for an unscattering atmosphere: $J(z) = \varepsilon_1(z)B(z)$, where ε_1 is atmospheric emissivity,

$$B(z) = \left(2hf^3/c^2\right)[\exp\{hf/(kT)\} - 1]^{-1},$$

h, c, and k are fundamental constants, $f = vc$ is frequency. It is known that $hf < kT$ in microwave range and Rayleigh-Jeans approximation $B(z) = 2kT/\lambda^2$ is valid. Here $k = 1.38 \times 10^{-23}$ JK^{-1} is the Boltzmann constant, λ is the wavelength.

Equation (8.10) can be solved by

$$T_j = \kappa T_e \tau(0, \infty) + \int_0^\infty T(u)Q(u)du, \tag{8.11}$$

where

$$\tau(a,b) = \exp[-\int_a^b \alpha(z)dz]$$

The weight function $Q(u) = \partial\tau(u,\infty)/\partial u$ describes the influence of various atmospheric layers on the receiving signal. Formula (8.11) converts expression $T_j = \kappa T_e$ into a whole. Examples of specific values are given in Table 8.15.

Thus, a set of vegetation-related parameters should be generated using specific time and space resolutions as the basis for the calculation of EMW attenuation by the vegetation layer. Available remote monitoring data should be used as input information for algorithms and models to synthesize the spatial-temporal distribution of the attenuation effects. In this context, the most important parameter is the dielectric constant. Some estimations of this characteristic are given in Table 8.14. For example, for the stem dielectric constant a representative value at the L band is found to be 27.3 − j8.4. The leaf dielectric value varies from 8.77 − j2.88 to 19.58 − j5.54. An average value of the leaf dielectric constant equals to 19.58 − j5.54. The values of the dielectric constant at X band equals to 14.9 − j4.9 and 20.0 − j9.7 for leaves and branches, respectively.

Detail and complete study of the EMW attenuation by the vegetation layer was realized in the framework of the ISTC Project #2059 "Screening Effect of Vegetation in Active (SAR) and Passive (Radiometric) Remote Sensing of the Earth's Covers and in Radio-Communication in UHF-, VHF-, and Microwave (X-, C-, S-, L-, and P-) bands" (Chukhlantsev 2006). The goals of this Project were the following:

- acquiring systematized data about electromagnetic waves attenuation by different types of vegetation covers, mainly in the meter wavelength band, but also in decimeter and centimeter wavelength bands (UHF, VHF, P-, L-, S-, C-, and X-bands) using theoretical and experimental investigations.
- developing advanced, more sophisticated, and generalized models, compared with the existing SAR and microwave radiometric models of different types of

Table 8.15 Brightness temperatures typical of some land covers

Land cover	λ = 0.8 cm		λ = 1.35 cm	
	$T_{j,min}$	$T_{j,max}$	$T_{j,min}$	$T_{j,max}$
Desert	251	261	245	270
Cuntivated steppe	255	268	257	277
Steppe covered by snow	246	261	241	260
Ice covered by snow	247	263	231	253
Ice	239	247	235	254
Sea surface	136	171	110	168

vegetation with due regard for the electromagnetic properties of vegetative elements to satisfy the needs in reliable assessment of vegetation impact (screening effect of vegetation) on soil condition sensing;
- developing methodologies of vegetation impact assessment in active (SAR) and passive (MR) remote sensing of the Earth's covers to reduce/eliminate/remove the impact of vegetation on the land surface observation;
- developing methodologies for assessing the impact of vegetation on the radiocommunication.

This project solved the following main tasks.

- Building up of the measuring complex to conduct investigations of the attenuation of electromagnetic waves in vegetation covers and in plant elements (leaves, sticks, etc.).
- Systematization and generalization of the existing theoretical and experimental data about attenuation of electromagnetic waves in vegetation.
- Acquisition of experimental characteristics of electromagnetic waves attenuation by various types of vegetative covers in the laboratory and field conditions in meter, decimeter and centimeter wavelength range.
- Conducting model investigations of the attenuation of electromagnetic waves in vegetation covers in meter, decimeter, and centimeter wavelength range.
- Elaboration of methodologies for the screening effect of vegetation assessment by conducting single wavelength measurements and by combining active and passive methods.
- Elaboration of methodologies for the screening effect assessment by conducting multi-wavelength measurements and by combining active and passive remote sensing systems.
- Elaboration of methodologies for the screening effect of vegetation assessment in radio communication.

Technical approach and methodology under the solution of these tasks were founded on the following ideas. Theoretical part included utilization of the methods of the radio waves propagation for the two basic types of vegetation models:

(1) vegetation in a form of inhomogeneous randomized media (stochastically inhomogeneous or irregular), in particular to discuss the vegetation model as a continuous medium with an effective value of dielectric permittivity; and
(2) vegetation as a community of discrete, separately located scatters; the determination of the attenuation cross-section of these elements.

In a continuous model, the vegetative cover was described as a homogeneous medium with effective value of dielectric permittivity. The effective values of permittivity were calculated in a quasi-static approach (the size of irregularities, disks, needles, strips, is much less compared with a wavelength) by deriving different relationship based on the theory of dielectric mixtures. The corrections for effective value of dielectric permittivity was made with due regard for the oscillating (wavy) nature of the process.

In a discrete model, the aggregated electromagnetic parameters of vegetation chosen for evaluation of vegetation behavior were the attenuation, cross-section and scattering cross-section of the vegetation elements. The vegetation elements were modeled by disks, strips, and cylinders. A spectral dependence of the attenuation and scattering cross-sections are determined. Total attenuation in the vegetative media was studied by using the radiation transfer theory.

An experimental part of the ISTC-2059 Project included laboratory and field measurements. A special laboratory complex was built up to conduct investigation of the attenuation by plant samples and plant elements in a wide range of frequencies. This complex includes two waveguide transmission systems operating in the frequency range of 0.2–8.0 GHz (Figs. 8.21 and 8.22). The radio attenuation

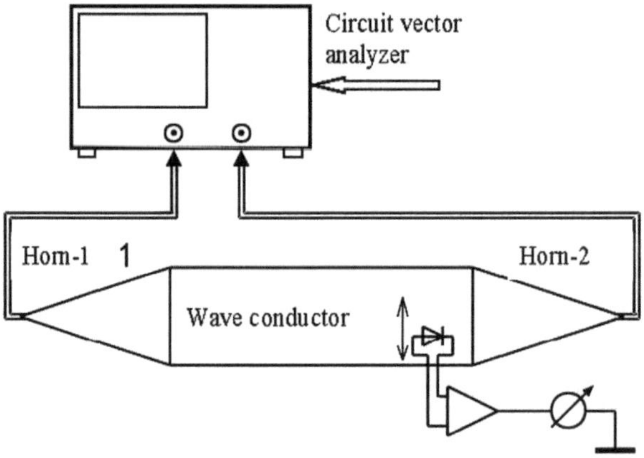

Fig. 8.21 Block diagram of super-wide band waveguide measuring system (Chukhlantsev 2006)

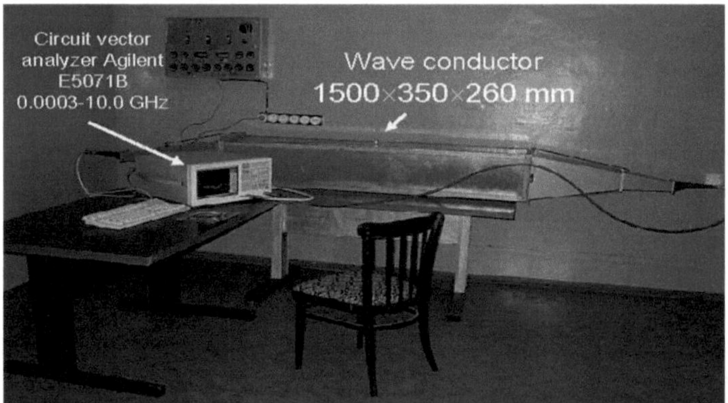

Fig. 8.22 General view of measuring system for the study of attenuation of electromagnetic waves by vegetation canopies in the 100–1,000 MHz frequency band (Chukhlantsev et al. 2003)

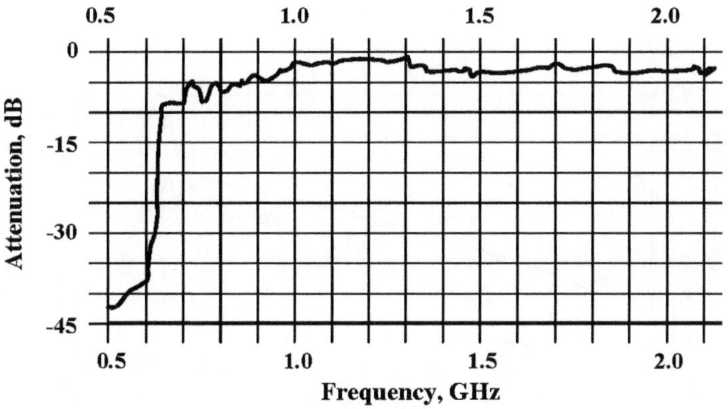

Fig. 8.23 Frequency dependence of attenuation coefficient for SHF channel of measuring system

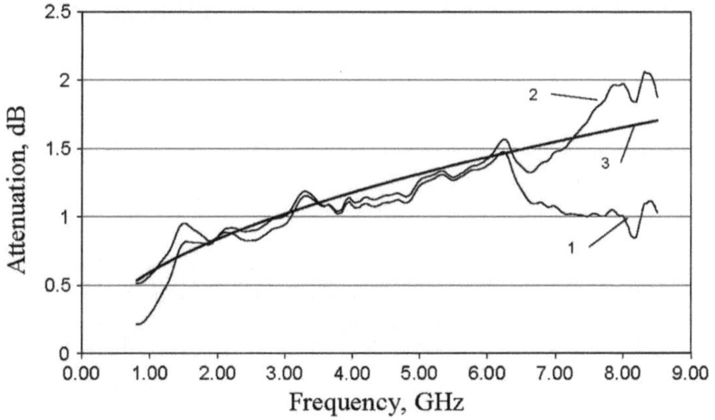

Fig. 8.24 Frequency dependence of the power attenuation in the pine branches under humidity equaled 62.7 %. *Notation 1* values of attenuation measured without correction; *2* corrected values of attenuation; and *3* approximation of *curve 2* by power function. Correction of measured values of attenuation was realized by means of neutralization of the power percolation effect from waveguide (Chukhlantsev 2006)

was studied by comparing the data with and without vegetation samples in the waveguides. Basic electrical characteristic of laboratory complex is shown in Fig. 8.23 It is clear that attenuation was less than 5 dB in the range of 0.8–2.15 GHz. The same dependence is observed in the range of 0.8–10.0 GHz. Some experimental results are given in Figs. 8.24, 8.25 and 8.26.

Experiments in field conditions were conducted at the territory of forests using the lowering unit (Figs. 8.27 and 8.28). The territory includes grassy and forested areas of different plant type and density. The generator is installed behind the trees. A receiver is located in different areas at different heights of tree crowns. The assessments of the attenuation were obtained by comparing the intensity of signals

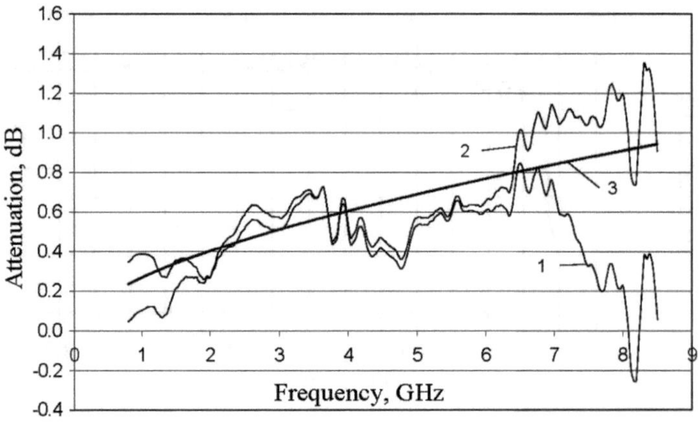

Fig. 8.25 Frequency dependence of the power attenuation in the pine branches under humidity equaled 17.1 %. *Notation 1* values of attenuation measured without correction; *2* corrected values of attenuation; and *3* approximation of curve 2 by power function. Correction of measured values of attenuation was realized by means of neutralization of the power percolation effect from waveguide (Chukhlantsev 2006)

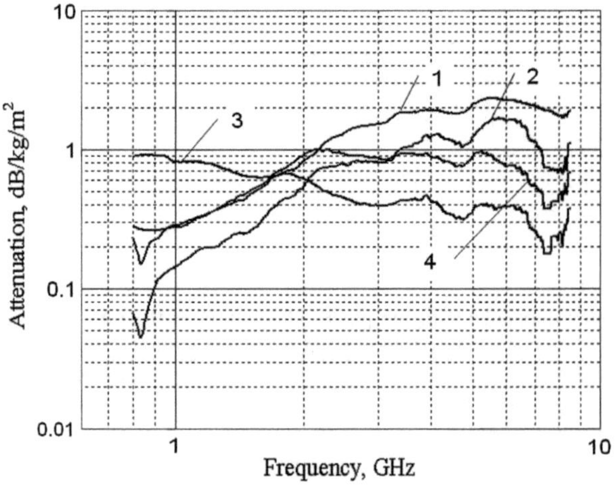

Fig. 8.26 Frequency dependence of electromagnetic wave attenuation in the aspen branches. *Notation 1* branches of 5 mm diameter with leaves; *2* branches of 5 mm diameter or less without leaves; *3* thick branches of 20–50 mm diameter; and *4* rough branches of 5–20 mm diameter

for different locations of the receiver off the tree. As a result, the database of the EMW attenuation was synthesized (Golovachev et al. 2004; Vinogradova et al. 2004).

In addition, the available microwave radiometric and SAR data are being processed and studied.

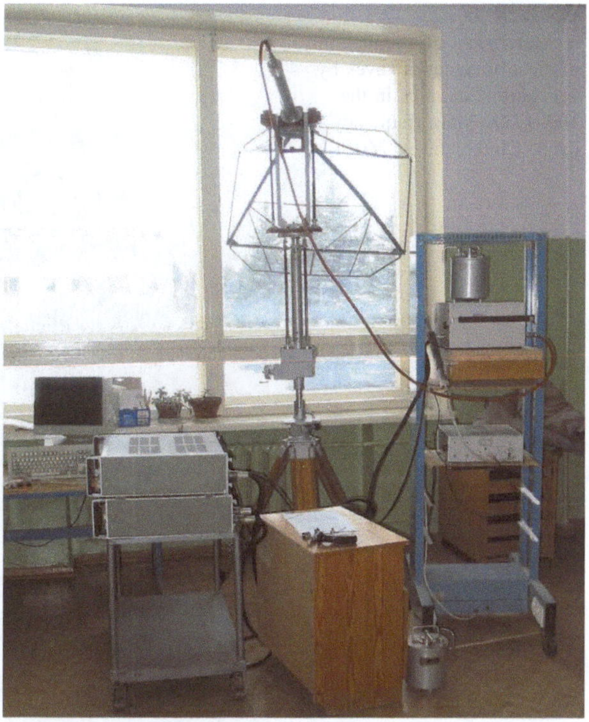

Fig. 8.27 Measuring system that was used for the assessment of electromagnetic wave attenuation in situ experiments (Chukhlantsev 2006)

The methodological part of the investigations also includes the following (Chukhlantsev 2002, 2006; Chukhlantsev et al. 2003):

- elaboration of new technologies and approaches for the screening effect assessment and its account for SAR and microwave radiometric remote sensing, including synchronous active and passive observations and radio-communication;
- determination of the optimal frequencies or a set of frequencies for remote sensing of the Earth's covers with due regard for the specifics of the areas of observation, terrain and vegetation type, climate, and current meteorological conditions;
- elaboration of new methodologies for multi-channel measurements of both, the properties of vegetation and of soil, by developing an advanced approach and adequate models for the system "soil-vegetation" sensing in absence of the reliable prior knowledge-based information about the investigated object;
- elaboration of an advanced approach for the typical Earth environments observation from space by optimizing the instrument set and developing approaches for artificial increase in a spatial resolution by conducting multi-wavelength measurements.

Fig. 8.28 In situ measurements of attenuation of electromagnetic waves by vegetation canopies in the 100–1,000 MHz frequency band (Chukhlantsev 2006)

8.7 Links Between Experiments, Algorithms, and Models

Field observations are the initial stage to determine the important properties of the vegetation cover. Many problems arise in providing the model descriptions of biological and physical processes, which operate on different timescales. The tasks of describing EMW attenuation by the vegetation cover have timescales ranging from less than a second to hours or days and also have various spatial scales. Consequently, the parametrical description of the EMW attenuation by the vegetation cover calls for an enormous number of experimental observations, the planning of which depends on the model type and the vegetation component under investigation. That is why the effective synthesis of the final results into given space-time scales may be realized by the application of GIMS-technology. This gives the opportunity to solve the problem of scaling the soil-vegetation-atmosphere model and to describe the radiative transfer processes on a scale from that of individual plants or small plots to large-scale biomes.

As mentioned above, the satellite data can be used to calculate many parameters of the vegetation cover, including unstressed stomatal conductance, photosynthetic capacity, fraction of photosynthetically active radiation absorbed by the green portion of the vegetation canopy, canopy reflectance, transpiration, and other important environmental characteristics. These data and the results of field experiments allow to generate model experiments.

The GIMS-based method uses a priori information to start the model experiment. Under this procedure, a set of algorithms operate over the whole data set and determine the input parameters of the model used. The model structure and its coefficients are changed depending on the fluctuation of the difference between the experimental and model estimations of the EMW attenuation effect. Usually, the central part of the model describes the fundamental processes within the SPF, such as the radiation balance, water circulation (evapotranspiration, water uptake by roots), photosynthesis, and mortality.

In a common case of environmental investigations, combined use of algorithms, computer models and remote sensing technologies is possible within the hybrid information-modeling system with the structure described by Soldatov (2011, 2012). This structure links experimental optical and microwave techniques with computer modeling supporting the correlations between different stages of simulation experiment.

The model-based analysis and optimization of computational experiments have specific features that certainly demand the determination of new design points. Unfortunately, in many scenarios of simulation experiments, the objective function, e.g., the response of survivability indicator to the impacts on the environmental system cannot be assessed exactly. Thus, the true quality of plan can be certainly obtained only by repeating the corresponding experiment and performing an estimation of the distribution of the responses. The GIMS-technology based on the evolutionary modeling and sequential analysis as follows from Fig. 1.5 optimizes the investigation process.

Complex engineering problems clearly cannot be resolved without using simulation experiments, which are usually brought about by numerical solution of a complex system of equations. In this case, the researcher runs into a conflict between his desire to enhance the accuracy of a simulation experiment and the limited capabilities of the used algorithms. The biggest difficulties arise when the investigator cannot use mathematical equations to parameterize the studied processes. These difficulties are overcome by the evolutionary computer technology that helps to design a new type of the model (Bukatova et al. 1991; René and Hassane 2005).

In common case, the GIMS-based approach provides the reproduction of a numerical simulation experiment, derived from a given set of quantitative models and algorithms independently from the complexity of problem for the study. Of course, the investigator has to compose the information about different parts of the simulation experiment including information about input and output data as well as about the models to use. All models used in the experiment must be identified, accessible, and fully described. The models required for the simulations must be provided with all governing equations, parameter values, and necessary conditions including initial state, boundary conditions, etc. The main step of the simulation experiment is a quality criteria definition in order to achieve the purpose of the experiment. Various examples of solution of such environmental problem, with the use of the GIMS-based approach are given in this book.

8.8 Microwave Model of Vegetation Cover

8.8.1 Two-Level Model of Vegetation Cover

The atmosphere-vegetation-soil system (AVSS) is commonly considered in Fig. 8.29. The microwave emission of the AVSS is formed by a combination of absorption and diffraction processes within the media and by repeated reflection upon their boundaries. Microwave emission models (MEMs) have been thoroughly studied in attempts to derive possible AVSS configurations by many authors (Ferrazzoli and Guerriero 1996; Phillips et al. 2009; Shutko 1997; Shutko et al. 2007; Krapivin et al. 2006b; Shutko and Krapivin 2011). The AVSS emissivity ability is usually described as

$$e = 1 - |R|^2 \tag{8.12}$$

where

$$R = \frac{R_{21}\exp[-2i\chi h] + R_{32}}{R_{21}R_{32} + \exp[-2i\chi h]};\ \chi = \chi_1 - i\chi_2 = \frac{2\pi}{\lambda}\left[\varepsilon_2 - \sin^2\theta\right]^{1/2};$$

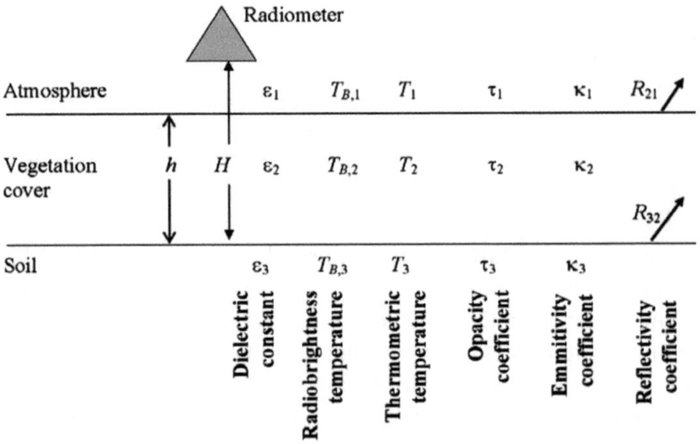

Fig. 8.29 Block-diagram of the model-based calculation of the shielding influence of the vegetation cover within the problem of microwave monitoring (Krapivin et al. 2006b)

θ is the angle between the direction of electromagnetic wave propagation in the atmosphere and the atmosphere/vegetation interface; R_{21} and R_{32} are the complex Fresnel coefficients for the atmosphere/vegetation and vegetation/soil interfaces, respectively; and h is the vegetation cover height.

The Fresnel coefficients for the interface between the ith and kth medium are defined by the expression:

$$\hat{R}_{ik} = |R_{ik}| \exp(-iq_{ik}) \tag{8.13}$$

where q_{ik} is the phase shift.

From (8.12) and (8.13), the following representation for the emissivity is given:

$$e = \frac{r - \hat{R}_{32}^2 \exp(-4\chi_2 h) - \hat{R}_{21}^2 - 2\kappa_1 \sin q_{21} \sin \xi}{r + \kappa_1 \cos(q_{21} + \xi)}$$

where

$$\xi = q_{32} + 2\chi_1 h;\ \kappa_1 = 2\hat{R}_{21}\hat{R}_{32} \exp(-2\chi_2 h);$$
$$\hat{R}_{ik} = |R_{ik}|;\ r = 1 + \hat{R}_{21}^2 \hat{R}_{32}^2 \exp(-4\chi_2 h);$$

$$\hat{R}_{ik} = \left\{ \frac{(\gamma_i - \gamma_k)^2 + (\delta_i - \delta_k)^2}{(\gamma_i + \gamma_k)^2 + (\delta_i + \delta_k)^2} \right\}^{1/2}; \tag{8.14}$$

$$tgq_{ik} = \frac{2(\delta_i \delta_k - \gamma_i \gamma_k)}{\gamma_i^2 - \gamma_k^2 + \delta_i^2 - \delta_k^2}; \tag{8.15}$$

$$\gamma_k = \begin{cases} \frac{A_k}{a_k} & \text{for horizontal polarization;} \\ \frac{A_k \varepsilon_k' + \beta_k \varepsilon_k''}{(\varepsilon_k')^2 + (\varepsilon_k'')^2} & \text{for vertical polarization;} \end{cases}$$

$$\delta_k = \begin{cases} \frac{\beta_k}{a_k} & \text{for horizontal polarization;} \\ \frac{A_k \varepsilon_k'' - \beta_k \varepsilon_k'}{(\varepsilon_k')^2 + (\varepsilon_k'')^2} & \text{for vertical polarization;} \end{cases}$$

$A_k = \sqrt{(a_k + b_k)/2}$; $\beta_k = \sqrt{(a_k - b_k)/2}$; $a_k = \sqrt{b_k^2 + (\varepsilon_k'')^2}$; $b_k = \sqrt{(\varepsilon_k')^2 - \sin\theta}$; $\varepsilon_2 = \varepsilon_2' - i\varepsilon_2''$ is the dielectric permittivity of the vegetation cover; $\varepsilon_3 = \varepsilon_3' - i\varepsilon_3''$ is the dielectric permittivity of the soil; and $\varepsilon_1 = \varepsilon_1' - i\varepsilon_1''$ is the dielectric permittivity of the air with $(\varepsilon_1' = 1,\ \varepsilon_1'' = 0)$.

Following Chukhlantsev and Shutko (1988), the Fresnel coefficients are determined as functions of the vegetation cover parameters. The dielectric constants of the vegetation and soil are input parameters to formulae (8.14) and (8.15).

Considering the vegetation and soil as double-component mixtures of dry matter and water, the dielectric constants can be determined by the following expressions:

$$\sqrt{|\varepsilon_2|} = \rho_P\sqrt{\varepsilon_B} + (1 - \rho_P)\sqrt{\varepsilon_W};\ \sqrt{|\varepsilon_3|} = \rho_B\sqrt{\varepsilon_B} + (1 - \rho_B)\sqrt{\varepsilon_W}; \tag{8.16}$$

where ρ_P and ρ_B are the relative volumetric concentrations of water in the plants and soil, respectively; and ε_B and ε_W are the dielectric constants of dry soil and water, respectively.

As was shown empirically by Engman and Chauhan, (1995) $\varepsilon_3' > \varepsilon_3''$ and both the real and imagery parts of the soil dielectric constant are increasing functions of the volumetric moisture content. These functions can be approximated by the following formulae:

$$\begin{aligned} \varepsilon_3' &\approx c_1 + \left(c_2\rho_B + c_3\rho_B^2\right)\exp(c_4 - c_5/\lambda), \\ \varepsilon_3'' &\approx \left(d_1\rho_B + d_2\rho_B^2\right)\exp(d_3 - d_4/\lambda), \end{aligned} \tag{8.17}$$

where λ is the wavelength (cm), and c_i and d_i are constants depending on the soil type. For example, soil that consists of 30.6 % sand, 55.9 % silt and 13.5 % clay is characterized by the dependencies (8.17) with $c_1 = 2.35$; $c_2 = 52.4$; $c_3 = 31.1$; $c_4 = 0.057$; $c_5 = 1.22$; $d_1 = 7.1$; $d_2 = 46.9$; $d_3 = 0.0097$ and $d_4 = 1.84$.

Thus, taking into account the dependencies of (8.17) it becomes possible to optimize the microwave range for passive remote sensing. Of course, a set of unsolved problems exists relating to the model corrections necessary in order to take into consideration the surface roughness and other obstacles distorting the brightness temperature T_B, as well as a set of special features which arises in the inverse tasks solution. A basis for future model refinements is the correlation between T_B, the atmospheric transmissivity for a radiometer at height H above the soil, the smooth surface reflectivity R and the thermometric temperatures of the SPF $T_{s\text{-}v}$ and atmosphere T_a. This correlation can be expressed by the Schmugge-Shutko formula (Schmugge 1990; Shutko 1986):

$$T_B = t(H)\left\lfloor RT_{sky} + (1 - R)T_{s-v}\right\rfloor + T_a \tag{8.18}$$

where T_{sky} is the contribution from the reflected sky brightness. Typical remote sensing applications use microwave wavelengths $\lambda \geq 1$ cm and in this case the atmospheric transmission approaches 99 % and $T_a + T_{sky} \leq 5$°K (Engman and Chauhan 1995).

The more precise correlation in (8.18) can be achieved considering the influence of the surface roughness on the soil reflectivity: $R' = R\,exp(-g \cdot \cos^2 \omega)$, where ω is the angle of incidence, and g is the roughness parameter ($g = 4\sigma^2k^2$, where σ is the root mean square height variation of the soil surface and $k = 2\pi/\lambda$).

Formulae (8.16) and (8.17) are simplified with the use of the correlation between the soil water content ρ_B and the vegetation water content ρ_P (DeWitt and Nutter 1988):

$$\rho_B = 78.9 - 78.4[1 - R' \exp(0.22\rho_P)]$$

For uniform media, the expression (8.18) is often represented in the following way:

$$T_B = \frac{\left(1 - \hat{R}_{21}^2\right)\left\{\left\lfloor T_3\left(1 - \hat{R}_{32}^2\right)\right\rfloor e^{-\tau_2} + T_2(1 - e^{-\tau_2})\left(1 + \hat{R}_{32}^2 e^{-\tau_2}\right)\right\}}{1 - \hat{R}_{21}^2 \hat{R}_{32}^2 \exp(-2\tau_2)}$$

where the opacity coefficient τ_2 can be evaluated by means of the formula (Kirdiashev et al. 1979):

where v is the vegetation volume density and ρ_1 is the specific density of the wet biomass. The coefficient $\hat{R}_{21}$ is estimated by the following expression:

$$\hat{R}_{21} \approx \alpha_2[1 - \exp(-\tau_2)].$$

There are many more simplified or more complicated models of the AVSS. For example, the model describing the forest medium, as sketched in Fig. 8.2, was developed by Ferrazzoli and Guerriero (1996) and is based on the Rayleigh-Jeans approximation of the electromagnetic properties of coniferous leaves (needles). The vegetation cover is represented as a two-layered structure. The first layer is the canopy composed of leaves and branches, represented by discs and cylinders, respectively. The second layer of the vegetation cover consists of trunks described by near-vertical cylinders. Such a model structure allows one to simulate the microwave forest emission.

Models for layered vegetation have been studied by many authors (Ferrazzoli and Guerriero 1996; Karam et al. 1992; Krapivin et al. 2006b). The main mathematical task involves solving the radiative transfer equation. Different models consider a wide frequency range for both deciduous and coniferous forests and account for the distributions of branch size and leaf orientation. The basic conclusion of these studies indicates that detailed knowledge of the forest structure is important for the adequacy of the model. For example, lichens in taiga forests and forest tundra are characterized by lower photosynthesis and can influence the attenuation index. The tasks arising here are the subject of future studies.

The problem of optimal electromagnetic range choice and synthesis of effective methodic for soil moisture measurement, certainly, needs additional study of formation conditions of the radiation fields in the atmosphere-soil-vegetation system. Some data for solution of this problem are given in Table 2.18.

8.8.2 Analytical Model of Vegetation Cover

To understand the dependence on the environmental parameters of the radio brightness temperature registered by a radiometer at height H above the soil, which is covered by plants, it is useful to consider simple parametrical descriptions of this dependence. A simplified microwave model can be written as the following integral:

$$T_B(H) = e_s T_s \exp\left[-\int_0^H \gamma(x)dx\right] + \int_0^H T(y)\gamma(y)\exp\left[-\int_0^y \gamma(x)dx\right]dy + \zeta(H) \tag{8.19}$$

where γ is the absorption coefficient, e_s is the soil emissivity coefficient, T_s is the soil temperature, T is the temperature of atmosphere, and ζ is the model precision. There exist different simplifications of equation (8.19). For example, various approximations for $T(y)$ and $\gamma(y)$ are often used:

$$T(y) = \begin{cases} T_2, & 0 \le y \le h; \\ T_1, & h < y \le H. \end{cases} \quad \text{or } T(y) = T_0 + T_1 y + \cdots + T_n y^n$$

and $\gamma(y) = \gamma_0 + \gamma_1 y$ or $\gamma(y) = \text{const}$.

For the satellite monitoring case it can be shown that

$$T_B = e_s T_s p_{sat} + \sum_{l=1}^{n-1} D_l(0) \tag{8.20}$$

where p_{sat} is the atmospheric transmittance between soil surface and satellite,

$$D_1 = T(h), D_2 = \frac{1}{\gamma(h)} \cdot \frac{dT(h)}{dh}, \ldots, D_l = \frac{1}{\gamma(h)} \cdot \frac{dD_{l-1}(h)}{dh}, \ldots .$$

Taking different analytical representations for the vertical profile $T(h)$ into consideration, formula (8.20) will calculate the brightness temperature T_B with an error equal to $D_n(0)$.

The more simplified microwave model (8.19) for the canopy-surface brightness temperature is given as:

$$T_B(\lambda, \theta) = e_2 T_2 + \hat{R}_{21}(\theta) T_{B,sky}(\theta_1, \theta_2) \tag{8.21}$$

where θ is the angle of incidence and $T_{B,sky}$ is the sky brightness temperature incident on the canopy level from a direction (θ_1, θ_2), with zenith and azimuth angles θ_1 and θ_2, respectively.

Model (8.21) simplifies the analysis of the influence of canopy temperature and wind speed, for example, on the variations of the brightness temperature. This analysis is possible with knowledge of the canopy roughness as a function of wind speed.

8.9 Land Cover Classifications

Remote sensing resources, image interpretation and classification tools provide land cover classification using remotely sensed data. There exist different options for conducting land cover classification including types of imagery, methods and algorithms, and classification schemes. The image classification process depends on the monitoring data types and quality. It is important to have the possibility to distinguish each class of vegetation on the map and to clearly define and distinct it from other classes. The most important problem is the forest ecosystem classification. The solution of the tasks arising here depends on the methods and algorithms for remote sensing data. Table 8.16 gives the list of basic indicative characteristics of forest ecosystems evaluated on the base of remote sensing of the forest. Table 8.17 enumerates the most known vegetative indexes used in the remote sensing monitoring. A vegetation index is a number that is generated by a combination of remote sensing bands and may have some relationship to the amount of vegetation in a given image pixel.

As far as the climate change and sustainable development problems are concerned, forest ecology plays more important role than other branches of ecology. Forests are studied at a number of organizational levels, from individual organisms

Table 8.16 List of basic indicative characteristics of the forest ecosystems evaluated on the remote sensing data

Characteristic	Sensors	Spatial resolution	Periodicity
Soil-plant formation type, area and age	Optical, microwave	1 km	5 years
Tree diameter, tree height, canopy biomass, parameters of the forest ecosystem models	Optical, microwave	0.5–30 m	5 years
Soil type and moisture	Optical, microwave	1–50 rm	Monthly
Forest canopy state: defoliation and color change	Optical	1–30 m	Monthly
Characteristics of the disturbing impacts on the forest: type of the influence, area and level of the damage, time of the event, velocity of the vegetation recovery	Optical, microwave	1–20 m	Depending on the impact type
Biophysical characteristics: LAI, NDVI, FPAR, chlorophyll concentration	Optical	0.25–1 km	Daily or monthly depending on the solved task
Phenological characteristics: snow cover timeline, depth and water content of the snow cover, vegetative period long time	Optical, microwave	1–50 km	Monthly
Surface characteristics: albedo and temperature	Optical, microwave	1–50 km	Daily

Table 8.17 The most known vegetative indexes evaluated in the framework of remote monitoring of the vegetation covers

Vegetative index	Description and model
NDVI	Normalized difference vegetation index. NDVI = (NIR − VIS)/(NIR + VIS) where NIR and VIS (or RED) stand for the spectral reflectance measurements acquired in the near-infrared and visible (red) regions, respectively
RVI	Ratio vegetation index. RVI = NIR/VIS
IPVI	Infrared percentage vegetation index. It is restricted to values between 0 and 1, which eliminates the need for storing a sign for the vegetation index values, and it eliminates the conceptual strangeness of negative values for vegetation indices. IPVI and NDVI are functionally equivalent and related to each other by the following equation: IPVI = NIR/(NIR + VIS) = (NDVI + 1)/2
NDII	Normalized difference infrared index. NDII = $(\rho_{819} - \rho_{1649})/(\rho_{819} + \rho_{1649})$ where ρ_{819} and ρ_{1649}) is surface reflectance measurements in ranges of 819 and 1,649 nm, respectively
WDVI	Weighed difference vegetation index. WDVI = NIR − S × VIS where S is the soil line slope
SAVI	Soil-adjusted vegetation index. SAVI = (NIR − VIS)/(NIR + VIS + L) where L is a correction factor which ranges from 0 for very high vegetation cover to 1 for very low vegetation cover. The most typically used value is 0.5 which is for intermediate vegetation cover
TSAVI	Transformed soil-adjusted vegetation index. TSAVI = $[s(\mathrm{NIR} - s \times \mathrm{VIS} - a)]/[a \times \mathrm{NIR} + \mathrm{VIS} - a \times s + \mathrm{X} \times (1 + s^2)]$ where s is the soil line slope, a is the soil line intersection, and X is the compensation factor reducing the soil noise. The soil line is a hypothetical line in spectral space that describes the variation in the spectrum of bare soil in the image. The line can be found by locating two or more patches of bare soil in the image having different reflectivities and finding the best fit line in spectral space
MSAVI	Modified soil-adjusted vegetation index. MSAVI = $(\mathrm{NIR} - \mathrm{VIS})(1 + L_0)/(\mathrm{NIR} + \mathrm{VIS} + L_0)$ where $L_0 = 1 - 2S \times \mathrm{NDVI} \times \mathrm{WDVI}$
MSAVI2	Modified soil-adjusted vegetation index-2. MSAVI2 = $(\mathrm{NIR} - \mathrm{VIS})(1 + L_1)/(\mathrm{NIR} + \mathrm{VIS} + L_1)$ where $L_1 = 1 - 0.5\{2\mathrm{NIR} + 1 - [(2\mathrm{NIR} + 1)^2 - 8(\mathrm{NIR} - \mathrm{VIS})^2]^{1/2}$
GEMI	Global environmental monitoring index. GEMI = E(1 − 0.25E)-(VIS-0.125)/(1-VIS) where E = {2(NIR-VIS) + 1.5NIR + 0.5VIS}/(NIR + VIS + 0.5)
ARVI	Atmospherically resistant vegetation index. ARVI = $(\mathrm{NIR} - R_b)/(\mathrm{NIR} + R_b)$ where R_b = RED − A(RED − BLUE), A is experimental parameter that characterizes the vegetation unity (usually A = 1 or A = 0.5 for small area of vegetation cover)
SARVI	Soil and atmospherically resistanf vegetation index. SARVI = $(\mathrm{NIR} - R_b)/[(1 + L)(\mathrm{NIR} + R_b)]$ where $R_b = \mathrm{VIS} - A(\mathrm{VIS} - \rho_{\mathrm{Blue}})$
GVI	Green vegetation index. GVI = −0.2848 × Band1 − 0.2435 × Band2 − 0.5436 × Band3 + 0.7243 × Band4 + + 0.0840 × Band5 − 1.1800 × Band7 where BandI (I = 1, 2, 3, 4, 5, 7) is I-th channel of the Landsat TM (Thematic Mapper). GVI = −0.29 × MSS4 − 0.56 × MSS5 + 0.60 × MSS6 + 0.49 × MSS7 where MSSI is I-th (I = 4, 5, 7) channel of the Landsat MSS (MultiSpectral Scanner)

(continued)

Table 8.17 (continued)

Vegetative index	Description and model
SWVI	Shortwave vegetation index. SWVI = (NIR − SWIR)/(NIR + SWIR) where SWIR is the surface reflectance in short-wave infrared channel (1.65 μm) of the Spot that is sensitive to vegetatiov water content
EVI	Enhanced vegetation index. EVI = 2(NIR − RED)/[(1 + L) (NIR + C_1RED − C_2BLUE + L)] where C_1 (≈6) and C_2 (≈7.5) are empirical correcting coefficients; RED and BLUE is surface reflectance in red (0.620 − 0.670 μm) and blue (0.459 − 0.479 μm) range, respectively
DVI	Difference vegetation index. DVI = NIR − RED
PVI	Perpendicular vegetation index. PVI = NIR × sinα-RED × cosα where α is the angle between the soil line and the NIR axis
PRI	Photochemical reflectance index. PRI is defined by the following equation using reflectance (ρ) at 531 and 570 nm wavelength: PRI = $(\rho_{531} - \rho_{570})/(\rho_{531} + \rho_{570})$ or PRI = $(\rho_{570} - \rho_{531})/(\rho_{531} + \rho_{570})$
TVI	Triangular vegetation index. TVI = $0.5[120(\rho_{750} - \rho_{550}) - 200(\rho_{670} - \rho 550)]$
TVI	Transformed vegetation index.TVI = 0.5 + $\mathrm{NDVI}^{1/2}$
CTVI	Corrected transformed vegetation index. CTVI = $(\mathrm{NDVI} + 0.5)/\lvert\mathrm{NDVI} + 0.5\rvert^{3/2}$
MTVI	Modified triangular vegetation index. MTVI = $[1.5(1.2 \times (\mathrm{NIR} - \rho_{Green}) - 2.5(\mathrm{VIS} - \rho_{Green})]/[(2\mathrm{NIR} + 1)^2 - (6\mathrm{NIR} - 5\mathrm{VIS}^{0.5} - 0.5]^{0.5}$
MCARI	Modified chlorophyll absorption in reflectance index. MCARI = $\rho_{700}[(\rho_{700} - \rho_{670}) - 0.2(\rho_{700} - \rho_{550})]/\rho_{670}$
MNDWI	Modified normalized difference water index. MNDWI = $(\rho_{Green} - \rho_{SWR1})/(\rho_{Green} + \rho_{SWR1})$ where ρ_{Green} and ρ_{SWR1} is surface reflectance value of Band2 (wavelength 0.52 − 0.60 μm) and Band5 (wavelength 1.55 − 1.75 μm) in the Landsat-7 ETM + sensor, respectively
LSWI	Land surface water index. LSWI = $(\mathrm{NIR} - \rho_{SWR1})/(\mathrm{NIR} - \rho_{SWR1})$

to the ecosystem. In common, forest ecology mainly concentrates on the level of the population, community or ecosystem. Present study of the forest ecosystems is based on not only in-situ observations but the satellite measurements are used as fundamental base for the forest ecology. In contrast to other land ecosystems, forests have specific features that consist in the following circumstances:

- trees can grow larger than other plant life-forms;
- there is the potential for a wide variety of forest structures;
- the infinite number of possible spatial arrangements of trees of varying size and species makes for a highly intricate and diverse micro-environment in which environmental variables such as solar radiation, temperature, relative humidity, and wind speed can vary considerably over large and small distances;
- an important proportion of a forest ecosystem's biomass is often underground, where soil structure, water quality and quantity, and levels of various soil nutrients can vary greatly; and

- forest trees are important regulators of hydrological processes, especially those involving groundwater hydrology and local evaporation and rainfall/snowfall patterns.

These and other features of the forest ecosystems and large areas occupied by them demand the use of coupled mathematical models to study the forest in their evolution. This chapter gives examples of such models and remote sensing technologies. In the future it is necessary to develop the proposed approaches mainly based on the GIMS technology. Such approach gives the possibility to study the structure of forest ecosystems in vertical and horizontal dimensions. The horizontal structure is usually described by how individual trees, shrubs, herbs, and openings or gaps are distributed. The vertical structure has several levels and its study using remote sensing means is possible provided that there are data about microwave radiation attenuation in the canopy.

Parallel with microwave monitoring of vegetation covers, optical sensors located on flying laboratory and satellites deliver spectral images of land covers with high spatial resolution. In this case, the classification of land covers is based on the difference of their spectral reflectance (Figs. 8.30 and 8.31). Moreover, spectral image recognition method described in Chap. 9 can be used for the land cover classifications.

Vegetation indices are widely used to assess qualitatively the biophysical and morphological characteristics of vegetation from remote sensing measurements. Sensitivity of each vegetation index to different vegetation characteristics covers restricted parametrical cluster of the atmosphere-soil-plant system. Different satellites deliver data for global and regional vegetation index imagery at various pixel size resolutions. The sites include boreal forests, temperate coniferous forests temperate deciduous forests, tropical rainforests, grasslands, savannas, and desert biomes. Differences and similarities in sensitivity to vegetation conditions are

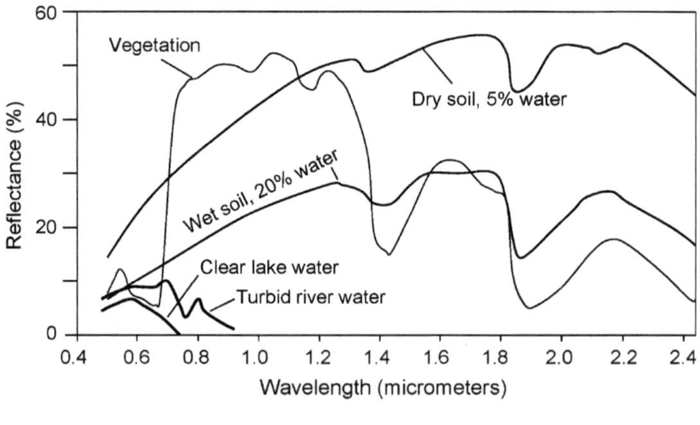

Fig. 8.30 Spectral reflectance of the land covers in the *red*, near infrared and short wave infrared bands. http://commons.wikimedia.org/wiki/File:Incoming_spectral_reflectance_from_different_objects_to_a_sensor_system.svg

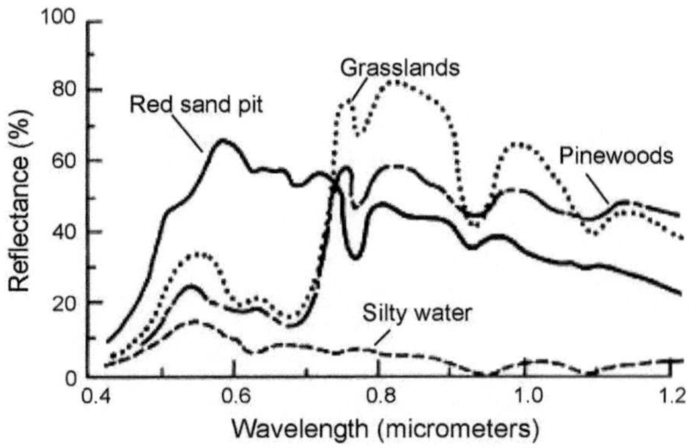

Fig. 8.31 Spectral reflectance of different objects. http://commons.wikimedia.org/wiki/File:Incoming_spectral_reflectance_from_different_objects_to_a_sensor_system.svg

registered by many vegetation indices. All vegetation indices show a qualitative relationship to variations in vegetation. However, there exist significant differences among the vegetation indices over desert, grassland, and forested biomes. For example, NDVI is sensitive to and responded primarily to the highly absorbing red reflectance band, while other indices such as SARVI are more responsive to variations in the near-infrared band. It is evident that different indices have their own advantages in retrieving vegetation information. It is very difficult to precisely attribute any vegetation index to any particular vegetation biophysical or morphological parameter. Therefore, creation of universal algorithm for the classification of vegetation characteristics is possible by using the model represented in Fig. 1.19 that realizes the GIMS technology. In this case, the vector $\Psi = \{y_i, i = 1, \ldots, m\}$ is synthesized, where y_i is separate vegetation index from Table 8.17 and $m \leq 25$ is the number of indices included to the vector Ψ. Vector Ψ gives a possibility to form cluster space that allows the recognition of different states of vegetation covers.

Chapter 9
Operational Diagnostics, Estimation of the Scale of Damage and Aftermath Reduction of the Stressful Natural Processes

9.1 Expert System for the Hydrophysical and Hydrochemical Investigations

9.1.1 Principal Matters

Risk of losses resulted from human factors, inadequate or failed internal processes and systems, or external events is closely connected with aftermath weakening of the damage reduction. This problem is important when stressful hydrophysical and hydrochemical processes occur. For example, the control and assessment of the state of the water environment require knowledge of numerous characteristics, which can be classified into:

(1) factors of the water chemistry;
(2) water quality; and
(3) dynamical characteristics.

The first group includes characteristics of the processes affecting the water quality and determining the dynamics of the parametric space that describes the complex state of the studied aqua-geosystem. Depending on the experimental conditions, these factors are connected with the spatial scale and can be point wise or of areal character.

To describe a set of factors and to determine their spatial significance, we denote the points on the aqua-geosystem's surface by latitude φ and longitude λ. Let us consider a rectangular system of coordinates with the axis z directed upwards in relation to the surface, so that the value z stating the depth at the point with coordinates (φ, λ). Then any measurement will be described by the parameter $\xi(\varphi, \lambda, z, t)$, which in a general case is a non-stationary random value. In the concrete experiment, the formation of the set $\Xi = \{\xi\}$ is connected with the factors shown in Fig. 9.1. In a general form, the scheme of the formation of the set Ξ is shown in Fig. 9.2. Depending on the spatial scales of the hydrophysical object under study,

V.F. Krapivin et al., *New Ecoinformatics Tools in Environmental Science*,
Environmental Earth Sciences, DOI 10.1007/978-3-319-13978-4_9

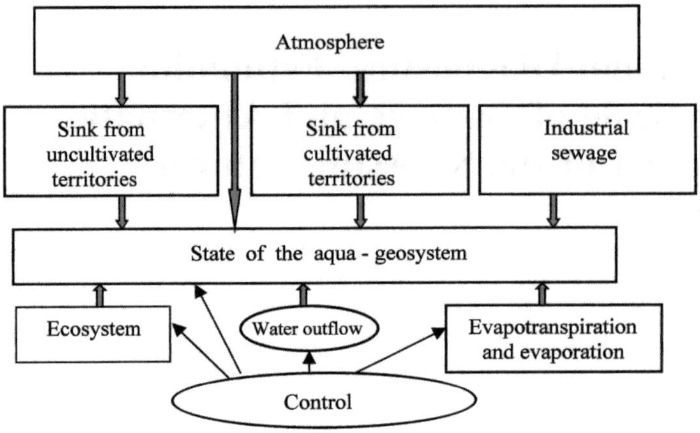

Fig. 9.1 Factors which form the water quality and should be taken into account in the hydrophysical experiment

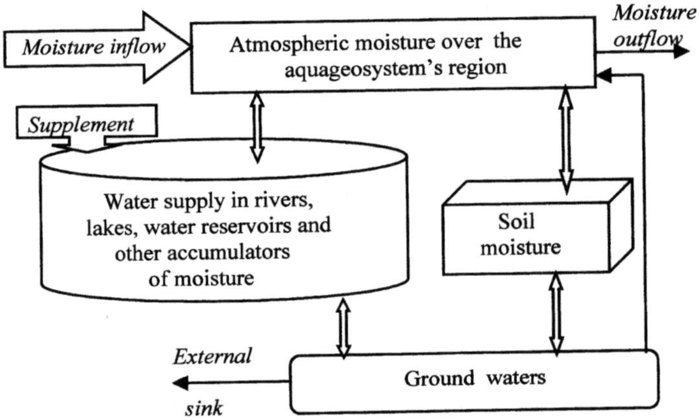

Fig. 9.2 Block-scheme of the model of the moisture cycle in the zone of the aqua-geosystem's functioning with the processes of evaporation, evapotranspiration, leakages, precipitation and anthropogenic use taken into account

a variety of possible schemes of the formation of water quality and other characteristics of the aqua-geosystem covers a wide spectrum of the physico-chemical interactions of the elements of the water basin ecosystem. The variety of all these schemes is an evidence of the complexity of assessing the water quality and the state of the aqua-geosystem. Therefore, an efficient monitoring of the aqua-geosystem depends on the adequate description of these interactions and the adequate list of parameters to be measured.

The parameters that directly determine the water quality are acidity, turbidity and the content of chemicals. In parallel, a set of criteria for the water quality exists depending on its application. In this regard, Table 9.1 exemplifies the sets of factors

Table 9.1 Some criteria and factors, the structure of substances and parameters that determine the water quality

(a) Factors	
Type of water pollution	*Factors*
Biological	Microorganisms favouring the fermentation of organic substances
Chemical	Toxic substances or changing the composition of water
Physical	Heating, radioactivity
(b) Elements	
Group of characteristics	*Elements*
Basic ions	Chloride $Cl-$, sulphate SO_4^{2-}, hydrocarbon HCO_3^-, carbon CO_3^{2-}, natrium Na^+, potassium K^+, magnesium Mg^{2+}, calcium Ca^{2+}
Dissolved gases	Oxygen O_2, carbon dioxide CO_2, nitrogen N_2, hydrogen sulfide H_2S
Biogenic elements	Nitrogen, phosphorus and silicon compounds
Microelements	*Zn, Cu, Pb, Ni, Co, Li, Rb, Cs. Be, Br, I, F,* etc.
Organic substances	Content of carbon, oxidizability, biochemical consumption of oxygen

that assess the water quality with different criteria. Abbasi and Abbasi (2012) describe water quality indices and show that only with the use of such indices is it feasible to assess, express, communicate, and monitor the overall quality of any water source. In general, the conditions of water quality include (http://water.epa.gov/type/rsl/monitoring/vms52.cfm)

- *Stream flow.* The stream flow, or discharge, is the volume of water that moves over a designated point over a fixed period of time. It is often expressed in cubic meters per second (m^3/s). The flow is affected by weather, increasing during rainstorms and decreasing during dry periods. It also changes during different seasons of the year, decreasing during the summer months when evaporation rates are high and shoreline vegetation is actively growing and removing water from the ground. August and September are usually the months of lowest flow for most streams and rivers in most of the country.
- *Dissolved oxygen and biochemical oxygen demand.* The stream system both produces and consumes oxygen. It gains oxygen from the atmosphere and from plants as a result of photosynthesis. Running water, because of its churning, dissolves more oxygen than still water, such as that in a reservoir behind a dam. Respiration by aquatic animals, decomposition, and various chemical reactions consume oxygen.
- *Temperature.* The rates of biological and chemical processes depend on temperature. Aquatic organisms from microbes to fish are dependent on certain temperature ranges for their optimal health. Temperature affects the oxygen content of the water (oxygen levels become lower as temperature increases); the rate of photosynthesis by aquatic plants; the metabolic rates of aquatic organisms; and the sensitivity of organisms to toxic wastes, parasites, and diseases.

- *Alkalinity (pH).* It affects many chemical and biological processes in the water. The pH scale measures the logarithmic concentration of hydrogen (H+) and hydroxide (OH−) ions, which form water (H+ + OH− = H_2O).
- *Turbidity.* Suspended materials including soil particles (clay, silt, and sand), algae, plankton, microbes, and other substances can affect the colour of the water and decrease the passage of light through it.
- *Phosphorus.* Phosphorus is essential nutrient for the plants and animals that makes up the aquatic food web.
- *Nitrates.* Nitrates are a form of nitrogen, which is found in several different forms in terrestrial and aquatic ecosystems. These forms of nitrogen include ammonia (NH_3), nitrates (NO_3), and nitrites (NO_2). Nitrates are essential plant nutrients, but in excess amounts they can cause significant water quality problems.
- *Total solids.* The concentration of total dissolved solids affects the water balance in the cells of aquatic organisms.
- *Conductivity.* It is a measure of the ability of water to pass an electrical current. Conductivity in streams and rivers is affected primarily by the geology of the area through which the water flows.
- *Total alkalinity.* Alkalinity is a measure of the capacity of water to neutralize acids. Alkalinity in streams is influenced by rocks and soils, salts, certain plant activities, and certain industrial wastewater discharges.
- *Fecal bacteria.* Presence of fecal bacteria in streams suggests that pathogenic microorganisms might also be present and that swimming and eating shellfish might be a health risk.

These parameters testify to the fact that it is very difficult to draw a distinction between hydrophysical and hydrochemical experiments. However, it is clear that the quality of water is, to a greater extent, hydrochemical function. In practice the aim is neither to draw this distinction not to substantiate a set of problems to be solved in the hydrophysical experiment. Apparently, speaking about data series of the hydrophysical experiment, one should not neglect the hydrochemical data. Ultimately, an assessment of the state of the aqua-geosystem requires a certain combination of such data and especially when studying the dynamic characteristics. The parameters of the water quality in the list mentioned above have various effects on the living organisms.

In particular, a change of $\xi(\varphi, \lambda, z, t)$ is connected with a consideration of correlations between all factors affecting the water quality. In the formation of the spatial image of the water object under study, it is necessary to take into account an interaction of climatic, biogeochemical, ecological and hydrological processes. This is possible by using a numerical modeling and, hence, developing new data series in the form of the models' coefficients.

The stages of the hydrophysical monitoring are connected with different levels of database and therefore require the respective algorithms of analysis and interpretation of measurement results. Moreover, an identification of a set of algorithms of data processing with the problems of the experiment cannot guarantee an adequate

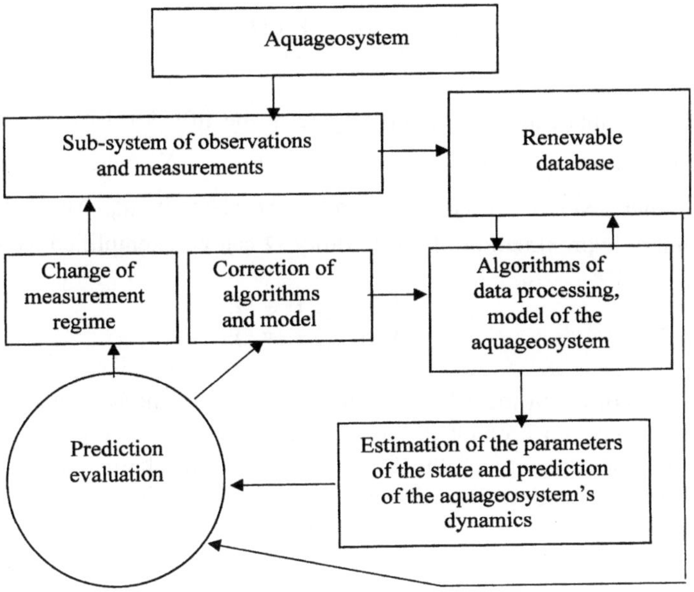

Fig. 9.3 Principal block-scheme of the adaptive regime of the hydrophysical experiment

complex of methods to solve these problems. A combination of the hydrophysical monitoring problems with the system of the automated data processing is only possible in an adaptive regime. It is shown schematically in Fig. 9.3.

The monitoring can be point wise and areal. In the first case the data processing is connected with the use of a number of algorithms providing an assessment of the required set of parameters without consideration of spatial scales of the process or object under study. In the second case the system of data processing requires special methods aimed at reducing their variability and forming a spatial image based on measurements episodic in time and fragmentary in space. In this case, it is essential to choose the scales of the cartographic grid and to correlate it with the problems to be solved. Only then can one guarantee a certain level of reliability in the assessment of the state of the aqua-geosystem.

Algorithms of formation and processing the spatial databases are approved in the GIS-technology. However, there exists a gap between traditional methods of GIS-technology and models as an instrument of prediction. To fill up the gap, the selection of the matrix structures-identifiers with the hierarchical topology both in space and in the elemental filling of the aqua-geosystem model is made. Each element of the identifier describes in the symbol space the image of an element (fragment) of the system under study. In other words, the territory Ω is presented in the form of the hierarchical spatial structure:

$$\Omega = \bigcup_{(i,j)} \Omega_{ij}, \; \Omega_{ij} = \bigcup_{s,l} \Omega_{i_s j_l} \; \text{etc.}$$

According to this scheme, the territory Ω at the fifth level of the database is represented by a set of matrices $A_k = \left\| a_{ij}^k \right\|$, $(k = 1, \ldots, N)$, where algorithms and data from other levels are used to determine the elements a_{ij}^k. An identifier as a simplest binary approximation of the territory Ω can be exemplified as follows:

$$a_{ij} = \begin{cases} 0, & (i,j) \notin \Omega; \\ \alpha, & (i,j) \in \Omega; \end{cases}$$

where the symbol α can identify the sets of elements of the aqua-geosystem supposed to be taken into account in the system of data processing. This semantics reduces the dimensionality of the fifth level of the database, but increases the probability of information distortion in the dialogue mode. In other words, the problem of an optimization of the fifth level of the database exists.

On the whole, models of hydrophysical and hydrochemical processes over a limited territory or on global scale permit to carry out a numerical experiment and to estimate the parameters of a hydrophysical object. Use of the numerical experiment in the hydrophysical studies has much in common with other fields of its application. A closed series of algorithms of analysis of the measured data is constructed and on its basis the fourth and fifth levels of the database are formed. An adaptive procedure of the use of the aqua-geosystem functioning model (according to the scheme in Fig. 9.3) makes it possible to start a dialogue mode of the estimation of many parameters, which cannot be reliably estimated from the measured data or which raise doubts.

By applying additional quantitative criteria of stability in the dialogue mode, it is possible to find permissible ranges of changes of some parameters. Schematically, it is shown in Fig. 9.4.

The most substantial link in the system of automatization of the hydrophysical experiment is the processing of the data on the hydrophysical factors, whose knowledge plays a certain role in assessing the complex state of the aqua-geosystem. These are factors, which introduce into the model some elements of instability and are the source of the aqua-geosystem's dynamics. Therefore, in all algorithms for the hydrophysical experiment, hydrodynamical equations serve for a reference point for strictness and accuracy. An adequate system of hydrodynamical equations describes changes of the components of the current velocity, pressure and density and contains five equations: equation of motion for three components, equation of continuity and equation of state. Since we are interested in the problem of pollutant transport in the hydrological network, these equations should be supplemented with equations of water balance in the territory.

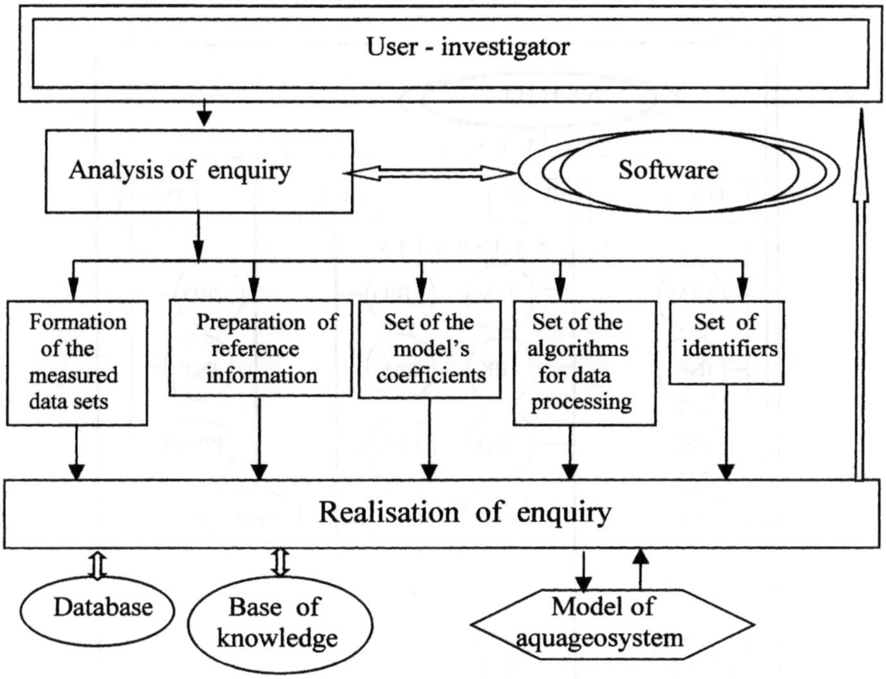

Fig. 9.4 Exemplary scheme of the dialogue regime of the use of algorithmic support of the hydrophysical experiment

9.1.2 Simulation System

The problem of synthesis of the simulation system for hydrophysical and hydrochemical investigations (SSHHI) in a given territory requires a classification of the elements of the aqua-geosystem. Analysis of the scheme in Fig. 9.5 shows that many fragments of SSHHI can be implemented with the help of GIS-technology. These are the formation of the spatial database, modeling the topographic structure, the computer cartography, and visualization. Other SSHHI fragments are accomplished in accordance with GIMS-technology.

The GIMS-technology recommends in the case of organization of studies in the territories with heterogeneous structures to form series of maps of different minuteness, as shown in Fig. 1.2. For any level of the spatial resolution $\Delta\varphi \times \Delta\lambda$, the database contains an information layer of the sets of identifiers controlled by the control unit and whose data are taken by the respective functions—units of the SSHHI. Here the scheme of cartographic modeling which is widely used in GIS, is realized.

As indicated above, the organization of any nature monitoring requires a solution of the problems of the spatial-temporal interpolation. This problem is solved by many methods applied in GIMS-technology. In SSHHI these methods are used to

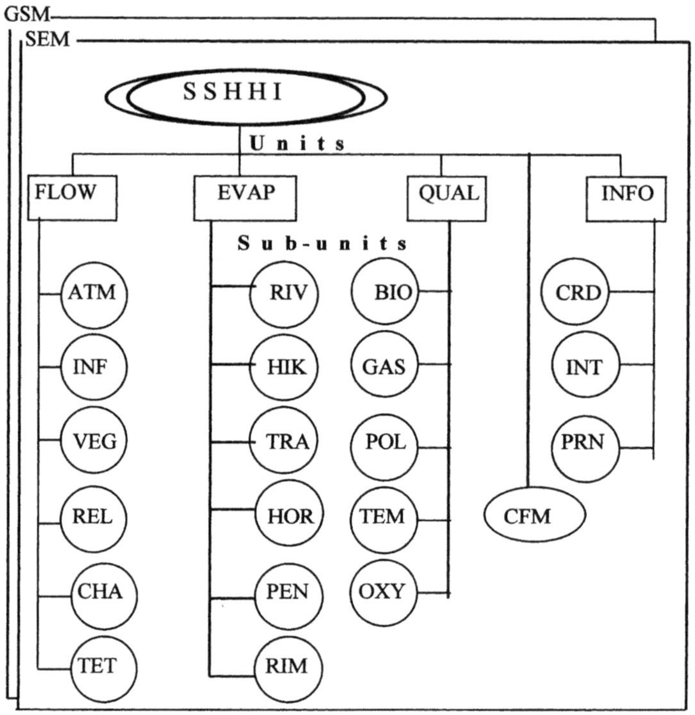

Fig. 9.5 Structure of the simulation system for hydrophysical and hydrochemical investigations (SSHHI). Notations are given in Table 9.2

integrate information from various sources. This procedure includes the following operations:

- integrating into a single system of spatial data from various sources, such as geographic maps, satellite photographs, remote measurements, ground observations, etc.;
- creation of the geometric description of the Earth surface within compatible topological structures;
- formation of models and software for transformations between the vector and expanded data, that is, between maps and images;
- provision of agreement between the series of hydrophysical and hydrochemical characteristics at each level of the spatial scale of the system;
- provision of co-ordination of information fluxes of spatial, linear and point wise data within one unit or for the whole model;
- synthesis of the scattered spatial information and formation of a representative image of either the aqua-geosystem or its fragments.

There are elements in the SSHHI database which, according to the scheme shown in Fig. 1.2, cover the whole set of information levels. As a result of the

Table 9.2 Decoded notations in the scheme in Fig. 9.5

Identifier of the SSHHI unit	Description of the unit
GSM	Global simulation model
SEM	System of ecological monitoring
SMMT	Simulation model of the moisture motion over the territory
SMPE	Simulation model of the process of moisture evaporation
SMWQ	Simulation model of water quality
PPWR	Realization of the procedure of predicting the state of the water regime and quality in a given territory
CIF	Control of information fluxes
MERS	Modelling the evaporation from the river surface
MESS	Modelling the evaporation from the soil surface
HM	Horton model
MT	Model of transpiration
MWTF	Model of the water temperature formation
MBAB	Modelling the biotic and abiotic balance of the water body
PM	Penman model
MATP	Model of atmospheric transport of pollutants
MI	Model of infiltration
MSPF	Model of the structure of soil-plant formations
MFRO	Model of the formation of the run-off with the topography of watershed and soil characteristics taken into account
MWMC	Model of water motion in the canal
MGE	Model of gas exchange on the border "water-atmosphere"
MPG	Model of pollutions generation
MFOL	Model of the formation of the oxygen level in water
CDM	Coordination of database with models
DI	Dialogue interface
FSOD	Formation of the structures of output data

simulation experiment, the computer memorizes at each time moment a set of schematic maps. Depending on the user's query, the FSOD unit forms an output file with the needed information structure. Means of computer visualization transform the symbol matrix into a schematic map, and the user obtains information that is easy to be analyzed.

To parameterize the hydrospheric part of the investigated site of the Earth surface and to create a universal schematic envelope of the system, cover the territory with the grid with a step $\Delta\varphi$ in latitude and $\Delta\lambda$ in longitude. The area of each cell will be, $\sigma_\Delta = \Delta\varphi \cdot \Delta\lambda \cdot k_\varphi \cdot k_\lambda$ where k_φ and k_λ are kilometers in 1° of latitude and longitude, respectively. A discrete numbering of the cells gives a totality of the sites $\Xi = \{\Omega_{ij}\}$, where each site Ω_{ij} has an area σ_{ij}. The coordinates of the center of the site Ω_{ij} are equal to:

$$\varphi_i = \varphi_1 + 0.5\varphi_i + \sum_{s=1}^{i} (s-1)\Delta\varphi_s,\ i = 1,\ldots,N;$$

$$\lambda_j = \lambda_1 + 0.5\lambda_j + \sum_{s=1}^{j} (s-1)\Delta\lambda_s,\ j = 1,\ldots,M;$$

The introduced grid of digitizing the territory enables one to create a universal procedure of referencing to a given object Ξ under study by using the following symbolic 2D arrays:

$$TERRITORY(I,J) = \begin{cases} 0 & \text{for} \quad \Omega_{IJ} \notin \Xi, \\ 1 & \text{for} \quad \Omega_{IJ} \in \Xi_S, \\ 2 & \text{for} \quad \Omega_{IJ} \in \Xi_W, \end{cases}$$

where $\Xi_S \cup \Xi_W = \Xi$, Ξ_S is a part of the land territory; Ξ_W is part of the territory under a hydrospheric object. The topological structure of the hydrospheric part of the territory is described by the identifier of the HYDROL(*I, J*) type in which the possible configurations of different types of hydrospheric elements are described. The HYDROL array permits to select a moving part of the water environment in the territory Ω and to more accurately consider thereby the variability of water chemistry, due to heterogeneous catchment areas.

To organize the numerical procedure of calculation of dynamic water flows between the cells $\Omega_{IJ,}$ introduce an array of the heights above sea level:

$$TOPOGR(I,J) = \begin{cases} 0 & \text{when} \quad \Omega_{IJ} \in \Xi_W, \\ \omega_{IJ} & \text{when} \quad \Omega_{IJ} \in \Xi_S, \end{cases}$$

where ω_{IJ} is the average height of the cell Ω_{IJ} above sea level in meters.

Each cell Ω_{IJ} is characterized by a certain relationship between water basins, wild and agricultural vegetation, urban constructions, and roads. The determination of this relationship is made by a set of the following 2D arrays: $\mathrm{PARTW}(I,J,\ \mathrm{TIME}) = S_{IJ}^1 \in [0,1]$ is the part Ω_{IJ} under water basins; $\mathrm{PARTA}(I,J,\mathrm{TIME}) = S_{IJ}^2 \in [0,1]$ is the part Ω_{IJ} under cultural vegetation; $\mathrm{PERTS}(I,J,\mathrm{TIME}) = \mathrm{S}_{IJ}^3 \in [0,1]$ is the part Ω_{IJ} under wild vegetation. The anthropogenic objects cover the remaining part of the territory: $S_{IJ}^4 = 1 - S_{IJ}^1 - S_{IJ}^2 - S_{IJ}^3$.

Each cell Ω_{IJ} is occupied by one of *n* types of vegetation and by one of *m* types of cultural vegetation. The schemes of this occupation are presented with 2D arrays: VEGETA (*I,J,*TIME) for agricultural vegetation, VEGETS (*I,J,*TIME) for wild vegetation and ANTROP (*I,J,*TIME) for anthropogenic landscapes.

Water can be polluted through the atmosphere, washing-out from soil and anthropogenic emissions directly into water reservoirs (common and industrial sewage, leakages and discharges from ships). Information about these processes is synthesized in the form of an identification scheme TYPEC(*I,J,*TIME). Based on the information contained in TYPEC array, the statistical mean situation of pollution in the absence of information is controlled.

The system is filled up with data by introducing the results of measurements provided either by the monitoring or by the services of control of observance of the norms of pollutants emissions into the environment by the anthropogenic systems. The system contains the TYPEP (*I*,*J*,TIME) array designed for introduction of more accurate information about the polluting productions. The TYPEP array includes productions characteristic of the territory, due to which the system provides the information network with average estimates of pollutants emissions in the case of the absence of other information. Information about the characteristic compositions of sewage in the territory of productions is used as a database.

Depending on requirements for the SSHHI efficiency, the corrective information can be obtained from satellites. For instance, information about the arrays TERRITORY, HYDROL, TOPOGR, PARTW, PERTS, VEGETA, VEGETS, and ANTROP can be obtained from satellites. The arrays TYPEC and TYPEP can be replenished and corrected by the regular use of the flying laboratory.

The structure of the SSHHI information base is determined both by possibilities of its formation and by need of practical solutions of the problems of the water quality control in a given territory. The gained experience in creation and use of such systems suggests a hierarchical structure of the database constructed by the principle of successive digitization of the territory. Following the scheme of hierarchical digitization of the space, one can, without damage to the system, constantly increase the minuteness of digitization, with the subject orientation of the territory Ω_{IJ} preserved. Thus the structure and functions of the SSHHI are adjusted to natural composition of data on various aspects of functioning of natural and anthropogenic landscapes. The SSHHI only synthesizes the available data and suggests data that should be obtained. Models in the SSHHI structure provide an integration of data in space and by subject areas, due to which an integrated image is formed of the territory with processes taking place in it.

The SSHHI has a set of formalized structures to control the monitoring regime. Spatial-temporal identifiers refer the SSHHI to the topology and morphology of the natural or anthropogenic object. The sets of the identifiers enumerated above parameterize the spatial and subject structure of the territory Ω, so that at each level of its spatial digitization a complete description is provided of an image of the system under study. By the space, the topology of this parameterization can be heterogeneous and non-uniform as well as excessive.

Following the hierarchy of the SSHHI menu the user, can implement the following operations:

- to ask for data on any identifier (array) and to correct any of its fragments;
- to ask for estimates of all or part of the parameters of simulation units and to correct them;
- to select the sets of parameters and identifiers for a more prompt access to them;
- to synthesize a symbolic schematic map of the distribution of the estimates of the environmental characteristics; and

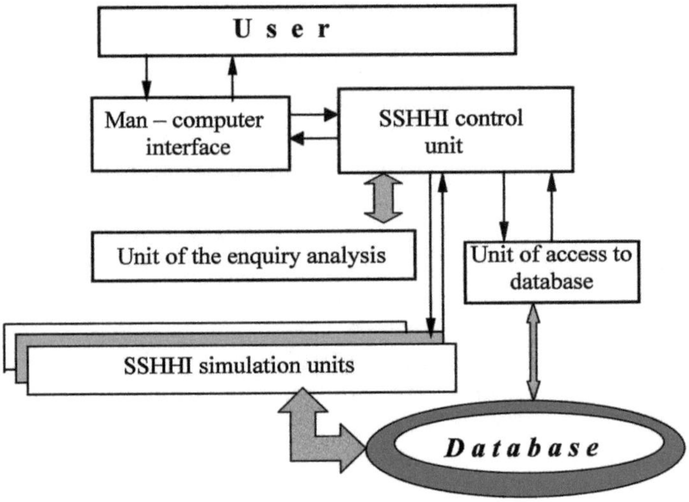

Fig. 9.6 SSHHI information units

- to predict the state of the environment down to a given depth or till accomplishing the a-priori formulated criterion of assessment of the state of the water environment.

Schematically, it is shown in Fig. 9.6. The user, through interface, sends permission at each step of the command dialogue, which is assessed in the unit of the query analysis, and from its response, the controlling unit realizes a chain of needed actions of the system. Via return channels of query, the resulting prediction is arranged in the needed form, which can change in each cycle of service. The final result is presented in the form of the protocol with enumerated characteristics of the water environment by objects and territories as well as in the form of schematic maps or digital information combined with the map.

9.2 Expert System for the Water Quality Control

9.2.1 Fresh Water Problems

The fresh water problems attract special attention, since, on the one hand, water is a key component of ecosystems and, on the other hand, about 1/3 of the global population is under the threat of chronic deficit of water in several decades. This and other facts have determined the declaration of the UN Fresh-Water Decade (2005–2015). The following facts characterize the present state of the global water resources (WCSDG 2004; WB 2009):

- The total volume of water resources on Earth constitutes about 1.4 billion km^3.
- The volume of fresh-water resources is about 35 million km^3 or 2.5 % of the total water volume.
- From these fresh-water resources, about 24 million km^3 or 68.9 % exist in the form of ice and constant snow cover in mountains, Antarctic and Arctic.
- About 8 million km^3 or 30.8 % are under ground in the form of ground waters (in shallow and deep basins of ground water down to 2,000 m deep). Soil moisture, bog water, and permafrost.
- In fresh-water lakes and rivers there are about 105 thousand km^3 or 0.3 % of global fresh-water supply.
- The general volume of fresh-water supply to be consumed by ecosystems and population constitutes about 2,000 thousand km^3, that is, less than 1 % of the global fresh-water supply.
- About 3,011 fresh-water biological species have included in the list of species, which either are under the threat or disappeared; 1,039 of them being fish. Four of the five existing species of river dolphins and two of the three existing species of sea-cows as well as about 40 species of fresh-water tortoises and more than 400 species of mollusks in fresh-water are under the threat.
- The annual ground waters diversion is evaluated at 600–700 km^3 or about 20 % of the global water diversion. About 1.5 billion people drink ground water.
- From estimates, 70 % of the global fresh water was spent in 2,000 on agriculture.
- The average per-capita consumption of water in developed countries is about 10 times greater than in developing countries. In developed countries this indicator varies between 500 and 800 ℓ/day, and in developing countries—from 60 to150 ℓ/day.
- Industry consumes about 20 % of the global fresh-water diversion. From 57 to 69 % of the global water diversion are spent on energy production at hydro and nuclear power stations, 30−40 %—in industrial processes, and 0.5−3 %—for needs of thermal energy.

The fresh-water problems are very broad and diverse. Table 9.3 (WCSDG 2004) presents a summary of the aspects of these problems on which the humans' life support depends in the context of the UN Millenium Declaradion (UNEP 2002, 2004).

On the whole, it should be stated that provision of sufficient pure water is of fundamental importance for achieving the goals of the socio-economic development and environmental protection. In this connection, of serious concern is an enhanced anthropogenic impact on the environment. So, for instance, the area of fresh-water wetlands that play an important role in natural purification of water and in the water cycle formation has almost doubled during the last 20 years. Meanwhile, according to the economic assessment of the role of wetlands functioning, their losses are equivalent to 20 thousand dollars per year. About 20 % of 10 thousand species of fresh-water fish are either on the verge of extinction or do not exist anymore. The number of large dams in the world has increased from 5,000 in 1950 to more than 45,000 now, which has led to negative ecological outcomes (Beder 2005; Nguyen Xuan Man et al. 2010).

Table 9.3 Various aspects of vital activity connected (directly of indirectly) with the fresh water problems

Millennium declaration	Problems to be resolved
Goal 1 To extirpate extreme poverty and hunger	Over the period 1990–2015, to halve the share of population with the income less than 1 dollar per day and to halve the share of the starving people
Goal 2 To provide universal elementary education	By the year 2015, to provide possibility for both boys and girls in the world to be elementarily educated
Goal 3 To promote equality of men and women and broaden the ability of women	To liquidate inequality of boys and girls in their access to elementary and secondary education preferably by 2005 and to all the levels of education by not later than 2015
Goal 4 To reduce infant mortality	To reduce by two thirds the mortality of children under 5 years over the period 1990–2015
Goal 5 Improve mothers' health	To reduce by three fourths the mothers' mortality over the period 1990–2015
Goal 6 To struggle against HIV/AIDS, malaria, and other diseases	By the year 2015, to stop the propagation of HIV/AIDS, malaria, and other basic diseases and start reducing its scale
Goal 7 To provide ecological stability	To take into account the principles of sustainable development in national strategies and programmes and reduce the loosing trend of ecological resources. By the year 2015, to halve the share of population without access to pure drinking water, to substantially improve the life of 100 millions of slums dwellers, as a minimum
Goal 8 To arrange global-scale partnership in the interest of development	To continue the formation of the undisguised, predictable and nondiscriminatory commercial and financial systems based on legal norms. To provide control, development and reduction of poverty on national and international levels. To satisfy special needs of the least developed countries having no access to the sea as well as small island developing countries (SIDC). In cooperation with the private practice sector, to take measures in order everybody could use the benefits of new technologies
Plan of realization of decisions of the Summit of Millennium	By the year 2015, to halve the share of population without access to safe drinking water partially because of want of money (as foreseen in the Millennium Declaration) and the share of population without access to base sanitation. By the year 2005, to work-out plans of a complex control of water resources and efficiency of water use

The geographical distribution of fresh-water resources is rather non-uniform: about half the global resources are the share of six countries (Brazil, Russia, Canada, Indonesia, China, Columbia), this non-uniformity being also characteristic of individual countries. So, for instance, China has 7 % of fresh-water resources (with a 21 % share of population with respect to global one), but most of the country is arid. Naturally, the countries with water deficit have to use ground waters, what brings forth a gradual decrease of their level. The data in Table 9.4 characterize the levels of per-capita water diversion in different countries as of 2011

Table 9.4 The annual per capita renewable freshwater resources

Country	Renewable freshwater resource per capita (m^3)	Country	Renewable freshwater resource per capita (m^3)
Greenland	10,522,275	Bosnia and Herzegovina	9,939
French Guiana	608,817	Estonia	9,205
Iceland	538,878	Guatemala	8,130
Guyana	315,678	United States	7,951
Suriname	236,836	Belarus	7,866
Congo	230,142	Lithuania	7,377
Papua New Guinea	121,791	Greece	6,465
Gabon	113,247	Mexico	4,353
Bhutan	106,292	Mauritania	3,546
Canada	8,3931	Luxembourg	3,421
Peru	66,339	Cuba	3,402
Liberia	61,159	Japan	3,328
Chile	54,868	Ukraine	3,034
Lao People's Dem. Rep.	53,752	France	3,003
Brazil	43,891	Italy	2,936
Panama	43,539	United Kingdom	2,683
Russian Federation	31,877	Uzbekistan	2,656
Sierra Leone	28,778	Afghanistan	2,389
Finland	20,737	Germany	2,285
Argentina	20,410	Belgium	1,882
Sweden	20,226	India	1,582
Australia	18,372	Egypt	1,065
Serbia	17,824	Kenya	779
Bulgaria	14,123	Algeria	417
Mongolia	13,176	Israel	237
Turkmenistan	12,067	China, Hong Kong SAR	213
Austria	10,075	Kuwait	7

(Krapivin and Varotsos 2007). These data visually show how the world water resources are irregularly distributed.

Besides, these data illustrate the existing socio-economic contrasts. About every fifth man in the developing world (they total about 1.1 billion) is subjected every day to the risk of falling ill for lack of good-quality drinking water. The main problem here is not the absence of water at all but adverse socio-economic conditions.

The major consumer of fresh water of rivers, lakes and underground sources is known to be agriculture (about 70 % on a global scale and up to 90 % in many developing countries). Since in the nearest future the enhancing use of irrigation will face a deficit of fresh-water resources, the efficient use of fresh water becomes

Table 9.5 Water expenditure (in litres) on productivity of 10 g of protein and provision of 500 food calories

Food	10 g of protein	500 calories
Potatoes	67	89
Peanut	90	210
Onions	118	221
Maize (grain)	130	130
Legumes	132	421
Wheat	135	219
Rice	204	251
Egg	244	963
Milk	250	758
Poultry	303	1,515
Pork	476	1,225
Beef	100	4,902

more and more important and can be substantially increased. In particular, this refers to the use of micro-irrigation (including drip irrigation), the scale of which remains rather limited.

A considerable potential of water economy is connected with food production. As seen, for instance, from Table 9.5, production of 10 g of protein in the form of beef requires water expenditure 5 times more than in the case of rice, and to obtain 500 food calories (this difference reaches a factor of 20). With an abundant meat diet, an average American needs 5.4 ℓ of water per day, whereas in the case of vegetarian diet this amount if halved.

A serious problem is the urban water supply and the implementation of measures on water economy. More urgent is still the problem of industrial use of fresh water whose share is 22 % of the globally used fresh-water resources (59 % in developed and 10 % in developing countries).

9.2.2 Water Quality Monitoring in the Estuary

An idea about the development of water quality diagnostics in the estuary was proposed by Krapivin et al. (1994a, b) in connection with environmental problems of South Vietnam. Development of industry and agriculture in Vietnam poses the problem of interial and coastal water pollution control and the synthesis of the monitoring systems with the functions of water quality assessment in the active regimes. Pollution levels in the Dong Nai and Sai Gon Rivers have been increasing since 2007, affecting the development of southern provincies. International Centre for Environmental Management and many scientific institutes of Vietnam Academy of Science and Technology develop environmental projects directed on the environment protection.

Vietnam has the coastal line equaled 3,260 km. There are many estuaries and lagoons on the seashore of which the development of anthropogenic systems depends on the sea water quality. At the same time these systems generate the pollutants and contaminate the sea. The pollutant sources in the Vietnam are grouped by following five classes (Bui Ta Long 1999; Bui Ta Long et al. 2010):

(1) agricultural sewage;
(2) domestic, municipal and industrial wastes;
(3) shipping;
(4) aquaculture; and
(5) coastal minings.

These sources are combined on the territories bordered to the South-China Sea. That is why the synthesis of environment monitoring system to assess the water quality in the estuary zone requires a systematic approach. A short description is given below about the expert system for ecological control of estuary zone (ESECEZ) developed in framework of collaborative agreement between Russian Academy of Sciences and the Vietnam Academy of Sciences and Technology on the ecoinformatics and remote sensing (Cao Van Phuong 2009a, b, 2012).

The expert system is based on the combination of modeling technology, algorithms of space-time reconstruction of measurement data and computer visualization software Block-diagram of expert system is given in Fig. 9.7 and Table 9.6.

The ESECEZ is oriented on the processes study of regulating and controlling the release of potential hazardous chemicals into the typical for the Vietnam estuary environment. The main functions of expert system can be realized automatically, but not the human components. Human activity can be displayed when it is required to explain the ecological situation or to make decisions about how to protect the environment.

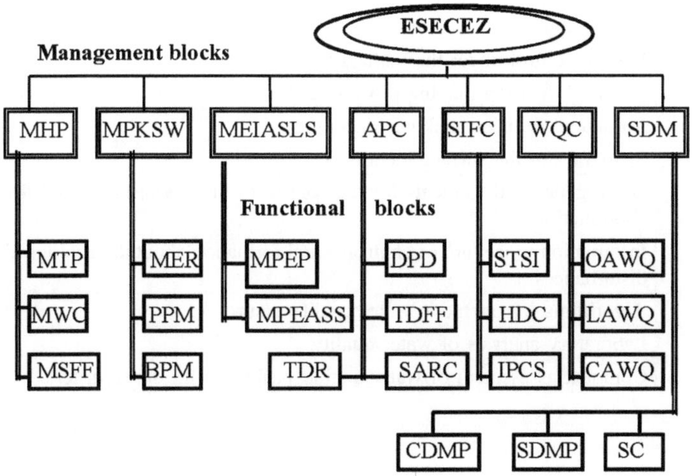

Fig. 9.7 Block-structure of the ESECEZ. Interpretation of abbreviations is given in the Table 9.6

Table 9.6 The abbreviations interpretation of Fig. 9.7 (Soldatov 2011)

Identifier	Description of the block function
ESECEZ	Expert system for ecological control of estuary zone
The first level blocks	
MHP	Modeling the hydrodynamics processes (Monin, krasitskyi 1985)
MPKSW	Modeling the pollutant kinetics in the sea waters
MEIASLS	Modeling the environmental interactions in the atmosphere-sea-land system
APC	An anthropogenic processes control
SIFC	Subsystem of information flows control
WQC	Water quality control
SDM	Subsystem of decision making
The second level blocks	
MTP	Model of the tide processes
MWC	Model of the waterflow from the coast
MSFF	Model of sewage flow formation
MER	Model of the ecosystem role in the processes of pollutants transformation
PPM	Physical processes model of interactions between the pollutant and sea water
BPM	Biological processes model of pollutant transformation in sea water
MPEP	Model of the pollutant exchange processes between the estuary zone and other territories (Krapivin 1995)
MPEASS	Model of the pollutant exchange in the atmosphere-sea system
DPD	Description of the passport data for the industrial and agricultural system of defined type, specific parameters of production, and activity scale. On the base of these data in the framework of the ESECEZ the control etalones are formed to describe the possible spectre of pollutants generated
TDFF	Technology of the database fragment forming based on the official information about the real volume of generated pollutants to the atmosphere and water area
TDR	Technology for the database renewal
SARC	Set of algorithms for the revealing of correlations in the anthropogenic system functioning
SC	System control under the realization of its functions
CDMP	Classical decision making procedure
SDMP	Sequential decision making procedure
STSI	Spatial tie of the surface identifiers
HDC	Heterogeneous documents for the control of the environment and their coordination
IPCS	Information procedure of control signal formation about the ecological condition disturbances
AOWQ	Algorithms for operative water quality control
LAWQ	Laboratory analysis of water quality
CAWQ	Complex assessment of water quality basing on the set of criteria

Total estuary aquatory is $\Omega = \{(\varphi, \lambda)\}$ where φ and λ is latitude and longitude, respectively. Water body of the estuary is $\Xi_{ijk} = \{(\varphi, \lambda, z),\ \varphi_i \le \varphi < \varphi_{i+1};\ \lambda_j \le \lambda < \lambda_{j+1};\ z_k \le z < z_{k+1}\}$ where z is water depth.

Construction of the ESECEZ is performed from the blocks which are connected by means of the parametrical inputs and outputs. There are the blocks having the same functional peculiarity. In framework of automatic regime a selection between these blocks is realized in the concordance with the database. Such a block is inserted, for which the database has the necessary parameters for the block input. A full set of blocks activated in framework of the ESECEZ consists of all the compatible blocks. Realization and control of this process is realized by the SC block. The clarification of this process is accompanied by the system response to request the necessary data. The ESECEZ has the working regime when the users can form the system structure in their own opinion.

The main ESECEZ function is risk assessment with the purpose of informing groups of user about the water quality in the estuary zone. There are various components of the risk assessment software system, each of which reflects a different approach to using the information contained in the system. The SDM block realizes such functions as the hazard identification, dose-response assessment, human exposure evaluation and risk characterization. These functions are conformed with the Vietnam water standards. The ESECEZ generates the signals which reflect the water quality in the site defined by system user. The CDMP and SDMP blocks rely on a risk assessment approach to make better regulatory decisions. The users can orient on their own specifical features and decide how to use the CDMP and SDMP blocks. In this way, the ESECEZ produces risk assessment of the highest possible quality. Moreover, it provides that each assessment is sensitive to site-specific conditions.

The WQC blocks are based on the sampling, laboratory analysis, data handling, data analysis, reporting, and information utilization. The SIFC blocks regulate the information flows between other blocks and provides the necessary data for analysis and interpretation.

Specifical features of the ESECEZ design are arisen to produce the water quality assessment in the real estuary zone including the anthropogenic systems. One of the most important and difficult tasks here is the simulation of pollutants kinetics in the sea water. This tasks are solved in framework of the MPKSW and MEIASLS blocks. Typical processes in the estuaries of Vietnam coastal line are taken into consideration. The DPD and TDR blocks realize the adaptation procedures to fit the ESECEZ for the individual estuary zone.

The process of the contaminants diffusion in the seawater depends on their state. For example, the dissolved fraction of the heavy metals (ψ) takes part in the biogeochemical processes more intensively than the suspended particles (e). But as suspended particles, the heavy metals fall out more rapidly to the sediment. A description of the entire spectrum of these processes in the framework of this study is impossible. Therefore, the PPM block describes the processes, which have estimates. The transport of heavy metals in the sea water includes absorption of the dissolved fraction ψ by plankton $\left(H_Z^{\psi}\right)$ and by nekton $\left(H_F^{\psi}\right)$, sedimentation of the solid fraction $\left(H_1^{e}\right)$, deposition with the detritus $\left(H_D^{\psi}\right)$, adsorption by detritophages from bottom sediments $\left(H_L^{e\psi}\right)$, and release from bottom sediments owing to the diffusion $\left(H_a^{e\psi}\right)$. As a result, the dynamic equations for the heavy metals become:

$$\frac{\partial e_w}{\partial t} + v_{\varphi}^{W}\frac{\partial e_w}{\partial_{\varphi}} + v_{\lambda}^{W}\frac{\partial e_w}{\partial_{\lambda}} + v_{z}^{W}\frac{\partial e_w}{\partial z} = \sum_{i=1}^{3}\alpha_2^{i}Q_{e\psi}^{i} - H_1^{e} + \alpha_1 H_a^{e\psi}$$

$$\frac{\partial \psi_w}{\partial t} + v_{\varphi}^{W}\frac{\partial \psi_w}{\partial \varphi} + v_{\lambda}^{W}\frac{\partial \psi_w}{\partial \lambda} + v_{z}^{W}\frac{\partial \psi_w}{\partial z} = (1-\alpha_1)H_a^{e\psi} + k_2^{W}\frac{\partial^2 \psi_W}{\partial z^2}$$
$$- H_Z^{\psi} - H_F^{\psi} - H_D^{\psi} - H_a^{\psi}$$

$$\frac{\partial e^*}{\partial t} = H_1^{e} - \alpha_1\left(H_L^{e\psi} + H_a^{e\psi}\right),$$

$$\frac{\partial \psi^*}{\partial t} = H_D^{\psi} - (1-\alpha_1)(H_L^{e\psi} + H_a^{e\psi})$$

where $e^*(e_w)$ and $\psi^*(\psi_w)$ are the concentrations of heavy metals in the bottom sediments (water) as solid and dissolved phases, respectively; H_a^{ψ} is the output of heavy metals from the sea to the atmosphere by evaporation and spray; $Q_{e\psi}^{i}$ is the input of heavy metals to the sea with river waters (i = 1), atmospheric deposition (i = 2) and ship's wastes (i = 3); α_2^{i} is the part of the suspended particles in the ith flow of heavy metals; $V = \left\{v_{\varphi}^{w}, v_{\lambda}^{w}, v_{z}^{w}\right\}$ is the flow velocity, and α_1 is the part of the solid fraction of heavy metals in the bottom sediments.

The MTP block simulates high and low tide processes that are subjected to the periodic law with time τ of the water mass moving across the estuary border Γ with velocity

$$V(\varphi,\lambda,t) = V_{\Gamma}(t) = V_{max}|\cos[\pi(t - t_{max})/\tau]|; (\varphi,\lambda)$$
$$\in \Gamma, t_{max} - \tau/2 \le t \le t_{max} + \tau/2;$$

where t_{max} is time when velocity of water flow across Γ reaches maximal value V_{max}.

Maximal change of the estuary depth is parameterized by the following formula:

$$z_p(\varphi,\lambda,t) = z_0(\varphi,\lambda,t) + \sigma_{\Gamma}V_{max}t/(2\sigma)$$

where σ_{Γ} is integral value of area for the estuary input section located along Γ, z_0 is the depth at the time of the tide finishing.

The MWC block realizes model of surface runoff and gives a possibility to assess the flux $E(\varphi, \lambda, \xi, \eta, d, t)$ of pollutant of type ξ at the state η with parameter d:

$$E(\varphi, \lambda, \xi, \eta, d, t) = S(\varphi, \lambda, \xi, \eta, d, t)Q(\varphi, \lambda, t)k(\xi, \eta, d, t)$$

where k is the washout coefficient, S is the pollutant concentration in the soil, Q is the runoff.

ESECEZ functioning needs the series environmental characteristics the measuring of which can be realized in framework of national hydrometeorological service and derived from satellite data. Islam et al. (2005) demonstrated a means of receiving monthly time series of SST data from NOAA satellites in the South-China Sea region.

9.3 Decision Making Under Dependence of Indicators

9.3.1 The Natural Disasters as a Dynamic Category of Environmental Phenomena

With the development of civilization the urgency of the problem of predicting the scale of expected changes in the environment, including climate, has grown. First of all, it is a question of the appearance and propagation of undesirable natural phenomena, which are dangerous (even lethal) for living beings and causing large-scale economic losses. Such phenomena have been called natural disasters. Disasters are distinguished between geophysical (earthquake, volcano and dry mass movement), meteorological (storm), hydrological (flood and wet mass movement), climatological (extreme temperature, drought and wildfire) and biological (epidemic, insect infestation and animal stampede). Economic losses and victims depend on the disaster subgroup.

In the course of long-term evolution, natural anomalies of various spatial and temporal scales are known to have played an important role in ecodynamics causing and activating the regulation mechanisms for natural systems. With the development of industry and the growth of population, such mechanisms have changed considerably and in some cases have become of threatening level. First of all, this is connected with the growth and propagation of the amplitude of anthropogenic environmental changes.

Numerous studies of last decades have shown that in some cases the frequency of occurrence of natural disasters and their scale have been growing, leading to an enhancement of the risk of large economic and human losses as well as to destruction of social infrastructure. So, for instance, in 2011, about 332 natural catastrophes happened over the globe, with victims amount to more than 30 thousand people and economic losses exceeding 360 billion dollars. In 2012, 357 natural triggered disasters were registered. This was both less than the average annual disaster frequency observed from 2002 to 2011 (394), and represented a decrease in associated human impacts of disasters in 2012, which were at their lowest level compared to previous years. However, natural disasters still killed a significant number, a total of 9,655 people were killed (annual average 2002 −2011:107,000) and 124.5 million people become victims worldwide (annual average 2002−2011:268 million). This list can be continued. Costliest of natural disasters is full of new losses. For example, disaster statistics of USA shows that average affected people per year consists about 0.9 millions and economic damage per year reaches US$18 billions.

The amount of damage depends strongly on the readiness of the territory to reduce the risk of losses and it changes substantially in time. Maximum losses are caused by floods and hurricanes. The spatial distribution of disasters is rather non-uniform. On the whole, the relative distribution of the number of catastrophes by their types and over continents is characterized by the following indicators: tropical storms—32 %, floods—32 %, earthquakes—12 %, droughts—10 %, and the share of other disasters constitutes 14 % (including Asia—38 %, America—26 %, Africa —14 %, Europe—14 %, and Oceania—8 %).

Natural disasters can be classified into various categories. Large-scale catastrophes are environmental phenomena leading to human victims and destruction of houses with the economic damage substantial for a given region. Hence, the scale of natural disasters depends on the level of economic development of the region, which determines the degree of population protection from natural disasters. Therefore, studies of phenomena connected with natural disasters should be accompanied by analysis of the level of poverty for a given region. The results of studies for the last 25 years show that in developing countries the dependence of losses due to natural disasters is much stronger than those in the economically developed regions. Bearing in mind that during the last decade the number and scale of natural disasters increased, on the average, 5 times and their danger 9 times, it becomes clear, what dangers may be expected for the population of these countries in the nearest future. There is no doubt that prediction of and warning about critical phenomena over the globe should be of concern for all countries independent of their economic development.

In mathematics, the theory of catastrophes has been well developed. But its application to the description of events and processes in the real environment requires the use of the methods of system analysis to substantiate the global model of the Nature-Society system (NSS) with the use of technical facilities of satellite monitoring. Solution of the problems is in the sphere of ecoinformatics, which provides a combination of analytical simple, semi-empirical and complex non-linear models of ecosystems with renewable global data sets. Many international and national programmes to study the environment could have recently raised the level of coordination in order to reach a needed level of efficiency. This coordination can be exemplified by the programmes Global Carbon Project (GCP) and Earth Observing System (EOS), within which most efficient information and technical facilities to evaluate and predict the NSS dynamic characteristics have been concentrated.

The development of constructive methods of prediction of natural disasters requires solution of a number of problems, including:

- Adaptation of methods of ecoinformatics as applied to the problem of diagnostics and prediction of natural disasters in all their diversity and scales.
- Formation of statistical characteristics of natural disasters in their historical aspects, separation of categories and determination of spatial and temporal scales of catastrophic changes in the environment. An analysis of the history of disasters is important for understanding the present dependences between crises in nature and society. Statistical characteristics of natural disasters in their dynamics enable one to formulate basic principles of mathematical theory of catastrophes and to determine the first-priority directions of studies.
- Development of concept and synthesis of the survivability model to use it for evaluation of the impact of natural disasters on the humans' habitat.
- Study of the laws of interactions between various elements and processes in the global NSS in correlation with the notion of biological complexity (biocomplexity) of ecosystems, considering it as a function of biological, physical, chemical, social, and behavioural interactions of sub-systems of the environment, including living organisms and their communities. The notion of biocomplexity correlates with the laws of biospheric functioning as a unity of the ecosystems of different scales (from local to global). In this connection, it is necessary to give a combined formalized description of biological, geochemical, geophysical, and anthropogenic factors and processes taking place at different levels of the spatial-temporal hierarchy of scales. It is important to assess the possibility to use biocomplexity in the form of an indicator of an approaching natural disaster.
- Study of correlations between survivability, biocomplexity, and evolution of NSS with the use of global modelling technology. Creation of the units for the global model, which describes the laws and trends in the environment that lead to an appearance of stress situations and are initiated by man's economic and political activity.
- Consideration of demographic conditions for the origin of natural disasters and revelation of mechanisms for regulation of the environment, which hamper the formation of these conditions.
- Assessment of the information content of the existing technical means of collection of data on the state of NSS sub-systems and available global data bases in order to allocate them in solving the problems of evaluation of conditions for the origin of stress situation in the environment.

As Walker (2003) noted, the definition of a natural disaster is rather vague, and its definition depends on many factors. Grigoryev and Kondratyev (2001) defined a natural disaster as an extreme and catastrophic situation in population viability caused by substantial unfavourable changes in the environment or "as a spasmodic change in the system in the form of its sudden response to smooth changes in external conditions". The amount of such critical situations in the environment increases. For instance, if before 1990 during the preceding 30 years, only in 1973 more than one thousand tornados were recorded on the territory of the USA, after 1990 this threshold was exceeded every year.

In general, natural disasters can arise from:

- weather patterns (storms, cyclones, hurricanes, floods, tornadoes, thunderstorms),
- other climatic conditions (droughts, bush fires, avalanches, colds, winterstorms),
- changes in the earth's crust (volcanoes, earthquakes, tsunami, tidal waves).

At present, natural disasters are floods, droughts, hurricanes, storms, tornados, tsunami. Volcanic eruptions, landslides, landslips, mud flows, snow avalanches, earthquakes, forest fires, dust storms, bitter frosts, heat, epidemics, locust invasions, and many other natural phenomena. In the future, this list can be extended due to appearance of new kinds of natural disasters, such as collisions with space bodies and anthropogenic ones—bioterrorism, nuclear catastrophes, sharp change of the Earth's magnetic field, plague, robots' invasion, etc. Therefore, it is important to develop effective quantitative technologies and criteria, which would reliably warn about possible dangerous catastrophic natural phenomenon.

Many experts associate the notion of a natural disaster with the notion of ecological safety, which appeared in connection with the necessity to evaluate the danger for human health on a given territory, constructions or possessions resulting from the changing environmental parameters. These changes can be both natural and anthropogenic. In the first case, the danger is caused by fluctuations in natural processes connected with changes of the synoptic situation, epidemics, or due to a natural disaster. In the second case, the danger occurs as a response of nature to man's activity. For instance, Gardner (2002), analyzing changes of the environment in the Himalayas, on the territory of India, came to the conclusion that such factors as deforestation and changes in vegetation cover, became the cause and intensifier of instability in this region characterized by a degradation of soil resources and increase of the consequences of environmental destruction due to water flows. Field and Raupach (2004) as well as Abrahamson (1989) connected a change in the pattern of the occurrence of natural disasters with the increase of instability of the carbon-climate-man system. According to Field et al. (2002), this instability during the nearest two decades can increase due to a change in many characteristics of the World Ocean's ecosystems. Milne (2004), analyzing the history of various large-scale disasters, has come to a pessimistic prognosis with respect to the fate of humankind, using the notion of "Doomsday".

In general, the ecological danger on a given territory results from a deviation of the parameters of the man's habitat beyond the limits, where after a long residence a living organism starts changing but not in the direction corresponding to the natural process of evolution. As a matter of fact, the notions of "ecological danger" or "ecological safety" are connected with the notions of stability, vitality, and integrity of the biosphere and its elements. Moreover, the Nature-Society System (NSS) being a self-organizing and self-structuring system and developing by the laws of evolution, creates within itself totalities of ecological niches whose degree of usability for population of a given territory is determined, as a rule, by natural criteria (totality of maximum permissible concentration, religious dogmas, national traditions, etc.).

Nevertheless, when considering the perspectives of life on the Earth, it is necessary to proceed from common criteria of evaluation of the levels of degradation of the environment, since in time, local and regional changes in the environment transform into global. The amplitudes of these changes are determined by mechanisms of NSS functioning, which ensure the optimal changes of its elements. Humankind deviates more and more from this optimality in its strategy of interaction with the surrounding abiotic, and biotic components of the natural environment. But at the same time, the human society as an NSS element, tries to understand the character of large-scale relationships with nature, applying many sciences and studying the cause-and-effect connections in this system. One of such cause-and-effect connections is a correlation of El Niño with the ozone layer. El Niño refers to natural phenomena directly related to natural disasters taking place in the equatorial sector of the East Pacific. It is a complex of interrelated variations of chemical and thermobaric parameters of the atmosphere and ocean. The anomalous character of natural processes called El Niño and observed mainly near the shore of Peru and Chili, is manifested via a sharp increase of water temperature, air pressure decrease in the Pacific and a substantial change of air fluxes direction. All this takes place over one of the most active parts of the world source of degassing—the East-Pacific elevation. The mechanism of El Niño occurrence is as follows. Hydrogen, rising from the ocean's bottom of the rift zone, reaches the ocean's surface, and due to heat release in the reaction with oxygen warms the water of the upper photic layer. As a result, the CO_2 solubility decreases, and its flux from the ocean to the atmosphere increases. From the estimates of Monin and Shishkov (1991), during the 1982−83 El Niño the atmosphere got 6,000 Tg CO_2. Besides, water evaporation increases, which, together with CO_2, enhances the greenhouse effect. The heating of water leads to the origin of typhoons, a decrease of atmospheric pressure, and breaking the standard trade-wind scheme of atmospheric dynamics. All these changes trigger feedbacks that restore the equilibrium of natural processes. One of the important regulators here is the ozone layer whose depletion during El Niño leads to an increase of the temperature gradient between the equatorial and southern c\sectors of the Pacific.

The humans' habitat is a complicated dynamic system. Its temporal stability is connected with the constancy of the structure, material composition and energy balance, as well as with stability of its response to the same external forcings. The system's stability can be broken due to the impact of both passive and active external forces. In other words, under present conditions, the nature N and human society H being a single planetary system and having hierarchical structures (|N|, |H|), interact, each with the aims of its own ($\underline{N}$, $\underline{H}$). From the formal point of view, this interaction can be considered as a random process $\eta(x, t)$ with the distribution law unknown, representing the level of tension in relationships between subsystems N and H or assessing the state of one of them. Here $x = \{x_1, \ldots, x_n\}$ is a set of identifying characteristics of subsystems N and H, which are the components of a possible indicator of the origin of natural disaster, that is, a deviation of $\eta(x, t)$ beyond the limits, where the state of the subsystem N threatens the subsystem H. It follows from this that the aims and behavior of subsystems N and H are functions of the indicator η depending on which their behavior can be antagonistic, indifferent or cooperative. The main goal of the subsystem H consists in reaching a high standard of living ensuring a long comfortable life. The goal and behavior of the subsystem N is determined by objective laws of co-evolution. In this sense, a division between N and H is conditional, and it can be interpreted as a division of a multitude of natural processes into controlled and non controlled. It is clear that with the growing population density, natural disasters will intensify the feeling of discomfort, affecting the social and cultural conditions in many regions.

Without dwelling upon philosophical aspects of this division, we shall consider the systems H and N symmetrical in a sense of their description given above, and open. Here the system H disposes of technologies, science, economic potential, industrial and agricultural productions, sociological arrangement, size of the population, etc. The process of interaction of the systems H and N leads to a change of η, whose level affects the structure of vectors $\underline{H}$ and $\overline{H}$. Really, there is a threshold η_{max}, beyond which humankind stops existing, and nature survives. The asymmetry of subsystems H and N in this sense causes a change of the goal and strategy of the system H. Apparently, under present conditions, interactions of these systems $\eta \rightarrow \eta_{max}$ take place rather rapidly, and therefore some elements of vector $\underline{H}$ can be attributed to the class of the cooperative. Since the present socio-economic structure of the world is represented with a totality of states, a country should be considered a functional element of the system H. The function $\eta(x, t)$ reflects the result of interaction of countries between themselves and with nature. A totality of the results of these interactions can be described by the matrix $B = \|b_{ij}\|$, each element of which with the symbolic sense of its own:

$$b_{ij} = \begin{cases} +\text{cooperative behavior;} \\ -\text{antagonistic relationships;} \\ 0\,\text{indifferent behavior.} \end{cases}$$

Many theories have been dedicated to studies of the laws of interactions of complicated systems of various origins. In the asymmetrical case considered here, it is a question of the survival of the system H and an attempt to find a means to assess the future dynamics of the system N. According to Podlazov (2001), the reflexive behavior of H will help humankind, eventually, to find a behavioral pattern "able to weight profits and danger, to understand principal limitations of our capabilities, and to feel new threats in due time". As Chernavsky (2004) notes, a human being is versatile, and knowledge of this synergetic capability will make it possible in future to describe the system H U N bearing in mind all social peculiarities in their variability, observing the boundaries of integral mentality of human society. Mechanisms of self-organization and self-regulation of natural systems determines the complexity of this method (Ivanov-Rostovtsev et al. 2001). Of course, deep semantic and philosophical notions of personal architectonics, which should have been taken into account when forming the model of vitality, at the present level of formalized description of intellect remain beyond feasibility of present ecoinformatics.

9.3.2 Decision Making and Indicators

The acceptance and documentation of decisions is an integral part of many tasks associated with artificial intelligence. In tasks of classification and recognition of images, as well as in tasks of choice and assessment, it is needed to take decisions and to validate the alternative ones (Shutko and Krapivin 2011). In suppressing number of cases, decision rules are made under the preposition about independence of indicators which describe recognized and analyzed objects. Under this these indicators are correlated and have probabilistic character. Thus, there exists the task of the developing the decision procedure, which could take into account of correlations and dependencies between different attributes.

Ensemble situations composed of S subsets of classes for the recognition is considered:

$$M = \{M_j\} \; (j = 1, \ldots, S) \tag{9.1}$$

In this case, "ensemble" is some population of the objects (situations) for which the notion of closeness or similarity is defined. Each ensemble is described by set of parameters or signs having dimension n. This implies that in common case description of given class or situation includes n components:

$$Q = \{Q_i\} \; (i = 1, \ldots, n) \tag{9.2}$$

Each component Q_i is characterized by specific degree of intensity or weight v_i. On the whole, recognized objects or processes are described by function that can consider as decision rule:

$$f_k = \{v_1, \ldots, v_I, \ldots, v_n\} \tag{9.3}$$

Method of etalons is commonly used to solve the recognition task when etalon base is created preliminary. Usually, in this case the decision rule is described by expression: $f_k \in M_i/R_f$ where R_f is some distance in space M.

Let us consider the case when the task of recognition of spectral images in hydrophysical experiment is solved. One of algorithms of multi-channel statistical analyzer (see Sect. 3.5.2) is the recognition of spectral images of polluted water solutions. Space of spectral images is formed during the learning procedure with the use of spectroellipsometric or spectrophotometer system (Mkrtchyan et al. 2011). This procedure is realized in laboratory conditions, when spectral images and chemical analysis are carried out at the same time. Figures 9.8, 9.9, 9.10 and Table 9.7 show examples of spectral images delivered by spectroellipsometric system (Fig. 9.11).

The problem of detection and identification of the pollution spots on the water surface, especially of oil spills, is being solved by many scientists. The recently developed technology of an adaptive identification of the environmental elements

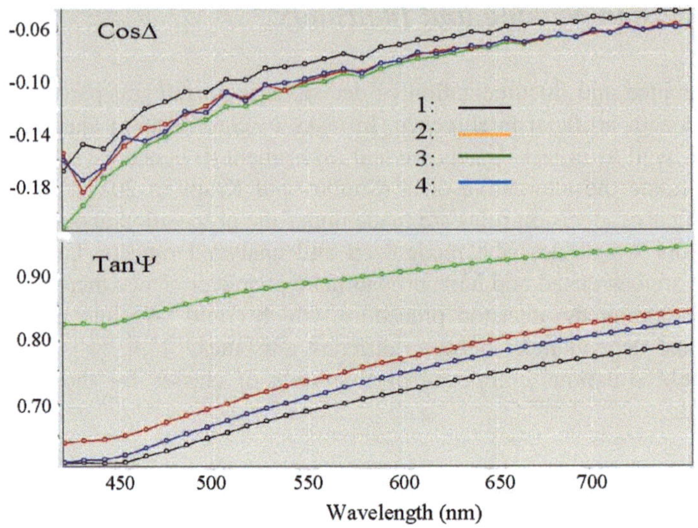

Fig. 9.8 Spectral images of bifidumbacterium water solution. *Notation* 1–5 doses of bifidumbacterium in 50 ml of water, 2–5 doses in 100 ml of water, 3–5 doses in 150 ml of water, 4–5 doses in 200 ml of water; CosΔ and TanΨ are spectroellipsometric parameters where Δ and Ψ are spectroellipsometric angles (Perov et al. 1994)

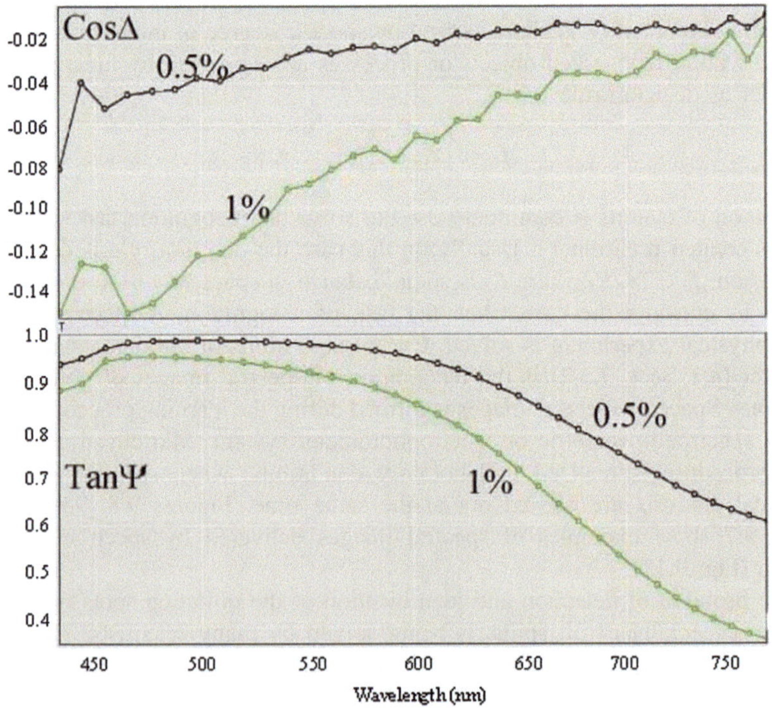

Fig. 9.9 Spectral images of water solution of Na + Cu + Zn + Mn + glucose

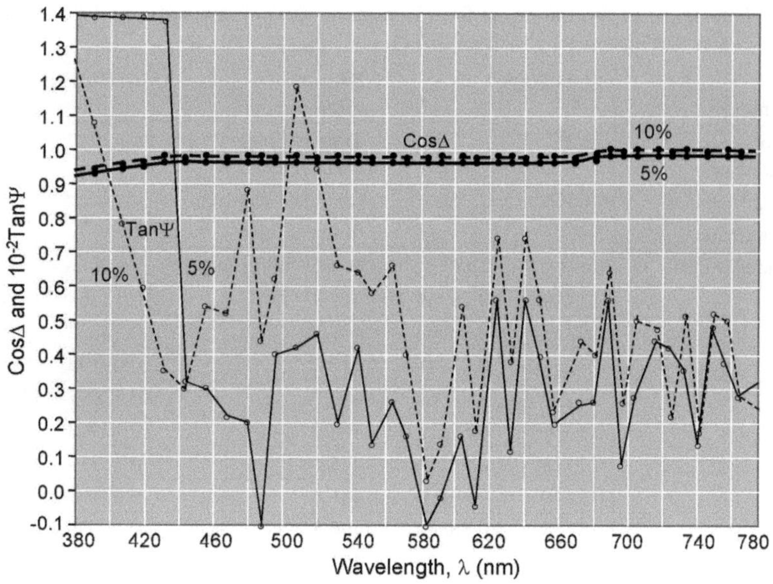

Fig. 9.10 Spectral characteristics of ammonia solution in the water. Solution concentration is shown on spectral curves

from measurements in the visible spectral region permits to synthesize an expert system for an adaptive identification of the environmental parameters (ESAIEP). The system's structure includes a compact multi-channel spectroellipsometer (MSP), information interface with computer, computer software (STW), and extending database (Mkrtchyan and Krapivin 2014).

The principle of the ESAIEP functioning is based on fixation of changes of the light flux at the MSP output and their transformation into a digital code. Further processing of these data with respect to their efficiency is determined by the STW composition containing various algorithms of recognition of 2D objects. The adaptability of the recognition procedure is determined by the level of accumulated knowledge about special features of intensity fluctuations and polarizing properties of the light reflected from the water surface. The STW includes the means that make it possible, in case of uncertain identification of the pollution spot, to make an expert decision based on the visual analysis of its spectral image. This procedure is carried out in the mode of dialogue with ESAIEP, and if decision is made, the operator can fix it in the database in the form of a standard for subsequent situations of an appearance of similar spots. Figure 9.11 characterizes general view of the ESAIEP.

Spectral image of water solution in optical range is *n*—volumetric vector that can be considered as element of cluster in the vector space describing concrete concentration of chemical element in water solution. Following to the signal

Table 9.7 Example of spectral image from base of the spectral standards, received by means of 35-channel spectroellipsometer (Soldatov 2011)

Date of measurement	14.04.2008		Date of measurement	14.04.2008	
Time of measurement	12:38		Time of measurement	12:40	
Chemical element	$CuSO_4$		Chemical element	$CuSO_4$	
Concentration	5 %		Concentration	10 %	
Wavelength (nm)	CosΔ	TgΨ	Wavelength (nm)	CosΔ	TgΨ
419.9	0.14495	0.79965	419.9	0.12868	0.78325
431.1	0.1438	0.83903	431.1	0.13372	0.82469
442.2	0.10568	0.87447	442.2	0.09193	0.86601
453.2	0.13294	0.88662	453.2	0.12097	0.88063
464.2	0.13363	0.89171	464.2	0.12963	0.88589
475	0.14282	0.89296	475	0.13339	0.88706
485.8	0.14405	0.89206	485.8	0.13707	0.8853
496.5	0.1448	0.88864	496.5	0.13698	0.87899
507.1	0.14249	0.88335	507.1	0.13936	0.86759
517.6	0.1364	0.87566	517.6	0.13176	0.85249
528	0.13757	0.86446	528	0.13611	0.83037
538.3	0.13057	0.84937	538.3	0.13048	0.8006
548.6	0.12315	0.82841	548.6	0.12313	0.76204
558.8	0.12031	0.80108	558.8	0.11616	0.71351
568.9	0.11322	0.76609	568.9	0.11295	0.65426
578.9	0.102	0.72329	578.9	0.1	0.58567
588.8	0.1098	0.67166	588.8	0.10753	0.50848
598.7	0.10002	0.61256	598.7	0.09581	0.42641
608.4	0.10565	0.54636	608.4	0.10102	0.34281
618.1	0.09871	0.47603	618.1	0.09216	0.26318
627.7	0.10558	0.40321	627.7	0.09899	0.1916
637.2	0.10169	0.33214	637.2	0.09465	0.13222
646.6	0.09366	0.26616	646.6	0.08989	0.08648
656	0.10029	0.2071	656	0.10013	0.05369
665.2	0.09585	0.15774	665.2	0.09004	0.03208
674.4	0.09909	0.11773	674.4	0.11971	0.01871
683.5	0.09421	0.08664	683.5	0.11268	0.01098
692.5	0.09696	0.06344	692.5	0.17604	0.00689
701.4	0.10772	0.04653	701.4	0.08982	0.0048
710.3	0.10526	0.03473	710.3	0.14121	0.004
719	0.10284	0.02651	719	0.16567	0.00382
727.7	0.12483	0.02099	727.7	0.2139	0.00402
736.3	0.13675	0.017357	736.3	0.27278	0.00444
744.8	0.14476	0.01522	744.8	0.27707	0.0051
753.2	0.17772	0.0141	753.2	0.26209	0.00592

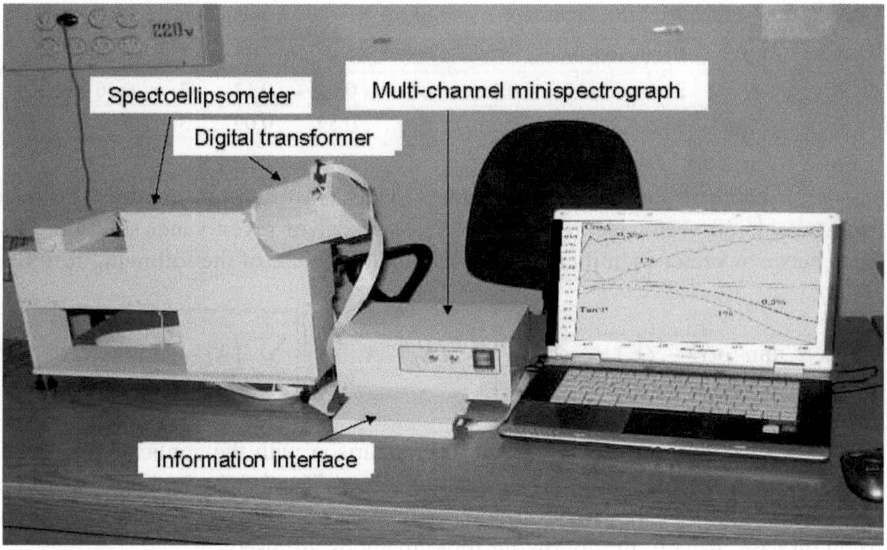

Fig. 9.11 General view of the multi-channel spectroellipsometric system for the water quality assessment (Krapivin et al. 2014)

recognition in radiotechnics, each optical spectrum can be represented as some standard:

$$Q_\Delta = \{A_{\Delta 1}, \ldots, A_{\Delta i}, \ldots, A_{\Delta n}\}; Q_\Psi = \{A_{\Psi 1}, \ldots, A_{\Psi i}, \ldots, A_{\Psi n}\}$$

Here Q_Δ is the light intensity on the photodetector output, Q_ψ is tangent of relative shift in phases of two orthogonal polarized components. An example of standard structure is given in Table 9.8. This structure is used as base of the decision making system (BDMS). Table 9.9 gives an example of standard etalon from this base.

An identification of spectral image for unknown water solution is realized by means of comparison his vector—identifier with elements of the BDMS. Spectral image of water solution can be represented by one or two vector-identifiers calculated with the use of rule described in Table 9.8. Final identification is realized by

Table 9.8 Structure of standard spectral image of water solution

Etalon number	A_1	A_2	A_3	A_4	A_5	A_6	A_7	A_8	A_9	B
1	A_{11}	A_{21}	A_{31}	A_{41}	A_{51}	A_{61}	A_{71}	A_{81}	A_{91}	B_1
……										
n	A_{1n}	A_{2n}	A_{3n}	A_{4n}	A_{5n}	A_{6n}	A_{7n}	A_{8n}	A_{9n}	B_n

Notation A_1 is the square occupied by spectral curve, A_2 is the maximal value of spectral curve, A_3 is the minimal value of spectral curve, A_4 is the distance between wavelengths with minimal and maximal values of spectral curve, respectively; A_5 is the maximal derivative of spectral curve; A_6 is the maximal value of second derivative of spectral curve; A_7 is the number of spectral curve maximums; A_8 is the average value of spectral curve; A_9 is the wavelength corresponded to average value of spectral curve; B is the chemical element concentration

Table 9.9 An example of the database element of the decision making system

Solution	Q	Vector-identifiers								
$CuSO_4$ 10 %	Q_Δ	21.6	0.17	0.1	143	0.67	0.12	2	0.16	0.21
	Q_Ψ	43.1	0.89	0.04	201	0.59	0.09	3	0.65	0.01

means of search in the BDMS of vector—identifiers which are minimal distance from considered vector-identifier $Q = \{X_1, \ldots, X_{n=}\}$ of given water solution. Distance between vector-identifiers is calculated with the use of the following formula:

$$\delta = \min_n \rho(Q - Q_n) = \frac{1}{2n} \min_i \left[\sum_{j=1}^{n} \left| X_j - A_j^i \right| + \sqrt{\sum_{j=1}^{n} \left(X_j - A_j^i \right)^2} \right] \qquad (9.4)$$

Use of Eq. (9.4) gives better result in comparison with the application of other known criteria of closeness between spectral curves. That is why in this case there is minimal risk to miss the situation with dangerous pollution of water reservoir. In common case, usually the following three methods are used:

- *Cluster analysis.* In this case two types of clusters are formed for CosΔ and TanΨ. Decision is made by weighted value of δ from Eq. (9.4).
- *Algorithm of discrepancy between spectra.* It is assessed average distance between the ordinates for both spectra and studied case and decision is made taking into account minimal value of this distance.
- *Algorithm of discrepancy between etalon vectors.* In this case decision made taking into considered minimal value of δ from Eq. (9.4).

Table 9.10 gives experimental results, which give possibility to compare these algorithms. Dependence of risk assessment as function of solution concentration is represented in Figs. 9.12 and 9.13. As it follows from these results, the risk to have

Table 9.10 Comparatively assessment of algorithms for recognition of spectral images of water solutions

Object for study	Identification algorithm and its error (%)		
	Cluster analysis	Discrepancy between spectra	Discrepancy between vector-etalons with the use (9.4)
$CuSO_4$	15	12	8
NaCl	17	11	7
$NaHCO_3$	16	10	5
NH_4OH	21	13	9
$ZnSO_4$	22	12	8
Potassium iodite	13	10	6
Na + Cu + Zn + Mn + glucose	18	9	9
Furaciline	23	11	5
Bifidumbacterium	14	10	4

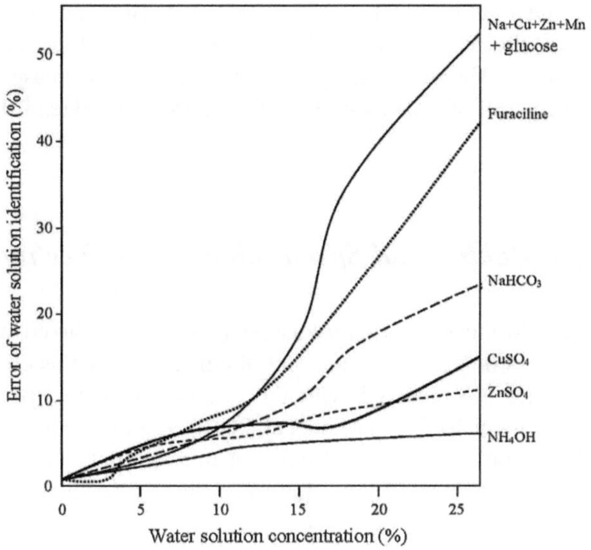

Fig. 9.12 Dependence of spectral image identification on the water solution concentration

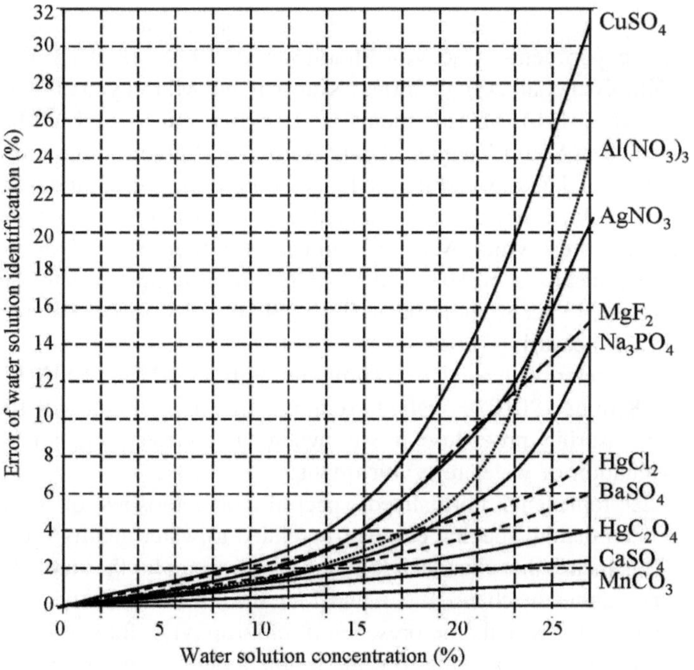

Fig. 9.13 Dependence of identification error on the liquid solution concentration

high error under the solution identification is reduced when algorithm of Table 9.8 is used. We see that the risk to have high error is growing with increase of chemical element concentration. The discrepancy between spectra is decreased with increase of chemicals concentration. In this case, it is necessary to extend the database of spectral etalons.

9.3.3 Decision Making and Spectrophotometric Technology

Many tasks related to the water quality assessment can be solved with the use of simple optical devices that deliver one spectral curve. Now let us consider the case when spectral images are received with the use of 8-channel spectrophotometer (Krapivin and Mkrtchyan 2007a, b; Man et al. 2010). According to the rule proposed by Bickel and Bo (2006) the sorting of the measurement data $\{x_i\}$ is accomplished by the following procedure. The value $R = (x_i - \overline{x})/\sigma$ is calculated, where $\sigma = \sqrt{\sum_{i=1}^{n} (x_i - \overline{x})^2/n}$, $\overline{x} = (1/n)\sum_{i=1}^{n} x_i$. Decision about the sample formation is made, depending on the value *R*. If $R > R_o$, measured value x_i is not included in the set $\{x_i\}$. If $R \le R_o$, value x_i is included in the set $\{x_i\}$ for the following analysis. Threshold value R_o is function of number of measurements (Table 2.22).

Both the Simulation System for Hydrophysical and Hydrochemical Investigations (SSHHI) (Fig. 9.5) and Expert System for Ecological Control of Estuary Zone (ESECEZ) (Fig. 9.7) were tested in the framework of joint Russian-Vietnamese hydrophysical experiments (Cao Van Phuong et al. 2009a, b; 2012; Bui Ta Long et al. 2010; Krapivin et al. 2008b, 2009a; Krapivin and Mkrtchyan 2006a, b, 2007d, 2008; Man et al. 2011b). Measurements were carried out to study different water reservoirs located in South Vietnam: Dong Nai River, Saigon River, coastal waters of South-China Sea, lagoons and flooded reservoirs of the fishing industry. Spectral characteristics of these objects are changed with high range. It says that characteristics of water reservoirs were dynamical and their operative control was implemented by means of SSHHI and ESECEZ. Measurements were obtained by means of 8-channel spectrophotometer (Figs. 9.14, 9.15, 9.16). Figures 9.17 and 9.18 explain the measuring process.

Spectrophotometer connected with computer with SSHHI and ESECEZ software according to Klimov (2010) is called Adaptive Identifier. Figures 9.16 and 9.17 explain the measuring procedure in-situ when an educated system is used for operative evaluation of water reservoir quality.

Figure 9.19 represents normalized spectral characteristics of studied water objects. It is seen that all spectral curves I/I_0 related to water quality intersect in the point (4,1). This means that channel 4 (510 nm) is invariant for the studied reservoirs. It should be noted that the channels 5, 6 and 7 are the most informative. Maximum of curve S for channel 7 reveals the presence of chlorophyll-a fluorescence on wavelength 690 nm and accompanying to it the pigments in red spectrum exists.

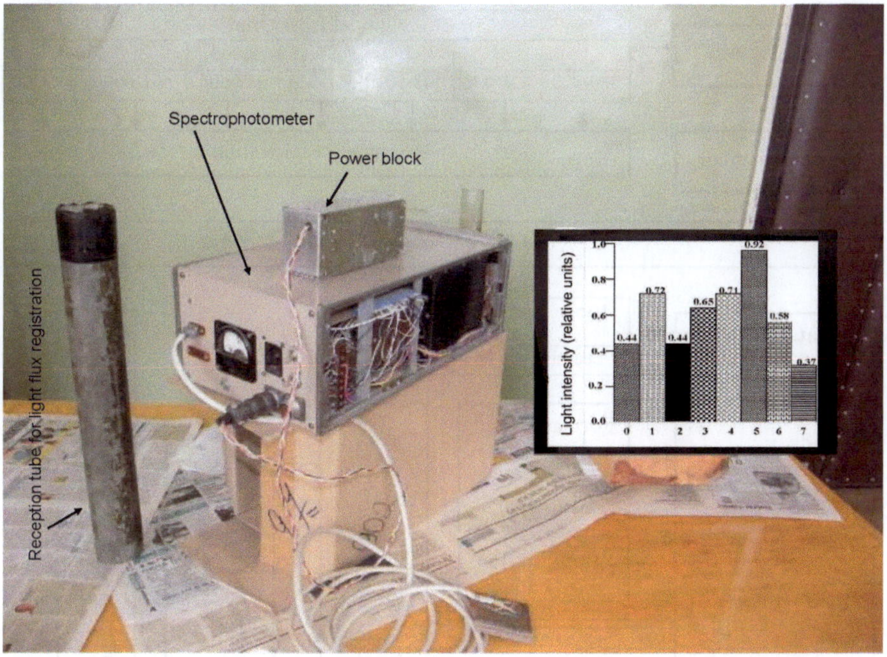

Fig. 9.14 General view of 8-channel spectrophotometer (Aleshin and Klimov 1992; Klimov et al. 1999)

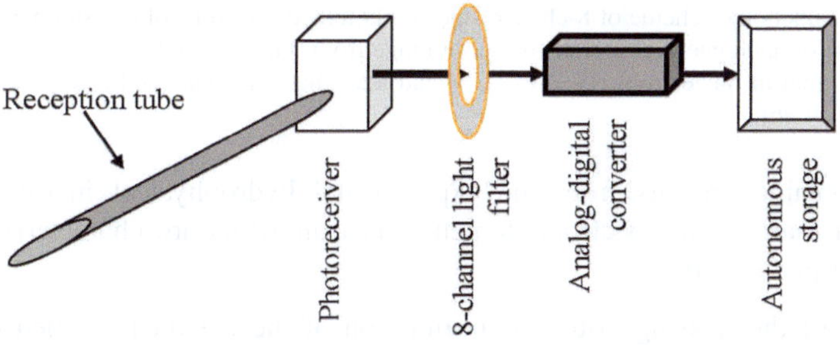

Fig. 9.15 Schematic diagram of the 8-channel spectrophotometer (Soldatov 2011)

Light intensity is attenuated by water layer according to the law $I(z) = I(0)\exp(-\beta z)$, where z is the depth (m), β is the attenuation coefficient (m^{-1}). One of the inaccurate methods for the assessment of water quality is based on the value of parameter β. This method does not give the precise assessment of water quality. Nevertheless, it shows how many pollutants there are into the water. Tables 9.11 and 9.12 characterize the optical parameters for a few reservoirs of South Vietnam.

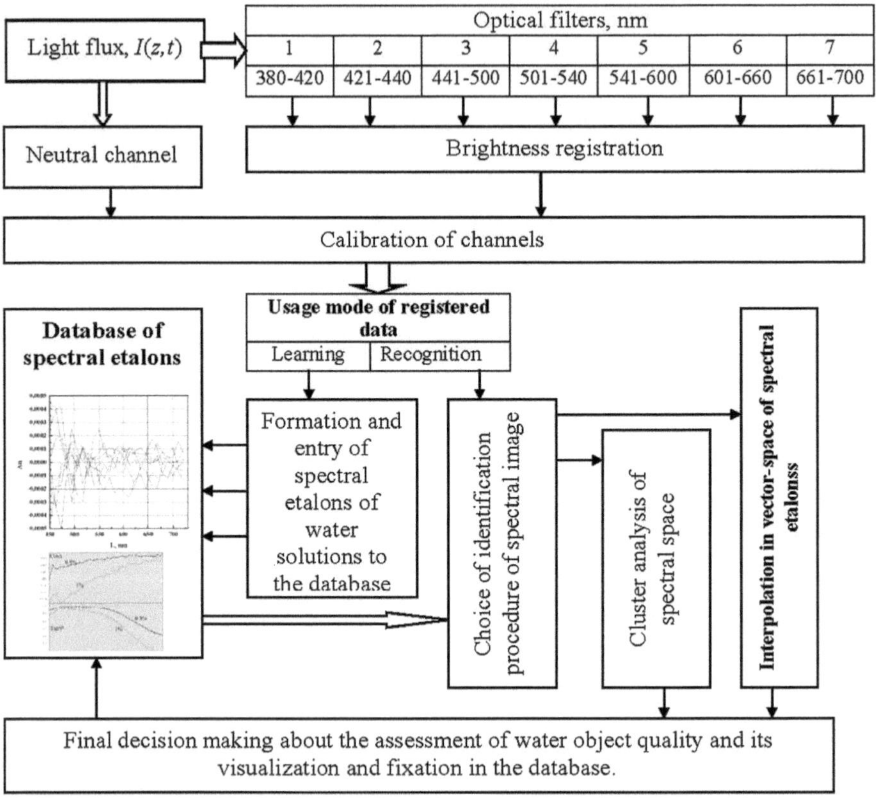

Fig. 9.16 Functional scheme of 8-channel spectrophotometer. Results of measurements are given by the scale of analogue-digital transformer in range from 0 to 2^{12}. Light signal is transformed in electrical signal in range from −5 V to +5 V and after that it is expressed as digital code in the interval [0, 4096]

The obtained results from the experimental hydrophysical investigations in South Vietnam using spectrophotometric measurements are characterized by the following parameters:

- Risks of the missing critical concentration of the chemical elements in water solution are varied from 0.12 to 0.31.
- Risks of the missing critical concentration of the oil products in water solution are varied from 0.09 to 0.22.
- Risks of the missing critical concentration of the biogenic salts in water solution are varied from 0.16 to 0.39.

These risks depend on the forms of spectral images and on the number of spectral channels. These risks are smaller for measurements obtained from the spectroellipsometer. That is why decrease of these risks can be achieved by means of coupled use of different devices and data processing methods. According to

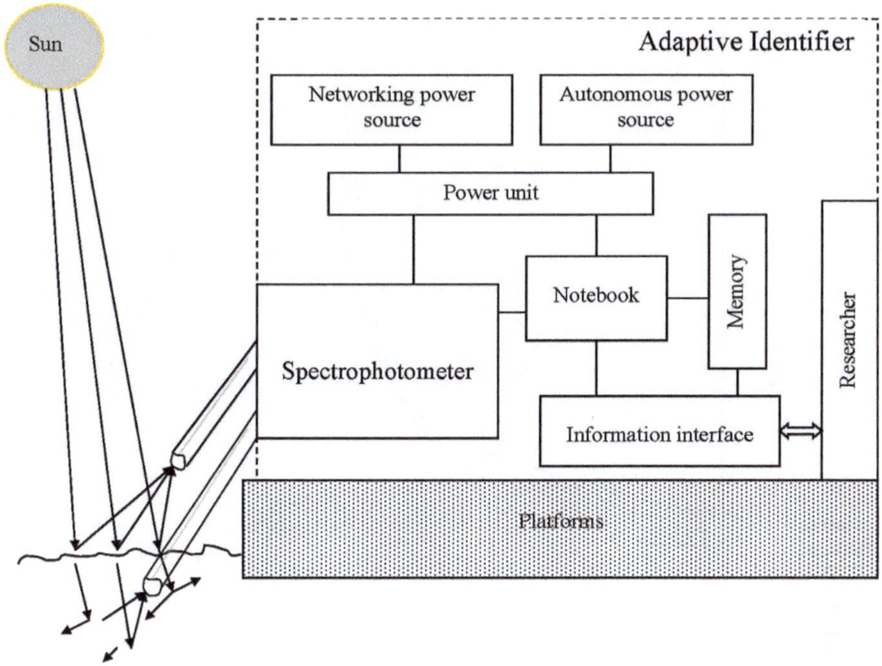

Fig. 9.17 General scheme and organization of the control system for monitoring of water bodies

Fig. 9.18 Preparation of adaptive identifier to measure water quality in site of water intake for Ho Chi Minh City (Dong Nai River)

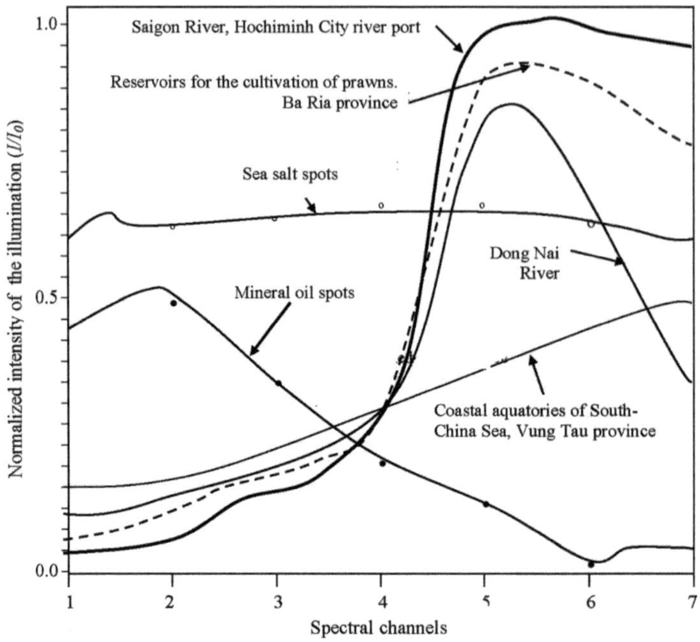

Fig. 9.19 Spectral characteristics of some reservoirs located in South Vietnam

Table 9.11 Evaluation of the integral attenuation of light β (m^{-1}) at the depth of 1 m under zenith angle of the Sun 75° for different reservoirs of South Vietnam

Substance	Water object				
	Dong Nai River	Saigon River	South-China Sea	Pond-1 in Ba Ria province	Pond-2 in Ba Ria province
Pure water	0.046	0.13	0.075	0.039	0.038
Oil	0.14	0.41	0.23	0.12	0.12
Dissolved organic matter	0.082	0.25	0.14	0.071	0.069
Weighed matter	0.094	0.28	0.15	0.079	0.077
Chlorophyll-a	0.051	0.15	0.083	0.043	0.041
Yellow matter	0.12	0.097	0.11	0.018	0.019
Biogenic salts	0.059	0.18	0.097	0.051	0.049

Soldatov (2011) the cluster analysis application provides the possibility to reduce these risks. The clusters that are formed in the space of the spectral images are very well separated. It enables to reduce these risks into the interval from 0.06 to 0.17.

The Adaptive Identifier is based on the employment of ESAIEP, ESECEZ and SSHHI functions which are used for the solution of the specific problems arising under the water quality diagnostics. The adaptive identifier allows the generation of

Table 9.12 Evaluation of the integral attenuation of light β (m^{-1}) at the depth of 1 m under zenith angle of the Sun 75° for different reservoirs of South Vietnam

Substance	Spectral channel						
	1	2	3	4	5	6	7
Pure water	0.121	0.109	0.078	0.117	0.121	0.134	0.038
Oil	0.431	0.482	0.391	0.217	0.189	0.009	0.074
Dissolved organic matter	0.033	0.027	0.015	0.183	0.309	0.295	0.177
Weighed matter	0.197	0.264	0.333	0.284	0.271	0.255	0.243
Chlorophyll-a	0.088	0.171	0.106	0.112	0.103	0.122	0.101
Yellow matter	0.123	0.112	0.109	0.065	0.064	0.063	0.059
Biogenic salts	0.172	0.181	0.164	0.183	0.179	0.159	0.147

Object for study is Saigon River in Ho Chi Minh City area

final product as the evaluation of physical-chemical characteristics of the water environment using models gives its spatial and dynamic distributions. The adaptive identifier takes into account both physical and ecological processes of the water characteristic transformations. In fact, the following equation is used to describe the dynamics of chemical substance B:

$$\frac{\partial B}{\partial t} + \vartheta_{\varphi}\frac{\partial B}{\partial \varphi} + \vartheta_{\psi}\frac{\partial B}{\partial \psi} + \vartheta_{z}\frac{\partial B}{\partial z} = Q_B - \sum_{i=1}^{n} H_B^i + \chi_{\varphi}\frac{\partial^2 B}{\partial \varphi^2} + \chi_{\psi}\frac{\partial^2 B}{\partial \psi^2} + \chi_{z}\frac{\partial^2 B}{\partial z^2} \tag{9.5}$$

where φ is the latitude, λ is the longitude, $v(v_{\varphi}, v_{\lambda}, v_z)$ is the current velocity, z is the water depth, χ is the turbulence factor, Q_B is the anthropogenic and natural sources of element B, H_B^i is the element B transformation process due to gravitational settling (i = 1), transfer to the food pyramid (i = 2), decomposition under the effect of microorganisms (i = 3), oxidizing process (i = 4), evaporation (i = 5), entry into neutral chemical compounds (i = 6), biosedimentation (i = 7) and other processes (i = 8).

Equation (9.5) is specified taking into consideration the condition on the uniform distribution of the water quality at the depth z within a limited volume. The entire water volume is divided into volumes $\Delta\varphi \times \Delta\lambda \times \Delta z$. The flow Q_B is divided into flows POL (*kk*,*i*), MARINE (*i*) and RUNS (*ss*,*i*) that characterize the inflow of the chemical element B_i derived from anthropogenic source *kk*-type, bordering water reservoir, and due to the coast drainage of *ss*-type, respectively. The functional RUNS (*ss*,*i*) is based on the data shown in the Table 9.13. The GIMS database contains series of the runoff coefficients depending on the soil groups and slopes. Considering this discretization procedure the Eq. (9.5) is changed by the system of the finite-difference equations that is solved by traditional methods (see Chap. 1).

Tables 9.14 and 9.15 show results of the water quality control. The Adaptive Identifier that is preliminary educated allows the physical-chemical characteristics assessment in their dynamics. The reconstruction error is no more than 7 %.

Table 9.13 List of runoff coefficients for different surfaces

Surface identifier (ss)	Character of surface	Runoff coefficient
Pavement		
0	Asphaltic and concrete	0.70–0.95
1	Brick	0.70–0.85
2	Roofs	0.75–0.95
Lawns, sandy soil		
3	Flat, 2 %	0.95–0.10
4	Average, 2–7 %	0.10–0.15
5	Steep, 7 %	0.15–0.20
Lawns, heavy soil		
6	Flat, 2 %	0.13–0.17
7	Average, 2–7 %	0.18–0.22
8	Steep, 7 %	0.25–0.35
9	Bare soil without vegetation	0.55–0.60
10	Soil covered by grass	0.36–0.45
11	Soil overgrown with shrubs	0.39–0.42
12	Forest plot (high trees)	0.28-0.37
13	Soil plot with many-storied houses	0.88–0.93
14	Soil plot with low-storied houses	0.79–0.86
15	Village territory	0.60–0.64
16	River	0
17	Reservoir	0

Table 9.14 Comparison of the assessments of physical-chemical characteristics of Sai Gon River delivered by the Adaptive Identifier with the laboratory analyses

Δx (km)	Sai Gon River, physical-chemical characteristics								
	Turbidity (NTU)			Suspended matters (mg/l)			Mineral oil (μg/l)		
	AI	L	Error (%)	AI	L	Error (%)	AI	L	Error (%)
0	68	66	3.0	51.5	50.2	2.6	0.42	0.38	5.3
1	73	69	5.8	49.7	52.9	7.0	0.51	0.44	5.9
5	87	84	3.6	70.3	65.7	7.0	0.55	0.48	4.6
10	89	92	3.3	88.6	90.4	2.0	0.72	0.65	7.6
15	87	91	4.4	81.3	77.7	4.6	0.61	0.55	0.9

The education of the Adaptive Identifier was realized 19 December, 2013 during 10–15 h of local time. The testing measurements were carried out on 20 and 21 December, 2013 during 10–15 h of local time. Final results of testing measurements are average values of the physical-chemical characteristics received during two days. *Notation* AI is the results delivered by the Adaptive Identifier, *L* is the data of laboratory analysis of water quality, Δx is the distance along the river from Ho Chi Minh City center

Table 9.15 The Adaptive Identifier effectiveness under the water salinity prognosis in the Nuoc Ngot Lagoon, realized from December 25, 2012 at $t_0 = 15$ h of local time

Checkpoint	Measured data ($^o/_{oo}$)	Prognosis and error (%)				
		t_0 + 7 days	t_0 + 14 days	t_0 + 21 days	t_0 + 1 month	t_0 + 1.5 month
1	30.2	32.0(5)	32.8(9)	26.4(10)	29.3(12)	34.4(14)
2	33.6	32.8(2)	34.4(10)	27.6(15)	26.4(19)	25.7(23)
3	33.6	32.4(5)	33.7(12)	37.5(17)	25.9(19)	26.2(22)
4	33.7	33.4(4)	28.3(12)	37.8(17)	26.6(20)	25.6(24)
5	33.8	33.4(4)	28.8(11)	27.6(17)	25.2(20)	26.5(22)
6	33.8	31.0(8)	30.9(12)	27.8(17)	27.5(20)	25.3(25)
7	30.7	31.4(3)	34.9(13)	27.9(17)	30.2(9)	34.5(19)
8	32.4	34.7(7)	30.9(9)	34.9(16)	26.6(19)	25.2(22)
9	32.5	33.0(6)	29.4(10)	28.6(18)	28.3(20)	25.1(23)
10	30.4	31.9(9)	33.2(12)	34.5(17)	33.2(10)	27.8(24)

9.3.4 Monitoring of Spatial Heterogeneous Water Systems

One of function of expert systems described in Chap. 8 is the spatial image reconstruction based on the data that are episodic in the time and fragmentary in the space. A specific case of the ESECEZ system is considered for the monitoring of Nuoc Ngot Lagoon located in South Vietnam (Binh Dinh province, Fig. 9.20). The Lagoon is located at the latitude and longitude coordinates of 14.15N and 109.183333W, respectively. This lagoon is used for fishing and it is located in the zone with growing intensive agriculture and industry. Its basic characteristics are given in Table 9.16.

Hydrophysical investigations in the lagoon zone were accomplishes in the frame of the Russian-Vietnamese collaboration in the ecoinformatics problems. Lagoon is divided in three levels by depth (upper, middle and near-bottom). The following characteristics were registered: direction and speed of currents, salinity, turbidity, phytoplankton biomass, zooplankton biomass, zoobenthos biomass. The scheme and the procedure of the in situ measurements are given in Fig. 9.18 and Table 9.17.

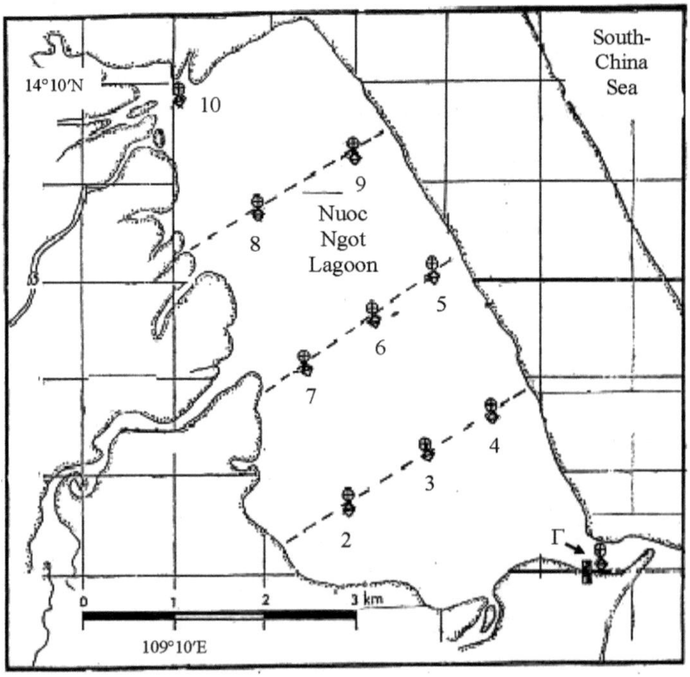

Fig. 9.20 Dam Nuoc Ngot Lagoon near Hung Lac (Vietnam)—14° 9′ 0″N; 109° 10′ 59.99″. Dislocation scheme for hydrophysical measurements. *Notation* 1−10 is points where water samples (⊕) are collected and hydrologic measurements are realized (◊); Γ is lagoon mouth and first point for water sample

Table 9.16 Parameters of the Nuoc Ngot Lagoon

Parameter	Parameter value
Area (km^2)	14.7
Wind speed (m/s)	
Average	1.9
Maximal	39.6
Air temperature (°C)	
Average	27.0
Minimal	15.8
Maximal	39.9
Relative humidity (mb)	
Average	27.9
Minimal	20.0
Maximal	32.7
Sun radiation (Kcal/cm^2)	144
Radiation balance (Kcal/cm^2)	92.5
Precipitation (mm/year)	
Average	1692.9
Minimal	778.0
Maximal	2587.0
River runoff (average annual, m^3/s)	5.6
Daily range of water flow rate across the boundary Γ with sea (m^3/s)	–380 to +470
Upper layer temperature (°C)	
Dry season	26
Rainy season	29
Water salinity (‰)	
Dry season	23.5
Rainy season	33
Mouths	2.5
Tide conditions	
Number of days with daily tide during month	18–22
Levels of tide (m)	0.3–2.0
Tide frequency (hours)	6–7
Distribution of depths (m)	
Average depth	1.6
Maximal depth	9.8

Figure 9.21 explains the tide process in lagoon. It is worth to note that the values $\grave{O}_1 = t_{lhw} - t_{shw}$ and $\grave{O}_2 = t_{shw} - t_{hwl}$ play significant role in the lagoon evolution ($T_1 \approx T_2 \approx 13$ h).

The use of ESECEZ system enables to realize the following simulation experiments:

Table 9.17 An evaluation of the ESECEZ system efficiency on the base of spatial reconstruction of water salinity in Nuoc Ngot Lagoon

Controlling sites	Water depth level	Measured water salinity (ppm)	Simulated water salinity (ppm)	Error (%)
2	Surface	30.1	33.1	10
	Middle	31.3	27.5	12
	Near-bottom	32.4	28.5	12
4	Surface	27.4	30.1	10
	Middle	30.2	33.2	10
	Near-bottom	32.2	28.0	13
6	Surface	24.9	28.4	14
	Middle	26.6	30.3	14
	Near-bottom	29.6	33.7	14
8	Surface	24.2	22.7	6
	Middle	26.8	22.5	16
	Near-bottom	28.4	26.4	7
10	Surface	28.5	25.6	10
	Middle	30.9	27.2	12
	Near-bottom	32.4	28.5	12

- Spatial structure of hydrologic and abiotic components for Nuoc Ngot Lagoon was reconstructed by means of the ESECEZ system on the basis of the in situ measurements in the control points 1, 3, 5, 7, 9 and the results obtained were compared with measurements in the control points 2, 4, 6, 8, 10.
- Particular results of prognosis of some characteristics of the lagoon were assessed taking into consideration of the in situ measurements in all control points during the prognosis time (Table 9.18).

General concept about the ESECEZ system capability to forecast the water quality in Nuoc Ngot Lagoon gives the results shown in Fig. 9.20. Simulation experiments with the ESECEZ system use show that all processes in the lagoon depend on its hydrophysical and physico-chemical interactions with South-China Sea where input data about anthropogenic impacts are available. In this case, the lagoon control can be realized using the ESECEZ system with input data obtained only in the lagoon mouth Γ (controlling site 1) with time intervals according to Fig. 9.22 (Table 9.19).

Figures 9.23 and 9.24 show phase portraits of the lagoon taking into consideration of its coordinates turbidity/salinity and salinity/water level variation. It was found that there exists a significant maximum of the dependence between the turbidity and salinity. This maximum obtains its minimal value in the mouth of the river flowed to the lagoon and its maximal value at the lagoon centre. The correlation between the salinity and the lagoon level variation has more complex character. Water salinity in the lagoon mouth weakly depends on the change of its

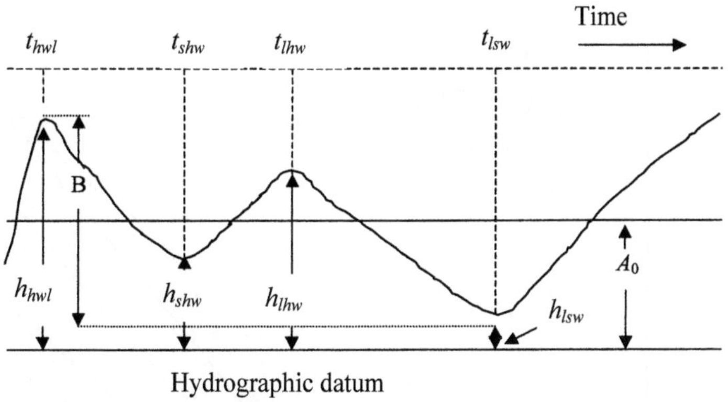

Fig. 9.21 Tide characteristics of Dam Nuoc Ngot Lagoon. *Notation* $t_{hwl}(h_{hwl})$ is the time (level) when absolute high water level is realized; $t_{shw}(h_{shw})$ is the time (level) when small high water is realized; $t_{lhw}(h_{lhw})$ is the time (level) when low high water takes place; $t_{lsw}(h_{lsw})$ is the time (level) when low small water takes place; B is the great tide level; A_0 is the mean tide level

Table 9.18 Effectiveness of the ESECEZ system to forecast water salinity in Nuoc Ngot Lagoon

Controlling sites	Measured water salinity ($^o/_{oo}$) (t_0)	Forecast and error (%)				
		t_0 + week	t_0 + two weeks	t_0 + three weeks	t_0 + month	t_0 + five weeks
1	30.2	32.0(5)	32.8(9)	26.4(10)	29.3(12)	34.4(14)
2	33.6	32.8(2)	34.4(10)	27.6(15)	26.4(19)	25.7(23)
3	33.6	32.4(5)	33.7(12)	37.5(17)	25.9(19)	26.2(22)
4	33.7	33.4(4)	28.3(12)	37.8(17)	26.6(20)	25.6(24)
5	33.8	33.4(4)	28.8(11)	27.6(17)	25.2(20)	26.5(22)
6	33.8	31.0(8)	30.9(12)	27.8(17)	27.5(20)	25.3(25)
7	30.7	31.4(3)	34.9(13)	27.9(17)	30.2(9)	34.5(19)
8	32.4	34.7(7)	30.9(9)	34.9(16)	26.6(19)	25.2(22)
9	32.5	33.0(6)	29.4(10)	28.6(18)	28.3(20)	25.1(23)
10	30.4	31.9(9)	33.2(12)	34.5(17)	33.2(10)	27.8(24)

depth. However, the water salinity in the river mouths flowed to the lagoon is sharply reduced with increase of its depth.

Table 9.20 gives some details about the precision of the ESECEZ system under the reconstruction of other characteristics in Nuoc Ngot Lagoon. The big discrepancy of the experimental and the simulated data is observed in the control sites 3 and 7 when PO_4^{3-} is assessed. Evidently, the main cause of this discrepancy lies in that washout processes for chemical elements from the bottom sediments are not

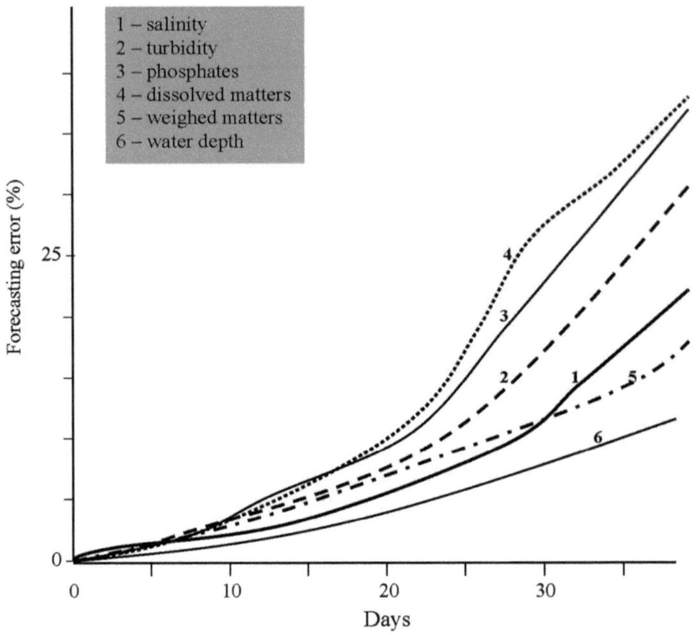

Fig. 9.22 Dependence of the forecast error on its depth in the time

Table 9.19 Comparison of the empirical and modeling results related to the evaluation of physical-chemical parameters of Nuoc Ngot Lagoon

Controlling site	Salinity ($^o/_{oo}$)		Turbidity (mg/l)		pH		PO_4^{-3}(mg/1)	
	M	E	M	E	M	E	M	E
1 (Г)	30.15	33.5	12.4	10	7.04	7.82	0.033	0.03
2	23.14	26	24.78	21	7.71	7.79	0.034	0.03
3	26.48	29.1	29.7	27	7.66	7.58	0.068	0.45
4	27.37	32.2	26.88	24	8.01	7.78	0.023	0.02
5	28.53	31.7	45.59	47	7.29	7.84	0.042	0.04
6	28.4	26.3	29.97	27	7.22	7.52	0.082	0.1
7	27.5	25.7	23.94	21	7.45	7.3	0.067	0.35
8	26.16	25.4	34.58	38	7.21	7.75	0.023	0.02
9	26.54	30.5	48.45	51	7.09	7.71	0.023	0.02
10	20.95	26.5	60.48	63	7.63	7	0.039	0.05
Average error (%)	10.7		11		6		27.4	

Notation M—modeling results, E—experimental measurements

taken into account in the ESECEZ system adequately. This follows from the fact that the in situ measurements in these control sites register anomalous rejections. This suggests that it is necessary to modify some ESECEZ items.

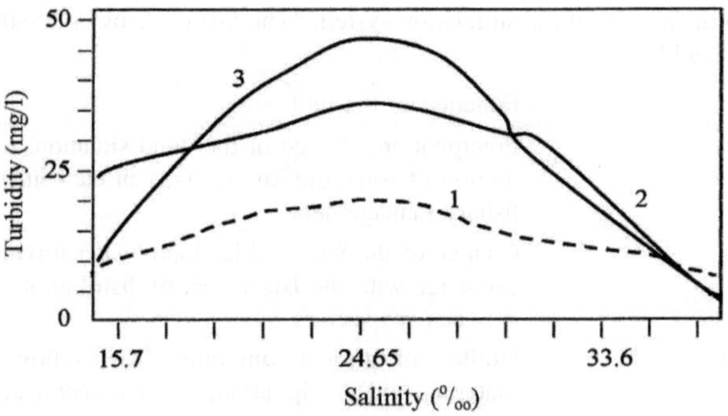

Fig. 9.23 Phase portrait of Dam Nuoc Ngot Lagoon. *Notation 1* river mouth, *2* boundary Γ, *3* lagoon centre

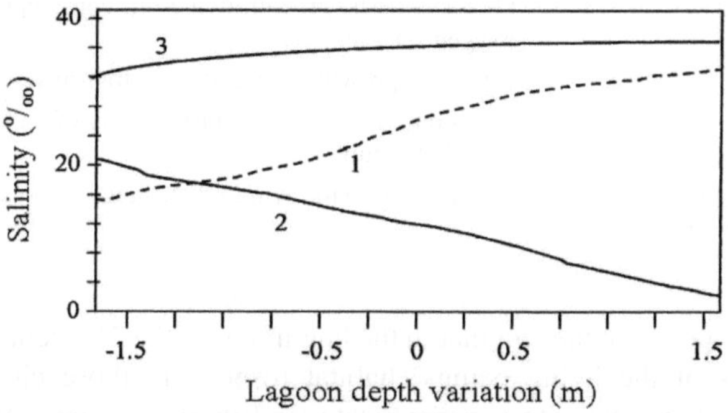

Fig. 9.24 Phase portrait of Dam Nuoc Ngot Lagoon in its section by plane salinity/depth variation. *Notation 1* average level on all aquatory; *2* average in the river mouths; and *3* boundary Γ

9.4 Risk Control and Sustainable Development

Assuming the definition of the sustainable development given by Kobayashi (2005) as a basis, we shall consider its different aspects connected with the human behavior strategy under conditions when the trend of NSS development depends on the decision making. The level of the risk of making decisions at this stage of humankind development can be evaluated by not only the results obtained in the nearest future, but also by the subsequent impacts. In that sense, the developed and currently used ineffective strategy of the risk control of the decisions made can lead to catastrophic consequences. In particular, this refers to various decisions obtained by international organizations on nature-use management, change of the Earth's covers, impacts on anthropogenic processes, etc. Under conditions of limited long-range forecasting methods, an erroneous decision can cause irreversible consequences of economic character with development of undesirable processes in the environment.

Table 9.20 The main units of the simulation system "The Sea of Azov" (Gorstko 1979; Gorstko and Ougolnitsky 1990)

Unit	Function of the unit
External factors	Precipitation, change of the wind situation, river run-off, motion of soils, outflow of nutrient elements and pollutants, fishery management
Water dynamics	Change of the water-table slope, water mixing, water exchange with the Black sea, re-distribution of ingredients between sea sectors
Nutrient elements	Outflow of nutrient compounds, their inflow with abrasive materials and precipitation, physico-chemical transformation of the compounds
Oxygen	Aeration, consumption of oxygen in biochemical processes
Water quality	Sea self-cleaning, propagation of pollutants over the water body
Phytoplankton	Photosynthesis, growth, dying-off, consumption of nutrients, oxygen release
Zooplankton	Nutrition, growth, dying-off, age transitions
Benthos	Nutrition, growth, metabolism, dying-off, change of the spatial structure
Nekton (goby, pike persch, bream, roach, etc.)	Nutrition, growth, dying-off, fishing-out, migration

It is of great importance the fact that in the beginning of the 21st century, most of the key indicators of the living beings' habitat (especially those characterizing the humans' safety), are close to a critical state, and in some regions to a threatening state.

The risk of the occurrence of unfavourable environmental conditions for people is mainly connected with the probability of a natural or technogenic catastrophe resulting from the decision made earlier. Catastrophes in general and natural disasters in particular, threaten the human life and therefore their prediction aimed at prevention is a useful element of science on the processes taking place in NSS. The risk evaluation of the occurrence of a natural disaster and its subsequent damage requires an interdisciplinary analysis of a huge information volume about various aspects of NSS functioning. First of all, the ecological safety, taking a broad view of this aspect of NSS, is closely connected with the socio-economic level of the development of society on a given territory. According to Vladimirov et al. (2000), for instance, for conditions in Russia, many indicators of the society development are in a critical state. So, compared to Germany, Russia spends about 50 times less finance to secure the ecological safety. Other indicators, such as nature protection expenditure and ecological losses on the territory of Russia deviate from the rated values towards degradation by factors 2.5 and 3.5, respectively. Moreover, most of the indicators of NSS development on the territory of Russia testify to a negative trend in preparedness of the Russian society to prevent natural disasters and to meet their consequences. While on the territory of many developed countries most of the natural disasters refer to the category of "nature whim", in many regions of Russia they cause a marked damage. For instance, a sudden snow fall on 13 October 2004 in Cheliabinsk resulting in a 40-cm snow layer caused a 24-hr paralysis of the urban life. Other example of catastrophic natural phenomenon is abundant snowfalls in Siberia during the first decade of March 2013 when numerous roads were paralyzed and twelve people were killed.

It is thought that global warming implies more snow for Northern Hemisphere and for Russian Siberia, in particular. The increase in snowfall in the Northern Hemisphere was first noted in the 1960s when satellite observation was first introduced. Snow cover in the Northern Hemisphere in winter is rising at a rate of 0.1 % per decade. For example, in December 2012 the area of snow cover in the Northern Hemisphere reached its higher level in the last 130 years (at almost 3 million km^2, or over 200,000 km^2 above the previous record in 1985). Nevertheless, the satellite monitoring data from 1966 to 2005 showed that the mean monthly snow-cover extent in the Northern Hemisphere was decreasing at a rate of 1.3 % per decade. For the calendar year of 2006 the average snow-cover extent was 24.9 million km^2, which is 0.6 million km^2 less than the 37-year average. In the Northern Hemisphere, spring and summer exhibited the strongest decreases in snow-cover extent. Satellite observations of snow-cover extent show a decreasing trend in the Northern Hemisphere for each month except November and December, with the most significant decreasing trends during May–August.

Undiscovered mysteries of many natural phenomena require a search of the ways of their formalized description, in order they could be predicted. Since all the processes in NSS are somehow interconnected, the desire to create a GMNSS with a broad set of functions is one of the possible ways to resolve the global problem of predicting emergency situations in the environment. Statistical mean dependences created for description of correlations between the frequency of occurrence of natural disasters and their consequences in many cases are a sufficiently effective means to make strategic decisions on the prevention of natural disasters. In that sense, the widely known law of Richter-Gutenberg (Vladimirov et al. 2000) for the USA territory depicts the following:

$$\log N = \begin{cases} 1.65 - 1.35F \text{ for floods,} \\ 1.93 - 1.39F \text{ in case of tornado,} \\ 0.45 - 0.58F \text{ for hurricanes,} \\ -0.55 - 0.41F \text{ in case of earthquakes.} \end{cases}$$

where F is the logarithm of the annual average number of victims for the last 100 years, N is the number of events.

Knowledge of the laws of occurrence of natural disasters, ability to predict catastrophic events and the presence of mechanisms for warning of disasters do not provide complete protection of people and their infrastructure from losses and destruction. These components should be requested by people with a certain level of understanding the risk and danger perception. Vladimirov et al. (2000)

introduced the notion of "safety culture" which reflects the code of conduct, morals, and emotional response to cataclysms.

Risk, as a measure of danger evaluated with numerous factors taken into account, can serve as a guiding line to resolve the problem of control of a set of potential factors that can disturb the human habitat and to change conditions for the functioning of the society. Risk is a more capacious notion compared to probability of a natural disaster. It includes a probability of an unfavourable event and the volume of the resulting losses. In other words, risk reflects a measure of danger of a natural phenomenon including estimates of the levels of adversity for different aspects. At the same time, risk has also a subjective constituent which is measured with the help of formal methods of decision making, taking into account an intuitive assessment of the situation and psychological norms of perception of the environment. Many problems are resolved at the level of acceptance of legislative acts, which reflect general principles of danger assessment.

On the whole, the notion of risk covers its various sections characterizing the components of probability of an unfavourable change of conditions of people's existence, including natural, economic, social, cultural, and religious factors. Each of these sections is a complex of non-linear functions of a multitude of NSS parameters, prediction of whose changes is the subject of the current global change science. Here of importance are economic levers of the control of the processes of NSS development. The relationship between economy, natural resources, and energy in many respects determines the direction of global processes development. In that sense, the main problem of humankind consists in a search of a compromise between increasing needs of the growing "consumption society" and nature having limited food and energy resources. On this way, humankind has turned out to be at the crossroads between the strategy of globalization and regional control. As seen from IEA (2013a, b), the level of risk to obtain an unfavourable direction of development for many poor countries remains sufficiently high due to the presence in the strategy of developed countries of an element of appropriation of excessive resources. Therefore, the problem of inclusion of a country or a region into the process of globalization is important from the viewpoint of evaluation of the risk of consequences. The index of globalization used for the evaluation of this risk and including variables, which characterize the level of economic integration, development of technologies, political aspects and openness of the society for information contacts does not foresee any free transmission of excessive resources from one country to another. As of the year 2004 62 countries with population constituting 84 % of the global size, covered 96 % of the global GDP. And it would be natural to render assistance to other countries to achieve the index of development corresponding to the annual energy consumption 5 t of oil equivalent per head. By 2020, the emerging economies will account for around 50 % of the world economy. By then, two-thirds of global GDP growth will be generated in those countries.

Many experts assess differently the risk of globalization strategy realization in developed countries. Lomborg (2001, 2004) explains how the "environmental myths" originate and propagate in order to achieve political goals by intimidating the population with coming catastrophes. As a result, a discussion of important problems of global ecodynamics follows a mistaken path without prospects. Lindsey (2001) discusses political and economic mechanisms of globalization control drawing attention to contradictoriness of the laws of economic development in the past and future. He gives a pattern of the present course of globalization as a favourable perspective of the future economic arrangement of the world. Callinicos (2001) analyzes and criticizes the so-called "third way" of humankind development which is propagated in the USA, England, and Germany with liberalism in free economy and possible national socialism. This way will lead humankind to still more contrasting economic differences with an uncontrolled increase of the social inequality. Criticizing the theoretical bases of a new market, Callinicos (2001) demonstrates the presence of the American imperial trend to political globalization. Anderson and Anderson (2001) attract attention to the fact that along with economy, in the process of globalization no less important are culture, politics, and biology, since people live in the world of open systems and, depending on interaction of these NSS components, many global processes can form. Tabb (2001) considers the process of globalization from the viewpoint of ethics whose coordination with economy is one of the factors of prevention of negative trends in the development of the modern society. Trade wars, capital concentration, displacement of investment processes and latent restriction of social rights—these are basic components of the present ideology of globalization.

9.5 Managing Natural Resources

9.5.1 Introduction

The problem of an optimal management of natural resources appears in the regions of the presence of elements of the anthropogenic forcing. In this connection, Khudoshina (1996) introduced the term "ecologo-economic system" (EES), which formalizes the problem of managing a totality of the nature components and economic activity and enables one to describe numerous contradictions between these components. The efficiency of the monitoring system depends on the correctness of the administrative decisions on determination of priorities, distribution and use of resources. It is also important to have a set of efficient indicators of the state of both individual elements of the environment (air, water, soil, flora/fauna, etc.) and the system on the whole. Such indicators enable one to assess the state of the ecological balance of the territory and to simplify thereby the problem of management of the ecosystems located on the territory.

Real problems of managing natural resources are multi-criterion and connected with analysis of a great number of factors, dependences between which often cannot be formalized. Khudoshina (1996) proposed a technique to solve such problems for an efficient management of the resources in the coastal regions. This technique is based on the method of the logico-informational modeling. This method, in turn, is based on the principles of simulation modeling and includes two types of models: dynamic and logico-informational. The dynamic model describes the temporal transformation for all characteristic parameters of the system, combines a group of models of the physical processes and determines the quantitative changes and relationships in the system. The logico-informational model, with the help of the Boolean algebra, assesses the quality of the system for various ranges of the values of its parameters. The logical-informational model is based on the notion of the functioning element characterized by the vector of state $X = (M_i, \{C_{ij}\}, \{P_{ik}\})$ where M_i is the power of the functional element i, C_{ij} is the indicator of the consumption by the i-th functioning element of the j-th resource per its unit power, P_{ik} is the production of the k-th resource per unit power of the functioning element. The state of the functioning element is described by a set of the X_{ij} indicators of its provision with the resources C_{ij}:

$$x_{ij} = \begin{cases} 1 & \text{if} \quad C_{ij} \in \left[a_{ij}^{H}, a_{ij}^{B}\right], \\ 0 & \text{if} \quad C_{ij} \notin \left[a_{ij}^{H}, a_{ij}^{B}\right] \end{cases}$$

In a general case, the estimates a_{ij}^{H} and a_{ij}^{B} are functions of many parameters and of M_i, in particular.

The functioning elements of EES include groups of the population of the region, objects of economic activity (transport, agriculture, industrial enterprises, municipal services, parks, beaches, sanatoria, ports and levees, etc.) and natural objects. Of course, in the synthesis of a simulation model it is impossible to reflect the interaction of all EES elements. Therefore it is necessary to select a first-priority group of these elements, attributing the rest of them to the background load. In analysis of EES in the coastal regions only those objects of economic activity are chosen which directly affect the marine environment. The scheme proposed by Khudoshina (1996) to parameterize the EES functioning in the zone of sea-land interaction is shown in Fig. 9.25.

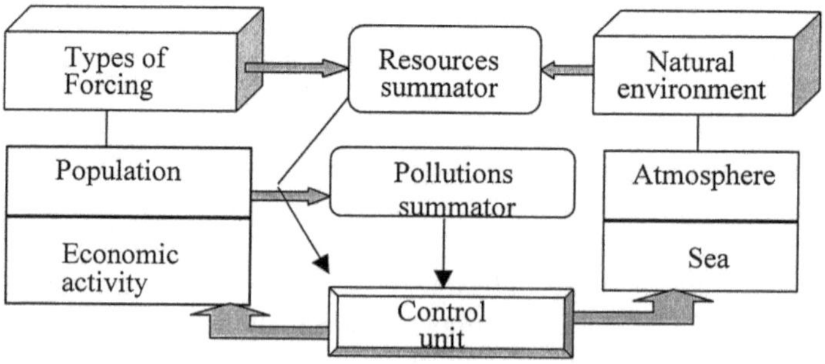

Fig. 9.25 Scheme of interaction of anthropogenic factors of land and marine environment

The functioning of the EES is determined by the volume of consumption and distribution of resources, which depends on population density. Hence, the prognostic estimate of the state of EES is connected with the accuracy of the demographic model and depends on the level of interaction with other EESs. It is necessary to bear in mind that all elements of the system consume and produce resources, some playing a progressive role, others (e.g., pollutions) being negative. Pollutions of the seas violate their main resource—the quality determined by a complex of indicators. The goal of the monitoring system is to find opportunely the critical situations when there is a possibility of rapid $x_{ij} \to 0$.

The EES complexity is determined by the number ζ of inputs and the number β of outputs: $L(\zeta,\beta)$. Assume the hypothesis of additivity of the quality of the EES complexity:

$$L = \sum_{i=1}^{K} L_i,\ L_i = \zeta_i^{\xi} \beta_i,$$

$\xi = (\zeta + 1)/\zeta$, that is, with an increase of the number of inputs of one sub-system by unity, the sub-system's complexity increases ξ times.

To decrease the model's dimensionality, divide the modelled system into sub-systems $E = \left\{E_k = \{\Xi_{kl}; l = 1, \ldots, L\};\ (k = 1, \ldots, K); E_{kj} \cap E_{ki} \text{ for } i \neq j\right\}$. Here $\Xi_{ck,l}$ are the lowest-level sub-systems. Also, reduce the number of the model's inputs. Introduce notations: X is the state of the system, $\{x\}$ is the space of the state of sub-systems, $\{x_M\}$ is the space of the state of the environment, x_j is the vector coordinate in the space of states: $X = \left\{x_1, \ldots, x_j, \ldots, x_M\right\}$. Each sub-system E_k has a totality of properties of the functioning: $\Omega = \{w_h\}, h = 1, \ldots, H$; where w is the regularity of the functioning E in certain conditions, a totality of partial descriptions $w_h = \left\{w_q; q = 1, \ldots, Q_h\right\}$. At an initial moment to a concrete realization w_q of the physical pattern corresponds the state of the sub-system $x_{c,t,0} = \left(x_{c1}, \ldots, x_{cy}\right)$ and the state of the environment $x_{m,t,0} = \left(x_{M1, \ldots}, M_d\right)$. Application of this procedure makes it possible to simplify the model's structure and to make its analysis easier.

9.5.2 *Assessment of the State of the Ecologo-Economic System*

The state of the modelled system based on the logico-informational modeling is assessed using the Boolean algebra. For economic and natural objects two possible states are considered: normal functioning and the working regime bug (refusal of the system's functioning). The function Y of the state of EES depends on the function of the state of its elements: $Y = Y_1 \forall \ldots \forall Y_K$. In other words, $Y = 0$ if $\exists Y_{i \in k} \in \{Y_{1,\ldots,} Y_K\}$, $Y_i = 0$.

Following the method discussed in item 8.7.2 we write:

$$\begin{aligned} Y &= F(Y_1, \ldots, Y_n), \\ Y_i &= F_i(Y_{i1}, \ldots, Y_{ini}), \\ &\ldots\ldots\ldots\ldots\ldots\ldots \\ Y_{ij} &= F_{ij}\left(Y_{ij1}, \ldots, Y_{inij}\right), \\ &\ldots\ldots\ldots\ldots\ldots\ldots \\ Y_{ij\ldots k} &= F_{ij\ldots k}\left(Y_{ij\ldots k1}, \ldots, Y_{ij\ldots nij\ldots k}\right) \end{aligned}$$

The EES functioning is assessed by the totality of the states of its elements. We have $Y_i = 0$ in case of scarcity of at least one of the resources used by the i-th element or its quantitative characteristics do not meet the assumed criteria. Here the interdependent pairs and chains of the system's elements are taken into account. The concrete structure of the model introduces certain corrections and special features into the logical hierarchy of division of the system into sub-systems and elements. Especially this is characteristic of the systems with the substantially heterogeneous structure. These systems include also the marine EES with the biological, economic and demographic units (Fig. 9.26).

Here pollutions play the role of a negative resource. Information about the quality of the marine environment and the state of the biota, that reflects the impact of economic activity on nature, enters the controlling unit of the model. This unit compares and analyzes the information about the state of EES on the whole. With

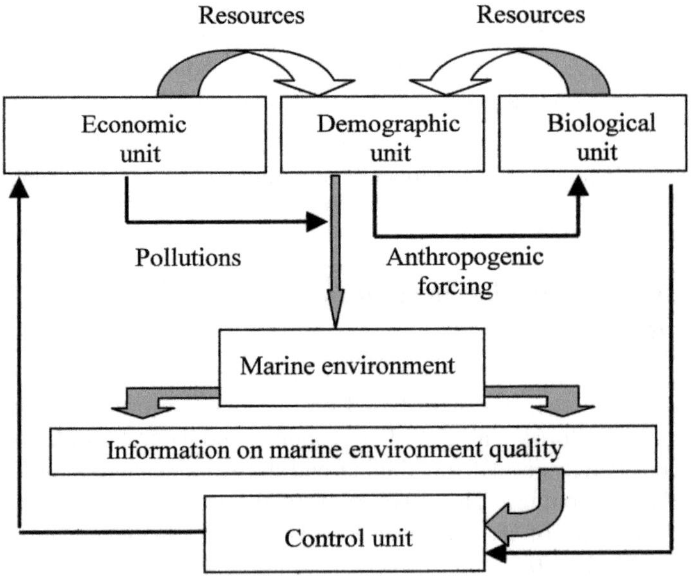

Fig. 9.26 Structural scheme of the model of the marine coastal ecotone (Khudoshina 1996)

respect to the biological unit of the marine EES, economic and demographic units simulate an external load in the form of pollutions, extraction of bioresources, transformation of the habitat. Information about changes of the quality of the marine environment, decrease of the size and diversity of populations of sea animals serves as a feedback in the model, which, of course, reflects on the economic activity of EES.

The demographic unit models the power and the kinds of the external forcings on the natural EES. The input characteristics of the unit reflect the need of population for economic and biological resources as well as the quality of the environment. The output parameters of this unit are really consumed resources, population size and an amount of pollutants emitted to the environment.

The economic unit describes complicated feedbacks between the EES elements, reflecting correlations between input and output parameters of the remaining units as functions of the economic strategy of the system. The biological unit reflects the composition of biotic communities and their trophic structure. The concrete model realization of all elements of the scheme in Fig. 9.26 is connected with the use of the models of partial processes described by many scientists (Krapivin and Kondratyev 2002; Sarancha 1995; Abakumov 1993; Polishchuk 1992; Ougolnitsky 1999; Ougolnitsky et al. 2000).

9.5.3 Simulation Model of the Ecologo-Economic System

The most widely used method of the ecologo-economic analysis is a simulation modeling which makes it possible to combine, for joint studied, the data of natural and social sciences and to create a computer version of the system of the EES management. This approach and its adequate application in concrete situations have been reflected in many studies accomplished by various scientists. For instance, Gorstko and Ougolnitsky (1990) described two approaches to study the EES using the numerical models. The first approach is based on the concepts of the optimization theory and consists in synthesis of the numerical model of the process. This model relates the internal (endogenis) and external (exogenis) factors of the EES functioning. The problem is to determine the management of the process of the EES functioning, which provide an extremum of some functional that characterizes the quality of management. In a general case, the solution of this problem is reduced to mathematical programming or game theory.

The second approach is connected with simulation modeling. Here a set of models of some functions of EES and expert parameterizations (scenarios) is created, which forms the basis of the simulation experiment on the search of an optimal management of the system. The simulation model is more complicated compared with the numerical model, it has a greater dimensionality, contains a multitude of non-linear direct couplings and feedbacks between the variables. As a rule, the simulation model has more flexible capabilities to assess the efficiency of the expert choice of the controlling functions to optimize the EES functioning.

It requires the development of extensive software for the computer search of the optimal or sub-optimal management. The structure of the simulation system is shown schematically in Fig. 9.27. The most important part of the simulation system is a unit of models of the object under study. This unit can include the duplicating elements with different adequacy and, which is especially important, various requirements for databases. The information retrieval system provides an agreement of the database with the set of partial problem-oriented models and adjusts the simulation model for the dialogue with the operator. The existing standard means of many programming languages make it possible to realize these procedures in the forms providing an accomplishment of a simulation experiment by the user without certain skills and knowledge in the field of informatics.

The simulation systems of the Sea of Okhotsk, the Aral Sea and the Sea of Azov can exemplify these simulation systems (Krapivin and Kondratyev 2002; Krapivin and Phillips 2001a). In the first case the anthropogenic factor is sufficiently small compared with the climatic impact on the ecosystem. On the contrary, the Aral Sea is subject to powerful anthropogenic forcings. In each case, in the synthesis of the simulation system it is necessary to take it into account, to fill the database and the base of knowledge with the needed reliable information. Following Gorstko (1979), consider the case of modelling the ecosystem of the Sea of Azov and the related problems. Its catchment basin covers the space 43°–54°N and 33°–46°E with an areas of about 0.5 million km^2 and population 35 million people. The Azov basin involves numerous anthropogenic forcings: developed agriculture with irrigation, heavy and light industry, oil and coal extraction, fishery, network of macadamized roads and railways, food industry, recreation zones. The simulation model of the Sea of Azov created by the Rostov scientists includes 150 components of its state divided into 16 units (Table 9.20). The model database includes extensive information obtained during many decades in the process of natural-scientific observations of the Sea of Azov ecosystem. The components of the vector of the sea state are functions

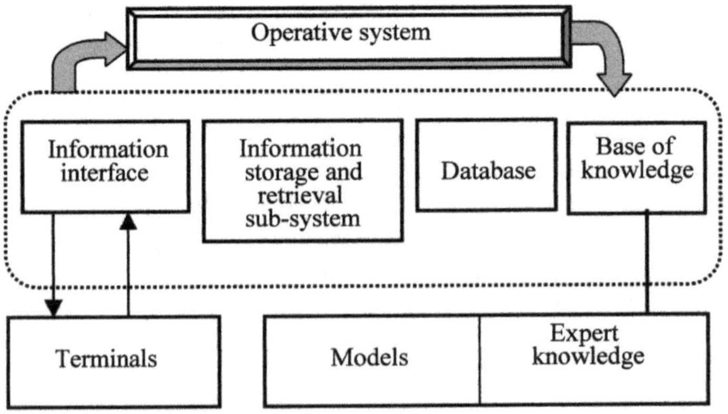

Fig. 9.27 The simulation system's structure (Gorstko and Ougolnitsky 1990)

of time and calculated in accordance with dependences included in the base of knowledge. For instance, the equation of the oxygen balance is written in the form:

$$dO_i(t)/dt = f_1 - f_2 - f_3 - f_4 - f_5,$$

where O_i is the concentration of oxygen in the water of the i-th section of the sea; f_1 is the input of oxygen into the i-th section; f_2 is the oxygen consumption by bottom sediments; f_3 is the oxygen expenditure on respiration of benthos animals; f_4 is the oxygen expenditure on oxidation of organic matter; f_5 is the loss of oxygen on decomposition of pollutants. The simulation procedure, which foresees a digitization of the sea area into non-intersecting zones (sections), provides a correlation of the functions O_i (t) through exchange fluxes between these zones.

Another example of the simulation model is the model of the aqua-geosystem of the Aral Sea basin. In contrast to the Sea of Azov, here the scales are greater and the information supply is smaller. There is also a considerable difference in the interaction of natural and anthropogenic factors. During the last decades, the remote monitoring systems have become the main source of observed data (Krapivin and Phillips 2001a).

The development of irrigation world, the respective complex of hydrotechnical measures, as well as secondary processes-consequences have substantially changed and continue to change the natural, economic and ecological situations in some regions of the Middle Asia. On this basis, the development and introduction of the simulation technology, as applied to the EES analysis in conditions of arid climate, have become one of the acceptable approaches to develop the management strategy and a complex of optimization measures in the NSS.

The conceptual models listed in Table 9.21 reflect the basic mechanisms and trends of the change of the natural situation in the Aral basin in 1960−1980 with a sufficiently negative result—ceasing inflow if river waters into the Aral Sea and degradation of the natural-economic situation in this region. The situations in other regions are sufficiently problematic (the Tashaus and Khoresm oases, Sarakamysh, the zone of Kara-Kum canal, etc.). It follows from these schemes, in particular, that the water component in the Middle Asia region has been and remains the leading object of studies, prediction and optimization(management) Table 9.22.

On the whole, the system of units of a general model—simulation of the current natural-economic situation in the region is as follows. It includes the units: water component, salts, ecological situation, economic load, macro- and meso-climatic background, mass- and energy-exchange at inputs—outputs, direct couplings and feedbacks (Kojima 1998).

Here is a general scheme of the distribution of water in the complex of aqua-geosystems "Amu Darya river—Kara Kum canal—Aral Sea—Sarakamysh". Receipt (input)—precipitation onto the territory (X) + the Amu Darya flowing off at the station of the Kara Kum Canal (R_0). Expenditure: (a) water scoop into canal (R_1), discharge of collector waters into Sarakamysh (R_2), watering (R_j), inflow to the Aral Sea (R_3), outflow from Syr Darya (R_S); (b) accumulation in water reservoirs, except Sarakamysh (R_{acc}—as the sum of dynamic, R_d, and non-operated

Table 9.21 Elements of the "Basin—Aral Sea" system's functioning before and after 1960

Inflow	Outflow		Result
Formation of the Amu-Darya and Syr-Darya run-offs and the rivers gravitating towards the Aral Sea basin in mountains	Evaporation from water bodies, natural geosystems, valleys and river deltas (lakes included)	Supply of ground waters	Inflow of river waters into the Aral Sea before 1960 varied around 54 m^3/year
	Irrevocable off-take for industry and municipal services	Off-take for irrigation; evaporation from cultural plants, water surface of irrigation systems, etc.	
Decrease of the inflow of river waters to the boundary of the region	Evaporation from the preserved few natural complexes and water basins; evaporation from hydroeconomic and recreation water reservoirs	Replenishment of sub-surface waters: irrevocable losses in deserts, water catchment of Uzboi, etc.	Irrevocable off-take for industry and municipal services
Formation of the river run-off			Inflow of river waters into the Aral Sea ceased in the 1980s
Filling-up the pores in soils and dead volumes of water reservoirs	Off-take for irrigation, evaporation	Return discharge into the rivers	
		Formation of collector and discharge waters in the irrigation systems	Discharge into lakes preserved in valleys and river deltas

Table 9.22 Statistics of large-scale natural disasters (Bondur et al. 2009)

Yeas	1950–1959	1960–1969	1970–1979	1980–1989	1990–1999	1992–2002	1998–2008
Number of natural disasters	20	27	47	63	91	70	83
Economical losses (bln dollars)	42.1	75.5	138.4	213.9	659.9	550.9	566.8

volumes, R_n) in the sub-bed and sub-channel lenses (U_O); (c) return (into the river, canal, Aral Sea), seepage and drainage waters (U_v), irrevocable water—in the zone of scattering in desert (U_2), infiltration to deep sub-ground horizons (U_g); (d) evaporation from the water surface (E_w), from irrigated crops and oases (E_O), from natural vegetation (E_v), from saline lands (E_1) and boundaries between facies (E_2); etc.

Introduced notations make it possible to write for each sub-system of the Aral Sea basin the following traditional balance relationships:

- For the zone of the Kara Kum canal

$$R_1 + X = R_j + R_d + U_0 - U_v - U_1 + U_2 + E_w + E_0 + E_1 + E_2$$

- For the zone of Amu Darya from the canal to Sarakamysh

$$R_0 + X = R_1 + R_j + U_0 - U_v - U_1 + E_w + E_0 + E_v = R_2 + R_3$$

- For Sarakamysh $R_2 + X = E_w$
- For the Aral Sea

$$X + R_3 + R_S = R_j + U_0 - U_1 - U_v + E_w + E_v + E_1 + E_2 + R_{akk} - U_w,$$

where U_w is the return sub-ground waters.

For the whole system, many scientists have studied relationships of the water balance, with a consideration of the synoptic situation and of dependences of its components both on each other and on other factors (Krapivin and Kondratyev 2002; Bortnik et al. 1994). The transition to dynamic equations requires additional information about the character of correlations between the elements of the environment, anthropogenic processes and hydrophysical constituents of the system. Since there are irrevocable losses in the system (U_g+U_2), there should exist the balancing branches of the hydrological cycle. Apparently, these branches exist in the flows of moisture in the region, which, by their character, play a dual role. So, for instance, Krapivin and Kondratyev (2002) considered a situation when a forced irrigation of saline lands and takyrs was carried out on the eastern coastline of the Caspian Sea at optimally chosen time moments. As it turned out, the resulting additional evaporation according to mean-statistical synoptic scenario has improved the hydrological situation in the Turan lowland, and with the forced precipitation it was possible to restore the level of the Aral Sea during 10–12 years. Of course, such calculations raise questions, but they show how to study and to search positive forcings on the natural-anthropogenic system.

9.6 Risk Control in Cases of Natural Disasters

An optimization of the risk of insurance from natural disasters becomes every year a more urgent problem, since economic losses increase and become practically poorly predictable. For instance, the insurance payments only for destructions from the hurricane Hugo in 1989 constituted 5 billion dollars exceeding by more than 50 % maximum losses from the earlier natural catastrophes. However, 3 years layer the insurance payment for losses from the hurricane Andrew exceeded the losses from the hurricane Hugo almost 4 times. Examples of comparison of unexpectedly large insurance losses grow in number.

The problem of risk control as a quantitative measure of NSS property to ensure the habitat safety on a given territory is a complicated mathematical problem of optimization which is successfully solved using logical-and-probabilistic methods (Solozhentsev 2004) based on construction, solution, and study of LP-function of risk and safety of the structural-complex systems. Boolean algebra and its combined use with methods of numerical modeling of processes taking place in NSS is a mathematical basis for these studies. The informative characteristic of such an approach to evaluation of the risk of the occurrence of a natural catastrophe is determined by accuracy of prediction of development of natural-anthropogenic processes on a given territory and reliability of produced scenarios, which substitute for individual elements of this forecast.

According to Solozhentsev (2004), to assess the efficiency of risk control, one should use discrete multitudes of parameters $\{Z_j\}$, which affect the efficiency, and the scale Y of the efficiency indicator. In this case, an introduction of the goal function $F = N_{1c} + \cdots + N_{kc}$, where N_{jc} is the number of correctly identified events on the scale Y in the class j, and maximization of F make it possible to determine the weights of factors affecting the efficiency. The complexity of this problem is determined by lack of reliable data on statistical characteristics of these factors. Appearing difficulties can be overcome with the use of methods of assessing the level of self-organization and self-regulation of natural systems (Ivanov-Rostovtsev et al. 2001) and synergetics (Chernavsky 2004). In particular, the use of D-Self theory gives a possibility to introduce a generalized characteristic of the NSS state in the form of non-linear function $\xi_0 = \sqrt{\xi_i \xi_i^*}$, where ξ_0, ξ_I, and ξ_i^* are NSS parameters which meet three axioms:

i. *discreteness* of elements within and beyond NSS with their interaction;
ii. *hierarchy* of elements within and beyond NSS;
iii. *contingency* between discrete elements of hierarchy of the structure within the system (ξ_i) and beyond the system (ξ_i^*) in the form of dependence $\xi_0 = \sqrt{\xi_i \xi_i^*}$.

Synthesis of the D-Self model as applied to NSS is an individual problem and is not considered here. We mention only that according to Ivanov-Rostovtsev et al. (2001), the method of modeling the evolutionary dynamics of natural systems based on D-Self technology has some features in common with the evolutionary modeling technology (Bukatova and Makrusev 2003). In both cases the evolution of the natural system is considered as a discrete process of a change of its states in some parametric space. In the case considered, the economic indicators and social infrastructure of the territory are made into a separate level of hierarchy of parameters and considered as indicators of the risk efficiency. Other characteristics of NSS functioning are considered as external parameters.

The frequency and intensity of natural disasters increase, which is well seen from data in Figs. 7.3, 7.4 and 7.5 and Tables 7.5 and 9.22. Respectively, economic damage from natural disasters grows (Tables 9.23, 9.24, 9.25, 9.26, 9.27, Figs. 9.28 and 9.29). Therefore, detailed analysis of relationships between parameters of natural disasters and regional NSS characteristics such as infrastructure, level of economic development, state of social order, level of people's education and attitude to religion are an important stage of development of a regional strategy to meet the consequences of natural disasters.

Losses from natural disasters between 1987 and 1997 reached the level 700 billion dollars, and this means that for poor countries, meting the consequences of natural disasters in the years to come will become an insoluble problem. For instance, in 1998, total economic losses from natural disasters constituted 65.6 billion dollars, 66 % of them falling on developing countries. It is clear that economic mechanisms of support to regions damaged by natural disasters should be optimized with the real state of NSS taken into account. From estimates of the World Bank, poverty is characterized by the following indicators:

- one third of population in developing countries are poor, 18 % of them being beggars;
- about 50 % of the poor and 50 % of beggars live in South Asia;
- poverty prevails in rural localities;
- basic means of subsistence for the poor are obtained from agriculture.

Development of infrastructure plays an important role in poverty reduction. But it is the infrastructure that is most damaged by natural disasters. So, in Asia almost 70 % of natural disasters fall on floods, each flood causing damage estimated recently at about 15 billion dollars, losses in agriculture contributing most. Therefore the problem consists in reducing the risk at the expense of infrastructure transformation. In developing countries the solution of this problem is connected with an optimization of processes of the use of land resources and working out strategies of ecological character. It is necessary to take into account the indicator of the growth of population density in developing countries. The quantitative and qualitative dynamism in the urban development leads to a super-proportional

Table 9.23 Characteristic of losses from some natural disasters in 2006 (Zanetti et al. 2007)

Natural disasters	Number of cases	Number of victims	Economic losses (million dollars)	
			Total	Insurance
Earthquakes	8	5,880	241.5	80
Floods	58	7,217	2970.6	984
Storms	47	4,600	24950,9	8,265
Droughts, wildfires, heat	5	2,259	362.3	120
Colds, frosts	12	1,617	4105.7	1,360
Hail	5	19	3103.4	1,028
Tsunami	1	19	3	1
Total	136	22,394	35737.4	11,838

Table 9.24 Natural disaster occurrence and impacts by the regions and subgroups (Guha-Sapir et al. 2013)

Region	Natural disaster subgroup									
	Climatological		Geophysical		Hydrological		Meteorological		Total	
	2012	*M*	2012	*M*	2012	*M*	2012	*M*	2012	*M*
Africa	16	14	0	3	30	46	11	9	57	72
Americas	12	14	6	7	26	41	35	39	79	95
Asia	12	12	23	22	71	82	39	39	145	156
Europe	45	17	3	2	16	23	1	14	65	56
Oceania	0	1	0	2	7	5	4	7	11	16
Global	85	59	32	36	150	197	90	102	357	394

Notation M is the average 2002–2011

Table 9.25 Average annual losses of global GDP due to floods (bln. dollars) (GAR 2011)

Region	1970–1979	1980–1989	1990–1999	2000–2009
Eastern Asia and countries of Pacific basin	2.8	5.1	10.2	21.5
Europe and Central Asia	2.2	2.7	2.7	3.1
Countries of Latin America and Caribean basin	2.5	3.2	3.9	5.4
Middle East and North Africa	0.3	0.4	0.6	0.9
OECD countries	24.1	32.8	43.5	52.9
Southern Asia	3.9	5.4	8.7	15.4
Countries of Africa southern Sahara	0.4	0.5	0.6	0.9
World	**36.2**	**50.0**	**70.2**	**100.1**

Table 9.26 Average annual losses in global GDP due to tropical cyclones (bln. dollars) (GAR 2011)

Region	1970−1979	1980−1989	1990−1999	2000−2009
Eastern Asia and countries of Pacific basin	16.0	25.3	39.5	90.2
Countries of Latin America and Caribean basin	2.3	4.9	3.7	24.3
Middle East and North Africa	0.0	0.0	0.0	0.0
OECD countries	506.6	665.1	1247.1	1455.0
Southern Asia	0.3	2.6	4.2	4.3
Countries of Africa southern Sahara	0.5	1.1	1.3	1.7
World	**525.7**	**699.0**	**1295.8**	**1576.5**

Table 9.27 Number of peoples affected by tropical cyclones (millions)

Region	1970–1979	1980–1989	1990–1999	2000–2009
Eastern Asia and countries of Pacific basin	36.6	42.2	44.3	53.7
Countries of Latin America and Caribean basin	1.1	1.6	1.2	5.2
Midle East and North Africa	0.0	0.0	0.0	0.1
OECD countries	26.2	27.2	39.7	53.2
Southern Asia	1.5	7.8	11.1	7.6
Countries of Africa southern Sahara	0.5	0.9	1.5	2.7
World	**65.9**	**79.8**	**97.8**	**122**

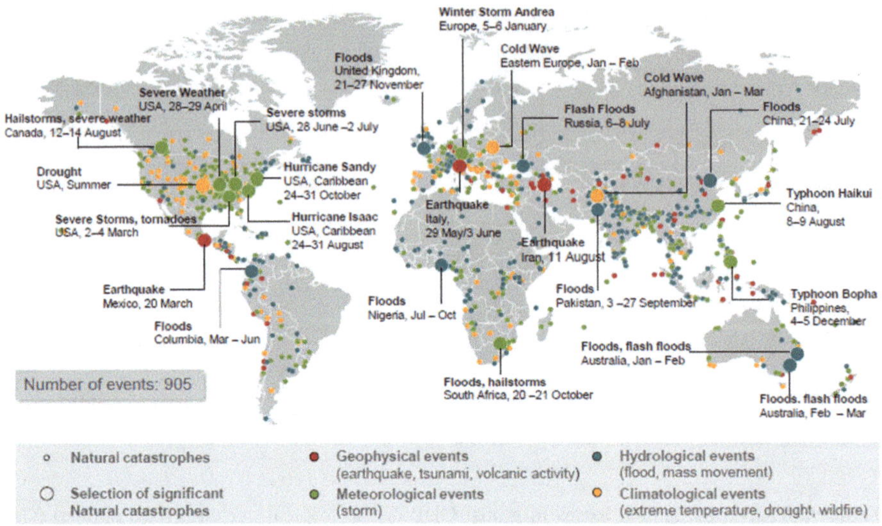

Fig. 9.28 World map of natural catastrophes 2012 (http://reliefweb.inf/map/world/natural-catastrophes-world-map-2012)

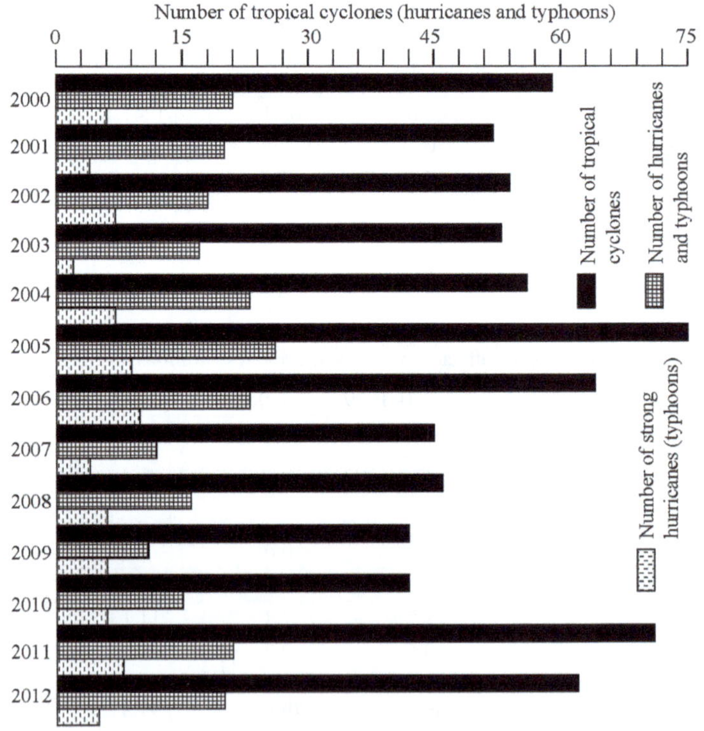

Fig. 9.29 Distribution of tropical cyclones at the beginning of 21-st century

increase of vulnerability of such territories. This trend of developing countries is provoked by either economic interests or socio-political factors of a given region, when a large number of people are concentrated over a small area, which drastically increases the risk of people's life in case of a natural disaster. On the other hand, in this situation, protective constructions can be easily built. Therefore the problem of the regional infrastructure optimization becomes more urgent in time (Gurjar and Leliveld 2005). To some extent, the solution of this problem is favored by the plan of putting into practice decisions of the World Summit on Sustainable Development (Johannesburg, South Africa) adopted at the 17th plenary meeting on 4 September 2002. This plan foresees a mobilization of technical and financial assistance to developing countries in order to balance the economic, social and ecological development of all regions of the world.

The ratio between economic losses in case of a natural disaster and real insurance arrangements for subsequent financial investment to liquidate consequences is one of the key problems in the formation of strategies of insurance companies. One of the successful attempts to formalize the appearing processes is the use of the models of catastrophes in assessment and control of the risk of extreme events (Grossi and Kunreuther 2005). The authors lay emphasis on the risk of natural disasters and discuss the urgent problems of controlling risk from terrorism. The goal of the study was to reduce the losses from potential dangers. The Pennsylvania University tested the proposed technology, carrying out several numerical experiments, which showed that the model approach to planning the financial risk from natural disasters makes it possible to optimize the insurance of extreme events. The necessity of this is seen from the data in Tables 9.27 and 9.28, which demonstrates a very non-uniform regional distribution of the ration of total losses from natural disasters. And if, in addition, one takes into account the distribution of losses by the types of events (Table 9.23), it becomes clear that the model optimization can substantially raise the effects of insurance of these events.

Flood is one of the most destructive and frequent natural events. Analysis of the consequences of numerous floods which occurred already in the 21st century shows that they markedly change the social and economic development of the regions. In connection with the real danger of floods for the society, the European Economic Community decided to finance the RIBAMOD (River Basin Modeling) project aimed at solving the problem of the hydrological risk control, covering the most important themes (Beven and Blazkova 1998):

- accumulation of databases on watershed basins, which can initiate floods;
- assessment of risks from decisions on protection and reduction of the consequences of floods;
- development of technologies of planning measures on flood prevention and operative interference into the flood control;
- planning of interactions of different groups of specialists when working on lessening the damage from flood;

Table 9.28 Natural disasters 2012 and their characteristics (Ferris et al. 2013)

Country	Disaster	Month	Fatalities
Phillipines	Typhoon bhopal/pablo	December	1,901
Europe	Cold wave	January–February	587
Pakistan	Flood	August–October	480
Nigeria	Flood	July–October	363
Iran	Earthquakes	August	306
Peru	Cold wave	June	252
Russia	Flash flood	July	171
Korea Dem. P. Rep.	Flood	July	169
USA, Caribean	Hurricane Sandy	October	141/220
Pakistan	Avalanche	April	135
China	Flood	April–May	132
India	Flood	June–July	120
Total 2012			9,330

- working out recommendations on accomplishing engineering and social measures whose realization during the flood can be hindered or impossible but necessary; and
- analysis of uncertainties which can appear with a threatening flood or its occurrence due to climatic or other natural processes.

Most of the countries that suffer floods have services for prevention of and protection from floods. The structure and equipment of these services are determined by the regional special features of the causes of floods. But regardless of this, there are general problems of assessment of the risk of life, infrastructure, and economy from flood. One of the most effective ways to solve these problems is to predict the hydrological processes at regional and global levels. Efficiency of such predictions depends on information content of the systems of nature monitoring.

A natural disaster always leads to economic losses. In poor countries the population becomes even poorer. The discussions of sustainable development acquire the meaning of people's survival, and problems of the environmental protection and search of NSS equilibrium at a regional level become of secondary importance (Freeman 2000). A natural disaster destroys the infrastructure as a key component of the economic growth of the region. For almost 18 % of the global population, poverty becomes critical, and with an occurrence of a natural disaster in their habitat a complicated social situation is always created, and the way out of this situation is only possible with an assistance of developed countries. Therefore a reduction of the risk of economic losses from natural disasters in developing regions is one of the first-priority problems of global ecodynamics (Krapivin and Varotsos 2007, 2008; Krapivin and Shutko 2012).

9.7 Social and Human Dimensions of Risk

Human society in its development has approached the state when correlation between processes in the environment and social organization of the society reaches critical values at which the behavior of even one man can drastically change the global processes. For instance, Tony Blair declared in November 2005 at the meeting of 20 countries in London that the Kyoto Protocol should be reconsidered and that science will find the ways to resolve the global warming problem. He also said that politically no country wants to sacrifice its economy to resolve the problem. Other ways of its solution should be found, which would not restrict the economic growth and keep a check on the processes of life improvement in the world. It is this statement of one man that opens up ways to search mechanisms of raising the living standards and poverty eradication.

According to Letz (2000), humankind has crossed the threshold of systems adaptation, which makes it possible to damp out deviations from acceptable values of NSS parameters and to preserve the habitat; therefore, immediate measures are needed on formation of a new view of the processes in the environment. Especially it concerns a technocratic paradigm which determines the decisions and respective mentality of national leaders. Therefore, the socio-political constituent of risk is very important in its evaluation and determination of the task for the society to prevent damage from potential consequences of the made decisions. Among the problems to be resolved by government authorities, the following are most important:

- reduction of the level of risk of the social vital activity due to public policy in the sphere of neutralization of negative processes in the environment;
- an increase of the socio-political stability of the society by formation of institutional and informative conditions for preparation of population to actions under extreme circumstances;
- an increase of mobilization capabilities of the society for preventive and adequate response to originating threats;
- development of the legal basis and respective structures to meet the consequences of extreme events and disasters of natural and technogenic origin.

As Churin (2005) noted, the social processes in the modern society have no purpose-oriented control, and therefore it is difficult to evaluate the risk of decision making. The interconnection of social and natural processes practically in all regions of the planet is determined by their interactivity. Therefore, regulation of social processes inevitably tells upon the functioning of natural systems, and this certainly leads to a necessity to develop legal mechanisms for transition to sustainable development.

A decision on the level of potential risk of possible change in the environment can be made on the basis of analysis of pre-history of such events and using the methods of prediction of natural events. As a rule, risk assessment methods are based on statistical processing of data on parameters of the processes whose interaction can initiate undesirable changes in the environmental characteristics. For instance, when assessing the risk of human health from the environmental pollution, to make a serious decision, a certain level of the informative description of the territory is needed:

- multi-year observations of concentration of chemical elements with their indicated allocation and characteristics;
- data on hydrology of the territory and synoptic characteristics;
- assessment of ecological consequences of pollution and their impact on human health;
- characteristics of the state of protective constructions and the level of development and technical equipment of services for monitoring, prevention and meeting extreme situations.

An efficiency of the systems of risk assessment depends on the form and kind of the applied procedure of decision making (Potapov et al. 2006; Garsey and McGlade 2006; George 2002). The most informative is when prediction procedures are combined with an environmental monitoring regime, which can foresee situations of decision-making in a real time on the basis of information accumulation before the decision was made or from analysis of database fragments with no time reference. Statistical analysis of several events that follow the functioning of the monitoring system can be carried out by numerous methods whose applicability in each case is determined by the totality of probabilistic parameters that characterize the phenomenon under study. However, non-stationary and parametric uncertainty in situations when each observation requires much effort and expense prompt a search for new methods of decision-making on the basis of observation data that are fragmentary both in time and in space.

With development of alternative methods of making statistical decisions, the problem of finding objective estimates of the perimeters of processes taking place in the environment has been substantiated anew. It is possible to consider and compare two approaches to this problem: classical approach based on a-priori restricted number of observations, and successive analysis based on the procedure of step-by-step decision making. The development of computer technologies facilitates realizing both approaches in the form of a single system of making statistical decisions. The classical and sequential procedures of making statistical decisions by the Neumann-Pearson (Neyman and Pearson 1933) and Wald (1947) methods were described in Chap. 3.

The notion of risk is closely connected with the notion of uncertainty of prediction of the development of events in the NSS undesirable for humans. By definition of Burgman (2005), risk is the chance of occurrence of an unfavorable event within some time interval. For the NSS such intervals are measured by centuries, and the society, first of all, wants to know the prospects for life improvement. However, the state of the current environmental science does not provide a guaranteed solution of this problem, and therefore it is necessary to search constructive technologies of risk assessment. One of the reliable and constructive approaches to the solution of this problem is an expansion of the GMNSS functions. In particular, the number of various uncertainties which hinder a reliable prediction of global ecodynamics is so large that without their purpose-oriented analysis and selection all their interactions cannot be formally taken into account in the global model. For instance, of millions of chemical compounds and species of living organisms, only thousands have been adequately studied.

One of the numerous factors of the evolutionary processes development in the NSS are sudden changes of the environmental characteristics that lead to stresses for living organisms with their possible death. In various periods of evolution, the scale and significance of individual factors had changed. The role of catastrophic processes taking place during 11,500 years in the present civilization development is being studied within two international ICSU programs: " Dark Nature—Rapid Natural Change and Human Responses" (started in 2004) and "The role of Holocene environmental catastrophes in human history" (2003–2007). The main goals of these projects are to:

- examine how past societies and communities reacted in the face of harmful change;
- explore the implications of rapid natural change for current environmental and public policies;
- focus on the inter-disciplinary investigation of Holocene geological catastrophes, which are of importance for civilizations and ecosystems.

A longer period of the Earth system's evolution has been considered by Condie (2005) who has analyzed the interaction of various components of the planet for the last 4 million years, selecting those which have affected the history of the land ecosystems, oceans, and the atmosphere. The current period is characterized by an increasing trend of anthropogenically caused natural disasters, such as floods, forest and peat-bog fires, deforestation, desertification, and epidemics. Of course, not always is it possible to distinguish between the causes of a natural disaster. But one fact is apparent—the present ecodynamics is followed by an increase of the number of extreme situations in the environment. Unfortunately, presence of statistics for natural disasters no allows their prognosis with high reliability. Events of the 1998 flooding in Bangladesh, the 2004 megathrust earthquake in Sumatra (Indonesia) and the 2013 typhoon Haiyan on Philippines when was killed over then 50,000, 230,000 and 6,000 peoples, respectively, are examples of the abrupt natural events.

The contribution of different types of disasters to human losses and the economic damage varies from year to year, but maximum victims result from earthquakes, floods, snow storms, and tsunami. For instance, of 700 natural disasters recorded in 1998, winter storms and floods occurred in 240 and 170 cases, respectively, and the resulting economic losses constituted 85 % of all losses. In Europe, in mid-November 1998, more than 215 people were frozen to death.

Natural disasters were during last years various. There have been 620 tornadoes reported in the U.S.A. in first six months of 2013 with 44 fatalities. 102 fatalities have been confirmed worldwide in first six months of 2013. Several wildfires cover numerous regions in Russia, U.S.A., Australia and Europe during summer 2012 and 2013. Intense floods were practically everywhere during last three years.

The significance and scale of critical situations in the environment depends on the state of many NSS components. As a rule, natural disasters are evaluated in the context of damage to human life and economic activity. However, it is well known that, on the other hand, natural disasters are an important factor of ecodynamics from local to global scales. As Lindenmayer et al. (2004) note, for many processes that determine the ecosystems dynamics, natural disasters are of the key importance as factors that regulate such processes. Therefore Wright and Erickson (2003) discussed the problem of taking into account the impact of natural disasters within the problems of the numerical modeling of the environmental change. Special attention has been recently paid to forest fires which, being an important component of ecodynamics, are mainly of anthropogenic origin and cause both material damage and humans' death.

Forest fires occurring regularly in various global regions have more and more become a factor of the ecosystem's dynamics, which is manifested through fire-induced emissions of greenhouse gases and aerosols to the atmosphere. According to available estimates, about 30 % of tropospheric ozone, carbon oxide and carbon dioxide contained in the atmosphere are determined by the contribution of forest fires. Aerosol emissions to the atmosphere connected with forest fires can substantially affect the microphysical and optical characteristics of the cloud cover, leading to climate change. So, satellite observations over Indonesia have demonstrated, for instance, that the presence of smoke in the atmosphere due to prolonged fires has led to precipitation suppression, which has further favored the development of fires. In this context, Ji and Stocker (2003) performed a statistical processing of the data of TRMM-satellite measuring rains in the tropics as well as of TOMS (total ozone mapping spectrometer) data for the period January 1998–December 2001 in order to analyze the laws of the annual course, intraseasonal and interannual variability of the number of forest fires on a global scale. During this period, there was a clear annual course of fires in south-Eastern Asia with a maximum in March, and in Africa and in North and South America—in August. An analysis of data revealed also the interannual variability of forest fires in Indonesia and Central America correlating with the El Niño/Southern Oscillation (ENSO) cycle in 1998−1999. In 1998, the boreal forests burned out over vast territories of Russia and North America. Fires covered the territory about 4.8 million ha in boreal forests of Canada and USA and 2.1 million ha in Russia.

There appears a clear correlation between variability of the atmospheric aerosol content and the above variations of the frequency and intensity of forest fires. An exception is the region of south-Western Australia where intensive fires recorded from the TRMM data were not followed by smoke layers formation (from the TOMS data). With the Australian region excluded, the coefficient of correlation between the number of fires and AI (from the TOMS data) constitutes 0.55. The statistical analysis of data using the calculated empirical orthogonal function (EOF) revealed a contrast between the Northern and Southern Hemispheres as well as the existence of the intercontinental transfer of aerosols originating from fires in Africa and America. The data of statistical analysis point to the presence of 25−60-day intra-seasonal variations superimposed onto the annual change of the number of fires and aerosol content. A similarity has been found out between the inter-seasonal variability of the number of fires and dynamics of the Julian-Madden oscillation (Kondratyev and Grigoryev 2004).

As McConnel (2004) has justly noted, so far it remains unknown why some forests have suffered fragmentation, degradation and loss of biodiversity, while other forests remain in a good state and even expand. There is no doubt that an increase of population size, further enhancement of the scale of market relations and an intensive development of various economic infrastructures belong to important factors of the observed deforestation. Factors acting against these processes are measures on nature protection favoring the preservation of forests. Finally, the forest cover dynamics is determined by a complicated and interactive totality of such factors as biogeophysical processes, increasing population density, market relations, various disturbing forcings (forest fires including) and institutional microstructures.

Forest fires affect the formation of global carbon cycle. Really, global scale of forest fires have recently become equivalent in area to the territory of Australia. The atmosphere receives almost 40 % of global CO_2 emissions, 90 % of forest fires being of anthropogenic origin. This means that the balance of natural factors is heavily broken, and the laws of natural evolution are strongly violated.

Among other natural phenomena affecting the environmental dynamics are volcanic eruptions, dust storms and thunderstorms. One of the consequences of a large-scale volcanic eruption is a powerful atmospheric pollution, which can lead to a climate change in some regions or even on a global scale. So, in particular, aerosols connected in many respects with emissions of sulphur dioxide during the 1991 Pinatubo eruption have led to a global decrease of temperature in the lower atmosphere in 1992 by about 0.5 °C. Also, it was noted that in 2 years the level of the global ozone content decreased by 4 % compared to the preceding 12-year period. And though during the last century there were no drastic climate changes due to volcanic eruptions, in the historical aspect such situations did take place. So, 73,500 years ago due to the Toba eruption on the Sumatra island, the global atmospheric temperature decreased by 3−5 °C. The area of glaciers increased and the Earth's population decreased (Oppenheimer 1996). As a result, 600 volcanoes existing now on Earth can, in principle, be a global threat to population, since they can seriously change climate and affect thereby further development of the NSS.

Dust storms occurring in places with dust and strong winds, like volcanoes, play a similar role in the environmental change. Dust propagates at great distances affecting negatively plants and soils as well as causing a strong pollution of the lower atmosphere for long time periods. In the regions with a thin vegetation cover and arid climate, dust storms take away the fertile soil layer damaging thereby the plants. Dust storms occur periodically, for instance, on the plains of Central and Western states of the USA during heavy hurricanes. Sometimes they last for several days, raise dust up to 1.5–1.8 km and partially to 5–6 km, and transfer it at hundreds and even thousands of kilometers towards the Atlantic Ocean. Catastrophic dust storms are connected with destruction of the soil-plant cover in steppes and partially-wooded steppes, as well as in the regions of intensive agriculture. Along with the short-duration impacts and marked damages, dust storms create a regular constituent of the global ecodynamics which can manifest itself in decades and even centuries.

Thunderstorms play a special role in the environmental change. Lightning discharges occurring practically in all latitudes, affect the photochemical reactions in the atmosphere and are a factor of fire risk. From available estimates, 1800 thunderstorms, on the average, occur at any moment on Earth, each being followed by 200 lightning flashes per hour (or 3.3 flashes per minute). From satellite observations, the global frequency of lightning flashes averages 22–65 flashes per s. Observations from Microlab-1 satellite gave 44 ± 5 flashes per s, which corresponds to 1.4 billion flashes per year. The use of Microlab-1 data has made it possible to draw global maps of the frequency of lightning discharges in different seasons. An analysis of these maps has shown that lightning discharges occur mainly over land, and the ratio between their number over land and ocean averages approximately 10:1. About 78 % of lightning discharges fall on the latitudinal band 30°N−70°S. The all-the-year-round regime of lightning discharges is most intensive in the river Congo basin where the average frequency of flashes reaches (in Rwanda) 80 flashes per km^2 per year, which corresponds to conditions in the Central Florida State (USA). The all-the-year-round intensive regime of lightning flashes is characteristic of the northern sector of the Pacific Ocean where under the influence of cold air advection over the warm ocean surface the atmosphere becomes unstable. In the eastern sector of the tropical Pacific and in the Indian Ocean, where the atmosphere is warmer, the lightning discharges are not that frequent. Maximum frequency of lightning occurrence in the Northern Hemisphere falls on summer, whereas in the tropics, lightning discharges are characterized by a semi-annual cycle.

In the northern regions, an important constituent of the controlling mechanism of evolution are strong frosts, whose impact on vegetation cover depends on plants' hardiness. Jönsson et al. (2004), with the Norwegian spruce *Picea abies* as an example have studied the response of boreal forests to temperature changes. It has been shown that sudden temperature variations cause changes in wood compactness, and that an expected climate change can change the vegetation cover.

During the last years, in connection with discussions of the causes of potential climate change, of key importance have been estimates of uncertainties of values which serve the basis for conclusions about climate change and measures to prevent it. Particularly important is the problem of assessing the levels of GHGs emissions to the atmosphere connected, first of all, with solution of the problem of global carbon cycle. It is evident that without a reliable verification of available estimates of emissions all discussions concerning the ecological benefit of various measured and respective expenditure are abstract (Nilsson et al. 2002). For instance, how can the fines for not meeting recommendations on GHGs emissions reduction be substantiated if it is impossible to prove that emissions in 2012 will differ from those in 1990? So far, in the course of discussions on the Kyoto Protocol problem, quantitative estimates of uncertainties of the levels of GHGs sinks have been ignored (especially it concerns the biosphere). Uncertainties of the estimates of total CO_2 fluxes are, however, very substantial (exceeding 100 % in Russia). Calculations of the errors in the estimates of total GHGs fluxes gave $\sim\pm 5-25$ %, whereas the levels of GHGs emissions reduction foreseen by the KP average about 5 %. An average global situation is illustrated, for instance, by the fact that uncertainties of the estimates of GHGs emissions due to the systems of energy production are approximately equal to uncertainties of the estimates of CO_2 assimilationby the biosphere and land.

In this situation, solution of the problem of the estimates uncertainties (first of all, the matter concerns reliable information on carbon cycles) and verification are particularly important. Solution of the verification problem requires a common opinion of its mechanisms, which is also of great importance from the financial point of view. Results of simulation modeling indicate, for instance, that if the confidence level of GHGs emissions reduction by 5.2 % is raised from 50 to 95 %, it will lead to an increase of expenditure on measures to reduce GHGs emissions by factors of 3–4. The main conclusion is that science should serve as the compass for recommendations on measures in the field of ecological policy. In this respect, the COP-6 failure can be a healing shock (Nilsson et al. 2002).

A serious uncertainty of the estimates of a potential anthropogenically induced climate change and its determining factors has brought forth a heated discussion on these problems both in scientific literature and in mass media. For instance, in the interview to a journalist from "Scientific American", Professor R. Lindzen from Massachusetts Institute of Technology, USA (Lindzen et al. 2001; Lindzen and Giannitsis 2003) said that his anxiety about speculative opinions of climate change emerged in 1988 when J. Hansen (Director of Goddard Institute for Space Studies, New York) announced in public that a global climate warming has resulted from an increase of CO_2 concentration due to emissions to the atmosphere caused by fossil fuel burning (Hansen and Sato 2001). This announcement has made R. Lindzen explain that climate science is only at an initial stage of development and, in particular, there is no consensus on the causes of climate change. In early 2001 he reported at the meeting of the US Cabinet on the problem of climate change. The fact that the global mean surface air temperature (SAT) increased by $\sim$0.5 °C during 100 years, and the atmospheric CO_2 concentration increased by $\sim$30 %, does not reflect a cause-and-effect feedback between CO_2 concentration increase and temperature increase. Lindzen et al. (2001) believe that the most reliable estimate of climatic sensitivity (SAT increase at a doubled CO_2 concentration) is 0.4 °C, and it means lack of a basis for anxiety about the catastrophic global climate change in future.

According to IPCC mandate, the problem posed before IPCC is to prepare an overview of "any climate change with time, both natural and anthropogenic". Pielke (2002) noted, however, that at least two climate-forming factors have turned out to be either unreliably considered or left out of account at all:

(1) impact on global climate of anthropogenic changes of land surface characteristics;
(2) biological impacts of the growing CO_2 concentration in the atmosphere (effect of fertilization including) (Tianhong et al. 2003).

If both these factors are really substantial, the conclusion suggests itself that an agreement of the results of the global climate numerical modeling is accidental (in fact, the matter concerns the global mean annual mean SAT).

In this connection, Pielke (2001a, b, 2002) discussed information which confirms the significance of both these climate-forming factors and expresses his opinion about possibilities to check the basis for this conclusion. For this purpose data can be used on the impact of anthropogenic changes of land surface characteristics on local, regional and global climate, which illustrate the fact that this impact is also should be taken into account, like the impact of CO_2 doubling in the atmosphere (as well as an increase of other GHGs concentrations). Also important is the fact that the atmosphere-surface interaction is characterized by the presence of various non-linear feedbacks, and therefore it can be impossible to predict a climate change for periods longer than a season.

As for potential biological impacts of CO_2 concentration growth, they are manifested through short-range (biophysical), middle-range (biogeochemical), and long-range (biogeographical) impacts of landscape-forming processes on weather and climate. The biophysical impact includes, for instance, the effect of transpiration on relationship of latent and sensible heat fluxes as surface heat balance components. Biogeochemical impacts include the effect of the growth of plants ("fertilization effect") on the leaf area index, from which evaporation takes place, on surface albedo, and carbon supply. One of manifestations of biogeographical impacts is a change in time of the species composition of plant communities. Numerical modeling results indicate that without consideration of biophysical/biogeochemical impacts, an assessment of climate change cannot be reliable (Bounoua et al. 2002).

Further development of climate models should contain, in particular, a consideration of the following aspects of climate formation:

(1) direct and indirect impacts of landscapes dynamics through biophysical, biogeochemical, and biogeographical processes;
(2) consideration of anthropogenic changes of land use on various (local, regional, and global) spatial-temporal scales; and
(3) assessment of possibilities of climate forecasts for longer periods than a season, bearing in mind the functioning of numerous non-linear feedbacks which determine the atmosphere-surface interaction.

The openness of these and other problems reduced the significance of the IPCC-2001 Report and the US National Report as containing only an assessment of global climate sensitivity to changes of some climate-forming factors.

In the long process of preparation of the three published already IPCC Reports (Houghton 2003, 2004), more and more complete analyses of climate-forming factors (e.g., apart from GHGs, atmospheric aerosols have been considered) and various feedbacks have been carried out. But nevertheless, the available very complicated numerical climate models cannot be considered sufficiently adequate from the viewpoint of consideration of all substantial climate-forming factors. A new step forward in the IPCC-2001 Report was a consideration of the forcing F of anthropogenic land use changes on climate limited, however, by a consideration of only the impact of land use dynamics, starting from 1,750, on surface albedo. The obtained estimates gave an average F value equal to $-0.2\ \mathrm{W{\cdot}m^{-2}}$ with an uncertainty interval $0/-0.4\ \mathrm{W{\cdot}m^{-2}}$. Thus such estimates are highly uncertain, and this concerns even the sign of F (a more complete consideration of biophysical, biogeochemical and biogeographical impacts of the land use evolution on climate has led to the conclusion that in this case $F > 0$).

In this connection, of importance is a new range of potential global warming by 2,100 assumed in the IPCC-2001 Report (1.4−5.8 °C) based only on the numerical modeling data (and therefore it will inevitably change in future). Also, the problem is that this new range cannot be directly compared with the earlier similar values. From the viewpoint of reliability of the estimates of future climate change, of importance is the use in the IPCC-2,001 Report of the term *projections* instead of *predictions*, since the latter implies that the factors left out of account will not be substantial in the future. It is unacceptability of the latter supposition that is the reason why none of the specialists in the field of numerical modeling will assert that the climate can be predicted 100 years in advance. This conclusion confirms the results of numerical modeling carried out in (Andronova and Schlesinger 2001).

The complexity of the problem of prognostic assessments of climate, especially of selecting the contribution of anthropogenic component is illustrated by the remaining inconsistency of analysis of the climatic impact of clouds. Tsushima and Manabe (2001) have analyzed the impact of cloud feedback (CF) on the formation of the annual change of global mean SAT using the data of space observations of the Earth's radiation budget (ERB), bearing in mind an assessment of adequacy of CF consideration in numerical climate models by comparing the calculated and observed change of the global mean SAT. It follows from the observed data that the global annual change of SAT is in phase with the annual change of SAT in the Northern Hemisphere, and its amplitude constitutes 3.3 K (this in-phase character is determined by concentration in the Northern Hemisphere of continents contributing most to the formation of the amplitude of the annual SAT change).

Analysis of the data of the ERB components observations for the period from February 1985 to February 1990 has shown that global mean values of both shortwave and longwave radiative forcings (SWRF and LWRF) depend weakly on the annual change of global mean SAT (the ERB data considered refer only to the band 60°N–60°S). Thus the cloud cover dynamics causes neither increase nor decrease in the annual change of SAT. The data on SWRF and optical properties of clouds show that not only albedo but an amount and top height of clouds depend weakly on SAT and hence, CF does not affect strongly the annual change of the global mean SAT.

It could be drawn from this conclusion that the impact of CF on the annual change of the global mean SAT or on the global warming is negligible. However, this speculative opinion is dangerous in view of strong complexity of the spatial field of SAT. Calculations with the use of three climate models which take into an interactive account the dynamics of the microphysical properties of clouds have led to the conclusion about a substantial increase of the cloud top albedo with increasing SAT, which disagrees with the data of observations. This situation reflects the prospects for comparison of assessment of the role of cloud feedbacks from the data of numerical modeling and observations from the viewpoint of models validation.

Lomborg (2001, 2004) is of course right rejecting apocalyptic predictions of global ecodynamics based on exaggerated fear of limited natural resources and the environmental state. Such opinions and assessments are confirmed, in particular, by the data of Holdren (2003), which characterize both real and potential global energy resources. The energy units are expressed here (in case of non-renewable energy sources) in TW-year, which is equivalent to 31 exa-J ($1\,\mathrm{TW} = 1\,\mathrm{TW\text{-}year} \cdot \mathrm{year}^{-1} = 31.5\,\mathrm{exa\text{-}J} \cdot \mathrm{year}^{-1}$). It should be added that in 2000 the global energy consumption constituted about 15 TW or 15 TW-year $\cdot$ year^{-1} with an assumed increase up to 60 TW-year $\cdot$ year^{-1} by 2100.

Despite the optimistic data by Holdren (2003), the absence of long-range prospects for development of the present consumption society illustrated by global ecodynamics estimates (Krapivin and Varotsos 2007, 2008) raises no doubts. Therefore at the World Summit on Sustainable Development held in Johannesburg in 2002, a necessity has been emphasized of accomplishing 10-year programmes in order to achieve stable production and consumption, which included the following recommendations (Starke 2004):

- developed countries should undertake the leading role in provision of stable production and consumption;
- these goals should be achieved on the basis of common but differentiated responsibility;
- the problem of stable production and consumption should play the key role;
- the young generation must take part in the solution of the problem of sustainable development;
- the "polluter pays" principle should be practiced;
- control of the complete cycle of products evolution from their production to consumption as well as waste in order to raise the production efficiency;
- support should be given to politics that favors the output of ecologically acceptable products and rendering of ecologically adequate services;
- to develop more ecological and effective methods of energy provision and liquidate energy subsidies;
- to support free-will initiatives of industry aimed at raising its social and ecological responsibility; and
- to study and introduce an experience of ecologically pure productions, especially in developing countries as well as in small- and medium-scale business.

Though the enumerated recommendations are rather declarative, they still clearly point to the necessity to change the paradigm of the socio-economic development (primarily in developed countries) from the consumption society to priorities of public and spiritual values. A concrete analysis of the ways of such development requires a participation of specialists in the field of social sciences. Some related opinions were expressed by Corcoran (2005) in the Introduction to *The Earth Charter in Action*. Therefore, the question whether a human can change climate needs further studies (Borisov 2005).

Finally, preliminary conclusions drawn from the above are as follows:

- the existing climate models cannot be used to make decisions and assess the risk of accomplishing the anthropogenic scenarios;
- the level of uncertainty of climate forecasts ca be reduced due to a broader consideration in global models of interactive bonds in the NSS and mechanisms of biotic regulation of the environment as well as improvement of the global monitoring system; and
- the use of hydrocarbon energy sources in the 21st century will not lead to a catastrophic climate change if the Earth covers are preserved and the World Ocean is protected from pollution.

9.8 Reducing Risks in Agriculture

9.8.1 Introduction

Uncertainty and risk are quintessential features of agricultural production. Main sources of uncertainty and risk are connected with the causes which influence on agricultural production:

- the amount and quality of output that will result from a given bundle of inputs are typically not known with certainty, i.e., the production function is stochastic;
- presence of uncontrollable elements, such as weather and absence reliable prognosis of many agronomy processes;
- price uncertainty is a standard attribute of agricultural activities;
- technological uncertainty is always the cause of excided risk; and
- policy uncertainty creates considerable risk for agricultural investments.

Unfortunately, basic rationality postulates in the agricultural risk are connected with economic models. Technological and nature-oriented risks are studied least of all. Decisions about environmental risks typically focus on the evaluation of consequences, and estimates of consequences are complicated by uncertainty. Some problems of environmental risks are discussed by Krapivin and Varotsos (2007, 2008). It is clear that agricultural risks are some fragment of policy decisions for sustainable management that are based on the reasonable certain knowledge base the effective use of which is possible in framework of GMNSS.

One of the main problems in framework of modern science is the estimation of the biosphere survivability in the conditions of the increasing anthropogenic impact. Such estimation can be realized by using the biosphere model. Well-known attempts in direction of the synthesis of a global model give unsatisfactory results.

Created global models have educational character and no give possibility for the real estimations. For the last years many investigators put a problem of creating of reliable and effective systems for control of the environment state in the global scale including agricultural sphere. Generally this problem includes elaboration of technical means for collection, storage and transfer of the data of the nature medium state and (from the other side) development of methods for processing these data. The existing means for collection of information about natural objects and processes give a possibility to form the data set, covering large territories up to the whole biosphere. Remote means of environmental monitoring become especially effective. In this context, Nitu et al. (2004) have formulated the basic model of the biosphere survivability and proposed a new view on the global modeling.

The behavior of any system is determined by the value which can take on different terms characterizing the state of the system. Upon interaction with an external medium and in particular with other systems, the values of these terms can vary in one way or another. For any technological or biological system it is always possible to show the field of change of the characteristic parameters, wherein the system can be considered to be functioning. Outside this field the system does not exist.

Thus one can substitute the complex behavior of a system by a description of the behavior depicting this system by a point in phase space of the characteristic parameters. If the change of any coordinate leads to the disappearance of the depicting point from the allowable field, the system is demolished (the organism as a whole perishes).

A system is defined by its structure and behavior. The behavior of such a system is aimed at providing uninterrupted functioning by means of a correspondingly organized structure and behavior. This characteristic of the complex system to actively withstand the hostile action of an external medium, we shall refer to as survivability (Nitu et al. 2000a, b, 2004; Krapivin and Shutko 2012). The key questions to be answered within the numerous investigations of global ecodynamics are:

- What are the level, interactions, and significance of "human dimension" (socio-economic factors) in the development of society and its role in global environmental changes?
- What are the present and possible future impacts of the global environment variability on economic development, what factors determine the capability of society to react to the occurring changes, what are possibilities to provide a sustainable development and to reduce the man's sensitivity to forcings?
- What are the possible methods of decision making in the interests of sustainable development under conditions of NSS complexity and high-level uncertainties regarding the global environment variability?
- What are the possible impacts of the global environmental changes on human health; what information about ecodynamics and socio-economic factors is needed to evaluate the respective cumulative risks for human health?

Most substantial directions of developments connected with answers to these questions include, in particular, the following:

(1) Regional problems of ecodynamics. Of primary importance here are studies of the problems of power supply, reconstruction of ecosystems, recovery of human health, and special phenomena such as droughts and forest fires al local and regional levels. In this connection, an integration of physics, chemistry and biology with respective socio-economic aspects taken into account should play a decisive role, which will help to make adequate decisions in ecological policy. Studies of the dependence of forest fires on climatic conditions can serve as an illustration of possible approaches to the solution of such problems, and the main goal is the prediction of anthropogenic wildfires.
(2) Analysis of economic efficiency. The complexity of quantitative assessments of the economic efficiency of one or another ecological policy is determined by an ambiguous choice of this policy with priorities taken into account. In this context, of great importance is the use of simulation models of ecodynamics, especially the models of changes in land supplies of carbon to assess the response of carbon supplies in vegetation and soil to specific features of land use and changes in land surface characteristics, to variations of the atmospheric CO_2 content, and to climate. Solution of such problems will provide a more reliable prediction of global trajectories of CO_2 emissions to the atmosphere.
(3) Possible consequences of global climate changes for forestry and agriculture over the country territory. Analysis of such changes with various scenarios of possible changes in global climate suggested the conclusion that in the case of several scenarios, the impact on forestry and agriculture in the country will be economically favorable. Partly, it is connected with the growth of forests production (with a CO_2 concentration increase) and determined by capability of forests to adapt for climate change. As for agriculture, according to available prognostic estimates for a long period, positive impact of the global warming on agriculture in the country will be less economically favorable than it follows from the earlier estimates.
(4) Impact of UV solar radiation on human health. Intensive developments in this direction are being carried out to study the impact of the lowering level of biologically active UV solar radiation due to decreasing total ozone content in the atmosphere on the agricultural production.

Talking risks assessments we mean at least 5-folds (5-levels) interpretation or definition:

(1) theory of different kinds of risks;
(2) experimental studies of risks in conditions of natural and man-made disasters;
(3) remote sensing application for risks determination (microwave radiometry allows to obtain by practically direct way the risks assessments);
(4) risks assessments in AIMS; and

(5) creation of state-of-the art of the technological complexes for strategic determination of risks of hunger throughout the World by developing aerospace complexes.

9.8.2 Practical Microwave Radiometric Risk Assessment of Agricultural Function

Advanced technology for the assessment of soil hydrological regimes for agricultural purposes using multichannel passive (radiometric) and active (SAR) microwave measurements were developed by many authors (Krapivin and Shutko 2012; Shutko 1986). Physical background of the technology, the equipment used, examples of experimental data, and prospects of the technology application were described by Shutko et al. (2007, 2010).

The problem of the Earth's environment monitoring and land use control has been growing in importance over the last 20–30 years. The successful solution of this problem requires research, development and application of different remote sensing technologies. Among these technologies there are radio-physical remote sensing methods and instruments that play an important role in the Earth's surface investigation. These methods and instruments are based either on the measurements of the parameters of natural radio-thermal electromagnetic radiation from the Earth's surface or on the measurements of the parameters of artificially radiated electromagnetic signals scattered by the Earth's terrain. In the first case, the remote sensing tools are called either Microwave Radiometers (MR) or Passive Microwave Radar (PMR); in the second case, the tools are called either Side-Looking Airborne Radar (SLAR) or Synthetic Aperture Radar (SAR), or Nadir Viewing Radar (NVR) with hyper-short (nanoseconds) length of pulses, etc. These instruments are being installed on board of aircraft and satellites and orbital stations and tested in many scientific campaigns.

Research team of the Kotelnikov's Institute of Radioengineering and Electronics of Russian Academy of Sciences and Radioengineering Corporation "Vega" of Russia has developed an advanced technology for the assessment of soil hydrological regimes using multispectral passive (radiometric) and active (SAR) microwave measurements (Gulyaev et al. 2009; Haarbrink et al. 2011; Shutko et al. 2006a, b). These technologies are aimed to measure:

- surface soil moisture;
- underground moistening at different depths (profiling);
- depth to a shallow water (down to 2 m in humid areas and down to 3−5 m in arid/dry areas, (to 10 m for active system);
- groundwater resources presence;
- contours and amount of water seepage through hydro technical constructions (levees); and
- biomass of vegetation above a water surface or wet ground.

Passive system includes airborne scanning multi-channel radiometer "Radius" (2.25, 5.5, 21, 43 cm) and a set of non-scanning radiometers (2.25, 6, 18, 21, 27 cm). Active system includes airborne multichannel SAR system (3.9, 23, 68, 254 cm). Successful tests of this equipment have been conducted in Russia, Ukraine, Uzbekistan, Turkmenia, Moldova, Bulgaria, Cuba, USA (Krapivin and Sutko 2012).

Microwave radiometry or passive microwave remote sensing is based on measurements of the natural electromagnetic radiation of objects and Earth covers in the millimeter to decimeter wavelength range. Inside this band, the land surface radiation is primarily a function of free water content in soil and also influenced by other parameters of the soil-canopy system, such as above ground vegetation biomass, soil density, water salinity and temperature of the system. Free water content in soil is dependent on rainfall rate, artificial watering, shallow groundwater and processes of soil drying of at interface soil-atmosphere. Hence microwave radiometry is an important tool for the assessment of soil hydrological regimes (Shutko 1986, 1992; Krapivin and Shutko 2012).

The measure of the intensity of radiation in the microwave band is referred to as a brightness temperature T_b, which is a product of emissivityκ, and thermodynamic temperature T_e, within the effectively emitting layer of the object: $T_b = \kappa T_e$. The emissivity is a function of dielectric permittivity/permeability of the object/surface of observation. For a land surface, the dielectric permittivity is first of all a function of soil moisture. The higher a soil moisture content, the higher the permittivity of soil, the lower the emissivity/intensity of radiation/brightness temperature of this piece of land. For a water surface, the dielectric permittivity is first of all a function of electric conductivity of a water solution that is dependent on a concentration of salts, acids, on a presence of oil films and many other chemical substances. For example, the higher salinity of water, the higher the dielectric permittivity of water solution, the lower the emissivity/intensity of radiation/brightness temperature of this water body (Gulyuev et al. 2005).

Within the 2 to 30 cm band, for air temperature t = 10−30 °C, the radiation characteristics of several surface types are shown in Table 2.13. Table 2.14 shows the sensitivity of radiation in the X-band (2−3 cm) and L-band (18−30 cm) to the changes in free water content in bare soil, soil density, salinity and temperature of the soil surface (Shutko 1986). These data show that the main parameter affecting the intensity of a bare soil radiation, practically independent of spectral band, is the soil moisture. Based on this sensitivity it is feasible to estimate the value of soil moisture without a priori data on the soil parameters. It has also been seen that even for rather high values of biomass, up to 2–3 kg/m^2, the plant canopy is still transparent in the decimeter wavelength range.

There are different sources of natural microwave radiation in microwaves, which can influence on accuracy of measurements: Galaxy, ionosphere, atmosphere (clouds), land. The minimum influenced by different sources of radio-noise is 2-cm to 21-cm wavelength band, which is window of transparency at microwaves. The centimeter and decimeter wavelength microwave radiometers are not influenced by the conditions of illumination, by presence of fog, smoke, clouds, and are very useful for measurements in different meteorological conditions (Shutko 1992).

Radar (SAR) survey or active microwave remote sensing is based on measurements of the parameters of artificially radiated electromagnetic signals scattered by the Earth's covers and objects. Due to high spatial resolution radar systems are very important instruments of remote sensing from space. Previously studies of scattered signal from soil and vegetation were conducted using on-ground and airborne scatterometers and radars. The results of these studies are the dependences of scattered signal on soil moisture, soil surface roughness, vegetation parameters, etc. (Chukhlantsev and Vinokurova 1991; Chukhlantsev 1992).

Theoretical models for electromagnetic waves scattering by bare soil are based on Kirchhoff approximation. Different expressions were obtained for the backscattering coefficient σ_s^0 but most of them may be written as

$$\sigma_s^0 = he^{-h}F(klSin\vartheta)R_sCos^2\vartheta + \sigma(h^2),\ h = 4k^2\sigma^2Cos^2\vartheta, \tag{9.24}$$

where σ is the standard deviation of rughness height, l is the roughness correlation length, k is the wave number in free space (2π/λ), ϑ is the nadir viewing angle, R is the soil reflectivity (Chukhlantsev 1992).

Analysis of formulas (9.24) showed that

- The sensitivity of the electromagnetic wave scattering by bare soils to reflectivity variations of bare soil surface doesn't depend on roughness parameters, wavelengths and observation angle; dynamic range of scattering by bare soils is no more 9 dB, when reflectivity varies from 0.05 (very dry soil) to 0.4 (very high soil moisture); 6–8 gradations of reflectivity may be obtained with an accuracy of soil backscatter measurements of 1.5 dB
- Maximal dynamic range of backscatter by bare soils is equal to about 18 dB, when standard deviation of roughness height σ varies from 0.5 to 4 cm.
- There are two independent parameters: the standard deviation of roughness and the reflectivity. It is obvious, to separate and to estimate the reflectivity (soil moisture) and roughness parameters, it is necessary to use measurements at minimum 2 frequencies. The best separation of these parameters is observed for observations at wavelengths satisfying the condition: $h(\lambda_1) \approx 1$, $h(\lambda_1) << 1$; that produces wavelengths 10 and 24 cm.

These conclusions were verified by experimental data. Quantitative and qualitative analysis showed that radar data may be good tool for soil moisture measurements. It is possible to obtain 6−8 gradations of soil moisture for bare soil and 3 −4 for soil covered by vegetation (Chukhlantsev 1992).

It has theoretically been shown and experimentally proven in investigations (Kutuza et al. 2000) and in the studies conducted by other teams that active and passive systems are sensitive to soil moisture changes and to changes in surface roughness, peculiarities of relief, vegetation biomass and structure, and angles of observation. In general, the terrain features, surface roughness, changes in relief, and vegetation characteristics influence the radar return more than natural emitted radiation, other conditions being equal.

The approximate ranges of radar return changes are as follows: 6–8 db for whole range of soil moisture changes; 10–18 db for whole possible scale of roughness changes; 8–10 db for vegetation characteristics changes; 6–8 db for changes in angle of observation, $30° < \Theta < 60°$ off nadir. Now, let the changes in radar returns caused by soil moisture changes be referred to as signal, and the changes in radar returns caused by the changes in other than moisture parameters referred to as noise. Then, the signal-to-noise (s/n) ratio calculations will result in a s/n = 0.33–0.8 for roughness, a s/n = 0.6–1.0 for vegetation, and a s/n = 0.75–1.33 for changes in angle.

Similarly, calculations of the signal-to-noise ratio for the naturally emitting radiation are higher: s/n = 2–5 for roughness, s/n = 1–3 for vegetation, s/n = 2–3 for changes in angle. This shows the benefits of using passive systems, in spite of a more coarse spatial resolution and a smaller swath, in combination with active systems.

Microwave remote sensing equipment was partly described in Chap. 2. An experimental usage of this equipment for solving agricultural tasks has shown high effectiveness and decreasing risks related with well-timed evaluation of basic characteristics of agricultural soils including soil salinity and its humidity, as well as the crop maturity. Figure 9.30 compares the different penetrative ability of L-, P-, and VHF-band radars to see a ground surface through vegetation cover. In L-wavelength band, the signal is mainly dependent on canopy characteristics. The P- band returns are much less affected by grass vegetation and a back-scattered. It clearly indicates the effect of a soil and water surface along with some effect of vegetation. The VHF-band image clearly demonstrates no effect of canopy on signal returns for grass and bushes. More detailed consideration of the raised problems are given in Chap. 2.

Comparison is given in Fig. 9.28 of a different penetrative ability of L-, P-, and VHF-band radars to "see" a ground surface through vegetation cover. In L-wavelength band, the signal is mainly dependent on canopy characteristics. The P- band returns are much less affected by grass vegetation and a back-scattered signal clearly indicates the effect of a soil and water surface along with some effect of

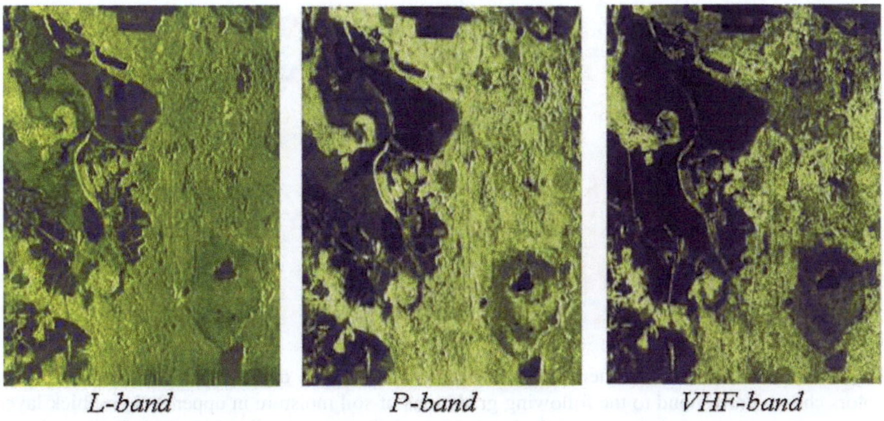

Fig. 9.30 Images obtained by the improved SAR System IMARC, Spas-Klepiki, Ryazanskaya Region (Gulyaev et al. 2005)

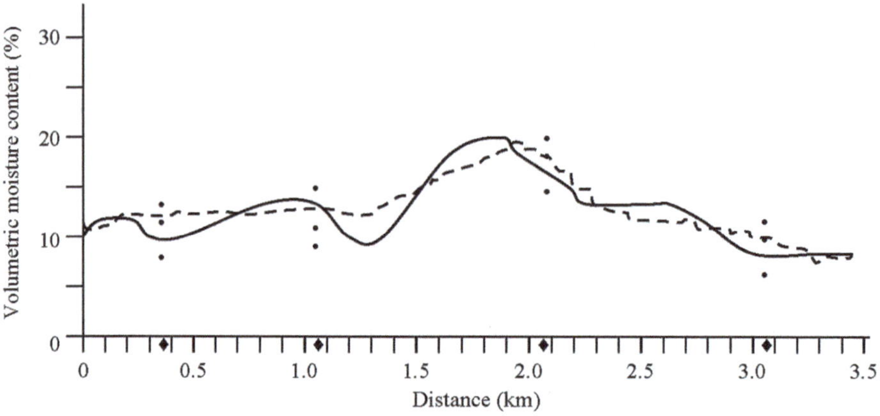

Fig. 9.31 Comparative analysis of results received by means of RIMS (*solid curve*), microwave monitoring (*dashed curve*) and in-situ measurements (dislocation of sites where measurements were carried out is marked by symbol ♦. Measurements were realized by means of flying laboratory Miramap on August 2, 2007 near village Nikolovo (Bulgaria) one day later the rich rainfall (Soldatov 2011)

vegetation. The VHF-band image clearly demonstrates no effect of canopy on signal returns for grass and bushes.

Calculation of the hydrological characteristics of agricultural fields based on data of microwave monitoring as input information for the RIMS gives possibility to increase the crops and to control the quality of soil resources. As Krapivin and Shutko (2012) repot, some experimental investigations of the RIMS were realized during spring of 2007 on the Bulgarian territory using Miramap flying laboratory equipped by microwave radiometers (Fig. 2.7). Some results are given in Figs. 2.14, 2.15, 9.31 and 9.32. The RIMS input information was quantized topographic maps

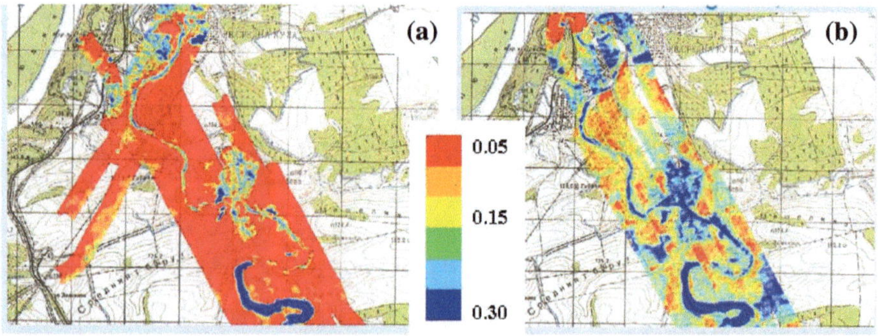

Fig. 9.32 Microwave assessments of surface soil moisture in dry (**a**) and wet (**b**) conditions. colors chosen correspond to the following gradations of soil moisture in upper 0–5 cm thick layer. measurements were carried out on July 31, 2007 (**a**) and on August 2, 2007 (**b**) in Rusenski Lom River region (Bulgaria). *Dark blue* sites correspond to the soil moisture content of 0.2–0.3 g/cm^3 or to open water

with spatial resolution 9 m × 9 m, meteorological data for the moment of the experiment realization, and systematic maps of soil-plant formations. Verification of the RIMS results was conducted on the basis of remote sensing and in situ data that was not the input of the RIMS. Disagreement between the RIMS results and observational data does not exceed 5 %.

9.9 Conclusion

Recent worst disasters, such as the earthquake and tsunami in Japan (2011), the earthquake in Haiti (January 2010), the oil spill in the Gulf of Mexico (April 2010), the floods in Pakistan (August 2010) and Australia (January 2011), the heat wave and wildfires in Russia (summer 2010), and typhoon Haiyan in Philippines (2013)

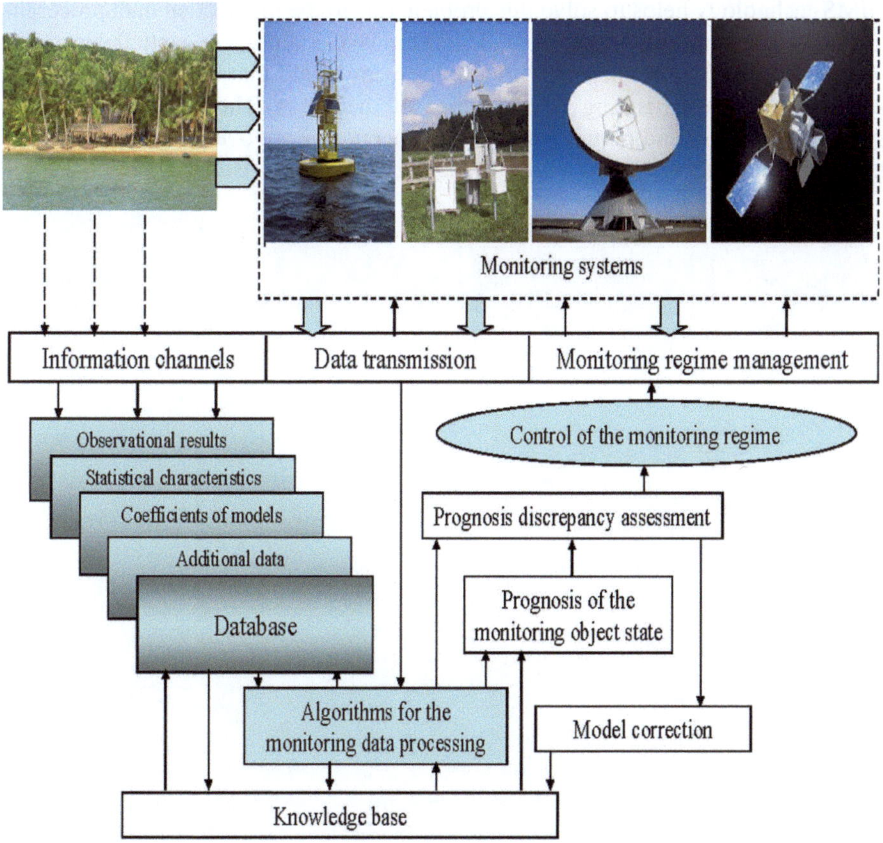

Fig. 9.33 Principal scheme for the GIMS functioning during its adaptation to the specific environmental sub-system to realize the monitoring procedures

are indicative of the increasing devastation and variety of disasters. This chapter draw attention to the fact that risk and disasters threaten development and security. The scale of damage and aftermath reduction of stressful natural processes is possible with the use of the GIMS technology.

Operational diagnostics of the environment is accomplished using the model that is synthesized for the concrete monitoring object. These models are stored in the GIMS knowledge base and than are selected automatically during the GIMS adaptation process according to the scheme represented in Fig. 9.33. This model describes the interactions and functioning of different spatial-temporal levels of whole of environmental processes that affect the monitoring object functioning. Model covers natural and anthropogenic processes that are specific for given territory during concrete time interval and the GIMS supports the model information inputs.

As it was mentioned above and was expressed by many experts (Krapivin and Shutko 2012; Shutko and Krapivin 2011; Kondratyev et al. 2004a; Hushon 1990) the synthesis of the model is complex problem to achieve the expected results. The GIMS-technology helps to solve this problem due to the intellectual data processing and the sorting through of models and algorithms. Certainly, conceptual description of the solved task is connected with fixed semantic field that is given in the GIMS knowledge basis depending on the series of meanings. The environmental subsystem complexity and partial parametrical uncertainty are overcome by the GIMS-technology due to the evolutionary modeling technology (Bukatova et al. 1991).

Chapter 10
Ecoinformatics Problems in the Future World

10.1 Problems, Perspectives and Waitings

As it was mentioned by Krapivin and Shutko (2012) that ecoinformatics is the science of information in ecology and environmental science. This book demonstrates different definitions, semantic structures, algorithms, methods, and models that help to parameterize various environmental processes in the nature-society system. One of main ecoinformatics objectives is to develop new effective informational technologies for the detection and prognosis of environmental factors that are significant for human life. Really ecoinformatics is an interdisciplinary study between ecology, GIS, mathematical modeling, and computer science. Ecoinformatics promotes novel exploitation and integration of ecological data and knowledge by means of state-of-the-state simulation technology. Ecoinformatics and GIMS technology are leading-edge technologies transforming the ways data that are collected and analyzed.

Environmental science such as ecoinformatics emerged practically at the same time in series publications in Russian Federation (Arsky et al. 1992; Krapivin and Potapov 2002) and United States (Kaeiva 2001; Brunt et al. 2002). Dengler et al. (2011) consider the ecoinformatics as ecological science that develops biodiversity informatics. The field of ecoinformatics provides concepts, methods and standards to guide management and analysis of ecological data with particular emphasis on exploration of co-occurrences of organisms and their linkage to environmental conditions and taxon attributes. Moreover, ecoinformatics develops new information-modeling technologies for future global change research with special focus on important environmental problems connected with human society survivability.

Nature-society system evolution depends on the numerous economical, social, natural and technological processes taking place both in regions and world. Coupled investigation of these processes is principal problem of its sustainable development. Effective instrument to solve regional and global problem of sustainable development is based on the combined use of knowledge from different sciences. One of such instruments is the GIMS-technology described in above chapters. It is evident

V.F. Krapivin et al., *New Ecoinformatics Tools in Environmental Science*,
Environmental Earth Sciences, DOI 10.1007/978-3-319-13978-4_10

that creating universal technology for global model of nature-society system is unlikely. Unfortunately, knowledge accumulated by the humanity does not cover all problems of global sustainable development that was discussed by many experts (Brucan 1984; Cozzi 2003; Cracknell 2009; Cracknell et al. 2009a, b; Krapivin and Varotsos 2007, 2008; Homer-Dixon 2008; Binenko et al. 2004; Degermendzhi 2009; Stoddard et al. 2003). Many of them have demonstrated the absence of constructive algorithms for existing data processing. This default lies mainly in the "dimension perdition", i.e. the weakness to take into consideration of interactions between all subsystems of the nature-society system.

Nevertheless, Degermendzhi (2009) and Degermendzhi et al. (2009) proposed new view on this problem and developed new directions in biophysical ecology that is based on small-scale models. This approach is based on simple scheme of the most obvious interactions between processes in the nature-society system. These interactions can have positive, negative or uncertain effects. Analogous simplified approach to the nature-society system modeling was proposed by Cracknell et al. (2009b) and Krapivin and Kelley (2009). It is clear that the choice of the effective model for global and regional ecological investigations requires simplified representations of interactions in the nature-society system. Basic difficulty arises in the choice of climate model.

Nowadays, the problems of globalization and sustainable development have been discussed in a vast scientific literature. Even more numerous are the concerns raised in the mass media. However, discussions of such problems contain many contradictions and disagreements. First of all, globalization and sustainable development problems are treated, as a rule, as independent and separate. Nevertheless, the priority of the globalization problem and subordinate importance of sustainable development problems as one of the most important aspects of the globalization processes raises few doubts.

A constructive view on the globalization and sustainable development problem was first broached by Russian scientist Kirill Ya. (Kondratyev et al. 1992, 1994, 1997). The development of globalization theory took a number of different directions that were often contradictory and colored by socio-political biases (Friedman 2005; Mander and Goldsmith 2006; Wijen et al. 2005). Global socio-political issues have a long history beginning with the era of colonialism. After 1917 there existed two periods: the pre- and post-Cold War eras. The post-Cold War era led to the increasing influence of what some people these days call quasi-governments (such as the International Monetary Fund or the World Bank). This delimitation served for a certain time as the cause of principal difference between the existing concepts of globalization process. At its most basic, there is nothing mysterious about globalization in spite of these differences. In fact, there exist two main conceptions of globalization:

(i) Globalization that brings prosperity to certain countries while impoverishing other countries.
(ii) Globalization as the objective process behind development of the nature-society system (NSS), taking into account existing global population dynamics.

In the first case, many experts around the world identify the globalization with "Americanization" of economics and culture (Friedman 2005). The rest of the world seems to be following the U.S.A. and leaving behind its own ways of life. "Americanization" is the contemporary term used for the influence that the U.S.A. has on the culture and economics of other countries, substituting their way of life by the American way. This process is managed by the many international structures under the control of the U.S.A. Significant social networks and international forums have been organized with the purpose of proliferating capitalist interests. Some of these forums date from earlier in the twentieth century, such as the International Chamber of Commerce, the International Organization of Employers, Center for Environmental Diplomacy, and the Bretton Woods Committee. They were created as a specific response to questions about globalization (Furth 1965; Devkota 2005). These and other organizations and forums continue to attend to the interests of developed countries in 21st century as well.

Reformists of global social democracy put forward the theory of neoliberal globalization which proposes the transformation of social ambience in the world. The countries of the Former Soviet Union have fully faithful negative consequences of this theory. Therefore, the question "what is globalization?" is becoming principal question today. Many experts see it as a primary economic phenomenon involving the increasing integration of national economic systems through the growth in international trade, investment, and capital flows. Other experts define globalization as social, cultural, political, and technological exchanges between countries. As the result two alternatives were formed: anti-globalization and pro-globalization.

The alternatives to globalization promoted by anti-globalists include different approaches to the dynamics of world structure:

- recognition of importance of free trade and investment for economic growth and development, but there also exist data that trade and investment have in fact increased poverty and inequality;
- sustainable development is best achieved through a single framework, integrating environmental protection and the promotion of economic growth and social equity;
- innovation is necessary for sustainable living;
- support of the accumulation of power at the local level;
- implementation of trade barriers to protect local production and renunciation of international trade unless goods or services cannot be produced locally;
- free trade usually benefits wealthy countries at the expense of poor countries; and
- environmental protection is the key to sustainable development and is imperative for economic development.

The argument of pro-globalization groups is based on the promotion of free trade as the key to eliminating poverty and ensuring effective development and that market instruments such as intellectual property rights are necessary to protect the environment and promote development. Supporters of globalization argue that it can be rolled back and point to the period between the First and Second World

Wars as evidence. The increase in world trade as a proportion of world GDP was proportionately greater between 1870 and 1914 that it has been since 1975. That expansion was stopped, not just by the First World War, but by the loss of support for free trade that followed. This led directly to the 1930 depression and indirectly to the Second World War. All these changes took place when the world population did not exceed 2 billion. At present time the situation is rapidly changing with the exponential growth of world population exceeding 6.6 billion. Therefore, globalization process is becoming an inevitable phenomenon of world life. It is for this reason that the list of the following key questions on globalization must be answered:

- Is there any alternative to globalization in the coming decades?
- What are the environmental impacts of globalization?
- How does globalization affect culture and religion?
- What is globalization and when did it start and how does it depend on population size?
- Who are the players in the interactions between nature and society?
- Why is there global inequality, and is it getting worse?
- What are the cost and benefits of free trade?
- What is the role of the internet and communications technology in globalization?
- Is globalization shifting power from nation states to undemocratic organizations?
- Is globalization resulting in industries in developed countries being undermined by industries in developing countries with inferior labor standards?
- What does globalization mean to separate countries like Australia?

Nevertheless, when considering the prospects for life on Earth, it is necessary to proceed from the human criteria of assessing the level of environmental degradation, since over time local and regional changes in the environment become global ones. The amplitude of these changes is determined by mechanisms that manage the NSS dynamics and structure (Gorshkov et al. 2000, 2004; Ehlers and Kraft 2005; Guista and Kambhampati 2006). Humankind deviates more and more from this optimality in the way it interacts with the surrounding inert environment. At the same time, human society as an element of the NSS tries to understand the character of large-scale relationships with nature, directing efforts of many sciences to study cause-and-effect feedbacks from this system. The basic item of a global human society is the country; that is considered as an NSS component (Kondratyev and Krapivin 2005a; Olenyev and Fedotov 2003; Wang et al. 2011).

The national safety in present-day conditions is estimated on the basis of many criteria, mainly military, economic, ecological, and social in character. Development of an efficient method of objective analysis of the problem of national security requires the use of the latest methods to collect and process the data on various aspects of the functioning of the world system. Such methods are provided by GIMS-technology (Kondratyev et al. 2002; Man et al. 2011a).

This discussion expounds ideas, methods and information technologies that can help to solve many problems of global ecodynamics . It is clear that formal modeling of nature-society system with consideration of different relationship between spatial and temporal scales does not give a desired result without the principles of sustainable development. Schematically it is shown in Figs. 4.26 and 8.9.

10.2 Global Survivability Problems

10.2.1 Sustainable Development and Ecoinformatics

The present ecological situation worldwide can be exemplified by growing instability, a civilization crisis, the global scale of which is expressed through deterioration of humans and animals habitats. The dominant features in global ecodynamics in the late 20th–early 21st century include the rapid growth in population (mainly in developing countries), urban population growth at the expense of rural population (the growing number of cities), broadening scales of dangerous diseases as HIV/AIDS, hepatitis, tuberculosis, etc. UN data predict global population in 2050 to reach 8.9 billion people. This means that the decrease in per capita consumption (expected as a result of the growing efficiency of technologies) will be cancelled out by the impact of population growth. For instance, if meat consumption by an average American increases by 20 % by 2050 (compared with 2000), then due to the growth of population size in the U.S.A. the total mean consumption will increase by 5 million tons. An important general fact is that despite the prevailing growth of population size in developing countries, their contribution to the impact on the environment will not necessarily exceed that reached in developed countries. For instance, U.S. population size will annually increase by about 3 million people and in India by 16 million people, but the impact of the U.S.A. on the environment (due to its higher per capita consumption) will be more substantial. This is illustrated, in particular, by estimates of the growing CO_2 emissions to the atmosphere: 15.7 million tC in the U.S.A., and 4.9 million tC in India.

The paradox of the contemporary situation is that despite all the unprecedented scale discussions on sustainable development, the world is continuing an unsustainable way of development. There are numerous ideas and principles to achieve sustainable development in different regions and in whole world. But all investigations in this area is restricted by solution of special problems connected with concrete region or ecosystem. Consideration of global nature-society system is possible with the GIMS technology application. In this case realization of global model for the parameterization of nature-society system dynamics needs the measures of key importance including the following:

- There is a necessity to realize that further development of the consumption society will lead to a global ecological catastrophe and the collapse of civilization. The solution of this problem can only be a refusal of the traditional paradigm of the consumption society and a radical change of life style towards spiritual values.

- To remove the present social contrasts between 'the south' and 'the north' there is an urgent need to help the developing countries in many ways, according to the recommendations which have been adopted by the UN.
- Since the level of understanding of the problem of social and environmental dynamics remains low, the support of interdisciplinary studies has particular importance. It would be important to organize the international scientific discussions on sustainable development problems in the context of "Rio+10", concerning first of all the problem of priorities and quantitative assessments of various threats for modern civilization.

Global change priorities are aimed at assessing changes in the NSS taking place in the past, those observed now, and possibly those in the future. A detailed realistic analysis of the prospects of civilization development can be realized with global model principal scheme of which is given in Figs. 1.18, 1.20 and 1.21.

The main priorities of studies of the NSS dynamics include:

- conceptual and analytical approaches to studies of interactions in the NSS;
- discovery of connections of the environment and socio-economic development with ecological policy;
- establishing connections between environmental changes taking place now and possible in the future;
- substantiation of the means that can be used to study these connections;
- observance of the laws of the international ecological rights in national environmental programmes; and
- revealing gaps in scientific developments and estimates.

These basic principles should be realized by taking the following four general requirements into account:

- Coordination of scientific developments, ecological policy and their accomplishment.
- Consideration of interaction of ecodynamics with socio-economic development.
- Examination of special features of the processes on different spatial scales.
- Reflection of special features of the processes on different spatial scales.

The following fact is the necessity to take into account an interaction of the key problems of ecodynamics determined by the presence of numerous feedbacks in the NSS and non-linearity due to which the "threshold effects" (Frankham 1995) can occur as well as synergism of technologies and ecological policy. Numerous illustrations of the urgency of taking into account different interactions are contained in the problems of global climate change.

An adequate analysis of the role of feedbacks and NSS non-linearity is seriously complicated by available fragmentary information. In this connection, the fact disappoints that the conception of biotic regulation of the environment developed in Russia, which could have served a conceptual basis for resolving the problems of global ecodynamics, has not been recognized yet (Gorshkov et al. 1998, 2000, 2002, 2004; Krapivin and Nazaryan 1995). Unfortunately, the biotic regulation conception remains "unnoticed" in the West, which has been illustrated, in particular, by recent polemics with respect to the Gaya conception (Ebel et al. 2004).

The biotic regulation conception is based on the following conclusions:

- The Earth is a unique planet of the Solar system, since on this planet, life exists in the form of biota—a totality of all living organisms, humans including. Important properties of life include: biological stability of species and their communities as well as a very rigid distribution of energy fluxes absorbed by biota among organisms of different sizes. Biota is responsible for the formation of environmental properties and their stability according to biotic needs. Only for this reason the long existence of biota on the Earth based on the principle of biotic regulation has become possible. Maintenance of environmental stability is one of the major goals of all living organisms.
- Like many other species, *homo sapiens* is one of the biotic species and therefore its important problem consists also in maintaining the global biospheric stability. Otherwise, sustainable development would be impossible. Very long ago, humans left their natural ecological niche and began to consume much more biospheric resources than it is permitted by requirements of ecological balance. After the beginning of industrial revolution, this process of breaking the natural balance has been accelerating under conditions of rapid growth of population size.
- Approximate estimates have shown that to provide the stable state of the biosphere, one can use not more than 1 % of its resources. At present, this share is close to 10 %. A similar situation exists with respect to the evolution of global biogeochemical cycles of matter. So, for instance, the completeness of global carbon cycle before the beginning of industrial revolution had been close to 0.01 % (biodiversity had played an important role in establishing the biospheric stability). By now, the completeness has decreased to 0.1 % and a threat of global ecological disaster becomes more and more pronounced. So the biosphere should be considered not a resource but a fundamental condition for life on Earth. At present, a major goal should be a restoration of the biosphere, already subjected to substantial disturbing forcings, and its maintenance in a state that would ensure the sustainable development. However a difficult problem is the presence of many insufficiently studied aspects of biospheric dynamics. In this context, creation of adequate observational systems and further improvement of numerical modelling methods play a decisive role. The first of these problems is especially urgent, since until now an adequate system of the global change monitoring not only does not exist but even has not been substantiated conceptually.
- A critically important feature of the present civilization development consists in inequality between developed and developing countries, which is manifested, first of all, through non-equivalent use of the existing resources. The per-capita consumption of the "golden billion" (this term is popular in the Russian-speaking world, referring to the relatively wealthy people in industrially developed nations, or the West) is incomparably higher than for the rest of the world. Therefore it is necessary to achieve an agreement on a new social order

based on the agreement on cooperation and partnership. An accomplishment of adequate systems of life support is only possible on the basis of respective international agreements.

- It becomes more and more clear that the present system of market economy does not provide the transition from the unstable to stable trajectory of development. The respective effort undertaken in the spheres such as education, legislation, and management cannot be considered sufficient. Non-governmental and religious organizations can play a more constructive role.
- To stimulate the transition to sustainable development, a complex accomplishment of a series of measures is needed based on understanding the present situation complexity.
  - It is extremely important to realize the fact that further development of the consumption society will lead to a global ecological disaster and civilization collapse. The only resolution of the problem consists in rejection of the traditional paradigm of the consumption society and in radical change of the way of life based on admission of priorities of spiritual values.
  - To overcome the North–South social contrasts, immediate measures are needed on rendering help to developing countries based on realization of the UN recommendations.

It is the biotic regulation conception that should serve the basis for the first priority of the UNEP Programme (UNEP 2006), which is defined as an interaction between man and the environment the conceptual approach to analysis of interactions. This problem can be resolved with the use of the UNEP DPIR (Drivers, Pressures, Impact, Response) model, which describes the cause-and-effect connections between ecological and socio-economic NSS components.

Concrete directions of the ecological development include:

- *Natural and anthropogenic environmental changes.*
  In this context, of key importance is a system classification and priorities of interaction between natural and anthropogenic changes of the environment. It should be emphasized that this unresolved problem is among the major reasons of the extremely contradictory character of the existing conclusions with respect to the nature of the present global climate change. The UNEP document has justly emphasized the priority of the problems of biogeochemical interactions and cycles. Even in the case of carbon cycle we are still far from an adequate understanding of the most substantial mechanisms of its formation.
- *Ecological factors and humans' well-being.*
  Both positive and negative ecological factors demand attention. Positive factors include ecosystems that perform functions of life support (water, food, recreations, and many others). Various factors of ecological stress (diseases, agricultural pests, natural disasters, etc.) are negative.
- *Anthropogenic impacts on the environment.*
  In this connection of special interest is an analysis of demographic dynamics and conditions of socio-economic development.

- *Ecological policy and its connections with ecodynamics.*
  In particular, bearing in mind the use of technologies, institutional measures and risk control. Quite separate are the problems of interaction of science and ecological policy. As for the concrete factors of ecodynamics, the UNEP Programme (2004, 2006) has justly emphasized the first priority of analysis and assessment of the role of interaction between problems such as biodiversity, climate change, degradation of soils fertility, fresh water, coastal and marine environment, local and regional air quality, ozone layer depletion, persistent organic pollutants (POPs), and heavy metals.

  Questions to be answered are as follows:

- What are the key interactions between various environmental changes and what factors determine them?
- How are interactions between various anthropogenic impacts realized and how can they be "separated" if needed?
- How do interactions between products provided by ecosystems and ecosystems' functioning manifest themselves?
- What are present interactions between impacts and responses to them and how can they be either re-grouped or removed in case of necessity to change interactions in the NSS?

As for the goals of the socio-economic development mainly during the period up to 2015 (compared to 1999 conditions), the main problems include (Dodds and Pippard 2005):

- Liquidation of extreme poverty and starvation (halving the number of people with the income less than 1 \$/day).
- Realization of universal elementary education.
- Provision of sexual equality and raising the role of women.
- Reduction of infant mortality (children below 5 years) by 2/3 by the year 2015.
- Improvement of mothers' health (reduction of mortality in childbirth by 3/4).
- Struggle against HIV/AIDS, malaria, and other diseases.
- Provision of ecodynamics stability (liquidation of decreasing trends of the environmental quality, halving the number of people without access to good-quality drinking water, improvement of living conditions for, at least, 100 million people living in slums).
- Development of cooperation in the interests of socio-economic development (with emphasis on the problem of the developing countries' debt).

Serious efforts to substantiate the priorities of the environmental studies in the 21st century have been undertaken by the NOAA scientists who developed the Strategic Plan (SP) for the period 2003–2007 and later (NOAA 2003, 2012; Lautenbacher 2005). Specific attention is given to the NOAA's life-saving information about hurricanes and other severe storms, as well as the data to monitor the Earth. The SP determines the priority of the four directions of developments for the NOAA:

1. Protection, reconstruction and control of the use of coastal and ocean resources based on understanding the laws of ecosystems' dynamics.
2. Understanding the laws of climate change in order to increase the ability of the society to respond to them (note that the document uses two synonymic notions: variability and change with respect to climate without their clear definition).
3. Provision of information on climate and water resources.
4. Maintenance of national commercial interests by provision of ecologically reliable information to ensure safe and efficient transportation. Of course, safe and efficient systems of transportation are one of the key components of economy. This concerns all kinds of transportation: land, sea, and air.

Six directions of developments are considered most substantial:

(1) Creation of the system of integral (complex) observations of the global environment, processing and analysis of observational data.
(2) Efficient resolution of the problem of propagation of knowledge about the environment.
(3) Accomplishment of the needed scientific studies.
(4) Provision of an efficient scientific cooperation.
(5) Resolution of national security problems.
(6) An adequate resolution of the organizational problems.

The NOAA strategic plan foresees an achievement of the following strategic goals for the first of the four directions of developments mentioned above:

- Increasing the number of coastal and marine ecosystems is not subjected to unfavourable changes and being in a stable condition.
- Raising the socio-economic importance of the marine environment and respective resources (sea food, recreations, tourism, etc.).
- Improving the living conditions for coastal and marine flora and fauna.
- Increasing the number of species protected from unfavourable impacts.
- Increase the number of cultivated species up to an optimal level.
- Improving the environmental quality in the coastal and marine regions.

An achievement of the enumerated goals should be based on the use of the ecosystem approach, and in this connection, on extending the scale of studies of various ecosystems dynamics within the framework of three strategic directions:

(a) Protection, reconstruction, and control of the use of resources of coastal and marine regions as well as Great Lakes.
(b) Reconstruction of species and their habitats to be protected.
(c) Expansion and maintenance of stable fishery.

To achieve the goals of the NOAA SP, of key importance is to create a complex system of the environmental monitoring which would provide the obtaining, processing, and analysis of the data on the ocean-atmosphere-land interactive system from local to global scales.

Assessing the NOAA SP on the whole, one should state that unfortunately it is too strongly directed towards meeting national interests. Even in this context it is impossible to adequately resolve the problems of ecodynamics without taking into account the interactivity of processes of local, regional and global scales. The section of the plan concerning climate problems causes a serious bewilderment. With an extreme urgency of these problems (mainly due to their strongly politicized character), of extreme importance is concrete understanding of limited means of numerical modeling and systems of climate observations as well as planning of efforts to overcome this limitedness.

Without dwelling upon concrete characteristic of the present global ecodynamics, we only emphasize: the absence of long-term perspectives of development of the modern consumption society illustrated by estimates of global ecodynamics raises no doubts. Therefore at the World Summit on Sustainable Development held in Johannesburg in 2002, the necessity was emphasized to accomplish 10 year programmes on reaching a stable production and consumption including the following recommendations (Southerton et al. 2004):

- Assuming by developed countries of the leading role in provision of stable production and consumption.
- Achieving these goals based on the general but differentiated responsibility.
- Imparting the key role to the problem of stable production and consumption.
- Laying emphasis on participation of young generation in resolution of the problems of sustainable development.
- Use of the "polluter pays" principle.
- Control of the cycle of products evolution from their production to consumption and waste to increase the productive efficiency.
- Backing the policy that favours both the output of ecologically acceptable products and ecologically adequate services.
- Development of more ecological and efficient methods of energy supply. An exception of energy subsidies.
- Support of free-will initiatives of industry in order to raise its social and ecological responsibility.
- Study and application of ecologically pure production especially in developing countries as well as in small and medium business.

Though the enumerated recommendations are rather declarative, still they demonstrate a necessity to change the paradigm of the socio-economic development (first of all, this refers to developed countries) from the consumption society to priorities of social and spiritual values. A concrete analysis of the ways of this development requires a participation of many specialists in the field of social sciences.

As has been mentioned above, the main aspect of global ecodynamics (in the context of biotic regulation of the environment) is the state of ecosystems on the planet. The development and realization of an efficient technology to assess the ecological safety on global scales may be possible with the organization of the International Centre on the Global Geoinformation Monitoring (ICGGM). This will make it possible to realize the mechanisms for the NSS coevolution. The basic mechanism of this combination will be new technologies of data processing based on the progress of the evolutionary informatics and global modeling. Here the point is about a realization of an approach developed by some authors (Bukatova et al. 1991; Krapivin and Varotsos 2007, 2008; Nitu et al. 2013c; Elinson 1983) to modeling the processes in conditions of inadequate a-priori information about their parameters and the presence of principally unavoidable information gaps.

The ICGGM performs the following functions:

- Data collection from national monitoring systems and international centers on environmental studies.
- Sorting, primary processing and accumulation of data on natural processes.
- Formation of knowledge bases for the processes in the environment.
- Simulation, mathematical and physical modelling of climatic, biospheric, cosmic, social and economic processes.
- Prediction of the environmental state and formation of a constantly renewed bank of scenarios of the anthropogenic activity.
- Answering inquiries from national and international agencies on protection of the environment.
- Substantiation of recommendations for national and international centres of the environmental monitoring.

Indicators characterizing the GIMS as the base subsystem of ICGGM are grouped by thematic principles of organization of its structure. They are specified in the process of GIMS exploitation and cover the key characteristics of the global topography, synoptic situation in the energy-active zones, content of dangerous atmospheric pollutants in characteristic latitudinal belts, as well as announcements about catastrophes.

A multitude of spatially irregular and fragmentary in time data of measurements of geographic, geochemical, ecological, biogeocenotic and synoptic characteristics serve as input parameters for the GIMS. Both in situ and remote sensing measurements are made in different wavelength regions and with different accuracies. Measurements agree with other GIMS units through algorithmic procedures of primary data processing. The volume of these data will decrease in the process of GIMS functioning. The GIMS input foresees also a possibility to receive signals from scenarios of the anthropogenic development of situations under study.

The GIMS model is demonstrated as a conceptual scheme in Fig.1.1. Relationships between input and output parameters are taken into account through a composition of information fluxes indicated here. The GIMS functions in the adaptive regime, and the final result of the system affects the input characteristics of its measuring section. The mathematical aspects of GIMS are shown in Fig. 1.22. Here all the biogeochemical and biogeocoenotic processes are described by the systems of balance equations. However, a considerable part of the different-to-parameterize processes is described using the method of the evolutionary modelling oriented towards a parameterization without formulas of very non-stationary processes. The GIMS functioning is presented in Figs. 1.5 and 1.7.

The global-scale socio-economic structure can be divided into m levels. According to Kondratyev (1998b), $m \geq 9$, and this structure has three main levels of the development of the regions: developed, developing and low-developed. A realization of the ICGGM project will accelerate the process of alignment of this structure, since the planning of the organizational structure of the human society will be accelerated, with the provision of the purposefulness of global processes for the benefit of humankind and without damage to nature and, which is most important, international mechanisms will be developed for the coordinated nature use by the population of the Earth. The ICGGM will be useful because people will not waste finances on realization of ecologically unacceptable projects and will preserve the balance with nature. With the scientific-technical progress, this profit will grow, since the transition to new kinds of resources is unavoidable (Gorshkov et al. 2004).

10.2.2 Sustainable Development Model

The biosphere is a complex unique system. Historically, the man was considered as an element of the biosphere. But at the present time the problem of co-evolution between the human society (*H*) and nature (*N*) has arisen. The influence of human activity on natural systems has reached the global scale, but it is possible to make the conditional division between anthropogenic and natural processes. The use of system analysis permits a more formal description of this division to be performed. Commonly, there are two interacting systems: *H* with the technologies, sciences, economics, sociology, agriculture, industry, etc.; and *N* with climatic, biogeocenotic, biogeochemical, hydrological, geophysical and other natural processes that are characterized by numerous direct couplings and feedbacks.

The solution of practical problems in investigating complex systems often involves evaluating their effectiveness and in particular their stability under indeterminate conditions of functioning. The theory of potential effectiveness of complex systems deals with the solution of these problems. A constructive mathematical apparatus has been developed within the framework of this theory by which it is possible to solve different problems of optimizing the structure and behavior of the functioning of *H* and *N* in certain situations.

The systems *H* and *N* are determined by their structures (number of elements and relations among them) and behaviour (responses to stimuli). The internal behavior of such a system is aimed at maintaining its uninterrupted functioning. The external behavior of the system is aimed at achieving a certain outside goal. Temporal stability of a complex system is a necessary property without which all its other properties become meaningless. It is connected with the structural stability of the material composition and energy balance of the complex system as well as with the regularity of its responses to the same external stimuli.

A breach in stability of a system may result from internal causes (the aging of its elements) or external causes associated with the unfavorable influence of the environment (an ill-intentioned enemy, in particular). Survivability of biological systems is determined by environmental conditions where man's interference with nature is one of the important factors. In connection with this and the prospects of constructing artificial biological systems, optimization problems also arise; an increase of productivity of a biological system being the main criterion of optimality.

For the global scale the problem of survivability of the interacting systems is complicated at the expense of the hierarchy of interaction levels. For a complete explanation of *H* and *N* systems their openness has to be taken into account. Commonly, let us consider the interaction of two open complex systems *H* and *N*, defined by their goals H_G and N_G, structures H_S and N_S and behavior H_B and N_B, respectively. It is suggested that the functioning of such systems should be described by equations of (*V, W*)-exchange. Namely, the interaction of an open system with the environment (or other system) is represented as a process whereby the system exchanges a certain quantity *V* of resources spent in exchange for a certain quantity *W* of resources consumed. The aim of the systems is the most advantageous (*V, W*)-exchange, i.e. it tries to get maximum *W* in exchange for minimum *V*. The *V* is a complex function of the structure and behavior of both systems:

$$V = V(W, H_S, N_S, H_G, N_G) = V(W, H, N) \tag{10.1}$$

As a result of their interaction the systems *H* and *N* get the following *(V, W)*-exchanges:

$$\begin{aligned} V_{H,\max} &= V_{H,\max}(W_H, H^*, N^*) = \max_{\{H_B,H_S\}} \min_{\{N_B,N_S\}} V_H(W_H, H, N) \\ V_{N,\max} &= V_{N,\max}(W_N, H^*, N^*) = \max_{\{N_B,N_S\}} \min_{\{H_B,H_S\}} V_N(W_N, H, N) \end{aligned} \tag{10.2}$$

where H^* and N^* are the optimal systems *H* and *N* respectively.

From (10.1) and (10.2) one can see that the value of (*V, W*)-exchange depends on the goal of the system and may vary within certain limits: $V_{1,\min} \le V_H \le V_{1,\max}$, $V_{2,\min} \le V_N \le V_{2,\max}$, where $V_{i,\min}$ ($i = 1, 2$) corresponds to the case, when both systems are most aggressive, and $V_{i,\max}$ ($i = 1, 2$) to the case, when they are most cautious. In brief, there is some spectrum of the interactions between *H* and *N*. For a formal description of these interactions we shall divide all the elements of both systems into three classes: the working (working), protective (defensive) and active elements, the latter designed to act on the environment. For short, we shall refer to the working elements of systems *H* and *N* as *a*-and *b*-elements, to the protective elements as R_a-and R_b-elements and to the active elements as C_a-and C_b-elements, respectively.

Let us assume that before interaction the systems H and N have certain limited energy resources (vital "substrates") V_a and V_b, where $V_a = \{V_{aj}, j = 1, \ldots, m_a\}$, $V_b = \{V_{bi}, i = 1, \ldots, m_b\}$. These substrates generate working elements in such a way, that the substrate V_{aj} (V_{bi}) can generate H_j (N_i) $a(b)$-elements of the $j(i)$-th type of values $a_j(b_i)$.

The protective and active elements of each system are generated by the working elements. First of all the protective $E^a_{Rm}(E^b_{Rm})$ and active $E^a_{Cm}(E^b_{Cm})$ substrates are created which, in their turn, generate R-and C-elements of the m-th type. These processes are described by the following dependencies:

$$E^a_{Rm} = E^a_{Rm}(V_a, H_1, \ldots, H_{m_a}) = \sum_{j=1}^{m_a} w^a_{mj} f^a_{jR}(V_{aj}, H_j),$$

$$E^b_{Rm} = E^b_{Rm}(V_b, N_1, \ldots, N_{m_b}) = \sum_{j=1}^{m_b} w^b_{mj} f^b_{jR}(V_{bj}, N_j),$$

$$E^a_{Cm} = E^a_{Cm}(V_a, H_1, \ldots, H_{m_a}) = \sum_{j=1}^{m_a} w'^a_{mj} f^a_{jC}(V_{aj}, H_j),$$

$$E^b_{Cm} = E^b_{Cm}(V_b, N_1, \ldots, N_{m_b}) = \sum_{j=1}^{m_b} w^b_{mj} f^b_{jC}(V_{bj}, N_j),$$

where $w^{a(b)}_{mj}, w'^{a(b)}_{mj}$ and $f^{a(b)}$ are certain present weights and functions respectively.

We shall assume that as a result of such hierarchical synthesis, elements in the systems H and N have at the beginning of the interaction ($t = 0$):

(1) m_j and n_j working elements of the j-th type with values a_j and b_j, respectively, where

$$\sum_{j=1}^{m_a} a_j H_j = M_a(0), \quad \sum_{j=1}^{m_b} b_j N_j = M_b(0); \tag{10.3}$$

(2) r_a and r_b types of protective elements, the m-th type having α_m and β_m elements, and:

$$\sum_{m=1}^{r_a} \alpha_m = M_{R_a}(0), \quad \sum_{m=1}^{r_b} \beta_m = M_{R_b}(0); \tag{10.4}$$

(3) s_a and s_b types of active elements the m-th type having v_m^a and v_m^b elements and:

$$\sum_{m=1}^{s_a} v_m^a = D_a(0), \sum_{m=1}^{s_b} v_m^b = D_b(0), \tag{10.5}$$

respectively.

In the discrete case, the change of the average number of system elements that have survived till the moment t_{i+1} will be described by the following relations:

$$H_s(t_{i+1}) = \max\{0, H_s(t_i) - \sigma_{hs}^n(t_i)p_{hs}^n(t_i)\}, \quad s = 1, \ldots, m_h; \tag{10.6}$$

$$\alpha_j(t_{i+1}) = \max\left\{0, \alpha_j(t_i) - \sigma_{Rj}^n(t_i)p_{Rj}^n(t_i)\right\}, \quad j = 1, \ldots, r_h \tag{10.7}$$

$$v_m^h(t_{i+1}) = \max\{0, v_m^h(t_i) - \sigma_{Cm}^n(t_i)p_{Cm}^n(t_i)\}, \quad m = 1, \ldots, s_h \tag{10.8}$$

$$N_l(t_{i+1}) = \max\{0, N_l(t_i) - \sigma_{nl}^h(t_i)p_{nl}^h(t_i)\}, \quad l = 1, \ldots, m_n \tag{10.9}$$

$$\beta_s(t_{i+1}) = \max\{0, \beta_s(t_i) - \sigma_{Rs}^h(t_i)p_{Rs}^h(t_i)\}, \quad s = 1, \ldots, r_n \tag{10.10}$$

$$v_m^n(t_{i+1}) = \max\{0, v_m^n(t_i) - \sigma_{Cm}^h(t_i)p_{Cm}^h(t_i)\}, \quad m = 1, \ldots, s_h \tag{10.11}$$

where the $\sigma_{\omega i}^{a(b)}(t)$ values characterize the external behavior of both these systems:

$$\overline{H}_e^{(i)} = \{\|\sigma_{bl}^a\|, \|\sigma_{Rs}^a\|\}, \overline{N}_e^{(i)} = \{\|\sigma_{as}^b\|, \left\|\sigma_{Rj}^b\right\|\};$$

and $p_{\omega i}^{a(b)}(t)$ are the respective probabilities of death of the elements as a result of their interaction.

The following limiting conditions should be taken into account here:

$$\sum_{i=0}^{T} \{\sigma_{Cs}^b(t_i)p_{Cs}^b(t_i) + \sum_{j=1}^{m_b} \sigma_{bj}^a(t_i) + \sum_{j=1}^{r_b} \sigma_{Rj}^a(t_i) + \sum_{j=1}^{s_b} \sigma_{Cj}^a(t_i)\} = v_l^h(0), \tag{10.12}$$

$$\sum_{i=0}^{T} \{\sigma_{Cs}^a(t_i)p_{Cs}^a(t_i) + \sum_{j=1}^{m_a} \sigma_{aj}^b(t_i) + \sum_{j=!}^{r_a} \sigma_{Rj}^b(t_i) + \sum_{j=1}^{s_a} \sigma_{Cj}^b(t_i)\} = v_l^n(0), \tag{10.13}$$

The stochastic solution of Eqs. (10.1)–(10.13) can be in the unrealized form in practice. There are many real situations when the realization of H^*or N^* system is impossible. Some tasks and algorithms were described by Krapivin and Klimov (1995, 1997).

10.2.3 Study of the Simple Survivability Model

Let us consider the interaction of the two systems in the frame of diagram shown in Fig. 10.1. At the beginning, the systems H and N have, respectively, $N_a(0)$ and $N_b(0)$ working elements, $N_{Ra}(0)$ and $N_{Rb}(0)$ protective elements, and $M_a(0)$ and $M_b(0)$ active agents for undertaking action against an external medium. In this case, we shall assume that the initial structures H_S and N_S of the systems are uniformly filled with elements. This means that at the time $t = 0$ in any sphere with a fixed radius ε, which is completely confined within system H, there are constant numbers of elements.

Specific characteristics of all elements can be depending on the solved task. For example, C_a and C_b element can be anthropogenic scenarios and diseases, respectively. Feedbacks and medicine can be as R_b and R_a elements, respectively. Elements a and b represent life conditions for people and hierarchy of ecological systems, respectively. Certainly, detailed interpretation of all types of accomplished elements depend on the spatial-temporal scale and knowledge base.

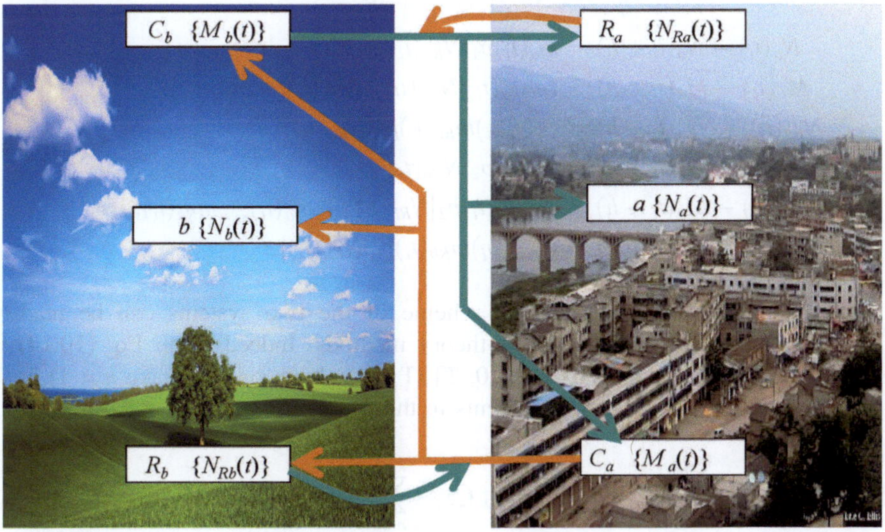

Fig. 10.1 Schematic diagram of the interaction of nature and humanity in the survivability problem. Structure of system elements may be changed and added

We shall consider that all elements of systems H and N, independently of their spatial location, are accessible to the same degree to the active agents of the external medium. The interaction of the systems consists in the situation that each system in a fixed interval of time [0, T] at discrete moments $t_i = ih$, $i = 0,1,2,\ldots, k$ (k = [T/h]), determines its behavior by a set of numbers:

$$H_B = \{m_b(t_i), m_{Rb}(t_i), \rho_i\}, N_B = \{m_a(t_i), m_{Ra}(t_i), r_i\}.$$

Here, m_a and m_b are parts of C_b- and C_a-elements aimed at destroying a-and b-elements, respectively; analogously, the parts $(1 - r_i)m_{Ra}$ and $(1 - \rho_i)m_{Rb}$ of C_b- and C_a-elements are directed towards the destruction of the corresponding protective elements. In the course of time portions of C_a- and C_b-elements uniformly fill the opposite system, and in such a manner the elements of the systems become with time uniformly depleted.

It is assumed that system $H(N)$ is put out of operation, if at the time $t_i \leq$ T, $N_a(t_i) \leq \theta_a N_a(0)$ ($N_b(t_i) \leq \theta_b N_b(0)$); that is, if more than $(1 - \theta_a)$-th ($(1 - \theta_b)$-th) portion of the initial amount of its working elements is out of operation, where $0 \leq \theta_a, \theta_b \leq 1$. On the other hand, when $N_a(\text{T}) > \theta_a N_a(0)$ ($N_b(\text{T}) > \theta_b N_b(0)$), then system $H(N)$ is considered to have been survived. The assignment of the values θ_a and θ_b is determined by the peculiarities of the system under consideration. It is clear that the smaller the number θ, the more survivable is the system.

Further, we shall assume that one of the elements of system H or N is put of operation by the action of one of the elements C_a or C_b with a probability $p_1[N_{Ra}(t)]$ or $p_2[N_{Rb}(t)]$, respectively. Consequently in the interval of time $[t_i, t_i + h]$ with step length h there occur on the average the following changes in the structure of the system:

$$\begin{aligned}
N_a(t_i) - N_a(t_i + h) &= m_a(t_i)p_1[N_{Ra}(t_i)],\\
M_a(t_i) - M_a(t_i + h) &= r_i m_{Ra}(t_i)[N_{Ra}(t_i) + m_b(t_i) + m_{Rb}(t_i)],\\
N_{Ra}(t_i) - N_{Ra}(t_i + h) &= (1 - r_i)m_{Ra}(t_i)p_1[N_{Ra}(t_i)],\\
N_b(t_i) - N_b(t_i + h) &= m_b(t_i)p_2[N_{Rb}(t_i)],\\
M_b(t_i) - M_b(t_i + h) &= \rho_i m_{Rb}(t_i)p_2[N_{Rb}(t_i)] + m_a(t_i) + m_{Ra}(t_i),\\
N_{Rb}(t_i) - N_{Rb}(t_i + h) &= (1 - \rho_i)m_{Rb}(t_i)p_2[N_{Rb}(t_i)]
\end{aligned} \tag{10.14}$$

The above-discussed interaction scheme of the two systems can be readily implemented with the aid of game-theory methods. Indeed, from Eq. (10.14) it follows that in the interval of time [0, T] (T = kh, here, one can take h = 1) both systems lose working a-and b-elements in the following amounts:

$$Q_1 = \sum_{n=0}^{k-1} m_a(n)p_1[N_{Ra}(n)],\ Q_2 = \sum_{n=0}^{k-1} m_b(n)p_2[N_{Rb}(n)], \tag{10.15}$$

In this case, system H with its behavior of

$$H_B(t) = \{m_b(t), m_{Rb}(t), \rho(t)\}$$

tends to minimize the function Q_1, which characterizes its losses, and to maximize Q_2, the losses of system N. On the other hand, the system N with its behavior $N_B(t) = \{m_a(t), m_{Ra}(t), r(t)\}$ tends to maximize function Q_1 and minimize function Q_2.

In practical problems, the win function is taken as such a characteristic of the antagonistic situation which most fully describes a given conflict. In this case, such a function is $Q = Q_1 - Q_2$. The maximizing participant in this case will be system N, the minimizing opponent, system H. In this manner the solution of the set problem on the optimal behavior of the two systems in an antagonistic situation becomes reduced to the solution of a game with a win function:

$$Q(H_B, N_B) = \sum_{n=0}^{i-1} \{m_a(n)p_1[N_{Ra}(n)] - m_b(n)p_2[N_{Rb}(n)]\} \tag{10.16}$$

where, according to Eq. (10.14),

$$\begin{aligned} N_{Ra}(n) &= N_{Ra}(0) - \sum_{i=0}^{n-1} (1 - r_i)m_{Ra}(i)p_1[N_{Ra}(i)], \\ N_{Rb}(n) &= N_{Rb}(0) - \sum_{i=0}^{n-1} (1 - \rho_i)m_{Rb}(i)p_{21}[N_{Rb}(i)] \end{aligned} \tag{10.17}$$

and conditions are imposed on the behavior of the system related to the limitations of C_a-and C_b-elements:

$$\begin{aligned} &\sum_{i=0}^{k-1} \{r_i m_{Ra}(i)p_1[N_{Ra}(i)] + m_b(i) + m_{Rb}(i)\} = M_a(0), \\ &\sum_{i=0}^{k-1} \{\rho_i m_{Rb}(i)p_2[N_{Rb}(i)] + m_a(i) + m_{Ra}(i)\} = M_b(0) \end{aligned} \tag{10.18}$$

Thus we have a k-step survival game. At the start of each step, system H provides a certain amount of its resources $u = \{m_b, m_{Rb}, \rho\}$ and N provides a certain amount of its resources $v = \{m_a, m_{Ra}, r\}$, so that limitations (10.18) are maintained. As a result of this distribution of resources, system N gets the advantage:

$$R(u, v, M_a, M_b) = m_a p_1[N_{Ra}] \tag{10.19}$$

However, this win of N and the loses of H are not counted on the basis of their initial resources and cannot be added to the remaining amounts of C-elements. After each step of the game Eq. (10.14), the change in the resources in the participants takes place, and as a result of the k-step process a total win of system N can be described by the equation:

$$\begin{aligned} Q_k &= Q_k[u_0, u_1, \ldots, u_{k-1}; v_0, v_1, \ldots, v_{k-1}; M_a(0), M_b(0)] \\ &= \sum_{n=0}^{k-1} \{m_a(n)p_1[N_{Ra}(n)] - m_b(n)p_2[N_{Rb}(n)]\} \end{aligned} \tag{10.20}$$

There are several methods with the aid of which an analysis can be made of this k-step process. One can consider this k-step game as a one-step game, in which case system H must select simultaneously a plurality of the vector $\{u_0, u_1, \ldots, u_{k-1}\}$, and system N, a plurality of the vector $\{v_0, v_1, \ldots, v_{k-1}\}$, where the selection of u_k and v_k depends on the previous values obtained according to Eq. (10.14).

For a solution of this quite complex problem, it is suggested to substitute for it a similar problem of multi-step optimization. The values of this game can be expressed as:

$$V_k = \max_F \min_G \{ \int Q_k dF(v_0, v_1, \ldots, v_{k-1}) dG(u_0, u_1, \ldots, u_{k-1}) \} = \min_G \max_F \{\ldots\} \tag{10.21}$$

where, the distribution functions F and G are determined on the boundaries of the complex form:

$$\begin{array}{ll} 0 \le v_0 \le M_b(0) & 0 \le u_0 \le M_a(0) \\ 0 \le v_1 \le M_b(1) & 0 \le u_1 \le M_a(1) \\ \ldots & \ldots \\ 0 \le v_{k-1} \le M_b(k-1) & 0 \le u_{k-1} \le M_a(k-1) \end{array} \tag{10.22}$$

By utilizing the optimality principle and taking into account the dependence $V_k = V_k[M_a(0), M_b(0)]$, we obtain the following functional equation:

$$\begin{aligned} & V_{n+1}[M_a(0), M_b(0)] \\ & = \max_F \min_G [\iint_{\substack{0 \le u \le M_a(n-1) \\ 0 \le v \le M_b(n-1)}} \{R(u,v) +_n [M_a(n-1), M_b(n-1)]\} dF(v) dG(u)] = \min_G \max_F [\ldots] \end{aligned} \tag{10.23}$$

where

$$V_1[M_a(0), M_b(0)] = \max_F \min_G [\iint_{\substack{0 \le u \le M_a(0) \\ 0 \le v \le M_b(0)}} R(u,v) dF(v) dG(u)] = \min_G \max_F [\ldots] \tag{10.24}$$

The solution of these functional equations is a difficult problem. Therefore, not without reason the expression curse of dimensionality was introduced. Next we shall present the solution of some specific cases. This task is solved in more detail by Nitu et al. (2000a). For the solution of the proposed problem, we shall begin with the case where in both systems the *C*- and *R*-elements are indistinguishable. The diagram of interaction between systems *H* and *N* can then be represented in such a form as shown in Fig. 10.2. By taking into account the designations used in this diagram, we obtain that the number of C_a-and C_b-elements that have reached the C_b-and C_a-elements of the opposite system in a one-step operation will be respectively: $\max\{0, a_C\text{-}N_{Rb}\}$ and $\max\{0, b_C\text{-}N_{Ra}\}$. Therefore, the numbers of C_a-and C_b-elements that are put out of operation on the average are $\max\{0, b_C\text{-}N_{Ra}\}$ $p_1[N_{Ra}]$ and $\max\{0, a_C\text{-}N_{Rb}\}p_2[N_{Rb}]$. Consequently, after one step of the game the following numbers of C_a-and C_b- elements remain in the system that are not put out of operation: $\max[0, M_a\text{-}\max(0, b_C\text{-}N_{Ra})p_1[N_{Ra}]]p_1[N_{Ra}]$, $\max[0, M_b\text{-}\max(0, a_C\text{-}N_{Rb})p_2[N_{Rb}]]p_2[N_{Rb}]$. Going over to the multi-step situation, we introduce the win function (2.16). The aim of system *N* is to destroy system *H*, or more precisely, to put out of operation all of the latter's *a*-elements. For this purpose the system *N* cannot use all C_b-elements, since it will then be left defenseless against the C_a-elements. An analogous situation holds true for the system *H*. Each system is compelled to release after each successive step of the game a certain number of working *a*-and *b*-elements. The magnitude of this payoff is proportional to the difference $M_b - b_C - N_{Rb} - (M_a - a_C - N_{Ra})$. Now the problems on both sides become clear. The system *N* must strive to provide such a number of C_b-elements in all directions so as to maintain the maximum value of this difference. In other

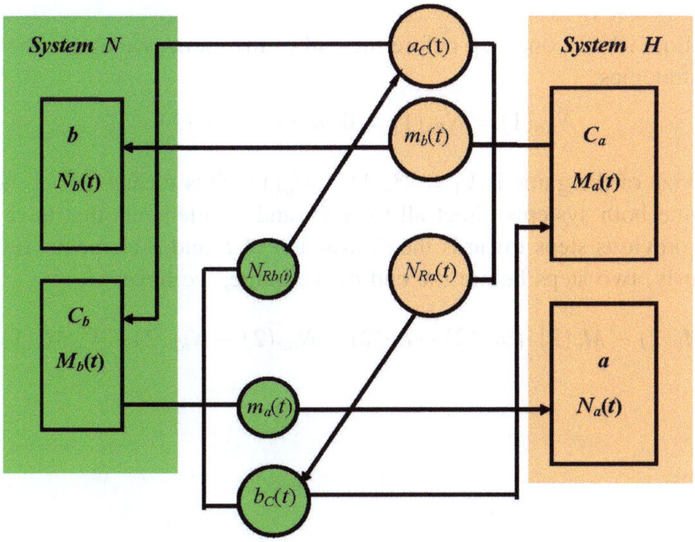

Fig. 10.2 Schematic diagram of the case of mutual interchangeability

words, it must provide the largest possible number of elements m_a for the destruction of a- elements and thus increase its winnings. However, one could also consider the presence of an analogous distribution of C_a- elements and provide sufficient protection for the system N through the maximum distribution of C_a-elements. An analogous situation exists for the system H. For $(n - 1)$ moves left before the end of the game, we have:

$$M_a(n-1) = \max\{0,\ M_a(n) - \max[0, b_C(n) - N_{Ra}(n)p_1][N_{Ra}(n)]\}p_1[N_{Ra}(n)],$$
$$M_b(n-1) = \max\{0,\ M_b(n) - \max[0,\ a_C(n) - N_{Rb}(n)]p_2[N_{Rb}(n)]\}p_2[N_{Rb}(n)] \tag{10.25}$$

The payoff for the entire game according to Eq. (10.20) will be:

$$Q_k = \sum_{n=1}^{k} \{M_b(n) - b_C(n) - N_{Rb}(n) - [M_a(n) - a_C(n) - N_{Ra}(n)]\} \tag{10.26}$$

The functional Eq. (10.23) will acquire the following form:

$$V_{n+1} = \max_{H_B} \min_{N_B}\{M_b(n+1) - b_C(n+1) - N_{Rb}(n+1) - [M_a(n+1) - a_C(n+1) - N_{Ra}(n+1)] + Q_k[M_a(n), M_b(n)]\} = \max_{H_B} \min_{N_B}\{\ldots\} \tag{10.27}$$

Since at the end of the game $Q_0 = 0$, we obtain from Eq. (10.26) for $k = 1$:

$$Q_1 = M_b(1) - b_C(1) - N_{Rb}(1) - M_a(1) + a_C(1) + N_{Ra}(1) \tag{10.28}$$

From Eq. (10.28), one step before the end of the game, we obtain the following optimal strategies:

$$N_{Rb}(1) = N_{Ra}(1) = 0,\ a_C(1) = b_C(1) = 0 \tag{10.29}$$

and the prize of the game is $V_1 = M_b(1) - M_a(1)$. This means that in the last step of the game both systems direct all their C_a-and C_b-elements that have remained from the previous steps towards the destruction of b-and a-elements, respectively. Analogously, two steps before the end of the game, we have:

$$Q_2 = M_b(2) - M_a(2) + a_C(2) - b_C(2) + N_{Ra}(2) - N_{Rb}(2) + V_1[M_a(1), M_b(1)] \tag{10.30}$$

where $V_1 = M_b(1) - M_a(1)$, and

$$M_a(1) = \max\{0, M_a(2) - \max[0, b_C(2) - N_{Ra}(2)]p_1[N_{Ra}(2)]\}p_1[N_{ra}(2)],$$
$$M_b(1) = \max\{0, M_b(2) - \max[0, a_C(2) - N_{Rb}(2)]p_2[N_{Rb}(2)]\}p_2[N_{Rb}(2)] \tag{10.31}$$

It is obvious that at this step of the game the participants have no pure strategies. Therefore the solution of this game is impossible in an analytic form, and it can be obtained in a concrete case only by the numerical method. The modelling of the game provides some understanding of the nature of its solution. Indeed with the aid of a computer it is possible either to construct a model of the game and to gather statistics or to solve the functional Eq. (10.23) numerically and with the aid of heuristic concepts to investigate the dependence of strategies on the initial conditions. Of course, such an approach for a short interval of modeling cannot give any significant information concerning the solution. Nevertheless, this is the only possible approach at the present time. The feeling of hopelessness in a specific situation should not deter us from seeking a solution by analytical methods. The importance of obtaining analytical solutions is obvious, since they have the advantage over the numerical solutions in that they make possible the detection of the general regularities of the optimal behavior of complex systems in antagonistic situations. The importance of the analytical solutions has been pointed out by Krapivin (1978), who has shown that a single numerical solution cannot replace an analytical solution in which the quantitative description of the phenomenon is most concentrated. In the case examined here, when the participants have no information concerning the action of the opponent in the process of the entire game, the solution of particular cases with the aid of a computer enables us to obtain the following quantitative description of optimal strategies.

Let $M_b(n)/M_a(n) = \delta_n$. Then if $\delta_n \gg 1$, system N in the initial stage of the game has a pure strategy, but for system H it is more advantageous to adopt a mixed strategy, using the tactics of deception. During the first steps system N destroys only C_a- elements and only during the last steps does it destroy the a-elements. By using the corresponding probability mechanism, system H must direct all its C_a-elements, with probability p_1 towards the destruction of the C_b-elements, with probability p_2 towards the defense of its own elements, and with probability $1 - p_1 - p_2$ towards the attack of the b-elements. In the case when $\delta_n \cong 1$, then at this stage of the game $p_1 + p_2 = 1$, $m_b(n) = 0$, and the behavior of the system becomes symmetrical. During the last steps of the game systems H and N, independent of the magnitude of δ_n, change over to the strategies: $m_a(t) \neq 0$, $m_b(t) \neq 0$, and $b_C = a_C = N_{Ra} = N_{Rb} = 0$, that is, to the destruction of working elements.

Let us consider a particular case where the systems have no protective elements, that is, $N_{Ra} = N_{Rb} = 0$, and therefore, $p_1 = p_2 = 1$. Then we obtain:

$$\begin{aligned} M_a(n-1) &= \max\{0, M_a(n) - b_C(n)\}, \\ M_b(n-1) &= \max\{0, M_b(n) - a_C(n)\} \end{aligned} \tag{10.32}$$

One step before the end of the game, the payoff, according to Eq. (10.31), is $V_1 = M_b(1) - M_a(1)$ and the optimal strategies $a_C^*(1) = b_C^*(1) = 0$; that is, at the last step the systems release all C-elements for the destruction of a-and b-elements respectively.

Next, by assuming $n = 2$, according to Eqs. (10.30) and (10.31), we obtain:

$$Q_2[a_C(2), b_C(2)] = \begin{cases} Q_{21}, & when \quad b_C(2) \geq M_a(2), a_C(2) \geq M_b(2); \\ Q_{22}, & when \quad b_C(2) < M_a(2), a_C(2) \geq M_b(2); \\ Q_{23}, & when \quad b_C(2) \geq M_a(2), a_C(2) < M_b(2); \\ Q_{24}, & when \quad b_C(2) < M_a(2), a_C(2) < M_b(2); \end{cases} \tag{10.33}$$

where $Q_{21} = M_b(2) - b_C(2) - M_a(2) + a_C(2)$, $Q_{22} = M_b(2) - 2M_a(2) + a_C(2)$, $Q_{23} = 2M_b(2) - b_C(2) - M_a(2)$ and $Q_{24} = 2M_b(2) - 2M_a(2)$.

The solution of the game with the win function (10.33) has the following form:

$$V_2 = 2M_b(2) - 2M_a(2),\ b_C^*(2) = M_a(2); a_C^*(2) = M_b(2) \tag{10.34}$$

Actually, if an account is made of the real situation, the optimal strategies of both systems two steps before the end of the game will be as follows:

$$a_C^*(2) = \min\{M_a(2), M_b(2)\}, b_C^*(2) = \min\{M_a(2), M_b(2) \tag{10.35}$$

Therefore at the penultimate step when $\delta_2 > 1$, the system N releases a portion of its force and the system H releases all its forces against the c-elements of the other system.

By reasoning in an analogous way, we obtain that n-steps before the end of the game the strategies of the systems H and N will be:

$$a_C^*(n) = \min\{M_a(n), M_b(n)\};\ b_C^*(n) = \min\{M_a(n), M_b(n)\}; \tag{10.36}$$

It is evident that if in the process of the game no replacement of C-elements takes place in the systems H and N, then it follows from Eqs. (10.35) and (10.36) that it makes sense to conduct the game in two steps; and the two following cases are distinguishable:

(1) When $M_a(0) = M_b(0)$, both systems release all their C-elements into battle against the C-elements of the other system, but in so doing the system themselves survive.
(2) When $M_a(0) > M_b(0)$, system N releases all the elements into battle against C_a-elements, and system H releases $M_b(0)$ of C_a-elements into battle against C_b-elements, and $M_a(0) - M_b(0)$ of C_a-elements into battle against b-elements.

Consequently system H survives in every case, while system N survives only when $M_a(0){-}M_b(0) < \theta_b N_b(0)$.

Now, the case of mutual indifference of C_a-and C_b-elements is considered. In this case, by supposing $r = \rho = 0$ in Eqs. (10.17) and (10.18), we obtain:

$$
\begin{aligned}
N_{Ra}(n) &= N_{Ra}(0) - \sum_{i=0}^{n-1} m_{Ra}(i)p_1[N_{Ra}(i)], \\
N_{Rb}(n) &= N_{Rb}(0) - \sum_{i=0}^{n-1} m_{Rb}(i)p_2[N_{Rb}(i)]
\end{aligned} \tag{10.37}
$$

and

$$
\sum_{n=0}^{k-1} [m_a(n) + m_{Ra}(n)] = M_b(0), \sum_{n=0}^{k-1} [m_b(n) + m_{Rb}(n)] = M_a(0) \tag{10.38}
$$

The solution of the game of the two systems, H and N, with the win function as shown in Eq. (10.16) and under conditions as expressed in Eqs. (10.37) and (10.38), is reduced to the problem of maximizing two functions:

$$
Q_1 = \sum_{n=0}^{k-1} m_a(n)p_1[N_{Ra}(n)] = \max_{Ra} \tag{10.39}
$$

$$
Q_2 = \sum_{n=0}^{k-1} m_b(n)p_2[N_{Rb}(n)] = \max_{Rb} \tag{10.40}
$$

From Eqs. (10.39) and (10.40) it is evident that when $p_1[N_{Ra}(n)] = p_2[N_{Rb}(n)] = 1$, the optimal strategies of both sides consist in the destruction of a- and b-elements. If the C-elements act independently of one another, and the probability of putting the C-elements out of operation by a single R-element is equal to a constant value D_a and D_b for H and N systems, respectively, then

$$
p_1[N_{Ra}(n)] = \exp[-d_a N_{Ra}(n)], p_2[N_{Rb}(n)] = \exp[-d_b N_{Rb}(n)] \tag{10.41}
$$

where, $d_a = -\theta_1 \ln(1 - D_a)$, $d_b = -\theta_2 \ln(1 - D_b)$ are the *effectiveness* coefficients of the R_a- and R_b-elements; and θ_1 and θ_2 are the *density* of the location of elements in the H and N system, respectively.

When $k = 1$, the optimal strategies of the behavior of the systems will be $m_a^*(0) = M_b(0)$ *and* $m_b^*(0) = M_a(0)$. This is natural, since in the last step it makes no sense to destroy protective elements. When $k = 2$, from Eqs. (10.37)–(10.40) we obtain the optimal strategies in the form:

$$\begin{aligned} m_a^*(0) &= m_b^*(0) = m_{Ra}^*(1) = m_{Rb}^*(1) = 0,\ m_a^*(1) = d_a^{-1}\exp[d_a N_{Ra}(0)]\,, \\ m_b^*(1) &= d_b^{-1}\exp[d_b N_{Rb}(0)],\ m_{Ra}(0) = M_b(0) - d_a^{-1}\exp[d_a N_{Ra}(0)]\,, \\ m_{Rb}(0) &= M_a(0) - d_b^{-1}\exp[b_b N_{Rb}(0)] \end{aligned} \tag{10.42}$$

The cost of the game is:

$$\begin{aligned} V_2 = {} & d_a^{-1}\exp\{-1 + d_a M_b(0)\exp[-d_a N_{Ra}(0)]\} \\ & - d_b^{-1}\exp\{-1 + d_b M_a(0)\exp[-d_b N_{Rb}(0)]\} \end{aligned} \tag{10.43}$$

From Eq. (10.43) it is evident that for the system N to destroy the system H it is necessary that the following inequality holds:

$$d_a M_b(0)\exp[-d_a N_{Ra}(0)] \geq l_n[d_a e(1-\theta_a)N_a(0)] \tag{10.44}$$

from which we have:

$$M_b(0) \geq d_a^{-1}\exp[d_a N_{Ra}(0)]\{1 + l_n[d_a(1-\theta_a)N_a(0)]\} \tag{10.45}$$

Similarly, we obtain the condition for system H:

$$M_a(0) \geq d_b^{-1}\exp[d_b N_{Rb}(0)]\{1 + l_n[d_b(1-\theta_b)N_b(0)]\} \tag{10.46}$$

Thus for the system N to destroy the system H it is necessary that the inequality (10.45) holds. In an analogous way, for the system H it is necessary that the inequality (10.46) holds. The number of a-and b-elements, as a rule, is a fixed number, as follows from considerations related to the work of the system. The number of R—elements, which perform protective functions, can be best selected when certain physical parameters in the problem are fixed and when a priori information concerning the amount of C-elements of the opposite system is available. For example, if the effectiveness of all R-elements is constant independent of their number, then in order for the system H to survive, the necessary number of protective elements must satisfy the inequality:

$$N_{ra}(0) \geq d_a M_b(0) l_n / \{d_a(1 + l_n[d_a(1-\theta_a)N_a(0)])\} \tag{10.47}$$

It appears clear that when the number of C_b-elements is constant and the survivability of the system H decreases, the necessary number of protective elements increases rapidly. In this case, the greater their effectiveness, the smaller the number of protective elements is required for carrying out one and the same task. With an increase in survivability the necessary number of protective elements can be decreased.

Now let the systems H and N have a fixed amount E_a and E_b of a certain substratum (for instance, energy) and be able to distribute it evenly between their protective elements, so that for each fraction of R_a-and R_b-elements there is $E_{1a} = E_a/N_{ra}(0)$ and $E_{1b} = E_b/N_{rb}(0)$ respectively. Then the efficiency coefficients of the protective elements must increase with the increase in the E_{1a} and E_{1b} portions, since in this case the probabilities of D_a and D_b increase. Therefore let $D_a = 1 - \exp\{-F_a F_{1a}^{\alpha}\}$, $D_a = 1 - \exp\left\{-F_b F_{1b}^{\beta}\right\}$, where, F_a, F_b, α and β are independent of the number of protective elements. This gives then the following expressions for the coefficients of effectiveness of the protective elements:(10.48)

$$\begin{aligned} d_a &= -\theta_a \ln(1 - D_a) = -\theta_a F_a E_{1a}^{\alpha} = G_a N_{Ra}^{-\alpha}(0), \\ d_b &= -\theta_b \ln(1 - D_b) = -\theta_b F_b E_{1b}^{\beta} = G_b N_{Rb}^{-\beta}(0), \end{aligned} \tag{10.48}$$

where, $G_a = \theta_a F_a E_a^{\alpha}$, $G_b = \theta_b F_b E_b^{\beta}$. From Eqs. (10.47) and (10.48) we obtain the following transcendental equations for the amount of the necessary protective elements in systems H and N:

$$N_{Ra}(0) = N_{Ra}^{\alpha}(0) G_a^{-1}\{\ln[G_a M_b(0)/f_1] - \alpha \ln N_{Ra}(0)\}, \tag{10.49}$$

$$N_{Rb}(0) = N_{Rb}^{\beta}(0) G_b^{-1}\{\ln[G_b M_a(0)/f_2] - \beta \ln N_{Rb}(0)\}, \tag{10.50}$$

where $f_1 = \ln[e G_a(1 - \theta_a) N_a(0) N_{Ra}^{-\alpha}]$, $f_2 = \ln[e G_b(1 - \theta_b) N_b(0) N_{Rb}^{-\beta}(0)]$

An analysis of Eqs. (10.49) and (10.50) reveals that in this case the number $N_{ra}(0)$ is very sensitive with respect to changes in the quantity $\ln[(1 - \theta_a)N_a(0)]$. And this is natural, since with the increase in the number of R_a-elements their effectiveness sharply decreases. It is clear that there exists a certain optimal level for the number of protective elements. This level is defined by the assigned survivability of the system. It is better to have a small number of R_a-elements of high effectiveness, than a large number of R_a-elements of low effectiveness. When $k = 3$, from Eqs. (10.37)–(10.39) we have:

$$Q_1 = \sum_{i=1}^{3} m_a(i-1) \exp[-\alpha N_{Ra}(i-1)], \tag{10.51}$$

where,

$$\begin{aligned} &N_{Ra}(1) = N_{Ra}(0) - m_{Ra}(0)\exp[-d_a N_{Ra}(0)], \\ &N_{ra}(2) = N_{ra}(0) - m_{Ra}(0)\exp[-d_a N_{Ra}(0)] - m_{Ra}(1)\exp[-d_a N_{Ra}(1)] \\ &m_a(1) + m_a(0) + m_a(2) + m_{Ra}(1) + m_{Ra}(0) + m_{Ra}(2) = M_b(0) \end{aligned} \tag{10.52}$$

From Eqs. (10.51) and (10.52) we obtain that the function Q_1 reaches its maximum value when $m_a^*(0) = m_{Ra}^*(2) = m_{Ra}^*(1) = 0$, with this maximum value being independent of the distribution of C_a-elements during the last two steps. In this case $Q_1^*(3) = Q_1^*(2)$. An analogous result is also obtained for system N.

Thus for an identical game, for instance, system N can destroy at the most $Q_1^*(2) = M_b(0)\exp[-d_a N_{Ra}(0)]$ of a-elements; when the number of steps is greater than 2, system N can destroy not more than $Q_1^*(2) = (ed_a)^{-1}\exp[d_a Q_1^*(1)]$ of a-elements. Therefore by comparing $Q_1^*(1)$ with $Q_1^*(2)$ we obtain that in the analyzed antagonistic situation the best strategy for both systems is the performance of all allowable operations in two steps. And that is, during the first step the action of a portion of each system's C-elements is to be set against the protective elements of the opposite system, and during the second step the action of the remaining force is to be set against the working elements of the opposite system. This conclusion completely agrees with the conclusions reached by Krapivin (1978), which were obtained by different methods.

In the above-examined models of the interaction of the two systems, it was assumed that the effectiveness of the protective elements does not change with respect to time. However, this assumption in many real situations must be withdrawn. Let us consider a case where both systems can vary the effectiveness of the protective elements from step to step, so that at each step the effectiveness is independent of the number of protective elements. Let the effectiveness of R_a-and R_b-elements be equal to d_{1a} and d_{1b} respectively during the first step in the two-step case. Similarly, during the second step the effectiveness acquires values d_{2a} and d_{2b} so that $d_{1a} + d_{2a} = 2d_a$, $d_{1b} + d_{2b} = 2d_b$; that is, the summed value of the effectiveness does not exceed a constant value. As a result, we obtain the following matrix game:

	$d_{1a} = d_{2a}$	$d_{1a} < d_{2a}$	$d_{1a} > d_{2a}$
$d_{1b} = d_{2b}$	Q_{11}	Q_{12}	Q_{13}
$d_{1b} < d_{2b}$	Q_{21}	Q_{22}	Q_{23}
$d_{1b} > d_{2b}$	Q_{31}	Q_{32}	Q_{33}

where,

$$\begin{aligned}
Q_{11} &= (ed_a)^{-1}\exp\{M_b(0)d_a\exp[-d_a N_{Ra}(0)]\} - (ed_b)^{-1}\exp\{M_a(0)d_b\exp[-d_b N_{Rb}(0)]\},\\
Q_{22} &= Q_{33} = Q_{23} = Q_{32} = (ed_{2a})^{-1}\exp\{(d_{1a} - d_{2a})N_{Ra}(0) + d_{2a}M_b(0)\exp[-d_{1a}N_{Ra}(0)]\}\\
&\quad - (ed_{2b})^{-1}\exp\{(d_{1b} - d_{2b})N_{Rb}(0) + d_{2b}M_a(0)\exp[-d_{1b}N_{Rb}(0)]\},\\
Q_{12} &= Q_{13} = (ed_{2a})^{-1}\exp\{(d_{1a} - d_{2a})N_{Ra}(0) + d_{2a}M_b(0)\exp[-d_{1a}N_{Ra}(0)]\}\\
&\quad - (ed_b)^{-1}\exp\{M_b(0)d_b\exp[-d_b N_{Rb}(0)]\},\\
Q_{21} &= Q_{31} = (ed_a)^{-1}\exp\{M_a(0)d_a\exp[-d_a N_{Ra}(0)]\}\\
&\quad - (ed_{2b})^{-1}\exp\{(d_{1b} - d_{2b})N_{Rb}(0) + d_{2b}M_b(0)\exp[-d_{1b}N_{Rb}(0)]\}
\end{aligned} \tag{10.53}$$

The matrix of this game has a saddle-shaped point in the case of following conditions: $M_a(0) \le N_{Rb}(0)\exp[d_b N_{Rb}(0)]$, $M_b(0) \le N_{Ra}(0)\exp[d_a N_{Ra}(0)]$ and that it takes place in the real system.

Thus for both systems it is advantageous during the first step to provide the protective elements with a small effectiveness and to increase their effectiveness during the second step. This is natural, since it is better to lose a greater number of protective elements during the first step, thus securing a reliable defense of the working elements during the second step of the game.

From Eq. (10.53) it follows that the loses of a-and b-elements by system H and N, respectively will amount to:

$$\begin{aligned} Q_{1^*} &= \exp\{-1 + (d_{1a} - d_{2a})N_{Ra}(0) + d_{2a}M_b(0)\exp[-d_{1a}N_{Ra}(0)]\}/d_{2a}, \\ Q_{2^*} &= \exp\{-1 + (d_{1b} - d_{2b})N_{Rb}(0) + d_{2b}M_a(0)\exp[-d_{1b}N_{Rb}]\}/d_{2b}. \end{aligned} \tag{10.54}$$

In order to determine the optimum value of the effectiveness of the R_a-and R_b-elements for each step of the game, we must find:

$$\min_{(d_{1a},d_{2a})} Q_{1^*}(d_{1a}, d_{2a}) \textit{ and } \min_{(d_{1b},d_{2b})} Q_{2^*}(d_{1b}, d_{2b})$$

From Eq. (10.54) we obtain:

$$\begin{aligned} \partial Q_{1^*}/\partial d_{1a} = {} & 1 + (2d_a - d_{1a})\{2N_{Ra}(0) - M_b(0) \\ & \times [1 + N_{Ra}(2d_a - d_{1a})]\exp[-d_{1a}N_{Ra}(0)]\} = 0, \end{aligned} \tag{10.55}$$

$$\begin{aligned} \partial Q_{2^*}/\partial d_{1b} = {} & 1 + (2d_b - d_{1b})\{2N_{Rb}(0) - M_a(0) \\ & \times [1 + N_{Rb}(0)(2d_b - d_{1b})]\exp[-d_{1b}N_{Rb}(0)]\}, \end{aligned} \tag{10.56}$$

From Eq. (10.55) it is evident that if the equation:

$$\{(2d_a - d_{1a})/(d_a - d_{1a})\}\exp[-d_{1a}N_{Ra}(0)] = 0.5N_{Ra}(0)/M_b(0) \tag{10.57}$$

has a real root $0 \le d^*_{1a} \le 2d_a$ then system H, by utilizing this root for its own optimum strategy, can guarantee on the average losses of elements not exceeding $Q^*_1 = [e(2d_a - d^*_{1a})]^{-1}$. If Eq. (10.57) does not have a root in the interval [0, $2d_a$], the optimum strategy is then determined either by the root of Eq. (10.55) or by $d^*_{1a} = 0$. In particular, when $M_b(0) = N_{ra}(0)$, then $d^*_{1a} = 0$. Similar calculations are realized for Eq. (10.56).

10.3 Concluding Remarks About Global Humanity Problems

The level of the anthropogenic impacts on the natural ecosystems has achieved global scale. The main problem of the present humanity is the sustainable development. The sustainable development refers to a mode of human development in which resource use aims to meet human needs, while preserving the environment so that these needs can be met not only in the present, but also for forthcoming generations. The sustainable development ties together concerns for the carrying capacity of natural systems with the social challenges faced by humanity. The concept of the sustainable development has in the past most often been broken out into three components: the environmental sustainability, the economic sustainability and the sociopolitical sustainability. More recently, it has been suggested that a more consistent analytical breakdown is to distinguish four domains, notably: the economic, the ecological, the political and the cultural sustainability. Unfortunately, solidarity does not exist between scientists who are studying the global environmental processes and other experts from economics, politics and other humanities. Nevertheless, the solution of global problems arising in the nature-society system evolution is possible only with the use of global ecoinformatics methods some of which were discussed in this book.

Undoubtedly, the global environmental problems derive from global environmental causes and effects. All the human causes of the global environmental change happen through a subset of proximate causes, which directly alter aspects of the environment in ways that have global effects. Main attention of most people is focused on the major economic problems. But the sustainable problems entail the more complex idea of humanity development. Fundamental attention should be paid to the nature evolution of which the principal aspect is the survivability of human civilization. A variety of existing point of views complicates search of stable strategies in the nature-society system management. Much has been written on this subject and a large number of scenarios have been worked out. But, the present trend of the interactions between human society and the environment remains mainly hopeless.

It is evident that the perspective evolution of the nature-society system depends on the coordinated strategies of practically all countries and the availability of the effective technology for global modeling and monitoring. Essential aspect is the proper definition of the sustainable development category. Chapin et al. (2004), Kobayashi (2005) and Wijen et al. (2005), Krapivin et al. (2009c) consider four categories of sustainability that refer to:

- *Nature*: (1) biodiversity, (2) global warming, (3) cycle of resources, (4) water, soil, and air, (5) ecological formation.
- *Economy*: (1) energy, (2) resources productivity, (3) food, (4) financial status, (5) international cooperation.

- *Society*: (1) security, (2) mobility, (3) sex characteristics, (4) traditions and culture, (5) financial fluxes.
- *Wellbeing*: (1) satisfaction with life, (2) science and education, (3) participation in social life, (4) health, (5) differences in life standards.

Consideration of these categories and search of optimal coordination between them can be the responsibility of international institutes in the frame of the Organization of United Nations. It is evident that these categories have sub-levels with the following more detailed structure:

- *Climate change*. The climate is changing. Many authors have agreed that climate change may be one of the greatest threats facing the planet. In this regard, the recent data show that temperature increases in various regions, and/or enhanced extremes in weather patterns. Research has shown that air pollutants from both anthropogenic and natural sources make clouds to reflect more of the sun's rays back into space. This leads to an effect known as global dimming whereby less heat and energy reaches the Earth. Therefore, solution of the sustainable development problems depends mainly on the precision of forecasting climate trend.
- *Conservation*. Natural resource conservation problems comprise a wide spectrum of administrative, scientific, social, political and educational aspects the solution of which is possible with global ecoinformatics.
- *Energy*. Energy is one of main elements of sustainable development. The Earth receives 174 petawatts of incoming solar radiation at the upper atmosphere. Approximately 30 % is reflected back to space, while the rest is absorbed by clouds, oceans and land masses. Other energy sources have terrestrial origin: coal, oil, gas, atomic and wind energy.
- *Environmental degradation*. Environmental degradation is a process through which the natural environment is compromised in some way, reducing biological diversity and the general health of the environment. This process can be entirely natural in origin, or it can be accelerated or caused by human activities.
- *Environmental health*. The environmental health and safety guidelines contain the performance levels and measures that are normally acceptable to International Finance Corporation and that are generally considered to be achievable in new facilities at reasonable costs by existing technology.
- *Electromagnetic fields*. An electromagnetic field is a physical field produced by electrically charged objects. It affects the behavior of charged objects in the vicinity of the field. The electromagnetic field extends indefinitely throughout space and describes the electromagnetic interaction. It is one of the four fundamental forces of nature (the others are gravitation, the weak interaction, and the strong interaction).
- *Genetic engineering*. Genetic engineering is the direct manipulation of an organism's genome using biotechnology. Genetic engineering techniques have been applied in numerous fields including research, agriculture, industrial biotechnology, and medicine.

- *Intensive farming*. Intensive farming or intensive agriculture is an agricultural production system characterized by a low fallow ratio and the high use of inputs such as capital, labour, or heavy use of pesticides and chemical fertilizers relative to land area. This is in contrast to many sorts of traditional agriculture in which the inputs per unit land are lower.
- *Land degradation*. Land degradation is a process in which the value of the biophysical environment is affected by one or more combinations of human-induced processes acting upon the land. Also environmental degradation is the gradual destruction or reduction of the quality and quantity of human activities or natural means example water causes soil erosion, wind, etc.
- *Soil*. Soil is a natural body consisting of layers that are primarily composed of minerals which differ from their parent materials in their texture, structure, consistency, color, chemical, biological and other characteristics. Soil forms a structure filled with pore spaces and can be thought of as a mixture of solids, water, and gases. Soil degradation is main problem of present science.
- *Land use*. Land use involves the management and modification of natural environment or wilderness into built environment such as fields, pastures, and settlements. Land use practices vary considerably across the world. Land use and land management practices have a major impact on natural resources including water, soil, nutrients, plants and animals.
- *Nanotechnology*. Nanotechnology is the manipulation of matter on an atomic and molecular scale. This new perspective scientific field can have unpredicted consequences for the environment and humans.
- *Nuclear issues*. Nuclear energy use is one of perspective way in the energy development, but there exist many problems connected with the safety.
- *Overpopulation*. Overpopulation is a generally undesirable condition where an organism's number exceeds the current carrying capacity of its habitat. The term often refers to the relationship between the human population and its environment, the Earth, or smaller geographical areas such as countries. Overpopulation can result from an increase in births, a decline in mortality rates, an increase in immigration, or an unsustainable biome and depletion of resources. It is possible for very sparsely populated areas to be overpopulated if the area has a meager or non-existent capability to sustain life.
- *Ozone depletion*. Ozone is very rare and important in the Earth's atmosphere. Ozone plays a vital role in the atmosphere. Ozone is mainly found in two regions of the Earth's atmosphere. Most ozone (about 90 %) resides in a layer that begins between 10 (near the poles) and 17 (over tropics) kilometers above the Earth's surface and extends up to about 50 km. This region of the atmosphere is called the stratosphere. The ozone in this region is commonly known as the ozone layer. The remaining ozone is in the lower region of the atmosphere, which is commonly called the troposphere. Stratospheric ozone plays a beneficial role by absorbing most of the biologically damaging ultraviolet sunlight, allowing only a small amount to reach the Earth's surface.
- *Pollution*. Pollution is the introduction of contaminants into the natural environment that causes adverse change. Pollution can take the form of chemical

substances or energy, such as noise, heat or light. Pollutants, or the components of pollution, can be either foreign substances/energies or naturally occurring contaminants.

- *Water quality*. Water quality refers to the chemical, physical and biological characteristics of water. It is a measure of the condition of water relative to the requirements of one or more biotic species and or to any human need or purpose. It is most frequently used by reference to a set of standards against which compliance can be assessed. The most common standards used to assess water quality relate to health of ecosystems, safety of human contact and drinking water.
- *Air quality*. The atmosphere is a complex dynamic natural gaseous system that is essential to support life on planet Earth. Air pollution is the introduction into the atmosphere of chemicals, particulates, or biological materials that cause discomfort, disease, or death to humans, damage other living organisms such as food crops, or damage the natural environment or built environment. Air pollution is a significant risk factor for multiple health conditions. The health effects caused by air pollution may include difficulty in breathing, wheezing, coughing and aggravation of existing respiratory and cardiac conditions.
- *Reservoirs*. The natural topography provides most of the basins of the reservoirs. Their exploitation is main problem of sustainability of many human life aspects. Many dammed river reservoirs and most bank-side reservoirs are used to provide the raw water feed to a water treatment plant which delivers drinking water through water mains. Main problem is in the protection of water quality.
- *Fishing*. Fishing is the activity of trying to catch fish. Fish are normally caught in the wild. Techniques for catching fish include hand gathering, spearing, netting, angling and trapping. There are many fishing techniques or methods for catching fish. The fishing industry includes any industry or activity concerned with taking, culturing, processing, preserving, storing, transporting. Sustainability problem of fishing include overfishing, by-catch, marine pollution, environmental effects of fishing, climate change and fish farming.
- *Logging*. In common usage, the term "logging" is used to indicate a range of forestry or silviculture activities. Logging is a dangerous occupation.
- *Toxins*. A toxin is a poisonous substance produced within living cells or organisms. Man-made substances created by artificial processes are thus excluded. Toxins can be products or byproducts of ordinary metabolism, such as lactic acid, and they must be broken down or excreted before building up to dangerous levels. Toxins can facilitate survival, as with snake venom that kills or immobilizes prey, or cyanide produced by some plants as a defense against being eaten. Bacterial toxins can sometimes be neutralized with antitoxins. Toxins may be dangerous to human health.
- *Waste*. There are many waste types defined by modern systems which support human life. Waste can contaminate surface water, groundwater, soil, and air which cause more problems for humans, other species, and ecosystems. Waste treatment and disposal produces significant green house gas emissions, notably methane, which is contributing significantly to global climate change.

Thus, solution of sustainable development problems needs the global model of the nature-society system evolution, which can give possibility to consider all the complex interactions listed above. This requires a participation of specialists in different areas of science. Unfortunately, global environmental problems are studied separately. Successful progress in the evolution of global humanity can be achieved by means of creation International Integrated System for Global Geoinformation Monitoring (IISGGM) the functional structure of which is based on the evolutionary neurocomputer technology. Such an accomplishment will take into account the architecture of a modern computer system with assess to a network of evolution —type tutorial servers. This will lead to such IISGGM structure that will simulate the MCBS system with capability of evaluating nature protection measures and other structural decisions on how human society can best interact with nature.

10.4 Mission to Mars: Reliable Method for Liquid Solutions Diagnostics

10.4.1 Introduction

The NASA Marshall Spaceflight Center, European Space Agency, China National Space Administration and Russian Space Agency have a long-term vision of sending a manned mission to Mars in the very near future. For this campaign many unresolved technical and social problems still remain, such as the creation of favorable conditions for human health during the flight and accommodation on this planet. In this context, negative health effects can be caused by both drinking water with distorted characteristics and medical products (e.g., solutions) with altered properties. Ensuring the reliable operational control of these vitally important elements is one of the main stages during the preparation for the manned mission to Mars. Such control of liquid solutions in the environment of Mars during ancient years will be an important stage in the exploration and life support.

Martian drinking water is concentrated in the frozen phase into the soil and as ice formations. Today, it is known that water on Mars exists almost exclusively as ice, with a small amount present in the atmosphere as vapour. Results from the recently published research suggest that the soil of the Martian surface contains water about 2 % by weight. Drinking water extracted from soil or surface ice needs the preliminary investigation and continuous monitoring in time. The implementation of this procedure is feasible by means of optical methods developed for the Earth conditions without chemical analysis (Krapivin and Shutko 2012).

Krapivin et al. (2013b, c) showed that the development of multi-channel polarization optical instruments and the use of spectro-ellipsometric technology are very important for real-time ecological monitoring of aquatic environment. It should be mentioned that efficient solution of this multi-parametric problem greatly depends on the precision and simplicity of ellipsometric devices. Recently, considerable successes achieved in this area (Krapivin and Shutko 2012; Mkrtchyan et al. 2013; Mkrtchyan and Krapivin 2014; Nitu et al. 2013a, b), notably:

- A technology of combined use of spectro-ellipsometry and algorithms of identification and recognition that allow the set-up of a standard integral complex consisting of instrumental, algorithmic, modular and software tools for the collection and processing of data about the quality of aquatic environment with forecasting and decision—making functions.
- A compact measuring—information multichannel spectro-ellipsometric system (device) for monitoring the quality of aquatic environment, that is based on the combined use of spectro-ellipsometry and training, classification, and identification algorithms. This spectro-ellipsometric system differs from the modern foreign analogues by the use of a new and very promising method of ellipsometric measurements, an original element base of polarization optics and a complex mathematical approach to estimate the quality of a water object subjected to anthropogenic influence. Unlike foreign analogues, the above-mentioned system has no rotating polarization elements. This allows one to increase the signal-to-noise ratio and the long-term stability of measurements, to simplify and reduce the price of multichannel spectro-ellipsometers. This system is trainable to the recognition of the pollutants of aquatic environment.

10.4.2 The Method Description

The detection and identification of water quality, especially of drinking water, is being addressed by many scientists (Sun et al. 2013). For example, the U.S. National Water-Quality Assessment Program (NAWQA) provides an understanding of water-quality conditions; whether conditions are getting better or worse over time; and how natural features and human activities affect those conditions. In the Earth's conditions, water quality assessment is defined depending on many environmental factors connected with the environmental management from pollution sinks and sources. Martian conditions make additional unknown amendments in the task of reliable evaluation of drinking water that can be extracted on the Mars. Therefore, method of the drinking water assessment does not depend on the environment. Such a method can be based on the spectral analysis of water mass in the optical range (Krapivin et al. 2008b).

The recently developed technology of an adaptive identification of the environmental elements from measurements in the visible spectral region permits to synthesize an expert system for an adaptive identification of the liquid solutions (ESAILS). The system's structure includes a compact multi-channel spectro-ellipsometer (MCS), information interface with computer (IIC), computer software (CS), and extending database (ED). The CS offers a number of algorithms to process the data fluxes from MCS and provides service functions for visualization and control of the regime of measurements. The ED consists of the sets of standard

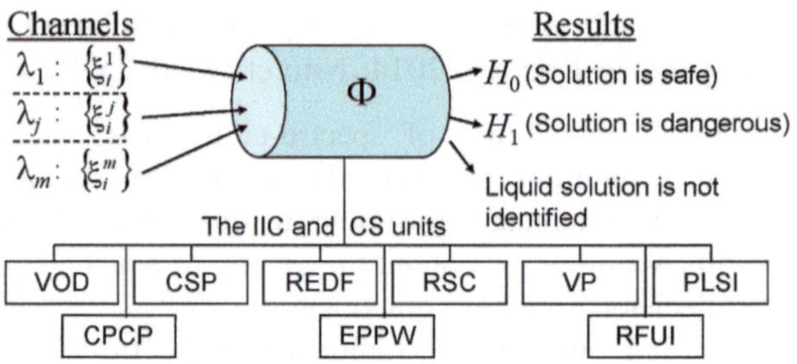

Fig. 10.3 A conceptual block-diagram of the procedure used to identify liquid solutions under the isolated conditions. Notation is given in Table 10.1

Table 10.1 Description of the IIC and CS as the ESAILS units schematically shown in Fig. 10.3

Unit	Characteristic of the unit
VOD	Visualization of the observed data (spectra, correlations, statistical parameters, empirical and theoretical functions of distribution)
CPCP	Calculation of parameters for the classical procedure of decision making by using the Neuman-Pearson criterion
CSP	Calculation of statistical parameters and other characteristics, formation of the vector space $\{u_i\}$
EPPW	Estimation of the parameters that will be used in the procedure of the Wald successive analysis in decision making
REDF	Reconstruction of the empirical distribution functions
RFUI	Reconstruction of the functions of the user's interface equipped with software to intervene in the working regime of the expert system at any stage of its functioning
RSC	Realization of statistical criterion to estimate the theoretical function of distribution
PLSI	Procedure of the liquid solution identification
VP	Visualization procedures

spectral images of the liquid solutions represented by points in the multi-dimensional vector space of indicators, pre-calculated on the basis of learning samples. In particular, identification algorithms (ID) carry out the task of liquid solution assessment.

The principal scheme of the CS unit providing the procedure of identification is a transformation Φ (Fig. 10.3, Table 10.1). The light intensity ξ_j^i recorded at the moment t_i in the channel λ_j enters the algorithm Φ, where two hypotheses H_0 and H_1 are identified. The ESAILS operator determines initial data $\mathbf{v}_i$, α and β and decides which parameters $\mathbf{u}_i = (u_1, \ldots, u_r)$ will be calculated from measurements of $\{\xi_j^i\}$. The service unit IIC makes it possible to form vector $\mathbf{u}_i$ from statistical characteristics of the series $\{\xi_j^i\}$ or to use direct measurements. The a-priori information characterizes the type of distribution $f_a(\mathbf{u}_i)$. The function

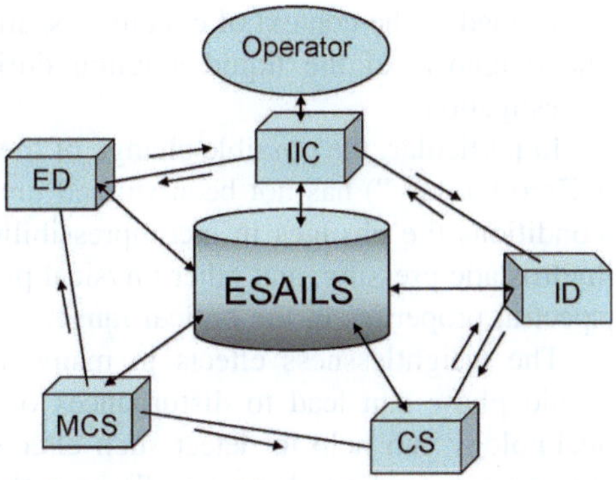

Fig. 10.4 The ESAILS model

$$Li = \sum_{i=1}^{\phi} \psi_j = f_{a1}\left(\mathbf{u}_i^j\right) / f_{a0}\left(\mathbf{u}_i^j\right)$$

is compared with its ultimate values $L_{i,min}$ and $L_{i,max}$ At the first stage, these values are chosen arbitrarily, but then they change till reaching a maximum accurate recognition of the hypotheses H_0 and H_1. We have $L_{i,min} \rightarrow L^*_{i,min}$ and $L_{i,max} \rightarrow L_{i,max}$. The values $L^*_{i,min}$ and $L_{i,max}$ are memorized in ED.

After the learning procedure, the functioning of the expert system is limited only by the volume of measurements fixed by the operator, proceeding from statistical reliability and the real-time regime. The operator has two possibilities to regulate this regime: establishing the volume of the series $\{\xi_j^i\}$ or fixing the time of their accumulation. Usually, the latter characteristic that is equal to 1 s is chosen. Fig. 10.4 explains this procedure. The operator is combined with the ESAILS units through the man-machine interface IIC, which provides the selective control of operations in all units.

A knowledge of spectrums allows the recognition of two alternative hypotheses: H_0 (liquid solution is normal) and H_1 (liquid solution is defective). The CS algorithms have more sweeping functions allowing the more detailed diagnostics of liquid solutions. These algorithms are based on cluster analysis and recognition of spectral images. Recognition of liquid solution structure or receive of content of chemical elements is possible after creation of spectral etalons in the ED. This stage can be realized in the Earth's conditions and on orbital space station taking into consideration knowledge about weightlessness effects on liquid solution.

10.4.3 Conclusion

The discussion above showed that the ESAILS technology version can carry out the task of liquid solution diagnostics in conditions where the use of chemical analysis is not feasible. Such situation occurs both on Earth and in the space. This version is performed in the context of concrete assumptions. It is shown that the actual use of the diagnosis of the liquid solution during the Mars mission requires detailed investigations.

In particular, the possible change of the liquid solution in the absence of gravity ("Zero Gravity") has not been studied until now. For instance, under zero gravity conditions the changes in incompressibility, elasticity, convective heat exchange, hydrostatic pressure, and other physical parameters of the solution can influence its spectral properties in the optical range.

The weightlessness effects in many natural processes including those in the liquid phase can lead to disturbances of living conditions. The above proposed technology can help to detect such effects well in advance. In particular, stop of engine can lead to change of flight path. We discussed above the fact that the operative spectral control of fuel solution can be realized by means of the ESAILS adapted to this area.

The ESAILS employment as a specific onboard device for space expeditions is possible considering the following actions:

- definition of liquid solutions that could be diagnosed by using the ESAILS tool;
- realization of the learning procedure in the Earth conditions to form database of spectral images for these liquid solutions and assessment of the ESAILS efficiency;
- realization of the recognition procedure for separate liquid solutions on the International Space Station and assessment of the weightlessness influence on spectral images of liquid solutions;
- the ESAILS modernization taking into account the assessments obtained; and
- production of the ESAILS as onboard device for space campaigns.

References

A Guide to World Resources (2000–2001) People and ecosystems: the frying web of life (2000). World Resources Institute, Washington, 389 pp

Aagard K (ed) (1998) Toward prediction of the arctic system: predicting future states of the arctic system on seasonal-to-century time scales by interacting observations, process research, modeling, and assessment. The Arctic Research Consortium of the United States (ARCUS), Fairbanks, 54 pp

Abakumov AI (1993) Control and optimization in the models of exploited populations. Dal'nauka, Vladivostok, 129 pp [in Russian]

Abbasi T, Abbasai SA (2012) Water quality indices. Springer, Berlin, 362 pp

Abrahamson DE (1989) Challenge of global warming. Island Press, Washington, 376 pp

Abu-Zaid A, Ahmadi G (1989) Chaos in a double diffusive convention model in the presence of noise. Appl Math Model 13(5):291–297

ACIA (2000) Arctic climate impact assessment (ACIA). The Assessment Steering Committee, Fairbanks, 35 pp

Adamenko VN, Kondratyev KYa (1990) Global climate changes and their empirical diagnostics. In: Izrael YA, Kalabin GV, Nikonov VV (eds) Anthropogenic impact on the nature of the North and its ecological implications. Kola Scientific Centre, Russian Academy of Sciences, Apatity, pp 17–34 [in Russian]

Agirre-Basurko E, Ibarra-Berastegi G, Madariaga I (2006) Regression and multilayer perception-based models to forecast hourly O_3 and NO_2 levels in the Bilbao area. Environ Model Softw 21 (4):430–446

Aibulatov NA (2000) Cold war echo in the Russian arctic seas. GEOS, Moscow, 307 pp [in Russian]

Alcamo J, Kreileman GJJ, Krol MS, Zuidema G (1994) Modeling the global society-biosphere-climate system: description and testing. Water Air Soil Pollut 76(1–2):1–35

Aleshin VA, Klimov VV (1992) An analysis of optical heterogeneity. Proceedings of the international symposium "ecoinformatics problems", Zvenigorod, Moscow Region, 14–18 Dec 1992. KIREE Publishing, Moscow, pp 202–204 [in Russian]

Alessio S, Longhetto A, Richiardone R (2004) Evolutionary spectral analysis of European climatic series. II Nuovo Cimento C 27(1):73–98

Alexandrov GA, Yamagata Y, Saigusa N, Oikawa T (2005) Long-term carbon exchange at the Takayama, Japan forest re-calibration TsuBiMo with eddy-covariance measurements at Takayama. Agric For Meteorol 134(1–4):135–142

Alexeev GV (1998) Arctic climate dynamics in the global environment. World climate research programme/world meteorological organisation, vol 908, Geneva, pp 11–14

V.F. Krapivin et al., *New Ecoinformatics Tools in Environmental Science*, Environmental Earth Sciences, DOI 10.1007/978-3-319-13978-4

Alexeev VV, Kryshev II, Sazhykina TG (1992) Physical and mathematical modeling of the ecosystem. Hydrometeoizdat, Sankt-Petersburg, 367 pp [in Russian]

Alimov AF, Dmitriyev VV, Florinskaya TM, Khovanov NV, Chistobayev AI (1999) Integral assessment of the ecological state and environmental quality of the urban territories. St. Petersburg Science Centre of RAS, St. Petersburg, 253 pp [in Russian]

Allen M (2002) Climate of twentieth century: detection of change and attribution of causes. Weather 57(8):296–303

Alley RB, Marotzke J, Nordhaus WD, Overpeck JT, Peteet DM, Pielke RA Jr, Stocker TF, Talley LD, Wallace JM (2003) Abrupt climate change. Science 299(5615):2005–2010

Allison I, Barry RG, Goodison BE (eds) (2001) Climate and cryosphere (CliC) project science and co-ordination plan. Version 1. WCRP-114, (WMO/TD no. 1053), Geneva, 73 pp

Aloyan AE (2004) Numerical modelling of minor gas constituents and aerosols in the atmosphere. Ecol Model 179:163–175

Alvarado E, Sandberg DV, Pickford SG (eds) (1998) Modeling large fires as extreme events. Northwest Sci 82:66–75

AMAP (2009) AMAP assessment 2009: radioactivity in the Arctic. Narayana Press, Oslo, 95 pp

Ananicheva MD (2005) Evolution of climatic snow line and equilibrium line altitudes in the North-Eastern Siberia mountains (20th century). Ice Clim News 6:2–6

Anderson W, Anderson WT (2001) All connected now: life in the first global civilization. Westview Press, New York, 320 pp

Andreev GA, Borodin LF, Zaentsev LV, Pubtsov SN (1984) On the remote determination of land cover temperature using the polarizational invariant of microwave radiothermal radiation. Res Earth Space 4:120–123 [in Russian]

Andronova NG, Schlesinger ME (2001) Objective estimation of the probability density function for climate sensitivity. J Geophys Res 106(D19):22, 605–22, 611

Aota M, Shirasawa K, Krapivin VF (1991a) Simulation model for study of the Okhotsk Sea. Proceedings of the international symposium "engineering ecology-91", Zvenigorod, Russia, 22–24 Oct 1991. IREE RAS Publisher, Moscow, pp 110–113

Aota M, Shirasawa K, KrapivinVF, Mkrtchyan FA (1991b) The system for data processing in Okhotsk Sea monitoring. Proceedings of the Sixth international symposium on Okhotsk Sea & Sea Ice, Mombetsu, Japan, 3 Feb 1991. Okhotsk Sea & Cold Ocean Research Association, Mombetsu, pp 317–318

Aota M, Shirasawa K, Krapivin VF, Mkrtchyan FA (1992) Simulation model of the Okhotsk Sea geoecosystem. Proceedings of the 7th international symposium on Okhotsk Sea & Sea Ice, Mombetsu, Japan, 2–5 Feb 1992. Okhotsk Sea & Cold Ocean Research Association, Mombetsu, pp 311–313

Aota M, Shirasawa K, Krapivin VF, Mkrtchyan FA (1993) A project of the Okhotsk Sea GIMS. Proceedings of the 8th international symposium on Okhotsk Sea & Sea Ice and ISY/Polar Ice extent workshop, Mombetsu, Japan, 1–5 Feb 1993. Okhotsk Sea & Cold Ocean Research Association, Mombetsu, pp 498–500

Archer D (2010) The global carbon cycle. Princeton University Press, Princeton, 224 pp

ARCSS (2003) ARCSS program. Proceedings of the arctic system science program all-hands workshop 2002. Arctic Research Consortium of the United States, Fairbanks, Alaska, 271 pp

ARCUS (1998) Toward an arctic system synthesis: results and recommendations. The Arctic Research Consortium of the United States (ARCUS), Fairbanks, Alaska, 165 pp

Armand NA, Krapivin VF, Mkrthyan FA (1987) Methods for data processing in the radiophysical investigation of the environment. Science Publishing House, Moscow, 270 pp [in Russian]

Arneth A, Kurbatova J, Kolle O, Shibistova OB, Lloyd J, Vygodskaya N, Schulze E-D (2002) Comparative ecosystem-atmosphere exchange of energy and mass in a European Russian and a central Siberian bog II. Interseasonal and interannual variability of CO_2 fluxes. Tellus 54B (5):514–530

Arrhenius S (1896) On the influence of carbonic acid in the air upon the temperature of the ground. Lond Edinb Dublin Philos Mag J Sci 41:237–276

Arsky YuM, Zakharov YF, Kalutskov VA, Kienko YuP, Kolosnitsin IV, Kuleshova ME, Linnik VG, Lopatin VN, Melnikova GL, Osipov YuB, Ostrovsky MA, Perelet RA, Puzatchenko YuG, Savransky VA, Sankovsky AG, Taskaeva AA (1992) Ecoinformatics. Hydrometeoizdat, Sankt-Petersburg, 520 pp [in Russian]

Artemov VT, Basharinov AE, Bulatnikov VL, Egorov ST, Mishenev VF (1974) Measurement of microwave radiation from land covers with airplane board in zones of forests and peatbogs. Invest Environ Means Cosm Means 4:150–159

Arutiunian VO, Aloyan AE, Chi J, Kuznetsov YuA (2004) Numerical modeling of regional transport of gas components taking into consideration of photochemical transformation. Proc RAS Phy Atmos Ocean 49(4):501–513 [in Russian]

Ashraf M, Öztürk M, Ahmad MSA, Aksoy A (eds) (2012) Crop production for agricultural improvement. Springer, London, 782 pp

Asrar G, Dozier J (1994) EOS: science strategy for the Earth observing system. AIP Press, Woodbury, 119 pp

Austin AT, Yahdjian L, Stark JM, Belnap J, Porporato A, Norton U, Ravetta DA, Schaeffer SM (2004) Water pulses and biogeochemical cycles in arid and semiarid ecosystems. Oecologia 141:221–235

Baker BB (2012) Common precepts of marine scientific research access in the Arctic. In: Wasum-Rainer S, Winkelmann I, Tiroch K (eds) Arctic science, international law and climate change, Heidelberg Beitrage zum ausländischen öffentlichen Recht und Völkerrecht 235:209–243

Ball JT (1988) An analysis of stomatal conductance. PhD Thesis, Stanford University, Stanford, 89 pp

Baltrenas P, Vasareviclus S, Petraltis E (2003) Numerical simulation of solid particle propagation in the atmosphere. Chem Petrol Eng 39(9–10):615–616

Bard S (1999) Global transport of anthropogenic contaminants and the consequences for the Arctic marine ecosystems. Mar Pollut Bull 38(5):356–379

Barenbaum AS (2002) Galaxy, solar system, the Earth: subordinate processes and evolution. Geos, Moscow, 393 pp [in Russian]

Barenbaum AS (2004) Mechanism for the formation of gas and oil accumulation. Annals of Acad Sci 399(6):1–4 [in Russian]

Bartsev SI, Degermendzhi AG, Erokhin DV (2003) Global generalized models of carbon dioxide dynamics. Probl Environ Nat Resour 12:11–28 [in Russian]

Basharinov AE, Zotova EN, Naumov MI, Chukhlantsev AA (1979) Radiation characteristics of plant covers in microwave range. Radiotechnics 34(5):16–20 [in Russian]

Bazhin NM (2000) Methane emission from a residual layer. Proceedings of the second international methane mitigation conference, 18–23 June 2000. Novosibirsk, Novosibirsk State University Publication, Novosibirsk, pp 231–236

Bazilevich NI, Rodin LE (1967) Diagrammatic map of productivity and biological cycle of basic terrestrial plant types. Proc All-Union Geogr Soc 99(3):190–194 [in Russian]

Beder S (2005) Setting the global agenda: corporate coalitions, coercion, and control. Earthscan, London, 224 pp

Beeby A, Brennan A-M (2003) First ecology: ecological principles and environmental issues. Oxford University Press, Oxford, 352 pp

Bekoryukov VI, Fedorov VV (1987) Empirical model of total ozone content above the Southern Hemisphere. Meteorol Hydrol (Moscow) 3:47–53 [in Russian]

Bellman R, Dreyfus S (1962) Applied dynamic programming. Princeton University Press, Princeton, 388 pp

Bellman R, Roth R (1966) A technique for the analysis of a broad class of biological systems. Bionics Symposium, Gordon and Breach, pp 725–737

Bengtsson L (1999) Climate modelling and prediction—achievements and challenges. WCRP/WMO Publ 954:59–73

Berdnikov SV, Dombrovsky YuA, Ostrovskaya AG, Prichodko MV, Titova LI, Tjutjunov JuV (1989) Simulation model of basic components of the Okhotsk Sea ecosystem. Mar Hydrophys J (Vladivostok) 3:52–57 [in Russian]

Berndt T, Boge O, Stratmann F (2004) Atmospheric particle formation from the ozonolysis of alkenes in the presence of SO_2. Atmos Environ 38(14):2145–2153

Beven K, Blazkova S (1998) Estimating changes in flood frequency under climate change by continuous simulation (with uncertainty). In: Bronstert P, Samuels RP (eds) Proceedings of the RIBAMOD final workshop, 26–27 Feb 1998. Wallingford, Balabanis, pp 269–279

Bichele I, Moldau X, Ross Yu (1980) Sub-model for the assimilation, distribution and vegetation growth under condition of water deficit. Tartu astrophysical observatory report A–5, Tartu, 22 pp [in Russian]

Bickel P, Li Bo (2006) Regularization in statistics. Test 15(2):271–344

Binenko VI, Khramov G, Yakovlev VV (2004) Emergency situations in the present world and problems of life safety. "Integratsiya" Publication, St-Petersburg, 398 pp [in Russian]

Binkowski FS, Roselle SJ (2003) Models-3 community multiscale air quality (CMAQ) model aerosol component, 1. Model description. J Geophys Res 108(D6):AAC3/1–AAC3/18

Bjorkstrom A (1979) A model of CO_2 in interaction between atmosphere, oceans and land biota. In: Bolin B (ed) Global carbon cycle, SCOPE 13. Wiley, Chichester, pp 403–457

Bobylev LP, Kondratyev KYa, Johannessen OM (eds) (2003) Arctic environment variability in the context of global change. Springer/Praxis, Chichester, 471 pp

Boden TA, Kaiser DP, Speranski RJ, Stoss FW (1994) Trend's 93: a compendium of data on global change. Carbon Dioxide Informational Analysis Center, Oak Ridge National Laboratory, Oak Ridge, 176 pp

Bodenbender J, Wassmann R, Papen H, Rennenberg H (1999) Temporal and spatial variation of sulfur-gas-transfer between coastal marine sediments and the atmosphere. Atmos Environ 33 (21):3487–3502

Bodri L, Čermák V (1999) Climate change of last millennium inferred from borehole temperatures: regional patterns of climate changes in the Czech Republic. Part III Glob Planet Change 21(4):225–235

Boehmer-Christiansen S (1997) Who is driving climate change policy? IEA Stud Educ 10:53–72

Boehmer-Christiansen S (2000) Who and how determines the policy concerning climate change? Trans Russ Geogr Soc 132(3):6–22 [in Russian]

Bogatyrev BG (1988) Modelling the transition processes on the border of vegetation zones in conditions of anthropogenic climate changes. PhD theses, Moscow State University, Moscow, 114 pp [in Russian]

Bogorodsky VV, Kondratyev KY, Rabinovitch YuI (1976) Microwave remote indication of the sea surface by the oil pollutants. Ann Main Geophys Observatory (Leningrad) 371:22–36 [in Russian]

Bolin B (1998) The WCRP and IPCC: research inputs to IPCC assessments and future needs. WCRP/WMO 904:27–36

Bolin B (1999) Global environmental change and the need for international research programmes WCRP/WMO. WMO/TD 954:11–14

Bolin R, Sukumar R (2000) Global perspective. In: Watson RT, Noble IR, Bolin B, Ravindranath NH, Verardo DJ, Dokken DJ (eds) Land use, land-use change, and forestry. Cambridge University Press, Cambridge, pp 23–51

Bonafoni S, Alimenti F, Angelucci G, Tasselli G (2011) Microwave radiometry imaging for forest fire detection: a simulation study. Prog Electromagnet Res 112:77–92

Bond-Lamberty B, Wang C, Gower ST (2003) Annual carbon flux from woody debris for a boreal black spruce fine chronosequence. J Geophys Res 108(3):WFX1/1–WFX1/10

Bondur VG, Krapivin VF (2014) Space monitoring of tropical cyclones. Scientific World Publisher, Moscow, 508 pp [in Russian]

Bondur VG, Pulinets SA, Kim GA (2008) On the role of the galactic cosmic rays in the tropical cyclogenesis on the Hurricane Katrina case. Proc Acad Sci 422(2):1–6 [in Russian]

Bondur VG, Krapivin VF, Savinikh VP (2009) Monitoring and prediction of the natural disasters. Scientific World Publisher, Moscow, 691 pp [in Russian]

Bondur VG, Krapivin VF, Potapov II (2010) Methods for the aerospace diagnostics of the forest ecosystems. Ecol Syst Devices 3:17–27 [in Russian]

Borisov PM (2005) Can man change climate? Science Publishing House, Moscow, 270 pp [in Russian]

Bornstein R (1999) Urban induced convergence zones and air pollution episodes. Proceedings of the international conference cum workshop on air quality management, 15–19 Nov 1999, Darussalam, Brunei, pp 51–54

Borodin LF (2008) Overcoming of information instability in the ecological monitoring systems. Problems Environ Nat Resour 5:80–87 [in Russian]

Borodin LF, Krapivin VF (1998) Remote measurements of the land cover characteristics. Problems Environ Nat Resour 7:38–54 [in Russian]

Borodin LF, Mitnik LM (1977) Remote indication of forest fires by means of microwave radiometry method. For Econ 6:123–131 [in Russian]

Borodin LF, Kirdiashev KP, Stakankin YuP, Chukhlantsev AA (1976) About the use of microwave radiometry to the investigation of forest fires. Radiotech Electron XI(9):1945–1950 [in Russian]

Borodin LF, Valendic EN, Veselov AP, Mironov AS, Stakanjin YP (1978) About anomaly in the spectre of passive radio-thermal radiation of peatbogs. Rep Russ Acad Sci 238(3):709–711 [in Russian]

Borodin LF, Vasenkov LV, Krapivin VF, Malyutin VI (1982) Solution of typical SHF-radiometric airborne tasks and determination of parameters for aquatories in radio astronomic range. Geodezy Aerophotogr 2:44–50 [in Russian]

Borodin LF, Mironov AS, Burkov VD, Krapivin VF, Potapov II, Shalaev VS (2008) Technologic measurement process of temperature anomalies in forest and forest-marsh complexes by means of microwave methods. Probl Environ Nat Resour 4:75–92 [in Russian]

Bortnik VN, Lopatina SA, Krapivin VF (1994) Simulation system for the study of hydrophysical fields in the Aral Sea. Meteorol Hydrol (Moscow) 9:102–108 [in Russian]

Bounoua L, Collatz GJ, Los SO, Sellers PJ, Dazlich DA, Tucker CJ, Randall DA (2000) Sensitivity of climate to changes in NDVI. J Clim 13:2277–2292

Bounoua L, Defries R, Collatz GJ, Sellers P, Khan H (2002) Effects of land cover conversion on surface climate. Clim Change 52(1–2):29–64

Bourke RH, Paquette RG, Blythe RF (1992) The Jan Mayen current of the Greenland Sea. J Geophys Res C 97(5):7241–50

Bouwman AF (1990) Global distribution of the major soils and land cover types. In: Bouwman AF (ed) Soils and the greenhouse effect. Wiley, Chichester, pp 33–59

Bowen DQ (2000) Calibration & correlation with the GRIP & GISP2 Greenland ice cores of radiocarbon ages from Paviland (Goat's Hole), Gower. In: Aldhouse-Green S (2000) (ed) Paviland Cave and the 'Red Lady'. Western Academic and Specialist Press, Bristol, pp 61–64. ISBN 0-9535418-1-9

Boysen M (ed) (2000) Biennial report 1998 & 1999. Potsdam Institute for Climate Impact Research, Potsdam, 130 pp

BP (2005) Putting energy in the spotlight: BP statistical review of world energy June 2005. BP Publisher, London, 44 pp

BP (2013) BP statistical review of world energy June 2013. BP Publisher, London, 45 pp

Braesicke P, Pyle JA (2004) Sensitivity of dynamics and ozone to different representations of SST's in the United model. Quart J Roy Meteorol Soc 130(601B):2033–2045

Bras RL (1990) Hydrology. Addison-Wesley, New York, 643 pp

Brasseur GP (2005) Aeronomy of the middle atmosphere: chemistry and physics of the stratosphere and mesosphere. Springer, Berlin, 644 pp

Brebbia CA (2004) Risk analysis IV. WIT Press, Southampton, 400 pp

Brebbia CA (ed) (2012) Modelling, monitoring and management of forest fires III. WIT Press, Southampton, 258 pp

Brennan MJ, Hennon CC, Khabb RD (2009) The operational use of QuikSCAT ocean surface vector winds at the National Hurricane Center. Weather Forecast 24:621–645

Bréon F-M (2004) POLDER level-2 products: data format and user manual. CNES NASDA, Tokyo, 37 pp

Broecker WS (2003) Does the trigger for abrupt climate change reside in the ocean or in the atmosphere? Science 300(5625):1519–1522

Brown RA, Kaufman CA, MacGorman DR (2002) Cloud-to-ground lightening associated with the evolution of a multicell storm. J Geophys Res 107(D19):ACL13/1–ACL13/13

Brucan S (1984) The global crisis. Int Stud Quart 28(1):97–109

Brunt JW, McCartney P, Baker K, Stafford SG (2002) The future of ecoinformatics in long term ecological research. In: Callaos N (ed) Proceedings of the 6th world multiconference on systemics, cybernetics and informatics: SCI 2002; 14–18 July 2002 Orlando, Florida, USA. Information systems development, II, vol 7. International Institute of Informatics and Systemics, Orlando, pp 367–372

Bui TL (1998) Application of GISM-technology to the numerical experiment problem of atmospheric pollution physics. Probl Environ Nat Resour 1:2–11 [in Russian]

Bui TL (1999) Data processing automatization into the systems of geophysical monitoring on the Vietnam territory. Doctorate dissertation, Institute of Radioengineering and Electronics, Russian Academy of Sciences, Moscow, 354 pp

Bui TL, Nitu C, Krapivin VF, Soldatov VY (2010) Urbanization processes and the environment. The Scientific Bulletin of Electrical Engineering Faculty, Valahia University of Targoviste (Romania), vol 10, issue 2/3, pp 13–16

Bukatova IL (1992) Evolutionary neurocomputer technology. KIREE Publisher, Moscow, 34 pp [in Russian]

Bukatova IL, Kipyatkov VA, Sharov AM (1991) Simulation-evolutionary technology of multi-channel processing of signals. Probl Radio Electron 3:5–26 [in Russian]

Bukatova IL, Makrusev VV (2003) Modern information technologies for the management. RCA Publisher, Moscow, 252 pp [in Russian]

Bukatova IL, Makrusev VV (2004a) Theory of integrity-evolutionary intellectual-actualization of social systems. Moscow Institute of State and Corporative Management Publishing, Moscow, 126 pp [in Russian]

Bukatova JL, Matveev MA (2004b) Integrity-evolutionary identification of neuron networks for image processing. Probl Environ Nat Resour 7:32–40 [in Russian]

Bulkeley H, Betsill MM (2003) Cities and climate change. Urban Sustainability and Global Environmental Governance, Routledge, London, 237 pp

Bunyard P (1999) Eradicating the Amazon rainforest will wreak havoc on climate. Ecologist 29 (2):81–84

Burgman M (2005) Risks and decisions for conservation and environmental management. Cambridge University Press, Cambridge, 488 pp

Burkov VD, Krapivin VF (2005) GIMS technology in the monitoring of vegetation covers. For Bull 4(40):117–121 [in Russian]

Burkov VD, Krapivin VF (2009) Ecoinformatics: algorithms, methods, and technologies. Moscow State University of Forest Publisher, Moscow, 430 pp [in Russian]

Burkov VD, Samartsev IE, Krapivin VF, Kovalev VI (2008) Application of technology of flexible information-modelling systems for synthesis of fiber optical devices and metrological systems in nano-technologies. Ecol Syst Devices 11:4–10 [in Russian]

Byakola T (2000) Technological options and policy measures for methane mitigation in Uganda—possibilities and limitations. Proceedings of the second international methane mitigation conference, 18–23 June 2000, Novosibirsk, Novosibirsk State University, Novosibirsk, pp 95–100

Cai W, Whetton PH (2002) Evidence for a time-varying pattern of greenhouse warming in the Pacific ocean. Geophys Res Lett 27(16):2577–2580

Cakaj S (2010) Modeling, simulation and optimization: tolerance and optimal control. In Tech, Shanghai, China, 312 pp
Callaghan J, Power SB (2011) Variability and decline in the number of severe tropical cyclones making land-fall over eastern Australia since the late nineteenth century. Clim Dyn 37(3–4):647–662
Callaghan TV, Johansson M, Key J, Prowse T, Ananichava M, Klepikov A (2011) Feedbacks and interactions: from the arctic cryosphere to the climate system. Ambio 40(1):75–86
Callinicos A (2001) Against the third way. Polity Press, London, 160 pp
Carvalho CM, Lopes HF (2007) Simulation-based sequential analysis of Markov switching stochastic volatility models. Comput Stat Data Anal 51:4526–4542
Caswell H (2006) Matrix population models. Sinauer Associates, Inc., Sunderland, 722 pp
Chang C-P, Zhang Y, Li T (2000) Interannual and interdecadal variations of the East Asian summer monsoon and tropical Pacific SSTs. Part 1: roles of the subtropical ridge. J Clim 13:4310–4325
Chapin FS III, Callaghan TV, Bergeron Y, Fukuda M, Johnstone JF, Juday G, Zimov S (2004) Global change and the boreal forest: thresholds, shifting states, or gradual change? Ambio 33:361–365
Charvat K, Alberts M, Harakova S (eds) (2010) INSPIRE, GMES and GEOSS activities, methods and tools towards a single information space in Europe for the environment. Wireless Information, London, 170 pp
Chen LJ, Cihlar J, Goulden ML (1999) Daily canopy photosynthesis model through temporal and spatial scaling for remote sensing applications. Ecol Model 124:99–119
Chen W, Chen J, Cihlar J (2000) An integrated terrestrial ecosystem carbon-budget model based on changes in disturbance, climate, and atmospheric chemistry. Ecol Model 135(1):55–79
Cheremisin MV (2013) Information processing delivered by the complex of active and passive sensors of the Earth remote sensing under aerospace monitoring. PhD thesis, Moscow University of Forest, Moscow, 137 pp [in Russian]
Chernavsky DS (ed) (2004) Recognition, autodiagnostics, thinking: synergetics and human science. Radiotekhnika Publisher, Moscow, 272 pp [in Russian]
Chin J (2011) Coral reefs. Roaring Brook Press, New York, 40 pp
Chobadian A, Goddard AJH, Gosman AD (1985) Numerical simulation of coastal internal boundary layer developments and a comparison with simple models. Air pollution modeling and its application. Plenum Press, New York, pp 343–358
Chock DP, Song Q, Hass H, Schell B, Ackerman I (2003) Comment on "Control of fossil-fuel particulate black carbon and organic matter, possibly the most effective method of slowing global warming" by M.Z. Jaconson. J Geophys Res 108(D24):ACH12/1–ACH12/3, ACH13/1–ACH13/4
Chou M-D, Chan PK, Wang M (2002) Aerososl radiative forcing derived from Sea-WiFS-retrieved aerosol optical properties. J Atmos Sci 59(3):748–757
Christensen JH (1997) The Danish Eulerian hemispheric model—a three-dimensional air pollution model used for the Arctic. Atmos Environ 31(24):4169–4191
Christian TJ, Kleiss B, Yokelson RJ, Holzinger R, Crutzen PJ, Hao WM, Saharjo BH, Ward DE (2003) Comprehensive laboratory measurements of biomass-burning emissions: 1. Emissions from Indonesian, African, and other fuels. J Geophys Res 108(D23):ACH3/1–ACH3/13
Christy JR, Spenser RW (2003) Reliability of satellite data sets. Science 301(5636):1046–1047
Chronis TG, Anagnostou EN (2003) Error analysis for a long-range lightning monitoring network of ground-based receivers in Europe. J Geophys Res 108(D24):ACL8/1–ACL8/10
Chukhlantsev AA (1986) Scattering and absorption of microwave radiation by elements of plants. Radio Eng Electron Phys 31:1095–1104 [in Russian]
Chukhlantsev AA (1992) Microwave emission and scattering from vegetation canopies. J Electromag Waves Appl 6(8):1043–1068
Chukhlantsev AA (2002) Microwave attenuation by coniferous branches. For News 1(21):110–112 [in Russian]

Chukhlantsev AA (2006) Microwave radiometry of vegetation canopies. Springer, Berlin, 287 pp
Chukhlantsev AA, Shutko AM (1988) Microwave radiometry of the earth's surface: effect of vegetation. Res Earth Space 2:67–72 [in Russian]
Chukhlantsev AA, Vinokurova SI (1991) On application of radar equipment for remote sensing of vegetation and soil. Res Earth Space 4:21–26 [in Russian]
Chukhlantsev AA, Shutko AM, Golovachev SP (2003) Attenuation of electromagnetic waves by vegetation covers. J Radio Electron 48(11):1285–1311 [in Russian]
Chukhlantsev AA, Golovachev SP, Krapivin VF, Shutko AM (2004) A remote sensing-based modeling system to study the Aral/Caspian water regime. In: Proceeding of the 25th ACRS, vol 1. 22–26 Nov 2004, Chiang Mai, Thailand, pp 506–511
Church J, White N (2006) A 20th century acceleration in global sea-level rise. Geophys Res Lett 33(L01602). doi:10.1029/2005GL024826
Churin GY (2005) Legal mechanisms for transition of society to sustainable development on global and regional levels. In: Troyan VN, Dementiev IA (eds). Sustainable development and ecological management, St-Petersburg State University Publisher, St.-Petersburg, pp 369–372 [in Russian]
Claussen M, Brovkin V, Ganopolski A, Kubatzki C, Petoukhov V, Rahmstorf S (1999) A new model for climate system analysis: outline of the model and application to paleoclimatice simulations. Environ Model Assess 4(4):209–216
Cocknell C, Koeberl C, Gilmour I (eds) (2006) Biological processes associated with impact events. Springer, Berlin, 376 pp
Coconea G (2000) Methane gas emissions from the Romanian natural gas transport system. In: Proceedings of the second international methane mitigation conference. 18–23 June 2000, Novosibirsk, Novosibirsk State University Publisher, Novosibirsk, pp 297–302
Collatz GJ, Berry JA, Farquhar JA, Pierce J (1990) The relationship between the rubisco reaction mechanism and models of leaf photosynthesis. Plant Cell Environ 13:219–225
Collatz GJ, Ball GJ, Grivet JT, Berry JA (1991) Physiological and environmental regulation of stomatal conductance, photosynthesis and transpiration: a model that includes a laminar boundary layer. Agric For Meteorol 54:107–136
Collatz GJ, Ribas-Carbo M, Berry JA (1992) Coupled photosynthesis stomatal conductance model for leaves of C_4 plants. Aust J Plant Physiol 19:519–538
Collins M, Senior CA (2002) Projections of future climate change. Weather 57(8):283–287
Collins WJ, Stevenson DS, Johnson CE, Derwent RG (1998) A simulation of long-range transport of CFCs in the troposphere using a 3-D global Lagrangian model with 6-hourly meteorological fields. Air Pollut Model Appl, XII, pp 227–235
Conant WC, Nenes A, Seinfeld JH (2002) Black carbon radiative heating effects on cloud microphysics ans implications for the aerosol indirect effect. 1. Extended Köhler theory. J Geophys Res 107(D21):AAC23/1–AAC23/9
Conant RT, Ryan MG, Ran GI, Gren A, Birge HE, Davidson EA, Eliasson PE, Evans SE, Frey SD, Giardina CP, Hopkins FM, Nen RH, Kirschbaum MUF, Lavallee JM, Feld JL, Parton WJ, Steinweg JM, Wallenstein MD, Wetterstedt JAM, Bradford MA (2011) Temperature and soil organic matter decomposition rates—synthesis of current knowledge and a way forward. Global Change Biol 17:3392–3404
Condie KC (2005) Earth as an evolving planetary system. Elsevier Academic, Burlington, 461 pp
Cook J, Highwood EJ (2004) Climate response to tropospheric absorbing aerosols in an intermediate general-circulation model. Quart J Roy Meteorol. Soc 130(296A):175–191
COPES (2004) The world climate research programme strategy 2005–2015. Coordinated observation and prediction of the Earth system (COPES). A discussion document. WMO, Geneva, p 31
Corcoran PP (ed) (2005) The Earth charter in action: toward a sustainable world. KIT Publisher, Amsterdam, 192 pp
Corliss JO (2002) Biodiversity and biocomplexity of the protista and an overview of their significant roles in maintenance of our biosphere. Acta Protozool 41:199–219

Cozzi L (2003) World energy outlook insights: global energy investment outlook. IEA, Paris, 20 pp
Cracknell AP (2009) Sustainabilityno hope! or sustainability no hope? In: Cracknell AP, Krapivin VF, Varotsos CA (eds) Global climatology and ecodynamics: anthropogenic changes to planet Earth. Springer/Praxis, Chichester, pp 479–511
Cracknell AP, Varotsos CA (2007) Fifty years after the first artificial satellite: from Sputnik 1 to ENVISAT. Int J Remote Sens 28:2071–2072
Cracknell AP, Krapivin VF, Varotsos CA (eds) (2009a) Global climatology and ecodynamics: anthropogenic changes to planet Earth. Springer/Praxis, Chichester, 518 pp
Cracknell AP, Krapivin VF, Varotsos CA (2009b) Sustainable development problems in the context of global ecoinformatics. In: Cracknell AP, Krapivin VF, Varotsos CA (eds) Global climatology and ecodynamics. Springer/Praxis, Chichester, pp 445–477
Crawford J, Davis D, Olson J, Chen G, Liu S, Fuelberg H, Hannan J, Kondo Y, Anderson B, Gregory G, Sachse G, Talbot R, Viggiano A, Heikes B, Snow J, Singh H, Blake D (2000) Evolution and chemical consequences of lightning-produced NO_x observed in the North Atlantic upper atmosphere. J Geophys Res 105(D15):19,795–19,809
Crichton M (2005) State of fear. Harper-Collins, London, 503 pp
Crowley TJ (2000) Causes of climate change over the past 1000 years. Science 289(5477):270–277
Daanen RP, Misra D, Epstein H, Walker D, Romanovsky V (2008) Simulating non-sorted circle development in arctic tundra ecosystems. J Geophys Res 113(G03S06):1–10
Dally JW, Riley WF, McConnell KG (1984) Instrumentation for engineering measurements. Wiley, New York, 576 pp
Dalton LA, Dougherty ER (2011) Bayesian minimum mean-square error estimation for classification error—part II: linear classification of Gaussian models. IEEE Trans Signal Process 59(1):130–144
Degermendzhi AG (1987) Structural laws of mixed cultures under modelling the water ecosystems. Doctor Dissertation Thesises. Institute of Biophysics of Russian Academy of Sciences, Krasnoyarsk, Russia. 511 pp (in Russian)
Degermendzhi AG (2009) New directions in biophysical ecology. In: Cracknell AP, Krapivin VF, Varotsos CA (eds) Global climatology and ecodynamics. Springer/Praxis, Chichester, pp 379–396
Degermendzhi AG, Bartsev SI, Gubanov VG, Erokhin DV, Shevirnogov AP (2009) Forecast of biosphere dynamics using small-scale models. In: Cracknell AP, Krapivin VF, Varotsos CA (eds) Global climatology and ecodynamics. Springer/Praxis, Chichester, pp 241–300
Del Frate F, Ferrazzoli P, Schiavon G (2003) Retrieving soil moisture and agricultural variables by microwave radiometry using neural networks. Remote Sens Environ 84(2):174–183
Delworth TL, Knutson TR (2000) Simulation of early 20th century global warming. Science 287 (5461):2246–2250
DeMaria M, Mainelli M, Shay LK, Knaff JA, Kaplan J (2005) Further improvements to the statistical hurricane intensity prediction scheme (SHIPS). Weather Forecast 20:531–543
Dementjeva TV (2000) Emission of gases from peat-bog ecosystems. Proceedings of the second international methane mitigation conference, 18–23 June 2000, Novosibirsk, Novosibirsk State University Publishing, Novosibirsk, pp 223–226
Demirchian KS, Demirchian KK, Danilevich YB, Kondratyev KY (2002) Global climate warming, energetics and geopolicy. Ann RAS Energ 3:221–235 [in Russian]
Demirchian KS, Kondratyev KY (2004) Global carbon cycle and climate. Trans Russ Geogr Soc 136(1):16–25 [in Russian]
Dengler J, Ewald J, Kühn I, Peet RK (2011) Ecoinformatics and global change—an overdue liaison. J Veg Sci 22:577–581
Despres A, Rancillac F, Bouville A (1986) First result of the data processing of the VIth European campaign on remote—sensing of air pollution. In: de Wispelaere C, Schiermeier FA, Gillani NV (eds) Air pollution modeling and its application V. Plenum Press, New York, pp 371–382
Dessler A (2011) Introduction to modern climate change. Cambridge University Press, Cambridge, 240 pp
Devkota SR (2005) Is strong sustainability operational? Ex Nepal. Sustain Dev 13(5):297–310

DeWitt DP, Nutter GD (eds) (1988) Theory and practice of radiation thermometry. Wiley, New York, 1137 pp
Diamond J (2005) Collapse: how societies choose to fall or succeed. Viking, London, 575 pp
Dickson R (1999) All change in the Arctic. Nature 397:389–391
Dimov I, Georgiev K, Ostromsky T, Zlatev Z (2004) Computational challenges in the numerical treatment of large air pollution models. Ecol Model 179(2):187–203
Dodds F, Pippard T (eds) (2005) Human and environmental security. An agenda for change. Earthscan, London, 270 pp
Dong J, Kaufmann RK, Myneni RB, Tucker CJ, Kauppi PE, Liski J, Buermann W, Alexeyev V, Hughes MK (2003) Remote sensing estimates of boreal and temperate forest woody biomass: carbon pools, sources, and sinks. Remote Sens Environ 84(3):393–410
Douglass DH, Blackman EG, Knox RS (2004) Temperature response of Earth to the annual solar irradiance cycle. Phys Lett A 323(3–4):315–322
Drewry D, Orheim O (eds) (2000) Arctic science summit week. 25–29 April 1999. Tromsø, norway. Joint-science day: marine climate of the arctic. Norsk Polarinstitutt, Intern Rapport, Tromsø, vol 3, 32 pp
Dulepova EP (2002) Comparative bioproductivity of macroecosystems of Far-East seas. Pacific Scientific-Research Center for Fishery Publisher, Vladivostok, 273 pp [in Russian]
Duncan BN, Martin RV, Staudt AC, Yevich R, Logan J (2003) Inter-annual and seasonal variability of biomass burning emissions by satellite observations. J Geophys Res 108(D2): ACH1/1–ACH1/11
Dunion JP, Velden CS (2004) The impact of the Saharan air layer on Atlantic tropical cyclone activity. Bull Am Meteorol Soc 85:353–365
Ebel A, Eckhardt S, Feldmann H, Forster C, Jäger H, Hermann J, James P, Kanter H-J, Kerschgens M, Kreipl S, Memmesheimer M, Scheel H-E, Stohl A, Trickl T, Vogelman H (2004) Atmospheric long-range transport and its impact on the trace-gas concentrations in the free troposphere over Central Europe (ATMOFAST). AFO 2000 Newsl 9:7–10
Efremov DF, Sapozhnikov AP (1997) Far-East forests biodiversity and succession dynamics. Proceedings of the IGBP siberian transect workshop "spatial-temporal dimension of high-latitude ecosystem change". 1–7 Sept 1997, Krasnoyarsk, V.N. Sukachev Institute of Forest of RAS, Krasnoyarsk, pp 13–14
Ehrenfreund P, Spaans M (2008) Cosmic carbon chemistry. ACS Symp Ser 981:232–245
Ehlers E, Kraft T (eds) (2005) Earth system science in the anthropocene. Emerging issues and problems. Springer, Heidelberg, 300 pp
EIA (2005) International energy outlook 2005. DOE/EIA-0484, Washington 194 pp
Elinson MI (1983) On the microchip building-block base for one class of cybernetic systems. Microelectronics 12(5):403–412 [in Russian]
Elsner JB, Jagger TH, Hodges RE (2010) Daily tropical cyclone intensity response to solar ultraviolet radiation. Geophys Res Lett 37(L09701). doi:10.1029/2010GL043091
Emerson S, Hedges J (2008) Chemical oceanography and the marine carbon cycle. Cambridge University Press, Cambridge, 461 pp
Emery WJ, Thompson RE (2001) Data analysis methods in physical oceanography. Elsevier, Berlin, 638 pp
Engman ET, Chauhan N (1995) Status of microwave soil moisture measurements with remote sensing. Remote Sens Environ 51(1):189–198
Enz R (ed) (2006) Natural catastrophes and man-made disasters 2005: high earthquake casualties, new dimension in windstorm losses. Swiss Reinsurance Company Publisher, Zurich, 40 pp
EPA (2001) Non-CO2 greenhouse gas emissions from developed countries-1990-2010. EPA-430-R-01-007, U.S. Environmental Protection Agency, Washington, 132 pp
EPA (2005) Proceedings of EPA Science Forum 2005: Collaborative Science for Environmental Solutions, 16–18 May 2005? Washington, US EPA, Washington, DC, 162 pp
EPA (2006) Inventory of US greenhouse gas emissions and sinks: 19902004 US. Environmental Protection Agency, Report No. EPA 430-R-06-002, Washington, 459 pp

Eriksson E (1963) The yearly circulation of sulfur in nature. J Geophys Res 68:401–403
Essam JW (1980) Percolation theory. Rep Prog Phys 43:833–912
Essex C, McKitrick R (2002) Taken by storm. The troubled science, policy and politics of global warming. Key Porter Books, Toronto, 320 pp
Everett JT, Fitzharris BB (eds) (2001) The regional impacts of climate change. Chapter 3: the arctic and antarctic. The IPCC Report—2001 (http://www.grida.no/climate/ipcc/regional/042.htm)
Faletau T, Leavai P, McFadzien D, Ronneberg E (2012) Strategies and on the ground options for climate change adaptation and disaster risk management in the Pacific. SPREP, Bangkok, pp 1–9
Fang H, Liang S (2003) Retrieving leaf area index with a neural network method: simulation and validation. IEEE Trans Geosci Remote Sens 41(9):2052–2062
Fang J, Chen A, Peng C, Zhao S, Ci L (2001) Changes in forest biomass carbon storage in China between 1949 and 1998. Science 292:2320–2322
FAO (2001) Global forest resources assessment 2000. Main report. FAO Forestry Paper, Rome, vol 140, 511 pp
Farquhar GD, von Caemmerer S, Berry JA (1980) A biochemical model of photosynthetic CO_2 fixation in leaves of C_3 species. Planta 149:78–90
Fasham MJR (ed) (2003) Ocean biogeochemistry. Springer, Berlin, 283 pp
Fedorov KN, Ginsburg AI (1988) The subsurface layer of the ocean. Gidrometeoizdat, Leningrad, 304 pp [in Russian]
Feichter J, Sausen R, Grassl H, Fiebig M (2003) Comment on "Control of fossil-fuel particulate black carbon and organic matter, possibly the most effective method of slowing global warming" by M.Z. Jacobson, J Geophys Res, 108(S24), ACH10/1–ACH10/2; ACH11/1–ACH11/4
Feldstein SB, Franzke C (2006) Are the North Atlantic oscillation and the Northern annular mode distinguishable? J Atmos Sci 63:2915–2930
Feller IC, Lovelock CE, Berger U, McKee KL, Joye SB, Ball MC (2010) Biocomplexity in mangrove ecosystems. Annu Rev Mar Sci 2:395–417
Ferm M, Hultberg H (1999) Dry deposition and internal circulation of nitrogen, sulfur and base cations to a coniferous forest. Atmos Environ 33(27):4421–4430
Ferrazzoli P, Guerriero L (1996) Passive microwave remote sensing of forests: a model investigation. IEEE Trans Geosci Remote Sens 34(2):433–443
Ferrazzoli P, Paloscia S, Pampaloni P (1992) Sensitivity of microwave measurements to vegetation biomass and soil moisture content: a case study. IEEE Trans Geosci Remote Sens 30(4):750–756
Ferris E, Petz D, Stark C (2013) The year of recurring disasters. A review of natural disasters in 2012. The Brookings Institution, London, 101 pp
Field CB (2001) Sharing the garden. Science 294(5551):2490–2491
Field CB, Raupach MR (eds) (2004) Global carbon cycle: integrating humans, climate, and the natural World. Island Press, Washington, 584 pp
Field CB, Raupach MK, MacKenzie SH (2004) The global carbon cycle: integrating humans, climate, and the natural World. Island Press, Washington, 568 pp
Field JG, Hempel G, Summerhayer CP (eds) (2002) Oceans 2020: science trends and the challenge of sustainability. Island Press, Washington, 296 pp
Filatov NN (2004) Climate of karelia: variability and impact on water objects and watersheds. Karel Sci. Centre of RAS, Petrozavodsk, 224 pp [in Russian]
Fink AH, Schrage JM, Kotthaus S (2010) On the potential causes of the nonstationary correlations between West African precipitation and Atlantic hurricane activity. J Clim 23(20):5437–5456
Flood J (1995) Indicators for the implementation and monitoring of Agenda. Habitat 1(5):13–14
Fogel LJ, Owens AJ, Walsh MJ (1966) Artificial intelligence through simulated evolution. Wiley, New York, 230 pp
Folland C, Frich P, Basnett T, Rayner N, Parker D, Horton B (2000) Uncertainties in climate datasets – a challenge for WMO. WMO Bull 49(1):59–68
Folland CK, Karl TR, Salinger MJ (2002) Observed climate variability and change. Weather 57 (3):269–278

Font J, Lagerloef GSE, Vine DL, Camps A (2002) Open issues for the soil moisture and ocean salinity (SMOS) satellite mission salinity retrieval. Proceedings of the first results workshop on EuroSTARRS, WISE, LOSAC compaigns, 4–6 Nov 2002, Touluse, France. European Space Agency, The Netherlands, pp 7–16

Fourie G, Djolov G, Syrakov D, Pienaar K, Prodanova M (2004) Modelling long-range transport and chemical transformation of pollutants in the Southern Africa region. Proceedings of 9th international conference on harmonisation within atmospheric dispersion modelling for regulatory purposes, 1–4 June 2004, Garmisch-Partenkirchen, Germany, pp 320–324

FRA (2011) The global forest resources assessment 2010. FAO, Rome, 340 pp

Frankham R (1995) Inbreeding and extinction: a threshold effect. Conserv Biol 9(4):792–799

Frederick KD (1994) Integrated assessment of the impacts of climate change on natural resources. An introductory essay. Clim Change 28(102):1–14

Freeman PK (2000) Infrastructure, natural disasters, and poverty. Proceedings of the euro-conference on global change and catastrophe risk management: flood risks in Europe. Luxemburg, Austria, 6–9 June 2000, IIASA, Luxemburg, pp 234–238

Friedi MA, Mclver DK, Hodges JCF, Zhang XY, Muchoney D, Strahler AH, Woodcock CE, Gopal S, Schneider A, Cooper A, Baccini A, Gao F, Schaaf C (2002) Global land cover mapping from MODIS: algorithms and early results. Remote Sens Environ 83(1–2):287–302

Frederick KD (1994) Integrated assessment of the impacts of climate change on natural resources. An Introductory essay. Clim Change 28(102):1–14

Friedman TL (2005) The world is flat: a brief history of the twenty-first century. Farrar, Straus and Giroux Publishing, New York, 496 pp

Friedrich R (ed) (2001) Generation and evaluation of emission data. GENEMIS Annual Report 2001, Stuttgart, Germany, 87 pp

Fuller M (1991) Forest fires: an introduction to wildland fire behavior, management, firefighting, and prevention. Wiley, New York, 238 pp

Fung I, John J, Lerner J, Matthews E, Prather M, Steele LP, Fraser PJ (1991) Three-dimensional model synthesis of the global methane cycle. J Geophys Res 96:13033–13065

Furiayev VV (1996) The role of fires in the process of forest formation. Science Publisher, Novosibirsk, 252 pp [in Russian]

Furth JH (1995) Professor James on the theory of monetary policy. J Econ 25(1–2):199–203

G8 (2003) Science and technology for sustainable development, a g8 action plan, 2 June 2003. Proc Natl Acad Sci (USA) 100(14):8059–8061

Gale J, Freund P (2000) Reducing methane emissions to combat global climate change: the role Russia can play. Proceedings of the second international methane mitigation conference, Novosibirsk, 18–23 June 2000. Novosibirsk State University Publisher, Novosibirsk, pp 73–80

Galitsky VV (1985) Horizontal structure and dynamics of an aged vegetation community: mathematical modeling. In: Svirezhev YM (ed) Mathematical modeling of biogeocenotic processes. Science Publishing House, Moscow, 1985, pp 59–69 [in Russian]

GAR (2011) Global assessment report on disaster risk reduction: revealing risk, redefining development. United Nations Information Centre, New York. http://www.preventionweb.net/english/hyogo/gar/2011/en/home/download.html

Garcia-Barrón L, Pita MF (2004) Stochastic analysis of time series of temperatures in the south-west of the Iberian Peninsula. Atmosfera 17(4):225–244

Gardner JS (2002) Natural hazards risk in the Kullu district, Himachal pradesh, India. Geogr Rev 92:172–177

Garsey E, McGlade J (eds) (2006) Complexity and co-evolution: continuity and change in socio-economic systems. Edward Elgar Publishing, Cheltenham, 219 pp

Gasiewski AJ, Kunkee DB (1994) Polarized microwave emission from water waves. Radio Sci 29 (6):1449–1466

Gbondo-Tugbawa SS, Driscoll CT (2002) Evaluation of the effect of future controls on sulfur dioxide and nitrogen oxide emissions on the acid-base status of a northern forest ecosystem. Atmos Environ 36(10):1631–1643

George RE (2002) Socioeconomic democracy: an advanced socioeconomic system (praeger studies on the 21st century). Praeger Publication, Santa Barbara, 328 pp

GCOM (2008) Global change observation mission. Earth Observation Research Center, Japan Aerospace Exploration Agency, Tokyo, 10 pp

Girnis PR, Orphan VJ, Hallam SJ, DeLong EF (2003) Growth and methane oxidation rates of anaerobic methanotrophic *archaea* in a continuous-flow bioreactor. Appl Environ Microbiol 69:5472–5482

Golitsin GS, Vasiliev AA (eds) (2001) Natural disasters of the Russia: hydrometeorological disasters. Kruk Publisher, Moscow, 295 pp [in Russian]

Golovachev SP, Krapivin VF, Lalaian MV, Shutko AM, Chukhlantsev AA (2004) Global mapping of microwave attenuation by the vegetation canopy. Probl Environ Nat Resour 1:59–69 [in Russian]

Goody RM, Jung YL (1989) Atmospheric radiation. Oxford University Press, New York, 519 pp

Goody RM (2002) Observing and thinking about the temperature. Annu Rev Environ 27:1–20

Gorshkov VG, Gorshkov VV, Makarieva AM (2000) Biotic regulation of the environment. Springer/Praxis Publishing, Chichester, 364 pp

Gorshkov VG, Makarieva AM, Mackey B, Gorshkov VV (2002) Biological theory and global change science. Global Change Newsl 48:11–14

Gorshkov VG, Esenin BK, Karibayeva KN, Kurochkina LY, Losev KS, Makarieva AM, Shukurov ED (2004) Scientific fundamental of strategic directions of the nature protection policy. Ecol Edu 1:29–31 [in Russian]

Gorstko AB (1979) Mathematical model of the Azov sea ecosystem. Znanie Press, Moscow, 64 pp [in Russian]

Gorstko AB, Ougolnitskiy GA (1990) Introduction to the modelling of ecologo-economic systems. Rostov-Don State University Press, Rostov-Don, 111 pp [in Russian]

Goto Y, Suzuki S (2013) Estimates of carbon emissions from forest fires in Japan, 1979–2008. Int J Wildland Fire 22(6):721–729

Graham B, Guyon P, Maenhaut W, Taylor PE, Ebert M, Matthias-Maser S, Mayol-Bracero OL, Gedoi RHM, Artaxo P, Meixner FX, Lima Moura MA, D'Almeida Rocha CHF, van Grieken R, Glovsky MM, Flagan RC, Andreae M (2003) Composition and diurnal variability of the natural Amazonian aerosol. J Geophys Res 108(D24):AAC5/1–AAC5/16

Grankov AG (1994) Passive radiometric diagnosis of vertical turbulent heat fluxes through the ocean surface from satellites. In: Donelan MA (ed) The air-sea interface. The University of Toronto Press, Toronto, pp 55–58

Grankov AG (2011) (ed) Elaboration of technologies for diagnosis of tropical hurricanes beginning in oceans with remote sensing methods. Final ISTC Report No. 3827, ISTC, Moscow, 99 pp

Grankov AG, Milshin AA (2004) Interconnection of radio-emission by the ocean-atmosphere system with heat and dynamic processes on the interface. Phizmatlit, Moscow, 166 pp [in Russian]

Grankov AG, Milshin AA (2009) Laboratory investigations of the response of water surface internal radiation on variations in enthalpy of its upper layer. Probl Environ Nat Resour 8:67–71 (in Ruaaian)

Grankov AG, Milshin AA (2010) Microwave radiation of the ocean-atmosphere: boundary heat and dynamic interaction. Springer, Berlin, 250 pp

Grankov AG, Soldatov VYu (2005) Evaluation of radiobrightness response function of the ocean-atmosphere system on variations in heat fluxes. Probl Environ Nat Resour 12:45–53 [in Russian]

Grankov AG, Milshin AA, Krapivin VF, Golovachev SP (2006) Analysis of the ocean and atmosphere roles in their heat interaction with use of satellite and vessel. Proceedings of the ISRS 2006 PORSEC, BEXCO Busan, Korea, 2–4 Nov 2006, pp 4–5

Grassl H (2000) Status and improvements of coupled general circulation models. Science 288:1991–1997

Gray WM, Neumann CJ, Tsui TL (1991) Assessment of the role of aircraft reconnaissance on tropical cyclone anlysis and forecasting. Bull Am Meteorol Soc 72:1867–1883

Griggs DJ, Noguer M (2002) Climate change 2001: the scientific basis. contribution of working group 1 to the third assessment report on the intergovernmental panel on climate change. Weather 57(8):267–269

Gregory D, Morcrette J-J, Jakob C, Beljaars ACM, Stockdale T (2000) Revision of convection, radiation and cloud schemes in the ECMWF Integrated Forecasting System. Q J Roy Meteorol Soc 126:1685–1710

Grigoryev AA, Kondratyev KY (2001) Ecodinamics and geopolitics. Ecological catastrophes, vol II. Research Centre for Ecological Safety Press, St-Petersburg, 688 pp [in Russian]

Grigoryev AA, Kondratyev KY (2004) Global urbanization, 1: general laws. Trans Russ Geogr Soc 136(4):1–8 [in Russian]

Grigoryev AA, Kondratyev KY (2005) Natural and anthropogenic forest fires: ecodynamics component and natural disasters. Trans Russ Geogr Soc 137(1):3–40 [in Russian]

Grossi P, Kunreuther H (eds) (2005) Catastrophe modeling: a new approach to managing risk, vol 25. Springer, New York, 252 pp

Guha A, Jacobs JM, Jackson TJ, Cosh MN, Hsu E-C, Judge J (2003) Soil moisture mapping using ESTAR under dry conditions from the Southern Greet Plains Experiment (SGP99). IEEE Trans Geosci Remote Sens 41(10):2392–2397

Guha-Sapir D, Hyois P, Below R (2013) Annual disaster statistical review 2012. CRED, Brussels, Belgium, 43 pp

Guista ML, Kambhampati US (eds) (2006) Critical perspectives on globalization. Edward Elgar, London, 656 pp

Gulyaev YuV, Krapivin VF, Bukatova IL (1987) On the way to evolutionary informatics. Bull USSR Acad Sci 11:53–61 [in Russian]

Gulyaev YV, Bukatova IL, Golubeva LN, Krapivin VF (1989) Evolution informatics and intellectual special processors. Preprint 6(507), KIREE Publisher, Moscow, 38 pp

Gulyaev YV, Shutko AM, Golovachev SP, Chukhlantsev AA, Krapivin VF, Pljushchev VA, Manakov VJ, Sidorov IA (2005) Advanced technology for assessment of soil hydrological regimes using multichannel passive (radiometric) and active (sar) microwave measurements. Proceedings of the 26-th ACRS-2005, 7–11 Nov 2005, Hanoi, Vietnam, AARS, Hanoi, pp 100–107

Gulyaev YV, Krapivin VF, Cherepenin VA, Chukhlantsev AA, Shutko AM, Potapov II (2009) Radioelectronics methods in the environmental investigations. Probl Environ Nat Resour 1:3–150 [in Russian]

Gurjar BR, Leliveld J (2005) New directions: megacities and global change. Atmos Environ 39 (2):391–393

Gutberlet J (2003) Cities, consumption and the generation of waste. Aviso 11:12–19

Gyde LH (2012) Definitions of forest, deforestation, afforestation, and reforestation. V.A. Forest Information Services. http://home.comcast.net/~gyde/DEFpaper.htm

Haarbrink R, Krapivin VF, Krisilov A, Krisilov V, Novitchikhin EP, Shutko A, Sidorov I (2011) Intelligent data processing in global monitoring for environment and security. ITHEA, Sofia-Kiev, 410 pp

Hackler JL, Houghton RA, Lawrence KT (1999) The US carbon budget: contributions from land-use change. Science 285(5427):574–577

Hall DK, Riggs GA, Salomonson VV (1995) Development of methods for mapping global snow cover using moderate resolution imaging spectroradiometer data. Remote Sens Environ 54:127–140

Hansell DA, Carlson CA (eds) (2002) Biogeochemistry of marine dissolved organic matter. Elsevier, Amsterdam, 469 pp

Hansen JE (1998) Book review of sir John Houghton's global warming: the complete briefing. J Atmos Chem 30:409–412

Hansen JE, Fung I, Lacis A, Rind D, Lebedeff S, Ruedy R, Russell G, Stone P (1988) Global climate changes as forecast by Goddard institute for space studies three-dimensional model. J Geophys Res 93:9341–9364
Hansen JE, Sato M (2001) Trends of measured climate forcing agents. Proc Nat Acad Sci (USA) 98(26):14,778–14,783
Hansen JE, Sato M, Ruedy R (1997) Radiative forcing and climate response. J Geophys Res 102:6831–6864
Hansen JE, Sato M, Lacis A, Ruedy R, Tegen L, Matthews E (1998) Climate forcing in the Industrial era. Proc Nat Acad Sci (USA) 95(22):12,753–12,758
Hansen JE, Ruedy R, Sato M, Lo K (2010) Global surface temperature change. Rev Geophys 48 (RG4004). doi:10.1029/2010RG000345
Hansen J, Ruedy R, Glascoe J, Sato M (1999) GISS analysis of surface temperature change. J Geophys Res 104(D24):30,997–31,022
Hansen J, Ruedy R, Sato M, Lo K (2013) NASA GISS surface temperature (gistemp) analysis. In trends: a compendium of data on global change. Carbon Dioxide Information Analysis Center, Oak Ridge National Laboratory, US Department of Energy, Oak Ridge, Tennessy, USA. doi: 10.3334/CDIAC/cli.001
Hansen MC, DeFries RS, Townshend JRG, Sohlberg R, Dimiceli C, Carroll M (2002) Towards an operational MODIS continuous field of percent tree cover algorithm: examples using AVHRR and MODIS data. Remote Sens Environ 83(1–2):303–319
Hanson RB, Ducklow HW, Field JG (2000) (eds) The changing ocean carbon cycle. Cambridge University Press, Cambridge, 505 pp
Haque CE (2005) Mitigation of natural hazards and disasters: international perspectives. Springer, Heidelberg, 200 pp
Hacenauer H (2012) Concepts within ecosystem modeling. In: Pötzelsberger E, Mäkelä A, Mohren G, Palahí M, Tomé M, Hasenauer H (eds) Modelling forest ecosystems—concepts, data and application. Proceedings of the COST FP0603 Spring School, 9–13th May 2011, Kaprun, Austria. University of Natural Resources and Life Sciences, Viena, 2012, pp 7–13
Heil P, Allison I (1999) The pattern and variability of Antarctic sea-ice drift in the Indian ocean and western pacific sectors. J Geophys Res [Oceans] 104(C7):15789–15802
Helfrich SR, McNamara D, Ramsay BH, Baldwin T, Kasheta T (2007) Enhancements to, and forthcoming developments in the interactive multisensor snow and ice mapping system (IMS). Hydrol Processes 21:1576–1586
Hellebrandt HJ, Kern J, Scholz V (2003) Long-term studies on greenhouse gas fluxes during cultivation of energy crops on sandy soils. Atmos Environ 37(12):1635–1644
Hodgkin AL, Huxley AF (1952) A quantitative description of membrane current and its application to conduction and extraction in nerve. J Physiol 117(4):500–544
Hoelzemann JJ, Schultz MG, Brasseur GP, Granier C, Simon M (2004) Global wildland fire emission model (GWEM): evaluating the use of global area burnt satellite data. J Geophys Res 109(14):D14504/1–D14504/18
Hoffmann WA (1998) Post-burn reproduction of woody plants in a neotropical savanna: the relative importance of sexual and vegetative reproduction. J Appl Ecol 35(3):422–433
Holdren JP (2003) Environmental change and human condition. Bull Am Acad Arts Sci (New York) 57(1):25–31
Holloway G, Sou T (2001) Is Arctic sea ice rapidly thinning? Ice Clim News 1:2–5
Homer-Dixon T (2008) The upside of down: catastrophe, creativity, and the renewal of civilization. Islandpress, Washington, 448 pp
Hong NS, Yen LB, Thao LH, Chien NV, Chihn LN, Son DC, Krapivin VF, Mkrtchyan FA (1992) On the problem of ecological monitoring in Vietnam. KIREE Publisher, Moscow, 46 pp [in Russian]
Hong NS, Yen LB, Son DT, Tinh LH, Luan NF, Son KT, Chi TM, Hanh ChK, Niat FH, Aleshin VA, Klimov VV, Krapivin VF, Mkrtchyan FA (1994) An investigation of the spatial-temporal structure of spectra for water objects of South Vietnam with adaptive identifier. In: Krapivin VF (ed) Ecoinformatics Problems. KIREE Publisher, Moscow, pp 128–135 [in Russian]

Houghton JT (2003) Global warming is now a weapon of mass destruction. The guardian, 28 July 2003
Houghton JT (2004) Global warming—the complete briefing. Cambridge University Press, Cambridge, 382 pp
Houghton JT, Ding Y, Griggs DJ, Noguer M, Van der Linden PJ, Dai X, Masskell K, Johnson CA (2001) Climate change 2001: the scientific basis. contribution of working group i to the third assessment report of the intergovernmental panel group on climate change. Cambridge University Press, Cambridge, 881 pp
Howard L (1820) The climate of London, deduced from meteorological observations, made at different places in the neighbourhood of the metropolis, vol 2. W. Phillips, London, 350 pp
Hrol VP (1993) Energetic balance of the North Polar area. Hydrometeo Press, St.-Petersburg, 168 pp [in Russian]
Huang S, Pollack HN, Shen P-Y (2000) Temperature trends over the past five centuries reconstructed from borehole temperatures. Nature 403:756–758
Hubbard KG, Lin X, Baker CB, Sun B (2004) Air temperature comparison between the MMTS and the USCRN temperature systems. J Atmos Oceanic Technol 21:1590–1597
Hughes TP, Baird AH, Bellwood DR, Card M, Connoly SR, Folke C, Grosberg R, Hoegh-Gulberg O, Jackson JBC, Kleypas J, Lough JM, Marshall P, Myström M, Palumbi SR, Pandolfi JM, Rosen BJ (2003) Climate change, human impacts, and the resilience of coral reefs. Science 301:529–533
Hulme M, Parry M (1998) Adapt of mitigate? Responding to climate change. Town Country Plan 67(2):50–51
van Hulst R (1979) The dynamics of vegetation: succession in model communities. Vegetatio 39 (2):85–96
Hushon JM (ed) (1990) Expert systems for environmental applications. Am Chem Soc, Washington, 232 pp
IASC (2001) IASC project catalogue. The International Arctic Science Committee, Oslo, 57 pp
Ichikawa A (2004) Global warming—the research challenges: a report of Japan's global warming initiative. Springer, Berlin, 161 pp
IEA (2005a) Energy policies of IEA countries. IEA Books, Paris, 588 pp
IEA (2005b) Key world energy statistics. IEA Books, Paris, 82 pp
IEA (2007a) Energy balances on non-OECD countries, 2004–2005. IEA Books, Paris, 482 pp
IEA (2007b) Key world energy statistics. IEA Books, Paris, 82 pp
IEA (2013a) World energy outlook 2012. www.worldenergyoutlook.org
IEA (2013b) World energy outlook 2012. IEA Books, Paris, 690 pp
IEA (2013c) Energy efficiency market report 2013—market trends and medium-term prospects. IEA Books, Paris, 279 pp
Inoue M, Isozaki H (2003) People and forest-policy and local reality in Southern Asia, the Russian far East, and Japan. Springer, Berlin, 376 pp
IPCC (2001) Third assessment report (volume 1) climate change 2001. The Scientific Basis, Cambridge University Press, Cambridge, 881 pp
IPCC (2005) IPCC workshop on new emission scenarios. Meeting report, (29 June-1 July 2005, Laxemburg, Austria). WMO/UNEP, Laxemburg, Austria. 47 pp
IPCC (2007) Climate change 2007: the physical science basis. WMO/UNEP, Geneva, 18 pp
Islam MM, Sado K, Soon CE (2005) Analysis of satellite derived sea surface temperature data for South China Sea and Java Sea. Asian J Geoinform 5(3):12–23
Ito A, Oikawa T (2002) A simulation model of the carbon cycle in land ecosystems (Sim-CYCLE): a description based on dry-matter production theory and plot-scale validation. Ecol Model 151(2–3):143–176
Ivachnenko AG, Peka PY, Vostrov NN (1984) Combined method for the water and oil fields modeling. Naukova Dumka, Kiev, Ukraina, 150 pp [in Russian]
Ivanov A (1978) An introduction to the oceanography. Mir, Moscow, 392 pp [in Russian]

Ivanov MV (1981) The global biogeochemical sulphur cycle. In: Likens GE (ed) Some perspectives of the major biogeochemical cycles. SCOPE, Luxembourg, pp 61–78

Ivanov MV, Freney JR (Eds) (1983) The global biogeochemical sulphur cycle. SCOPE 19. Wiley, Chichester, 470 pp

Ivanov-Rostovtsev AG, Kolotilo LG, Tarasiuk YF, Sherstiankin PP (2001) Self-organization and self-regulation of natural systems. Russian Geographical Society Publisher, St.-Petersburg, 216 pp [in Russian]

Jacobson MC (2000) Earth system science: from biogeochemical cycles to global change. Elsevier Academic, London, 527 pp

Jackson P (1999) Introduction to expert systems. Addison-Wesley, Michigan, 542 pp

Jackson TJ, Hsu AY, Shutko A, Tishchenko Yu, Petrenko B, Kutuza B, Armand N (2002) Priroda microwave radiometer observations in the Southern great plains 1997 hydrology experiment. Int J Remote Sens 23(2):231–248

Jacobson MZ (2002) Control of fossil-fuel particulate black carbon and organic matter, possibly the most effective method of slowing global warming. J Geophys Res 107(D19):4410–4419

Jacobson MZ (2003a) Reply to comment by J.F. Penner on "Control of fossil-fuel particulate black carbon and organic matter, possibly the most effective method of slowing global warming". J Geophys Res 108(D24):ACH15/1–ACH15/4

Jacobson MZ (2003b) Reply to comment by J. Feichter et al. on "Control of fossil-fuel particulate black carbon and organic matter, possibly the most effective method of slowing global warming". J Geophys Res 108(D24):ACH11/1–ACH11/4

Jacobson MZ (2003c) Reply to comment by D.P. Chock et al. on "Control of fossil-fuel particulate black carbon and organic matter, possibly the most effective method of slowing global warming". J Geophys Res 108(D24):ACH13/1–ACH13/4

Jagovkina SV, Karol IL, Zubov VA, Lagun VE, Reshemikov AI, Rosanov EV (2000) An estimation of the gas deposits leakage input into the total methane flux from the Weast Siberian region. Proceedings of the second international methane mitigation conference, 18–23 June 2000, Novosibirsk, Novosibirsk State University Publisher, Novosibirsk, pp 263–267

Ji Y, Stocker E (2003) Seasonal, intraseasonal, and interannual variability of global land fires and their effects on atmospheric aerosol distribution. J Geophys Res D 107(23):ACH10/1–ACY10/11

Johnson DE, Ulyatt MI (2000) Variations in the proportion of methane of total greenhause gas emissions from US and NZ dairy production systems. Proceedings of the second international methane mitigation conference, Novosibirsk 18–23 June 2000. Novosibirsk State University Publisher, Novosibirsk, pp 249–254

Johnson EA (1996) Fire and vegetation dynamics: studies from the North American boreal forest (Cambridge studies in ecology). Cambridge University Press, Cambridge, 144 pp

Johnson EA, Miyanishi K (eds) (2001) Forest fires: behavior and ecological effects. Academic Press, Washington, 594 pp

Jönsson AM, Linderson M-L, Stjernquist I, Schlyter P, Bärring L (2004) Climate change and the effect of temperature backlashes causing frost damage in *Picea abies*. Global Planet Change 44 (1–4):195–207

Jordan MI, Sejnowski TJ (2001) Graphical model: foundations of neural computation. MIT Press, Massachusetts, 421 pp

Justice CO, Bailey GB, Maiden ME, Rassol SI, Strebel DE, Tarpley JD (1995) Recent data and information system initiatives for remotely sensed measurements of the land surface. Remote Sens Environ 51(1):235–244

Kabanov MV, Matvienko GG, Tikhomirov AA, Soldatkin NP, Klyuev VV, Filinov VN, Ketkovitch AA (2000) Optical ecological monitoring. In: Klyuev VV (ed) Ecological diagnostics. Znanie, Moscow, pp 152–247 [in Russian]

Kaevitser VI, Krapivin VF, Soldatov VY (2013) A new information-modeling technology for monitoring environment in the Okhotsk Sea. Proceedings of the 28-th international symposium on Okhotsk Sea & Sea Ice, 17–21, Feb 2013, Mombetsu, Hokkaido, Japan. The Okhotsk Sea & Cold Ocean Research Association, Mombetsu, Hokkaido, Japan, pp 295–299

Kaiser J (2000) Soaking up carbon in forests and fields. Science 290(5493):922
Kalabin GV (2000) Ecodynamics of anthropogenic environment province of the North. Kola Scientific Centre, Russian Academy of Science, Apatity, 294 pp [in Russian]
Kalb D, Pansters W, Siebers H (2004) Globalization and development: themes and concepts in current research. Springer, Berlin, 203 pp
Kaminski T, Rayner PJ, Heimann M, Enting I (2001) On aggregation errors in atmospheric transport inversions. J Geophys Res 106(D5):4703–4715
Karam MA, Fung AK, Lang RH, Chauhan NS (1992) A microwave scattering model for layered vegetation. IEEE Trans Geosci Remote Sens 30(4):767–784
Karasulu B, Korukoglu S (2012) Moving object detection and tracking by using annealed background subtraction method in videos: performance optimization. Expert Syst Appl 39 (1):33–43
Kareiva P (2001) Ecoinformatics: facilitating access to existing data sets. Trends Ecol Evol 16:226–231
Kasting JF, Canfield DE (2012) The global oxygen cycle. In: Knoll AH, Canfield DE, Konhauser KO (eds) Fundamentals of geobiology. Blackwell Publishing Ltd., Oxford, pp 93–104
Keeling CD, Bacastow RB (1977) Impact of industrial gases on climate. In: Keeling CD, Bacastow RB (eds) Energy and climate. National Academy of Sciences, Washington, pp 72–95
Keigwin LD, Boyle EA (2000) Detecting holocene changes in thermohaline circulation. PNAS 97 (4):1343–1346
Keller M, Rocha H, Trumbora S, Kruijt B (2001) Investigating the carbon cycle of the Amazon forests. Glob Change Newsl 45:14–19
Kelley JJ (1987) Carbon dioxide in the Arctic environment. J Earth Sci 35(2):341–354
Kelley JJ, Krapivin VF, Popovich PR (1999a) Problems of the Arctic environment monitoring. Probl Environ Nat Resour 6:32–40 [in Russian]
Kelley JJ, Krapivin VF, Popovich PR (1999b) Problems of the Arctic monitoring. Probl Environ Nat Resour 1:2–6 [in Russian]
Kelley JJ, Krapivin VF, Vilkova LP (1992a) A conception of model for the estimation of arctic ecosystems role in global carbon budget. Proceedings of international symposium 'problems of ecoinformatics', Zvenigorod, Russia, 12–18 Dec 1992. KIREE Publisher, Moscow, pp 19–20
Kelley JJ, Rochon GL, Novoselova OA, Krapivin VF, Mkrtchyan FA (1992b) To the global geo-eco-information monitoring. Proceedings of international symposium 'problems of ecoinformatics', Zvenigorod, Russia, 12–18 Dec 1992. KIREE Publisher, Moscow, pp 3–7
Kelley RE, Chang AT, Tsang L, Foster JL (2003) A prototype AMSR-E global snow area and snow depth algorithm. IEEE Tran Geosci Remote Sens 41(2):230–242
Kelly RE, Drake NA, Barr SL (2005) Spatial modelling of the terrestrial environment. Wiley, New York, 276 pp
Kerr RA (2000) Dueling models: future US climate uncertain. Science 288:2113
Khalil MAK, Rasmussen RA, Ren L, Wang MX, Shearer MJ, Dalluge RW, Duan C-L (2000) Methane emissions from rice fields. Proceedings of the second international methane mitigation conference, Novosibirsk, 18–23 June 2000. Novosibirsk State University Publisher, Novosibirsk, pp 13–30
Khudoshina MY (1996) Automatisation of investigations on the control of resources of sea coastal regions. Ph.D. theses, Institute of Radioengineering and Electronics, Moscow, 268 pp [in Russian]
Kiehl JT, Gent PR (2004) The community climate system model, version 2. J Clim 17:3666–3682
Kirdiashev KP, Chukhlantsev AA, Shutko AM (1979) Microwave radiation of thew Earth's surface in the presence of vegetation cover. Radio Eng Electron Phys Engl Transl 24:256–264
Kirilenko AP (1990) Numerical modeling of the production process and of the water cycle of the forest ecosystems. Ph.D. theses, Computing Center of RAS, Moscow, 151 pp [in Russian]
Kiseleva SV (1990) Study of the processes of CO_2 transport under different conditions of the water-air interface. Ph.D. theses, Moscow State University Publishing, Moscow, 171 pp [in Russian]

Klimov VV (2010) Software for the experiments of optical sensing of water objects basing on the adaptive identifier. Proceedings of the ninth international symposium on ecoinformatics problems, Moscow, 9–11 Dec 2010. The Moscow Sciences Engineering A.S. Popov Society for Radio, Electronics and Communication, Moscow, 2010, pp 161–165 [in Russian]
Klien E (2008) GIGAS—GEOS, INSPIRE, and GMES an action in support. Institute for Computer Graphics Research, Darmstandt (Germany), 13 pp
Klimov VV, Kovalev VI, Krapivin VF, Mkrtchyan FA, Nitu C (1999) An expert system to diagnose pollutant spills on the water surface. Proceedings of International Conference on Control Systems and Computer Science, Bucharest, Romania, 3–5 May 1999. Politechnik University of Bucharest, Bucharest, pp 277–283
Klyuev VV (ed) (2000) Ecological diagnostics. Znanie, Moscow. 495 pp [in Russian]
Knutson TR, VcBride JL, Chan J, Emanuel K, Holland G, Landsea C, Held I, Kossin JP, Srivastava AK, Sugi M (2010) Tropical cyclones and climate change. Nat Geosci 3:157–163
Knuuttila S, Svendsen LM, Staaf H, Kotilainen P, Boutrup S, Pyhala M, Durkin M (2011) Baltic Sea pollution load compilation (PLC-5). Baltic Sea environment proceedings No. 128, Helsinki Commission, Helsinki, 220 pp
Kobayashi K (2005) Toward a sustainable Japan—corporations at work article series No. 28: a key player in information and culture (Toppan Printing Co.). Japan for sustainability—newsletter 036(August 31):1–11
Kohler MA, Richards MA (1962) Multicapacity basin accounting for predicting runoff from storm precipitation. J Geophys Res 67(13):5187–5197
Koike T (2004) The coordinated enhanced observing period—an initial step for integral global water cycle observation. WMO Bull 53(2):2–8
Kojima T (1998) The carbon dioxide problem. Integrated energy and environmental policies for the 21st century. Gordon and Breach, London, 240 pp
Kolber ZS, Plumley FG, Lang AS, Beatty JT, Blankenship RE, van Dover CL, Vetriani C, Koblizek M, Rathgeber C, Falkowski PG (2001) Contribution of aerobic photoheterotrophic bacteria to the carbon cycle in the ocean. Science 292(5526):2492–2495
Koller D, Friedman N (2009) Probabilistic graphical models: principles and techniques. MIT Press, London, 1280 pp
Kondratyev KYa (1982) The world climate research programme: state, perspectives and the role of space-borne observational means. Progress in science and engineering: meteorology and climatology, vol 8. ARISTI Publication, Moscow, 274 pp [in Russian]
Kondratyev KYa (1990) The biosphere and climate. Atmósfere 3:165–182
Kondratyev KYa (1992) Global climate. Nauka Publication, St-Petersburg, 359 pp [in Russian]
Kondratyev KYa (1998a) Ecological risk: real and hypothetical. Bull Russ Geogr Soc 130(3):13–24 [in Russian]
Kondratyev KYa (1998b) Multidimensional global change. Wiley, Chichester, 771 pp
Kondratyev KYa (1999) Ecodynamics and geopolicy, vol 1. Global problems. St.-Petersurg Science. Center of RAS, St.-Petersburg, 1036 pp [in Russian]
Kondratyev KYa (2000a) Global changes on the border of two millennia. Herald Russ Acad Sci 70 (9):788–796 [in Russian]
Kondratyev KYa (2000b) Research of the Earth from space: scientific plane EOS. Res Earth Space 3:82–91 [in Russian]
Kondratyev KYa (2001) Key issues of global change at the end of the second millennium. In: Tolba MK (ed) Our fragile world: challenges and opportunities for sustainable development, vol 1. Eolls Public Co. Ltd., Oxford, pp 147–165
Kondratyev KYa (2002) Global climate change: reality, hypotheses, and fiction. Res Earth Space 1:3–23 [in Russian]
Kondratyev KYa (2004a) Uncertainties in observational data and numerical climate modeling. Meteorol Hydrol 4:103–119 [in Russian]
Kondratyev KYa (2004b) Priorities of global climatology. Trans Russ Geog Soc 136(2):3–25 [in Russian]

Kondratyev KYa (2005) Key aspects of the global climate change problem. Trans Russ Geog Soc 5:92–99 [in Russian]
Kondratyev KYa, Cracknell AP (1998) Observing global climate change. Taylor & Francis, London, 562 pp
Kondratyev KYa, Fedchenko PP (1992) Precise structure of the sun's spectre and its role in the biosphere evolution. Propo Press, St.-Petersburg, 40 pp [in Russian]
Kondratyev KYa, Grigoryev AA (2004) Forest fires as a global ecodynamics component. Opt Atmos Ocean 136(4):279–292 [in Russian]
Kondratyev KYa, Johannessen OM (1993) Arctic and climate. Propo Press, St. Petersburg, 139 pp [in Russian]
Kondratyev KYa, Krapivin VF (2001a) An expert system to control the greenhouse effect. Probl Environ Nat Resour 6:23–39 [in Russian]
Kondratyev KYa, Krapivin VF (2001b) Biocomplexity and the global ecological monitoring. Res Earth Space 1:1–8 [in Russian]
Kondratyev KYa, Krapivin VF (2001c) Land ecosystems and their global dynamics. Probl Environ Nat Rersour 5:24–38 [in Russian]
Kondratyev KYa, Krapivin VF (2001d) Global dynamics of basic land ecosystems. Res Earth Space 4:3–12 [in Russian]
Kondratyev KYa, Krapivin VF (2003a) Global change: real and possible in the future. Res Earth Space 4:1–10 [in Russian]
Kondratyev KYa, Krapivin VF (2003b) Global carbon cycle and climate. Res Earth Space 1:3–15 [in Russian]
Kondratyev KYa, Krapivin VF (2004a) Global carbon cycle: state, problems and perspectives. Res Earth Space 3:12–21 [in Russian]
Kondratyev KYa, KrapivinVF (2004b) Modelling the global carbon cycle. Physics-Mathematics Publication, Moscow, 335 pp [in Russian]
Kondratyev KYa, Krapivin VF (2005a) Civilization development and its ecological limitations: numerical modelling and monitoring. Res Earth Space 4:3–23 [in Russian]
Kondratyev KYa, Krapivin VF (2005b) Monitoring and prediction of natural disasters. Il Nuovo Cimento 27(6):657–672
Kondratyev KY, Krapivin VF (2006) On the Earth' heat balance. Energy Econ Eng Ecol 3:7–13 [in Russian]
Kondratyev KYa, Moskalenko NI (1984) The greenhouse effect of the atmosphere and climate. Progress in science and engineering: meteorology and climatology, vol 12, ARISTI Publication, Moscow, 262 pp [in Russian]
Kondratyev KYa, Varotsos CA (1996) Global total ozone dynamics—Impact on surface solar ultraviolet radiation variability and ecosystems 1. Global ozone dynamics and environmental safety. Environ Sci Pollut Res 3(3):153–157
Kondratyev KYa, Varotsos CA (2000) Atmospheric ozone variability: implications for climate change, human health, and ecosystems. Springer/Praxis, Chichester, 758 pp
Kondratyev KYa, Fedchenko PP, Kozoderov VV, Topchiev AG (1990a) Biosphere. Methods and results of remote sensing. Science Publication, Moscow, 223 pp [in Russian]
Kondratyev KYa, Krapivin VF Phyllips GW (1990b) High latitude environmental pollution problems. Research Centre for Ecological Safety, Russian Academy of Sciences, St.-Petersburg, 279 pp
Kondratyev KYa, Ortner J, Preining O (1992) Priorities of global ecology now and in the next century. Space Policy 8(1):39–48
Kondratyev KYa, Moreno-Pena F, Galindo I (1994) Global change: environment and society. Universidad de Coloma, Mexico, 147 pp
Kondratyev KYa, Donchenko VK, Losev KS, Frolov AK (1996) Ecology-economics-politics. St-Petersburg Science Centre of Russian Academy Science, St-Petersburg, 828 pp [in Russian]
Kondratyev KYa, Moreno-Pena F, Galindo I (1997) Sustainable development and population dynamics. Universidad de Coloma, Mexico, 128 pp

Kondratyev KYa, Krapivin VF, Pshenin ES (2000) Conception of regional geoinformation monitoring Res Earth Space 6:1–8 [in Russian]
Kondratyev KYa, Krapivin VF, Borodin LF (2001) Reconstruction of two-dimensional images in the monitoring problem for the Aral-Caspyi aquageosystem. Res Earth Space 5:19–26 [in Russian]
Kondratyev KYa, Krapivin VF, Phillips GW (2002) Global environmental change: modelling and monitoring. Springer, Berlin, 316 pp
Kondratyev KYa, Krapivin VF, Phillips GW (2003a) Arctic Basin pollution dynamics. In: Bobylev LP, Kondratyev KY, Johannesses OM (eds) Arctic environment variability in the context of global change. Springer/Praxis, Chichester, pp 309–362 (2003)
Kondratyev KYa, Krapivin VF, Savinykh VP (2003b) Perspectives of civilization development: multidimensional analysis. Logos, Moscow, 574 pp [in Russian]
Kondratyev KYa, Krapivin VF, Varotsos CA (2003c) Global carbon cycle and climate change. Springer/Praxis, Chichester, 372 pp
Kondratyev KYa, Losev KS, Ananicheva MD, Chesnokova IV (2003d) Scientific-natural bases of life sustainability. Institute of Geography Publication, Moscow, 340 pp [in Russian]
Kondratyev KYa, Losev KS, Ananicheva MD, Chesnokova I (2003e) Stability of life on earth. Springer-Praxis, Chichester, 152 pp
Kondratyev KYa, Krapivin VF, Savinykh VP, Varotsos CA (2004a) Global ecodynamics: a multidimensional analysis. Springer/Praxis, Chichester, 658 pp
Kondratyev KYa, Losev KS, Ananicheva MD, Chesnokova IV (2004b) Stability of life on earth. Springer/Praxis, Chichester, 162 pp
Kondratyev KYa, Ivlev LS, Krapivin VF, Varotsos CA (2006a) Atmospheric aerosol properties: formation, processes and impacts. Springer/Praxis, Chichester, 572 p
Kondratyev KYa, Krapivin VF, Varotsos CA (2006b) Natural disasters as interactive components of global ecodynamics. Springer/Praxis, Chichester, 574 pp
Konrad M, Joachim S (2013) Agroecology. Springer, London, 330 pp
Korgenevsky AV, Krapivin VF, Cherepenin VA (1989) Modelling the global processes of magnetosphere. In: Novichikhin EP (ed) Methods of informatics in radiophysical investigations of environment. Science Publication House, Moscow, pp 25–43 [in Russian]
Kosaka Yu, Xie S-P (2013) Recent warming hiatus tied to equatorial Pacific surface cooling. Nature 501:403–407
Kovalev VI, Rukavishnikov AI, Perov PI, Rossukanyi NM (1999) Elaboration of optical methods and equipment for the control of technology and parameters of semiconductor structures of nano- and microelectronics. Radiotech Electron 44(11):1404–1408 [in Russian]
Kram P (1999) Application of the forest-soil-water model. Ecol Model 120(1):9–30
Krapivin VF (1969) On approximate solutions of initial value problems for integro-differential equation with quasilinear differential operator and generalized Volterra operator. Casopis pro pestovani matematiky (Praha) 94:21–33
Krapivin VF (1972) Theory-game methods for the synthesis of complex systems in conflict states. Soviet Radio Press, Moscow, 192 pp [in Russian]
Krapivin VF (1978) On theory of the complex systems survivability. Science Publication House, Moscow, 248 pp [in Russian]
Krapivin VF (1993) Mathematical model for global ecological investigations. Ecol Model 67:103–127
Krapivin VF (1995) Simulation model for the investigation of pollution dynamics in the Arctic basin. Oceanol (Moscow) 35(3):366–375 [in Russian]
Krapivin VF (1996) The estimation of the Peruvian current ecosystem by a mathematical model of biosphere. Ecol Model 91:1–14
Krapivin VF (2000a) Biospheric oxygen balance and its modeling. Probl Environ Nat Resour 10:15–25 [in Russian]
Krapivin VF (2000b) Radiowave ecological monitoring. In: Klyuev VV (ed) Ecological diagnostics. Znanie, Moscow, pp 295–311 [in Russian]

Krapivin VF (2000c) Global nitrogen cycle modeling. Probl Environ Nat Resour 10:2–15 [in Russian]

Krapivin VF (2000d) The greenhouse effect and the global biogeochemical cycle of carbon dioxide. Probl Environ Nat Resour 2:2–18 [in Russian]

Krapivin VF (2008a) Biocomplexity related to the ecosystem survivability. Proceedings of the eighth international symposium "ecoinformatics problems", Moscow, 16–17 Dec 2008. The Moscow Sciences Engineering A.S. Popov Society for Radio, Electronics and Communication, Moscow, pp 12–18

Krapivin VF (2008b) Resources and global ecodynamics. Proceedings of the eighth international symposium "ecoinformatics problems" Moscow, 16–17 Dec 2008. The Moscow Sciences Engineering A.S. Popov Society for Radio, Electronics and Communication, Moscow, pp 4–11

Krapivin VF (2010) Tropical hurricanes as display of the climatic processes. Proceedings of the IX international symposium "ecoinformatics problems", 9–11 Dec 2010, Moscow. The Moscow Sciences Engineering A.S. Popov Society for Radio, Electronics and Communication, Moscow, pp 4–8

Krapivin VF, Kelley JJ (2009) Model based method for the assessment of global change in the nature-society system. In: Cracknell AP, Krapivin VF, Varotsos CA (eds) Global climatology and ecodynamics. Springer/Praxis, Chichester, pp 133–184

Krapivin VF, Klimov VV (1995) Valuation of convergence of 'physical mixture' strategies in the matrix games. Theory Control Syst 6:209–217 [in Russian]

Krapivin VF, Klimov VV (1997) Stable strategies in the games with gain functions M(x-y). Methods Cybern Inform Technol 2:36–45 [in Russian]

Krapivin VF, Kondratyev KY (2002) Global changes of the environment. St.-Petersburg University Publisher, St.-Petersburg, 724 pp [in Russian]

Krapivin VF, Mkrtchyan FA (1991) Applications in study of environment. Proceedings of the 8th international conference on control systems and computer science, Bucharest, 22–25 May 1991. Polytechnical Institute, Bucharest, pp 49–56

Krapivin VF, Mkrtchyan FA (2006a) Expert system for the operative environmental diagnostics. Proceedings of the 21st international symposium on Okhotsk Sea & Sea Ice, Mombetsu, Hokkaido, Japan, 19–24 Feb 2006. Okhotsk Sea & Cold Ocean Research Association, Mombetsu, Hokkaido, Japan, pp 128–131

Krapivin VF, Mkrtchyan FA (2006b) Expert system for the operative environmental diagnostics. Proceedings of the international symposium on geoinformatics for spatial-infrastructure development in earth and allied sciences, Hochiminh City (Vietnam), 9–11 Nov 2006. Institute for Environment and Resources, Hochiminh City, Vietnam, pp 293–298

Krapivin VF, Mkrtchyan FA (2006c) Microwave radiometry technology for the nature-society system biocomplexity assessment. Proceedings international symposium on geoinformatics for spatial-infrastructure and applied sciences, Hochiminh City, Vietnam, 9–11 Nov 2006. Institute for Environment and Resources, Hochiminh City, Vietnam, pp 187–192

Krapivin VF, Mkrtchyan FA (2007a) Expert system for the operative environmental diagnostics. Proceedings IGARSS 2007 symposium, Barselona, Spain, 23–27 July 2007. IEEE, Barselona, Spain, pp 397–402

Krapivin VF, Mkrtchyan FA (2007b) Expert system for the operative environmental diagnostics. Abstracts of the PICES XVI annual meetings. The changing North Pacific: previous patterns, future projections, and ecosystem impacts, Victoria, Canada, 26 Oct–2 Nov 2007. PICES, Victoria, Canada, pp 108

Krapivin VF, Mkrtchyan FA (2007c) Microwave radiometry technology for the nature-society system biocomplexity assessment (by the example of Okhotsk Sea). Proceedings of the 22nd international symposium on Okhotsk Sea & Sea Ice, Mombetsu, Hokkaido, Japan, 18–23 Feb 2007. Okhotsk Sea & Cold Ocean Research Association, Mombetsu, pp 33–37

Krapivin VF, Mkrtchyan FA (2007d) Microwave radiometry technology for the nature-society system biocomplexity assessment. Proceedings URBAN/URS 2007 symposium, Paris, France, 11–13 April 2007. URBAN-URS, Paris, France, pp 733–738

Krapivin VF, Mkrtchyan FA (2008) About simulation model of Okhotsk Sea ecosystem. Proceedings of the 23th international symposium on Okhotsk Sea and Sea Ice, Mombetsu, Hokkaido, Japan, 16–22 Feb 2008. The Okhotsk Sea & Cold Ocean Research Association, Mombetsu, Hokkaido, Japan, pp 67–70

Krapivin VF, Mkrtchyan FA (2009a) An adaptive technology for nature/society system biocomplexity assessment. Proceedings of the PICES-2009, Jeju, Republic of Korea, 23 Oct–1 Nov 2009. North Pacific Marine Science Organization (PICES), Jeju, Republic of Korea, pp 19–20

Krapivin VF, Mkrtchyan FA (2009b) Microwave radiometry technology for the nature-society system biocomplexity assessment (by the example of Okhotsk Sea). Proceedings of the international symposium of mathematical modelling of tropical hurricane beginning and searching its indicators-precursors, Hochiminh City, Vietnam, 16 Dec 2009. Institute of Applied Mechanics and Informatics, Hochiminh City, pp 57–64

Krapivin VF, Mkrtchyan FA (2009c) Remote sensing radiometry technology for the Okhotsk Sea ecosystem biocomplexity assessment. In: Kashiwai M, Kantakov G (eds) Proceedings of the fourth workshop on the Okhotsk Sea and adjacent areas. PICES Scientific Report, 2009, No. 36, pp 219–224

Krapivin VF, Mkrtchyan FA (2010) Development of the simulation model of pollutants propagation in the Arctic Basin. Abstracts of PICES annual meeting "North Pacific ecosystems today, and challenges in understanding and forecasting change", Portland, O.R., USA, 22–31 Oct 2010. PICES, Portland, pp 25

Krapivin VF, Mkrtchyan FA (2012) Spatial simulation model of dynamics of the Arctic Basin polution. Proceedings of the 27th international symposium on Okhotsk Sea & Sea Ice, Mombetsu, Hokkaido, Japan, 19–24 Feb 2012. Okhotsk Sea & Cold Ocean Research Association, Mombetsu, pp 65–68

Krapivin VF, Nazaryan NA (1995) Theory games approach to the biosphere survivability. KIREE Publisher, Moscow, 28 pp [in Russian]

Krapivin VF, Nazarian NA (1997) A numerical model to study the global sulfur cycle. Numer Model 9(8):36–50 [in Russian]

Krapivin VF, Nitu C (2009) Flexible information technology for the diagnostics of the ocean-atmosphere system instability. Proceedings of the international conference on control systems and computer science (CSCS-17), vol 2, Bucharest, Romania, 26–29 May 2009. University "Politechnica", Bucharest, pp 277–280

Krapivin VF, Nitu C (2011) Carbon cycle and climate. Proceedings of the international conference on control systems and computer science, vol 2, Bucharest, Romania, 25 May 2011. University "Politechnica", Bucharest, pp 145–148

Krapivin VF, Phillips GW (2001a) A remote sensing-based expert system to study the Aral-Caspian aquageosystem water regime. Remote Sens Environ 75:201–215

Krapivin VF, Phillips GW (2001b) Application of a global model to the study of arctic basin pollution: radionuclides, heavy metals and oil hydrocarbons. Environ Model Softw 16:1–17

Krapivin VF, Potapov II (2001) Algorithms for the reconstruction of two-dimensional image fragments in the monitoring problems. Probl Environ Nat Res 6:16–23 [in Russian]

Krapivin VF, Potapov II (2002) Methods of ecoinformatics. ARISTI Publisher, Moscow, 496 pp [in Russian]

Krapivin VF, Potapov II (2006a) Habitat and monitoring of disaster situations. Environ Probl Nat Res 5:80–86 [in Russian]

Krapivin VF, Potapov II (2006b) Technology for the detection and diagnosis of zones where tropical hurricanes are began using the radio-sensing and mathematical modeling. Probl Environ Nat Res 12:3–11 [in Russian]

Krapivin VF, Potapov II (2007) What happen with the biosphere? Probl Environ Nat Res 5:3–15 [in Russian]

Krapivin VF, Shutko AM (1989) Observation and prognosis of the state of environmental resources, ecological and meteorological situations by geoinformational monitoring system. Proceedings of the fourth international symposium on Okhotsk Sea & Sea Ice, Mombetsu, Japan, 5–7 Feb 1989. Okhotsk Sea & Cold Ocean Research Association, Mombetsu, Japan, pp 1–5

Krapivin VF, Shutko AM (2002) Investigations in the field of microwave monitoring of land covers. Probl Environ Nat Res 4:44–53 [in Russian]

Krapivin VF, Shutko AM (2012) Information technologies for remote monitoring of the environment. Springer/Praxis, Chichester, 498 pp

Krapivin VF, Soldatov VY (2009a) Biocomplexity problem related to the Okhotsk Sea ecosystem. Proceedings of the 24th international symposium on Okhotsk Sea and Sea Ice, Mombetsu, Hokkaido, Japan, 15–20 Feb 2009. The Okhotsk Sea & Cold Ocean Research Association, Mombetsu, Hokkaido, Japan, pp 143–146

Krapivin VF, Soldatov VY (2009b) Indicators of the phase states for the ocean-atmosphere system. Proceedings of the international symposium of mathematical modelling of tropical hurricane beginning and searching its indicators-precursors, Hochiminh City, Vietnam, 16 Dec 2009. Institute of Applied Mechanics and Informatics, Hochiminh City, Vietnam, pp 94–104

Krapivin VF, Thu NV (2001) Assessment of physical parameters of ocean surface basing on microwave monitoring data. Probl Environ Nat Res 12:30–37 [in Russian]

Krapivin VF, Varotsos CA (2007) Globalization and sustainable development: environmental agendas. Springer/Praxis, Chichester, 304 pp

Krapivin VF, Varotsos CA (2008) Biogeochemical cycles in globalization and sustainable development. Springer, Chichester, 562 pp

Krapivin VF, Vilkova LP (1990) Model estimation of excess CO_2 distribution in biosphere structure. Ecol Model 50:57–78

Krapivin VF, Svirezhev YM, Tarko AM (1982). Mathematical modeling of the global biosphere processes. Science Publisher House, Moscow, 272 pp [in Russian]

Krapivin VF, Shutko AM, Strelkov GM, Loskutov VS (1990) The modeling of ecological situation in upper sea water layer for climate conditions Northern region. Proceedings of the fifth international symposium on Okhotsk Sea & Sea Ice, Mombetsu, Hokkaido, Japan, 4–6 Feb 1990. Okhotsk Sea & Cold Ocean Research Association, Mombetsu, Japan, pp 243–247

Krapivin VF, Mkrtchyan FA, Nitu C, Petrache G, Tertisco M, Graur A (1991) Multidirectional data processing automatization for remote study of environment. Proceedings of the 8th international conference on control systems and computer science, vol 2, Bucharest, Romania, 22–25 May 1991. Bucharest Polytechnical Institute, Bucharest, Romania, pp 57–62

Krapivin VF, Mkrtchyan FA, Graur A, Nitu C (1994a) Identification and pattern recognition in ecology. Stefan cel Mare Suceava, Sectiunea Electrica, 1, 23–27

Krapivin VF, Mkrtchyan FA, Pham HH, Pham S (1994b) Formation of the databases for Vietnam basing on the geoinformation monitoring. In: Krapivin VF (ed) Ecoinformatics problems. Institute of Radioengineering and Electronics of Russian Academy of Sciences, Moscow, pp 69–72 [in Russian]

Krapivin VF, Bui TL, Rochon GL, Hicks DR (1996) A global simulation model as a method for estimation of the role of regional area in global change. Proceedings of Second Hochiminh City Conference on Mechanics, Hochiminh City, Vietnam, 24–25 Sept 1996. Institute of Applied Mechanics, Hochiminh City, pp 68–69

Krapivin VF, Cherepenin VA, Nazaryan NA, Phillips GW, Tsang FY (1997a) Simulation model for radionuclides transport in the Angara-Yenisey river system. Probl Environ Nat Res 2:41–58

Krapivin VF, Bui TL, Dean C, Nguyen MN, Rochon GL, Hicks DR (1997b) System of survey and simulation for air pollution over large industrial regions. Proceedings of the IASTED international conference "Modelling, simulation and optimization" (MSO'97), Singapore, 11–13 Aug 1997. IASTED/Acta Press, Anaheim, pp 307–311

Krapivin VF, CherepeninVA, Phillips GW, August RA, Pautkin AY, Harper MJ, Tsang FY (1998a) An application of modeling technology to the study of radionuclear pollutants and heavy metals dynamics in the Angara-Yenisey river system. Ecol Model, 111:121–134

Krapivin VF, Michalev M, Mkrtchyan FA, Nitu C, Shutko AM (1998b) A project of intelligent information technology and its application to the agricultural systems monitoring. Proceedings of the Third international symposium on ecoinformatics problems, Moscow, 8–9 Dec 1998. Institute of Radioengineering and Electronics Publisher, Moscow, pp 6–12

Krapivin VF, Graur A, Nitu C, Oancea D (1998c) A new approach for geo-eco-information monitoring. Proceedings of the international conference D&AS, Suceava, Romania, 21–22 May 1998. "Stefan cel Mare" University of Suceava, Suceava, Romania, pp 20–25

Krapivin VF, Mkrtchyan FA, Klimov VV, Kovalev VI, Rukavishnikov AI (2001) Adaptive identifier—a new approach to ecological monitoring of the aquatic environment. New technologies for the 21st century, vol 5. pp 30–32

Krapivin VF, Mkrtchyan FA, Bui DT (2005a) Microwave radiometry technology for the nature-society system biocomplexity assessment. Proceedings of the 26th Asian conference on remote sensing (ACRS2005), Hanoi,Vietnam, AARS, Hanoi, Vietnam, pp 43–44

Krapivin VF, Mkrtchyan FA, Ivanov MV (2005b) Expert system for the operative environmental diagnostics. Proceedings of the 26th Asian conference on remote sensing (ACRS2005), Hanoi, Vietnam, 7–11 Nov 2005. AARS, Hanoi, Vietnam, 58 pp

Krapivin VF, Nitu C, Soldatov VY (2006a) Computer simulation of acid rains. Proceedings of the 8th international conference on development and application systems, Suceava, Romania, 25–27 May 2006. The "Stefan cel Mare" University, Suceava, pp 270–276

Krapivin VF, Shutko AM, Chukhlantsev AA, Golovachev SP, Phillips GW (2006b) GIMS-based method vegetation microwave monitoring. Environ Model Softw 21:330–345

Krapivin VF, Mkrtchyan FA, Nitu C (2007a) Simulation model of Okhotsk Sea ecosystem. Proceedings of the international symposium on engineering ecology, Moscow, 4–6 Dec 2007. KIREE Publisher, Moscow, pp 172–178

Krapivin VF, Nitu C, Soldatov VY (2007b) An adaptive geoinformation technology based approach to the monitoring and prediction of the natural disasters. Proceedings of the 16th international conference on control systems and computer science, Bucharest, Romania, 22–25 May 2007. Printech, Bucharest, Romania, pp 450–455

Krapivin VF, Phillips GW, Nitu C (2007c) A remote sensing—based modeling system to study the Aral-Caspian water regime. Proceedings of the 16th international conference on control systems and computer science, Bucharest, Romania, 22–25 May 2007. Printech, Bucharest, Romania, pp 524–528

Krapivin VF, Burkov VD, Bukatova IL, Potapov II (2008a) Technology for the synthesize of information-measuring systems of new generation. Probl Environ Nat Res 10:3–14 [in Russian]

Krapivin VF, Mkrtchyan FA, Kovalev VI, Klimov VV (2008b) An adaptive system to identify the spots of pollutants on the water surface. Proceedings of the eighth international symposium on ecoinformatics problems, Moscow, 16–17 December 2008. The Moscow Sciences Engineering A.S. Popov Society for Radio, Electronics and Communication, Moscow, pp 35–46

Krapivin VF, Mkrtchyan FA, Nguyen XM, Bui TL (2009a) Urbanization process and the environment. Proceedings of the international symposium on engineering ecology, Moscow, 8–9 Dec 2009. The Russian Sciences Engineering A.S. Popov Society for Radio, Electronics and Communication, Moscow, pp 22–26

Krapivin VF, Nitu C, Soldatov VY (2009b) Direct and inverse problems of the microwave monitoring of the environment. J Control Eng Appl Inform CEAI-2009 11(1):56–61

Krapivin VF, Nitu C, Soldatov VY (2009c) Resources and global ecodynamics. Proceedings of the international conference on control systems and computer science (CSCS-17), Bucharest, Romania, 26–29 May 2009. University "Politechnica" of Bucharest, pp 369–372

Krapivin VF, Shutko AM, Soldatov VY, Novichikhin EP, Stepanov VN (2009d) Creation problems of geo-ecological information-modelling system for the Sea of Azov. In: Stepanov VN (ed) Economic-ecological problems of Sea of Azov. Fenix Publisher, Odessa, Ukraine, pp 327–343 [in Russian]

Krapivin VF, Mkrtchyan FA, Shutko AM (2010a) Remote sensing methods of the soil-plant formations. Papers of the workshop on remote sensing methods for change detection and process modeling, Cologne, Germany, 18–19 Nov 2010. University of Cologne, Germany, pp 31–36

Krapivin VF, Nitu C, Soldatov VY (2010b) Synthesis of geoecological information-modeling systems, vol 10(1/12). The Scientific Bulletin of Electrical Engineering Faculty, Valahia University of Targoviste, Romania, pp 101–105

Krapivin VF, Mkrtchyan FA, Shutko AM (2011) Microwave monitoring of soil moisture as global water cycle component. Proceedings of the international symposium on engineering ecology, Moscow, 6–8 Dec 2011. The Russian Sciences Engineering A.S. Popov Society for Radio, Electronics and Communication, Moscow, pp 8–12

Krapivin VF, Shutko AM, Nitu C (2012a) The GIMS-based research remote sensing platforms vol V.XVII(2). Bulletin AGIR, Bucharest, pp 1224–1228

Krapivin VF, Soldatov VY, Varotsos CA, Cracknell AP (2012b) An adaptive information technology for the operative diagnostics of the tropical cyclones; solar-terrestrial coupling mechanisms. J Atmos Solar Terr Phys 89:83–89

Krapivin VF, Potapov II, Shalaev VS, Burkov VD (2013a) About the role of forest ecosystems in climate change. Probl Environ Nat Res 3:13–30 [in Russian]

Krapivin VF, Soldatov VY, Potapov II (2013b) Expert system for water environment diagnostics. Ecol Syst Devices 1:4–13 [in Russian]

Krapivin VF, Soldatov VY, Potapov II (2013c) Adaptive information-modeling system for hydrophysical investigations. Ecol Syst Devices 2:29–39 [in Russian]

Krapivin VF, Shutko AM, Potapov II, Soldatov VY, Burkov VD, Shalaev VS (2013d) Microwave monitoring of soil moisture as the element of regional water balance. Probl Environ Nat Res 3:33–52 [in Russian]

Krapivin VF, Varotsos CA, Soldatov VY (2014) Mission to Mars. Reliable method for liquid solutions diagnostics. Frontiers in environmental science: environmental informatics. Science 2 (21). doi:10.3389/fenvs.2014.00021

Krimmer RW, Lake FK (2001) The role of indigenous burning in land management. J Forest 99 (11):36–41

Kristjánsson JE (2002) Studies of the aerosol indirect effect from sulfate and black carbon aerosols. J Geophys Res 107(D15):AAC1/1–AAC1/19

Krylov YM (1966) Spectral methods for investigation and calculation of wind-generated waves. Hydrometeoizdat, Leningrad, 255 pp [in Russian]

Krylov YuM (1981) Metodic instructions for the estimation of wind and wave conditions under design of seaports. Soyuzmorniiproject, Moscow, 35 pp [in Russian]

Krylov YuM, Kuznetsov VV, Strekalov SS (1981) Systems of wind-generated waves. Ann Sov Union Acad Sci 208(4):958–961 [in Russian]

Kuck LR Jr, Smith T, Balsley BB, Helmig D, Conway TJ, Tans PP, Davis K, Jensen ML, Bognar JA, Arrieta RV, Rodriques R, Birks J (2000) Measurements of landscape-scale fluxes of carbon dioxide in the Peruvian Amazon by vertical profiling through the atmospheric boundary layer. J Geophys Res 105(D17):22, 137–22, 146

Kurbatsky NP (1964) On the forest fire in the region of Tungus Meteorit. Forest Econ 2:59–61 [in Russian]

Kurz C, Grewe V (2002) Lightning and thunderstorms. Part 1: observational data and model results. Meteorol Z 11(6):379–392

Kutuza BG (1977) Measurement of the atmosphere radio-brightness polarization during rainfall on the wavelength 2.25 cm. In: Kondratyev KY (ed) Radiophysical investigations of the atmosphere. Hydrometeoizdat, Leningrad, pp 201–204 [in Russian]

Kutuza BG, Zagorin G, Hornbostel A, Schroth A (1998) Physical modeling of passive polarimetric microwave observations of the atmosphere with respect to the third stokes parameter. Radio Sci 33(3):677–695

Kutuza B, Shutko A, Plushchev V, Ramsey E, Logan B, DeLoach S, Haldin A, Novichikhin E, Sidorov I, Manakov V, Nelson G (2000) Advantages of synchronous multi-spectral SAR and microwave radiometric observations of land covers from aircraft platforms. Proceedings EUSAR'2000, 3rd European conference on synthetic aperture radar, Munich, Germany, 23–25 May 2000. German Aerospace Center, Frankfurt, pp 663–666

Kuzmin PO (1957) Hydrological investigations of land waters. Int Assoc Sci Hydrol 3:468–478

KWES (2012) Key world energy statistics. OECD/IEA, Paris, 80 pp

Lai CT, Katul G, Oren R, Ellsworth D, Schäfer K (2000) Modeling CO_2 and water vapor turbulent flux distributions within a forest canopy. J Geophys Res 105(D21):26, 333–26, 351

Lai TL (2001) Sequential analysis: some classical problems and new challenges. Stat Sinica 11:303–408

Lal R, Follett RF, Stewart BA (eds) (1998) Soil processes and the carbon cycle. Springer/CRC Press, Berlin, 609 pp

Langmann B (2000) Numerical modelling of regional scale transport and photochemistry directly together with meteorological processes. Atmos Environ 34(21):3585–3598

Laudon KC (2011) Essentials of management information systems. Prentice Hall, New Jersey, 640 pp

Laudon KC, Traver CG (2011) Management information systems. Prentice Hall, New Jersey, 640 pp

Lautenbacher CC (ed) (2005) New Priorities for the 21st Century-NOAA's strategic plan undated for FY 2006- FY 2011, NOAA, Washington, 24 pp

Lawford R (2004) Earth observations: a renewed opportunity area for GEWEX. GEWEX News 14 (3):2

Le C (1977) Architecture of the 20th century. Progress Publisher, Moscow, 304 pp [in Russian]

Legendre L, Krapivin VF (1992) Model for vertical structure of phytoplancton community in Arctic regions. Proceedings of the seventh international symposium on Okhotsk Sea & Sea Ice. Mombetsu, Hokkaido, Japan, 2–9 Feb 1992. Okhotsk Sea and Cold Ocean Research Association, Mombetsu, Hokkaido, Japan, pp 314–316

Legendre P, Legendre L (1998) Numerical ecology. Elsevier, Amsterdam, 853 pp

Lemke P, Jacobi H-W (eds) (2012) Arctic climate change. The ACSYS decade and beyond. Atmospheric and oceanographic sciences library, vol 43(XIV). Springer, Berlin, 464 pp

Leonov AV, Mogilnicova TA, Pischalnik VM, Zenkin OV (2007) Development of microalgae in Okhotsk Sea under winter conditions and modeling of their annual dynamics in the Aniva bay. Water Resour 34(2):204–215 [in Russian]

Letz SE (2000) Social, political, legal aspects of extreme situations. In: Makarov IM (ed) Risk management: risk, sustainable development, Synergetics. Science Publishers, Moscow, pp 331–365 [in Russian]

Levinson DH, Waple AM (2004) State of the climate in 2003. Bull Am Meteorol Soc 85(6):1–72

Levitus S, Antonov JI, Boyer TP, Stephens C (2000) Warming of the world ocean. Science 287:2225–2229

Levitus S, Antonov JI, Wang J, Delworth TL, Dixon KW, Broccoli AJ (2001) Anthropogenic warming of Earth's climate system. Science 292(5515):267–270

Li Y, Zhao X (2012) An empirical study of the impact of human activity on long-term temperature change in China: a perspective from energy consumption. J Geophys Res 117:D17117. doi:10.1029/2012JD018132

Liao K-J, Tagaris E, Manomaiphiboon K, Wang C, Woo J-H, Amar P, He S, Russell AG (2009) Quantification of the impact of climate uncertainty on regional air quality. Atmos Chem Phys 9 (3):865–878

Lieth H (1985) A dynamic model of the global carbon flux through the biosphere and its relations to climatic and soil parameters. Int J Biometeorol 29(2):17–31

Linacre E, Geerts B (1997) Climates and weather explained. Routledge, New York, 432 pp

Lindenmayer DB, Foster DR, Franklin JF, Hunter ML, Noss RF, Schmiegelow FA, Perry D (2004) Salvage harvesting policies after natural disturbance. Science 303(5662):1303

Lindsey B (2001) Against the dead hand: the uncertain struggle for global capitalism. Wiley, New York, 368 pp

Lindzen RS, Chou M-D, Hou AY (2001) Does the earth have an adaptive infrared iris? Bull Am Meteorol Soc 82:417–432
Lindzen RS, Giannitsis C (2003) Reconciling observations of global temperature change. Geophys Res Lett 29(10):X1–X3
Liou Y-A, Pavelyev A, Igarashi K, Hocke K, Huang C-Y (2003) Analytic method for observation of the gravity waves using radio occultation data. Geophys Res Lett 30(20):1-1–1-5. doi:10.1029/2003GL017818. (ASC)
Liou Y-A, Pavelyev AG, Matyugov SS, Yakovlev OI, Wickert J (2010) Radio occultation method for remote sensing of the atmosphere and ionosphere. InTech Publisher, Rijeka, 169 pp
Liping G, Erda L, Zhongpei L (2000) Methane emission flux and mitigation options and its relationship with N_2O emission from paddy soils. Proceedings of the second international methane mitigation conference, Novosibirsk, 18–23 June 2000. Novosibirsk State University Publisher, Novosibirsk, pp 217–222
Lischke H, Lotter AF, Fischlin A (2002) Untangling a Halocene pollen record with forest model simulations and independent climate data. Ecol Model 150(1–2):1–21
Liu K, Fearn ML (1993) Lake-sediment record of late Holocene hurricane activities from coastal Alabama. Geology 21:793–796
Logofet DO (2002) Matrix population models: construction, analysis and interpretation. Ecol Model 148(3):307–310
Lohmann U (2004) Can anthropogenic aerosols decrease the snowfall rate? J Atmos Sci 61 (10):2457–2468
Lohmann G, Sirocko F (2004) Paleoclimatic research within DEKLIM. Pages News 112(12):6–7
Lohmann G, Butzin M, Dima M, Grosfeld K, Knorr G, Könnecke L, Romanova V, Schubert S, Zech S (2004) Climate transitions: forcing and feedback mechanisms of glacial-interglacial and recent climate change. Pages News 12(2):21–22
Lomborg B (2001) The sceptical environmentalist: measuring the real state of the world. Cambridge University Press, Cambridge, 496 pp
Lomborg B (ed) (2004) Global crisis, global solutions. Cambridge University Press, Cambridge, 670 pp
Lowd RB (1977) The Design of a neural events analyzer. Thayer School of Engineering, Hanover, 206 pp
Lu F, Li Z (2003) A model of ecosystem health and its application. Ecol Model 170(1):55–59
Luus KA, Kelly REJ, Lin JC, Humphreys ER, Lafreur PM, Oechel WC (2013) Modeling the influence of snow cover on low Arctic net ecosystem exchange. Environ Res Lett 8(3):1–10
Lyapunov AA (1971) On the construction of the mathematical model of balance relationships in the ecosystem of the ocean tropical waters. In: Vinogradov ME (ed) Functioning of pelagic communities of the tropical areas of the ocean. Nauka, Moscow, pp 13–24 [in Russian]
Maguire DJ, Goodchild MF, Rhind DW (eds) (1991) Geographical information systems, vol 1–2. Longman Scientific & Technical, New York, 649 and 447 pp
Mahlman JD (1998) Science and nonscience concerning human-caused climate warming, vol 23. Annual Review of Energy and the Environment, Palo Alto, pp 83–105
Majorovicz H, Safanda J, Skinner W (2002) East to west retardation in the onset of the recent warming across Canada inferred from inversions of temperature logs. J Geophys Res 107 (B10):ETG6/11–ETG6/12
Maki M, Ishiahra M, Tamura M (2004) Estimation of leaf water status to monitor the risk of forest fires by using remotely sensed data. Remote Sens Environ 90:440–450
Man NX, Tuet DV, Klimov VV, Krapivin VF, Mkrtchyan FA, Soldatov VY (2010) An experience of hydrophysical experiments on the Vietnamise territory. Proceedings of IX international symposium on ecoinformatics problems, Moscow, 11–13 Dec 2010. Moscow Science Popov Society for Radio, Electronics and Communication, Moscow, pp 167–176
Man NX, Krapivin VF, Mkrtchyan FA, Potapov II (2011a) GIMS-technology in the remote monitoring of forest ecosystems. Probl Environ Nat Resour 3:3–13 [in Russian]

Man NX, Krapivin VF, Mkrtchyan FA, Klimov VV, Kovalev VI, Rukovishnikov AI, Soldatov VY (2011b) Measurements and identification of the water solutions basing on the spectroellipsometer. Probl Environ Nat Resour 2:3–22 [in Russian]

Mander J, Goldsmith E (eds) (2006) The case against the global economy. IFG Publisher, San Francisco, 560 pp

Mansell ER, Mac Gorman DR, Ziegler CL, Starka JM (2002) Simulated three-dimensional branches lightning in a numerical thunderstorm model. J Geophys Res 107(D9):ACL2/1–ACL2/14

Manzini E, Steil B, Brühl C, Giorgetta MA, Krüger K (2003) A new interactive chemistry-climate model: 2. Sensitivity of the middle atmosphere to ozone depletion and increase in greenhouse gases and implications for recent stratospheric cooling. J Geophys Res 108(D14):ACL10/1–ACL10/22

Marhaba TF, Van D, Lippincott RL (2000) Rapid identification of dissolved organic matter in water by spectral fluorescent signatures. Water Res 34(14):3543–3550

Marquardt D (1963) An algorithm foe least-squares estimation of nonlinear parameters. SIAM J Appl Math 11(2):431–441

Marshall SJ (2011) The cryosphere. Princeton University Press, USA, 304 pp

Marshall TP (2001) Birth of the fujita scale. Storm Track 24(3):6–10

Marshall TC, Stolzenburg M (2002) Electrical energy constraints on lightning. J Geophys Res 107 (D7):ACL1/1–ACL1/13

Marti YY (1971) Biological productivity of the Peruvian current area. In: Vinogradoc ME (ed) Principles of the biological productivity of the ocean and its utilization. Nauka, Moscow, pp 72–80 [in Russian]

Matishov DG, Matishov GG (2001) Radiational ecological oceanology. Kola Science Centre, Russian Academy of Sciences, Apatity, 419 pp [in Russian]

Matthew C, Walter SHF (2010) The volume of Earth's ocean. Oceanography 23(2):112–114. doi:10.5670/oceanog.2010.51

Matusevich NN (1956) Foundations of the sea astronomy. CANO Press, Moscow, 548 pp [in Russian]

Mayers JC (2004) London's wettest summer and wettest year—1903. Weather 59(10):274–278

Mazur II, Ivanov OP (2004) Dangerous natural processes. Economics, Moscow, 702 pp [in Russian]

McBean GA (1998) The climate agenda: the role of the WCRP as the research thrust, vol 904. WCRP/WMO Publisher, pp 11–18

McCauley LL, Meier MF (eds) (1991) Arctic system science. Land/Atmosphere/Ice Interactions. ARCUS, Alaska, 48 pp

McCloy KR, Lucht W (2004) Comparative evaluation of seasonal patterns in long time series of satellite image data and simulations of a global vegetation model. IEEE Trans Geosci Remote Sens 42(1):140–153

McConnel WJ (2004) Forest cover change—tales of the unexpected. Global Change Newsl 57:8–11

McGillis WR, Edson JB, Hare JE, Fairrall CW (2001) Direct covariance air-sea CO_2 fluxes. J Geophys Res 106(C8):16,729–16,746

MCGuire AD, Hayes DJ, Kicklighter DW, Manizza M, Zhuang Q, Chen M, Follows MJ, Gurney KR, MCclelland JW, Melillo JM, Peterson BJ, Prinn RG (2010) An analysis of the carbon balance of the Arctic Basin from 1997 to 2006. Tellus 62B:455–474

McIntyre AD (1999) The environment and the oil companies. Mar Pollut Bull 3(3):155–156

McIntyre S, McKitrick R (2003) Corrections to the Mann et al (1998), proxy data base and northern Hemisphere temperature series. Energy Environ 14(6):751–757

McKitrick R (2010) A critical review of global surface temperature data products. http://ssrn.com/abstract=1653928

McNulty SG (2002) Hurricane impacts on US forest carbon sequestration. Environ Pollut 116: S17–S24

Meaden GT (1985) A study of tornadoes in Britain. TORRO HQ, Wiltshire, 131 pp

Meadows M (2000) United Kingdom methane emissions: trends, projections and mitigation options. Proceedings of the second international. Methane mitigation conference, Novosibirsk, 18–23 June 2000. Novosibirsk State University Publisher, Novosibirsk, pp 37–44

Mearns LO, Bogardi L, Giorgi F, Matyasovsky I, Palecki M (1999) Comparison of climate change scenarios generated from regional climate model experiments and statistical downscaling. J Geophys Res 104(D6):6603–6621

Mebust MR, Eder BK, Binkowski FS, Roselle SJ (2003) Models-3 community multiscale air quality (CMAQ) model aerosol component. 2. Model evaluation. J Geophys Res 108(D6): AAC4/1–AAC4/18

Mees A, Vincent TL, Teo KL, Judd K (1997) Control and chaos (mathematical modelling). Birkhäuser, Boston, 333 pp

Mehdizadeh F, Rifal H (2004) Modeling point source plumes at high altitudes using a modified Gaussian model. Atmos Environ 38(6):821–831

Melentiev VV, Jochannessen OM, Kondratyev KY, Bobilev LP, Tichomirov AI (1998) An experience of satellite-based radiolocation diagnostics of the ice-lake cover: ecology and history. Res Earth Space 2:91–101 [in Russian]

Melillo J, Field CB, Moldan B (eds) (2003) Interactions of the main biogeochemical cycles: global change and human impacts. SCOPE series, vol 61. Island Press, Washington, 320 pp

Mendelsohn R, Rosenberg NJ (1994) Framework for integrated assessment of global warming impacts. Climatic Change 28(1–2):15–44

Meriakri VV (1992) Spectroscopy of the millimeter and sub-millimeter range. Bull Moscov State Univ Phys Astron 33(3):86–94 [in Russian]

Metternicht GI (1998) Fuzzy classification of JERS-1 SAR data: an evaluation of its performance for soil salinity mapping. Ecol Model 111(1):61–74

Miki M, Rakov VA, Rambo KJ, Schnetzer GH, Uman MA (2002) Electric fields near triggered lightning channels measured with Pockels sensors. J Geophys Res 107(D16):ACL2/1–ACL2/11

Mikolajewicz U, Gröger M, Maier-Reimer F, Schurgers G, Vizcaino M, Winguth A (2004) Climcyc: modeling of the last glacial cycle: response of climate and vegetation to insolation forcing between 132–112 ka BP. Pages News 12(2):24–25

Milne A (2004) Doomsday: the science of catastrophic events. Praeger Publisher, Westport, 194 pp

Mintzer IM (1987) A matter of degrees: the potential for controlling the greenhouse effect. World Resources Institute Research Report No. 15, Washington, 70 pp

Mirkin BM (1986) Vegetation communities: what are they? Science Publisher House, Moscow, 160 pp [in Russian]

Mitnik LM (1977) Physical foundations of environment remote sensing. Polytechnical Institute Press, Leningrad, 56 pp [in Russian]

Mkrtchyan FA, Krapivin VF (2011) GIMS-technology in monitoring marine ecosystems. Proceedings of the 26th international symposium on 20–25 Feb 2011. Okhotsk Sea & Sea Ice, Mombetsu, Hokkaido, Japan, pp 163–166

Mkrtchyan FA, Krapivin VF (2014) Spectroellipsometry opportunities in the monitoring aquatic ecosystems. Proceedings of the 29-th international symposium on Okhotsk Sea and Sea Ice, Mombetsu, Hokkaido, Japan, 16–19 Feb 2014. The Okhotsk Sea and Cold Ocean Research Association, Mombetsu, Hokkaido, Japan, pp 154–156

Mkrtchyan FA, Krapivin VF, Kovalev VI, Klimov VV, Rukovishnikov AI, Golovachev SP (2004) An adaptive spectroellipsometric technology for the ecological monitoring of the aquatic environment. Proceeding of 25-th ACRS, Chiang-Mai, Thailand, 22–26 November 2004. AARS, Bangkok, pp 13–15

Mkrtchyan FA, Krapivin VF, Kovalev VI, Klimov VV (2011) An adaptive multi-cannel spectroellipsometer for ecological monitoring. Proceedings of the ISPRS hannover workshop 2011 "High-resolution Earth imaging for geospatial information", Hannover, Germany, 14–17 June 2011. ISPRS, Hannover, Germany, pp 143–147

Mkrtchyan FA, Krapivin VF, Klimov VV, Kovalev VI (2013) Hardware-software system of the water environment monitoring with use of microwave radiometry and spectroellipsometry means. Proceedings of the 28-th International Symposium on Okhotsk Sea & Sea Ice, Mombetsu, Hokkaido, Japan, 17–21 Feb 2013. The Okhotsk Sea & Cold Ocean Research Association, Mombetsu, Hokkaido, Japan, pp 104–109

Mo Q, Helsdon JH, Winn WP (2002) Aircraft observations of the creation of lower positive charges in thunderstorms. J Geophys Res 107(D22):ACL4/1–ACL4/15

Moberg Å (2006) Environmental systems analysis tools for decision-making. Royal Institute of Technology, Stockholm, 56 pp

Mohr T, Bridge G (2003) The evolution of an integrated global Earth observing system. Res Earth Space 1:64–73

Mon J, Flury M, Harsh JB (2006) A quantitative structure–activity relationships (QSAR) analysis of triarylmethane dye tracers. J Hydrol 316(1–4):84–97

Monin AS, Obukhov AM (1954) Spatial characteristics of turbulence in the surface layer of the atmosphere. Ann USSR Acad Sci 93:223–226 [in Russian]

Monin AS, Shishkov YuA (1991) The warming dilemmas in the 20th century. In: Monin AS (ed) Man and chaos. Hydrometeoizdat, St. Petersburg, pp 47–49 [in Russian]

Morén L, Påsse T (2001) Climate and shoreline in Sweden during Weichsel and the next 150,000 years. Swedish Nuclear Fuel and Waste Management Co., Stokholm, 70 pp

Morgan MR (2008) Implications of climate change on Eastern Canadian waters. Can Nav Rev 3 (4):19–22

Morgan J, Codispoti L (eds) (1995) Department of defense Arctic nuclear waste assessment program—FY's 1993–1994. Office of Naval Research, ONR 322-95-5, Arlington, pp 15–30

Morice CP, Kennedy JJ, Rayner NA, Jones PD (2012) Quantifying uncertainties in global and regional temperature change using an ensemble of observational estimates: the HadCRUT4 dataset. J Geophys Res 117(D8):2–22

Moritz RE, Bitz CM, Steig EJ (2002) Dynamics of recent climate change in the Arctic. Science 297:1497–1502

Moron V, Vautard R, Ghil M (1998) Trends, interdecadal and interannual oscillations in global sea-surface temperatures. Clim Dyn 7–8:545–569

Morris J (1997) Introduction: climate change—prevention or adaptation? IEA Stud Educ 10:13–37

Mourant AE (1964) The international biological programme. Eugenics Rev 55(4):201–202

Mu Mu, Wansuo Duan, Jiacheng Wang (2002) The predictability problems in numerical weather and climate prediction. Adv Atmos Sci 19(2):191–204

Muller R, Peter T (1992) The numerical modeling of the sedimentation of polar stratospheric cloud particles. Ber Bunsenges Phys Chem 96:353–361

Mur RK, Fen AK (1979) Radiolocation determination of the wind parameters above sea. Proc Eng Inst Electron Radiotechnics (EIER) 67(11):40–63

Murata A, Takizawa T (2002) Impact of a coccolithophorid bloom on the CO_2 system in surface waters of the eastern Bering Sea shelf. Geophys Res Lett 29(11):1547. doi:10.1029/2001GL013906

Murphy KP (2012) Machine learning: a probabilistic perspective (adaptive computation and machine learning series). The MIT Press, Cambridge, 1104 pp

Mustard J (2001) Spatial discretization of discrete dynamical systems. Deakin University, Victoria, 368 pp

Myers TS, Atkinson IM (2008) The semantic reef: a hypothesis-based, eco-informatics platform to support automated knowledge discovery for remotely monitored reef systems. Proceedings of the 11th international coral reef symposium, Ft. Lauderdale, Florida, 7–11 July 2008. pp 524–528

Myerson RB (1997) Game theory: analysis of conflicts. Harvard University Press, Cambridge, 600 pp

Najam A, Sagar A (1998) Avoiding a cop-out: Moving towards systematic decision-making under the climate convention. Clim Change 39(4):III–IX

Nakamura H, Lin G, Yamagata T (1997) Decadal climate variability in the North Pacific during the recent decades. Bull Am Meteor Soc 78:2215–2225

Nazaretyan AP (2004) Anthropogenic crisises: hypothesis of technician-humanitarian balance. Herald Russ Acad Sci 74(4):319–330 [in Russian]

Neale M, Cardon LR (1992) Methodology for genetic studies of twins and families. Springer, Berlin, 496 pp

Nefedova EI (1994) Mathematical modeling of the global carbon cycle in the atmosphere-ocean system. Ph.D. theses, Computing Center RAS, Moscow, 112 pp [in Russian]

Nefedova EI, Tarko AM (1993) Study of the global carbon cycle using the zonal model in the atmosphere-ocean system. Ann Russ Acad Sci 333(5):645–647 [in Russian]

Nelepo BA, Korotaev GK, Suetin VS, Terechin YV (1985) Investigations of ocean from space. Naukova Dumka, Kiev, 185 pp [in Russian]

Neyman J, Pearson ES (1933) On the problem of the most efficient tests of statistical hypotheses. Philos Trans R Soc A Math Phys Eng Sci 231(694–706):289–337

Newland JA, De Luca TH (2000) Influence of fire on native nitrogen-fixing plants and soil nitrogen status in ponderosa pine-Douglas-fir forests in western Montana. Can J For Res 30 (2):274–282

Nguyen XM, Dao VT, Klimov VV, Krapivin VF, Mkrtchyan FA, Soldatov VY (2010) Results of hydrophysical experiments on the Vietnamese territory. Proceedings of the IX international symposium on ecoinformatics problems, Moscow, 9–11 Dec 2010. A.S. Popov Society for Radio, Electronics and Communication, pp 12–18

Nicholls RJ, Cazenave A (2010) Sea-level sea-level rise and its impact on coastal zones. Sci Mag 328(5985):1517–1520

Nilsson S, Jonas M, Obersteiner M (2002) COP-6: a healing shock? Editorial Essay Clim Change 52(1–2):25–28

Nitu C, Krapivin VF (2011) Study of the CO_2 greenhouse effect. Proceedings of the international conference on control systems and computer science, vol 2, Bucharest, 25 May 2011. University "Politechnika", Bucharest, Romania, pp 149–152

Nitu C, Krapivin VF, Bruno A (2000a) Intelligent techniques in ecology. Printech, Bucharest, 150 pp

Nitu C, Krapivin VF, Bruno A (2000b) System modelling in ecology. Printech, Bucharest, 260 pp

Nitu C, Krapivin VF, Pruteanu E (2004) Ecoinformatics: intelligent systems in ecology. Magic Print, Onesti, 411 pp

Nitu C, Krapivin VF, Soldatov VY (2007) The GIMS-based method for vegetation microwave monitoring. Proceedings of the 16th international conference on control systems and computer science, Bucharest, Romania, 22–25 May 2007. Printech, Bucharest, Romania, pp 444–449

Nitu CI, Krapivin VF, Dobrescu AS (2010) Application of a global model to the study of Arctic Basin pollution. Sci Bull Electr Eng Fac Valahia Univ Targoviste Rom 10(1/12):111–114

Nitu C, Krapivin VF, Soldatov VY, Dobrescu A (2013a) A device to measure the geophysical and hydrophysical parameters. Proceedings of the 19th international conference on control systems and computer science—CSCS19, 29–31 May 2013, Bucharest, Romania, pp 281–284

Nitu C, Krapivin VF, Soldatov VY, Anda D (2013b) A device to measure the geophysical and hydrophysical parameters. 2013 Proceedings of the 19th international conference on control systems and computer science—CSCS19, 29–31 May 2013, Bucharest, Romania, pp 281–284

Nitu C, Krapivin VF, Varotsos CA, Cracknell AP, Soldatov VY, A Dumitrascu (2013c) An effective tool for the tropical cyclones monitoring. Proceedings of the 19th international conference on control systems and computer science—CSCS19, 29–31 May 2013, Bucharest, Romania, pp 276–280

NOAA (2003) New Priorities for the 21st century. NOAA's strategic plan for FY 2003—FY 2008 and beyond. NOAA, Washington, 16 pp

NOAA (2012) Implementing the NOAA next generation strategic plan: annual guidance memorandum. NOAA, Washington, 5 pp

Norris K, McCulloch N (2003) Demographic models and the management of endangered species: a case study of the critically endangered Seychelles Magpie Robin. J Appl Ecol 40(5):890–899

Novelli PC, Masarie KA, Lang PM, Hall BD, Myers RC, Elkins JW (2003) Reanalysis of tropospheric CO trends: effects of the 1997–1998 wildfires. J Geophys Res 108(D15):4464. doi:10.1029/2002JD003031

NRC (2013) Assessment of intraseasonal to interannual climate prediction and predictability. National research council. Committee on assessment of intraseasonal to interannual climate prediction and predictability. The National Academies Press, Washington, 190 pp

Obermayer K, Sejnowski TJ (2001) Self-organizing map formation: foundations of neural computation. MIT Press, Massachusetts, 440 pp

Olenyev VV, Fedotov AP (2003) Globalistics on the verge of the 21st century. Probl Philos 4:18–30 [in Russian]

Oppenheimer C (1996) Volcanism. Geography 81(1):65–81

Orheim O (ed) (1999) Proceedings of the international symposium on polar aspects of global change. Trømso, Norway, 24–28 Aug 1998. (1999) Polar Research, 18(2), 404 pp

Ougolnitsky GA (1999) Management of the ecologo-economic systems. High School Book Publisher, Moscow, 132 pp [in Russian]

Ougolnitsky GA, Demianenko YM, Dubrov DV (2000) Methods of the hierarchical optimization in the problems of forestry. In: Ougolnitsky GA (ed) Computer modeling: ecology. High School Book Publisher, Moscow, pp 79–89 [in Russian]

Overland JE, Adams JM (2001) On the temporal character and regionality of the Arctic Oscillation. Geophys Res Lett 28(14):2811–2814

Pachauri RK, Reisinger A (eds) (2007) Climate change 2007: synthesis report. Contribution of working groups I, II and III to the fourth assessment report of the intergovernmental panel on climate change. IPCC, Geneva, Switzerland, 104 pp

Palutikof JP, Goodess CM, Watkins SJ, Burgess PE (1999) Long-term climate change. Prog Environ Sci 1(1):89–96

Pampaloni P, Paloscia S (eds) (2000) Microwave radiometry and remote sensing of the Earth's surface and atmosphere. VSP Publisher, Florence, 549 pp

Pan H (2005) The cost efficiency of Kyoto flexible mechanisms: a top-down study with the GEM-E3 world model. Environ Model Softw 20(11):1401–1411

Panikov NS, Dedysh SN (2000) Cold season CH_4 and CO_2 emission from boreal peat bogs (West Siberia): winter fluxes and thaw activation dynamics. Glob Biogeochem Cycvles 14(4):1071–1080

PARCS (2000) PARCS develops two updated research goals. Witness Arct 9(1):8

Parin NV (1971) On distributional pattern of midwater fishes of the peruvian current zone. In: Zenkevich LA (ed) Oceanographic investigation in the peruvian current area, vol 89. Nauka, Moscow, pp 81–86 [in Russian]

Park KH, Thompson AG, Marinier M, Clark K, Wagner-Riddle C (2006) Greenhouse gas emissions from stored liquid swine manure in a cold climate. Atmos Environ 40:618–627

Parkinson CL (1994) Spatial patterns in the length of the sea ice season in the Southern Ocean. J Geophys Res 99:16327–16339

Parkinson CL, Comiso JC, Zwally HJ, Cavaliery DJ, Gloersen P, Campbell WJ (1987) Arctic sea ice, 1873–1976: satellite passive-microwave observations. NASA SP-489, GPO, Washington, 296 pp

Parson EA, Fisher-Vander K (1997) Integrated assessment models of global climate change. Ann Rev Energ Environ 22:589–628

Paulsen CA (1970) The mathematical representation of wind and temperature profiles in the unstable atmospheric surphace layer. J Appl Meteor 9:857–861

Pavelyev AG, Liou Y-A, Huang C-Y, Reigber C, Wickert J, Igarashi K, Hocke K (2002) Radio holographic method for the study of the ionosphere, atmosphere and terrestrial surface from space using GPS occultation signals. GPS Solutions 6:101–108

Pavelyev AG, Tsuda T, Igarashi K, Liou Y-A, Hocke K (2003) Wave structures in the electron density profile in the ionospheric D and E-layers observed by radio holography analysis of the GPS/MET radio occultation data. J Atmos Solar Terr Phys 65(1):59–70

Pavelyev AG, Matyugov SS, Yakovlev OI (2010) Global satellite monitoring of the atmosphere and ionosphere. J Commun Technol Electron 53(9):1021–1033

Payne JR, McNabb GD, Clayton JR (1991) Oil-weathering behavior in Arctic environments. Polar Res 10(2):631–662

Peng MS, Fu B, Li T, Stevens DE (2012) Developing versus nondeveloping disturbances for tropical cyclone formation. Part 1: North Atlantic. Mon Weather Rev 140(4):1047–1066

Penman HL (1948) Natural evaporation from open water, bare soil and grass. Proc R Soc Lond 193:120–145

Penner JE (2003) Comment on "Control of fossil-fuel particulate black carbon and organic matter, possibly the most effective method of slowing global warming" by M. Z. Jacobson. J Geophys Res 108(D21):ACH14/1–ACH14/5; ACH5/1–ACH5/4

Perona G, Brebbia CA (2010) Modelling, monitoring and management of forest fires II. WIT Press, Washington, 304 pp

Perov PI, Kovalev VI, Rukovishnikov AI, Rossukanyi NM, Johnson WH (1994) New high precision and high speed automatic ellipsometer with polarization switching for in situ control in semiconductor device technologies. Int J Electron 76(5):797–803

Perry I (2001) Marine ecosystem and climate interactions in the past, present, and future: report of a workshop between GLOBEC, PAGES and CLIVAR. Newsletter 7(1):6–8

Pervaniuk VS (2001) A spatial numerical model of the global biogeochemical cycles of carbon and nitrogen in the atmosphere-ocean system. Ph.D. theses, Computing Center RAS, 118 pp (in Russsian)

Peterson A, Krantz WB (2008) Differential frost heave model for patterned ground formation: Corroboration with observations along a North American arctic transect. J Geophys Res 113 (GE):1–2

Petty GW (1995) The status of satellite-based rainfall estimation over land. Remote Sens Environ 51(1):125–137

Phillips GW, August RA, Cherepenin VA, Harper MJ, King SE, Krapivin VF, Pautkin AY, Tsang FY (1997) Radionuclear pollutants in the Angara and Yenisey rivers of Siberia. Radioprotection-Colloques 32:299–304

Phillips GW, Nitu C, Chukhlantsev AA, Golovachev SP, Krapivin VF, Shutko AM (2004) A new technology of vegetation microwave monitoring. Proceedings of the 7th international conference on development and application systems, Suceava, Romania, 27–29 May 2004, pp 70–74

Phillips GW, Nitu C, Chukhlantsev AA, Golovachev SP, Krapivin VF, Shutko AM (2009) A new technology of vegetation microwave monitoring. Sci Bull Electr Eng Fac Valahia Univ Targoviste Rom 2(11):95–98

Picket EE (ed) (1987) Atmospheric pollution. Springer, Berlin, 286 pp

Pielke RA Jr (1999) Hurricane forecasting. Science 284:1123

Pielke RA Sr (2001a) Carbon sequestration—the need for an integrated climate system approach. Bull Am Meteorol Soc 82(11):20–21

Pielke RA Sr (2001b) Earth system modeling—an integrated assessment tool for environmental studies. In: Matsuno T, Kida H (eds) Present and future of modeling global environmental change: toward integrated modeling. Terra Scientific Publisher, Company, Tokyo, pp 311–337

Pielke RA Sr (2002) Overlooked issues in the US national climate and IPCC assessments. Clim Change 52(1–2):1–11

Plotnikov VV (1996) Long-term prognosis of Okhotsk Sea ice conditions with the consideration of large-scale atmospheric processes. Meteorol Hydrol 12:93–100 [in Russian]

Plumlee GS, Foreman WT, Griffin DW, Lovelace JK, Meeker GP, Demas CR (2006) Characterization of flood sediments from hurricanes Katrina and Rita and potential implications for human health and the environment. Science and the storms: the USGS response to the hurricanes of 2005. US Geological Survey, Denver, pp 245–256

Podlazov AV (2001) Self-organizing criticism and risk analysis. Proc High Sch Appl Nonlinear Dyn 9(1):49–88 [in Russian]

Polishchuk YM (1992) Simulation modelling of the systems with natural components. Science Publishing House, Novosibirsk, 228 pp [in Russian]

Polishchuk VY, Polishchuk YM (2013) Geosimulation modeling of fields for the thermokarst lakes in the permafrost zones. Yugorsky State University, Khanty Mansiysk, 128 pp
Pollak RA (1990) Two-sex demographic models. J Polit Econ 98(2):399–420
Potapov II, Krapivin VF, Soldatov VY (2006) Risk assessment in the monitoring regime. Ecol Syst Devices 8:11–17 [in Russian]
Power HC (2000) Estimating atmospheric turbidity from climate data. Ecol Model 135(1):125–134
Preller RH, Cheng A (1999) Modeling the transport of radioactive contaminants in the Arctic. Mar Pollut Bull 38(2):71–91
Prestrud P, Overland JE, Walsh JE, Wang M, Barry RG, Armstrong R, Callaghan T, Cherry J, Gearheard S, Nolin A, Russell D, Zöckler C, Gerland S, Aars J, Bracegirdle T, Carmack E, Hop H, Hovelsrud GK, Kovacs K, Perovich DK, Richter-Menge J, Rybråten S, Strøm H, Turner J, Bentley CR, Thomas RH, Velicogna I, Zemp M, Haeberli W, Bajracharya S, Chinn TJ, Hagen JO, Huggel C, Kääb A, Kaltenborn BP, Karki M, Kaser G, Kotloyakov VM, Lambrechts C, Li Z, Monia BF, Mool P, Nellemann C, Novikov V, Osipova GB, Rivera A, Shrestha B, Svoboda F, Tsvetkov DG, Yao T, Church JA, Nicholls R, Hay JE, Gornitz V, Romanovsky VE, Gruber S, Instanes A, Jin H, Marchenko SS, Smith SL, Trombotto D, Walter KM, Prowse TD, Bonsal BR, Duguay CR, Hessen DO, Vuglinsky VS, Eamer J, Young O, Cochran P, Dixit K, Nakalevu T, Pungowiyi C (2007) Global outlook for ice & snow. UNEP, Norway, 235 pp
Quaas J, Bouher O, Dufresne J-L, Le Treut H (2004) Impacts of greenhouse gases and aerosol direct and indirect effects on clouds and radiation in atmospheric GCM simulations of the 1930–1989 period. Clim Dyn 23:779–789
Raaschou-Nielsen O, Hertel O, Vignati E, Berkowitcz R, Jensen SS, Larsen VB, Lohse C (2000) Evaluation of an air pollution model with respect to use in epidemiologic studies; comparison with measured levels of nitrogen dioxide and benzene. J Exp Anal Environ Epidemiol 10:4–14
Radakov DV (1972) Flocking of fishes as an ecological phenomenon. Nauka, Moscow, 174 pp [in Russian]
Rahimzadeh-Bajgiran P, Omasa K, Shimizu Y (2012) Comparative evaluation of the vegetation dryness index (VDI), the temperature vegetation dryness index (TVDI) and the improved TVDI (iTVDI) for water stress detection in semi-arid regions of Iran. ISPRS J Photogrammetry Remote Sens 68:1–12
RAISE (2001) RAISE plan focuses on ship-based research in Russia. Witness Arct 9(1):7
Rakov VA, Tuni WG (2003) Lightning electric field intensity at high latitudes: inferences for production of elves. J Geophys Res 108(D20):ACL8/1–ACL8/16
Ram B (2010) Numerical methods. Pearson Education India, Delhi, 505 pp
Ramad F (1981) Principles of applied ecology. Hydrometeoizdat, Leningrad, 544 pp [in Russian]
Randall DA, Wood RA, Bony S, Colman R, Fichefet T, Fyfe J, Kattsov V, Pitman A, Shukla J, Srinivasan J, Stouffer RJ, Sumi A, Taylor KE (2007) Cilmate models and their evaluation. In: Solomon S, Qin D, Manning M, Chen Z, Marquis M, Averyt KB, Tignor M, Miller HL (eds) Climate change 2007: the physical science basis. Contribution of working group I to the fourth assessment report of the intergovernmental panel on climate change. Cambridge University Press, Cambridge, New York, 1008 pp
Randel WJ, Wu F, Rios WR (2003) Thermal variability of the tropical tropopause region derived from GPS/MET observations. J Geophys Res 108(D1):7/1–7/12
Rashid R, Seizov P (2012) Assessment of carbon stocks as an ecosystem service at Rusenski Lom nature park. Geosogt EOOD, WWF Danube-carpathian programme, Viena, Austria, http://www.panda.org/dcpo, 20 pp
Rasmussen CE, Williams CKI (2006) Gaussian processes for machine learning. MIT Press, Cambridge, 266 pp
Rastetter EB, Aber JD, Peters DPC, Ojima DS, Burke IC (2003) Using mechanistic models to scale ecological processes across space and time. BioScience 53(1):1–9

Rawlins MA, McDonald KC, Frolking S, Lammers RB, Fahnestock M, Kimball JS, Vörösmarty CJ (2005) Remote sensing of snow thaw at the pan-Arctic scale using the SeaWinds scatterometer. J Hydrol 312:294–311
Remizov LT (1985) Natural radionoise. Nauka Publishing, Moscow, 196 pp [in Russian]
René D, Hassane A (2005) Discrete, continuous, and hybrid petri nets. Springer, Berlin, 526 pp
Report (2001) Report from the Arctic climate impact assessment modeling and scenarios workshop. Stockholm, Sweden, 29–31 Jan 2001. The ACIA Secretariat, Fairbanks, 28 pp
Reutov EA, Shutko AM (1986) Prior-knowledge-based soil-moisture determination by microwave radiometry. Sov J Remote Sens 5(1):100–125 [in Russian]
Rheinheimer G (1998) Pollution in the Baltic Sea. Naturwissenschaften 85(7):318–329
Riedlinger SH, Preller RH (1991) The development of a coupled ice-ocean model for forecasting ice conditions in the Arctic. J Geophys Res 96:16955–16977
Riseborough D, Shiklomanov N, Etzelmüller B, Gruber S, Marchenko S (2008) Recent advances in permafrost modeling. Permafrost Periglac Process 19(2):137–156
Rochon GL, Krapivin VF, Watson M, Fauria S, Tsang FY, Fernandez M (1996) Remote characterization of the landsea interface: a case study of the Viet-Nam/South China Sea coastal zone. Proceedings of second HoChiMinh City Conference on Mechanics, Hochiminh City, Vietnam, 24–25 September 1996. Institute of Applied Mechanics, Hochiminh City, pp 72–74
Rodean HC (1996) Stochastic lagrangian models of turbulent diffusion. American Meteorological Society, Washington, 84 pp
Rovinsky FY, Chernogaeva GM, Paramonov SG (1995) A role of river flow and atmospheric transport in the pollution of Russian northern seas. Meteorol Hydrol (Moscow) 9:22–29 [in Russian]
Rudels B, Larsson A-M, Sehistendt P-I (1991) Stratification and water mass formation in the Arctic ocean. Polar Res 10(1):19–31
Russel S, Norvig P (2009) Artificial intelligence: a modern approach. Prentice Hall, New Jersey, 1152 pp
Russo G, Eva C, Palau C, Caneva A, Saechini A (2000) The recent abrupt increase in the precipitation rate, as seen in an ultra-centennial series of precipitation. II Nuovo Cimento C 23 (1):39–51
Safai PD, Rao PSP, Momin GA, All K, Chate DM, Praveen PS (2004) Chemical composition of precipitation during 1984–2002 at Pune, India. Atmos Environ 38(12):1705–1714
Santer BD, Wigley TML, Boyle JS, Gaffen DJ, Hnilo JJ, Nychka D, Parker DE, Taylor KE (2000) Statistical significance of trends and trend differences in layer average atmospheric temperature time series. J Geophys Res 105(D6):7337–7356
Santer BD, Sausen R, Wigley TML, Boyle JS, Achuta RK, Doutriaux C, Hansen JE, Meehl GA, Roeckner E, Ruedy R, Schmidt G, Taylor KE (2003) Behavior of tropopause height and atmospheric temperature in models, re-analysis, and observations: decadal changes. J Geophys Res 108(D1):1/1–1/22
Sarancha DA (1995) Biomodelling. Computer Centre of Russian Academy of Science Press, Moscow, 138 pp [in Russian]
Sarkisyan AS (2000) Synthesis of the observational data and the results of modeling as a perspective direction in the investigation of oceans, seas, and lakes. Phys Atmos Ocean (Moscow) 2:202–210 [in Russian]
Sasaki Y, Asanuma I, Muneyama K, Naito G, Suzuki T (1987a) A simplified microwave model and its application to the determination of some oceanic environmental parametric values, using multichannel microwave radiometric observations. IEEE Trans Geosci Remote Sens GE-25(3):384–392
Sasaki Y, Asanuma I, Muneyama K, Naito G, Suzuki T (1987b) The dependence of sea-surface microwave emission on wind speed, frequency, incidence angle, and polarization over the frequency range from 1 to 40 GHz. IEEE Trans Geosci Remote Sens GE-25(2):138–146
Sasaki Y, Asanuma I, Muneyama K, Naito G, Suzuki T (1988) Microwave emission and reflection from the wind-roughened sea surface at 6.7 and GHz. IEEE Trans Geosci Remote Sens 26(6):860–868

Sasaki Y, Naito G, Wilheit TT (1989) Rainfall rate dependence of brightness temperature of microwave emission from rain. Proceedings of annual spring meeting of oceanographical society of Japan, Tokyo, 5–9 April 1989. Oceanographic Society of Japan, Tokyo, pp 65–66
Savin AI, Bondur VG (2000) Scientific principles for the creation and diversification of global aerospace systems. Opt Atmos Ocean (Moscow) 1:46–63 [in Russian]
Savinikh VP, Krapivin VF, Potapov II (2007) Information technologies in the ecological monitoring systems. Geodezkartizdat, Moscow, 388 pp [in Russian]
Savorskij VP (1992) Spatial-temporal structure of radio-thermal field in the atmosphere-ocean system. Ph.D. theses, Institute of Radioengineering and Electronics, Russian Academy of Sciences, Moscow, 199 pp [in Russian]
Schimel DS (1995) Terrestrial biogeochemical cycles: global estimates with remote sensing. Remote Sens Environ 51(1):49–56
Schlesinger ME, Ramankutty N, Andronova N (2000) Temperature oscillations in the North Atlantic. Science 289(5479):547–548
Schmugge T (1990) Measurements of surface soil moisture and temperature. In: Hobbs RJ et al (eds) Remote sensing of biosphere functioning. Springer, New York, pp 31–62
Schneider EK (2006) Stochastic forcing of surface climate. COLA Technical Report 224, Calverton, Maryland, 34 pp ftp://grads.iges.org/pub/ctr/ctr_224.pdf
Schneider EK, Fennessy MJ, Kinter JL (2009) A statistical-dynamical estimate of winter ENSO teleconnections in a future climate. J Clim 22(24):6624–6638
Schrank WE (2007) The ACIA, climate change and fisheries. Mar Policy 31:5–18
Schulte RPO (2003) Analysis of the production stability of mixed grasslands II: a mathematical framework for the qualification of production stability in grassland ecosystems. Ecol Model 159(1):71–99
Schulte RPO, Lantinga EA, Struik PC (2003) Analysis of the production stability of mixed grasslands I: A conceptual framework for the qualification of production stability in grassland ecosystems. Ecol Model 159(1):43–69
Schulze E-D, Heimann M, Harrison S, Holland E, Lloyd J, Prentice IC, Schimel DS (2001) Global biogeochemical cycles in the climate system. Academic Press, New York, 350 pp
Scorer RS (1997) Dynamics of meteorology and climate. Wiley, New York, 350 pp
SEARCH (2001a) SEARCH develops implementation framework. Witness Arctic 9(1):3
SEARCH (2001b) SEARCH research opportunities emerging (Winter 2000/2001). Witness the Arctic 8(2):8
Sellers PJ (1985) Canopy reflectance, photosynthesis and transpiration. Int J Remote Sens 6:1335–1372
Sellers PJ (1987) Canopy reflectance, photosynthesis and transpiration. II: the role of biophysics in the linearity of their interdependence. Remote Sens Environ 21:143–183
Sellers PJ, Meeson BW, Hall FG, Asrar G, Murphy RE, Schiffer RA, Bretherton FP, Dickinson RE, Ellingson RG, Field CB, Huemmrich KF, Justice CO, Melack JM, Roulet NT, Schimel DS, Try PD (1995) Remote sensing of the land surface for studies of global change: models—algorithms—experiments. Remote Sens Environ 51(1):3–26
Sellers PJ, Bounoua L, Collatz GL, Randall DA, Dazlich DA, Los SO, Berry JA, Fung I, Tucker CJ, Field CB, Jensen TG (1996) Comparison of radiative and physiological effects of doubled atmospheric CO_2 on climate. Science 271:1402–1406
Sellers PJ, Dickinson RE, Randall DA, Betts AK, Hall FG, Berry JA, Collatz GJ, Denning AS, Mooney HA, Nobre CA, Sato N, Field CB, Henderson-Sellers A (1997) Modeling the exchanges of energy, water, and carbon between continents and the atmosphere. Science 275:502–509
Semenov SM (2004) Greenhouse gases and the present climate of the Earth. Meteorology and Hydrology Center Publisher, Moscow, 175 pp [in Russian]
Seppelt R, Voinov A (2002) Optimization methodology for land use pattern using spatially explicit landscape models. Ecol Model 151(2–3):125–142
Seppelt R, Voinov A (2003) Optimization methodology for Landuse patterns: evaluation based on multiscale habitat pattern comparison. Ecol Model 168(3):217–231

Sergin SYa (ed) (1974) Study of the Glaciers-Ocean-Atmosphere system. USSR Far-East Science Centre Publication, Vladivostok, 138 pp [in Russian]

Shahramantyan MA (2003) New information technologies in tasks of the provision of national safety of Russia (Natural-Technogenic Aspects). Russian Ministry for Excessive Situations Publisher, Moscow, 358 pp [in Russian]

Shapovalov SM (ed) (2010) Physical, geological, and biological researches of oceans and seas. Scientific World Publishing, Moscow, 628 pp [in Russian]

Sharkov EA (2003) Passive microwave remote sensing of the Earth: physical foundations. Springer/Praxis, Berlin, 613 pp

Shaver GR, Rastetter EB, Salman V, Street LE, van de Weg MJ, Rocha A, van Vijk MT, Williams M (2013) Pan-Arctic modeling of net ecosystem exchange CO_2. Philos Trans Roy Soc B 368 (1624):1–13 doi:10.1098/rstb.2012.0485

Shevchenko G, Rabinovich A, Thomson R (2004) Sea-ice drift on the northeastern shelf of Sakhalin Island. J Phys Oceanogr 34(11):2470–2491

Shultze E-D, Heimann M, Harrison S, Holland E, Lloyd J, Prentice IC, Schimel DS (2001) Global biogeochemical cycles in the climate system. Academic Press, New York, 350 pp

Shutko AM (1986) Microwave radiometry of water surface and grounds. Science Publishing House, Moscow, 204 pp [in Russian]

Shutko AM (1987) Remote sensing of waters and lands via microwave radiometry (The principles of method, problems feasible for solving, economic use). Pontificiae Acad Scientiarum Scripta Varia 68:413–441

Shutko AM (1992) Soil-vegetation characteristics at microwave wavelengths. In: Mather P (ed) Understanding the terrestrial environment. The role of the earth observations from space. Taylor & Francis, London, pp 53–66

Shutko AM (1997) Remote sensing of soil moisture and moisture related parameters by means of microwave radiometry. In: Sorooshian S, Gupta H, Rodda J (eds) Global environmental change and land surface processes in hydrology: the trials and tribulations of modeling and measuring. NATO ASI Ser Ser I Glob Environ Changes 46:263–273

Shutko AM, Krapivin VF (eds) (2011) Operational diagnostics, estimation of the scale of damage and aftermath reduction of stressful natural processes. Prof. Marina Drinova's Academic Publisher, Sofia, Bulgaria, 287 pp [in Russian]

Shutko AM, Krapivin VF, Mkrtchyan FA, Reutov EA, Novichkhin EP, Leonidov VA, Mishanin VG, Tsankov NS (1994) Econo-ecological estimates of the effectiveness of utilizing remotely sensed data and GIS information for soil moisture and moisture related parameters determination (Geoinformation Monitoring System Approach - GIMS). Proceedings of the ICID, May 7–9, 1994, Varna, Bulgaria. ICID, Varna, Bulgaria, pp 185–189

Shutko AM, Reutov EA, Golovachev SP (1995) Estimation of soil moisture profiles and root zone moisture content by means of microwave radiometry and a priory information. In: Choudhury B, Kerr Y, Njoku E, Pampaloni P (eds) Proceedings of the ESA/NASA international workshop on passive microwave remote sensing of land-atmosphere interaction, January 11–15, 1993 Saint-Larry, France, Utrecht Publisher, The Netherlands, pp 461–474

Shutko A, Gulyaev Yu, Chusov I, Sidorov I, Novichikhin E, Golovachev S, Krapivin V, Haldin A, Nazarov L, Tishchenko Yu, Chukhlantsev A, Marechek S, Haarbrink R, Dachev TS, Kancheva R, Petkov D, Semkova J, Yanev T, Kolionsky N, Coleman T, Archer F, Pampaloni P, Paloscia S, Krissilov A, Kosteniuk B, Baryshnikov I, Carmona A (2006a) Microwave radiometric remote sensing for emergency mapping of the areas with water seepage through levees and of the zones with dangerously high groundwater level. In: Marcal A (ed) Proceedings of the 25th symposium of the european association of remote sensing laboratories, June 6–11, 2005, Porto, Portugal. Millpress Science Publishers, Rotterdam, The Netherlands, pp 305–314 (www.fc.up.pt/earsel2005/index.html)

Shutko A, Haldin A, Krapivin V, Novichikhin E, Tishchenko Yu, Haarbrink R, Georgiev G, Kancheva R, Nikolov H, Coleman T, Archer F, Pampaloni P, Paloscia S, Krissilov A, Carmona A (2006b) Microwave radiometry in monitoring and emergency mapping of water seepage and dangerously hugh groundwaters. In: Proceedings of the 2-nd Microwave & Radar week in Poland. Workshop on space technologies, STW-2006, May 23, 2006, Krakow, Poland. Polish Academy of Sciences, Krakow, Poland, pp 79–84

Shutko AM, Haldin A, Krapivin V, Novitchikhin E, Sidorov I, Tishchenko Yu, Haarbrink R, Georgiev G, Kancheva R, Nikolov H, Coleman T, Archer F, Pampaloni P, Paloscia S, Krisilov A, Carmona A (2007) Microwave radiometry in monitoring and emergency mapping of water seepage and dangerously high groundwaters. J Telecommun Inform Technol 1:76–82

Shutko AM, Krapivin VF, Haarbrink RB, Sidorov IA, Novichikhin EP, Archer F, Krisilov AD (2010) Practical microwave radiometric risk assessment. Professor Marin Drinov Academic Publishing House, Sofia, Bulgaria, 88 pp

Silver D (1998) Tropical rain forest. McGraw-Hill, New York, 48 pp

Sims DA, Gamon JA (2003) Estimation of vegetation water content and photosynthetic tissue area from spectral reflectance: a comparison of indices based on liquid water and chrolophyll absorption features. Remote Sens Environ 84(4):526–537

Singh VP, Singh P, Haritashya UK (eds) (2011) Encyclopedia of snow, ice and glaciers vol XLVI. Springer, Berlin, 1253 pp

Sirocko F, Cubash U, Kaspar F, von Storch H, Widmann M, Litt T, Kühl N, Mangini A, Pachur H-J, Claussen N, Kubatzki C, Junge FW, Böttger T, Krbetschek M, Degering D (2004) Climate change at the very end of a warm stage: first results from the last glacial inception at 117,000 yr BR. Pages News 12(2):18–20

Sirotenko OD (2000) The future of agriculture in Russia in connection with expected climate change. Probl Ecol Monit Ecosyst Model XVII:258–274 [in Russian]

Singer SF (2008) Nature, Not human activity, rules the climate. Report of the nongovernmental international panel on climate change. Heartland Institute Publisher, Chicago, 30 pp

Sinik N, Loncar E, Vidic S (1985) The use of field data in average wet deposition modeling. Air pollution modeling and its application. Plenum Press, New York, pp 155–161

Sivasakthivel T, Reddt KKSK (2011) Ozone layer depletion and its effects: a review. Int J Environ Sci Dev 2(1):30–37

Skou N (2008) Microwave radiometry and radiometers for ocean applications. In: Barale V, Gade M (eds) Remote sensing of the European seas. Springer, London, pp 202–222

Smith HL (2008) Mondone dynamical systems: an introduction to the theory of competitive and coopperative systems. American Mathematical Society, Washington, 174 pp

Solanki SK, Usoskin IG, Kromer B, Schorsler M, Beer J (2004) Unusual activity of the Sun during recent decades compared to the previous 11,000 years. Nature 431(7012):1084–1087

Soldatov VY (2006) Microwave diagnostics of the ocean-atmosphere system. Probl Environ Nat Res 12:28–35 [in Russian]

Soldatov VYu (2007) Diagnostics of the ocean-atmosphere system by means of percolation model. Probl Enviro Nat Res 5:52–64 [in Russian]

Soldatov VY (2009a) A sequential analysis method for the prediction of typhoon. In: Proceedings of the international symposium on engineering ecology – 2009, December 8–9, 2009, Moscow. The Russian Sciences Engineering A.S. Popov Society for Radio, Electronics and Communication, Moscow, pp 105–109

Soldatov VY (2009b) An information technology for the operative diagnostics of the ocean-atmosphere system. In: Proceedings of the international symposium on mathematical modelling of process of tropical hurricane beginning and searching its indicator Precursors, December 16, 2009, Hochiminh City, Vietnam. Institute of Applied Mechanics, Hochiminh City, Vietnam, pp 65–85

Soldatov VY (2009c) Possibilities of monitoring systems to detect tropical hurricanes. In: Proceedings of the international symposium on mathematical modelling of process of tropical hurricane beginning and searching its indicator precursors, December 16, 2009, Hochiminh City, Vietnam. Institute of Applied Mechanics, Hochiminh City, Vietnam, pp 41–45

Soldatov VY (2010) Multi-functional information-modeling system for the hydrophysical experiment. In: Proceedings of the IX international symposium on ecoinformatics problems, December 9–11, 2010, Moscow. The Russian Sciences Engineering A.S. Popov Society for Radio, Electronics and Communication, Moscow, pp 96–100 [in Russian]

Soldatov VY (2011) Multi-functional information-modeling system for the hydrophysical experiments. PhD Theses, KIREE, Moscow, 145 pp. [in Russian]

Soldatov VY (2012) Some questions of diagnostics of physical phenomena and processes in hydrophisical systems. Ecol Syst Devices 3:41 [in Russian]

Soldatov VY, Krapivin VF, Borodin LF, Startsev AA (2008) Direct and inverse problems of the microwave monitoring of the environment. In: Proceedings of the eighth international symposium on ecoinformatics problems, December 16–17, 2008, Moscow. The Moscow Sciences Engineering A.S. Popov Society for Radio, Electronics and Communication, Moscow, pp 19–25

Soldatov VY, Nitu C, Krapivin VF (2010a) Diagnosis of transition processes in the ocean-atmosphere system. In: Proceedings of the international conference DAS (Data Application System), May 27–29, 2010, Suceava, Romania. University Stefan cel Mare Suceava, Romania, pp 36–39

Soldatov VY, Nitu C, Krapivin VF (2010a) Diagnosis of transition processes in the ocean-atmosphere system. Control Eng Appl Inform 12(2):22–29

Soldatov VY, Nitu C, Krapivin VF (2011b) Information-modeling technology for the operative diagnostics of the ocean-atmosphere system. Sci Bull Electr Enfineering Fac Valahia Univ Targoviste Romania 2:82–89

Solimini D (ed) (1995) Microwave radiometry and remote sensing of the environment. VSP BV, The Netherlands, 539 pp

Solozhentsev ED (2004) The scenario logical-probabilistic risk control in business and engineering. Business-Press Publisher, St.-Petersburg, 416 pp [in Russian]

Somes NLG (1999) Numerical simulation of wet land hydrodynamics. Environ Int 25(6–7):773–779

Soon W, Postmentier E, Baliunas S (2000) Climate hypersensitivity to solar forcing? Ann Geophys 18:583–588

Southerton D, Chapells H, van Vliet B (eds) (2004) Sustainable consumption. The implications of changing infrastructures of provision. Edward Elgar Publishing Ltd, London, 192 pp

Spedicato E (1991) A catastrophical scenario for discontinuities in human history. J New Engl Antiquities Res Assoc 26:1–14

Sportisse B (2000) Box models versus Eulerian models in air pollution modeling. Atmos Environ 35(1):173–178

Starke L (ed) (2004) State of the World-2004. Progress towards a sustainable society. Earthscan, London, 246 pp

Stauffer D, Aharony A (1994) Introduction to percolation theory. Plenum Press, New York, 215 pp

Steffen W, Sanderson A, Tyson PD, Jäger J, Matson PA, Moore B III, Oldfield F, Richardson K, Schellnhuber H-J, Turner BL II, Wasson RJ (2004) Global change and the Earth system—a planet under pressure. Springer, New York, 336 pp

Stevenson DS, Johnson CE, Collins WJ, Derwent RG, Edwards JM (2000) Future estimates of tropospheric ozone radiative forcing and methane turnover—the impact of climate change. Geophys Res Lett 27(14):2073–2076

Stevenson DS, Johnson CE, Collins WJ, Derwent RG (2003) The atmospheric sulphur cycle and the role of volcanic SO_2. Geol Soc London Spec Publ 213:295–305

Stoddard JL, Kahl JS, Deviney FA, DeWalle DR, Driscoll CT, Herlihy AT, Kellogg JH, Mudoch PS, Webb JR, Webster KE (2003) Response of surface water chemistry to the clean air act amendments of 1990. U.S. Environmental Protection Agency, EPA/620/R-03/001, Corvallis, Oregon, 92 pp

Stoll-Kleemann S, O'Riordan T (2004) The possible role of public private partnerships. IHDP Update 3:6–8

Stone RS, Dutton EG, Harris SM, Longenecker D (2002) Earlier spring snowmelt in northern Alaska as an indicator of climate change. J Geophys Res 107(D10):ACL10/1–ACL10/15

Straub CP (ed) (1989) Practical handbook of environmental control. CRC Press, Boca Raton, 537 pp
Strelkov GM (1995) About the possibility to determine the content of admixtured gases CO and N_2O using the method of active sounding of the atmosphere in the millimeter range. Res Earth Space 4:3–7 [in Russian]
Strelkov GM (1996) Active sounding of the water wapours within the mesopause with the use of millimeter and sub-millimeter waves. Res Earth Space 14:169–177 [in Russian]
Strelkov GM, Soldatova IV (1989) Radio-impulse propagation within the line of the oxygen absorption. Radiotechnics (Moscow) 3:16–17 [in Russian]
Strong AE, Kearns EJ, Gjovig KK (2000) Sea surface temperature signals from satellites update. Geophys Res Lett 27(11):1667–1670
Sud YC, Walker GK, Kim J-H, Liston GE, Sellers PJ, Lau WK-M (1996) Biogeophysical consequences of a tropical deforestation scenario: A GCM simulation study. J Clim 9 (12):3225–3247
Sudarshana P, Nageswara M, Soneji JR (2012) Tropical forests. InTech, New York, 388 pp
Sukov AI, Soldatov VY (2007) Multi-channel sequential procedure of statistical decision making in the detection task of tropical hurricanes. Proceedings of the Russian sciences engineering A. S. Popov society for radio, electronics and communication. Ser Eng Ecol 4:35–41
Sukov AI, Soldatov VY, Krapivin VF, Cracknell AP, Varotsos CA (2008a) A sequential analysis method for the prediction of tropical hurricanes. Int J Remote Sens 29(9):2787–2798
Sukov AI, Soldatov VYu, Nitu C, Krapivin VF (2008b) Direct and inverse problems of the microwave monitoring of the environment. In: Proceedings of the international world energy system conference, June 30-July 2, 2008, Iasi, Romania. Technical University Gheorghe Asachi, Iasi, Romania, pp 123–129
Sun R, Wang ZZ, Chen L, Wang WW (2013) Assessment of surface water quality at large watershed scale: land-use, anthropogenic, and administrative impacts. JAWRA J Am Water Res Assess 49(4):741–752
Suni T, Berninger F, Markkanen T, Keronen P, Rannik Ü, Vesala T (2003) Interannual variability and timing of growing—season CO_2 exchange in a boreal forest. J Geophys Res 108(9):ACL2/1–ACL2/8
Suzuki A (1992) Results of collection fishes, and tropical to temperate migrant fishes comes into Okhotsk Sea coast during 1988 to 1991 in Northern Hokkaido Japan. In: Proceedings of the seventh international symposium on Okhotsk Sea and Sea Ice, February 2–5, 1992, Mombetsu, Hokkaido (Japan). Okhotsk Sea & Cold Ocean Res. Assoc., Mombetsu, pp 225–231
Syvorotkin VL (2002) Deep degassing of the earth and global catastrophes. Geoinformcenter Publisher, Moscow, 250 pp [in Russian]
Szathmáry E, Griesemer J (2003) The principles of life. Oxford University Press, Oxford, 220 pp
Tabb WK (2001) The amoral elephant: globalization and the struggle for social justice in the twenty-first century. Monthly Review Press, New York, 288 pp
Takemura T, Nozawa T, Emori S, Nakajima TY, Nakajima T (2005) Simulation of climate response to aerosol direct and indirect effects with aerosol transport-radiation model. J Geophys Res 110(D02202). doi:10.1029/2004JD005029
Tang J, Yang R, Kafatos M (2005) Data mining for tropical cyclone intensity prediction. In: Proceedings of the sixth conference on coastal atmospheric and oceanic prediction and processes, January 9—13, 2005, San Diego C.A. George Mason University, Fairfax. https://ams.confex.com/ams/pdfpapers/84949.pdf
Tarko AM (2001) Investigation of global biosphere processes with the aid of a global spatial carbon dioxide cycle model. In: Sixth international carbon dioxide conference, extended abstracts, vol 2. Tokohu University, Sendai, pp 899–902
Tarko AM (2003) Analysis of global and regional changes in biogeochemical carbon cycle: a spatially distributed model. Interim Report, IR-03-041, IIASA, Laxenburg, Austria, 28 pp
Terziev FS (ed) (1992) Hydrometeorology and hydrochemistry of USSR seas. vol 1: Barents Sea. Gidrometeoizdat, St-Petersburg, 182 pp [in Russian]

Terziev FS, Zatuchnoy BM, Gershanovitch DE (1993) Okhotsk Sea. Gidrometeopress, St-Petersburg, 167 pp [in Russian]
Tett SFB, Stott PA, Allen MR, Ingram WJ, Mitchell JFB (1999) Causes of twentieth-century temperature change near the Earth's surface. Nature 399:569–572
Thiessen KM, Thorne MC, Maul PR, Prohl G, Wheater HS (1999) Modelling radionuclide distribution and transport in the environment. Environ Pollut 100(1–3):151–177
Tianhong L, Yanxin S, An X (2003) Integration of large scale fertilizing models with GIS using minimum unit. Environ Model 18(3):221–229
Timmermann A, Oberhuber J, Bacher A, Esch M, Latif M, Roeckner E (1999) Increased El Nińo frequency in a climate model forced by future greenhouse warming. Nature 398:684–687
Topics G (2003) Annual review: natural catastrophes 2002. Munich Re Group, München, 52 pp
Topics G (2004) Annual review: natural catastrophes 2003. Munich Re Group, München, 56 pp
Topics G (2005) Annual review: natural catastrophes 2004. Munich Re Group, München, 60 pp
Topics G (2004) Annual review: natural catastrophes 2003. Munich Re Group, München, 56 pp
Topics G (2005) Annual review: natural catastrophes 2004. Munich Re Group, München, 60 pp
Topics G (2006) Annual review:nnatural catastrophes 2005. Munich Re Group, München, 56 pp
Topics G (2007) Annual review: natural catastrophes 2006. Munich Re Group, München, 54 pp
Topics G (2008) Natural catastrophes 2007: analyses, assessments, positions. Munich Re Group, München, 54 pp
Toth F, Bruckner T, Füssel H-M, Helm C, Hooss G, Leimbach M, van Minnen J, Petschel-Held G, Schellnhuber H-J, Tothne-Hizsnyik E (2000) ICLIPS—Integrierte Abschätzung von Klimaschutzstrategien: Methodisch-naturwissenschaftliche Aspekte [in German]. Research report 01 LK 9605/0, Federal Ministry for Science and Research, Bonn, 115 pp
Tricomi F (1960) Integral equations. Foreign Literature Publisher, Moscow, 299 pp [in Russian]
Tsegaye TD, Metzl R, Wang X, Schamschula M, Tadesse W, Clendenon D, Golson K, Coleman TL, Archer F, Schaefer G (2005) Near real time variability of soil moisture and temperature under different land use and cover: the Alabama Mesonet. J Environ Hydrol 13:1–13
Tsushima Y, Manabe S (2001). Influence of cloud feedback on annual variation global mean surface temperature. J Geophys Res 106(D19):22646–22655
Turetsky MR, Wieder RK, Vitt DH, Evans RJ, Scott KD (2007) The disappearance of relict permafrost in boreal north America: effects on peatland carbon storage and fluxes. Glob Change Biol 13:1922–1934
Ugolnitsky GA (1999) Management of the ecologo-economic systems. High School Book Publisher, Moscow, 132 pp [in Russian]
UN (2012) World urbanization prospects. The 2011 revision. United Nations Publication, New York, 34 pp
UNEP (2002) Global environmental outlook 3. Past, present and future perspectives. UNEP, Nairoby, 20 pp
UNEP (2004) UNEP science initiative. Environmental change and human needs: assessing Interlinkages. An input to the fourth global environment outlook (GEO-4). Concept Paper, Nairoby, 36 pp
UNEP (2006) Convention on biological diversity. In: Proceedings of the conference on the parties to the convention on biological diversity, Curitiba, Brazil, 20–31 Mar 2006, UNEP/CBD/COP, New York, 374 pp
UNIT TWO (2008) The arctic and biocomplexity—a global and local science. TERC, University of Tsukuba, Japan, 65 pp
Vakulenko NV, Kotliakov VM, Monin AS, Sonechkin DM (2004) Evidence of the leading role of temperature variations with respect to variations of greenhouse gas concentration from data of ice core from the station "Vostok". Ann RAS 397(5):663–667 [in Russian]
Valette-Silver NJ, Hameedi MJ, Efurd DW, Robertson A (1999) Status of the contamination in sediments and biota from the western Beaufort Sea (Alaska). Mar Pollut Bull 38(8):702–722
Vallis GK, Thoumi G (2012) The cryosphere-princeton primers in climate: a book review. Princeton University Press, Princeton, 288 pp

Van Cao P, Khue NB, Cuong DM, Tan NN, Krapivin VF, Mkrtchyan FA (2009a) Thunderstorms as interactive component of global ecodynamics. In: Proceedings of the Popov A.S. Russian society for radiotechnics, electronics, and communication. Series: "Engineering ecology", vol 5, pp 9–14

Van Cao P, Khue NB, Cuong DM, Tan NN, Krapivin VF, Mkrtchyan FA, Soldatov VY (2009b) Synthesis of geoecological information-modeling systems. In: Proceedings of the Popov A.S. Russian society for radiotechnics, electronics, and communication. Series: "Engineering ecology", vol 5, pp 18–22

Van Cao P, Nguyen BK, Krapivin VF, Mkrtchyan FA (2012) An adaptive information technology for the operative diagnostics of the ocean-atmosphere system. Binh Duong Univ J Sci Technol 9(4):50–67

Varotsos CA (2005) Airborne measurements of aerosol, ozone, and solar ultraviolet irradiance in the troposphere. J Geophys Res Atmos 110(D9), Art No. D09202. doi:10.1029/2004JD005397

Varotsos CA, Krapivin VF, Soldatov VYu (2014) Modeling the carbon and nitrogen cycles. Front Environ Sci Air Pollut 2(8):1−6. doi:10.3389/fenvs.2014.00008

Vasilevich VI (1983) Essays on theoretical phytocenology. Science Publishing House, Leningrad, 226 pp [in Russian]

Verba VS, Gulyaev YV, Shutko AM, Krapivin VF (2013) SHF-radiometry of the land and water surfaces: about theory and practice. Marina Drinova Academic Publishing House, Sofia, 296 pp [in Russian]

Vermeer M, Rahmstor S (2009) Global sea level linked to global temperature. PNAS 106 (51):21527–21532

Vernadsky VI (1944) Several words about noosphere. Prog Present Biol 18(2):49–93 [in Russian]

Vinogradov ME (1975) Ecosystems of the Pacific Ocean Pelagial. Nauka, Moscow, 408 pp [in Russian]

Vinogradov ME, Gitelzon II, Sorokin JI (1970) The vertical structure of a pelagic community in the tropical ocean. Mar Biol 6(4):187–194

Vinogradov ME, Krapivin VF, Fleishman BS, Shushkina EA (1975) The use of the mathematical model for analysing the behaviour of ecosystem of oceanic pelagic zone. Oceanology (Moscow) 10(2):313–320 [in Russian]

Vinogradov ME, Volkov AF, Semenova TN (eds) (1996) Hyperiid amphipods (amphipoda, hiperiidea) of the world oceans, vol XXVII. Science Publishers, Lebanon, 632 pp

Vinogradov GM, Marasaeva EF, Larionov VV, Matishov GG (2001) Communities of zooplankton for the ice covered aquatories of Barents and Kara Seas in the winter-summer period of 2000. Ann Russ Acad Sci 376(6):815–817 [in Russian]

Vinogradova TO, Golovachev SP, Grankov AG, Milshin AA, Priadko AV, Chukhlantsev AA, Shelobanova NK, Shutko AM (2004) Spectral measurements of microwave attenuation by crowns of a fir and chestnut. Probl Environ Nat Resour 1:69–76 [in Russian]

Vladimirov VA, Vorobyev YL, Salov SS, Faleye MI, Arkhipova NI., Kapustin MA, Kashchenko SA, Kosiachenko SA, Kuznetsov IV, Kul'ba VV, Malinetsky GG, Makhutov NA, Pisarenko VF, Podlazov AV, Posashkov SA, Potapov AB, Shnirman MG (2000) Risk management: risk. Sustainable development. Science Publishing, Moscow, 431 pp [in Russian]

Vörösmarty CJ, Hinzman LD, Peterson BJ, Bromwich DH, Hamilton LC, Morison J, Romanovsky VE, Sturm M, Webb RS (2001) The hydrological cycle and its role in arctic and global environmental change: a rationale and strategy for synthesis study. Arctic Research Consortium of the US, Fairbanks, 84 pp

Wald A (1947) Sequential analysis. Wiley, New York, 224 pp

Wald A (2004) Sequential analysis. Dover Publications, Mineola, 224 pp

Walker DA, Bockheim JG, Chapin FS, Eugster W, King JY, McFadden J, Michaelson GJ, Nelson FE, Oechel WC, Ping CL, Reeburgh WS, Regli S, Shiklomanov NI, Vourlitis GL (1998) Energy and trace-gas fluxes across a soil-pH boundary in the Arctic. Nature 394:469–472

Walker DA, Epstein HE, Gould WA, Kelley AM, Kade AN, Knudson JA, Krantz WB, Michaelson G, Peterson RA, Ping C-L, Raynolds MK, Romanovsky VE, Shur Yu (2004) Frost-boil ecosystems: complex interactions between landforms, soils, vegetation, and climate. Permafrost Pelagic Process 15:171–188

Walker DA, Epstein HE, Welker JM (2008) Introduction to special section on biocomplexity of Arctic tundra. J Geophys Res 113(G03800):1–5

Walker DA, Bhatt US, Epstein HE, Bieniek PA, Comiso JC, Frost GV, Pinzon J, Raynolds MK, Tucker CJ (2012) Changing Arctic tundra vegetation biomass and greenness. Bull Am Meteorol Soc 93(7):138–139

Walker G (2003) Snowball Earth: the story of the great global catastrophe that spawned life as we know it. Crown Publishers, New York, 269 pp

Wallace JM (1998) Observed climatic variability: spatial structure. In: Anderson DLT, Willebrand J (eds) Decadal climate variability: dynamics and predictability. Springer, Berlin, pp 31–81

Wallace JM, Thompson DWJ (2002) Annual modes and climate prediction. Phys Today 55(2):28–33

Wanders N, Karssenberg D, Bierkens M, Parinussa R, de Jeu R, van Dam J, de Jong S (2012) Observation uncertainty of satellite soil moisture products determined with physically-based modeling. Remote Sens Environ 127:341–356

Wang H, Christiansen JH (1986) A reentry plume fumigation model. In: De Wispelaere C, Schiermeier FA, Gillani NV (eds) Air pollution modeling and its application (vol V). Plenum Press, New York, pp 565–579

Wang C, Bond-Lamberty B, Gower ST (2003) Soil surface CO_2 flux in a boreal black spruce fire chronosequence. J Geophys Res 108(3):WFX5/1–WFX5/8

Wang LK, Shammas NK, Hung Y-T (eds) (2011) Advanced biological treatment processes, vol 9. Humana Press, New York, 737 pp

Wang W, Davis KJ, Cook BD, Bakwin PS, Yi C, Butler MP, Ricciuto DM (2005) Surface layer CO_2 budget and advective contributions to measurements of net ecosystem-atmosphere exchange of CO_2. Agric For Meteorol 135(1–4):202–214

Wania F, Hoff JT, Jai CQ, Mackay D (1998) The effects of snow and ice on the environmental behavior of hydrophobic organic chemicals. Environ Pollut 102(1):25–41

Waple AM, Lawrimore JH (eds) (2003) State of the climate in 2002. Bull Am Meteorol Soc 84(6): S1–S68

Watanabe S, Hajima T, Sudo K, Nagashima T, Takemura T, Okajima H, Nozawa T, Kawase H, Abe M, Yokohata T, Ise T, Sato H, Kato E, Takata K, Emori S, Kawamiya M (2011) MIROC-ESM 2010: model description and basic results of CMIP5-20c3m experiments. Geosci Model Dev 4:845–872

Watson RT, Noble IR, Bolin B, Ravindranath NH, Verardo DJ, Dokken DJ (eds) (2000) Land use, land-use change, and forrestry. Cambridge University Press, Cambridge, 377 pp

Wauben WMF, Van Velthoven PFJ, Kelder H (1997) 3d chemistry transport model study of changes in atmospheric ozone due to aircraft—NO emissions. Atmos Environ 31(12):1819–1836

WB (2009) World development report 2010: development and climate change. World Bank, New York 424 pp

WCSDG (2004) A fair globalization: creating possibilities for all. Report of the world commission on the social aspects of globalization. International Trade Organization, Geneva, 190 pp

Weber GR (1992) Global warming. The rest of the story. Dr. Boettiger Verlags GmbH, Wiesbaden, 188 pp

Webster PJ, Holland GJ, Curry JA, Chang H-R (2005) Changes in tropical cyclone number, duration, and intensity in a warming environment. Science 309:1844–1846

Weeks SJ, Pitcher GC, Bernard S (2004) Satellite monitoring of the evolution of a coccolithophorid bloom in the southern Benguela upwelling system. Oceanography 17 (1):83–89

Weinkle J, Maue R, Pielke R (2012) Historical global tropical cyclone landfalls. J Climate 25:4729–4735

Weis RF, Jahnke RA, Keeling CD (1982) Seasonal effects of temperature and salinity of the partial pressure of CO_2 in seawater. Nature 300:511–513
Weller G, Lange M (eds) (1999) Impacts of global climate change in the Arctic region. Intern. Arctic Science Com., Fairbanks, 59 pp
WEO (2002) World energy outlook. OECD/IEA, Paris, 533 pp
WEO (2004) World energy outlook. OECD/IEA, Paris, 550 pp
WEO (2005) Middle East and North Africa Insights. IEA, London, 600 pp
WEO (2006) Energy market reform; energy policy; energy projections. IEA, London, 600 pp
WEO (2013) World energy outlook 2013. OECD/IEA, Paris, 523 pp
Widmann M, Jones JM, von Storch H (2004) Reconstruction of large-scale atmospheric circulation and data assimilation in paleoclimatology. PAGES News 12(2):12–13
Wielgolaski FE (ed) (1997) Polar and alpine tundra. Elsevier, New York, 930 pp
Wigley TML, Schimel DS (2005) The carbon cycle. Cambridge University Press, Cambridge, 312 pp
Wijen F, Zoeteman K, Pieters J (eds) (2005) A handbook of globalization and environmental policy. National government interventions in a global Arena. Edward Elgar Publishing. Ltd., London, 768 pp
Williams RG, Follows MJ (2011) Ocean dynamics and the carbon cycle. Massachusetts Institute of Technology, Massachusetts, 434 pp
Wirtz KW (2000) Second order up-scaling: theory and an exercise with a complex photosynthesis model. Ecological Modelling 126(1):59–71
WMO (2000) World climate research programme. International programme for antarctic buoys (IPAB). Report on the Third Meeting of Progrsamme Participants, Fairbanks, p 57
Wong S (2005) Suppression of deep convection over the tropical North Atlantic by the Saharan Air Layer. Geophys Res Lett 32(9), L09808:1–5
Wood R, Mechoso CR, Bretherton CS, Weller RA, Huebert B, Straneo F, Albrecht BA, Coe H, Allen G, Vaughan G, Daum P, Fairall C, Chand D, Gallardo Klenner L, Garreaud R, Grados C, Covert DS, Bates TS, Krejci R, Russell LM, de Szoeke S, Brewer A, Yuter SE, Springston SR, Chaigneau A, Toniazzo T, Minnis P, Palikonda R, Abel SJ, Brown WOJ, Williams S, Fochesatto J, Brioude J, Bower KN (2011) The VAMOS ocean-cloud-atmosphere-land study regional experiment (VOCALS-REx): goals, platforms, and field operations. Atmos Chem Phys 11:627–654
Woodcock A (1999a) Global warming—a natural event? Weather 54(5):162–163
Woodcock A (1999b) Global warming: the debate heats up. Weather 55(4):143–144
WRI (2011) Decision making in a changing climate. World Resources Institute, Washington, 175 pp
WRI (2012) Making big ideas happen. World Resources Institute 2011–2012 Annual report, Washington, 50 pp
Wright J (2003) Environmental chemistry. Rutledge, London, 432 pp
Wuerthner G (ed) (2006) Wildfires: a century of failed forest policy. Island Press, Washington, 350 pp
Xu S, Jaffe P, Mauzerall DL (2007) A process-based model for methane emission from flooded rice paddy systems. Ecol Model 205:475–491
Yakimov SP (1996) Algorithms for the forest fire dangerous assessment basing on the remote sensing data. Ph.D. theses, Krasnoyarsk State of Technical University, Krasnoyarsk, 155 pp [in Russian]
Yakovlev OI (2001) Space radio science. Taylor and Francis Group, London, 320 pp
Yakovlev OI, Paveliev AG, Matyugov SS (2009) The earth satellite monitoring. Book House Publishing, Moscow. 206 pp [in Russian]
Yamagata T, Behera SK, Luo JJ, Masson S, Jury MR, Rao SA (2004) Coupled ocean-atmosphere variability in the tropical Indian ocean. In: Wang C, Xie X-P, Carton JA (eds) Earth climate: the ocean-atmosphere interactions. Springer, Berlin, pp 189–212
Yan Q, Blum RS (2001) Distributed signal detection under the Neyman-Pearson criterion. IEEE Trans Inf Theory 47(4):1368–1377

Yohe G, Malone E, Brenkert A, Schlesinger M, Meij H, Xing X, Lee G (2006) A synthetic assessment of the global distribution of vulnerability to climate change from the IPCCP perspective that reflects exposure and adaptive capacity. CIESIN, NewYork, 17 pp

Yoshioka H (2004) Vegetation isoline equations for an atmosphere-canopy-soil system. IEEE Trans Geosci Remote Sens 42(1):166–175

Yu R, Zhang M, Cess RD (1999) Analysis of the atmosphere energy budget: a consistency study of available data sets. J Geophys Res 104(D8):9655–9661

Yue C, Ciais P, Luyssaert S, Cadule P, Harden J, Randerson J, Bellassen V, Wang T, Piao SL, Poulter B, Viovy N (2013) Simulating boreal forest carbon dynamics after stand-replacing fire disturbance: insights from a global process-based vegetation model. Biogeosciences 10:8233–8252

Zagorin GK (1999) Polarizational characteristics (stock's parameters) for the own and scattered microwave radiation in the rains. PhD theses, Institute of Radioengineering and Electronics, Russian Academy of Sciences, Moscow, 178 pp [in Russian]

Zagorin GK, Kutuza BG (1998) Special features of the transportation of polarizational heat microwave radiation in the clouds and precipitation. Radiotechnics (Moscow) 10:566–575 [in Russian]

Zaikov GE, Maslov SA, Rubailo VL (1991) Acid rains and the environment. Chemistry Publications, Moscow, 139 pp [in Russian]

Zaitsev SI (1988) Laboratory modeling of the processes of the ocean-atmosphere CO_2 exchange. PhD theses, Moscow State University, Moscow, 170 pp [in Russian]

Zanetti A, Schwarz S, Lindemuth S (2007) Natural catastrophes and man-made disasters in 2006: low insured losses. Sigma 2:1–40

Zhan X, Sohlberg RA, Townshend JRG, DiMiceli C, Carroll ML, Eastman JC, Hansen MC, DeFries RS (2002) Detection of land cover changes using MODIS 250 m data. Remote Sens Environ 83(1–2):336–350

Zhang X, Helsdon JH Jr, Farley RD (2003a) Numerical modeling of lightning-produced NO_x using an explicit lightning scheme. 1. Two-dimensional simulation as a "proof of concept". J Geophys Res 108(D18):4579. doi:10.1029/2002JD003224

Zhang X, Helsdon JH Jr, Farley RD (2003b) Numerical modeling of lightning-produced NO_x using an explicit lightning scheme. 2. Three-dimensional simulation and expanded chemistry. J Geophys Res 108(D18):4580. doi:10.1029/2002JD003225

Zhou X, Chen JCL (2006) Ensemble forecasting of tropical cyclone motion using a baroclinic model. Adv Atmos Sci 23(3):342–354

Zhu C, Anderson G (2002) Environmental applications of geochemical modeling. Cambridge University Press, Cambridge, 298 pp

Zillman JW (2000) The challenges ahead. WMO Bull 49(1):8–13

Zlatev Z, Chrisensen J, Hov O (1992) A Eulerian air pollution model for Europe with nonlinear chemistry. J Atmos Chem 15:1–37

Zuev VV (2000) Destruction of the ozone layer: an ecological catastrophe or the greatest mystery of the century? In: Proceedings of the second symposium on the ecology of Siberian rivers and the Arctic, Tomsk, 24–26 Nov 2000, Tomsk University Press, Tomsk, Russia, 36040, pp [in Russian]

Index

V.F. Krapivin et al., *New Ecoinformatics Tools in Environmental Science*, Environmental Earth Sciences, DOI 10.1007/978-3-319-13978-4

D

E

K

L

M

R

Y

Z

MIX
Papier aus verantwortungsvollen Quellen
Paper from responsible sources
FSC® C105338

If you have any concerns about our products,
you can contact us on
ProductSafety@springernature.com

In case Publisher is established outside the EU,
the EU authorized representative is:
Springer Nature Customer Service Center GmbH
Europaplatz 3, 69115 Heidelberg, Germany

Printed by Libri Plureos GmbH
in Hamburg, Germany